U0191918

nature

The Living Record of Science

《自然》学科经典系列

总顾问：李政道（Tsung-Dao Lee）

英方总主编：Sir John Maddox
Sir Philip Campbell
中方总主编：路甬祥

《自然》百年物理经典 II

（英汉对照）

英方主编：Philip Ball　中方主编：赵忠贤

外语教学与研究出版社 · 麦克米伦教育 · 自然科研

FOREIGN LANGUAGE TEACHING AND RESEARCH PRESS · MACMILLAN EDUCATION · NATURE RESEARCH

北京 BEIJING

图书在版编目（CIP）数据

《自然》百年物理经典. Ⅱ：英汉对照 /（英）菲利普·鲍尔（Philip Ball），赵忠贤主编. -- 北京：外语教学与研究出版社，2020.9
（《自然》学科经典系列）
ISBN 978-7-5213-1946-0

Ⅰ. ①自… Ⅱ. ①菲… ②赵… Ⅲ. ①物理学－文集－英、汉 Ⅳ. ①O4-53

中国版本图书馆 CIP 数据核字（2020）第 126023 号

出 版 人　徐建忠
项目统筹　章思英
项目负责　刘晓楠　王丽霞
责任编辑　王丽霞
责任校对　黄小斌
封面设计　孙莉明　高　蕾
版式设计　孙莉明
出版发行　外语教学与研究出版社
社　　址　北京市西三环北路 19 号（100089）
网　　址　http://www.fltrp.com
印　　刷　北京华联印刷有限公司
开　　本　787×1092　1/16
印　　张　56.5
版　　次　2020 年 8 月第 1 版　2020 年 8 月第 1 次印刷
书　　号　ISBN 978-7-5213-1946-0
定　　价　568.00 元

购书咨询：（010）88819926　电子邮箱：club@fltrp.com
外研书店：https://waiyants.tmall.com
凡印刷、装订质量问题，请联系我社印制部
联系电话：（010）61207896　电子邮箱：zhijian@fltrp.com
凡侵权、盗版书籍线索，请联系我社法律事务部
举报电话：（010）88817519　电子邮箱：banquan@fltrp.com
物料号：319460001

《自然》学科经典系列

（英汉对照）

《自然》百年物理经典

（英汉对照）

英方主编：Philip Ball　　中方主编：赵忠贤

审稿专家（以姓氏笔画为序）

于　贵	于　渌	马宇蒨	王　琛	王乃彦	邓祖淦	厉光烈
石锦卫	朱永生	朱道本	刘　纯	刘京国	刘朝阳	江丕栋
杜江峰	李　淼	李芝芬	李兴中	李军刚	肖伟科	何香涛
狄增如	汪长征	沈宝莲	宋心琦	张元仲	张泽渤	张焕乔
陆朝阳	陈　方	尚仁成	郑东宁	赵见高	郝　伟	夏海鸿
顾镇南	郭建栋	陶宏杰	曹　俊	曹庆宏	葛墨林	韩汝珊
鲍重光	蔡荣根	翟天瑞	熊秉衡			

翻译工作组稿人（以姓氏笔画为序）

王耀杨	刘　明	何　铭	沈乃澂	郭红锋	黄小斌	蔡　迪

翻译人员（以姓氏笔画为序）

王　锋	王　静	王耀杨	牛慧冲	邓铭瑞	史春晖	刘　霞
刘东亮	安宇森	孙惠南	李　琦	李世媛	李忠伟	吴　彦
何　钧	沈乃澂	金世超	周　杰	孟　洁	胡雪兰	姜　克
姜　薇	钱　磊	高如丽	黄　娆	崔　宁	葛聆沨	韩　然
韩少卿	曾红芳					

校对人员（以姓氏笔画为序）

于平蓉	于同旭	马　荣	马晨晨	王　羽	王帅帅	王阳兰
王丽霞	王晓萌	王晓蕾	王赛儿	元旭津	牛慧冲	公　晗
甘秋玲	田胜聪	史未卿	丛　岚	冯　翀	吕秋莎	朱　玥
乔萌萌	刘　明	刘子怡	刘本琼	刘晓楠	齐文静	闫　妍
许梅梅	孙　娟	孙瑞静	杜赛赛	李　芳	李　娟	李　琦
李　景	李　渝	李世媛	李红菊	李若男	李盎然	杨　茜
杨学良	吴　茜	邱彩玉	何　钧	何　铭	何　敏	何思源
邹伯夏	沈乃澂	张　帆	张　敏	张向东	张亦卓	张美月
张竞凤	张梦璇	张琦玮	张媛媛	陈　雄	陈思原	陈露芸
范艳璇	罗小青	周玉凤	郑　琪	郑婧澜	郑期彤	宗伟凯
赵凤轩	胡海霞	侯鉴璇	顾海成	钱　磊	徐　玲	徐秋燕
郭晓博	黄小斌	黄雪嫚	曹则贤	崔天明	梁　瑜	葛　越
葛云霄	葛聆沨	董静娟	韩少卿	曾红芳	曾芃斐	蔡　迪
蔡军茹	Eric Leher（澳）					

Contents
目　录

1940

1945

1947

1948

1949

1956

1958

1960

1965

Volume II

Physical Evidence for the Division of Heavy Nuclei under Neutron Bombardment

O. R. Frisch

Editor's Note

Otto Hahn and Fritz Strassmann in 1938 had found convincing chemical evidence for the fission of uranium nuclei. After bombarding uranium nuclei with neutrons, they had found traces of barium nuclei, with atomic weight 56, suggesting a roughly equal splitting of the nuclei. Here Otto Frisch of the Institute of Theoretical Physics in Copenhagen gives clinching physical evidence of the process. Calculations suggested that the fission fragments should have energies of perhaps 200 million electronvolts, and so should create significant ionization in a chamber. In experiments, Frisch had detected such ionization events and, from the level of ionization, estimated the atomic weight of the fragments as about 70, not too far from 56, and certainly less than uranium's 92.

FROM chemical evidence, Hahn and Strassmann[1] conclude that radioactive barium nuclei (atomic number $Z = 56$) are produced when uranium ($Z = 92$) is bombarded by neutrons. It has been pointed out[2] that this might be explained as a result of a "fission" of the uranium nucleus, similar to the division of a droplet into two. The energy liberated in such processes was estimated to be about 200 Mev., both from mass defect considerations and from the repulsion of the two nuclei resulting from the "fission" process.

If this picture is correct, one would expect fast-moving nuclei, of atomic number about 40–50 and atomic weight 100–150, and up to 100 Mev. energy, to emerge from a layer of uranium bombarded with neutrons. In spite of their high energy, these nuclei should have a range, in air, of a few millimetres only, on account of their high effective charge (estimated to be about 20), which implies very dense ionization. Each such particle should produce a total of about three million ion pairs.

By means of a uranium-lined ionization chamber, connected to a linear amplifier, I have succeeded in demonstrating the occurrence of such bursts of ionization. The amplifier was connected to a thyratron which was biased so as to count only pulses corresponding to at least 5×10^5 ion pairs. About fifteen particles a minute were recorded when 300 mgm. of radium, mixed with beryllium, was placed one centimetre from the uranium lining. No pulses at all were recorded during repeated check runs of several hours total duration when either the neutron source or the uranium lining was removed. With the neutron source at a distance of four centimetres from the uranium lining, surrounding the source with paraffin wax enhanced the effect by a factor of two.

中子轰击导致重核分裂的物理证据

弗里施

编者按

1938 年奥托·哈恩和弗里茨·施特拉斯曼发现了铀核裂变的可信的化学证据。他们在使用中子轰击铀核之后，发现了原子序数为 56 的钡原子核，这表明铀核几乎是对等地分裂为两半。这篇文章中哥本哈根理论物理研究所的奥托·弗里施给出了这个过程的明确的物理证据。计算表明裂变碎片应该具有大约 200 兆电子伏的能量，因此应该在电离室中产生显著的电离反应。弗里施在实验中已探测到了这种相关的电离情况，从电离的量级他估算出产生的碎片的原子序数应为 70 左右，与 56 相差不远，但明显小于铀的原子序数 92。

根据化学证据，哈恩和施特拉斯曼[1]作出了当中子轰击铀（原子序数 $Z = 92$）时会产生放射性钡核（原子序数 $Z = 56$）的结论。此前我们已经指出[2]，这种现象可以解释为铀核的“裂变”，就像一个液滴一分为二那样。无论是根据质量亏损，还是根据“裂变”过程中产生的两核之间的排斥反应，都可估算出这一铀核“裂变”过程中释放出来的能量大约为 200 兆电子伏。

如果上述描述是正确的，那就可以预期，用中子轰击铀原子层时可以发出高速运动的原子核，其原子序数和原子量分别处于 40~50 和 100~150 的范围内，能量上达到 100 兆电子伏。尽管这些核的能量很高，但它们在空气中只有几毫米的射程，这是因为它们具有高的有效电荷（估计为 20），而这意味着具有极为密集的电离作用，每个这样的粒子会产生总计约 300 万个离子对。

利用一个与线性放大器连接的铀衬电离室，我成功证实了这种电离脉冲的出现。放大器与一个有偏置的闸流管相连，以便能对那些相当于至少 5×10^5 个离子对的脉冲计数。把 300 毫克混有铍的镭放置在距离铀衬一厘米处时，每分钟能记录到大约 15 个粒子。在重复进行的长达几个小时的检验测量中，无论是移走中子源还是移走铀衬，都根本记录不到任何脉冲。而当中子源距离铀衬 4 厘米远时，用石蜡包裹中子源却能使效应增加一倍。

It was checked that the number of pulses depended linearly on the strength of the neutron source; this was done in order to exclude the possibility that the pulses are produced by accidental summation of smaller pulses. When the amplifier was connected to an oscillograph, the large pulses could be seen very distinctly on the background of much smaller pulses due to the alpha particles of the uranium.

By varying the bias of the thyratron, the maximum size of pulses was found to correspond to at least two million ion pairs, or an energy loss of 70 Mev. of the particle within the chamber. Since the longest path of a particle in the chamber was three centimetres and the chamber was filled with hydrogen at atmospheric pressure, the particles must ionize so heavily, in spite of their energy of at least 70 Mev., that they can make two million ion pairs on a path equivalent to 0.8 cm. of air or less. From this it can be estimated that the ionizing particles must have an atomic weight of at least about seventy, assuming a reasonable connexion between atomic weight and effective charge. This seems to be conclusive physical evidence for the breaking up of uranium nuclei into parts of comparable size, as indicated by the experiments of Hahn and Strassmann.

Experiments with thorium instead of uranium gave quite similar results, except that surrounding the neutron source with paraffin did not enhance, but slightly diminished, the effect. This gives evidence in favour of the suggestion[2] that also in the case of thorium, some, if not all, of the activities produced by neutron bombardment[3] should be ascribed to light elements. It should be remembered that no enhancement by paraffin has been found for the activities produced in thorium[3] (except for one which is isotopic with thorium and is almost certainly produced by simple capture of the neutron).

Prof. Meitner has suggested another interesting experiment. If a metal plate is placed close to a uranium layer bombarded with neutrons, one would expect an active deposit of the light atoms emitted in the "fission" of the uranium to form on the plate. We hope to carry out such experiments, using the powerful source of neutrons which our high-tension apparatus will soon be able to provide.

(**143**, 276; 1939)

Otto Robert Frisch: Institute of Theoretical Physics, University, Copenhagen, Jan. 16.

References:

1. Hahn, O., and Strassmann, F., *Naturwiss.*, 27, 11 (1939).
2. Meitner, L., and Frisch, O. R., *Nature* [143, 239 (1939)].
3. See Meitner, L., Strassmann, F., and Hahn, O., *Z. Phys.*, **109**, 538 (1938).

现已证明脉冲数目与中子源的强度具有线性关系；这是为了排除脉冲源于偶然的较小脉冲累积而产生的可能性。当把放大器与示波器相连时，可以看到在铀发射的 α 粒子产生的非常小的脉冲背景之上有非常明显的大脉冲。

通过调节闸流管的偏压，可以发现最大脉冲幅度对应于至少 200 万个离子对，或者说相当于粒子在电离室内损失了 70 兆电子伏的能量。由于粒子在电离室内的最长路径只有 3 厘米，并且电离室内充满了一个大气压的氢气，因此，尽管粒子具有至少 70 兆电子伏的能量，但它们必定会发生强烈的电离以至于在相当于 0.8 厘米空气的路径上产生了 200 万个离子对。根据这一结果，假设原子量与有效电荷之间具有合理关联的话，我们就可以估算出电离粒子的原子序数应该至少是 70。这对铀核分裂成大小相当的两个部分似乎是个决定性的物理证据，正如哈恩和施特拉斯曼的实验结果所示。

用钍代替铀所做的实验给出了相当类似的结果，只是用石蜡包裹中子源后效应并没有增强，反而是略有减弱。先前我们曾提出 [2] 在钍的实验中用中子轰击所产生的某些（如果不是全部的话）放射性 [3] 应该归属为轻元素，上述实验结果为这一观点提供了有力的证据。同时我们还应该记住，在钍产生的放射性实验 [3] 中用石蜡包裹中子源后效应并没有增强（只有一次例外，那是因为使用了钍的某种同位素，并且几乎可以确定是由于单纯的中子俘获而产生的）。

迈特纳教授曾提出过另外一个有趣的实验。如果将一块金属板置于用中子轰击的铀层附近，就可以预期金属板上会形成铀“裂变”过程中所发射出的轻原子的放射性沉积物。我们希望实现诸如利用高压装置便能提供更强中子源的实验。

（王耀杨 翻译；张焕乔 鲍重光 审稿）

The Fundamental Length Introduced by the Theory of the Mesotron (Meson)*

H. J. Bhabha

Editor's Note

By 1939, physicists understood that the framework in which they discussed the behaviour of particles such as electrons and protons was in some sense incomplete. There was particular concern over the particles, called mesotrons, which were intermediate in mass between electrons and protons, but which carried electrical charge. Homi J. Bhabha was an Indian graduate student at Cambridge and Bristol. He was among the first to appreciate the wider implications of mesons. In the 1960s Bhabha returned to India and became the head of the Indian Atomic Energy Commission. He was killed in an air crash in 1966.

IT is well known that the vector theory of the meson[1] contains a fundamental length in the interaction of mesons with protons and neutrons determined by the fact that the mass of the meson appears explicitly in the denominator of some of the interaction terms. This circumstance has the result that in those elementary processes in which the momentum change is large compared with mc, m being the mass of the meson, the interaction becomes very large, leading to Heisenberg explosions, and to greater divergences in some second-order effects than is the case in radiation theory. This has led Heitler[2] and others to the view that the meson theory in its present form is quite incorrect for meson energies larger than about mc^2, and Heisenberg[3] to the position that quantum mechanics is competent to deal accurately with only those elementary processes in which the condition[4] due to Wataghin,

$$\left| \left(\frac{E_1 - E_2}{c} \right)^2 - (\boldsymbol{p}_1 - \boldsymbol{p}_2)^2 \right| \ll \left(\frac{\hbar}{r_0} \right)^2, \tag{1}$$

is satisfied, E and p being the initial and final energy and momenta of a particle concerned in the process, and r_0 a fundamental length of the order $\hbar/mc$. The purpose of this note is to bring forward an argument which, it seems to me, shows first that the limitation of quantum mechanics by the condition (1), if true, cannot be based on the explosions as derivable from the theory of the meson, and secondly, to throw some doubt on (1) itself as a limit to the correctness of quantum mechanics.

---

* The name "mesotron" has been suggested by Anderson and Nedder-meyer (*Nature*, **142**, 874; 1938) for the new particle found in cosmic radiation with a mass intermediate between that of the electron and proton. It is felt that the "tr" in this word is redundant, since it does not belong to the Greek root "meso" for middle; the "tr" in neutron and electron belong, of course, to the roots "neutr" and "electra". In these circumstances, it seems better to follow the suggestion of Bohr and to use electron to denote particles of electronic mass independently of their charge, and negaton and positon to differentiate between the sign of the charge. It would therefore be more logical and also shorter to call the new particle a meson instead of a mesotron.

介子理论中的基本长度*

巴巴

编者按

1939 年，物理学家认识到，讨论电子和质子等粒子行为的理论框架在某种意义上是不够完善的，尤其是对于一种叫介子的粒子，它的质量介于电子和质子之间而且携带电荷。霍米·巴巴是剑桥大学和布里斯托尔大学的印度籍研究生。他也是第一批对介子具有的更广泛含义进行认真思考的学者之一。20 世纪 60 年代巴巴回到印度并成为印度原子能委员会的主席。1966 年他在一次空难中不幸去世。

众所周知，介子矢量理论[1]包含介子与质子和中子相互作用的基本长度，它是由介子质量出现在一些相互作用项的分母中的事实所决定的。这种情况使得在一些动量变化大于 mc（m 为介子质量）的基本过程中相互作用变得非常大，从而导致了海森堡爆炸，同时也使一些二阶效应与辐射理论所得结果相比更加发散。这使海特勒[2]和其他一些人注意到介子理论的现有形式因给出介子能量大于 mc^2 而是完全错误的，海森堡[3]也指出，量子力学仅可以完美地解决那些满足瓦塔金条件[4]的基本过程，该条件如下：

$$\left|\left(\frac{E_1-E_2}{c}\right)^2-(\overline{p}_1-\overline{p}_2)^2\right| \ll \left(\frac{\hbar}{r_0}\right)^2 \tag{1}$$

式中 E 和 p 在这个过程中分别代表一个粒子初态和终态的能量和动量，r_0 为与 $\hbar/mc$ 具有同样量级的基本长度。我这篇短文的目的是要引出一个论点，首先该论点指出量子力学的应用范围如果真的仅限于条件 (1) 满足的情况，那么就不可能出现介子理论所预言的爆炸现象；其次，我对条件 (1) 本身作为量子力学正确性的限制条件表示质疑。

---

* "mesotron" 是安德森和尼德–迈耶（《自然》，142，874；1938）为在宇宙线中发现的新粒子所取的名字，该粒子的质量介于电子和质子之间。大家感觉 "tr" 在这个单词中是多余的，因为它并不属于前面表示 "中间" 的意思的希腊词根 "meso"，而中子和电子中的 "tr" 是属于词根 "neutr" 和 "electra" 的，在这种情况下最好接受玻尔的建议，即用电子表示与所带电荷正负无关的带电粒子，而用负电子和正电子来区分所带电荷的正负号。因此把这种新粒子命名为 "meson" 比将其称为 "mesotron" 更符合逻辑、也更简洁。

The argument runs as follows. Let us consider uncharged mesons[5] for simplicity, since this changes nothing essential to the argument, and consider the Hamiltonian given in A (49)[1]. The interaction (58 *a*) in this contains terms which become very large when the momentum change of the meson becomes large compared to mc in a suitable Lorenz frame. These terms, which lead to explosions, are due *as much to the transverse meson waves as to the longitudinal ones*, even in the limit when the proton may be considered to be moving non-relativistically. Further, the critical momentum above which explosions begin to appear becomes lower the smaller mc, and becomes vanishingly small when $mc \rightarrow 0$.

On the other hand, *the exact quantized equations of motion for the meson field derivable from this Hamiltonian* (A (14) and (15), with the appropriate simplifications for a neutral meson), namely,

$$G_{\mu\nu} = -\left(\frac{\partial}{\partial\chi^{\mu}}U_{\nu} - \frac{\partial}{\partial\chi^{\nu}}U_{\mu}\right) + \frac{g_2}{\hbar c}\psi + \gamma^{\mu}\gamma^{\nu}\tau_3\psi$$

$$\frac{\partial}{\partial\chi_{\mu}}b_{\mu\nu} = \frac{m^2c^2}{\hbar^2}U_{\nu} + \frac{g_1}{\hbar c}\psi + \gamma^{\nu}\tau_3\psi$$

go over continuously into the Maxwell equations when $m \rightarrow 0$. But one knows from electrodynamics that although there are circumstances in which the emission of a large number of quanta may be more probable than the emission of a single quantum, as in the so-called "infra-red catastrophe", this in no way sets a limit to the accuracy of quantum mechanics *and does not interfere with the calculation of less probable processes by the methods of perturbation theory*. Moreover, it is just those processes where the emission of a large number of quanta is very probable which can be calculated classically.

In view of the above circumstances, we must conclude that the appearance of the fundamental length determined by the mass of the meson in the interaction term in no way sets a limit to the accuracy of quantum mechanics. For example, in the collision of two protons with energy very large compared to mc^2, the probability becomes large for the simultaneous emission of a large number of mesons, which is the analogue of the "infra-red catastrophe" for quanta of finite rest mass, and hence quantum mechanics is none the less competent to deal with it. It can similarly be shown that *we can calculate the production of large explosions to a high degree of accuracy by treating the meson field quantities classically*, that is, as non-quantized magnitudes, for since mesons satisfy Einstein-Bose statistics, the meson field becomes a classical one just in the case where we are dealing with a large number of mesons.

Hence if a fundamental length r_0 exists which limits the applicability of present quantum mechanics to the cases satisfying (1), this length r_0 has nothing to do with the mass of the meson or the appearance of explosions. Quantum mechanics in its present form cannot be strictly valid since it leads to divergent results connected with the self-energies of point charges; but these limitations are probably due to the fact that it is not the quantization of the correct classical equations for point charges, and not to the existence of a fundamental length r_0. These equations have only recently been given by Dirac[6] and their quantization has not yet appeared.

这个论点具体如下：为了简单起见，我们考虑不带电荷的介子[5]的情况，因为带电情况的变化对这个论题没有本质上的影响，另外我们还用到 A(49) 式[1]中给出的哈密顿量。在适当的洛伦兹框架下，当介子动量的变化大到可以和 mc 相比拟时，表示相互作用的 (58 *a*) 式中的一些项就会变得非常大。即使在质子运动可能被认为是非相对论性的极限情况下，这些可引发爆炸的项也是一样由**横向介子波和纵向介子波**引起的。此外，爆炸开始出现时的临界动量在 mc 变小时也会变小，当 mc 趋于 0 时，临界动量会变得几乎为 0。

另一方面，**由哈密顿量**（A(14) 和 (15)，并在中性介子的假设下作适当简化）**导出的严格介子场量子化运动方程**，即：

$$G_{\mu\nu} = -\left(\frac{\partial}{\partial \chi^{\mu}} U_{\nu} - \frac{\partial}{\partial \chi^{\nu}} U_{\mu}\right) + \frac{g_2}{\hbar c}\psi + \gamma^{\mu}\gamma^{\nu}\tau_3\psi$$

$$\frac{\partial}{\partial \chi_{\mu}} b_{\mu\nu} = \frac{m^2 c^2}{\hbar^2} U_{\nu} + \frac{g_1}{\hbar c}\psi + \gamma^{\nu}\tau_3\psi$$

在 m 趋于 0 时会连续地过渡到麦克斯韦方程。但是根据电动力学，我们知道正如所谓的“红外灾难”，虽然发射大量量子的概率要大于发射单个量子的概率，但这决不会限制量子力学的适用范围，**也不会干预应用微扰理论计算小概率的过程**。而且正是那些可以用经典方法计算的过程有可能发射大量量子。

我们可以由此得出结论，取决于相互作用项中介子质量的基本长度绝不可能对量子力学的精确性有所限制。举例来讲，当两个能量大到足以和 mc^2 相比拟的质子发生碰撞的时候，同时发射大量介子的可能性就会变大，这类似于与具有有限静止质量的量子相关的“红外灾难”，因此量子力学仍然可以解决这类问题。同样我们可以指出，**用经典方法处理介子场量能够精确地计算出大爆炸的结果**，也就是说，既然介子满足爱因斯坦 – 玻色统计，那么当我们处理包括大量介子的问题时，对于非量子化的场量，介子场就变成了一个经典场。

因此，如果基本长度 r_0 的出现使现有的量子力学理论只能应用于满足式 (1) 的情况，那么这个长度 r_0 就与介子质量或爆炸的出现无关。量子力学的现有形式不可能是严格正确的，因为它在涉及点电荷自能问题时会产生发散的结果；但是，这种局限性可能是由于正确的经典点电荷方程没有量子化，而不是因为引入了基本长度 r_0。最近只有狄拉克[6]给出了这些方程，不过尚未得到这些方程的量子化形式。

Accordingly, we might expect that very fast protons (or neutrons) would produce explosions consisting of mesons of momenta roughly *mc*, while mesons with energy much larger than the proton rest-energy would *not* do so, and their scattering by protons would also decrease with increasing energy, in analogy with the Compton effect.

It can be shown that the classical retarded meson field and potentials due to the world line of a classically moving proton or neutron can be written as the sum of two parts. The first part has exactly the form that the corresponding electromagnetic quantities would have for a point charge and point dipole (represented by a six-vector) moving along a classical world line, and does not contain the mass of the meson. The second part has no singularity at any point of space including the world line of the proton, and goes to zero as the mass of the meson $m \to 0$. The meson singularities are therefore *identical* with the electromagnetic singularities, and it is possible to eliminate these to the same degree and in the same way as has been done by Dirac[6] for the electromagnetic field of a point charge.

The detailed calculations will be published elsewhere.

(**143**, 276-277; 1939)

H. J. Bhabha: Gonville and Caius College, Cambridge, Dec. 17.

References:

1. Kemmer, *Nature*, **141**, 116 (1938); *Proc. Roy. Soc.*, A, **166**, 127 (1938). Fröhlich, Heitler and Kemmer, *Proc. Roy. Soc.*, A, **166**, 154(1938). Bhabha, *Nature*, **141**, 117 (1938); *Proc. Roy. Soc.*, A, **166**, 501 (1938), referred to above as A; Yukawa, Sakata, and Taketani, *Proc. Phys. Math. Soc. Japan*. **20**, 319 (1938). Stueckelberg, *Helv. Phys. Acta*, **11**, 299 (1938).
2. Heitler, *Proc. Roy. Soc.*, A, **166** (1938).
3. Heisenberg, *Z. Phys.*, **110**, 251 (1938).
4. Wataghin, *Z. Phys.*, **66**, 650 (1931); **73**, 126 (1931).
5. Kemmer, *Proc. Camb. Phil.* Soc., **34**, 354 (1938).
6. Dirac, *Proc. Roy. Soc.*, A, **167**, 148 (1938). See also Pryce, *Proc. Roy, Soc.*, A, **168**, 389 (1938).

因此，我们可以预测快速运动的质子（或中子）会引发爆炸，放出动量大约为mc的介子，而当介子能量远大于质子的静止能量时则**不**可能引发爆炸，而且能量越大，质子对介子的散射越弱，这与康普顿效应类似。

这表明，由经典运动的质子或中子的世界线所决定的经典的推迟介子场和势能可以写成两部分之和。第一部分的具体形式为一个点电荷和点偶极子（用一个六维矢量描述）沿经典的世界线运动所对应的电磁场量，而不包含介子质量。第二部分在包括质子世界线在内的任何空间点上都不会出现奇异点，而且当介子质量m趋于0时它也变为0。因此，介子的奇异点与电磁场的奇异点是**一样的**，可以参照狄拉克[6]对点电荷电磁场的处理方法以同样的方式、在同一程度上消除这些奇异点。

详细的计算结果将发表在其他地方。

（胡雪兰 翻译；厉光烈 审稿）

Energy Obtained by Transmutation

Editor's Note

This editorial comments on the possibility of deriving energy, perhaps in explosive form, from nuclear transmutation. The likelihood, it suggests, is remote. Meanwhile, recent experiments had demonstrated the artificial transmutation of most elements, including gold, though without any repercussions for world financial stability. The goal of harnessing nuclear energy seemed far off, partly because significant energy can be released only in processes involving the heaviest and rare elements. Yet the first atomic weapon was detonated only six years later, made possible by the enrichment of vast quantities of uranium with the easily fissionable isotope ^{235}U, which constitutes only 0.7% of the element naturally.

MR. Robert D. Potter, of "Science Service", Washington, D.C., points out that the confirmation of the artificial breakdown of uranium announced in New York (see also *Nature*, Feb. 11, p. 233) is in the direct succession of experiments carried out in recent years on the transmutation of the elements. For centuries, alchemists had dreamed of transmuting base metals into gold. It was imagined that enormous wealth would be at hand for the discoverer of this transmutation, and dire forecasts of the effects of this discovery were made, such as a complete revolution on the financial pattern of the world. We know that this transmutation has now been achieved for most of the known chemical elements. Transmutation's biggest result is the theoretical incentive it has provided for further physical researches. In a similar way, the dream of releasing the large amounts of energy locked inside atoms has been in the minds of men for many years. When the most efficient transformation of energy takes place in the atom of uranium so that a neutron can slip into it, the energy released is only one fifteenth of that required to bring it about. In fact, neutrons are so easily absorbed by all atomic nuclei that many of the neutrons produced with such poor efficiency will only go into atoms other than uranium. There need be little fear of an explosion in Nature due to uranium. The very heavy elements, in which such an energy release can be secured, occur only in very small amounts in the Earth's crust. The release of atomic energy can only be achieved by direct experiment with this end in view and with elaborate laboratory apparatus.

(**143**, 328; 1939)

嬗变产生的能量

编者按

这篇社论就从核嬗变中获取能量（或许以爆炸的形式）的可能性进行了评论。文中提出，这种可能性还很遥远。同时，最近的实验已经证实了包括金在内的大多数元素可实现人工嬗变，而这并未对世界金融的稳定造成影响。利用核能的想法似乎遥不可及，部分原因是只有最重且稀有的那些元素的嬗变过程才能释放出巨大的能量。然而仅在六年后，通过大量天然铀富集获得的铀–235使得第一颗原子弹爆炸成功，而这种易裂变的同位素铀–235仅占天然铀元素的0.7%。

华盛顿特区科学服务社的罗伯特·波特先生指出，纽约宣布的铀的人工嬗变（也见《自然》2月11日，第233页）是近年来对元素嬗变问题所进行的实验的直接延续。几个世纪以来，炼金术士都梦想将普通金属转变成金子。有人认为发现了这种嬗变，巨大的财富就会唾手可得，有人还对这个发现的结果做出了极端的预测，例如这将是一次世界金融格局的彻底变革。我们知道，现在对于大多数已知的化学元素，这类嬗变都已实现。嬗变的最大成效是，它为进一步的物理研究提供了理论发展的动力。同样，将禁锢在原子内部的大量能量释放出来，一直是人们心中多年以来的梦想。当铀原子中发生最有效的能量转化，并使一个中子进入其中时，所释放的能量仅是促使该过程发生所需能量的1/15。实际上，中子很容易被各种原子核吸收，以至于许多如此低效的中子只能进入其他原子而不能进入铀。不必过于担心自然界的铀会引起爆炸。能够确保这种能量释放得非常重的元素在地壳中的含量很少。原子能量的释放只能用直接的实验来实现，为达到这样的目的而做的实验是要在精心制作的实验装置中进行的。

（沈乃澂 翻译；朱永生 审稿）

Cause, Purpose and Economy of Natural Laws*: Minimum Principles in Physics

M. Born

Editor's Note

In the 1920s, Max Born, based at the University of Göttingen, had been one of the founders of quantum mechanics. The German government's decision that Jewish people could no longer teach at German universities made it necessary for him to leave Germany; after a brief stay at Cambridge he settled in Edinburgh where he and his family remained for several decades. Born was an exceptionally lucid lecturer. This general lecture on the laws of physics is a model of its kind.

WITHOUT claiming to be a classical scholar, I think that the earliest reference in literature to the problems which I wish to treat here is contained in Virgil's "Aeneid", Book I, line 369, in the words "taurino quantum possent circumdare tergo."

When Dido landed at the site of the citadel of Carthage, she opened negotiations with the inhabitants for some land and was offered for her money only as much as she could surround with a bull's hide. But the astute woman cut the bull's hide into narrow strips, joined them end-to-end and with this long string encompassed a considerable piece of land, the nucleus of her kingdom. To do this she had evidently to solve a mathematical question—the celebrated "problem of Dido": to find a closed curve of given circumference having maximum area. We do not know how she solved it, by trial, by reasoning, or by intuition. In any event, the correct answer is not difficult to guess: it is the circle. But the mathematical proof of this fact has only been attained by modern mathematical methods.

In saying that the first appearance of this kind of problem in literature is that quoted above I am not, of course, suggesting that problems of minima and maxima had never occurred before in the life of mankind. In fact, nearly every application of reason to a definite practical purpose is more or less an attempt to solve such a problem: to get the greatest effect from a given effort, or, putting it the other way round, to get a desired effect with the smallest effort. We see from this double formulation of the same problem that there is no essential distinction between maximum and minimum; we can speak shortly of an "extremum", and "extremal" problems.

It was during Isaac Newton's lifetime, at the end of the seventeenth century, that geometrical and mechanical problems of extremals began to interest mathematicians,

* Substance of the Friday evening discourse delivered at the Royal Institution on February 10.

自然规律的起源、宗旨和经济性*：物理学的极小原理

玻恩

编者按

20世纪20年代，在哥廷根大学工作的马克斯·玻恩已经是量子力学的奠基人之一了。后来，德国政府关于犹太人不能再在德国大学任教的决定使他不得不离开德国；在剑桥短暂停留后，他最终定居爱丁堡，并和其家人在那里住了几十年。玻恩是一位思路极为清晰的讲演者，这篇关于物理定律的讲演就是其中的一个典范。

我并不是要表明自己是一名古典主义学者，不过我认为有关这里我要研究的问题的最早记录恐怕要追溯到维吉尔的《埃涅阿斯纪》，即第一册第369行所写的“用公牛皮围一块尽可能大的地”。

当狄多来到迦太基古城时，她与当地的土著人谈判，用一定量的金币换取一头公牛的牛皮能够围起来的土地。这个精明的女人把公牛皮割成细长条并将它们首尾相连，最终圈得一片相当大的土地，这就是她王国的核心。为了做到这点，显然她得解决这样一个数学问题——著名的“狄多问题”：如何使给定周长的闭合曲线所围成的面积达到最大。我们并不知道她是如何解决这个问题的，可能用试验或推理，也可能是靠直觉。无论如何，我们不难猜到这个问题的正确答案是：一个圆。但直到现代数学方法的出现才最终从数学方面给出了证明。

上面提到的内容是第一次在文字记载出现这类问题的相关记录，当然我不是指在此记录之前的人类历史中从未出现过这类求极小值和极大值的问题。实际上，每次应用推理达到某个明确实际的目标时，我们都或多或少在试图解决这样的问题：在一定的努力的基础上得到最大的效果，换句话说，即以最小的代价得到预期的效果。从以上对同一个问题的两种表述方式中我们看到，求极大值与极小值的问题之间并没有本质的区别；我们可以简单地称之为“极值”或“极值函数”问题。

在17世纪末的艾萨克·牛顿时代，极值函数的几何问题和力学问题开始引起数学家们的兴趣，在牛顿去世（1727年）后不久，自然界的意义或经济性等形而上学

---

*2月10日周五晚上玻恩在英国皇家研究院发表演说的主要内容。

and shortly after Newton's death (1727) the metaphysical idea of purpose or economy in Nature was linked up with them.

One of the simplest examples is the optical "law of reflection" which can be expressed as a minimum principle: the beam of light from a point P_1 to another point P_2 selects just that reflecting point Q which makes the total path P_1Q+QP_2 as short as possible. The light behaves as if each beam had a tendency to contract, and the French philosopher, Fermat, has shown that all the laws of geometrical optics can be reduced to the same principle. Light moves like a tired messenger boy who has to reach definite destinations and carefully chooses the shortest way possible. Are we to consider this interpretation as accidental, or are we to see in it a deeper metaphysical significance? Before we can form a judgment, we must learn more about the facts and consider other cases.

The straight line is the shortest connexion between two points in space. But if we travel on our earth, we can never go exactly in a straight line since the earth's surface is not plane. The best we can do is to follow a great circle, which is the curve in which the sphere is intersected by a plane passing through the centre. The globe, however, is not an exact sphere, but is slightly flattened at the poles and bulges at the equator. What, then, about the shortest line on such a surface?

Gauss hit on this problem when occupied with a geodetic triangulation of his country, the electorate of Hanover. He attacked the problem from the most general point of view and investigated the shortest lines on arbitrary surfaces. But in remembrance of his starting point he called these lines "geodesics". They are in many ways of fundamental importance for physics.

Let us consider a point P on a surface and all curves through P which have the same direction at P. It is evident that there is among them a "straightest curve", that is, one with the smallest curvature. Hence the geodesic can be characterized by two somewhat different minimum properties: one which can be called a "local" or "differential" property, namely, to be as little curved as possible at a given point for a given direction; and the other, which can be called "total" or "integral", namely, to be the shortest path between two points on the surface.

This dualism between "local" and "total" laws appears not only in this simple geometrical problem, but also has a much wider application in physics. It lies at the root of the old controversy whether forces act directly at a distance (as assumed in Newton's theory of gravitation and the older forms of the electric and magnetic theories), or whether they act only from point to point (as in Faraday's and Maxwell's theory of electromagnetism and all modern field theories).

There seems to be no objection to extremal laws of the local type; but those of the integral type make our modern mind feel uneasy: although we understand that the particle may choose at a given instant to proceed on the straightest path, we cannot see how it can

的思想逐渐与这些极值问题联系起来了。

最简单的一个例子是光学中的“反射定律”，这个定律可以表述成极小原理：从某一点 P_1 传播到另一点 P_2 的光束所选择的反射点 Q 要满足总路程 $P_1Q + QP_2$ 为最短。光的行为性质表现为每束光都具有缩短路程的趋势，法国哲学家费马已经指出，几何光学的所有规律都可以归纳为这一原理。光的运动就像一个疲劳的信使，他会在到达确定的目的地之前谨慎地选择尽可能最短的路程。我们如何理解以上的解释呢？认为这是一种偶然的情况，还是我们可以从中发现更深刻的形而上学方面的意义？在我们做出判断之前，我们必须了解更多的事实，并考虑其他更多的情况。

空间中两点之间最短的连线是直线。但如果我们在地球上运动时，我们永远不能精确地沿直线行进，这是因为地球的表面不是平面。而最好的选择是沿着大圆的路径，这个大圆是大圆所在的球与某一穿过球心的平面相交的曲线。然而地球并不是一个精确的球体，它的两极稍微有些扁平、赤道略微凸起。那么在这样的球面上，最短的路线是什么样子的呢？

高斯利用大地三角测量方法对他所在的德国汉诺威选区进行测绘时也碰到了这个问题。他从最普遍的观点出发来处理问题，并研究了任意表面上最短的线。为了纪念他处理这个问题的出发点，他仍将这些最短的线称为“测地线”。对于物理学来说，“测地线”在许多方面都具有重要意义。

我们考虑球体表面上的一点 P，且通过 P 的所有曲线在 P 处方向相同。显然，在它们之中有一条“最直的曲线”，即曲率最小的曲线。因此，测地线可以用两种稍微不同的最小性原理来表征：第一种可被称为“局部的”或“微分的”性质，即经过某个定点并给定方向的曲线，其曲率最小；第二种被称为“整体的”或“积分的”性质，即表面上的两点之间路程最短。

“局部的”和“整体的”定律之间的二元论不仅出现在这类简单的几何问题中，而且在物理学中有着更加广泛的应用。它在早期争论物理学的基础时就已经出现，即力的作用是直接的超距作用（如牛顿引力理论和电磁理论早期形式中的假设），还是只能是点到点的作用（如法拉第和麦克斯韦的电磁学理论及所有的现代场论）。

人们对局部形式的极值定律似乎并无反对意见；但以我们现在的思维来说，我们还是不太能自然地接受总体形式的极值定律：虽然我们知道粒子将在给定的瞬间

quickly compare all possible motions to a distant position and choose the shortest one—this sounds altogether too metaphysical.

But before we follow out this line of thought, we must convince ourselves that minimum properties appear in all parts of physics, and that they are not only correct but also very useful and suggestive formulations of physical laws.

One field in which a minimum principle is of unquestionable utility is statics, the doctrine of the equilibrium of all kinds of systems under any forces. The centre of gravity tends to descend as far as possible; to find the configuration of stable equilibrium, one has only to look for the minimum of the height of the centre of gravity. This height, multiplied by the force of gravity, is called potential energy.

A chain hanging from both ends assumes a definite shape, which is determined by the condition that the height of the centre of gravity is a minimum. If the chain has very many links, we get a curve called the catenary. It can readily be shown by means of a heavy chain, the centre of gravity of which is marked by a construction of light levers, that disturbance of the equilibrium of the chain in any arbitrary way causes the centre of gravity of the chain to rise.

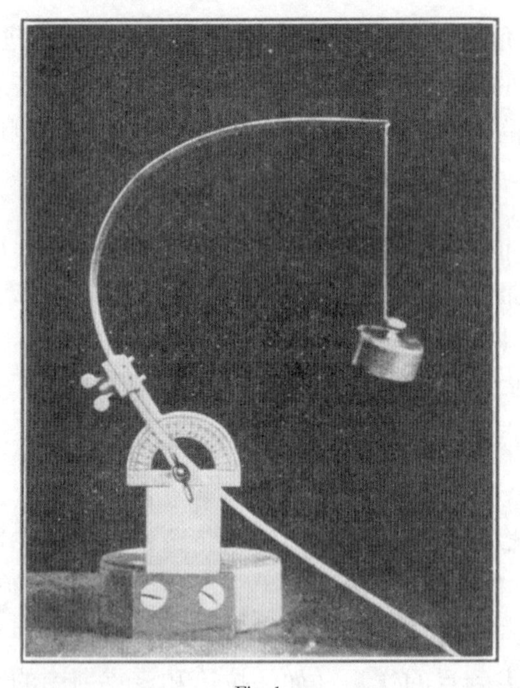

Fig. 1

Fig. 1 illustrates an example where gravity competes with another force, elasticity. A steel tape is clamped at one end and carries a weight at the other. This weight is pulled downwards by gravity, while the tape tries to resist bending in virtue of its elasticity. This

沿直线行进，但是我们并不明白它是如何迅速地对到达远处某个地方所有可能的路线进行对比并从中选择最短的一条的——这些听起来似乎都很形而上学。

但我们在探究以上想法之前，我们自己必须深信，在物理学的所有方面都存在最小性原理，它们不仅是正确的，而且是物理定律中很有用且具有启发性的公式。

毫无疑问，静力学，即各类系统在力的作用下最终都将处于平衡态的学说，成功应用了极小原理。重心趋向于处在尽量低的位置；为了得到稳定的平衡态结构，我们只能去寻求重心高度的极小值。重力乘以这个高度将得到所谓的势能。

两端悬挂的一条链将呈现出确定的形状，这个形状是在重心高度取极小值的条件下得到的。如果这条链有很多链环，我们将得到被称为悬链线的曲线形状。这很容易通过一条重的链条表示出来，它的重心是用光杠杆的测量方法进行标记的，用任何方式对链条的平衡位置进行扰动都会导致链条重心的上升。

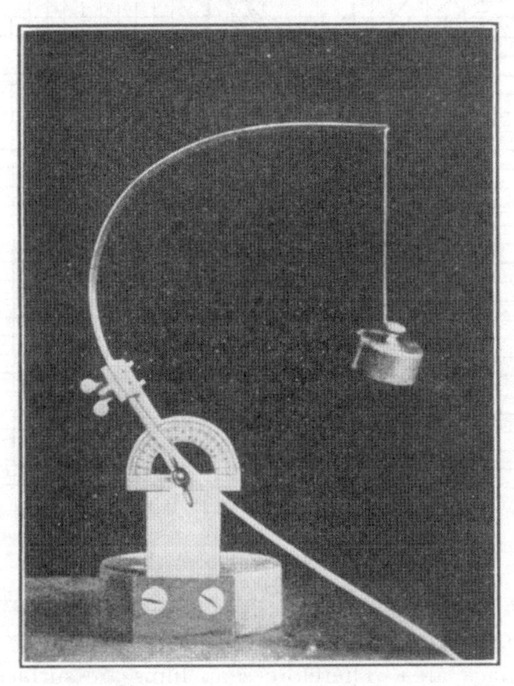

图 1

图 1 描述了一个重力与另一种力即弹力相互竞争的例子。一根钢尺一端被夹紧固定住，另一端悬挂着一个砝码。这个砝码被重力往下拉，而钢尺由于具有弹性而

elastic force also has a potential energy; for a definite amount of work must be done to bend the tape into a given curved shape. Now there are definite positions of equilibrium in which the total energy, that of gravitation plus that of elasticity, is as small as possible. Generally there are two such positions. Changing the clamping angle carefully, a position is found when a jump suddenly occurs from one position to another on the opposite side. This instability is determined by the condition of minimum energy. We can summarize the facts connected with the limits of stability by drawing a graph, Fig. 2, not of the elastic lines themselves, but by plotting the angle of inclination against the distance from the free end. We obtain wave-shaped curves, all starting horizontally from the line representing the end carrying the weight. These curves have an envelope which separates the regions in which one or several curves are going through each point, and calculation shows that this envelope is just the limit of stability. We shall return to this example later when discussing the minimum principles of dynamics.

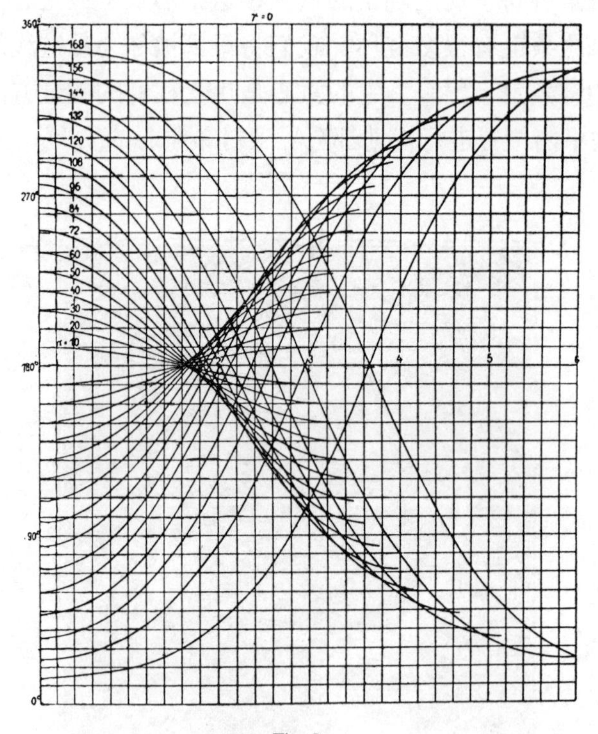

Fig. 2

Another example of the statical principle of minimum energy is provided by soap bubbles. Soap films have the property of contracting as much as possible; the potential energy is proportional to the surface area. Therefore soap films are surfaces of smallest area, or minimal surfaces. Nature is an expert mathematician, quickly finding the solution.

These experiments are not merely pretty toys without serious background. They have been chosen only for the sake of illustration. The real importance of the principle of minimum energy can scarcely be exaggerated. All engineering constructions are based on it, and also

抵制弯曲。因为必须做一定量的功才能使钢尺弯曲成一定形状的弧线，所以弹力也有相应的势能。事实上有一些确定的平衡位置，其总的能量即引力和弹力相应的势能之和取最小值。通常存在两个这样的平衡位置。慢慢地改变钢尺被夹的角度，我们可以找到这样一个位置，该位置上的钢尺可能突然朝反方向弯曲成相反的形状。这种不稳定性是由能量极小值的条件所决定的。我们总结了与稳定性极限相关的事实并用图形（图 2）表示出来，图 2 所画的并不是弹性曲线本身，而是倾斜角与自由端到固定点距离两者的关系曲线。我们得到了波形曲线，这些曲线都是一端挂着砝码的曲线沿水平方向的变化曲线。这些曲线具有一个包络面，它把通过同一个点的一条或几条曲线与其他曲线分隔开来，相关的计算表明，包络面就是稳定性的极限条件。后面讨论到动力学的极小原理时我们将回来继续研究这个例子。

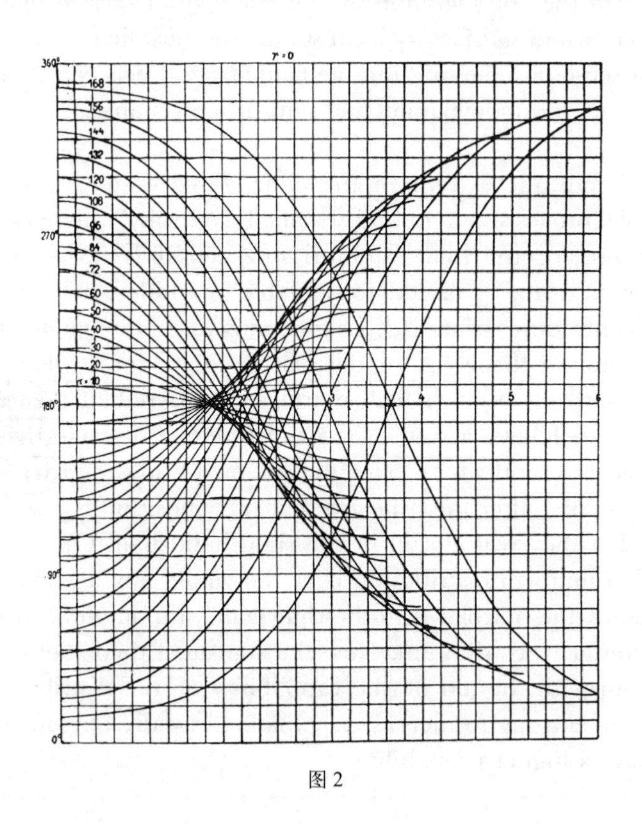

图 2

另一个证明极小能量静态原理的例子是肥皂泡。肥皂泡具有尽可能收缩变小的性质；其势能与表面面积成正比。因此，肥皂泡具有最小范围的表面，或者说极小的表面。自然界是一位专业的数学家，它很快就找到了这个问题的解决办法。

这些实验并不是不能应用于实际的玩具模型。选择它们只是为了便于说明。我们绝对没有夸大极小能量原理的实际重要性。它是所有工程建筑的基础，也是物理

all structural problems in physics and chemistry.

Models of crystal lattices provide examples of this. A crystal is a regular arrangement of atoms of definite kinds in space. Now the models representing the crystal lattices of two chemically similar compounds, namely, common salt (NaCl) and caesium chloride (CsCl) are different in structure. Why are they different? Because, as the caesium atoms are much larger than the sodium atoms, the potential energy differs in the two cases and its minimum is attained for different configurations.

Considerations of this kind, more or less quantitative, enable us to understand a great number of facts about the internal structure of solid matter.

Before we proceed to the consideration of minimum principles in dynamics, where the situation is not as clear and satisfactory as in statics, we must first mention another part of physics which in a sense occupies an intermediate position between statics and dynamics. It is the theory of heat, thermodynamics and statistical mechanics.

There is a very important extremum principle, discovered by Lord Kelvin, which governs irreversible processes: a quantity called "entropy" increases in the process and has a maximum for the final equilibrium state. It is not easy to describe this miraculous entropy in terms of directly observable quantities, such as volume, pressure, temperature, concentration, heat. But its meaning is immediately obvious from the point of view of atomic theory. A model will facilitate the explanation. Take a flat box like a little billiard table into which marbles can be put. If I carefully place them in the right-hand half, I have a state of partial order; if I shake the box they spread out over the whole box and attain a configuration of lower order. If I throw some marbles into the box one after the other so that their position is purely accidental, it is very improbable that they will all fall in the right-hand half. One can easily calculate the probability of a uniform distribution over the whole box as compared with one in which the majority of the marbles is in the right half; and one finds overwhelming odds in favour of the uniform distribution. Now the statistical theory of heat interprets the entropy of a system with the aid of the probability of the distribution of the atoms, and the tendency of entropy to increase is explained by the obvious fact that states of higher disorder have a higher probability.

Let us now come back to the minimum principles of dynamics.

The first problem of this kind—first both in historical order and in order of simplicity—was formulated at the end of the seventeenth century by Johann Bernoulli of Basle, one of a great family which produced many famous scholars and especially mathematicians. It is the problem of the curve of quickest descent or brachistochrone: given two points at different levels, not in the same vertical line, to determine a connecting curve in such a way that the time taken by a body to slide without friction under the action of gravity

学和化学中所有结构问题的基础。

晶格模型提供了这类例子。晶体是确定种类的原子在空间中规则排列形成的。由两种化学性质类似的化合物即普通的盐（NaCl）和氯化铯（CsCl）各自组成的晶格的模型在结构上是不同的。为什么二者会存在这样的差别呢？因为铯原子的尺寸远大于钠原子，所以两种情况下的势能不一样，而势能取极小值时对应的原子构型也不一样。

以上这类问题的考虑或多或少能从定量的角度帮助我们了解很多固体物质内部结构的情况。

由于动力学中的极小原理并不像静力学中的那样清楚和令人满意，因此，在开始考虑动力学的极小原理之前，我们有必要先提到物理学的另一部分，它在某种意义上来说是介于静力学和动力学之间的。这就是热学、热力学和统计力学的理论。

开尔文勋爵发现了控制不可逆过程的很重要的极值原理：被称为“熵”的物理量在不可逆过程中是不断增大的，并在最终的平衡态下达到最大值。很难用直接的可观测量，如体积、压力、温度、浓度和热量等来描述这个不可思议的熵。但是我们可以很容易地从原子理论的角度得到熵的物理意义。以下这个模型将帮助我进一步解释熵的意义。取一个如同台球桌那样扁平的可放入弹球的盒子。如果我们小心地把这些弹球摆在盒子的右半部分，这时得到的状态是部分有序；如果我们摇动盒子使这些弹球分布在整个盒子的范围内，得到的是一个更低序的状态。如果我将这些弹球一个接一个地扔进盒子里，则每个弹球的位置都是随机的，就不太可能出现所有的弹球都落到右半部分的情况了。我们可以很容易地计算出弹球在整个盒子内均匀分布的概率，并把这个概率与大部分弹球都位于右半部分的概率进行比较；人们发现均匀分布的情况存在压倒性的优势。热学的统计理论借助原子的分布概率解释了系统中的熵，而我们可以由越无序的状态出现的概率越大这个明显的事实来解释熵增加的趋势。

现在我们回来继续讨论动力学极小原理。

17 世纪末巴塞尔的约翰·伯努利阐述了我们要讨论的这类问题中的一个，这个问题不仅在历史顺序上是最早出现的，在这类问题中也是最简单的，伯努利家族是一个大家族，并且产生了许多著名的学者，尤其是数学家。这是一个关于最快速下降曲线或者说捷线的问题：给定两个不同水平高度上的点，它们不在同一竖直线上，为了确定这样一条曲线，使物体在重力作用下从较高点沿着该曲线无摩擦地滑到较

from the higher point to the lower is a minimum—compared of course with all possible curves through the two points. With a model having three paths, namely, a straight line, an are of a circle and an intermediate curve known as a cycloid, it can be shown that it is not the straight line on which a rolling ball arrives first, nor the steep descent of the circular arc, but the cycloid. If you were to try with any other curves, you would always find the same result, for the cycloid has been constructed according to the theoretical calculation.

The determination of this brachistochronic property of the cycloid was a very satisfactory piece of mathematics. It attracted much attention, and there is no philosopher of this period who did not test his analytical powers by solving similar extremal problems. Another member of the Bernoulli family, Daniel Bernoulli, developed at the beginning of the eighteenth century the minimum principle of statics and applied it to the catenary and the elastics line. Encouraged by these successes, he raised the question whether it was possible to characterize the orbit, and even the motion in the orbit, of a body subject to given forces—for example, a planet—by a minimum property of the real motion as compared with all other imagined or virtual motions. He put this question to the foremost mathematician of his time, Leonard Euler, who was very much interested in it and found, in the autumn of 1743, a solution which he explained with the help of various examples in a book published in 1744. It is the basis of the *principle of least action* which has played so prominent a part in physics right up to the present time.

The history of this principle, however, is an amazing tangle of controversies, quarrels over priority and other unpleasant matters. Maupertuis, in the same year, 1744, presented a paper to the Paris Academy of Sciences in which he substituted for Fermat's optical principle of the shortest light path, which we have already discussed, a rather arbitrary hypothesis, and extended the latter, in 1746, to all kinds of motions. He never gave a satisfactory proof of his principle (which is not surprising as it is incorrect), but defended it by metaphysical arguments based on the economy of Nature. He was violently attacked by Chevalier d'Arcy in Paris, Samuel König from Bern and others, who showed that if maupertuis's principle were true, thrifty Nature would be forced in certain circumstances to spend not a minimum but a maximum of action! Euler, whose principle is quite correct, behaved rather strangely; he did not claim his own rights, but even expressed his admiration for Maupertuis's principle, which he declared to be more general. The reasons for this attitude are difficult to trace. One of them seems to have been the publication by König of a fragment of an alleged letter of Leibniz in which the principle was enunciated. The genuineness of this letter could never be proved, and it seems probable that it was a forgery designed to weaken Maupertuis's position. This may have brought Euler over to the side of Maupertuis, who was at this time president of the Berlin Academy and a special favourite of the king, Frederic II, later known as the Great. The dispute was now carried over into the sphere of the court of Sanssouci and even into the arena of politics. Voltaire, friend of Frederic, who heartily disliked the haughty president of the Academy,

低点的过程中时间最短，我们当然要对通过这两点的所有可能曲线进行比较。使用一个具有三条路径的模型，三条路径分别为直线、圆的一段弧线和被称为摆线的中间曲线，可以看到，最先到达较低点的并不是沿着直线滚动的小球，也不是沿着圆弧的陡峭弧度滚动的小球，而是沿着摆线滚动的小球。如果你尝试其他任何曲线，你将会发现结果总是一样的，因为摆线就是按照理论计算所得出的时间最短的轨迹。

确定摆线的捷线过程在数学处理方面是完美、无可挑剔的。这类极值问题受到了广泛的关注，在那个时期，没有一个哲学家不是通过解决类似的极值问题来检验自己的分析结果的。18 世纪初，伯努利家族的另一个成员丹尼尔·伯努利发展了静力学的极小原理，并成功地把该原理应用到悬链线和弹性线的情况中。在以上这些成就的鼓舞下，他提出这样一个问题：通过真实运动情况的极小性质（与其他所有想象的或可能实现的运动情况相比较），是否可以描述出给定作用力情况下物体的运动轨迹，甚至是轨迹中物体的运动情况呢？例如描述一颗行星的运动情况。他向当时最有名的数学家伦纳德·欧拉提出了这个问题，欧拉对此很感兴趣并于 1743 年的秋天给出了这个问题的答案，之后，他在 1744 年出版的书中借助各类例子对此作了进一步的解释。这就是迄今为止在物理学中一直扮演着重要角色的**最小作用量原理**的基础。

与这个最小作用量原理有关的是一段充满混乱的不可思议的历史，其中出现了发现原理的优先权之争和其他不愉快的事件。同样是在 1744 年，莫佩尔蒂向巴黎科学院提交了一篇论文，文中他把我们前面已经讨论过的费马提出的最小光程的光学原理替换成一个他自己主观的假设，并于 1746 年把他的假设推广到所有的运动情况中。他并没有对他的原理给出满意的证明（这并不奇怪，因为他的原理本身就是不正确的），而是在自然经济性的基础上以形而上学的论据进行论证。他受到了巴黎的达西爵士、伯尔尼的塞缪尔·柯尼希和其他人的猛烈攻击，他们指出，如果莫佩尔蒂的主张是正确的话，那么一向经济节俭的自然在某些情况下将被迫做最大的而不是最小的功！欧拉提出的原理是正确的，但他的行为颇为奇怪；他非但没有声明自己发现该原理的优先权，反而对莫佩尔蒂的原理表示赞赏，并宣称这个原理是更普遍的。我们很难找到欧拉持这种态度的原因。其中的一个原因似乎是由于柯尼希公开了所谓的莱布尼茨信件的一部分，该部分对最小作用量原理也进行了阐述。这封信的真实性从未得到证实，似乎很有可能是伪造的，目的是用来削弱莫佩尔蒂的地位。这也许是使欧拉站在莫佩尔蒂这一边的原因，当时的莫佩尔蒂是柏林学院的院长并且特别受当时的国王弗雷德里克二世，也就是后来所谓的弗雷德里克大帝的喜爱。这场关于最小作用量原理的争辩不但在无忧宫的范围内展开了，后来甚至进入了政界的活动舞台。弗雷德里克的朋友伏尔泰非常讨厌傲慢的柏林学院院长莫

took the side of the "underdog", König, and wrote a caustic pamphlet, "Dr. Akakia", against Maupertuis. But the king, although he thoroughly enjoyed Voltaire's witty satire, could not sacrifice his grand president and was compelled to defend Maupertuis. This led at last to the disruption of their friendship and to Voltaire's flight from Berlin, as described in many biographies of Frederic and of Voltaire.

The curse of confusion has rested for a long period on the principle of least action. Lagrange, whose work was the culmination of the development of Newton's dynamics, gives an unsatisfactory formulation of the principle. Jacobi restricts it in such a way that the minimum condition determines the orbit correctly; the motion in the orbit must be found with the help of the energy equation. This was an important step. But the spell was at last broken by the great Irishman, Sir William Rowan Hamilton, whose principle is mathematically absolutely correct, simple and general. At the same time, it put an end to the interpretation of the principle expressing the economy of Nature. For there is, by a kind of fortunate mathematical coincidence, a statical problem for which the statical minimum principle for the potential energy coincides formally with Hamilton's principle of least action for the pendulum; this statical problem is the loaded steel tape considered before. The curves representing the angle of inclination of the elastic line as a function of the distance from the free end are exactly the same lines as those representing the angle of deflection of the pendulum as a function of time (see Fig. 2).

Now we have seen that only those regions of the graph which are simply covered by the lines correspond to a real minimum, a stable configuration of the elastic line. There are other regions, those beyond the envelope, where two or more lines pass through a given point. Only one of those lines corresponds to a real minimum. But both represent possible motions of the pendulum. Although the conditions at the ends of the elastic tape do not correspond exactly to those at the ends of the time interval in Hamilton's principle, there is this fact in common: if the length of the tape or the corresponding time interval in Hamilton's principle for the pendulum exceed a certain limit, there is more than one possible solution, and not each of them can correspond to a true minimum, though to a possible motion. In this way we come to the conclusion that the actual motion is not in every case distinguished by a genuine extremal property of action but by a less obvious mathematical property called "stationary" configuration.

Thus the interpretation in terms of economy breaks down. We may regard the idea of finding purpose and economy in natural laws as an absurd piece of anthropomorphism, a relic of a time when metaphysical thinking dominated science.

The importance of Hamilton's principle lies in a different direction altogether. It is not Nature that is economical, but science. All our knowledge starts with collecting facts; but proceeds by summarizing numerous facts by simple laws and these again by more general laws. This process is very obvious in physics. We may recall, for example, Maxwell's electromagnetic theory of light, by which optics became a branch of general electrodynamics. The minimum principles are a very powerful means to this end of

佩尔蒂，因此，他站到了“受压迫者”柯尼希的一边，并写了《阿卡基亚博士》这本小册子来讽刺莫佩尔蒂。虽然弗雷德里克非常欣赏伏尔泰机智的讽刺风格，但是为了不让自己尊贵的院长的利益受到损害，他不得不替莫佩尔蒂辩护。这最终导致了弗雷德里克和伏尔泰的友谊破灭，随后伏尔泰逃离柏林，这些在许多关于弗雷德里克或伏尔泰的传记中都有所描述。

如同受到了诅咒一样，人们在最小作用量原理上困惑了很长时间。拉格朗日的工作把牛顿力学发展到了极致，但是他给出的最小作用量原理的形式并不令人满意。雅可比给拉格朗日的形式加上了限制条件使得极小值条件可以给出物体正确的运动轨迹；但必须知道能量方程才能得到轨迹中具体的运动情况。这是极小原理发展历史上很重要的一步。最终，对这个原理的诅咒被伟大的爱尔兰人威廉·罗恩·哈密顿爵士打破了，他所表述的该原理在数学上是完全正确、简单且普适的。同时，他也结束了用自然的经济性来解释该原理的时代。由于幸运的数学巧合，势能的静态极小原理这个静力学问题与摆的哈密顿最小作用量原理在形式上完全符合；先前我们讨论的一端悬挂砝码的钢尺就是这种静力学情况之一。钢尺弯曲的倾斜角是其自由端到固定点距离的函数，而摆的偏转角是时间的函数，但是表示钢尺倾斜角的曲线与表示摆偏转角的曲线却是严格一致的（见图 2）。

现在我们已经看到，图中只有那些被曲线简单覆盖的区域对应于实际的极小值范围，即对应弹性尺形变的稳定构型。而包络面以外的其他区域中存在两条或多条曲线穿过同一个给定点的情况。其中只有一条对应实际的极小值。以上两种对应实际极小值的情况都是摆可能的运动情况。虽然弹性钢尺端点的限制条件与哈密顿原理中每个时间间隔内端点的限制条件并不严格一致，但是在以下情况中它们有一样的规律：如果钢尺的长度或相应于哈密顿原理中钟摆模型的时间间隔超过一定的极限，将不止有一种可能的解存在，虽然这些解都对应于可能的运动情况，但并不是每一个解都存在真正的极小值与之对应。在这种情况下，我们得到的结论是：实际的运动情况是否存在并不总是以真正的作用量的极值性质作为判断依据的，而是以相对不明显的数学性质即所谓的“稳态”构型作为判断依据的。

因此从自然的经济性解释极小原理失败了。我们认为这种在自然规律中寻找目的和经济性的想法是一种可笑的拟人论的想法，它是形而上学的思维主导科学的时代留下的残余。

总而言之，哈密顿的极小原理的重要性体现在完全不同的方面。具有经济性的并不是自然而是科学。我们所有的知识都是从收集到的事实经验出发的；但我们进一步要做的是把这些无数的事实经验总结发展成为简单的规律，并把这些简单规律再次总结发展成为更加普适的规律。在物理学中这个过程是显而易见的。我们可以回想一下，例如在麦克斯韦关于光的电磁理论中，光学变成了广义电动力学的一个

unification. The ideal would be to condense all laws into a single law, a universal formula, the existence of which was postulated more than a century ago by the great French astronomer, Laplace.

If we follow the Viennese philosopher, Ernst Mach, we must consider economy of thought as the only justification of science. I do not share this view—there are other aspects and justifications of science. But I do not deny that economy of thought and condensation of the results are very important, and I consider Laplace's universal formula as a legitimate ideal. There is no question that the Hamiltonian principle is the adequate formulation of this tendency. It would be the universal formula if only the correct expressions for the potential energy of all forces were known. Nineteenth century thinkers believed more or less explicitly in this programme, and it was successful in an amazing degree.

By choosing a proper expression for the potential energy, nearly all phenomena could be described, including not only the dynamics of rigid and elastic bodies, but also that of fluids and gases, as well as electricity and magnetism, together with electronics and optics. The culmination of this development was Einstein's theory of relativity, by which the abstract principle of action regained a simple geometrical interpretation. The motions of the planets can be considered as "geodesics" in the four-dimensional space formed by adding time to our common space. Einstein's law of gravitation, which contains Newton's law as a limiting case, can also be derived from an extremum principle in which the quantity that is an extremum can be interpreted as the total curvature of the space-time world.

We call this period of physics which ends with the theory of relativity the classical period, in contrast to the recent period which is dominated by quantum theory. The new quantum mechanics assumes that all laws of physics are of statistical character. The fundamental quantity is a wave function which obeys laws similar to those of an acoustical or optical wave; it is not, however, an observable quantity, but determines indirectly the probability of observable processes. The point which interests us here is the fact that even this abstract wave function of quantum mechanics satisfies an extremum principle of the Hamiltonian type.

We are still far from knowing Laplace's universal formula, but we may be convinced that it will have the form of an extremal principle, not because Nature has a will or purpose or economy, but because the mechanism of our thinking has no other way of condensing a complicated structure of laws into a short expression.

(**143**, 357-361; 1939)

分支。在达到理论大统一的过程中，极小原理是一个强有力的手段。最理想的情况是把所有的规律浓缩成一个规律，一个普适的公式，而早在一个多世纪之前，法国伟大的天文学家拉普拉斯就已经提出过这种猜想了。

如果按照维也纳哲学家恩斯特·马赫的观点，则我们应该把是否具有经济性作为判断科学是否合理的唯一条件。我并不赞同这种观点——我认为还有其他判断科学合理性的方面和办法。这里我并不是认为思维的经济性和对结果的归纳不重要，而且我也认为拉普拉斯提到的普适的公式是一种合理的想法。毫无疑问，哈密顿原理是一个大家认可的倾向于成为普适规律的原理。如果已知所有作用力势能的正确表达式，哈密顿原理就将是一个普适的原理。19 世纪，几乎所有的思想家们都明确相信这个原理是普适的，而这个原理也确实取得了惊人的成功。

如果我们选择合适的势能表达式，哈密顿原理就可以描述几乎所有的现象，不仅包括刚体、弹性体的动力学，也包括流体、气体的动力学以及电学、磁学、电子学和光学。爱因斯坦的相对论是这个理论发展的极致，其中抽象的作用量原理又可以用简单的几何学进行解释了。在通常的空间中加入时间这个维度后，则我们可以认为行星的运动轨迹是这种四维空间中的“测地线”。牛顿运动定律作为一种极限情况被包含在爱因斯坦的引力定律当中，而我们也可以从极值原理导出后者，其中，极值原理中的极值这个量可以被解释为时空世界总的曲率。

我们把相对论出现之前的时期称为经典物理学的时期。与之相对的是最近以量子理论为主导的时期。新的量子力学假定所有的物理规律都具有统计特性。其中最基本的物理量是一个与声波和光波遵循类似规律的波函数；然而，它并不是一个可观测量，而是间接决定了可观测过程的概率。这里我们感兴趣的方面是，即便是量子力学中这个抽象的波函数也是满足哈密顿形式的极值原理的。

虽然要得到拉普拉斯所谓的普适公式还有很长的路要走，但我们深信最终我们将得到极值原理普适的形式，这并非出于自然界的愿望、宗旨或经济性，而是因为以我们现在的思维方式还没有其他办法能将复杂的结构规律浓缩成简单的表达式。

（沈乃澂 翻译；葛墨林 审稿）

The "Failure" of Quantum Theory at Short Ranges and High Energies

A. S. Eddington

Editor's Note

Arthur Eddington was by now the most distinguished astronomer in Britain. He had turned his attention to the problems thrown up by the new quantum mechanics. He quickly replied to Bhabha's claim that quantum mechanics could not accommodate the behaviour of atomic particles at high energy with an account of how he proposed to deal theoretically with nuclei of all kinds, concluding that quantum mechanics was not in need of revision. Both Bhabha and Eddington were, unfortunately, unaware of the complications that would arise from the introduction of mesons into their developing theory.

IN Dr. Bhabha's letter in *Nature* of February 18, reference is made to a breakdown of quantum theory at high energies and short intervals. This seems to be widely interpreted as setting a limit to the validity of our present physical conceptions. Some indeed would associate it with a fundamental discontinuity of structure of space and time. I have no occasion to criticize Dr. Bhabha's letter which, so far as it goes, is opposed to the more extreme interpretations. But I venture to suggest that an unnecessary mystery is being made of what is really an elementary point of relativity theory. In short, we know why the present theory has got into difficulties, and we know what must be taken into account if it is to get out of them.

Relativity theory begins with a denial of absolute motion. Every observed velocity dx/dt is a relative velocity of two physical objects. Likewise the "x" of which velocity is the time-derivative is a relative displacement of two physical objects. Both objects are connected with the space-time frame in the way pointed out by Heisenberg; namely, they are not locatable as points (or, in four dimensions, worldlines) but have an uncertainty of position and momentum.

Usually dx/dt and x are assigned to one of the objects (here called the object particle), the other being regarded as a reference body. In precise formulae, the reference body must evidently be a particle. The reference particle is then the physical "origin" from which the observable co-ordinate x of the object particle is measured. Current quantum theory has repeated the pre-relativity mistake of paying insufficient attention to the definition of the physical reference system to which its exact formulae are intended to apply. It enunciates formulae involving x and $\partial/\partial x$, but omits to specify the standard deviation (uncertainty) of position and momentum of the origin from which x is measured. Clearly, the formulae cannot be true for an arbitrary standard deviation; if true at all, they must hold for a

短程和高能中量子理论的“失败”

爱丁顿

编者按

阿瑟·爱丁顿是英国迄今为止最杰出的天文学家。后来他将注意力转向了新量子力学所引发的问题上。巴巴声称量子力学不适用于高能原子粒子的行为，爱丁顿随即对这一观点作出了回应。他阐述了理论上如何描述所有的原子核，进而认为量子力学没必要修正。不过遗憾的是巴巴和爱丁顿都没有意识到，当在他们的理论中引进介子后，相应会出现的一些新的问题。

在 2 月 18 日《自然》杂志上巴巴博士的文章中，提到了量子理论在短程和高能中存在问题。这似乎可以广泛的解释为需要对我们现有的物理概念的有效性做出限定。某些问题确实与基本的时间和空间结构的不连续性相关。现在严苛地质疑巴巴博士的文章还为时过早，就该文而言，它与更极端的解释相反。但我要冒昧地提出一个小的疑问，即究竟什么才真正是相对论的基本点。简言之，我们知道为什么现有的理论已经处于困境，并且我们也知道如果要克服这些困难必须要考虑什么。

相对论以否认绝对运动为出发点。每一个我们观测到的速度 dx/dt 都是两个物理对象的相对速度。同样地，“x”是两个物理对象之间的相对位移，而速度是“x”对时间的一阶导数。两个物体都与海森堡所指出的时空框架相联系；即它们并不是像三维空间中的点（或四维时空中的世界线）那样有确定的位置，它们的位置和动量具有不确定性。

通常，dx/dt 和 x 赋予其中一个物体（本文称之为实物粒子），另一个则被视为是参照物。显然，在精确的公式中，参照物必须是一个粒子。这个参照粒子即为物理“原点”，我们可以由此测量实物粒子的可观测坐标 x。目前的量子理论已重复了相对论前时期的错误，即对精确公式所对应的物理参考系的定义不够重视。该理论虽然清楚地表达了包含 x 和 $\partial/\partial x$ 的公式，但却忽略了对 x 所参照的原点的位置和动量的标准偏差（不确定度）的测定。显然，公式并不可能对任意的一个标准偏差都是正确的；如果公式正确，它必然对应于特定位置的标准偏差 ε，并且这个标准偏

particular standard deviation of position ε which ought to have been specified. Since there is no explicit reference, the actual value of ε must be implicit in the empirical constants (such as h/mc) of the quantum formulae.

When two object particles with co-ordinates x_1 and x_2 are considered, the displacement ξ_0 of one relative to the other is observable independently of any origin of co-ordinates, being the original observable called "x" in our second paragraph. This must not be confused with the co-ordinate difference $\xi = x_2 - x_1$, which introduces twice over the uncertainty of the physical origin from which x_1 and x_2 are measured. We have (in the notation of the theory of errors)

$$\xi = \xi_0 \pm \varepsilon\sqrt{2}\,.$$

The failure of current theory is due especially to its omission to distinguish the two observables ξ and ξ_0.

The physical origin has uncertainty both of position and momentum; for if either were zero the other would be infinite, and the physical origin would not approximate to a geometrical origin with a definite world-line. An energy m_0c^2 corresponding to the mean square of the uncertain momentum is therefore associated with the origin. Except in two-body problems (in which one object particle is used as physical origin for the other) the practice is to treat all the object particles symmetrically; the physical origin must then be an additional *virtual* particle, that is, a particle inserted in the system as part of the apparatus of measurement, but not counted among the object particles, and only taken into account as representing the disturbance of the system which the carrying out of a measurement implies. The energy m_0c^2 belongs to the physical origin contemplated as a virtual particle, and gives it a proper mass m_0. In order that quantum equations may be definite, the uncertainty constants ε and m_0 of the physical origin must have standard values.

Naturally physicists who have neglected the uncertainty ε of the origin will find that their equations break down at distances of order ε. Nothing has gone wrong with space; it is the theorists who have failed to apply their own principles in relating the observable physical system to space. To state their failure summarily: there are two recognized principles of observability, namely, the quantum principle that an observable *object* has an uncertainty relation to the geometrical space-time frame, and the relativity principle that an observable *quantity* relates to two observable objects. Current theory recognizes these principles separately but not in combination; and in dealing with co-ordinates and momenta it pays attention only to the uncertainty at the object-particle end of the relation.

The remedy is obvious. I do not say that the application of the remedy is an easy matter; but, if it is clearly the thing most worth doing, that will not deter anyone. An astronomer, unable to solve his own "problem of three bodies", can only admire the success with which physicists tackle the more numerous closely interacting particles of the nucleus. But I think the advance would not be less rapid or less substantial if they gave up using the wrong formulae. My own work[1] (chiefly concerned with the momentum uncertainty m_0) has been

差也应当已确定。由于不存在明确的参照系，ε 的实际值必然隐含在量子公式的经验常数（例如 h/mc）中。

当考虑分别位于坐标 x_1 和 x_2 中的两个实物粒子时，一个粒子相对于另一个粒子的位移 ξ_0 是可观察的，这与任何坐标原点无关，也就是本文第二段中称为“x”的原始可观测量。一定不要把它与坐标差 $\xi = x_2 - x_1$ 相混淆，坐标差引入了测量 x_1 和 x_2 时所参照的物理原点两倍的不确定度。我们得到（用误差理论的符号表示）

$$\xi = \xi_0 \pm \varepsilon\sqrt{2}$$

目前理论的失败主要是忽略了两个可观测量 ξ 和 ξ_0 的区别所致。

物理原点具有位置和动量两种不确定度；因为如果其中任意一个物理量为零，那么另一个物理量必将是无穷大，那么物理原点就不会以有限的世界线接近几何原点。因此，存在一个对应于不确定动量均方值的能量 m_0c^2 与原点相联系。除了在二体问题中（其中一个实物粒子作为另一个的物理原点），实际上对所有的物体粒子都作无差别处理；我们用一个附加的**虚**粒子作为物理原点，即把一个粒子作为测量装置的一部分插入到系统中，但不计入实物粒子，而仅将其作为测量时所包含的系统扰动来考虑。能量 m_0c^2 对应于虚粒子考虑的物理原点，并给出其固有质量 m_0。为了可以确定量子方程，物理原点的不确定量 ε 和 m_0 必须具有标准值。

当然，忽略了原点不确定度 ε 的物理学家将会发现，他们的方程在量级为 ε 时会失效。空间本身没有问题；问题出在那些理论工作者身上，他们无法成功地应用自己的理论将可观测的物理系统与空间相联系。他们的失败可以概括地表述为：存在两个公认的关于可观测性的原理，即量子理论和相对性原理，在前者中可观测的**物体**对几何时空框架具有不确定关系，在后者中可观测的**量**与两个可观测的物体相关。目前的理论认为，这些原理相互独立无法结合；当我们处理坐标和动量时，仅需注意其关系中实物粒子端的不确定度。

补救措施是明显的。我并不是说完成修正是一件简单的事情；但是，如果它是一件最值得去做的事情，那就不会使任何人畏缩不前。一位不能解决自己“三体问题”的天文学家，只能羡慕物理学家处理原子核中大量存在短程相互作用的粒子时所取得的成就。但我认为，如果他们放弃使用错误的公式，不会使进展速度变缓，或使进展的实质性减弱。我自己的工作 [1]（主要是关于动量不确定量 m_0）一直局限

confined to extra-nuclear problems; no insuperable difficulty has appeared, tough progress has not been easy.

I turn now to the actual values of ε and m_0. A geometrical frame of space-time is not a physical reference system, since its exactitude is incompatible with observability. (To assign infinite mass to the frame, so as to make both its position and velocity exact, would introduce infinite curvature, and defeat its use in a different way.) We turn it into a physical frame by associating with it uncertainty constants, namely, a particular length ε and a particular mass m_0. (In mathematical treatment we should assign to the frame a wave function describing a probability distribution corresponding to these constants.) These "put the scale into" physical systems, and all other natural lengths and masses will stand in a definite numerical ratio to them. Since we observe only relative scale, the arbitrariness of the initial choice of ε and m_0 is eliminated. From extra-nuclear investigations I have found that m_0 is 10/136 of the mass of a hydrogen atom; and ε, which is very simply related to certain magnitudes calculated in cosmological theory[2], has the value 1.10×10^{-13} cm. (For technical reasons, the constant more usually given is $k_0 = 2\varepsilon = 2.20\times10^{-13}$ cm.)

That the virtual particle of mass m_0, originally studied in extra-nuclear theory, is the particle now used in nuclear theory under the name of "mesotron", or "meson", admits, I think, of no doubt. But whether the usual assumption is correct, that nuclear mesotrons are the same as the actual mesotrons observed in a Wilson chamber, I have no means of judging. Since m_0 and ε are conjugate, it is optional which we treat as the more fundamental; but I would point out that to proceed *via* the mesotron mass is a roundabout way of getting at the range of nuclear forces, since the range is an immediate manifestation of the uncertainty of position of the origin. The nuclear force between two protons comes from an energy-singularity occurring when two protons coincide—a sink which (in uniform distribution) just compensates the Coulomb energy occurring when they do not coincide. The condition of coincidence $\xi = 0$ gives $\xi_0 = \pm\,\varepsilon\sqrt{2}$, which (for the calculated value of ε) agrees exactly with the range of force found by Breit, Condon and Present in their discussion of the scattering of protons.

(**143**, 432-433; 1939)

A. S. Eddington: Observatory, Cambridge, Feb. 27.

References:

1. Eddington, "Relativity Theory of Protons and Electrons", Chapters XI, XII.

2. Eddington, *Proc. Roy. Soc.*, A, **162**, 155 (1937).

于核外问题；在此过程中虽然尚未出现不能克服的困难，但也是举步维艰。

现在我回到 ε 与 m_0 的实际值问题的讨论。时空的几何框架并不是物理参考系，因为其精确性与可观测性互不相容。（设框架质量无限大，以使其位置和速度两者的值均是精确的，但这将引出无限曲率，使得该框架以另外一种方式失效。）我们将它转化为具有不确定性常数的物理框架，即其中的不确定性常数就是特定的长度 ε 和特定的质量 m_0。（在数学处理中，我们应赋予框架一个波函数来描述与这些常数相应的概率分布。）对物理系统“进行标度变换”，即所有其他自然长度和质量将以与其成确定比率的数值形式出现。由于我们仅观测相对标度，排除了 ε 和 m_0 最初选择的任意性。根据核外研究我发现，m_0 是氢原子质量的 10/136；ε 与宇宙论 [2] 中计算得到的确定量值有非常简单的关系，数值为 1.10×10^{-13} 厘米 (出于技术方面的考虑，通常将常数取为 $k_0 = 2\varepsilon = 2.20 \times 10^{-13}$ 厘米)。

最初在核外理论中研究的质量为 m_0 的虚粒子，正是现在核理论中的“介子”，我认为这种观点是毋庸置疑的。但通常的假设是否正确，即核介子是否与威尔逊云室中观测到的实际介子相同，我还无法做出判断。由于 m_0 与 ε 是共轭的，因此选择哪个作为基本量是任意的；但我要指出的是，由介子的质量出发确定核力的作用范围是一种不直接的方式，因为此范围是原点位置不确定性的直接表现。在两个质子之间的核力来自两个质子重合时产生的能量奇点——一个下陷，它（在均匀分布中）刚好可以补偿两个质子不重合时产生的库仑能。在刚好符合 $\xi=0$ 的条件时，有 $\xi_0=\pm\varepsilon\sqrt{2}$，这（在 ξ 的计算值方面）与布莱特、康登和普雷森特在他们对质子散射的论述中建立的力的范围精确地一致。

（沈乃澂 翻译；鲍重光 李军刚 审稿）

Liberation of Neutrons in the Nuclear Explosion of Uranium

H. von Halban, jun. *et al.*

Editor's Note

Recent experiments had shown that considerable energy was released in the explosion of uranium or thorium nuclei, triggered by impacting neutrons. The possibility of an energy-releasing chain reaction depended on whether such processes might release further neutrons. Here Hans von Halban, Frédéric Joliot and Lew Kowarski report experiments suggesting a positive answer. They sent neutrons into two substances, ammonium nitrate and uranyl nitrate, and measured the number of detected slow neutrons as a function of distance. A discrepancy between the two substances clearly seemed attributable to the presence of uranium and to additional neutrons being created in fission events. They could say little about the energy of the neutrons, but these observations were a step towards sustained nuclear fission.

RECENT experiments[1,2] have revealed the existence of a new kind of nuclear reaction: neutron bombardment of uranium and thorium leads to an explosion of the nucleus, which splits up into particles of inferior charge and weight, a considerable amount of energy being liberated in this process. Assuming a partition into two particles only, so that the nuclear mass and charge of uranium have to be distributed between two lighter nuclei, the latter contain considerably more neutrons than the heaviest stable isotopes with the same nuclear charges. (A splitting into, for example, ^{98}Rb and ^{141}Cs means an excess of 11 neutrons in the first, and of 8 neutrons in the second of these two nuclei.) There seem to be two possibilities of getting rid of this neutron excess. By the emission of a β-ray, a neutron is transformed into a proton, thus reducing the neutron excess by two units; in the example given above, five and four successive β-activities respectively would be needed to restore the neutron-proton stability ratio. In fact, the explosion products have been observed to be β-active and several periods have been recorded, so that a part, at least, of the neutron excess is certainly disposed of in this way. Another possible process is the direct liberation of neutrons, taking place either as a part of the explosion itself, or as an "evaporation" from the resulting nuclei which would be formed in an excited state.

In order to find some evidence of this second phenomenon, we studied the density distribution of the thermal neutrons produced by the slowing down of photo-neutrons from a Ra γ-Be source in a 1.6 molar solution of uranyl nitrate and in a 1.6 molar solution of ammonium nitrate (the hydrogen contents of these two solutions differ by only 2 percent). Plotting Ir^2 as a function of r (where r is the distance between the source and a given point, and I is the local density of thermal neutrons at the same point, measured by the activity induced in a dysprosium detector), a curve is obtained the area of which is

铀核爆炸时中子的释放

冯·哈尔班等

编者按

最近的实验显示由中子碰撞触发的铀核或钍核爆炸释放出相当多的能量。释放能量的链式反应发生的可能性依赖于这个过程是否释放更多的中子。本文中，汉斯·冯·哈尔班、弗雷德里克·约里奥和卢·科瓦尔斯基称实验给出了肯定答案。他们用中子轰击硝酸铵和硝酸双氧铀两种物质，对观测到的慢中子数目随距离的变化进行测量。在两种物质中产生的差异无疑是因为铀核的存在和在裂变中产生的额外中子。关于中子能量他们能讲得很少，但是这些观测是迈向持续核裂变研究的一步。

最近的实验[1,2]已经揭示了存在一种新的核反应：用中子轰击铀和钍会引发核爆炸，核分裂为电荷和质量较小的粒子，同时在这个过程中释放了相当大的能量。假设一次分裂仅产生两个粒子，则铀核的质量和电荷必定在两个更轻的核之间分配，与带有相同核电荷的最重的稳定同位素相比，两个轻核含有较多的中子。（例如，分裂为 ^{98}Rb 和 ^{141}Cs 意味着在两个核子中，第一个核子多出了 11 个中子，第二核子多出了 8 个中子。）似乎存在着两种可能性来消除这些过剩中子。通过发射 β 射线，一个中子转变为一个质子，这样就减少了两个单位的过剩中子；在上述给出的例子中，分别需要相继 5 次和 4 次 β 放射来恢复中子－质子的稳定率。实际上，已观测到的爆炸产物是具有 β 活性的，并且已经记录到了几个活性周期。因此，至少有一部分过剩的中子必定是通过这种方式去除的。另一种可能的过程是中子以部分爆炸产物的形式被直接释放，或从处于激发态的生成核中“蒸发”掉。

为了找到这类次级现象的某些证据，我们研究了 Ra γ-Be 光中子在 1.6 摩尔硝酸双氧铀溶液和 1.6 摩尔硝酸铵溶液中的减速所产生的热中子的密度分布（这两种溶液中的氢含量只相差 2%）。将 Ir^2 作为 r 的函数绘图（其中 r 是源点与给定点之间的距离，I 是热中子在相同点的局部密度，通过镝探测器中感生到的放射性进行测量），图中所得曲线的面积与 $Q\cdot\tau$ 成正比，Q 是每秒由中子源发射出的或在溶液中形

proportional to $Q \cdot \tau$, Q being the number of neutrons per second emitted by the source or formed in the solution and τ the mean time a neutron spends in the solution before being captured[3,4]. Any additional nuclei, which do not produce neutrons, brought into the solution, will increase the chances of capture and therefore decrease τ and the area. If, however, these dissolved nuclei are neutron-producing, Q will be greater and the area of the curve will tend to increase. Evidence of neutron production, as indicated by an actual increase of the area, will only be obtained if the gain through Q (neutron production) is greater than the loss through τ (neutron capture). This loss can anyway be studied separately, since it has been shown[5] that the introduction of nuclei which act merely by capture or by increasing the hydrogen content of the solution can affect the shape of the density curve only in a characteristic way: the modified curve can always be brought to coincide with the primitive curve by multiplying all abscissae by a suitable factor and all ordinates by another factor.

The accompanying graph shows the two curves obtained. At small distances from the source the neutron density is greater in the ammonium solution than in the uranyl solution; at distances greater than 13 cm., the reverse is true. In other words, the decrease of the neutron density with the distance is appreciably slower in the uranyl solution.

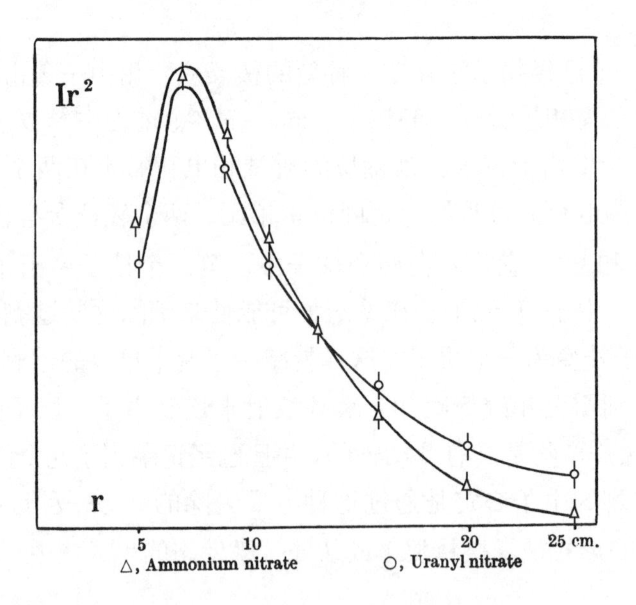

The observed difference must be ascribed to the presence of uranium. Since the two curves cannot be brought to coincide by the transformation mentioned above, the uranium nuclei do not act by capture only; an *elastic* diffusion by uranium nuclei would have an opposite effect: it would "contract" the abscissae, instead of stretching them. The density excess, shown by the uranyl curve beyond 13 cm., must therefore be considered as a proof of neutron production due to an interaction between the primary neutrons and the uranium nuclei. A reaction of the well-known $(n,2n)$ type is excluded because our primary neutrons are too slow for such a reaction (90 percent of Ra+Be photo-neutrons

成的中子数，τ是中子被俘获之前在溶液中度过的平均时间[3,4]。将任何其他不产生中子的原子核放入溶液中将会增加中子被俘获的概率，因此时间τ和面积也会相应减小。然而，如果溶液中的这些核是产生中子的，那么Q将更大，曲线的面积也将趋于增大。由面积的实际增大所表明中子生成的证据将仅在以下情况中才能获得，即由Q（中子生成）所造成的增益比由时间τ（中子俘获）所导致的损耗大的情况。我们可以通过各种方式来分别研究这项损耗，因为其已经表明[5]，仅仅是通过俘获，或通过增加溶液中的氢含量而引入的核只通过一种特定的方式来影响密度曲线的形状，那么将原始曲线的所有横坐标乘上一个合适的因子，而将纵坐标乘上另一个合适的因子，修正的曲线总能与其相吻合。

附图显示了得到的两条曲线。在与源距离较近的位置，硝酸铵溶液中的中子密度大于硝酸双氧铀溶液中的密度；但在距离大于13厘米的位置时，情况则相反。换言之，在硝酸双氧铀溶液中中子密度随距离的减小明显较为缓慢。

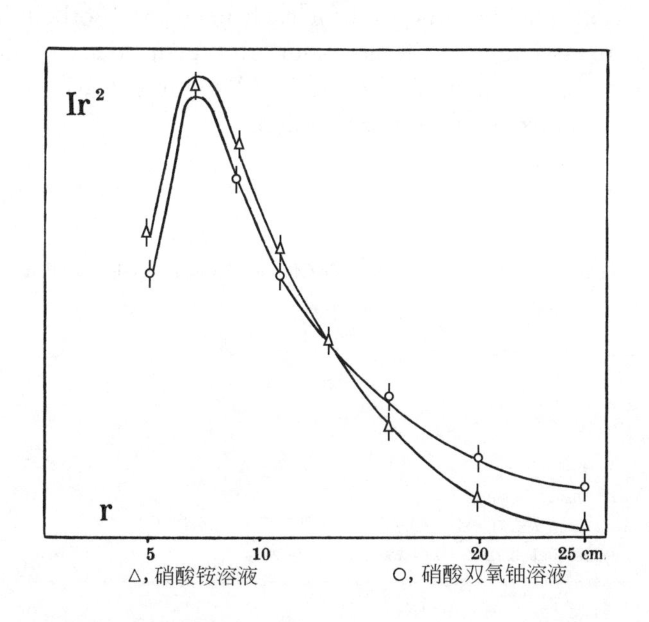

△，硝酸铵溶液　　○，硝酸双氧铀溶液

观测到这样的差别一定是由铀的存在所致。因为这两条曲线不能通过上述修正而重合，所以铀核不仅仅起俘获的作用；由铀核所产生的**弹性**漫射将会起到相反的效果：它会“缩短”横坐标，而不是延长横坐标。因此，当距离超过13厘米时硝酸双氧铀曲线所示的密度超量，必定会被当作中子是由初级中子与铀核之间的相互作用的产物的证据。熟知的$(n,2n)$型的反应是被排除在外的，这是由于对于这类反应而言我们的初级中子的能量太小了（90%的Ra+Be光中子的能量小于0.5兆电子伏，

have energies smaller than 0.5 Mev. and the remaining 10 percent are slower than 1 Mev.).

The degree of precision of the experiment does not permit us to attribute any significance to the small increase of the area in the uranyl curve (as compared to the ammonium curve), which we obtain by extrapolating the curves towards greater distances. In any event, an inferior limit for the cross-section for the production of a neutron can be obtained by assuming that the density excess due to this production is equal throughout the whole curve to the excess observed at r = 25 cm.; this limit, certainly inferior to the actual value, is 6×10^{-25} cm.[2].

Our measurements yield no information on the energy of the neutrons produced. If, among these neutrons, some possess and energy superior to 2 Mev., one might hope to detect them by a (n,p) process, for example, by the $^{32}S(n,p)^{32}P$ reaction. An experiment of this kind, Ra γ-Be still being used as the primary neutron source, is under way.

The interest of the phenomenon observed as a step towards the production of exo-energetic transmutation chains is evident. However, in order to establish such a chain, more than one neutron must be produced for each neutron absorbed. This seems to be the case, since the cross-section for the liberation of a neutron seems to be greater than the cross-section for the production of an explosion. Experiments with solutions of varying concentration will give information on this question.

(**143**, 470-471; 1939)

H. von Halban, jun., F. Joliot and L. Kowarski: Laboratoire de Chimie Nucléaire, Collège de France, Pairs, March 8.

References:

1. Joliot, F., *C.R.*, **208**, 341 (1939).
2. Frisch, O. R., *Nature*, **143**, 276 (1939).
3. Amaldi, E., and Fermi, E., *Phys. Rev.* **50**, 899 (1936).
4. Amaldi, E., Hafstad, L., and Tuve, M., *Phys. Rev.*, **51**, 896 (1937).
5. Frisch, O. R., von Halban, jun., H., and Koch, J., *Danske Videnskab. Kab.*, **15**, 10 (1938).

其余的 10% 小于 1 兆电子伏）。

实验的精度不允许我们对硝酸双氧铀曲线中面积的微小增加（与硝酸铵曲线比较）赋予任何含义，这样的增加是我们将曲线向距离更大的方向外推时得到的。在任何情况下，假设由于中子产生造成的密度增量等于在 r=25 厘米处观测到的整个曲线的增量，则可得到中子产生的截面下限；这个极限肯定在实际值之下，为 6×10^{-25} 平方厘米。

我们的测量还无法给出关于所产生中子的能量的信息。如果在这些中子之中，某些中子具有的能量超过 2 兆电子伏，我们有望通过一个（n,p）过程来探测它们，例如，通过 $^{32}S(n,p)^{32}P$ 反应。初级中子源仍然为 Ra γ-Be 的这种实验正在进行之中。

我们所观测的现象使实现产生外能变换链向前迈进了一步，这个意义是很明显的。然而，为了建立这样的一个链条，每吸收一个中子，都必须再产生一个以上的中子。看起来事实就是如此，因为释放一个中子的截面似乎大于产生一次爆炸的截面。采用各种不同浓度的溶液进行的实验将给出该问题的相关信息。

（沈乃澂 翻译；王乃彦 审稿）

Products of the Fission of the Uranium Nucleus

L. Meitner and O. R. Frisch

Editor's Note

Lise Meitner and Otto Frisch here report new experiments probing the products of uranium fission experiments. Earlier work suggested that fission fragments should emerge with energies of several hundred million electron volts. Here the researchers sent neutrons into a sample of uranium hydroxide and attempted to collect the fission fragments 1 mm away, either in a paper surface or in water. They found evidence for a range of different fission fragments. It seemed most unlikely that the mere absorption of a neutron could give a uranium nucleus enough kinetic energy to reach their collecting surfaces. This new technique offered a route to the more detailed examination of the nuclear fragments created in fission processes.

O. Hahn and F. Strassmann[1] have discovered a new type of nuclear reaction, the splitting into two smaller nuclei of the nuclei of uranium and thorium under neutron bombardment. Thus they demonstrated the production of nuclei of barium, lanthanum, strontium, yttrium, and, more recently, of xenon and caesium.

It can be shown by simple considerations that this type of nuclear reaction may be described in an essentially classical way like the fission of a liquid drop, and that the fission products must fly apart with kinetic energies of the order of hundred million electron-volts each[2]. Evidence for these high energies was first given by O. R. Frisch[3] and almost simultaneously by a number of other investigators[4].

The possibility of making use of these high energies in order to collect the fission products in the same way as one collects the active deposit from alpha-recoil has been pointed out by L. Meitner (see ref. 3). In the meantime, F. Joliot has independently made experiments of this type[5]. We have now carried out some experiments, using the recently completed high-tension equipment of the Institute of Theoretical Physics, Copenhagen.

A thin layer of uranium hydroxide, placed at a distance of 1 mm. from a collecting surface, was exposed to neutron bombardment. The neutrons were produced by bombarding lithium or beryllium targets with deuterons of energies up to 800 kilovolts. In the first experiments, a piece of paper was used as a collecting surface (after making sure that the paper did not get active by itself under neutron bombardment). About two minutes after interrupting the irradiation, the paper was placed near a Geiger-Müller counter with aluminium walls of 0.1 mm. thickness. We found a well-measurable activity which decayed first quickly (about two minutes half-value period) and then more slowly. No attempt was made to analyse the slow decay in view of the large number of periods to be expected.

铀核的裂变产物

迈特纳，弗里施

编者按

莉泽·迈特纳和奥托·弗里施在这篇文章中报道了关于探测铀核裂变产物的新实验。早期的研究认为裂变碎片的产生会伴随有几百兆电子伏的能量出现。本文作者用中子轰击氢氧化铀样品，尝试在1毫米远的纸张表面或者水中收集裂变碎片。他们发现了一系列不同的裂变碎片存在的证据。可以看出，仅靠吸收一个中子似乎不能提供铀原子核足够的动能使其到达他们的收集表面。这项新技术提供了一个途径去更仔细地检测裂变过程中产生的原子核碎片。

哈恩和施特拉斯曼[1]发现了一种新型核反应，即在中子轰击下，铀核和钍核分裂为两个更小的核。以此他们证明了钡核、镧核、锶核、钇核以及最近又发现的氙核和铯核等裂变产物。

可以简单地以一种本质上经典的方式，比如类似于液滴分裂的形式，来描述这类核反应，并且分裂的产物必定以上百兆电子伏的动能飞离[2]。弗里施[3]首先给出这些高能反应的证明，几乎是在同一时间其他多位研究者也给出了证明[4]。

迈特纳已经指出（见参考文献3），像人们从α反冲中收集放射性沉积物那样，利用这些高能量来收集裂变产物是可能的。在此期间，约里奥已独立地进行了这类实验[5]。我们利用哥本哈根理论物理研究所最近制成的高压装置，现在已经开展了一些实验。

将一薄层氢氧化铀放在距收集表面1毫米的位置，并用中子对其进行轰击。中子是用能量高达800千电子伏的氘核轰击锂靶或铍靶而产生的。在第一个实验中，使用一张纸作为收集表面（已经确定纸在经过中子轰击后纸本身不具有放射性）。在辐照中断大约两分钟后，将纸放在具有0.1毫米厚铝壁的盖革–米勒计数器的附近。我们发现放射性可以很好地被测量到，并且开始时放射性衰变得很快（约两分钟的半衰期），然后会较为缓慢。鉴于估计其中存在数目繁多的周期，我们并不打算对缓慢的衰变进行分析。

The considerable intensity, however, of the collected activity encouraged us to try to get further information by chemical separations. The simplest experiment was to apply the chemical methods which have been developed in order to separate the "transuranium" elements from uranium and elements immediately below it[6]. The methods had to be slightly modified on account of the absence of uranium in our samples and in view of the light element activities discovered by Hahn and Strassmann[1].

In these experiments, the collecting surface was water, contained in a shallow trough of paraffin wax. After irradiation (of about one hour) a small sample of the water was evaporated on a piece of aluminium foil; its activity was found to decay to zero. It was checked in other ways, too, that the water was not contaminated by uranium. To the rest of the water we added 150 mgm. barium chloride, 15 mgm. lanthanum nitrate, 15 mgm. platinum chloride and enough hydrochloric acid to get an acid concentration of 7 percent. Then the platinum was precipitated with hydrogen sulphide, in the usual way; the precipitate was carefully rinsed and dried and then placed near our counter.

The results of three such experiments were found to be in mutual agreement. The decay of the activity was in one case followed for 28 hours. For comparison, a sample of uranium irradiated for one hour was treated chemically in the same way. The two decay curves were in perfect agreement with one another and with an old curve obtained by Hahn, Meitner and Strassmann under the same conditions. In the accompanying diagram the circles represent our recoil experiment while the full line represents the uranium precipitate. A comparison of the activity (within the first hour after irradiation) of the precipitate and of the evaporated sample showed that the precipitate contained about two thirds of the total activity collected in the water. After about two hours, however, the evaporated sample was found to decay considerably more slowly than the precipitate, presumably on account of the more long-lived fission products found by Hahn and Strassmann[1].

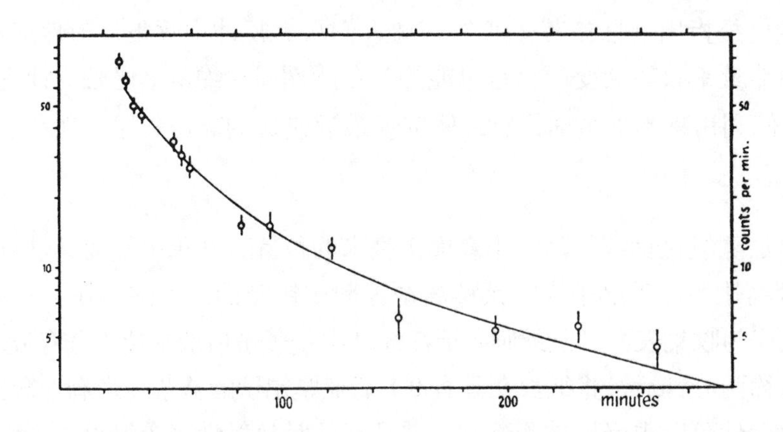

From these results, it can be concluded that the "transuranium" nuclei originate by fission of the uranium nucleus. Mere capture of a neutron would give so little kinetic energy to the nucleus that only a vanishing fraction of these nuclei could reach the water surface.

然而，收集到的放射性有相当大的强度，这促使我们试图通过化学分离的方法得到更进一步的信息。最简单的实验是采用一种化学方法，最初发展这种方法是为了把所谓的“超铀”元素从铀及紧邻它的较低的元素中分离出来[6]。考虑到在我们的样品中不存在铀，并鉴于哈恩和施特拉斯曼[1]发现的轻元素的放射性，这种实验方法必须稍做改进。

这些实验均以石蜡浅槽中的水作收集表面。在（大约一小时的）辐照后，少量样品水被蒸发到一片铝箔上；并且其放射性衰变为零。用其他方法也可证明这些水中并不含铀。在剩余的水中，我们加入 150 毫克氯化钡、15 毫克硝酸镧、15 毫克氯化铂和足够的盐酸，以便得到 7% 的酸浓度。采用通常的方法，用硫化氢将铂析出；析出的沉淀物经过精心漂洗和烘干，然后放在我们的计数器附近。

我们发现三个这样的实验得到的结果是相互一致的。在其中一种情况下，对放射性物质的衰变跟踪了 28 小时。作为对照，对经过一个小时辐照的铀样品，按同样方式进行化学处理。结果显示，这两条衰变曲线完全一致，并且它们与之前由哈恩、迈特纳和施特拉斯曼在相同条件下得到的曲线也完全符合。在附图中，圆圈表示我们的反冲实验，而实线表示铀的沉淀物。沉淀物与经过蒸发样品的放射性（辐照后的第一小时内）的对比表明，沉淀物大约含有水中收集到的总放射性的 2/3。然而，在大约两小时后，我们发现经过蒸发的样品比沉淀样品的衰变要更加缓慢，这大概是由哈恩和施特拉斯曼[1]所发现的更长寿命的裂变产物所致。

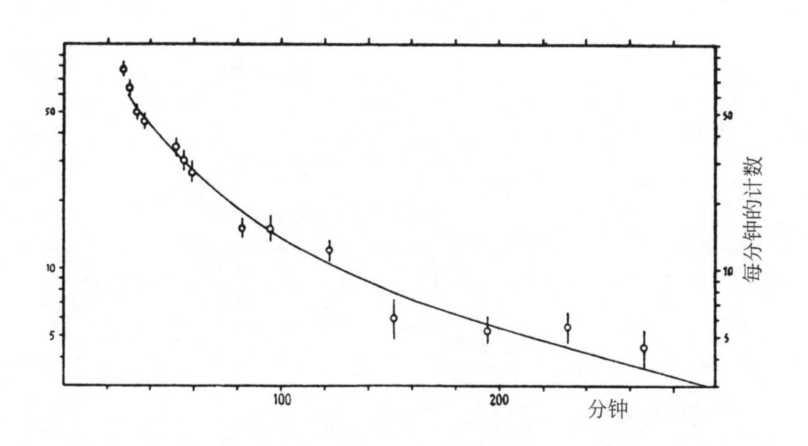

根据这些结果我们可以断定，所谓“超铀”核是由铀核裂变产生的。仅仅俘获一个中子只能为核提供极少的动能，因此这些核中没有一个能够达到水的表面。因

So it appears that the "transuranium" periods, too, will have to be ascribed to elements considerably lighter than uranium.

In conclusion, we wish to thank Dr. T. Bjerge, Dr. J. Koch and K. J. Brostrøm for putting the high-tension plant at our disposal and for kind help with the irradiations. We are also grateful to Prof. N. Bohr for the hospitality extended to us at the Institute of Theoretical Physics, Copenhagen.

(**143**, 471-472; 1939)

Lise Meitner: Physical Institute, Academy of Sciences, Stockholm.

Otto Robert Frisch: Institute of Theoretical Physics, University, Copenhagen, March 6.

References:

1. Hahn, O., and Strassmann, F., *Naturwiss.*, 27, 11, 89, and 163 (1939).
2. Meitner, L., and Frisch, O. R., *Nature*, 143, 239 (1939). Bohr, N., *Nature*, 143, 330 (1939).
3. Frisch, O. R., *Nature*, 143, 276 (1939).
4. Fowler, R. D., and Dodson, R. W., *Nature*, 143, 233 (1939). Jentschke, W., and Prankl, F., *Naturwiss.*, 27, 134 (1939).
5. Joliot, F., *C. R.*, **208**, 341 (1939).
6. Hahn, O., Meitner, L., and Strassmann, F., *Chem. Ber.*, **69**, 905 (1936); and **70**, 1374 (1937).

而所谓“超铀”周期似乎也必然是由比铀轻得多的元素所致。

最后，我们要感谢伯格博士、科赫博士和布罗斯特伦，他们将高压设备交给我们自由使用并好心地对实验中的辐照提供了帮助。我们同样对玻尔教授在哥本哈根理论物理研究所对我们的热情款待表示感激。

（沈乃澂 翻译；张焕乔 审稿）

New Products of the Fission of the Thorium Nucleus

L. Meitner

Editor's Note

In late 1938, German physicists Otto Hahn and Fritz Strassmann in Berlin discovered nuclear fission: the splitting of large atomic nuclei such as uranium into smaller fragments, with an immense release of energy. Production of neutrons during uranium fission meant that the process could be self-sustaining: in reactors and in a bomb. Hahn and Strassmann had been working with the Austrian physicist Lise Meitner, but as a Jew she had been forced to flee from the Nazi regime that summer. Here Meitner, now in Stockholm, extends that work by showing that the element thorium can also undergo fission, and that both uranium and thorium fission lead to similar end products. Hahn was later awarded the Nobel prize for the discovery of fission, but it is now acknowledged that Meitner's contributions were vital to understanding the process.

IN a preceding communication[1] it has been shown that the "transuranium" elements are found among the fission products of uranium under neutron bombardment. Consequently they must be lower elements, probably partly somewhere near tellurium[2], partly (the complementary fission fragments) in the region of ruthenium.

It was then natural to carry out similar experiments with thorium in order to see whether the fission of thorium gave rise to elements with chemical properties similar to those of the "transuranium" elements, that is, elements which can be precipitated with hydrogen sulphide from a strong hydrochloric acid solution. No search for such elements had previously been carried out with thorium, since the formation of elements beyond uranium, from thorium, was not to be expected. So far, chemical analysis of the radioactive bodies produced in thorium by neutron bombardment has revealed (apart from a thorium and a protactinium isotope resulting from pure neutron capture) only products which, on the basis of their chemical properties, had originally to be assigned to radium and actinium isotopes[3] and which have more recently been identified with barium and lanthanum isotopes[4].

A "thick" layer of thorium oxide on a glass plate was irradiated by neutrons obtained by bombarding a lithium target with deuterons of 800 kv. from the high-tension equipment of the Institute of Theoretical Physics, Copenhagen. In order to collect the recoil nuclei, a water surface was used in the way previously described[1]. After irradiation (of about one hour) a fraction (2 c.c.) of the water was evaporated without chemical separation; with the rest (9 c.c.) the usual hydrogen sulphide precipitation was carried out in exactly the same way as in the uranium recoil experiments[1].

钍核裂变的新产物

迈特纳

编者按

1938 年末，德国物理学家奥托·哈恩和弗里茨·施特拉斯曼在柏林发现了核裂变：大原子核（例如铀核）分裂成较小的碎片，并释放出大量能量。铀裂变过程中产生了中子，意味着无论在反应堆中还是在原子弹中，该过程都是可以自持的。哈恩和施特拉斯曼一直以来都与奥地利物理学家莉泽·迈特纳保持合作，但迈特纳作为犹太人，被迫于 1938 年夏天逃离纳粹统治区。本文中，已经来到了斯德哥尔摩的迈特纳延续了上述工作。她在实验中展示了钍裂变，并发现铀核和钍核裂变都产生相似的最终产物。后来，哈恩因发现裂变而获得了诺贝尔奖。但现在人们认识到，迈特纳的贡献对于理解裂变过程至关重要。

在此前的一篇通讯[1]中我们曾指出，在中子轰击下产生的铀的裂变产物中，发现了“超铀”元素。因此它们一定是序数较小的元素，可能部分位于碲附近[2]，部分（互补的裂变碎片）则位于钌的附近区域。

于是很自然地我就使用钍来进行类似的实验，目的是看看钍的裂变是否也能产生与“超铀”元素化学性质类似的元素，所谓的“超铀”元素就是那些可以从浓的盐酸溶液中利用硫化氢沉淀下来的元素。此前从未利用钍对这些元素进行过探寻，因为人们从未期望用钍来制得铀后面的元素。迄今为止，对于那些通过中子轰击在钍中产生的放射性物质所做的化学分析表明（除去由单纯的中子俘获所产生的一种钍的同位素和一种镤的同位素之外），只存在这种产物：根据其化学性质，它们最初曾被归为镭和锕的同位素[3]，但最近已确认其为是钡和镧的同位素[4]。

利用哥本哈根理论物理研究所的高压设备，以 800 千电子伏的氘核轰击锂靶，用所得的中子辐照玻璃板上的氧化钍“厚”涂层。按照以前曾描述过的方法[1]，利用水的表面来收集反冲核。经过辐照（大约 1 个小时）后，一部分（2 立方厘米）水蒸发但不发生化学分离；对剩余部分（9 立方厘米）进行常规的硫化氢沉淀，所用方法与铀反冲实验[1]中的完全一致。

The precipitate showed a clearly measurable activity, the decay of which was distinctly different from the decay of the analogous uranium products. After an initial decrease with a half-value period of about 40 minutes, an activity remained which was followed for almost two days and showed a single decay period of 14–15 hours. The evaporated sample showed first a much faster decay which then gradually became slower; after two days, the activity had vanished.

A second experiment with 2.5 hours' irradiation gave the same result, with correspondingly greater intensity, especially of the longer period. In this experiment, after the sulphide precipitation, the filtrate was neutralized and the barium and lanthanum[1] were precipitated as carbonates. The carbonate precipitate decayed first with about 20 minutes and then with 4 hours half-value period, in agreement with the periods already known[3,4]. Analysis of the decay of the sulphide precipitate showed again the presence of a substance of half-period about 40 minutes and of one of 14.5 hours, which was followed for nearly three days and found to decay to zero. The initial intensities of the two periods were about equal, in spite of the short duration of bombardment. The possible existence of very short or very long additional periods can, of course, not be excluded on the basis of this experiment. The two periods observed are quite different from those of the chemically analogous uranium products.

Several check experiments without neutrons showed that there was no contamination of the water by thorium B or C, which might have resulted from radioactive recoil or traces of emanating thoron. As a further protective measure, the thorium layer was sealed, in the second experiment, by a celluloid membrane of 0.3 mm. stopping power.

From the evidence given above, one can conclude that some of the fission products of thorium show a chemical behaviour similar to that of the "transuranium" elements. This is a further indication that essentially the same chemical elements are produced in the fission of uranium and thorium.

In conclusion, I wish to express my thanks to Dr. T. Bjerge, Dr. J. Koch, and K. J. Brostrøm for kind help in the irradiations with the high-tension tube. I am especially grateful to Prof. N. Bohr for the opportunity to carry out these experiments and for the facilities kindly put at my disposal at the Institute of Theoretical Physics, Copenhagen.

(**143**, 637; 1939)

Lise Meitner: Physical Institute, Academy of Sciences, Stockholm, March 26.

References:

1. Meitner, L., and Frisch, O. R. *Nature*, 143, 471 (1939).
2. Abelson, P., *Phys. Rev.*, 55, 418 (1939). Feather, N., and Bretscher. E., *Nature*, 143, 516 (1939).
3. Meitner, L., Hahn, O., and Strassmann, F., *Phys.*, Z. **109**, 538 (1938).
4. Hahn, O., and Strassmann, F., *Naturwiss.*, 27, 89 (1939).

沉淀表现出明显的、可测量到的放射性，其衰变显著区别于类似的铀产物的衰变。在经历半衰期约 40 分钟的初始衰减后，在随后几乎两天仍保持放射性并且表现出单一的 14~15 小时的半衰期。蒸发的样品最初衰变很快，随后便逐渐减慢；两天后，放射性消失。

第二次实验经过了 2.5 小时的辐射，得到了相同的结果，但相应的强度更大一些，尤其是对于更长的周期。在这次实验中，在硫化物沉淀后滤液呈中性，钡和镧[1]以碳酸盐的形式沉淀。这些碳酸盐最初的半衰期为 20 分钟，随后变为 4 小时，这与已知的半衰期[3,4]相一致。对硫化物沉淀的衰变进行分析，再次表明一种具有 40 分钟半衰期和一种具有 14.5 小时半衰期的物质存在，后者持续大约三天，然后消失。尽管轰击持续时间很短，然而两种半衰期的初始强度大致上是相等的。当然，根据这个实验并不能排除存在额外的极短或极长半衰期的可能性。不过我观测到的这两个半衰期与化学性质类似的铀的产物的半衰期全然不同。

几次无中子辐照的检查性实验表明，水中不存在钍 B 或钍 C（钍 B 或钍 C 可能由放射性物质反冲或者极少量钍射气的发射而产生）。为了保险起见，在第二次实验中用阻止本领为 0.3 毫米的赛璐珞膜将钍层密封。

根据上面给出的证据可以得出这样的结论，钍的某些裂变产物表现出与“超铀”元素类似的化学特性。这进一步表明在铀和钍的裂变中实质上产生了同样的化学元素。

最后，我要向伯格博士、科赫博士和布罗斯特伦表示感谢，他们对于高压管辐射实验给予了热心帮助。我还要特别感谢玻尔教授，他使我有机会进行这些实验并慷慨地为我提供了哥本哈根理论物理研究所的实验设备。

（王耀杨 翻译；朱永生 审稿）

Number of Neutrons Liberated in the Nuclear Fission of Uranium

H. von Halban, jun. *et al.*

Editor's Note

Experiments had confirmed that uranium nuclei could be fissioned—induced to split into two large nuclei—when bombarded by a neutron, and that secondary neutrons were given off in the process. But physicists had little idea how many neutrons were produced in such a fission event. This was important to the possibility of a self-sustaining fission chain-reaction. Here physicist Hans von Halban, then working with Frédéric Joliot-Curie and Lew Kowarski in Paris, estimates experimentally how many neutrons are released. The answer—3.5 on average—suggested that a chain reaction might be possible in uranium if the likelihood for a neutron to trigger a fission event were sufficiently high. After the German invasion in 1940, von Halban fled France for England.

RECENT experiments have shown that neutrons are liberated in the nuclear fission of uranium induced by slow neutron bombardment: secondary neutrons have been observed which show spatial[1], energetic[2] or temporal[3] properties different from those which primary neutrons possess or may acquire. Such observations give no information on the mean number of neutrons produced per nucleus split; this number ν may be very small (less than 1) and the result of the experiment will still be positive.

We are now able to give information on the value of ν. Let us consider the curve representing the density distribution of neutrons slowed down in an aqueous solution surrounding a primary neutron source[1]; the area S of this curve is proportional to $Q\cdot\tau$, Q being the number of neutrons per second emitted by the source or formed in the solution, and τ the mean time a neutron spends in the solution before being captured. Assuming that the solution contains only nuclei which absorb neutrons according to the $1/v$ law (the only exception to this rule will presently be dealt with), τ is proportional to $1/\Sigma c_i\sigma_i$, where c_i is the concentration (atom grams per litre) of an absorbing nucleus, σ_i its cross-section for the capture of neutrons of velocity 1 and the index i is extended to all kinds of neutron-absorbing reactions attributable to nuclei present in the solution. Substituting the symbol A_i for $c_i\sigma_i$ and A_{tot} for ΣA_i, we have identically:

$$\frac{\Delta S}{S}=\frac{\Delta Q}{Q}-\frac{\Delta A_{tot}}{A_{tot}}, \tag{1}$$

neglecting all terms of higher orders, such as those containing $(\Delta Q)^2$, $\Delta Q\cdot\Delta A_{tot}$, etc.

Let the symbol Δ stand for the differences observed between the two solutions (uranyl and ammonium) used in our previous experiment[1]. Neglecting ΔA_{tot} before A_{tot} introduces an ambiguity in the definition of A_{tot} (uranyl *vs.* ammonium value) which is numerically

铀核裂变时释放的中子数量

冯·哈尔班等

编者按

实验已证实被中子轰击时铀核可以分裂为两个大核，并且裂变过程中会释放出次级中子。但是关于一个裂变事件中释放出多少中子，物理学家对此了解很少。但是这直接关系到自持裂变链式反应实现的可能性。物理学家汉斯·冯·哈尔班当时与费雷德里克·约里奥–居里及卢·科瓦尔斯基在巴黎一起工作，基于实验他对释放中子数进行了估测。答案是——平均 3.5 个——这意味着如果一个中子诱发一个裂变事件的可能性足够高，铀就可以发生链式反应。1940 年德国入侵后，冯·哈尔班从法国逃亡到英国。

最近的实验表明，慢中子轰击铀核引发核裂变可以释放出中子：次级中子在空间[1]、能量[2]、时间[3]上的性质都与初级中子所拥有或者可能获得的性质不同。这些实验没有给出每次核分裂所产生的中子的平均数量；虽然这个数量 ν 可能非常小（小于 1），但是实验测量仍然可以得出正的结果。

现在我们可以给出 ν 值的一些信息。让我们仔细思考一下在包围着初级中子源[1]的水溶液中减速的中子的密度分布曲线；曲线的面积 S 和 $Q\cdot\tau$ 成正比，Q 是每秒由中子源发射出的或在溶液中形成的中子数，τ 是中子被俘获之前在溶液中度过的平均时间。假设溶液中只有依照 $1/v$ 定律吸收中子的原子核（这个法则的唯一例外将在不久后处理），τ 正比于 $1/\Sigma c_i\sigma_i$，其中 c_i 是吸收中子的原子核浓度（克原子每升），σ_i 是所有速度为 1 的中子的俘获截面，指标 i 遍历溶液中原子核所致的各种中子吸收反应。用 A_i 代替 $c_i\sigma_i$，用 A_{tot} 代替 ΣA_i，我们将得到：

$$\frac{\Delta S}{S}=\frac{\Delta Q}{Q}-\frac{\Delta A_{tot}}{A_{tot}} \tag{1}$$

上式忽略了所有包含诸如 $(\Delta Q)^2$、$\Delta Q\cdot\Delta A_{tot}$ 等的高阶项。

符号 Δ 代表我们在以前实验所使用的两种溶液（硝酸双氧铀和硝酸铵）中观察到的差值[1]。在 A_{tot} 引入使 A_{tot} 定义不明确的因素之前，ΔA_{tot} 可以忽略不计，这（硝酸双氧铀对硝酸铵的值）在数值上并不重要，并且可以采用算数平均值

unimportant and can be reduced by adopting the arithmetical mean (A_{tot} (amm.) $+\Delta A_{tot}$) /2.

In the quantity ΔA_{tot} the uranium nuclei are represented by several separate terms standing for the different modes of neutron capture (see below); let A_f be the term for the capture leading to fission. Every neutron has the probability A_f/A_{tot} of causing a fission and, since one individual fission process liberates ν neutrons on the average, the total number ΔQ of neutrons thus created is $Q \cdot \frac{A_f}{A_{tot}} \cdot \nu$, and the equation (1) can be rewritten as follows:

$$\nu = \frac{\Delta S}{S} \cdot \frac{A_{tot}}{A_f} + \frac{\Delta A_{tot}}{A_f}. \tag{2}$$

Let us estimate the values of all quantities necessary to calculate ν according to this formula. The area variation $\Delta S/S$ can be read from the graph given in our previous letter with an error of less than 20 percent (due to the uncertainties of inter- and extrapolation; in order to facilitate the latter, we added to the curves a further experimental point for r = 29 cm.). The value of A_{tot} for the ammonium solution can be easily calculated from the known concentrations and capture cross-sections (hydrogen, nitrogen and oxygen). A_f is equal to c_U (1.6 in our experiment), multiplied by the value of σ_f given in a recent paper by Anderson *et al.*[4]. ΔA_{tot} contains a term expressing the small difference of the hydrogen content between the two solutions; and three terms relative to uranium, namely, the fission term A_f, already dealt with, the thermal capture term A_{ct} which can be calculated by using a recently found value for σ_{ct}[5] and finally the resonance capture term A_r which requires some explanation.

Our reasoning assumed that all neutrons introduced into the solution spend practically all their life, and are absorbed, in the thermal state. This is true in so far as the $1/v$ law is valid for absorption of neutrons in all nuclei concerned; and, therefore, not wholly true for uranium, which shows a pronounced resonance capture of neutrons of about 25 volts[6]. A certain proportion of neutrons entering the solution is bound to come within this resonance band and to be absorbed by resonance; therefore, it will never reach the thermal state. This proportion depends on the width of the resonance band and on the concentration c_U; its value in our system of symbols is equal to A_r/A_{tot} and was numerically determined by an experiment reported elsewhere[5].

Putting all numerical values in the formula (2) (with 10^{-24} cm.^{2} as the unit of cross-section), that is: $\Delta S/S = 0.05 \pm 0.01$; $A_{tot} = 36 \pm 3$; $A_f = 1.6 \times 2 = 3.2$; $\Delta A_{tot} = 8.7 \pm 1.4$ decomposable into $\Delta A_H = 1.2 \pm 0.1$, $A_{ct} = 1.6 \times (1.3 \pm 0.45) = 2.1 \pm 0.7$, $A_r = 6.4 \pm 1.1$ and $A_f = 3.2$, we find:

$$\nu = 3.5 \pm 0.7.$$

We were not able to allow for an error in A_f, since the value of σ_f given by Anderson *et al.* contains no indication of probable error. Any error in σ_f will affect $\nu - 1$ in an inversely proportional way; in any case ν will remain greater than 1.

The interest of the phenomenon discussed here as a means of producing a chain of

$(A_{tot}$ (铵) $+\Delta A_{tot})/2$ 来进行简化。

在 ΔA_{tot} 这个量中，铀原子核由代表不同中子俘获模式的分离项所描述（如下）；设 A_f 为引发裂变的俘获项。每个中子都有 A_f/A_{tot} 的引起裂变的概率，而且因为一次单独的裂变过程平均释放 ν 个中子，这样产生的中子总数 $\Delta Q = Q \cdot \frac{A_f}{A_{tot}} \cdot \nu$，则式(1) 可以写成如下形式：

$$\nu = \frac{\Delta S}{S} \cdot \frac{A_{tot}}{A_f} + \frac{\Delta A_{tot}}{A_f} \tag{2}$$

让我们估算一下用上式来计算 ν 所必需的量。面积变化量 $\Delta S/S$ 可以从我们以前文章的图中得知，其误差小于 20%（误差是由插值和外推的不确定性造成；为了便于外推，我们在曲线 $r = 29$ 厘米处增加了一个实验点）。利用已知的浓度和俘获截面（氢、氮和氧），可以很容易地计算出铵溶液中的 A_{tot} 值。A_f 等于 c_U（在我们实验中值是 1.6）乘以安德森等人[4] 在近期文章中给出的 σ_f 值。ΔA_{tot} 中包含了一个表示两种溶液中关于氢含量微小差异的项，以及三个与铀相关的项，即前面已经讨论过的裂变项 A_f，可以用最近得到的 σ_{ct} 值来计算的热俘获项 A_{ct}[5]，还有共振俘获项 A_r，最后这一项需要一些解释。

我们合理地假设：所有进入溶液的中子，几乎全部寿命都处于热态，并且在这个态被吸收。只要保证所涉及的原子核对中子的吸收满足 $1/v$ 定律，这个假设就是成立的；然而，对于 25 电子伏的中子有显著中子共振俘获的铀[6]，这个假设并不完全成立。因为以一定比例进入溶液的中子会被束缚在这种共振带中，并被共振吸收；因此，它们就不会到达热态。这个比例依赖于共振带宽和浓度 c_U；在我们的符号系统中，其数值等于 A_r/A_{tot}，我们已经有实验确定了它的数值并已在别处发表[5]。

把上述所有的数值都代入公式（2）（以 10^{-24} 平方厘米作为截面的单位），即：$\Delta S/S = 0.05 \pm 0.01$；$A_{tot} = 36 \pm 3$；$A_f = 1.6 \times 2 = 3.2$；$\Delta A_{tot} = 8.7 \pm 1.4$，可以分解为 $\Delta A_H = 1.2 \pm 0.1$；$A_{ct} = 1.6 \times (1.3 \pm 0.45) = 2.1 \pm 0.7$，$A_r = 6.4 \pm 1.1$，$A_f = 3.2$，我们得到：

$$\nu = 3.5 \pm 0.7$$

我们无法推算 A_f 的误差，因为安德森等人给出的 σ_f 值并不包含任何可能的误差值。σ_f 中的任何误差都会对 $\nu - 1$ 产生一个反比例的影响；但无论怎样，ν 都会大于 1。

本文讨论的现象可以作为一种产生链式核反应的方法，我们在以前的文章中

nuclear reactions was already mentioned in our previous letter. Some further conclusions can now be drawn from the results reported here. Let us imagine a medium containing only uranium and nuclei the total neutron absorption of which, as compared to that of uranium, may be neglected (containing, for example, only some hydrogen for slowing down purposes). In such a medium, if $\frac{A_f}{A_{tot}} \cdot v > 1$ (A_{tot} includes now only uranium terms), the fission chain will perpetuate itself and break up only after reaching the walls limiting the medium. Our experimental results show that this condition will most probably be satisfied (the quantity $\frac{A_f}{A_{tot}} \cdot v - 1$, though positive, will be, however, small), especially if one keeps in view that the term A_r, because of the self-reversal of the resonance absorption line, increases much more slowly than the other uranium terms when the uranium content of the medium is increased.

(**143**, 680; 1939)

H.von Halban, jun., F. Joliot and L. Kowarski: Laboratoire de Chimie Nucléaire, College de France, Paris, April 7.

References:

1. von Halban, jun., H., Joliot, F., Kowarski, L., *Nature*, 143, 470 (1939).
2. Dodé, M., von Halban, jun., H., Joliot, F., Kowarski, L., *C.R.*, **208**, 995 (1939).
3. Roberts, R., Meyer, R., Wang, P., *Phys. Rev.*, 55, 510 (1939).
4. Anderson, H., Booth, E., Dunning, J., Fermi, E., Glasoe, G., Slack, F., *Phys. Rev.*, 55, 511 (1939).
5. von Halban, jun., H., Kowarski, L., Savitch, P., *C.R.* (in the Press).
6. Meither, L., Hahn, O., Strassmann, F., *Z. Phys.*, **106**, 249 (1937).

已经提到过它的重要性。这里报道的结果可以得到一些更进一步的结论。设想一种介质中只含有铀和一些与铀相比对中子的吸收可以忽略不计的原子核（比如，只含有一些用于减速的氢）。在这种介质中，如果 $\frac{A_f}{A_{tot}}\cdot\nu>1$（$A_{tot}$ 此处只含有铀的项），链式裂变反应将一直继续下去，直到到达介质的边界为止。我们的实验结果表明，这种条件非常有可能实现（$\frac{A_f}{A_{tot}}\cdot\nu-1$ 的值虽然为正，但是会非常小），特别是考虑到当介质中的铀含量增加时，由于共振吸收线的自蚀，A_r 要比铀的其他项增加得缓慢得多。

（王静 翻译；夏海鸿 审稿）

Control of the Chain Reaction Involved in Fission of the Uranium Nucleus

F. Adler and H. von Halban, jun.

Editor's Note

Because roughly 3.5 neutrons are typically released in a uranium fission event, physicists now understood that chain reactions were probably possible. There is thus a considerable danger, as Felix Adler and Hans von Halban here note, that a sample of uranium might explode. But they also suggest a means for controlling such reactions. Into a system of highly concentrated uranium one might insert materials known to absorb neutrons efficiently, such as cadmium. In so doing, the number of further fission events triggered by one such event could be tuned close to 1, so that the reaction would grow only slowly. This suggestion ultimately pointed the way to the "moderator rods" used to control reactions in modern nuclear reactors.

IT has recently been shown that the number[1] of neutrons liberated[2] in the nuclear fission of a uranium nucleus is sufficiently high to make the realization of a self-perpetuating reaction chain seem possible. The danger that a system containing uranium in high concentration might explode, once the chain is started, is considerable. It is therefore useful to point out a mechanism which gives the possibility of controlling the development of such a chain.

We form an expression which is characteristic for the behaviour of the chain:

$$\nu'' = \frac{A_f}{A}\nu\,(1 - \alpha) \tag{1}$$

A_f being the product of the cross-section for nuclear fission for a thermal neutron of the uranium nucleus with the concentration of the uranium; A_i the product of the absorption cross-section for thermal neutrons of the nucleus of kind i multiplied with its concentration; A the sum of all A_i's, which is to be taken over all kinds of nuclei present in the solution; ν is the average number of neutrons liberated in one fission, α the average probability for a neutron to diffuse out of the system before being absorbed.

The energy liberated by the chain will be

$$E = NF, \tag{2}$$

F being the energy liberated in one fission and N the number of fissions produced by the chain. We have

$$N = \nu'' + \nu''^2 + \nu''^3 + \cdots \tag{3}$$

The chain gives thus a quantity of energy, which is increasing rapidly with time, if ν''

铀核裂变中链式反应的控制

阿德勒，冯·哈尔班

编者按

由于平均每个铀核裂变事件中大约释放 3.5 个中子，物理学家此时已明白链式反应是可以发生的。正如费利克斯·阿德勒和汉斯·冯·哈尔班在本文中所说的，铀样品可能会爆炸，并且会产生相当大的危险。不过他们也提出了一个控制这种反应的方法。对于包含高浓缩铀的系统，人们可以往里面注入已知的可以有效吸收中子的材料，例如镉。通过这种做法，可以将一个裂变事例触发的次级裂变的数目控制在 1 左右，因此整个反应可以缓慢地进行。这个建议从根本上提供了现代核反应堆中反应的“控制棒”的制作途径。

最近实验表明，一个铀核在核裂变时释放[2]出的中子的数目[1]似乎高到足以使自持反应链的实现成为可能。一个含有高浓度铀的体系一旦开始链式反应就可能会爆炸，这种危险是不容忽视的。因此，提出一种可能能够控制这种链式反应发展的机制将是有益的。

我们构建了一个描述链式反应行为的特征表达式：

$$\nu'' = \frac{A_f}{A}\nu(1-\alpha) \tag{1}$$

其中 A_f 是热中子铀核裂变截面与铀浓度的乘积；A_i 是 i 类核的热中子吸收截面与该类核的浓度的乘积；A 是所有 A_i 之和，此求和遍及存在于溶液中的所有类型的核；ν 是一次裂变中释放出中子的平均数目，α 是一个中子在被吸收前扩散到体系外的平均概率。

链式反应释放的能量为：

$$E = NF \tag{2}$$

其中 F 是一次裂变释放出的能量，N 是由链式反应所产生的裂变数目。我们有

$$N = \nu'' + \nu''^2 + \nu''^3 + \cdots \tag{3}$$

于是链式反应给出了能量的值，当 ν'' 大于 1 时，该值随时间的增加快速增加。

is greater than 1. Let us consider the case of a chain which is due to fission produced by thermal neutrons; that is, a chain propagating itself in a system containing sufficient hydrogen for the slowing down of the neutrons.

If the cross-sections for capture or fission of all nuclei present follow the $1/v$ law, ν'' will not depend on the velocity of the neutrons and therefore not on the temperature of the system (since α will in practice be small and since it depends in the first place on the distance necessary for slowing down the neutron; the temperature has, of course, an effect, although it will be very small).

Let us, however, introduce an absorbent, such as cadmium, the cross-section of which does not depend on the neutron energy in the thermal region. We will have, instead of (1),

$$\nu'' = \frac{A_f}{A' + A_c}(1 - \alpha), \tag{4}$$

where A' is the sum of all A_i's following the $1/v$ law and A_c is a constant, the term due to the newly added absorbent. ν'' will now decrease with increasing temperature. At a temperature, which will be characteristic for the composition and the geometrical constants of the system, ν'' will become smaller than unity and the system will stabilize itself somewhere near this temperature; the equilibrium being determined by the fact that the amount of energy given out per unit of time by the system (in the form of heat and nuclear radiation) is equal to the energy produced by the system. Similar questions have been discussed by F. Perrin[3].

Added in proof: In the case of a chain propagating itself by thermal neutrons, the time necessary for the slowing down and for the absorption of a neutron, that is, its mean life, is of the order of 10^{-4} sec. If one makes ν'' as small as 1,007, it needs 100 times the mean life of a neutron or about 10^{-2} sec. to double the number of neutrons, and with that the energy liberated per unit of time. It is therefore possible to control the development of the chain by a periodical interaction of absorbers which break up the chains by entering the system.

(**143**, 793; 1939)

F. Adler and H. von Halban, jun.: Laboratoire de Chimie Nucléaire, Collège de France, Paris.

References:

1. von Halban, jun., H., Joliot, F., and Kowarski, L., *Nature*, **143**, 470 (1939).
2. von Halban, jun., H., Joliot, F., and Kowarski, L., *Nature*, **143**, 680 (1939). Roberts, R., Meyer, R., and Wang, P., *Phys. Rev.*, **55**, 510 (1939). Haenny, C., and Rosenberg, A., *C.R.*, **208**, 898 (1939). Szilard and Zinn (private communication). Huber and Buldinger (private communication).
3. Perrin, F., *C.R.*, in the Press.

让我们来考虑由热中子引发的裂变所导致的一个链式反应；即一条在含有足够用来慢化中子的氢的体系中的自持增殖链。

如果所有核的俘获或裂变截面都遵循 $1/v$ 定律，那么 ν″ 就不会依赖于中子的速度，因此也就不依赖于系统的温度（因为 α 实际上会是很小的，而且它首先取决于慢化中子所必需的距离；当然，温度也会产生一定的影响，尽管这种影响很小）。

不过，我们可以引入一种吸收剂，比如镉，它的截面不依赖于热区的中子的能量。我们将用下面的式子来代替式（1）：

$$\nu'' = \frac{A_f}{A' + A_c}(1 - \alpha) \tag{4}$$

其中 A' 是所有遵循 $1/v$ 定律的 A_i 的和，而 A_c 是一个常数，它取决于新加入的吸收剂。此时的 ν″ 将会随着温度的增加而减小。在系统的组成和几何常数的特征温度下，ν″ 会变得比 1 还小，而系统则会在该温度附近的某个温度下达到自稳定；平衡是由下面的事实所决定的，即单位时间内系统释放出的能量大小（以热或者核辐射的形式）等于系统产生的能量。佩兰曾经讨论过类似的问题[3]。

附加说明：在热中子所导致的自持增殖链的情况中，对中子慢化和吸收所需要的时间，即中子的平均寿命，为 10^{-4} 秒的数量级。如果使 ν″ 小到 1,007，就需要 100 倍于中子平均寿命或者说大约 10^{-2} 秒的时间来使中子数量加倍，同时单位时间释放的能量也随之翻倍。于是，有可能通过利用吸收剂周期性的相互作用（它会进入系统破坏链式反应）来控制链式反应的发展。

（王耀杨 翻译；张焕乔 审稿）

Emission of Neutrons Accompanying the Fission of Uranium Nuclei

J. Rotblat

Editor's Note

The Polish physicist Joseph Rotblat here supplies further evidence on the release of neutrons in the fissioning of uranium nuclei, indirect evidence for which had already been reported. Using a new technique, he suggests that the number of secondary neutrons produced in a fission event might be as high as six. Rotblat went on to work on the Manhattan Project which led to the atomic bomb, but after the war became a fierce critic of nuclear weapons and worked tirelessly on international arms control. In 1957, with the philosopher Bertrand Russell, Rotblat founded the highly influential Pugwash Conferences, which bring together scientists and public figures in an effort to reduce the proliferation of nuclear weapons.

THE experiments of Halban, Joliot, and Kowarski described in *Nature*[1] give an indirect proof of the neutron multiplication accompanying the fission of uranium nuclei after neutron capture. It can be deduced from these experiments that the additional neutrons are, on the average, faster than the photo-neutrons from radium C-beryllium used as active primary radiation. This conclusion is confirmed in a subsequent note[2] by the same authors in which they find that the neutrons contributed by uranium are able to produce the endo-energetic reaction S(*n*, *p*)P.

I have made some experiments on the same problem by a different method. Any increase of neutron effects observed when neutrons are allowed to pass through a given medium may be due: (*a*) to the inelastic scattering of neutrons, (*b*) to the reaction (*n*,2*n*) or, exceptionally (*n*,3*n*); (*c*) to an unknown cause, as in the case of uranium to the fission of its nuclei. In the experiments of Halban, Joliot and Kowarski, effects (*a*) and (*b*) are excluded owing to the integrating method adopted and to the low energy of the primary neutrons. But if these effects are present, it is possible to estimate their importance relatively to effect (*c*) by comparing uranium with substances in which only effect (*a*) or only (*a*) and (*b*) are possible. I have used aluminium and copper as comparison substances of the first and of the second type respectively.

A radon plus beryllium source was placed in the cylindrical axial hole of a cylinder of aluminium, of 2.2 cm. diameter and 5 cm. height, or, alternatively, of a cylindrical double-walled vessel of identical dimensions filled with uranium oxide (U_3O_8) or copper oxide. The mass of aluminium was 40 gm., of uranium oxide 49 gm. and of copper oxide 42 gm., and the thickness of walls could be neglected. The number of absorbing or scattering uranium nuclei was therefore 9.2 times smaller than the corresponding number

伴随铀核裂变的中子发射

罗特布拉特

编者按

波兰物理学家约瑟夫·罗特布拉特在本文中提供了关于铀核裂变过程中中子释放的进一步证据，此前已经报道过相关的间接证据。他使用了一项新技术，并提出一个裂变事件中产生的次级中子数目可能高达6个。罗特布拉特持续从事研制原子弹的曼哈顿项目工作，但是战后他成了核武器的强烈批评者，并且坚持不懈地致力于国际核武器控制。1957年，罗特布拉特和哲学家伯特兰·罗素一起发起了具有极大影响力的帕格沃什会议，这次会议上科学家和社会名人一起努力以实现减少核武器的扩散。

哈尔班、约里奥和科瓦尔斯基在《自然》中所描述的实验[1]为在中子俘获之后伴随着铀核裂变的中子倍增提供了间接证据。从这些实验可以推知，一般来说，这些新产生的中子比来自作为放射性初级辐射的镭C–铍源的光中子更快。这一结论在这几位作者随后的一篇文章[2]中得到证实，他们发现由铀贡献的中子能够产生吸能反应 S(*n*, *p*)P。

我对同样的问题用不同方法做了一些实验。在令中子通过一种给定介质的实验中，所观察到的中子效应的任何增长都可能是由于以下原因造成的：(*a*) 中子的非弹性散射；(*b*)（*n*,2*n*）反应，个别时候是（*n*,3*n*）反应；(*c*) 未知原因，就像在铀核裂变的情况中那样。在哈尔班、约里奥和科瓦尔斯基的实验中，由于集成方法的采用和初级中子的能量较低，故可以排除效应（*a*）和（*b*）。但是如果这些效应存在的话，就有可能通过将铀与只可能有（*a*）效应或只有（*a*）和（*b*）效应的物质进行比较，进而估计出它们相对于效应（*c*）的重要性。我曾经使用铝和铜分别作为第一类和第二类对比物质。

把氡–铍放射源置于一个铝质圆柱体（直径为2.2厘米，高为5.0厘米）轴向的孔中，或者，也可以采用同样尺寸的装填有氧化铀（U_3O_8）或氧化铜的圆柱形双层壁容器。铝的质量为40克，氧化铀质量为49克，氧化铜质量为42克，器壁厚度是可忽略的。结果吸收或散射的铀核数目是相应的铝核数目的1/9.2，是铜核数目的

of aluminium nuclei and 3.3 times smaller than the number of copper nuclei. One would expect, therefore, that the effect (*a*) due to aluminium, and effects (*a*) and (*b*) due to copper would be at least of the same importance and probably larger than the same effects due to uranium. The number and quality of neutrons issuing from these substances were compared by measuring the activation of a silver foil surrounding the cylinders in two cases: first, when no appreciable scattering of neutrons took place outside the cylinders, and secondly, when the neutrons were scattered back by a cylindrical sheet of paraffin wax of 6 mm. thickness. The results are given below, the figures being the total numbers of counts of a Geiger-Müller counter in corresponding series.

	Aluminium	Uranium	Increase (%)	Copper	Uranium	Increase (%)
No paraffin	9,089	9,325	2.6	4,810	4,869	1.2
With paraffin	10,285	10,775	4.8	10,049	10,292	2.4

It can be inferred from these data that uranium gives off, in fact, more neutrons than aluminium or copper. The increase is larger relatively to aluminium than to copper, which must be attributed to the reaction $(n, 2n)$ occurring in this last element. In both cases, the increase is larger when the neutrons are slowed down by a small quantity of paraffin, which shows that the additional neutrons from uranium are, on the average, slower than the bulk of primary neutrons emitted by the source. As, from other evidence, they appear to be faster than the radium C-beryllium neutrons, we can estimate that their *average* energy must be of the order of 1 Mev.

Owing to the small number of uranium nuclei acting as absorbers or scatterers, it seems very unlikely that the apparent excess of neutrons given off by these nuclei should be due to some trivial cause like the inelastic scattering or the reaction $(n, 2n)$. It probably represents the neutron shower' accompanying the fission of an activated uranium nucleus. Assuming the cross-section for this process produced by the neutrons from radon plus beryllium equal to 5×10^{-25} cm.[2], I calculate that the number of neutrons emitted in a single fission is equal to 6.

(**143**, 852; 1939)

J. Rotblat: Miroslaw Kernbaum Radiological Laboratory of the Scientific Society of Warsaw, Warsaw, April 8.

References:

1. *Nature*, **143**, 470 (1939).
2. *C.R.* **208**, 995 (1939).

1/3.3。由此我们预期，铝产生的效应（*a*）与铜产生的效应（*a*）和（*b*）至少是同样重要的，甚至可能比铀产生的同种效应更大。我们将这些物质所发出的中子的数目和性质进行比较，比较的方法是分别在下述两种情况下测量包裹在圆柱体周围的银箔的放射性：第一种情况是在圆柱体外部没有发生明显的中子散射时，第二种情况是中子经 6 毫米厚的圆柱形石蜡层散射而返回时。下面数据列出了该实验的结果，数字表示用盖革–米勒计数器在相应的序列中得到的总计数。

	铝	铀	增量（%）	铜	铀	增量（%）
无石蜡	9,089	9,325	2.6	4,810	4,869	1.2
有石蜡	10,285	10,775	4.8	10,049	10,292	2.4

根据这些数据可以推测，事实上铀比铝或者铜发射出更多的中子。相对于铝的增量比相对于铜的增量要大，这必定是因为后一种元素中发生了（*n*,2*n*）反应。在铝和铜两种情况中，当中子被少量石蜡慢化时这种增量是比较大的，这表明从铀所产生的新的中子平均来说比放射源发射出的大量初级中子慢。由于在其他的证据中，它们似乎比镭 C–铍中子快，因此我们可以估算出它们的**平均**能量必定具有 1 兆电子伏的数量级。

由于作为吸收体或散射体的铀核数目较小，所以由这些核发射出的中子的明显过剩似乎不可能归结为某些细微的原因，比如非弹性散射或（*n*,2*n*）反应。它可能代表了伴随着一个放射性铀核的裂变而产生的中子簇射。假定由氡–铍源的中子产生这个过程的截面等于 5×10^{-25} 平方厘米，我计算出单独一次裂变发射出的中子数等于 6。

（王耀杨 翻译；张焕乔 审稿）

Fission of Heavy Nuclei: a New Type of Nuclear Disintegration

N. Feather

Editor's Note

Physicist Norman Feather offers a review of advances in the understanding of nuclear fission reactions, and the possibility of a controlled chain reaction. In retrospect, he notes, the first evidence for fission in uranium nuclei came in experiments by Enrico Fermi and colleagues in 1934, though five years passed before the physics could be clarified. It was now clear that each such fission gave off several neutrons on average, and may well suffice to drive a sustained chain reaction. However, experiments also showed that slow "thermal" neutrons are most effective in stimulating fission events. No experiments on chain reactions had at that point been carried out, however.

THE first indication that the transmutation of heavy nuclei could be effected in a laboratory experiment was obtained by Fermi in March 1934. Curie and Joliot had just discovered that short-lived radioactive species are produced as the result of α-particle bombardment of certain light elements, and Fermi, accepting the appearance of such "induced" radioactivity as proof of transmutation, very soon showed that the nuclei of almost all elements, even those of highest atomic weight, undergo transformation when neutrons are used. From his early experiments Fermi concluded that in general the neutron is simply captured by the nucleus, and he went on to show that this process of capture is usually more efficient—and sometimes very much more efficient—when the neutron is moving with a small ("thermal") velocity, before the collision, than when its energy is large. Eventually he found that negative electrons were emitted in the disintegration of the radioactive products obtained in all these capture transformations, and thus the final result of the combined process of neutron capture and β disintegration was in every case shown to be the production of a nucleus having both mass and charge numbers greater by one unit than the mass and charge numbers of the nucleus bombarded.

In June 1934[1], Fermi and his collaborators obtained negative electron activities from thorium and uranium under neutron bombardment, and they quite naturally supposed, on the basis of their previous investigations, that in the latter case a nuclear species of atomic number $Z = 93$ must remain after disintegration of the unstable product first formed. Further examination showed that not one but several distinct radioelements were produced as a result of the bombardment of uranium, and rough chemical tests proved that one of these, at least (half-value period $\tau = 13$ minutes), was not attributable to an element of atomic number $Z = 92, 91, 90, 89, 88, 83$ or 82. It seemed quite clear, then, that after the uranium nucleus had captured a neutron, not less than two β transformations followed, and consequently that a new element, for which Z was not less than 94, was ultimately

重核裂变：一种新型的核蜕变

费瑟

编者按

物理学家诺曼·费瑟根据他对核裂变以及可控链式反应的可能性的理解写了一篇综述性文章。他指出实验上首次观察到铀核裂变的证据是在1934年由恩里科·费米及其同事发现的，然而5年后裂变的物理原理才被阐释。现在人们知道每个这样的裂变反应平均释放几个中子，并且有可能驱动持续的链式反应。但实验也表明慢的"热"中子对于激发裂变是非常有效的。不过关于这一点，还没有链式反应实验予以证实。

在实验室所进行的实验中，费米于1934年3月首次发现影响重核嬗变的迹象。居里和约里奥刚刚发现α粒子轰击某些轻元素时会产生短寿命的放射性核素，费米便认识到这种"诱发的"放射性的出现可以作为嬗变的证据，并且很快指出，几乎所有元素的核——即使是那些具有最大原子量的核——在用中子轰击时都会发生转变。费米根据自己的早期实验推断，中子通常只是被核俘获，他接着指出，当撞击前的中子以慢的（"热"）速度运动时，这个俘获过程一般会比它具有高能量时更为有效——而且有时会高效很多。最后他发现，在所有这些俘获转变中生成的放射性产物，它们的分裂都会释放出负电子。因此，在任何情况下，兼有中子俘获与β蜕变的组合过程的最终结果都显示出，生成的核的质量和电荷数都比被轰击核大一个单位。

1934年6月[1]，费米及其合作者通过用中子轰击钍和铀，得到了负电子放射性。于是，基于之前的研究，他们很自然地认为，在后一情况中，最初形成的不稳定产物蜕变之后一定会有一种原子序数$Z=93$的核素保存下来。进一步的检测表明，轰击铀的结果不是生成了一种放射性元素，而是生成了几种截然不同的放射性元素，而粗略的化学检验证明，其中至少有一种（半衰期$\tau=13$分钟）不能归结为原子序数$Z=92$、91、90、89、88、83或82的元素。于是，看起来非常明确，铀核在俘获了一个中子后至少会接着发生两次β转变，因而最终会生成一种Z不小于94的

produced. These conclusions at once gave rise to much argument amongst chemists as to what the chief chemical properties of these hypothetical transuranic elements might be, but a great deal of careful research in many physical laboratories throughout the world only served to strengthen the assumption that, apart from any preconceived ideas about chemical behaviour, such elements were certainly formed when uranium was bombarded by neutrons.

By May 1937, Meitner, Hahn and Strassmann[2] had recognized the existence of nine separate species arising from this transformation, and had suggested genetic relations between them which supposed that every element of atomic number between (and including) 92 and 97 was represented, either as intermediate or end product of a series. They identified three of these nine products (τ = 10 sec., 40 sec., 23 min.) as unstable isotopes of uranium, and the remainder, which could all be obtained by precipitation with platinum as sulphide in acid solution, as the transuranic elements already mentioned. The scheme presented certain difficulties, it was admitted, (no evidence for the final return of the unstable nucleus within the ordinary range, Z less than 93, could be found, and awkward questions concerning isomeric forms were raised in an acute form) but it was the best that could be done.

Then, in October 1937, Curie and Savitch[3] discovered a tenth activity of about $3\frac{1}{2}$ hours period, and at once proceeded to investigate the chemical nature of the element to which it belonged. They showed that this radioelement did not separate with the platinum precipitate and soon discovered that it bore a close resemblance to lanthanum in chemical behaviour. At first the most plausible suggestion appeared to be that the new species was an isotope of actinium (Z = 89). Then two very disturbing facts were encountered. First, a body of the same period and almost identical properties was among the active products formed in the disintegration of thorium by neutrons, and, secondly, it was found possible to separate actinium almost completely from the new body by a lanthanum fractionation. In September 1938, Curie and Savitch[4] wrote, "On the whole, the properties of R 3.5 h. are those of lanthanum, from which it appears that until now it has not been separated".

The work of Curie and Savitch immediately prompted a further search for activities belonging to elements of atomic number less than 92 produced in the uranium transformation, and, as a result, Hahn and Strassmann[5] discovered two other lanthanum-like and three barium-like products. These workers believed that they had demonstranted the production of each of the former species from one of the latter, but at that time they still inclined to the view that actinium and radium isotopes were really in question. However, in January of this year, they reported[6] that fractionation of the new bodies with mesothorium and barium (or lanthanum) invariably concentrated the neutron-produced activity with the lighter carrier and resulted in a complete separation of mesothorium 1 (radium), or mesothorium 2 (actinium). The conclusion now appeared inescapable that active isotopes of barium and lanthanum were among the products of the bombardment of uranium with neutrons.

新元素。这些结论立刻在化学家中间引起了很多争论，内容是关于这些假想的超铀元素可能具备的主要化学性质是什么的问题。然而除去任何预先的关于化学行为的想法，在全世界多所物理实验室中所进行的大量细致的研究却只能用来支持下面的假设，即这些元素确实是在中子轰击铀时形成的。

到 1937 年 5 月时，迈特纳、哈恩和施特拉斯曼[2]已经认识到存在着 9 种由这类转变所产生的独立核素，并且提出了它们之间的亲缘关系，这种关系假定原子序数介于（且包括）92 和 97 之间的每一种元素不是作为某一序列的中间产物就是作为其最终产物而出现。他们鉴别出这 9 种产物中的 3 种（半衰期 τ=10 秒、40 秒、23 分钟）是铀的不稳定同位素，而其余的几种都可以通过在酸溶液中与铂以硫化物的形式沉淀而得到，它们被视为是前面提到的超铀元素。虽然这一方案存在着某些难点，但人们还是接受了，（没有能够找到最终收回的不稳定核处于 Z 小于 93 的常规范围内的证据，这就尖锐地提出了有关异构形式的棘手问题），而这也是当时所能做到的方案中最好的一个。

然后，1937 年 10 月，居里和萨维奇[3]发现了半衰期为 3.5 小时的第 10 种放射性，并立即着手研究该放射性所属元素的化学本质。他们指出这种放射性元素不能用铂沉淀来分离，并且很快发现它的化学行为与镧极为相似。最初看来最有道理的观点似乎是，这一新核素是锕（$Z=89$）的某种同位素。但是后来遇到两个令人极为困扰的事实。第一，在中子轰击所导致的钍的分裂中形成的放射性产物中，存在着某种具有相同周期和几乎同样性质的物质，第二，发现通过镧分级分离的方法，从这种新物质中有可能将锕几乎完全分离出来。1938 年 9 月居里和萨维奇[4]写到，“大体上讲，半衰期 3.5 小时就是镧的性质，而到目前为止，似乎仍不能把这种新元素与镧分离开来”。

居里和萨维奇的工作直接促进了在铀转变过程中产生的原子序数小于 92 的元素的放射性的进一步研究。而其结果是，哈恩和施特拉斯曼[5]发现了另外两种类镧产物和三种类钡产物。这些研究者相信，他们已经论证了所有类镧产物都是由类钡产物中的一种产生的，但是在那时，他们仍然倾向于这样的观点，即锕和镭的同位素有待研究。不过，今年 1 月他们报道了[6]含有新钍和钡（或镧）的新物质的组分，总是可以通过较轻载体聚集由中子引起的放射性，从而导致新钍 1（镭）或新钍 2（锕）的完全分离。这时得到如下结论似乎是顺理成章的，即在铀经中子轰击所得产物之中存在着钡和镧的放射性同位素。

At this stage, Meitner and Frisch[7] discussed the problem on the Bohr theory of heavy nuclei, making particular use of the essentially classical "water-drop model" of the highly condensed system of particles of which such a nucleus is constituted. They concluded, "It seems therefore possible that the uranium nucleus has only small stability of form, and may, after neutron capture, divide itself into two nuclei of roughly equal size. ... These two nuclei ... should gain a total kinetic energy of *c.* 200 Mev. ... This amount of energy may actually be expected to be available from the difference in packing fraction between uranium and the elements in the middle of the periodic system". Then Frisch[8] obtained direct evidence for the projection of fission fragments with approximately the energy predicted, being able to detect the production of large bursts of ionization in a uranium-lined ionization chamber which was irradiated by neutrons. Similar results were obtained when thorium was substituted for uranium in the chamber, and it was concluded that some of the activities previously ascribed to isotopes of radium and actinium, in this case also resulted from fission of the nucleus under neutron bombardment.

The investigations begun by Meitner and Frisch were rapidly followed by many others in physical laboratories both in Europe and in the United States: the confirmation of the findings of Curie and Savitch by Hahn and Strassmann had indicated quite clearly to many workers that something new was involved. In Paris, in Berkeley, in Washington, New York and Baltimore, direct proof of the fission of uranium and thorium was obtained within the space of a few days. Now, some three and a half months after the original announcement, so much has been published that rigorous selection is necessary in any report on the subject. For the remainder of this survey, therefore, only the most interesting features of the new phenomenon can possibly be included.

Perhaps the first such feature concerns the radioelements of the platinum precipitate, of which the previously supposed transuranic nature was now in question. Before the fission process was discovered, both in Berkeley and in Cambridge projects had already been formed of investigating these elements for natural X-radiations, in the hope of being able to deduce the atomic number from the energy of the radiations (natural *L*-radiations) as determined by the method of critical absorption. This problem became much simpler once the presence of medium-heavy elements was suspected, since *K*-radiations could be looked for, instead of the more complex radiations of the *L*-series. Almost at once, Abelson[9] and Feather and Bretscher[10] found evidence for the natural *K*-radiations of iodine from the long-lived activities of the platinum precipitate, and, guided by this observation, were able to identify chemically as tellurium and iodine two products previously described as eka-iridium and eka-platinum, respectively. Then several workers found other of the so-called transuranic activities in the products collected by recoil from bombarded uranium. Observations concerning the rates of decay of these recoil activities[11]—and the results of chemical tests[12]—left little doubt that almost all the previous assignments had been seriously in error. At the present time, one might justly say that it cannot definitely be maintained, concerning any of the activities separable from uranium, that it does not arise in a process of fission of the uranium nucleus.

在这个阶段，迈特纳和弗里施[7]对玻尔的重核理论中存在的问题进行了讨论，尤其是应用了本质上经典的“液滴模型”来描述构成这种核的粒子所组成的高密度系统。他们得出结论，“由此看来，可能铀核只具有较低稳定性的形式，并且可能在俘获中子之后将自身分裂成两个大小基本相同的核。……这两个核……应该获得大约200兆电子伏的总动能。……根据铀和位于周期表中部的元素之间聚集率的差异，实际上这部分能量预计是可以得到的”。接着，弗里施[8]获得了裂变碎片基本上是以预计能量发射的直接证据，在用中子照射以铀填衬的电离室时，检测到有强烈的电离脉冲出现。在用钍代替电离室中的铀的情况下，得到了类似的结果，从而得出结论，某些从前被归结为来源于镭和锕的同位素的放射性，在这种情况下也是由中子轰击下发生的核裂变所致。

由迈特纳和弗里施所开创的研究很快便引发了欧洲及美国物理实验室的众多研究人员的追随：哈恩和施特拉斯曼对居里与萨维奇的发现进行的证实已经向很多研究者非常明确地指出，这涉及了某些新的事物。在巴黎、伯克利、华盛顿、纽约和巴尔的摩，短短几天的时间内就得到了有关铀和钍裂变的直接证据。现在，在最初宣布之后大约三个半月的时间内，已经发表了大量的文章，这样就有必要对关于这一主题的报道进行严格的筛选。因此，在这篇综述的其余部分中将只能包括有关新现象的最值得关注的特征。

也许第一个值得关注的特征就是关于铂沉淀物中的放射性元素，以前曾认为其所具有的超铀性质现在遭到了质疑。在发现裂变过程之前，对这些元素的天然X辐射研究项目在伯克利和剑桥已经展开，希望能够用通过临界吸收方法测定的辐射能量(天然的L辐射)来推断原子序数。一旦假设存在中重元素，这个问题就变得非常简单了，因为这样就可以去寻找K–辐射，而不再需要寻找较为复杂的L序列的辐射。埃布尔森[9]以及费瑟和布雷切尔[10]几乎立刻从铂沉淀物的长寿命放射性中发现了碘的天然K辐射的证据，并且在这一观测结果的指引下，将两种最初分别被描述为类铼和类铂的产物，用化学方法确定为碲和碘。接着，在收集到的被轰击的铀的反冲产物中，几位研究者发现了其他一些所谓的超铀放射性。这些反冲产物放射性衰变速率的观测结果[11]及化学检测的结果[12]毫无疑问地表明，此前几乎所有的认定结果都是严重错误的。目前，对于任何可以从铀中分离出的放射性，我们都可以公正地说，认为它不是在铀核的裂变过程中产生的这一观点肯定是得不到支持的。

There is, however, one important activity, of 23 minutes half-value period, which is not separable from uranium and for which, in consequence, the fission process cannot be assumed to be responsible. This non-separable activity arises in a process of resonance capture of neutrons of about 25 ev. energy, and the fact that negative electrons are involved must clearly indicate that a species for which $Z = 93$ is formed as the result of the disintegration. Yet, in spite of much careful investigation, no radioactivity of any kind has been discovered which is unquestionably due to the transformation of this species. Furthermore, in respect of the parent species (the uranium isotope for which $\tau = 23$ min.), this is clearly a quasi-stable modification of the body which undergoes fission in the majority of cases, and Meitner and Frisch and Bohr[13] have discussed this aspect of the phenomenon. They have pointed out that there is nothing intrinsically incomprehensible in the occurrence of resonance capture (emission of γ-radiation) rather than fission in certain circumstances. In any event, division of the nucleus into two fragments must be preceded by the concentration of the available energy in a type of nuclear motion of large deformation, and this concentration may be very unlikely if the original state of the system, after capture of the neutron, is one of considerable symmetry. Or if, as Bohr has assumed, the effect of thermal neutrons in producing fission in the case of uranium (such neutrons are quite ineffective when thorium is bombarded) is ascribed to capture by the rare isotope ^{235}U, and the resonance effect and the fission process due to fast neutrons are ascribed to ^{238}U, it may even be that, at the resonance energy, the compound nucleus is not formed with sufficient energy of excitation for the neutralization of the small stability of form which it naturally possesses. As regards the whole of this question, more definite conclusions must clearly await further experiments: Joliot[14] has reported a variation in the relative proportions of the different fission products as the energy of the neutrons is altered—and this might be held to favour the suggestions of Bohr—but Bjerge, Brostrøm and Koch[15] have failed to establish any difference in the decay of the products obtained with high-energy neutrons and thermal neutrons, respectively.

Hitherto, the process of fission has been spoken of without any precise statement regarding the nature of the fragments which result from the primary act of division of the nucleus, and in fact very little exact knowledge is as yet available on this point. Determinations of the range of the fission products provide some information. In the first place, they indicate that (with uranium and fast neutrons) the process occupies less than 5×10^{-13} sec. from the time of capture of the neutron (the forwards range is slightly greater than the backwards range, showing that very little of its original momentum is lost before the compound nucleus divides[16]), and, secondly, they appear to favour a very small number of competing primary processes, rather than a large number of possibilities[17]. On the other hand, chemical investigation reveals such a wealth of active products[18] (with atomic numbers lying between 35 (Br) and 57 (La), if not more widely distributed) that some adequate explanation of their complexity must certainly be found. It would appear that the discovery that very frequently neutrons are emitted almost instantaneously by the original products of fission[19] already provides a basis for such explanation (nuclei, first formed, presumably, in states of high excitation, emit either neutrons or quanta of radiation in passing to longer-lived states). Also, even after these states have decayed with

但是，存在着一种半衰期为23分钟的重要的放射性，它是不能从铀中分离出来的，因此，不能认为它是在裂变过程中产生的。这种不可分离的放射性来源于能量约为25电子伏的中子共振俘获的过程，而有负电子参与的事实必然明确表明，一种 $Z = 93$ 的核素的形成是由蜕变所导致的。可是，尽管已进行了很多细致的研究，却仍没有发现任何一种无疑是来自于该核素的转变过程的放射性。此外，关于母体核素（相应的铀的同位素半衰期 τ=23分钟），很明显它是该物质的一种准稳态变体，在大多数情况下会发生裂变，迈特纳、弗里施和玻尔[13]已经就这种现象对这个方面的问题进行了讨论。他们已经指出，在某些情况下发生共振俘获（γ射线的发射）而不是裂变，本质上并不存在任何不可理解的问题。无论如何，核分裂成两块碎片之前，必定要先通过核的巨大形变的运动形式聚集可用的能量，而如果在俘获中子之后，系统的初始状态仍然具有相当大的对称性，那么出现这种聚集的可能性就非常小了。或者如果，就像玻尔假设的那样，热中子对引起铀的裂变的影响（在轰击钍时这样的中子所产生的效果是相当弱的）是由稀有同位素铀–235的俘获所造成的，而由快中子导致的共振效应和裂变过程是由铀–238所造成的，那么甚至可以说，由于复合核天然具有形式的低稳定性导致的抵消作用，使得它在共振能量下形成时不会具有足量的激发能。就这个问题整体而言，要得到更为确定的结论显然必须等待进一步的实验研究：约里奥[14]曾报道过中子能量发生改变时会发生不同裂变产物相对比例的变化，而这有可能被认为是对玻尔的提议的支持。但是对于分别由高能中子和热中子得到的产物，伯格、布罗斯特伦和科赫[15]却没有从它们的衰变过程中发现任何差别。

到目前为止，谈及的裂变过程仍未给出任何关于裂变碎片性质的精确陈述，这些碎片是在核分裂第一阶段产生的，而实际上至今对这一点的确切的认识也只是极少量的。对裂变产物范围进行测定可以提供一些信息。首先，它们意味着（利用铀和快中子）该过程从俘获中子时刻（向前的范围略大于向后的，这表明在复合核分裂之前，初始动量几乎没有损失[16]）算起持续时间不足 5×10^{-13} 秒。其次，它们似乎倾向于极少量的竞争性初级过程，而不具有该过程大量出现的可能[17]。另一方面，化学研究显示出如此丰富的放射性产物[18]（如果没有更广泛的分布，其原子序数介于35（Br）到57（La）之间），这样就必定可以对它们的复杂性找到一些充分的解释。看来，裂变初始产物经常几乎是瞬间发射出中子[19]，这一现象的发现，已经为这种解释提供了基础（据推测，最初以高激发态形成的核，在到达寿命较长态的过程中会发射出中子或者量子辐射）。实验也表明，甚至当这些态发生β衰变之后，有时产物核仍然具有足够的激发能，为了到达更稳定态，相对于辐射跃迁而言“蒸发”中子

β-emission, experiment shows that occasionally product nuclei result, still with sufficient energy of excitation for the "evaporation" of neutrons to be a possible alternative to radiative transitions leading to more stable states. In this connexion, Roberts, Meyer, and Wang[20] and Booth, Dunning and Slack[21] have reported delayed neutron periods of about 12 sec. and 45 sec., when uranium is bombarded, whilst a similar feature has also been established in the case of thorium.

The frequency of the neutron-evaporation process accompanying fission, and the energies of the neutrons so produced, have been studied by many workers, but so far most exhaustively by Joliot[22] and his colleagues, and by Fermi[23] and others in New York. The general result appears to be that, for each process of fission with uranium, at least two neutrons, having a mean energy of the order of 10^6 ev., eventually evaporate from the residual fragments. Since neutrons of less than this energy are still capable of producing fission on their own account (probably in ^{235}U, as already suggested), the possibility of a cumulative process of exothermic disintegration has to be considered. Clearly, if the probability of removal of neutrons in processes other than those which result in fission is sufficiently reduced, the latter process must eventually build up in any solid substance containing uranium. Direct experiments on this aspect of the matter have not yet been reported in the scientific literature, but at this stage it may be pointed out that, even in pure uranium, it is well known that a non-fission capture process takes place (*v. sup.*), whilst the unlimited generation of energy in the solid material would ultimately increase the energy of the "thermal" neutrons until their efficiency as agents for fission was greatly reduced. Already several attempts have been made[24] to calculate the course of the phenomenon using existing data, but the assumptions upon which they have been based have generally been so severely idealized that no confidence in numerical values is at present likely to result.

(**143**, 877-879; 1939)

N. Feather: Cavendish Laboratory, Cambridge.

References:

1. *Nature*, **133**, 898 (1934).
2. *Z. Phys.*, **108**, 249 (1937).
3. *J. Phys.*, **8**, 385 (1937).
4. *J. Phys.*, **9**, 355 (1938).
5. *Naturwiss.*, **26**, 755 (1938).
6. *Naturwiss.*, **27**, 11 (1939).
7. *Nature*, **143**, 239 (1939).
8. *Nature*, **143**, 276 (1939).
9. *Phys. Rev.*, **55**, 418 (1939).
10. *Nature*, **143**, 516 (1939).
11. Meitner and Frisch, *Nature*, **143**, 471 (1939). Glasoe and Steigman *Bull. Amer. Phys. Soc.*, (2), 14, 20 (1939).
12. Bretscher and Cook, *Nature*, **143**, 559 (1939).

仍是一个可能的选择。关于这一点，罗伯茨、迈耶和王[20]以及布思、邓宁和斯莱克[21]都曾报道，在轰击铀时缓发中子的周期约为12秒和45秒，而对于钍而言，类似的特征也得到了证实。

对于伴随裂变的中子蒸发过程的频率，以及这种过程中产生的中子的能量，已经有很多研究者进行过研究，但是到目前为止最为详尽的当属约里奥[22]和他的同事，以及费米[23]和另外一些来自纽约的研究者。看来一般性的结论是，对于铀的每一次裂变过程，至少有两个具有10^6电子伏数量级的平均能量的中子，最终会从残存碎片中蒸发。由于能量比这还低的中子仍然可以靠自己来产生裂变（如前所述，可能是在铀–235中），因此必须要对可能出现的放热分裂累积过程加以考虑。很明显，除了中子导致裂变的可能性之外，如果过程中消去中子的可能性充分减小，那么对任何含铀的固体物质，裂变过程最终都一定会逐渐累积。在科学文献中还没有关于该问题的直接实验的报道，但是目前我们可以指出，即使是在纯铀中，众所周知会发生非裂变俘获过程（见上），同时在固体材料中无限制产生的能量最终将会增加“热”中子的能量，直到它们引发裂变的效能大大减弱为止。已经有人利用现有数据进行过几次计算该现象过程的尝试[24]，但是一般来说他们所依赖的前提过于理想化，因而这些数值目前还不大可能获得认可。

（王耀杨 翻译；夏海鸿 审稿）

13. *Phys. Rev.*, **55**, 418 (1939).
14. *J. Phys.*, **10**, 159 (1939).
15. *Nature*, **143**, 794 (1939).
16. Feather, *Nature*, **143**, 597 (1939).
17. McMillan, *Phys. Rev.*, **55**, 510 (1939).
18. Heyn, Aten and Bakker, *Nature*, **143**, 516 (1939). Abelson, *phys. Rev.*, **55**, 670 (1939).
19. v. Halban, Joliot and Kowarski, *Nature*, **143**, 470 (1939).
20. *Phys. Rev.*, **55**, 510 (1939).
21. *Bull. Amer. Phys. Soc.*, (2), **14**, 19 (1939).
22. v. Halban, Joliot and Kowarski, *Nature*, **143**, 680 (1939). Dodé, v. Halban, Joliot and Kowarski, *C.R.*, **208**, 995 (1939).
23. Anderson, Fermi and Hanstein, *Phys. Rev.*, **55**, 797 (1939). Szilard and Zinn, *Phys. Rev.*, **55**, 799 (1939).
24. For example, Perrin, *C.R.*, **208**, 1394 (1939).

Energy of Neutrons Liberated in the Nuclear Fission of Uranium Induced by Thermal Neutrons

H. von Halban, jun. *et al.*

Editor's Note

The fissioning of a uranium nucleus, physicists now knew, releases further neutrons, which might in principle trigger splitting in other nuclei. But researchers knew little about the energy of the released neutrons. Here Hans von Halban and colleagues in Paris clarify this matter. In a series of experiments, they used an ionization chamber to probe the distribution of energies of "fast" neutrons—that is, those having energy above 1.5 million electronvolts (1.5 MeV). The number of neutrons declines at higher energies, but some neutrons can carry away as much as 11 MeV. Knowledge of the distribution of neutron energies played an important role in the engineering of nuclear reactors, as slow neutrons trigger fission events more effectively than fast ones.

IT has been shown that *fast* neutrons are liberated in the process of nuclear fission induced in uranium by primary *thermal* neutrons. Two different methods of detection have been used: in the first method[1], the primary and (if any) secondary neutrons are absorbed in a medium in which an endo-energetic reaction can take place, leading to the formation of an easily detectable radioactive nucleus. If the energy threshold is situated above the maximum energy of the primary neutrons, any positive results observed must be ascribed to the secondary neutrons. In the second method[2], elastic collisions of fast neutrons with heavier nuclei are observed by means of an ionization chamber filled with a gas at atmospheric pressure and connected to a linear amplifier. In order to study separately the effect due to the primary thermal neutrons, the experiment is performed with, and without, a cadmium shield between the source and the uranium mass.

The first method having shown us that fast secondary neutrons are produced with energies of at least 2 Mev. (sufficient to transform ^{32}S into radioactive ^{32}P in detectable quantities), we sought to ascertain, by the second method, whether neutrons of energy notably higher than 2 Mev. are also present in the secondary radiation. In our experiment, the oxygen-filled ionization chamber was placed in a nearly cubical box (9 cm. × 9 cm. × 8 cm.) containing uranium oxide and surrounded by a thick layer of paraffin wax. The source (300 mgm. RaY + Be), surrounded by a lead shield (5 cm. in the direction of the chamber) was buried in the wax. In order to absorb thermal neutrons, the uranium box could be screened on all sides with a cadmium foil. The pulses were recorded either in the presence or in the absence of this foil and the part of the effect (projection of oxygen nuclei by fast neutrons liberated in the uranium) due to thermal neutrons could thus be evaluated.

在热中子诱发的铀核裂变中释放出的中子的能量

冯·哈尔班等

编者按

物理学家现在已经知道，原则上铀核裂变释放的中子可以诱发其他核的分裂。但是研究人员关于释放中子的能量知之甚微。这里汉斯·冯·哈尔班和他巴黎的同事阐述清楚了这个问题。他们在一系列的实验中采用电离室来探测“快”中子的能量分布——即那些能量高于 1.5 兆电子伏的中子。中子数目随着能量升高而降低，但是一些中子携带能量多达 11 兆电子伏。中子能量分布的认知在核反应堆工程技术中有重要影响，因为慢中子比快中子能够更有效地诱发裂变。

目前实验已表明，**快**中子是从初级**热**中子诱发的铀核裂变过程中释放出来的。已经使用了两种不同的检测方法：在第一种方法中[1]，初级中子和次级中子（如果有的话）在一种可以发生吸能反应的介质中被吸收，以致形成了一种易于检测的放射性核。如果能量阈值高于初级中子的最大能量，那么所观测的一切肯定的结果都必定要归结为次级中子。在第二种方法中[2]，快中子与较重核的弹性碰撞是借助电离室进行观测的，这个电离室充满着一个大气压的气体并与线性放大器相连。为了单独研究由初级热中子所产生的影响，我们在源与铀块之间，分别在加入和没有加入镉屏蔽层的两种情况下进行实验。

第一种方法向我们表明，产生的次级快中子的能量至少为 2 兆电子伏（足以将可检测量的 ^{32}S 转变为放射性的 ^{32}P），我们设法通过第二种方法来确定，能量明显高于 2 兆电子伏的中子是否也会在次级辐射中出现。在我们的实验里，将充满氧气的电离室置于装有氧化铀且四周用厚的固体石蜡层包裹的近似立方（9 厘米 ×9 厘米 ×8 厘米）的盒子中，将用铅屏蔽层包裹（在电离室的方向上的厚度为 5 厘米）的源（300 毫克镭铍源）埋入石蜡之中。为了吸收热中子，我们可以用镉箔将铀盒的所有面都遮挡起来。在镉箔存在或者不存在的两种情况下记录脉冲，并且由此可以估算出受到热中子影响的部分（铀核所释放的快中子导致氧核的反冲）。

In view of the large number of of accidental pulses due to the strong γ-radiation emitted by the source, only nuclei recoiling with at least 1.5 Mev. could be taken into consideration. The distribution curve shows that the frequency of pulses observed falls off rapidly between 1.5 Mev. and 2.5 Mev.; between 2.5 Mev. and 3.7 Mev. the frequency decreases much more slowly, pulses observed in this second region being, however, very rare. The total number of pulses recorded is small (with cadmium: 84 pulses in 90 minutes; without cadmium: 161 pulses in 90 minutes); but it appears clearly that recoils with energy of about 2.5 Mev. are notably more frequent in the absence of cadmium and, therefore, that *neutrons possessing an energy of at least* 11 Mev. *are liberated in uranium irradiated with thermal neutrons.*

The high energy of these fast neutrons shows that their parent nuclei are in a highly excited state at the moment of their liberation, which is probably simultaneous with the fission. In this way a non-negligible fraction of the fission energy is disposed of; a further fraction is carried off by the β- and γ-rays afterwards emitted by the nuclei produced in the fission. The remainder available as kinetic energy for these recoiling nuclei is therefore considerably smaller than the total amount of energy liberated in the fission process (about 200 Mev.).

(**143**, 939; 1939)

H. von Halban, jun., F. Joliot and L.Kowarski: Laboratoire de Chimie Nucléaire, Collège de France, Paris, May 20.

References:

1. Dodé, M., von Halban, jun., H., Joliot, F., and Kowarski, L., *C.R.*, **208**, 995 (1939).
2. Szilard, L., and Zinn, W., *Phys. Rev.*, 55, 799 (1939).

鉴于由源发出强γ辐射所导致随机脉冲的数目很大，我们只考虑反冲能量至少为1.5兆电子伏的核。分布曲线显示，观测到的脉冲频率在能量1.5兆电子伏到2.5兆电子伏之间快速下降；在能量2.5兆电子伏到3.7兆电子伏之间，频率的下降减慢了很多，不过，在这第二个能量区域所观测到的脉冲非常稀少。记录到的总的脉冲数很少（使用镉时：90分钟里有84个脉冲；不使用镉时：90分钟里有161个脉冲）；但是似乎很明确的是，在没有镉时能量约为2.5兆电子伏的反冲明显出现得更为频繁，因此，**用热中子辐照铀时释放出至少具有11兆电子伏能量的中子**。

这些快中子所具有的高能量表明，在它们释出时其母核处于高激发态，释出与裂变可能是同时发生的。这样，裂变能中不容忽视的一部分就被转移走了；更大的一部分则被在裂变中生成的核通过发射β射线和γ射线带走了。因此，能够作为反冲核的动能的剩余能量就明显少于裂变过程中释放出的能量总量（约为200兆电子伏）。

（王耀杨 翻译；夏海鸿 审稿）

Nuclear Reactions in the Continuous Energy Region

N. Bohr *et al.*

Editor's Note

The chief interest of this Letter to *Nature* is its authors: Niels Bohr at Copenhagen was the acknowledged father of quantum mechanics, Rudolf Peierls was a graduate research worker at Copenhagen who eventually migrated to Britain and was the designer of the thermal diffusion process for separating isotopes, and George Placzek was a close colleague of the other two. The point of their letter was to emphasise the utility of an idea due to Bohr that when nuclei collide with each other they first form a "compound nucleus" which can then split up in several different ways.

IT is typical for nuclear reactions initiated by collisions or radiation that they may, to a large extent, be considered as taking place in two steps: the formation of a highly excited compound system and its subsequent disintegration or radiative transition to a less excited state. We denote by A, B, ... the possible alternative products of the reaction, specified by the nature, internal quantum state, and spin direction both of the emitted particle or photon and of the residual nucleus and the orbital momentum. Further, we call P_A, P_B ... the probabilities, per unit time, of transitions to A, B, ... respectively, from the compound state.

The cross-section of the reaction $A \rightarrow B$ is then evidently

$$\sigma_B^A = \sigma^A \frac{P_B}{P_A + P_B + \dots}, \tag{1}$$

where σ^A is the cross-section for a collision in which, starting from the state A, a compound nucleus is produced. This formula implies, of course, that we are dealing with energies for which the compound nucleus can actually exist, that is, that we are either in a region of continuous energy values or, if the levels are discrete, that we are at optimum resonance. Moreover, it is assumed that all possible reactions, including scattering, proceed by way of the compound state, neglecting, in particular, the influence of the so-called "potential scattering", where the particle is deflected without actually getting into close interaction with the individual constituents of the original nucleus.

On these assumptions a very general conservation theorem of wave mechanics[1] yields the relation

$$\sigma^A = \frac{\lambda^2}{\pi}(2l + 1)\frac{P_A}{P_A + P_B + \dots}, \tag{2}$$

连续能量区域的核反应

玻尔等

编者按

这篇寄给《自然》的通讯最令人感兴趣的是他的作者：哥本哈根的尼尔斯·玻尔，他是众所周知的量子力学之父，鲁道夫·佩尔斯为哥本哈根的研究生，最后移居英国，是应用热扩散过程来分离同位素方法的设计者，乔治·普拉切克是上述两位作者的亲密同事。他们的这篇文章的要点是强调玻尔的关于核碰撞的观点，他们认为，原子核在发生碰撞时首先形成“复合核”，这种复合核可以通过多种方式分裂。

通常认为，由碰撞或者辐射引发的核反应在很大程度上分为两步发生：首先形成一个处于较高激发态的复合体系，然后通过衰变或辐射跃迁到较低的激发态。我们用 A，B，……表示反应中可能出现的两类不同的产物，并通过发射出的粒子或光子及残余核的性质、内在量子态、自旋方向及轨道角动量来决定。此外，我们定义 P_A，P_B，……为单位时间内由复合态分别跃迁到 A，B，……态上的概率。

显然，$A \to B$ 反应的有效截面为：

$$\sigma_B^A = \sigma^A \frac{P_B}{P_A + P_B + \cdots} \tag{1}$$

式中 σ^A 为从 A 状态发生碰撞产生一个复合核的有效截面。这个公式表示，我们讨论的能量是这个复合核能够真正存在的能量，也就是说，我们或者有一个能量连续的区域，或者如果能级是分立的，我们将得到最适合的共振态。而且，假定所有可能发生的反应，包括散射，都是通过复合状态进行的，特别忽略了被称作“势散射”的影响，在“势散射”中，粒子的轨迹发生偏斜，因而未能与原始原子核中的某个粒子发生近距离的相互作用。

基于这些假定，根据波动力学 [1] 普适的守恒定理给出以下关系式：

$$\sigma^A = \frac{\lambda^2}{\pi}(2l+1)\frac{P_A}{P_A + P_B + \cdots} \tag{2}$$

where λ is the wave-length of the incident particle and l is the angular momentum.

In the case of discrete levels, (1) and (2) give the same cross-section as the usual dispersion formula, if one applies it to the centre of a resonance level and neglects the influence of all other levels. In this case we have for each resonance level a well-defined quantum state of the compound nucleus, and its properties, in particular the probabilities P_A, P_B, ... then cannot depend on the kind of collision by which it has been formed, that is, they would be the same if we had started from the fragments B, or C, ... instead of A.

In the case of the continuum, however, where there are many quantum states with energies that are indistinguishable within the life-time of the compound nucleus, the actual state of the system is a superposition of several quantum states and its properties depend on their phase relations, and hence on the process by which the compound nucleus has been produced.

This dependence is made particularly obvious if we consider the formula

$$\overline{\sigma^A} = \frac{\hbar}{2}\rho\lambda^2(2l+1)P_A^0, \tag{3}$$

for the mean value of the cross-section over an interval containing many levels, which follows from the well-known considerations of detailed balancing. Here ρ is the density per unit energy of levels (of suitable angular momentum and symmetry) of the compound nucleus. P_A^0 is the probability for process A in statistical equilibrium and thus refers to a micro-canonical ensemble of compound states built up from the fragments A, B ... respectively, with proper statistical weights.

In the case of discrete levels, where formula (3) can also be derived directly from the dispersion formula, P_A^0 is simply an average over the individual levels of the probability P_A, which in this case is well defined.

In the continuum, (3) must be identical with (2), since the cross-section does not vary appreciably over an energy interval containing many levels, and hence, comparing (2) and (3)

$$\frac{P_A^{(A)}}{P_A^0} = \frac{\pi}{2}\hbar\rho\,(P_A^{(A)} + P_B^{(A)} + \cdots) = \frac{\pi}{2}\frac{\Gamma^{(A)}}{d}, \tag{4}$$

where the superscript A has been added to the probabilities occurring in (1) in order to show explicitly the dependence on the mode of formation, and where $\Gamma^{(A)}$ is the total energy width of the compound state concerned and $d = \frac{1}{\rho}$ the average level distance. In the continuum, where $\Gamma^{(A)} \gg d$, the probability $P_A^{(A)}$ of re-emitting the incident particle without change of state of the nucleus will thus be much larger than the probability of the same process in a compound nucleus produced in other ways.

式中 λ 为入射粒子的波长，l 为其角动量。

假如人们应用色散公式于一个共振能级的中心而且忽略了所有其他能级的影响的话，对于分立的能级，由式 (1) 和式 (2) 给出的有效截面与通常色散公式得到的结果相同。在这种情况下，我们将为复合核中的每一个共振能级确定一个明确的量子态，同时它的特性，尤其是概率 P_A，P_B，……并不依赖于通过什么类型的碰撞而形成，也就是说反应从阶段 B 或者 C 及其他碎片开始与从 A 开始得到的结果是一样的。

然而，对于能量连续的情况来说，在复合核的寿命期内存在着很多不易区分能量的量子态，系统的实际状态就是多个量子态的叠加，系统的性质取决于它们的相位关系，因此取决于复合核的形成过程。

如果我们考虑以下的公式，这种相关性就会变得非常明显：

$$\overline{\sigma^A} = \frac{\hbar}{2}\rho\lambda^2(2l+1)P_A^0 \tag{3}$$

对于在一个包含若干能级的区间中截面的平均值，它服从著名的细致平衡原理。式中 ρ 在这里表示复合核中能级（具有适当的角动量和对称性）的每单位能量上的密度，P_A^0 为 A 过程的统计平衡概率，因此引出一个由各自具有特有的统计权重的碎片 A，B，……构成的复合状态的微正则系综。

对于分立能级的情况，式 (3) 也可以直接由色散公式导出，在这种情况下我们可很容易的定义 P_A^0 为对概率 P_A 的单个能级上的简单的平均值。

对于能量连续的情况，式 (3) 和式 (2) 必须一致，因为有效截面在一个包含多个能级的能量区间中的变化并不明显。因此，比较式 (2) 和式 (3)，我们会得出：

$$\frac{P_A^{(A)}}{P_A^0} = \frac{\pi}{2}\hbar\rho\,(P_A^{(A)} + P_B^{(A)} + \cdots) = \frac{\pi}{2}\frac{\Gamma^{(A)}}{d} \tag{4}$$

其中，在式 (1) 中的概率上添加上标 A 是为了明确地表示其与形成模式的相关性，$\Gamma^{(A)}$ 是对应复合态的总能级宽度，$d=\frac{1}{\rho}$ 表示能级的平均间距。在能量连续的体系中，当 $\Gamma^{(A)} \gg d$ 时，在不改变原子核状态的条件下再次发射入射粒子的概率 $P_A^{(A)}$ 会比在以其他方式形成的复合核中发生同样过程的概率大得多。

While the arguments used so far are of a very general character, more detailed considerations of the mechanism of nuclear excitation are required for a discussion of the dependence $P_B^{(A)}$ of the mode A of the compound nucleus provided A=B.

One can think of cases in which such a dependence must obviously be expected; in fact, if a large system be hit by a fast particle, the energy of excitation might be localized in the neighbourhood of the point of impact, and the escape of fast particles from this neighbourhood may be more probable than in statistical equilibrium. Further, if the system had modes of vibration very loosely coupled, the excitation of one of them, for example by radiation, would be unlikely to lead to the excitation of a state of vibration made up of very different normal modes, even though the state may be quite strongly represented in statistical equilibrium.

In actual nuclei, however, the motion cannot be described in terms of loosely coupled vibrations, nor would one expect localization of the excitation energy to be of importance in nuclear reactions of moderate energy. If we suppose that there are no other special circumstances which would lead to a dependence of $P_B^{(A)}$ on A, it is thus a reasonable idealization to assume that, even in the continuum, all $P_A^{(A)}$ are equal to P_B^0, except, of course, for A=B, where we have seen in (4) that the phases are necessarily such as to favour the re-emission of the incident particle.

A typical case of a reaction in the continuum is the nuclear photo-effect in heavy elements, produced by γ-rays of about 17 mv. In the first experiments of Bothe and Gentner, there seemed to be marked differences between the cross-sections of different elements, but the continuation of their investigations[2] indicated that these differences can be accounted for by the different radioactive properties of the residual nuclei, and that the cross-sections of all heavy nuclei for photo-effect are of the order of 5×10^{-26} cm.^{2}.

In previous discussions, based on formulae (1) and (2), where the distinction between $P_A^{(A)}$ and P_A^0 was not clearly recognized, it was found difficult, however, to account for photo-effect cross-sections of this magnitude. In fact, if one estimates the probability of neutron escape P_B at about 10^{17} sec.$^{-1}$, one should have for P_A 10^{15} sec.$^{-1}$ and as long as this was taken as P_A^0 it seemed much too large, since it evidently must be much smaller then the total radiation probability, estimated at about 10^{15}, which included transitions to many more final levels besides the ground state.

We see now, however, that $P^{(A)}$ is here considerably larger than P_A^0, since the level distance at the high excitations concerned is probably of the order of 1 volt, whereas the level width corresponding to the above value of P_B is about 100 volts. From (4), or more directly from (3), P_A^0 is thus seen to be only about 10^{13} sec.$^{-1}$, which would appear quite reasonable.

(**144**, 200-201; 1939)

前面的论述针对的都是总体上的特征，要进一步考量核激发机制需要讨论当 $A=B$ 时，$P_B^{(A)}$ 对复合核中模式 A 的依赖性。

人们可以设想一些这种依赖性肯定存在的情况，实际上，如果一个很大的系统被一个快速粒子撞击，激发的能量也许只局限于碰撞点附近，快速粒子逃离该区域的可能性比到达统计平衡时的概率大。而且，如果这个系统具有松散耦合的振动模式，它们当中一个被激发，比如通过辐射，很难导致一个差异很大的普通模式组成的振动态被激发，即使这种状态在统计平衡中占有很大的优势。

然而，在实际的原子核中，我们不能用松散耦合的振动模式来描述运动，也不能期望激发能的局域化对中等能量的核反应的重要性。如果我们假定没有其他的特殊情况可以导致 $P_B^{(A)}$ 对 A 的依赖性，那么除去 $A=B$ 的情况以外，理想化地假定所有 $P_A^{(A)}$ 都等于 P_B^0 是合理的，即使是在能量连续的体系中，正如我们在式 (4) 看到的那样，该状态必然更有利于产生入射粒子的再发射过程。

在能量连续体系中的一个典型反应是重元素原子的光电效应，由 17 毫伏的 γ 射线引发。博特和根特纳在最初的实验中发现，不同元素的有效截面明显不同，但是他们后来的研究 [2] 又表明这种不同可以用残余核子的放射性差异来解释，而且所有重元素在光电效应中的有效截面均为 5×10^{-26} 平方厘米的量级。

在前面基于式 (1) 和 (2) 的讨论中，由于没有清楚地认识到 $P_A^{(A)}$ 和 P_A^0 之间的差别，我们发现要解析这种大小的光电效应的有效截面是有困难的。事实上，如果人们估算中子逃逸的概率 P_B 约为每秒 10^{17}，那么 P_A 应为每秒 10^{15}，但如果认为 P_A^0 也具有同样的数值，那就太大了，因为它显然必须远远小于整个辐射的概率，整个辐射的概率约为 10^{15}，其中包括向基态以及更多其他最终能级的跃迁过程。

然而，我们现在得出，在本文 $P^{(A)}$ 要远大于 P_A^0，因为在高激发态中能级间距的数量级大约为 1 电子伏，但是对于上面的 P_B 值得到的能级宽度大约为 100 电子伏。从式 (4)，或者更直接地从式 (3) 中看出，P_A^0 的合理值只有每秒 10^{13} 左右。

（胡雪兰 翻译；王乃彦 审稿）

N. Bohr, R. Peierls and G. Placzek: Institute of Theoretical Physics, Copenhagen, July 4.

References:

1. The details of this and of the other arguments of this note will be published in the *Proceedings of the Copenhagen Academy*.

2. Bothe, W., and Gentner, W., *Z. Phys.*, **106**, 236 (1937); **112**, 45 (1939).

The Scattering by Uranium Nuclei of Fast Neutrons and the Possible Neutron Emission Resulting from Fission

The Scattering by Uranium Nuclei of Fast Neutrons and the Possible Neutron Emission Resulting from Fission

L. Goldstein *et al.*

Editor's Note

Here Goldstein, Rogazinski and Walen describe experiments measuring how neutrons interact with uranium nuclei. They used neutrons from a polonium-beryllium source to irradiate samples of lead oxide and uranium oxide, and detected neutrons with an ionization chamber. They are able to estimate the "cross-section"—the "size" of the nuclei as seen by the neutrons—for neutrons that are scattered with or without a change in energy. As they noted, their cross-section was somewhat higher than previous estimates, implying that neutrons can travel a little less far than thought in uranium. This suggested, in turn, that a critical mass for a chain reaction might also be smaller than previously suspected.

THE work to be described concerns only fast neutrons, and its object is the study of their scattering by uranium and the possible neutron emission which accompanies the fission of the nucleus.

The experiments were performed with a polonium plus beryllium source equivalent to 3 mC. of radon plus beryllium. An ionization chamber surrounded with 2.5 cm. lead, filled with hydrogen at a pressure of 35 atm., was used as a neutron detector. The insulated electrode was connected to a compensated electrometer valve[1], the grid leak being 10^{11} ohms and the sensitivity 1.2×10^{-15} amp./div. on the scale.

We have employed two experimental arrangements in which the source was placed (1) between the chamber and the substance used as scatterer, the nature and the thickness of which were variable; (2) in the centre of a cube of 16 cm. side, alternately filled with uranium oxide (specific gravity, d=4.0) and lead oxide (compressed to d=3.8).

The first type of experiment gave us the total scattering cross-section, which is, as can be shown, $\sigma_t=\sigma_e+k_i\sigma_i$; for uranium oxide $\sigma_t=\sigma_e+k_i\sigma_i+k_r\nu_r\sigma_r$, where σ_e, σ_i, σ_r are respectively the average cross-sections of elastic and inelastic scattering and of fission; ν_r is the average number of neutrons produced per fission; k_i and k_r are the average efficiency factors of the chamber for the neutrons having undergone an inelastic collision or for the

铀核的快中子散射与可能源于裂变的中子发射

戈尔德施泰因等

编者按

本文戈尔德施泰因、罗格兹尼斯基和瓦伦描述了测量中子与铀核如何相互作用的实验。他们使用钋－铍源的中子辐照氧化铅和氧化铀的样品，并用电离室探测中子。他们可以估算出有能量变化和无能量变化两种实验情况下散射中子的“截面”，即中子与核碰撞的“尺寸”。正如他们所说的，他们计算得出的截面比先前估算的稍高，这意味着中子在铀中的行程比想象的稍小。反过来说，这也表明了链式反应的临界质量值也可能小于原先估算的值。

本文所描述的内容仅涉及快中子，目标是研究铀的快中子散射以及核裂变过程中可能伴随的中子发射。

实验是用一个钋－铍源（相当于 3 毫居氡－铍源）完成的。我们用 2.5 厘米厚的铅板包围电离室来作为中子探测器，室内充满了 35 个大气压的氢气。绝缘电极与补偿静电计电子管相连接[1]，栅漏电阻为 10^{11} 欧姆，标度灵敏度为每分度 1.2×10^{-15} 安培。

我们使用了两套实验装置：(1) 实验源位于电离室和作为散射体的物质之间，其中散射体物质的性质和厚度是可变的；(2) 实验源位于边长为 16 厘米的立方体中心，其中交替填充氧化铀（比重 $d = 4.0$）和氧化铅（压缩后比重 $d = 3.8$）。

第一种实验给出了总的散射截面，可以表示为 $\sigma_t = \sigma_e + k_i\sigma_i$；对于氧化铀，散射截面为 $\sigma_t = \sigma_e + k_i\sigma_i + k_r v_r \sigma_r$，其中，$\sigma_e$、$\sigma_i$、$\sigma_r$ 分别是弹性散射、非弹性散射及裂变的平均截面；v_r 是每次裂变产生的平均中子数；k_i 和 k_r 分别是电离室对发生非弹性碰撞的中子和裂变产生的中子的平均效率因子；直接入射中子的效率因子取为 1，

neutrons resulting from fission. The efficiency for the direct neutrons was taken to be unity, $k=1$. For neutrons elastically scattered by nuclei of sufficiently high mass, $k_e=k=1$. We have calculated k, taking into account the size of the chamber, the cross-section for proton projection, etc. The spectrum of polonium plus beryllium neutrons has been considered[2] to contain 50 percent of neutrons of W_n less than 10^5 ev. We thus obtain:

$10^{-6} W_n$	0.1	0.5	3	5	10	ev.
k	0.3	1	1.9	1.7	1.2	

In view of a possible extrapolation that would give $\sigma_e + k_i \sigma_i$ for uranium, we have in the same way experimented with scattering by lead oxide, lead, copper and zinc.

The results of the first experiment were as follows:

Substance	Cu	Zn	Pb	PbO_2	UO_2	$(O)_{calc.}$	$(U)_{calc.}$
$\sigma_t \times 10^{-24}$ cm.^{2} (±10%)	2.2	2.3	5.4	9.5	14.4	2	10.3

The values for uranium and oxygen are calculated on the assumption of the additivity of the cross-sections in lead oxide and uranium oxide.

The second experiment gives us, in the first approximation, the absorption coefficient $(1 - k_i)\sigma_i + (1 - k_r v_r)\sigma_r$, the value of σ_e being only as a correction term in the determination of the mean free path λ and the average distance L travelled by the neutrons before they escape from the whole mass, which is supposed spherical, the radius being r and $L = r\left(1 + \frac{1}{2}\frac{r}{\lambda}\right)$. This experiment, taking into account the results of the previous experiments, gives for lead, $(1 - k_i)\sigma_i \simeq 2 \times 10^{-24}$ cm.^{2}. Assuming that σ_i can reach 30 percent of σ_e[3], this gives $k_i (\simeq) 0$.

With the exception of uranium, for which one must consider not only σ_i, but also $v_r \sigma_r$, it is probable that σ_t is not very different from σ_e because of the small value of k_i.

In the case of uranium, however, we have,

$$(1 - k_i)\sigma_i + (1 - k_r v_r)\sigma_r \simeq 0.9 \times 10^{-24} \text{ cm.}^2 \qquad (1),$$

or, by adding to σ_t, thus eliminating k_i and k_r,

$$\sigma_e + \sigma_i + \sigma_r \simeq 11.2 \times 10^{-24} \text{ cm.}^2. \qquad (2).$$

If it is supposed that each fission produces radioelements, the cross-section measured by Joliot, and by Anderson *et al.*[4] would be identical with σ_r, which they found to be $\sigma_r \simeq 10^{-25}$ cm.^{2}. In this case we see that $(\sigma_e + \sigma_i)$ is much greater ($\simeq 11.1 \times 10^{-24}$ cm.^{2})

即 k=1。对于由质量足够高的原子核弹性散射的中子，$k_e = k$ =1。我们在考虑了电离室的尺寸、质子投射截面等值后计算了 k 的值。现在认为[2]，钋 – 铍源的中子能谱包含了 W_n 小于 10^5 电子伏的中子的 50%。因此，我们得到：

$10^{-6}W_n$	0.1	0.5	3	5	10	电子伏
k	0.3	1	1.9	1.7	1.2	

鉴于一种可行的外推法可以给出铀的 $\sigma_e + k_i\sigma_i$ 值，我们以相同的方法用氧化铅、铅、铜和锌与中子发生散射进行实验。

第一种实验结果如下：

物质	Cu	Zn	Pb	PbO_2	UO_2	(O)$_{计算值}$	(U)$_{计算值}$
$\sigma_t \times 10^{-24}$ 平方厘米 (±10%)	2.2	2.3	5.4	9.5	14.4	2	10.3

假定氧化铅和氧化铀截面具有可加性的前提下，我们计算了铀和氧的截面值。

第二种实验我们在一阶近似下给出了吸收系数 $(1 - k_i)\sigma_i + (1 - k_r v_r)\sigma_r$ 的值，σ_e 的值只是作为中子从假设为球形的块状物体逃逸前的测定平均自由程 λ 和中子穿行平均距离 L 的修正项，球块的半径是 r，且 $L = r\,(1 + \frac{1}{2}\frac{r}{\lambda})$。考虑到以前实验的结果，这个实验给出了铅的 $(1 - k_i)\sigma_i \simeq 2 \times 10^{-24}$ 平方厘米。假定 σ_i 的值能达到 σ_e 值[3]的 30%，则有 $k_i(\simeq)0$。

对于铀我们不仅要考虑 σ_i，还要考虑 $v_r\sigma_r$，而对于其他物质很有可能由于 k_i 的值很小，σ_t 与 σ_e 的值相差并不大。

对于铀的情况，我们有

$$(1 - k_i)\sigma_i + (1 - k_r v_r)\sigma_r \simeq 0.9 \times 10^{-24} \text{ 平方厘米} \quad (1)$$

或者，在式子两边加上 σ_t 将消掉 k_i 和 k_r，得到

$$\sigma_e + \sigma_i + \sigma_r \simeq 11.2 \times 10^{-24} \text{ 平方厘米} \quad (2)$$

如果假定每次裂变都产生放射性元素，约里奥和安德森等人[4]测得的截面值则与 σ_r 一致，他们测得的结果是 $\sigma_r \simeq 10^{-25}$ 平方厘米。在这种情况下，我们看到，

than that given by an extrapolation ($\simeq 6 \times 10^{-24}$ cm.^{2})

On the other hand, it results from (1) that, if the value of σ_i is comparable to that of the next elements (1 to 2×10^{-24} cm.^{2}), v_r can, with plausible assumptions as to the coefficients k_i and k_r, take variable values, for example, from 1 to 5, or even more.

One can see that, so long as σ_i is not determined separately, the experiments of the type described do not allow us to determine v_r and σ_r (characteristics of the fission), or to conclude that neutrons are liberated; or *a fortiori*, to form a conclusion as to the possibility of chain reactions, contrary to the results of similar experiments[5].

The only suitable case for showing with certainty, by means of an ionization chamber, the production of neutrons, would be that in which, by the use of a sufficient quantity of uranium, the chain mechanism would give multiplication of neutrons, if such chain is realizable[6].

In conclusion, it results from these experiments with neutrons of polonium plus beryllium that the sum of the cross-sections $\sigma_e + \sigma_i + \sigma_r$ for the uranium nucleus is $(11.2\pm1.5)10^{-24}$ cm.^{2}. This value implies a mean path in uranium much shorter than that usually admitted; this suggests that smaller masses than those hitherto expected might be used to show chain fission.

(**144**, 201-202; 1939)

La. Goldstein, A. Rogozinski and R. J. Walen: Laboratoire Curie, Institut du Radium, Paris, 5, July 13.

References:

1. Rogozinski, A., *C.R.*, **208**, 427 (1939).
2. Auger, P., *J. Phys. Radium*, **4**, 719 (1933).
3. Seaborg, G. F., Gibson, G. E., and Graham, D. C., *Phys, Rev.*, **52**, 408 (1937).
4. Joliot, F., *J. Phys. Radium*, **10**, 159 (1939). Anderson, H. L., Booth, E. T., Dunning, J. R., Fermi, E., Glasoe, G. N., and Slack, F. G., *Phys. Rev.*, **55**, 511 (1939).
5. Haenny, C., and Rosenberg, A., *C.R.*, **208**, 898 (1939).
6. v. Halban, H., Joliot, F., and Kowarski, L., *Nature*, **143**, 680 (1939). Perrin, F., *C.R.*, **208**, 1394, 1573 (1939).

$(\sigma_e + \sigma_i)$ 的值 ($\simeq 11.1 \times 10^{-24}$ 平方厘米) 远大于外推法所给出的值 ($\simeq 6 \times 10^{-24}$ 平方厘米)。

另一方面，根据（1）式的结果可知，如果 σ_i 的值与接下来 σ_r 的值（1×10^{-24}～2×10^{-24} 平方厘米）相近的话，在 k_i 和 k_r 系数运用合理的假定下，ν_r的值是可变的，例如可以从 1 取到 5 甚至更大的值。

可以看到，只要不能独立地测定 σ_i 值，在这类实验中我们就不能测定 ν_r 和 σ_r（分裂的特征）的值，也就不能断定是否释放了中子；更不用说得到可能发生链式反应的结论了，这与类似实验的结果 [5] 相反。

通过使用电离室，唯一可以确定的合理情况是，对于中子的产生，如果链式反应可以实现的话 [6]，通过使用足量的铀，链式反应就可以产生出中子的倍增效应。

综上所述，通过运用钋－铍中子源的这些实验可知，对铀核截面求和，即 $\sigma_e + \sigma_i + \sigma_r = (11.2 \pm 1.5) \times 10^{-24}$ 平方厘米。这个值表明铀中的平均自由程远小于通常公认的值；而这也意味着可以用比现在预期的更小的质量来实现链式核裂变。

（沈乃澂 翻译；张焕乔 审稿）

Recent Experimental Results in Nuclear Isomerism

B. Pontecorvo

Editor's Note

The idea that two atomic nuclei might be identical in their physical composition and mass but nevertheless might differ in their radioactive properties goes back to 1917 and to Frederick Soddy, one of the pioneers of radioactive studies. It is equivalent to saying that isomeric nuclei are capable of decaying by more than one route. With the discovery of artificial radioactivity, pairs of isomeric nuclei became more common. Bruno Pontecorvo, who migrated from France to Britain in 1940, had made a special study of isomeric nuclei and published this account of their properties a few months before the Second World War began.

THE hypothesis that two atomic nuclei indistinguishable in respect of atomic and mass number could nevertheless have different radioactive properties (the hypothesis of nuclear isomerism) was put forward for the first time by Soddy[1] in 1917. In 1921 uranium Z was discovered by Hahn[2]; by studying the chemical and radioactive properties of this element, Hahn deduced that uranium Z and uranium X_2 are isomeric nuclei. The problem of uranium Z has been taken up recently by Feather and Bretscher (*Proc. Roy. Soc.*, **165**, 542; 1938). It should be noted that, for many years, uranium Z and uranium X_2 were the only known example of an isomeric pair.

After the discovery of artificial radioactivity, the study of isomerism received considerable impetus on account of the experimental material assembled in the course of research on artificial radio-elements. The first *certain* example of an isomeric pair to which it has been possible to attribute a mass number (A = 80) in the domain of the artificial radio-elements was furnished[3] by the study of the radioactivity produced in bromine by neutrons (slow and fast) and by γ-rays of great energy.

Then, as the experimental material on artificial radio-elements has increased, the number of pairs of nuclei which are undoubtedly isomeric has grown to such an extent that it is not possible to quote here all the investigations which have been published on the question. More than thirty such pairs are known and there is no doubt that the number still unknown is much greater. We can say, now, *that nuclear isomerism is by no means an exceptional phenomenon.*

It is natural to think that the physical difference between two isomeric nuclei is connected with two states of different excitation of the same nucleus (let us say ground state and first excited state). But in this case, how could the upper state be metastable, that is,

核同质异能性的最新实验结果

庞蒂科夫

编者按

两个原子核可以具有相同的物理组成和质量但却有不同的放射性质，这一想法最初是由放射性研究的开拓者之一——弗雷德里克·索迪于 1917 年提出的。这一观点还可以理解为，同质异能核衰变的方式不止一种。随着人工放射性的发现，同质异能核对变得更为普遍。布鲁诺·庞蒂科夫对同质异能核进行了一番专门研究，并在第二次世界大战前几个月发表了这篇关于同质异能核性质的文章，1940 年他从法国移居英国。

原子序数和质量数都相同的两个原子核却可能具有不同的放射性质（即核同质异能性的假说），这一假说最早是由索迪[1]于 1917 年提出的。1921 年，哈恩[2]发现了铀 Z；通过对该元素化学性质和放射性质的研究，哈恩推测铀 Z 与铀 X_2 是同质异能核。最近，费瑟和布雷切尔重新提出了铀 Z 的问题（《皇家学会会刊》，第 165 卷，第 542 页，1938 年）。值得人们注意的是，很多年以来铀 Z 和铀 X_2 是仅有的已经知道的同质异能核对。

发现人工放射性以后，科研工作者在对人工放射性元素的研究过程中积累的大量实验材料为同质异能现象的研究起到了相当大的推动作用。通过研究（慢和快）中子和高能 γ 射线轰击溴所产生的放射性可为人工放射性元素的领域中增添一个质量数为 80 的新元素，该元素也是首次得到的**确定的**同质异能对[3]。

于是，随着关于人工放射性元素实验材料的增加，已确定为同质异能核对的数目大量增加，以至于我在此处不能一一引述所有关于这一问题的已发表的研究结果。目前已知道超过 30 对这样的核，并且毫无疑问的是，尚未发现的同质异能核对比这数目要多得多。现在我们可以说，**核的同质异能性绝对不是一种个别现象**。

我们可以很自然地想到，两个同质异能核之间的物理差异与相同原子核的两个不同激发态有关（比如说基态和第一激发态）。但是，在这种情况下，高能态是如何成为亚稳态的，或者说，它怎么能维持这么长时间（在某些情况下甚至超过一天）呢？

how could it live for any length of time (greater than one day, in some cases)? By what mechanism would it be preserved from destruction in a very short time by the emission of an electromagnetic radiation? Weiszäcker has answered this question[4].

According to Weiszäcker's *hypothesis*, nuclear isomerism may be explained by assuming that *the lowest excited state of the nucleus has an angular momentum differing by several units from that of the ground state.* Selection rules may then be invoked to weaken considerably the probability per unit of time of the transition from the upper to the ground state of the nucleus. Of course, experiments which make it possible to test the truth of Weiszäcker's hypothesis are of great interest.

One of the most important points is the study of the γ-radiation eventually emitted in the transition from one isomeric state to another: I say "eventually emitted" because, the nucleus being radioactive, the upper state corresponding to one of the isomeric forms may be destroyed by an ordinary β transition. The γ-ray–β-ray branching ratio will depend on the relative life-times for the two modes of decay. The first researches made to observe this γ-radiation failed. But it should be noted[5] that the considerable difficulty in detecting this radiation might be due to the fact that transitions between isomeric nuclei can be strongly converted: in this case electrons of small energy would be emitted and not γ-rays.

The very complete theory of the internal conversion of radiations emitted in the transitions between isomeric states[6], given by Hebb and Uhlenbeck, Dancoff and Morrison, has shown that these radiations must have internal conversion coefficients of approximately 1. Since these calculations are based on Weiszäcker's hypothesis, it can be concluded that *experiments which prove that the conversion coefficient in question is very high would indicate, to a certain extent, that Weiszäcker's hypothesis is correct.*

Indeed, in the case of the isomerism of radio-rhodium, Pontecorvo[7] has observed a radiation of low-energy electrons, which he interpreted as an electron line emitted in the transition from the metastable state to the ground state of the nucleus.*

At the present time, after a number of recent experiments, *there is no longer any doubt as to the fact that these transitions are generally strongly converted.* In particular, in the cases of isomeric nuclei of radio-bromine[8] and of element 43[9], strong lines of conversion electrons have been photographed in the Wilson chamber or in the magnetic spectrograph. Of course, the internal conversion is accompanied by emission of X-rays: as a rule, the analysis of these rays is an invaluable test in the interpretation of these phenomena[9,10].

It is interesting to find possible genetic relations between isomeric states of the same nucleus (β-radioactive): in this direction an extremely brilliant method has been described

* Note added in proof. A similar conclusion was independently obtained by Roussinow and Yusephovitch [*C. R. Acad. Sci. U.R.S.S.*, **20**, 9 (1938)] who observed a soft electron radiation in the case of isomeric forms of radio-bromine.

是什么机制使它们免于在很短时间内通过发射电磁辐射而崩溃呢？魏茨克回答了上述问题[4]。

根据魏茨克的**假说**，核的同质异能现象可以这样解释：即假定**核的最低激发态具有与基态相差若干单位的角动量**。这样，根据选择规则，原子核在单位时间内从上述激发态跃迁到基态的概率就会明显减少。当然，能够检验魏茨克假说真实性的实验才是最令人感兴趣的。

最重要的问题之一是，对于在从一种同质异能态到另一种同质异能态的跃迁过程中最终发射出的γ辐射的研究。我之所以要说“最终发射”，是因为原子核具有放射性，相应于同质异能形式之一的较高能态可能会被一次普通的β跃迁破坏。γ射线–β射线分支比将取决于两种衰变方式的相对寿命。虽然最初观测这种γ辐射的研究未能成功，不过值得注意的是[5]，探测这种辐射时所遇到的相当大的困难可能是因为同质异能核之间的跃迁可以发生强烈的转换，即在这种情况下，发射出来的是低能电子而不是γ射线。

关于在同质异能态之间的跃迁过程中发射出的辐射的内转换问题，赫布与乌伦贝克，以及丹克夫与莫里森给出了非常完整的理论[6]，该理论指出，这些辐射必须具有接近于1的内转换系数。这些计算都是基于魏茨克的假说，因此可以得出下面的结论：**如果能够找到证明其所涉及的内转换系数的确很高的实验，就在一定程度上意味着魏茨克的假说是正确的**。

确实，庞蒂科夫[7]在研究放射性铑时曾观测到低能电子辐射，他将其解释为原子核从亚稳态跃迁到基态时发射出来的电子线。*

在近期的大量实验之后，**人们目前对于这些跃迁普遍发生强烈转换这一事实再没有任何怀疑**。特别是在放射性溴[8]和43号元素[9]的同质异能核的实验研究中，人们已经在威尔逊云室和磁谱仪中拍下了转换电子的强线的照片。当然，内转换伴随着X射线的发射，因此，通常情况下在解释上述现象时[9,10]，对X射线的分析就是极其重要的检验标准。

寻找同种原子核（β放射性）的同质异能态之间可能具有的遗传关系是很有意思的：在这个研究方向上，塞格雷、哈尔福德和西博格曾描述过一种极为出色的方

* 附加说明：卢西诺和约瑟夫维奇独立地得到了一个类似的结论 [《苏联科学院院刊》**20**, 9(1938)]，他们在研究放射性溴的同质异能现象时观测到了软性电子辐射。

by Segré, Halford and Seaborg[11], who have succeeded in *separating*, one from the other, the two isomeric forms of radio-bromine. The principle of their method is as follows. Suppose the element, of which the isomeric states are being studied, can give compounds suitable for the application of the Szilard-Chalmers method of concentration. When the isomer in the upper state decays to the lower state, there is a γ-ray emission: corresponding recoil may be sufficient to knock the decayed atom out of the compound. The daughter activity can then be separated, as in the Szilard-Chalmers method.

This method, which has been successfully applied in several cases[12], can then be used (*a*) to separate known isomers in some cases; (*b*) to discover the existence of isomeric pairs, still unknown, in the study of artificial radioactivity.

Moreover, it has given a striking new proof that the transitions between isomeric states are strongly converted: in effect, the recoil due to the γ-emission is not sufficient to knock the decayed atom out of the compound, while the recoil of a conversion electron *can* be sufficient.

So far we have discussed radioactive isomers: the isomerism in this case, implies a difference in the life-times of the isomers. It has been noticed by Pontecorvo[5] that β-stable nuclei with a metastable excited state ought not to be very rare and should be revealed by the study of the radiation emitted by this metastable state. These nuclei are interesting for the understanding of nuclear isomerism, because the radiation corresponding to the transition from one isomeric state to the other is not troubled by the presence of β- or γ-rays. It should be possible to obtain a β-stable nucleus in a metastable state, after a nuclear transmutation or a radioactive disintegration.

Dodé and Pontecorvo[13], by bombardment of cadmium with fast neutrons, have obtained an activity (T=50 min.) which chemical proofs have shown to be due to an isotope of cadmium. On the other hand, there is no question of a reaction of simple neutron capture or of an $n,2n$ reaction. They interpreted the soft radiation emitted by cadmium (50 min.) as proceeding from a metastable state of an isotope of cadmium; the reaction of excitation without capture by fast neutrons (reaction n,n), having a considerable cross-section (some 10^{-24} cm.^{2}), it is not impossible, indeed, that part of the nuclei so excited might fall into a metastable state.

Segré and Seaborg[9] have observed a metastable state of element 43, decaying (only a line spectrum of electrons) with a 6-hour period into the ground state, which is stable or perhaps radioactive with a long life: the 6-hour activity is daughter of a β-radioactive molybdenum.

A very interesting case has been observed and studied thoroughly by Goldhaber, Szilard and Hill[14]. They have obtained by the n,n reaction already quoted, a metastable state (^{115}In*) of ^{115}In, decaying with a period of 4.1 hours; moreover, the same state can be obtained after the disintegration of a radio-cadmium (T=2.5 days). The radiation emitted

法[11]，他们成功地将放射性溴的两种同质异能形式彼此分离开来。他们所用方法的原理是，假定所要研究其同质异能状态的元素能够形成某些适用于齐拉特－查默斯富集方法的化合物。当处在较高能态的同质异能核衰变到较低能态时，就会放射出 γ 射线，而由此引起的反冲作用可能足以将衰变后的原子撞出化合物。于是就可以分离出子体放射性，如同在齐拉特－查默斯方法所描述的那样。

于是，这种已经成功应用于若干实例[12]的方法就可以用于：(*a*) 分离某些研究中的已知同质异能核；(*b*) 在人工放射性研究中，发现未知的同质异能核对的存在。

此外，这种方法对于同质异能态之间的跃迁发生强烈转换这一事实给出了一个令人惊奇的新证据：实际上，由于 γ 辐射而产生的反冲是不足以将衰变后的原子撞出化合物的，但是转换电子引起的反冲却**能够**做到这一点。

到目前为止我们讨论了放射性的同质异能素：在这些情况下出现的同质异能现象意味着同质异能素寿命之间的差异。庞蒂科夫[5]已经注意到，具有亚稳激发态的 β 稳定核应该不是很稀有，并且，对这种亚稳态所发射的辐射进行研究应该会有助于对它的了解。这些核对于理解核的同质异能现象是有帮助的，因为对应于从一个同质异能态到另一个同质异能态的跃迁的辐射并没有受到 β 射线或 γ 射线的干扰。在经过一次核嬗变或者一次放射性衰变之后，可能会得到一个处于亚稳态的 β 稳定核。

多德和庞蒂科夫[13]通过用快中子轰击镉的方法获得了一种放射性（T = 50 分钟），化学证据表明这种放射性源自镉的一种同位素。另外，无疑同时发生了单纯的中子俘获反应或者（n,2n）反应。他们将镉（50 分钟）发射的软辐射解释为镉的一种同位素的某一亚稳态的辐射；事实上，对具有相当可观截面（大约为 10^{-24} 平方厘米）的不伴随快中子俘获的（n,n）激发反应，激发的那部分核落入亚稳态是可能的。

塞格雷和西博格[9]曾观测到 43 号元素的亚稳态，它以 6 小时的周期衰变（只有电子的线型能谱）到基态，基态是稳定的，或者可能具有长寿命的放射性。因此，周期为 6 小时的放射性是一种 β 放射性钼的子体。

戈德哈伯、齐拉特和希尔[14]曾观测到一种非常有趣的情况，并对其进行了充分的研究。他们通过上面提到的（n,n）反应得到了 ^{115}In 的一个半衰期为 4.1 小时的亚稳态（^{115}In*），此外，一种放射性镉（T = 2.5 天）衰变后也能得到同样的态。^{115}In*

by $^{115}In^*$ has not yet been sufficiently studied; its properties are of the greatest interest both for the understanding of the nuclear isomerism and for that of isobaric pairs. In effect, ^{115}In and ^{115}Sn are one of the rare cases of stable neighbouring isobaric nuclei: $^{115}In^*$ might then decay into ^{115}Sn (β-emission) or into ^{115}In, or into both together.

The same metastable state of ^{115}In has been obtained also by irradiating indium with 5.8 Mev. protons (reaction *p,p*), by Barnes and Aradine[15]. Nevertheless, it is not yet clear whether the mechanism of nuclear excitation is that discussed by Weisskopf, that is, excitation by the action of the electric field of the proton[16].

In all these cases and in others studied more recently[17], the metastable states of stable isotopes are obtained from *nuclear transmutations*.

Lazard and Pontecorvo[18] have tried a new method, by which *it would be impossible to transmute the nucleus* and, consequently, to obtain artificial radio-elements, the presence of which may interfere with the investigation.

This method consists of irradiating the target with a continuous spectrum of X-rays, the energy of which is less than the dissociation energy of the nuclei. Suppose the radiated nuclei have a metastable state; the X-rays may excite higher levels of the nuclei; a part of the nuclei thus excited can fall into the metastable state, and it is the radiation from this state which can be observed. The maximum energy of the continuous spectrum utilized was 1,850 kilovolts: indium gives an activity of approximately 4 hours period, which is obviously due to the same metastable state $^{115}In^*$, of which we have already spoken. Similar results on the stable nuclear fluorescence of indium were obtained by Collins and others[19].

There is no doubt that new isomers of β-stable nuclei will be discovered, in the course of research undertaken in different laboratories; systematic research on the radiations emitted by metastable states will certainly be very useful for the understanding of nuclear isomerism.

In conclusion, we may remark that it is very probable, on account of the great number of known isomers, that the radiative transitions of life-times between, say, 10^{-7} sec. and 1 sec. are much more frequent than is generally supposed. These transitions, on the other hand, are strongly converted[6]. We would expect[20], consequently, that *transitions of this kind, with conversion coefficients approximately* 1, *may be found frequently.*

Indeed, in the radiation emitted in the capture of slow neutrons by gadolinium, a strong component of soft electrons has been observed by Amaldi and Rasetti[21] (life-time less than 10^{-3} sec.). Soft electronic components have also been observed in the capture of slow neutrons by other nuclei[22]. On the other hand, these strongly converted transitions may play a considerable part in the interpretation of γ- and X-spectra emitted by certain natural radio-elements[20].

发射出的辐射还没被充分的研究；它的性质对理解核的同质异能现象与同量异位对来说都是非常重要的。实际上，^{115}In 与 ^{115}Sn 是稳定的相邻同量异位核的罕见实例之一：^{115}In* 可以衰变成 ^{115}Sn（β 辐射），或者衰变成 ^{115}In，或者同时衰变成这两种元素。

巴恩斯与阿拉丁[15]通过用 5.8 兆电子伏的质子照射铟元素（反应 *p,p*），也得到了 ^{115}In 的同一亚稳态。不过，我们目前仍不清楚，核激发机制是否如魏斯科普夫所讨论的那样，是由质子电场的作用而导致的激发[16]。

在上述各情况以及一些更为近期的研究[17]中，稳定同位素的亚稳态都是通过**核嬗变**得到的。

拉扎德和庞蒂科夫[18]尝试了一种新方法，该方法使**核嬗变不可能发生**从而获得人工放射性元素，而人工放射性元素的出现有可能干扰实验研究。

这种方法要使用具有连续谱的 X 射线照射靶子，所用 X 射线的能量低于核的离解能。假定被照射的核具有一个亚稳态；X 射线可以激发核的较高能级；由此而被激发的一部分核就可能落入亚稳态，可以观测到的正是来自这个态的辐射。所用连续谱的最大能量为 1,850 千电子伏，铟原子产生了周期约为 4 小时的放射性，这一放射性显然是来自我们曾提到过的 ^{115}In* 亚稳态。柯林斯与其他一些研究人员[19]从铟的稳定核荧光现象上得到了类似结果。

毫无疑问，在不同实验室所进行的研究工作中，一定会发现 β 稳定核的新同质异能素；关于亚稳态发射辐射的系统性研究对于理解核同质异能现象无疑是大有助益的。

总而言之，根据大量已知的同质异能素，我们可以说，具有介于诸如寿命在 10^{-7} 秒到 1 秒之间的放射性跃迁，很有可能比我们一般所认为的更为频繁。另一方面，这些跃迁发生了强烈的转换[6]。由此，我们预期[20]**会频繁地发现这种转换系数近乎 1 的跃迁**。

实际上，在钆俘获慢中子的过程所发出的辐射中，阿马尔迪和拉塞蒂[21]已经观测到了强的软电子成分（寿命不足 10^{-3} 秒）。在其他核的慢中子俘获过程中也观测到了软电子成分[22]。此外，这些强烈转换的跃迁对于解释某些天然放射性元素发射的 γ 光谱和 X 光谱可能也会起到相当重要的作用[20]。

(**144**, 212-213; 1939)

B. Pontecorvo: Laboratory of Nuclear Chemistry, College de France, Paris.

References:

1. Soddy, *Proc. Roy. Inst.*, **22**, 117 (1917).
2. Hahn, *Ber. dtsch. Chem. Ges.*, B, **54**, 1131 (1921).
3. Kourtchatow, Myssowsky, Roussinow, *C.R.*, **200**, 1201 (1935). Amaldi, d'Agostino, Fermi, Pontecorvo, Segré, *Ric. Scient.*, **6**, 581 (1935). Amaldi and Fermi, *Phys. Rev.*, **50**, 899 (1936). Bothe and Gentner, *Z. Phys.*, **106**, 236 (1937).
4. Weiszäcker, *Naturwiss.*, **24**, 813(1936).
5. Pontecorvo, *Congrès du Palais de la Decouverte*, Paris, 1937, p. 118.
6. Hebb and Uhlenbeck, *Physica*, **5**, 605 (1938). Dancoff and Morrison, *Phys. Rev.*, **55**, 122 (1939).
7. Pontecorvo, *Phys. Rev.*, **54**, 542 (1938).
8. Valley and McCreary, *Phys. Rev.*, **55**, 666 (1939). Siday, *Nature*, **143**, 681 (1939).
9. Seaborg and Segré, *Phys. Rev.*, **55**, 808 (1939). Kalbfell, *Phys. Rev.*, **54**, 543 (1938).
10. Alvarez, *Phys. Rev.*, **54**, 486 (1938). Roussinow, Yusephovitch, *Phys. Rev.*, **55**, 979 (1939). Siday, ref. 8. Walke, Williams and Evans, *Proc. Roy. Soc.*, A, **171**, 360 (1939).
11. Segré, Halford and Seaborg, *Phys. Rev.*, **55**, 321 (1939).
12. De Vault and Libby, *Phys. Rev.*, **55**, 322 (1939). Le Roux, Lu and Sugden, *Nature*, **143**, 517 (1939). Seaborg and Kennedy, *Phys. Rev.*, **55**, 410 (1939).
13. Dodé and Pontecorvo, *C.R.*, **207**, 287 (1938).
14. Goldhaber, Hill and Szilard, *Phys. Rev.*, **55**, 46 (1939).
15. Barnes and Aradine, *Phys. Rev.*, **55**, 50(1939).
16. Weisskopf, *Phys. Rev.*, **53**, 1018 (1938).
17. Delsasso, Ridenour, Sherr and White, *Phys. Rev.*, **55**, 113.
18. Pontecorvo and Lazard, *C.R.*, **208**, 99(1939).
19. Collins, Waldman, Stubblefield and Goldhaber, *Phys. Rev.*, **55**, 507 (1939).
20. Pontecorvo, *C.R.*, **207**, 230 (1938).
21. Amaldi and Rasetti, *Ric. Scient.*, **10**, 115 (1939).
22. Hoffman and Bacher, *Phys. Rev.*, **54**, 644 (1938). Pontecorvo, *C.R.*, **207**, 856 (1938).

（王耀杨 翻译；厉光烈 审稿）

Cosmic Ray Ionization Bursts

H. Carmichael and C. N. Chou

Editor's Note

Since the beginning of the twentieth century, the study of cosmic rays had preoccupied a great many physicists for a number of reasons. First, they appeared to consist of fast-moving particles whose identity was for a long time unknown. The energy of these particles was much higher than could be created artificially in the laboratory, and they seemed to offer clues to the nature of phenomena and matter at great distances in the Universe. This paper is based on observations carried out under 30 metres of London clay on the disused part of a platform in Holborn Station on London's Underground railway network. The object of the experiment was to measure the numbers of showers of cosmic-ray particles created within electrically sensitive ionization chambers. Curve *C* represents the results of an experiment at the surface of the Earth, *A* an experiment running for 150 hours at the Underground station itself and *B* an experiment lasting 350 hours. The bends in the experimental curves *A* and *C* are taken to indicate a change in the mechanism of the production of showers of cosmic-ray particles.

COSMIC ray ionization bursts produced by showers of ten or more particles in a small ionization chamber (volume 1 litre) have been recorded at sea-level in Cambridge and, thanks to the hospitality of Prof. P. M. S. Blackett and Dr. H. J. J. Braddick, under 30 m. of clay in London. The data discussed here, curves *A* (150 hours) and *B* (350 hours), are from runs with no lead or other dense shower-producing material above the chamber (the wall of the chamber was of duralumin, 1.2 cm. thick, so as to avoid as much as possible showers produced by cascade multiplication in the chamber itself). Curve *C* (500 hours) is the result of similar observations at sea-level with a large ionization chamber (volume 175 litres, wall-thickness 0.3 cm. of steel), most of which were published by one of us[1] in 1936, when the method of recording was also described. The curves show the number of showers per hour in which *N* or more particles intersected the ionization chamber; the number of shower particles *N* was estimated on the basis of a specific ionization in normal argon of 90 ion-pairs per cm.

The ionization bursts obtained at sea-level with the big ionization chamber (curve *C*) involve much larger numbers of shower particles than those obtained with the small chamber (curve *A*), and also each curve has a remarkable change of slope at a rate of occurrence about 0.16 per hour. The curves, however, can be superposed fairly closely if the size of the showers in the small chamber is multiplied by 10. We deduce from this approximate proportionality of the *size* of these showers (as distinct from their rate of occurrence) to the areas of the thin-walled chambers used to observe them that they are mostly *extensive showers*, originating in the atmosphere, of the same type as those found by

宇宙射线的“电离暴”

卡迈克尔，周长宁

编者按

从20世纪初开始，由于各种原因，很多物理学家都对宇宙射线的研究给予了关注。首先，它们似乎是由快速运动的粒子组成，但究竟是什么粒子，这在很长一段时间里都是未知的。这些粒子的能量大大地超过了在实验室里人工所能产生出来的能量，并且它们似乎能够为宇宙中远距离的现象和物质性质的研究提供线索。本论文是基于伦敦地面以下30米处的观察结果，该地是伦敦地铁网络霍尔本车站一个废弃不用的站台。实验的目的是采用对电荷灵敏的电离室测量宇宙射线粒子簇射的数目。*C*曲线代表地球表面的实验结果，*A*曲线代表一个在地下车站为时150个小时的实验结果，*B*曲线代表为时350小时的实验。*A*和*C*实验结果曲线的弯曲则用来显示宇宙射线粒子簇射产生机制的变化。

把一个小电离室（容量为1升）放在剑桥的海平面上，对其中10个或更多粒子的簇射产生的宇宙射线电离暴进行了记录，并且多亏了布莱克特教授和布拉迪克博士的热心帮助，我们在伦敦地面以下30米的地方也完成了同样的实验。在本文要讨论的数据中，*A*曲线（150小时）和*B*曲线（350小时）是在小电离室上部没有覆盖铅或者其他密度较大的引发簇射的物质时得到的（电离室内壁由硬铝制造，厚1.2厘米，以尽可能避免在电离室内由连续倍增效应产生的簇射）。*C*曲线（500小时）是我们在海平面上用大电离室（容量为175升，壁厚0.3厘米，钢制）重复类似实验得到的结果，我们中的一人[1]已于1936年发表了大部分数据，当时他也描述了记录的方法。从这些曲线中可以得到每小时的簇射数量，即有*N*个或是更多的粒子横穿电离室；簇射粒子数*N*是在每厘米90个离子对的普通氩气的特定离子化过程中得到的。

在海平面上的大电离室观察到的离子爆发过程（*C*曲线）所产生的簇射粒子数量要比在小电离室里（*A*曲线）看到的数量多很多，而这两条曲线的斜率都在每小时观测到的簇射线为0.16处出现明显变化。如果将小电离室里的簇射量大小乘以10，那么*A*曲线和*C*曲线会很好地重叠在一起。我们根据簇射量（这一点可以从发生率中很明显地看到）与所用薄壁电离室的横截面面积大致成正比这一现象可以推断出，所发生的簇射多半是**广延簇射**，它们产生于空气中并且与俄歇及其同事[2]用

Auger and his co-workers[2] with counters. We should not expect to find exactly the ratio of the areas (approximately 1:30) because narrow showers or condensations of rays of cross-section smaller than the area of the large chamber tend to increase disproportionately the bursts in the small chamber.

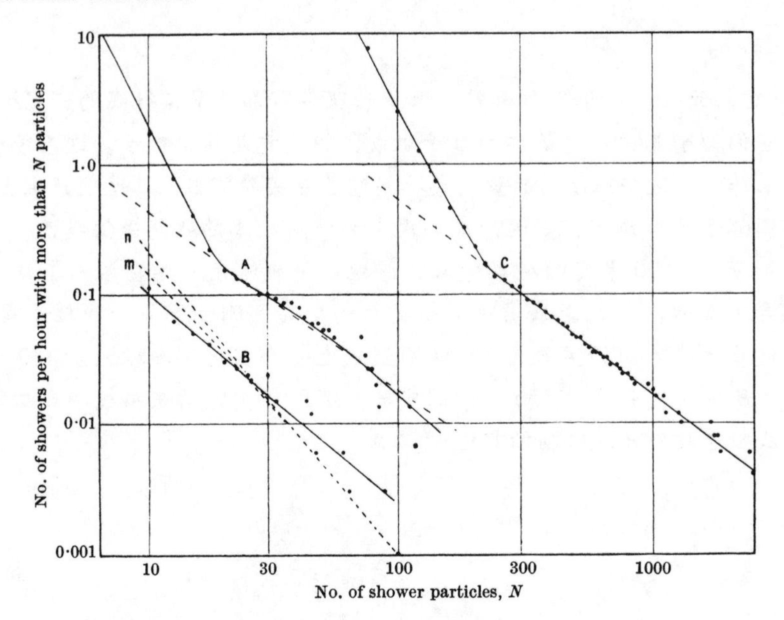

The bursts which are found underground (curve *B*) must be produced by the penetrating component of the cosmic rays. We have calculated the distribution curve to be expected in the tube station for cascade showers produced by electrons "knocked on" by mesons using the data given in the paper by Bhabha[4] in which, however, the meson was assumed to have spin $\frac{1}{2}$. We adopted, following Euler and Heisenberg[3], an exponent 1.87 for the integral energy distribution of the mesons originating in the atmosphere. The calculation shows (curve *m*) that the showers resulting from this process alone are nearly sufficient to account for the bursts recorded underground, if we assume that the cross-sectional area of the showers underground is not much greater than the area of the chamber (actually more of the larger bursts are found than are given by this calculation, but the theoretical implications of this discrepancy will be discussed later).

At sea-level a similar calculation (curve *n*) gives much fewer bursts than are observed even if we suppose that the showers are so narrow that all the shower particles in any one burst can intersect the small chamber: but we already know that most of the showers observed at sea-level are at least wider than the large chamber. We therefore conclude that a negligible number of the extensive showers observed at sea-level is produced by electrons knocked-on by mesons. It would seem also that an insufficient number of such showers can be produced by the spontaneous decay of the meson, a process which might be invoked to explain the steeper parts of the sea-level curves.

计数器测量到的簇射属于同一类型。我们不能指望由此得到精确的面积之比（大约是 1∶30），因为当窄簇射或射线凝聚效应的横截面比大电离室的面积小时更容易引起非均匀的离子爆发。

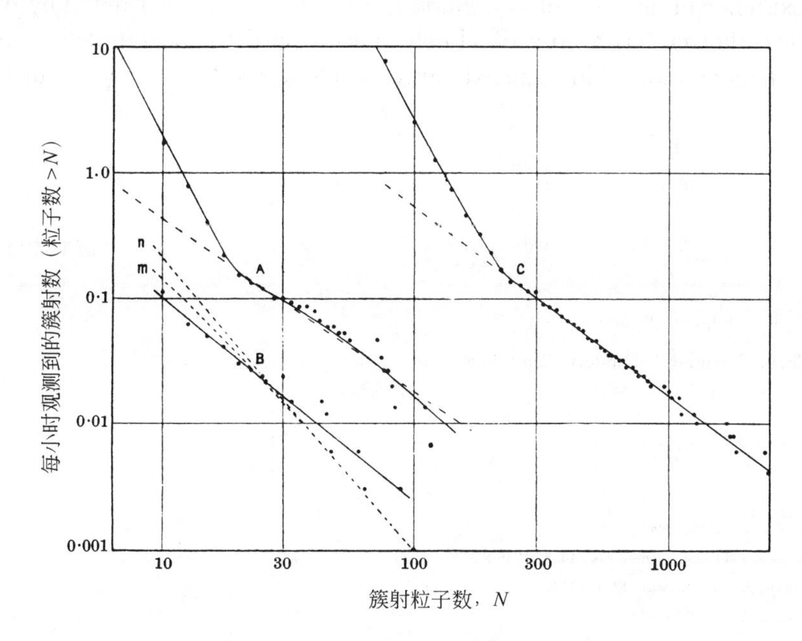

在地面以下测量到的爆发（*B* 曲线）肯定是由穿越了的宇宙射线产生的。利用霍米 · 巴巴[4]在其论文里给出的数据，假设介子的自旋为$\frac{1}{2}$，我们已经计算出由电子被介子撞击而产生的级联簇射在地铁站里的预期曲线分布。根据欧拉和海森堡[3]的方法，我们认为大气中介子累积能量分布曲线的指数为 1.87。如果我们假设在地面以下的实验中，簇射区域的横截面不比电离室大多少，那么从计算结果中可以看到（*m* 曲线）：仅由介子和电子的碰撞过程所产生的簇射就足以解释在地面以下记录的爆发了（实际上测量到的大型爆发数值要多于计算值，对这种差异的理论解释将在以后讨论）。

在海平面上用上述方法得到的计算值（*n* 曲线）比观察到爆发的数值低很多，即使我们假设簇射范围非常窄，以至于每次爆发中的所有簇射粒子都能够横穿小电离室，但事实上我们已经知道，在海平面上所观察到的大部分簇射范围至少都比大电离室的范围大。因此，我们得出结论：在海平面上所观察到的少量爆发簇射是由介子撞击电子产生的。这种簇射的量不够大，可能是由介子的自发衰退造成的，这也许可以作为由海平面上的观测数据得到的 *n* 曲线有一个斜率较大的部分的原因。

We think, therefore, that nearly all the showers which produce bursts at sea-level in our thin-walled ionization chambers originate very high in the atmosphere, and that the dual nature of the distribution curve at sea-level (as indicated by the change of slope) is due, not to the existence of showers of two kinds, but to the fact (already noted by Auger[2]) that each extensive shower has a core of closely spaced particles surrounded by a relatively wide fringe of much more thinly spaced particles able to produce bursts of small size.

(**144**, 325-326; 1939)

Hugh Carmichael: St. John's College, Cambridge.
Chang-Ning Chou: Cavendish Laboratory, Cambridge.

References:

1. Carmichael, *Proc. Roy. Soc.*, A, **154**, 223 (1936).
2. Auger, Maze, Ehrenfest and Freon, *J. Phys. et Rad.*, **10**, 39(1939).
3. Euler and Heisenberg, *Er. exak. Naturwiss.*, 17, 1 (1938).
4. Bhabha, *Proc. Roy. Soc.*, A, **164**, 257 (1938).

因此，我们认为，几乎所有能在位于海平面上的薄壁电离室内产生爆发的簇射都来自于大气层中非常高的地方，而海平面上分布曲线所表现出的双重特性（由斜率的变化可以看出）并不是由于存在两类簇射，而是由于以下原因（这一点俄歇[2]也注意到了），即每次爆发簇射都有一个由密排粒子组成的核心，其周围是由排列非常松散的粒子组成的广大边界地区，而在这些边界地区就会产生规模较小的爆发。

（李忠伟 翻译；马宇蒨 审稿）

Measurements of the Velocity of Light

Editor's Note

Commenting on the earlier proposal of M. E. J. Gheury de Bray, Lord Kitchener asserted that observations suggesting a very slight slowing of the speed of light over time were probably due to error. Here Gheury de Bray responds. Although some techniques for measuring the speed of light could be criticized, he says, this cannot be said of his toothed-wheel method. He also notes that Simon Newcomb and Albert Michelson had independently reported similar results, which differed by just the amount expected if the speed of light were slowly changing. All the same, there seems now no evidence of such an effect, although the possibility of changes in physical constants over time is still debated.

M. E. J. Gheury de Bray suggests[1] that measurements of the velocity of light show that it is changing according to a formula:

$$c = 299{,}900 - 3.855\ T \qquad (1)$$

T being measured in years from 1900.

Assuming that Planck's constant and the energy E given out by an excited atom remain unchanged, the relation

$$E\lambda = ch \qquad (2)$$

shows that a change in c must be accompanied by a proportional change in λ. But the wave-length of the red calcium line has remained constant to within one part in five million for thirty years, only 0.5 percent of the change required by (1).

It is probable that there are unsuspected systematic errors in the determinations of c.

Kitchener

* * *

The decrease of velocity of light deduced from the red-shifts is of the order of 1 km./sec. in 6,000 years, or 1 in 600,000,000 for thirty years, which is quite consistent with the apparent constancy of the wave-length mentioned by Lord Kitchener. Our observations, being affected by unsuspected systematic errors, and covering only a third of a century, give what is probably a greatly exaggerated rate of slowing down. The remarkable fact is that *all* the determinations are unanimous in indicating the existence of such a variation, and the red-shifts, if interpreted so as to escape from fantastic results, confirm it.

光速的测定

编者按

基奇纳勋爵在评论谷瑞·德布雷早前的猜想时断言，观测结果所显示的光速随时间的轻微变慢很可能是由于误差造成的。谷瑞·德布雷在这篇文章中予以回应。他说，尽管测量光速所用的一些技术值得怀疑，但他所使用的齿轮方法不在此列。他也指出西蒙·纽科姆和阿尔伯特·迈克尔逊都曾独立地报道过类似的结果，区别仅仅在于光速减慢的变化速率不同。尽管如此现在看来似乎仍没有证据支持这个效应的存在，虽然对物理常数随时间变化的可能性仍然在探讨之中。

谷瑞·德布雷指出[1]，光速的测定结果表明它依照如下公式而变化：

$$c = 299{,}900 - 3.855\ T \tag{1}$$

式中的 T 是自 1900 年以来的年份。

假定普朗克常数和受激发原子释放出的能量 E 保持恒定，则关系

$$E\lambda = ch \tag{2}$$

表明光速 c 变化的同时必定伴随着 λ 与之成比例的变化。但是我们观测到红色钙谱线的波长在 30 年中的变化小于五百万分之一，而这只有公式（1）中变化量的 0.5%。

这可能是在光速 c 的测量过程中存在的未知的系统误差。

基奇纳

*　*　*

根据红移推断，光速在 6,000 年中降低了 1 千米 / 秒的数量级，或者说在 30 年中降低了 1/600,000,000，这与基奇纳勋爵所谈到的波长的表观不变性颇为一致。由于受未知的系统误差的影响，以及观测仅持续了 1/3 个世纪，所以我们观测到的结果很可能远远夸大了光速减慢的速率。不过显而易见的是，**所有**测量结果都一致地表明这种变化的存在，而且，如果不对红移进行荒诞怪异的解释的话，那么红移就证实了这一点。

The Table 1 of the communication referred to cannot be dismissed on the ground of "unsuspected systematic errors". It is admitted that the method of the revolving mirror may suffer from physical bias, but no such reproach can be levelled against the toothed wheel method, which is only open to objections of a physiological nature. These can be readily overcome by substituting for the observer's eye a photo-electric sensitive device. While France and the United States share between them practically the whole initiative in the measurement of c, Great Britain has only to her credit a conspicuous failure. Is there in this country no one who can redeem it from this position and settle this question, which lies at the basis of physical science, considered in its broadest aspect?

Two observations, of Newcomb (1882.7 : 299,860) and of Michelson (1882.8 : 299,853) agree so closely that, if we consider that they were made by different observers, working independently with different instruments and different techniques, in different places, they must be extremely accurate, despite their large probable errors. It is significant that the second in date gives a lesser value of c.

M. E. J. Gheury de Bray

(**144**, 945; 1939)

Kitchener: Trinity College, Cambridge.

M. E. J. Gheury de Bray: 49, Great Thrift, Petts Wood, Nov. 2.

Reference:

1. *Nature*, 144, 285 (1939).

不能由于存在“未知的系统误差”而把我之前提到的表 1（见本书第 689 页）排除在考虑的范围之外。我们知道，旋转镜方法可能会受到物理偏置的影响，但是齿轮方法就没有受到这类影响，这种齿轮方法存在的唯一问题是研究者的生理特征的影响。不过通过用光电敏感装置代替观测者的眼睛，我们就可以很容易地克服这些困难。实际上，当法国和美国共同占据了测定光速 c 值的全部主动性时，留给英国的只是赤裸裸的失败。难道这个国家就没有人能挽回这种局面并且从最广泛的视角来解决这个物理科学的基础问题吗？

纽科姆（1882.7 : 299,860）和迈克尔逊（1882.8 : 299,853）的观测结果惊人一致，如果考虑到它们是由不同的观测者用不同的装置和技术方法在不同的地点各自独立地得到的，我们甚至可以断言这些结果是极为精确的，哪怕存在着较大的或然误差。重要的是，观测日期靠后的迈克尔逊给出了较小的 c 值。

谷瑞 · 德布雷

（王耀杨 翻译；张元仲 审稿）

The Evolution of the Stars

F. Hoyle and R. A. Lyttleton

Editor's Note

Astronomers Fred Hoyle and Raymond Lyttleton review how recent advances in nuclear theory were fundamentally changing astronomers' understanding of stellar evolution. It was now possible, as George Gamow had recently argued, to probe the nuclear chemistry likely to be important at the densities and temperatures prevailing in stars. But Hoyle and Lyttleton think Gamow went rather too far in suggesting most problems of stellar evolution were now solved. In Gamow's view, for example, all red giant stars must be of very recent origin, which seemed most unlikely. Hoyle and Lyttleton were correct: physicists then understood only the rudiments of nuclear chemistry, and knew nothing of nuclear fusion, a key process driving all stellar activity and evolution.

PROF. G. Gamow has recently discussed in *Nature*[1] the consequences of recent developments in nuclear theory on the problem of stellar evolution. In the light of the exact knowledge that is now available of a large number of nuclear reactions, it is possible to decide which processes rise to importance at the densities and temperatures prevailing in the stars, and on this basis trustworthy estimates have been given for the rate of liberation of subatomic energy. These results, which are the outcome of laboratory investigations, furnish the mathematical theory of internal constitution of the stars with a new equation that enables the luminosity of a star to be calculated by direct methods. The information so obtained has been utilized to attempt to resolve the many paradoxes and discrepancies encountered in discussing the general problem of stellar evolution. All this recent work has been authoritatively summarized by Gamow with great clarity and understanding in the article referred to above. It is therefore with some surprise that we find that Prof. Gamow concludes his article with the impression that these new developments practically solve the problems of stellar evolution. This seems to us to be so far from being the case that some further discussion of the claims of nuclear theory as the main factor in stellar evolution would be desirable.

In the first place, it should be noticed that the application of nuclear theory in its present state inevitably leads to a result at least as embarrassing as any of the questions that it might possibly resolve. For the theory maintains that no synthesis of atomic nuclei from hydrogen is possible within the stars except for the very light elements. This would imply that the stars can no longer be regarded as the building place of the heavy elements, which must have formed before they became part of the star—if indeed they ever were formed. Now although such a conclusion does not itself constitute a logical contradiction, it seems to us to present such overwhelming difficulty that it is much more reasonable to conclude that the basis of nuclear theory is in need of revision rather than that the heavy

恒星的演化

霍伊尔，利特尔顿

编者按

天文学家弗雷德·霍伊尔和雷蒙德·利特尔顿回顾了核物理理论的近期进展是如何从根本上改变天文学家对恒星演化的理解的。如乔治·伽莫夫近期所指出的，现在有可能证明核化学在恒星当时的密度和温度下是有重要作用的。伽莫夫认为恒星演化的大多数问题现在都已经解决了，但霍伊尔和利特尔顿对此并不认同。例如，伽莫夫认为所有红巨星一定是在新近才产生的，但这似乎是最不可能的事情。霍伊尔和利特尔顿是对的：那时的大多数物理学家只知道核化学方面的初级知识，但对核聚变这一驱动恒星活动和演化的关键过程却一无所知。

乔治·伽莫夫教授最近在《自然》[1] 杂志上讨论了核物理理论的最新进展在恒星演化问题上所取得的结果。根据现有可用大量核反应得到的准确知识，人们有可能确定在恒星当时密度和温度下哪些过程具有重要的作用，并且在此基础上，人们得到了对亚原子能量释放率的可靠估计。这些实验研究得到的结果为计算恒星内部结构的数学理论提供了一个新的方程，由此可以采用直接的方法计算恒星的光度。我们利用所获得的信息尝试解决恒星演化的普遍问题中遇到的许多矛盾和差异。在上面所引用的那篇文章中，伽莫夫以十分清晰易懂的方式对近来的所有工作进行了权威性的总结。令人有些惊讶的是伽莫夫教授在他的文章中得出的结论给人这样的印象：这些新的进展从实际上解决了恒星演化的多种问题。而在我们看来情况远非如此，核物理理论在恒星演化中作为主要因素的论断还需要进一步讨论。

首先，我们应该注意到目前核物理理论的应用和它有可能解决的任何问题一样，会不可避免地导致一种同样令人为难的结果。因为该理论认为，在恒星内部氢仅能合成一些非常轻的元素，而不能合成其他的原子核。这意味着恒星不能再被当成是重元素产生的地方，如果这些重元素确实曾经产生过，那么它们必须在变成恒星的一部分之前就已经产生了。现在尽管这个结论自身在逻辑上并不矛盾，但是目前在我们看来它却造成了非常巨大的困难，所以更为合理的结论是：核物理理论的基础所需要的是修正而非重元素不是由合成产生的。另一方面，许多研究者似乎已经认

elements were not formed by synthesis. On the other hand, many investigators seem to have accepted the former result as satisfactory, and in particular Gamow has proceeded to make it the basis of a theory of the red giant stars.

Secondly, Prof. Gamow apparently regards the conclusion that various classes of stars should be of totally different ages as a natural one. Thus the fact that the present theory leads to a life-time for certain massive stars of order 10^{-3} the life-time of the Sun is not regarded as a difficulty at all; indeed it is merely supposed that this is the case and the theory remains unquestioned. In point of fact, it is an essential part of the theory as proposed by Gamow that all red giant stars are considered as of very recent formation, since the presence of lithium, etc., is required to enable them to radiate with their supposed low internal temperatures.

Now if all the stars could be regarded as single autonomous bodies, it would be difficult to dispute the validity of these views by direct means. But it so happens that the frequent occurrence of binary systems enables a simple test of the theory to be made, for in the case of binary stars both components must be of comparable age. It is immediately clear that the well-known difficulty concerning the relative emission per unit mass of the components of doublestars must remain in any theory that appeals only to the internal properties of the stars, although by making very artificial assumptions in Gamow's theory some of these discrepancies might be avoided. For example, it would require as a general result that the less massive components of binary systems form with a hydrogen content differing systematically from that of their companions. In certain cases this would lead to even more dubious initial conditions: thus, whilst Sirius must have formed with high hydrogen content, it would have to be assumed (according to the more generally accepted theory of degenerate matter) that the companion formed almost solely from heavy elements for it to have been practically exhausted during the whole of its existence. Such a solution of this difficulty could scarcely be regarded as satisfactory even if there were no other objection to the theory described by Gamow.

It should be particularly noticed that the foregoing suggestion assumes that the components have not evolved by fission, for this latter process (even if it were dynamically satisfactory) would clearly lead to two stars of closely similar compositions. We have been able to show, however, that to produce close binary systems, periods of order 5×10^{10} years are necessary[2]. Thus the existence of spectroscopic binary systems in which one component is a red giant or any highly luminous star presents an immediate contradiction of the theory given by Gamow. Moreover, there seems to be a general tendency for the mass of binary systems to increase with decreasing separation.

While on the subject of fission, it is perhaps only proper to point out here that although the mathematical investigations of the development of rotating fluid masses clearly demonstrate that binary stars cannot be generated by fission[3], many astronomers do not yet seem to have realized the physical significance of the mathematical results. As a consequence of this there are still many who "believe" in the fission theory. But as will

同前一个结论是令人满意的，特别是伽莫夫，他已着手把它作为了红巨星理论的基础。

其次，伽莫夫教授显然认为各种类型的恒星有完全不同的年龄是很自然的事情。因此，由现有理论得出的某些大质量恒星的寿命仅仅是太阳寿命的 10^{-3} 倍的结论完全没有被认为是一件困难的事情；事实上大家认为这就是事实且现有理论是毋庸置疑的。其实正如伽莫夫所提出的，所有红巨星形成于最近这样一个观点是现有理论里必不可少的一部分，这是因为需要锂等元素的存在，以使其可以在假定的内部低温中发出辐射。

现在，如果所有恒星都可以被看作是独立的天体，那么可能很难用直接的方法对这些观点的正确性提出质疑。但幸亏多次发现的双星系统使我们可以对该理论进行简单的检验，因为在此情况下，两颗成员星应该有同等的年龄。尽管伽莫夫理论中人为的假设避免了某些矛盾，但很显然，对于任何单纯只研究恒星内部性质的理论来说，著名的关于双星中成员星的单位质量相对辐射的困难仍然存在。例如，作为一个一般性结果人们可能会假设双星系统中质量较小的成员星在形成的时候氢含量系统性地和它们另外的成员星不同。在某些情况下，这可能会让初始条件变得更加可疑：所以，虽然天狼星在形成时氢含量可能很高，但我们不得不假设（考虑到简并态物质的理论已被更为广泛地接受了）其成员星几乎只由重元素组成，以使其在存在的整个期间耗尽主要的核燃料。即使对于伽莫夫提出的理论没有其他反对的理由，但是对于上述困难，这样的解决方案也是绝不可能令人满意的。

应该特别注意，上述理论假设了成员星不是通过分裂演化而来的，因为分裂演化（尽管在动力学上是合理的）会产生两颗组分相似的恒星。然而，我们已经能够证明，必须有 5×10^{10} 年量级的时间 [2] 才能产生密近双星系统。于是分光双星中，若其中一颗成员星是红巨星或者任一颗高亮度的星，就必然同伽莫夫所提出的理论相矛盾。另外，似乎存在一个普遍趋势，即双星系统的质量随间距的减小而增加。

在恒星分裂的问题上，或许在这里指出最为合适，即尽管转动流体演化的数学研究结果清楚地证明了双星不可能由分裂产生 [3]，但是许多天文学家似乎还并没有意识到这些数学结果在物理方面上的重要性。因此仍然还有很多人“相信”分裂理论。但是正如人们将要看到的，即使求助于分裂理论也不能挽救伽莫夫的红巨星理论，

be seen, even an appeal to fission could not save Gamow's theory of the red giant stars, in addition to which it would raise afresh the difficulty of the relative emissions of the components in binary stars. Even if some process of break-up of a single star led to a binary system, it is evident that the components must have similar chemical compositions, while in close binary systems consisting of a red giant star and a class *B* star, Gamow's theory would require the red star to be the less massive component on account of the mass luminosity relation. But observation shows that in such pairs the giant star tends to be the more massive component. This is the case, for example, in the three stars υ Sagittarius, *V V* Cephei and ζ Aurigae. Thus it seems that no matter from what angle we approach the questions raised by the observed properties of binary systems, the paradoxes already recognized by astronomers must remain in one form or another.

However, quite apart from the foregoing objections to the constructive portion of Gamow's article, it is very noticeable that no reference is made to the wide class of dynamical features that is associated with the stars. This, of course, is the direct result of attending only to the internal physical properties of the stars; but the dynamical features we have in mind are altogether too marked to remain unaccounted for in a satisfactory theory. Thus such questions as the formation of individual stars, and of binary and multiple systems, together with the general increase of mass with decreasing separation, and the observed approximation to equipartition of energy among the stars seem to present the real key to any theory of stellar evolution. An internal theory can give no explanation for the correlation between peculiar velocity and spectral class or the observed tendency for massive stars to lie in the galactic plane, features that must be related to the previous history of the stars.

It has been customary during recent years for investigators on stellar evolution to devote attention to internal constitution with little or no regard for the dynamical features. It appears that Prof. Gamow has followed essentially in this tradition and therefore confined his article to the modifications effected by the introduction of modern nuclear theory. Thus, in dealing with the properties of variable stars, no attempt is made to account for the three distinct periodicity groups comprised by stars of periods of order half a day, four days and 300 days. These variables also show a marked preference as regards spectral class, the first being largely of classes *B* and *A*, the second of *F* and *G* and the third of class *M*. Moreover, the two short-period groups exhibit a most remarkable property in that none of them, out of more than two hundred available examples, possesses a close companion, whereas about one star in five of normal stars of similar spectral classes does possess a close companion. On the other hand, long-period variables appear to possess a normal complement of companions. The first and third types are stars of moderate luminosity and show no pronounced galactic concentration, whereas the variables of intermediate period, the Cepheids, are strongly concentrated to the galactic plane and are among the most luminous known stars. Thus it is clear that very remarkable dynamical properties are intimately connected with even the different types of variability, and therefore that purely internal considerations are most unlikely to prove capable of elucidating the nature of the connexion.

此外双星成员星的相对辐射问题将再次出现。就算单星分裂的某一阶段会产生双星系统，但可以肯定的是成员星之间仍应该有相似的化学组成，对于一个由一颗红巨星和一颗 B 型星组成的密近双星系统，伽莫夫的理论认为基于质光关系，红巨星的质量相对较小。但观测结果表明在这种双星系统中，红巨星的质量往往相对较大。例如在人马座 υ、仙王座 $V\,V$、御夫座 ζ 这三个双星系统中，情况均是如此。这样看来，似乎无论我们从什么角度来处理双星系统观测性质上的问题，这个已被天文学家意识到的矛盾总会以某种形式存在。

然而，除了前面对伽莫夫文章理论构建部分提出的异议外，很明显，还没有文献提到与恒星相关的广泛动力学特征。当然，这是只关注恒星内部物理性质的直接结果；但是，总的来说我们考虑的动力学性质是个很重要的问题，为此任何一个理论要令人信服都应该对此给出说明。于是，诸如单个恒星、双星以及聚星系统的形成，双星质量普遍随间距的减小而增加，以及观测到类似在恒星之间近似均分能量，类似这样的问题似乎是所有恒星演化理论的真正关键。一个恒星内部结构的理论不能解释本动速度和光谱类型之间的关系，也不能解释为什么观测到的大质量恒星都倾向位于银道面上，而这些特征必然和恒星早先的历史有关。

近年来，致力于恒星演化研究的科学家们都习惯性地只关注恒星内部组成而很少或者不关心其动力学特征。伽莫夫教授似乎也基本上遵循了这个传统，因此他的文章局限于引入现代核物理理论加以改进。因此，在处理变星性质的时候，他没有试图解释由周期为 0.5 天、4 天和 300 天量级的恒星组成的三个不同的周期组。这些不同周期的变星也明确显示出了与光谱型相关的顺序，第一类主要是 B 和 A 型星，第二类是 F 型星和 G 型星，而第三类是 M 型星。此外，两个短周期的组还显示出一个最显著的特性：在已有的两百多个例子中，没有一个有密近的伴星。而在具有类似光谱型的一般双星中，每 5 对就有一个具有密近的伴星。另一方面，长周期的变星有正常的伴星。第一类和第三类中等亮度的恒星并没有显著的向银道面聚拢，而中等周期的变星，如造父变星，却都位于已知最明亮的恒星之中并且都高度聚集在银道面上。这清晰地表示，明显的动力学性质甚至与不同类型的光变有着紧密的联系，因此只考虑恒星的内部结构是不大可能诠释这些联系的本质的。

From these and many other dynamical qualities associated with various types of stars, it appears to us that Prof. Gamow has over-estimated the importance of nuclear theory in the problem of stellar evolution. Indeed, in our opinion nuclear physics has very little to add to the results already conjectured by astrophysicists, and can merely serve to confirm these conjectures, a typical instance being the mass-luminosity relation itself. Finally, we wish to point out that although the present article consists largely of criticism, we have discussed elsewhere a number of the questions raised[2], and it has been found that purely dynamical considerations may be sufficient to provide a natural explanation of many of the difficulties mentioned in this article.

(**144**, 1019-1020; 1939)

F. Hoyle and R. A. Lyttleton: St. John's College, Cambridge.

References:

1. *Nature*, **144**, 575, 620 (Sept. 30 and Oct. 7, 1939).
2. *Proc. Camb. Phil. Soc.*, (4), **35** (1939).
3. *Mon. Not. Roy. Astr. Soc.*, **98**, 646 (1938).

考虑到上述这些情况以及其他与各种类型恒星相联系的动力学性质，在我们看来伽莫夫教授高估了核物理理论在恒星演化问题中的重要性。事实上，我们认为核物理对天体物理学家所推测出的结果几乎没有什么补充，只能用其来验证这些推断，一个典型的例子是质光关系本身。最后，我们想指出，尽管本文主要是批评，但是我们已经在另外一篇文章中对所提出的许多问题[2]进行了讨论，并且我们发现从纯粹的动力学角度考虑可能足以为本文中提到的这些困难提供一个合理的解释。

（钱磊 翻译；何香涛 审稿）

Radioactive Gases Evolved in Uranium Fission

L. Wertenstein

Editor's Note

The discovery of nuclear fission and its release of enormous energy posed the possibility of putting that energy to use, either in weapons or in industry. But scientists still knew very little of many fundamental processes which might influence its successful engineering. Here physicist Ludwik Wertenstein, writing from a Poland recently occupied by Nazi Germany, reports that the fission process produces radioactive gases. He had detected the radioactivity induced into several gases circulating about a sample of uranium bombarded by neutrons, and found traces of essentially two distinct radioactive components, with half-lives of about 30 seconds and four minutes. This was merely the beginning of detailed investigation required to bring nuclear energy into engineering practice.

IN this letter, a brief account is given of the preliminary results of an investigation of the radioactive gases evolved in the fission of uranium nuclei. This investigation was commenced this summer but was interrupted by the outbreak of the War. Even if the results so far obtained are not more complete than those obtained in the meantime by other investigators[1], it may still be of interest to describe the method employed, which follows somewhat different lines.

The radioactive gases evolved in the fission process were carried by a circulating stream of an inactive gas through two Geiger-Müller counters placed in succession. The time lag of the arrival of the gas in the two counters could be varied within wide limits by means of a system of capillary glass tubes of various bores placed in parallel and fitted with stopcocks. The magnitude of the lag was ascertained by separate experiments in which actinon or thoron was circulated. As a carrier gas, acetone vapour at a pressure of 10 mm. mercury was used because of its favourable properties for the working of the counters. The circulation was kept up by means of a Vollmer glass pump.

The uranium vessel, containing about 30 gm. of uranium oxide (U_2O_3), was surrounded by paraffin and provided with a well to allow introduction or removal of the neutron source, which consisted of about 30 mgm. of radon + beryllium. In the experiments, the counting rate of both counters was recorded for a period of time immediately following the commencement or cessation of irradiation. Typical results, using a time lag between the counters of 15 sec., are shown in the accompanying figure, from which it may be seen that the curves obtained for increase and for decay are almost complementary, and indicate several periods, of which some are of the order of a minute, while others are evidently much longer, giving rise to a residual activity almost constant within the time of the experiment.

铀核裂变时放出的放射性气体

维腾斯坦

编者按

核裂变的发现及其释放的巨大能量显示了将这种能量用于武器和工业的可能性。但是科学家对它的许多基本的过程仍了解很少，这可能影响到对它的成功工程化。物理学家卢德维克·维腾斯坦从最近被纳粹德国占领的波兰写出的信中指出，裂变过程产生放射性气体。他探测到在中子轰击铀样品周围几种气体循环引入的放射性，并发现两种截然不同的放射性组分的踪迹，它们的半衰期分别为30秒钟和4分钟。这仅仅是将核能引入工程应用详细研究的必要开端。

这封信对铀核裂变中放出放射性气体研究的初步结果进行了简要地说明。这项工作原计划于今年夏天开始着手，但是因二战的爆发而中断。尽管到目前为止我们所得到的结果并不比其他研究者在此期间所得到的结果[1]更为完善，但我们采用了一种稍为不同的研究方法，而描述该方法可能也是很有趣的。

裂变过程中放出的放射性气体依次流经放置好的两套盖革–米勒计数器，被其中无放射性气体环流携带出来。利用一套平行放置且装有活栓的不同孔径的毛细玻璃管构成的系统，可以对两套计数器内气体到达的时间延迟在很宽的限度内进行调节。时间延迟的长短可以通过锕射气或钍射气循环的分立实验来确定。实验中我们使用10毫米汞柱压强的丙酮作为载气，因为它具有有利于计数器正常运转的性质。气体循环是通过瓦尔莫玻璃泵来维持的。

将盛有大约30克氧化铀（U_2O_3）的容器用石蜡包围起来，并提供一个很容易引入和取走中子源的井，中子源由大约30毫克氡和铍组成。实验过程中，在照射开始或停止之后的一段时间，记录两台计数器的计数率。附图中显示了两计数器间的时间延迟为15秒钟时得到的典型结果，从图中我们可以看到，所得到的增长和衰减曲线几乎是互补的，并且展现出若干个周期（其中一些的数量级为分钟，另外一些则明显要长很多），这些在实验期间引起了几乎恒定的剩余放射性。

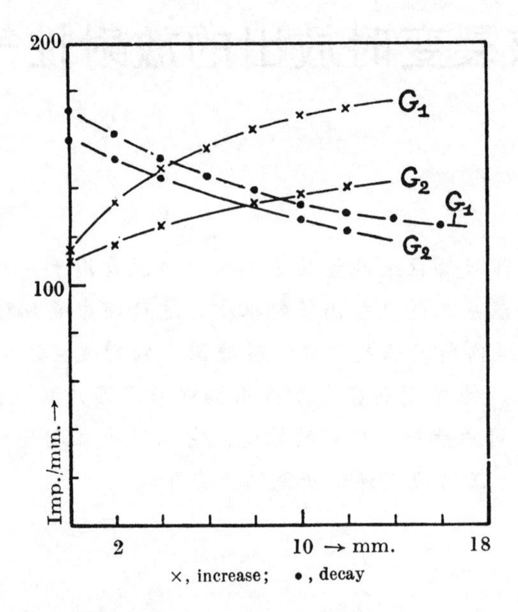

The shorter periods were estimated from a decay curve obtained in a separate experiment, in which the flow of gas was stopped at the same time that the source was removed, and the records of the two counters at any subsequent time simply added. This gave two periods of about 30 sec. and 4 min., the ratio between the rates of production of the corresponding gases being estimated as 1.82. (Glasoe and Steigman (*loc. cit.*) find two gaseous products of uranium fission of periods 30 sec. and 5 min., of which the first transforms into a product of 3 min. period. It is possible that the period of 4 min. found in our experiments results from a combination of the periods 3 min. and 5 min.)

A test of these estimates was further obtained by calculating from them the values of the ratio between the counting rates of the two counters for steady flow with the various circulating periods obtained by inserting different capillary tubes in the circuit, and comparing the calculated values with the values measured 12 min. after the beginning of the irradiation. This interval of time is not quite sufficient for complete equilibrium to be established, but, on the other hand, it is sufficiently short compared to the periods of the long-lived transformation products to ensure that the values obtained do not depend on the formation of these products to any appreciable extent.

Assuming that the radioactive gas consists only of the two short-period components of decay constants λ_1, λ_2 produced at the rate of n_1, n_2 atoms per sec., the counting rate in any of the counters can readily be shown to be given by

$$G = n_1 e^{-\lambda_1\tau}\frac{1-e^{-\lambda_1\theta}}{1-e^{-\lambda_1 T}} + n_2 e^{-\lambda_2\tau}\frac{1-e^{-\lambda_2\theta}}{1-e^{-\lambda_2 T}},$$

where τ is the time taken by the circulating gas to travel from the uranium container to the counter in question, θ the time taken by the gas to flow through the counter, and T the period of the whole circulation process.

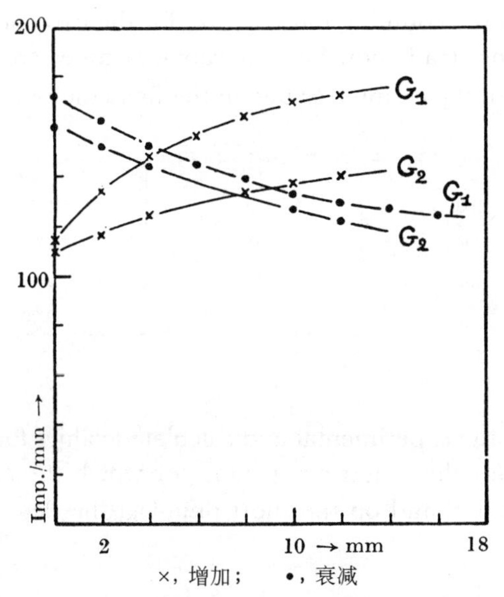

×，增加；　•，衰减

我们根据分立实验中得到的衰减曲线估算了较短的周期。实验中，气体的流动随着移开中子源而同时停止，在此之后对两台计数器的记录结果简单相加。这给出了约为 30 秒钟和 4 分钟的两个周期，产生相应气体的速率的比值的估算值为 1.82。（格拉索和斯泰格曼（见上述参考文献）发现铀裂变的两种气体产物周期分别为 30 秒钟和 5 分钟，其中第一种转变为一种周期为 3 分钟的产物。我们在实验中发现的周期为 4 分钟的产物有可能是 3 分钟和 5 分钟周期结合的结果。）

对上述估算的检验可以通过如下方式获得：通过在环路中插入不同的毛细管，获得各种环流周期，对具有这些周期的稳定流动计算两台计数器计数率的比值，并将计算所得的数值与辐射开始 12 分钟后测得的数值相比较。这一时间间隔对于完全建立平衡来说不是很充足，不过从另一方面来说，它与长寿命衰变产物的周期相比却已足够短，这足以确定所获得的值在任何可观程度上都不依赖于这些产物的形成。

假设放射性气体仅由衰变常数分别为 λ_1 和 λ_2、生成速率分别为每秒 n_1 和 n_2 个原子的两种短周期成分组成，其任一台计数器中的计数率都可以很容易由下列式子给出：

$$G = n_1 e^{-\lambda_1 \tau} \frac{1 - e^{-\lambda_1 \theta}}{1 - e^{-\lambda_1 T}} + n_2 e^{-\lambda_2 \tau} \frac{1 - e^{-\lambda_2 \theta}}{1 - e^{-\lambda_2 T}}$$

式中，τ 表示循环气体从盛铀容器到达我们正在讨论的计数器所需的时间，θ 表示气体流过计数器所需的时间，T 表示整个循环过程的周期。

In the table are recorded the counting rates, G_1, G_2, for the two counters measured 12 min. after the beginning of the irradiation for four capillary tubes corresponding to values of the time lag $\tau_2-\tau_1$ between the counters stated in the first column.

$\tau_2-\tau_1$	G_1	G_2	$G_1 : G_2$	$(G_1 : G_2)$ calc.
15 sec.	108	63	1.72	1.31
50 sec.	94	45	2.09	2.01
2 min.	52	24	2.15	2.10
3 min.	42	21	2.00	2.01

The agreement between the experimental and calculated values for the ratio $G_1 : G_2$ is seen to be very good except for the fastest circulation, for which the calculated value depends mostly on the shorter period and on the short time-lag, the determinations of which are the least precise.

(**144**, 1045-1046; 1939)

L. Wertenstein: Miroslaw Kernbaum Radiological Laboratory, Warsaw Society of Sciences, At Turczynck, near Warsaw, Sept. 24.

Reference:

1. Hahn, O., Strassmann, F., *Naturwiss.*, 27, 163 (1939). Heyn F., Aten, A., and Bakker, C., *Nature*, 143, 516 and 679 (1939). Glasoe and Steigman, *Phys. Rev.*, 55, 982 (1939).

下面的表格中记录了两台计数器的计数率 G_1 和 G_2，其中，G_1 和 G_2 是在辐照开始 12 分钟后，两台计数器对相应于表中第一栏所列出的两个计数器之间的时间延迟 $\tau_2-\tau_1$ 的四个毛细管进行计数所得结果。

$\tau_2-\tau_1$	G_1	G_2	$G_1:G_2$	（$G_1:G_2$）计算值
15 秒钟	108	63	1.72	1.31
50 秒钟	94	45	2.09	2.01
2 分钟	52	24	2.15	2.10
3 分钟	42	21	2.00	2.01

我们可以看到，除了对于最快的循环之外，实验值和计算值之比 $G_1:G_2$ 符合得相当好，这是因为对于最快的循环计算值主要依赖于较短的周期和短时间延迟，所以由此确定的结果精确性最差。

（王耀杨 翻译；张焕乔 审稿）

Scattering of Mesons and the Magnetic Moments of Proton and Neutron

W. Heitler

Editor's Note

Walter Heinrich Heitler was a German scientist who had emigrated to Britain in the 1930s. The particles of matter called mesons had not been observed in practice but only their existence inferred; their role was supposed to mediate the forces between particles of nuclear matter, neutrons and protons for example. This highly technical note suggests that other properties of the nuclear particles such as their magnetic moment could be calculated from their functions as mediators of the nuclear force.

THE meson theory in its present form exhibits a number of difficulties which are connected with the particular way in which the conservation of charge and the spin enter the interaction between mesons and the nuclear particles. The expression for the anomalous magnetic moment of the proton and neutron[1], for example, diverges as $\int_0^\infty dk$. The cross-section for the scattering of mesons by a nucleus is found to be very much larger than experiments permit and increases rapidly with increasing energy. This would be incompatible with the high penetrating power of cosmic ray mesons. The cross-section for scattering of a longitudinal meson (rest mass μ) with energy ε (momentum p/c) is, according to the present theory, given by[2]

$$\varphi = 4\pi\left(\frac{g^2}{\mu c^2}\right)^2 \frac{p^4}{\varepsilon^2 (\mu c^2)^2} \tag{1}$$

From the analogy of mesons with light quanta, one would expect a cross-section of the order $(g^2/Mc^2)^2$, where M is the mass of the proton, and no increase with energy for $\varepsilon > \mu c^2$. The experiments by J. G. Wilson[3] have shown that the scattering cross-section even for an energy so low as a few times μc^2 (10^8 ev.) is smaller than (1) by an order of magnitude and does not increase with energy.

As can be seen from the computation of (1), both difficulties are due to the fact that the conservation of charge forbids a number of transitions which could occur if mesons were neutral particles[2]. A neutral meson could, for example, be *emitted and absorbed* by a proton. A positive meson can *only be emitted by a proton but not absorbed.* The cross-section for scattering would be of the right order of magnitude if we allow a positive meson also to be absorbed and a negative one to be emitted by a proton, etc., or, in other words, if "proton states" with charges $-e$ and $+2e$ existed[4]. The introduction of those particles meets, however, with the following difficulties. First, particles of this nature are not observed and are unlikely to have escaped observation if they occur in heavy nuclei. Secondly, if a proton were capable

介子散射与质子和中子的磁矩

海特勒

编者按

瓦尔特·海因里希·海特勒是一位德国科学家，他于20世纪30年代移居英国。名为介子的物质粒子实际上并没有观察到，它们的存在只是推测；其作用被认为是核物质的粒子（比如中子和质子）之间作用力的媒介。这篇专业性很强的短文指出核内粒子的其他一些性质，如磁矩，也能被视为核力的介质，我们可以通过它们的功能函数计算得到它们的值。

现有形式的介子理论呈现若干困难，这是与电荷守恒和自旋引入介子与核粒子之间相互作用的特殊方式相联系的。例如，质子和中子的反常磁矩表达式[1]作为$\int_0^\infty dk$是发散的。我们发现介子被核散射的截面比实验允许值大很多，而且其随着能量的增加而迅速增加。这种现象与宇宙线介子具有高穿透本领是相矛盾的。根据现有的理论，能量为ε（动量为p/c）的纵向介子（静止质量为μ）的散射截面为[2]

$$\varphi = 4\pi\left(\frac{g^2}{\mu c^2}\right)^2 \frac{p^4}{\varepsilon^2(\mu c^2)^2} \tag{1}$$

通过将介子与光量子类比，我们可以预计散射截面的量级为$(g^2/Mc^2)^2$，其中M为质子的质量，并且当能量$\varepsilon>\mu c^2$时不再增加。由威尔逊[3]的实验表明，即使在能量低至几倍μc^2（10^8电子伏）时散射截面的量级比表达式（1）的值小，而且其值不随能量的增加而增加。

从对表达式（1）的计算我们可以看出，前面所述的两种困难都源于电荷守恒禁止了一些跃迁这个事实，而如果介子是中性粒子这些跃迁就会发生[2]。例如，一个中性介子可以被一个质子**发射和吸收**。而一个正介子**只能被一个质子发射却不能被其吸收**。如果我们允许一个正介子也可以被一个质子吸收且一个负介子可以被一个质子发射等等，或者换句话讲，如果带有$-e$和$+2e$电荷的“质子态”存在[4]，那么将得出数量级正确的散射截面。但是这些粒子的引入会遇到下面的困难。首先，具有这些性质的粒子并没有被观测到，而如果它们在重核中出现时它们是不太可能不

of emitting also a negative meson, the negative meson would give a contribution to the anomalous magnetic moment of the proton of opposite sign and—all other quantities being equal—of the same value as the contribution from the positive meson. Thus, there would be no anomalous magnetic moment at all.

These difficulties can be overcome if we assume that the rest mass of the new particles is considerably *higher* than that of the proton, say, by 25–50 electron masses (see below). The particles would then be extremely unstable and would not play any part in the structure of heavy nuclei. Denoting the mass difference between the new particles and the proton by ΔM, the cross-section for the scattering of a longitudinal meson becomes, for $\varepsilon < Mc^2$, assuming $\Delta M \ll \mu$,

$$\varphi = 4\pi\left(\frac{g^2}{\mu c^2}\right)^2\left\{\frac{1}{3}\left(\frac{\mu}{M}\right)^2\left(\frac{\mu c^2}{\varepsilon}\right)^4+\left(\frac{\Delta M}{\mu}\right)^2\frac{p^4}{\varepsilon^4}\right\} \tag{2}$$

This expression does *not* increase with energy for $\varepsilon > \mu c^2$. If p is approximately μc^2, (2) is smaller by a factor $(\Delta M/\mu)^2$ than (1). A value $\Delta M/\mu$ of approximately 1/5 would be sufficient to bring (2) into harmony with the experimental requirements. For ε greater than Mc^2, φ will probably decrease owing to the relativistic features of the proton.

Similar considerations must be applied to the spin. The spin contributes also to the high scattering cross-section (for transverse mesons). This can be avoided if we introduce also “higher spin states”, for example, heavy particles with spin s of 3/2. Transitions from the normal proton state, s equal to 1/2, to these higher states under the influence of a meson field can be included in the theory in a very simple manner. In the present theory the spin-dependent interaction between a meson field φ and a nuclear particle is f/λ (σ curl φ), where σ is the spin matrix and has only matrix elements for $\Delta s = 0$. σ has to be extended in such a way as to include transitions $\Delta s = 1$. This can be done if we replace σ by the matrix of a dipole moment $\boldsymbol{r}/r$. For transitions $\Delta s = 0$, the matrix elements of $\boldsymbol{r}/r$ and of σ are identical. If this is done, the cross-section for the scattering of transverse mesons also is small and of the order of magnitude (2). The physical significance of $\boldsymbol{r}/r$ is that of the *intrinsic* magnetic dipole moment of the proton, which is a characteristic feature of the meson theory ($\boldsymbol{r}$ is not, of course, the spacial position of the proton).

As a further result of our new assumptions, the anomalous magnetic moments diverge only *logarithmically*. Such a divergence can, at the present stage of the theory of the meson, scarcely be considered as a very serious difficulty. The relativistic features of the proton have so far been neglected, and it may well be that the logarithmic divergence would disappear if they are taken into account properly. As an upper limit for the validity of the meson theory in the form proposed above, we therefore take the rest energy of the proton. The anomalous magnetic moment of the proton then becomes (in units of the nuclear magneton μ_0)

$$\frac{\mu}{\mu_0}=\frac{16}{3\pi}\frac{M\Delta M}{\mu^2}\frac{f^2}{\hbar c}\left(\log\frac{2E_m}{\mu c^2}-\frac{4}{3}\right),\quad E_m \sim Mc^2 \tag{3}$$

被观测到的。其次，如果一个质子也能够发射一个负介子，那么负介子会对异号质子的反常磁矩有一定的贡献，并且其他所有的量都是相同的话，这些贡献与来自正介子的贡献有相同的值，因此，根本就不会存在反常磁矩。

如果假定这些新粒子的静止质量远远**大于**质子的质量，比如说为电子质量的 25 ~ 50 倍（见下文），我们就能够克服这些困难。那么，这些粒子将会极其不稳定，并且在重核结构中不会起任何作用。用 ΔM 表示新粒子与质子之间的质量差，当能量满足 $\varepsilon < Mc^2$ 时，假定 $\Delta M \ll \mu$，纵向介子的散射截面可以表示为：

$$\varphi = 4\pi\left(\frac{g^2}{\mu c^2}\right)^2\left\{\frac{1}{3}\left(\frac{\mu}{M}\right)^2\left(\frac{\mu c^2}{\varepsilon}\right)^4+\left(\frac{\Delta M}{\mu}\right)^2\frac{p^4}{\varepsilon^4}\right\} \tag{2}$$

然而当能量满足 $\varepsilon>\mu c^2$ 时，这个表达式的值**不会**随能量增加。如果 p 趋近于 μc^2，表达式（2）比表达式（1）小 $(\Delta M/\mu)^2$ 倍。一个趋于 1/5 的 $\Delta M/\mu$ 值，足以使表达式（2）与实验要求相吻合。如果能量 ε 大于 Mc^2，φ 可能会因为质子的相对论特征而减小。

对于自旋问题必须采取相似的分析思路。自旋同样会对高散射截面（对于横向介子）有贡献。如果我们同样引入“较高自旋态”，比如，具有自旋 s 为 3/2 的重粒子，就可以避开这个问题。在介子场的影响下，自旋 s 为 1/2 的正常质子态跃迁到较高能态，可以以非常简单的方式包含于理论中。在现有的理论中，介子场 φ 与核粒子之间与自旋相关的相互作用为 f/λ(σ 与 φ 做旋度计算)，其中 σ 为自旋矩阵，且仅当 $\Delta s = 0$ 时才有矩阵元。必须对 σ 矩阵进行扩展，这样才能够包含 $\Delta s = 1$ 的跃迁。如果我们用偶极矩 $\vec{r}/r$ 矩阵来替代 σ，那么就可以满足上面的要求。对于 $\Delta s = 0$ 的跃迁，$\vec{r}/r$ 的矩阵元和 σ 的相同。假若这些得以实现，在 $\vec{r}/r$ 替代 σ 后，横向介子的散射截面仍然很小，同表达式（2）中的散射截面具有一样的数量级。$\vec{r}/r$ 的物理意义为质子的**内禀**磁偶极矩，这是介子理论的显著特点（当然，$\vec{r}$ 并不是质子的空间位置）。

作为新的假定所得到的进一步的结果是，反常磁矩仅仅是**对数**上的发散。从现阶段的介子理论看来，这种发散几乎可以不被当作一个很严重的困难。到目前为止，质子的相对论效应都被忽略了，如果对此进行适当的考虑的话，目前存在的对数发散也会消失。作为上述形式的介子理论有效性的上限，我们采用了质子的静止能量。因此质子的反常磁矩表示为（以核子的磁矩 μ_0 为单位）：

$$\frac{\mu}{\mu_0} = \frac{16}{3\pi}\frac{M\Delta M}{\mu^2}\frac{f^2}{\hbar c}\left(\log\frac{2E_m}{\mu c^2}-\frac{4}{3}\right),\quad E_m \sim Mc^2 \tag{3}$$

This is of the right order of magnitude when $\Delta M/\mu$ is approximately 1/5, which is the value assumed above.

(**145**, 29-30; 1940)

W. Heitler: H. H. Wills Physical Laboratory, University, Bristol, 8, Nov. 28.

References:

1. Fröhlich, Heitler and Kemmer, *Proc. Roy. Soc.*, A, **166**, 154 (1938). Yukawa, Sakata and Taketani, *Proc. Math. Phys. Soc. Japan*, **20**, 319 (1938).
2. Heitler, *Proc. Roy. Soc.*, A, **166**, 529 (1938), and Report of the Eighth Solvay Conference, Brussels, in the press, where the reasons for the high cross-section are analysed.
3. Wilson, *Proc. Roy. Soc.*, in the press. I am very much indebted to Dr. Wilson for having sent me his MS. before publication.
4. This possibility was first mentioned to me by Bhabha in a private discussion in connexion with his classical theory for neutral mesons. The whole problem was very much clarified in discussions with Bhabha, Fröhlich and Kemmer.

当 $\Delta M/\mu$ 趋于 1/5 时这是正确的数量级，而 1/5 正是上面所假定的值。

（胡雪兰 翻译；张焕乔 审稿）

Evidence for Transformation of Mesotrons into Electrons

E. J. Williams and G. E. Roberts

Editor's Note

If seeing is believing, then Figure 2*b* of this paper is a complete demonstration that mesons, thought to mediate nuclear forces between particles such as neutrons and protons, themselves decay into electrons. The tracks exhibited in this paper were made in a cloud chamber, a device filled with nearly super-saturated water vapour and air which can be expanded quickly, thus forming droplets of water wherever there are free electric charges in the chamber. The first author of this paper, Evan J. Williams, was one of the most talented British physicists of his generation. He died at the age of 43. (Since 1940, several different kinds of mesons have been discovered; that described in this paper is now called a π-meson.)

ONE of the outstanding questions regarding the mesotron is that of its ultimate fate. Certain properties of this particle are remarkably like those of the hypothetical particle assumed by Yukawa in his theory of nuclear forces and β-disintegration, and this has led to the view that the two may be identical. Within a rather large experimental error they have the same mass, and both are unstable in the free state, having an average life of the order of 10^{-6} seconds. The disappearance of the particle of Yukawa's theory at the end of its life takes place through its transformation into an electron and a neutrino, and it is regarding this that hitherto there has been no evidence of a parallel between it and the mesotron of cosmic rays. In fact, existing experimental evidence has rather gone to show that mesotrons suffer at the end of their life some other fate than befalls the Yukawa particle.

With the object of obtaining information on this crucial point we constructed a large cloud-chamber (24 in. diameter, 20 in. deep) which, with its large sensitive period and volume, might catch a cosmic ray mesotron coming to the end of its range in the gas of the chamber. A recent photograph taken with this shows a mesotron track terminating in the gas as desired. From its end there emerges a fast electron track, the kinetic energy of which is very much greater than the kinetic energy of the mesotron, but is comparable with its mass energy. This indicates that the mesotron transforms into an electron, in which case the remarkable parallel between the mesotron and the Yukawa particle is taken one stage further. In terms of Yukawa's theory, the phenomenon observed may be described as a disintegration of the mesotron with the emission of an electron, thus constituting the most elementary form of β-disintegration.

Fig. 1 is a reproduction of one of the photographs of the stereoscopic pair. The dense

介子向电子转化的证明

威廉姆斯，罗伯茨

编者按

如果说眼见为实，那么本文图 2*b* 就是作为诸如中子和质子这样的粒子之间核力媒介的介子衰变成电子的实证。本文所展现的径迹是在云室中获得的，那是一个充满近乎过饱和水蒸气和空气并且其中的气体可迅速扩张的装置，这样在云室中只要是存在自由电荷的地方就会形成很多小水滴。本文的第一作者埃文·威廉姆斯是同时代人中最有才华的英国物理学家之一，享年 43 岁。（自 1940 年以来，已经发现了几种不同类型的介子，本文中描述的介子现在被称为 π 介子。）

关于介子的一个最突出的问题是它的最终命运。这种粒子的某些性质与汤川秀树在他的核力和 β 衰变理论中假设的粒子的性质是明显一致的，这使得人们认为这两种粒子是同一种粒子。在相当大的实验误差范围内，它们具有相同的质量，而且在自由态时都是不稳定的，平均寿命的数量级为 10^{-6} 秒。汤川秀树理论中的粒子在其寿命结束时转化成电子和中微子，但至今都没有发现在这一点上它与宇宙线中的介子有相似之处的证据。事实上，有实验证据表明这种粒子在其寿命终结时会遭受另一种命运——一种不同于汤川秀树粒子的命运。

为了得到与这个至关重要的问题有关的信息，我们构建了一个大云室（直径为 24 英寸，深度为 20 英寸），由于其具有高灵敏周期和庞大的体积，因此可以在云室气体中捕捉到即将到达其行程尽头的宇宙线介子。最近拍摄的相关照片中显示出了那些我们所期望得到的在气体中终结的介子径迹。其终结处出现快电子的径迹，快电子的动能明显大于介子的动能，但是质量和能量则与介子的相当。这表明，介子会转化成为电子。在这种情况下介子与汤川秀树粒子之间的相同之处又更近了一步。根据汤川秀树的理论，观察到的这种现象可以这样描述，介子在衰变中发射了一个电子，从而构成了 β 衰变的一种最基本的形式。

图 1 为其中一张立体像对照片的复制片。密集的径迹 *AF* 是介子的，而从其终

track *AF* is that of the mesotron, and the faint track *FG* leaving its end, near the bottom of the chamber, is that of the fast electron. It will be noticed that the latter is comparable in density with the tracks of other fast particles which happened to traverse the chamber in the same region. Fig. 2 is a larger reproduction of the stereoscopic pair, showing only the end portion of the mesotron track and the emergent electron (2*a* is not in as good focus as 2*b*). Fig. 3 is a heavily exposed reproduction of the last few millimetres of the mesotron track to bring out its shape though the electron track is thereby nearly lost, and Fig. 4 is an enlargement of the δ-track at *E* to show more clearly its initial direction. The tracks in the present reproductions are much less distinct than in the original negatives and photographic prints, and this particularly applies to the fast tracks (including *FG*) and the short δ-tracks, of which there are at least six obvious ones to be seen between *C* and *F* on the original negative.

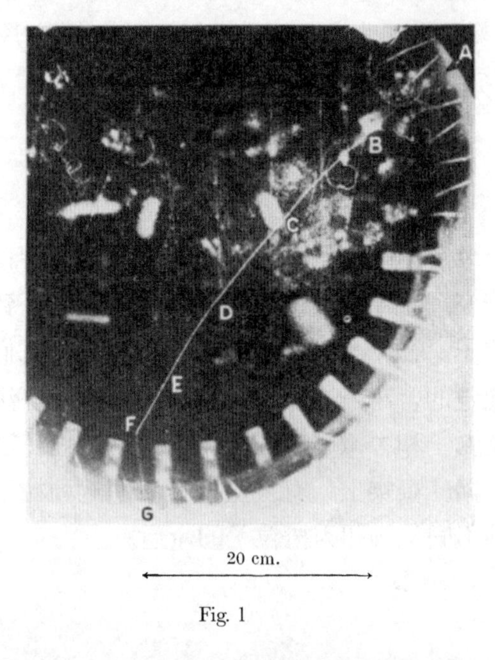

Fig. 1

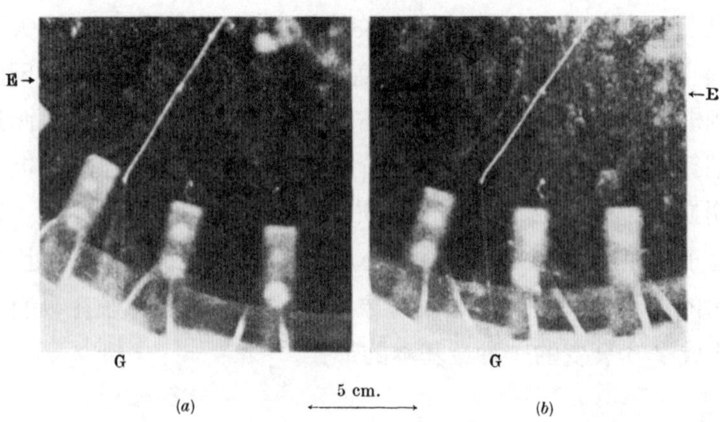

Fig. 2. *a* and *b* are arranged for stereoscopic observation with the naked eye, when usually the left eye sees the right-hand picture.

端远去的那条模糊的、靠近云室底部的径迹 FG 则是快电子的。我们将会注意到，后者与碰巧在同一区域穿过云室的其他快速粒子的径迹在密度上是相当的。图 2 是立体像对照片的一个更大的复制片，仅仅显示了介子径迹的末端部分和出射的电子（2*a* 不如 2*b* 的聚焦效果好）。图 3 为介子径迹最后几毫米的一个严重曝光的复制片，以此来显示介子的径迹形状，尽管这样基本上会看不到电子的径迹。图 4 是对 E 处 δ 径迹的放大，是为了更清楚地显示出其初始方向。在这里给出的复制片中的径迹不如原底片和用底片冲洗出来的照片中那么清晰，尤其是对于快电子的径迹（包括 FG）和短的 δ 径迹，在原始底片中的 C 和 F 之间至少能够看到 6 条它们这样的清晰径迹。

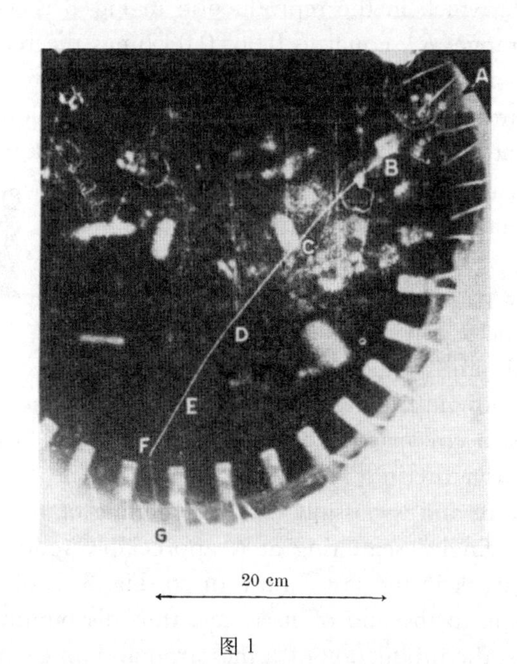

图 1

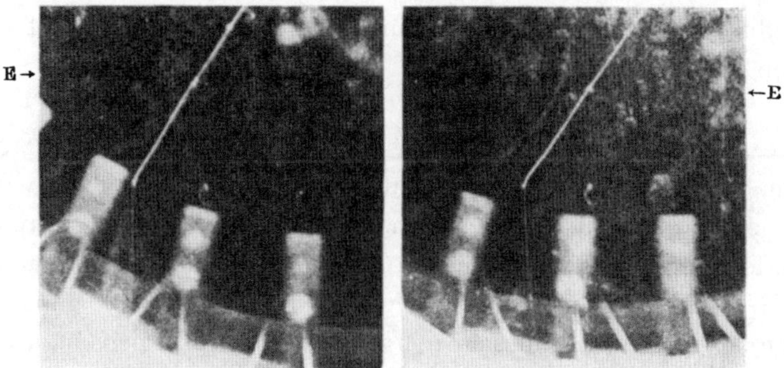

E→ ←E

G G

(*a*) 5 cm (*b*)

图 2. *a* 和 *b* 是为用肉眼进行立体观察所准备的，通常用左眼看着右手边的图片

That the dense track is that of a mesotron follows from its range and curvature, and from the δ-tracks. An accurate estimate of the mass from the curvature is not possible because the scattering which the particle suffers interferes appreciably with the curvature due to the magnetic field. The straightness of *FG* and of neighbouring fast tracks shows that there was no appreciable distortion from air motion. The radius of curvature, ρ, at *B*, measured over *AC* (~20 cm.), is 70 cm., giving $H\rho = 1{,}180 \times 70 = 8.3 \times 10^4$. The range beyond *B* is 33 cm. in the chamber, corresponding to 41 cm. of normal air. These data give a mass, μ, of $(250\pm70)\ m$, where m represents electronic mass. This is of the same order as previous estimates of the mass of the mesotron, and is sufficiently far removed from the mass of the proton (1,840 m) to establish the particle as a mesotron. The number and range of the δ-tracks also indicate mesotronic mass, and rule out a proton. In particular the long δ-track at *E*, which in the reproduction in Fig. 3 is seen to be directed nearly forward, has a path-range, R', equal to 0.06 ± 0.03 times the remaining range, R, of the heavy particle. This is roughly the range that would be expected for a secondary electron knocked nearly forward by a mesotron with the observed remaining range. It is, however, at least five times greater than the range of the longest δ-track that could be produced by a proton. The latter is approximately $(2^{3.4}/1{,}840)\ R = 0.006\ R$. Regarding the "scattering" of the track, while it is more pronounced than the *average* effect expected for a mesotron, it is more compatible with the latter than with a proton or any other known particle. The natural "curvature" of cloud-tracks due to multiple and single scattering is discussed by one of us in a paper now in the press (*Physical Review*). It is there shown that towards the end of its range—last 5 cm. or so—the natural curvature of a mesotron track may well exceed its magnetic curvature in a field of 1,200 gauss. (The "kink" at *D* contributes little to the average curvature and is possibly more a thinning of the track on one side than a true deflection. The "single" scattering at *C* appreciably reduces the overall curvature.) The bending of the track in the last 5 mm. or so (Fig. 3) is of interest. It indicates that the mesotron has come to the end of its range, thus discounting the possibility that the photograph represents the production of a mesotron and an electron by a neutral particle. Against this supposition are also the facts that the long δ-track at *E* is initially directed forward, and that the δ-tracks are more numerous in the lower half of the track. Both indicate motion of the mesotron towards *F*.

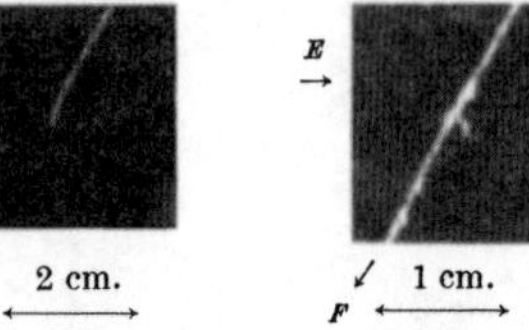

Fig. 3　　Fig. 4

The curvature of the electron track, *FG*, is very small. Actually there is detectable (Fig. 2*b*), a small curvature in a direction indicating a positive charge, which is also the direction of the curvature of the mesotron. The photograph thus represents a positive mesotron transforming into a positive electron. So far as it can be estimated, the radius of curvature of *FG* is 200 cm. ± 50 percent, which in the field of 1,180 gauss (neglecting any distortion due to air-motion) indicates an energy of 70 Mev. ± 50 percent. Taking $\mu = 200\ m$, and assuming that a neutrino takes half the energy, the energy of the electron would be $100\ mc^2 = 50$ Mev.

The large energy of the electron shows, quite apart from Yukawa's theory, that mass

密集的径迹是介子的，这是根据该径迹的范围和曲率，以及δ径迹分析得出的。根据曲率是不可能精确估算出粒子质量的，因为粒子受到的散射对由磁场引起的曲率有明显的干扰。*FG*和近邻的快电子径迹的平直度表明，不存在明显的由空气运动而导致的畸变。在*AC*段（约20厘米）上测量到的*B*处的曲率半径ρ为70厘米，它使得$H\rho=1{,}180\times70=8.3\times10^4$。在云室中*B*以后的行程范围为33厘米，对应于普通空气中的41厘米。这些数据可以给出质量μ为(250±70) *m*，其中*m*为电子的质量。这同先前估算的介子质量具有相同的量级，而且与质子质量（1,840 *m*）相差甚远，进而可以确定这一粒子为介子。δ径迹的数量和范围同样也表明了介子的质量，从而排除了质子的可能。特别是图3的复制片中*E*处的长δ径迹看上去几乎一直向前，这个直的径迹的行程长度*R'*，为普通重粒子的残留径迹*R*的0.06±0.03倍。这就是介子在观测到的持续范围内被近似向前敲出的次级电子的粗略范围。然而，这至少比质子可以产生的最长δ径迹范围大五倍，后者大约是$(2^{3.4}/1{,}840)R = 0.006R$。至于径迹的"散开"，虽然它比对介子预期的**平均**效应要更为明显，但是与质子或其他任何已知粒子相比，它同介子的更为符合。由多重和单个散射引起的云迹的自身"曲率"已被我们中的一位在一篇即将出版的文章中进行了讨论（《物理学评论》）。该文章指出，对于介子行程末端，大约是最后5厘米，其径迹自身的曲率会超过它在1,200高斯的磁场中的磁曲率。（*D*处的"纽结"对平均曲率几乎没有作用，与真正的偏转相比更有可能的情况是径迹一边的稀疏。*C*处的"单"散射明显使整个曲率减小了。）径迹的最后大约5毫米处的弯曲（图3）是非常有价值的。它表明了介子已经到达了其行程的末端，这就使照片表示由一个中性粒子产生一个介子和一个电子的可能性不可置信。*E*处的长δ径迹最初是一直向前的事实，以及在径迹下半段δ径迹的数量更多的事实，也驳斥了这一假说。这两者均表明介子是朝向*F*方向运动的。

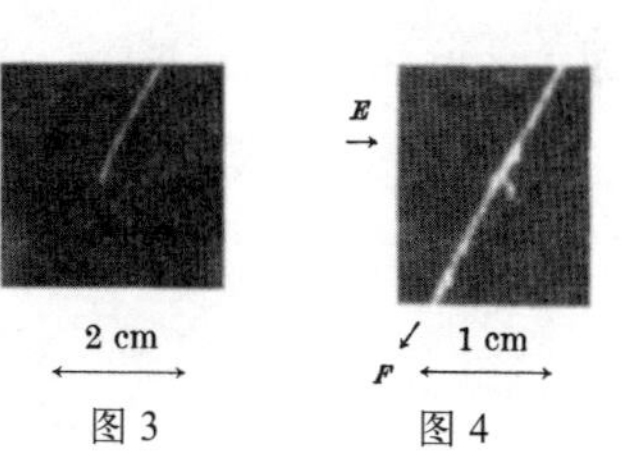

图3　　图4

电子径迹*FG*的曲率是很小的。实际上可以发现（图2*b*）在一个表明是正电荷的方向上有一个较小的曲率，这个方向也就是介子曲率的方向。因此，照片所表示的是一个正介子正在向一个正电子转化的过程。到目前为止可以估算得出，*FG*的曲率半径为200厘米±50%，这表明在1,180高斯的磁场中（忽略所有由空气运动造成的扭曲）能量为70兆电子伏±50%。设μ=200 *m*，并且假定一个中微子具有一半的能量，那么电子的能量将会是100 mc^2 =50兆电子伏。

完全抛开汤川秀树理论，高的电子能量表明，质量已经湮没了。对于介子，即

has been annihilated—for the mesotron, even if we suppose it has disintegrated before "stopping", has certainly less than 4 Mev. of kinetic energy. Actually, the large bending of the end of the mesotron track indicates (as already pointed out) that E is the normal end of its range, where it has reached too low a velocity to ionize further. In this connexion it is of interest that an upper limit to the lifetime, τ, of this mesotron, since its entry into the chamber, can be set from the fact that the electron track starts from a point certainly not more than 0.4 mm. from the end of the mesotron track. Assuming the mesotron, after it ceases to ionize, to diffuse with gas-kinetic free path (10^{-5} cm.) and thermal velocity (10^6 cm./sec.) this gives an upper limit to τ of $(0.04^2/10^{-5}\times10^6)\sim2\times10^{-4}$ seconds. Actually it is likely that a mesotron, when it stops ionizing, has a velocity of at least 10^7 cm./sec., and a free path considerably greater than gas-kinetic values, so that τ must be much less than the above limit. The average value of τ deduced from the anomalous absorption of cosmic ray mesotrons is of the order of 10^{-6} seconds.

(**145**, 102-103; 1940)

E. J. Williams and G. E. Roberts: University College of Wales, Aberystwyth, Dec. 21.

使我们假定它在“停止”之前已经衰变，它所具有的动能也小于 4 兆电子伏。实际上，介子径迹末端的大弯曲表明（正如以前所指出的）E 是它的行程范围的常规终点，到达此处时它所具有的速度太小，以致不能发生进一步的电离。关于这个问题值得注意的是，可以根据电子径迹从介子径迹末端不超过 0.4 毫米处算起这一事实判断出介子进入云室后的寿命上限 τ。假定介子不再电离后，沿着气体分子运动自由程（10^{-5} 厘米）以热速度（每秒 10^{6} 厘米）扩散，这就给定了寿命的上限 τ 为 $(0.04^{2}/10^{-5}\times10^{6})\sim2\times10^{-4}$ 秒。事实上，很有可能的是介子在停止电离后，其速度至少为每秒 10^{7} 厘米，而其自由程比气体分子运动的值要大很多，因此寿命上限 τ 肯定会远小于上述的值。由宇宙射线介子的反常吸收推断出 τ 平均值的量级为 10^{-6} 秒。

（胡雪兰 翻译；厉光烈 审稿）

Capture Cross-sections for Thermal Neutrons in Thorium, Lead and Uranium 238

L. Meitner

Editor's Note

Physicists had not yet measured the "capture cross-section"—the tendency to become captured—of thermal neutrons in the nuclei of ^{238}U, an important parameter for the possibility of a nuclear chain reaction. A small capture cross-section—fewer neutrons captured—would make the process more likely, as this leaves others to trigger fission events. Here Lise Meitner measures this cross-section in thorium and then uses it to estimate the corresponding value for uranium. Using a relationship derived by Niels Bohr linking the ratio of beta decay events in thorium and uranium, she reports a value very close to that reported independently by Enrico Fermi and Carl Anderson.

EXPERIMENTS on the processes arising in thorium under neutron bombardment have shown that nuclear fission is induced only by fast neutrons of energies of about 2 Mev. or more. There exists also a radiative capture process producing an isotope of thorium (Th 233) of 26 min. half-life; this process has a resonance character with a large contribution from thermal neutrons[1]. So far, the capture cross-section of thermal neutrons in thorium has not been measured. The following experiments were carried out in order to determine this cross-section.

As the neutron source available was not very strong (100 mgm. Ra+Be), all dimensions had to be kept as small as possible. On the other hand, in order to obtain high accuracy of measurement, one had to use an absorbing thorium layer of reasonable thickness. By the kindness of Prof. Coster, I obtained a sample of metallic thorium of more than 99 percent purity. Dysprosium of the highest purity, also kindly given me by Prof. Coster, was used as detector. The thorium was almost exactly prismatic in form (1.2 cm. × 1.2 cm. × 2.96 cm.). The dysprosium was a thin layer (15.7 mgm./cm.^{2} Dy) of rectangular form, 1.0 cm. × 2.7 cm., and its upper face was covered with 2 μ "Cellophane". For the absorption measurement the thorium was placed directly on the dysprosium with or without a cadmium screen, so that the neutrons impinging normally had to go through 1.2 cm. thickness corresponding to 13.4 gm. thorium.

The experimental arrangement was as follows. In a plate of paraffin wax of 3.8 cm. thickness, there was cut out a rectangular cavity of 1.3 cm. depth, the bottom and sides of which were covered with cadmium of 0.5 mm. thickness so that thermal neutrons could not enter except from above. This plate was put between two other plates of paraffin wax, forming in this way a block of 11.5 cm. height and about 25 cm. × 25 cm. area. The dysprosium was placed on the cadmium-shielded bottom of the cavity. The upper paraffin plate contained the neutron source 3.3 cm. below the surface in such a way that the source just touched the upper edges of the cadmium screened cavity.

钍、铅和铀–238中热中子的俘获截面

迈特纳

编者按

物理学家还没有测量出铀–238对热中子的“俘获截面”，该截面表征热中子被俘获的倾向，它是核链式反应可能性的一个重要参数。小的俘获截面表明有很少中子被俘获，这使得更有可能发生剩下的中子引发裂变的事件。本文中莉泽·迈特纳测量了钍的热中子俘获截面，然后用它估算铀中的相应值。根据尼尔斯·玻尔导出的钍核和铀核β衰变事例比值之间的关系，她宣称这个值与恩里科·费米和卡尔·安德森所报道的值非常接近。

在用中子轰击钍的实验过程中，我们发现，只有能量大于或等于大约2兆电子伏的快中子才能诱发核裂变。这里也存在一个辐射俘获过程，产生了一个半衰期为26分钟的钍的同位素钍–233；而且这一过程具有源自热中子[1]贡献的共振特征。迄今为止，热中子在钍元素中的俘获截面还没能被成功测量。为了测量出这个截面，我们做了下面的实验。

由于可利用的中子源并不是特别的强（100毫克的镭和铍），所以全部实验材料的尺寸都要保持尽量的小。另一方面，为了得到高精度的测量结果，我们必须使用厚度合适的吸收钍片。在科斯特教授的慷慨帮助下，我得到了一个纯度高于99%的金属钍的样品。科斯特教授还提供了纯度最高的镝，我们把它用作探测器。钍样品的形状是一个规则的棱柱形（1.2厘米×1.2厘米×2.96厘米）。镝为一个矩形（1.0厘米×2.7厘米）的薄层（15.7毫克每平方厘米镝），上表面覆盖了一层2微米的“玻璃纸”。为了进行吸收测量，我们把钍直接放在带镉屏或者不带镉屏的镝片上，这样中子轰击时一般就必须通过厚度为1.2厘米的钍片，其相当于13.4克的钍。

具体的实验做法如下。在一个3.8厘米厚的石蜡平板上，挖出一个1.3厘米深的矩形空腔，底部和侧面都覆盖有0.5毫米厚的镉，这样使得热中子只能从上面进入。把这个石蜡板放在另外两个石蜡板的中间，用这种方法组成一个高11.5厘米、底面为25厘米×25厘米的长方体。镝放在被镉屏蔽了的空腔的底部。最上面的石蜡平板里含有一个中子源，中子源距离上表面3.3厘米，这样放置后，中子源就刚好接触到了被镉屏蔽的空腔的上边界。

The activity of the dysprosium detector was measured with a Geiger-Müller counter with 0.1 mm. aluminium walls connected to an amplifier. The dysprosium, the half-life of which was carefully determined and found to be 156 ± 3 min., was in all experiments irradiated up to saturation, and the decay of the activity was followed for several hours in order to increase the accuracy of measurements. All measurements were referred to a uranium standard. The contribution from thermal neutrons was determined by carrying out the irradiation with and without 0.5 mm. cadmium directly over the exposed face of the detector. With cadmium screens on all faces of the detector, the observed activity is due to neutrons faster than thermal neutrons. Under the experimental conditions used here, it amounted to 9 percent of the activity obtained without cadmium on the exposed face.

For the determination of the capture cross-section for thermal neutrons in thorium, one has to consider the different kinds of interaction of neutrons with the thorium nucleus. The fission cross-section of fast neutrons is so small as to be negligible. The same holds for the radiative capture cross-section of fast neutrons. Therefore for fast neutrons one has to take into account the scattering cross-section only. Because of the arrangement used—the absorber being put directly on a detector of nearly equal size—one would expect that practically all the scattered neutrons would be efficient in the irradiation, that is, the scattering cross-section would not enter into these measurements. Experiment confirmed this expectation. When the detector was screened on both faces by cadmium, the measurements of the activity with and without thorium (or with and without lead) gave the same values within the experimental error of 2–3 percent. Further, in order to test the influence of inelastic scattering, the cadmium was placed by turns either directly on the exposed face of the detector (with the thorium put upon that), or between the neutron source and the thorium absorber. No difference could be detected. Thus the inelastic scattering does not give rise to thermal neutrons in any observable quantity, a result to be expected.

These results suggest that, under the conditions actually used, the scattering of thermal neutrons too will be negligible, and thus the decrease in activity (of about 28 percent) caused by the thorium absorber is due to radiative capture processes only. To obtain a direct proof the absorption in metallic lead was measured. The lead absorber had practically the same dimensions as the thorium absorber, but the thickness of the cast lead prism of density 10.6 was kept a little smaller (1.10 cm.) in order to have the same number of absorbing nuclei per $cm.^2$.

Of course, in determining the cross-sections, the angular distribution of the thermal neutrons was taken into account and obliquity corrections (angles up to nearly 70° were involved) were made according to the data given by Frisch[2].

The total cross-section for thermal neutrons in lead was found to be $\sigma_{Pb}^{th} = 2.5 \pm 0.2 \times 10^{-24}$ $cm.^2$. This value is in very good agreement with the value of 2.3×10^{-24} $cm.^2$ obtained by Fleischmann[3] from γ-ray measurements. Thus one can be sure that for thorium too the radiative capture cross-section alone enters into the measurements. The value obtained is $\sigma_{Pb}^{th} = 6.0 \pm 0.3 \times 10^{-24}$ $cm.^2$. This cross-section can be used to evaluate the capture cross-

镝探测器的放射性是由盖革–米勒计数器进行测量的，计数器的壁是厚度为 0.1 毫米的铝箔，并与一个放大器相连。镝的半衰期被精确地测定为 156±3 分钟，在所有的实验中，镝都会被强烈照射达到饱和，为了提高测量的精度，对镝放射性强度的衰减要跟踪数个小时。所有的测量都参照着铀的标准进行。对于探测器的被照表面直接覆盖和不覆盖 0.5 毫米镉这两种情形分别进行辐照测量，可以测定热中子的贡献。在探测器的所有面都有镉层屏蔽的时候，探测到的放射性来源于比热中子快的中子。在这里所使用的实验条件下，其放射性仅为无镉片屏蔽探测器被照表面情形下的 9%。

为了测定钍的热中子俘获截面，我们必须考虑到中子与钍核之间各种各样的相互作用。快中子的裂变截面很小几乎可以忽略。快中子的辐射俘获截面同样如此。因此，对于快中子来说，我们只需要考虑它的散射截面。由于实验安排中吸收体直接放在尺寸几乎相等的探测器的上面，可以认为实际上所有的散射中子在辐照中均是有效的，也就是说，这个散射截面在测量中不起作用。实验证实了这一推测。当探测器的两个面全被镉遮盖时，在实验误差 2%～3% 允许的范围内，实验中测得的有钍和没有钍时（或者有铅和没有铅）的放射性是一样的。为了进一步测定非弹性散射的影响，我们把镉先后放在探测器的被照表面（钍放在该表面之上）和中子源与钍吸收器的中间，分别对它进行了研究。结果显示没有任何的差别，因此，正如人们预测的那样，非弹性散射并不产生可观测到的热中子。

这些研究结果表明，在实际使用的条件下，热中子的散射也是可以忽略的，因此由钍吸收体导致的放射性的减少（28% 左右）仅仅是由辐射俘获过程引起的。为了得到更直接的证明，我们对在金属铅中的吸收进行了测量。铅吸收体的大小实际上和钍吸收体相同，但是为了使单位面积上吸收体原子核的数量相同，密度为 10.6 的棱柱形铅铸件的厚度要小一些（1.10 厘米）。

当然，在测量截面的过程中，热中子的角分布也被考虑在内，而且根据弗里施[2]提供的数据进行了倾斜度校正（最大角度接近 70°）。

经计算，铅的热中子总截面为 $\sigma_{\mathrm{Pb}}^{th}$=（2.5±0.2）$\times 10^{-24}$ 平方厘米。这与弗莱施曼[3]在 γ 射线测量中得到的结果 2.3×10^{-24} 平方厘米是非常吻合的。因此我们可以肯定，对于钍来说，在测量中也仅仅是辐射俘获截面起了作用。所测得的结果为 $\sigma_{\mathrm{Th}}^{th}$=（6.0±0.3）$\times 10^{-24}$ 平方厘米，这个值可以用来估算铀–238 的俘获截面。在这种

section of ^{238}U. In this isotope, as Bohr[4] has emphasized, thermal neutrons do not produce fission processes. Thus when equal small quantities of uranium and thorium are subjected to neutron bombardment under identical conditions, ^{239}U and ^{233}Th respectively being produced, it is clear that if on account of their nearly equal half-life the efficiency of the β-rays is assumed to be approximately the same, the β-ray activities due to thermal neutrons (corrected for equal numbers of nuclei) must be proportional to the respective cross-sections:

$$\frac{T^{th}_{U(239)}}{T^{th}_{Th(233)}} = \frac{\sigma^{th}_{U(238)} \cdot \frac{1}{238}}{\sigma^{th}_{Th} \cdot \frac{1}{232}}$$

From earlier measurements carried out in Dahlem, I find for this ratio the value 1/4.15. Using the above value for the cross-section of thorium, the cross-section for uranium 238 is

$$\sigma^{th}_{U(238)} = 1.5 \pm 0.2 \times 10^{-24} \text{ cm.}^2.$$

Anderson and Fermi[5], measuring directly the β-ray intensity of ^{239}U due to a known number of thermal neutrons, found

$$\sigma^{th}_{U(238)} = 1.2 \times 10^{-24} \text{ cm.}^2.$$

Considering the possibility of fairly large errors in this type of measurement, the agreement is very good.

I wish to express my gratitude to the Academy of Sciences for a grant and in particular to Prof. Siegbahn for the facilities kindly put at my disposal.

(**145**, 422-423; 1940)

Lise Meitner: Forskningsinstitutet för Fysik, Stockholm, Feb.1.

References:

1. Meitner, L., Hahn, O., and Strassmann, F., *Z. Phys.*, **109**, 538 (1938).
2. Frisch, O. R., *Kgl. Dansk Vid. Selskab. Math. Phys. Medd.*, **14**, No. 5 (1936).
3. Fleischmann, R., and Bothe, W., *Ergeb. exact. Naturwiss.*, **16**, 37 (1937).
4. Bohr, N., *Phys. Rev.*, **55**, 418 (1939).
5. Anderson, H. L., and Fermi, E., *Phys. Rev.*, **55**, 1106 (1939).

同位素中，正如玻尔[4]强调的，热中子不会诱发裂变。因此，当同样少量的铀和钍在相同的条件下受到中子的轰击时，结果就会相应地产生铀–239和钍–233。显然，考虑到它们的半衰期几乎都是相同的，如果假设产生β射线的效率也近似相等，那么由热中子（修正为相同数量的原子核）引起的β射线的放射性一定正比于相应的截面：

$$\frac{T^{th}_{U(239)}}{T^{th}_{Th(233)}}=\frac{\sigma^{th}_{U(238)}\cdot\frac{1}{238}}{\sigma^{th}_{Th}\cdot\frac{1}{232}}$$

在达勒姆早期的测量中，我找到了这个比值为1/4.15。应用上面钍俘获截面的值，得出铀–238的俘获截面应为：

$$\sigma^{th}_{U(238)}=（1.5\pm0.2）\times10^{-24}\text{ 平方厘米}$$

安德森和费米[5]通过直接测量由已知数量的热中子引起的铀–239的β射线强度，得出：

$$\sigma^{th}_{U(238)}=1.2\times10^{-24}\text{ 平方厘米}$$

考虑到这种类型的测量存在较大的误差，我们认为结果的一致性还是很好的。

在此对科学院的资金支持表示衷心的感谢，尤其要感谢西格巴恩教授，他慷慨地为我提供了实验设备。

（胡雪兰 翻译；朱永生 审稿）

The Mass Centre in Relativity

M. Born and K. Fuchs

Editor's Note

Part of the interest of this paper is its authorship. Max Born was a German émigré to Britain, awarded a Nobel Prize in 1954 for his work on quantum mechanics, while Klaus Fuchs was also a German exile working as a physicist in Britain. Fuchs afterwards joined the Manhattan Project at Los Alamos, New Mexico, and was later convicted by the British government of espionage on behalf of the Soviet Union; he working in Berlin after serving a prison sentence. The point of their paper was to show that, despite appearances, relativity can deal well with the centre of mass of a collection of particles.

THE question whether there exists in relativity mechanics a theorem analogous to the classical law for the motion of the mass centre (conservation of total momentum) has, as far as we can see, never found a satisfactory answer. Eddington[1] has taken this fact as the starting point for a general attack against the usual application of wave mechanics to fast-moving particles without contributing himself anything positive to the question. The reason why this problem has never been seriously treated seems to be this.

In classical mechanics the internal potential energy depends on the simultaneous relative positions of the particles; therefore one can separate the relative motion from the translatory motion of the centre. In relativity, however, all forces are retarded, the interaction does not depend on simultaneous relative positions and the separation of the relative motion from the translation of the whole system loses its meaning.

Quantum mechanics circumvents this problem by considering interactions as produced by emission and reabsorption of other particles. We were induced to reconsider this problem by its bearing on a relativistic and "reciprocal" formulation of second quantization. Without touching this question, we shall state here some simple results concerning free particles. It is clear that in this case there must exist a "rest system" Σ°, that is, a Lorentz frame in which the total momentum vanishes. The problem is to describe the relative motion in an invariant way.

We start by bringing the classical derivation into a form permitting generalization. If $\mathbf{r}_1$, $\mathbf{r}_2$ are the position vectors, $\mathbf{P}_1$, $\mathbf{P}_2$ the momenta of two particles, we form the vector of relative position and that of total momentum

$$\rho = \mathbf{r}_1 - \mathbf{r}_2, \qquad \mathbf{P} = \mathbf{P}_1 + \mathbf{P}_2 \tag{1}$$

and determine their canonical conjugate variables' components of the vectors π and $\mathbf{R}$. A simple calculation shows that these are not uniquely determined but have the form

相对论中的质心

玻恩，富克斯

编者按

本文作者是这篇文章令人感兴趣的原因之一。马克斯·玻恩是流亡到英国的德国人，1954 年因为他在量子力学方面的工作获得了诺贝尔奖，而克劳斯·富克斯也是从德国流亡至英国并在英国工作的物理学家。后来富克斯加入了在美国新墨西哥州洛斯阿拉莫斯实施的“曼哈顿计划”，之后被英国政府以苏联间谍的名义定罪；刑满之后他回到柏林工作。他们这篇文章的观点认为，相对论并不像表现的那样，实际上它可以很好地处理一个粒子集合的质心问题。

在相对论力学中是否存在一个类似于经典质心运动定律（总动量守恒）的定理，关于这个问题，就我们所知尚未找到一个满意的答案。爱丁顿[1]以这个事实作为出发点，对通常将波动力学应用到快速运动粒子的做法进行了一般性的抨击，而他自己却没有对此问题做出任何积极的贡献。这似乎就是此问题从未被认真对待的原因。

在经典力学中，内部势能取决于粒子的瞬时相对位置；因此，我们可以将相对运动从质心的平移运动中分离出来。然而在相对论中，所有的力都是迟滞的，相互作用并不取决于瞬时相对位置，所以把相对运动从整个系统的平移运动中分离出来是没有意义的。

量子力学把相互作用当作是由其他粒子的发射和重吸收所引起的，以此来规避这个问题。这使得我们需要通过一个相对论的和“倒易的”二次量子化公式来重新考虑这个问题。暂不考虑这个问题，我们在此将给出一些关于自由粒子的简单结果。显然，在这种情况下，必然存在一个“静止系统”Σ°，即一个总动量等于零的洛伦兹参考系。具体来说，这个问题就是要以一种不变量的形式对相对运动进行描述。

我们先把经典推导变为允许进行推广的形式。如果 $\vec{r_1}$、$\vec{r_2}$ 是两个粒子的位置矢量，$\vec{p_1}$、$\vec{p_2}$ 是它们的动量，我们可以得出相对位置的矢量和总动量的矢量为：

$$\vec{\rho}=\vec{r_1}-\vec{r_2},\qquad \vec{P}=\vec{p_1}+\vec{p_2} \tag{1}$$

我们还可以确定其正则共轭变量，即矢量 $\vec{\pi}$ 和 $\vec{R}$。经简单的计算表明，它们并不是唯一确定的，而是具有下列形式：

$$\pi = (1 - a)\,\mathbf{p}_1 - a\mathbf{p}_2, \qquad \mathbf{R} = a\mathbf{r}_1 + (1 - a)\,\mathbf{r}_2, \tag{2}$$

where a is an arbitrary constant. Hence another condition must be added.

We postulate that the kinetic energy $p_1^2/2m_1 + p_2^2/2m_2$ assumes the form $P^2/2m + \pi^2/2\mu$. This condition leads to a determination of the three constants a, m, μ, namely,

$$a = \frac{m_1}{m_1+m_2}, \quad m = m_1 + m_2, \quad \mu = \frac{m_1 m_2}{m_1 + m_2},$$

which introduced into (2) give the usual expressions for relative momentum and centre of mass.

In relativity, the energies E_1, E_2 of two free particles are given by

$$E_1^2 = m_1^2 + p_1^2, \quad E_2^2 = m_2^2 + p_2^2. \tag{3}$$

We consider now the 4-vectors $\mathbf{P}_+ = \mathbf{P}_1 + \mathbf{P}_2$, $E_+ = E_1 + E_2$ and $\mathbf{P}_- = \mathbf{P}_1 - \mathbf{P}_2$, $E_- = E_1 - E_2$. A simple calculation leads to

$$E^2{}_+ = m^2{}_+ + p^2{}_+ + \pi^2, \quad E^2{}_- = m^2{}_- + p^2{}_- - \pi^2; \tag{4}$$

here $m_+ = m_1 + m_2$, $m_- = m_1 - m_2$

and
$$\pi = 2m_1m_2 \sinh \Gamma/2, \tag{5}$$

where Γ is the "angular distance" of the two 4-vectors, given by

$$m_1m_2 \cosh \Gamma = E_1\, E_2 - \mathbf{P}_1\mathbf{P}_2. \tag{6}$$

Γ is invariant, hence π is invariant also. π has a simple meaning in the case of equal masses. In the rest system Σ°, where $\mathring{P} = \mathring{p}_+ = \mathring{p}_1 + \mathring{p}_2 = 0$, we have $m_1 - m_2 = m_- = 0$, and $\mathring{E}_1 - \mathring{E}_2 = \mathring{E}_- = 0$; hence $\pi^2 = (\mathring{p}_-)^2$. This shows that π is the length of the vector $\boldsymbol{\pi}$ representing relative momentum.

For different masses π can be described as the relative momentum in that Lorentz frame (which always exists) in which

$$E_1 - E_2 = \pm\,(m_1 - m_2).$$

The first equation (4) can now be written

$$E^2 = M^2 + P^2, \quad M^2 = \mu^2 + \pi^2, \tag{7}$$

Where $\mu = m_1 + m_2$ is the sum of the rest masses, M the total internal energy, which represents also the rest mass of the whole system, and $\mathbf{P} = \mathbf{P}_+ = \mathbf{P}_1 + \mathbf{P}_2$ the total momentum.

$$\vec{\pi} = (1-a)\,\vec{p}_1 - a\vec{p}_2, \qquad \vec{R} = a\vec{r}_1 + (1-a)\,\vec{r}_2 \tag{2}$$

式中 a 是一个任意常数。因此，必须加入另一个条件。

我们假定，动能 $p_1^2/2m_1 + p_2^2/2m_2$ 满足 $P^2/2m + \pi^2/2\mu$ 的形式。根据这个条件可以确定三个常数 a、m 和 μ，即

$$a = \frac{m_1}{m_1+m_2}, \quad m = m_1 + m_2, \quad \mu = \frac{m_1 m_2}{m_1+m_2},$$

将它们代入式 (2) 可得相对动量和质心的通常表达式。

在相对论中，两个自由粒子的能量 E_1、E_2 如下式：

$$E_1^2 = m_1^2 + p_1^2, \quad E_2^2 = m_2^2 + p_2^2 \tag{3}$$

我们现在考虑 4 矢量 $\vec{p}_+ = \vec{p}_1 + \vec{p}_2$，$E_+ = E_1 + E_2$ 以及 $\vec{p}_- = \vec{p}_1 - \vec{p}_2$，$E_- = E_1 - E_2$。简单的计算可导出：

$$E^2_+ = m^2_+ + p^2_+ + \pi^2, \quad E^2_- = m^2_- + p^2_- - \pi^2 \tag{4}$$

式中，$m_+ = m_1 + m_2$，$m_- = m_1 - m_2$

且

$$\pi = 2m_1m_2 \sinh \Gamma/2 \tag{5}$$

其中 Γ 是这两个 4 矢量的“角距离”，如下式：

$$m_1m_2 \cosh \Gamma = E_1 E_2 - \vec{p}_1\vec{p}_2 \tag{6}$$

Γ 是不变量，因此 π 也是不变量。在两个自由粒子的质量相等的情况下，π 的含义是简单的。在静止系 Σ° 中，$\mathring{P} = \mathring{p}_+ = \mathring{p}_1 + \mathring{p}_2 = 0$，我们已知 $m_1 - m_2 = m_- = 0$，及 $\mathring{E}_1 - \mathring{E}_2 = \mathring{E}_- = 0$；因此 $\pi^2 = (\mathring{p}_-)^2$。这表明 π 是表示相对动量的矢量 $\vec{\pi}$ 的长度。

对于两个自由粒子的质量不相同的情况，在使下式成立的洛伦兹系（它总是存在的）中，$\vec{\pi}$ 是相对动量。

$$E_1 - E_2 = \pm\,(m_1 - m_2)$$

(4) 中的第一个方程现可写为：

$$E^2 = M^2 + P^2, \quad M^2 = \mu^2 + \pi^2, \tag{7}$$

式中，$\mu = m_1 + m_2$ 是静止质量之和，M 是总内能，它也代表整个系统的静止质量，$\vec{P} = \vec{p}_+ = \vec{p}_1 + \vec{p}_2$ 是总动量。

Taking the components of $\mathbf{P}$ and π as new canonical momenta, one can determine the conjugate coordinates, $\mathbf{R}$ and ρ. They are linear in $\mathbf{r}_1$, $\mathbf{r}_2$; the coefficients are, however, not constants but functions of $\mathbf{P}_1$, $\mathbf{P}_2$.

For small $\mathbf{P}_1$, $\mathbf{P}_2$ the formulae reduce to the classical ones.

It is interesting to remark that in relativity there exists a "reciprocal"[2] theorem obtained by interchanging coordinates and momenta.

(**145**, 587; 1940)

Max Born and Klaus Fuchs: Department of Applied Mathematics, University of Edinburgh.

References:

1. Eddington, A., *Proc. Camb. Phil. Soc.*, 35, 186 (1939).
2. Born, M., *Proc. Roy. Soc. Edinburgh*, (ii), 59, 219(1939).

取 $\vec{P}$ 和 $\vec{\pi}$ 的分量为新的正则动量，我们可以确定共轭坐标 $\vec{R}$ 和 $\vec{\rho}$。它们对于 $\vec{r_1}$，$\vec{r_2}$ 是线性的；然而系数不是常数，而是 $\vec{p}_1$、$\vec{p}_2$ 的函数。

当 $\vec{p}_1$、$\vec{p}_2$ 很小时，该公式可简化为经典形式。

值得注意的是，在相对论中存在着“倒易”[2] 定理，它是通过交换坐标和动量得到的。

（沈乃澂 翻译；张元仲 审稿）

The Theory of Nuclear Forces*

R. Peierls

Editor's Note

Rudolph Peierls offers a summary of the emerging theory of nuclear forces. These forces could be most easily probed in two-particle interactions involving neutrons and protons. Scattering experiments had established the range of the force as about 1.2×10^{-13} cm, and supported the notion that the nuclear forces between neutrons and protons were identical, ignoring electrostatic differences. Nuclear forces could also depend strongly on direction in relation to the orientation of quantum-mechanical spins of the particles. The best theory proposed so far was that of Japanese physicist Hideki Yukawa, which viewed the nuclear force as originating from a nuclear field associated with a new quantum particle having mass of around 300 electron masses—later identified with the pion.

THE forces between the constituents of a nucleus are "short-range" forces, which have no appreciable effect over distances of more than a few times 10^{-13} cm. Hence it is impossible to find the laws of force by extrapolation from large-scale observations (as in the case of Coulomb's law) and the only possible lines of attack are either, to deduce the law of force from direct observations on the properties of nuclei, or to derive them from some simpler, and more general, laws.

Using the direct experimental approach, it is best to start from phenomena which involve only two particles, since only these permit an unambiguous theoretical interpretation. This group consists of the properties of the deuteron, including its disintegration, and the scattering of protons and neutrons by hydrogen.

Practically all experiments are carried out in conditions in which the de Broglie wave-length of the particles is greater than the range of the forces, and this has the effect that one need consider only those states of motion in which the motion of one particle with respect to the other has no angular momentum. (In all other states of motion the centrifugal force prevents the particles from approaching sufficiently close for any interaction to take place.) Moreover, so long as the wave-length is greater than the range of the force, the effect of the field of force is largely independent of the exact variation of force with distance within the range, and depends approximately only on one constant, which may be called the "strength" of the field of force. (In the case of a potential of the type of a potential well, this strength is approximately proportional to the depth times the square of the width.)[1]

---

* Based on lectures given at the Royal Institution on January 17, 24 and 31.

核力的理论*

佩尔斯

编者按

鲁道夫·佩尔斯概述了现有的关于核力的理论。这种力可以很容易地在含有质子和中子的两粒子相互作用中被探测到。散射实验确定这种力的作用距离大约为 1.2×10^{-13} 厘米，同时在忽略静电力差别的前提下，实验也支持了作用于中子之间、质子之间和中子－质子之间的核力是相同的这一观点。核力在很大程度上还由方向决定，该方向与粒子的量子力学自旋取向有关。目前关于核力的最好理论是由日本物理学家汤川秀树提出的，这一理论认为核力源于一种新的质量约为 300 个电子质量的量子粒子所产生的核场，这一粒子后来被定义为 π 介子。

核的各组分之间的力是“短程力”，当各组分之间的距离超过几个 10^{-13} 厘米时，这个力的作用效果就不明显了。因此，利用大尺度的观测结果（如库仑定律适用的情况）进行推导来得到力的定律是不可能的，因此仅有的两种处理方法是，要么对核的性质进行直接观测来推导微观尺度下力的定律，要么从一些更简单、更一般的定律去导出它们。

利用直接的实验方法，最好从只包含两个粒子的现象出发，因为只有这类情况才有可能作出清晰的理论解释。这样的组合具有氘核的性质，包括氘核的蜕变，以及质子和中子在氢核上的散射。

实际上，所有的相关实验都是在粒子的德布罗意波长大于力程的条件下进行的，在这种情况下，我们只需要考虑相对于另一个粒子而言角动量为 0 的粒子的运动状态。(在所有其他运动状态下，离心力使粒子间保持一定的距离从而不能产生相互作用。) 此外，只要波长大于力程，在力程的距离范围内，力场的影响在很大程度上与力的精确变化无关，而几乎仅仅取决于一个常数，我们称之为力场的“强度”。(对于具有势阱形式的势能来说，这种强度近似正比于势阱深度与势阱宽度平方的乘积。)[1]

* 基于 1 月 17 日、24 日和 31 日在英国皇家研究院所做的演讲。

In the experiments on proton-neutron interaction two possible cases are discussed, according to whether the spins of the two colliding particles are parallel or opposite (in the case of like particles, Pauli's principle ensures that, so long as there is no orbital angular momentum, the spins are necessarily opposite) and the strength of the interaction may be different for these two cases. For the case of a proton and a neutron with parallel spin; the strength constant can be obtained from the binding energy of the deuteron; for those with opposite spin, it follows from the scattering of neutrons by protons, since in this scattering the effect of the neutrons with parallel spin can be allowed for once their strength constant is known. The very high cross-section for slow neutrons is an indication that the strength constant of the interaction for neutrons with opposite spin is either just sufficient or just insufficient to give rise to a bound state of the deuteron with no resultant spin[2]. The choice between these alternatives can be made by means of the scattering of neutrons in *para-* and *ortho*-hydrogen where, because of interference, the results depend on the phase of the wave scattered by each proton. The result is that the force is just insufficient to give a bound state[3].

For the proton-proton force, the strength can be derived from measurements of the scattering of protons in hydrogen.

Once these strength constants are known it is possible to calculate all other observable quantities, namely, the energy variation of the proton-proton and neutron-proton scattering, the photo-effect of the deuteron, etc., to the degree of approximation in which the ratio between the range of the forces and the de Broglie wave-length is negligible. The fact that these calculations give approximate agreement with observations serves as a check on the initial hypothesis of short-range forces.

Actually, all results of this simple theory have to be supplemented by correction terms involving the range. If the experimental data were accurate enough, it should thus be possible to estimate the range of the forces. The use of these correction terms requires greater accuracy than is at present available in the data on neutron-proton scattering, and the determination can only be carried out in the case of the proton-proton scattering, in which more accurate data have been obtained. The result of this is that the "range" for this interaction is about 1.2×10^{-13} cm. The experimental data are even accurate enough to yield a certain amount of information on the actual dependence of the force on distance[4].

More information on these points would be obtained by measuring the neutron-proton scattering at medium energies with higher accuracy, or by observations at higher energies including, in particular, the angular distribution of the scattered particles.

An important contribution to the problem was the discovery by Rabi and others of a quadripole moment of the deuteron[5]. This proves that the neutron and proton have a tendency to have their spins in the direction of the line joining them, rather than at right angles to it. This implies obviously that the forces are not central, but that the potential energy depends on the angle between the spin direction and the line joining

在质子–中子相互作用的实验中，对两个碰撞粒子的自旋呈平行和反平行这两种可能的情况分别进行了讨论（对于同类粒子，只要轨道角动量为0，泡利不相容原理将确保两个粒子的自旋方向必然是相反的），这两种情况中质子–中子相互作用的强度有可能不同。对于自旋方向相同的质子和中子，强度常数可以由氘核的结合能得到；而对于那些自旋方向相反的质子和中子，强度常数可以通过中子对质子的散射来确定，在这种散射中，一旦获知它们的强度常数，我们就可以考虑自旋平行的中子的效应了。慢中子的截面非常高，这表明与自旋相反的中子间相互作用的强度常数或者刚好能够又或者刚好不够形成总自旋为0的氘核束缚态[2]。我们可以通过**仲**氢和**正**氢中的中子散射对以上两种可能的结果做出判断，由于中子散射中干涉现象的存在，这个结果取决于每一个质子的散射波的相位。结果是这个力刚好不足以形成束缚态[3]。

对于质子–质子之间的作用力，可以通过测量氢核上的质子散射来导出该作用力的强度。

一旦知道这些强度常数，在力程与德布罗意波长之比可以忽略不计的近似程度下，我们就有可能计算得到所有其他的可观测量，即质子–质子散射和中子–质子散射的能量变化量、氘核的光效应等。这些计算结果与观测结果近似一致，这一点可用来检验与短程力有关的最初假设。

实际上，所有这些简单理论的结果都必须加上与力程有关的修正项。只有实验数据足够准确，才有可能对力程进行估计。考虑这些修正项时所需数据的精确度要比目前在中子–质子散射中获得的数据的精确度更高，因此只有在质子–质子散射情况下，得到了更加精确的数据以后，我们才能确定力程的值。结果是，这种相互作用的“力程”约为 1.2×10^{-13} 厘米。在实验数据更加准确的情况下，我们将可得到一些关于力与距离的明确关系的信息[4]。

以更加精确的方式对中等能量的中子–质子散射进行测量，或者通过在更高能量情况下的观测，特别是对散射粒子的角分布的观测，我们都将获得更多关于这些问题的信息。

拉比及其他研究人员发现的氘核的四极矩对此问题有重要的贡献[5]。这证明了中子和质子的自旋方向倾向于与它们的连线方向一致，而不是与连线方向垂直。这显然意味着，中子与质子之间的力并不是中心力，而且它们之间的势能与自旋方向和两个粒子的连线方向之间的夹角有关。尽管上面提到的所有计算都是在中心力的

the two particles. Although all the calculations referred to above were carried out on the assumptions of central forces, the results remain practically unchanged. For opposite spins, which is the only case of interest for proton-proton interaction at low energies, the force must still be central, since the resultant spin is zero and hence does not set up a preferential direction; for parallel spin, the fact that the force is not central means that the angular momentum due to the motion of the particles is no longer constant but fluctuates. Hence the motion in the state of lowest energy is no longer one with zero orbital angular momentum; it can be shown that the state of motion is a mixture of a state with zero angular momentum and one with two units (*S* and *D* states in spectroscopic notation). From the electrical quadripole moment one can estimate that the contribution of the *D* state amounts to only a few percent[6].

However, an appreciable non-central force is required to produce even this small effect, since in the *D* state a very strong centrifugal force has to be overcome. It seems probable, therefore, that the non-central forces must represent an appreciable fraction of the total force. Yet, again owing to the effect of the centrifugal force, no appreciable influence of this *D* state should be expected on the other observable phenomena, except possibly on some finer features which have not yet been thoroughly investigated. Thus the approximate agreement with the experimental evidence, and the estimate of the range remain practically unaffected by the correction terms.

The data on the strength and the estimate of the range give the same answer, within the experimental error, for the proton-proton and proton-neutron interactions for opposite spins, and the suggestion has therefore been made that these two forces are exactly the same, except for the electrostatic interaction between the protons ("charge-independence hypothesis")[7].

From the fact that the charge of stable light nuclei is about half their mass, and that only a small change is produced in the energy of a light nucleus if in its constituents the number of neutrons and the number of protons are interchanged, has further led to the belief that the nuclear forces are symmetric in proton and neutron, that is, the neutron-neutron interaction is exactly the same as the proton-proton interaction, except for the effect of the electric forces[1].

Turning now to the evidence from nuclei containing more than two particles, this is usually discussed on the basis of the assumption that the forces are additive; that is, that the interaction force between two particles is not affected by the simultaneous presence of a third particle. Beyond its simplicity and the fact that it holds in the atom, this assumption is not founded on any evidence, and, in fact, there are certain theoretical arguments against it[8]. However, without this assumption the available evidence is insufficient to draw any conclusions at all, and one must therefore use the assumption of additivity as a working hypothesis, which may at a later stage be disproved. Moreover, most of the theoretical work on the nuclei of weight 3 and 4 was done before the discovery of the quadripole moment of the deuteron, and hence central forces were used. It is likely that

假设前提下进行的，但实际上结果并没有因为该假设成立与否而发生变化。在低能量的质子–质子相互作用中，我们唯一感兴趣的是自旋相反的情况，其中的作用力一定仍是中心力，这是因为该情况下的合自旋为零，没有产生任何的偏向；对于自旋平行的情况，作用力不是中心力意味着由粒子运动产生的角动量不再是常数，而是有涨落的。因此，粒子在最低能态中的运动不再是轨道角动量为 0 的状态，研究发现此运动状态是一个包含角动量等于 0 和 2 两种情况（用光谱学符号表示为 S 态和 D 态）的混合态，根据电四极矩我们能估计出 D 态的贡献只占几个百分点[6]。

然而，即使 D 态产生的效应很小，我们仍需要一个可观的非中心力来产生这个效应，因为 D 态中需要克服一个很强的离心力。因此，非中心力在合力中占有相当大的比重是很有可能的。然而，也是由于离心力的作用，除了一些可能尚未彻底研究过的某些较细微的特性外，预计 D 态不会对其他可观测的现象产生明显的影响。因此，修正项并不会影响理论与实验现象的近似吻合以及力程的估计结果。

在实验误差范围内，对于自旋相反的质子–质子相互作用和质子–中子相互作用，作用强度数据和力程的估算结果给出了同样的答案。由此可知，除了质子间的静电相互作用外，这两种相互作用力是完全相同的（“电荷独立性假说”）[7]。

稳定轻核的电荷数约为其质量数的一半，并且核组分中的中子数与质子数相互交换时，轻核的能量只会发生微小的变化，上述两个事实使我们更加相信质子和中子间的核力是对称的，也就是说，除了电场力的影响以外，中子–中子相互作用与质子–质子相互作用是完全相同的[1]。

下面来讨论含有两个以上粒子的核的相关证据，这些研究通常是以力具有可加性这一假设为基础的，即两个粒子之间的相互作用力不受同时存在的第三个粒子的影响。除了假设的简单性及其在原子中成立的事实以外，这个假设的建立没有任何根据，实际上，甚至存在某些理论论据认为该假设是不成立的[8]。然而，如果没有这个假设，现有证据是根本不足以得出任何结论的。因此，我们必须把可加性的假设作为有效的假说来使用，当然后面的研究有可能会证明该假设是不正确的。此外，在发现氘核的四极矩以前，人们就已经完成了大部分质量数为 3 和 4 的核的理论工作，因此当时的相关工作都采用了中心力的假设。实际上，这些问题中力对方向的依赖

the directional dependence of the force will not be negligible in these problems, but its precise effect has yet to be investigated.

So far as the calculations go, they show that the mass defects of the nuclei of mass 3 and 4 are very sensitive to the range of the force, and that the range required to give the right values is roughly of the same order of magnitude as that obtained from scattering experiments[1,9]. Quantitative agreement, however, was not obtained. The laws of force used in these calculations were usually restricted by adopting the charge-independence hypothesis, and also by assuming that the range of the force was the same for parallel and for opposite spins. Whether the remaining discrepancy is due to these restrictions, or to the assumption of central forces, or whether it represents a failure of the additivity, remains to be seen.

For nuclei beyond ^{4}He, the observed binding energy ceases to rise rapidly with the number of particles. This fact, often briefly referred to as "saturation", obviously means that the constituent particles of a large nucleus do not *all* attract each other with the same forces with which the constituents of a helium nucleus attract each other. This may be due to several reasons. For one thing, the additivity of the force might fail, and the attractive force between two particles may depend on the number of other particles in the immediate neighbourhood. This dependence might be such as to ensure that the total potential energy per particle always remained below a certain saturation value. A possible description on these lines has been put forward by Teller, Critchfield and Wigner[10], but the idea has not yet been pursued very far.

Another possibility is that the attraction may turn into repulsion at very close approach, in analogy with the forces between the atoms of a liquefied inert gas. In this case the repulsion might ensure that any one particle can, within the range of its forces, be surrounded only by a small number of others, and hence the binding energy per particle is again limited. This possibility has not been fully explored, but it has the disadvantage that, owing to the wave mechanical penetrability of potential barriers, an extremely strong repulsion would have to be assumed to make this explanation possible.

The most attractive explanation is no doubt that the forces are "exchange forces" which depend on the symmetry of the wave function describing the motion of the particles, like the valency forces between the constituents of organic molecules. On this idea the neutron is trivalent, capable of forming a "bond" with one other neutron and two protons, and correspondingly for the proton. This would give a very natural explanation of the helium nucleus as a saturated structure[1].

The directional properties of the forces, which must be inferred from the existence of the quadripole moment of the deuteron, may have a bearing on this question, since in a large nucleus different particles must necessarily be arranged in different directions relative to a given particle, and the forces may quite well be attractive for some pairs of particles and repulsive for others, if the directional dependence of the forces is strong[6]. This effect will

性很有可能是不可忽略的，但是它的确切的效应还有待研究。

现在的计算结果表明质量数为 3 和 4 的核的质量亏损与力程密切相关，得出正确结果所需的力程与散射实验得到的力程在量级上大致相等[1,9]。然而，在定量结果上并没有取得一致。上述计算中用到的作用力的相关定律通常受到以下两个条件的限定，即电荷独立性假说以及自旋平行和自旋反平行两种情况中力程相同的假设。至于定量结果的不一致是由这些限制引起的，还是由中心力的假设引起的，或者它是否代表了可加性的失效，这些还有待研究。

对于质量数超过 ^{4}He 的核，观测到的结合能不再随着粒子数的增加而快速增长。我们通常把这种现象简单称之为“饱和”，这显然意味着，大核内粒子之间彼此吸引的力，并不**都**与氦核内粒子间相互吸引的力相同。这可能是由多种原因造成的。一方面，力的可加性可能失效，两个粒子之间的吸引力可能与邻近的其他粒子的数目有关。这种关系可能使每个粒子的总势能始终保持在一定的饱和值以下。特勒、克里奇菲尔德和维格纳[10]已对此提出了一种可能的描述，但该想法还没有得到深入研究。

另一种可能性是两个粒子在非常接近的情况下吸引力有可能转变为排斥力，这与液化惰性气体中原子之间的力相类似。在这种情况下，斥力可以确保任何一个粒子在其力程范围内，只被少量的几个粒子所包围，因而每个粒子的结合能仍是有限的。这种可能性还没有得到充分的研究，但是根据势垒在波动力学下的穿透原理，这种可能性的缺点是，为了确保这种解释可能是正确的，必须假定斥力的强度非常大。

而最具吸引力的解释无疑是，这些力是与描述粒子运动的波函数的对称性有关的“交换力”，类似于有机分子组分之间的化合价力。按照这种观点，中子是三价的，它能够与另外一个中子及两个质子形成一个“键”，相应地，质子也具有类似的性质。这将为具有饱和结构的氦核给出一个很自然的解释[1]。

必须从氘核中四极矩的存在推出粒子间作用力的方向性，这或许与以上论述的问题有关，因为在一个大原子核内，不同的粒子相对于某个给定粒子的排列方向必定都不一样，而如果这些粒子间作用力的方向依赖性很强的话[6]，那么这些作用力很有可能对某些粒子对而言是吸引力，而对其他粒子对而言则是排斥力。这种效应

give rise to some kind of saturation; whether this saturation is of the right kind, and in particular whether it leads naturally to the α-particle as a stable structure, remains to be seen.

Of the attempts to derive the nuclear forces from a more elementary phenomenon, the most interesting is the meson theory, which is based on an idea of Yukawa[11]. Yukawa supposes that the nuclear forces are due to a "nuclear field" in the same manner in which the electromagnetic forces are caused by the electromagnetic field. This field must, however, satisfy different field equations in order to produce short-range forces, and if this requirement is coupled with that of the principle of relativity, there is only one possible type of wave equation, and the law of interaction is limited to laws of a particular type, of which the simplest has the potential:

$$V = g \cdot e^{-kr}/r, \tag{1}$$

where k and g are constants, and r is the distance between the particles. (Coulomb's law is a special case of this with $k = 0$). In quantum theory, just as the electromagnetic field is associated with light quanta, the "nuclear field" will be associated with a new type of particle; the fact that the wave equation differs from that of light indicates that the rest mass of these particles is not zero, like that of light quanta, but has the finite value $hk/2\pi c$, where k is the same constant as in (1). In order to obtain a range of the right order of magnitude this mass has to be assumed to be a few hundred times that of the electron. The subsequent discovery of "mesons" of just such a mass in cosmic rays has very much increased our confidence in the "meson theory" of nuclear forces.

If we try to fit a law of the form of (1) to the observations on proton-proton scattering, we obtain very good agreement, but we have to choose a value of k which corresponds to a meson mass of about 300 electron masses. This is almost certainly higher than the mass of the mesons found in cosmic rays. The origin of this discrepancy is not clear.

On the meson theory of nuclear forces, mesons are capable of being absorbed and emitted in nuclear reactions provided sufficient energy is available for their creation, and they may be exchanged (that is, temporarily created by one particle and absorbed by another) even if the available energy is insufficient to liberate them. If this view is taken, conservation of angular momentum in the process requires the mesons, like light quanta, to have integral spins. (The electron, which has a half-integral spin, can only be absorbed or produced in pairs.) Zero spin would make the nuclear forces repulsive when they should be attractive, and the most likely assumption seems that of a spin of one unit, as for the photon[12].

This assumption fixes the equations of the meson field completely, but not its interaction with the proton and neutron. This interaction is governed by two terms which, by analogy with the electric charge and magnetic moment, one may term the "meson charge" and the "meson moment" of the heavy particles. The law (1), in particular, is obtained if the heavy

将导致某种饱和；但这种饱和是否是上述的那种饱和现象，特别是，能否自然地解释 α 粒子是一种稳定的结构，这些还有待观察。

在所有尝试从更加基本的现象出发来推导核力的理论中，最令人关注的是基于汤川秀树[11]关于介子理论的一个构想。汤川秀树认为，核力是由“核场”引起的，这与电磁力是由电磁场引起的相类似。但是，为了产生短程力，这个场必须满足不同的场方程，而如果这个要求与相对性原理相结合，则只可能存在一种形式的波动方程，相互作用的定律也被限制在某种特定类型的定律之中，其中最简单的势能表达形式如下所示：

$$V = g \cdot e^{-kr}/r \tag{1}$$

式中 k 和 g 是常数，r 是粒子之间的距离。（库仑定律是上式中 k=0 时的一种特殊情况。）在量子理论中，正如电磁场是与光量子相联系的，“核场”也将与一种新型的粒子相关联；其波动方程与光的波动方程不同，这一事实表明，与光量子不一样，这些粒子的静止质量不为 0，而具有有限值 $hk/2\pi c$，其中 k 是式 (1) 中的常数。为了得到合适的数量级范围，必须假设这个粒子的质量是电子质量的几百倍。随后在宇宙射线中发现的“介子”正好具有这样的质量，这大大增强了我们对核力的“介子理论”的信心。

假如我们试图用一个具有式 (1) 形式的定律来拟合观测到的质子 – 质子散射，将会获得吻合得很好的结果，但为此我们必须选取一个 k 值，该值对应的介子质量约为电子质量的 300 倍。几乎可以肯定的是，这比在宇宙射线中发现的介子质量要高。而这种差异的根源尚不清楚。

在核力的介子理论中，当系统提供形成介子所需的足够能量时，介子将能够在核反应中被吸收和发射，但即使系统提供的能量并不足以释放出介子时，介子仍可以发生交换现象（即在极短的时间内介子由一个粒子产生后马上被另一个粒子吸收）。如果采用这种观点，该过程中角动量守恒要求介子如光量子一样，具有整数自旋。（具有半整数自旋的电子只能成对地被吸收或发射。）假设自旋为 0 的话将使原本应该为引力的核力表现为斥力，因此最有可能的假设似乎应该是，介子如同光子一样，具有一个单位的自旋[12]。

这个假设使介子场方程完全确定下来了，但介子场与质子和中子的相互作用还不确定。这类相互作用是由类似于电荷和磁矩的两个因素决定的，我们称之为重粒子的“介子电荷”和“介子矩”。特别地，假设重粒子只具有介子电荷 g，而不具有介子矩时可得到定律 (1)。这个定律在细节上与实验结果并不一致，因为它既没

particles have only a meson charge g, but no meson moment. This law does not agree with experiment in detail, since it gives neither a spin dependence of the force (as required to explain the properties of the neutron-proton scattering) nor a directional dependence that would account for the quadripole moment of the deuteron. The introduction of a "meson moment" would help to give the right dependence[12], but the force would then increase so rapidly at short distances that the proton and the neutron would fall into another, producing an infinite binding energy. Probably this result should not be taken too seriously, since the methods of quantum theory are likely to fail for too close approach, but it would mean in any event that the quantitative study of nuclear forces would have to be abandoned until an exact description of this failure of quantum theory is available.

It has also been suggested that two kinds of mesons might exist, of which one has the spin one, the other zero spin, and with such properties that the singular terms in the interaction energy just cancel. In the crude approximation to which such calculations are usually carried out, the directional dependence of the forces would then also just cancel, together with the infinities, but it is possible that finer effects would give a sufficiently large directional variation[13].

Lastly, there arises the question as to the electric charge of the meson. The mesons observed in cosmic rays are charged, and if one of them is emitted by one nuclear particle and absorbed by another, this will involve an exchange of charge, thus ensuring that the forces are of the exchange type, as required by the most widely accepted explanation of the saturation of nuclear forces. Such an exchange will, however, be possible only if one of the heavy particles is a proton and the other a neutron, but not for two like particles. In order to account for the equally strong forces between like particles, one has to assume the existence of neutral mesons in addition to the charged ones[14]. There is certain independent evidence for the existence of neutral mesons in cosmic rays[15]. If we take the view that the saturation of the nuclear forces is due to their directional dependence and not to their exchange character, it is, in fact, possible to attribute *all* nuclear forces to neutral mesons[6]. This would have the advantage of removing the discrepancy between the mass of the charged mesons from cosmic rays, and the range of the forces as determined from proton-proton scattering. It would mean, on the other hand, that the particles, the discovery of which was hailed as a confirmation of Yukawa's theory, had actually no connexion with the particles postulated by Yukawa.

This summary of the present state of the theory[16] closes with a number of questions to which the answer is not known. But the fact that it is possible to ask these questions at all is a sign of the rapid progress that has been made in this field in the last few years.

(**145**, 687-690; 1940)

Rudolph Peierls: University of Birmingham.

有给出自旋与力的关系（在解释中子–质子散射的性质时需要用到），也没有给出可用来解释氘核四极矩的方向依赖关系式。引入“介子矩”将有助于给出正确的依赖关系[12]，但在距离很近时，作用力将会迅速增大，使质子和中子落入到另一个核子之中，形成一个无限大的结合能。也许我们不该过于认真地看待这个结果，因为在粒子距离非常接近的情况下量子理论的方法可能是失效的，但这也可能意味着，在对量子理论失效的原因进行准确的说明以前，在任何情况下都不得不放弃对核力做定量研究的努力。

也有人提出，可能存在两类介子，其中一类自旋为 1，而另一类自旋为 0，在这种假设性质下，相互作用能中的奇异项恰好得以消除。在粗略的近似下，通常会进行这类计算，从而表示力的方向依赖性的项连同无限大的项一起，也刚好得以消除，但可能出现的情况却是，更加细微的效应将会使力在方向上出现足够大的变化[13]。

最后，出现了与介子电荷有关的问题。在宇宙射线中观测到的介子是带电的，而如果其中的一个介子是由一个核粒子发射并被另一个核粒子吸收，那么这就将涉及电荷的交换，因此，如同最为广泛接受的关于核力饱和的解释所要求的那样，这些力应该保证是交换型的力。然而，这样的一个交换仅可能在重粒子中的质子和中子之间发生，而不可能出现在两个相同类型的粒子之间。为了解释同类粒子间大小相同的力，就必须假定除带电介子外，还存在中性介子[14]。而宇宙射线中存在着中性介子，这是有独立可靠的证据的[15]。如果我们接受这样的观点，即认为核力的饱和是由核力对方向的依赖性所致而不是由其交换特性所致，那么实际上就可以认为所有的核力**都**是由中性介子引起的[6]。这将会带来的好处是，消除了宇宙射线中带电介子的质量与质子–质子散射过程所确定的力程之间的不一致。另一方面，它也将意味着，这个被认为确证了汤川秀树理论的粒子的发现，实际上与汤川秀树假设的粒子无关。

这篇有关核力理论现况[16]的综述，以许多尚未解决的难题作为结尾。但是，能提出这些问题本身就标志着最近几年这一领域取得了迅速的进展。

（沈乃澂 翻译；厉光烈 审稿）

References:

1. Bethe and Bacher, *Rev. Mod. Phys.*, **8**, 82 (1936).
2. Simons, *Phys. Rev.*, **55**, 793 (1939).
3. Brickwedde, Dunning, Hoge and Manley, *Phys. Rev.*, **54**. 266 (1938).
4. Hoisington, Share and Breit, *Phys. Rev.*, **56**, 884 (1939).
5. Kellogg, Rabi, Ramsay and Zacharias, *Phys. Rev.*, **55**, 318 (1939).
6. Bethe, *Phys. Rev.*, **55**, 1261 (1939).
7. Breit, Condon and Present, *Phys. Rev.*, **50**, 825 (1936).
8. Primakoff and Holstein, *Phys. Rev.*, **55**, 1218 (1939).
9. Rarita and Present, *Phys. Rev.*, **51**, 788 (1937); Rarita and Slawsky, *Phys. Rev.*, **54**, 1053 (1938).
10. Wigner, Critchfield and Teller, *Phys. Rev.*, **56**, 530 (1939).
11. Yukawa, *Proc. Phys. Math. Soc. Japan*, **17**, 48 (1935).
12. Fröhlich, Heitler and Kemmer, *Proc. Roy. Soc.*, A, **166**, 154 (1938).
13. Möller and Rosenfeld, *Nature*, **144**, 241 (1939); **144**, 476 (1939).
14. Kemmer, *Proc. Cam. Phil. Soc.*, **34**, 354 (1938).
15. Arley and Heitler, *Nature*, **142**, 158 (1938).
16. Cf. also Peierls, R., *Ann. Rep. Chem. Soc.*, in the press.

Excited States of Stable Nuclei

C. F. Powell *et al.*

Editor's Note

The idea that the stable nuclei of many atoms could exist in excited energetic states was first canvassed by a group of scientists who had used the newly built cyclotron at the University of Liverpool to bombard atomic nuclei such as neon. This paper shows that neon does indeed have an excited state lying some 2.5 million electron volts above that of the ground state, but that oxygen appears not to have an excited state within the range explored. Of the authors, Cecil Frank Powell and James Chadwick won Nobel Prizes for their work. Alan Nunn May, a lecturer at King's College London, was compromised by his association with Klaus Fuchs, afterwards shown to be a Soviet spy.

WE have recently carried out some experiments on the scattering of protons by light elements, using the proton beam provided by the Liverpool cyclotron and detecting the scattered particles by the photographic method.

A proton beam of about 10^{-8} amperes, with a divergence of one degree, is defined by a system of stops, and emerges from an attachment to the vacuum tank of the cyclotron into the "camera" through a mica window covering a hole $\frac{1}{8}$ in. in diameter. In the camera this narrow proton beam passes down the axis of a tube, which is interrupted for a length of 3 mm. to allow the scattered particles to emerge. A flat photographic plate is placed so that its surface is parallel to the axis of the beam and at a distance of 1 cm. from it. The protons scattered by the gas with which the camera is filled emerge through the interruption and enter the plate at a small glancing angle. This arrangement has the advantage that a single plate can contain the information for determining the probability of scattering from about 15° to 150°, providing for each angle regions containing a suitable number of tracks for counting purposes. At the same time, the energy of the scattered particle can be determined from the length of its track in the photographic emulsion.

Once the difficulties of defining the beam in the stray field of the magnet had been overcome, we found that suitable exposures could be obtained for some six to eight different scattering gases per day.

We have taken plates of the scattering from eleven elements which could be obtained either as elementary gases or in the form of suitable gaseous compounds. The energy of the incident particles at the point of scattering was about 4 million volts. The plates for hydrogen, deuterium and helium are satisfactory, and work is proceeding on these; but we wish to direct attention here to the inelastic scattering which accompanies the elastic

稳定核的激发态

鲍威尔等

编者按

一些科学家利用利物浦大学新建的回旋加速器轰击原子核（如氖），从而首次对激发能态下多种原子的稳定核能够存在的观点进行了探讨。这篇论文表明，氖确实存在一个激发态，且该激发态能量比基态能量大约高 2.5 兆电子伏，但是在实验研究范围内氧似乎没有激发态。本文作者中的塞西尔·弗兰克·鲍威尔和詹姆斯·查德威克凭借他们的研究成果获得了诺贝尔奖。另一位作者艾伦·农·梅是伦敦国王学院的一位讲师，他因与克劳斯·富克斯（后被发现是苏联间谍）有交往而受到了牵连。

最近，我们利用一些轻元素进行了质子散射的实验，实验使用利物浦的回旋加速器来产生质子束，并用照相的方法来检测散射粒子。

我们把一束强度为 10^{-8} 安培、分散度为 1° 的质子束，约束于一个静止的系统中，然后将其从回旋加速器的真空罐的附加装置中射出，穿过覆盖了云母窗的直径为 1/8 英寸的小洞后进入“照相机”。在照相机中，这条狭窄的质子束沿着一个长管轴向穿行，这一长管有个 3 毫米长的断口，散射粒子能从这个断口中逃逸出来。在与质子束的轴向相平行的方向上放置一块扁平的照相底板，底板与质子束的轴相距 1 厘米。被充满于照相机中的气体所散射的质子能够穿过长管的断口，并以很小的掠射角射到底板上。这样安置的好处是，假如每个角度区域中都包括适当数量的能用于计数的轨迹的话，那么只用一块底板就能获得用以确定 15° 到 150° 范围内散射概率的相关信息。与此同时，我们还能通过照相乳胶中轨迹的长度来确定散射粒子的能量。

如果能够克服在磁体的杂散场中约束质子束的困难，那么我们一天就能得到大约 6~8 种不同气体散射质子的合适的曝光照片。

我们已经得到了 11 种元素散射质子的底板，实验中采用的这些元素有的是单质气体的形式，有的是适当的气态化合物的形式。入射粒子在散射点的能量约为 4 兆电子伏。由氢、氘和氦得到的底板是令人满意的，后续处理的工作还在进行中；但是在这里，我们想把注意力放在较重元素中与弹性散射伴随出现的非弹性散射上，

scattering from the heavier elements, and gives information about the excited states of the stable nuclei in a particularly direct way.

The distributions in energy of the protons scattered through 90° from the gases oxygen and neon are shown in Fig. 1. In oxygen a single peak appears, corresponding in energy to protons elastically scattered through 90°. With neon, in addition to the elastically scattered group, there is a peak at lower energy which we attribute to inelastic scattering from ^{20}Ne. This view is supported by the fact, deduced from the analysis of the neutron spectrum of fluorine under deuteron bombardment, that ^{20}Ne has an excited state of 1.4 Mev., for the difference in energy of the two groups of scattered particles is just of this amount.

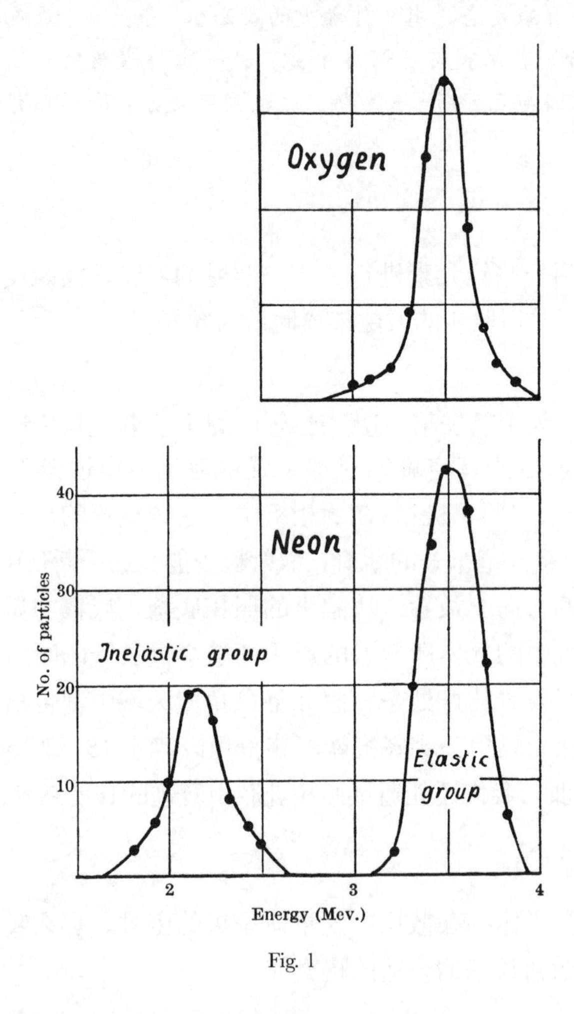

Fig. 1

We have examined the variation of the probability of scattering with angle for elastic and inelastic scattering in neon, and the results are shown in Fig. 2. It will be seen that in the range from 40° to 80° the elastic scattering follows very closely that expected from pure Coulomb scattering. In contrast with this, the inelastically scattered particles are distributed spherically symmetrically about the centre of mass of the system, to within the present accuracy of the measurements. This suggests that the inelastically scattered protons have

这种非弹性散射以特别直接的方式给出了关于稳定核激发态的信息。

图 1 显示了在氧气和氖气中散射后偏转了 90° 的质子的能量分布。在氧气中散射的结果是，只有一个单峰，其对应着发生弹性散射后偏转了 90° 的质子的能量。而在氖气中散射的结果是，除了弹性散射组之外，在能量较低的位置上还有一个峰，我们将其解释为由 ^{20}Ne 导致的非弹性散射。这一观点是有事实依据的，因为，对氟在氘核轰击下得到的中子能谱进行的分析表明 ^{20}Ne 具有一个能量为 1.4 兆电子伏的激发态，而我们的实验结果中两组散射粒子的能量差刚好就是这个数值。

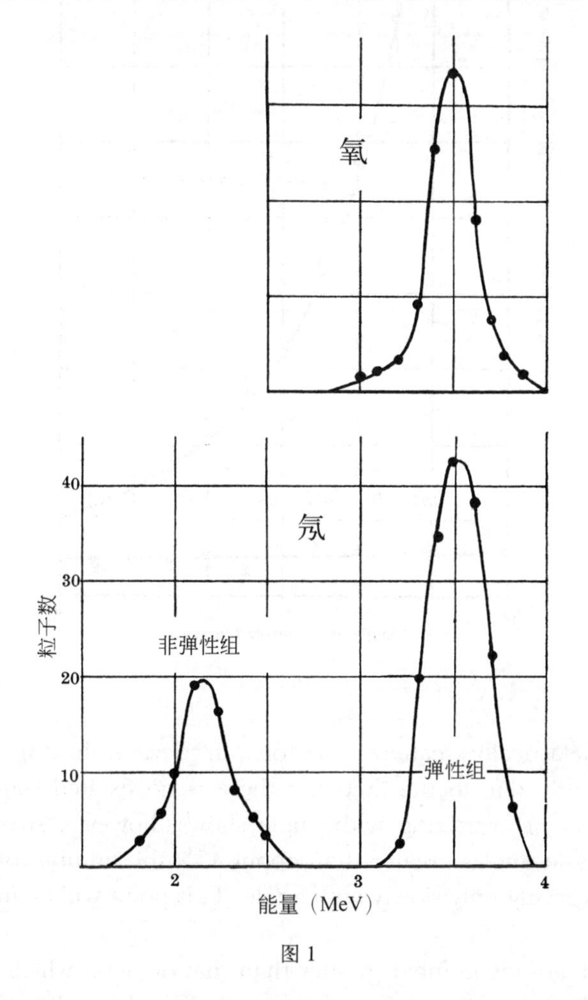

图 1

我们还研究了氖实验中弹性散射和非弹性散射的散射概率随角度的变化，结果如图 2 所示。从图中我们可以看到，在角度为 40°~80° 的范围内，弹性散射的结果与完全由库仑散射而得到的预期结果几乎完全吻合。但与此相反的是，在目前观测所能达到的精度范围内，非弹性散射粒子是以系统质心为中心呈球对称分布的。这

been "evaporated" from the compound nucleus formed in a close collision of an incident proton and a ^{20}Ne nucleus.

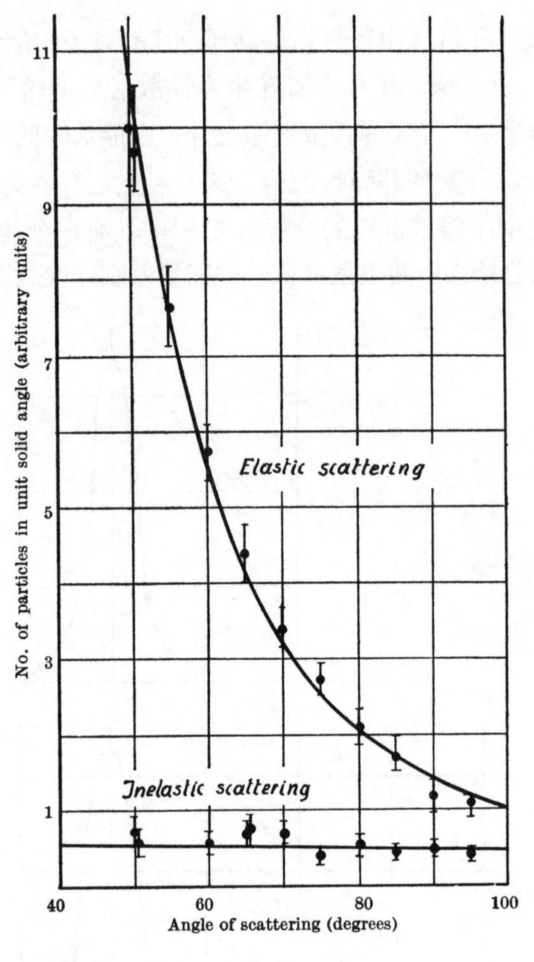

Fig. 2

The absence of inelastically scattered protons in oxygen in the conditions of our experiments is evidently due to the fact that there is no excited state of oxygen below 4 Mev. The variation of scattering with angle shows, however, strong anomalies from Rutherford scattering at angles greater than about 45°, the number of scattered particles per unit solid angle varying only slowly with angle. This point will be investigated further.

With the elements of atomic number greater than that of neon which we have examined, such as chlorine and argon, the ratio of the number of inelastically to elastically scattered particles is very much smaller than in the case of neon, corresponding to the decreasing probability of the protons entering the nucleus with increasing nuclear charge. It is therefore evident that it will be desirable to continue the experiments with protons of higher energy in order that the higher excited states of the light elements may become accessible to investigation, and to obtain results for elements of higher atomic number. It

意味着，一个入射质子与一个 ^{20}Ne 核发生近程碰撞后形成了复合核，其中非弹性散射质子就是从这个复合核中“蒸发”出来的。

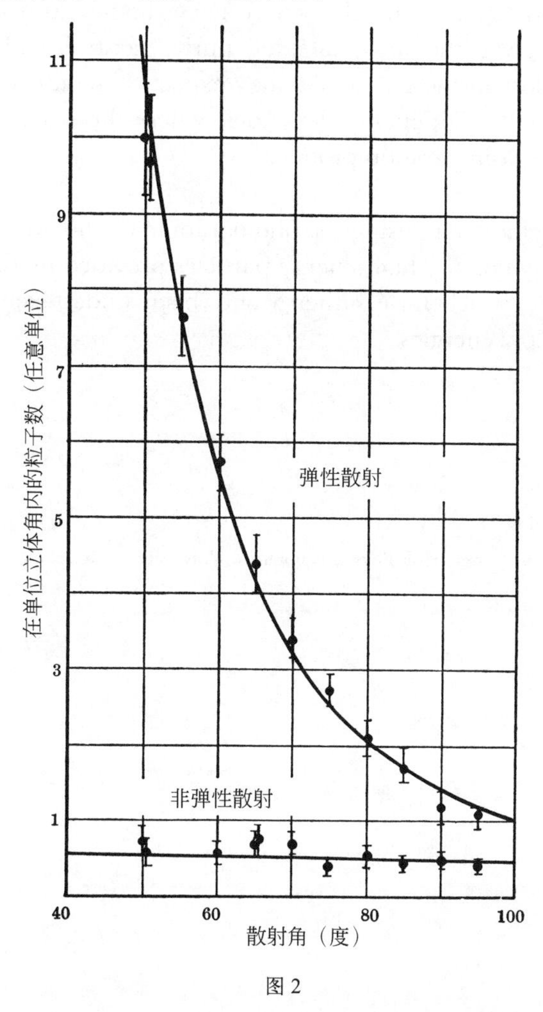

图 2

在我们的实验条件下，氧的实验中没有出现非弹性散射质子，这显然是因为氧不存在比基态能量高出 4 兆电子伏以内的激发态。不过，依据散射随角度的变化关系显示，在大于 45° 时散射结果严重背离了卢瑟福散射，且随着角度的变化，单位立体角中散射粒子的数目只发生缓慢的变化。在后面的工作中我们将对此进行深入的研究。

对于我们研究过的原子序数比氖大的元素，比如氯和氩，非弹性散射粒子数与弹性散射粒子数的比值远远小于氖实验中得到的结果，这与核电荷增大导致质子入核概率减小是相一致的。因此，要研究轻元素的更高能激发态，或者要得到较大原子序数元素的质子散射结果，就必须用更高能量的质子来进行实验。不过，从我们

is clear from our experience, however, that the method is very powerful, the plates being obtained with an exposure of a few minutes and the analysis of the energy distribution of the scattered protons being complete within a few hours. Also the use of what are essentially gas targets gives the advantage of purity control and absence of effects associated with energy loss in the target. We may expect these advantages in experiments of a similar character with high-energy deuterons, where the scattered primary particles may be accompanied by disintegration products.

In general, we may conclude that, using the photographic method of detection, it becomes possible to take advantage of the high-energy particles provided by the cyclotron to make experiments of the kind which have hitherto only been undertaken with direct current generators at relatively low energies.

(**145**, 893-894; 1940)

C. F. Powell: Wills Physical Laboratory, University of Bristol.

A. N. May: King's College, London.

J. Chadwick and T. G. Pickavance: George Holt Physics Laboratory, University of Liverpool.

的实验中可以清楚地看出，我们用的这种方法是非常强大的，短短的几分钟就可以得到曝光底板，而对散射质子能量分布的分析结果也能在短短的几个小时之内完成。另外，至关重要的是，气体靶标的使用有利于纯度控制以及与靶标中的能量损失相关的各种效应的消除。我们可以预期，在使用高能氘核进行的类似实验中这些优点也将会得以体现，只是在这些实验中被散射的入射粒子可能会与衰变产物混在一起。

总之，我们可以确定的是，如果使用照相法来进行检测，那么此前那些只能依靠直流发生器在相对较低的能量水平上进行的实验，现在就可以利用由回旋加速器产生的高能粒子来进行了。

（王耀杨 翻译；汪长征 审稿）

Considerations Concerning the Fundaments of Theoretical Physics*

A. Einstein

Editor's Note

Five years before his death, Einstein had been driven to the view that quantum mechanics was a valid description of quantum phenomena but that as a consequence physics had been robbed of its explicit and deterministic foundations. He concluded that "it is open to every man to choose the direction of his striving; and also every man may draw comfort from Lessing's fine saying that the search for truth is more precious than its possession".

WHAT we call physics comprises that group of natural sciences which base their concepts on measurements; and the concepts and propositions of which lend themselves to mathematical formulation. Its realm is accordingly defined as that part of the sum total of our knowledge which is capable of being expressed in mathematical terms. With the progress of science, the realm of physics has so expanded that it seems to be limited only by the limitations of the method itself. The larger part of physical research is devoted to the development of the various branches of physics, in each of which the object is the theoretical understanding of more or less restricted fields of experience, and in each of which the laws and concepts remain as closely as possible related to experience. It is this department of science with its ever-growing specialization, which has revolutionized practical life in the last centuries.

On the other hand, from the very beginning there has always been present the attempt to find a unifying theoretical basis for all these single sciences, consisting of a minimum of concepts and fundamental relationships, from which all the concepts and relationships of the single disciplines might be derived by logical process. This is what we mean by the search for a foundation of the whole of physics.

It is clear that the word foundation in this connexion does not mean something analogous in all respects to the foundations of a building. Logically considered, of course, the various single laws of physics rest upon this foundation. But whereas a building may be seriously damaged by a heavy storm or spring flood, and yet its foundations remain intact, in science the logical foundation is always in greater peril from new experiences or new knowledge than are the branch disciplines with their closer experimental contacts.

The first attempt to lay a uniform theoretical foundation in physics was the work of

* An address, slightly abridged, delivered at the Eighth American Scientific Congress at Washington on May 15.

关于理论物理基础的思考*

爱因斯坦

编者按

爱因斯坦在他生命最后的五年中坚持认为，虽然量子力学对量子现象的描述是合理的，但却摒弃了物理学的明确性和确定性的基础。他做出了如下的结论：“世事无绝对，每个人都可以选择自己奋斗的方向；而每个人也都可以从莱辛的一句精辟的名言中得到安慰：追求真理比拥有真理更为可贵。”

我们所说的物理学由这样一类自然科学组成，它们的概念的定义是以测量为依据，而且其中的概念和命题也都可用数学公式表示出来。因此可以说，所有能用数学语言来表达的那部分知识，都可划归为物理学的范畴。因此随着科学的进步，物理学的范畴似乎已经扩展到只受研究方法自身所限制的程度。物理学研究的大部分工作是致力于发展物理学的各个分支，目标是对一定领域内的实践经验给予理论解释，而且其中的定律和概念都要尽可能与实践经验紧密联系。正是这样一门科学，随着它研究发展的不断专业细化，在过去的几个世纪里彻底改变了人们的现实生活方式。

另一方面，人们从一开始就一直尝试寻找一个能把所有独立的科学统一起来的理论基础，这个统一的理论由最少的概念和基本关系组成，在此基础之上各个独立分支的所有概念和关系都可以通过逻辑推导产生。这就是我们探求整个物理学基础的用意所在。

显而易见这个基础并不完全类似于一座建筑的基础。当然，从逻辑上讲，各条物理定律确实都是建立在这一基础之上的。然而建筑物可能在风暴或洪水过后被严重毁坏，而其基础却完好无损；但在科学领域中，相比于与实验联系较为紧密的各个具体学科而言，这一逻辑基础总是由于新经验或新知识而处于危险的境地。

建立物理学统一理论基础的第一次尝试是牛顿的工作。在他的理论中，所有的

* 这是一篇经过删节的演讲，发表于5月15日在华盛顿举行的第八届美国科学大会。

Newton. In his system everything is reduced to the following concepts:

(1) Mass points with invariable mass;
(2) action at a distance between any pair of mass points;
(3) law of motion for the mass point.

There was not, strictly speaking, any all-embracing foundation, because an explicit law was formulated only for the actions-at-a-distance of gravitation; while for other actions-at-a-distance nothing was established *a priori* except the law of equality of *actio* and *reactio*. Moreover, Newton himself fully realized that time and space were essential elements as physically effective factors of his system, if only by implication.

This Newtonian basis proved eminently fruitful and was regarded as final up to the end of the nineteenth century. It not only gave results for the movements of the heavenly bodies down to the most minute details, but also furnished a theory of the mechanics of discrete and continuous masses, a simple explanation of the principle of the conservation of energy, and a complete and brilliant theory of heat. The explanation of the facts of electrodynamics within the Newtonian system was more forced; least convincing of all, from the very beginning, was the theory of light.

It is not surprising that Newton would not listen to a wave theory of light; for such a theory was most unsuited to his theoretical foundation. The assumption that space was filled with a medium consisting of material points that propagated light waves without exhibiting any other mechanical properties must have seemed to him quite artificial. The strongest empirical arguments for the wave nature of light, fixed speeds of propagation, interference, diffraction, polarization, were either unknown or else not known in any well-ordered synthesis. He was justified in sticking to his corpuscular theory of light. During the nineteenth century the dispute was settled in favour of the wave theory. Yet no serious doubt of the mechanical foundation of physics arose, in the first place because nobody knew where to find a foundation of another sort. Only slowly, under the irresistible pressure of facts, there developed a new foundation of field-physics.

From Newton's time on, the theory of action-at-a-distance was constantly found artificial. Efforts were not lacking to explain gravitation by a kinetic theory, that is, on the basis of collision forces of hypothetical mass particles. But the attempts were superficial and bore no fruit. The strange part played by space (or the inertial system) within the mechanical foundation was also clearly recognized, and criticized with especial clarity by Ernst Mach.

The great change was brought about by Faraday, Maxwell and Hertz—as a matter of fact half-unconsciously and against their will. All three of them, throughout their lives, considered themselves adherents of the mechanical theory. Hertz had found the simplest form for the equations of the electromagnetic field, and declared that any theory leading to these equations was Maxwellian theory. Yet towards the end of his short life he wrote a paper in which he presented as the foundation of physics a mechanical theory freed from

内容都可归结为如下的几个概念：

（1）具有恒定质量的质点；
（2）任何两个质点之间存在超距作用；
（3）质点的运动定律。

严格说来，这里并没有什么绝对的基础，因为这些明确的定律只有在引力的超距作用下才成立；而对于别的超距作用，除了作用力和反作用力相等这条定律以外，我们不能先验地得到其他任何定律。此外，牛顿自己也充分地意识到，空间和时间是他的理论体系中的根本有效因子，不过他并未明说。

牛顿的基本理论被证明是卓有成效的，并且直到 19 世纪末，它一直被视为是物理学的终极理论。它不但解释了天体运动以及其中最详细的细节，而且也建立了描述离散物质和连续物质的力学理论，对能量守恒原理给予了简单的解释且给出了一套完整而天才的热学理论。在牛顿的理论体系中，对电动力学的事实规律的解释是比较牵强的；而到目前为止，用牛顿的理论体系进行解释的所有理论中，最令人难以信服的是对光的理论的解释。

牛顿不相信光的波动论，这不足为奇；因为这个理论与他的理论基础最不相容。假定空间里充满着一种由质点组成的媒质，这些质点传递着光波但却不显示出任何力学性质，牛顿认为这个假定人为的痕迹很明显。光具有波动性的最有力的经验论据，如不变的传播速度、干涉、衍射、偏振等，这些在当时要么还不清楚，要么就是还没有被整理总结出来。牛顿坚持认为他的光的粒子论是有道理的。到了 19 世纪，这场争论才以波动论的胜利而告终。但是当时人们并没有严重质疑物理学建立的力学基础，这主要是因为没有找到另一种更好的基础。在不可抗拒的事实的压力之下，一种新的场物理学基础才慢慢发展起来。

从牛顿时代起，所谓的超距作用的理论就一直是难以令人信服的。试图根据假想质点发生碰撞的动力学理论来解释引力而做的努力并不少。但是所做的这些尝试都流于表面，没有得出满意的结果。另外，人们显然意识到了空间（或者惯性系）在以力学为基础的理论中有奇特的作用，而恩斯特·马赫也对此做了清楚详细的论述。

法拉第、麦克斯韦和赫兹带来了物理学的伟大变革，事实上这一变革是他们在半无意的情况下做出的，并且同他们最初的意愿相悖。自始至终，这三位物理学家都认为自己是力学理论的拥护者。赫兹发现了电磁场方程最简单的形式，并且宣称任何能够导出这些方程的理论都是麦克斯韦理论。可是就在他短暂的一生结束之前，他写了一篇论文，其中他提出了一种摆脱了力这个概念的力学理论并将其作为物理

the force-concept.

For us, who took in Faraday's ideas so to speak with our mother's milk, it is hard to appreciate their greatness and audacity. Faraday must have grasped with unerring instinct the artificial nature of all attempts to refer electromagnetic phenomena to actions-at-a-distance between electric particles reacting on each other. All these electric particles together seemed to create in the surrounding space spatial states, today called fields, which he conceived as states of mechanical stress in a space-filling medium, similar to the states of stress in an elastically distended body. For at that time this was the only way one could conceive of states that were apparently continuously distributed in space. The peculiar type of mechanical interpretation of these fields remained in the background—a sort of placation of the scientific conscience in view of the mechanical tradition of Faraday's time.

With the help of these new field concepts, Faraday succeeded in forming a qualitative concept of the whole complex of electromagnetic effects discovered by him and his predecessors. The precise formulation of the time-space laws of those fields was the work of Maxwell. Imagine his feelings when the differential equations he had formulated proved to him that electromagnetic fields spread in the form of polarized waves and with the speed of light! At that thrilling moment he surely never guessed that the nature of light, apparently so completely solved, would continue to baffle succeeding generations. Meantime, it took physicists some decades to grasp the full significance of Maxwell's discovery, so bold was the leap that his genius forced upon the conceptions of his fellow-workers. Only after Hertz had demonstrated experimentally the existence of Maxwell's electromagnetic waves did resistance to the new theory break down.

But if the electromagnetic field could exist as a wave independent of the material source, then the electrostatic interaction could no longer be explained as action-at-a-distance; and what was true for electrical action could not be denied for gravitation. Everywhere Newton's actions-at-a-distance gave way to fields spreading with finite velocity.

Of Newton's foundation there now remained only the material mass points subject to the law of motion. But J. J. Thomson pointed out that an electrically charged body in motion must, according to Maxwell's theory, possess a magnetic field the energy of which acted precisely as does an increase of kinetic energy to the body. If, then, a part of kinetic energy consists of field energy, might that not then be true of the whole of the kinetic energy? Perhaps the basic property of matter, its inertia, could be explained within the field theory? The question led to the problem of an interpretation of matter in terms of field theory, the solution of which would furnish an explanation of the atomic structure of matter. It was soon realized that Maxwell's theory could not accomplish such a programme. Since then many men of science have zealously sought to complete the field theory by some generalization that should comprise a theory of matter; but so far such efforts have not been crowned with success.

学的基础。

对于我们来说，接受法拉第的理论如同我们吸吮母亲的乳汁一样自然，我们很难体会到这些物理学家们的伟大和过人的胆识。对于一切试图把电磁现象归结为相互作用的带电粒子之间存在超距作用的做法，法拉第以准确无误的直觉看出了这些做法是人为不客观的。所有这些带电粒子似乎共同在其周围空间中产生了一些空间态，现在我们称之为场，法拉第设想这些空间态是当机械协强作用于充满空间的介质时出现的状态，类似于弹性膨胀体受到协强作用时的状态。因为在那个时候，为了想象这些在空间中明显是连续分布的状态，这是唯一可行的方法。这种对场的特殊的力学解释只属于那个特定的历史背景——从法拉第时代传统的力学理论的角度来看，这是对当时科学意识的一种妥协和退让。

借助这些新的场的概念，法拉第成功地为他和他的先辈们所发现的复杂的电磁效应提出了一整套定性的理论概念。麦克斯韦的工作则是为这些场的时空律推导出严密的公式。当麦克斯韦建立的微分方程证明了电磁场是以偏振波的形式，并且以光速在传播着的时候，想象当时的他该有怎样的感觉呀！在那激动人心的时刻，他肯定不会想到，光的本质这个表面看来已经被完美解决了的问题，仍会继续困惑以后的好几代人。同时，物理学家们也花了几十年的时间才完全领会了麦克斯韦伟大发现的全部意义，由此可见，他的同事们要在观念上作出多么勇敢的飞跃才能接受其天才般的智慧啊。直到赫兹在实验中证实了麦克斯韦电磁波的存在以后，对这个新理论的抵制才彻底消除。

如果电磁场可以独立于介质源而以波的形式存在，那么，静电相互作用就不能再解释成超距作用了；而在电的相互作用情况中是正确的东西，在引力中就也可能是正确的。渐渐地，在物理学的各个方面，牛顿的超距作用都被以有限速度传播的场所替代了。

牛顿的基础理论中，只有质点的运动定律这个概念直到现在仍然保留着。但是汤姆森指出：依照麦克斯韦的理论，运动着的带电体必定具有磁场，磁场的能量正好是带电体增加的那部分动能。那么如果动能的一部分是由场能组成的，那么全部动能不也可能是这样的吗？或许作为物质基本性质的惯性，是否也能在场论中得到解释？这个问题导致了利用场论对物质进行解释的困难，该困难的解决应该会给物质原子结构提供一种解释。人们很快意识到，麦克斯韦理论不能完成这个任务。从那时起，许多科学工作者就充满热情地试图完成将包含物质的理论推广到场论当中；但是到目前为止，这种努力并没有取得圆满成功。

For several decades most physicists clung to the conviction that a mechanical substructure would be found for Maxwell's theory. But the unsatisfactory results of their efforts led to gradual acceptance of the new field concepts as irreducible fundamentals—in other words, physicists resigned themselves to giving up the idea of a mechanical foundation.

Thus physicists held to a field-theory programme. But it could not be called a foundation, since nobody could tell whether a consistent field theory could ever explain on one hand gravitation, on the other hand the elementary components of matter. In this state of affairs it was necessary to think of material particles as mass points subject to Newton's laws of motion. This was the procedure of Lorentz in creating his electron theory and the theory of the electromagnetic phenomena of moving bodies.

Such was the point at which fundamental conceptions had arrived at the turn of the century. Immense progress was made in the theoretical penetration and understanding of whole groups of new phenomena; but the establishment of a unified foundation for physics seemed remote indeed; and this state of things has even been aggravated by subsequent developments.

The development during the present century is characterized by two theoretical systems essentially independent of each other: the theory of relativity and the quantum theory. The two systems do not directly contradict each other; but they seem little adapted to fusion into one unified theory.

The theory of relativity arose out of efforts to improve, with reference to logical economy, the foundation of physics as it existed at the turn of the century. The so-called special or restricted relativity theory is based on the fact that Maxwell's equations (and thus the law of propagation of light in empty space) are converted into equations of the same form, when they undergo Lorentz transformation. This formal property of the Maxwell equations is supplemented by our fairly secure empirical knowledge that the laws of physics are the same with respect to all inertial systems. This leads to the result that the Lorentz transformation—applied to space and time coordinates—must govern the transition from one inertial system to any other. The content of the restricted relativity theory can accordingly be summarized in one sentence: all natural laws must be so conditioned that they are co-variant with respect to Lorentz transformations. From this it follows that the simultaneity of two distant events is not an invariant concept and that the dimensions of rigid bodies and the speed of clocks depend upon their state of motion.

A further consequence was a modification of Newton's law of motion in cases where the speed of a given body was not small compared with the speed of light. There followed also the principle of the equivalence of mass and energy, with the laws of conservation of mass and energy becoming one and the same. Once it was shown that simultaneity was relative and depended on the frame of reference, every possibility of retaining actions-at-a-distance within the foundation of physics disappeared, since that concept presupposed the absolute character of simultaneity (it must be possible to state the location of the two

几十年以来，大多数物理学家都坚信能为麦克斯韦理论找到更基本的力学基础。但是由于他们的努力没有得到令人满意的结果，人们逐渐将场这个新的概念作为物理学不可约化的基础——换句话说，物理学家不得已放弃了力学是物理学基础的想法。

因此，后来物理学家都转而支持场论的研究体系。但是并不能把场论称为基础，因为尚不确定是否有一个统一的场论，既能解释引力，又能解释物质的基本组成成分。在这种情况下，就有必要把物质粒子看作是服从牛顿运动定律的质点。这是洛伦兹在创立其电子论的过程以及研究与运动物体电磁行为有关的理论中都采用了的假设。

以上这些就是我们在世纪之交物理学基本概念的由来。当时对所有新奇现象的理解和理论方面的突破都有了极大的进展；但是建立物理学统一的基础的希望似乎仍然很渺茫；而随后的进展更是加重了这一困难。

二十世纪科学发展的标志是本质上各自独立的两个理论体系：相对论和量子论。这两个体系彼此没有直接的矛盾；但是它们似乎很难融合成一个统一的理论。

考虑到逻辑上的便利，科学家们对物理学基础的探究导致了相对论在世纪之交的诞生。所谓狭义的或者有限制的相对论所根据的事实是：在洛伦兹变换下，麦克斯韦方程在形式上没有变化（因而光在真空中传播的定律也一样是保持不变的）。另外，这一麦克斯韦方程组的形式上的特征可以用我们已知的十分可靠的经验来进行补充，该经验就是：在所有惯性系中，物理定律都是相同的。这导致了一个惯性系到其他任何惯性系的变换必须满足应用于时空坐标的洛伦兹变换。因此，狭义相对论的内容可以总结成一句话：一切自然规律都是有条件限制的，即它们都得具有洛伦兹协变性。由此得知，两个异地事件发生的同时性并非是绝对的，刚体的尺寸和时钟的快慢都同它们的运动状态有关。

相对论的另一个结果是，在物体的速率并非远小于光速的情况下，牛顿运动定律必须做相应的修正。我们还得到了质能相当性原理，即质量守恒定律和能量守恒定律可以合并成同一个定律。注意，一旦指明了同时性是相对的，并且同参照系有关，在物理学的基础中保留超距作用的任何可能性都没有了，因为这一概念是以同时性的绝对性作为前提的（即必须能够指明“在同一时刻”两个相互作用的质点

interacting mass points "at the same time").

The general theory of relativity owes its origin to the attempt to explain a fact known since Galileo's and Newton's time but hitherto eluding all theoretical interpretation: the inertia and the weight of a body, in themselves two entirely distinct things, are measured by one and the same constant, the mass. From this correspondence follows that it is impossible to discover by experiment whether a given system of co-ordinates is accelerated, or whether its motion is straight and uniform, and the observed effects are due to a gravitational field (this is the equivalence principle of the general relativity theory). It shatters the concepts of the inertial system, as soon as gravitation enters in. It may be remarked here that the inertial system is a weak point of the Galilean–Newtonian mechanics. For there is presupposed a mysterious property of physical space, conditioning the kind of co-ordination systems for which the law of inertia and the Newtonian law of motion hold good.

These difficulties can be avoided by the following postulate: natural laws are to be formulated in such a way that their form is identical for co-ordinate systems of any kind of states of motion. To accomplish this is the task of the general theory of relativity. On the other hand, we deduce from the restricted theory the existence of a Riemannian metric within the time-space continuum, which, according to the equivalence principle, describes both the gravitational field and the metric properties of space. Assuming that the field equations of gravitation are of the second differential order, the field law is clearly determined.

Aside from this result, the theory frees field physics from the disability it suffered from, in common with the Newtonian mechanics, of ascribing to space those independent physical properties which heretofore had been concealed by the use of an inertial system. But it cannot be claimed that those parts of the general relativity theory which can today be regarded as final have furnished physics with a complete and satisfactory foundation. In the first place, the total field appears in it to be composed of two logically unconnected parts, the gravitational and the electromagnetic. In the second place, this theory, like the earlier field theories, has not yet supplied an explanation of the atomistic structure of matter. This failure has probably some connexion with the fact that so far it has contributed nothing to the understanding of quantum phenomena.

In 1900, in the course of a purely theoretical investigation, Max Planck made a remarkable discovery: the law of radiation of bodies as a function of temperature could not be derived solely from the laws of Maxwellian electrodynamics. To arrive at results consistent with the relevant experiments, radiation of a given frequency had to be treated as though it consisted of energy atoms of the individual energy $h\nu$, where h is Planck's universal constant. During the years following it was shown that light was everywhere produced and absorbed in such energy quanta. In particular, Niels Bohr was able largely to understand the structure of the atom, on the assumption that atoms can have only discrete energy values, and that the discontinuous transitions between them are connected

所处的位置)。

广义相对论源自对一个从伽利略和牛顿时代起就已经出现的问题进行解释的尝试,迄今为止这个问题也无法用现有的任何理论进行解释:物体的惯性和重量本身是两种完全不同的概念,但却用同一个物理量(质量)进行量度。由此可以推论,我们不可能通过实验判断给定的坐标系是否在加速,或是否在做匀速直线运动,而观察到的结果最终是由引力场所决定的(这就是广义相对论的等效原理)。一旦引入了引力场,惯性系的概念就失效了。这里我们注意到,惯性系是伽利略-牛顿力学的弱点。因为引入惯性系即相当于我们预先假定物理空间具有一种神秘的性质,使得惯性定律和牛顿的运动定律在这种坐标系中仍然有效。

为了避免出现上面的困难,我们可以做以下假设:在处于任何一种运动状态的坐标系中,自然规律的表达形式都是完全一样的。完善这一假设正是广义相对论的任务。另一方面,我们从狭义相对论推出时空连续区中存在黎曼度规,依照等效原理,它既描述了引力场,也描述了空间的度规性质。假定引力的场方程是二阶的微分方程,则我们可以明确地得到与场有关的定律。

除了以上的结果,相对论还使场物理学摆脱了牛顿力学中把独立的物理性质归咎于空间而出现的问题,而这个问题在相对论出现之前一直是通过引入惯性系而被隐藏起来的。但是现在我们还不能断言,当今广义相对论中被视作终极理论的那些部分是否为物理学提供了完整且令人满意的物理学基础。因为首先,相对论中所有的场都由逻辑上毫无关系的两个部分,即引力部分和电磁部分所组成。其次,跟以前的场论一样,相对论直到现在还未能对物质的原子结构给予解释。相对论的这种局限性也许同它至今对理解量子现象尚无贡献这一事实有一定的关系。

1900 年,在纯理论研究的过程中,马克斯·普朗克得到了一个不同寻常的发现:仅从麦克斯韦电动力学定律出发,不能推导出作为温度的函数的物体辐射定律。为了得到同相关实验一致的结果,必须把那些具有一定频率的辐射看作是由一些单个能量为 $h\nu$ 的能量子所组成,此处 h 是普朗克普适常数。随后几年,事实证明光始终都是以这样的能量子的形式产生或被吸收的。特别是尼尔斯·玻尔假定原子只能具有分立的能量值,而它们之间不连续的跃迁都同这种能量子的发射或者吸收有关,据此他能够基本上推测出原子的结构。这有助于人们理解气体状态下的元素及其化

with the emission or absorption of such an energy quantum. This threw some light on the fact that in their gaseous state elements and their compounds radiate and absorb only light of certain sharply defined frequencies. All this was quite inexplicable within the frame of the theories hitherto existing. It was clear that, at least in the field of atomistic phenomena, the character of everything that happens is determined by discrete states and by apparently discontinuous transitions between them, Planck's constant h having a decisive role.

The next step was taken by de Broglie. He asked himself how the discrete states could be understood by the aid of the current concepts, and hit on a parallel with stationary waves, as for example in the case of the fundamental frequencies of organ pipes and strings in acoustics. True, wave actions of the kind here required were unknown; but they could be constructed, and their mathematical laws formulated, employing Planck's constant h. De Broglie conceived an electron revolving about the atomic nucleus as being connected with such a hypothetical wave train, and made intelligible to some extent the discrete character of Bohr's "permitted" paths by the stationary character of the corresponding waves.

Now in mechanics the motion of material points is determined by the forces or fields of force acting upon them. Hence it was to be expected that those fields of force would also influence de Broglie's wave fields in an analogous way. Erwin Schrödinger showed how this influence was to be taken into account, re-interpreting by an ingenious method certain formulations of classical mechanics. He even succeeded in expanding the wave mechanical theory to a point where, without the introduction of any additional hypotheses, it became applicable to any mechanical system consisting of an arbitrary number of mass points, that is to say, possessing an arbitrary number of degrees of freedom. This was possible because a mechanical system consisting of n mass points is mathematically equivalent, to a considerable degree, to one single mass point moving in a space of $3n$ dimensions.

On the basis of this theory there was obtained a surprisingly good representation of an immense variety of facts which otherwise appeared entirely incomprehensible. But on one point, curiously enough, there was failure: it proved impossible to associate with these Schrödinger waves definite motions of the mass points—and that, after all, had been the original purpose of the whole construction.

The difficulty appeared insurmountable, until it was overcome by Born in a way as simple as it was unexpected. The de Broglie–Schrödinger wave fields were not to be interpreted as a mathematical description of how an event actually takes place in time and space, though, of course, they have reference to such an event. Rather they are a mathematical description of what we can actually know about the system. They serve only to make statistical statements and predictions of the results of all measurements which we can carry out upon the system.

Let me illustrate these general features of quantum mechanics by means of a simple example: we shall consider a mass point kept inside a restricted region G by forces of finite

合物只能辐射和吸收具有某些特定频率的光这一事实。然而所有这些在现存的理论框架内都还无法解释。显然，至少在原子现象的领域里，所发生的每件事情的特征均由分立的状态以及分立的状态之间的不连续的跃迁决定，普朗克常数 h 在这里起着决定性的作用。

进一步的工作是由德布罗意完成的。他向自己提出这样一个问题：如何借助现有的物理概念来理解这些分立的状态呢？他突然想到可以类比驻波，把分立的状态理解为如声学中管风琴和弦的基频一样的情况。虽然这里需要的这样一种波动行为还是未知的；但是用上述普朗克常数 h 应该可以构造出这种波动作用并且写出相应的数学表达式。德布罗意设想电子绕原子核的旋转是同某种有待证实的波列有关的，并且认为通过相应波的驻波特征能在一定程度上理解玻尔的"允许"轨道的分立特征。

既然力学中质点的运动是由作用在质点上的力或力场所决定的，那么我们可以预测，那些力场也以类似的方式影响着德布罗意所谓的波场。欧文·薛定谔向我们展示了如何考虑以上力场对波场的影响，并用巧妙的方法重新解释了经典力学中的某些公式。薛定谔甚至不需要引入任何额外的假设就成功地将波动力学理论拓展至含有任意个质点（换句话说就是具有任意个自由度）的任何力学体系。这是可能的，因为从数学的角度来看，由 n 个质点所组成的力学系统在一定程度上可以看作是在 $3n$ 维空间里运动着的单个质点。

根据以上由薛定谔建立的波动力学的理论，许多用别的理论好像完全无法理解的事实都意外地得到了完美的解释。但奇怪的是，以上理论有一个缺点：它无法将薛定谔波与质点确定的运动联系起来，而毕竟这是当初构造这整个理论的最初目的。

上面提到的困难似乎是不可克服的，直到玻恩用一个意想不到的简单的方法解决了该困难。德布罗意–薛定谔波场不能被理解为时空中确实发生的事件的数学描述形式，尽管该形式肯定是同这样的事件有关系的。说得恰当些，它们实际上是我们能够从体系中获知的事物的一种数学描述。它们只能用来对这个体系所进行的一切测量结果进行统计上的说明和预测。

让我举个简单的例子来说明量子力学这些普遍的特点：我们考查一个质点，它

strength. If the kinetic energy of the mass point is below a certain limit, then the mass point, according to classical mechanics, can never leave the region *G*. But according to quantum mechanics, the mass point, after a period not immediately predictable, is able to leave the region *G*, in an unpredictable direction, and escape into surrounding space. This case, according to Gamow, is a simplified model of radioactive disintegration.

The quantum theoretical treatment of this case is as follows: at the time t_0 we have a Schrödinger wave system entirely inside *G*. But from the time t_0 onwards, the waves leave the interior of *G* in all directions, in such a way that the amplitude of the outgoing wave is small compared to the initial amplitude of the wave system inside *G*. The farther these outside waves spread, the more the amplitude of the waves inside *G* diminishes, and correspondingly the intensity of the later waves issuing from *G*. Only after infinite time has passed is the wave supply inside *G* exhausted, while the outside wave has spread over an ever-increasing space.

But what has this wave process to do with the first object of our interest, the particle originally enclosed in *G*? To answer this question, we must imagine some arrangement which will permit us to carry out measurements on the particle. For example, let us imagine somewhere in the surrounding space a screen so made that the particle sticks to it on coming into contact with it. Then from the intensity of the waves hitting the screen at some point, we draw conclusions as to the probability of the particle hitting the screen there at that time. As soon as the particle has hit any particular point of the screen, the whole wave field loses all its physical meaning; its only purpose was to make probability predictions as to the place and time of the particle hitting the screen (or, for example, its momentum at the time when it hits the screen).

All other cases are analogous. The aim of the theory is to determine the probability of the results of measurement upon a system at a given time. On the other hand, it makes no attempt to give a mathematical representation of what is actually present or goes on in space and time. On this point the quantum theory of today differs fundamentally from all previous theories of physics, mechanistic as well as field theories. Instead of a model description of actual space-time events, it gives the probability distributions for possible measurements as functions of time.

The new theoretical conception owes its origin not to any flight of fancy but to the compelling force of the facts of experience. All attempts to represent the particle and wave features displayed in the phenomena of light and matter, by direct recourse to a space-time model, have so far ended in failure; and Heisenberg has convincingly shown, from an empirical point of view, that any decision as to a rigorously deterministic structure of Nature is definitely ruled out, because of the atomistic structure of our experimental apparatus. Thus it is probably out of the question that any future knowledge can compel physics again to relinquish our present statistical theoretical foundation in favour of a deterministic one which would deal directly with physical reality. Logically, the problem seems to offer two possibilities, between which we are in principle given a choice. In the

被有限大小的力束缚在一个有限的区域 G 内。如果质点的动能小于某一极限，那么根据经典力学，质点就永远不可能离开 G 这个区域。可是根据量子力学，经过一段无法直接预测的时间之后，质点却可能沿某个不确定的方向离开区域 G 而跑到周围的空间里去。按照伽莫夫的观点，这就是放射性蜕变的一个简化模型。

用量子理论处理以上情况如下所示：在 t_0 时刻，薛定谔波系统完全处于 G 的区域内。但是从时间 t_0 以后，这些波沿着所有可能的方向离开 G 的内部区域，在这个过程中，往外传播的波的振幅要小于 G 区域内波系统最初的振幅。往外传播的波扩散得越远，由 G 发出的处于 G 区域内的波的振幅就越小，相应地，其强度也越小。因而只有经过无限长的时间之后由 G 发出的波才会耗尽，与此同时传播到 G 区域外的波则已持续不断地扩散到了更大的空间。

但是这种波动过程同我们刚才提到的最初的对象，即原来被包围在 G 内的粒子有什么关系呢？要回答这个问题，我们必须设想能对粒子进行测量的某种装置。例如，想象周围空间的某个位置上有这样一个屏幕，粒子一旦与它接触就会被粘住。于是，根据波射到屏上某个点的强度，我们就可断定粒子在那时射到屏上这一点的概率。但是一旦粒子被射到屏上的点都是特定的点时，则整个波场立即失去了所有的物理意义；这种做法唯一的目的就是对粒子射到屏上的位置和时间（又或者如，粒子射到屏上时的动量）的概率做出相应的预测。

所有其他的例子也都是类似的。量子理论的目的就是要确定某一时刻体系测量结果的概率。但是另一方面，它并没有试图对时空中真实存在或正在进行的事物作出相应的数学描述。在这一点上，今天的量子理论同以前所有的物理学理论，不管是力学还是场论，都有本质的区别。量子理论不是就真实时空中的事件进行模型化的描述，而是对可能的测量结果给出概率分布随时间的变化。

必须承认，这个新理论的概念并不是来源于任何异想天开的想法，而是在经验事实的强制下产生的。光和物质在现象中都显示出了粒子性和波动性，所有试图借助时空模型来融合这两种性质的做法到目前为止都以失败告终。而且海森堡也给出了令人信服的观点：从经验的角度来看，由于我们的实验仪器的结构是由原子组成的，所以一定不可能出现任何与自然界严格的决定论理论结构有关的结果。因此，虽然决定论可以直接处理物理实在，但是即使是未来进一步发展的知识也不太可能使物理学放弃现在统计性的理论基础而以决定论取而代之。从逻辑上来看，这个问题似乎给出了两种可能性，原则上我们可以在这两种可能性中做出选择。最终，作

end, the choice will be made according to which kind of description yields the formulation of the simplest foundation, logically speaking. At the present, we are quite without any deterministic theory directly describing the events themselves and in consonance with the facts.

For the time being, we have to admit that we do not possess any general theoretical basis for physics which can be regarded as its logical foundation. The field theory, so far, has failed in the molecular sphere. It is agreed on all sides that the only principle which could serve as the basis of quantum theory would be one that constituted a translation of the field theory into the scheme of quantum statistics. Whether this will actually come about in a satisfactory manner, nobody can venture to say.

Some physicists, among them myself, cannot believe that we must abandon, actually and for ever, the idea of direct representation of physical reality in space and time; or that we must accept the view that events in Nature are analogous to a game of chance. It is open to every man to choose the direction of his striving; and also every man may draw comfort from Lessing's fine saying, that the search for truth is more precious than its possession.

(**145**, 920-924; 1940)

Albert Einstein: For. Mem. R. S., Institute of Advanced Study, Princeton University.

为选择依据的是，所选择的描述方式要尽可能得到逻辑上最简单的物理学基础所对应的公式。目前，还完全没有任何一种决定论性的理论既能直接描述事件本身又能同事实相符合。

目前我们还不得不承认，暂时还没有任何全面的物理学的基本理论可被作为物理学的逻辑基础。到现在为止，场论已经被证明不适用于分子领域。人们从各个方面进行考虑，都认为唯一可能作为量子力学根基的原理应是一种能够把场论和量子统计学体系对应统一起来的理论。至于实际上这个原理能否以一种令人满意的方式出现，现在谁也不敢断言。

有些物理学家，包括我自己在内，都不能相信：我们现在甚至将来都必须永远放弃在时空中直接表示物理实在的想法；或者我们必须接受这样的观点，即自然界中发生的事件就像碰运气的赌博一样。每个人都可以自由选择自己奋斗的方向；而每个人也都可以从莱辛的一句精辟的名言中得到安慰：追求真理比拥有真理更为可贵。

（沈乃澂 翻译；葛墨林 审稿）

Causality or Indeterminism?

H. T. H. Piaggio

Editor's Note

John von Neumann had recently claimed to prove that no theory could go beyond the current quantum theory in giving a causal account of quantum physics. Thus while some physicists, such as Einstein, suspected that quantum theory might only be an approximate theory, statistical in character because it left out details of some deeper level of physical reality, von Neumann claimed that this was impossible. Henry Piaggio here reviews the arguments for and against this claim, concluding that while the balance of evidence seemed to weigh against the possibility of deterministic laws, they could not be ruled out. Many years later, von Neumann's argument would be dismantled by John Bell, who helped revitalize interest in deterministic quantum theories.

A short article published in *Nature* of July 22, 1944[1], entitled "Collapse of Determinism", contained a brief statement of von Neumann's claim to have demonstrated that the results of the quantum theory cannot be obtained by averaging any exact causal laws. If one may judge from the number of communications referring to this point which have been submitted to the Editors, many regard this claim with suspicion and desire a more detailed discussion of the grounds on which it is based. Mr. W. W. Barkas[2] suggested that the existence of statistical regularity when large numbers of events are considered is incompatible with indeterminism, and that if the final result of the behaviour of a million photons were fixed, the behaviour of the first 999,000 must influence the other 1,000. Prof. (now Sir Edmund) Whittaker[2] replied that it might be profitable to consider the behaviour of tossed coins. He asked, in particular, whether the statistical regularity for this case, calculated by the ordinary theory of probability, involves the assumption of "crypto-determinism" (that is, real determinism hidden by lack of detailed information) or merely the assumption of symmetry. This reply produced further letters, too numerous for the Editors to publish in full, and I have been asked to give a connected account of the points raised. I shall start with the experimental evidence concerning coin-tossing, and contrast it with the theoretical discussion. After this I shall touch upon similar considerations for the kinetic theory of gases. Finally, and most important, I shall give some details of von Neumann's supposed disproof of causality, and give the arguments for and against it.

Buffon, the French naturalist, tossed a coin until he obtained 2,048 heads. The results were quoted by De Morgan[3], who gave also an account of three similar experiments by his own pupils or correspondents. The arrangement by which the last toss ended with a head gave a small advantage to heads, but too small to make any significant difference. Much more extensive experiments, on somewhat different lines, were carried out by W. S. Jevons[4], who took "a handful of ten coins, usually shillings", and tossed the ten together. He made two series of 1,024 such tossings of ten coins, so that in each series 10,240 coins were tossed.

因果律还是非决定论？

皮亚焦

编者按

约翰·冯·诺伊曼最近宣称，他将证明在对量子物理给出的因果解释方面没有任何一个理论能够超过当今的量子理论。虽然一些物理学家，例如爱因斯坦，认为量子理论也许仅仅是一个近似理论，具有统计特征，因为它忽视了物理现实中某些更深层次的细节，但冯·诺伊曼却不这样认为。在这篇文章中，亨利·皮亚焦对这两种不同的观点进行了评论，最后得出结论：虽然证据不利于决定论，但也不能将其完全排除。许多年以后，约翰·贝尔又推翻了冯·诺伊曼的观点，这使得人们开始再次关注决定论的量子理论。

发表在 1944 年 7 月 22 日《自然》上的一篇题为《决定论的崩溃》的短文[1]，简单地描述了冯·诺伊曼的观点，即不能通过对一些精确的因果规律取平均来得到量子理论的结果。如果仅从编辑部收到的关注这一观点的信函的数量来看，我们发现很多人都对这一观点表示怀疑，并且希望能就这一观点建立的基础进行更详细的讨论。巴卡斯[2]先生认为，当考虑大量的事件时，统计规律与非决定论是不相符的，他还提到，如果 100 万个光子的最终行为结果是确定的，那么，前 999,000 个光子的行为必定会影响另外 1,000 个光子的行为。惠特克[2]教授（现在是埃德蒙爵士）对此的回应是：我们也许可以分析一下抛掷硬币的行为。他特别问道，对于抛掷硬币来说，根据一般的概率理论计算出来的统计规律是涉及了“隐秘决定论”（即由于缺乏详细信息而隐藏了真正的决定论）的假设，还是仅与对称性假设有关。这一回应又引发了很多评论文章，因其数量太多致使编辑部很难将其全部发表，而且我也被要求针对这些观点给出有关的解释。我将从抛掷硬币的实验结果开始谈起，并将它与理论值进行比较。之后，我也将对气体的动力学理论作相似的分析。最后，也是最为重要的，我将详细解释冯·诺伊曼对于因果律的假设反驳，并且给出支持和反对它的理由。

法国博物学家布丰不停地抛掷一枚硬币，直到他得到 2,048 次正面朝上为止。德·摩尔根[3]引证了这一结果，并介绍了由他的学生或相关人员完成的 3 个与之相似的实验。这种规定以正面朝上为结束的抛掷实验让正面朝上的结果略占优势，但是，这个优势太小以至于可以忽略不计。杰文斯[4]以稍为不同的方式进行了更为广泛的实验，他以“10 个硬币（通常是面额为 1 先令的硬币）作为一组”，并一起抛掷。他做了两个系列的实验，各含 1,024 次这种 10 个一组的抛掷，因此每个系列中

The results of these six experiments were as follows:

No. of heads	2,048	2,048	2,048	2,048	5,222	5,131
No. of tails	1,992	2,044	2,020	2,069	5,018	5,109
Total	4,040	4,092	4,068	4,117	10,240	10,240
Excess of heads over mean	28	2	14	−10.5	102	11
Proportion of heads	0.5069	0.5005	0.5034	0.4974	0.5100	0.5011
Excess over mean	0.0069	0.0005	0.0034	−0.0026	0.0100	0.0011

If we examine these results, we see that it is easy to misinterpret the meaning of "statistical regularity". It is certainly not true, as some correspondents seemed to think, that the numbers of heads and tails are bound to be equal. In fact, the divergence from the mean actually increased from a maximum of 28 in the first four experiments, each based on roughly 4,096 tosses, to a maximum of 102 in the last pair, each based on 10,240 tosses. This is quite in accordance with theory, which, assuming that the probability of a head in one toss is 0.5, deduces that for a large number n of tosses, it is as likely as not that the divergence from the expected mean $n/2$ will exceed $0.3372\sqrt{n}$, but it is almost certain (99.73 percent probability) that it will be less than $1.5\sqrt{n}$. For $n = 4{,}096$ the "as-likely-as-not divergence" is, to the nearest integer, 22, and the "scarcely-ever divergence" is 96. For $n = 10{,}240$, the corresponding numbers are 34 and 152. Thus the actual divergences, though larger and more one-sided than some might have expected, are quite compatible with the theory. But the phrase "statistical regularity" really refers to the *proportion* of heads, which, according to theory, should be very nearly 0.5, with an "as-likely-as-not divergence" of $0.3372\sqrt{n}$ and a "scarcely-ever divergence" of $1.5/\sqrt{n}$. Both these divergences diminish indefinitely as n increases. We may also estimate the theoretical divergence of the proportion by its root-mean-square or "standard deviation". This has the value $0.5/\sqrt{n}$, a result which we shall contrast later with Heisenberg's Principle of Uncertainty.

We now come to an important criticism of the theory of probability on which the above calculations are based. As pointed out by Lieut.-Colonel E. Gold[5], there is an assumption of symmetry, not only in the two faces of the coin, but also in the actions of the hand that tosses the coin. When the hand was replaced by a machine, such as that devised by J. Horzelski[6], the absence of this symmetry was manifest. By a certain adjustment of the pressure actuating a lever, he obtained 98 heads out of 100 tosses. He then slightly altered the pressure, keeping the head, as before, initially upwards on the machine, and obtained only one head in the next 100 tosses. In this case the tossing mechanism is not a hidden parameter, but is visible and definite, whereas in the usual tossing it is indefinite and, so far as we can manage it, symmetrically distributed. It is possible that the excess of heads in Jevons's experiments was due to some slight lack of symmetry in his tossing conditions. Whether this was so or not, it appears obvious that *the description of reality given by the theory of probability in coin-tossing is not complete.*

It is therefore erroneous to suppose that the properties of a perfectly normal distribution must necessarily correspond exactly with physical reality, however useful they may be in

都抛掷了 10,240 个硬币。上述 6 个实验的统计结果如下：

正面朝上次数	2,048	2,048	2,048	2,048	5,222	5,131
反面朝上次数	1,992	2,044	2,020	2,069	5,018	5,109
总次数	4,040	4,092	4,068	4,117	10,240	10,240
正面朝上次数超过平均数的量	28	2	14	–10.5	102	11
正面朝上的比例	0.5069	0.5005	0.5034	0.4974	0.5100	0.5011
正面朝上的比例超过 50% 的量	0.0069	0.0005	0.0034	–0.0026	0.0100	0.0011

如果我们查看这些统计数字，就很容易曲解“统计规律”的含义。有些人认为，正面朝上和反面朝上的次数一定会是相等的，这显然是不正确的。实际上，前 4 个实验都抛掷了约 4,096 次，正面朝上的次数超过平均数的最大值为 28，而最后 2 个实验各抛掷了 10,240 次，其最大值为 102，因此，偏离平均数的差值实际上是在增加的。这一点同理论吻合得非常好，理论认为，在一次抛掷硬币的过程中，正面朝上的概率为 0.5，据此推断，如果抛掷 n（一个很大的数字）次硬币，那么偏离预期平均值 $n/2$ 的差值很可能会超过 $0.3372\sqrt{n}$，但是几乎可以肯定（99.73% 的可能性）该值会小于 $1.5\sqrt{n}$。当 $n = 4,096$ 时，“很可能的偏差”最接近整数 22，而“几乎不可能出现的偏差”是 96。当 $n = 10,240$ 时，相应的值为 34 和 152。因此，尽管实际的偏差要比一些人预期的大且不均衡，但与这个理论还是非常一致的。但是，根据理论，“统计规律”这个短语真正的意义指的是正面朝上出现的**比例**应该非常接近 0.5，其“很可能的偏差”为 $0.3372\sqrt{n}$，“几乎不可能出现的偏差”为 $1.5/\sqrt{n}$。随着 n 的增加，这两个偏差都会无限地减少。我们也可以根据这个比例的均方根或“标准偏差”来估算其理论偏差。得到的值为 $0.5/\sqrt{n}$，稍后我们会将这一结果与海森堡的测不准原理进行比较。

现在我们来讨论一下上面计算所依据的概率理论的一个重要的不妥之处。正如陆军中校戈尔德[5]指出的，对称性的假设不仅存在于硬币本身的正反两面，而且还存在于抛掷硬币时手的动作。当用机器取代手时，比如霍莱斯基[6]设计的机器，实验结果就出现了明显的不对称性。通过对驱动杠杆的压力进行适当调节，在抛掷 100 次硬币后，他得到了 98 次正面朝上的结果。接着，他轻微地改变了压力，初始时硬币在机器上同样保持正面朝上，但在接下来的 100 次抛掷中，只得到了一次正面朝上的结果。在这种情况下，抛掷机制不是一个隐藏的参数，而是可见并且确定的参数，而在一般的抛掷情况下，这个参数是不确定的，而且在我们所能控制的范围内，它就是对称分布的。在杰文斯的实验中，正面朝上的情况太多很可能是因为抛掷情况缺乏对称性。不管情况是否如此，很明显，**在抛掷硬币的实验中，概率理论给出的事实描述是不完善的。**

因此，认为完美正态分布的性质必定严格符合物理现实是错误的，不过通常情况下这些性质可以为事实提供一个很好的近似。我们不能仅仅通过宣称抛掷细节搅

giving a good approximation to the facts. We cannot disprove the existence of the details of the projection merely by claiming that they upset the purity of the normal distribution. It is rather the very purity of that distribution which goes beyond the physical facts, and so is not a complete description of reality. Similar considerations apply to the kinetic theory of gases, but in this case the symmetry assumed in the theory[7] is a much closer approximation to the actual facts. But it is only an approximation, and here, as elsewhere in classical physics, pure statistical aggregates do not exist.

This brings us to the question whether such aggregates exist in non-classical physics, in particular in quantum mechanics. We shall examine von Neumann's arguments, using for this purpose not only his well-known treatise "Mathematische Grundlagen der Quantenmechanik" (1932), but also the shorter account, in English, that he gave in Warsaw[8] in 1937, and the discussion that ensued. The starting point is an analysis of the qualitative laws obeyed by the mathematical "hypermaximal projective" operators which correspond to the physical quantities occurring in quantum mechanics. Everything is said to be based on six laws, of which two seem more important than the rest. One is the principle of superposition, extended to quantities not necessarily simultaneously measurable. The other may be called the principle of exact functional correspondence; for example, if an operator represents a physical quantity, then the square of the operator represents the square of the quantity.

In my opinion, however, the emphasis on these simple laws conceals the fact that other conditions of greater importance are imposed by the definition of "hypermaximal projective" operators. This definition implies some characteristic results of quantum mechanics, and the simple laws are merely the final requirements. Von Neumann shows that aggregates are of two kinds, "pure" and "mixed". The essential property of a pure aggregate is that it cannot be regarded as a mixture of two other non-identical aggregates. The qualitative laws of quantum mechanics show that the aggregates concerned must be pure, whereas all aggregates based upon causal laws, such as tossed coins or gas particles, must be mixed. Hence, he concludes, causality is incompatible with quantum mechanics, and the process of averaging causal laws, as applied in the kinetic theory of gases, cannot possibly be extended to quantum mechanics. There is no need, he says, to go more deeply into the details of a supposed system which is governed by further conditions ("hidden parameters") in addition to the wave functions. These hidden parameters would upset the qualitative laws of quantum mechanics. He admits that quantum mechanics in its present form is certainly defective, and, in spite of its great success in explaining physical phenomena, may possibly, in the long run, turn out to be false. But this is true of every theory; we can never say that it is proved by experiment, but only that it is the best summing up of experiment at present known.

Von Neumann therefore concludes that there is at present no reason and no excuse for supposing the existence of causality in quantum mechanics. This conclusion is described by Whittaker[9] as not only novel and unexpected, but also almost incredible, yet he endorses it with the exultant declaration "the bonds of necessity have been broken; for

乱了正态分布的纯正性，从而否认这些细节的存在。事实上，正是因为分布太过纯正，超出了物理的现实，因此，它才不是对现实的完整描述。同样的道理也适用于气体的动力学理论,但在这种情况下,该理论中[7]的对称性假设更接近于实际的事实。但仅仅只是接近而已，就像在经典物理学的其他地方一样，在这里纯粹的统计集合是不存在的。

这使我们遇到这样一个问题，即这样的集合是否也存在于非经典物理学中，特别是量子力学之中呢？为了得到解答，我们将查证冯·诺伊曼的观点，其中不仅涉及他的著作《量子力学的数学基础》（1932 年），还涉及他于 1937 年在华沙[8]用英文写的简短解释和对此所作的进一步的讨论。其起始点是对数学的“超大投影”算符所遵循的定性法则的分析，该算符与量子力学中出现的物理量相对应。一切都基于六条法则，其中有两条似乎尤为重要。一条是叠加原理，它适用于不必同时测量的物理量。另一条是严格的函数对应性原理；比如说，如果一个算符代表一个物理量，那么算符的平方就代表这个物理量的平方。

然而在我看来，过分注重这些简单的法则使我们忽视了通过“超大投影”算符的定义所带来的其他一些更为重要的条件。这个定义暗含了量子力学的一些特定结果，并且这些简单的法则仅仅是最后的必要条件。冯·诺伊曼认为集合分为“纯集合”和“混合集合”两种。纯集合的主要性质是它不能被表示为两个不同集合的混合。量子力学的定性法则表明与其有关的集合必然是纯集合，而如抛掷硬币或气体粒子等所有服从因果论的集合都必然是混合的。因此他得出结论：因果论与量子力学是不兼容的，并且在气体动力学理论中用到的平均因果论的方法不能应用于量子力学。他认为，对于一个受波函数控制以外还受其他条件（“隐参量”）控制的设定系统，没有必要去深究细节。这些隐参量将会打乱量子力学的定性法则。他承认当时的量子力学的确是有缺陷的，尽管它在解释物理现象方面取得了巨大的成功，但以后仍然有可能被证明是错误的。不过，每一个理论都是这样的；我们永远不可能声称用实验证明了它，只能说目前这是对实验的最好总结。

因此，冯·诺伊曼得出结论：目前在量子力学中没有任何理由能够推测出因果律的存在。惠特克[9]认为这个结论不仅是新奇且出人意料的，而且几乎是令人难以置信的，他还用兴奋的语言对其表示认同，“必然性的联系已被打破；对于某几类现

certain classes of phenomena, crypto-determinism is definitely disproved".

Other comments have been more sceptical. At the Warsaw conference, the president, C. Bialobrzeski, after hearing von Neumann, admitted the validity of the argument that it was impossible to fit causality into the framework of quantum mechanics, but expressed a doubt as to the logical coherence of that framework. In his opinion it is deficient because it does not take account of irreversible changes, and also because, in certain conditions of measurement, the indeterminism of the final state disappears, and the assumptions of discontinuity and indeterminism do not correspond to reality. He thought it necessary to introduce a new postulate concerning measurement. At a later meeting of the same conference a letter from Heisenberg said that the quantum theory, in its present form, could not yet give a logically coherent account of nuclear physics or of cosmic rays.

Many critics are suspicious of purely abstract arguments which make no reference to experiment. Of course, such experiments as those of Davisson and Germer on electron diffraction and of Condon and Gurney on radioactivity, though excellent as illustrations of the Uncertainty Principle, yet have no value in deciding whether this uncertainty is due only to lack of detailed knowledge, or to true indeterminism. On a somewhat different plane is the argument of Whittaker[9], who, though a supporter of von Neumann, illustrates his argument by a reference to the passage of plane-polarized light passing through a Nicol prism, and shows that the phenomena cannot be explained by causal laws governing any hidden parameters attributed to the photons. However, H. Pelzer[10] gives two models in which hidden parameters, attributed at least partly to the Nicol prism, can exist and obey causal laws. From this he infers that the arguments of Whittaker and von Neumann are incomplete, even though he agrees with their conclusion that quantum phenomena are truly indeterminate.

My own criticism of von Neumann is founded upon a paper by A. Einstein, B. Podolsky and N. Rosen[11]. By considering the problem of making predictions concerning a system on the basis of measurements made on another system which had previously interacted with it, they conclude that *the description of reality as given by a wave function is not complete*. As a wave function is a mathematical way of representing a probability distribution, this conclusion is almost exactly the same as that which I enunciated concerning coin-tossing. I therefore, with great diffidence, offer the opinion that the existence of causality has *not* been disproved. It is true that Einstein's opinion has been rejected by N. Bohr[12], but there are other grounds for supporting it. In fact, the postulate of quantum mechanics that electrons cannot be distinguished from one another appears, at least to me, not to be a statement that Nature is incomprehensible, but merely that quantum mechanics is incomplete. However, I do not wish to insist that there is no difference between coin-tossing and quantum mechanics. One striking difference is that in coin-tossing the standard deviation of the proportion of heads depends only upon the number of tosses, and can be diminished indefinitely; but in quantum mechanics the Principle of Uncertainty gives for the product of the standard deviations of the momentum and displacement a minimum value, namely, $h/4\pi$. The occurrence of Planck's constant in this result seems to show

象，隐秘决定论将完全被推翻”。

其他评论则对此持有更多的怀疑。在华沙会议上，会议主席比亚洛布尔泽斯基在听取了冯·诺伊曼的演讲后，承认现在确实不可能将因果律纳入量子力学的理论框架中，但他对这一理论框架的逻辑连贯性提出了质疑。他认为，由于没有考虑到不可逆变化，也由于在一定的测量条件下终态的不确定性消失了，而且不连续性和不确定性的假想与现实并不相符，所以这是有缺陷的。他认为有必要根据测量提出一个新的假定。在会议后期的一个会上，海森堡在一封来信中说，当前的量子理论还无法为核物理学或宇宙射线给出一个逻辑一致的解释。

许多评论家都质疑这些没有实验作为依托的完全抽象的理论观点。当然，尽管像戴维孙和革末在电子衍射方面的实验以及康登和格尼在放射性方面的实验都对测不准原理进行了完美的阐释，但是人们依旧无法说明这个不确定性究竟只是因为缺乏具体知识还是确实无法确定。惠特克[9]从一个稍微不同的角度阐述了自己的观点，惠特克是冯·诺伊曼的支持者，但是他通过引用一篇平面偏振光通过尼科尔棱镜的文章来表明，这种现象不能用控制着由光子引起的一些隐参量的因果律来解释。但是佩尔泽[10]给出了两个模型，其中的隐参量能够存在并遵循因果律，而这些隐参量至少部分来源于尼科尔棱镜。从这点他得出推断：惠特克和冯·诺伊曼的观点是不完整的，尽管他也认同他们关于量子现象确实不可确定的结论。

我对冯·诺伊曼的批判基于一篇由爱因斯坦、波多尔斯基和罗森[11]共同撰写的论文。他们基于对曾经与该系统发生过相互作用的另一系统的测量来考虑对该系统进行预测的问题，他们得出的结论是：**仅仅用波函数来描述现实世界是不完整的**。因为波函数只是描述概率分布的一种数学方法，这个结论和我对于抛掷硬币实验所得的结论几乎完全一样。因此我非常不自信地提出：因果律的存在还**没有**完全被否认。爱因斯坦的观点确实已经被玻尔[12]推翻，但还是存在一些其他依据来支持它。事实上，量子力学中关于电子与电子间无法区分的假设出现了，至少对我来说，这并不表明大自然是不可理解的，它仅仅说明量子力学是不完整的。但我并非想说量子力学与抛掷硬币没有任何区别。两者之间一个很明显的不同是在抛掷硬币中正面朝上的比例的标准偏差仅取决于抛掷的次数，并且可以无限减小；但是在量子力学中测不准原理给出了动量与位移的标准偏差的乘积的最小值，即 $h/4\pi$。这个结果中普朗克常数的出现似乎说明有一些新的东西尚待挖掘。我觉得如果冯·诺伊曼的结论

that there is something essentially new. I should find it easier to accept von Neumann's conclusions if his arguments, instead of being purely qualitative, contained this constant. Perhaps it is really concealed somewhere in the background, like a hidden parameter!

To conclude, I will quote the opinion expressed by Bertrand Russell[13] in 1936, that at present there is no decisive reason in favour of complete determinism (causality) in physics, but that there is no reason against it, and that it is theoretically impossible that there should be any such reason. But Russell does not mention von Neumann's arguments. My own conclusion is that the balance of the present evidence is rather against complete causality, but that the question is still unsettled.

(**155**, 289-290; 1945)

H. T. H. Piggio: University College, Nottingham.

References:

1. *Nature.* **154**, 122 (1944).
2. *Nature,* **154**, 676 (1944).
3. "Budget of Paradoxes", 170 (1872).
4. "Principles of Science", 238 (1874); or 2nd ed., 208 (1877).
5. *Nature,* **155**, 111 (1945).
6. *Nature,* **155**, 111 (1945).
7. Preston, "Theory of Heat", 4th ed., 782 (1929).
8. "New Theories in Physics", 30-45.
9. *Proc. Phys. Soc.*, **55**, 459 (1943).
10. *Proc. Phys. Soc.*, **53**, 195 (1944).
11. *Phys. Rev.*, **47**, 777 (1935).
12. *Phys. Rev.*, **48**, 696 (1935).
13. *Proc. Univ. of Durham Phil.* Soc., **9**, 228 (1936).

中也包含这个常数，而不仅仅是一些完全定性的分析的话，我们将更容易接受他的结论。当然，也许在其背景理论中也存在像隐参量一样的某个常数！

综上所述，我将引用伯特兰·罗素[13]在1936年表述的观点：目前尚没有强有力的理由支持物理学中的完全决定论（因果律），但也没有任何理由反对它，并且理论上也不会存在这种理由。但是罗素没有提到冯·诺伊曼的观点。我的结论是现在的证据不利于完全的因果律，但是这个问题依旧没有被彻底解决。

（刘霞 翻译；赵见高 审稿）

Processes Involving Charged Mesons

C. M. G. Lattes *et al.*

Editor's Note

César Lattes and colleagues from the University of Bristol here report recent results from cosmic-ray experiments suggesting the existence of hitherto unknown particles. They had found that some particles having masses small compared to the proton could cause nuclear disintegrations involving the emission of several heavy particles. They referred to the particles as mesons, this being the accepted term for any particle of mass intermediate to the electron and proton. Lattes and colleagues detected the particles from their unusual tracks on photographic plates, showing occasional decays of these mesons into secondary mesons. The primary mesons were later identified as the positive and negative charged pi mesons (pions), which would be created in the laboratory in a few years.

IN recent investigations with the photographic method[1,2], it has been shown that slow charged particles of small mass, present as a component of the cosmic radiation at high altitudes, can enter nuclei and produce disintegrations with the emission of heavy particles. It is convenient to apply the term "meson" to any particle with a mass intermediate between that of a proton and an electron. In continuing our experiments we have found evidence of mesons which, at the end of their range, produce secondary mesons. We have also observed transmutations in which slow mesons are ejected from disintegrating nuclei. Several features of these processes remain to be elucidated, but we present the following account of the experiments because the results appear to bear closely on the important problem of developing a satisfactory meson theory of nuclear forces.

In identifying the tracks of mesons we employ the method of grain-counting. The method allows us, in principle[3], to determine the mass of a particle which comes to the end of its range in the emulsion, provided that we are correct in assuming that its charge is of magnitude $|e|$. We define the "grain-density" in a track as the number of grains per unit length of the trajectory. Knowing the range-energy curve for the emulsion[4], we can make observations on the tracks of fast protons to determine a calibration curve showing the relation between the grain-density in a track and the rate of loss of energy of the particle producing it. With this curve, the observed distribution of grains along the track of a meson allows us to deduce the total loss of energy of the particle in the emulsion. The energy taken in conjunction with the observed range of the particle then gives a measure of its mass.

We have found that the above method gives satisfactory results when, in test experiments, it is applied to the determination of the mass of protons by observations on plates

涉及带电介子的过程

拉特斯等

编者按

布里斯托尔大学的塞萨尔·拉特斯及其同事们在这篇文章中报告说，近期宇宙射线的实验结果表明存在迄今未知的粒子。他们发现一些质量比质子小的粒子可以引起核嬗变，并且在此过程中会发射出一些重粒子，他们称其为介子。介子这一专有名词是指任意一个质量介于电子和质子之间的粒子，这已得到公认。拉特斯及其同事们在照相底片上不寻常的粒子径迹中观测到了介子，同时这些照相底片还显示出这些介子有时会衰变为次级介子。随后人们将初级介子确定为正负 π 介子，几年后在实验室中生成了这些粒子。

近期，我们利用照相法进行的研究[1,2]显示，作为高纬度处宇宙辐射的组分，小质量的带电慢粒子可以进入核并引起核嬗变同时发射出重粒子。对于任何质量介于质子和电子之间的粒子，用“介子”一词来表示都是合适的。通过连续的实验我们发现了有关介子的实验证据，它们在射程末端产生了次级介子。与此同时，我们还观测到了嬗变的核发射出慢介子的嬗变过程。这些过程的某些特性虽然还有待阐明，但是由于其实验结果似乎与发展一套令人满意的核力介子理论这样的重大问题有密切关系，我们将介绍以下的实验内容。

在鉴别介子的径迹时我们应用了颗粒计数法。假设一种粒子的电荷量为 $|e|$，如果我们的假设是正确的，那么当该粒子的射程末端终止于乳胶内时，颗粒计数法从原理上[3]允许我们测定该粒子的质量。我们定义径迹的“颗粒密度”为单位径迹长度中的颗粒数。知道了对于乳胶的射程–能量曲线[4]，我们就可以观测快质子的径迹，以测定描述径迹的颗粒密度与粒子能量损失率间关系的标定曲线。利用这条曲线，所观测到的沿着介子径迹的颗粒分布能够使我们推导出乳胶中粒子的总能量损失。利用粒子被带走的能量以及观测到的粒子射程可以对粒子的质量进行测量。

在测试实验中，对受粒子辐射后立即显影的照片进行观测，并用上述方法对质子质量进行测定，我们发现上述方法给出了令人满意的结果。各条径迹颗粒计数观

developed immediately after exposure. The errors in the observed values, based on grain-counts along individual tracks, are only a little greater than those corresponding to the statistical fluctuations associated with the finite number of grains in a track. As we have previously emphasized, however, serious errors arise when the method is applied to the plates exposed for several weeks to the cosmic rays[2]. These errors are due mainly to the fading of the latent image in the time elapsing between the passage of the particle and the development of the plate.

We have attempted to allow for fading by determining a calibration curve for each individual plate by grain-counts on the tracks of a number of protons, chosen at random from those originating in "stars". Such a calibration curve corresponds to an average value of the fading of the tracks in the plate. While we thus obtain improved mean values for the mass of particles of the same type, as shown by test measurements on the tracks of protons other than those used in making the calibration, the individual values are subject to wide variations. In no case, however, have mass determinations by grain-counts of particles, judged to be protons from the frequency of the small-angle scattering, given values exceeding 2,400 m_e or less than 1,300 m_e.

In these circumstances it is not possible to place serious reliance on the masses of individual mesons determined by grain-counts; and we employ the method, in the present experiments, only to distinguish the track of a meson from that of a proton. In searching a plate, an experienced observer quickly learns to recognize the track of a meson by inspection, provided that its range in the emulsion exceeds 100 μ. Nevertheless, we regard it as established that a particular track was produced by a meson only if both the grain-density and the frequency of the Coulomb scattering correspond to the values characteristic of a particle of small mass. We have considered the possibility that as a result of a rare combination of circumstances we might, in spite of the above precautions, wrongly attribute the track of a proton to a meson of mass less than 400 m_e. It is difficult to give a numerical estimate of the probability of making such an error, but we believe it to be very small.

Secondary Mesons

We have now made an analysis of the tracks of sixty-five mesons which come to the end of their range in the emulsion. Of these, forty show no evidence for the production of a secondary particle. The remaining twenty-five lead to the production of secondary particles. Fifteen of them produce disintegrations with the emission of two or more heavy particles, and from each of the remaining ten we observe a single secondary particle. Of these latter events, the secondary particle is in four cases a hydrogen or heavier nucleus; in four other cases the identification is uncertain, and in the last two cases it is a second meson.

Fig. 1 is a reproduction of a mosaic of photomicrographs which shows that a particle, m_1, has come to the end of its range in the emulsion. The frequent points of scattering and

测值的误差，只是略大于一条径迹中有限的颗粒数的统计波动对应的误差。然而，正如我们先前所强调的那样，在将这种方法用于被宇宙射线[2]辐射数周的照片时，会产生严重的误差。这些误差主要是由于在粒子穿过照片和照片显影之间的这段时间内发生的潜影衰退。

通过随机地选取来自“星”状的质子，对若干条质子径迹进行颗粒计数，来测定每张照片的标定曲线，我们试图以此顾及潜影衰退。这样一个标定曲线对应于照片中径迹衰退的一个平均值。于是，我们得到改进后同类粒子质量的平均值，对标定曲线的测定中没有用到的那些质子的径迹进行的测试性测量显示，单条径迹的颗粒计数测量值变化很大。不过，根据小角度散射的发生频率而判定其为质子的粒子，由其颗粒计数确定的粒子质量都不会超过 2,400 m_e 或小于 1,300 m_e。

在这些情况下，不可能太认真地相信由颗粒计数测定的单个介子质量；因为在目前的实验中我们采用这种方法，仅是为了把介子与质子的径迹区分开来。在观察一张照片时，如果乳胶中粒子的射程超过 100 微米，有经验的观测者通过观察很快就能学会识别出介子的径迹。然而，只有在颗粒密度和库仑散射的发生频率都与小质量粒子的特征值相对应的情形下，我们才确定这一特别的径迹是由介子产生的。尽管采取了上述的防范措施，我们仍然考虑了由于罕见的综合环境条件而将质子径迹误认为是质量小于 400 m_e 的介子径迹的可能性。虽然很难给出发生这类误差的概率估计值，但我们相信这个数值是很小的。

次级介子

现在，我们已经对 65 个介子（射程末端落在乳胶之内）的径迹做了分析，其中 40 个介子没有产生次级粒子的迹象。剩余的 25 个介子则产生了次级粒子，其中的 15 个介子发生嬗变并放射出两个或更多个重粒子，我们观测到剩余的 10 个介子中每个介子都产生一个单独的次级粒子。而在后面提到的 10 个事例当中，经我们研究发现其中有 4 个事例的次级粒子是氢或是更重的核；另 4 个事例产生的次级粒子的种类还不能确定，最后 2 个事例产生的则是次级介子。

图 1 是显微相片中的图像的再现，显示了粒子 m_1 在乳胶中到达了射程的末端。由靠近射程末端处散射点的频繁出现和颗粒密度的快速变化，我们推算出该径迹是

the rapid change of grain-density towards the end of the range show that the track was produced by a meson. It will be seen from the figure that the track of a second particle, m_2, starts from the point where the first one ends, and that the second track also has all the characteristics of that of a particle of small mass. A similar event is shown in Fig. 2. In each case the chance that the observation corresponds to a chance juxtaposition of two tracks from unrelated events is less than 1 in 10^9.

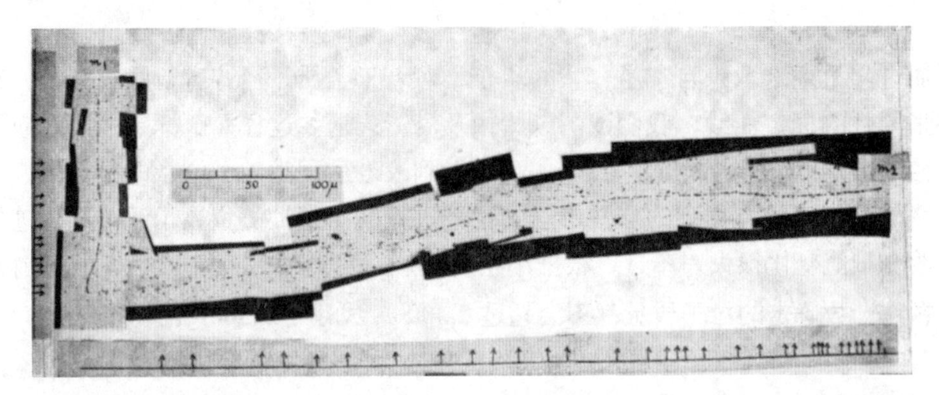

Fig. 1. Observation by Mrs. I. Roberts. Photomicrograph with Cooke × 45 "fluorite" objective. Ilford "Nuclear Research", boron-loaded *C*2 emulsion. m_1 is the primary and m_2 the secondary meson. The arrows, in this and the following photographs, indicate points where changes in direction greater than 2° occur, as observed under the microscope. All the photographs are completely unretouched

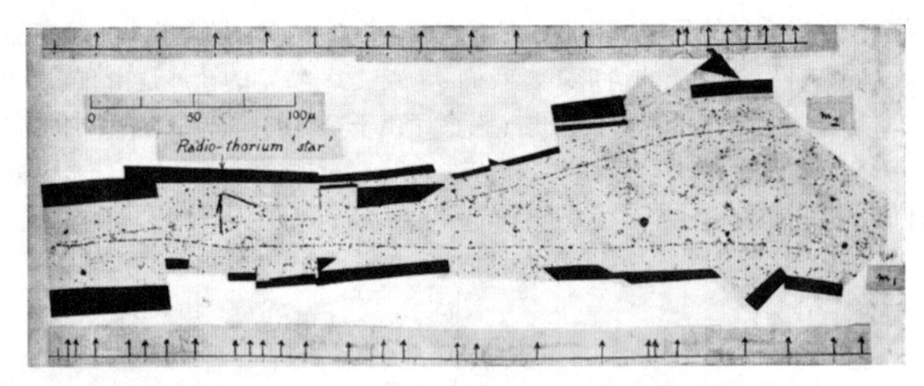

Fig. 2. Observation by Miss M. Kurz. Cooke × 45 "fluorite" objective. Ilford "Nuclear Research" emulsion, type *C*2, boron-loaded. The secondary meson, m_2, leaves the emulsion

Grain-counts indicate that the masses of the primary particles in Figs. 1 and 2 are 350±80 and 330±50 m_e, respectively; and of the secondary particle in Fig. 1, 330±50 m_e, the limits of error corresponding only to the standard deviations associated with the finite numbers of grains in the different tracks. All these values are deduced from calibration curves corresponding to an average value of the fading in the plate, and they will be too high if the track was produced late in the exposure, and too low if early. We may assume, however, that the two-component tracks in each event were produced in quick succession and were therefore subject to the same degree of fading. In these circumstances the measurements indicate that if there is a difference in mass between a primary and a secondary meson, it is

由介子产生的。如图 1 所示，一个次级粒子 m_2 的径迹起始于第一个粒子 m_1 的终点，并且它的径迹也具有小质量粒子径迹的所有特征。图 2 显示了一个类似的事例。在这两张图的情形中，我们所观测到的图像是由两个不相关的事例产生的两条径迹造成的偶然毗邻的可能性不到十亿分之一。

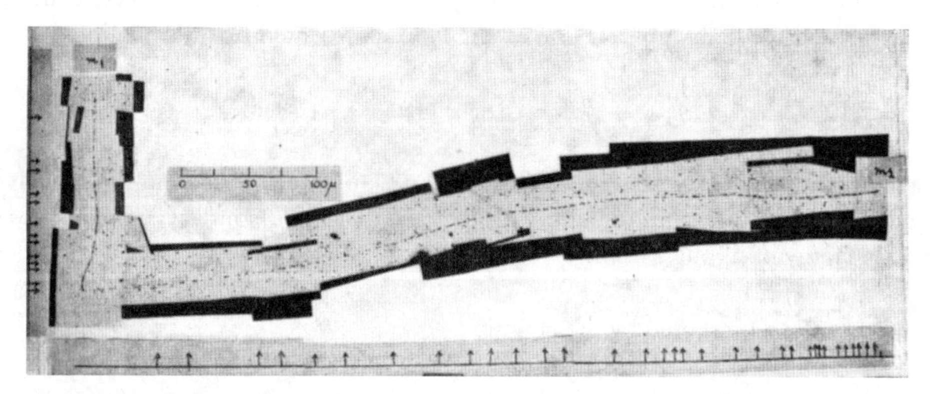

图 1. 由罗伯茨夫人观测。用库克 45 倍“萤石”物镜得到的显微照片。伊尔福载硼 C2“核”乳胶。m_1 是初级介子，m_2 是次级介子。这张和以下各张照片中的箭头，指示出显微镜观测下方向变化大于 2°的点，所有的相片全部未经修改。

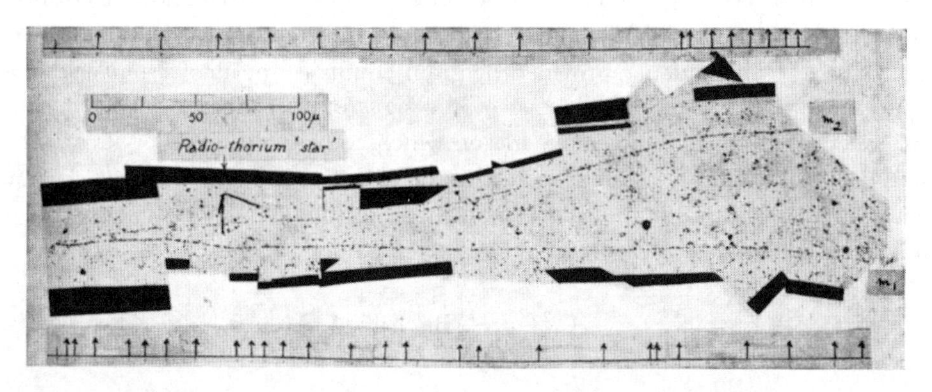

图 2. 由库尔茨小姐观测。库克 45 倍“萤石”物镜。伊尔福载硼 C2“核”乳胶。次级介子 m_2 射离了乳胶。

颗粒计数表明，图 1 和图 2 中初始粒子的质量分别为 350±80 m_e 和 330±50 m_e；图 1 中次级粒子的质量为 330±50 m_e，误差限仅相应于与不同径迹中有限数量的颗粒关联的标准偏差。所有这些值都是从照片中衰退的平均值所对应的标定曲线导出的，如果径迹在照射后期产生，该数值将偏高；而如果该径迹在照射早期产生，此数值就会太低。然而，我们可以假定，每个事例中的双组元径迹是迅速接连产生的，因此衰退程度相同。在这类情况下，测量表明，如果初级介子与次级介子之间存在质量差，其量级不可能大于 100 m_e。图 2 提供的证据并不充分，因为次级粒子穿出

unlikely that it is of magnitude greater than 100 m_e. The evidence provided by Fig. 2 is not so complete because the secondary particle passes out of the emulsion, but the variation in the grain density in the track indicates that it was then near the end of its range. We conclude that the secondary mesons were ejected with nearly equal energy.

We have attempted to interpret these two events in terms of an interaction of the primary meson with a nucleus in the emulsion which leads to the ejection of a second meson of the same mass as the first. Any reaction of the type represented by the equations

$$A_Z^N+\mu_{-1}^0 \rightarrow B_{Z-2}^N+\mu_{+1}^0 \text{ or } A_Z^N+\mu_{+1}^0 \rightarrow C_{Z+2}^N+\mu_{-1}^0, \quad (1)$$

in which A represents any stable nucleus known to be present in the emulsion, involves an absorption of energy, in contradiction with the fact that the secondary meson is observed to have an energy of about 2 MeV.

A second process, represented by the equation

$$\mathrm{Ag}_7+\mu_{-1}^0 \rightarrow X_Z+Y_{45-Z}+\mu_{+1}^0, \quad (2)$$

in which X and Y represent two nuclei of approximately equal charge number, may be energetically possible, but the chance of it occurring in conditions where the total energy of the two recoiling nuclei is of the order of only a few million electron-volts is remote. It is therefore possible that our photographs indicate the existence of mesons of different mass[5,6,7]. The evidence provided by grain counts is not inconsistent with such an assumption. We have no direct evidence of the signs of the charges carried by the two mesons, except that the one secondary meson which comes to the end of its range in the emulsion does not lead to a disintegration with the emission of heavy particles. If, however, we assume that the transmutation corresponds to the interaction of the primary meson with a light nucleus, of a type represented by the equation

$$\mathrm{C}_6^{12}+\mu_{-1}^0 \rightarrow \mathrm{Be}_4^{12}+\mu_{+1}^0, \quad (3)$$

the difference in mass of the two mesons must be of the order of 60 m_e, according to estimates of the mass of the beryllium nucleus.

The only meson theory, to our knowledge, which assumes the existence of mesons of different mass is that of Schwinger[8]. It is visualized[9] that a negative vector meson should have a very short life and should lead to the production of a pseudo-scalar meson of the same charge but lower mass, together with a quantum of radiation. It will therefore be of great interest to determine whether the secondary meson, in transmutations of the type we have observed, is always emitted with the same energy. If this is so, we must assume that we are dealing with a more fundamental type of process than one involving particular nuclei such as is represented in equation (3). If, as an example of such a process, we assume that the momentum of the secondary meson appearing in our experiments is equal and opposite to that of an emitted

了乳胶，但径迹颗粒密度的变化表明，它接近射程末端。由此我们得出结论，次级介子是以几乎相等的能量发射的。

我们试图用初级介子与乳胶中的核相互作用来解释这两个事例，由这种相互作用导致了一个与初级介子质量相同的次级介子的发射。方程（1）表示的任何类型的反应都涉及能量的吸收，

$$A_Z^N+\mu_{-1}^0 \to B_{Z-2}^N+\mu_{+1}^0 \text{ 或者 } A_Z^N+\mu_{+1}^0 \to C_{Z+2}^N+\mu_{-1}^0 \tag{1}$$

式中 A 表示乳胶中存在的已知的任何稳定核，但这与观测到的次级介子的能量约为 2 兆电子伏的事实相矛盾。

第二个过程用以下方程表示

$$\mathrm{Ag}_7+\mu_{-1}^0 \to X_Z+Y_{45-Z}+\mu_{+1}^0 \tag{2}$$

式中 X 和 Y 表示电荷数近似相等的两个核，这个反应按能量守恒可能发生，但在两个反冲核的总能量仅为几百万电子伏特量级的条件下发生的可能性很小。因此我们的照片显示的可能是不同质量的介子 [5,6,7] 的存在。由颗粒计数提供的证据与这个猜想不矛盾。除了知道在乳胶中到达射程末端的一个次级介子并不引发伴随重粒子发射的嬗变外，我们还没有两个介子带电荷符号的直接证据。然而，如果我们假设，嬗变对应于初级介子与轻核的相互作用用方程（3）表示，

$$\mathrm{C}_6^{12}+\mu_{-1}^0 \to \mathrm{Be}_4^{12}+\mu_{+1}^0 \tag{3}$$

那么，根据对铍核的质量估计，2 个介子的质量差一定是 60 m_e 的量级。

我们所知道的，认为存在不同质量介子的唯一介子理论是施温格模型 [8]。该模型设想 [9] 一个负矢量介子应具有很短的寿命，并应产生具有相同电荷但质量较小的赝标量介子，同时伴随量子辐射。因此，在我们已观测的嬗变类型中测定次级介子是否总释放相同的能量将具有很大意义。如果事实果真如此，我们必定认为，我们正在处理一个比公式（3）所表示的含有特殊核的过程更为基本的过程。作为这类过程的一个例子，如果我们假定在我们实验中出现的次级介子与发射的光子的动量大

photon, the total release of energy in the transmutation is of the order of 25 MeV.

In recent communications[10,11] very radical conclusions have been drawn from the results of observations on the delayed coincidences produced by positive and negative mesons in interactions with light and heavy nuclei[12,13]. It is assumed that a negative meson, at the end of its range, falls into a *K* orbit around a nucleus. In the case of a heavy nucleus, it is then captured, giving rise to a disintegration with the emission of heavy particles. With a light nucleus, on the other hand, it is regarded as suffering β-decay before being captured, so that, like a positive meson, it can produce a delayed coincidence. The conclusion is drawn that the nuclear forces are smaller by several orders of magnitude than has been assumed hitherto. Since our observations indicate a new mode of decay of mesons, it is possible that they may contribute to the solution of these difficulties.

Emission of Mesons from Nuclei

Fig. 3 shows a mosaic of photomicrographs of a disintegration in which six tracks can be distinguished radiating from a common centre. The letters at the edge of the mosaic indicate whether a particular track passes out of the surface of the emulsion, *s*, into the glass, *g*, or ends in the emulsion, *e*. The grain-density in tracks *a* and *c* indicate that the time between the occurrence of the disintegration and the development of the plate was sufficiently short to avoid serious fading of the latent image.

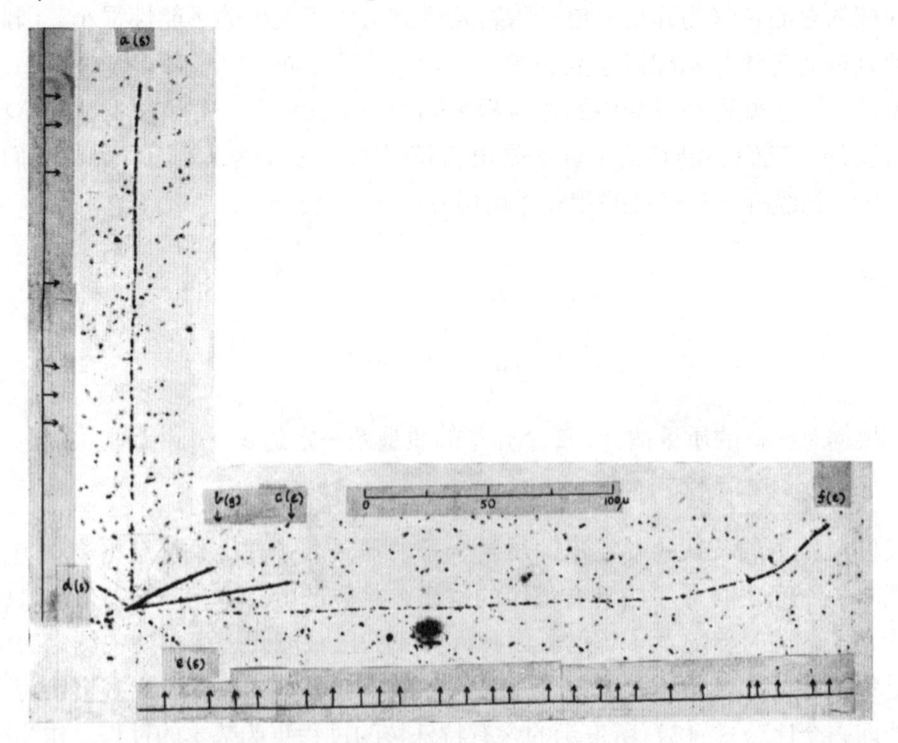

Fig. 3. Observation by Mrs. I. Roberts. Photomicrograph with Cooke × 45 "fluorite" objective. Ilford "Nuclear Research", boron-loaded *C*2 emulsion. The track (*b*) dips steeply and its apparent grain density is greater than the true value through foreshortening. Both (*b*) and (*c*) were probably produced by α-particles

小相等而方向相反，则在嬗变中释放的总能量是25兆电子伏的量级。

在最近的报道中[10,11]，从对正负介子与轻核和重核的相互作用[12,13]产生的延迟符合的观测结果中已得出基本性结论。该结论认为，一个负介子在其射程终端落入了核的K轨道中。在重核情况下，负介子会被俘获而导致嬗变及发射重粒子。另一方面，在轻核情况下，负介子被认为在被俘获前先发生β衰变，因此像一个正介子一样，可以产生一个延迟符合。由此得出的结论如下，核力要比到目前为止认为的数值小几个数量级。鉴于我们的观测表明了介子衰变的一种新模式，其可能有助于解决这些困难。

由核发射的介子

图3显示了一次嬗变的显微照相图像，其中可以分辨出从同一中心辐射出的6条径迹。图像边缘的字母表示特定的径迹是否穿透乳胶表面*s*，进入玻璃*g*，或终止于乳胶*e*中。径迹*a*和*c*的颗粒密度表明，嬗变的发生与照片显影之间的时间间隔足够短，可以避免严重的潜影衰退。

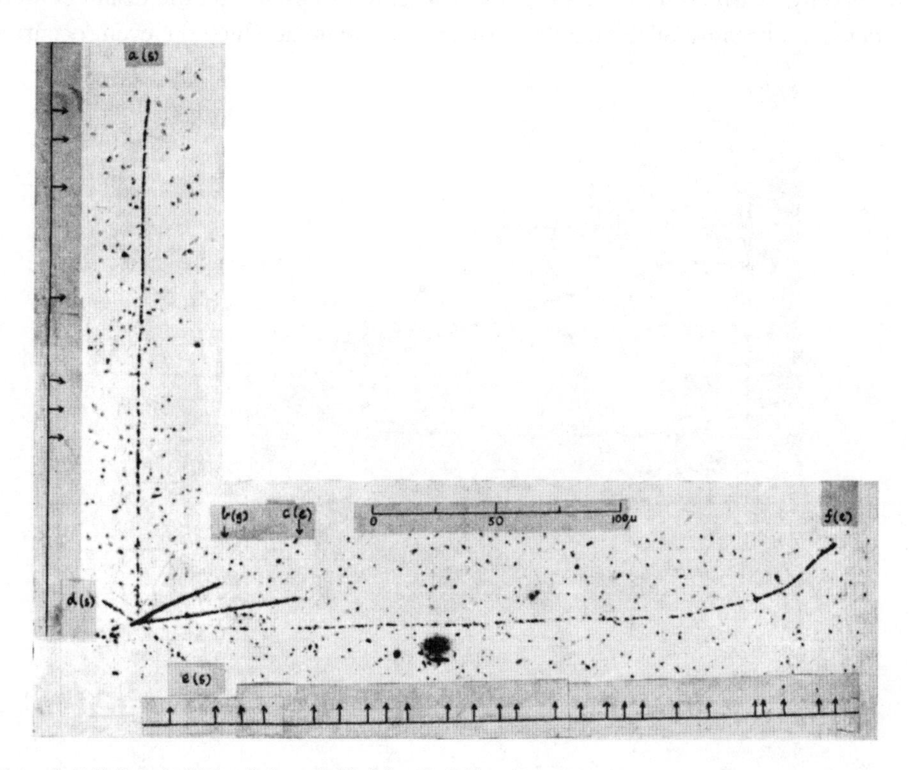

图3. 由罗伯茨夫人观测。库克45倍的“萤石”物镜下的显微相片。伊尔福载硼C2“核”乳胶。径迹(*b*)陡峭地下沉，其表观颗粒密度大于用透视缩小法得到的真值。(*b*)和(*c*)可能都是由α粒子产生的。

The track marked *f* suffers frequent changes in direction due to scattering, and there is a very rapid change in the grain-density in moving along the trajectory. These two features, taken together, make it certain that the track was produced by a light particle, and grain counts give an estimate for the mass of 375±70 m_e[14].

We have now observed a total of 1,600 disintegration "stars", in each of which three or more charged particles are ejected from a nucleus. Of these, 170 correspond to the liberation of an amount of energy equal to, or greater than, that in the "star" represented in Fig. 4, but only in two cases can we identify an emitted particle as a meson. We cannot conclude, however, that the emission of mesons in such disintegrations is so rare as these figures suggest. If a meson is emitted with an energy greater than 5 MeV., it is likely to escape detection in the conditions of our experiments. Mr. D. H. Perkins, of the Imperial College of Science and Technology, has shown that, in the B_1 emulsion, the grain-density in the track of a meson becomes very small at energies greater than 2 MeV., and we must anticipate a similar result in the C_2 emulsion at higher energies. Our observations are therefore not inconsistent with the view that the ejection of mesons is a common feature of the disintegration of nuclei by primary particles of great energy, and that the present instance, in which the velocity of ejection has been exceptionally low so that an identification of the particle has been possible, is a rare example. It is possible that the example of meson production recently described[15] is due to a similar process, produced by a primary particle of higher energy, in which some of the heavier fragments emitted on the disintegration have escaped detection because of the depth inside the lead plate at which the event occurred.

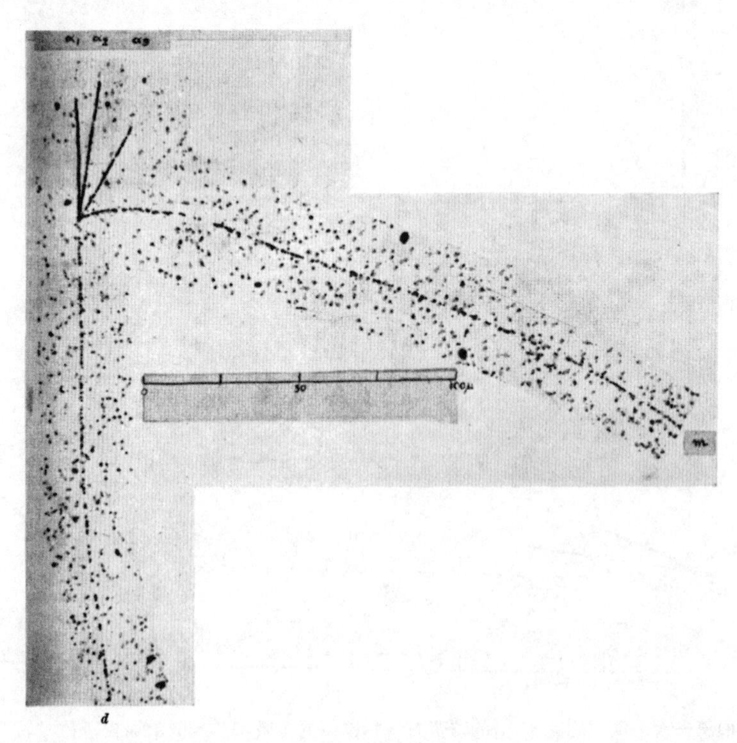

Fig. 4. Observation by Mrs. I. Roberts. Cooke × 95 achromatic objective. Ilford "Nuclear Research" emulsion, type *C*2, lithium-loaded

标记 f 的径迹由于散射而频繁地改变方向，并且在沿着径迹的运动中，颗粒密度也快速变化。这两个特性一起确定了径迹是由轻粒子产生的，颗粒计数给出的质量估计值是 $375 \pm 70\ m_e$[14]。

我们现在已观测到总共 1,600 个“星”状嬗变，每一个都会从一个核中发射出 3 个或更多个的带电粒子。其中的 170 个能量释放大于或等于图 4 表示的“星”状嬗变中的能量释放，但仅在两个嬗变事例中我们可以判定发射粒子是介子。然而，我们并不能认为，这类嬗变中的介子发射是如这些图中所示的如此罕见。如果发射介子的能量大于 5 兆电子伏，在我们的实验条件下似乎很难观测到它。帝国理工学院的珀金斯先生已指出，在 B_1 乳胶中，当能量大于 2 兆电子伏时介子径迹的颗粒密度变得很小，在 C_2 乳胶中更高的能量下我们可以预测到类似的结果。因此，我们的观测与以下观点不矛盾：介子的发射是高能初级粒子核嬗变的共同特性；而目前的事例即粒子的发射速度特别低以至于能够鉴别粒子的种类，是一个罕有的例子。最近描述的介子产生的例子 [15] 是高能初级粒子由类似过程产生的，其中，由于事例发生时铅片的深度，没有观测到嬗变中发射的一些重质量碎片。

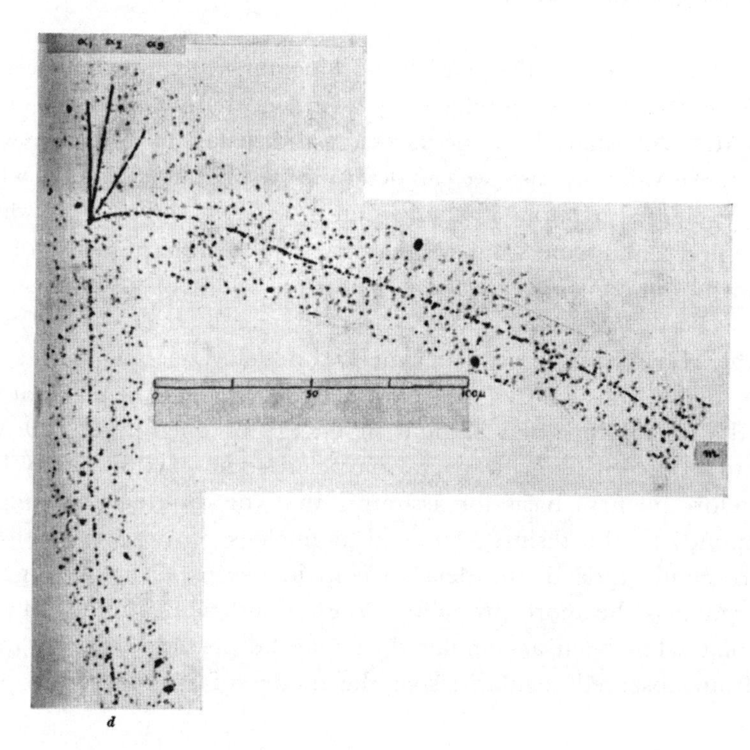

图 4. 由罗伯茨夫人观测。库克 95 倍消色差物镜。伊尔福载锂 C2“核”乳胶。

The disintegration shown in Fig. 3 may be the representative of a type, common in the high atmosphere with particles of great energy. In the present instance the energy of the primary particle must have been of at least 200 MeV., and, if its mass was equal to or less than that of a proton, it would not have been recorded by the emulsion.

Disintegrations Produced by Mesons

The observation of the transmutations of nuclei by charged mesons has led to the suggestion of a method for determining the mass of these particles based on observations of the total energy released in the disintegration[1,2]. In attempting to apply the method, we meet the difficulty of identifying the particular type of nucleus undergoing disintegration and of taking account of any ejected neutrons which will not be recorded by the emulsion. A photograph of such a disintegration which, at first sight, appears to allow us to draw definite conclusions, is shown in Fig. 4. In the photograph, the tracks of four heavy particles can be distinguished, of which the short tracks α_1, α_2 and α_3 end in the emulsion; α_1 and α_2 were certainly produced by α-particles, and grain-counts show that α_3 is due to a proton. The observations are therefore consistent with the equation

$$N_7^{14}+\mu_{-1}^{0} \rightarrow 2He_2^4+H_1^1+H_1^1+4n_0^1; \tag{4}$$

or, less probably, to a similar equation involving the emission of a deuteron or a triton in addition to the particles of short range.

Grain-counts on the track of the particle of long range, *d*, which passes out of the emulsion, indicate that if it was produced by a proton, the initial energy of the particle was about 15 MeV. Alternatively, if the particle was a deuteron, its energy was 30 MeV.; or, if a triton, 45 MeV. In any case, we can determine the minimum energy which must be attributed to the emitted neutrons if momentum is to be conserved in the disintegration. As a result, we find a minimum value for the mass of the primary meson of 240 m_e. The value determined by grain-counts is also 240±50 m_e.

In view of the recent results of experiments on delayed coincidences, referred to previously[12,13], such results must, for the present, be accepted with great reserve. We must expect the liberation of an amount of energy of magnitude 100 MeV. in any nucleus to lead to the ejection of several particles, some of which may be neutrons. There is therefore no firm basis for assuming that the disintegration represented in Fig. 4 corresponds to the disintegration of a nucleus of nitrogen rather than one of silver or bromine. Indeed, the delayed coincidence experiments suggest that the second assumption is the more probable. When a sufficient number of observations with loaded plates has been accumulated, it may be possible to draw more definite conclusions from observed regularities in the modes of disintegration of particular types of nuclei.

图 3 所示的衰变可能是高能粒子在高层大气中一类具有代表性的常见过程。在当前的例子中，初级粒子的能量至少已达到 200 兆电子伏，如果其质量小于或等于质子的质量，将不会在乳胶中被记录到。

介子导致的嬗变

基于对嬗变中释放的总能量的观测，通过对带电介子导致的核嬗变的观测，我们提出了测定这些粒子质量的方法[1,2]。尝试采用这个方法时，在鉴别嬗变核的种类和考虑不能被乳胶记录的发射中子问题上我们遇到了困难。图 4 所示的这类嬗变的一张照片初看起来似乎允许我们做出确定的结论。在照片中，我们可以分辨出四个重粒子的径迹，其中的短径迹 α_1、α_2 和 α_3 终止于乳胶中；α_1 和 α_2 无疑是由 α 粒子产生的，颗粒计数表明，α_3 是由质子产生的。因此，观测与以下方程一致

$$N_7^{14}+\mu_{-1}^0 \rightarrow 2He_2^4+H_1^1+H_1^1+4n_0^1 \tag{4}$$

或者与发射若干短程粒子加上一个氘核或氚核的类似方程一致，但是后者的可能性较小。

穿透乳胶的长程粒子径迹 d 的颗粒计数表明，如果它是由质子产生的，那么粒子的初始能量约为 15 兆电子伏。如果粒子是氘，那么其能量为 30 兆电子伏；或者，如果是氚核，则能量为 45 兆电子伏。无论如何，在嬗变中如果动量守恒，我们可以确定来自发射的中子的最小能量。结果，我们发现初级介子质量的最小值为 240 m_e。颗粒计数测定的值也是 240 ± 50 m_e。

鉴于最近的关于延迟符合的实验结果，并参考以前的结果[12,13]，目前而言，必须以相当保留的态度采纳上述结论。我们可以预测，在任何核中释放出 100 兆电子伏量级的能量将导致若干个粒子的发射，其中有些可能是中子。因此，认为图 4 中所示的衰变是氮核嬗变而不是银或溴的嬗变的假设并没有牢固的依据。实际上，延迟符合的实验表明第二种假设具有更大的可能性。当用乳胶片积累足够数量的观测时，可能会从各类核嬗变中观测到的规律中得出更明确的结论。

A detailed account of the experiments will be published elsewhere.

(**159**, 694-697; 1947)

C. M. G. Lattes, H. Muirhead, G. P. S. Occhialini and C. F. Powell: H. H. Wills Physical Laboratory, University of Bristol.

References:

1. Perkins, D. H., *Nature*, **159**, 126 (1947).
2. Occhialini and Powell, *Nature*, **159**, 186 (1947).
3. Bose and Choudhuri, *Nature*, **148**, 259 (1941); **149**, 302 (1942).
4. Lattes, Fowler and Cuer, *Nature*, **159**, 301 (1947).
5. Hughes, *Phys. Rev.*, **69**, 371 (1946).
6. Leprince-Ringuet and L'Héritier, *J. Phys. et Rad.*, 7, 65 (1946).
7. Bethe, *Phys. Rev.*, **70**, 821 (1947).
8. Schwinger, *Phys. Rev.*, **61**, 387 (1942).
9. Wentzel, *Rev. Mod. Phys.*, **19**, 4 (1947).
10. Fermi, Teller and Weisskopf, *Phys. Rev.*, **71**, 314 (1947).
11. Wheeler, *Phys. Rev.*, **71**, 320 (1947).
12. Conversi, Pancini and Piccioni, *Phys. Rev.*, **71**, 209 (1947).
13. Sigurgeirsson and Yamakawa, *Phys. Rev.*, **71**, 319 (1947).
14. Zhdanov, *Akad. Nauk. Odtel. Bull. Ser. Phys.*, **3**, 734 (1938); **4**, 272 (1939). *C.R.* (*U.S.S.R.*), **28**, 110 (1940).
15. Rochester, Butler and Runcorn, *Nature*, **159**, 227 (1947).

实验的详细情况即将在别处发表。

（沈乃澂 翻译；朱永生 审稿）

A Floating Magnet

V. Arkadiev

Editor's Note

Russian physicist V. Arkadiev here reports his observation of a levitation phenomenon achieved with the aid of a superconductor. He notes that a diamagnetic substance (in essence, a conventionally "non-magnetic" one) pulled toward a magnetic surface by gravity can hover over it. For example, a linear magnet one centimetre long could hover about one centimetre above a copper sphere some 20 metres in diameter. To prevent a magnet falling onto a diamagnetic sphere the size of the Earth, however, requires a superconductor beneath the magnet. Arkadiev demonstrates the effect with a small magnet hovering over a concave lead disk held over liquid helium. Today, with liquid nitrogen and high-temperature superconductors, Arkadiev's levitation is a common trick in the physics classroom.

BY assuming that diamagnetic bodies are pushed out of a magnetic field, it may be shown that a diamagnetic particle attracted to a magnet by gravitational forces will take up a position in space in the equatorial plane of the straight magnet at a certain distance from the latter. The "satellite" can vibrate elastically about the point of equilibrium, describing a certain curve. The period of vibration in the radial and meridional directions is close to the period of the Kepler rotation of a magnetically indifferent satellite about a body of the same mass. Several identical particles arrange themselves around the magnet. Such a combination of bodies is in the nature of a static planetary system as distinct from the Kepler dynamic planetary system.

However, systems thus formed can only be of small dimensions. The orbit of the outermost bodies can be no larger than several metres, and in the case of small magnetized iron meteorites amounts to several millimetres.

Computation shows that in space a straight magnet keeps at a certain distance from a large diamagnetic body. Thus a magnet 1 cm. long will take up a position at a distance of 1 cm. from the surface of a copper sphere about 20 m. in diameter. Diameters of 300 m. and 3,000 m. respectively would be necessary for bismuth and carbon spheres. To prevent a magnet falling on to a diamagnetic sphere the size of the Earth, the sphere must consist of the strongest diamagnetic substance, or be a superconductor. In this case, however, it is sufficient that the superconductor is placed only under the magnet itself.

The approach of a magnet to the surface of a superconductive semispace is accompanied by the appearance of the magnetic image of this magnet within the superconductor. In the case of a common steel magnet, this may lead to demagnetization, while a ferro-nickel-

悬浮的磁体

阿卡迪耶夫

编者按

俄罗斯物理学家阿卡迪耶夫在本文中报道了他借助超导体对悬浮现象所做的观察。他注意到，在重力的作用下被拉向磁体表面的抗磁物质（其实就是传统的“非磁性”物质）可以停留在其上方。譬如说，一个1厘米长的条状磁体可以飘浮在直径约为20米的铜球上方约1厘米处。然而，为了避免磁体落到具有地球尺寸大小的抗磁性球体上，需要在磁体下方放置一个超导体。阿卡迪耶夫通过一个在置于液氦中的凹面铅盘上方悬浮的小磁体来演示上述效应。如今，利用液氮和高温超导体技术，阿卡迪耶夫悬浮已成为物理课堂中一个常用的演示。

假设磁场排斥抗磁体，我们也许可以看到，当抗磁粒子因为重力作用而落向磁体时，其将会处于条形磁体中心面所在空间的某个位置上，与条形磁体相距一定距离。通常这个“卫星”可以围绕平衡点作弹性振动，得到某一曲线轨迹。而径向和经向的振动周期接近于与磁性无关的卫星围绕具有相同质量的物体作开普勒转动的周期。若有几个这样的粒子则会围绕磁体取一定的排列。这种物体的组合在性质上是静态行星系，与开普勒动力学的行星系有所不同。

然而，由此形成的系统只能是小尺寸的。最远物体的轨道可能不会大于几米，对于小块磁化了的铁陨石，其量级仅为几毫米。

计算表明，条形磁体与大块抗磁体间可在空间中保持一定的距离。因此1厘米长的磁体将与直径约为20米的铜球表面保持1厘米的距离。而对铋球和碳球而言，则其直径分别需要达到300米和3,000米才能实现上述距离。为了防止磁体落到具有地球尺寸大小的抗磁球体上，该球体必须由最强的抗磁物质组成，或是一个超导体。然而，在这种情况下，只要在磁体下方放置一个超导体就能实现上述情况。

磁体靠近超导半空间表面时在超导体内部产生该磁体的映像。对一个普通的钢磁体而言，这将导致退磁；而对一个含有铁–镍–铝的钢磁体，在这种力的作用下将被排斥在超导半空间的水平面以外，即使没有支撑物也能悬浮（“漂浮”）在该水平

aluminium steel magnet will be repelled from the horizontal surface of the semispace with such force that it will hang suspended ("float") over the latter without any support. Thus one of the cases of a static planetary system may be reproduced in the laboratory. The Earth, screened by a superconductor in the neighbourhood of a magnet, repels the latter with the same force as it is attracted owing to universal gravitation. The accompanying photograph shows a magnet, 4 mm. × 4 mm. × 10 mm. in dimensions, floating above a concave lead disk 40 mm. in diameter in a Dewar vessel over liquid helium.

The experimental test of these views was possible through the kindness of Prof. P. L. Kapitza, in the Institute of Physical Problems, Moscow.

The lower the coercive force of the magnet, the smaller the magnet itself must be. Carbon steel magnets, for example, can "float" when they have the dimensions of 0.5 mm. × 9 mm. By scattering microscopically small magnets over the surface of a body, it is possible to reveal superconductive inclusions directly, since the magnetic particles will roll to the spots where there is no superconductivity.

(**160**, 330; 1947)

V. Arkadiev: Maxwell Laboratory, Physical Department, University, Moscow, March 25.

面上方。因此，一种静态行星系中的情况可能在实验室中重现。但被与磁体相邻的超导体所屏蔽的地球，将以与万有引力相同大小的力把磁体推开。附图的照片展示了尺寸为 4 毫米 × 4 毫米 × 10 毫米的磁体，悬浮在充有液氦的杜瓦瓶内直径为 40 毫米的凹面铅盘的上方。

在莫斯科物理问题研究所卡皮查教授的热心帮助下，检验这些观点的实验才得以顺利开展。

磁体的抗磁力越低，则磁体本身必然越小。例如，碳钢磁体在其尺寸小至 0.5 毫米 × 9 毫米时才可能“浮起”。将微观尺度中的小磁体散在物体表面上，就有可能直接揭示出超导杂质的存在，因为磁性粒子将会滚到没有超导电性的地方。

（沈乃澂 翻译；赵见高 审稿）

Observations on the Tracks of Slow Mesons in Photographic Emulsions*

C. M. G. Lattes *et al.*

Editor's Note

Physicists in the late 1940s had adopted the term "meson" to refer to any particle having mass between that of an electron and a proton. Here the Brazilian physicist Cesar Lattes and colleagues reported on their observations of several hundred meson-like particles detected in photographic plates exposed at high altitudes, and presumably having origins as cosmic rays. Their observations led them to suggest that some of the initial mesons, when striking the photographic emulsions, created secondary mesons of a new kind. It soon became clear that this new particle—the pion—takes part in nuclear interactions. It was initially predicted to mediate such interactions by Hideki Yukawa in 1935, a prediction for which he won the Nobel Prize in 1949.

INTRODUCTION. In recent experiments, it has been shown that charged mesons, brought to rest in photographic emulsions, sometimes lead to the production of secondary mesons. We have now extended these observations by examining plates exposed in the Bolivian Andes at a height of 5,500 m., and have found, in all, forty examples of the process leading to the production of secondary mesons. In eleven of these, the secondary particle is brought to rest in the emulsion so that its range can be determined. In Part 1 of this article, the measurements made on these tracks are described, and it is shown that they provide evidence for the existence of mesons of different mass; In Part 2, we present further evidence on the production of mesons, which allows us to show that many of the observed mesons are locally generated in the "explosive" disintegration of nuclei, and to discuss the relationship of the different types of mesons observed in photographic plates to the penetrating component of the cosmic radiation investigated in experiments with Wilson chambers and counters.

Part 1. Existence of Mesons of Different Mass

As in the previous communications[1], we refer to any particle with a mass intermediate between that of a proton and an electron as a meson. It may be emphasized that, in

---

* This article contains a summary of the main features of a number of lectures given, one at Manchester on June 18 and four at the Conference on Cosmic Rays and Nuclear Physics, organised by Prof. W. Heitler, at the Dublin Institute of Advanced Studies, July 5-12. A complete account of the observations, and of the conclusions which follow from them, will be published elsewhere.

感光乳胶中慢介子径迹的观测*

拉特斯等

编者按

20 世纪 40 年代后期，物理学家采用“介子”来称谓所有质量介于电子和质子间的粒子。在这篇论文中，巴西物理学家塞萨尔·拉特斯和他的同事们报道了在高海拔处受照射的照片中探测到几百个类介子粒子的观测结果，并推测其源自宇宙射线。根据上述观测结果，他们提出，初级介子进入感光乳胶时，有一部分产生了一种新的次级介子。不久他们就弄清楚了这种新粒子（即 π 介子）参与了核相互作用。早在 1935 年汤川秀树首先预言了 π 介子是核作用的媒介，由此他获得了 1949 年的诺贝尔奖。

引言　最近实验显示，终止于感光乳胶中的带电介子有时会导致次级介子的产生。我们现已通过检查置于玻利维亚安第斯山脉海拔 5,500 米处的受照射的底片来延伸这些观测，现已发现共计有 40 个导致次级介子产生的事例。其中 11 个事例的次级粒子终止于乳胶中，因此其射程是可以确定的。本文的第一部分描述了对这些粒子径迹的测量，同时也证明了不同质量介子的存在。第二部分中（编者注：本书未收录第二部分），我们进一步证明了介子的产生，这使得我们可以得出，许多可观测到的介子是以“爆炸”式的核嬗变在局部区域产生的，同时也使我们可以对威尔逊云室和计数器实验中研究的宇宙射线的穿透成分与受照的感光底片中观测到的不同类型介子之间的关系进行探讨。

第一部分．不同质量介子的存在

如之前的报道 [1] 所述，我们把所有质量介于电子与质子之间的粒子称为介子。在这里需要强调的是，使用这个术语并不意味着我们认为相应的粒子必然与核子有

* 这篇论文包含一些已做的演讲主要内容的总结，其中一篇是 6 月 18 日在曼彻斯特发表的，还有 4 篇是于 7 月 5 日～12 日在都柏林高等研究所由海特勒教授组织的宇宙射线和核物理会议上发表的。关于这些观测及由此导出的结论的完整描述将在别处发表。

using this term, we do not imply that the corresponding particle necessarily has a strong interaction with nucleons, or that it is closely associated with the forces responsible for the cohesion of nuclei.

We have now observed a total of 644 meson tracks which end in the emulsion of our plates. 451 of these were found, in plates of various types, exposed at an altitude of 2,800 m. at the Observatory of the Pic du Midi, in the Pyrenees; and 193 in similar plates exposed at 5,500 m. at Chacaltaya in the Bolivian Andes. The 451 tracks in the plates exposed at an altitude of 2,800 m. were observed in the examination of 5 c.c. emulsion. This corresponds to the arrival of about 1.5 mesons per c.c. per day, a figure which represents a lower limit, for the tracks of some mesons may be lost through fading, and through failure to observe tracks of very short range. The true number will thus be somewhat higher. In any event, the value is of the same order of magnitude as that we should expect to observe in delayed coincidence experiments at a height of 2,800 m., basing our estimates on the observations obtained in similar experiments at sea-level, and making reasonable assumptions about the increase in the number of slow mesons with altitude. It is therefore certain that the mesons we observe are a common constituent of the cosmic radiation.

Photomicrographs of two of the new examples of secondary mesons, Nos. III and IV, are shown in Figs. 1 and 2. Table 1 gives details of the characteristics of all events of this type observed up to the time of writing, in which the secondary particle comes to the end of its range in the emulsion.

Table 1

Event No.	Range in emulsion in microns of Primary meson	Range in emulsion in microns of Secondary meson
I	133	613
II	84	565
III	1,040	621
IV	133	591
V	117	638
VI	49	595
VII	460	616
VIII	900	610
IX	239	666
X	256	637
XI	81	590

Mean range 614±8 μ. Straggling coefficient $\sqrt{\Sigma\Delta_i^2/n}$ = 4.3 percent, where $\Delta_i = R_i - \overline{R}$, R_i being the range of a secondary meson, and $\overline{R}$ the mean value for n particles of this type.

强相互作用，或是与核的凝聚力密切相关。

我们现已观测到终止于底片乳胶中的介子径迹总共为644条，其中451条发现于比利牛斯山脉的米迪山峰海拔2,800米处天文台受照射的不同类型的底片中；另外193条发现于玻利维亚安第斯山脉恰卡尔塔亚山海拔5,500米处受照射的类似底片中。海拔2,800米处受照底片中的451条径迹是在5厘米^{3}的乳胶检测中观测到的，这相当于每天每立方厘米乳胶中到达约1.5个介子。该数据是一个下限，因为某些介子的径迹可能由于潜影衰退和射程极短难以观测而丢失。因此，真正的数值将会比该数值要稍高一些。基于我们在海平面上类似实验观测的估计，以及慢介子数目随着海拔上升而增大的合理猜想，总是可以得出，这个数值与我们在2,800米高度延迟符合实验的观测中所预期的数值具有相同量级。因此可以确定，我们观测到的介子是宇宙射线的一般组分。

图1和图2分别是次级介子的两个新事例III和IV的显微照片。表1给出了直至本文执笔时观测到的所有这类事例的详细特性，其中次级粒子射程末端终止于乳胶中。

表1

事例编号	初级介子在乳胶中的射程	次级介子在乳胶中的射程
I	133	613
II	84	565
III	1,040	621
IV	133	591
V	117	638
VI	49	595
VII	460	616
VIII	900	610
IX	239	666
X	256	637
XI	81	590

平均射程：614±8微米。离散系数：$\sqrt{\Sigma\Delta_i^2/n} = 4.3\%$，式中，$\Delta_i = R_i - \overline{R}$，$R_i$是次级介子的射程，$\overline{R}$是$n$个这类粒子的射程平均值。

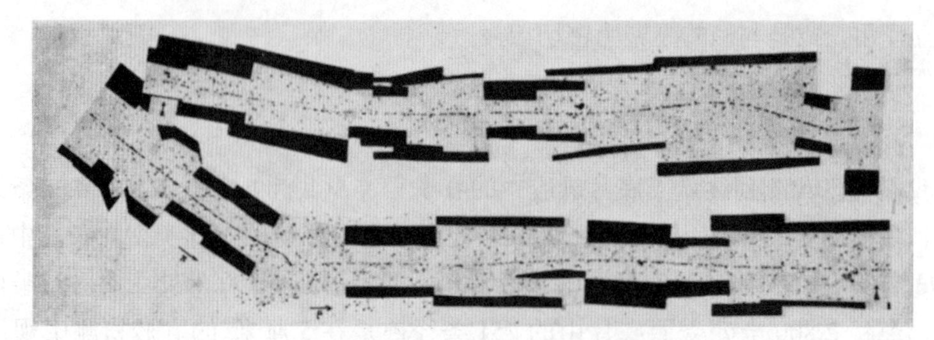

Fig. 1. Observation by Mrs. I. Powell. Cooke × 95 achromatic objective; *C*2 Ilford Nuclear Research emulsion loaded with boron. The track of the μ-meson is given in two parts, the point of junction being indicated by *a* and an arrow

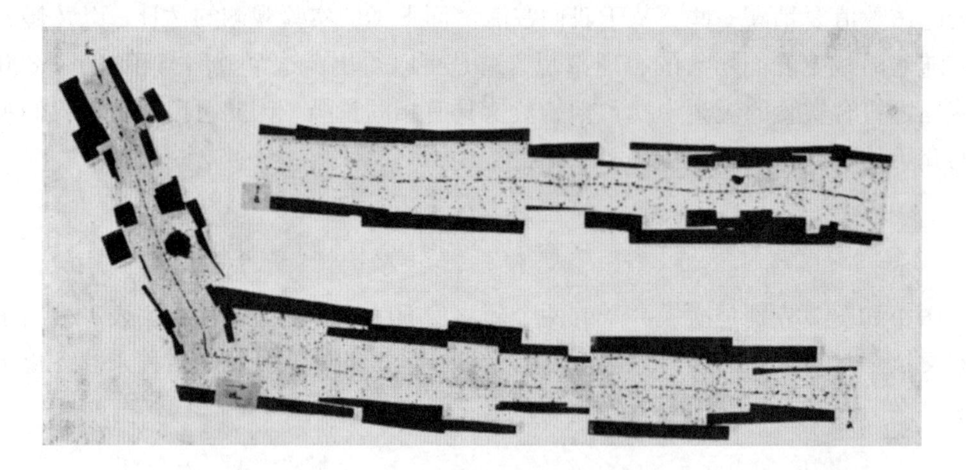

Fig. 2. Cooke × 95 achromatic objective. *C*2 Ilford Nuclear Research emulsion loaded with boron

The distribution in range of the secondary particles is shown in Fig. 3. The values refer to the lengths of the projections of the actual trajectories of the particles on a plane parallel to the surface of the emulsion. The true ranges cannot, however, be very different from the values given, for each track is inclined at only a small angle to the plane of the emulsion over the greater part of its length. In addition to the results for the secondary mesons which stop in the emulsion, and which are represented in Fig. 3 by black squares, the length of a number of tracks from the same process, which pass out of the emulsion when near the end of their range, are represented by open squares.

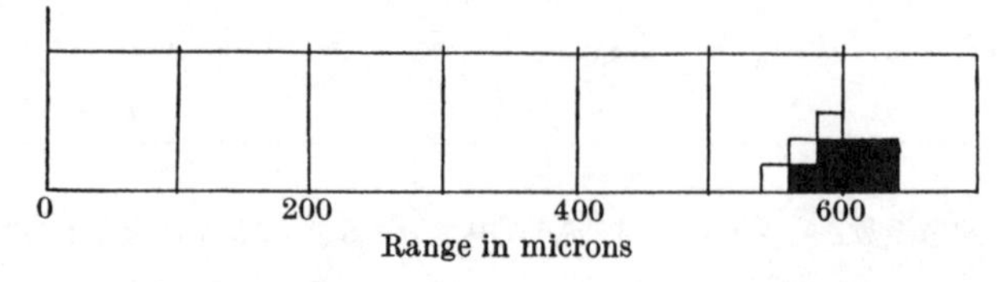

Fig. 3. Distribution in range of ten secondary mesons. Those marked ■ stop in the emulsion; the three marked □ leave the emulsion when near the end of their range. Mean range of secondary mesons, 606 microns. The results for events Nos.VIII to XI are not included in the figure

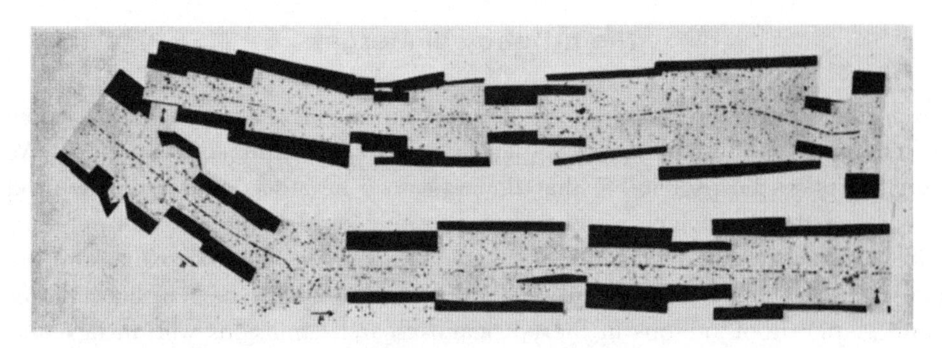

图 1. 由鲍威尔夫人观测。库克 95 倍消色差物镜；*C*2 伊尔福载硼核乳胶。μ 介子（编者注：现在粒子物理认为“μ 介子”不是介子，而是一种轻子，因此正确的称谓应为“μ 子”。考虑到历史原因，文中仍保留“μ 介子”的译法）的径迹由两部分给出，连接点用 *a* 和箭头表示。

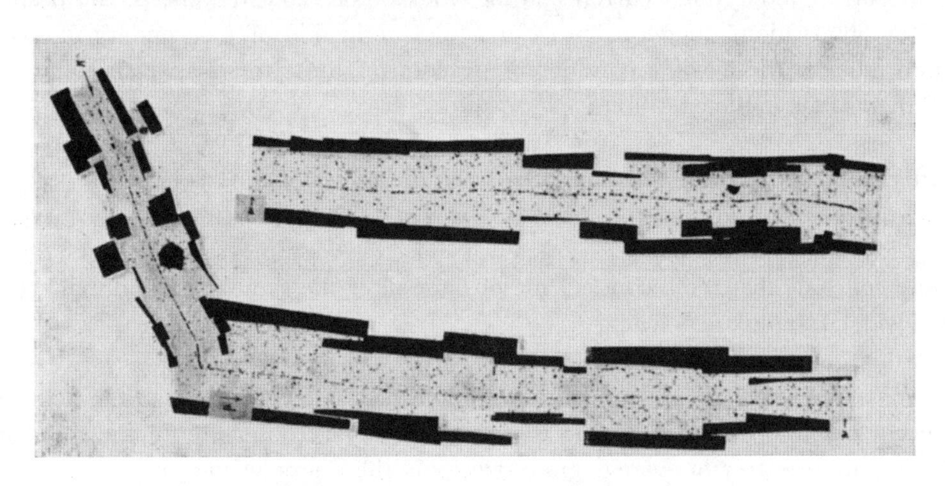

图 2. 库克 95 倍消色差物镜。*C*2 伊尔福载硼核乳胶。

图 3 是次级粒子射程的分布，这些值指的是粒子实际径迹在平行于乳胶表面的平面上的投影长度。不过真实的射程不会与给出的数值有很大的差异，因为每个径迹在大部分长度中相对于乳胶平面只倾斜很小的角度。图 3 中除了用黑方块表示出终止于乳胶中的次级粒子的结果之外，还用白方块表示出产生于相同过程的、于射程末端附近穿出乳胶的一些径迹的长度。

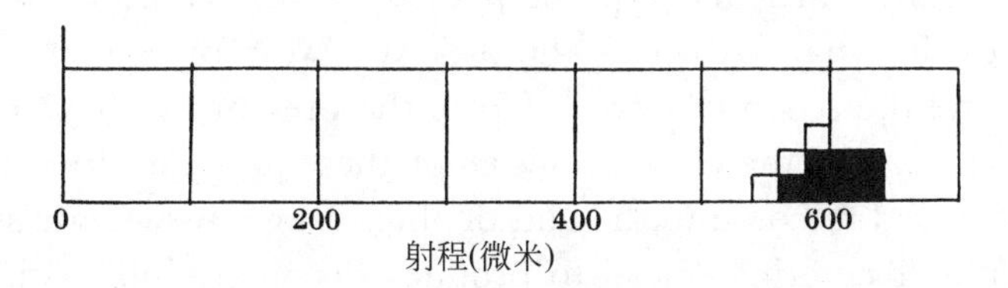

图 3. 10 个次级介子的射程分布。■表示 7 个次级介子在乳胶中停止；3 个□表示 3 个次级介子在接近射程末端时离开乳胶。次级介子的平均射程为 606 微米。事例 VIII~XI 的结果未包含在本图内。

The μ-Decay of Mesons

Two important conclusions follow from these measurements. Our observations show that the directions of ejection of the secondary mesons are orientated at random. We can therefore calculate the probability that the trajectory of a secondary meson, produced in a process of the type which we observe, will remain within the emulsion, of thickness 50 μ, for a distance greater than 500 μ. If we assume, as a first approximation, that the trajectories are rectilinear, we obtain a value for the probability of 1 in 20. The marked Coulomb scattering of mesons in the Nuclear Research emulsions will, in fact, increase the probability of "escape". The six events which we observe in plates exposed at 2,800 m., in which the secondary particle remains in the emulsion for a distance greater than 500 μ, therefore correspond to the occurrence in the emulsion of 120 ± 50 events of this particular type. Our observations, therefore, prove that the production of a secondary meson is a common mode of decay of a considerable fraction of those mesons which come to the end of their range in the emulsion.

Second, there is remarkable consistency between the values of the range of the secondary mesons, the variation among the individual values being similar to that to be expected from "straggling", if the particles are always ejected with the same velocity. We can therefore conclude that the secondary mesons are all of the same mass and that they are emitted with constant kinetic energy.

If mesons of lower range are sometimes emitted in an alternative type of process, they must occur much less frequently than those which we have observed; for the geometrical conditions, and the greater average grain-density in the tracks, would provide much more favourable conditions for their detection. In fact, we have found no such mesons of shorter range. We cannot, however, be certain that mesons of greater range are not sometimes produced. Both the lower ionization in the beginning of the trajectory, and the even more unfavourable conditions of detection associated with the greater lengths of the tracks, would make such a group, or groups, difficult to observe. Because of the large fraction of the mesons which, as we have seen, can be attributed to the observed process, it is reasonable to assume that alternative modes of decay, if they exist, are much less frequent than that which we have observed. There is, therefore, good evidence for the production of a single homogeneous group of secondary mesons, constant in mass and kinetic energy. This strongly suggests a fundamental process, and not one involving an interaction of a primary meson with a particular type of nucleus in the emulsion. It is convenient to refer to this process in what follows as the μ-decay. We represent the primary mesons by the symbol π, and the secondary by μ. Up to the present, we have no evidence from which to deduce the sign of the electric charge of these particles. In every case in which they have been observed to come to the end of their range in the emulsion, the particles appear to stop without entering nuclei to produce disintegrations with the emission of heavy particles.

Knowing the range-energy relation for protons in the emulsion, the energy of ejection

介子的 μ 衰变

从以上测量中得到了两个重要的结论。我们的观测表明，次级介子的发射方向是随机的。因此，我们可以计算出我们所观测的那类过程中产生的次级介子径迹停留在厚度为 50 微米乳胶内且射程大于 500 微米的概率。作为一个初步近似，如果我们假定径迹是直线，则我们得出的概率为 1/20。实际上，核乳胶中介子的显著的库仑散射将增加“逃逸”的概率。我们在海拔 2,800 米处受照底片中观测到的 6 个事例中的次级粒子在乳胶中的射程大于 500 微米，因而这 6 个事例相当于乳胶中出现 120±50 个这类特殊形式的事例。因此我们的观测证明，对于射程终止于乳胶内的介子，相当大一部分的衰变模式通常是产生次级介子。

其次，次级介子射程值之间存在显著的一致性，如果粒子总以相同的速度发射，各个数值之间的变化类似于“离散”预期的值。因此我们可以得出，次级介子是质量相同的粒子，它们会以恒定的动能发射。

如果较短射程的介子偶尔以另一种类型的方式发射，它们出现的频率必定比我们观测到的要低得多；因为几何条件以及径迹中较大的平均颗粒密度将会为探测它们提供更为有利的条件。实际上，我们尚未发现这类较短程的介子。然而，我们不能肯定是否有时候会产生更长射程的介子。长程介子径迹开始处的电离能力较低，加之与长径迹相关的更不利的探测条件，这两个因素将使这一类或这些类长程介子难以观测。如我们所见，由于大部分介子产生于我们已观测到的过程，因此有理由假设，如果存在其他模式的衰变，其出现频率要远低于我们已观测到的过程。因此，有力的证据证明，只产生了质量和动能为常数的一类次级介子。这就强有力地提出了一个基本过程，但这个过程并不涉及初级粒子与乳胶中的特殊类型核的相互作用。为方便起见，这一过程在后面的陈述中被称为 μ 衰变。我们用 π 表示初级介子，μ 表示次级介子。至今，我们尚无可以推断出这些粒子的电荷符号的证据。在粒子射程末端终止于乳胶中的事例中，粒子停止时似乎并未进入核并产生发射重粒子的嬗变。

知道了乳胶中质子的射程–能量关系后，如果设定了粒子的某个质量值，次级介

of the secondary mesons can be deduced from their observed range, if a value of the mass of the particles is assumed. The values thus calculated for various masses are shown in Table 2.

Table 2

Mass in m_e	100	150	200	250	300
Energy in MeV.	3.0	3.6	4.1	4.5	4.85

No established range-energy relation is available for protons of energies above 13 MeV., and it has therefore been necessary to rely on an extrapolation of the relation established for low energies. We estimate that the energies given in Table 2 are correct to within 10 percent.

Evidence of a Difference in Mass of π- and μ-Mesons

It has been pointed out[1] that it is difficult to account for the μ-decay in terms of an interaction of the primary meson with the nucleus of an atom in the emulsion leading to the production of an energetic meson of the same mass as the first. It was therefore suggested that the observations indicate the existence of mesons of different mass. Since the argument in support of this view relied entirely on the principle of the conservation of energy, a search was made for processes which were capable of yielding the necessary release of energy, irrespective of their plausibility on other grounds. Dr. F. C. Frank has re-examined such possibilities in much more detail, and his conclusions are given in an article to follow. His analysis shows that it is very difficult to account for our observations, either in terms of a nuclear disintegration, or of a "building-up" process in which, through an assumed combination of a negative meson with a hydrogen nucleus, protons are enabled to enter stable nuclei of the light elements with the release of binding energy. We have now found it possible to reinforce this general argument for the existence of mesons of different mass with evidence based on grain-counts.

We have emphasized repeatedly[1] that it is necessary to observe great caution in drawing conclusions about the mass of particles from grain-counts. The main source of error in such determinations arises from the fugitive nature of the latent image produced in the silver halide granules by the passage of fast particles. In the case of the μ-decay process, however, an important simplification occurs. It is reasonable to assume that the two meson tracks are formed in quick succession, and are subject to the same degree of fading. Secondly, the complete double track in such an event is contained in a very small volume of the emulsion, and the processing conditions are therefore identical for both tracks, apart from the variation of the degree of development with depth. These features ensure that we are provided with very favourable conditions in which to determine the ratio of the masses of the π- and μ-mesons, in some of these events.

In determining the grain density in a track, we count the number of individual grains in successive intervals of length 50 μ along the trajectory, the observation being made with

子的发射能量便可以根据其观测到的射程导出。据此，表 2 列出了不同质量对应的介子能量的计算值。

表 2

质量 (m_e)	100	150	200	250	300
能量（兆电子伏）	3.0	3.6	4.1	4.5	4.85

对于能量大于 13 兆电子伏的质子尚未确立射程–能量关系，因此必须依靠已经确立的低能关系外推。我们估计，表 2 中给出的能量在 10% 以内是正确的。

π 介子和 μ 介子质量差的证据

我们已指出[1]，用初级介子与乳胶中原子核的相互作用导致一个与初级介子质量相同的高能介子的产生难以解释 μ 衰变。因此可以认为，已有的观测表明存在不同质量的介子。由于支持这个观点的论据完全依赖于能量守恒定律，因此无论它们在其他理论背景中的合理性如何，都对能够产生必要的能量释放的若干过程进行了探索。弗兰克博士更加仔细地重新检验了这种可能性，并在随后发表的一篇论文中给出了他的结论。他的分析指出，无论用核嬗变或用“聚集”过程都很难解释我们的观测，其中“聚集”过程是指，假设负介子能够与氢核结合，质子可以进入轻元素的稳定核内并释放结合能。我们现已发现，基于颗粒计数的证据，可能会巩固存在不同质量介子的总论点。

我们已反复强调[1]，由颗粒计数推出粒子质量的结论时，一定要十分地谨慎。这类测定中的误差主要来自由快速粒子通过卤化银颗粒所产生的潜影的易变性。然而，在 μ 衰变过程的情况下，出现了重要的简化现象。我们可以合理地假设，两个介子的径迹是快速接连产生的，并遭到相同程度的潜像衰退。其次，在这类事例中完全的双径迹包含于很小的乳胶体积内，因此除了显影的程度在深度上的变化之外，两条径迹的处理条件是一样的。这些特点保证了我们是在非常有利的条件下，利用这一类事例来测定 π 介子和 μ 介子的质量比。

测定径迹的颗粒密度时，我们沿着径迹对每 50 微米长度间隔内的颗粒计数依次进行了统计，观测是用大放大倍数（×2,000）的光学装置，并达到了最高可能的分

optical equipment giving large magnification (×2,000), and the highest available resolving power. Typical results for protons and mesons are shown in Fig. 4. These results were obtained from observations on the tracks in a single plate, and it will be seen that there is satisfactory resolution between the curves for particles of different types. The "spread" in the results for different particles of the same type can be attributed to the different degrees of fading associated with the different times of passage of the particles through the emulsion during an exposure of six weeks.

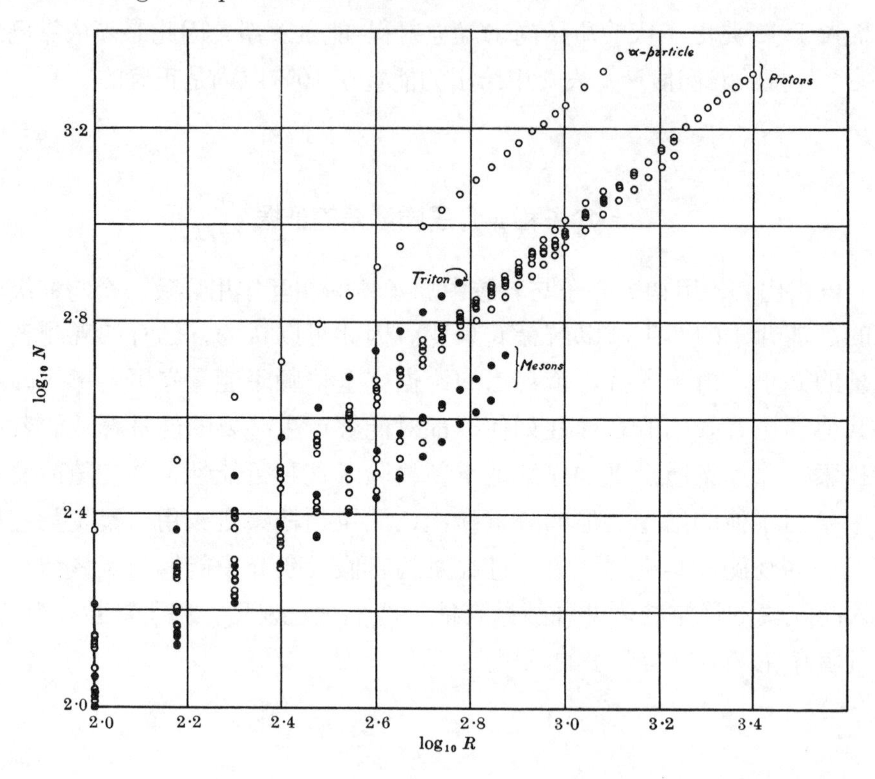

Fig. 4. N is total number of grains in track of residual range R (scale-divisions). 1 scale-division=0.85 microns

Applying these methods to the examples of the μ-decay process, in which the secondary mesons come to the end of their range in the emulsion, it is found that in every case the line representing the observations on the primary meson lies above that for the secondary particle. We can therefore conclude that there is a significant difference in the grain-density in the tracks of the primary and secondary mesons, and therefore a difference in the mass of the particles. This conclusion depends, of course, on the assumption that the π- and μ-particles carry equal charges. The grain-density at the ends of the tracks, of particles of both types, are consistent with the view that the charges are of magnitude $|e|$.

A more precise comparison of the masses of the π- and μ-mesons can only be made in those cases in which the length of the track of the primary meson in the emulsion is of the order of 600 μ. The probability of such a favourable event is rather small, and the only examples we have hitherto observed are those listed as Nos. III and VIII in Table 1.

辨率。图 4 是质子和介子的典型结果。这些结果是从单张底片的径迹观测中获得的，我们看到在不同类型的粒子曲线间得到了满意的分辨率。结果中相同类型的不同粒子的“发散”可归因于，在 6 周的受照期间内，与粒子穿过乳胶的时间相关的不同程度的潜影衰退。

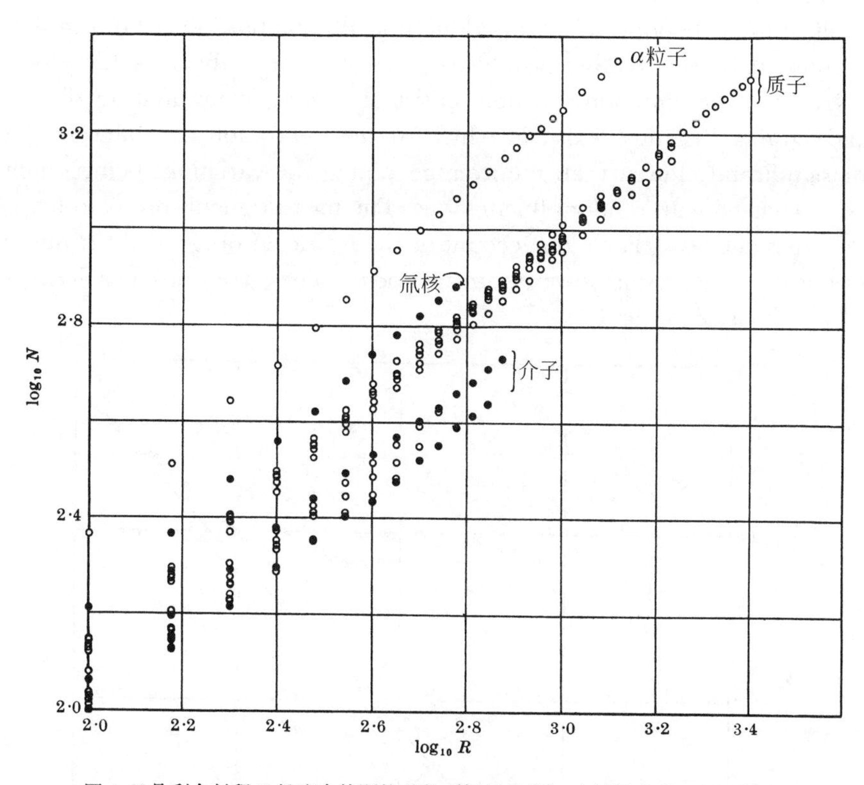

图 4. N 是剩余射程 R 径迹中的颗粒总数（标尺分度）。1 标尺分度 = 0.85 微米

将这些方法应用于那些次级介子射程末端终止于乳胶内的 μ 衰变过程，我们发现，在所有这些事例中表示初级介子观测的线位于次级粒子之上。因此我们可以认为，初级介子与次级介子径迹中的颗粒密度存在显著差异，从而可知它们的质量也是不同的。当然，这个结论依赖于 π 介子和 μ 介子带有相等电荷的假定。两类粒子在径迹末端的颗粒密度与电荷量为 $|e|$ 的观点是一致的。

对 μ 介子和 π 介子质量更精密的比较只能在下述情况下进行，即乳胶中初级介子的径迹长度为 600 微米量级的情况。这样理想的事例的出现概率是很小的，表 1 中的 III 和 VIII 列出了我们迄今为止观测到的仅有的例子。第一个这样事例的部分

A mosaic of micrographs of a part only of the first of these events is reproduced in Fig. 1, for the length of the track of the μ-meson in the emulsion exceeds 1,000 μ. The logarithms of the numbers of grains in the tracks of the primary and secondary mesons in this event are plotted against the logarithm of the residual range in Fig. 5. By comparing the residual ranges at which the grain-densities in the two tracks have the same value, we can deduce the ratio of the masses. We thus obtain the result m_π/m_μ=2.0. Similar measurements on event No. VIII give the value 1.8. In considering the significance which can be attached to this result, it must be noticed that in addition to the standard deviations in the number of grains counted, there are other possible sources of error. Difficulties arise, for example, from the fact that the emulsions do not consist of a completely uniform distribution of silver halide grains. "Islands" exist, in which the concentration of grains is significantly higher, or significantly lower, than the average values, the variations being much greater than those associated with random fluctuations. The measurements on the other examples of μ-decay are much less reliable on account of the restricted range of the π-mesons in the emulsion; but they give results lower than the above values. We think it unlikely, however, that the true ratio is as low as 1.5.

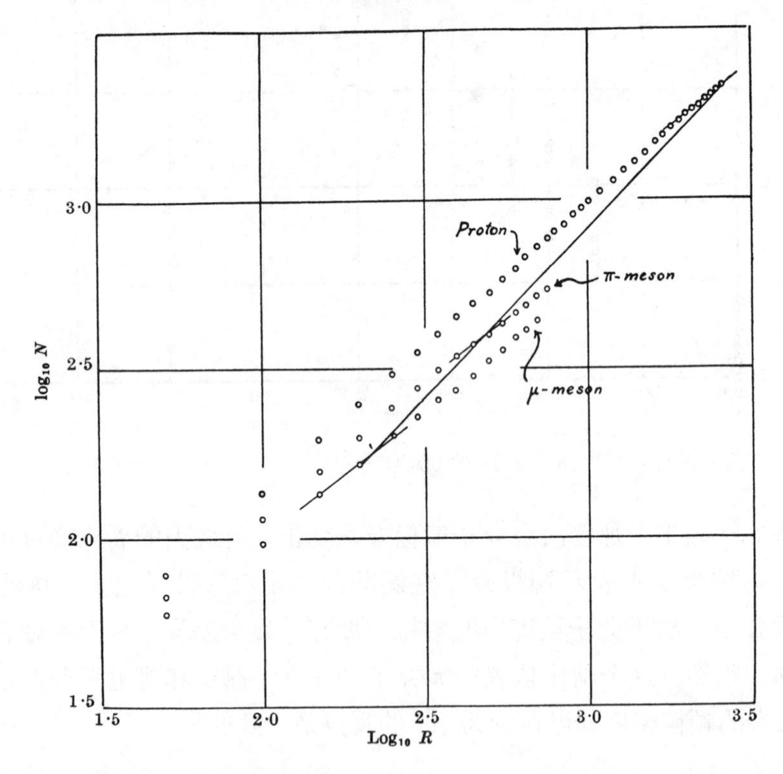

Fig. 5. N is total number of grains in track of residual range R (scale-divisions). 1 scale-division=0.85 microns The 45°-line cuts the curves of the mesons and proton in the region of the same grain density

The above result has an important bearing on the interpretation of the μ-decay process. Let us assume that it corresponds to the spontaneous decay of the heavier π-meson, in which the momentum of the μ-meson is equal and opposite to that of an emitted photon. For any assumed value of the mass of the μ-meson, we can calculate the energy of

显微照片的图像重现于图 1，其中乳胶中的 μ 介子径迹长度超过 1,000 微米。图 5 是这一事例中初级和次级介子径迹颗粒数的对数与剩余射程的对数的关系图。通过比较在两个径迹中具有相同颗粒密度的剩余射程，我们可以导出质量比。据此，我们获得的结果为 $m_\pi/m_\mu = 2.0$。对事例 VIII 的类似测量给出的值为 1.8。在考虑这项结果的意义时必须注意到，除颗粒计数的标准偏差外，还存在其他可能的误差来源。例如，由于乳胶不是由完全均匀分布的卤化银颗粒构成而导致的误差。其中存在一些“岛屿”，其颗粒密度明显高于或低于平均值，其变化量比随机涨落要大得多。考虑到乳胶中 π 介子的有限射程，μ 衰变的其他事例的测量不太可靠；不过它们给出的结果低于上述数值。尽管如此，我们认为真正的比值不可能低到 1.5。

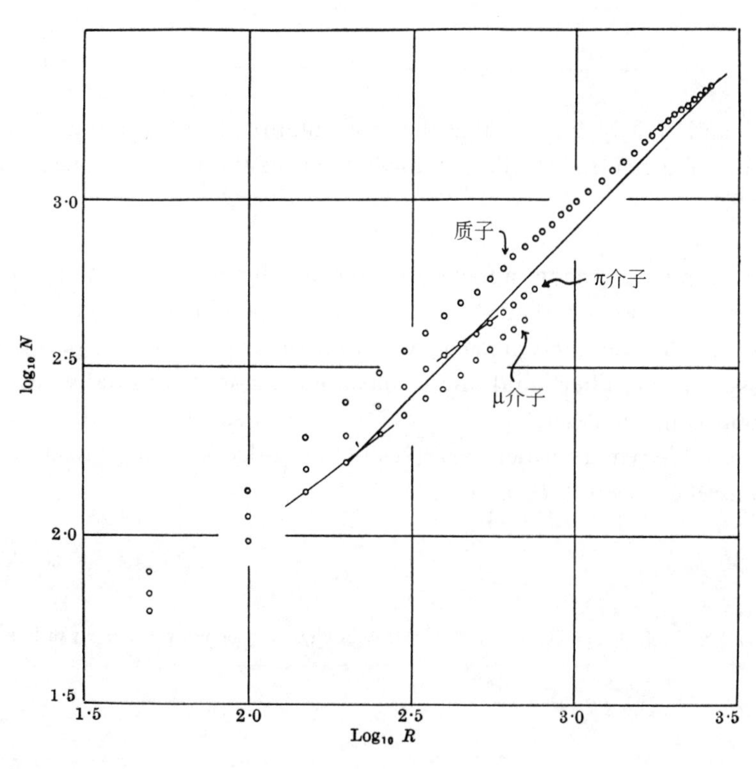

图 5. N 是剩余射程 R 的径迹中的颗粒总数（标尺分度）。1 标尺分度 = 0.85 微米。45°线在相同的颗粒密度区域处与介子和质子曲线相交。

上述结果对 μ 衰变过程的解释具有重要的意义。我们假定，它相应于较重的 π 介子的自发衰变，其中 μ 介子的动量与发射光子的动量数值相等但方向相反。对 μ 介子质量的任何假定值，我们均能通过粒子的观测射程值计算其发射能量值，从

ejection of the particle from its observed range, and thus determine its momentum. The momentum, and hence the energy of the emitted photon, is thus defined; the mass of the π-meson follows from the relation

$$c^2 m_\pi = c^2 m_\mu + E_\mu + h\nu.$$

It can thus be shown that the ratio m_π/m_μ is less than 1.45 for any assumed value of m_μ in the range from 100 to 300 m_e, m_e being the mass of the electron (see Table 3). A similar result is obtained if it is assumed that a particle of low mass, such as an electron or a neutrino, is ejected in the opposite direction to the μ-meson.

Table 3

Assumed mass m_μ	E(MeV.)	$h\nu$(MeV.)	m_π	m_π/m_μ±3 percent
100 m	3.0	17	140 m_e	1.40
150	3.6	23	203	1.35
200	4.1	29	264	1.32
250	4.5	34	325	1.30
300	4.85	39	387	1.29

On the other hand, if it is assumed that the momentum balance in the μ-decay is obtained by the emission of a neutral particle of mass equal to the μ-meson mass, the calculated ratio is about 2.1:1.

Our preliminary measurements appear to indicate, therefore, that the emission of the secondary meson cannot be regarded as due to a spontaneous decay of the primary particle, in which the momentum balance is provided by a photon, or by a particle of small rest-mass. On the other hand, the results are consistent with the view that a neutral particle of approximately the same rest-mass as the μ-meson is emitted. A final conclusion may become possible when further examples of the μ-decay, giving favourable conditions for grain-counts, have been discovered.

(**160**, 453-456; 1947)

C. M. G. Lattes, G. P. S. Occhialini and C. F. Powell: H. H. Wills Physical Laboratory, University of Bristol.

Reference:

1. *Nature*, 159, 93, 186, 694 (1947).

而确定其动量。据此可以确定发射光子的动量和能量；π 介子的质量遵从以下关系：

$$c^2 m_\pi = c^2 m_\mu + E_\mu + h\nu$$

因此可以表明，对质量 m_μ 在 100 m_e~300 m_e 区间内的任意假定值，m_π/m_μ 比值都小于 1.45，其中 m_e 是电子质量（见表 3）。如果假定一个小质量粒子，例如电子或中微子，与 μ 介子以相反方向射出，亦会得到类似结果。

表 3

假定质量 m_μ	E（兆电子伏）	$h\nu$（兆电子伏）	m_π	m_π/m_μ ±3%
100 m	3.0	17	140 m_e	1.40
150	3.6	23	203	1.35
200	4.1	29	264	1.32
250	4.5	34	325	1.30
300	4.85	39	387	1.29

另一方面，如果假定在 μ 衰变中的动量平衡是通过发射一个与 μ 介子质量相等的中性粒子而达到的，则计算出的比值约为 2.1∶1。

因此，我们的初步测量表明，不能将次级介子的发射归因于初级粒子的自发衰变，其中一个光子或静止质量很小的粒子提供了动量平衡。另一方面，观测结果与发射了一个与 μ 介子质量近似相等的中性粒子的观点相一致。当发现了更多的能够为颗粒计数提供更有利条件的 μ 衰变事例时，才有可能做出最终的结论。

（沈乃澂 翻译；朱永生 审稿）

Evidence for the Existence of New Unstable Elementary Particles

G. D. Rochester and C. C. Butler

Editor's Note

By 1947, physicists knew of the positron and the neutrino, and had recently discovered the neutral pion. A tidy picture of the particle zoo seemed to be emerging, except for some minor confusion concerning the muon. But here physicists George Rochester and Clifford Butler reported shocking evidence for yet further heavy particles of an unknown kind. They analysed photographs of energetic particles penetrating a cloud chamber, and in two cases noted V-like patterns apparently showing one incident massive particle spontaneously disintegrating into two others. These processes would later be identified as involving two new particles, the theta and tau particles: just two of a host of new particles to be discovered over the coming decades.

AMONG some fifty counter-controlled cloud-chamber photographs of penetrating showers which we have obtained during the past year as part of an investigation of the nature of penetrating particles occurring in cosmic ray showers under lead, there are two photographs containing forked tracks of a very striking character. These photographs have been selected from five thousand photographs taken in an effective time of operation of 1,500 hours. On the basis of the analysis given below we believe that one of the forked tracks, shown in Fig. 1 (tracks *a* and *b*), represents the spontaneous transformation in the gas of the chamber of a new type of uncharged elementary particle into lighter charged particles, and that the other, shown in Fig. 2 (tracks *a* and *b*), represents similarly the transformation of a new type of charged particle into two light particles, one of which is charged and the other uncharged.

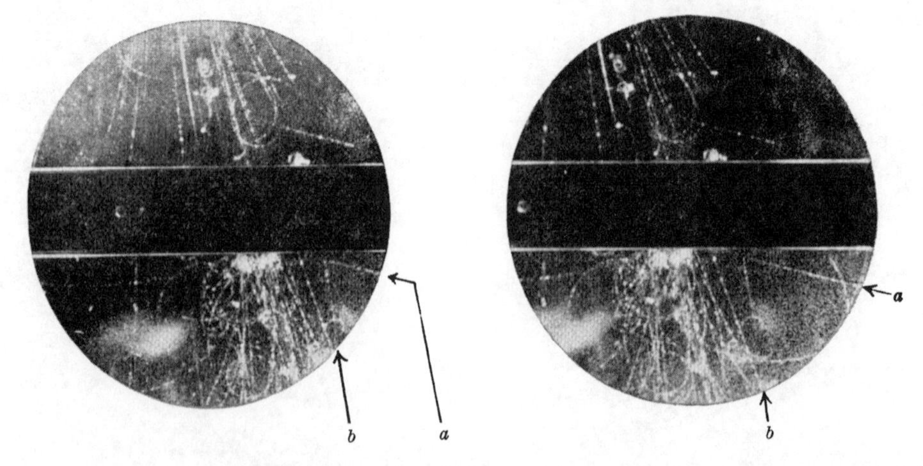

Fig. 1. Stereoscopic photographs showing an unusual fork (*a b*) in the gas. The direction of the magnetic field is such that a positive particle coming downwards is deviated in an anticlockwise direction

新的不稳定基本粒子存在的证据

罗切斯特，布特勒

编者按

到 1947 年，物理学家们已经知道正电子和中微子，并刚刚发现了中性介子。除了关于 μ 介子存在一些小争议外，一个关于粒子园的令人满意的图景似乎已经成形了；但是在本文中物理学家乔治·罗切斯特和克利福德·布特勒报道了一种未知重粒子存在的证据，令人震惊。他们分析了穿透云室的高能粒子的照片，并在其中两张照片中注意到 V 形径迹，这显然表明一个有质量的入射粒子自发衰变为另外两种粒子。后来证实这些衰变过程与两种新粒子有关，分别是 θ 粒子和 τ 粒子，而它们仅仅是后来几十年中发现的大量新粒子中的两种。

在过去的一年中，我们得到了约 50 张由计数器控制的贯穿簇射的云室照片，这些射线流是在对宇宙射线辐射铅板时，穿透铅板的粒子，照片是在研究穿透粒子的自然特征的实验中得到的，其中有两张具有显著的叉状特征。这些照片是从经过 1,500 个小时的有效工作时间中拍摄得到的 5,000 张照片中筛选出来的。在以下分析的基础上，我们相信其中图 1 中的一条叉状径迹（*a*、*b* 径迹）代表了云室气体中的自发转变过程：一种新的不带电的基本粒子转化为更轻的带电粒子。而图 2 中的另一条径迹（*a*、*b* 径迹）代表了相似的转变过程：一种新的带电粒子转化为两种轻粒子，其中一种粒子带电，另一种粒子不带电。

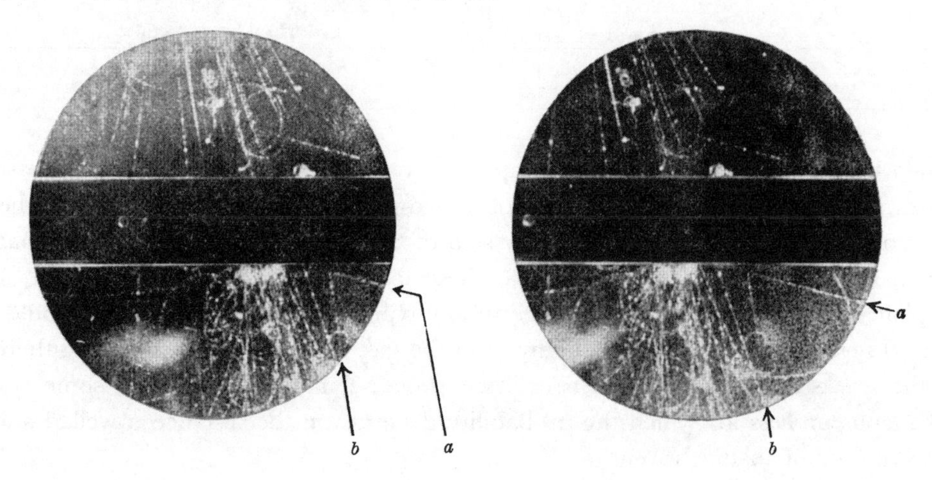

图 1. 立体照片显示了云室气体中一条不同寻常的叉状径迹 (*a b*)。磁场的方向使得向下运动带正电的粒子沿着逆时针方向发生偏转。

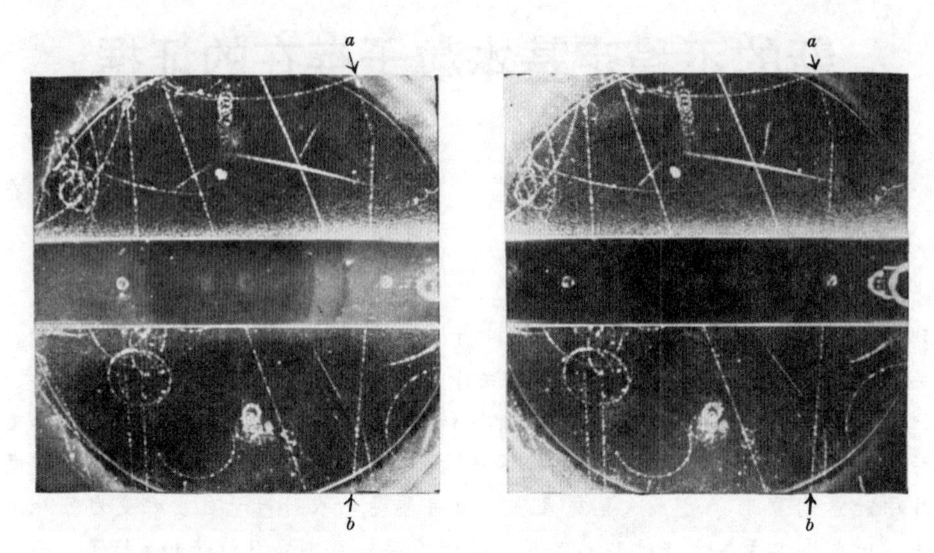

Fig. 2. Stereoscopic photographs showing an unusual fork (*a b*). The direction of the magnetic field is such that a positive particle coming downwards is deviated in a clockwise direction

The experimental data for the two forks are given in Table 1; H is the value of the magnetic field, α the angle between the tracks, p and Δp the measured momentum and the estimated error. The signs of the particles are given in the last column of the table, a plus sign indicating that the particle is positive if moving down in the chamber. Careful re-projection of the stereoscopic photographs has shown that each pair of tracks is copunctal. Moreover, both tracks occur in the middle of the chamber in a region of uniform illumination, the presence of background fog surrounding the tracks indicating good condensation conditions.

Table 1. Experimental data

Photograph	H (gauss)	α (deg.)	Track	p (eV./c.)	Δp (eV./c.)	Sign
1	3,500	66.6	*a*	3.4×10^{8}	1.0×10^{8}	+
			b	3.5×10^{8}	1.5×10^{8}	−
2	7,200	161.1	*a*	6.0×10^{8}	3.0×10^{8}	+
			b	7.7×10^{8}	1.0×10^{8}	+

Though the two forks differ in many important respects, they have at least two essential features in common: first, each consists of a two-pronged fork with the apex in the gas; and secondly, in neither case is there any sign of a track due to a third ionizing particle. Further, very few events at all similar to these forks have been observed in the 3 cm. lead plate, whereas if the forks were due to any type of collision process one would have expected several hundred times as many as in the gas. This argument indicates, therefore, that the tracks cannot be due to a collision process but must be due to some type of spontaneous process for which the probability depends on the distance travelled and not on the amount of matter traversed.

This conclusion can be supported by detailed arguments. For example, if either forked

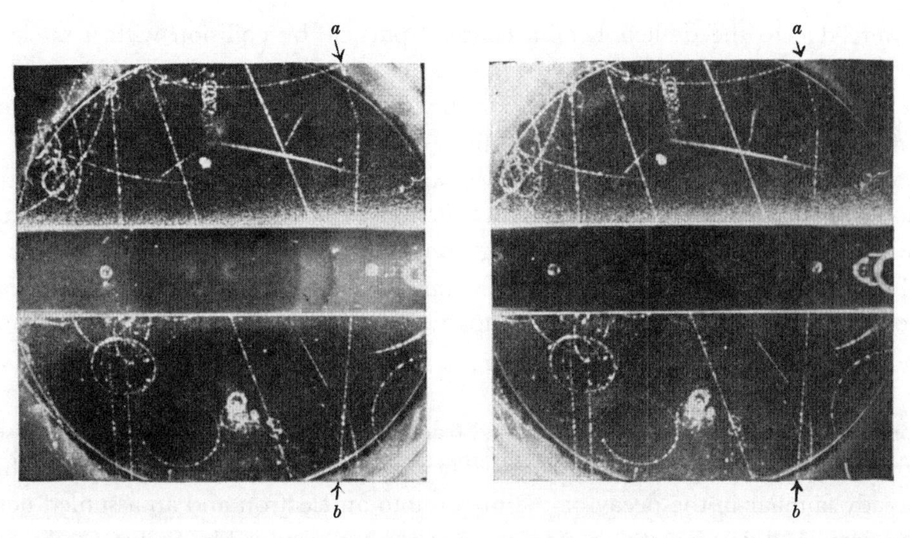

图 2. 立体照片显示了云室气体中一个不同寻常的叉状径迹 (a b)。磁场的方向使得向下运动带正电的粒子沿着顺时针方向发生偏转。

这两条叉状径迹的实验数据如表 1 所示。H 为磁场的强度值，α 为叉状径迹的夹角，p 和 Δp 为动量的测量值和估计的误差值。在表格的最后一列给出这些粒子的符号，其中正号表示如果粒子穿过云室向下运动，则粒子带正电。对立体照片进行细致的重新投影发现，每一对径迹都是共点的。而且，每对中的两条径迹均出现在具有相同亮度的云室的中间区域，径迹周围雾化背景的存在表示该情况具有良好的凝聚状态。

表 1. 实验数据

照片	H（高斯）	α（度）	径迹	p（电子伏/光速）	Δp（电子伏/光速）	符号
1	3,500	66.6	a	3.4×10^8	1.0×10^8	+
			b	3.5×10^8	1.5×10^8	−
2	7,200	161.1	a	6.0×10^8	3.0×10^8	+
			b	7.7×10^8	1.0×10^8	+

尽管这两条叉状径迹在很多重要的方面是不一样的，但它们具有至少两个相同的基本特征：第一，每一个都由两条叉状径迹组成，且尖端在气体中；第二，没有任何迹象表明任何一条径迹的符号是由第三个离子化的粒子引起的。另外，在 3 厘米厚的铅板的观测实验中几乎没有出现与这种叉状径迹相类似的情况。然而，如果叉状径迹是由任何一种类型的碰撞过程产生的，则预期在气体中能探测到的分叉应多几百倍。因此，以上讨论表明，这些径迹并不是由碰撞过程产生的，而是由某种自发过程引起的，而这种过程发生的概率依赖于粒子走过的路程而不是横向穿过粒子的数量。

下面更细节的讨论将支持以上的结论。例如，如果叉状径迹的任意一支是由于

track were due to the deflexion of a charged particle by collision with a nucleus, the transfer of momentum would be so large as to produce an easily visible recoil track. Then, again, the attempt to account for Fig. 2 by a collision process meets with the difficulty that the incident particle is deflected through 19° in a single collision in the gas and only 2.4° in traversing 3 cm. of lead—a most unlikely event. One specific collision process, that of electron pair production by a high-energy photon in the field of the nucleus, can be excluded on two grounds: the observed angle between the tracks would only be a fraction of a degree, for example, 0.1° for Fig. 1, and a large amount of electronic component should have accompanied the photon, as in each case a lead plate is close above the fork.

We conclude, therefore, that the two forked tracks do not represent collision processes, but do represent spontaneous transformations. They represent a type of process with which we are already familiar in the decay of the meson into an electron and an assumed neutrino, and the presumed decay of the heavy meson recently discovered by Lattes, Occhialini and Powell[1].

The Masses of the Incident Particles

Let us assume that a particle of mass M and initial momentum P is transformed spontaneously into two particles of masses m_1 and m_2, momenta p_1 and p_2 at angles of θ and φ with the direction of the incident particle. Then the following relations must hold:

$$\sqrt{M^2c^4+P^2c^2}=\sqrt{m_1^2c^4+P_1^2c^2}+\sqrt{m_2^2c^4+P_2^2c^2} \tag{1}$$

$$P=p_1\cos\theta+p_2\cos\varphi \tag{2}$$

$$p_1\sin\theta=p_2\sin\varphi. \tag{3}$$

These general relations may be used to obtain the mass of the incident particle as a function of the assumed messes of the secondary particles.

The value of M must be greater than that obtained by taking the rest masses of the secondary particles as small compared with their momenta; thus the minimum value $M_{\min}$ is given by the following equation:

$$M_{\min}c^2=c\sqrt{(p_1+p_2)^2-P^2}. \tag{4}$$

Applying this equation to the forked track of Fig. 1, after calculating P from the observed values of p_1 and p_2, it is found that $M_{\min}$ is $(770\pm200)m$, where m is the mass of the electron. The application of equation (4) to the forked track of Fig. 2, however, after calculating p_2 from the observed values of P and p_1, shows that $M_{\min}=(1{,}700\pm150)m$. This

带电粒子与核粒子碰撞发生偏转而产生的，动量的转移将足以产生一个清晰易辨的反冲径迹。这时，试图用碰撞过程来解释图 2 再次遇到了困难，即入射粒子在气体中的单个碰撞过程中偏离了 19°，而在穿透 3 厘米铅板的过程中仅偏离了 2.4°，这几乎是不可能的。基于以下两点考虑，我们可以排除在核子场中由高能光子产生电子对这种特有的碰撞过程：被观测到的叉状径迹之间的角度仅仅是 1 度的几分之一，例如图 1 中叉状径迹的角度只有 0.1°；而且，在每次实验中铅板紧挨着叉状径迹的上方都应该有大量的电子伴随着光子出现，但事实上却没有发现大量电子的出现。

因此，我们得出结论：这两条叉状径迹与碰撞过程无关，而是由自发转变过程产生的。它们代表了我们已经熟悉的介子衰变成电子和一种假定的中微子的过程，而且，最近拉特斯、奥恰利尼和鲍威尔 [1] 已经观测到这种假想的重介子的衰变过程。

入射粒子的质量

让我们假定质量为 M、初始动量为 P 的一个粒子自发转化为两个粒子，这两个粒子的质量分别为 m_1 和 m_2，动量分别为 p_1 和 p_2，与入射粒子的夹角分别为 θ 和 φ，则该过程应该满足以下关系式：

$$\sqrt{M^2c^4+P^2c^2}=\sqrt{m_1^2c^4+P_1^2c^2}+\sqrt{m_2^2c^4+P_2^2c^2} \tag{1}$$

$$P=p_1\cos\theta+p_2\cos\varphi \tag{2}$$

$$p_1\sin\theta=p_2\sin\varphi \tag{3}$$

这些普适的关系式可以用来求解入射粒子的质量，求得的解为假定的次级粒子质量的函数。

假设次级粒子的静止质量相对于其动量来说很小，则真实的 M 值大小应该比在该假设极限情况下得到的解要大，因此 M 的最小值 $M_{\min}$ 可由下面的公式给出：

$$M_{\min}c^2=c\sqrt{(p_1+p_2)^2-P^2} \tag{4}$$

通过测得的 p_1 和 p_2 值计算得出 P 之后，将以上公式用于求解图 1 中的叉状径迹，计算得到 $M_{\min}$ 为 $(770\pm200)m$，其中 m 为电子的质量。然而由观测得到的 P 和 p_1 值得出 p_2 后，将等式（4）用于求解图 2 中的叉状径迹，计算得到 $M_{\min}=(1,700\pm150)m$。这一质量值将对应于质量为两倍质量最小值的入射粒子发生电离，这与观测到的电

value of the mass would require an ionization for the incident particle of twice minimum, which is inconsistent with the observed ionization. We are therefore justified in assuming that the real value of P is greater than the observed value which, as indicated in Table 1, has a large error. If larger values of P are assumed, then M_{min} is reduced in value. The lowest value of M_{min} is $(980\pm150)m$ if P is 14.5×10^8 eV./c. Beyond this value of P the mass increases slowly with increasing momentum. No choice of incident momentum will bring the mass of the incident particle below $980m$.

In the special case where the incident particle disintegrates transversely into two particles of equal mass m_0, giving a symmetrical fork, equation (1) reduces to the following expression,

$$\frac{M}{m} = \frac{2m_0}{m}\left(1 + \frac{p^2c^2}{m_0^2c^4} \cdot \sin^2\theta\right)^{1/2}, \tag{5}$$

where p is the momentum of each of the secondary particles. Some typical results for different assumed secondary particles, calculated from equation (5), are given in Table 2. On the reasonable assumption that the secondary particles are light or heavy mesons, that is, with masses of $200m$ or $400m$, we find that the incident particle in each photograph has a mass of the order of $1,000m$.

Table 2. Mass of incident particle as a function of mass of secondary particle

Photograph	Assumed secondary particle m_0/m	Momentum of observed secondary particle (eV./c.)	Incident particle M/m
1	0	$3.5\times10^8\pm1.0\times10^8$	770 ± 200
	200	"	870 ± 200
	400	"	$1,110\pm150$
	1,837	"	$3,750\pm\ 50$
2	0	$7.7\times10^8\pm1.0\times10^8$	980 ± 150
	200	"	$1,080\pm100$
	400	"	$1,280\pm100$
	1,837	"	$3,820\pm\ 50$

Upper values of the masses of the incident particles may also be obtained from the values of the ionization and the momenta. Thus for each of the observed particles in Fig. 1, the ionization is indistinguishable from that of a very fast particle. We conclude, therefore, that $\beta=v/c \geqslant 0.7$. Since the momentum of the incident particle may be found from the observed momenta of the secondary particles, we can apply equation (1) to calculate M. In this way we find $M/m \leqslant 1,600$. Again, since the ionization of the incident particle in Fig. 2 is light, $\beta \geqslant 0.7$, from which it can be shown that $M/m \leqslant 1,200$. This last result, however, must be taken with caution because of the uncertainty in the measured value of the momentum of the incident particle.

One further general comment may be made. This is that the observation of two spontaneous disintegrations in such a small number of penetrating showers suggests that

离情况不一致。因此，我们有理由认为动量 P 的真实值大于测量值，正如表 1 中给出的那样，存在很大的误差。如果假定动量 P 的值更大，则 M_{min} 将会减小。如果 P 为 14.5×10^8 电子伏 / 光速，则 M_{min} 的最小值为 $(980\pm150)m$。当动量 P 超过这一值后，入射粒子的质量随着动量的增加而缓慢地增加。无论动量取何值，入射粒子的质量都不会低于 $980m$。

在一个入射粒子分解为两个质量均为 m_0 的粒子的特定过程中，产生了一个对称的叉状径迹，而等式（1）将简化成以下表达式：

$$\frac{M}{m}=\frac{2m_0}{m}\left(1+\frac{p^2c^2}{m_0^2c^4}\cdot\sin^2\theta\right)^{1/2} \tag{5}$$

式中 p 为每个次级粒子的动量。对于不同的假定的次级粒子，由等式（5）计算得到的特定结果如表 2 所示。在次级粒子分别为质量 $200m$ 的轻介子或质量 $400m$ 的重介子的合理假定下，我们发现，每一张照片中入射粒子的质量都在 $1,000m$ 的量级上。

表 2. 入射粒子质量与次级粒子质量的函数关系

照片	假定的次级粒子 m_0/m	探测到的次级粒子动量（电子伏 / 光速）	入射粒子 M/m
1	0	$3.5\times10^8\pm1.0\times10^8$	770 ± 200
	200	”	870 ± 200
	400	”	$1,110\pm150$
	1,837	”	$3,750\pm50$
2	0	$7.7\times10^8\pm1.0\times10^8$	980 ± 150
	200	”	$1,080\pm100$
	400	”	$1,280\pm100$
	1,837	”	$3,820\pm50$

上表中入射粒子的质量值也可以通过电离值和动量值计算得到。因此对于图 1 中观测到的每一种粒子，其电离情况与快粒子的电离情况是难以辨别的。因此，我们得出结论 $\beta=v/c\geqslant0.7$。既然入射粒子的动量可以由探测到的次级粒子的动量得出，我们就可以用公式（1）计算得出入射粒子的质量 M。按这种方法分析，我们得到 $M/m\leqslant1,600$。此外，由于图 2 中入射粒子的电离结果是轻粒子，$\beta\geqslant0.7$，由此得到 $M/m\leqslant1,200$。然而，我们必须谨慎处理最后的结果，因为入射粒子动量的测量值具有不确定性。

下面进行更深一步的全面讨论。即在这种少量的贯穿簇射中观测到的两个自发

the life-time of the unstable particles is much less than the life-time of the ordinary meson. An approximate value of this life-time may be derived as follows. The probability of an unstable particle of life-time τ_0 decaying in a short distance D is given by

$$p = \frac{D(1-\beta^2)^{1/2}}{\tau_0 c\beta}. \qquad (6)$$

Since the total number of penetrating particles in the penetrating showers so far observed is certainly less than 50, we must assume that the number of our new unstable particles is unlikely to have been greater than 50. Since one particle of each type has been observed to decay, we can therefore put $p \approx 0.02$. Setting $D \approx 30$ cm., and $\beta = 0.7$, we find from equation (6) that $\tau_0 = 5.0\times10^{-8}$ sec.

We shall now discuss possible alternative explanations of the two forks.

Photograph 1. We must examine the alternative possibility of Photograph 1 representing the spontaneous disintegration of a charged particle, coming up from below the chamber, into a charged and an uncharged particle. If we apply the argument which led to equation (4) to this process, it is readily seen that the incident particle would have a minimum mass of 1,280*m*. Thus the photograph cannot be explained by the decay of a back-scattered ordinary meson. Bearing in mind the general direction of the other particles in the shower, it is thought that assumption of the disintegration of a neutral particle moving downwards into a pair of particles of about equal mass is more probable. Further, it can be stated with some confidence that the observed ionizing particles are unlikely to be protons because the ionization of a proton of momentum 3.5×10^8 eV./c. would be more than four times the observed ionization.

Photograph 2. In this case we must examine the possibility of the photograph representing the spontaneous decay of a neutral particle coming from the right-hand side of the chamber into two charged particles. The result of applying equation (4) to this process is to show that the minimum mass of the neutral particle would be about 3,000*m*. In view of the fact that the direction of the neutral particle would have to very different from the direction of the main part of the shower, it is thought that the original assumption of the decay of a charged particle into a charged penetrating particle and an assumed neutral particle is the more probable.

We conclude from all the evidence that Photograph 1 represents the decay of a neutral particle, the mass of which is unlikely to be less than 770*m* or greater than 1,600*m*, into the two observed charged particles. Similarly, Photograph 2 represents the disintegration of a charged particle of mass greater than 980*m* and less than that of a proton into an observed penetrating particle and a neutral particle. It may be noted that no neutral particle of mass 1,000*m* has yet been observed; a charged particle of mass 990*m* ± 12 percent has, however, been observed by Leprince-Ringuet and L'héritier[2].

衰变表明，不稳定粒子的寿命要比普通介子的寿命短得多。下面的推导中将给出这个寿命的近似值。寿命为 τ_0 的不稳定粒子在短距离 D 范围内发生衰变的概率为：

$$p = \frac{D(1-\beta^2)^{1/2}}{\tau_0 c\beta} \tag{6}$$

目前为止探测到的贯穿簇射中贯穿粒子的总数量小于 50，我们必须假设这种新的不稳定粒子的数目不可能大于 50。由于观测发现每一种类型的粒子中都有一个粒子发生衰变，因此我们可以令 $p \approx 0.02$。取 $D \approx 30$ 厘米以及 $\beta = 0.7$，通过等式（6）我们可以得出 $\tau_0 = 5.0 \times 10^{-8}$ 秒。

下面我们将讨论这两种叉状径迹另一种可能的解释。

照片 1　我们必须研究照片 1 所代表的另一种可能性：从云室的下方向上运动的带电粒子发生自发衰变，成为一个带电粒子和一个不带电粒子。如果我们将导出等式（4）的讨论应用到这一过程中，将会很容易地发现，入射粒子质量的最小值为 1,280m。因此，不能用普通的背散射介子的衰变来解释这张照片。考虑到簇射中其他粒子的一般方向，我们认为一个中性粒子向下移动，转化为具有相同质量的一对粒子这种假定的可能性比较大。进一步讲，我们有理由认为探测到的电离粒子不可能是质子，因为动量为 3.5×10^8 电子伏 / 光速的质子发生电离的结果将是实际探测到的电离结果的四倍多。

照片 2　在这种情况中，我们必须研究照片所代表的一种可能性：来自云室右边的一个中性粒子自发衰变为两个带电粒子。在这一过程中应用等式（4）得到的结果表明，中性粒子质量的最小值大约为 3,000m。鉴于实际情况为中性粒子的方向应该不同于簇射中大部分粒子的方向，因此认为一个带电粒子衰变为一个带电的贯穿粒子和一个中性粒子这种最初的假设可能性更大。

通过所有的证据我们推断，照片 1 代表了一个质量介于 770m 和 1,600m 之间的中性粒子衰变为两个观测到的带电粒子的过程。类似的，照片 2 代表了质量大于 980m 但小于质子质量的带电粒子衰变为观测到的贯穿粒子和中性粒子的过程。这里还必须声明，目前我们还没有探测到质量为 1,000m 的中性粒子。然而，勒普兰斯–兰盖和莱里捷 [2] 已经探测到一个质量为 990$m \pm 12\%$ 的带电粒子。

Peculiar cloud-chamber photographs taken by Jánossy, Rochester and Broadbent[3] and by Daudin[4] may be other examples of Photograph 2.

It is a pleasure to record our thanks to Prof. P. M. S. Blackett for the keen interest he has taken in this investigation and for the benefit of numerous stimulating discussions. We also wish to acknowledge the help given us by Prof. L. Rosenfeld, Mr. J. Hamilton and Mr. H. Y. Tzu of the Department of Theoretical Physics, University of Manchester. We are indebted to Mr. S. K. Runcorn for his assistance in running the cloud chamber in the early stages of the work.

(**160**, 855-857; 1947)

G. D. Rochester and C. C. Butler: Physical Laboratories, University, Manchester.

References:

1. Lattes, C. M. G., Occhialini, G. P. S., and Powell, C. F., *Nature*, **160**, 453, 486 (1947).
2. Leprince-Ringuet, L., and L'héritier, M., *J. Phys. Radium*.(Sér. 8), 7, 66, 69 (1946). Bethe, H. A., *Phys. Rev.*, **70**, 821 (1946).
3. Jánossy, L., Rochester, G. D., and Broadbent, D., *Nature*, **155**, 1 42 (1945). (Fig. 2. Track at lower left-hand side of the photograph.)
4. Daudin, J., *Annales de Physique*, 11e Série, **19** (Avril-Juin), 1944 (Planche IV, Cliché 16).

由亚诺希、罗切斯特、布罗德本特[3]以及多丹[4]得到的特殊的云室照片可能是照片 2 中的其他情况。

在这里感谢布莱克特教授对本项研究的极大关注及许多受益匪浅的讨论，感谢曼彻斯特大学理论物理学院罗森菲尔德教授、哈密顿先生和滋先生为我们提供的帮助，最后我们还要感谢朗科恩先生在本项研究工作的初期帮助我们操控云室。

（胡雪兰 翻译；尚仁成 审稿）

A New Microscopic Principle

D. Gabor

Editor's Note

Dennis Gabor was an employee of a large electrical manufacturer when be produced this paper, ostensibly directed at the improvement of the electron microscope but which has now become the principle underlying the technique called holography, by means of which all the information needed to construct a three-dimensional view can be stored on a two-dimensional surface, and which is widely used in devices such as bank and credit cards. Gabor moved to Imperial College London in the 1950s and received a Nobel Prize in 1971.

IT is known that the spherical aberration of electron lenses sets a limit to the resolving power of electron microscopes at about 5 A. Suggestions for the correction of objectives have been made; but these are difficult in themselves, and the prospects of improvement are further aggravated by the fact that the resolution limit is proportional to the fourth root of the spherical aberration. Thus an improvement of the resolution by one decimal would require a correction of the objective to four decimals, a practically hopeless task.

The new microscopic principle described below offers a way around this difficulty, as it allows one to dispense altogether with electron objectives. Micrographs are obtained in a two-step process, by electronic analysis, followed by optical synthesis, as in Sir Lawrence Bragg's "X-ray microscope". But while the "X-ray microscope" is applicable only in very special cases, where the phases are known beforehand, the new principle provides a complete record of amplitudes *and* phases in one diagram, and is applicable to a very general class of objects.

Fig. 1 is a broad explanation of the principle. The object is illuminated by an electron beam brought to a fine focus, from which it diverges at a semi-angle α. Sufficient coherence is assured if the nominal or Gaussian diameter of the focus is less than the resolution limit, $\lambda/2 \sin \alpha$. The physical diameter, determined by diffraction and spherical aberration of the illuminating system, can be much larger. The object is a small distance behind (or in front of) the point focus, followed by a photographic plate at a large multiple of this distance. Thus the arrangement is similar to an electron shadow microscope; but it is used in a range in which the shadow microscope is useless, as it produces images very dissimilar to the original. The object is preferably smaller than the area which is illuminated in the object plane, and it must be mounted on a support which transmits an appreciable part of the primary wave. The photographic record is produced by the interference of the primary wave with the coherent part of the secondary wave emitted

一种新的显微原理

盖伯

编者按

丹尼斯·盖伯在完成这篇论文时还是一家大型电子制造企业的一名雇员，表面上这篇论文指导了电子显微镜的改进，而如今，这篇论文业已成为全息摄影技术的原理。这项技术可以将构造三维影像所需的全部信息储存于二维表面中，并已广泛应用于银行和信用卡等设备中。20 世纪 50 年代盖伯进入伦敦帝国学院，并于 1971 年获得诺贝尔奖。

我们已经知道，电子透镜的球差决定了电子显微镜的分辨率极限约为 5 埃。虽然人们已经提出建议对物镜进行校正，但是这些都很难实现，而分辨率极限正比于球差的四次方根这一事实使得改进分辨率的希望更为渺茫。也就是说，要将分辨率提高 10 倍，需要将物镜的校正提高一万倍，这几乎是一项不可能完成的任务。

下面所描述的新的显微原理提供的方法可绕过上述困境，因为它完全无需使用电子物镜。可通过两步处理获得显微图像，即电子分析和随后的光学合成，就像劳伦斯·布拉格爵士的“X 射线显微镜”那样。但是“X 射线显微镜”只适用于那些预先知道相位的非常特殊的情况，而新原理则在一张图像中提供了振幅**和**相位的完整记录，因而适用于极其一般的各类物体。

图 1 是对该原理的一个概括性解释。用一束向细微焦点聚焦后并以半角 α 发散的电子束照明物体。如果焦点的标称直径或高斯直径小于分辨率极限值 $\lambda/2\sin\alpha$，就能确保充分的相干性。由照明系统的衍射和球差所决定的物理直径可能还要大得多。物体位于焦点后方（或前方），与焦点相隔很小的一段距离，在其后面比这段距离大很多倍的位置放置照相干板。因而这样的排列方式与电子阴影显微镜类似；但是它可以在阴影显微镜力不能及的范围使用，因为它会产生与初始图像截然不同的图像。物体最好小于物平面中的照明面积，而且必须被固定在一个可透射相当一部分初级波的支架上。通过初级波与物体发出的次级波中的相干部分之间的干涉产生照相记录结果。可以看到，至少在图像靠外的区域中会出现干涉极大，其位置非常

by the object. It can be shown that, at least in the outer parts of the diagram, interference maxima will arise very nearly where the phases of the primary and of the secondary wave have coincided, as illustrated in Fig. 1.

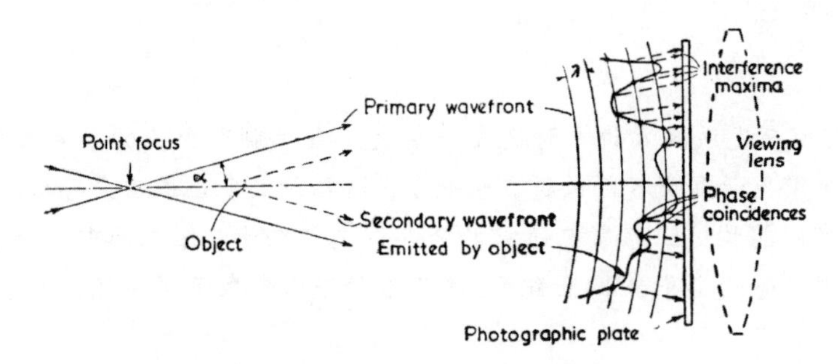

Fig. 1. Interference between homocentric illuminating wave and the secondary wave emitted by a small object

If this photograph is developed by reversal, or printed, the loci of maximum transmission will indicate the regions in which the primary wave had the same phase as the modified wave, and the variations of the transmission in these loci will be approximately proportional to the intensity of the modified wave. Thus, if one illuminates the photographic record with an optical imitation of the electronic wave, only that part of the primary wave will be strongly transmitted which imitates the modified wave both in phases and in amplitudes. It can be shown that the "masking" of the regions outside the loci of maximum transmission has only a small distorting effect. One must expect that looking through such a properly processed diagram one will see behind it the original object, as if it were in place.

The principle was tested in an optical model, in which the interference diagram was produced by monochromatic light instead of by electrons. The print was replaced in the apparatus, backed by a viewing lens which admitted about $\sin \alpha = 0.04$, and the image formed was observed and ultimately photographed through a microscope. It can be seen in Fig. 2 that the reconstruction, though imperfect, achieves the separation of some letters which could just be separated in direct observation of the object through the same optical system. The resolution is markedly imperfect only in the centre, where the circular frame creates a disturbance. Other imperfections of the reconstruction are chiefly due to defects in the microscope objectives used for the production of the point focus, and for observation.

It is a striking property of these diagrams that they constitute records of three-dimensional as well as of plane objects. One plane after another of extended objects can be observed in the microscope, just as if the object were really in position.

接近于图 1 所示的初级波相位和次级波相位达到一致的位置。

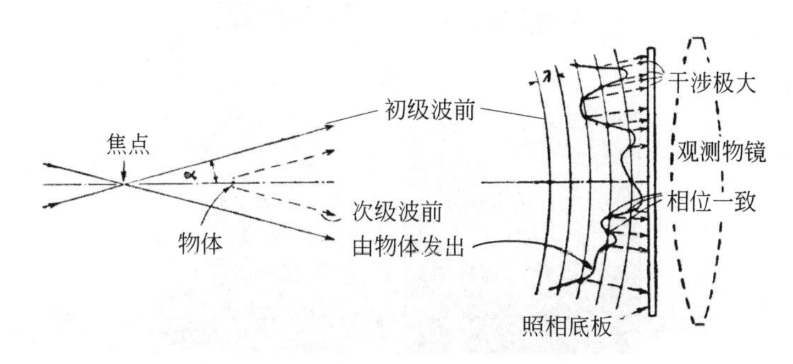

图 1. 共心照明波与微小物体发出的次级波之间的干涉

如果将照片进行反转处理或者翻印出来，最强的透射位点就代表了初级波（译者注:“初级波”的一部分被用作物体的照明光；另一部分，即其未被物体散射的部分被用作参考光）与修正波（译者注:“修正波”是“初级波”照射在物体上，被物体散射而发生改变的波，也就是“物波”）具有相同相位的区域，而这些位置上透射率的变化将近似正比于修正波的强度变化。因此，如果用电子波的光学模拟去照射照相记录，只有能够同时模拟修正波的相位和振幅的部分初级波才会被强烈透射。可以看出，对最强透射位置之外区域的“掩蔽”只会产生小的扭曲效应。应该可以预期，通过仔细查看这类经过适当处理的图像，我们将会看到它背后的原物体，好像它就在眼前一样。

可用一个光学模型来检验这一原理，即用单色光而不是电子产生干涉图像。将装置中的照片放回原处，背后加一面观测物镜，它允许的入射半角近似为 $\sin\alpha = 0.04$，通过一台显微镜观察所形成的图像并最终进行拍照。在图 2 中可以看到，虽然重现影像并不完美，但是利用相同的光学系统直接观察物体时刚好可以区分的那些字母，在重现影像中也能够被区分。只有中心部分的分辨率明显不足，那是圆框造成了干扰。重现影像的其他不足主要是由用以产生焦点和进行观测的显微镜物镜的瑕疵导致的。

这些图像具有一个引人注目的特性，即它们能像记录平面物体一样地记录三维物体。可以在显微镜中逐一地连续观察延伸物体的不同平面，就好像它们真的在那个位置一样。

Fig. 2. (*a*) Original micrograph, 1.4 mm. diameter. (*b*) Micrograph, directly photographed through the same optical system which is used for the reconstruction (*d*). Ap. 0.04. (*c*) Interference diagram, obtained by projecting the micrograph on a photographic plate with a beam diverging from a point focus. The letters have become illegible by diffraction. (*d*) Reconstruction of the original by optical synthesis from the diagram at the left. To be compared with (*b*). The letters have again become legible

Racking the microscope through and beyond the point focus, one finds a second image of the original object, in central-symmetrical position with respect to the point focus. The explanation is, briefly, that the photographic diagram cannot distinguish positive and negative phase shifts with respect to the primary wave, and this second image corresponds to the same phase shifts as the original, but with reversed sign.

If the principle is applied to electron microscopy, the dimensions in the optical synthetizer ought to be scaled up in the ratio of light waves to electron waves, that is, about 100,000 times. One must provide an illuminating system which is an exact optical imitation of the electronic condenser lens, including its spherical aberration. To avoid scaling-up the diagram, one has to introduce a further lens, with a focal length equal to the distance of the object from the photographic plate in the electronic device, in such a position that the plate appears at infinity when viewed from the optical space of the point focus. Work on the new instrument, which may be called the "electron interference microscope", will now be taken in hand.

I wish to thank Mr. I. Williams for assistance in the experiments, and Mr. L. J. Davies, director of research of the British Thomson-Houston Company, for permission to publish this note.

(**161**, 777-778; 1948)

D. Gabor: Research Laboratory, British Thomson-Houston Co., Ltd., Rugby.

图 2. (*a*) 原始显微图，直径为 1.4 毫米。(*b*) 显微图，通过直接照相得到，与重现影像 (*d*) 使用相同的光学系统。Ap. 0.04。(*c*) 干涉图，利用从一个焦点发散出来的射线束将显微图投影于照相干板上得到。字母由于衍射而变得模糊。(*d*) 利用光学合成从左侧图像得到的重现影像。以 (*b*) 为参照，字母再度由模糊变得清晰。

编者注：图中所使用的三个人名 Huygens（惠更斯）、Young（杨）、Fresnel（菲涅耳）代表证明光的波动性的三位重要科学家。

将显微镜经过焦点向另一侧移动，我们会在相对于该焦点中心对称的位置发现原物体的第二个像。对此现象的简要解释是，照相图像无法区分初级波的正相移与负相移，这第二个像与原物体的相移相同，但是符号相反。

如果将该原理应用于电子显微镜，那么在光学合成仪器中的尺度应该按光波对电子波的比值（即大约 100,000 倍）同比例增加。同时必须提供一个照明系统，它是电子聚焦透镜及其球差的严格光学模拟。为避免图像按比例增大，还必须再引入一个透镜，其焦距等于从电子装置中的照相干板到物体之间的距离，其放置位置应使得从焦点的光学距离看来干板出现在无穷远处。对于这种我们不妨称为“电子干涉显微镜”的新型装置的研究即将展开。

我要感谢威廉姆斯先生在实验中给予的协助，感谢英国汤姆森–休斯敦电气公司研究主管戴维斯先生对这篇短文的公开发表给予允准。

（王耀杨 翻译；熊秉衡 审稿）

Observations with Electron-sensitive Plates Exposed to Cosmic Radiation

R. Brown *et al.*

Editor's Note

Inspired by recent evidence of new particles having a mass approximately 1,000 times that of the electron, physicist Cecil Powell and colleagues here report on high-altitude cloud-chamber experiments on cosmic rays: high-energy particles coming from space. They analyse one track from a particle having unit charge, called a "*k*-particle" (later a kaon), which entered the chamber and subsequently decayed into two other particles. Estimating the *k*-particle's mass by counting photographic grains, they found a value of roughly 1,080 times the electron mass. They interpret its decay as a process involving the creation of two particles called pions. These discoveries led towards the discovery, within a decade, that interactions involving the weak nuclear force do not respect mirror symmetry.

ONE of the first events found in the examination of electron-sensitive plates exposed at the Jungfraujoch is represented in the mosaic of photomicrographs shown in Fig. 8. There are two centres, *A* and *B*, from which the tracks of charged particles diverge, and these are joined by a common track, *t*. Because of the short duration of the exposure, and the small number of disintegrations occurring in the plate, the chance that the observation corresponds to a fortuitous juxtaposition of the tracks of unrelated events is very small—of the order 1 in 10^7. It is therefore reasonable to exclude it as a serious possibility. Further observations in support of this assumption are presented in a later paragraph.

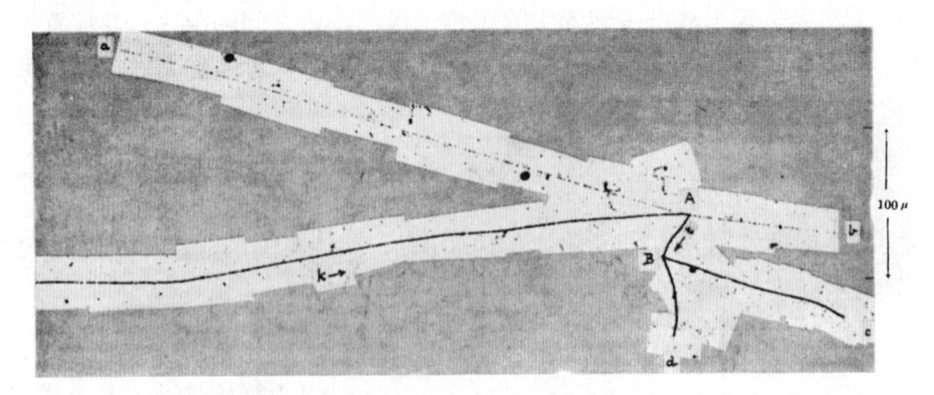

Observer: Mrs. W. J. van der Merwe

Fig. 8

在宇宙辐射下曝光的电子敏感底片的观测

布朗等

编者按

最近新发现了一种大约具有1,000倍电子质量的粒子，受到其实验证据的启发，物理学家塞西尔·鲍威尔和同事们在本文中报道了高海拔云室宇宙射线实验：来自太空的高能粒子。他们分析了一个具有单位电荷的“*k*粒子”（后来称为K介子）的径迹，这个粒子进入云室后衰变为另外两个粒子。通过计算照相颗粒他们估算了这个*k*粒子的质量，结果发现其质量大约是电子质量的1,080倍。他们把这种衰变解释为一个产生两个π介子的过程。这些结果促使人们在之后十年内有了新的发现，即包含弱核力相互作用并不遵守镜面对称。

在检查少女峰上曝光的电子敏感底片时，我们最初的重要发现之一就是如图8中显微照片的拼嵌图所表示的记录。图中存在两个中心A和B，带电粒子的径迹从这两点开始分裂，并且以常见的径迹t相连接。由于曝光的时间很短，并且仅有少量的蜕变反映在底片上，因此观测结果中对应于无关事件径迹偶然交叉重叠的可能性是非常小的，仅为1×10^{-7}量级，因此我们可以将这种重要的可能性合理地排除。后面我们将会给出进一步支持上述假设的观测结果。

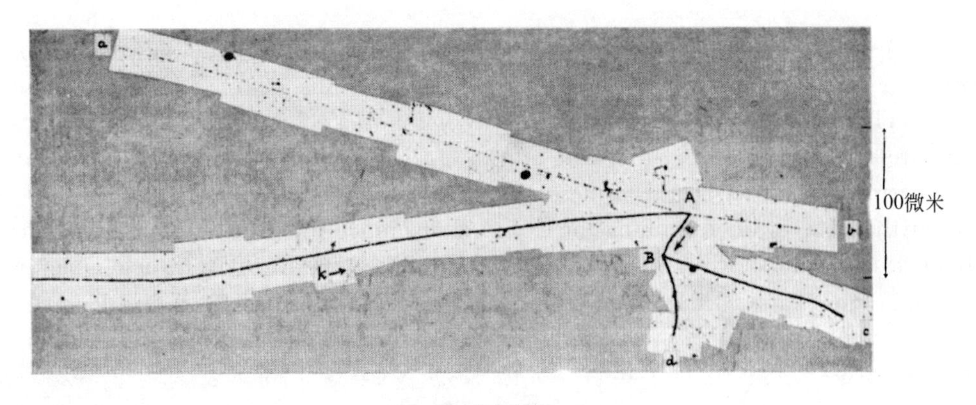

观测者：范德梅伟

图8

An inspection of the track k shows that the particle producing it approached the centre of disintegration A. The range of the particle in the emulsion exceeds 3,000 μ, and there is continuous increase in the grain-density along the track in approaching A. Near A, the grain-density is indistinguishable from that of particles of charge e, recorded in the same plate, near the end of their range.

The evidence for the direction of motion of the particle based on grain-counts is supported by observations on the small-angle deviations in the track due to Coulomb scattering. These deviations are most frequent near A, and the scattering is less marked at points remote from it.

From these observations, it is reasonable to conclude that the particle k approached the point A; that it carried the elementary electronic charge and that it had reached, or was near, the end of its range at the point A. We therefore assume that the particle k initiated the train of events represented by the tracks radiating from A and B. It follows that the particle producing track t originated in star A, and produced the disintegration B. In order to analyse the event, we first attempted to determine the mass of the particle k.

Mass Determinations by Grain-Counts

About a year ago, experiments were made in this Laboratory to determine the ratio, m_π/m_μ, of the masses of π- and μ-mesons, by the method of grain-counting[5], and by studying the small-angle scattering of the particles in their passage through the emulsion[4]. The values obtained by the two methods were $m_\pi/m_\mu = 1.65 \pm 0.11$, and $m_\pi/m_\mu = 1.35 \pm 0.10$*, respectively. Recent experiments at Berkeley[6] suggest that the true value is 1.33 ± 0.02, a result which throws serious doubt on the reliability of the method based on grain-counts. Because of the advantage of this method, and of the important conclusions which have been based on it, experiments were made to determine the conditions in which reliable results can be obtained.

In the first experiments[5], the two most serious experimental difficulties arose from the fading of the latent image and from the variation of the degree of development with depth. This made it necessary to work only with tracks formed contemporaneously; to compare the grain-density along the tracks of the π- and μ-mesons of the same pair. As a result, the tracks of the π-mesons available for measurement were, in most cases, shorter than 400 μ. In continuing the experiments, much more favourable conditions were obtained by using short exposures, so that the effects of fading were negligible; and by developing the plates by the method employed by Dilworth, Occhialini and Payne[7], which

* For the following reasons, the limits of error quoted above, in the determination of m_π/m_μ by observations on scattering, are less than those given in ref. 4. Previously, values for the mass of the different types of mesons, classified phenomenologically, were given separately. It is now known, however, that at least the majority of the σ-mesons are π^--particles; and the ρ-mesons, μ^+- and μ^--particles. The different results can therefore be combined to give a value for m_π/m_μ with a greater statistical weight.

对径迹 k 的检验表明，产生它的粒子到达蜕变中心 A 的附近。该粒子在乳胶中的射程超过 3,000 微米，在接近 A 的过程中，径迹的颗粒密度逐渐增加。在 A 附近，其颗粒密度与同在该底片上记录的电荷为 e 的粒子接近射程终端处的颗粒密度无法区分。

观测发现库仑散射造成了径迹的小角度偏离，这支持了基于颗粒计数得到的关于粒子运动方向的证据。这些偏离在靠近 A 时出现得最为频繁，而散射在远离 A 的各点处比较不明显。

根据这些观测我们有理由认为，粒子 k 向 A 点靠近；它携带着基本电子电荷，并在 A 点附近达到或接近其射程的终端。因此我们假设，粒子 k 引发了一连串由 A 和 B 的辐射径迹所代表的事件。由此可知，该粒子产生了源自 A 点并在 B 点产生蜕变的径迹 t。为了分析该事件，我们首先尝试了对粒子 k 的质量进行确定。

根据颗粒计数所做的质量测定

约在一年前，本实验室采用颗粒计数的方法[5]以及研究粒子通过乳胶时小角度散射的方法[4]，进行了测定 π 介子和 μ 介子（编者注：现在粒子物理认为“μ 介子”不是介子，而是一种轻子，因此正确的称谓应为“μ 子”。考虑到历史原因，文中仍保留“μ 介子”的译法。）的质量比 m_π/m_μ 的实验。用这两类方法得到的结果分别为 $m_\pi/m_\mu=1.65\pm0.11$ 和 $m_\pi/m_\mu=1.35\pm0.10$*。最近在伯克利进行的实验[6]表明，其真值为 1.33 ± 0.02，这个结果使人们对颗粒计数方法的可靠性产生了严重的怀疑。但是考虑到该方法的优点，以及据此得到的一些重要结论，我们进行了一些实验来确定在何种条件下才能获得可靠的结果。

在首次实验中[5]，两个最为严重的实验难点来自潜影的褪色和显像度随深度的变化。若要对同一对 π 介子和 μ 介子径迹的颗粒数密度进行对比，就必须研究同时形成的径迹。然而这样一来，可以用于测量的 π 介子的径迹大多小于 400 微米。在后续的实验中，通过运用短期曝光的方法获得了更多良好条件，这使得褪色效应可以被忽略；与此同时，还采用了迪尔沃思、奥恰利尼和佩恩[7]所使用的冲洗底片的

* 基于以下理由，上文所引用的由散射观测给出的 m_π/m_μ 值的误差极限要比参考文献 4 中的小。以前，基于现象分类的不同介子的质量是分别给出的。然而现在已知，至少 σ 介子中大部分是 π^- 粒子，ρ 介子中大部分是 μ^+ 和 μ^- 粒子，将不同的结果综合起来可得一个更大统计权重下的 m_π/m_μ 数值。

gives a nearly uniform degree of development with depth.

In the plates obtained by these methods, it is legitimate to compare the grain-density in the tracks of unrelated particles. Further, it is now known that at least the majority, and possibly all, the mesons which produce "stars" are π^--particles[6,8]; and that most of the ρ-mesons are μ^+- and μ^--particles. In determining m_π and m_μ, we have therefore made measurements on the tracks of π^+- and π^--, μ^+- and μ^--particles, of length greater than 1,000 μ, comparing the results with those of similar measurements made on the tracks of protons. In these conditions, we have found $m_\pi/m_\mu = 1.33 \pm 0.05$. A detailed account of the observations will be published elsewhere; but, for the purpose of the present paper, it is sufficient to note that the results appear to be in good accord with those obtained by other methods. We conclude that, using the Ilford *C*2 emulsion in the new conditions, reliable information can be obtained.

We have seen that the conditions of uniform development and absence of fading have been achieved in the present experiments with the new Kodak emulsions, and we therefore attempted to measure the mass of particles by similar methods to those employed with the Ilford plates. The results obtained in observations on the tracks of four protons and four μ-particles, occurring in the same plate, are represented in Fig. 9. In this figure, the number of grains per unit length in the tracks is plotted for different values of the residual range; and the mean values, for tracks of the same type, are indicated by the full lines. The ratio of the masses of the two types of particles can be deduced by making a comparison of the values of the residual range at which the grain densities have the same value. The result thus obtained is $m_\mu = 220 \pm 20\ m_e$.

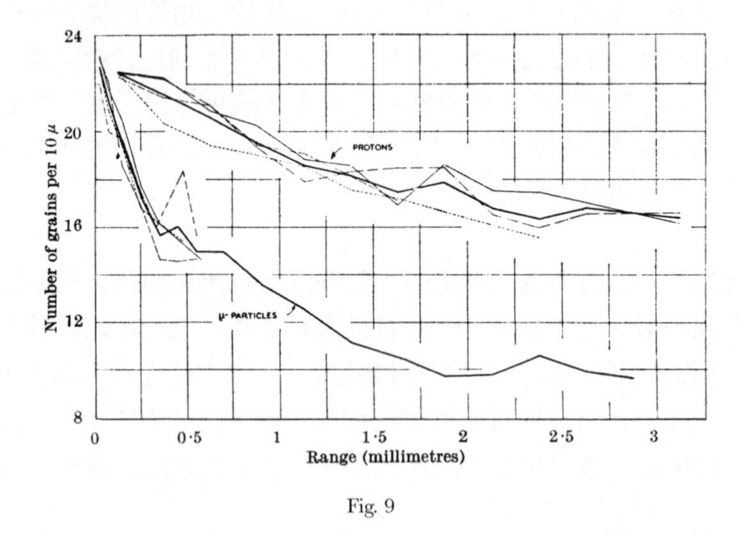

Fig. 9

Using similar methods, we have made estimates of the mass of the particle, *k*, and the measurements are represented in Fig. 10. This figure shows the mean values of the grain-density in the tracks of the four μ-mesons and four protons, together with the corresponding results for the particle *k*. All the tracks under consideration occurred in the same plate.

方法，这种方法获得了与深度几乎一致的显像度。

用上述方法得到的底片来比较不相关粒子径迹的颗粒密度是合理的。此外，现在知道，产生“星”点的介子中至少多数甚至可能全部是 π^- 粒子 [6,8]；而大多数 ρ介子是 μ^+ 粒子和 μ^- 粒子。因此在测定 m_π 和 m_μ 的实验中，我们对长度大于 1,000 微米的 π^+ 和 π^-、μ^+ 和 μ^- 粒子的径迹进行了测量，并将其与用类似方法对质子径迹的测量结果做了比较。在上述条件下，我们得出 m_π/m_μ=1.33 ± 0.05。观测的详细计算将另行发表；考虑到本文的目的，这里只需指出该结果与用其他方法获得的结果非常一致就足够了。由此我们认为，在新的条件下，用伊尔福 $C2$ 乳胶能够获得可靠的信息。

我们已经知道，目前实验中所用的新柯达乳胶已可以同时满足均匀显像和不变色的条件，因此我们试图用类似方法并使用伊尔福胶片对粒子的质量进行测量。图 9 显示了在同一胶片中四个质子和四个 μ 粒子的径迹的观测结果。在图中，用径迹中单位长度的颗粒数对不同残余射程值作图；同一类型径迹的平均值用粗线表示。两类粒子的质量比可通过比较颗粒密度相同时所对应的残余射程值导出。由此得到的结果为 $m_\mu = 220 \pm 20\ m_e$。

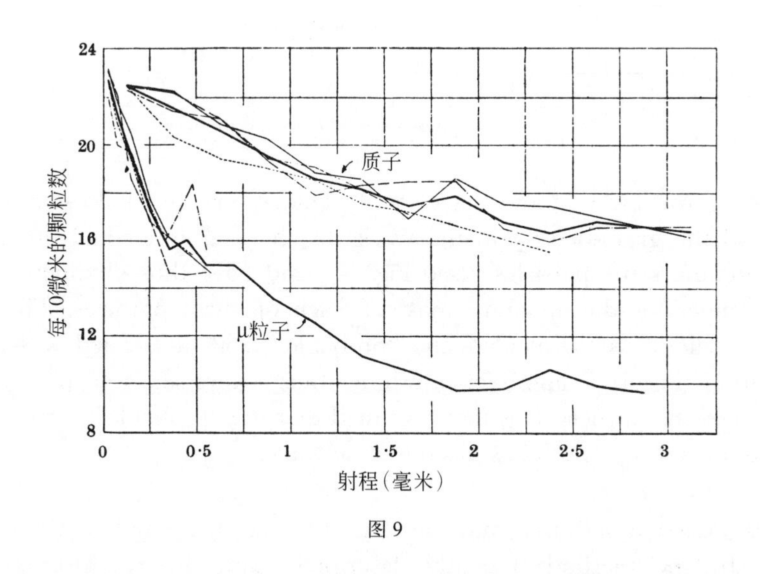

图 9

用类似的方法，我们已估算了粒子 k 的质量值，测量结果如图 10 所示。这个图给出了四个 μ 介子和四个质子的径迹中的颗粒密度的平均值，同时也给出了粒子 k 的颗粒密度。上述所有径迹均在同一张底片中获得。

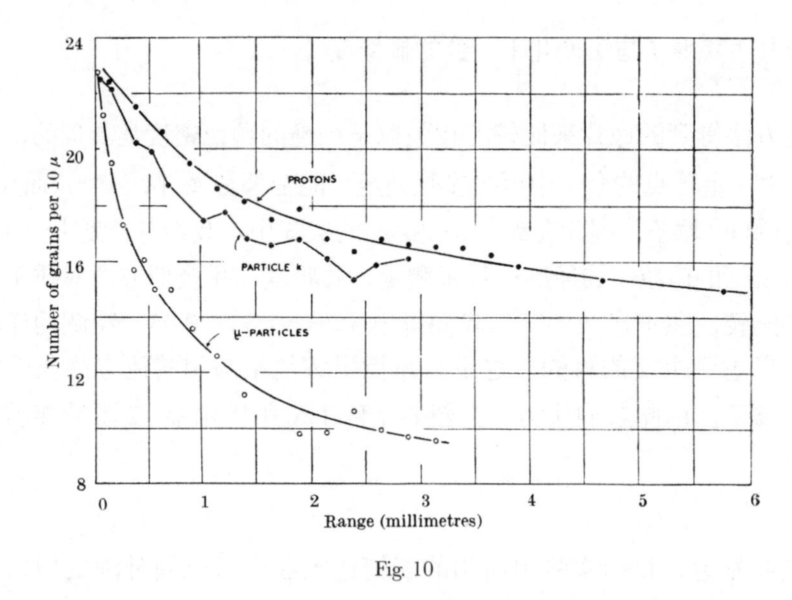

Fig. 10

Table 1 shows the values of the mass of the particle, *k*, as determined from these results, by making a comparison of the grain-density in the track of the particle with the mean curve for protons. The values thus obtained are all independent and the mean is m_k=1,080±160 m_e.

Table 1. Determination of the ratio, m_P/m_k, of the mass of a proton to that of particle, *k*, by grain-counting

	Individual independent values						
m_P/m_k	1.77	1.88	1.49	1.64	2.17	1.79	1.32
	1.71	1.66	1.27	1.69	2.13	1.55	
	Mean value: 1.70; m_k =1,080 ± 160 m_e						

The limits of error given above have been deduced in the following manner: We have compared the grain-density in the tracks of the four individual protons with the mean curve for the same particles—(see Fig. 9)—and have thus obtained a number of independent values for the apparent mass of each of these particles. The distribution in these values allows us to calculate the "probable error" associated with the mass as determined from the observations on any one track, expressed as a percentage of the apparent mass of the particle. It is then assumed that the "probable" percentage error in the calculated mass of the particle *k* has the same value.

We have also determined the mass m_k by studying the small-angle scattering of the particle, by the methods recently described[4], and the result thus obtained is m_k = 1,800 ± 400 m_e. If the true mass of the particle is 1,080 m_e, the chance that the value obtained by observations on scattering shall be equal to, or greater than, 1,800 m_e is one in four. Because of the large statistical fluctuations associated with the observations in the scattering experiments, we give more weight to the measurements by grain-counting.

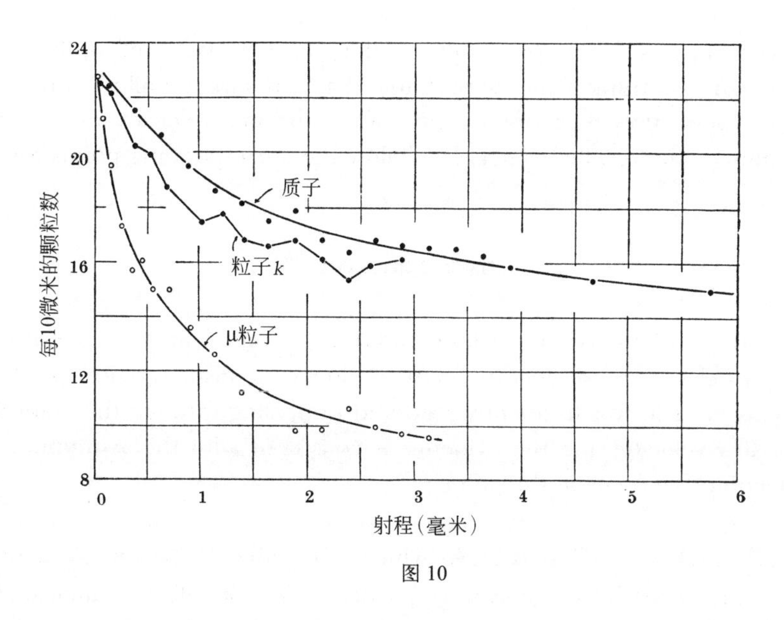

图 10

将粒子 k 径迹中的颗粒密度与质子平均曲线进行对比，表 1 给出了比较结果以及由此测定的 k 粒子的质量值。这些值都是独立的，平均值是 m_k =1,080 ±160 m_e。

表 1. 利用颗粒计数测得的质子质量与粒子 k 质量的比值 m_P/m_k

	各自独立的值						
m_p/m_k	1.77	1.88	1.49	1.64	2.17	1.79	1.32
	1.71	1.66	1.27	1.69	2.13	1.55	
	平均值：1.70; m_k =1,080 ± 160 m_e						

上述给出的误差极限由下述方式导出：我们已对四个单独质子的径迹中的颗粒密度与它们的径迹颗粒密度的平均曲线（见图 9）进行了比较，并由此得到了每个粒子表观质量的若干独立值。根据这些值的分布，我们可以计算出在依据每个单独径迹测定质量的过程中所产生的“或然误差”,并将其表示为粒子表观质量的百分比。然后我们假设计算出的粒子 k 的质量具有相同的“或然”百分比误差。

我们也用最近介绍的方法[4]，即通过研究粒子的小角度散射确定了质量 m_k，由此得到的结果是 m_k=1,800±400 m_e。如果粒子质量的真值是 1,080 m_e，通过散射观测得到的值等于或大于 1,800 m_e 的概率是 1/4。由于散射实验的观测有较大的统计涨落，因此我们给颗粒计数的测量以更大的权重。根据这些观测，似乎可以肯定，m_k 的真

It appears certain, from these observations, that the true value of m_k lies between 700 and 1,800 m_e, and we think it highly probable that it is substantially less than that of the proton. Thus every individual point representing the grain-density in the track *k*, at a particular value of the residual range, lies below the corresponding points for each of the four protons.

Disintegration "*B*"

The tracks, *c* and *d*, of the two particles emitted from point *B* are characteristic of protons or heavier particles, and we regard them as due to a disintegration produced by the particle *t*. This particle was frequently scattered in passing through the emulsion and was therefore of low velocity; and the evidence is consistent with the assumption that it had reached the end of its range at the point *B*.

The only known slow charged particle which is capable of producing a disintegration of the type represented by star *B* is a π^--particle[6,8]. We therefore assume that a negative meson of mass 286 m_e was created at the point *A*, and reached the end of its range to produce the disintegration *B*.

Transmutation "*A*"

In order to interpret the transmutation *A*, we first made a detailed examination of the tracks of the emitted particles. Of the two tracks *a* and *b*, the former has a length in the emulsion of more than 2,000 μ, and ends in the surface, whereas *b* ends in the glass and is 116 μ long. The grain-densities in the two tracks are equal to within the limits defined by the statistical fluctuations. The average grain-density in the long track *a* is 49.0 grains per 100 μ; that is, 2.17 times the value characteristic of minimum ionization for a particle of charge *e*. Unless we admit the existence of fractional values of the electronic charge, we must conclude that the particles producing the tracks *a* and *b* both carried charges of magnitude *e*.

In order to determine the possible values for the energy of the particles producing tracks *a* and *b*, we have calculated the variation with energy of the specific ionization of a particle of charge *e*, from the formula of Halpern and Hall[9], assuming the atomic composition of the emulsion to be identical with that of the Ilford *C*2 plates. This formula is a modification of that of Bloch[10]; it applies to particles moving in a solid medium and gives results in good agreement with experiment for particles of low energy. The results are shown in Fig. 11, where the specific ionization is plotted as a function of the quantity E/m, where E is the energy and m the mass of the particle, both quantities being measured in MeV. From Fig. 11, we have determined the possible values of the energy of the particles, *a*, *b*, corresponding to the observed grain-density in the tracks, assuming them to be protons, π-mesons, μ-mesons or electrons. The resulting values are tabulated in Table 2.

值位于 700 m_e 和 1,800 m_e 之间，我们认为它极可能远小于质子的质量。因此，在某一特定的残余射程值处，径迹 k 中每一个表示颗粒密度的点，都位于四个质子中相应的点之下。

“B”蜕变

由点 B 发射的两个粒子的径迹 c 和 d 具有质子或较重粒子的特性，我们认为这源自于粒子 l 产生的蜕变。该粒子在通过乳胶时频繁地散射，因此速度很低；这个现象与其在 B 点达到射程终点的假设是一致的。

目前已知能够产生具有星点 B 特征的蜕变的慢带电粒子只有 π^- 粒子 [6,8]。因此我们假定，在点 A 产生了质量为 286 m_e 的负介子，在达到其射程终点时产生 B 蜕变。

“A”嬗变

为了解释 A 嬗变，我们首先对发射粒子的径迹做了详细的检查。在径迹 a 和 b 中，前者的径迹在乳胶中的长度大于 2,000 微米，并在表面终止，而径迹 b 的长度是 116 微米且在玻璃中终止。两个径迹的颗粒密度在统计涨落所确定的极限之内相等。在长径迹 a 中，平均颗粒密度是每 100 微米 49.0 个颗粒，该密度是电荷为 e 的粒子最小电离特征值的 2.17 倍。除非我们允许存在电荷的分数值，否则我们必然会得出，产生径迹 a 和 b 的粒子的带电量均为 e 的量级。

为了测定产生径迹 a 和 b 的粒子能量的可能值，我们假定乳胶的原子组分与伊尔福 C2 胶片相同，根据哈尔彭和霍尔的公式 [9] 计算了电荷为 e 的粒子的比电离值随能量的变化。这个公式是布洛赫公式 [10] 的修正形式；该公式适用于在固体介质中运动的粒子，并给出了与低能粒子的实验非常符合的结果。结果示于图 11 中，将比电离值作为量 E/m 的函数绘图，E 和 m 分别为粒子的能量和质量，两个量均以兆电子伏为单位测量。根据图 11，假定粒子是质子、π 介子、μ 介子或电子，我们确定了对应于径迹中观测到的颗粒密度的粒子 a、b 能量的可能值。所得到的结果列于表 2 中。

Table 2. Values of the energy and momentum of the particle producing track *a*, as deduced from the observed grain-density and scattering, making various assumptions concerning the mass of the particle

Assumed particle		proton	π-meson	μ-meson	electron
Energy in MeV.	(*a*) below minimum ionization	235±95	37±13	27±11	0.13±0.05
	(*b*) above minimum ionization				>1,000
Momentum MeV./*c*	(*a*) below minimum ionization	700±160	109±22	80±15	0.4±0.1
	(*b*) above minimum ionization				>1,000
	(*c*) from scattering observations	245±40	113±18	100±16	68±11
	(*d*) from momentum balance	98±5	98±5	98±5	98±5

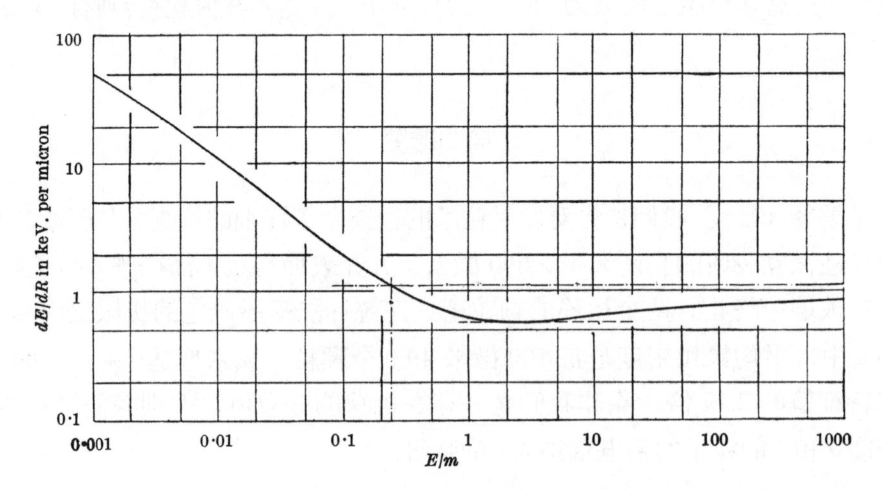

Fig. 11. Variation of the rate of loss of energy of a particle of charge $|e|$ as a function of the quantity E/m, where E is the kinetic energy and m the mass of the particle, both quantities being measured in MeV.

There are two possible interpretations of the transmutation produced at *A* by the particle *k*. We can assume, either that the particle was captured by a nucleus, or that it decayed spontaneously. From the measured values of the mass of the particle, it would be possible, from the point of view of the conservation of mass and energy, to admit that, at the end of its range in the emulsion, it was captured by a nucleus and led to the ejection of two energetic protons and a π^--particle. It appears almost certain, however, that the release in a nucleus of such a large amount of energy would lead to the "evaporation" of many nucleons, a process commonly observed in plates exposed to the cosmic radiation; and that two protons of great energy would be only two components of a "many-pronged" star. (It may be noticed that we cannot assume that the particle *k* was captured by one of the rare nuclei of heavy hydrogen, present in the gelatine. In such an interaction, the algebraic sum

表 2. 基于有关粒子质量的各类假设，根据观测颗粒密度和散射导出的产生径迹 a 的粒子的能量和动量值

假设的粒子		质子	π 介子	μ 介子	电子
能量（兆电子伏）	(a) 在最小电离作用之下	235±95	37±13	27±11	0.13±0.05
	(b) 在最小电离作用之上				> 1,000
动量（兆电子伏 / 光速）	(a) 在最小电离作用之下	700±160	109±22	80±15	0.4±0.1
	(b) 在最小电离作用之上				> 1,000
	(c) 根据散射观测	245±40	113±18	100±16	68±11
	(d) 根据动量平衡	98±5	98±5	98±5	98±5

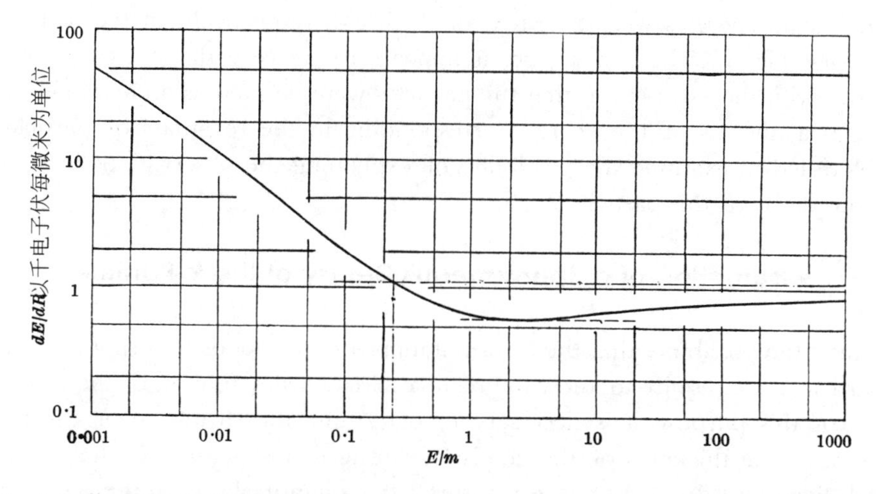

图11. 电荷为 $|e|$ 的粒子能量损失的速率随 E/m 变化的函数。其中 E 是动能，m 是粒子质量，两个量的测量单位都是兆电子伏。

粒子 k 在 A 点产生嬗变有两种可能的解释。我们可以假定，粒子或被核俘获，或产生自发衰变。从粒子质量的测量值来看，根据质量和能量守恒的观点，可能的情况是它在乳胶中射程的终点处被核俘获，并导致两个高能质子和一个 π^- 粒子的发射。然而几乎同样确定的是，如此大量的能量在核中释放将导致很多核子的“蒸发”，这个过程在对宇宙辐射曝光的底片中很常见；还可以基本确定的是两个高能质子将是“多股”星仅有的两个组分。（可以注意到，我们不可以假定粒子 k 是被存在于凝胶体中的某个稀有的重氢核所俘获。在这种相互作用中，两个初始粒子电荷的代数

of the charges on the two initial particles is 0 or $2e$, whereas that of the product particles is e or $3e$.) We shall see later that there are other objections to the hypothesis that the tracks *a* and *b* were produced by protons, or heavier nuclei of charge *e*.

It follows from the above considerations that if we are to describe the transmutation in terms of particles of which the existence is already established, we must attribute the tracks *a* and *b* either to electrons, to μ-mesons or to π-mesons. Considering the first of these possibilities, we must assume the electrons to have had an energy value greater than that corresponding to minimum ionization, namely, greater than 1,000 MeV.; for with the alternative lower value corresponding to the observed ionization, 300 keV., the particle would have had a range in the emulsion of only about 100 μ, and would have been frequently scattered. The assumption that the particles *a* and *b* were electrons is therefore inconsistent with the conservation of energy and can be rejected. We are left with the alternatives that the tracks were produced either by π- or by μ-mesons.

If the particles *a* and *b* were mesons, we must assume, in order to conserve mass-energy, that their kinetic energies were 27 MeV. or 37 MeV., respectively, in the case of μ- or π-mesons (see Fig. 11). In either case, it appears to be very difficult to reconcile the observations with the assumption that the particles were emitted as a consequence of the liberation in a nucleus of the energy corresponding to the rest-mass of particle *k*. We are therefore led to examine the possibility of explaining the observations in terms of a spontaneous decay of this particle.

Assumption of a Spontaneous Decay of the *k*-Particle

In examining the possibility that the transmutation *A* corresponds to a spontaneous decay of the particle *k*, we require to know the relative directions of motion of the three ejected particles. For this purpose it is necessary to determine the shrinkage of the emulsion; the ratio, *S*, of the thickness of the emulsion during exposure to that after it had been developed, fixed and dried. We have measured this quantity by examining the tracks of α-particles, produced in the emulsion by uncontrolled radioactive contamination. Among such "stars", it is possible to identify some, due to an original atom of radiothorium, from which an α-particle of thorium C′ was emitted. The shrinkage has been measured by determining the lengths of the projection of the corresponding tracks on the surface of the emulsion, and their apparent angles of "dip". The value of the "shrinkage" thus found is $S = 2.7 \pm 0.1$. Knowing the value of *S*, the original orientation of a track in the emulsion, before processing, can be determined, in favourable cases, with a precision of the order of 1°, by observing the apparent angle of "dip" of the particle, and the direction of its projection on the plane defined by the surface of the emulsion. Using these methods, the original directions of motion of the three particles *a*, *b* and *t* were found to be coplanar. The departure of the direction of motion of any one particle from the plane defined by the other two is less than 4°. The error in this determination is largely due to the fact that track *t* is of short range, and the particle producing it was of low velocity, and frequently scattered.

和是 0 或 $2e$，而产生的粒子的电荷代数和是 e 或 $3e$。）对于径迹 a 和 b 是由质子或带电荷 e 的较重核产生的这一假说，稍后我们将介绍其他一些反对意见。

根据上述考虑，如果我们要用已确定存在的粒子来描述嬗变，那么我们必须将径迹 a 和 b 认为是由电子、μ 介子或 π 介子这三者之一所产生的。如果是电子，则我们必须假设电子的能量大于相应的最小电离能，即大于 1,000 兆电子伏；对应于观测到的较低的电离入射能量 300 千电子伏，粒子在乳胶中仅有约 100 微米的射程，并且会被频繁地散射。因此认为粒子 a 和 b 是电子的假设与能量守恒不相符，可以被排除。我们只能在剩下的 π 介子或 μ 介子中择一作为产生径迹的粒子。

如果粒子 a 和 b 是介子，那么为了使质能守恒，我们必须假定在 μ 介子或 π 介子的情况下（见图 11），其动能分别为 27 兆电子伏和 37 兆电子伏。如果假设粒子的发射是由于核内释放了相应于粒子 k 静止质量的能量，那么无论是上述哪种介子，观测结果都很难与该假设相一致。因此，我们只能尝试用这个粒子的自发衰变来对观测结果进行解释了。

k 粒子自发衰变的假定

在确定 A 嬗变是否对应于粒子 k 的自发衰变时，我们需要知道三个射出粒子的相对运动方向。为此必须确定乳胶的收缩，即在曝光时乳胶的厚度与经过显影、定影和烘干后的乳胶的厚度比值 S。我们已经通过检验在不可控制的放射性沾染下乳胶中产生的 α 粒子的径迹测量了这个量。在这类“星”中，可以确定其中一些情况是由放射性钍的天然原子产生的，一个钍 C′ 发射一个 α 粒子。通过测定乳胶表面上相应径迹投影的长度以及其“下陷”的表观角度，已测得了乳胶的收缩。由此获得的“收缩”值为 $S = 2.7 \pm 0.1$。知道了 S 的值，那么通过观测粒子“下陷”的表观角度以及由乳胶表面确定的平面投影方向，便可以测定乳胶被处理前其中的径迹的原始取向，在较好的情况下，精度可达 1° 量级。采用这些方法，获得的三个粒子 a、b、t 运动的原始方向是共面的。任何一个粒子的运动方向与另两个粒子所确定的平面之间的偏离小于 4°。在这项测定中的误差主要来源于径迹 t 是短程的，产生它的粒子是低速的并被频繁地散射。

The values of the angles between the directions of motion of the particles in the common plane are shown in Fig. 12. The observed coplanarity makes it legitimate to assume that the three particles arise as a result of the spontaneous decay of the *k*-particle at the end of its range in the emulsion, and that they are the only product of its disintegration; that no neutral particles, which would escape observation, are emitted. It follows that the vector sum of the momenta of the three particles must be assumed to be equal to zero.

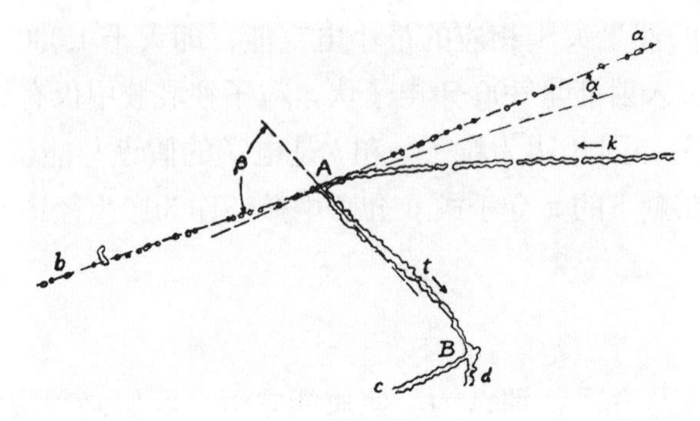

Fig. 12. Facsimile drawing of the event shown in Fig. 8, made with the projection microscope. The actual angles α and β, measured in the common plane of the three tracks, *a*, *b* and *t*, are: $\angle\alpha=9.8°$; $\angle\beta=76.6°$

If we are correct in attributing the track *t* to a π^--particle, it follows from the observed range, 45 μ, that the kinetic energy of ejection was 1.04 MeV. The corresponding value of the momentum of the particle is 17.5 MeV./*c*. From the observed directions of motion, the momenta of the particles giving tracks *a* and *b* are then found to be 98±5 and 104±5 MeV./*c*, respectively. These values are to be compared with those corresponding to electrons or mesons listed in Table 2, which have been deduced from the observed grain-density in the tracks. We have seen that the values given in Table 2 for the momenta of the two particles, if they are assumed to be electrons, are many times too large. It follows that there is a wide departure from a momentum balance if the tracks *a* and *b* are assumed to be due to either electrons or protons. Further, the values of the momenta, as deduced from observations on the scattering of the particles, are inconsistent with those obtained from grain-counts, if the particles are assumed to have been either electrons or protons (see Table 2).

The agreement between the sets of values for mesons, however, is most remarkable, and gives strong support for the assumption of a spontaneous decay of the *k*-particle. Only a very rare combination of unrelated features, including the co-planarity of the tracks and the directions of motion of the particles in the common plane, the range of the particle *t*, and the specific ionization of the particles producing tracks *a* and *b*, could produce such an agreement between the estimated values of the momenta, if the result is fortuitous.

The values of the momenta of the particles producing tracks *a* and *b*, as determined by the three different methods, are consistent, within the errors of measurements, with the assumption of a spontaneous decay of the *k*-particle whether the product particles are assumed to be μ-mesons or π-mesons. We can apply a further test by calculating the values of the rest-mass of the particle *k* which corresponds to the two different assumptions, and the results are tabulated in Table 3.

共面的粒子运动方向之间的角度值示于图 12 中。根据观测到的共面性可以合理地得出以下假设：三个粒子来自粒子 k 在乳胶的射程终端的自发衰变，并且它们是蜕变仅有的产物，整个过程没有中性粒子发射（中性粒子无法观测）。由此，必然会得出三个粒子动量的矢量和等于零的结论。

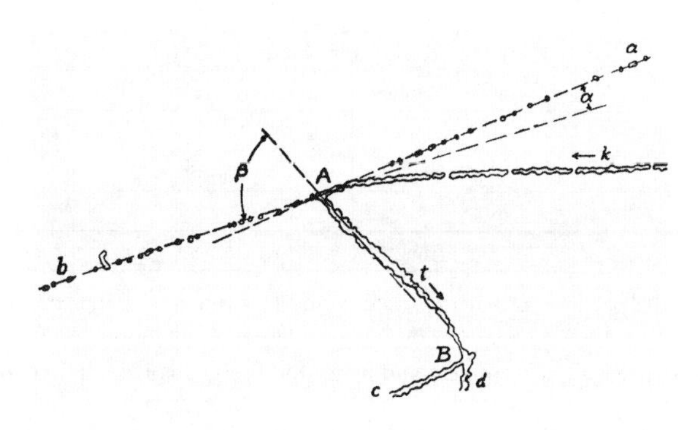

图 12. 用投影显微镜作的图 8 中示出的事件的复制图。在三个径迹 a、b、t 的共面中测得的实际角 α 和 β 分别是：$\angle\alpha = 9.8°$，$\angle\beta = 76.6°$。

如果我们将径迹 t 归因于 π^- 粒子是正确的，则根据观测到 45 微米的射程可得，发射的动能是 1.04 兆电子伏。该粒子相应的动量值是 17.5 兆电子伏 / 光速。根据观测到的运动方向，可得产生径迹 a 和 b 的粒子的动量分别为 98 ± 5 兆电子伏 / 光速和 104 ± 5 兆电子伏 / 光速。这些值将与表 2 中列出的根据观测的径迹中颗粒密度而导出的相应的电子或介子的数值相比较。我们已看到表 2 中列出的两个粒子的动量值，如果假定它们是电子，则这些值显然已大出很多倍。因此如果假定径迹 a 和 b 是由电子或质子产生的，则与动量平衡存在很大的偏差。而且无论粒子是电子还是质子(见表 2)，根据粒子观测到的散射而导出的动量值与从颗粒密度得出的值也是不一致的。

然而，介子的数值之间的一致性是最为突出的，这有力地支持了粒子 k 的自发衰变的假设。如果结果是偶然的，那么须同时具备以下几个不相关的特征才能得到与估算的动量值吻合很好的情况，这些特征包括：在同一平面中粒子的径迹和运动方向的共面性、粒子 t 的射程以及产生径迹 a 和 b 的粒子电离比值，而这种情况是非常罕见的。

假设粒子 k 发生自发衰变，则无论产生的粒子是 μ 介子还是 π 介子，用三类不同方法测定的产生径迹 a 和 b 的粒子的动量值，在测量误差允许的范围内都是一致的。我们可以通过计算两种不同假设下的粒子 k 的静止质量来进行进一步的测试，结果列于表 3 中。

Table 3. Estimates of the mass of particle k based on total release of mass and energy, for two assumed modes of decay

(i) $k \rightarrow \pi^- + \pi + \pi$			
	Track "a"	Track "b"	Track "t"
Particle	π	π	π^-
Rest-mass (m_e)	286	286	286
Energy in m_e	61	64	2
Total = m_k = 985 m_e			
(ii) $k \rightarrow \pi^- + \mu + \mu$			
Particle	μ	μ	π^-
Rest-mass (m_e)	212	212	286
Energy in m_e	76	81	2
Total = m_k = 869 m_e			

In calculating the energy of the particles producing tracks a and b, it is assumed that the particle producing track t is a π^--particle, of momentum 17.5 MeV./c; knowing the relative directions of motion of the three ejected particles, the momenta of the other two particles are determined, and hence the energies corresponding to any assumed mass.

It will be seen from Table 3 that the assumption of two μ-mesons corresponds to a rest mass of the k-particle of 869 m_e; and for two π-mesons, 985 m_e. The assumption of different particles, one π- and one μ-meson, gives an intermediate value of approximately 925 m_e. In view of the error in the direct determination of m_k, the results are not decisive.

If the transmutation is to be interpreted in terms of particles of which the existence is already established, we are left with four possibilities for the nature of the particles producing tracks a and b. These are indicated schematically in Table 4.

Table 4. Comparison of the observed and calculated values of the grain-density in track b, for various assumptions regarding the nature of the particles producing the tracks a and b

	Length of track in microns	Number of grains	Grain-density	Assumed particles			
				1	2	3	4
Track a	2,100	1,025	49±1.5	π	μ	π	μ
Track b	116	59	51±6	π	μ	μ	π
Calculated grain-density in b				45	45	34	64

Values of the grain-density are given in grains per 100 μ.

For the following reasons, case 3, Table 4, is the most improbable. If track a is that of a π-meson, we can calculate the momentum and the grain-density to be expected in track b. We thus obtain the value of 34 grains per micron instead of 51.0±6.0 as observed. For case 4, on the other hand, if a is a μ-meson, the calculated grain-density for track b is 64, a value which differs from that observed by an amount only twice that corresponding to the standard deviations. The observed grain-densities agree best with the assumption that the two particles are of the same type.

Observations on the scattering of the particle producing track a are in better accord

表 3. 对两类假定的衰变方式，基于总的质量和能量释放对粒子 k 的质量估算

(i) $k \to \pi^- + \pi + \pi$			
	径迹“a”	径迹“b”	径迹“t”
粒子	π	π	π^-
静止质量（m_e）	286	286	286
能量（m_e）	61	64	2
总质量 = m_k = 985 m_e			
(ii) $k \to \pi^- + \mu + \mu$			
粒子	μ	μ	π^-
静止质量（m_e）	212	212	286
能量（m_e）	76	81	2
总质量 = m_k = 869 m_e			

在计算产生径迹 a 和 b 的粒子的能量时，假设产生径迹 t 的是一个 π^- 粒子，动量为 17.5 兆电子伏 / 光速；已知三个出射粒子的相对运动方向，则另两个粒子的动量可确定，因此可以得出对应于任一假定质量的能量值。

由表 3 可见，两个 μ 介子的假设对应于粒子 k 的静止质量为 869 m_e；两个 π 介子对应 985 m_e。不同粒子，即一个 μ 介子和一个 π 介子的假设，给出的中间值近似为 925 m_e。考虑到直接测定 m_k 时产生的误差，上述结果无法给出决定性判据。

如果用已确定存在的粒子来解释嬗变，考虑到产生径迹 a 和 b 的粒子所具有的特征，那么就只剩下四种可能性，结果列于表 4 中。

表 4. 根据产生径迹 a 和 b 的粒子的特征而对粒子种类做出的各类假设，并对径迹 b 的颗粒密度的观测值与计算值进行了比较

	径迹的长度（微米）	颗粒数	颗粒密度	假设的粒子			
				1	2	3	4
径迹 a	2,100	1,025	49 ± 1.5	π	μ	π	μ
径迹 b	116	59	51 ± 6	π	μ	μ	π
径迹 b 的颗粒密度的计算值				45	45	34	64

颗粒密度的值指每 100 微米的颗粒数。

由于下述的理由，表 4 中第三种情况是最不可能的。因为如果径迹 a 是一个 π 介子形成的，我们可以计算出径迹 b 中预期的动量值和颗粒密度。由此我们获得的颗粒密度是每微米 34 个颗粒，而不是观测到的 51.0 ± 6.0。反之，对于第四种情况，如果 a 是一个 μ 介子，计算的径迹 b 的颗粒密度为 64，这个值与观测量之间的差别仅为相应标准偏差的 2 倍。观测的颗粒密度与在假定两个粒子是同类型的情况下计算得到的颗粒密度值吻合得最好。

产生径迹 a 的粒子的散射观测更符合 π 介子的假设，而不是 μ 介子（见表 2）；

with the assumption that it is a π-meson rather than a μ-meson (see Table 2); but the results are again indecisive. We may sum up this evidence, and that provided by the mass determinations by grain-counting, by saying that there is some support for the view that the three product-particles are π-mesons; but that the alternative possibilities of one π- and two μ-, or two π- and one μ-meson cannot be excluded.

Chance Juxtaposition of Unrelated Events

In the light of the analysis made in the preceding sections, we can now return to the original assumption that the event is not to be regarded as a fortuitous juxtaposition of tracks. The accuracy of the determination of the mass of the particle *k* does not allow us to exclude the possibility that it has a mass as great as that of a proton, although the observations by grain-counts render it very improbable. Suppose then that a proton, unrelated to the particles producing the other tracks, came to the end of its range at *A*. Even with this assumption, the event is still difficult to explain in conventional terms. Many examples of π^--particles ejected from stars have been observed in this Laboratory[8], but in the present instance the existence of a nuclear interaction in which two protons of great energy are emitted, unaccompanied by slow protons and α-particles, would remain to be explained. A similar difficulty is met if we assume that a particle producing one of the tracks, *a* or *b*, approached *A* and produced the transmutation.

If, alternatively, the tracks *c* and *d*, diverging from star *B*, represent an unrelated disintegration—produced, for example, by a γ-ray—we could then assume track *t* to be that of a proton. We are then left with the difficulties associated with the features peculiar to star *A*, which must now be assumed to have been produced by a slow, charged particle; difficulties which have already been discussed in a previous paragraph. These considerations give further support to the original assumption, that all the tracks shown in the mosaic represent a succession of associated processes.

Relation of the Present Results to Other Observations

If a particle with the elementary electronic charge suffers a spontaneous decay, the law of the conservation of charge demands that the number of emitted particles of charge *e* shall be odd. From this point of view, the sign of the charge of the original particle can have been either positive or negative. If the particles producing tracks *a* and *b* form a pair of opposite sign, then the original *k*-particle was negative. The only other alternative is that they were both positively charged, in which case the *k*-particle was also positive. It is therefore possible that our observations correspond to a mode of decay of positive particles of mass approximately 900 m_e, and that the observation by Leprince-Ringuet[2] demonstrates the fate of the corresponding negative particles—nuclear capture with the production of a "star" and the ejection of a π^--particle.

Rochester and Butler[2] have published an expansion-chamber photograph which appears to be due to the spontaneous decay of a neutral particle of mass approximately 900 m_e

但结果也是非决定性的。我们可以将这个证据与用颗粒密度的质量测定所提供的证据综合在一起，认为这是对产生的三个粒子都是 π 介子的观点的支持；但一个 π 介子和两个 μ 介子的组合或两个 π 介子和一个 μ 介子的组合的可能性也不能排除。

无关事件的机遇并置

按照前几节的分析，我们现在回到最初的假设，即事件并不是径迹偶然的交叉重叠。粒子 k 质量测定的准确度不允许我们排除其质量大如质子的可能性，虽然颗粒计数的观测结果证明这种情况不可能。假设质子与产生其他径迹的粒子无关，在 A 处到达射程终点。即使采用这个假设，事件仍然很难用传统观点来解释。本实验室[8]观测到许多从星发射出 π^- 粒子的例子，但在目前的例子中，发射两个高能质子，并且不伴随产生慢质子和 α 粒子，这种核相互作用的情况是否存在仍然需要进一步证实。而如果我们假设产生径迹 a 或 b 的粒子靠近 A，并产生嬗变，也会遇到类似的困难。

另一种可能性是，如果从星 B 散射的径迹 c 和 d 代表一个无关的蜕变（例如是由 γ 射线所导致的），那么我们可以假定径迹 t 是质子的径迹。我们可以将困难与星 A 独有的特点相联系，并必须假定它是由低速带电粒子产生的，而这样假设遇到的困难已在以前的段落中进行了讨论。这些分析给原始假设以进一步的支持，拼嵌图中所有的径迹代表了一个不断延续发生的相关过程。

目前结果与其他观测的关系

如果带有基本电子电荷的粒子发生自发衰变，根据电荷守恒定律可知，其发射的电荷为 e 的粒子的数量应为奇数。根据这个观点，原始粒子的电荷符号不是正号就是负号。如果产生径迹 a 和 b 的粒子的电荷符号相反，则原始的 k 粒子带负电。仅有的另一种可能是，它们两者都带正电荷，在这种情况下，k 粒子也是正的。因此可能的情况是，我们的观测对应于质量近似为 900 m_e 的带正电粒子的衰变模式，而由勒普兰斯·兰盖[2]的观测则表明了相应的负电粒子的情况，即核俘获产生“星”并发射 π^- 粒子。

罗切斯特和布特勒[2]已发表了一张膨胀室的照片，它记录的可能是一个质量近似为 900 m_e 的中性粒子自发衰变形成了一对电荷相反的静止质量近似为 300 m_e 的粒

into a pair of oppositely charged particles of rest-mass approximately 300 m_e. We have therefore considered the possibility that the decay process suggested by the present results can be regarded as taking place in two stages: the emission of a π^--particle of low energy, followed by the spontaneous decay of the resulting neutral particle. On this view, however, it would be necessary to assume that the neutral particle has a life-time of the order of 10^{-14} sec. Otherwise, in recoiling from the π-particle, it would move away from the original point of decay, and the two charged particles into which it became transformed would originate from a point separated from the beginning of the track of the π^--particle. It follows that we cannot identify such a postulated unstable neutral particle with that for which evidence is provided in the experiments of Rochester and Butler.

Finally, we have considered the possible relations of the present results to the particles of mass approximately 800 m_e referred to as τ-mesons, evidence for which has been recently reported by Bradt and Peters[2]. It is a remarkable feature of their experiments that their τ-mesons give rise to no recorded secondary particles at the end of their range. It appears to be possible that these particles also decay with the emission of three fast mesons, but that the transmutation usually takes place with a more equal partition of kinetic energy than in the case we have observed. It would then follow that in the Ilford *C*2 emulsion the disintegration products would commonly escape observation. If this view is correct, we must regard the event we have observed as representing a rare example of a common mode of decay of these mesons; an example which, by chance, has allowed a detailed analysis to be carried out. If so, the τ-meson of Bradt and Peters, when recorded by electron-sensitive emulsions, should show the tracks of three particles, of low specific ionization, and of which the directions of motion are co-planar.

We have pleasure in thanking Prof. von Muralt and members of the staff of the Jungfraujoch Forschungsstation for hospitality and assistance in obtaining the exposures; Dr. E. R. Davies and Dr. W. E. Berriman, of Messrs. Kodak, Ltd., for special photographic plates; Miss C. Dilworth and Dr. G. P. S. Occhialini for advice on development; Mr. W. O. Lock and Mr. J. H. Davies for assistance in making observations on the scattering of particles in the emulsion; and to the team of microscope observers of this Laboratory. We are indebted to Prof. N. F. Mott and other colleagues for a number of discussions on the processes associated with the capture of negative mesons by nuclei.

Note added in proof. Since completing this article, we have been informed by Dr. Peters that, in Ilford *C*2 emulsions exposed at 90,000 feet, he and Dr. Bradt have observed three events with the following characteristics. A particle, which they judge to be similar in mass to their τ-mesons, appears to come to rest and to lead to the emission of a particle of smaller mass, which, at the end of its range, produces a nuclear disintegration. The ranges of the secondary particles, in the three cases, are 20, 25 and 45 μ, respectively. The authors were not aware of our results when they suggested to us that their observations may correspond to the spontaneous decay of heavy mesons. According to their description, these events are precisely similar to those we should expect to observe in *C*2 emulsions as a result of the

子。因此我们认为，该结果反映的衰变过程可能分为两个步骤：一个低能 π^- 粒子的发射和随后产生的中性粒子的自发衰变。然而，根据这个观点，必须假设，中性粒子的寿命为 10^{-14} 秒的量级。否则中性粒子在 π 粒子的反冲中，可能会离开衰变的原点，因此嬗变产生的两个带电粒子将从与 π^- 粒子径迹的起点不同的点起源。因此，根据罗切斯特和布特勒的实验中提供的证据，我们并不能确认存在这类假定的不稳定的中性粒子。

最后，我们考虑了目前的结果与质量近似为 800 m_e 的 τ 介子（编者注：现在粒子物理认为“τ 介子”不是介子，而是一种轻子，因此正确的称谓是 τ 子。考虑到历史原因，文中仍保留“τ 介子”的译法。）的可能关系，关于 τ 介子的证据最近已由布拉特和彼得斯 [2] 做了报道。他们实验的一个显著特征就是，在 τ 介子的射程内没有记录到次级粒子。因此很可能是 τ 介子在衰变的过程中伴随三个快介子的发射，但是这种嬗变通常发生在比我们所观测到的动能更加均分的情况下。由此可知，在伊尔福 *C*2 乳胶中，蜕变产物通常会观测不到。如果这个观点是正确的，我们必须认为我们已观测到的事件是这些介子衰变的一般模式中的一个稀有例子，一个碰巧允许我们进行详细分析的例子。如果是这样，那么布拉特和彼得斯的 τ 介子在被用电子敏感的乳胶记录时，应该会显示低电离比值的三个粒子的径迹，且运动方向是共面的。

诚挚感谢冯·穆拉尔特教授和少女峰研究站的工作人员，感谢他们的热情和在曝光过程中对我们的帮助；感谢柯达公司的戴维斯博士和贝里曼博士提供的特殊的照相底片；感谢迪尔沃斯女士和奥恰利尼博士关于显影的建议；感谢洛克先生和戴维斯先生在观察感光乳液中的粒子散射时所给予的协助；感谢本实验室的显微镜观测小组。此外，我们还特别感谢莫脱教授和其他同事关于核子捕捉负电介子的过程而进行的大量讨论。

附加说明：本文完成以后，彼得斯博士又给我们提供了一些新的信息：将伊尔福 *C*2 乳胶置于海拔 90,000 英尺处曝光，他和布拉特博士观测到三个具有下述特性的事件。一个质量与他们的 τ 介子相当的粒子在其射程的终端趋于静止，并产生一个质量更小的粒子，该粒子在其射程的终端激发了一个核蜕变。在三类情况下，次级粒子的射程分别为 20 微米、25 微米和 45 微米。当他们向我们提出他们的观测可能与重介子的自发衰变相符合时，作者并不知道我们的结果。按照他们的描述，这些事件非常类似于我们预期在 *C*2 乳胶中观测的我们所假设的这类重粒子的自发衰变

spontaneous decay of heavy particles of the type we have postulated; for any particles of low specific ionization will not be recorded by the Ilford plates. The observations of Peters and Bradt appear, therefore, to give further support for the assumption that the present observations are not due to a chance juxtaposition of tracks; and they suggest that it will be possible, in the near future, to find similar examples suitable for making a detailed analysis.

(**163**, 82-87; 1949)

R. Brown, U. Camerini, P. H. Fowler, H. Muirhead and C. F. Powell: H. H. Wills Physical Laboratory, University of Bristol.

D. M. Ritson: Clarendon Laboratory, Oxford.

References:

1. Berriman, *Nature*, 162, 992 (1948).
2. Leprince-Ringuet, *C.R.*, **226**, 1897 (1948). Rochester and Butler, *Nature*, **160**, 855 (1947). Bradt and Peters, Report to the Bristol Symposium, 1948 (in the press). Alichanian. Alichanov and Weissenberg, *J. Exp. and Theoret. Phys., U.S.S.R.*, **18**, 301 (1948); and other references.
3. Camerini, Muirhead, Powell and Ritson, *Nature*, **162**, 433 (1948).
4. Goldschmidt-Clermont, King, Muirhead and Ritson, *Proc. Phys. Soc.*, **61**, 138 (1948).
5. Lattes, Occhialini and Powell, *Proc. Phys. Soc.*, **61**, 173 (1948).
6. Serber, Report of Solvay Conference for 1948.
7. Dilworth, Occhialini and Payne, *Nature*, **162**, 102 (1948).
8. Occhialini and Powell, *Nature*, **162**, 168 (1948).
9. Halpern and Hall, *Phys. Rev.*, **73**, 477 (1948).
10. Livingston and Bethe, *Rev. Mod. Phys.*, **9**, 263 (1937).
11. Camerini and Lattes (private communication); see also Powell and Occhialini, "Nuclear Physics in Photographs", 112 (Oxford, 1947).

结果；伊尔福 *C*2 胶片不能记录任何低电离比值的粒子。因此，彼得斯和布拉特的观测结果进一步支持了下述假设，即目前的观测并不是偶然的径迹交叉重叠；他们表示在不久的将来有可能发现适合做详细分析的类似实例。

（沈乃澂 翻译；尚仁成 审稿）

Correlation between Photons in Two Coherent Beams of Light

R. Hanbury-Brown and R. Q. Twiss

Editor's Note

In classical interferometry, phase differences between two interacting light beams probe differences in their travelled paths. In the early 1950s, Robert Hanbury-Brown and Richard Twiss proposed that a new means of interferometry, using only the intensity (not the phase) of light, could be used to estimate the angular size of distant stars. Here they report a laboratory demonstration of the effect. The result illustrated the quantum" bunching" of photons—the tendency for photons in two separate, coherent beams to arrive together at two detectors. The Hanbury-Brown Twiss effect now refers generally to any correlation or anti-correlation in the intensities of signals measured by two detectors from a beam of particles. Intensity interferometry has become an important technique in nuclear and particle physics.

IN an earlier paper[1], we have described a new type of interferometer which has been used to measure the angular diameter of radio stars[2]. In this instrument the signals from two aerials A_1 and A_2 (Fig. 1*a*) are detected independently and the correlation between the low-frequency outputs of the detectors is recorded. The relative phases of the two radio signals are therefore lost, and only the correlation in their intensity fluctuations is measured; so that the principle differs radically from that of the familiar Michelson interferometer where the signals are combined before detection and where their relative phase must be preserved.

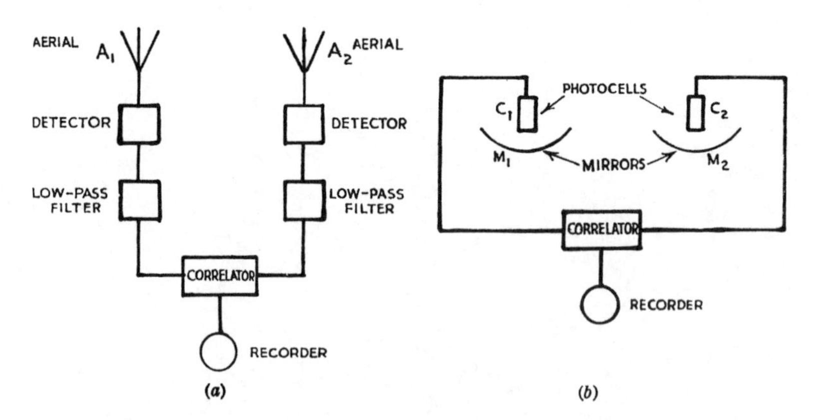

Fig. 1. A new type of radio interferometer (*a*), together with its analogue (*b*) at optical wave-lengths

This new system was developed for use with very long base-lines, and experimentally it has proved to be largely free of the effects of ionospheric scintillation[2]. These advantages led us to suggest[1] that the principle might be applied to the measurement of the angular

两个相干光束中光子间的相关性

汉伯里–布朗，特威斯

编者按

在经典的干涉技术中，由两个相互作用光束之间的相位差可以探查出它们传播路径的不同。20 世纪 50 年代早期，罗伯特·汉伯里–布朗和理查德·特威斯提出了一种新的干涉测量法，此方法仅仅利用光强（非相位）就可以估测遥远恒星的角大小。本文中他们报道了对这种效应所做的实验论证。结果阐释了光子的量子“聚束”效应——两束独立的相干光束中的光子有同时到达两个探测器的趋势。现在，汉伯里–布朗–特威斯效应一般是指使用两个探测器测量出的从一个粒子束发出的信号强度之间的任何相关性或者反相关性。强度干涉法已经成为核物理以及粒子物理中一项重要的技术。

在一篇早期的论文中[1]，我们描述了一台新型的干涉仪，它已被用于测量射电星的角直径[2]。用这台仪器分别检测来自 A_1 和 A_2 两个天线 (如图 1*a* 所示) 的信号，并记录检测器低频输出之间的相关系数。因此我们没有测量到两个射电信号的相对相位，而仅测量到它们强度起伏的相关系数；这种原理与常见的迈克尔逊干涉仪的原理有根本上的差异，因为后者的信号是在检测前合成的，从而必然保留它们的相对相位。

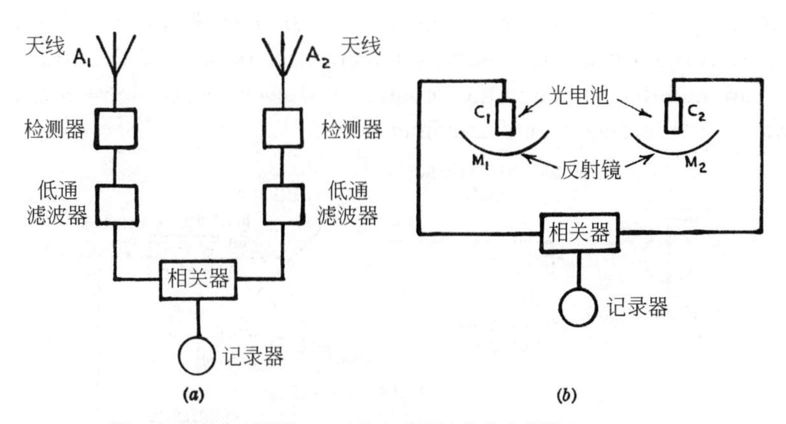

图 1. 新型射电干涉仪 (*a*) 及其在光学波段的类似设备 (*b*)

开发这套新系统是为了利用甚长基线，实验证明它在很大程度上不受电离层闪烁效应的影响[2]。鉴于这套系统的优点，我们提出[1]从原理上它可以用于测量目视

diameter of visual stars. Thus one could replace the two aerials by two mirrors M_1, M_2 (Fig. 1*b*) and the radio-frequency detectors by photoelectric cells C_1, C_2, and measure, as a function of the separation of the mirrors, the correlation between the fluctuations in the currents from the cells when illuminated by a star.

It is, of course, essential to the operation of such a system that the time of arrival of photons at the two photocathodes should be correlated when the light beams incident upon the two mirrors are coherent. However, so far as we know, this fundamental effect has never been directly observed with light, and indeed its very existence has been questioned. Furthermore, it was by no means certain that the correlation would be fully preserved in the process of photoelectric emission. For these reasons a laboratory experiment was carried out as described below.

The apparatus is shown in outline in Fig. 2. A light source was formed by a small rectangular aperture, 0.13 mm. × 0.15 mm. in cross-section, on which the image of a high-pressure mercury are was focused. The 4,358 A. line was isolated by a system of filters, and the beam was divided by the half-silvered mirror M to illuminate the cathodes of the photomultipliers C_1, C_2. The two cathodes were at a distance of 2.65 m. from the source and their areas were limited by identical rectangular apertures O_1, O_2, 9.0 mm. × 8.5 mm. in cross-section. (It can be shown that for this type of instrument the two cathodes need not be located at precisely equal distances from the source. In the present case their distances were adjusted to be roughly equal to an accuracy of about 1 cm.) In order that the degree of coherence of the two light beams might be varied at will, the photomultiplier C_1, was mounted on a horizontal slide which could be traversed normal to the incident light. The two cathode apertures, as viewed from the source, could thus be superimposed or separated by any amount up to about three times their own width. The fluctuations in the output currents from the photomultipliers were amplified over the band 3–27 Mc./s. and multiplied together in a linear mixer. The average value of the product, which was recorded on the revolution counter of an integrating motor, gave a measure of the correlation in the fluctuations. To obtain a significant result it was necessary to integrate for periods of the order of one hour, so very great care had to be taken in the design of the electronic equipment to eliminate the effects of drift, of interference and of amplifier noise.

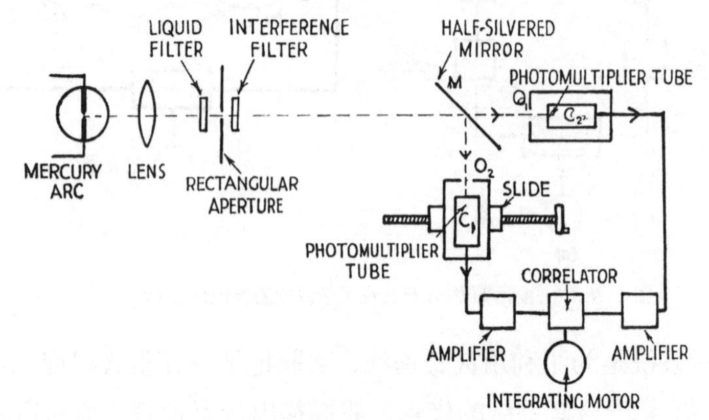

Fig. 2. Simplified diagram of the apparatus

恒星的角直径。因此我们可以用 M_1 和 M_2 两面反射镜来代替 A_1 和 A_2 两根天线（如图 1*b* 所示），用光电池 C_1 和 C_2 作为射频检测器，将测量到的由恒星照射导致的电池中电流起伏间的相关系数作为反射镜间距的函数。

当然，当入射到两个反射镜上的光束有相干性时，光子到达两个光电阴极的时间应该是相关的，这对于这套系统的运行是很有必要的。然而，据我们所知，目前尚未用光直接观测到这个基本效应，况且此效应存在与否还存在疑问。此外，在光电发射过程中，无法确定是否完全保留着这种相关性。为此，我们开展了下列实验。

图 2 展示了此套装置的概略图。一个截面为 0.13 毫米 × 0.15 毫米的小矩形孔径形成光源，高压汞灯在此截面上聚焦成像。我们通过滤波系统将 4,358 埃的谱线分离出来，光束通过半透半反分束镜 M 被分成两束，分别照射在光电倍增管 C_1 和 C_2 的阴极上。这两个阴极与光源相距 2.65 米，它们的面积受到截面为 9.0 毫米 × 8.5 毫米的相同矩形孔径 O_1 和 O_2 的限制。（可以证明这类仪器的两个阴极与光源之间的距离并不需要精确相等。在目前的实验中，它们的距离只是大致相等，精度约为 1 厘米。）为了使两束光束的相干度可以任意改变，光电倍增管 C_1 装在水平轨道上，这样光电倍增管 C_1 可以在垂直于入射光的方向上移动。因此从光源看来，两个阴极孔径能叠加或分离到最大可达约为其自身宽度 3 倍的任何量值。在 3 兆周 / 秒 ~ 27 兆周 / 秒波段内来自光电倍增管的输出电流的起伏被放大，并在线性混频器中倍增。在积分电机的旋转计数器上记录乘积的平均值，并对起伏相关系数进行度量。为了得到有意义的结果，必须要对量级 1 小时的周期进行积分，所以我们非常精心地设计了电子装置，从而消除由漂移、干涉和放大器噪声所造成的影响。

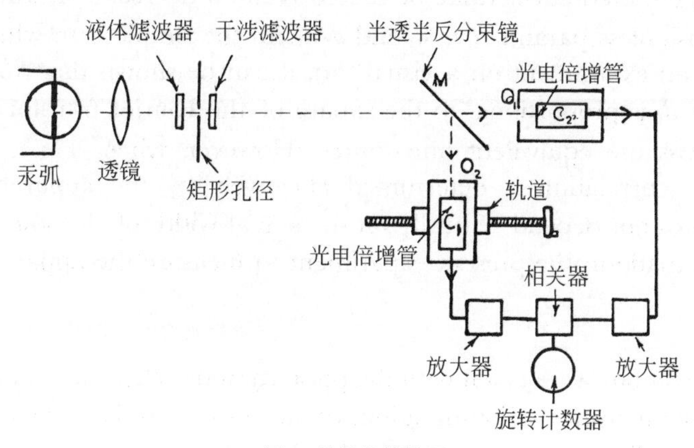

图 2. 装置的简化图

Assuming that the probability of emission of a photoelectron is proportional to the square of the amplitude of the incident light, one can use classical electromagnetic wave theory to calculate the correlation between the fluctuations in the current from the two cathodes. On this assumption it can be shown that, with the two cathodes superimposed, the correlation $S(0)$ is given by:

$$S(0) = A \cdot T \cdot b_v \mathrm{f}\left(\frac{a_1 \theta_1 \pi}{\lambda_0}\right) \cdot \mathrm{f}\left(\frac{a_2 \theta_2 \pi}{\lambda_0}\right) \int \alpha^2(v) \cdot n_0^2(v) \cdot \mathrm{d}v \tag{1}$$

It can also be shown that the associated root-mean-square fluctuations N are given by:

$$N = A \cdot T \cdot \frac{2m}{m-1} \cdot b_v (b_v T)^{-\frac{1}{2}} \int \alpha(v) \cdot n_0(v) \cdot \mathrm{d}v \tag{2}$$

where A is a constant of proportionality depending on the amplifier gain, etc.; T is the time of observation; $\alpha(v)$ is the quantum efficiency of the photocathodes at a frequency v; $n_0(v)$ is the number of quanta incident on a photocathode per second, per cycle bandwidth; b_v is the bandwidth of the amplifiers; $m/(m-1)$ is the familiar excess noise introduced by secondary multiplication; a_1, a_2 are the horizontal and vertical dimensions of the photocathode apertures; θ_1, θ_2 are the angular dimensions of the source as viewed from the photocathodes; and λ_0 is the mean wave-length of the light. The integrals are taken over the complete optical spectrum and the phototubes are assumed to be identical. The factor $\mathrm{f}\left(\frac{a\theta\pi}{\lambda_0}\right)$ is determined by the dimensionless parameter η defined by

$$\eta = a\theta/\lambda_0 \tag{3}$$

which is a measure of the degree to which the light is coherent over a photocathode. When $\eta \ll 1$, as for a point source, $\mathrm{f}(\eta)$ is effectively unity; however, in the laboratory experiment it proved convenient to make η_1, η_2 of the order of unity in order to increase the light incident on the cathodes and thereby improve the ratio of signal to noise. The corresponding values of $\mathrm{f}(\eta_1)$, $\mathrm{f}(\eta_2)$ were 0.62 and 0.69 respectively.

When the centres of the cathodes, as viewed from the source, are displaced horizontally by a distance d, the theoretical value of the correlation decreases in a manner dependent upon the dimensionless parameters, η_1 and d/a_1. In the simple case where $\eta_1 \ll 1$, which would apply to an experiment on a visual star, it can be shown that $S(d)$, the correlation as a function of d, is proportional to the square of the Fourier transform of the intensity distribution across the equivalent line source. However, when $\eta \geqslant 1$, as in the present experiment, the correlation is determined effectively by the apparent overlap of the cathodes and does not depend critically on the actual width of the source. For this reason no attempt was made in the present experiment to measure the apparent angular size of the source.

The initial observations were taken with the photocathodes effectively superimposed ($d = 0$) and with varying intensities of illumination. In all cases a positive correlation was observed which completely disappeared, as expected, when the separation of the photocathodes

假设发射一个光电子的概率与入射光的振幅的平方成正比，我们可以用经典电磁波理论计算来自两个阴极的电流起伏之间的相关系数。根据这个假设我们可以证明，两个阴极叠加，相关系数 $S(0)$ 由下式表示：

$$S(0) = A \cdot T \cdot b_v \mathrm{f}\left(\frac{a_1\theta_1\pi}{\lambda_0}\right) \cdot \mathrm{f}\left(\frac{a_2\theta_2\pi}{\lambda_0}\right)\int \alpha^2(v) \cdot n_0^2(v) \cdot \mathrm{d}v \tag{1}$$

同样地，相应的方均根涨落 N 由下式给出：

$$N = A \cdot T \cdot \frac{2m}{m-1} \cdot b_v(b_v T)^{-\frac{1}{2}}\int \alpha(v) \cdot n_0(v) \cdot \mathrm{d}v \tag{2}$$

式中 A 是与放大器增益等有关的比例常数；T 是观测时间；$\alpha(v)$ 是光电阴极在频率 v 处的量子效率；$n_0(v)$ 是每秒钟每周带宽入射到光电阴极上的量子数；b_v 是放大器的带宽；$m/(m-1)$ 是次级倍增引入的常见过量噪声；a_1 和 a_2 是光电阴极孔径的水平尺寸和垂直尺寸；θ_1 和 θ_2 是从光电阴极来看光源的角尺寸；λ_0 是光的平均波长。在假定光电管完全相同的情况下，对整个光谱积分。因子 $\mathrm{f}\left(\frac{a\theta\pi}{\lambda_0}\right)$ 由无量纲参数 η 决定，η 的定义为：

$$\eta = a\theta/\lambda_0 \tag{3}$$

这是光在光电阴极范围内相干度的度量。当 $\eta \ll 1$ 时，对于点光源，$\mathrm{f}(\eta)$ 可有效地认定为单位 1；然而，在实验室中做实验时，可方便地证明：设置 η_1 和 η_2 为单位 1 的数量级，以增加阴极上的入射光，从而提高信噪比。此时对应的 $\mathrm{f}(\eta_1)$ 和 $\mathrm{f}(\eta_2)$ 的值分别是 0.62 和 0.69。

当从源处看阴极的中心水平方向上位移距离为 d 时，相关系数的理论值会降低，它的多少与无量纲参量 η_1 和 d/a_1 有关。在简单情况下 $\eta_1 \ll 1$，将其用于可视恒星的实验上，可以证明相关系数 $S(d)$ 作为 d 的函数与等效线源上强度分布的傅里叶变换的平方成正比。然而，当 $\eta \geqslant 1$ 时，正如在目前的实验中一样，相关系数是由阴极的表观叠加有效决定的，而并不严格地依赖于源处的实际宽度。因此，在目前的实验中我们并不试图测量源的视角大小。

最初的实验观测是在光电阴极有效地叠加 ($d=0$) 以及改变照明强度下进行的。在我们所预期的所有情况下，当光电阴极相距很远时，观测到相关系数为正的情

was large. In these first experiments the quantum efficiency of the photocathodes was too low to give a satisfactory ratio of signal to noise. However, when an improved type of photomultiplier became available with an appreciably higher quantum efficiency, it was possible to make a quantitative test of the theory.

A set of four runs, each of 90 min. duration, was made with the cathodes superimposed ($d=0$), the counter readings being recorded at 5-min. intervals. From these readings an estimate was made of N_e, the root mean square deviation in the final reading $S(0)$ of the counter, and the observed values of $S_e(0)/N_e$ are shown in column 2 of Table 1. The results are given as a ratio in order to eliminate the factor A in equations (1) and (2), which is affected by changes in the gain of the equipment. For each run the factor

$$\frac{m-1}{m}\int \alpha^2(\nu)\, n_0^2(\nu) d\nu \Big/ \int \alpha(\nu)\, n_0(\nu)\, d\nu$$

was determined from measurements of the spectrum of the incident light and of the d.c. current, gain and output noise of the photomultipliers; the corresponding theoretical values of $S(0)/N$ are shown in the second column of Table 1. In a typical case, the photomultiplier gain was 3×10^5, the output current was 140 μamp., the quantum efficiency $\alpha(\nu_0)$ was of the order of 15 percent and $n_0(\nu_0)$ was of the order of 3×10^{-3}. After each run a comparison run was taken with the centres of the photocathodes, as viewed from the source, separated by twice their width ($d = 2a$), in which position the theoretical correlation is virtually zero. The ratio of $S_e(d)$, the counter reading after 90 minutes, to N_e, the root mean square deviation, is shown in the third column of Table 1.

Table 1. Comparison between the Theoretical and Experimental Values of the Correlation

	Cathodes superimposed ($d = 0$)		Cathodes separated ($d = 2a = 1.8$ cm.)	
	Experimental ratio of correlation to r.m.s. deviation $S_e(0)/N_e$	Theoretical ratio of correlation to r.m.s. deviation $S(0)/N$	Experimental ratio of correlation to r.m.s. deviation $S_e(d)/N_e$	Theoretical ratio of correlation to r.m.s. deviation $S(d)/N$
1	+7.4	+8.4	−0.4	~0
2	+6.6	+8.0	+0.5	~0
3	+7.6	+8.4	+1.7	~0
4	+4.2	+5.2	−0.3	~0

The results shown in Table 1 confirm that correlation is observed when the cathodes are superimposed but not when they are widely separated. However, it may be noted that the correlations observed with $d=0$ are consistently lower than those predicted theoretically. The discrepancy may not be significant but, if it is real, it was possibly caused by defects in the optical system. In particular, the image of the arc showed striations due to imperfections in the glass bulb of the lamp; this implies that unwanted differential phase-shifts were being introduced which would tend to reduce the observed correlation.

况完全消失。在第一批实验中，光电阴极的量子效率太低，以致没有得到令人满意的信噪比。然而，当我们采用具有较高量子效率的改进型光电倍增管时，就能进行这个理论的定量检验。

我们所做的观测 4 次为一组、每次持续时间为 90 分钟，是在阴极叠加 ($d=0$) 时进行的。计数器以 5 分钟为时间间隔进行记录。根据这些读数，我们对 N_e 做出估测，其中 N_e 为计数器最终读数 $S(0)$ 的均方根偏差，将 $S_e(0)/N_e$ 的观测值列入表 1 的第 1 列内。为了消除 (1) 式和 (2) 式中的因子 A，结果以比值的形式给出，此比值受装置增益变化的影响。对每次测量的因子如下式所示：

$$\frac{m-1}{m}\int\alpha^2(v)n_0^2(v)\mathrm{d}v\Big/\int\alpha(v)n_0(v)\mathrm{d}v$$

这是由入射光的光谱、直流电流、增益和光电倍增管的输出噪声所决定的；表 1 的第 2 列内示出了 $S(0)/N$ 相应的理论值。在典型情况下，光电倍增管的增益是 3×10^5，输出电流是 140 微安，量子效率 $\alpha(v_0)$ 的数量级为 15%，$n_0(v_0)$ 的数量级为 3×10^{-3}。在每次测量之后，在光电阴极中心做一次对比测量，从源处看两阴极之间的距离为自身宽度的两倍 ($d=2a$)，在光电阴极中心时理论修正实际上为零。表 1 的第 3 列内列出的是 $S_e(d)$ 与 N_e 的比值，其中 $S_e(d)$ 是 90 分钟后的计数器读数，N_e 是均方根偏差。

表 1. 相关系数的理论值和实验值之间的比较

	阴极叠加 ($d=0$)		阴极分离 ($d=2a=1.8$ 厘米)	
	相关系数与均方根偏差的实验比 $S_e(0)/N_e$	相关系数与均方根偏差的理论比 $S(0)/N$	相关系数与均方根偏差的实验比 $S_e(d)/N_e$	相关系数与均方根偏差的理论比 $S(d)/N$
1	+7.4	+8.4	–0.4	~0
2	+6.6	+8.0	+0.5	~0
3	+7.6	+8.4	+1.7	~0
4	+4.2	+5.2	–0.3	~0

表 1 中所示的结果证实了在阴极叠加时可观测到相关性，但当阴极相距甚远时未观测到相关性。然而可以注意到的是，距离 $d=0$ 时实验所观测到的相关系数比理论上预料的数值一致偏低。这个差异可能并不重要，但如果事实如此，那么它可能是由光学系统中的缺陷引起的。尤其是灯的玻璃泡中的不完善性引起弧光的像呈现辉纹，这意味着引入了多余的微分相移，它们往往会降低观测到的相关系数。

This experiment shows beyond question that the photons in two coherent beams of light are correlated, and that this correlation is preserved in the process of photoelectric emission. Furthermore, the quantitative results are in fair agreement with those predicted by classical electromagnetic wave theory and the correspondence principle. It follows that the fundamental principle of the interferometer represented in Fig. 1*b* is sound, and it is proposed to examine in further detail its application to visual astronomy. The basic mathematical theory together with a description of the electronic apparatus used in the laboratory experiment will be given later.

We thank the Director of Jodrell Bank for making available the necessary facilities, the Superintendent of the Services Electronics Research Laboratory for the loan of equipment, and Mr. J. Rodda, of the Ediswan Co., for the use of two experimental phototubes. One of us wishes to thank the Admiralty for permission to submit this communication for publication.

(177, 27-29; 1956)

R. Hanbury-Brown: University of Manchester, Jodrell Bank Experimental Station.
R. Q. Twiss: Services Electronics Research Laboratory, Baldock.

References:

1. Hanbury Brown, R., and Twiss, R. Q., *Phil. Mag.*, 45, 663 (1954).
2. Jennison, R. C., and Das Gupta, M. K., *Phil. Mag.* (in the press).

这项实验表明，两束相干光的两个光子之间的相关性是毫无疑问的，这种相关性存在于光电发射的过程中。此外，定量结果与经典电磁波理论及相应原理的预料吻合得相当好。图 1*b* 所示干涉仪的基本原理是正确的，建议在目视天文学的应用中进行进一步详细的检验。基本数学理论与用于实验室的电子学装置的描述将在以后一起给出。

我们感谢焦德雷尔班克天文台台长允许我们使用所需的实验设备，同时对电子服务研究实验室的负责人所提供的实验设备表示感谢，还要对爱迪斯旺公司的罗达先生所提供的两个光电管表示感谢。我们当中的一位作者对英国海军部同意将此文予以发表表示感谢。

（沈乃澂 翻译；熊秉衡 审稿）

The Neutrino

F. Reines and C. L. Cowan

Editor's Note

The existence of the subatomic particle called the neutrino had been inferred in the early 1930s chiefly as a result of work by Enrico Fermi in Rome and Wolfgang Pauli in Zurich from the phenomenon of radioactive beta-decay: it was supposed to be a particle with insignificant mass and no electric charge. Frederick Reines and Clyde L. Cowan at the Los Alamos Scientific Laboratory, acknowledging that nuclear reactors should be powerful sources of neutrinos, here described an experiment to demonstrate the existence of neutrinos by their interaction with protons (provided by a large tank of water) and their individual conversion into photons which were to be measured by an array of photoscintillators. The use of photoscintillators for detecting neutrinos produced in analogous nuclear reactions has become standard practice. In the 1980s, however, it became clear that there are three different kinds of neutrinos and that even those studied by Reines and Cowan may have a small mass.

EACH new discovery of natural science broadens our knowledge and deepens our understanding of the physical universe; but at times these advances raise new and even more fundamental questions than those which they answer. Such was the case with the discovery and investigation of the radioactive process termed "beta decay". In this process an atomic nucleus spontaneously emits either a negative or positive electron, and in so doing it becomes a different element with the same mass number but with a nuclear charge different from that of the parent element by one electronic charge. As might be expected, intensive investigation of this interesting alchemy of Nature has shed much light on problems concerning the atomic nucleus. A new question arose at the beginning, however, when it was found that accompanying beta decay there was an unaccountable loss of energy from the decaying nucleus[1], and that one could do nothing to the apparatus in which the decay occurred to trap this lost energy[2]. One possible explanation was that the conservation laws (upon which the entire structure of modern science is built) were not valid when applied to regions of subatomic dimensions. Another novel explanation, but one which would maintain the integrity of the conservation laws, was a proposal by Wolfgang Pauli in 1933 which hypothesized a new and fundamental particle[3] to account for the loss of energy from the nucleus. This particle would be emitted by the nucleus simultaneously with the electron, would carry with it no electric charge, but would carry the missing energy and momentum—escaping from the laboratory equipment without detection.

The concept of this ghostly particle was used by Enrico Fermi (who named it the "neutrino") to build his quantitative theory of nuclear beta decay[4]. As is well known, the theory, with but little modification, has enjoyed increasing success in application to nuclear

中微子

莱因斯，考恩

编者按

20世纪30年代初期，罗马的恩里科·费米和苏黎世的沃尔夫冈·泡利根据放射性β衰变现象得到的研究结果，推测出被称为中微子的亚原子粒子的存在，他们认为该粒子质量极小且不带电荷。洛斯阿拉莫斯科学实验室的弗雷德里克·莱因斯和克莱德·考恩认为，核反应堆是强大的中微子源，本文中他们描述了一个证明中微子存在的实验，实验中质子（由一大箱水提供）与中微子相互作用，使中微子转换为光子，并用光闪烁体阵列对光子进行测量。如今用光闪烁体检测这类核反应中产生的中微子已成为标准方法。然而，直到20世纪80年代人们才逐渐清楚存在三种不同类型的中微子以及莱因斯和考恩研究的那些中微子也可能具有微小的质量。

自然科学的每项新发现都扩展了我们的知识，加深了我们对物质世界的了解；但是与它们所解决的问题相比，这些进展有时会引出一些新的甚至更基本的问题。放射性“β衰变”过程的发现和研究就是这样一个例子。在这个过程中，原子核自发地发射一个负电子或正电子，这样它就变成了一个与母元素质量数相同但相差一个电子电荷的不同元素。正如所预料的那样，对自然界这一有趣的神奇过程的深入研究阐明了关于原子核的一些问题。然而随即出现的一个新问题是：人们发现伴随β衰变，衰变核[1]中有无法解释的能量损失；并且人们无法通过改善发生衰变的装置来捕获这些损失的能量[2]。一种可能的解释是，守恒定律（现代科学的整个结构建立在这个定律之上）应用到亚原子尺度时不再成立。沃尔夫冈·泡利于1933年提出另一种新奇的但能继续遵守守恒定律的解释，他假设存在一类新的基本粒子[3]对应于损失的核能量。核在发射电子的同时发射出这类粒子，该粒子不带电荷，但是带有丢失的那部分能量和动量，即实验装置未检测到的那部分能量和动量。

恩里科·费米在构建他的核β衰变定量理论[4]时，采用了这个如幽灵般的粒子的概念（他称之为“中微子”）。众所周知，这个理论作了微小修正后应用到核问题

problems and has itself constituted one of the most convincing arguments in favour of the acceptance of Pauli's proposal. Many additional experimental tests have been devised, however, which have served to strengthen the neutrino hypothesis; and also to provide information as to its properties. The very characteristic of the particle which makes the proposal plausible—its ability to carry off energy and momentum without detection—has limited these tests to the measurement of the observable details of the decay process itself: the energy spectra, momentum vectors and energy states associated with the emitted electron and with the recoiling daughter nucleus[5]. So, for example, an upper limit has been set on the rest mass of the neutrino equal to 1/500 of the rest mass of the electron by careful measurement of the beta-energy spectrum from tritium decay near its end point[6], and it is commonly assumed that the neutrino rest mass is identically zero.

While there is no theoretical reason for the expectation of a finite neutrino rest mass, there is some expectation for a small but finite neutrino magnetic moment of perhaps as much as 10^{-10} Bohr magneton based on a consideration of possible virtual states in which the neutrino may exist effectively dissociated into other particles[7]. An upper limit of 2×10^{-9} electron Bohr magneton has been set on the magnetic moment by calculations concerning the maximum assignable heat transfer to the Earth by neutrinos from the Sun[8]. We have recently obtained an improved upper limit of 10^{-9} electron Bohr magneton using a large scintillation detector near a fission reactor at the Savannah River Plant of the United States Atomic Energy Commission. The counting rate of single pulses in an energy range of 0.1–0.3 MeV. in 370 gallons of liquid scintillator was observed, and all changes due to reactor power changes were assigned to possible electron recoils in the liquid through magnetic moment interaction with neutrinos. It is hoped that this limit may be further improved by lowering the gamma-ray and neutron background at the detector.

The Pauli-Fermi theory not only requires the neutrino to carry energy and linear momentum from beta-decaying nuclei but also angular momentum, or "spin". The simplest of beta-decay processes, the decay of the free neutron[9], illustrates this:

$$n^0 \rightarrow p^+ + \beta^- + \nu_- \tag{1}$$

As the neutron, proton and beta particle all carry half-integral spin, it is necessary to assign a spin quantum number of 1/2 to the neutrino to balance the angular momenta of equation (1), where any two of the three product particles must be oriented with spin vectors antiparallel. As all four of the particles in equation (1) are, therefore, fermions and should obey the Dirac relativistic Wave equations for spin 1/2 particles, there are presumably antiparticles corresponding to each, of which as yet only the anti-electron (or positron) and the antiproton have been identified. The antiparticle corresponding to the neutrino in equation (1) may be obtained by rearrangement of the terms in the following manner:

$$p^+ \rightarrow n^0 + \beta^+ + \nu_+ \tag{2}$$

中已经取得了越来越多的成功,并成为支持泡利观点的最令人信服的论据之一。此外，还设计了许多别的实验测试，这些测试巩固了中微子假设，并提供了中微子性质的信息。这种粒子具有带走能量和动量而不被检测到的本领，这表明之前提出的解释可能是正确的。也正是这一特性限制了这些测试只能测量衰变过程本身可观测的细节：与发射的电子和反冲子核[5]相关的能谱、动量矢量和能态。例如，通过仔细测量接近端点的氚衰变的β能谱[6]，得到中微子静止质量的上限为电子的静止质量的1/500，通常可以假定中微子的静止质量恒等于零。

然而预测中微子确定的静止质量并无理论上的根据，但考虑到中微子可能存在的虚态（中微子在这种虚态中可以分裂为其他粒子[7]），可以预测中微子具有很小且确定的磁矩，大小可能为 10^{-10} 玻尔磁子。通过计算中微子从太阳[8]到地球所传送的最大可转让热量，得到磁矩的上限可取为 2×10^{-9} 电子玻尔磁子。最近我们在美国原子能委员会萨凡纳河工厂的裂变反应堆附近使用大的闪烁探测器，得到改进上限值为 10^{-9} 电子玻尔磁子。在 370 加仑的液体闪烁器内，观测到了 0.1 兆电子伏 ~0.3 兆电子伏能量区间内单脉冲的计数率，通过与中微子的磁矩相互作用，由反应堆功率变化引起的所有变化都被归因于液体中可能的电子反冲。我们希望通过降低探测器中 γ 射线和中子的背景，可以进一步改进这个上限。

泡利–费米理论不仅要求中微子从β衰变核带走能量和线性动量，而且还要求带走角动量,即“自旋”。最简单的β衰变过程就是自由中子的衰变[9],可用下式说明：

$$n^0 \rightarrow p^+ + \beta^- + \nu_- \tag{1}$$

因中子、质子和β粒子都带有半整数的自旋，为了式 (1) 中的角动量平衡（式中三个产物粒子中的任何两个必须自旋反平行），必须赋予中微子一个量子数为 1/2 的自旋。因为式 (1) 中四个粒子全都是费米子，所以应遵从自旋为 1/2 的粒子的狄拉克相对论波动方程。每个粒子都可能存在相应的反粒子，但只有反电子（或称为正电子）和反质子已被确认。式 (1) 的中微子相应的反粒子可以通过把这几项重排而获得：

$$p^+ \rightarrow n^0 + \beta^+ + \nu_+ \tag{2}$$

This process is observed in positron decay of proton-rich radioactive nuclides where the proton and daughter neutron are both constituent nucleons. Further rearrangement results in the reaction:

$$\beta^- + p^+ \rightarrow n^0 + \nu_+ \tag{3}$$

This is descriptive of the capture of an electron from one of the inner atomic shells by a nuclear proton and is equivalent to equation (2). The question of the identity of the neutrino, ν_+, appearing in equations (2) and (3) with the neutrino, ν_-, appearing in equation (1) thus arises. With no finite mass or magnetic moment yet measured for either of the neutrinos, one is under no compulsion to assume that they are not in fact identical. The rule of algebraic conservation of fermions, which states that fermions are produced or disappear in particle-antiparticle pairs, requires the ν_-, of equation (1) to be named "antineutrino", since it is emitted with a negative electron. The identity or non-identity of the neutrino, ν_+, and the antineutrino, ν_-, although of no observable significance in single beta decay, should be amenable to test by measurement of the decay constant for double beta decay of certain shielded isotopes. This process was studied theoretically by M. Goeppert-Mayer[10] for the case in which neutrinos are not identical with antineutrinos and by Furry[11] for the case in which the two neutrinos are identical, as proposed by Majorana[12]. Double beta decay is typified by the possible decay of neodymium-150:

$$^{150}\text{Nd} \rightarrow {}^{150}\text{Sm} + 2\beta^- + 2\nu_- \text{(Dirac-Mayer)} \tag{4a}$$
$$^{150}\text{Nd} \rightarrow {}^{150}\text{Sm} + 2\beta^- \text{(Majorana-Furry)} \tag{4b}$$

If the neutrino and antineutrino are identical, then the virtual emission of one neutrino and its immediate re-absorption by the nucleus are equivalent to the real emission of two neutrinos, and equation (4*b*) is applicable. This cancellation is not possible if the neutrino and antineutrino differ. The half-lives for processes such as equation (4) have been shown by Primakoff[13] and by Konopinski[14] to be quite different in the two cases, of the order 10^{19} years for equation (4*a*) and 10^{15}years for equation (4*b*), where 5.4 MeV. is available for the decay. Furthermore, a line spectrum for the total energy of the two beta particles is to be expected for the Majorana-Furry case (equation 4*b*).

That a decay period consistent with equation (4*b*) does not exist has been shown for a number of shielded isotopes[15], first by Kalkstein and Libby, then by Fireman and Schwartzer for tin-124; by Awschalom for calcium-48; and our associates and us for neodymium-150. In the neodymium-150 experiment, a lower limit of 4×10^{18} years (corresponding to one standard deviation in the background) was set on the mean life against Majorana-Furry decay. This limit is to be compared with a reasonable value on this hypothesis of 1.3×10^{15} years and one calculated for identical neutrinos (using most severe assumptions) to be 6×10^{17} years. The conclusion remains that the neutrino and antineutrino are distinct particles with an as yet undetected "difference". This conclusion is further supported by the negative results of an experiment recently reported by R. Davis[16] employing the reaction:

这个过程是在丰质子放射性核素的正电子衰变中观测到的，其中质子和中子都是核子的组成成分。进一步的重排引起了如下反应：

$$\beta^- + p^+ \rightarrow n^0 + \nu_+ \tag{3}$$

与式 (2) 类似，这是一个核内质子从原子内壳层捕获电子的描述。这导致了一个问题，即式 (2) 和式 (3) 中的中微子 ν_+ 与式 (1) 中的中微子 ν_- 是否相同。目前二者的质量和磁矩都还没有确定的测量值，我们更倾向于认为它们实际上是相同的。费米子代数上的守恒法则表示，费米子是通过粒子–反粒子对成对产生或消失的，由于式 (1) 的 ν_- 是伴随一个负电子发射的，因此应被称为“反中微子”。尽管中微子 ν_+ 和反中微子 ν_- 是否相同这个问题在单一的 β 衰变中并无重要观测结果，但却易于通过测量某些屏蔽同位素的双 β 衰变的衰变常数来检验。梅耶夫人 [10] 在理论上对这个过程做了研究，认为中微子和反中微子并不相同，而弗里 [11] 的研究则认为马约拉纳 [12] 的观点是正确的，即两类中微子是相同的。以钕-150 的可能衰变表征双 β 衰变：

$$^{150}\mathrm{Nd} \rightarrow {}^{150}\mathrm{Sm} + 2\beta^- + 2\nu_- \text{（狄拉克–梅耶夫人）} \tag{4a}$$

$$^{150}\mathrm{Nd} \rightarrow {}^{150}\mathrm{Sm} + 2\beta^- \text{（马约拉纳–弗里）} \tag{4b}$$

如果中微子和反中微子是相同的，则一个中微子的虚发射和它立即被原子核重新吸收等效于两个中微子的实发射，可用式 (4*b*) 来表示。如果中微子和反中微子并不相同，便不可能存在这种抵消。对于式 (4) 所表达过程的半衰期，普里马科夫 [13] 和科诺平斯基 [14] 在两种情况下给出的结果完全不同，其中式 (4*a*) 的量级为 10^{19} 年，式 (4*b*) 的量级为 10^{15} 年，衰变可用的能量为 5.4 兆电子伏。此外，对于马约拉纳–弗里 (式 4*b*) 的情况，可以预期得到两个 β 粒子总能量的线谱。

一些屏蔽同位素表明，与式 (4*b*) 相一致的衰变周期并不存在 [15]。卡尔克施泰因和利比以及法尔曼和施瓦策先后在锡-124 中发现了这一点，阿沙洛姆在钙-48 中以及我们和我们的同事在铷-150 中也发现了这一点。在钕-150 的实验中，马约拉纳–弗里衰变的平均寿命下限被设置为 4×10^{18} 年（对应于背景中的一倍标准偏差）。这个极限可以与基于合理假设的值 1.3×10^{15} 年以及相同中微子情况下计算（用最严格的假设）的值 6×10^{17} 年进行比较。这个结论依旧认为中微子和反中微子是不同的粒子，只是至今尚未检测到它们之间的“差异”。戴维斯 [16] 最近报道的实验负结果进一步支持了这个结论，他采用以下反应：

$$^{37}Cl + \nu_{+} \rightarrow {}^{37}A + \beta^{-} \quad (5)$$

The chlorine target was supplied by 1,000 gallons of carbon tetrachloride placed near a large reactor, and the liquid was tested for the presence of argon-37. Fission fragments, being rich in neutrons, should emit only the antineutrino, ν_{-}.

While careful reasoning from experimental evidence gathered about all terms in the beta-decay process—except the neutrino—may support the inference that a neutrino exists, its reality can only be demonstrated conclusively by a direct observation of the neutrino itself. If the neutrino is a real particle carrying the missing energy and momentum from the site of a beta decay, then the discovery of these missing items at some other place would demonstrate its reality. Thus, if negative beta decays as in equation (1) could be associated at another location with the inverse reaction:

$$\nu_{-} + p^{+} \rightarrow \beta^{+} + n^{0} \quad (6)$$

which is observed to occur at the predicted rate, the case would be closed. An expression for this reaction cross-section has been obtained by application of the principle of detailed balancing to equation (1), knowing the decay constant and electron energy spectrum for the beta decay of free neutrons:

$$\sigma = \left(\frac{G^2}{2r}\right)\left(\frac{\hbar}{mc}\right)^2\left(\frac{p}{mc}\right)^2\left(\frac{1}{v/c}\right)^2(\text{cm.}^2) \quad (7)$$

where σ is the cross-section in cm.^{2}; G^2 ($=44\times10^{-24}$) is the dimensionless lumped beta-coupling constant based on neutron decay[9]; and p, m and v are the momentum, mass and speed of the emitted positron, respectively, c is the speed of light, and $2\pi\hbar$ is Planck's constant, all in c.g.s. units. For neutrinos of 3-MeV. energy incident on free protons, this cross-section is 10^{-43} cm.^{2}. Explicit solution of equation (6) for the cross-section as a function of the neutrino energy yields:

$$\sigma = 1.0 \times 10^{-44} \times (E - a)\sqrt{(E - a)^2 - 1}(\text{cm.}^2) \quad (8)$$

where $a+1$ ($=3.53$) is the threshold for the reaction and E is the neutrino energy, both in units of $m_e c^2$. The threshold for a proton bound in a nucleus is higher by an amount equal to the energy difference between the target and daughter nuclei. It is interesting to note that the penetrability of matter is given by equation (8) to be infinite for neutrinos with low energies ($E<a+1$) and is very large for neutrinos of only a few MeV., the mean free path for absorption being measured in the latter case in terms comparable to the radius of the universe.

Equation (6) may be employed in an experiment in which a large number of hydrogen atoms are provided as targets for an intense neutrino flux and are watched by a detector capable of recording the simultaneous production of a positron and a neutron. Such a

$$^{37}Cl + \nu_+ \rightarrow {}^{37}A + \beta^- \tag{5}$$

将 1,000 加仑的四氯化碳放置在大反应堆的附近作为氯靶，然后检验液体中是否存在氩-37。富含中子的裂变碎片应仅发射反中微子 ν_-。

根据 β 衰变过程各方面聚集的实验证据（除中微子外）的审慎推理可以支持存在中微子这一结论，而是否真实存在只能通过对中微子进行直接观测才能证明。如果中微子是一个真实的粒子，且携带 β 衰变位置处丢失的能量和动量，那么若在另一地点发现这些丢失的能量和动量便可以证明它真实存在。因此，如果式 (1) 中的负 β 衰变能与另一地点的下列逆反应（观测发现该逆反应以预期的速率发生）相关：

$$\nu_- + p^+ \rightarrow \beta^+ + n^0 \tag{6}$$

那么这个问题就解决了。已知自由中子 β 衰变的衰变常数和电子能谱，将细致平衡原理应用于式 (1)，可以得到这个反应截面的表达式：

$$\sigma = \left(\frac{G^2}{2r}\right)\left(\frac{\hbar}{mc}\right)^2\left(\frac{p}{mc}\right)^2\left(\frac{1}{v/c}\right)^2(\text{厘米}^2) \tag{7}$$

式中截面 σ 的单位为厘米^2；G^2(值为 44×10^{-24}) 是基于中子衰变 [9] 的无量纲集总 β 耦合常数；p、m 和 v 分别是发射的正电子的动量、质量和速度，c 是光速，$2\pi\hbar$ 是普朗克常数，所有量都采用厘米 · 克 · 秒制单位。对于入射到自由质子上能量为 3 兆电子伏的中微子而言，这个截面是 10^{-43} 厘米^2。式 (6) 中截面的精确解是中微子能量的函数：

$$\sigma = 1.0 \times 10^{-44} \times (E - a)\sqrt{(E - a)^2 - 1}(\text{厘米}^2) \tag{8}$$

式中 $a+1$ (值为 3.53) 是反应的阈值，E 是中微子能量，两者都以 m_ec^2 为单位。束缚在核内的质子的阈值会随着靶与子核之间的能量差的增加而增高。有趣的是，我们注意到由式 (8) 可知，低能 ($E<a+1$) 的中微子对物质的穿透力无限大，仅有几兆电子伏的中微子的穿透力都很大，测得的吸收平均自由程可与宇宙的半径相比。

式 (6) 可以应用到如下实验：大量的氢原子作为强中微子流的靶，并用一个能够记录同时产生的正电子和中子的探测器进行监测。多兆瓦反应堆中的裂变碎片的

direct experiment is made possible by the availability of high beta-decay rates of fission fragments in multi-megawatt reactors and advances in detection techniques through the use of liquid scintillators. An estimate of the neutrino flux available from large reactors shows that a few protons should undergo reaction (6) per hour in 50 litres of water placed near the reactor. The problem, then, is to observe these events with reasonable efficiency against the background of reactor neutrons and gamma-rays, natural radioactivity and cosmic rays. In an experiment conducted at the Hanford Plant of the Atomic Energy Commission by us[17] in 1953, an attempt was made in this direction. The target protons were supplied by 300 litres of liquid scintillator (toluene plus trace amounts of terphenyl, and alpha-naphtha-phenyloxayole in which cadmium propionate was dissolved). A delayed coincidence-rate of pairs of pulses, the first of each pair being assignable to the positron and the second to a neutron capture in cadmium, of 0.4 ± 0.2 counts per minute was observed, in agreement with the predicted rate, and with a large reduction in the backgrounds mentioned above. The signal-to-total-background ratio, however, was still very low (1/20), rendering further testing of the signal impractical and leaving the results tentative. On the basis of the Hanford experience it was felt that the detection problem was soluble in a definitive manner, and a second experiment was designed[18] with the view of further reduction of backgrounds and providing means for checking each term of equation (6) independently.

Fig. 1 is a schematic diagram of the detection scheme employed in this experiment. The sequence of events pictured is as follows: a neutrino from the decay of a fission fragment in a reactor causes a target proton to be changed into a neutron with the simultaneous emission of a positron. The positron is captured by an electron in the target water, emitting two 0.51-MeV. annihilation gamma-rays, which are detected simultaneously by counters I and II. The neutron moderates and diffuses for several microseconds and is finally captured by the cadmium giving a few gamma rays (totalling 9 MeV.), which are again detected by I and II. Thus we have a prompt coincidence followed in several microseconds by a second prompt coincidence, providing a very distinctive sequence of events.

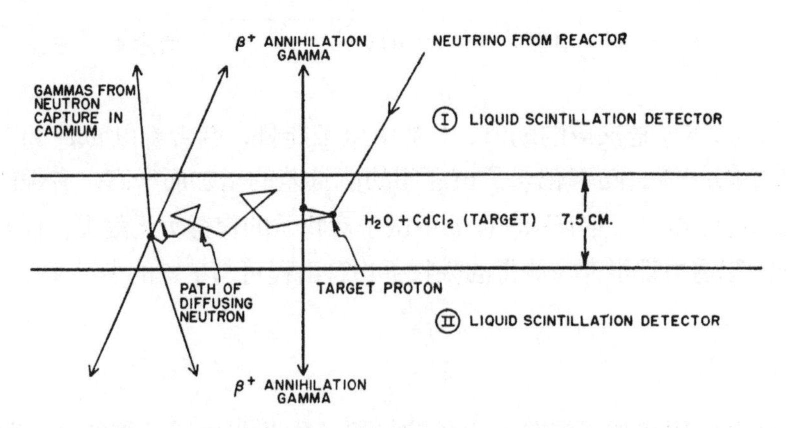

Fig. 1. Schematic diagram of neutrino detector

高 β 衰变速率的可用性，以及通过使用液体闪烁器而带来的检测技术的改进，最终使得这类直接实验得以实施。对来自大反应堆的中微子流的估计表明，反应堆附近放置的 50 升水中每小时有几个质子发生反应 (6)。问题是如何排除反应堆中的中子和 γ 射线、天然放射性和宇宙射线的背景的影响，合理有效地观测这些结果。1953 年，我们[17]在原子能委员会的汉福特场地进行的实验中，对这方面进行了尝试。我们用 300 升的液体闪烁器（内含甲苯和微量三联苯的混合物以及溶解了丙酸镉的α–挥发油–苯基）提供靶质子。观测到脉冲对的延迟速率为每分钟 0.4 ± 0.2 个计数（每对的第一个可归为正电子，第二个可归为在镉中俘获的中子），该观测速率与预测速率相符，而且前面提到的背景影响也大大减小。然而，信号与整个背景的比仍然很低 (1/20)，因而进一步测试还不现实，结果也只是暂时性的。在汉福特经验的基础上，我们认为可以以一种确定的方式解决检测问题，综合考虑了进一步减小背景因素的目的之后，设计了第二个实验[18]，该实验可以独立检验式 (6) 中的每一项。

图 1 是实验采用的检测方案的示意图。图示事件的发生顺序如下：从反应堆的裂变碎片中衰变得到的中微子使靶质子转变为中子，同时发射出一个正电子。正电子被靶水中的电子俘获，并发射两束 0.51 兆电子伏的湮没 γ 射线，被计数器 I 和 II 同时检测到。中子在几微秒内减速和扩散，最终被镉俘获并发射一些 γ 射线（总能量为 9 兆电子伏），这些射线再次被计数器 I 和 II 检测到。因此，我们得到一个瞬间符合，几微秒后，又得到第二个瞬间符合，从而形成了一个非常独特的事件序列。

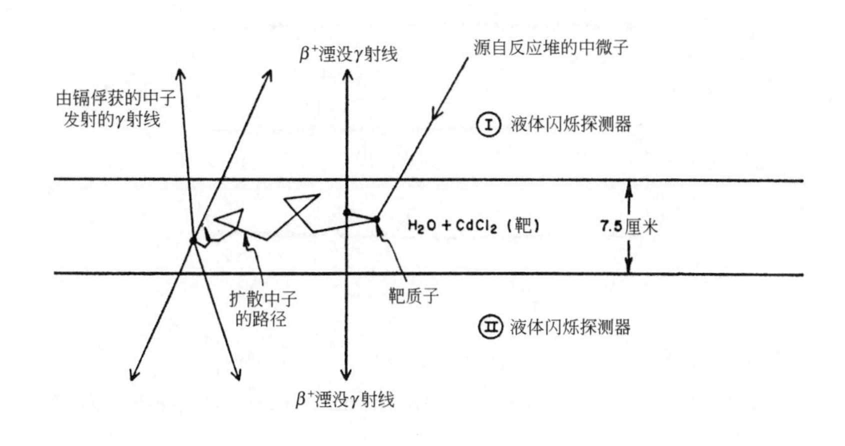

图 1. 中微子探测器示意图

The over-all size of the equipment was set by the number of events expected per hour per litre of water, and the detection efficiency one could hope to achieve. A primary factor in the design geometry and detection efficiency was the absorption of the positron annihilation radiation by the target water itself. Experimentation and calculations showed that an optimum water thickness was 7.5 cm. Since the over-all efficiency dictated a target volume of about 200 litres to yield several counts per hour, two target tanks were used, each measuring 1.9 m. ×1.3 m. × 0.07 m. The depth of the liquid scintillation detector (61 cm.) was such as to absorb the cadmium-capture gamma-rays with good efficiency and transmit the resultant light to the ends of the detector with minimal loss. The scintillating liquid (triethylbenzene, terphenyl and POPOP wave-length shifter) were viewed from the ends of each detector tank by 110 5-in. Dumont photomultiplier tubes, a number determined primarily by the amount of light emitted in a scintillation. The complete detector consisted of a "club sandwich" arrangement employing two target tanks between three detector tanks, comprising two essentially independent triads which used the centre detector tank in common. The entire detector was encased in a lead-paraffin shield and located deep underground near one of the Savannah River Plant production reactors of the United States Atomic Energy Commission. Signals from the detectors were transmitted via coaxial cables to an electronics trailer located outside the reactor building. The pulses were analysed by pulse-height and time-coincidence circuits and, when acceptable, were recorded photographically as traces on triple-beam oscilloscopes. Fig. 2 is a record of an event in the bottom triad. The entire system was calibrated using a plutonium-beryllium neutron source and a dissolved copper-64 positron source in the target tanks; and standardized pulsers were used to check for stability of the electronics external to the detector itself. The response of the detector to cosmic ray μ-mesons was also employed as a check on its performance. After running for 1,371 hr., including both reactor-up and reactor-down time, it was observed[19] that:

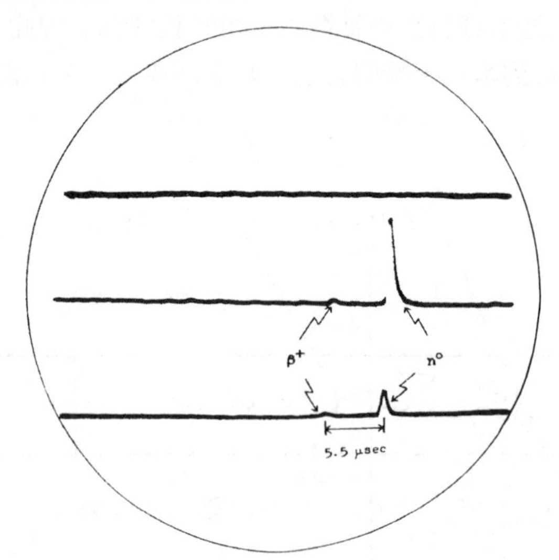

Fig. 2. A characteristic record. Each of the three oscilloscope traces shown corresponds to a detector tank. The event recorded occurred in the bottom triad. First seen in coincidence are the "positron" annihilation gamma-ray pulses in each tank followed in 5.5 μsec. by the larger "neutron" pulses. The amplification was chosen in this case to enable measurement of the neutron pulses. A second oscilloscope with higher amplification was operated in parallel to enable measurement of the positron pulses

根据预测的每小时每升水中的事件数以及我们希望得到的探测效率来设置实验装置的总尺寸。设计几何形状和探测效率的一个主要影响因素是靶水本身对正电子湮灭辐射的吸收。实验和计算表明，水的最佳厚度为 7.5 厘米。由于整个效率需要约 200 升的靶以保证每小时产生几个计数，因此需要两个尺寸为 1.9 米 × 1.3 米 × 0.07 米的靶箱。液体闪烁探测器的深度 (61 厘米) 可以保证高效率地吸收镉俘获的 γ 射线，并以最小的损耗传输 γ 射线产生的光到探测器终端。用 110 个 5 英寸的杜蒙特光电倍增管从每个探测箱的端部观察闪烁液体（三乙基苯、三联苯和 POPOP 波长变换剂），其中光电倍增管的数量主要取决于闪烁体中发射光子的数量。完整的检测器是一个“总汇三明治”式的组合，即在三个探测箱之间放置两个靶箱，从而构成了两个基本独立但共同使用中心探测箱的三件套。整个探测器装在一个铅–石蜡保护箱内，并放置在深层地下，位于美国原子能委员会的萨凡纳河工厂的一个生产堆附近。检测信号通过同轴电缆传输到位于反应堆建筑外的电子仪器拖车上。通过脉冲–高度以及时间符合电路来分析脉冲，当接收到脉冲时，用照相方法将脉冲记录为三踪示波器上的径迹。图 2 是三件套底部一个事件的记录结果。用靶箱中的钚–铍中子源和溶解的铜-64 正电子源校准整个系统；用标准化的脉冲发生器检测探测器以外的电子设备的稳定性。还通过探测器对宇宙射线中 μ 介子的响应检验它自身的性能。在运行 1,371 小时后（包括反应堆功率增长的时间和反应堆功率下降的时间），观测结果 [19] 如下：

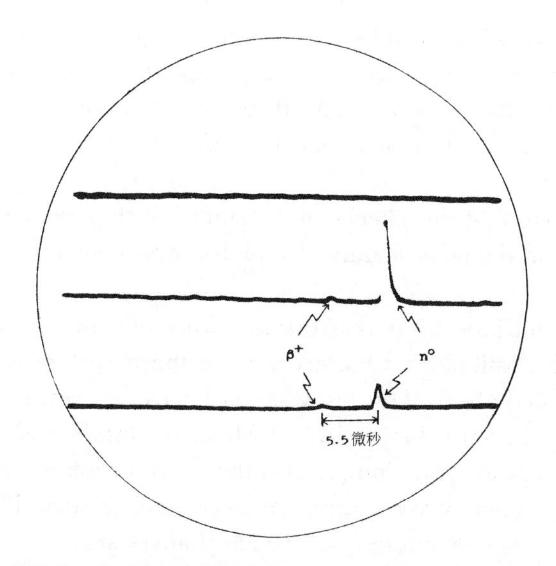

图2. 一次特征记录。显示的三条示波器径迹分制对应三个探测箱。记录的这个事件发生在底部的三件套。在符合中首先见到的是每个探测箱中的“正电子”湮没γ射线脉冲，5.5微秒后，出现较大的“中子”脉冲。在这种情况下，选用能测量中子脉冲的放大倍数。并行运行的第二台示波器的放大倍数更大，用以测量正电子脉冲。

(1) A signal dependent upon reactor-power, 2.88±0.22 counts/hr. in agreement with the predicted[20] cross-section (6×10^{-44} cm.[2]), was measured with a signal-to-reactor associated accidental background in excess of 20/1. The signal-to-reactor independent background ratio was 3/1.

(2) Dilution of the light water solution in the target tank with heavy water to yield a proton density of one-half normal caused the reactor signal to drop to one-half its former rate. The efficiency of neutron detection measured with the plutonium-beryllium source was unchanged.

(3) The first pulse of the pair was shown to be positron annihilation radiation by subjecting it to a number of tests: its spectrum agreed with the spectrum of positron annihilation radiation from copper-64 dissolved in the water, and it was absorbed in the expected manner by thin lead sheets inserted between the target tank and one detector.

(4) The second pulse of the pair was identified as due to the capture in cadmium of a neutron born simultaneously with the positron by virtue of its capture-time distribution as compared both with calculations and observations with a neutron source. The second pulse spectrum was consistent with that of cadmium-capture gamma-rays, and removal of the cadmium resulted in disappearance of the reactor signal.

(5) Reactor-associated radiations such as neutrons and gamma-rays were ruled out as the source of the signal by two kinds of experiment. In the first, a strong americium-beryllium neutron source was placed outside the detector shield and was not only found very inefficient in producing acceptable delayed coincidences but was also found to produce a first-pulse spectrum which was unlike the required signal in that it was monotonically decreasing with increasing energy. In the second experiment an additional shield, which provided an attenuation factor of at least 10 for reactor neutrons and gamma-rays, was observed to cause no change in the reactor signal outside the statistical fluctuations quoted in (1).

Completion of the term-by-term checks of equation (6) thus demonstrated that the free neutrino is observable in the near vicinity of a high-power fission reactor.

The availability of neutrinos from reactors in sufficiently intense fluxes has opened a number of interesting possibilities. One arises from the use of heavy water to dilute the proton target as described above. This test was valid because the threshold for the neutrino interaction with the deuteron is higher by 2.2 MeV., the binding energy of the deuteron, than the threshold energy for equation (6), and the cross-section is for this reason an order of magnitude smaller; other considerations reduce it still further. The neutrino-deuteron interaction is itself, however, of interest, as two alternatives arise:

$$\nu_{-} + D \rightarrow \beta^{+} + n + n \qquad (8a)$$
$$\nu_{-} + D \rightarrow \beta^{+} + n_2 \qquad (8b)$$

(1) 取决于反应堆功率的信号，为 2.88±0.22 个计数 / 小时，是用超过 20/1 的信号–反应堆偶然背景比测量的，与预计[20]的截面 (6×10^{-44} 厘米^2) 相一致。信号–反应堆的独立背景比为 3/1。

(2) 用重水稀释靶箱内的轻水溶液从而使质子密度为正常情况下的一半，这样可使反应堆信号的速率下降到其以前的一半。而用钚–铍源测量的中子检测效率不变。

(3) 通过一些实验表明，这对脉冲中的第一个脉冲是正电子湮没辐射形成的：其光谱与水中溶解的铜-64 的正电子湮没辐射谱一致，并以预期的方式被插放在靶箱和探测器间的薄铅板吸收。

(4) 这对脉冲中的第二个脉冲是由与正电子同时产生的中子经延迟后在镉中俘获产生的，这可由其俘获时间分布与中子源的计算和观测结果相比较来确定。第二个脉冲的谱与镉俘获 γ 射线谱一致，移除镉将导致反应堆的信号也随之消失。

(5) 对于与反应堆有关的辐射，例如中子和 γ 射线，有两类实验可将它们排除在信号来源之外。在第一类实验中，在检测器保护箱外放置一个强镅–铍中子源，我们发现它不仅在产生可接收的延迟符合方面效率很低，而且它产生的第一个脉冲谱并不像要求的信号那样随着能量的增高单调下降。在第二类实验中，一个附加的保护箱为反应堆的中子和 γ 射线提供了一个至少为 10 的衰减因子，观测发现除了影响到式 (1) 统计涨落外，它并不引起反应堆的信号变化。

因此，对式 (6) 的逐项检测结果表明，在高能裂变反应堆附近能观测到自由中微子。

来自反应堆的高通量中微子的可用性带来了许多有趣的可能性。其中一个源于前面提到的用重水稀释质子靶。这个试验是成立的，因为中微子与氘核相互作用的阈能比式 (6) 的阈能高 2.2 兆电子伏，这个值正是氘核的结合能，因此截面小一个量级；如果考虑到其他因素，其值可能还会进一步减小。然而中微子–氘的相互作用本身就很有意义，得到下面两个反应：

$$\nu_- + D \rightarrow \beta^+ + n + n \tag{8a}$$

$$\nu_- + D \rightarrow \beta^+ + n_2 \tag{8b}$$

where n_2 is the bound state of the bineutron[21], as yet unobserved. If reaction (8*a*) were observed to occur, then a careful measurement of its rate relative to the rate of reaction (6) and a knowledge of the fission neutrino spectrum should enable a direct determination of the ratio of the Fermi and Gamow-Teller coupling constants in beta decay. This follows from the fact that the coupling constant in equation (6) includes a mixture of both types, whereas in (8*a*) it is composed of the Gamow-Teller constant alone. If, on the other hand, equation (8*b*) were observed, not only would these considerations hold, but also the existence of a bound state of the bineutron, which would necessarily be a singlet state (antiparallel spins) because of the Pauli exclusion principle, would bear directly on the question of the dependence of nuclear forces on charge. This follows because the singlet state of the (n, p) system is known to be unbound. As the two neutrons in equation (8*a*) can possess only a few kilovolts of energy when produced by fission-fragment neutrinos, and as they leave the event in antiparallel spin states, the conditions seem favourable for the formation of bineutrons, even if the binding energy were only tens of kilovolts.

Since the proposal of the neutrino hypothesis by Pauli and its success in Fermi's theory of nuclear beta decay, the particle has been called upon to play similar parts in the observed decay of a number of different mesons[22]. The question arises as to the identity of these neutrino-like particles with the neutrino of nucleon decay. It is to be noted that in nuclear beta decay the initial and final nuclei both quite obviously interact strongly with nuclei. This is not the case in (π, μ) decay, where the emission of a "neutrino" converts the interaction of the heavy particles with nuclei from strong to weak. Furthermore, despite the apparent equality of the nuclear beta-decay matrix elements with those associated with (μ, β) decay, both the initial and final products of the latter interact weakly with nuclei.

The neutrino is the smallest bit of material reality ever conceived of by man; the largest is the universe. To attempt to understand something of one in terms of the other is to attempt to span the dimension in which lie all manifestations of natural law. Yet even now, despite our shadowy knowledge of these limits, problems arise to try the imagination in such an attempt. If nuclear reactions played a part in a cataclysmic birth of the universe as we assume, what fraction of the primordial energy was quickly drained into the irreversible neutrino field? Are these neutrinos—untouched by anything from almost the beginning of time—trapped by the common gravitational field of the universe, and if so, what is their present density, their energy spectrum and angular distribution? Do neutrinos and antineutrinos exist in equal numbers? If the neutrino has zero rest mass, is it to be considered with "matter" particles in discussing its gravitational potential, or with electromagnetic radiation? The problem of detecting these cosmic end-products of all nuclear energy generation processes and the measurement of their characteristics presents a great challenge to the physics of today.

式中 n_2 是双中子[21]的束缚态，不过目前尚未观测到。如果观测到反应 (8*a*) 发生，则精确测量其反应速率与式 (6) 反应速率之比，并应用裂变中微子谱的知识，将能直接测定在 β 衰变中费米常数与伽莫夫 – 特勒耦合常数的比值。这是因为：式 (6) 中的耦合常数包含上述两个常数，而式 (8*a*) 仅包含伽莫夫 – 特勒常数。另一方面，如果观测到式 (8*b*) 表示的反应，则不仅这些考虑成立，而且双中子束缚态 (根据泡利不相容原理这个束缚态必然是一个单重态，即反平行自旋) 的存在将直接影响到核力是否依赖于电荷这一问题。这是因为已经知道 (n, p) 体系的单重态是非束缚的。由于式 (8*a*) 中由裂变碎片中的中微子产生的两个中子仅具有几千伏的能量，而且事件结束时它们处于自旋反平行态，因此条件似乎对双中子的形成十分有利，即使结合能仅几十千伏。

自从泡利提出中微子假设并且在费米的核 β 衰变理论上取得成功后，这种粒子已经被认为在观测到的许多不同介子[22]的衰变中发挥着类似的作用。然而存在一个问题，即这些类中微子粒子与核子衰变的中微子是否是同一种粒子。我们注意到在核 β 衰变中，最初的核和最终的核均与原子核有明显的强相互作用。而在 (π, μ) 衰变中并不是这种情况，其中“中微子”的发射使重粒子与核的相互作用从强变弱。此外，尽管核 β 衰变矩阵元与那些 (μ, β) 衰变矩阵元明显相等，但是后者的最初产物和最终产物与核只有很弱的相互作用。

人们曾认为，中微子是最小的物质实体；最大的是宇宙。试图用一事物来理解另一事物，就是希望横跨自然定律所有表现形式的尺度。然而，尽管现在我们对这些极限的知识还是模糊不清，但仍在试图通过想象解决这些问题。如果核反应在我们假定的宇宙的剧烈的诞生过程中发挥作用，那么有多少比重的初始能量快速地流入不可逆的中微子场中？这些中微子几乎从时间一开始就不与任何物质接触，那么它们会被宇宙普通的引力场捕获吗？如果答案是肯定的，那么它们现在的密度有多大，它们的能谱和角度分布又如何？中微子和反中微子以相等数量存在吗？如果中微子的静止质量为零，那么在讨论其引力势时，它是作为“物质”粒子还是电磁辐射来考虑？所有核能产生过程的宇宙最终产物的探测及其特性的测量，对当今的物理学无疑是一个重大的挑战。

The known properties of the neutrino are summarized below.

Properties of the Neutrino

Spin	$1/2\hbar$.
Mass	<1/500 electron mass, if any.
Charge	0.
Magnetic moment	$<10^{-9}$ Bohr magneton
Cross-section for reaction	$\nu_{-}+p^{+}\rightarrow\beta^{+}+n^{0}$ at 3 MeV. $=10^{-43}$ cm.^{2}. Neutrino ν_{+} not identical with antineutrino ν_{-}

Our work and that of our associates reported in this paper were supported by the United States Atomic Energy Commission.

(**178**, 446-449; 1956)

Frederick Reines and Clyde L. Cowan, jun.: University of California, Los Alamos Scientific Laboratory, Los Alamos, New Mexico.

References:

1. Chadwick discovered that the beta spectrum was continuous. L. Meitner suggested in 1922 that a quantized nucleus should not be expected to emit a continuous spectrum, and Ellis found non-conservation of energy from experiments on the emitted electron. Chadwick, J., *Verh. Deutsch. Phys. Ges.*, **16**, 383 (1914). Ellis, C. D., Internat. Conf. on Phys., **16**, 209 (1934).
2. Ellis and Wooster, *Proc. Roy. Soc.*, A, **117**, 109 (1927). Chadwick, J., and Lea, D. E., *Proc. Camb. Phil. Soc.*, **30**, 59 (1934); Nahmias, M. E., *Proc. Camb. Phil. Soc.*, **31**, 99 (1935). Wu, C. S., *Phys. Rev.*, **59**, 481 (1941).
3. Pauli, W., in "Rapports du Septième Conseil de Physique Solvay", Brussels, 1933 (Gauthier-Villars, Paris, 1934).
4. Fermi, E., *Z. Phys.*, **88**, 161 (1934).
5. We do not attempt here to describe the many beautiful and difficult, recoil experiments in which recoils of neutrino-emitting nuclei (~8-200 eV.) have been measured. A summary can be found in an article by O. Kofoed-Hansen in Siegbahn's "Beta and Gamma-Ray Spectroscopy" (Interscience Publishers, Inc., New York, 1955).
6. Langer, L. M., and Moffat, R. J. D., *Phys. Rev.*, **88**, 689 (1952). Hamilton, Alford and Gross, *Phys. Rev.*, **92**, 1521 (1953). This question is treated in detail in an article by C. S. Wu in Siegbahn (*op. cit.*). We quote Dr. Wu's most conservatively estimated limit.
7. Houtermans, F. G., and Thirring, W., *Helv. Phys. Acta*, **27**, 81 (1954). H. A. Bethe has given the relationship between the recoil electron spectrum and the energy and magnetic moment of a neutrino in *Proc. Camb. Phil. Soc.*, **31**, 108 (1935).
8. Crane, H. R., *Revs. Mod. Phys.*, **20**, 278 (1948). This article also summarizes neutrino detection attempts to 1948. The status of the neutrino in 1936 is given by H. A. Bethe and R. F. Bacher. *Revs. Mod. Phys.*, **8**, 82 (1936).
9. Snell, A. H., and Miller, L. C., *Phys. Rev.*, **74**, 1714 A (1948). Snell, A. H., Pleasanton, F., and McCord, R. V., *Phys. Rev.*, **78**, 310 (1950). Robson, J. M., *Phys. Rev.*, **78**, 311 (1950); **83**, 349 (1951).
10. Goeppert-Mayer, M., *Phys. Rev.*, **48**, 512 (1935).
11. Furry, W. H., *Phys. Rev.*, **56**, 1184 (1939).
12. Majorana, E., *Nuovo Cimento*, **14**, 171 (1937).
13. Primakoff, H., *Phys. Rev.*, **85**, 888 (1952).
14. Konopinski, E. J., Los Alamos Report *LAMS* 1949 (1955).
15. Kalkstein, M. I., and Libby, W. F., *Phys. Rev.*, **85**, 368 (1952). Fireman, E. L., and Schwartzer, D., *Phys. Rev.*, **86**, 451 (1952). Awschalom, M., *Phys. Rev.*, **101**, 1041 (1956). Cowan, jun., C. L., Harrison, F. B., Langer, L. M., and Reines, F., *Nuovo Cimento*, **3**, 649 (1956).
16. Davis, jun., R., Contributed Paper, American Physical Society, Washington, D. C., Meeting, 1956. This experiment was originally suggested by Pontecorvo and considered by Alvarez in a report *UCRL*-328 (1949).
17. Reines, F., and Cowan, jun., C. L., *Phys. Rev.*, **90**, 492 (1953); **92**, 830 (1953).
18. Cowan, jun., C. L., and Reines, F., Invited Paper, American Physical Society, New York Meeting, January 1954.
19. Cowan, jun., C. L., and Reines, F., Postdeadline Paper, American Physical Society, New Haven Meeting, June 1956. Cowan, Reines, Harrison, Kruse and McGuire, *Science*, **124**, 103 (1956).
20. The neutrino spectrum was deduced from the spectrum of beta-radiation from fission fragments as measured by C. O. Muehlhause at the Brookhaven National Laboratory. Dr. Muehlhause kindly communicated his results to us in advance of publication.
21. The evidence for and against the existence of a "bineutron", also called "dineutron", is discussed by B. T. Feld in his article on the neutron in the volume edited by E. Segrè entitled "Experimental Nuclear Physics", 2 (John Wiley and Sons, Inc., New York, 1953).
22. Oneda, S., and Wakasa, A., discuss the question of classes of interactions between the elementary particles in *Nuclear Phys.*, **1**, 445 (1956).

中微子的已知性质概括如下：

中微子的性质

自旋	$1/2\hbar$
质量	即使有，< 1/500 电子质量
电荷	0
磁矩	$< 10^{-9}$ 玻尔磁子
反应截面	在 3 兆电子伏时，$\nu_{-}+p^{+} \rightarrow \beta^{+}+n^{0}$ 截面为 10^{-43} 厘米^2。中微子 ν_{+} 与反中微子 ν_{-} 不相同

我们的工作以及本文中涉及的我们同事的工作得到美国原子能委员会的支持。

（沈乃澂 翻译；尚仁成 审稿）

Production of High Temperatures and Nuclear Reactions in a Gas Discharge

P. C. Thonemann *et al.*

Editor's Note

The first hydrogen bomb, which derived energy from nuclear fusion rather than fission, was detonated in 1952. Two years later, physicists began trying to bring controlled nuclear fusion into the laboratory. Here Peter Thonemann and colleagues report on initial experiments with a device called the ZETA, a toroidal chamber 3 metres in diameter holding a dilute plasma. They were attempting to heat that plasma to temperatures approaching those in the Sun by using powerful bursts of current. The team describe encouraging initial results, but would later find that numerous plasma instabilities foiled the scheme's ultimate success. Today, more than 50 years later, a wide variety of plasma instabilities still put practical fusion energy out of reach.

Introduction

THE basic conditions which must be established before a thermonuclear reactor is possible are, first, the containment of a high-temperature gas so that it is isolated from the walls of the surrounding vessel, and second, the attainment of temperatures sufficiently high for nuclear reactions to take place between the light elements. These two conditions are interdependent. Poor containment results in energy losses so large that gas temperatures much exceeding 10^6 °K. are unattainable.

The experimental apparatus described was designed to study the containment of ionized hydrogen (or deuterium) by the magnetic field associated with the current flowing in the gas and to reach temperatures sufficiently high for nuclear reactions to be detectable using deuterium. This apparatus, known as ZETA, was built at the Atomic Energy Research Establishment Harwell.

The principles underlying the containment of a high-temperature gas by means of the "pinch effect" have been discussed by a number of authors[1-3].

The constricted gas discharge formed by passing a high current through a low-pressure gas, while isolating the gas from the tube walls for short periods of time, rapidly develops instabilities and distortions which lead to bombardment of the walls by electrons and positive ions. The consequent cooling and recombination make it impossible to maintain high temperatures except for a transient period[4-8]. These instabilities can be suppressed by the combination of an axial magnetic field parallel to the direction of the discharge current and by the fields produced by eddy currents induced in the surrounding metal walls when the current channel changes its position.

气体放电中产生的高温和核反应

索恩曼等

编者按

1952 年第一颗氢弹爆炸，其能量来自核聚变而不是核裂变。两年后，物理学家开始尝试将可控的核聚变引入实验室。本文中彼得·索恩曼及其同事报道了最初的实验，在实验中他们使用一个叫 ZETA 的装置，这个装置是一个直径为 3 米并装有稀薄等离子体的环形腔。他们试图通过强大的电流爆发加热等离子体使其温度接近太阳的温度。这个小组描述了令人振奋的初步结果，但不久发现大量的等离子体的不稳定性阻碍了课题的最终成功。五十多年后的今天，多种等离子体的不稳定性仍然阻碍着聚变能的应用。

引　言

在成功建立热核反应堆前必须具备的基本条件是：第一，约束高温气体，确保将其与周围的容器壁隔离开；第二，达到足够高的温度，使轻元素间能发生核反应。这两个条件是相互依存的。不好的约束条件会使能量损失太大，以致气体温度难以达到远大于 10^6 K 的温度。

设计的实验装置通过与气体中电流相关的磁场来研究电离氢(或氘)的约束情况，该装置能达到可以用氘检测是否发生核反应所需的足够高的温度。这台在哈威尔原子能研究中心建立的装置被称为 ZETA。

许多作者 [1-3] 已讨论了用“箍缩效应”方法约束高温气体的原理。

短时间内将气体与管壁隔离开时，在低压气体中通过强电流产生的压缩气体放电会很快变得不稳定并发生变形，这会导致电子或正离子轰击壁。除非在过渡周期 [4-8]，否则随后的冷却和复合将使等离子体不可能保持高温。这些不稳定性可以通过平行于放电电流方向的轴向磁场的叠加和电流通道改变位置时在周围金属壁感生的涡旋电流产生的磁场来加以抑制。

Theoretical studies of discharge stability in the presence of an axial magnetic field have been published[9-12], together with experimental evidence for stability in a straight discharge tube[13].

The preliminary results reported in this article show that relatively long-time stability can be achieved in a toroidal metal-walled tube. The neutron yield and the kinetic ion temperatures have been measured over a limited range of conditions. The nuclear reaction-rates observed are not inconsistent with those expected from a thermonuclear process.

Apparatus

ZETA is a ring-shaped discharge tube of aluminium, 1-m. bore and 3-m. mean diameter, containing gas at low pressure. The gas, usually at a pressure of about 10^{-4} mm. of mercury, is made weakly conducting by a radio-frequency discharge. The toroidal ionized gas plasma forms the secondary of a large iron-cored pulse transformer. A condenser bank, storing up to a maximum of 5×10^5 joules, is discharged into the primary of the transformer, and produces a unidirectional current pulse in the gas up to a maximum of 200,000 amp. The current pulse in the gas lasts for about 4 msec. and is repeated every 10 sec. A steady axial magnetic field is generated by current-carrying coils wound on the torus. This field can be varied from zero to 400 gauss.

Fig. 1 shows the discharge tube assembled with the transformer. A vacuum spectrograph can be seen connected to the torus body.

Fig. 1. Photograph of ZETA. The apparatus is enclosed in a room with concrete walls 3 ft. thick for radiation shielding

Electrical Characteristics

Typical current and voltage oscillograms are shown in Fig. 2. The top trace is the "voltage

轴向磁场存在时放电稳定度的理论研究[9-12]，以及在直线放电管中稳定性的实验证据[13]现已发表。

本文中报道的初步结果表明，在环形的金属壁管中能实现较长时间的稳定。在有限范围的条件下，已测量了中子产生和离子动力学温度。观测到的核反应速率并没有与热核过程预期的结果相矛盾。

装　置

ZETA 是一个铝的环形放电管，管径为 1 米，平均直径为 3 米，内装有低压气体。通常气体的压强约为 10^{-4} 毫米汞柱，射频放电使其具有弱的导电性。环形电离的气体等离子体形成了一个大铁芯的脉冲变压器的次级。电容器储备了最大为 5×10^5 焦的能量，它放电到变压器的初级，在气体中产生一个单向的最大值为 200,000 安的电流脉冲。在气体中的电流脉冲持续约 4 毫秒，每 10 秒重复一次。稳定的轴向磁场是由绕在环上的载流线圈产生的。磁场从 0 变到 400 高斯。

图 1 显示了装有变压器的放电管。可以看到环流器主体与一台真空光谱仪连接。

图 1. ZETA 的照片。为了屏蔽辐射，装置密封在 3 英尺厚的混凝土墙的室内。

电学特性

图 2 是电流和电压的特征波形图。上面的轨迹是用放电管包围的铁芯环绕的回

per turn" measured by a loop around the iron core enclosed by the discharge tube. The length of the discharge path is approximately 1,000 cm. so that the initial electric field at the boundary of the ionized gas is about 2 V. per cm. The primary winding is short-circuited when the voltage per turn is zero, thus preventing the charge on the condensers reversing. The second trace shows the current flowing in the gas, which persists for about 2 msec. after the transformer primary is short-circuited.

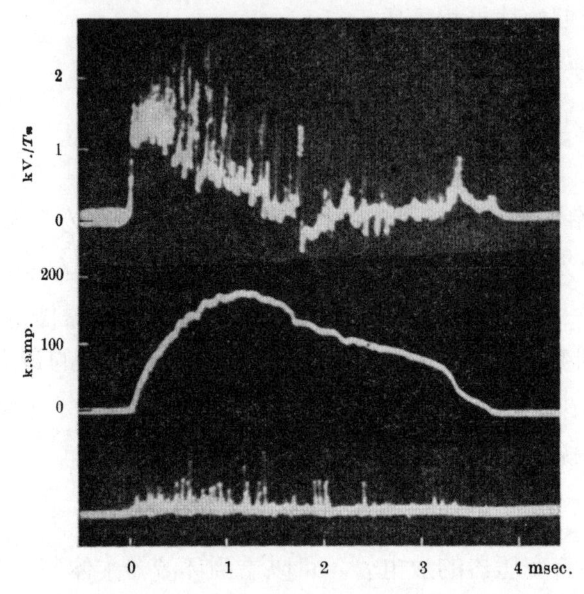

Fig. 2 Oscillograph recordings of the voltage per turn of the transformer, and the secondary current I_s. The lower trace shows the pulses produced by proton recoil in a scintillation neutron counter. Conditions: gas, deuterium +5 percent nitrogen + 10 percent oxygen; pressure, 0.13×10^{-3} mm. mercury; axial field, 160 gauss

As the current decreases the plasma expands until it reaches the walls. It is then cooled. A sudden increase in resistance accompanies this process and the consequent increase in $|dI/dt|$ produces a severe voltage transient which may rise to tens of kilovolts. Destructive voltage transients of this type are suppressed by the addition of 5 percent nitrogen without affecting the neutron yield.

Stability

Measurements with magnetic field-probes and Langmuir probes, together with streak photographs of the current channel taken through a slit in the vacuum vessel, show the current channel to be quasi-stable and clear of the walls for the greater part of the current pulse. Fig. 3 is a reproduction of a streak picture of a helium discharge. The limit of the black area represents the internal diameter of the tube. The light recorded is that of the spark lines of impurities and of (He II) 4,686 A. Streak pictures taken of discharges in deuterium are difficult to interpret as the light is emitted by neutral atoms and by impurity ions released into the current channel from the walls.

路测量的“每匝的电压”。放电路程的长度近似为 1,000 厘米，因此，在电离气体边界的初始电场约为 2 伏/厘米。当每匝的电压为零时，初级线圈被短路，因此避免了电容器的电荷反向。第二个轨迹显示了气体中的电流，它在变压器初级线圈短路后持续约 2 毫秒。

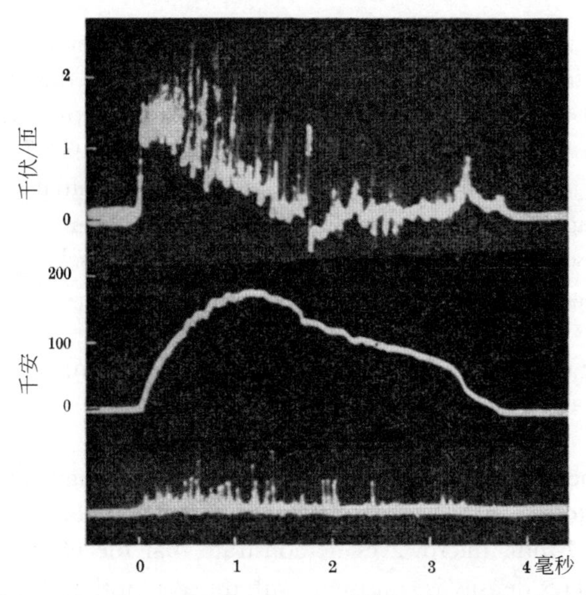

图 2. 示波器记录的变压器每匝电压以及次级电流 I_s。下方的径迹显示的是在闪烁中子计数器中质子反冲产生的脉冲。条件：气体，氘 +5% 氮 +10% 氧；气压，0.13×10^{-3} 毫米汞柱；轴向场为 160 高斯。

当电流减小时，等离子体扩展直到它到达器壁，然后冷却。伴随这一过程，电阻突然增大，$|dI/dt|$ 相继增大，产生瞬间急剧增长的电压，它可以增高到几十千伏。通过附加不影响中子产生的 5% 的氮可抑制这类破坏性电压瞬变。

稳 定 度

由磁场探针和朗缪尔探针进行的测量与通过真空容器的狭缝拍摄的电流通道的条纹照片一起可知，电流通道是准稳定的并且显示大部分的电流脉冲没有接触管壁。图 3 是氦放电的条纹图的重现。黑色区域的边界代表了管的内径。记录的光是杂质和 4,686 埃 (He II) 的火花放电谱线。氘放电的条纹图难以解释为光是由中性原子和从器壁进入电流通道的杂质离子所发射的。

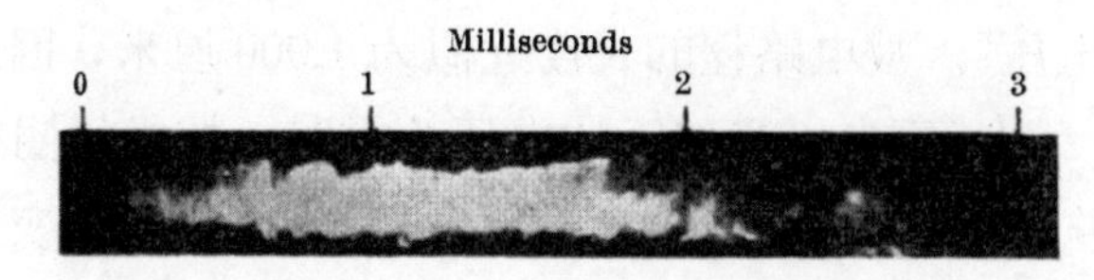

Fig. 3. Streak picture of a helium discharge. Conditions: initial gas pressure 0.25×10^{-3} mm. mercury; axial field 160 gauss; peak current, 130 k.amp. The tube walls lie at the boundary of the dark region

The centre of the current channel is displaced towards the outer wall due to the tendency of the ring current to expand. This expansion is opposed by eddy currents in the metal walls, which are of 1-in. thick aluminium. Measurements of the internal magnetic fields in the plasma are reproducible and show that the axial magnetic field, B_z , is trapped in the gas. On the axis it increases to approximately ten times the initial value. In general, the resultant lines of magnetic force due to the B_θ and B_z components are helical and vary in pitch over the cross-section of the plasma. The stability of a discharge with this magnetic-field configuration has not been treated theoretically.

The presence of the magnetic field-probe, which is 1 in. in diameter, greatly increases the discharge resistance and reduces the production of neutrons.

The diameter of the current channel estimated from the magnetic-field measurements and the streak photographs is between 20 and 40 cm. at peak current. Transmission measurements with 4-mm. microwaves demonstrate that the electron density is greater than 6×10^{13} cm.$^{-3}$. This density is consistent with the assumption that all the gas present is ionized and contained in the current channel.

High-Energy Radiations

Neutron emission arising from the D-D reaction is observed for gas currents in deuterium in excess of 84 k.amp. Emission occurs for a period of about 1 msec., centred about the peak current.

Table 1 shows the average number of neutrons emitted per pulse as the peak current is increased.

Table 1

Conditions: gas D_2 + 5 percent N_2. Pressure: 0.12×10^{-3} mm. B_z= 160 gauss

Current (k.amp.)	Total neutron yield per pulse	T_c (calc.)
84	0.4×10^4	2.4×10^4 °K.
117	3.1	2.9
126	9.2	3.3
135	14.2	3.6
141	26.5	3.8
150	41.6	4.0
177	108	4.5
178	125	4.6
187	134	4.65

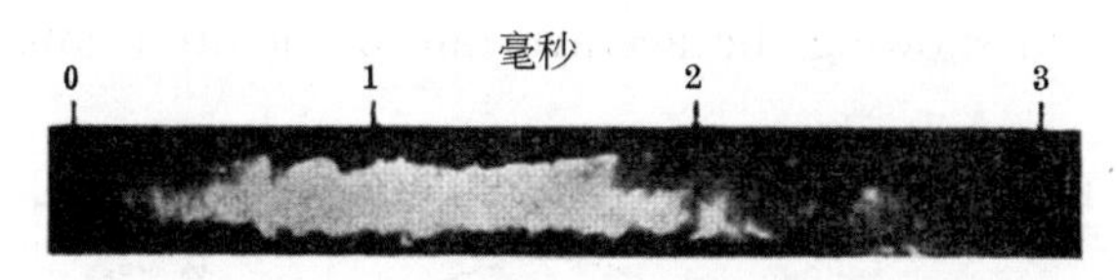

图3.氦放电的条纹图。条件：初始气压为0.25×10^{-3}毫米汞柱；轴向磁场为160高斯；峰值电流为130千安。管壁位于黑色区边缘。

电流通道的中心会由于环形电流扩张的趋势而移向外壁。这种扩张受到1英寸厚铝金属壁中涡旋电流的抵制。等离子体内的磁场测量是可重复的，测量还显示轴向磁场B_z局限于气体中。在轴上，它增大到约初始值的十倍。通常由B_θ和B_z分量产生的磁力线是螺旋形的，螺旋间距随等离子体截面变化。理论上尚未对具有这类形状的磁场的放电稳定度做过处理。

磁场探针直径为1英寸，极大地增加了放电电阻并减少了中子的产生。

从磁场测量和条纹照片估计的电流通道在峰值电流处的直径在20厘米~40厘米之间。通过使用4毫米微波的透射测量表明，电子密度大于6×10^{13}厘米$^{-3}$。这个密度与所有气体是电离的并包含在电流通道中的假设一致。

高能辐射

在氘的气体电流超过84千安时观测D-D反应产生的中子发射，发现发射集中在峰值电流附近，时间约为1毫秒。

表1显示了在峰值电流增高时每个脉冲发射的平均中子数。

表1

条件：气体D_2+5% N_2；气压：0.12×10^{-3}毫米汞柱；$B_z = 160$高斯

电流（千安）	每个脉冲产生的总中子数	T_c（计算值）
84	0.4×10^4	2.4×10^4 K
117	3.1	2.9
126	9.2	3.3
135	14.2	3.6
141	26.5	3.8
150	41.6	4.0
177	108	4.5
178	125	4.6
187	134	4.65

Fig. 4 is a histogram showing the average rate of neutron emission during the current pulse.

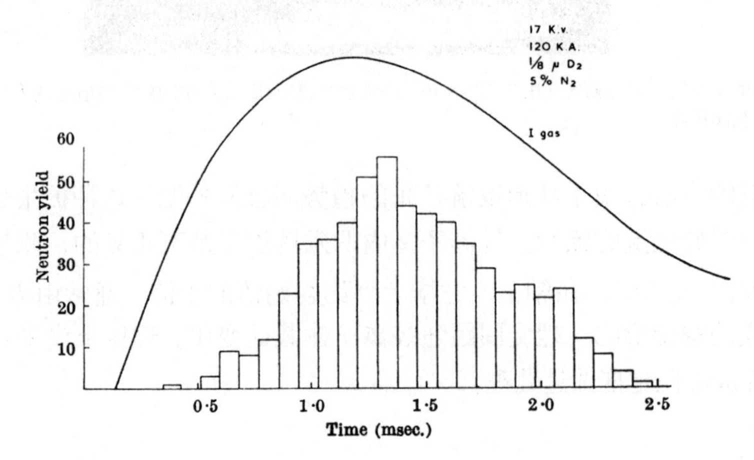

Fig. 4 Histogram showing the number of neutrons counted at various times during the current pulse

The third column of Table 1 gives the temperatures required to produce the observed neutron yields assuming a thermonuclear process. In calculating these figures, it has been assumed that the current channel is 20 cm. in diameter, emits neutrons uniformly for a period of 1 msec. and all the deuterium initially present is contained in the current channel. Since the reaction rate is an extremely sensitive function of temperature, variations in these parameters do not greatly affect the calculated temperature T_c. A comparison is made of the calculated and spectroscopically observed temperatures in Fig. 6.

Within the pressure-range investigated ($0.8–10.0\times10^{-4}$ mm.) the neutron yield decreased with increasing pressure. Neutrons were observed when 25 percent of the gas initially present was nitrogen, but the yield was much reduced.

The results obtained with a directional neutron counter moved around the torus showed that, within a factor of two, the neutron emission was uniform and did not arise from localized sources.

No correlation is found between the time of neutron emission and the voltage fluctuations during the current pulse. However, neutrons are produced at the large voltage transient at the end of the pulse, but these can be eliminated by the addition of nitrogen gas.

X-rays are observed towards the beginning of the current pulse. Their average energy lies in the range 20–30 kV. and on the average some 10^5 quanta per pulse are emitted by the whole tube. The number and energy of the X-ray quanta are insensitive to gas pressure and current, but increase in intensity as the axial magnetic field is increased.

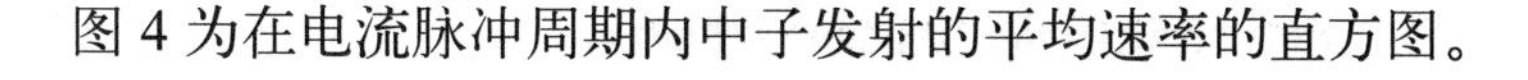

图 4 为在电流脉冲周期内中子发射的平均速率的直方图。

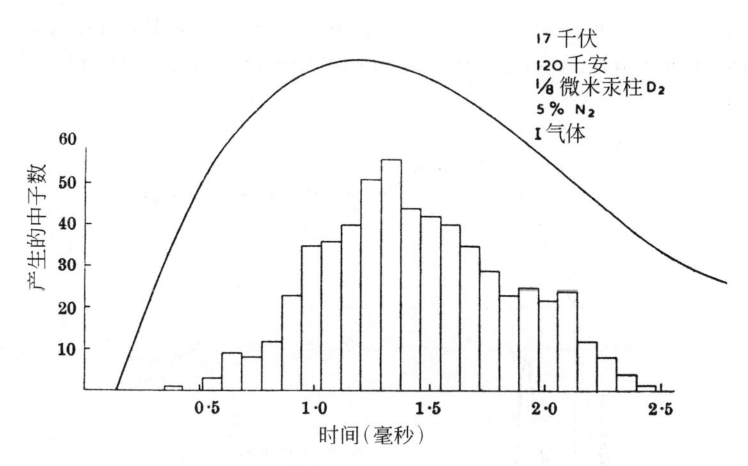

图 4. 电流脉冲周期内不同时间的中子计数的直方图

表 1 的第三列给出了假定的热核过程产生观测到的中子数所需的温度。在计算这些值时假定，电流通道的直径是 20 厘米，在 1 毫秒期间内均匀地发射中子，并且所有最初存在的氘均在电流通道内。由于反应速率是一个对温度极其灵敏的函数，这些参量的变化并不会对计算的温度 T_c 产生很大的影响。在图 6 中对计算的温度与光谱仪上观测到的温度做了对比。

在所研究的气压范围内 (0.8×10^{-4} 毫米汞柱 ~ 10.0×10^{-4} 毫米汞柱)，随着压强增高，产生的中子将会减少。最初存在的气体中 25% 是氮气时，观测到了中子，但中子产额大大减少。

用沿环流器运动的定向中子计数器得到的结果表明，在最大值与最小值相差不到两倍的意义上，中子发射是均匀的，并不是由局域源产生的。

在中子发射时间和电流脉冲期间的电压起伏之间并没有发现相关性。然而，中子是在脉冲末端的大电压瞬态时产生的，但这可以通过添加氮气来消除。

在电流脉冲起始时观测到 X 射线，其平均能量在 20 千伏 ~ 30 千伏范围内，整个管平均每个脉冲发射 10^5 个量子。X 射线量子的数量和能量对于气压和电流并不敏感，但是强度随轴向磁场的增大而增强。

Spectroscopic Observations

Both arc and spark line intensities vary greatly over the period of the pulse. In general, emission lines of normal atoms and ions up to three times ionized have a maximum intensity before peak current. Fig. 5 shows the intensity variation of (He II) 4,686 A. and (O V) 2,781 A.

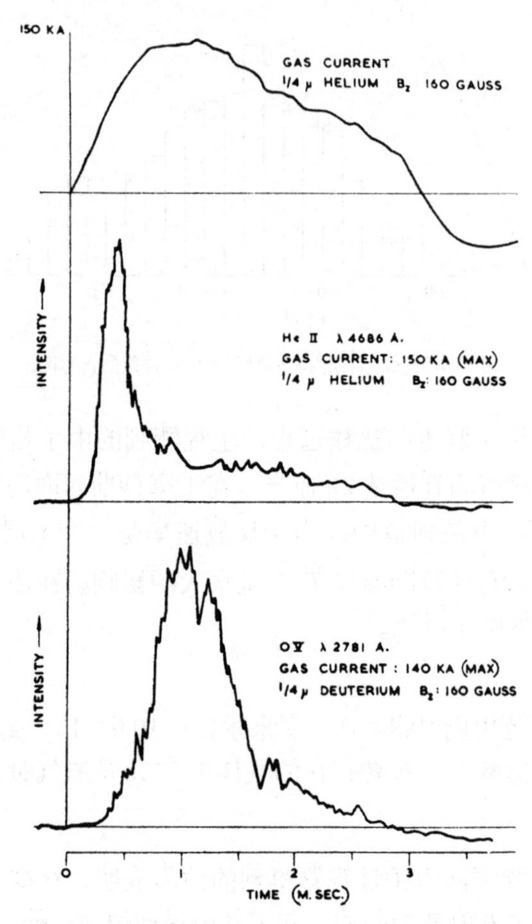

Fig. 5. The two lower traces are photomultiplier records of the intensity variation of two selected spark lines during the current pulse

The Doppler broadening of spark lines emitted in a radial direction is used for estimating the kinetic ion temperature of deuterium and neon discharges. Small quantities of oxygen and nitrogen introduced into deuterium discharges provide spark lines in a convenient part of the spectrum. The breadth of the lines is of the order of 1 A. and can be measured with a quartz spectrograph having a dispersion of 20 A./mm. Calculations show that both Stark and Zeeman effects make a negligible contribution to the line-breadth. Mass motion may contribute to the observed broadening to an appreciable extent, but both probe and streak records show no evidence of gross motion. The contribution to the line breadths of small-scale instabilities and turbulence remains to be measured.

光谱观测

弧光和火花放电谱线的强度在脉冲周期内变化都很大。通常正常的原子和三次电离的离子的发射谱线在峰值电流前有一个最大强度。图 5 显示了 4,686 埃 (He II) 和 2,781 埃 (O V) 的谱线强度变化。

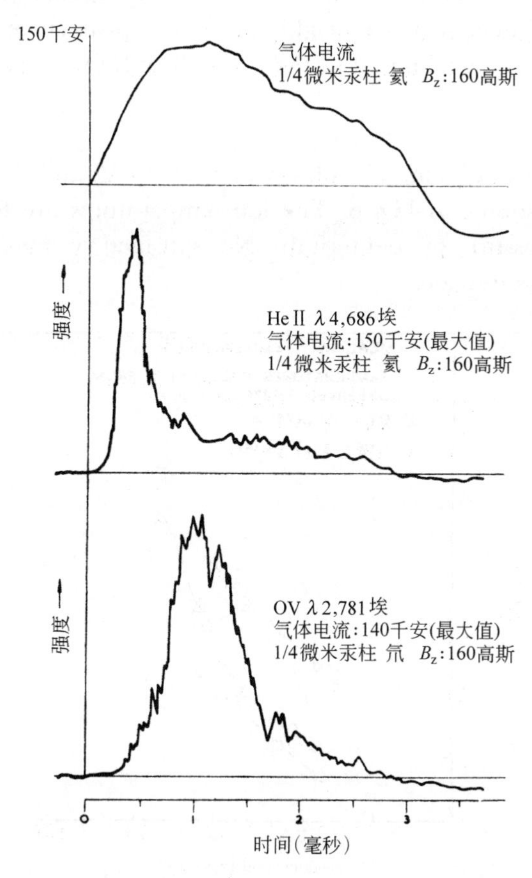

图 5. 图中下方的两条径迹是光电倍增管记录的在电流脉冲期间选择的火花放电的两个谱线强度的变化

径向发射的火花放电谱线的多普勒展宽可用来估计氘和氖放电的运动学离子温度。引入到氘放电中的少量氧和氮，在谱线适当部分提供火花放电谱线。谱线宽度是 1 埃的量级，可以用色散为 20 埃 / 毫米的石英光谱仪测量。计算表明，斯塔克效应和塞曼效应对线宽的影响可忽略不计。集体运动可能有助于谱线增宽变显著，但探针和条纹记录均没有证据显示总体运动。小范围的不稳定性及扰动对谱线宽度的影响尚有待测量。

Some 300 emission lines have been identified in the wave-length range 400–2,500 A. The most prominent are those oxygen, nitrogen, aluminium and carbon. Strong lines of (O VI) are recorded. In this wave-length range more than 400 lines remain unidentified.

The (O V) line has been used for Doppler breadth determination without time resolution as it has a maximum intensity in the neighbourhood of peak current. For (N IV) 3,479 A. this is not so, and light was admitted to the spectrograph for a period of 1 msec. centred about peak current.

The kinetic temperatures obtained by observation of (O V) and (N IV) lines as a function of peak current are shown in Fig. 6. The ion temperatures are found to decrease with increasing initial pressure of deuterium. No satisfactory measurement of electron temperature has yet been made.

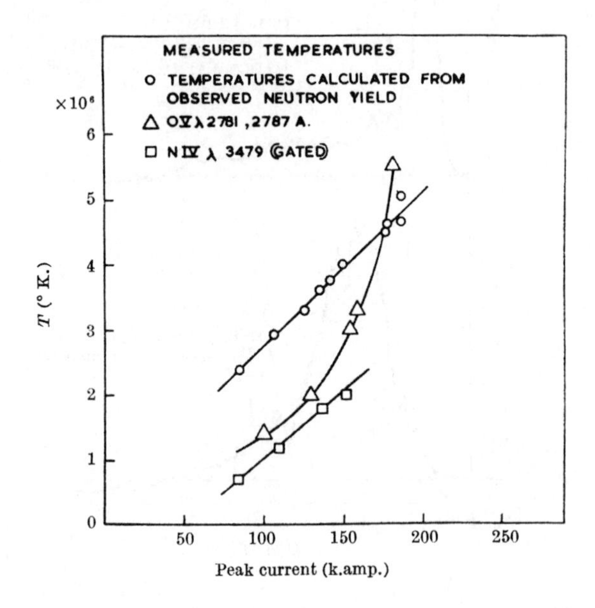

Fig. 6. Ion temperature as a function of peak current determined from the Doppler broadening of (O V) and (N IV). Conditions: initial gas pressure, 0.13×10^{-3} mm. deuterium and 5 percent nitrogen; B_z=160 gauss. The temperature of the deuterium gas, estimated from the observed neutron yield, is shown for comparison

Conclusion

These preliminary results demonstrate that it is possible to produce a stable highly ionized plasma isolated from the walls of a toroidal tube. Hydrogen gas has been maintained in a state of virtually complete ionization with a particle density lying between 10^{13} and 10^{14} per cm.^{3}, for times of milliseconds. The mean energy of the ions in the plasma is certainly of the order of 300 eV., and there are many indications that the electron temperature is of the same order. The containment time and the high electrical conductivity are both adequate for the detailed study of magnetohydrodynamical processes.

在波长为400埃~2,500埃的范围内,约300条发射谱线已被确认。最显著的是氧、氮、铝和碳的发射谱线。其中，记录了(O VI)的强谱线。在这个波长区间内，还有400多条谱线有待确认。

由于(O V)谱线在峰值电流附近有最大的强度，它已被用于测定无时间分辨的多普勒宽度。(N IV) 3,479埃并不是这样，光在峰值电流中心1毫秒范围内被放入到光谱仪中。

图6显示了根据(O V)和(N IV)谱线的观测得出的动力学温度，它是峰值电流的函数。发现离子温度随氘的最初压强的增高而降低。对电子温度尚未得到满意的测量结果。

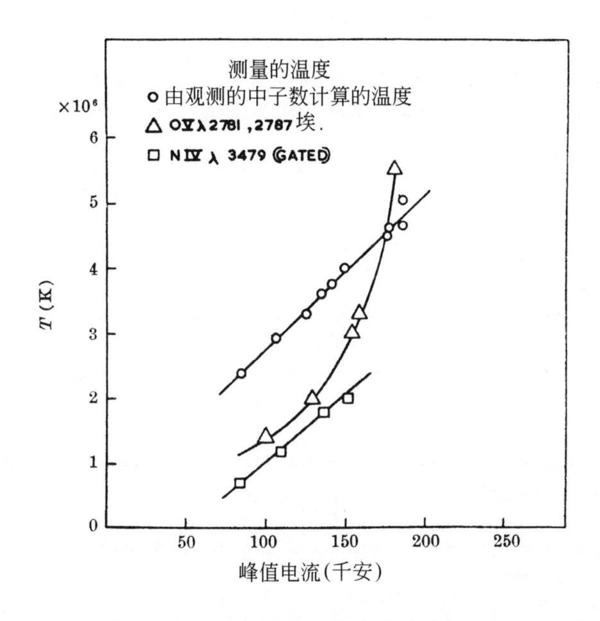

图6. 根据(O V)和(N IV)的多普勒展宽确定的离子温度作为峰值电流的函数。条件：最初气压为0.13×10^{-3}毫米汞柱的氘和5%的氮；B_z=160高斯。作为对比，图中还给出了根据观测的中子产生估计的氘气体的温度。

结　论

这些初步结果表明,可以产生与螺旋管壁隔离的稳定的高度电离等离子体。在几毫秒时间内，氢气体维持完全电离的状态，其粒子密度在每立方厘米10^{13}至10^{14}之间。等离子体内的离子平均能量量级为300电子伏，许多事实表明，电子的温度处于相同量级。约束时间和高电导对于详尽研究磁流体力学过程都是合适的。

To identify a thermonuclear process it is necessary to show that random collisions in the gas between deuterium ions are responsible for the nuclear reactions. In principle, this can be done by calculating the velocity distribution of the reacting deuterium ions from an exact determination of both the energy and direction of emission of the neutrons. The neutron flux so far obtained is insufficient to attain the desired accuracy of measurement.

Investigations leading up to the present results have been constantly encouraged and supported by Sir John Cockcroft and the late Lord Cherwell. The theoretical investigations have been directed by Dr. W. B. Thompson, of the Theoretical Physics Division.

A major part of the engineering design and construction of ZETA was done by the Metropolitan-Vickers Electrical Co., Ltd.

(**181**, 217-220; 1958)

P. C. Thonemann, E. P. Butt, R. Carruthers, A. N. Dellis, D. W. Fry, A. Gibson, G. N. Harding, D. J. Lees, R. W. P. McWhirter, R. S. Pease, S. A. Ramsden and S. Ward: Atomic Energy Research Establishment, Harwell.

References:

1. Post, B. F., *Rev. Mod. Phys.*, **28**, 338 (1956).
2. Pease, R. S., *Proc. Phys. Soc.*, B, **70**, 11 (1957).
3. Burkhart, L. C., Dunaway, B. E., Mather, J. W., Phillips, J. A., Sawyer, G. A., Stratton, T. F., Stovall, E. J., and Tuck, J. L., *J. App. Phys.*, **28**, 519 (1957).
4. Kurchatov, I. V., *Atomnaya Energiya*, **1**, No. 3, 65 (1956). English translation, *J. Nuclear Energy*, **4**, 198 (1957).
5. Artaimovitch, L. A., Andrianov, A. M., Bazilevakaya, O. A., Prokhorov, YU. G. and Filippov, N. V., *Atomnaya Energiya*, **1**, No. 3, 76 (1956). English translation, *J. Nuclear Energy*, **4**, 203 (1957).
6. Colgate, S. } Proc. 3rd Int. Conf. on Ionization Phenomena in Gases, Venice (1957).
7. Berglund, S., Nilsson, R., Ohlin, P., Siegbahn, K., Sundstrom, T., and Svennerstedt, S., *Nuclear Instruments*, **1**, 233 (1957). } Proc. 3rd Int. Conf. on Ionization Phenomena in Gases, Venice (1957).
8. Curran, S., and Allen, K. W. } Proc. 3rd Int. Conf. on Ionization Phenomena in Gases, Venice (1957).
9. Shrafranov, V. D., *Atomnaya Energiya*, **1**, No. 5, 38 (1956). English translation, *J. Nuclear Energy*, **5**, 86 (1957).
10. Tayler, R. J., *Proc. Phys. Soc.*, B, **70**, 1049 (1957).
11. Bosenbluth, M. } Proc. 3rd Int. Conf. on Ionization Phenomena in Gases, Venice (1957).
12. Bickerton, R. J. } Proc. 3rd Int. Conf. on Ionization Phenomena in Gases, Venice (1957).
13. Bezba chenko, A. L., Golovin, I. N., Ivanov D. P., Kirillov, V. D., Yavlinsky, N. A., *Atomnaya Energiya*, **1**, No. 5, 26 (1956). English translation, *J. Nuclear Energy*, **5**, 71 (1957).

为了确认热核过程，必须说明核反应是由气体中氘离子之间随机碰撞引起的。原则上，这一点可以通过计算反应氘粒子的速率分布得到，速率分布又可以通过中子能量和发射方向的精确测定而得到。至今得出的中子通量还不足以获得所要求的测量精确度。

为获得现有结果所做的研究得到了约翰·考克饶夫爵士和已故的彻韦尔勋爵不断的鼓励和支持。理论研究得到了理论物理部的汤普森博士的指导。

工程设计的主要部分和ZETA的构建是由茂伟电气公司完成的。

（沈乃澂 翻译；尚仁成 审稿）

Co-operative Phenomena in Hot Plasmas

L. Spitzer, jun.

Editor's Note

The new experiments producing high-temperature plasmas in the ZETA device at Harwell not only shed light on the potential for controlled fusion energy, but also provided an opportunity for studying plasma physics in extreme conditions. Here physicist Lyman Spitzer reports on a mystery evident in the ZETA data: an anomalously fast rise in the temperature of positive ions, which in theory should gain energy more slowly through collisions with hot electrons. These findings indicated that physicists still had much to learn about the rich dynamics within plasmas, and the many collective instabilities which drive it. This theme of the overwhelming collective complexity of dynamical behaviour would be a common one in plasma physics for the next half century.

THE interesting results described by Thonemann and his co-workers in the preceding article provide not only an important step forward in the controlled release of thermonuclear energy, but also a challenging problem in the dynamics of fully ionized gases or plasmas. The spectroscopic line profiles and the neutron counts provide incontrovertible evidence for the acceleration of the positive ions. However, theory would seem to indicate that electron-ion collisions are inadequate to explain the observed rate of heating, and some unknown mechanism would appear to be involved.

The rate at which the positive-ion temperature increases, as a result of electron-ion collisions, has been given elsewhere[1]. It has been suggested, both in Great Britain and in the United States, that this process might be inadequate to explain the rate at which positive ions gain energy in certain electrical discharges. A relatively simple procedure, due to Stix[2], may be used to set an upper limit to the rate at which positive ions are heated by electron-ion collisions. In this method the rate of heating is made a maximum by setting the electron temperature equal to three times the ion temperature, T_i, and setting the electron density equal to a constant, n_e, corresponding to complete ionization and concentration within the discharge channel of all the gas initially present in the tube. On this basis (with $\ln \Lambda$ set equal to 15, see ref. 1) T_i is given by the relation

$$T_i^{3/2} = 1.71 \times 10^{-2} n_e t \tag{1}$$

where T_i is in degrees K., n_e in electrons per cm.^{3}, and t in seconds.

To determine the density in the ZETA experiments, we compute the radius, r_d, of the discharge channel on the assumption that the flux of the axial magnetic field, B_z, is held constant in the gas during the contraction, and that the field, B_θ, due to the current equals the compressed axial field at the boundary of the discharge. As a result of the high

热等离子体中的协同现象

小斯皮策

编者按

在哈威尔的ZETA设备上产生高温等离子体的新实验不仅为可控核聚变的研究带来一线光明，同时也为我们提供了一个研究极端条件下等离子体物理的机会。本文中物理学家赖曼·斯皮策指出了ZETA数据中一个谜团：正离子温度异常地快速上升，而理论上来说，由于它与热电子相互碰撞，因此获得能量的速率应该更加缓慢。这些发现说明，物理学家们对于等离子体内丰富的动力学现象以及导致这种现象的众多集体不稳定性仍需进一步的探索。在此后的半个世纪的时间里，动力学行为这个错综复杂且极具挑战性的问题成为等离子体物理中一个很常见的研究主题。

索恩曼及其同事们在先前的文章中阐述了一个很有意思的结果，这个结果不仅说明我们在热核能量的可控释放中迈出了重要的一步，而且还提出了一个极具挑战性的有关完全电离的气体（即等离子体）的动力学问题。光谱线剖面和中子数提供了与正离子加速相矛盾的证据。然而，理论显示仅靠电子–离子碰撞并不足以解释观测到的升温速率，其中应该包含某些未知的机制。

在之前的一篇文章[1]中，我曾提到正离子温度升高的速率是由电子–离子碰撞导致的。然而在英国和美国都有人提出，这个碰撞过程不足以解释正离子在释放电荷后获得能量的速率。斯蒂克斯[2]提出了一个相对简单的过程，可以用来确定电子–离子碰撞所导致的正离子升温的速率上限。在这种方法中，将电子温度取为离子温度 T_i 的三倍，设定电子密度为一常数 n_e，相当于完全电离并且管内最初存在的所有气体都浓缩在放电通道内，在这些条件下可以得到加热速率的最大值。据此 ($\ln\Lambda$ 取为15，见参考文献1)，T_i 由下列关系式给出：

$$T_i^{3/2} = 1.71 \times 10^{-2} n_e t \tag{1}$$

式中，T_i 的单位是K，n_e 的单位是每立方厘米的电子数，t 的单位是秒。

为了测定ZETA实验中的密度，我们在下面两个假设的基础上计算了放电通道的半径 r_d。一是假定在收缩过程中轴向磁场 B_z 方向上的通量保持不变，二是假定由电流产生的磁场 B_θ 等于放电界面上被压缩的轴向磁场。由于ZETA实现了高温，因

temperatures achieved in ZETA, appreciable leakage of axial flux out of the discharge does not appear to be possible. Neglect of the finite gas pressure decreases the computed channel radius, increases n_e and again makes the rate of increase of T_i a maximum. On these assumptions, values of r_d have been computed for two currents, and an initial B_z of 160 gauss, and are listed in the second column of Table 1.

Table 1

I (amp.)	r_d (cm.)	n_e (cm.$^{-3}$)	T_i (°K.)	t (theor.)	t (obs.)
126,000	16	8.6×10^{13}	3.3×10^{6}	4.0×10^{-3}	1.0×10^{-3}
140,000	14	1.1×10^{14}	2.5×10^{6}	2.0×10^{-3}	1.0×10^{-3}

These values of r_d are consistent with those cited by the Harwell group. The values of n_e in the third column correspond to complete ionization of all the deuterium initially present in the torus, at a pressure of 1/8μ, and its concentration in the current channel. The value of T_i for the lower current in column 4 is taken from Table 1 in the article by Thonemann *et al.*, and is based on the neutron yield. The value of T_i at the higher current is the temperature obtained from the Doppler width of the O V triplet, as shown in Fig. 6 of the Harwell article. The ion temperature obtained from the neutron yield under this condition is about 50 percent greater, but no information is available on the time at which the neutrons appear in this case. In the fifth column are given the values of the time, in seconds, required for the positive ions to reach this temperature, computed from equation 1.

For comparison, the final column lists the observed times, in seconds, at which the positive-ion energies reach the values corresponding to the fourth column. For the lesser current, this is the time at which the neutron yield reaches half its peak value as shown in Fig. 4 of the paper from Harwell. For the greater value of current, t (obs.) is set equal to the time at which the O V radiation reaches its peak intensity, as shown in Fig. 5 of the Harwell paper. The simplifications made have tended to reduce t (theor.), and the correct value may be about an order of magnitude greater than given in Table 1. Thus the discrepancy seems real. There does not appear to be any simple model, based on a quiescent plasma, which is consistent with the observed rapid heating of the positive ions.

Following a suggestion by the Harwell group as to the importance of non-thermal heating processes, Stix[2] in 1956 arrived at conclusions similar to the above from experimental results obtained at Project Matterhorn. A discharge was produced in helium gas in a stainless-steel race-track tube of 10 cm. diameter and 240 cm. axial length, with an initial pressure of 0.63μ and an externally produced axial field of 19,000 gauss. The magnetic field was arranged so that intersection of the outer lines of force with material walls restricted the discharge to a channel of 5 cm. diameter. A loop voltage of 300 V. was applied around an iron transformer threading the race-track, and a maximum current of 8,000 amp. observed; since this current produces only a minor perturbation in the magnetic field, there was no pinching of the discharge. Time-resolved spectroscopic profiles of the He II line, λ 4,686, indicated that the kinetic temperature of these ions increased to 1.2×10^6 degrees K. in 1.5×10^{-4} sec., as compared to a theoretical maximum

此基本不会出现放电造成的轴向磁通量的明显泄漏。忽略有限的气压使得计算出的通道半径变小，并使 n_e 增大，最后使 T_i 的增长率达到最大值。基于这些假设，对两个电流强度下的 r_d 值进行计算，最初的 B_z 为 160 高斯，计算结果详见表 1 的第二列。

表 1

I(安)	r_d(厘米)	n_e(厘米$^{-3}$)	T_i (K)	t(理论值)	t(观测值)
126,000	16	8.6×10^{13}	3.3×10^{6}	4.0×10^{-3}	1.0×10^{-3}
140,000	14	1.1×10^{14}	2.5×10^{6}	2.0×10^{-3}	1.0×10^{-3}

这些 r_d 的值与哈威尔小组引用的数值是一致的。在第三列中 n_e 的值对应的是最初存在于环内的所有氘在气压为 1/8 微米汞柱时完全电离并浓缩在电流通道内的情况。第四列中较低电流下的 T_i 值取自索恩曼等人的论文中的表 1，它是根据中子产额而得出的。较高电流下的 T_i 值是根据哈威尔小组的论文中图 6 所示的 O V 三重线的多普勒宽度而得出的。在这种条件下，根据中子产额得到的离子温度升高了约 50%，但还无法确定在这种情况下中子是什么时候开始产生的。在第五列中给出了以秒为单位的时间的值，这是根据式 (1) 计算出的正离子达到这个温度所需要的时间。

为了进行比较，最后一列给出了以秒为单位的观测时间，在这个时间内正离子能量达到第四列的温度所对应的值。对于较小电流的情况，这是中子产额达到其峰值一半时所用的时间，如哈威尔小组的论文中图 4 所示。对于较大电流的情况，t(观测值) 取 O V 辐射值达到其峰值强度时所用的时间，如哈威尔小组的论文中图 5 所示。这种简化常常使得 t(理论值) 偏小，修正值可能会比表 1 给出的值大约大一个数量级。因此差异看起来确实存在。似乎并没有能和观测到的阳离子的快速加热相吻合的基于静等离子体的任何简单模型。

1956 年，按哈威尔小组关于非热效应的加热过程的重要性所提出的建议，斯蒂克斯 [2] 得到了与马特峰计划的实验结果相类似的结论。在直径 10 厘米、轴长 240 厘米充满氦气的不锈钢环形管中进行放电，其初始气压为 0.63 微米汞柱，外部产生的轴向磁场为 19,000 高斯。磁场的分布使得外部的磁力线与材料壁交叉从而将放电限制在直径为 5 厘米的通道内。将 300 伏回路电压加在穿过粒子轨道的铁变压器周围，观测到的最大电流为 8,000 安；因为电流在磁场中只产生了较小的扰动，所以并不存在放电的箍缩。λ 为 4,686 埃的氦 II 线的时间分辨光谱图表明，这些离子的动力学温度在 1.5×10^{-4} 秒内增高到 1.2×10^{6}K，而理论上在这一时间内所能达到的最大值

value of 0.8×10^6 degrees in this same time-interval.

It has been known since the work by Langmuir[3] that electrons in a conventional gas discharge approach a Maxwellian distribution much more rapidly than can be explained by inter-particle collisions. Recent research by Gabor and his collaborators[4] has shown that oscillations generated in the plasma sheath are responsible for much of this effect, but the detailed mechanism is still unexplained. Possibly the high ion energies observed in ZETA represent a phenomenon related to Langmuir's paradox. To analyse the possible processes involved, such as oscillations, shocks, hydromagnetic turbulences, etc., it would be helpful to obtain information on the extent to which the positive-ion velocities are thermalized, that is, on how nearly the distribution function is isotropic and Maxwellian. Evidently detailed experimental investigations of these cooperative effects in hot plasmas will be of great interest in basic physics.

(**181**, 221-222; 1958)

Lyman Spitzer: Princeton University.

References:

1. Spitzer, L., "Physics of Fully Ionized Gases" (Interscience Publishers, 1956, Section 5.3).
2. Stix, T., Talk at Berkeley, California (February 20-23, 1957).
3. Langmuir, I., *Phys. Rev.*, **26**, 585 (1925); *Z. Phys.*, **46**, 271 (1928).
4. Gabor, D., Ash, E. A., and Dracott, D., *Nature*, **176**, 916 (1955).

是 0.8×10^6 K。

依据朗缪尔[3]的工作我们知道，在传统的气体放电中，电子以极快的速率达到麦克斯韦分布而单纯用粒子间的相互碰撞无法对此进行解释。最近盖伯及其同事们[4]的研究表明，这种效应主要是由等离子体鞘层中产生的振荡造成的，但具体的机制还无法解释。在 ZETA 中观测到的高离子能也许代表了一种与朗缪尔悖论相关的现象。为了分析可能包含的过程（例如振荡、冲击、磁流体湍流等），有必要收集与正离子速度被热能化的程度有关的信息，即分布函数在多大程度上接近于各向同性和麦克斯韦分布。显然，对热等离子体中协同效应的详细情况的实验研究，将是基础物理学科中一个非常有趣的课题。

（沈乃澂 翻译；尚仁成 审稿）

Stimulated Optical Radiation in Ruby

T. H. Maiman

Editor's Note

Albert Einstein identified the phenomenon of the stimulated emission of radiation early in the 20th century—how radiation of a particular frequency could stimulate more of the same to be emitted in an excited medium. Here Theodore Maiman of the Hughes Research Laboratories in California reported the first observation of the effect in a cube of ruby with ends coated with reflective silver. These mirrors create an "optical cavity" within which light bounces back and forth to stimulate emission from chromium ions in an electronically excited state. This "light amplification by stimulated emission of radiation" was the first basic demonstration of the principles of laser action.

SCHAWLOW and Townes[1] have proposed a technique for the generation of very monochromatic radiation in the infra-red optical region of the spectrum using an alkali vapour as the active medium. Javan[2] and Sanders[3] have discussed proposals involving electron-excited gaseous systems. In this laboratory an optical pumping technique has been successfully applied to a fluorescent solid resulting in the attainment of negative temperatures and stimulated optical emission at a wave-length of 6,943 Å.; the active material used was ruby (chromium in corundum).

A simplified energy-level diagram for triply ionized chromium in this crystal is shown in Fig. 1. When this material is irradiated with energy at a wave-length of about 5,500 Å., chromium ions are excited to the $^4F_2$ state and then quickly lose some of their excitation energy through non-radiative transitions to the ^2E state[4]. This state then slowly decays by spontaneously emitting a sharp doublet the components of which at 300°K. are at 6,943 Å. and 6,929 Å.(Fig. 2*a*). Under very intense excitation the population of this metastable state (^2E) can become greater than that of the ground-state; this is the condition for negative temperatures and consequently amplification via stimulated emission.

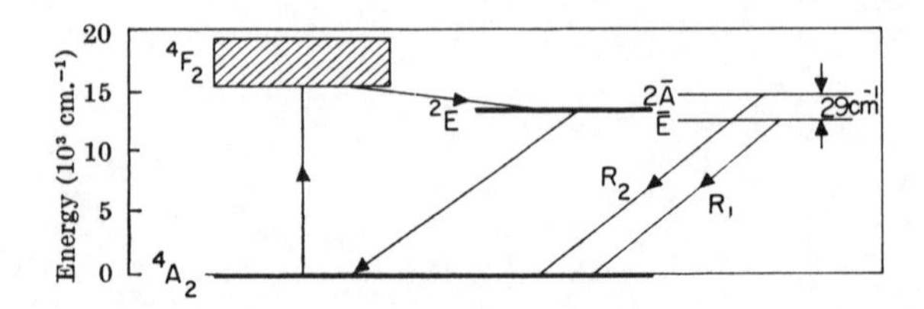

Fig. 1. Energy-level diagram of Cr^{3+} in corundum, showing pertinent processes

红宝石中的受激光辐射

梅曼

编者按

在 20 世纪初，阿尔伯特·爱因斯坦指出存在受激辐射跃迁现象，即一种特定频率的辐射是怎样从受激发介质中激发出更多同样频率的辐射的。这篇文章中，加利福尼亚州休斯研究实验室的西奥多·梅曼利用一块红宝石首次观察到了这种现象。这块红宝石的两端镀着反射型银膜，因此而形成的镜子构成了一个“光学谐振腔”，在这个“光学谐振腔”中光束反复地将处于激发态的镉离子激发并辐射出来。“受激辐射的光放大”是关于激光原理的第一次基本说明。

肖洛和汤斯[1]提出了一种以碱金属蒸气作为工作介质产生单色性极高的红外光辐射的技术。贾范[2]和桑德斯[3]也讨论了在电子激发气体中产生这种辐射的若干方案。在本实验室中，我们已经成功地将光泵浦技术用于固体荧光材料中，并以此获得了负温度及波长为 6,943 埃的受激光辐射，其中所用的工作介质是红宝石（含铬的刚玉）。

图 1 给出的是该晶体中三价铬离子的能级结构简图。当我们用波长约为 5,500 埃的能量照射此材料时，铬离子被激发到 $^4F_2$ 态，然后通过非辐射跃迁迅速损失部分激发能而跃迁到 ^2E 态[4]。随后这个能态通过自发辐射而缓慢衰变，辐射产生了明锐的双线，它们在温度为 300 K 时的波长分别是 6,943 埃和 6,929 埃（图 2*a* 所示）。在高强度的激发条件下，处于亚稳态（^2E）的粒子数将多于处于基态的粒子数，这就是产生负温度的条件，并最终导致受激辐射的光放大。

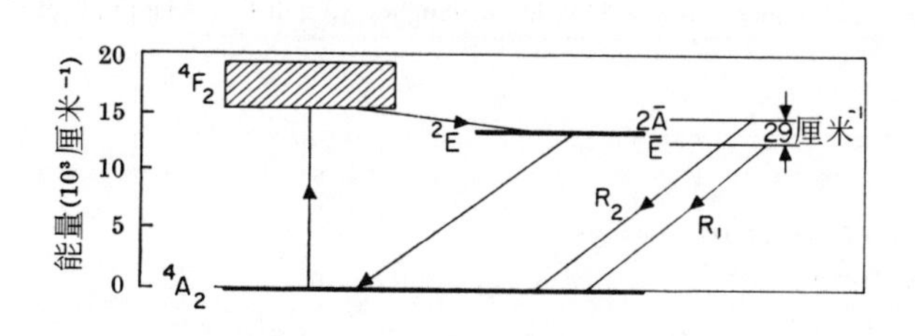

图 1. 刚玉中 Cr^{3+} 的能级图及相关的跃迁过程

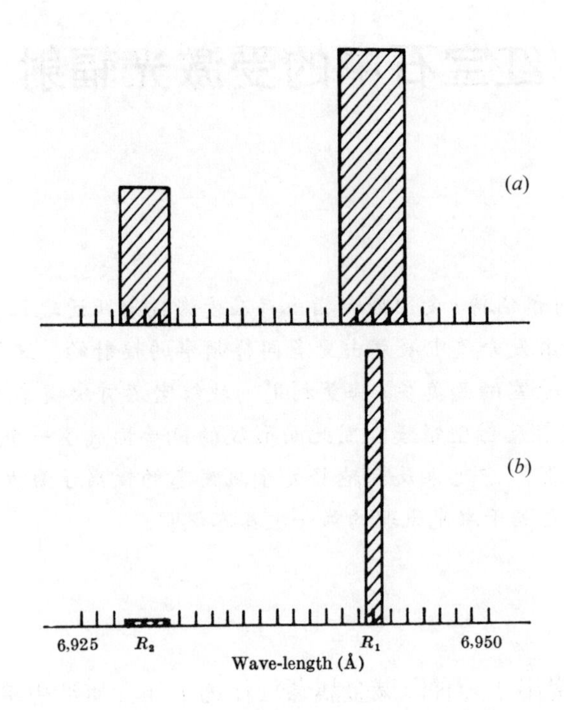

Fig. 2. Emission spectrum of ruby: *a*, low-power excitation; *b*, high-power excitation

To demonstrate the above effect a ruby crystal of 1-cm. dimensions coated on two parallel faces with silver was irradiated by a high-power flash lamp; the emission spectrum obtained under thesse conditions is shown is Fig. 2*b*. These results can be explained on the basis that negative temperatures were produced and regenerative amplification ensued. I expect, in principle, a considerably greater ($\sim 10^8$) reduction in line width when mode selection techniques are used[1].

I gratefully acknowledge helpful discussions with G. Birnbaum, R. W. Hellwarth, L. C. Levitt, and R. A. Satten and am indebted to I. J. D'Haenens and C. K. Asawa for technical assistance in obtaining the measurements.

(**187**, 493-494; 1960)

T. H. Maiman: Hughes Research Laboratories, A Division of Hughes Aircraft Co., Malibu, California.

References:

1. Schawlow, A. L., and Townes, C. H., *Phys. Rev.*, 112, 1940 (1958).
2. Javan, A., *Phys. Rev. Letters*, 3, 87 (1959).
3. Sanders, J. H., *Phys. Rev. Letters*, 3, 86 (1959).
4. Maiman, T. H., *Phys. Rev. Letters*, 4, 564 (1960).

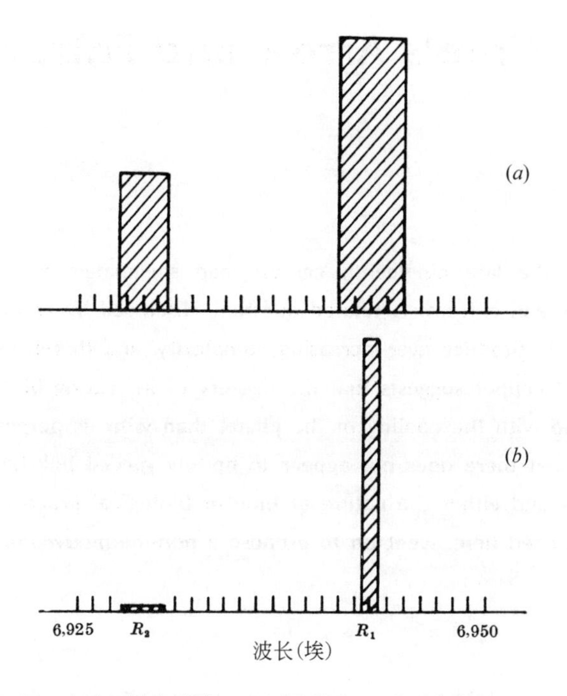

图 2. 红宝石的发射谱：*a*, 低功率激发；*b*, 高功率激发

为了演示上述效应，我们用高功率的闪光灯照射一块尺寸为 1 厘米且两个平行表面都镀了银膜的红宝石晶体，在这种条件下得到的发射谱如图 2*b* 所示。这些结果可以用负温度的产生以及随后的再生放大来解释。我预期，原则上当采用模式选择技术时 [1] 谱线宽度将会有显著的（约 10^8）减小。

我非常感谢伯恩鲍姆、赫尔沃思、莱维特和萨滕与我进行了非常有益的讨论，也感谢黑内斯和浅轮等人在测量中提供的技术帮助。

（沈乃澂 翻译；尚仁成 审稿）

Time's Arrow and Entropy

K. Popper

Editor's Note

Physicists since the late nineteenth century had speculated over the links between time, thermodynamics and entropy. A related question, discussed by Erwin Schrödinger, is how life on Earth seems to produce ever-increasing complexity, and therefore lower entropy. Here the philosopher Karl Popper suggests that the mystery of increasing biological order may actually have more to do with the cooling of the planet than with its perpetual warming by the Sun. Popper argues that there does not appear to be any special link between the second law of thermodynamics and either the nature of time or biological processes on Earth. Physicist Ilya Prigogine, mentioned here, went on to propose a non-thermodynamic, quantum origin for the "arrow of time".

SEVERAL years ago[1] I suggested that we should distinguish between two essentially different ways in which energy can be degraded or dissipated: "Dissipation in the form of increasing disorder (entropy increase) is one of them, and dissipation by expansion without increase of disorder is the other. For an increase of disorder, walls of some kind are essential: a sufficiently thin gas expanding in a 'vessel without walls' (that is, the universe) does not increase its disorder." Reasons for this view were given in the place cited.

In order to explain this a little more precisely, I shall here introduce, following Prigogine[2], the term "system" to denote the (energy and material) "contents of a well-defined geometrical volume of macroscopic dimensions" (so that, for example, an organism enclosed by its skin, or our solar system as enclosed by a sphere round the Sun with a radius of 10^5 light seconds, would be a "system"); and I shall speak of the "exterior" of a system X as a region of space (leaving it open whether or not this is in its turn a geometrically well-defined "system") of which X forms a part.

Following Prigogine, I shall distinguish between (materially or at least energetically) "open" and "closed" systems. (An energetically closed system is called "isolated".) Moreover, I shall call a system X "essentially open" if it is part of a system Y such that all geometrically convex systems of which Y is a part are (at least energetically) open. (This definition makes it possible even for an isolated system to be essentially open.)

I further call X "essentially open towards a cooler exterior" if X is enclosed by some convex system Y such that: (*a*) all elements of any sequence Z_i of convex systems of which Y is a part are essentially open and of a lower average temperature than Y, and that (*b*)

时间之箭和熵

波普尔

编者按

从19世纪末期起，物理学家们便开始思索时间、热力学和熵之间的相互联系。埃尔温·薛定谔曾讨论过一个与此相关的问题，即地球上的生命是如何不断地增加其复杂性，从而使熵值降低的。本文中，哲学家卡尔·波普尔认为生物有序性逐渐增加的奥秘实际上更多地在于地球自身的冷却，而不是由于太阳给地球的持续加温。波普尔认为热力学第二定律无论是与时间的本质还是与地球上的生物过程似乎都不存在任何特别的联系。本文还提到了物理学家伊利亚·普里高津进一步提出的“时间之箭”的非热力学的量子起源。

几年前[1]我曾指出，我们应该区分两种完全不同的使能量降低或耗散的途径：“一种是以增加无序性（熵增加）的形式进行耗散，另一种是不增加无序性，通过膨胀产生的耗散。某种形式的壁的存在对于无序性的增加是必不可少的：足够稀薄的气体在一个‘没有器壁的容器’（即宇宙）中进行膨胀并不会增加其无序性。”在前述引文中已给出了此观点的理由。

为了对此进行更为精确的解释，在本文中我将引入普里高津[2]定义的术语“系统”来表示“在宏观维度内给定几何体积之中的内含物”（能量和物质）（因此，如皮肤包被的有机体，或环绕太阳的半径为10^5光秒的球面之内的太阳系都属于此“系统”）；我将系统X的“外界”说成是一个空间区域（让它开放，无论这样反过来会不会是一个具有明确几何意义的系统），系统X形成了其中的一部分。

按照普里高津的观点，我将（从物质或至少从能量的角度）区分“开放”系统和“封闭”系统。（所谓的“孤立系统”就是指一个能量封闭的系统。）此外，如果系统X是系统Y的一部分，我就称系统X是“基本开放的”，这样包含系统Y的所有在几何上凸出的系统（至少在能量上）是开放的。（这个定义甚至使得一个孤立系统有可能是基本开放的。）

另外，如果系统X被某些凸出系统Y所包围并满足以下条件：(a) 包含Y的凸出系统的任意序列Z_i的所有元素都是基本开放的，并且它们的平均温度比系统Y的

for every such system Z_i there is a system Z_j which encloses Z_i and which is not of a higher average temperature than Z_i.

The terminology here introduced makes it possible to clarify a number of points in connexion with the second law of thermodynamics which seem in urgent need of clarification.

Again following Prigogine[3], we can split the change of entropy dS_X in any system X into two parts: dS_{Xe}, or the flow of entropy due to interaction with the exterior of X, and dS_{Xi}, the contribution to the change of entropy due to changes inside the system X. We have, of course:

$$dS_X = dS_{Xe} + dS_{Xi} \tag{1}$$

and we can express the second law by:

$$dS_{Xi} \geqslant 0 \tag{2}$$

For an energetically closed (or "isolated") system X, for which by definition $dS_{Xe} = 0$, expression (2) formulates the classical statement that entropy never decreases. But if X is open towards a cooler exterior:

$$dS_{Xe} < 0 \tag{3}$$

holds, and the question whether its total entropy increases or decreases depends, of course, on both its entropy production dS_{Xi} and its entropy loss dS_{Xe}.

The fact that entropy can decrease in an open system X does not, of course, conflict with the second law as given by expression (2). But the second law is often formulated in a different way; for example, it is said that "if we … expand our system to include all the energy exchange, it would be found that in the larger system the entropy had increased. For example, to measure the entropy change taking place in living organisms as a whole, it would be necessary to include in our system the Sun and some additional portion of the universe, as well as the Earth itself"[4]. Thus it is suggested that for sufficiently large systems X of our universe, $dS_X \geqslant 0$, so that the entropy always increases.

Yet, so far as our knowledge of the Universe goes, the precise opposite appears to be the case. With very few and short-lived exceptions, the entropy in almost all known regions (of sufficient size) of our universe either remains constant or decreases, although energy is dissipated (by escaping from the system in question). This is so, at any rate, if we assume that the law of conservation of energy is valid; and it is also so if we assume the "steady state" theory of the expanding universe. (It is not so on the assumption of a finite and non-expanding universe with non-zero energy density.)

要低；(*b*) 对于每一个这样的系统 Z_i，都有一个包含 Z_i 在内的且温度不比 Z_i 高的系统 Z_j 存在，那么我称系统 X“对更冷的外界基本开放”。

本文中引入的术语有可能阐明与热力学第二定律有关的许多观点，该定律似乎亟待阐明。

再次按照普里高津的观点[3]，我们可将任何系统 X 熵值的变化 $\mathrm{d}S_X$ 分成两部分：由于与系统 X 的外界相互作用引起的熵的流动 $\mathrm{d}S_{Xe}$，以及由于系统 X 内部的变化引起的熵的变化 $\mathrm{d}S_{Xi}$。由此，我们可得：

$$\mathrm{d}S_X = \mathrm{d}S_{Xe} + \mathrm{d}S_{Xi} \tag{1}$$

我们可用下式表示第二定律：

$$\mathrm{d}S_{Xi} \geqslant 0 \tag{2}$$

对于能量封闭的 (或“孤立的”) 系统 X，我们通过定义可知 $\mathrm{d}S_{Xe} = 0$，式 (2) 变成了经典表述：熵永不减少。但是，如果系统 X 对更冷的外界开放，则有：

$$\mathrm{d}S_{Xe} < 0 \tag{3}$$

那么总熵是增加还是减少便自然取决于增加的熵 $\mathrm{d}S_{Xi}$ 和减少的熵 $\mathrm{d}S_{Xe}$ 分别有多少。

事实上，开放系统 X 中熵的减少与式 (2) 所表述的第二定律并不相矛盾，但第二定律通常用另一种方式来表述。例如，它可表述为“如果我们……扩展我们的系统至其包含所有的能量交换，就会发现在更大的系统中熵增加了。又如，要测量整个生命有机体发生的熵变化时，在这个系统中有必要包含太阳以及宇宙某些其他的部分，还有地球自身”[4]。因此，如果 X 是宇宙中足够大的系统，则 $\mathrm{d}S_X \geqslant 0$，那么熵总是增加的。

然而，就我们对宇宙的认知不断地变化而言，完全相反的情况似乎也存在。除了极少的短暂的例外，宇宙中几乎所有已知（足够大的）地区的熵都是保持不变或减少的，虽然能量（从当前系统中逃逸出）耗散了。如果我们假设能量守恒定律成立，那么无论怎样都会有上述结论；如果我们假设膨胀宇宙的“稳态”理论也成立，那么上述结论仍然成立。(但是如果假设具有非零能量密度的宇宙是有限的和非膨胀的，那上述结论则不成立。)

In order to see this, all that is needed is to be clear about the empirical fact that in our universe we know only essentially open systems, and only systems *X* which, so far as they produce entropy at all, are essentially open towards a cooler exterior. (This is true even of all so-called "closed" or "isolated" systems.) But for all such systems, one of the following cases must hold: (*a*) they are (practically) stationary, like the solar system and most stars known to us, in which case their entropy production (practically) equals their entropy loss, at least temporarily; or (*b*) they are losing temperature, and thereby entropy; or (*c*) they are producing more entropy than they lose, in which case they are in process of getting hotter, a process which, whether energy conservation is assumed or the steady state theory, can be only a comparatively rare and short-lived temporary process. (Even if the system in question should be one that collects matter from its environment until its gravitational field becomes so strong as to encapsulate and separate off the system from the rest of the universe, it would thereby presumably become stationary.) All we know about the universe points to (*a*) and (*b*) as being by far the most frequent and important cases: in almost all sufficiently large systems known to us, entropy production seems to be equalled, or even exceeded, by entropy loss through heat radiation.

This may be explained by the conjecture that every entropy-producing region is open towards some large (perhaps infinite) sinks of energy—regions the energy capacity or heat capacity of which, at least for heat in the form of radiation, is infinite (or approximately so for all practical purposes). The existence of such sinks seems to be strongly indicated by the darkness of the night sky. (We might represent this conjecture by the model of an infinite universe with zero energy density; or by that of an energy-conserving expanding—and therefore cooling and entropy-destroying—universe which tends towards zero energy density; of by that of an expanding steady-state universe with constant temperature, and entropy production equalled by entropy escape.)

So there do not seem to be theoretical or empirical reasons to attribute to expression (2) any cosmic significance or to connect "time's arrow" with that expression; especially since the equality sign in expression (2) may hold, for almost all cosmical regions (and especially for regions empty of matter). Moreover, we have good reason to interpret expression (2) as a statistical law; while the "arrow" of time, or the "flow" of time, does not seem to be of a stochastic character: nothing suggests that it is subject to statistical fluctuation, or connected with a law of large numbers.

As for the evolution of life, this seems to be connected, if at all, with a cooling rather than a heating process on Earth (or perhaps with periodic temperature fluctuations); that is, with increasing order and decreasing entropy. Yet it does not seem that "feeding on neg-entropy" has much to do with the preservation of life, as has been suggested, for example, by Schroedinger[5]. For during the incubation of birds' eggs entropy rather than neg-entropy is supplied to them, though they are in a period of increasing organization; and while in an organism dying of heat or of fever entropy may increase, if it dies of cold—say, by deep-freezing—its entropy certainly decreases.

(**207**, 233-234; 1965)

为了证明这一点，需要弄清这样一个经验事实，即在我们的宇宙中，我们只知道基本开放的系统，只知道系统 X（只要它们还能产生熵）对更冷的外界基本开放。(即使对于所谓的“封闭的”或“孤立的”系统也是成立的。）但对上述所有的系统，下列情况之一必须成立：(*a*) 它们（实际上）是定态的，像太阳系和大多数我们已知的恒星，在这种情况下，至少暂时它们的熵产生（实际上）等于其熵减少；或者 (*b*) 它们的温度正在降低，因此熵在减少；或者 (*c*) 它们正在增加的熵大于其减少的熵，在这种情况下，它们处于变热的过程中，无论我们假定能量是守恒的抑或是稳态理论，这个过程只能是一个比较罕见和短暂的瞬间过程。(尽管讨论的系统应该从外界环境中收集物质直到引力场增强到可以容纳这些物质并将系统和外部的宇宙分割开来，但是仍然可以假定这个系统是定态的。）我们对宇宙的所知显示出 (*a*) 和 (*b*) 是迄今为止最常见和最重要的两种情况：在几乎所有足够大的已知系统中，通过热辐射减少的熵可以抵消甚至超过产生的熵。

这可以用下述推测来解释，即每一个正在产生熵的区域都是在向某些大的（也许是无限大的）能量库开放的——它们的能量容量或热容量（至少对辐射形式的热容量）是无限的（或实际上就是近似于无限）。夜晚天空的黑暗似乎强有力地表明这些能量库的存在。(我们可以用能量密度为零的无限宇宙模型来表达这种推测；或用能量守恒的膨胀宇宙模型——其温度和熵都在减少——即用一种能量密度趋于零的宇宙模型表达这种推测；抑或用具有恒定温度的、熵的产生与熵的逃逸相等的膨胀稳态宇宙模型来表达这种推测。)

因此本文似乎并没有理论依据或经验依据赋予式 (2) 任何宇宙的意义，或把“时间之箭”同其联系起来；尤其因为式 (2) 的等号对几乎所有的宇宙区域（特别是对没有物质的真空）都能成立。此外，我们有很好的理由可以将式 (2) 解释为一种统计学定律；而时间之“箭”或者时间之“流”似乎并不具有随机的特性：没有证据说明它受统计性涨落支配或和大数定律相关。

至于生物进化，似乎是（如果有的话）与地球上的冷却过程而不是与升温过程有联系（或者可能同周期性的温度涨落相联系）；也就是说，其与有序性增加和熵减少的过程有联系。然而，薛定谔[5]提出似乎没有显示生命的维系与“摄入负熵”有很多联系。在鸟蛋孵化期间，尽管它们处于有机体增加的时期，但它们摄入的是正熵而不是负熵；在死于发热或高烧的有机体中熵可能增加，如果死于寒冷——比如深度冷冻——它们的熵肯定要减少。

（沈乃澂 翻译；张元仲 审稿）

Karl Popper: University of London.

References:

1. Popper, K. R., *Nature*, **178**, 381 (1956); 177, 538 (1956); **179**, 1296 (1957); **181**, 402 (1958); *Brit. J. Phil. Sci.*, **8**, 151 (1957).
2. Prigogine, I., *Introduction to Thermodynamics of Irreversible Processes*, 3 (1955).
3. Prigogine, I., *Introduction to Thermodynamics of Irreversible Processes*, 16 (1955).
4. Blum, Harold F., *Time's Arrow and Evolution*, 15 (1935). (Similar statements are to be found, for example, on pages 16, 24, 33, 201.) Compare also Planck, M., *A Survey of Physics*, 17, 27 (1925).
5. Schroedinger, E., *What is Life?* 72 (1944).

Opening Electrical Contact: Boiling Metal or High-density Plasma?

F. Llewellyn Jones and M. J. Price

Editor's Note

The opening of an electrical contact may not appear to hide rich physics. Yet physicist Frank Llewellyn Jones and colleagues here report that the problem is more complex than usually thought. As a closed contact begins to open, the current flowing through it must follow an ever more constricted path. This will necessarily cause great heating, and perhaps even boiling of the metal. Indeed, their simple calculations suggest that small volumes of metal of the order of 10^{-12} cm^3 should be lost upon each opening, eventually damaging the contact. The researchers also show a series of high-speed photographs documenting the formation of tiny bridges of metal between contacts, which are key to understanding the process.

THE processes occurring at the opening of a low-voltage (~4 V) electrical contact have considerable fundamental physical interest as well as having practical importance in the field of electronic and communication engineering. It is well known[1] that, starting with the electrodes closely pressed together in the fully closed position, the opening process leads to a constriction of the current stream lines, which can produce intense local heating and melting of the penultimate microscopic region of contact. The maximum temperature in the contact is related to the potential difference by the ψ, θ theorem:

$$\psi = \left[2 \int_{\theta}^{\theta m} \frac{\lambda}{\kappa} d\theta \right]^{1/2} \tag{1}$$

where ψ = a generalized potential equal to the electrical potential in the absence of thermo-electric effects, θ = temperature, λ = thermal conductivity and κ = electrical conductivity. Thus, on gradual separation of the electrodes the constriction resistance increases and the temperature rises up to and past the melting-point of the metal. On continuing the withdrawal the molten volume thus increases and gets drawn out into a microscopic bridge of molten metal joining the solid electrodes; the contacts finally separate and the circuit opens only when this bridge is broken. The rupture process, however, can be very complicated and lead to transfer of metal from one electrode to the other, a process which, when continually repeated, can lead to the "pip" and "crater" formation which renders the contacts useless after some time. There is evidence[1,2] to show that the matter transferred per operation (~10^{-12} cm^3 in a 5-amp circuit) is related to the size of the molten metal bridge (width ~10^{-4} cm/amp), so that the stability, growth and final rupture of the bridge are a matter of importance, not only from practical

断电接触：沸腾的金属还是高密度等离子体？

卢埃林–琼斯，普赖斯

编者按

电接触的断开中似乎不会藏有高深的物理意义。然而，在本文中物理学家弗兰克·卢埃林–琼斯及其同事们指出：这个问题比人们通常想象的要复杂得多。在一个闭合电路的接触即将断开的时候，通过它的电流一定会经过一个更加狭窄的路径。这种情况必然会产生大量的热，甚至导致金属的沸腾。确实，他们通过简单的计算表明在每次断开电路时都会有体积为 10^{-12} 厘米^3 数量级的小块金属消失，最终使这个电接触毁坏。研究者们也展示了一系列电接触间形成的微小金属桥的高速照相图片，这是理解此过程的关键。

低压下（约为 4 伏特）电接触断开的过程有重要的基础物理学意义，同时对电子工程学和通信工程领域也有重要的实用价值。众所周知 [1]，初始状态是两个电极紧压在一起处于完全导通的状态，断开过程中会导致电流线的收缩，这会导致在最后的微观接触区域产生很强的局部加热和熔化。在接触区的最高温度与电势差有关，可以通过 ψ，θ 定理表示，如下式：

$$\psi = \left[2 \int_{\theta}^{\theta m} \frac{\lambda}{\kappa} \, \mathrm{d}\theta \right]^{1/2} \tag{1}$$

式中 ψ 为广义势（在没有热电效应情况下就等于电势），θ 为温度，λ 为热导率，κ 为电导率。因此，随着电极的逐渐分离，收缩电阻增大，温度升高并超过了金属的熔点。继续分离，熔化的体积随之增加，这种熔融金属逐渐被拉成连接着两个固体电极的微观桥；电接触最终会断开，但只有这个微观的桥断开时整个电路才会断开。然而，这个断裂过程可能是非常复杂的并且会导致金属从一个电极转移到另一个电极上，当不断重复这个过程时，便会产生“点”和“坑”状形态，以致过一段时间后触点失效。有证据 [1,2] 表明，每次操作过程中物质的转移量（在电流为 5 安培的电路中约 10^{-12} 厘米^3）与熔融金属桥的尺寸（宽度约为 10^{-4} 厘米 / 安培）有关，因此桥的稳定性、增长性以及最终的断裂都是具有重要意义的，这不仅是对实际应用而言，而且对研究熔融状态和高温状态下金属的一些物理性质也有重要意义。

considerations, but also from the point of view of the physical properties of metals in the molten state and at high temperatures.

In the first place, an important condition of equilibrium, at least in the earlier stages, is that which depends on the application of surface tension forces. The shapes of the bridges would then be surfaces of revolution satisfying the equation:

$$\Delta p = T\left(\frac{1}{R_1} + \frac{1}{R_2}\right) \tag{2}$$

and these are unduloids, catenoids or nodoids according as Δp is positive, zero or negative respectively[3]. Photographs of static microscopic bridges have indeed confirmed that these stable shapes can be attained[1]. In the later stages of opening Δp will be negative, and experiment has established that the final stable shape is usually the nodoid. The ψ, θ theorem shows that the hottest region of the microscopic molten metal bridge between like electrodes will probably be the narrow neck and, at first sight, it might appear that this is the region at which the bridge is most likely to break. However, detailed investigation of this final process raises some important problems in the physics of metals at high temperatures, and, in particular, near their boiling points.

Mechanisms of Break

It can be seen at once from the ψ, θ theorem that the mechanism of rupture of the molten metal bridge involves the physical properties of the metal, not at any one temperature, but over a wide range of temperatures up to boiling-point, and a number of different processes of rupture are possible.

In the first place, continued separation of the electrodes and the drawing out of the bridge increases the contact resistance R_c; consequently, the contact voltage V_c ($=R_cI_c$) for a given circuit current I_c continually rises. Inspection of the ψ, θ theorem shows that the maximum temperature θ_m correspondingly increases; in fact, it is readily seen by using the Wiedermann-Franz law that, when V_c exceeds about 1.5 V, the corresponding value of θ_m from (1) exceeds the boiling-point of any known metal. Thus, this process of rise of maximum temperature may well continue until θ_m reaches the boiling-point θ_b so breaking the molten metal bridge. The voltage V_b at which this occurs is called the boiling voltage, and is related to θ_b by equation (1). Further analysis[4] of the ψ, θ theorem shows that, since the relationship between ψ and θ depends on the variation of λ/κ with θ, thermal instability can occur for certain functional relationships of λ/κ with θ, in which case a sudden rise of θ_m up to the boiling point can, in certain circumstances, take place. Measurements of the contact current, potential difference and maximum temperature, and its location, enable a determination to be made of the transport properties of metals such as their thermal and electrical conductivities and Thomson coefficients and their dependence on temperature to be determined for the molten state at high temperatures[1].

首先，至少是在较早期阶段，达到平衡的一个重要条件取决于表面张力的应用。桥的形状将由满足式（2）的旋转曲面构成：

$$\Delta p = T\left(\frac{1}{R_1} + \frac{1}{R_2}\right) \tag{2}$$

当 Δp 分别为正值、零或负值[3]时，形态依次是类波状、类链状或类结节状。静态微观桥的照片的确已证实这些稳定形态是可以达到的[1]。在断开的较后期阶段，Δp 将是负值，实验已证实最终的稳定形态通常是类结节状。ψ，θ 定理表明，相同电极间的微观熔融金属桥中最热的区域可能位于窄口处，初看之下，这里非常有可能就是金属桥即将断裂的地方。然而，有关这个最终过程的详细研究引发出了一系列高温下金属物理的重要问题，尤其在接近沸点时。

断裂的机制

从 ψ，θ 定理可以一目了然地看出，熔融金属桥的断裂机制涉及许多金属的物理性质，但断裂并不是发生在某一特定温度下，而是在逼近沸点的较大温度范围内都会发生，许多不同的断裂过程都是有可能的。

首先，电极的连续分离和桥的拉长增大了接触电阻 R_c；结果，对于给定的回路电流 I_c，其接触电压 V_c $(=R_cI_c)$ 连续增高。对 ψ，θ 定理的分析表明，最高温度 θ_m 相应地增高；实际上，用威德曼–弗朗兹定律很容易看出，当接触电压 V_c 超过约 1.5 伏特时，由式（1）得出的 θ_m 对应值超出了任何已知金属的沸点。因此使最高温度上升的过程会持续下去直至 θ_m 达到沸点 θ_b，从而使熔融金属桥断裂。此时的电压 V_b 称为沸点电压，它通过式（1）与 θ_b 相关。对 ψ，θ 定理做进一步分析[4]可知，由于 ψ 和 θ 的关系取决于 λ/κ 随 θ 的变化，因此 λ/κ 和 θ 之间具有特定的函数关系时可以产生热的不稳定性，在这种情况下，θ_m 在特定的环境中可能突然上升至沸点。通过测量接触电流、电势差和最高温度及其位置，便能得出金属的输运性质，例如，金属的热导率、电导率和汤姆孙系数以及这些性质在高温熔融态下与温度之间的关系[1]。

These thermal effects, however, are not the only processes which may sever the bridge. For example, the electromagnetic pinch effect might well, with large currents, so constrict the bridge at the narrowest, hottest and therefore weakest point as to rupture it there. Again, the known variation of surface tension with temperature is an important factor influencing the stability of the bridge, and the consequences of this can only be neutralized by the influence of surface impurities or compensating internal viscous motions in the bridge. Restriction due to the size and geometry of the actual electrodes in a given practical contact might well prevent the continued formation of the stable nodoid form, and instability could result.

The shape and volume of the final bridge and its mechanism of rupture are very important from the practical point of view on account of their relation to the rate of matter transfer on rupture. For example, suppose that a thermo-electric effect displaced the hottest section of the molten metal bridge towards one electrode, then it follows that the rupture at that particular section could have the effect of producing net transfer from one electrode to the other. The amount transferred may then only be that of the hottest region in the neck. On the other hand, if, in a different process of rupture, the molten metal bridge disintegrated as a whole and was transferred to one or other electrode (say, by mechanical splashing of minute droplets), then in this case the matter transfer could be relatively high and this, of course, could have serious practical effects.

Thus, the precise processes of the actual opening of the circuit are a matter of considerable practical and theoretical importance, and these phenomena have been under investigation at Swansea for some time. The relationships of the size of the molten metal bridge to the current and to the matter transferred per operation and the local self-inductance have been investigated[1,5-10].

Accurate measurement of the amount of metal of a given electrode ($\sim 10^{-12}$ g) actually transferred on the rupture of the molten metal bridge was found possible using the radioactive tracer technique, and a very rapid variation with local inductance, particularly in the range 10^{-6} to 10^{-8} H, was found for a number of metals and, particularly so, for platinum[1,6-9]. Further, optical examination of the crater formed on bridge rupture indicated that in some metals the whole volume of the molten metal bridge took part in the transfer. The shape and volume of the microscopic bridge were determined from the geometry of the melting isothermals in each electrode after rupture.

Facts such as these are difficult to reconcile with the picture of matter transfer occurring as the result of a mechanical splashing of small droplets formed from the disintegrating molten metal bridge. On the contrary, they are more consistent with the view that the transfer may well be ionic, the motions of the ions being determined by the oscillatory electric field between the electrodes after the bridge has broken. For reasons such as these an alternative view was put forward based on the production of a micro-plasma, possibly initially formed at the broken neck of the molten metal bridge[11]. Consequently, in recent years effort has been directed to finding direct evidence for the existence of

然而，这些热效应并不是导致桥断裂的唯一过程。例如，电磁箍缩效应可产生很大的电流，因此在最窄、最热因而最薄弱的点使桥收缩，并导致其断裂。此外，已知的表面张力随温度而变化也是影响桥稳定性的一个重要因素，这些结果只能被表面杂质的影响或对桥内部黏滞运动的补偿所抵消。在特定的实际电接触中，由于真实电极尺寸和几何形状的限制会有效地阻止稳定类结节状桥的连续形成，从而使其具有了不稳定性。

从实际的观点看，考虑到最终桥的形态和尺寸及其断裂机制与桥断裂时物质转移的速率有关，所以它们都是非常重要的。例如，假定热电效应会将熔融金属桥的最热部分移向某个电极，则将引起在某个特定位置上发生断裂并导致从一个电极净转移到另一个电极的效果。那么传递的量只是颈部最热区域的量。另一方面，在另一不同断裂过程中，熔融金属桥作为整体分解，并转移到一个或另一个电极（譬如说，通过微滴的机械喷溅），那么在这种情况下，物质转移量可以相当高，当然，这也可能具有重大的实际影响。

因此，电路实际断开的详细过程是一个具有重要的应用性和理论性的课题，我们在斯旺西大学已对此现象做了一段时间的研究。研究人员也对熔融金属桥的尺寸与电流、每次操作中转移的物质以及局部自感之间的关系开展了研究[1,5-10]。

采用放射性示踪技术，能够精确测量出在熔态金属桥断裂时实际转移到给定电极的金属量(约为 10^{-12} 克)。人们已发现在许多金属中，特别是铂[1,6-9]中，局部电感(尤其是在 10^{-6} 亨 ~ 10^{-8} 亨之间）有很快的变化。并且，通过对由桥断裂形成的坑状结构进行的光学检测发现，在一些金属中，整个熔融金属桥的体积都参与了转移。这个微观桥的形态和大小取决于各个电极断裂后熔融等温线的几何形状。

这些情况很难与由熔融金属桥分解时产生的微滴机械喷溅而导致的物质转移图像相吻合。相反，这些情况与如下观点更为一致，即这种物质的转移是呈离子态的，而这种离子的运动取决于熔融金属桥断裂后在电极间振荡的电场。由于上述原因，人们提出了另一种基于微等离子体产物的观点，这一微等离子体很有可能最初形成于熔融金属桥的断裂颈上[11]。因此，近年来我们努力寻找这类等离子体存在的直接证据。可喜的是，由于熔融金属桥发展的最后一步的变化速度，使得等离子体扩张

such a plasma. It will be appreciated that, owing to the speed of events in the final stages of the development of the molten metal bridge, extensive expansion of the plasma to a size which can readily be seen may not take place. In fact, particularly in the presence of a high-pressure ambient atmosphere, a plasma of metal vapour might well be severely restricted in size throughout its short life.

Photography of the Microscopic Molten Metal Bridge

Early attempts to photograph the development of the exploding bridge were confined to cases in which it would be expected that surface tension would be the dominant controlling force and large stable bridges obtainable. Such photographs have been previously obtained for large iron bridges in air[1]. In appropriate circumstances the oxide film on the surface would enable a constant surface tension to be set up over the whole surface, and thus produce stability in accordance with (2).

A number of standard ciné films (25 f.p.s.) were taken of the formation, development and final rupture of the bridge, and many thousands of frames were examined in the hope of finding an illustration of the actual rupture. One or two frames were found which showed that the actual rupture process might not be a simple parting of the nodoid, and this indicated that it was necessary to use high-speed photography if rupture was to be examined in more detail. There were considerable difficulties in the high-speed photography of the molten bridge mainly on account of the small area to be photographed ($\leqslant 10^{-4}$ cm^2), the low luminosity for metals other than platinum and tungsten, and the difficulty associated with the synchronization of the camera and the phenomenon to be photographed. A certain degree of elusiveness of the bridge at all stages of its life also made photography difficult. However, an optical system incorporating a high-speed camera was designed and constructed to examine the development of a molten metal bridge for time-intervals down to about 10 μsec. In this way a large number of metals were investigated under varying conditions of ambient atmosphere and pressure. Some preliminary photographs thus obtained were shown at conferences at Oklahoma[5], at Graz[9] and at Berlin[10]. Sets of later photographs giving a succession of frames extending over a total time of a few milliseconds covering in some detail various phases of the rupture of the microscopic molten metal bridge are given here. The results for iron are of particular interest in that they illustrate the three different aspects of the rupture process discussed here, and these are given in Figs. 1, 2 and 3.

Fig. 1 shows a series of photographs at a rate of 10^4 frames per sec and deals with iron in air. Doubtless on account of consequent oxidation affecting the surface tension, well-shaped stable bridges were formed after a number of operations, and these are in accordance with the theoretical prediction of the stable nodoid. Stability could be controlled by surface tension forces, and the fact that the bridge ruptured when the hottest section boiled is clearly indicated by the photographs, which show the two white-hot separate parts of the bridge after rupture.

到能易于看到的尺寸的情况并没有发生。实际上，尤其是在高压环境的气氛下，在金属蒸气等离子体的短寿命内，它的大小受到了严格的限制。

微观熔融金属桥的拍摄

早期对于拍摄桥断裂过程的尝试受到一些条件的限制，因为人们认为表面张力会成为首要的控制力，从而能获得大而稳定的桥。这样的照片最先是在空气[1]中对大的铁金属桥拍摄而得。在适宜的环境下，表面上的氧化膜将在整个表面范围内产生恒定的表面张力，因此得到如式（2）中的稳定性。

我们采用许多标准的电影胶卷（25 帧 / 秒）对桥的形成、发展和最终的断裂过程进行拍摄，我们对成千上万个镜头进行分析，希望能够发现桥实际断裂时的景象，最终找到了一两个镜头，显示出实际的断裂过程可能不是一个简单的类结节体的分开，并且这也表明如果想对断裂进行更细致地分析，就必须使用高速照相机。利用高速照相机拍摄熔融金属桥面临相当多困难，这主要是因为拍摄的目标范围非常小(小于等于 10^{-4} 厘米^2)，除铂金属和钨金属外，其他金属的亮度都很低，控制好现象发生与拍摄的同步也很困难。此外桥在寿命中不同阶段持续的时间不好把握，这同样为拍摄带来了困难。尽管如此，还是设计和制造了一个结合了高速照相机的光学系统，它能够以约为 10 微秒的时间间隔对熔融金属桥的断裂过程进行拍摄。利用这种方法，我们在不同的气氛和气压下对大量的金属进行了研究。通过这种方式获得的一些初步的照片，在俄克拉荷马州[5]、格拉茨[9]和柏林[10]的会议上进行了展示。这里给出几组后来得到的照片，照片拍摄到一段连续的镜头，覆盖了几毫秒内微观熔融金属桥断裂时不同阶段的一些细节。铁金属的结果格外有趣，因为它们说明了本文中讨论的在不同条件下断裂过程的三种状态，结果如图 1、图 2 和图 3 中所示。

图 1 示出了以每秒 10^4 帧拍摄到的一系列照片，其中金属铁是在空气中处理的。毫无疑问的是随后的氧化反应会影响到表面张力，在经过多次操作后才最终形成形状完好的稳定桥，并且这与稳定的类结节型桥的理论预测是相一致的。稳定性受到表面张力的控制，事实上照片清楚显示出，桥是在最热部分沸腾时断裂的，并且在断裂后形成两个白热分离的部分。

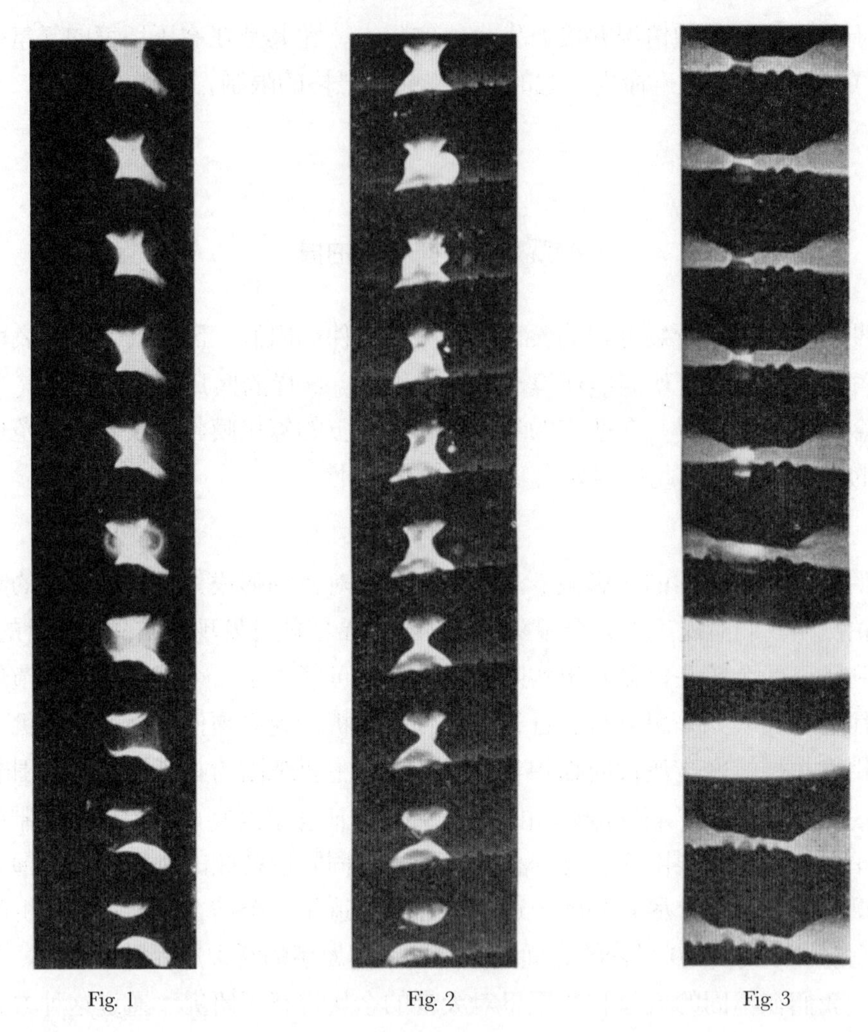

Fig. 1 Fig. 2 Fig. 3

Fig. 1. Material, iron; atmosphere, air; current, 30 amp; circuit E.M.F., 6 V; polarity, top electrode negative; magnification, ×6; framing rate, 5,000 f.p.s.

Fig. 2. Material, iron; atmosphere, air; current, 30 amp; circuit E.M.F., 6 V; polarity, top electrode negative; magnification, ×6; framing rate, 2,000 f.p.s.

Fig. 3. Material, iron; atmosphere, vacuum; current, 60 amp; circuit E.M.F., 6 V; polarity, top electrode positive; magnification, ×19.5; framing rate, 7,000 f.p.s.

Fig. 2 illustrates a less stable condition in which the degree of oxidation was such that the surface tension could not be maintained constant over the surface. In such cases stability can only be produced by internal viscous motion, and this is consistent with the effect illustrated by the rapid change from frame to frame in the location of the hottest region of the bridge surface.

In order to minimize the effect of surface tension forces in establishing stability, the iron surfaces were cleaned by a glow-discharge treatment in hydrogen and the contact was also operated in a vacuum in order to avoid further oxidation. The development of the

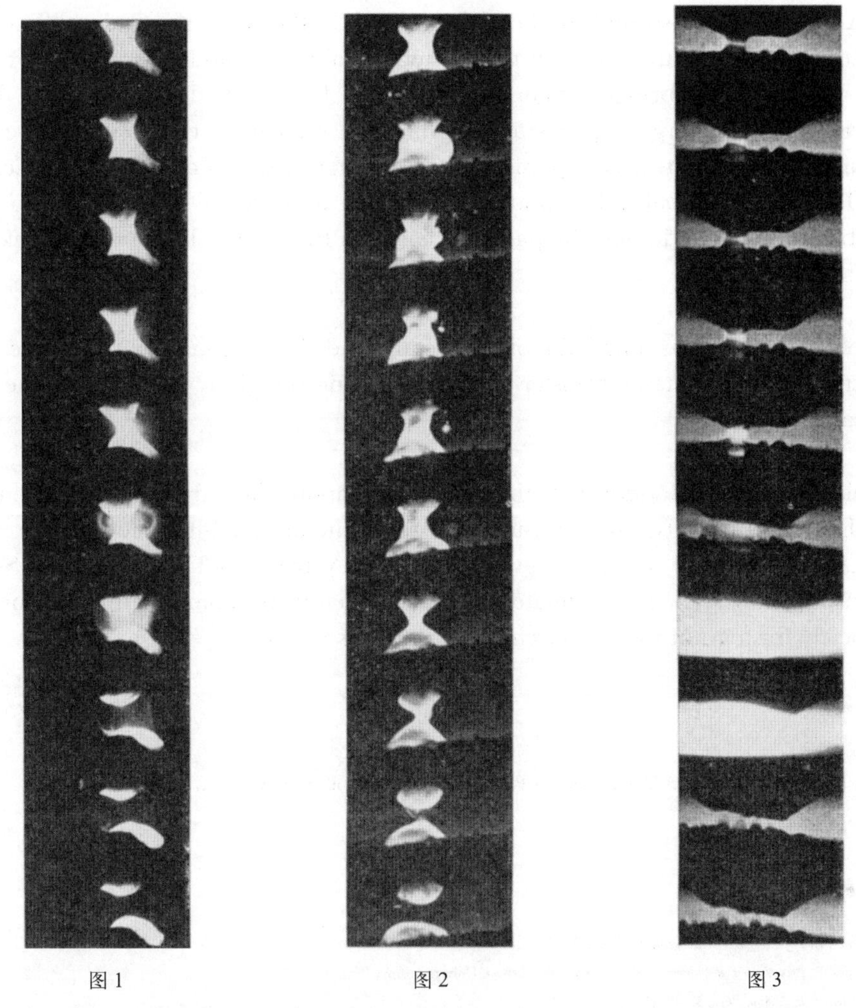

图 1　　图 2　　图 3

图 1. 材料：铁；气氛：空气；电流：30 安培；回路电动势：6 伏特；极性：顶部电极为负极；放大倍数：×6；帧速率：5,000 帧 / 秒。

图 2. 材料：铁；气氛：空气；电流：30 安培；回路电动势：6 伏特；极性：顶部电极为负极；放大倍数：×6；帧速率：2,000 帧 / 秒。

图 3. 材料：铁；气氛：真空；电流：60 安培；回路电动势：6 伏特；极性：顶部电极为正极；放大倍数：×19.5；帧速率：7,000 帧 / 秒。

图 2 示出一个稍不稳定的条件，该条件下氧化的程度使表面张力不能在表面保持不变。这种情况下，稳定性仅靠内部的黏滞运动产生，并且这与用桥表面最热区域处帧－帧之间的快速变化说明的效应是一致的。

在稳定性的建立过程中为了使表面张力的效应最小，我们利用在氢中辉光放电处理法对铁的表面进行了清洗，而且为了避免进一步的氧化，接触在真空中进行。

bridge in this case is illustrated by the series in Fig. 3. This is dramatically different from the two series in Figs. 1 and 2 in that the molten metal bridge no longer severs over a very small section while the bridge remains in the liquid form. On the contrary, in this case an electric micro-discharge appears to have been formed from the completely exploded bridge, producing a high-density plasma of great brightness the duration of which is less than 100 μsec and probably less than 10 μsec. On account of the inability to expand appreciably in these conditions, the particle density in the micro-plasma must be extremely high.

Similar effects have been found for other metals, and the radio-tracer technique enables the directions of the metallic transfer occurring during this short existence of the plasma to be measured.

We thank Mr. Ieuan Maddock of the Atomic Weapons Research Establishment, U.K. Atomic Energy Authority, Aldermaston, for the loan of a high-speed camera and for advice. One of us (M. J. P.) is also grateful for the award of a Department of Scientific and Industrial Research postgraduate research studentship. Thanks are also due to the Royal Society for a grant for the purchase of precious metals.

(**207**, 255-257; 1965)

F. Llewellyn Jones and M. J. Price: Department of Physics, University College of Swansea, University of Wales.

References:

1. Llewellyn Jones, F., *Physics of Electrical Contacts* (Clarendon Press, 1957).
2. Llewellyn Jones, F., *Proc. Intern. Res. Symp. Electric Contact Phenomena*, University of Maine (1961).
3. Davidson, P. M., *Brit. J. App. Phys.*, **5**, 189 (1954).
4. Greenwood, J. A., and Williamson, J. B. P., *Proc. Roy. Soc.*, A, **240**, 13 (1958).
5. Llewellyn Jones, F., *Proc. Twelfth Intern. Conf. Electromagnetic Relays*, Oklahoma State Univ. (1964).
6. Price, M. J., *Proc. Thirteenth Intern. Conf. Electromagnetic Relays*, Oklahoma State Univ. (1965).
7. Jones, C. R., Hopkins, M. R., and Llewellyn Jones, F., *Brit. J. App. Phys.*, **12**, 485 (1961).
8. Jones, C. H., and Hopkins, M. R., *Brit. J. App. Phys.*, **14**, 137 (1963).
9. Llewellyn Jones, F., *Proc. Intern. Symp. Electrical Contacts*, Technische Hochschule, Graz, Austria (1964).
10. Hopkins, M. R., *Proc. Intern. Symp. Electrical Contacts*, Technische Hochschule, Graz, Austria (1964).
11. Llewellyn Jones, F., *Proc. Third Intern. Conf. Ionization Phenomena in Gases*, Venice, **2**, 620 (1958).
12. Llewellyn Jones, F., *Proc. Intern. Conf. Electrical Contacts*, Deutsche Akademie der Wissenschaften zu Berlin (1964); also *Elektrie*, **3**, 129 (1965).

在这种情况下建立的桥通过图 3 中的系列照片进行了说明。这与图 1 和图 2 的两个系列照片有很大差别，图 3 中当熔融金属桥以液体的形式存在时，其不会以很小的截面分离。相反，在这种情况下，在完全分解的桥上已出现电的微放电，在小于 100 微秒或可能小于 10 微秒的时间内产生了高亮度的高密度等离子体。由于在这些情况下无法明显地扩展，因此微等离子体中的粒子密度必然是极高的。

在其他金属中已发现了类似效应，放射性示踪技术使在存在等离子体的这段极短时间内对金属转移方向进行测定成为可能。

我们感谢位于奥尔德马斯顿的英国原子能管理局原子武器研究中心的爱恩·马多克先生借给我们一台高速照相机并提供了一些建议；感谢科学及工业研究署颁发给我们之间的一员（普赖斯）的研究生研究奖学金，也同样感谢皇家学会同意为给我们购买一些贵金属。

（沈乃澂 翻译；赵见高 审稿）

Structural Basis of Neutron and Proton Magic Numbers in Atomic Nuclei

L. Pauling

Editor's Note

Physicists by the mid 1960s had noted the surprising stability of atomic nuclei having certain "magic numbers" of either protons or neutrons, including 2, 8, 20, 50, 82 and 126. Here American chemist Linus Pauling offers an explanation. These numbers do not correspond to atomic shells being completely filled with fermions (particles with spin 1/2, like protons and neutrons). However, they do appear to correspond to closed shells of nucleons which achieve efficient packing in space, possibly with an extra halo of alpha particles. For example, the magic number 50 arises as 8 neutrons or protons in a closed-shell core, and another 42 in an outer halo. These magic-number nuclei have a higher average binding energy than other nuclei.

IN 1933 Elsasser[1] pointed out that some of the properties of atomic nuclei correspond to greater stability for certain numbers of neutrons and protons (given the name magic numbers) than for other numbers; the magic numbers for both N (neutron number) and Z (proton number) are 2, 8, 20, 50, 82 and 126. (Less-pronounced effects are observed also for N or Z equal to 6, 14, 28, 40, and some larger numbers. The set of magic numbers is often assumed to include 28.)

The magic numbers do not have the values ($2n^2$) for completed shells of fermions (with all states with total quantum number n, azimuthal quantum number $l \leqslant n-1$, occupied by pairs), which are 2, 8, 18, 32, 50, ..., nor the values for certain shells and sub-shells that lead to maximum stability for electrons in atoms, which are 2, 10, 18, 36, 54 and 86.

It was discovered by Mayer[2] and by Haxel, Jensen and Suess[3] that the magic numbers can be accounted for by use of the sub-sub-shells corresponding to spin-orbit coupling of individual nucleons; that is, to the values of $j = l + 1/2$ and $l - 1/2$ for the two sub-sub-shells of each sub-shell. For example, they[4] assign to N or Z = 50 the configuration $(1s1/2)^2(1p3/2)^4(1p1/2)^2(1d5/2)^6(2s1/2)^2(1d3/2)^4(1f7/2)^8(2p3/2)^4(1f5/2)^6(2p1/2)^2(1g9/2)^{10}$, which may be written more briefly as $1s^2 1p^6 1d^{10} 2s^2 2p^6 1f^{14}(1g9/2)^{10}$.

The evidence for spin-orbit coupling and for the Mayer–Jensen shell model is convincing. It is, however, difficult to understand, on the basis of their arguments, why the six magic numbers should be outstanding among the many numbers corresponding to the completion of spin-orbit sub-sub-shells, which (for the Mayer–Jensen sequence[4] of energy-levels) are 2, 6, 8, 14, 16, 20, 28, 32, 38, 40, 50, 56, 64, 68, 70, 82, 92, 100, 106, 110, 112, 126, 136, 142, ...

中子和质子在原子核中的幻数的结构基础

鲍林

编者按

在20世纪60年代中期，物理学家注意到，那些出奇稳定的原子核都含有一些特定“幻数”的质子或中子，这些“幻数”包括2、8、20、50、82和126。本文中，美国化学家莱纳斯·鲍林提出了一种解释：他认为这些幻数并不对应于那些完全被费米子（自旋为1/2的粒子，例如质子和中子）填满的原子壳层。而是看起来对应于在空间中被有效堆积的核子的闭壳层，可能还带有一圈外加的α粒子晕。例如，幻数50就意味着在闭壳层核心内有8个中子或质子，另外42个在外层晕中。与其他原子核相比，这些幻数核具有更高的平均结合能。

艾尔萨瑟[1]于1933年指出，较之于其他数，具有某些特定中子数和质子数（将其命名为幻数）的原子核的一些性质更为稳定。对于N（中子数）和Z（质子数），这些幻数是2、8、20、50、82和126。（对于N或Z等于6、14、28、40以及某些更大的数，也能注意到不太明显的效应。通常认为幻数序列包括28。）

幻数既不是与费米子满壳层（或者说费米子成对填满主量子数为n、角量子数为$l \leqslant n-1$的所有状态）有关的数值$(2n^2)$，即2、8、18、32、50……也不是与某些可导致电子在原子中有最大稳定性的壳层和子壳层有关的数值：2、10、18、36、54和86。

梅耶夫人[2]以及哈克塞尔、延森和苏斯[3]都发现，幻数可以利用对应于独立核子的自旋–轨道耦合的支壳层来进行解释；也就是说，对应于某一子壳层的总角动量分别为$j=l+1/2$和$l-1/2$的两个支壳层。例如，他们[4]将N或Z为50的组态标记为$(1s1/2)^2\ (1p3/2)^4\ (1p1/2)^2\ (1d5/2)^6\ (2s1/2)^2\ (1d3/2)^4\ (1f7/2)^8\ (2p3/2)^4\ (1f5/2)^6\ (2p1/2)^2\ (1g9/2)^{10}$，或更简略地写成$1s^2\ 1p^6\ 1d^{10}\ 2s^2\ 2p^6\ 1f^{14}\ (1g9/2)^{10}$。

有关自旋–轨道耦合和梅耶夫人–延森壳层模型的证据是令人信服的。然而，对于他们的论据，让人难以理解的是，为什么这6个幻数在可以填满自旋–轨道耦合的支壳层的数字集合中是如此与众不同，这个集合（对于能级的梅耶夫人–延森序列[4]）是2、6、8、14、16、20、28、32、38、40、50、56、64、68、70、82、92、100、106、110、112、126、136、142……

In the course of developing a theory of nuclear structure based on the assumption of closest packing of clusters of nucleons[5], I have found that the magic numbers have a very simple structural significance: 2 and 8 correspond to the closed shells $1s^2$ and $1s^2 1p^6$, and the others to a closed-shell core with an outer layer (the mantle of the nucleus) containing the number of spherons (helions[6], He^4, tritons, H^3, or dineutrons) required to surround the core in closest packing.

Triangular (icosahedral) closest packing, as found, for example, in the intermetallic compound[7] $Mg_{32}(Al,Zn)_{49}$, involves the sequence 1, 12, 32, 72 of spheres in successive layers. These numbers are approximated by the equation $n_0 = (n_i^{1/3} + 1.30)^3 - n_i$, in which n_0 is the number of spheres in an outer layer and n_i is the number in the core. (The form of this equation corresponds to assigning equal effective volumes to the spheres, and the value of the constant reflects the nubbling of the surface and the packing of outer spheres into pockets of the core.) This equation can be applied to obtain the number of spherons in the successive layers in a nucleus, and thus to obtain the sequence of nucleonic energy-levels. Sub-shells (with given value of l) occurring once (as $1s$, $1p$, etc.) are assigned to the mantle of spherons, those occurring twice ($1s$ and $2s$, for example) to the mantle and next inner layer, and so on. Thus I interpret the configuration for N or Z = 50, given above, as representing 8 neutrons or protons in the core ($1s^2 1p^6$) and 42 in the outer layer ($2s^2 2p^6 1d^{10} 1f^{14} (1g9/2)^{10}$).

The application of the packing equation leads to a sequence of levels essentially as given by Mayer and Jensen, but often with sub-sub-shells for different layers being filled over overlapping ranges of values of N or Z. For example, the $3s1/2$, $2d3/2$, and $1h11/2$ sub-sub-shells all begin to be occupied at about N or Z = 60 and are all completed at about N or Z = 82.

The configurations found in this way for the magic numbers are given in Table 1 and Fig. 1. Each of the first two represents a completed shell. The third (20) has a completed shell as core and another as mantle. Each of the others has a core of a completed shell or two completed shells, with a mantle that is required by the packing to include a sub-sub-shell $(1g9/2)^{10}$ for 50, $(1h11/2)^{12}$ for 82, and $(1i13/2)^{14}$ for 126. Until 184 is reached, there are no other values of N or Z for which the packing equation leads to a core consisting of layers that are completed shells.

Table 1. Nucleon configurations for magic numbers

N or Z	Mantle	Core or outer core	Inner core
2	$1s^2$		
8	$1s^2 1p^6$		
20	$2s^2 1p^6 1d^{10}$	$1s^2$	
50	$2s^2 2p^6 1d^{10} 1f^{14} (1g9/2)^{10}$	$1s^2 1p^6$	
82	$3s^2 2p^6 2d^{10} 1f^{14} 1g^{18} (1h11/2)^{12}$	$2s^2 1p^6 1d^{10}$	$1s^2$
126	$3s^2 3p^6 2d^{10} 2f^{14} 1g^{18} 1h^{22} (1i13/2)^{14}$	$2s^2 2p^6 1d^{10} 1f^{14}$	$1s^2 1p^6$

在构建基于核子集团最密堆积假设的核结构理论[5]的过程中，我已发现，幻数具有非常简单的结构意义：2 和 8 对应于闭壳层组态 $1s^2$ 和 $1s^21p^6$，而其他的则对应于带有一个外层（称为核的幔壳层）的闭壳层核，其中外层内含有一定数量的球子（氦核[6]、He[4]、氚核、H[3] 或双中子），这些球子以堆积的方式围绕在核心周围。

例如，在金属间化合物[7] $Mg_{32}(Al,Zn)_{49}$ 中发现的（二十面体的）三角形最密堆积各外层中包含的球子数按顺序依次为 1、12、32、72。这些数可用方程 $n_0=(n_i^{1/3}+1.30)^3-n_i$ 近似表示，其中 n_0 是在一个外（壳）层中球子的数目，n_i 是核中球子的数目。（这个方程的形式相当于为球子赋予等量的有效体积，常数值反映了表面的小块和外部球子的堆积进入了核心的内部。）这个方程可用来计算核内相继外层中球子的数目，进而获得核子能级的序列数。子壳层（具有特定的角量子数 l）一旦出现（如组态 $1s1p$ 等）就分配给球子的幔壳层，那些出现二次的（如 $1s$ 和 $2s$）就分配给幔壳层和下一个内层，等等。因此，对于上述给定的 N 或 Z = 50 的组态，我的解释是：在核心 $(1s^21p^6)$ 中有 8 个中子或质子，另外 42 个中子或质子在外层 $(2s^22p^61d^{10}1f^{14}(1g9/2)^{10})$ 里。

堆积方程的应用基本上导出了梅耶夫人和延森给出的能级序列，但通常是在 N 或 Z 值有重叠的范围内填满不同外层的支壳层。例如，支壳层 $3s1/2$，$2d3/2$ 和 $1h11/2$ 都在 N 或 Z 约为 60 时开始填充，直到 N 或 Z 约为 82 时才都被填满。

在表 1 和图 1 中分别给出了将这种方法应用于幻数后得到的组态。前两个幻数都表示满壳层，第三个幻数（20）则有一个满壳层作为核心和另一个作为幔壳层。其他每一个幻数都包含一个或两个满壳层作为核心，至于它们的幔壳层，在幻数是 50 的情况下，堆积要求其包含组态为 $(1g9/2)^{10}$ 的支壳层，在幻数为 82 的状态下，包含 $(1h11/2)^{12}$ 的支壳层，而幻数 126 则包含 $(1i13/2)^{14}$ 的支壳层，一直至幻数 184，不存在 N 或 Z 的其他数值，由堆积方程导出的仅包含满壳层的核。

表 1. 对于各种幻数的核子组态

N 或 Z	幔壳层	核心或外层核心	内层核心
2	$1s^2$		
8	$1s^21p^6$		
20	$2s^21p^61d^{10}$	$1s^2$	
50	$2s^22p^61d^{10}1f^{14}(1g9/2)^{10}$	$1s^21p^6$	
82	$3s^22p^62d^{10}1f^{14}1g^{18}(1h11/2)^{12}$	$2s^21p^61d^{10}$	$1s^2$
126	$3s^23p^62d^{10}2f^{14}1g^{18}1h^{22}(1i13/2)^{14}$	$2s^22p^61d^{10}1f^{14}$	$1s^21p^6$

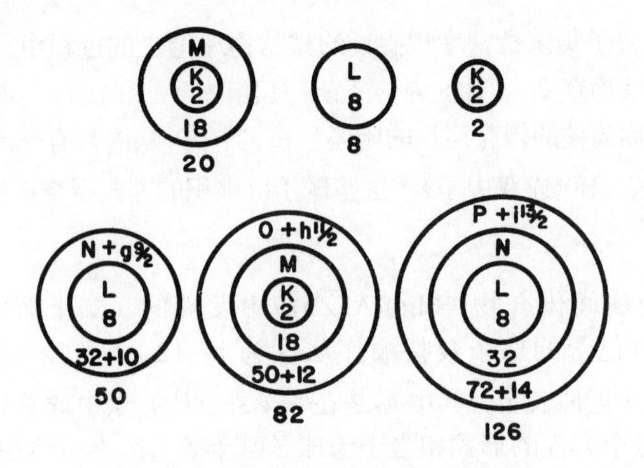

Fig. 1. The magic-number structures of atomic nuclei

I conclude that the stability that characterizes the magic numbers results from the completion of shells for a single layer (2, 8) or two layers (20) of spherons, or, for the larger magic numbers (50, 82, 126), for the core layers, the mantle having a completed shell plus a completed sub-sub-shell.

(**208**, 174; 1965)

L. Pauling: Big Sur, California.

References:

1. Elsasser, W. M., *J. Phys. et Radium*, 4, 549 (1933); 5, 389, 635 (1934).
2. Mayer, M. Goeppert, *Phys. Rev.*, 75, 1969 (1949).
3. Haxel, O., Jensen, J. H. D., and Suess, H. E., *Phys. Rev.*, 75, 1766 (1949); *Z. Physik*, **128**, 295 (1950).
4. Mayer, M. Goeppert, and Jensen, J. H. D., *Elementary Theory of Nuclear Shell Structure*, 58 (John Wiley and Sons, Inc., New York and London, 1955).
5. This theory is described in papers to be published in *Phys. Rev. Letters*, *Proc. U. S. Nat. Acad. Sci.*, and *Science*.
6. Pauling, L., *Nature*, **201**, 61 (1964): Proposal of the name helion for the α-particle.
7. Bergman, G., Waugh, J. L. T., and Pauling, L., *Nature*, **169**, 1057 (1952); *Acta Cryst.*, **10**, 254 (1957). Pauling, L., *The Nature of the Chemical Bond*, third ed., 427 (Cornell University Press, Ithaca, New York, 1960).

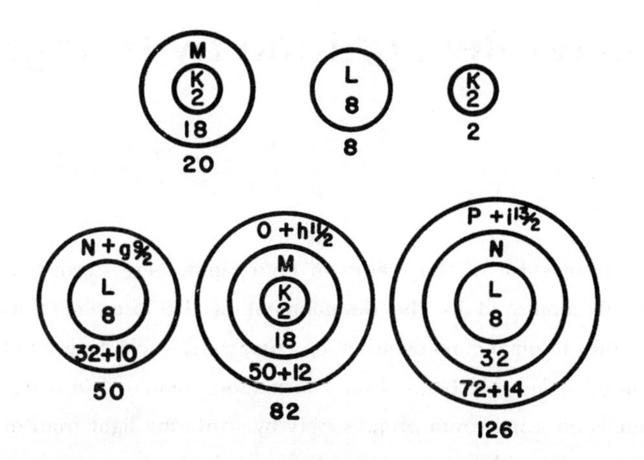

图 1. 原子核的幻数结构

我的结论是，幻数所体现的稳定性来自由球子组成的单层（2、8）或者双层（20）满壳层，或者对于更大的幻数（50、82、126）和具有核心的层，则幔壳层是由一个满壳层加上一个填满的支壳层构成。

（沈乃澂 翻译；厉光烈 审稿）

Character Recognition by Holography

D. Gabor

Editor's Note

Dennis Gabor, the inventor of the theory of holography, here suggests an application of the technique, recently improved by the development of the laser, to a long-standing problem in engineering: the automatic recognition of characters, such as printed letters. Holography produces an image of an object based on information contained in a scattered coherent light wave. If one builds up a hologram progressively by scattering light from many possible variants of a single character, then illumination of a character similar to any such variant can be made to produce a visual code easily readable by a machine. Gabor's basic idea, with many modifications, is now commonly used in pattern recognition techniques based on holography.

WAVE-FRONT reconstruction or holography, on which the first report[1] was published in Nature seventeen years ago, had a powerful renaissance in the past years. E. N. Leith and J. Upatnieks[2], G. W. Stroke[3] and others have greatly improved the original method, and showed that it was possible to reconstruct complicated two- and three-dimensional objects, with half-tones, in previously unattainable perfection. The revival of holography owed much of its impulse to the invention of the laser, which made it possible to produce holograms with interferences of the order of 10,000, and thus to make full use of the information capacity of fine-grain photographic plates.

I wish to show that it has now become possible to harness holography for the solution of one of the most urgent problems of computers and other date-processing devices; the recognition of characters with many variants.

Wave-front reconstruction contains a principle which has not yet been fully exploited. Expressed in a general form: two coherent waves are made to fall simultaneously on a photographic plate, one coming from an object A, the other from an object B. The photograph links these together in such a way that if the hologram is illuminated by A alone, B will appear too, and vice versa. So far this principle has been applied in the form that A was the object of interest and B a light source, usually a simple one, and in the reconstruction the hologram was illuminated by B. I now propose to turn this around. Let A be a character, such as a printed or hand-written letter or numeral, which can be read by human beings but not by a machine, and let B be a combination of point-sources, forming a code-word which can be read by a machine. Produce the hologram by combining A and B. When A, or a character sufficiently close to it is presented to the hologram, with the original illumination, the code-word B will flash out. This means that the hologram can act as a translator, or coding device.

利用全息术的字符识别

盖伯

编者按

全息术理论的发明者丹尼斯·盖伯在本文中提出了该技术（最近因激光器的发展而有所改进）的一项应用，即应用于工程技术中长期存在的一个问题：字符（例如印刷字体）自动识别。全息术产生物体的像建立在包含有相干光波散射信息的基础之上。如果将同一个字符各种可能的变形所散射的光递增地制作出一张全息图，那么与任何一个变形相似的字符的照明就能够产生一个便于机器读出的视觉编码。盖伯的基本创意经过多次改进后，现在已经普遍用在基于全息术的图形识别技术中。

17 年前《自然》首次报道[1]了波前重建，即全息术，近年来这项技术又再次兴起，势头强劲。利思和乌帕特尼克斯[2]、斯托克[3]以及其他一些人已经对原来的方法做了重大的改进，并表明有可能利用半色调重建复杂的二维物体和三维物体，并达到过去不能实现的完美程度。激光器的发明在很大程度上推动了全息术的复苏，因为它使利用量级为 10,000 的干涉产生全息图成为可能，这样也可以充分利用细粒照相干板上的信息容量。

我要说明的是，现在利用全息术有望解决计算机和其他数据处理装置中最为紧迫的问题之一——对具有多种变形字符的识别。

波前重建包含了一个尚未被充分应用的原理。通常表现为：两束相干波同时照射在一张照相干板上，其中一束来自物体 A，另一束来自物体 B。如果全息图单独用 A 照明，B 也会出现，反之亦然，全息图就是通过这样的方式将它们联系在一起的。至今，这个原理已经通过下述方式加以利用：若 A 是被关注的物体，B 是光源（通常是简单的一个单个的光源），在重建中，全息图用 B 照明。现在我要对此进行一些改动。设 A 是一个字符，例如一个印刷的或手写的字母或数字，它可以被人读出，但不能被机器识别，再设 B 是点光源的一个组合，它形成了一个可被机器识别的编码。将 A 和 B 混合便可产生全息图。当 A 或一个与它十分相似的字符呈现在全息图前，并使用原来的照明光照亮时，编码 B 将立刻浮现出来。这表明全息图起到了转换器或编码装置的作用。

The interest of this principle is in the enormous recognition capacity which can be stored in a single hologram, and which one might not perhaps suspect at first sight. I wish to show that with N characters to be discriminated, each with M variants, the product $M{\cdot}N$ can be made of the order of a thousand or even more.

Fig. 1 shows the optics for producing the master hologram, and for using it in the read-out. The recording medium is assumed to be a transparency, such as a microfilm; but reflecting media can also be used. The hologram is built up by repeated exposures in what may be called "layers". These are not, of course, physically separated in the emulsion. Each layer corresponds to one of the N characters to be discriminated, with all its M variants, and is marked with one code-word. The layer contains the part-holograms, to be called "engrams" of the variants side by side, with little overlap. Each engram is produced with one direction of illumination, and as the photographic plate is arranged in the rear focal plane of a lens viewing the character, it is a "Fourier-hologram". This has the advantage that the hologram is translation-invariant, that is to say, independent of the position of the character so long as this appears alone in the window. An engram need not occupy much more area in a photographic plate than would be needed for a good record of the corresponding character, but as a cautious example we will assume 120 engrams, each with a diameter of about 5 mm, on a photographic plate of 50 mm × 50mm. This is sufficient to record without overlap 30 variants, each in four or six "identical" engrams. Fig. 1 shows how this is achieved.

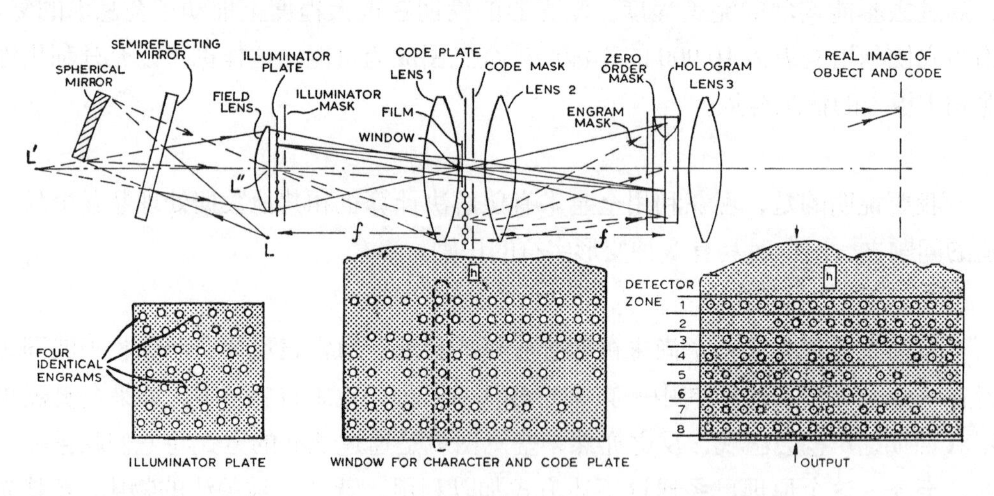

Fig. 1. Apparatus for producing a coding hologram and using it for the read-out.

The light of a laser issues from a point L, and a beam splitter consisting of a spherical mirror and a semi-reflecting mirror produces of this two images L' and L''. The first of those serves the illuminator; the second, in the centre hole of the illuminator plate, serves the code plate. The illuminator plate, backed by a field lens, consists of a plastic plate embossed with, say, 120 lenticules, and is black outside the lenticules. These produce 120 point sources which illuminate the window containing the character through a lens 1, which removes the illuminator points into star space. The point sources correspond one-

这个原理的价值在于它所具有的巨大的识别容量，并且能存储在单个全息图中，对此人们最初也许不会察觉得到。我要说明的是，要对 N 个字符进行辨别，每个有 M 种变形，其乘积 $M \cdot N$ 能够达到上千甚至更高的量级。

制作主全息图并利用它进行读取的光学装置如图 1 所示。假定记录介质是透射型的，例如显微胶片，不过也能够使用反射型介质。通过在“层”（姑且这样命名）上重复曝光而产生全息图。当然，这并不是说乳胶中存在物理上的分离。每一层都对应 N 个字符中的一个字符，每个都具有 M 种变形的待辨别字符，并用一个编码字标记。这个层包含了部分全息图，称为各个变形的“忆迹”，它们并排相连并略微重叠。每个忆迹是用一个方向上的光束照射产生的，又因为照相干板位于观测字符的透镜的后焦平面上，因此是一个“傅里叶全息图”。其优点在于全息图具有平移不变性，换言之，字符只需在窗口中单独出现，而其所处的位置是无关紧要的。与良好地记录一个字符所需要的面积相比，相应的一个忆迹并不需要占用照相干板更大的面积。而作为一个严谨的例子，我们将假设在 50 毫米 × 50 毫米的照相干板上有 120 个直径约为 5 毫米的忆迹，这就足够在无须重叠的方式下记录 30 个变形，其中每个有 4 个或 6 个“同样的”忆迹。图 1 显示了这是如何实现的。

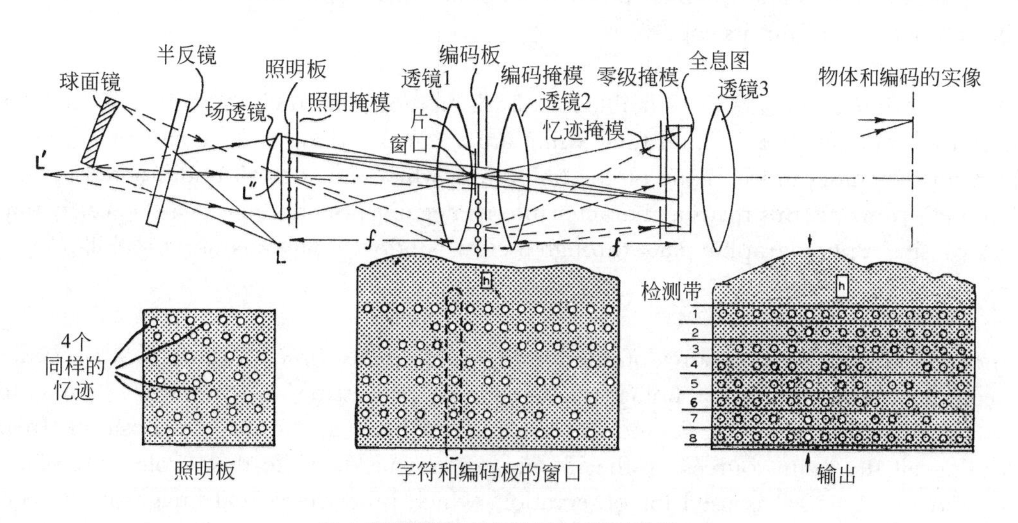

图 1. 制作编码全息图并将它读出的装置

激光器的光束从点 L 射出，通过一个包含有一个球面镜和一个半反镜的分束器后产生两个像 L' 和 L''。其中第一个像被用作照明光；位于照明板中心孔处的第二个像被用作编码板。照明板在场透镜的背后，其由一片有 120 个微透镜的模压塑料板构成，在微透镜外为黑色。这些微透镜形成了 120 个点光源，它们穿过透镜 1 照射在包含字符的窗口上，使照明点移入了星空。点光源与忆迹一一对应。它们的排列稍有随机性是有利的。当任何一个变形发生时，一次均可获得 4 个同样的忆迹。

by-one to the engrams. There is a certain advantage in randomizing them slightly. Four identical engrams are taken at a time of any one variant. These are spaced out, as far as possible, to increase the resolving power of the hologram. They are selected with a mask, and a different mask with four holes is used for every variant.

The point source L'' in the centre of the illuminator plate illuminates the code-plate, through the same lens 1, which serves in this area as a field lens. The code-plate, like the illuminator, is an embossed plastic plate, which contains the code-word in the form of groups of luminous points, arranged in one or several arrays. It is advantageous to use self-checking codes, in which every word has the same number of code-points. In the example there are six positions, of which two remain dark and four light up. This code has $6{\cdot}5/1{\cdot}2 = 15$ words. Two more positions have been added. There do not contribute to the discrimination of characters, but improve the signal-to-noise ratio, as eight points have to light up for every valid character.

In the making of the master hologram all engrams in one layer are marked, that is to say, exposed simultaneously, with one distinctive code-word, which is selected by a mask. But as each code-word illuminates the whole area of the hologram, a further mask must be used near the plane of the photographic plate, which cuts out the light except in the area of the engrams which are made at any one time. This makes it possible to observe the rule of optimum illumination, which postulates about equal light sums on any engram from the character and from its code-word.

Black-on-white letters are less suitable for discrimination than their negatives, because they have too much in common; all their white area. But this disadvantage can be eliminated by a further mask, in the plane of the hologram, which cuts out all undiffracted light. By Babinet's principle this turns a character into its negative. Such a mask can be easily made by exposing a photographic plate through a clear window simultaneously to all illuminator points.

After $M{\cdot}N$ successive exposures of the photographic plate, which add up to a convenient medium density, the master hologram is made by processing and printing it, preferably with an overall gamma of 2, and the print is put back in the original position. In the reading all the point-sources of the illuminator are used, while the whole code-plate is covered up. A lens 3 is used for observation, which produces a real image of the code-plate. If now the recording medium is dragged across the window, whenever a character or a variant appears in it, its code-word will flash up. It is advantageous to arrange in the image plane a mask, which is a replica of the code-plate, with very fine holes, so as to exclude all but the signalling light. This mask, too, can be made photographically.

A method of reading the code-words is to sum up all the light which appears in one zone, corresponding to one position in the code, and guide it to a separate photoelectric detector. Each detector is fitted with a level discriminator, so as to reject spurious signals below a certain level. This method is simple; but it has only moderate discriminating

为了提高全息图的分辨率，这些忆迹要尽可能地相互远离。用掩模对它们进行挑选，每个变形使用一个不同的带有 4 个孔的掩模。

位于照明板中心的点光源 L'' 同样通过透镜 1 照明编码板，透镜 1 在这一区域作为场透镜使用。与照明器类似，编码板是一块模压塑料板。它包含的编码字以一群照明点的形式存在，它们排成一个或几个阵列。使用自检编码具有一定的优点，因为其中每个字都具有相同数目的编码点。在这个例子中有 6 个位置，其中 2 个始终为暗，另外 4 个被照亮。这个编码具有的字数为 $(6\times5)/(1\times2)=15$。已经又增加了 2 个位置。这对字符的分辨率并没有贡献，但提高了信噪比，因为每个有效的字符必须有 8 个点被照亮。

在主全息图的制作过程中，用一个独特的编码字来标记一个层中所有的忆迹，换言之，它们是同时曝光的，而每个编码字的选择取决于掩模。但由于每个编码字都照明了全息图的整个区域，因此必须在靠近照相干板的平面附近再使用一个掩模，它遮去了在任何时刻获得的忆迹区域以外的光。这样就有可能观测到最佳照明的规律，最佳照明要求对于任一忆迹，来自字符和其编码字的光的总和都大致相等。

相比于其负片，白底黑字的字母不太适合于鉴别，因为它们之间共同的部分即它们所有的白色区域太多。但这一不利条件可以通过在全息图平面中再加一个掩模来消除，这个掩模遮去了所有的非衍射光。根据巴比涅原理，这样可使一个字符转换为它的负片。令所有的照明点通过一个清晰的窗口同时对照相干板曝光就可以很容易地制作出这类掩模。

照相干板经 $M\cdot N$ 次连续曝光后，增至一个适合介质的光密度，再通过处理和洗印就制成了主全息图，总伽马值最好为 2，并将照片放回原始位置。在读取过程中使用了照明光的所有点源，而整个编码板是被遮盖住的。透镜 3 用于观测，它产生了一个编码板的实像。这时如果记录介质被拖过窗口，只要有字符或其变形在其中出现，其编码字便会立刻显现出来。在像平面放置一个掩模是有好处的，这个掩模是编码板的复制品，具有很细的孔，因此可以排除信号光外的所有光。这类掩模也能通过照相而制得。

读取编码字的方法是：将与编码中某一位置相对应的一个带中出现的所有的光相加，并将它导入一个单独的光电探测器。每个探测器与一水平鉴别器相配，以排除低于一定水平的乱真信号。这种方法简单，但只具有一般的鉴别能力，因为如果

power, because if the characters are not clearly distinct, some light might show up in the same zone in the code-words of other characters. One can reduce this by making the code-words of characters which are not clearly distinct as different as possible. But the maximum of discrimination is achieved by a somewhat more complicated apparatus. In this the image of the code-plate is projected on the screen of an image camera. The code-words flash up at intervals corresponding to the time allotted to each letter, during 10-30 percent of this period. In the time between flashes all code positions are scanned word by word, and points above a certain level of intensity are transferred to a memory organ, such as a core store. But unless the full number of points appear in a word, the record is erased. If the full number is counted, the code-word is transferred to the computer.

The great discriminating power of the holographic method stems from its high angular resolution. Assume, for example, $N = 35$, $M = 30$, $M{\cdot}N = 1050$. The group of four engrams corresponding to the character presented to the reader receives 1/30 of the light, and can diffract about 1/35 of it, altogether about 10^{-3} of the total. (Not counting, of course, in black-on-white records, the undiffracted light which goes into the zero order.) Of the diffracted light, under the proper conditions, that is to say, when the engrams were taken with about equal light sums from the letter and from the code-word, one-quarter will go into the reconstruction of the code-word. One half appears in the object, another quarter goes into the "twin" image of the code-word, which, however, is washed out by intermodulation with the character, and is useless for recognition. But the useful quarter is concentrated in extremely small solid angles. For example, if four or six identical engrams are spaced out by about 25 mm, the solid angle in which the major part of the light corresponding to a code-point is concentrated will be of the order 10^{-8}. Let the light of, say, 10^{-4} of the total be distributed among ten code-points, this means that 10^{-5} of the light appears in one code-point, in a solid angle which is perhaps 10^{-6} of the solid angle covered by the whole code; a concentration of the order ten. Moreover, this estimate is somewhat pessimistic, because it takes no account of the confirmation of the character by the engrams of slightly different variants in the same layer.

In conclusion, there is good reason to believe that a single hologram may discriminate between all the numerals and the letters of the alphabet, each with 30 variants.

(**208**, 422-423; 1965)

D. Gabor: Department of Electrical Engineering, Imperial College of Science and Technology, London.

References:

1. Gabor, D., *Nature*, **161**, 777 (1948); *Proc. Roy. Soc.*, A, **197**, 475 (1949); *Proc. Phys. Soc.*, B, **64**, 244 (1951).
2. Leith, E. N., and Upatnieks, J., *J. Opt. Soc. Amer.*, **53**, 1377 (1963); **54**, 1295 (1964); **55**, 569 (1965).
3. Stroke, G. W., *Optics of Coherent and Non-coherent Electromagnetic Radiations, Univ. Michigan* (1965), with Falconer, D. G., *Physics Letters*, **13**, 306 (1964); **15**, 283 (1965).

有些字符没有明显的区别，那么其他字符的编码字的光可能会出现在同一带内。对于那些没有明显区别的字符，可以通过制作尽可能不同的编码字来减少这种情况。但通过更加复杂的装置可以达到最大的鉴别力。在这个装置中，编码板的像投影在相机的屏上。编码字每隔一段时间闪现，占这个周期的10%~30%，而闪现的时间间隔与分配给每个字母的时间相对应。在闪现间隔的时间内，所有的编码位置被逐一扫描，那些高于一定强度水平的点被传送到一个存储元件上，例如磁心存储器。但是直到一个字中全部的点都出现，这个记录才会被清除。如果点的总数被记录下来，那么这个编码字就会被传送到计算机中。

全息照相方法的高辨别能力来自其角度的高分辨率。例如假定N=35，M=30，则$M{\cdot}N$=1050。提交给读出器的字符所对应的4个忆迹的组接收到光的1/30，其中约1/35会发生衍射，因此全部的光约占总数的$1/10^3$。（当然，没有将白底黑字的记录计算在内，因为其非衍射光的量级接近于零。）也就是说，在适当的条件下，当忆迹从字符中和从编码字中接收到的光近乎相等时，衍射光的1/4将进入编码字的重建中去。此外，一半出现在物体中，另外的1/4进入编码字的“孪生”像中，然而，由于这部分与字符交互调制，并且对于识别没有帮助，所以被消除了。但是有用的1/4集中在极小的立体角中。例如，如果4个或6个同样的忆迹彼此之间的距离间隔约为25毫米，那么与一个编码点相对应的光将大部分集中在量级为10^{-8}的立体角内。也就是说，占总量$1/10^4$的光分布在10个编码点上，这意味着一个编码点上出现的光为$1/10^5$，并且集中在一个可能为布满全部编码的10^{-6}的立体角中，汇集度为10。此外，这种估计有些悲观，因为它忽略了在同一层中略有差异的变形忆迹对字符的证实。

总之，我们完全有理由相信，单个全息图可以在所有的数字和字母表中的字母之间进行辨别，其中每个有30个变形。

（沈乃澂 翻译；熊秉衡 审稿）

A Dense Packing of Hard Spheres with Five-fold Symmetry

B. G. Bagley

Editor's Note

Crystals are forbidden from having five-fold symmetry on geometric grounds: it is impossible to pack pentagons without gaps. The discovery in 1984 of a metal alloy with apparent ten-fold symmetry seemed to challenge that idea, but this so-called quasicrystal proved to lack true crystallinity. This paper from B. G. Bagley describes, two decades earlier, another way to create five-fold symmetry from a dense, infinitely extended packing of spheres. The packing is not periodic in three dimensions, however, but has a definite centre. The following year, an example of Bagley's scheme was reported for virus particles. Bagley also cites five-fold-symmetric clusters proposed by Desmond Bernal to exist in simple liquids, which were later invoked as possible nuclei of incipient quasicrystals.

SUPPOSE a plane of hard spheres is constructed such that the spheres form concentric pentagons with an odd number of balls per pentagon side. A second plane of hard spheres is now constructed such that the spheres form concentric pentagons with an even number of spheres per pentagon side. If this second plane is placed in intimate contact with the first, with their five-fold axes coincident, there results a layer which, within the plane of the layer, can be continuously packed to infinity (Fig. 1). Identical layers can then be stacked one on another, with their five-fold axes coincident, to give an infinite packing along the five-fold axis. An infinite structure can thus be constructed the nucleus of which is a pentagonal dipyramid of seven spheres.

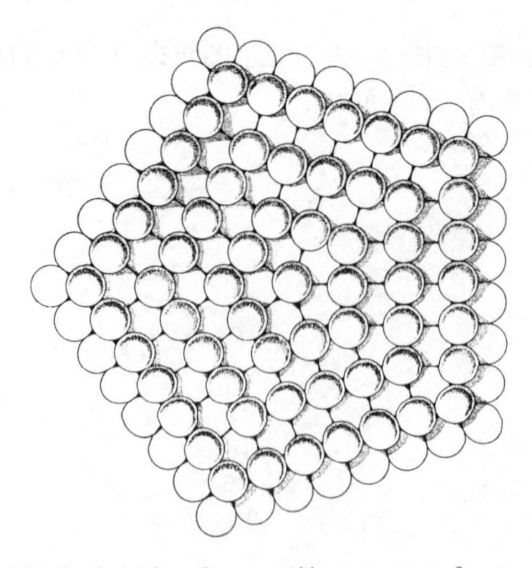

Fig. 1. A layer of hard spheres based on a packing sequence of concentric pentagons.

具有五重对称性的硬球的密堆积

巴格利

编者按

根据几何学原理，晶体不能具有五重对称性：做五边形堆积而不产生空隙是不可能的。1984 年人们发现一种金属合金具有明显的十重对称性，这一发现似乎是对上述观点的挑战，但是后来发现这个所谓的准晶体缺乏真正的晶体性质。巴格利的这篇文章发表于此前 20 年，文中他介绍了另外一种利用无限外延密堆积小球而构建出五重对称结构的方法。然而，这种堆积在三维空间中不具有周期性，但是却具有一个明确的中心。第二年，人们在病毒粒子中发现了一个巴格利方案的实例。巴格利还引用了德斯蒙德·贝尔纳提出的、存在于简单液体中的五重对称性团簇结构，而后来这被认为是初期准晶体的晶核。

假定如此构建一个硬球的平面，使若干硬球形成许多同心五边形，且每个五边形边上的硬球数为奇数。而第二个硬球的平面虽也由硬球构成同心五边形，但每个五边形边上的硬球数为偶数。如果将第二个硬球平面与第一个硬球平面紧密接触，且使它们的五重轴一致，这样就形成了一个在层的平面内可以无限连续堆积的硬球层 (如图 1 所示)，那么相同的这些层在五重轴一致的方向上可以连续地堆积，这样就可以沿着五重轴的方向形成一个无限堆积。一个无限堆积的结构也因此被构建出来，这个结构的核是由 7 个球构成的五边形双棱锥体。

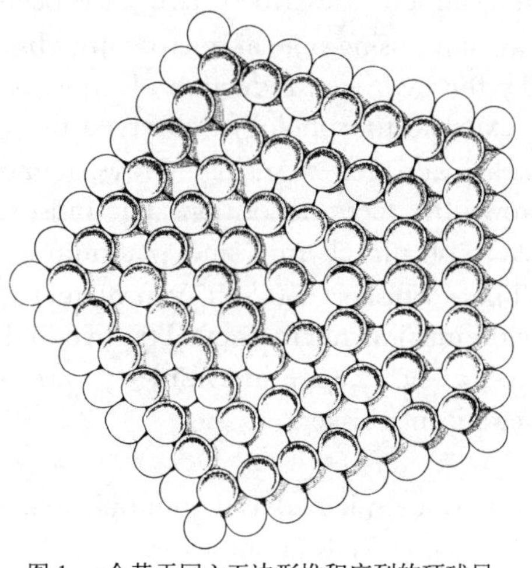

图 1. 一个基于同心五边形堆积序列的硬球层

Following the foregoing packing sequence with polygons other than the pentagon results in other, well-known structures. The same sequence with squares yields cubic close packing[1] and, with hexagons, primitive hexagonal. A difficulty arises when attempts are made to apply the exact sequence to triangles, because concentric triangles with an even (or odd) number of spheres per side cannot be made coplanar. It is important that, of these polygons, only the pentagon cannot form a regular tessellation and therefore, although it can be packed to infinity, it has a unique axis, the single five-fold rotation axis.

An alternative way of generating the same pentagonal structure is as follows: Construct n ($n = 1, 2, 3, \cdots \infty$) pentagonal pyramidal shells of hard spheres such that each face is an equilateral triangle of side length n (spheres). If shell 1 is placed in the cavity of shell 2 there results a pentagonal dipyramid of seven atoms. Likewise when shell 3 is placed on the structure there results a pentagonal dipyramid of twenty-three spheres. In fact, as each subsequent shell is placed on the growing structure there always results a pentagonal dipyramid bounded by close-packed planes, each face of which is an equilateral triangle with n (shell number) spheres to a side. This pentagonal dipyramid consists of five distorted tetrahedral the edges parallel to the five-fold axis being expanded by 5.15 percent. Within each tetrahedron the structure is body-centred orthorhombic with cell dimensions chosen such that the pentagonal dipyramid faces will be close packed and two adjacent tetrahedra will be joined by a coincidence boundary. These conditions yield a body-centred orthorhombic cell with dimensions (diameter of sphere=1.000), a=1.000, $b=\cot 36^{\circ}=1.3764$, $c=(2^2-\csc^2 36^{\circ})^{1/2}=1.0515$. Thus this pentagonal structure has a density independent of position of 0.72357. This density is slightly lower than that for close packing (0.74048), but higher than body-centred cubic (0.68017) or icosahedral shell packing (0.68818) (ref. 2). The co-ordination is 10 at a distance of 1.000 and 2 at a distance of 1.052. This structure is an example of G_3^1 type symmetry, that is, a one-dimensionally periodic group in three dimensions, and its symmetry group is $5mP2ml$ (Niggli's[3] nomenclature).

Structures which have the symmetry described here have been observed experimentally. Gedwill, Altstetter and Wayman[4], using optical microscopy, observed five-fold symmetry in cobalt crystals produced by the hydrogen reduction of cobaltous bromide. Wentorf[5], also using optical microscopy (external morphology), observed five-fold symmetry in synthetic diamonds. Ogburn, Paretzkin and Peiser[6], using X-rays, found pentagonal symmetry in copper [110] dendrites grown by electrodeposition. The most striking examples, however, are the sub-micron whiskers of nickel, iron and platinum grown from the vapour by Melmed and Hayward[7]. These whiskers, 50-200 Å in diameter, had a five-fold rotational symmetry observed by field emission microscopy. The five-fold symmetry was found not to be limited to the surface, as no change in symmetry was observed in the continuous reduction in length of several iron whiskers.

In all these cases the structure was explained as a quintuple twin ((111) twinning plane) with five face-centred cubic individual crystals about a common [110] axis, the 7°20′ difference between 5×70°32′ and 360° being made up with lattice strain or imperfections. It is

接着如果将上述五边形的堆积序列换成其他多边形堆积序列，那么我们就可以得到另外一些大家所熟知的结构。例如，正方形按相同序列堆积便可得到立方密堆积[1]，六边形按相同序列堆积则可得到简单的六方密堆积。但是，人们在试图做一个精确的三角形序列时出现了困难，因为在共面上不能形成每边具有偶数（或奇数）个球的同心的三角形。重要的是，在这些多边形中，只有五边形不能形成规则的镶嵌，因此，尽管它能无限堆积，却只具有一个特殊的轴，即唯一的五重旋转轴。

产生相同的五边形结构的另一种方法如下：构建 n（$n = 1, 2, 3, \cdots \infty$）的五边形棱锥的硬球壳层，使它的每个面都是一个边长为 n（球）的等边三角形。如果将壳层 1 置于壳层 2 的腔体内，结果就形成了一个具有 7 个原子的五边形双棱锥体。同样，将壳层 3 置于上述结构上时，它就形成了一个 23 个球的五边形双棱锥体。实际上，当每个次级壳层置于这种生长的结构上，总可形成一个以密堆积平面为界面的五边形双棱锥体，其中每个面都是边长为 n 个（壳层数）球的等边三角形。这种五边形双棱锥体是由五个畸变的四面体组成的，它的边平行于向外扩展 5.15% 的五重轴。每个四面体内都是体心正交晶结构，具有的单胞尺度能使五边形双棱锥体的面可以呈密堆积，且两个相邻的四面体通过重合边界连接在一起。这些条件所产生的体心正交单胞的尺度是（球的直径 =1.000）：a=1.000、b=cot36°=1.3764、$c=(2^2-\csc^2 36°)^{1/2}$=1.0515。因此，这个五边形结构具有的密度值为 0.72357，它与位置无关。这个密度比密堆积密度（0.74048）稍低，但高于体心立方堆积（0.68017），或二十面体的壳层堆积（0.68818）（参考文献 2）。在距离为 1.000 时，配位数为 10，在距离为 1.052 时，配位数就为 2。这种结构是 G_3^1 型对称性的一个例子，也就是说，在三维中的一维的周期群，其对称群为 *5mP2ml*（尼格利[3]的命名法）。

本文描述的对称性结构已在实验中观测到。格德威尔、阿尔特施泰特和韦曼[4]用光学显微镜观测到了由氢还原溴化钴所产生的钴晶体的五重对称性。温托夫[5]也用光学显微镜（外形貌学）观测到了合成金刚石的五重对称性。奥格本、帕雷茨金和派泽[6]用 X 射线发现了通过电沉积生长的铜 [110] 枝晶的五角对称性。然而，最显著的例子是梅尔梅德和海沃德[7]利用蒸汽生长出镍、铁和铂的亚微米晶须。这些晶须的直径为 50 埃 ~200 埃，通过场发射显微镜观测到其具有五重旋转对称性。这种五重对称性不只限于表面，因为在一些铁晶须的连续剥离的过程中，并没有观测到对称性的变化。

在所有这些情况下，该结构都解释为五重孪晶（(111) 孪晶平面），即在共同的 [110] 轴的周围有 5 个面心立方的晶体，5×70°32′与 360°之间的 7°20′之差是点阵应变或欠完美性引起的。然而，不同的是孪晶机制未必能产生梅尔梅德和海沃德[7]

unlikely, however, that a twinning mechanism could generate a structure having the small size (50–200 Å) and atomic perfection (at the five-fold axis) of Melmed and Hayward's[7] whiskers. On the other hand, it appears that the formation of a pentagonal dipyramid nucleus and its subsequent growth is a more probable and simpler mechanism for the formation of this structure. Furthermore, if a twinning mechanism were responsible for the five-fold symmetry one would expect [110] to be an observed whisker orientation in normal, non-pentagonal, whiskers. This is indeed the case for nickel and platinum, but the observed orientation for face-centred cubic iron is [100] (ref. 8).

It is also to be noted that the pentagonal nucleus for the structure described here has the same form as one of the configurations which has been proposed as an important element of liquid structure by Bernal[9,10]. It is evident from the foregoing discussion that crystallization can occur by the growth of such a configuration.

I thank Profs. F. C. Frank, C. S. Smith, and D. Turnbull for their advice, and the Xerox Corporation for a fellowship. This work was supported in part by the Office of Naval Research under contract *Nonr* 1866 (50), and by the Division of Engineering and Applied Physics, Harvard University.

(**208**, 674-675; 1965)

B. G. Bagley: Division of Engineering and Applied Physics, Harvard University, Cambridge, Massachusetts.

References:

1. Coxeter, H. S. M., *Illinois J. Math.*, 2, 746 (1958).
2. Mackay, A. L., *Acta Cryst.*, 15, 916 (1962).
3. Niggli, A., *Zeit. Krist.*, 111, 288 (1959).
4. Gedwill, M. A., Altstetter, C. J., and Wayman, C. M., *J. App. Phys.*, 35, 2266 (1964).
5. Wentorf, R. H., jun., in *The Art and Science of Growing Crystals*, edit. by Gilman, J. J., 192 (Wiley, New York, 1963).
6. Ogburn, F., Paretzkin, B., and Peiser, H. S., *Acta Cryst.*, 17, 774 (1964).
7. Melmed, A. J., and Hayward, D. O., *J. Chem. Phys.*, 31, 545 (1959).
8. Melmed, A. J., and Gomer, R., *J. Chem. Phys.*, 34, 1802 (1961).
9. Bernal, J. D., *Nature*, 183, 141 (1959).
10. Bernal, J. D., *Nature*, 185, 68 (1960).

晶须的小尺寸（50埃～200埃）结构以及在五重轴上的原子完美性。另一方面，五边形双棱锥体晶核的形成以及其随后的生长，也许是这种结构形成的更可能且更简单的机制。而且，如果孪晶机制是造成五重对称性的原因，可以预期在通常的、非五边形的晶须中法向取向应为 [110]。这确实符合镍和铂的情况，但在面心立方的铁中观测到的方向却是 [100]（参考文献 8）。

还应注意到，本文所述结构的五边形核与由贝尔纳[9,10]提出的液体结构的重要组态之一具有相同的形式。从上述讨论中我们可以明显得出，这类组态在生长的过程中能够产生结晶。

我要感谢弗兰克教授、史密斯教授和特恩布尔教授提供的意见，同时也感谢施乐公司提供的科研经费。这项研究的部分工作得到了海军研究办公室（合同为 *Nonr* 1866（50））以及哈佛大学工程与应用物理学院的支持。

（沈乃澂 翻译；赵见高 审稿）

Reconstruction of Phase Objects by Holography

D. Gabor *et al.*

Editor's Note

In 1948, Hungarian physicist Dennis Gabor reported in *Nature* a kind of three-dimensional imaging process which he called holography. He developed the idea initially in connection with the optics of electron beams, although he later generalized it to light. Holography relies on the interaction of the phases of reflected rays, but as Gabor and his coworkers say here, that phase information might appear to be lost in photographic images which merely record brightness of the beams. They show, however, a way to reconstruct both the phases and the amplitudes of the beams using optical holography, and thus to obtain all the available information about the object from which the rays are scattered.

THE principle of wave-front-reconstruction imaging, first described by one of us in 1948 (refs.1–3), has recently resulted in spectacular advances, notably in the form of three-dimensional "lensless" photography and imaging, with both macroscopic and microscopic objects[3-8]. Excellent images have been obtained in a number of variations of the basic method, notably when the objects used were "half-tone" intensity objects, or transparencies, rather than "phase" objects.

"Phase" objects may be of a primary interest in a number of holographic applications, notably in microscopy, and in several other applications, where phase rather than amplitude variations in the light field may be predominantly characteristic of the physical phenomena under investigation, for example, in work with wind tunnels and in acoustical applications.

As holograms are recorded on photographic emulsions which register only intensities, not phases, one might easily believe that a hologram is not a full substitute for a real object. Indeed, the total wave-front which issues from a hologram in the reconstruction cannot be the same as the original wave-front emitted by the object, because one half of the information is missing. In order to obtain the total information one requires two holograms, which are in sine-cosine relation to one another. Two such "complementary" or "quadrature" holograms have been used in the "total reconstruction microscope" by one of us[9]. Adding up the wave-fronts issuing from two such holograms, one obtains in the reconstruction the original wave-front, and nothing else, except a uniform background.

However, somewhat paradoxically, the original wave-front is also contained in the modified wave-front diffracted in the reconstruction by a single hologram. The incompleteness

用全息术实现的相位物体的重建

盖伯等

编者按

匈牙利物理学家丹尼斯·盖伯于1948年在《自然》上发表了一篇关于一种三维成像过程的报道，他把这种过程称作全息术。他提出的这个想法最初与电子束光学有关，但是后来他将其推广至光学。全息术依赖于反射光线相位的相互作用，但是正如盖伯及其同事们在本文中所说，在利用照相术得到的影像中相位信息似乎会丢失，这些影像记录到的只是光束的亮度。然而他们展示了一种能够通过光学全息术同时重建光束相位和振幅的方法，并通过这种方法获得了有关散射光线的物体的全部有效信息。

1948年，笔者之一首次描述了波前重建成像的原理（参考文献1~3），最近这个原理产生了惊人的进展，在宏观和微观物体的三维“无透镜”照相术和成像中尤其明显[3-8]。通过对基本方法做许多改进，人们已经获得了极好的图像，当所用的物体不是“相位”物体而是“半色调”强度物体或透明物体时，效果更为显著。

在许多全息的应用中，尤其是在显微术以及其他几种应用中，“相位”物体可能是首选，其中所研究物理现象的主要特征可能是光场的相位变化，而非振幅变化，例如在风道的研究和声学的应用中。

由于全息图是用照相感光乳胶记录的，且该乳胶记录的只是强度，而非相位，因此人们或许很容易相信全息图并不是一个实物的完全替代。实际上，由重建得到的全息图产生的总的波前并不等同于由物体发射的原始的波前，因为有一半的信息丢失了。为了获得全部信息，我们需要两个彼此之间是正弦–余弦关系的全息图。这样两个“互补的”或“正交的”全息图已被笔者之一[9]用于“完全重建显微术”中。把这样的两个全息图产生的波前相加，我们就得到了重建过程中的原波前，除此以外仅有均匀的背景。

然而，有些自相矛盾的是，原始的波前也包含在修正波前中（修正波前是单个全息图在重建中通过衍射得到的）。实际上这种不完整性本身表明，波前中混合了一

shows itself in the fact that this wave-front is mixed up with an additional wave, which appears to issue from a "conjugate object". But the two partial wave-fronts can be separated by various methods, the simplest of which is using a skew reference beam at an angle to the plate, in the taking of the hologram. The information-theoretical paradox that an incomplete record contains the full information in a retrievable form is explained by the fact that there is a loss of one half of the definition. This, however, is as good as unnoticeable in almost all present-day applications of holography.

Consequently, since one of the two (or more) waves diffracted by the hologram in the reconstruction process contains information on both the amplitude and the phase distribution in the object, both the phase and the amplitude information may be extracted from the reconstructed wave-front, for example, by suitable "filtering" of the aerial reconstructed images, or diffraction patterns, before the final image is recorded on a photographic film. In essence, the image-forming wave-fronts reconstructed from holograms are indistinguishable from the wave-front which would be obtained from an ideal lens or mirror looking directly at the object, when it is possible to form an image in the ordinary "one-step" imaging, for example, in microscopy. It has therefore been clear to us for some time that we may display the phase in the holographically reconstructed images of "phase" objects, when necessary, with the aid of any one of the several well-known methods[10,11] (for example, phase contrast, interferometry, Foucault or Schlieren methods) used to display the phase variations in the form of amplitude variations, as used in microscopy and other phase measuring applications in optics.

Because of the great present interest in holography, and because some of our recent advances seem to indicate a good likelihood that high resolutions may indeed be attainable in microscopy at very short wave-lengths (for example, X-rays), it may be of interest to demonstrate that phase-preserving imaging and "phase-contrast" image reconstruction may indeed be readily achieved in holography, using single holograms.

As one example of the "phase-preserving" reconstruction of the image of a phase object, we have used the arrangement shown in Fig. 1. The phase object used is shown (barely visible) in Fig. 2*a*, and an enlarged transmission two-beam interferogram of the object is shown in Fig. 2*b*. The phase object was formed by photographing the word "phase" and the letter "φ " on a Kodak 649*F* plate, and by bleaching the emulsion (using Kodak chromium intensifier as the bleacher). It is well known that photographic emulsions will shrink with the density of the exposure. (Typical emulsion shrinkage with exposure factors, 1, 2, 3, 4, is shown in Fig. 2*c*.) The phase object was recorded by projecting an image on to the plate deliberately slightly out of focus, in order to avoid steep gradients at the edges of the letters. (The amount of shrinkage shown in Fig. 2*b* was achieved in a 20-sec exposure, with a 75 W bulb, at *f*/11, in 1:1 imaging in the enlarger, and suitable bleaching.)

个似乎由“共轭物体”产生的附加波。但是这两个部分波前可用多种方法分离，其中最简单的方法是在记录全息图的过程中，使用与感光板呈一定角度的偏斜参考光束。一个不完整的记录包含并可以恢复得到完整的信息，这种信息－理论的矛盾可用精确度损失了一半的事实来解释。然而，在目前所有全息图的应用中这一点几乎没有引起人们的注意。

因此，既然在重建过程中通过全息图衍射得到的两个（或多个）波中，有一个包含了物体中振幅和相位分布的信息，那么相位和振幅的信息或许都可从重建的波前中提取出，例如，在感光胶片上记录下最终图像之前，对存在于空中的重建图像或衍射图样进行适当的“滤波”。本质上，从全息图重建形成的图像，其波前与从一个直接对准物体的理想透镜或镜子得到的波前几乎完全相同，在这种情况下像是通过通常的“一步”成像形成的，例如显微镜中的成像。因此，不久之前我们已经清楚，我们可以在“相位”物体的全息重建图像中显示相位，必要时，可从人们熟知的方法[10,11]（例如相衬法、干涉测量法、傅科法或纹影法）中任意选取一种以振幅变化的形式来呈现相位变化，就像在显微术中和光学中其他相位测量应用中所使用的方法一样。

由于目前人们对全息术有很大的兴趣，并且我们最新取得的一些进展似乎表明在波长很短 (例如 X 射线) 的情况下可使显微镜得到非常高的分辨率，因此似乎有必要给出以下说明：利用全息术，使用单个全息图或许就可轻易实现保持相位的成像过程及“相衬”图像重建。

我们使用图 1 所示的装置，作为一个相位物体图像的“相位保持”重建的例子。所用的相位物体如图 2*a* 所示 (勉强可见)，图 2*b* 显示的是物体放大的透射型双光束干涉图。将单词“phase”和字母“φ”拍摄在一张柯达 649*F* 干板上，并漂白感光乳胶（使用柯达铬增强剂作为漂白剂）从而制得了相位物体。众所周知，感光乳胶将随着曝光强度的增加而收缩。（图 2*c* 为在曝光因子为 1、2、3、4 时，典型的感光乳胶的收缩。）为了避免字母边缘斜度过大，相位物体是通过将像投影在故意稍稍偏离焦点的胶片上来记录的。（图 2*b* 中所示的收缩量是用 75 瓦的灯泡在 *f*/11 和在 20 秒的曝光下以 1:1 在放大机内成像并作适当漂白而获得的。）

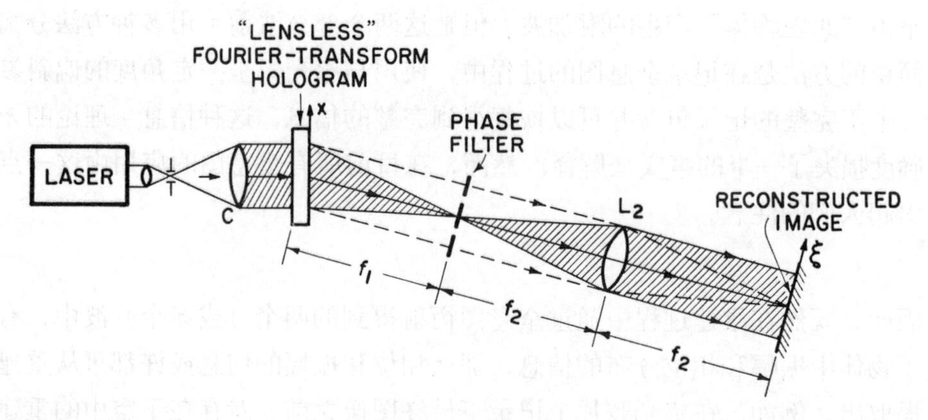

Fig. 1. Modified Fourier-transform holographic image-reconstruction arrangement, permitting "phase-contrast" detection and imaging of phase objects. The distance f_1 is equal to the distance of the object, respectively point-reference, from the hologram in the "lensless" recording of the Fourier-transform hologram[12,13]; "Conventional" Fourier-transform reconstruction of the images from the Fourier-transform holograms is obtained in the absence of the "phase filter". (The geometrical magnification obtained in Fourier-transform holographic imaging is equal to the ratio f_2/f_1. An additional magnification factor equal to λ_2/λ_1 is obtained, when the reconstructing wave-length λ_2 exceeds the recording wave-length λ_1. In this work, f_1=415 mm, f_2=600 mm, $\lambda_2=\lambda_1$=6,328 Å)

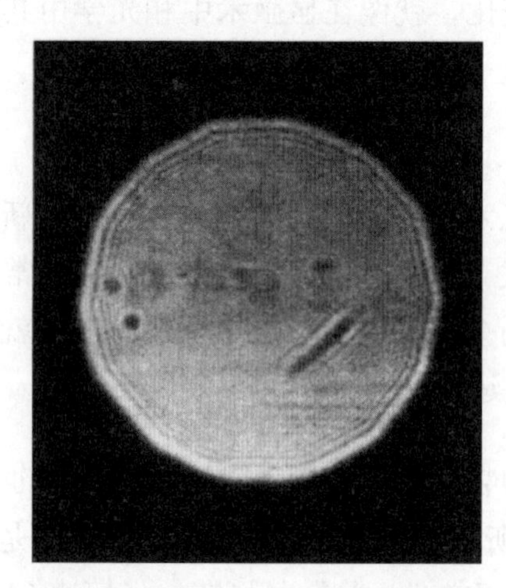

Fig. 2*a*. Direct (not holographic) image of phase object used in this work, showing degree to which a "pure" phase object was obtained by suitable bleaching of a Kodak 649*F* emulsion (see text, and Fig. 2*b*). The object is the word "phase" and the letter "φ" (the word phase being 20 mm long). The slight contrast detectable is due to some slight defocusing, and some residual absorption in the plate

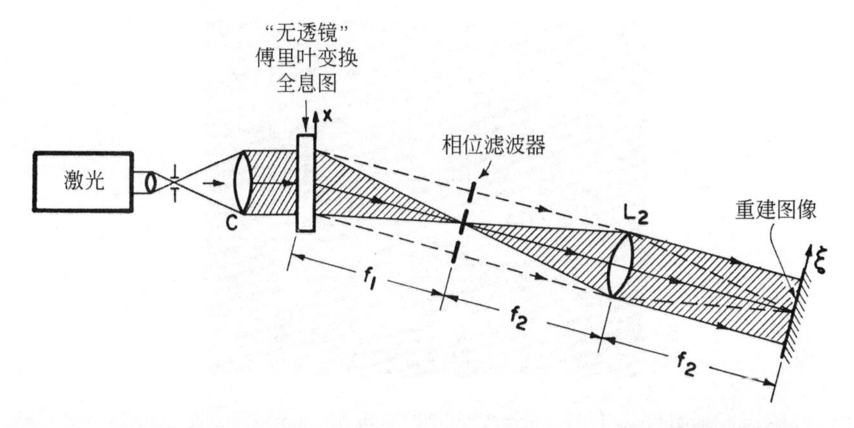

图 1. 改进型的傅里叶变换全息像重建装置，可以进行相位物体的“相衬”检测和相位物体的成像。在傅里叶变换全息图的“无透镜”记录中，距离 f_1 分别等于物体和点参考光源到全息图间的距离[12,13]。从傅里叶变换全息图得到的像的“传统的”傅里叶变换重建是在无“相位滤波器”的情况下得到的。(在傅里叶变换全息成像过程中获得的几何放大倍率等于 f_2/f_1。当重建波长 λ_2 超过记录波长 λ_1 时，得到的附加放大倍数因子等于 λ_2/λ_1。在本文中，f_1 = 415 毫米，f_2= 600 毫米，λ_2 =λ_1 = 6,328 埃。)

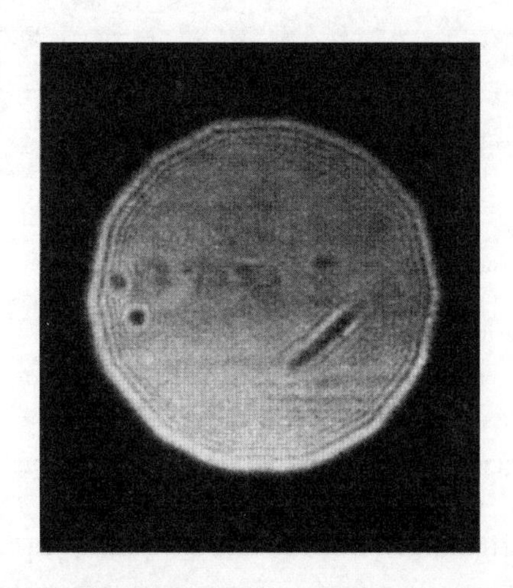

图 2*a*. 在本研究工作中使用的相位物体的直接 (不是全息的) 图像，这是用柯达 649*F* 感光乳胶 (见正文和图 2*b*) 进行适当漂白得到的“纯”相位物体的程度。该物体是单词“phase”和字母“φ” (单词“phase”长 20 毫米)。可以检测到的微弱衬度是由于轻微的散焦以及干板上一些剩余吸收而产生的。

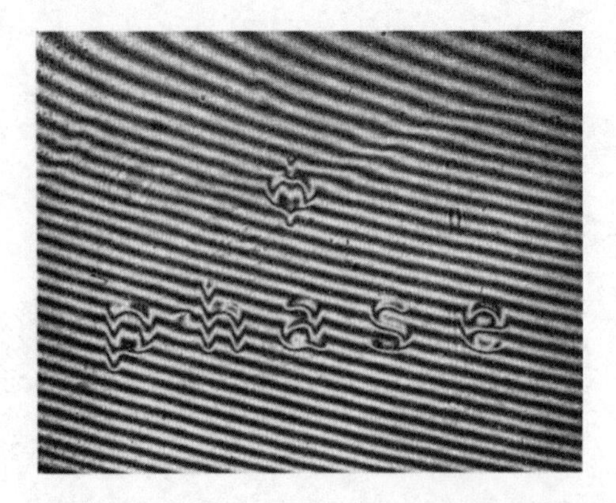

Fig. 2*b*. Two-beam single-pass interferogram (6,328 Å) of the phase object used in this work. The hologram of this phase object was recorded in the "lensless" Fourier-transform hologram recording arrangement according to ref. 12

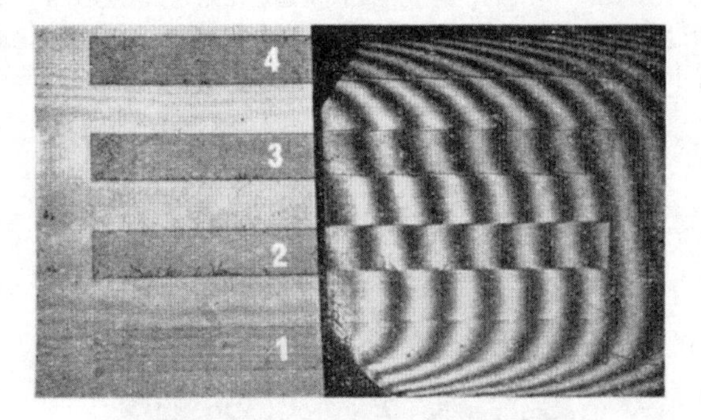

Fig. 2*c*. Two-beam single-pass interferogram (6,328 Å), illustrating the amount of emulsion shrinkage and corresponding phase variation achievable in a Kodak 649*F* emulsion with four different exposures (in ratios ×1, ×2, ×3, ×4), and suitable bleaching with Kodak chromium image intensifier, used to obtain almost pure "phase objects", with minimum residual absorption

The hologram of the phase object was recorded in the "lensless Fourier-transform" hologram recording arrangement, first described by one of us[12,13], in which the spherical waves originating from the various object points are made to interfere with a "single" spherical reference wave, originating from a source "point" in the mean plane next to the object. A reference wave of a radius f_1=415 mm was used in the recording. (We may note that the "lensless" recording of the hologram permits storage of the information about the phase distribution in the object without introducing any other optical elements between the object and the emulsion, thus avoiding any extraneous scattering, which might reduce the fidelity and sensitivity of the method.)

The images reconstructed from the "lensless Fourier-transform" hologram are shown in Figs. 3, 4 and 5. Fig. 3 shows a Fourier transform reconstruction, without filtering, obtained by simply projecting a plane monochromatic (6,328 Å) wave through the hologram, and by recording one of the side-band images in the focal pane of a f_2=600 mm lens. Unlike the reconstruction-imaging of intensity or amplitude objects, the reconstructed image of the phase object in Fig. 3 shows no amplitude contrast, because of the "pure" phase

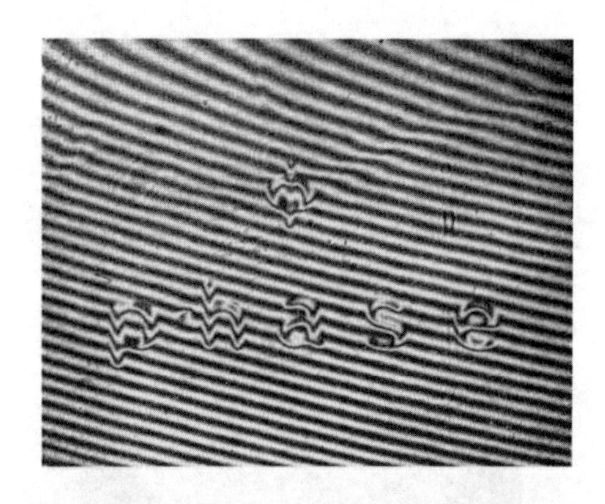

图 2*b*. 本研究工作中所使用的相位物体的双光束单通干涉图（6,328 埃）。这个相位物体的全息图是用参考文献 12 中记录的“无透镜”傅里叶变换全息图记录装置进行记录的。

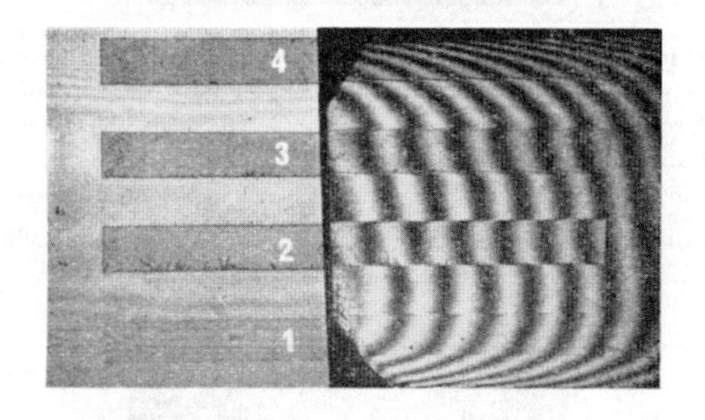

图 2*c*. 双光束单通干涉图（6,328 埃），它显示了感光乳胶的收缩程度以及在 4 种不同曝光量下（比例为 ×1、×2、×3、×4）用柯达 649*F* 感光乳胶得到的相应的相位变化，并用柯达铬像增强剂作了适当的漂白，以获得残余吸收量最小的近乎纯的“相位物体”。

相位物体的全息图是用“无透镜傅里叶变换”全息图记录装置记录的，笔者之一[12,13]首次对这种装置进行了描述，其中，令各个物点发出的球面波与一个“单个”球面参考波相干，而这个参考波来自与物体相邻的平均平面中的“点”源。记录中所用的参考波的半径 f_1 = 415 毫米。（我们可以注意到，在物体与感光乳胶之间不引入任何其他光学元件的情况下，全息图的“无透镜”记录可以对物体的相位分布信息进行存储，这避免了任何程度上的外来散射，而这种散射可能会降低所用方法的保真度和灵敏度。）

图 3、图 4 和图 5 是通过“无透镜傅里叶变换”全息图得到的重建图像。图 3 是一个傅里叶变换重建图像，其未经滤波，仅是用单色平面波 (6,328 埃) 透过全息图，并在一个 f_2=600 毫米的透镜的焦平面上记录一个边带像而得。与强度或振幅的重建图像不同，图 3 中相位物体的重建图像并没有表现出振幅衬度，这是由于物体具有

nature of the object.

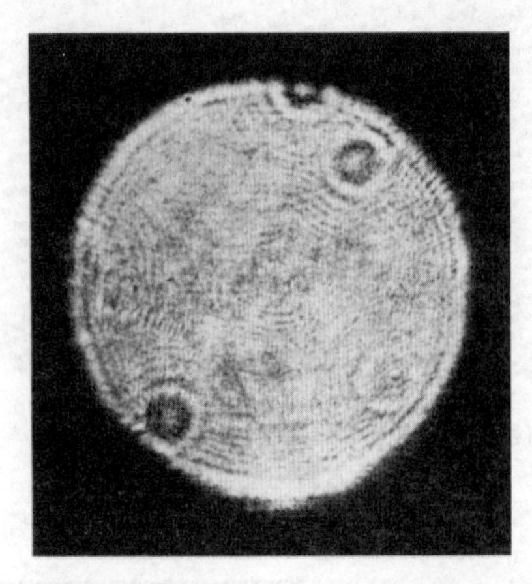

Fig. 3. Reconstructed image, obtained by Fourier-transform reconstruction in the arrangement of Fig. 1, without the use of any phase-contrast enhancement. (The "lensless" Fourier-transform recording of the hologram of the object was obtained according to ref. 12.) The image shown here is characterized by an almost complete absence of any amplitude contrast in the phase-portions of the image (the various interference effects are spurious, and are caused mainly by imperfectly clean reconstructing optics). It may be noted that excellent imaging is obtained under similar conditions when the original objects are amplitude or intensity objects[12,13], rather than pure phase objects

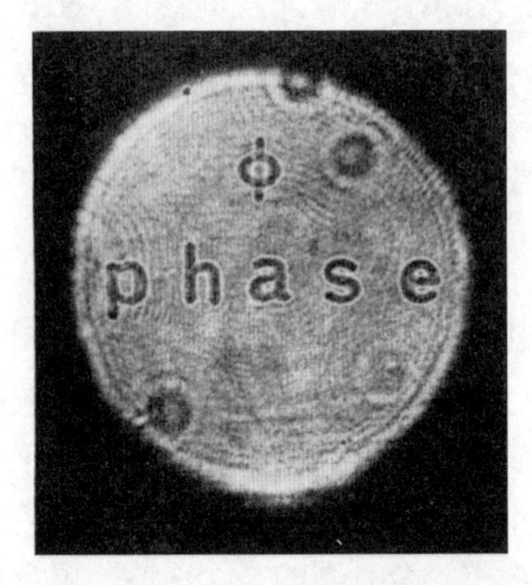

Fig. 4*a*. Reconstructed image of phase object, obtained with "phase-contrast" enhancement by defocusing (here towards the L_2 lens of Fig. 1 by $-f_2/4$, with f_2=600 mm). The length of the word in the object was 20 mm (in the image, it is 30 mm, because of the f_2/f_1 magnification (see Fig. 1))

“纯”相位性质。

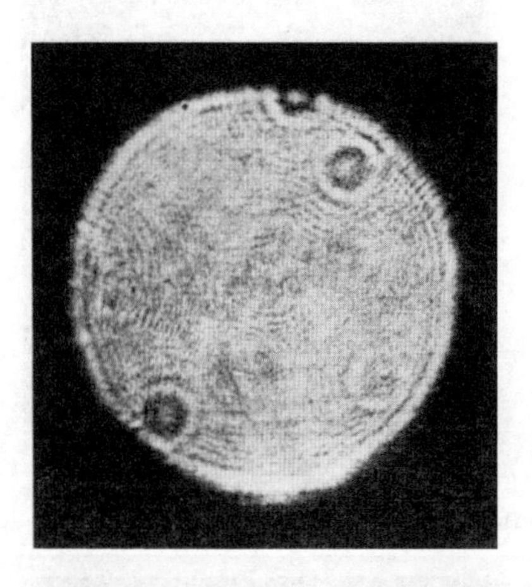

图 3. 用图 1 装置中的傅里叶变换重建得到的重建图像，没有做任何相衬增强处理。(根据参考文献 12 得到了物体的“无透镜”傅里叶变换全息图的记录。）这个像的特征是，在像的相位部分几乎没有任何振幅衬度。(各种各样的干涉效应都是干扰性的，这主要是由于重建光学元件不够干净而造成的。) 可以注意到，当原物体是振幅或强度物体 [12,13] 而不是纯相位物体时，在相似条件下可以获得极好的像。

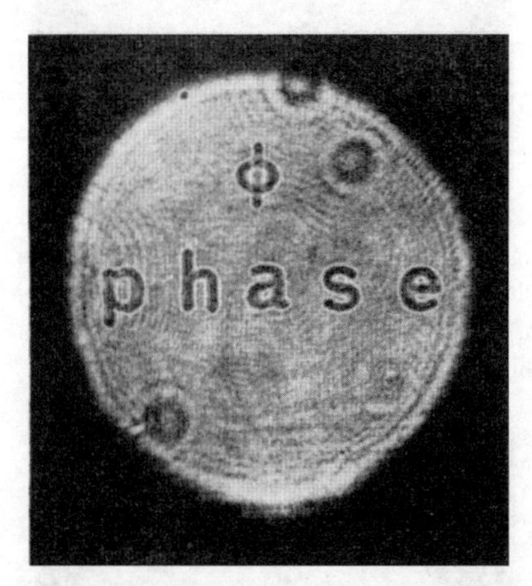

图 4*a*. 通过散焦增强“相衬”得到的相位物体的重建图像（与图 1 中透镜 L_2 的距离变化为 $-f_2/4$，其中 f_2= 600 毫米）。物体中单词的长度是 20 毫米（由于放大倍数为 f_2/f_1，因此像中长度是 30 毫米（见图 1））。

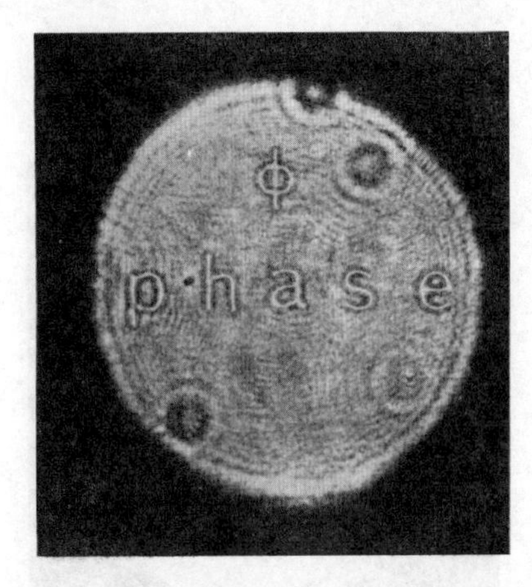

Fig. 4*b*. Reconstructed image of phase object, with "phase-contrast" enhancement, obtained by defocusing (here, out of focus, away from the lens L_2 by $+f_1/4$)

Fig. 4*c*. Reconstructed image of phase object, with phase-contrast enhancement obtained by using a phase-contrast filter (the corner of one of the rectangular phase strips, shown in Fig. 2*c*) in the arrangement of Fig. 1. (Here, the interference effects are spurious, and due to some imperfect cleanliness in the reconstruction optics)

Fig. 4*d*. Reconstructed image of phase object, with phase-contrast enhancement, obtained with a Foucault knife-edge used in the phase-filter plane of Fig. 1

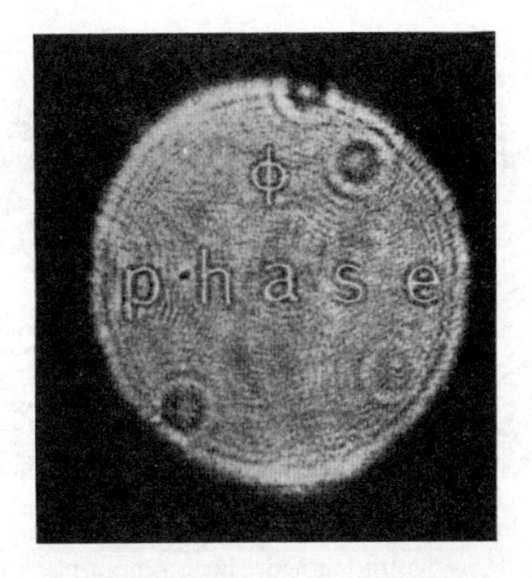

图 4*b*. 利用散焦增强“相衬”得到的相位物体的重建图像（此时散焦，与透镜 L_2 距离变化为 $+f_1/4$）。

图 4*c*. 在图 1 的装置中使用相衬滤波器（图 2*c* 中所示的其中一个矩形相位条的角）增强相衬而得到的相位物体的重建图像（这里的干涉效应是干扰性的，这是由重建光学器件不够干净造成的）。

图 4*d*. 在图 1 的相位滤波板中使用一个傅科刀口边沿增强相衬而得到的相位物体的重建图像。

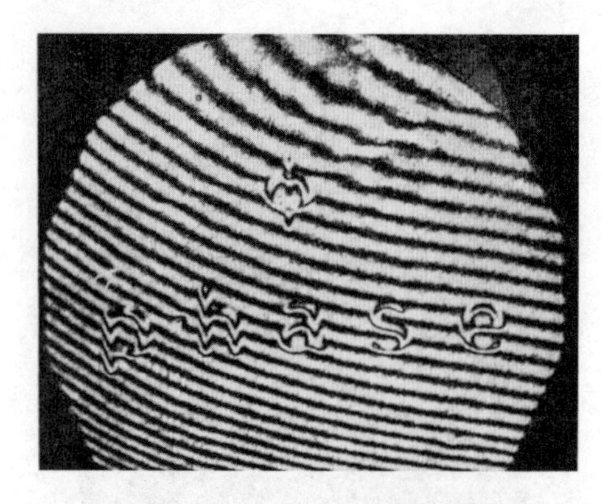

Fig. 5. "Hologram of the hologram." Two-beam single-pass interferogram (6,328 Å laser light) of the interference pattern between the reconstructed aerial image and a plane reference beam. Comparison of the image-interferogram shown here with the similarly obtained object-interferogram of Fig. 2*b* demonstrates that phase-distribution in the object is indeed preserved and completely reconstructed in Fourier-transform wavefront-reconstruction imaging, using a modified "phase-contrast" enhancing arrangement, such as that illustrated in Fig. 1. (We may note that phase preservation and phase-enhancing reconstruction apply to holograms recorded at one wave-length, and reconstructed in a second wave-length, for example, when λ_1 is in the X-ray domain and λ_2 in the visible-light laser domain)

Figs. 4 and 5 show well-contrasted images of the phase object, obtained by a number of the well-known phase filtering or phase-contrast methods, in which the phase variations are made visible in the form of amplitude (that is, intensity) variations in the image.

Figs. 4*a* and *b* show the reconstructed images, obtained by a "defocusing" phase-contrast enhancement, in a Fourier-transform reconstruction arrangement, as in the sharply focused Fig. 3, but now by recording the images together slightly ($\pm 1/4\, f_2$ at $f/24$) out of focus, with respect to the in-focus image of Fig. 3.

Fig. 4*c* shows an in-focus image, in which the phase-contrast enhancement was obtained in the filtering arrangement shown in Fig. 1, with the help of the corner of a phase-contrast filter (also recorded photographically, and bleached, similarly to the object recording already described here).

Fig. 4*d* shows an in-focus image, with "phase-contrast" enhancement obtained in the filtering arrangement of Fig. 1, with the help of a Foucault knife-edge filter (at right angles to the word "phase").

Fig. 5 shows a two-beam interferogram of the reconstructed aerial image, formed by interference of the aerial image with a plane wave (in a suitable beam-splitting arrangement): by comparing the interferogram of the phase object (Fig. 2*b*), the degree of phase preservation and of fidelity of "phase-preserving" reconstruction may be readily assessed.

It is clear from a comparison of the images and interferogram of the image, of Fig. 4

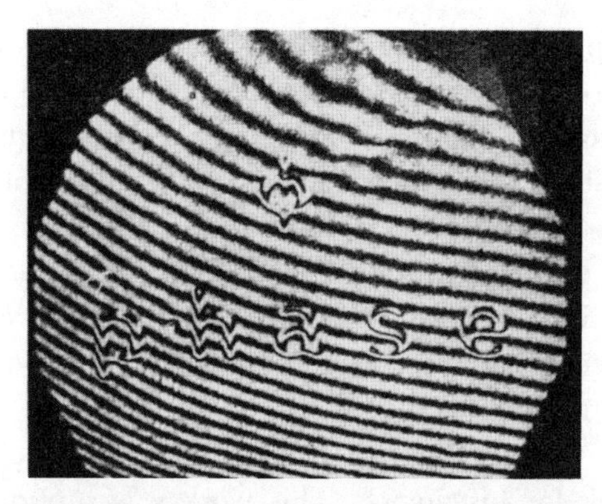

图 5.“全息图的全息图”。重建的空间像与平面参考光之间的干涉图样的双光束单通干涉图（激光波长为 6,328 埃）。将这里所示的像干涉图与通过相似方法得到的图 2*b* 中的物体干涉图进行比较，可以证明物体中的相位分布确实得到了保存，并在傅里叶变换波前重建中完全实现了重建。其中使用的是改进型的“相衬”增强装置，如图 1 所示。（我们可以注意到，相位保存和相位增强重建用作全息图记录时使用的是一个波长，而用于全息图重建时使用的是第二个波长。例如，当 λ_1 属于 X 射线波段时，λ_2 属于可见光激光波段。）

图 4 和图 5 是用人们熟知的相位滤波或相衬法得到的一些衬度很好的相位物体的像，其中相位变化是通过图像中的振幅（即强度）变化来体现的。

图 4*a* 和图 4*b* 是通过“散焦”相衬增强得到的重建图像，使用的傅里叶变换重建装置与经过严格对焦的图 3 所用的装置相同，但是相对于图 3 的准确对焦像，这里都是略微 ($f/24$ 处 $\pm 1/4f_2$) 偏离焦点对像进行记录的。

图 4*c* 是一个准确对焦的像，用图 1 所示的滤波装置中借助相衬滤波器的角来增强相衬（同样用照片做了记录，并进行了漂白，类似于前文描述的对物体所做的记录）。

图 4*d* 是在图 1 所示的滤波装置中借助傅科刀口滤波器（与单词“phase”垂直）增强“相衬”得到的一个准确对焦的像。

图 5 是一幅重建空间像的双光束干涉图，它由空间像与一个平面波（通过适当的光束分束装置）的干涉所形成：通过比较相位物体的干涉图（图 2*b*），可以很容易地对相位保持的程度和“相位保持”重建的保真度进行评估。

将图 4 和图 5 的像和像的干涉图与图 2 的相位物体和物体–干涉图对比，可以

and of Fig. 5, with the phase object and object-interferogram of Fig. 2, that the phase-distribution in the phase object was not only retrievably recorded in the hologram, but also that the phase in the image of the phase object can be readily displayed as an amplitude (respectively intensity) in the reconstructed image, with the aid of phase-contrast or other image-filtering methods, including interferometry of various types. We may note that there is some indication that holograms of phase objects, recorded with a 1/1 ratio of reference/diffracted field intensity (rather than the about 5/1 used here), appear to display some noticeable "phase-contrast" enhancement simply in the focus of the "conventional" Fourier-transform reconstruction arrangement.

We thank R. C. Restrick for his advice. This work was supported in part by the U.S. National Science Foundation and the U.S. Office of Naval Research.

(**208**, 1159-1162; 1965)

D. Gabor: F.R.S., Imperial College of Science and Technology, London.

G. W. Stroke, D. Brumm, A. Funkhouser and A. Labeyrid: University of Michigan, Ann Arbor, Michigan.

References:

1. Gabor, D., *Nature*, **161**, 777 (1948).
2. Gabor, D., *Proc. Roy. Soc.*, A, **197**, 475 (1949).
3. Gabor, D., *Proc. Phys. Soc.*, B, 244 (1951).
4. Leith, E. N., and Upatnieks, J., *J. Opt. Soc. Amer.*, **53**, 1377 (1963).
5. Leith, E. N., and Upatnieks, J., *J. Opt. Soc. Amer.*, **54**, 1295 (1964).
6. Leith, E. N., Upatnieks, J., and Haines, K. A., *J. Opt. Soc. Amer.*, **55**, 981 (1965).
7. Stroke, G. W., *Optics of Coherent and Non-Coherent Electromagnetic Radiations* (Univ. of Michigan, first ed., May 1964; second ed., March 1965).
8. Stroke, G. W., and Falconer, D. G., *Phys. Letters*, **13**, 306 (1964); **15**, 283 (1965); with Funkhouser, A., *Pyhys. Letters*, **16**, 272 (1965); with Restrick, R., Funkhouser, A., and Brumm, D., *Phys. Letters*, **18**, 274 (1965); *App. Phys. Letters*, 7,178 (1965).
9. Gabor, D., with Goss, W. P., British Patent No. 727, 893/1955, application date July 6, 1951.
10. Zernike, F., *Physica, Haag*, **1**, 43 (1934); *Physik. Z.*, **36**, 848 (1935); *Z. Techn. Physik.*, **16**, 454 (1935).
11. Francon, M., *Le Contraste de Phase en Optique et en Microscopie* (*Revue d'Oplique*, Paris, 1950).
12. Stroke, G. W., *App. Phys. Letters*, **6**, 201 (1965).
13. Stroke, G. W., Brumm, D., and Funkhouser, A., *J. Opt. Soc. Amer.*, 55, 1327 (1965).

清晰地看出，记录在全息图内相位物体的相位分布不仅可以获取，而且相位物体中像的相位也能容易地借助于相衬法或其他图像–滤波方法，包括各种类型的干涉测量方法，以重建图像的振幅（各自强度）来表示。我们可能会注意到一些迹象，就是以 1/1 的参考 / 衍射场强比（而不是本文中所用的大约 5/1 的场强比）记录的相位物体的全息图，似乎可简单地在“传统的”傅里叶变换重建装置的焦点处表现出某些明显的“相衬”增强。

我们感谢雷斯特里克提出的建议。本项研究工作部分由美国国家科学基金会及美国海军研究办公室支持。

（沈乃澂 翻译；熊秉衡 审稿）

Ball Lightning as an Optical Illusion

E. Argyle

Editor's Note

For centuries, eyewitness accounts have attested to an elusive atmospheric phenomenon known as ball lightning: the creation during thunder storms of spherical balls of luminous energy which float through the air, persist for several seconds before abruptly disappearing, and range in size from that of a small stone to several feet in diameter. The phenomenon has been controversial, because it is difficult to study scientifically. Several reported incidents in 1970 here led Edward Argyle to suggest that ball lightning is an optical illusion due to visual afterimages. Argyle's explanation remains one of many possibilities, and controversy over the causes or even the reality of ball lightning continues today.

DURING the past year there have been numerous publications on ball lightning[1-9], many attempting to account for the formation, properties and behaviour of lightning balls. None have questioned the reality of the phenomenon, in spite of the lack of progress toward an understanding of these baffling objects. Serious doubt about the existence of ball lightning was expressed by Humphreys[10] in 1936, and more recently by Schonland[11]. Both regarded the phenomenon as probably an optical illusion. Now that Altschuler *et al.*[8] have invoked nuclear reactions to account for the lightning ball it seems appropriate to re-examine the possibility of finding an explanation in the physiology of vision.

The phenomenon of visual afterimages in very complex, but is predominant effect—the negative afterimage—is well known. Of less frequent occurrence is the positive afterimage, which results from the observation of a light source which is very bright relative to the surround. In this communication I shall assume that a spherical region of ionized air and white hot particles is sometimes generated at the ground end of a lightning stroke by partial vaporization of Earth, vegetation or metal, and that the luminous intensity within the sphere is high enough to create a positive afterimage in the eye of an observer. In his comprehensive survey of reports on ball lightning Rayle[12] made the interesting discovery that "the number of persons reporting ball lightning observations is 44% of the number reporting observation of ordinary lightning impact points". This can be understood by assuming that about half of all strokes to ground generate a high luminosity ball at the impact point. Indeed, this sort of assumption is required whether ball lightning is real or illusory.

The behaviour and apparent properties of the positive afterimage are strikingly similar to those of ball lightning. Its shape will be the same as that of the exciting source, and it well commonly be described as a ball. The apparent size of the positive afterimage will be the same as that of the exciting source only if it is "projected" by the observer to the distance

球状闪电是一种视错觉

阿盖尔

编者按

几个世纪以来的目击证据已证明有一种被称为球状闪电的神秘大气现象是存在的：在遭遇雷暴天气时会形成一些漂浮于空气中的具有光能的球状物，短暂停留几秒钟然后突然消失，它们的大小从一块小石头尺寸到几英尺直径不等。因为很难通过科学方式来研究这一现象，所以人们对此一直争论不休。1970 年报道的几桩事件使爱德华·阿盖尔联想到球状闪电可能是由视觉余像引起的错觉。阿盖尔的解释只是多种可能性中的一种，人们对球状闪电的成因甚至球状闪电的真实性的争论直到现在也没有停止。

去年发表的关于球状闪电的文章[1-9]数不胜数，其中有很多文章在试着解释闪电球的形成过程、属性特征和活动特点。尽管人们在理解这些莫名其妙的现象方面并未取得进展，但却从来没有人质疑过这一现象的真实性。1936 年，汉弗莱斯[10]对球状闪电的存在提出了强烈的质疑，最近舍恩兰德[11]也表达了同样的观点。他们都认为这一现象可能是一种视错觉。既然阿尔特舒勒等人[8]都已经引入了核反应来解释这个闪电球，那么人们似乎值得重新审视在视觉生理学中寻找一种解释的可能性。

虽然视觉余像现象非常复杂，但它最主要的效应——负余像是大家所熟知的。正余像的出现频率略低，它是在我们观察一个比周围环境亮得多的光源时产生的。在这篇文章中，我将要假设大地的水蒸气、植物或金属有时能够被闪电靠地面的一端击中，从而形成了由电离气体以及高温粒子组成的球状区域，并且从这个球内发出的光的强度足够高，以至于能在观察者的眼中形成一个正余像。雷勒[12]在他的报告中全面考查了有关球状闪电的情况，他发现了一个有趣的现象："报告观察到球状闪电的人数是报告看到一般闪电击中某个点的人数的 44%"。这一现象可以通过假设有大约半数的击中地面的闪电在击中点处产生了高光度的球来解释。事实上，无论球状闪电是真实的还是虚幻的，这类假设都是必不可少的。

正余像效应的行为和表观特性与球状闪电非常相似。正余像的形状总是和激发源的形状保持一致，人们通常会用一个球来描述正余像的形状。只有在正余像被观察者"投射"到与他和激发源之间的距离一样远时，所看到的正余像才会和激发源

of the source. If the source is outdoors, for example, but the observer focuses on the window through which it was seen, the apparent size will be reduced. Thus the relatively small size of indoor lightning balls is accounted for. Rayle[12] pointed out that linear size and distance were one of the few pairs of parameters that showed a significant correlation, but he did not go on to infer the implied limited range of angular diameters.

The colour of positive afterimages depends more on the brightness of the source than on its colour. There is no conflict with the yellows, oranges and occasional other colours reported for ball lightning. The degree to which the size, shape, brightness and colour of the lightning ball remains constant throughout its lifetime of a few seconds is the despair of both the theoretician[8,13-16] who seeks a plausible physical mechanism for it, and the experimenter[3] who tries to reproduce it. On the other hand, the approximate constancy of these features is characteristic of a positive afterimage.

Being a cone effect, the positive afterimage will usually be formed near the centre of the retina, and the observer, wishing to examine the supposed object, will attempt the impossible task of centring it exactly. The result is a linear drift of the projected positive afterimage across the observer's visual background, as the eye muscles try to correct the centring error. In natural circumstances this feedback motion will be combined with varying amounts of voluntary adjustment as the observer strives to make sense out of the motion.

Passage through physical surfaces such as a glass or metal screen is possible for positive afterimages and is reported for lightning balls[3,4]. In neither case is the surface burned or damaged, nor is the size, shape, colour or brightness of the ball altered by the penetration.

Positive afterimages last 2–10 s, depending on circumstances, and most lightning balls are reported to have a duration in the same range[12]. Positive afterimages disappear rather suddenly, as do lightning balls. Positive afterimages generate no sound but the typical observer finds it easy to imagine "suitable" accompanying sounds, if he has any relevant preconceptions. The ease with which most people "hear" appropriate sounds while observing natural phenomena has been documented by Beals[17], who found that most of the inhabitants of northern Canada claimed to have heard the aurora borealis. The spectrum of sounds reported was persuasively narrow and differed only moderately from observer to observer. Apparently it is no more maladaptive to hear the rustling of the northern lights than it is to hear the implosive pop of a disappearing lightning ball. Curiously, the disappearance of indoor lightning balls is reported often to be violent but seldom to be damaging.

The psychological principle for imagining odour is the same as for sound but fewer individuals have strong odour expectations and in only a few cases are odours reported. An appropriate odour would be one the observer associates with electrical apparatus, if he has some familiarity with it; or with deeper metaphysical concomitants of danger and mystery, such as fire and brimstone, if his associations tend to the supernatural. Such are

的大小相同。举例来说，如果激发源在户外，而观测者把注意力集中在窗户上，通过窗户看外面的激发源，那么他看到的尺寸会小于实际的大小。这样就可以解释在室内观察到的闪电球为什么相对较小了。雷勒[12]指出线性尺寸和距离是能表现出显著相关性的为数不多的参数中的一对，但是他没有继续提到这意味着角直径存在极限范围。

正余像的颜色主要取决于激发源的亮度，而非激发源的颜色。在关于球状闪电的报告中提到过黄色、橙色，偶尔还出现过其他颜色，这些颜色都是合理的。在几秒钟的生命期内，闪电球的尺寸、形状、亮度以及颜色能在一定程度上保持稳定的现象使想要寻找一个合理的机制来解释球状闪电的理论家[8,13-16]和想要再现球状闪电现象的实验家[3]感到无计可施。而另一方面，这些特征的近似稳定不变又恰好是正余像效应所特有的。

正余像属于视椎细胞效应中的一种，它通常会在视网膜中心附近成像，而想要看清这一假想目标的观察者，会试图使其更精确地聚焦在视网膜的中心，尽管这是不可能做到的。如果观察者的眼部肌肉试图校正这一偏离中心的误差，就会导致正余像投影在观察者的视觉背景中产生线性漂移。在自然环境中，当观察者努力弄清这种反馈运动的意义时，反馈运动将会与不同程度的自发调整动作结合在一起。

正余像穿过诸如玻璃屏或金属屏这样的物质表面是有可能的，这已在闪电球的报道中得到了证实[3,4]。不论在哪种情况下，物质表面都不会被烧焦或破坏，而且余像的尺寸、形状、颜色或亮度也不会因为穿过过程而发生变化。

正余像可以持续 2 秒 ~10 秒，持续时间长短与周围环境有关，据报道大多数闪电球的持续时间具有同样的量级[12]。与闪电球一样，正余像的消失会很突然。正余像在产生时是没有声音的，但是假如普通的观察者事先对该现象的产生有所预见，那么他会发现想象出“合适”的伴随音并不是一件很难的事情。比尔斯[17]已经证明：在观察自然现象时，多数人能轻而易举地“听到”适当的声音，他发现大多数居住在加拿大北部的居民曾声称听到过北极光的声音。报道的声音频谱有很高的吻合度，在不同观察者所描述的声音之间并没有太大的差别。显然，如果能够听到一个正在消失的闪电球发出的爆裂声，那么听到北极光的沙沙声也就是很自然的事了。奇怪的是，报道称人们在室内看到的闪电球在消失时通常是很猛烈的，但是几乎不会造成破坏。

根据心理学原理，在听觉和嗅觉上的幻觉是一样的，但是很少有人对气味有强烈的预期，因此关于球状闪电有气味的报道只有为数不多的几个例子。一种适当的气味可以让观察者联想到他所熟悉的某种电气设备；或者更玄妙地和危险、神秘联系起来。比如，如果他的联想朝超自然的方向延伸，他会联想到地狱里的磨难。以

the reported odours of ball lightning[18].

In bright daylight positive afterimages are a rare occurrence. But in subdued light, as during a thunderstorm, especially if the observer is indoors, a positive afterimage is readily formed by any source as intense as a frosted light bulb. Considering that lightning is primarily an outdoor phenomenon it is remarkable how many lightning balls are seen indoors. This strange fact can be understood if positive afterimages are involved.

That the unsuspecting observer of a chance natural phenomenon can be misled by a positive afterimage is more than speculation. Two of the eyewitnesses to the passage of the Revelstoke bolide[19] reported independently that the meteor had landed on low ground in the middle distance, and had bounced several times before going out. It is more likely that these observers each followed a positive afterimage down to zero elevation after the bolide passed behind distant mountains. At ground level a conflict between positive afterimage drift and physical anticipations led to an unstable situation in which the positive afterimage oscillated briefly before disappearing.

There are a few reports which indicate the release of large amounts of energy from the lightning ball. In the most famous of these cases, described by Goodlet[20], water in a rain barrel was heated by a lightning ball. If ball lightning is an optical illusion it will be necessary to categorize this and similar reports as unreliable. To do so would not be unreasonable in view of the many observations on record.

Final resolution of the question of the nature of ball lightning will no doubt depend more on the outcome of further experimentation than on the collection of more observer reports. Fortunately, the question of the reality of ball lightning is amenable to laboratory investigation. I have found it easy to simulate the lightning ball by using light bulbs, strobe-flash lamps and photographic flash bulbs as intense sources for the generation of drifting positive afterimages. This qualitative work requires verification, however. It is hoped that a laboratory equipped for psychovisual studies will take up the problem and report on the degree to which descriptions of "artificial" ball lightning resemble those of the natural phenomenon that are recorded in the scientific literature.

(**230**, 179-180; 1971)

Edward Argyle: Dominion Radio Astrophysical Observatory, Penticton, British Columbia.

Received December 7; revised December 20, 1970.

References:

1. Lowke, J. J., Uman, M. A., and Lieberman, R. W., *J. Geophys. Res.*, 74, 6887 (1969).
2. Jennison, R. C., *Nature*, **224**, 895 (1969).

上谈到的是对球状闪电气味的报道[18]。

在白天日光非常强烈的情况下，正余像是十分罕见的。但是，当光线暗淡的时候，比如恰好遇到雷暴天气，尤其是在观察者处于室内的情况下，任何一个亮度类似于磨砂灯泡的光源都很容易使正余像得以形成。因为闪电主要是发生在户外的现象，所以在室内看到那么多闪电球就会显得不同寻常。如果用正余像来解释，这一奇怪现象就不言自明了。

一个确信自己看到了自然界中偶尔发生的现象的观察者，很有可能是被正余像欺骗了，这一论点绝不仅仅是一个推测。两名看到了雷夫尔斯托克火流星下落的目击者[19]分别报告称流星落在了不远处的低地上，并且在地面上弹了几下才熄灭。很可能这两位观察者的视线在火流星落到远处的山峰后面之后，就跟随着正余像降到了海拔为零的地方。在地面上，正余像漂移与物理预期之间的偏差导致了一个不稳定的状态，处于这种状态下的正余像在消失之前会发生短暂的振动。

有几份报告指出，大量的能量会从闪电球中释放出来。最著名的例子要数古德利特[20]所描述的雨桶中的水被闪电球烧热的现象。如果球状闪电是一种视错觉，那么就应当将这个报告以及一些类似的报告归为不实的报告。这样做并非不讲道理，因为有很多观察记录可以证明这一点。

毫无疑问，要最终解决球状闪电的本质问题还需要更多的实验结果来验证，而非更大量地收集观察者的报告。幸运的是，通过实验研究就可以揭开球状闪电的真相之谜。我发现很容易用灯泡、频闪闪光灯和摄影闪光灯作为强光源模拟闪电球以产生漂移的正余像。不过，这种定性研究还需要得到进一步的证实。我们希望有一个配备心理视觉研究设备的实验室能够接受这一课题；并且希望在科学文献中对“人造”球状闪电的报道量接近于对自然现象中球状闪电的报道量。

（孟洁 翻译；肖伟科 审稿）

3. Powell, J. R., and Finklestein, D., *Amer. Sci.*, **58**, 262 (1970).
4. Covington, A. E., *Nature*, **226**, 252 (1970).
5. Bromley, K. A., *Nature*, **226**, 253 (1970).
6. Felsher, M., *Nature*, **227**, 982 (1970).
7. Hill, E. L., *Amer. Sci.*, **58**, 479 (1970).
8. Altschuler, M. D., House, L. L., and Hildner, E., *Nature*, **228**, 545 (1970).
9. Zimmerman, P. D., *Nature*, **228**, 853 (1970).
10. Humphreys, W. J., *Proc. Amer. Phil. Soc.*, **76**, 613 (1936).
11. Schonland, B., *The Flight of Thunderbolts* (Clarendon Press, Oxford, 1964).
12. Rayle, W. D., *NASA Technical Note TN D-3188* (Washington, DC, 1966).
13. Bruce, C. E. R., *Nature*, **202**, 996 (1964).
14. Singer, S., *Nature*, **198**, 745 (1963).
15. Hill, E. L., *J. Geophys. Res.*, **65**, 1947 (1960).
16. Wooding, E. R., *Nature*, **199**, 272 (1963).
17. Beals, C. S., *J. Roy. Astron. Soc. Canad.*, **27**, 184 (1933).
18. Barry, J. D., *J. Atmos. Terr. Phys.*, **29**, 1095 (1967).
19. Folinsbee, R. E., Douglas, J. A. V., and Maxwell, J. A., *Geochim. Cosmochim. Acta*, **31**, 1625 (1967).
20. Goodlet, B. L., *J. Inst. Elec. Eng.*, **81**, 1 (1937).

Synchrotron Radiation as a Source for X-ray Diffraction

Synchrotron Radiation as a Source for X-ray Diffraction

G. Rosenbaum *et al.*

Editor's Note

The new DESY synchrotron accelerator in Hamburg, Germany, had recently become operational. Here biologist Gerd Rosenbaum of the Max Planck Institute for Medical Research and his colleagues demonstrate the potential usefulness of this accelerator as an intense source of X-rays for imaging in biology. Relativistic electrons travelling around on DESY's circular path naturally emitted X-rays in beams roughly 100 times brighter than any produced by then standard X-ray sources. The researchers show that this source could be used to produce images of biological specimens that are far clearer than those using the best conventional sources. Synchrotron X-ray sources have now become indispensable for imaging in biology.

WHEN an electron is accelerated it emits radiation. At the very high energies used in DESY, the emitted radiation is confined to a narrow cone about the instantaneous direction of motion of the electron. Thus the synchrotron radiates tangentially. Synchrotron radiation is polychromatic, with a peak in the X-ray region for an electron energy of 7.5 GeV (see ref. 1 for the original theoretical description and refs. 2–4 for experimental details).

The DESY synchrotron uses bursts of 50 pulses/s and each 10 ms pulse contains 6×10^{10} electrons (10 mA average beam current). The injection energy is relatively low and the electrons are accelerated up to 7.5 GeV in the 10 ms.

Most of the X-radiation is emitted during the last 3 ms of each pulse: little radiation is produced at the lower electron energies, and so the time averaged intensity at 1.5 Å is about 20% of the peak value.

We have evaluated the spectral luminance (that is, the power in photons per second radiated per unit area, solid angle, and wavelength interval) of both the synchrotron and a fine-focus rotating anode X-ray tube (see Table 2). The values are 2×10^{22} (time averaged) and 3×10^{20} photons s^{-1} $sterad^{-1}$ cm^{-2} $Å^{-1}$ respectively at 1.54 Å, showing clearly that the synchrotron is, relative to present X-ray tubes, a very bright source. The actual advantage to be gained in a diffraction experiment depends critically on the optical system necessary to focus and monochromate the radiation. Three types of focusing monochromators used in normal X-ray diffraction can be used: bent glass mirrors, quartz monochromators and graphite monochromators.

作为X射线衍射光源的同步辐射

罗森鲍姆等

编者按

德国汉堡的新 DESY（Deutsches Elektronen Synchrotron：德国电子同步加速器）最近开始运行。在本文中，马克斯·普朗克医学研究所的生物学家格尔德·罗森鲍姆及其同事们证明了这台加速器作为强 X 射线源在生物成像方面的潜在用途。在 DESY 环路中运动的相对论性电子自然发射的 X 射线束要比当时的标准 X 射线源所产生的射线束亮 100 倍左右。研究人员发现：使用这个源产生的生物样品的像比使用最好的传统 X 射线源要清晰得多。现在，同步加速器 X 射线源已经成为不可或缺的生物成像设备。

当一个电子被加速时，它会发出辐射。在 DESY 所用的非常高的能量下，发出的辐射被限于电子瞬时运动方向周围的窄椎体之内。因此同步辐射是沿切线方向发出的。同步辐射是多波长的辐射，对于能量为 7.5 GeV 的电子，其同步辐射的峰值落在 X 射线范围内（见参考文献 1 中对原始理论的描述及参考文献 2~4 中对实验细节的介绍）。

DESY 同步加速器采用每秒 50 个脉冲的爆丛，每 10 ms 脉冲含有 6×10^{10} 个电子（平均束流为 10 mA）。注入能量相对较低，电子在 10 ms 内被加速到 7.5 GeV。

大多数 X 辐射是在每个脉冲的最后 3 ms 内发射的：极少量的辐射来自较低的电子能量，所以在 1.5 Å 处的时间平均强度约为峰值的 20%。

我们已经估算出同步加速器和细焦旋转阳极 X 射线管的谱线亮度（即以单位面积、单位立体角和单位波长间隔内每秒发射的光子数表示的功率）（见表 2），在 1.54 Å 处分别为 2×10^{22}（时间平均）个和 3×10^{20} 个光子 s^{-1} $sterad^{-1}$ cm^{-2} $Å^{-1}$。这清楚地表明：与目前的 X 射线管相比，同步加速器是一种亮度非常高的源。衍射实验能从中得到多大的实际好处在很大程度上取决于使辐射聚焦和单色化所必需的光学系统。在标准 X 射线衍射中可以应用三种类型的聚焦单色器：弯曲玻璃镜、石英单色器和石墨单色器。

A preliminary investigation of the properties of bent quartz monochromators[5] used with synchrotron radiation is reported here. We have chosen quartz because of its suitable elastic and optical properties which allow it to be used asymmetrically cut and bent to form an accurate focusing monochromator, with a comparatively large numerical aperture. It also behaves substantially as a perfect crystal with a reflectivity near unity in a narrow angular range. We predict that it should be possible to focus the synchrotron radiation down to a point ($200 \times 200\ \mu m^2$) with a Berreman[6] monochromator to give a total flux of 10^{10} photons s^{-1} at 1.5 Å, which is higher than the flux available from other known X-ray sources (Table 2); also the beam is well collimated. The flux density, the important parameter when using film, is comparatively even higher because of the small focus.

Because of its large mosaic spread (300 times greater than that of quartz) a graphite monochromator might seem advantageous for our application. When used with the white radiation from the synchrotron, however, the mosaic spread of graphite would produce a highly divergent reflected beam with considerable wavelength inhomogeneity, thus restricting us to small monochromator-film distances for reasonable spot diameters on the film. For these short distances it would not then be possible to collect radiation from a large area of graphite by focusing. Alternatively, for larger film distances the reflected beam would require collimation, which would again reduce the expected intensity. We do not, therefore, expect graphite to give more intensity than quartz. Furthermore, the optical and mechanical properties of graphite are much less convenient.

Experimental Details

All experiments took place in the F41 (synchrotron radiation) group bunker at DESY in Hamburg (Fig. 1 and 2). The experimental area can only be entered when the main beam shutter between the synchrotron and the bunker is closed so that all the experiments had to be done by remote control. The quartz crystal was mounted in the vacuum pipe leading to the synchrotron ring. The reflected beam came out through a beryllium window (0.5 mm thick) of diameter 1.5 cm. A rotating disk containing a slot was used as an attenuator. This and a lead shutter were mounted near the window. The rotating disk was arranged to run synchronously with the synchrotron. A film holder was mounted about 120 cm from the quartz crystal on a table movable by remote control. Intensities were recorded on Ilford Industrial G film. The monochromator (Steeg and Reuter) consisted of a slab of quartz ($45 \times 13 \times 0.3\ mm^3$) with the face containing the long axis cut at about 8° to the 1011 planes. The slab was bent by two sets of pins. Before mounting the crystal in the beam, the curvature was pre-adjusted to the required radius with laser light. The final position of the focus was determined by through-focal photographs. The best focal line had a width of 180 μm and represented the image of the radiating electron beam in the synchrotron. (The total effective source size, including the betatron and synchrotron oscillations, was about 4 mm.) Photographs were also taken close to the monochromator, where the reflected beam was wide, to evaluate the total reflected flux. Experiments with aluminium filters were made to estimate the strength of the higher harmonics in the quartz reflected radiation.

本文报道了对用于同步辐射的弯曲石英单色器[5]性质进行的初步研究。我们选择石英是因为它具有合适的弹性和光学性质，因而可以对它进行非对称切割，并能使之弯曲形成一个具有相当大数值孔径的准确聚焦单色器。实质上它还是一个在窄角度范围内折射率接近于1的理想晶体。我们预言使用贝雷曼[6]单色器应该可以将同步辐射聚焦到一个点（$200 \times 200\ \mu m^2$）上，在1.5 Å处给出的总通量为每秒10^{10}个光子，高于从其他已知X射线源得到的通量（表2）；束的准直也很好。因为焦距小，所以通量密度，这个在使用胶片进行测量时的一个重要参数，比起其他X射线源还会更高一些。

由于石墨有很大的嵌镶度（是石英的300倍），因而石墨单色器或许会有利于我们的应用。然而，当我们使用来自同步加速器的白辐射时，石墨的嵌镶度将导致产生波长相当不均匀的高度发散反射束，因此为了在胶片上形成一个大小合适的光斑，单色器–胶片的距离必须非常短。对于这么短的距离，通过聚焦来收集来自大面积石墨的辐射是不可能的。另一方面，在单色器–胶片距离较大的情况下，反射束需要准直，这也将降低强度的预期值。因此，我们不认为石墨能给出比石英更高的强度。此外，石墨的光学和力学性质在使用上也很不方便。

实验的详细过程

全部实验都是在汉堡DESY的F41(同步辐射)小组的掩体中进行的(图1和图2)。只有当同步加速器和掩体之间的主射束光闸关闭的时候，人才能进入实验区，所以整个实验必须通过遥控进行。石英晶体被放在通向同步加速器环的真空管内。反射束从一个直径为1.5 cm的铍窗（厚0.5 mm）中射出。用一个带有狭缝的旋转圆盘作为衰减器。这个衰减器和一个铅闸装在铍窗附近。使旋转圆盘与同步加速器同步运转。胶片的支架装在一张能通过遥控移动的实验台上，支架距石英晶体约120 cm。在伊尔福工业G胶片上记录强度。单色器（施特格和罗伊特）是由一个石英片（$45 \times 13 \times 0.3\ mm^3$）构成的，其包含长轴的面的切割角度为相对1011面约成8°夹角。石英片的弯曲是通过两组销子实现的。在把晶体安装到光束中之前，先要用激光将曲率半径调整至所需的数值。用经过聚焦的照片确定焦点的最终位置。最佳聚焦线的宽度为180 μm，这代表了同步加速器内辐射电子束的像。（包括电子回旋加速器和同步加速器振荡在内的总有效源尺寸约为4 mm。）照相也是在靠近单色器处进行的，在这里反射束很宽，便于估计总的反射通量。用铝滤波器所作的实验来估计在石英反射的辐射中高次谐波的强度。

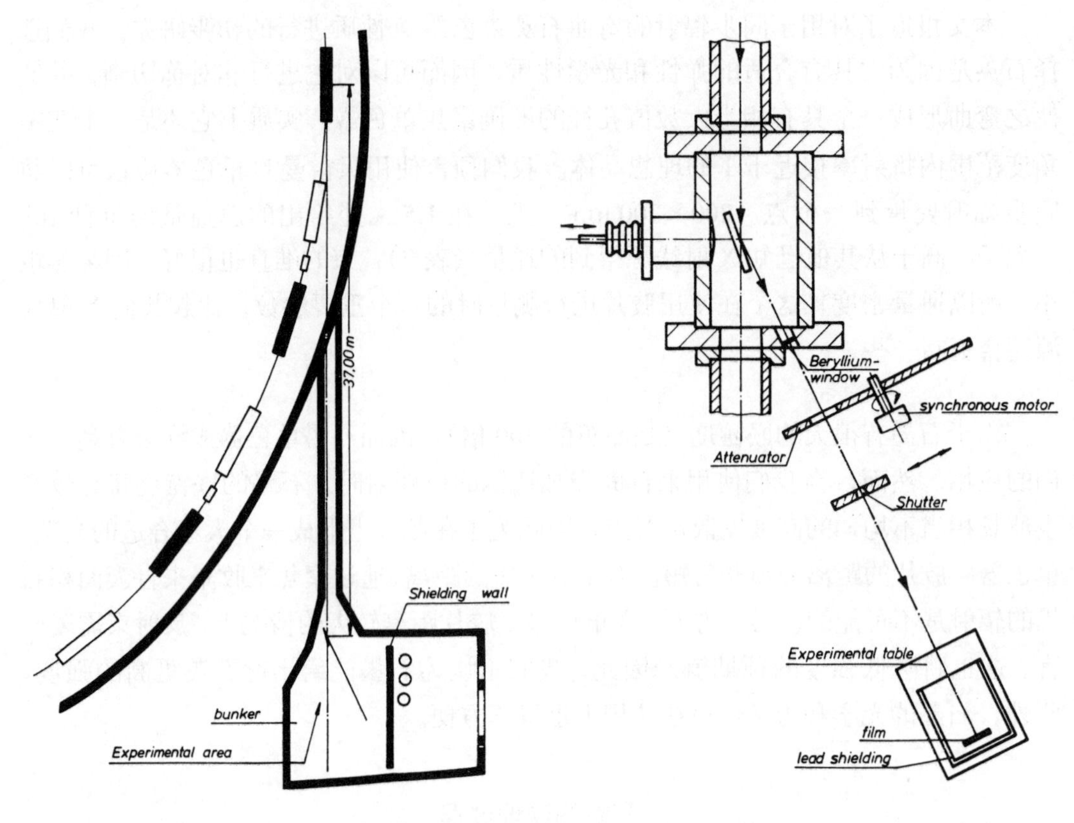

Fig. 1. The F41 bunker at DESY and its position with respect to the synchrotron.

Fig. 2. Monochromator housing and the experimental set-up.

With a source-to-monochromator distance of about 40 m the crystal, if set exactly for one wavelength (for example, in the Johann arrangement[5]), would give a focus at 10 m. The white radiation fortunately allowed us to relax this condition and obtain a more practical focal length (1.5 m) at the expense only of very little wavelength inhomogeneity (Table 1). Furthermore, the angular adjustment of the quartz monochromator was not critical. The central wavelength of the reflected beam was determined by measuring the angle between incident and reflected beam. The position and size of the Be-window limited our observations to Bragg angles of $13\pm1°$ (that is, 1.5 ± 0.15 Å).

Finally, a simple camera was constructed (specimen-film distance 40 cm), and a photograph (Fig. 3*a*) was taken of the equatorial reflexions from a 2 mm strip of the longitudinal flight muscle from the giant water bug *Lethocerus maximus*[7]. The entrance aperture of the camera was approximately 2 mm × 2 mm. A helium-filled tube minimized air absorption in the space between the radiation-pipe window and the camera. The exposure was 15 min with the synchrotron running at 5 GeV. On one side of the direct beam a large area of parasitic scattering is visible apparently resulting from the quartz and from the steel pins used to bend the quartz. Fortunately, the camera entrance-slits were not symmetrically positioned, so that a clear view of one side of the diffraction pattern

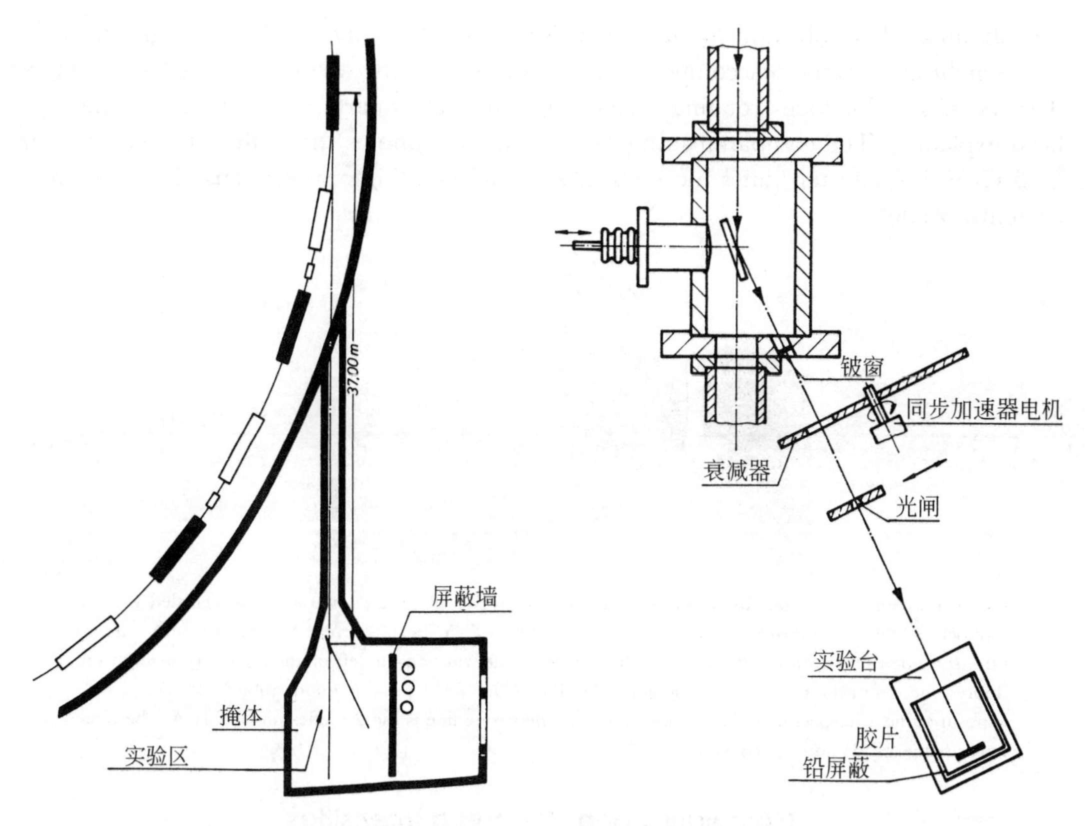

图1. DESY中的F41掩体及其相对于同步加速器的位置

图 2. 单色器的构造和实验装置

如果源到单色器的距离约为 40 m，并且晶体对一个波长精确地设定（例如在约翰式聚焦系统中[5]），那么给出的焦点将在 10 m 处。所幸的是，白辐射使我们可以放宽这个条件，从而在波长不均匀性损失很小的情况下得到一个更实际的焦距（1.5 m）（表 1）。此外，对石英单色器的角度调整并不是很重要。反射束的中心波长是通过测量入射束与反射束之间的夹角来确定的。铍窗的位置和尺寸把我们的观测范围限制在布拉格角为 13° ±1° 之内（即 1.5 Å ± 0.15 Å）。

最后要构造的是一台简易的照相机（样品 – 胶片的距离为 40 cm），拍摄的照片（图 3*a*）为巨型水虫纵向飞行肌的一条 2 mm 带状区域[7]的赤道反射。照相机的进口孔径约为 2 mm × 2 mm。用一只充有氦气的管使辐射 – 管窗与照相机之间的空间内空气吸收最小。在 5 GeV 运行的同步加速器的曝光时间为 15 min。在直射光束的一侧，明显可以看到由石英和用于使石英弯曲的钢质定位销造成的大面积寄生散射。所幸的是，照相机的进口狭缝并不是对称定位的，因此可获得一侧衍射条纹的清晰图像。与采用传统 X 射线源（图 3*b*，埃利奥特细焦旋转阳极管和弯曲石英单色器）

was obtained. The substantially greater width of the "20" line on the photograph made with synchrotron radiation, compared with that made using a conventional X-ray source (Fig. 3*b*, Elliott fine-focus rotating anode tube and bent quartz monochromator) has not been explained. The comparative intensity of the two photos shows that the synchrotron (at 5 GeV) is about ten times more effective than one of the most intense X-ray sources currently available.

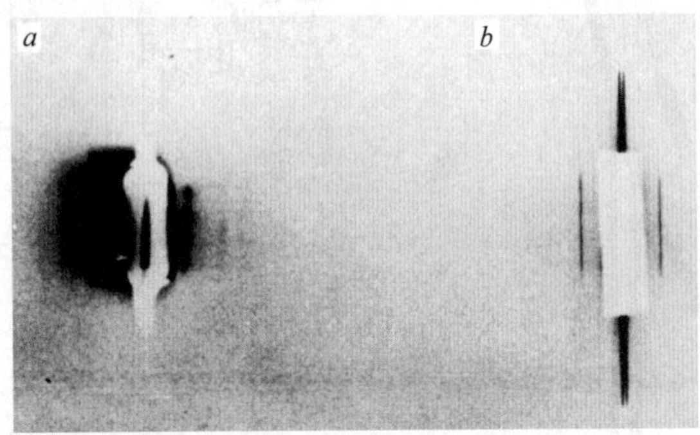

Fig. 3. Equatorial reflexions from dorsolongitudinal flight muscle of *Lethocerus maximus* recorded with: *a*, monochromated synchrotron radiation; electron energy 5 GeV, beam current 8 mA, exposure time 15 min, specimen film distance 40 cm; note the parasitic scattering on the left of the backstop arising from fluorescence from the monochromator holder; *b*, Elliott fine-focus rotating anode tube at 40 kV, 15 mA, exposure time 1 h, specimen film distance 36 cm. The strong line is the 20 reflexion (*d*=231 Å); the weak lines are the 21, 31 and 32 reflexions.

Calculated and Observed Intensities

Using the theory of Schwinger[1] and a programme written by Klucker, DESY group F41, we have calculated the intensities at 1.5 Å wavelength and at the harmonics of 1.5 Å: when the synchrotron runs at 7.5 GeV the second and third harmonics are twice as intense (photons/s) as the 1.5 Å radiation.

We have measured photographically the instantaneous intensity of the reflected beam passing the disk attenuator at the eighth ms of each synchrotron acceleration cycle. The contribution of higher orders has been estimated from measurements made through aluminium filters of various known thicknesses, and we have adopted values for the absorption coefficients[8]. The sensitivity of Ilford Industrial G film at 1.5 Å has been extrapolated from the calibrated value at 1.54 Å (ref. 9). The experimental conditions and data are summarized in Table 1.

Table 1. Data for Quartz Monochromator in Synchrotron Radiation Beam

Synchrotron	7.5 GeV, 10 mA beam current
Electron beam diameter	approximately 4 mm (=effective X-ray source diameter)
Distance	37 m from synchrotron to monochromator
Cross-fire of the incident beam	approximately 10^{-4} rad
Polarization	85% at 1.5 Å in the eighth ms of the cycle, polarized in the plane of the synchrotron
Be-window	0.5 mm (96 mg cm^{-2})

相比，用同步辐射制成的照片具有相对较大的“20”线宽度，这一现象目前尚未得到解释。从两张照片的相对强度可以看出，同步辐射（5 GeV）比现有的最强 X 射线源要有效约 10 倍。

图 3. 拍摄巨型水虫背部纵向飞行肌的赤道反射，采用以下两种方式记录。*a*，单色的同步辐射；电子能量：5 GeV，束流：8 mA，曝光时间：15 min，样品与胶片的距离：40 cm；注意在托架左侧会出现由单色器支架的荧光引起的寄生散射。*b*，40 kV 和 15 mA 的埃利奥特细焦旋转阳极管，曝光时间：1 h，样品与胶片的距离：36 cm。强线是 20 反射（d=231 Å）；弱线是 21、31 和 32 反射。

计算强度和观测强度

根据施温格 [1] 的理论和由 DESY 的 F41 小组成员克卢克编写的程序，我们计算了 1.5 Å 波长处和 1.5 Å 谐波处的强度：当同步加速器在 7.5 GeV 运行时，二次和三次谐波的强度（每秒的光子数）是 1.5 Å 辐射的两倍。

我们用照相方法测量了在每个同步加速器加速周期第 8 个 ms 时通过圆盘衰减器的反射束的瞬时强度。从对不同已知厚度的铝滤波器所作的测量中估计出了高阶的贡献，我们还用到了一些吸收系数数据 [8]。伊尔福工业 G 胶片在 1.5 Å 处的灵敏度是由在 1.54 Å 处的校准值外推得到的（参考文献 9）。实验条件和数据汇总在表 1 中。

表 1. 同步辐射束中石英单色器的数据

同步加速器	7.5 GeV, 10 mA 束流
电子束直径	约 4 mm（= 有效X射线源直径）
距离	同步加速器距单色器 37 m
入射束的交叉发射	约 10^{-4} 弧度 (rad)
偏振	对于在同步加速器平面内的偏振：在周期的第8个 ms 内 1.5 Å 处为 85%
铍窗	0.5 mm (96 mg · cm^{-2})

Continued

Crystal	quartz cut at σ=8°30′ to the 1011 planes, dimensions 45×13×0.3 mm³
Bender	pins: outer pair 40.5 mm inner pair 39.5 mm radius of curvature of crystal, 9 m
Wavelength	1.53 Å (θ = 13°15′)
Wavelength spread	$\Delta\lambda = 3\times10^{-3}$ Å (due to deviation from Johann focusing and to finite source size)
Focus	1.5 m from crystal, line focus 180 μm wide
Angular aperture of reflected beam	horizontal: 2 mrad (convergence) vertical: 3–4 mrad (divergence)
Measured flux in line focus	1.8×10^9 photons s^{-1} mm^{-1} (of focal length) (at the eighth ms of the cycle)

The ratio of the intensity at 1.5 Å, evaluated as indicated above, to the calculated incident intensity per unit wavelength interval is an "integrated band pass" which was found to be

$$\int R(\lambda)d\lambda = 0.7 \times 10^{-4}\ \text{Å}$$

Transforming the wavelength into an angle using Bragg's law we find an integrated reflectivity

$$\int R(\theta)d\theta = R_{int} = 1.0 \times 10^{-5}\ \text{rad}$$

for a quartz crystal cut at 8°30′ to the 1011 plane.

Quartz behaves essentially as a perfect dynamical diffractor[10]. Renninger[11] has calculated the reflectivity of a perfect quartz crystal (without corrections for absorption) to be

$$R_{int} = 4.4\times10^{-5}\ \text{rad}$$

and Brogren[12] measured an integrated reflectivity of

$$R_{int} = 3.9\times10^{-5}\ \text{rad}$$

for a polished quartz crystal cut parallel to the $10\bar{1}1$ planes. The case of an asymmetrically cut perfect crystal with absorption is treated in the Darwin–Prins theory. Using Zachariasen's formulae[13] we have calculated an integrated reflectivity of

$$R_{int} = 1.45\times10^{-5}\ \text{rad}$$

for a quartz crystal cut at 8°30′ to the $10\bar{1}1$ planes (λ = 1.5 Å) which agrees with our experimental value.

We emphasize that the aim of our experiments was not to make quantitative measurements of the reflectivity of quartz but to show that quartz is a suitable material

续表

晶体	石英切割到相对1011面为 σ=8°30′ 的夹角，尺寸为 $45 \times 13 \times 0.3\ mm^3$
弯曲装置	定位销：外对 40.5 mm 内对 39.5 mm 晶体的曲率半径为 9 m
波长	1.53 Å (θ = 13°15′)
波长展开度	$\Delta\lambda = 3 \times 10^{-3}$ Å（源自约翰式聚焦的偏差以及有限源的尺寸）
焦点	距晶体 1.5 m，线聚焦的宽度为 180 μm
反射束的孔径张角	水平：2 mrad（会聚） 垂直：3 mrad~4 mrad（发散）
在线聚焦时测量的通量	（焦距处）1.8×10^9 个光子 $s^{-1}\ mm^{-1}$（在周期的第8个ms内）

由上述方法估计得到的在 1.5 Å 处的强度与单位波长间隔入射强度的计算值之比是一个“积分带通”，可由下式表示：

$$\int R(\lambda)\mathrm{d}\lambda = 0.7 \times 10^{-4}\ \text{Å}$$

用布拉格定律将波长转换为角度，我们得到了以相对 1011 面 8°30′夹角切割的石英晶体的积分反射率：

$$\int R(\theta)\mathrm{d}\theta = R_{int} = 1.0 \times 10^{-5}\ \text{rad}$$

石英在本质上是个理想的动态衍射器 [10]。伦宁格 [11] 已经计算出了一个理想石英晶体的反射率（未作吸收校正）：

$$R_{int} = 4.4 \times 10^{-5}\ \text{rad}$$

布罗格伦 [12] 通过测量得到了一个经平行于 $10\bar{1}1$ 面切割的抛光石英晶体的积分反射率：

$$R_{int} = 3.9 \times 10^{-5}\ \text{rad}$$

这类具有吸收的非对称切割理想晶体的情况按照达尔文 – 普林斯理论来处理。利用扎卡里亚森公式 [13]，我们计算了以相对 $10\bar{1}1$ 面 8°30′ 夹角切割的石英晶体的积分反射率（λ = 1.5 Å）：

$$R_{int} = 1.45 \times 10^{-5}\ \text{rad}$$

这与我们通过实验得到的数值是一致的。

需要强调的是：我们的实验目的并不是定量测量石英的反射率，而是要说明就

for the construction of a focusing monochromator for synchrotron radiation, and to check that there was no large disparity between the observed and calculated flux of monochromated synchrotron radiation. Our results show that the monochromator has properties which can be accurately predicted. We have emphasized neither the accurate determination of the attenuation ratio of the rotating disk nor the speed of the shutter. Moreover, the evaluation of the contribution from higher harmonics may be inaccurate. We estimate that the error in our result may amount to 50%. Furthermore, the state of the surface of the quartz crystal is difficult to control, although it has a considerable influence on the actual shape and height of the reflectivity curve[14,15].

Estimated Intensities for Various Configurations

We intend to set up a Berreman monochromator[6] to give a point-focused beam from a quartz crystal ground so as to give the required curvature in one plane and bent to the corresponding curvature in the second. There seem to be no theoretical reasons why this should not produce foci of similar dimensions to those that we have obtained with a simple bent crystal, especially as the geometry of the synchrotron beam relaxes some of the stringent conditions which the radii of curvature of the crystal must otherwise satisfy.

The estimated performance of such an arrangement for each of three typical configurations used in biological applications of X-ray diffraction is shown in Table 2, and the performance is compared with a "conventional" fine-focus rotating anode tube. The calculated intensities are based on the effective band pass give above, 0.7×10^{-4} Å.

Table 2. Biological Applications

Specimen	Elliott fine-focus X-ray tube*	DESY synchrotron with Berreman point-focusing monochromator‡
Single crystal	Standard collimator, 0.5 mm diameter	
	A = 12.5 cm	D = 1 m
a=0.5 mm	d = 0.7 mm	d = 120 μm
b=0.5 mm	$P = 10^9$ photons s^{-1}	$P = 4\times10^9$ photons s^{-1}
L=7.5 cm	$I = 2\times10^9$ photons s^{-1} mm^{-2}	$I = 2.5\times10^{11}$ photons s^{-1} mm^{-2}
Tobacco mosaic virus gel	Double-crystal focusing monochromator†	
		D = 0.8 m
a = 0.6 mm	d = 80 μm	d = 100 μm
b = 1 mm	$P = 10^7$ photons s^{-1}	$P = 3\times10^9$ photons s^{-1}
L = 12 cm	$I = 2\times10^9$ photons s^{-1} mm^{-2}	$I = 3\times10^{11}$ photons s^{-1} mm^{-2}
Insect muscle	Double-crystal focusing monochromator†	
		D = 1.5 (3) m
a = 3 mm	d = 100 μm	d = 180 (350) μm
b = 0.3 mm	$P = 5\times10^5$ photons s^{-1}	$P = 5\times10^8$ (2×10^9) photons s^{-1}
L = 40 cm	$I = 5\times10^7$ photons s^{-1} mm^{-2}	$I = 1.5\times10^{10}$ photons s^{-1} mm^{-2}

a, Width of specimen; *b*, height of specimen; *L*, specimen film distance; *A*, anode specimen distance; *D*, focal length, that is, monochromator film distance; *d*, spot or focus diameter on film; *P*, X-ray power reaching the specimen; and *I*, flux density at the focus.

构成应用于同步辐射中的聚焦单色器而言，石英是一种很适合的材料，并证实了单色同步辐射的观测通量与计算通量之间并无很大的差异。我们得到的结果表明，单色器的性质是可以准确预言出来的。我们既没有强调对旋转圆盘衰减比的准确测定，也没有强调对光闸速率的准确测定。此外，对高次谐波贡献的估计也可能是不准确的。我们估计，在我们所得结果中的误差也许会达到 50%。此外，虽然石英晶体的表面状态对反射率曲线的实际形状和高度影响很大[14,15]，但很难对它进行控制。

对三类样品的强度估计

为了从石英晶体底部给出一束点聚焦辐射，我们打算装配一台贝雷曼单色器[6]，以便在一个平面上形成所需的曲率，并在第二个平面上也弯曲到相应的曲率。似乎无法用理论来说明为什么这样做得到的焦距不能与用简单弯曲晶体得到的焦距有类似的大小，尤其是在同步加速器束流的几何形状不那么严格时，晶体的曲率半径本应该满足的这时不满足了。

以这样一种方式用 X 射线衍射研究生物学中的三种典型样品，将估测的性能列于表 2，并把 DESY 的性能与“传统的”细焦旋转阳极管的性能进行了比较。强度的计算值是以前面给出的有效带通值 0.7×10^{-4} Å 为基础的。

表 2. 在生物学中的应用

样品	埃利奥特细焦X射线管*	配备贝雷曼点聚焦单色仪的DESY同步加速器‡
单晶	标准准直仪，直径为 0.5 mm	
	$A = 12.5$ cm	$D = 1$ m
$a = 0.5$ mm	$d = 0.7$ mm	$d = 120$ μm
$b = 0.5$ mm	$P = 10^9$ 个光子 s^{-1}	$P = 4 \times 10^9$ 个光子 s^{-1}
$L = 7.5$ cm	$I = 2 \times 10^9$ 个光子 s^{-1} mm^{-2}	$I = 2.5 \times 10^{11}$ 个光子 s^{-1} mm^{-2}
烟草花叶病毒凝胶体	双晶聚焦单色器†	
		$D = 0.8$ m
$a = 0.6$ mm	$d = 80$ μm	$d = 100$ μm
$b = 1$ mm	$P = 10^7$ 个光子 s^{-1}	$P = 3 \times 10^9$ 个光子 s^{-1}
$L = 12$ cm	$I = 2 \times 10^9$ 个光子 s^{-1} mm^{-2}	$I = 3 \times 10^{11}$ 个光子 s^{-1} mm^{-2}
昆虫的肌肉	双晶聚焦单色器†	
		$D = 1.5$ (3) m
$a = 3$ mm	$d = 100$ μm	$d = 180$ (350) μm
$b = 0.3$ mm	$P = 5 \times 10^5$ 个光子 s^{-1}	$P = 5 \times 10^8$ (2×10^9) 个光子 s^{-1}
$L = 40$ cm	$I = 5 \times 10^7$ 个光子 s^{-1} mm^{-2}	$I = 1.5 \times 10^{10}$ 个光子 s^{-1} mm^{-2}

a，样品宽度；b，样品高度；L，样品与胶片的距离；A，阳极与样品的距离；D，焦距，即单色器与胶片的距离；d，在胶片上的光斑或焦点直径；P，到达样品上的X射线功率；I，焦点处的通量密度。

* Loaded with 40 kV, 50 mA into a 0.2×2 mm^2 electron focus at the anode in the first case, and 40 kV, 15 mA into a 0.14×0.7 mm^2 focus in the other two cases. This set is the most powerful fine-focus X-ray tube currently available.

† The setting of this Johann-type[5] monochromator is optimized for each type of specimen.

‡ Conditions of the synchrotron are as in Table 1, computed for 1.5 Å radiation.

The tube values were calculated from measurements made with Ilford Industrial G film and a rotating disk attenuator on an Elliot fine-focus rotating anode tube used with single and double focusing quartz monochromators.

Higher Intensities and Longer Wavelengths

Some possible methods of obtaining higher intensities and utilizing the continuous spectrum are as follows. (*a*) According to current plans, the DESY synchrotron current will be raised from 10 mA to 50 mA. Also the electrons will be kept at the maximum energy for 1 or 2 ms, giving overall a six-fold improvement. (*b*) Sakisaka[14] suggests that both the height and width of the rocking curve of quartz can be increased appreciably by gentle grinding. A gain of 2 or 3 should be possible without affecting the size of the focus. (*c*) For special applications, where only pulses of X-rays can be used, the synchrotron is a very advantageous source if the experiment can be synchronized with the periodic maximum emission from the synchrotron. The integrated reflectivity of quartz increases approximately linearly with wavelength up to 3–4 Å (ref. 12). The intensity of the synchrotron radiation decreases, however, in the wavelength range 1.5–4.5 Å, approximately as the inverse of wavelength. The reflected intensity is thus roughly independent of wavelength. Previously, long wavelength experiments were avoided because of the low conversion efficiency of the anode materials involved.

We thank the Direktorium of DESY for facilities; Dr. R. Haensel and group F41 for advice; Drs. U. W. Arndt and H. G. Mannherz (who prepared the muscle specimen) and Dr. J. Barrington Leigh for the use of his calculations for the Berreman monochromator. The equipment was constructed in the workshops of DESY and the Max-Planck-Institut, Heidelberg. G. R. and J. W. have EMBO short term fellow-ships.

(**230**, 434-437; 1971)

G. Rosenbaum and K. C. Holmes: Max-Planck-Institut für Medizinische Forschung, Heidelberg.

J. Witz: Laboratoire des Virus des Plantes, Institut de Botanique de la Faculté des Sciences de Strasbourg, Strasbourg.

Received March 3, 1971.

References:

1. Schwinger, J., *Phys. Rev.*, **12**, 1912 (1949).
2. Godwin, R. P., *Springer Tracts in Modern Phys.* (edit. by Höhler, G.), **51**, 1 (1969).
3. Haensel, R., and Kunz, C., *Z. Angew. Phys.*, **23**, 276 (1967).

* 对于第一个样品，将40 kV和50 mA加载到阳极处 0.2 × 2 mm^2 电子焦点上；对于另外两个样品，将 40 kV 和 15 mA 加载到 0.14 × 0.7 mm^2 焦点上。这个装置是目前可用的最有效的细焦X射线管。

† 这种约翰式[5]单色器装置对每一种类型的样品都是最佳的。

‡ 同步加速器的条件如表1所列，以1.5 Å 的辐射进行计算。

X 射线管的数据是根据测量伊尔福工业 G 胶片和与单、双聚焦石英单色器协同使用的埃利奥特细焦旋转阳极管上的旋转圆盘衰减器所得到的结果计算出来的。

更高的强度和更长的波长

几个可能获得更高强度及连续光谱的方法是：(*a*) 按照目前的设计，可以将 DESY 同步加速器的电流从 10 mA 升至 50 mA。电子也将在 1 ms ~ 2 ms 内保持在最大能量，这样一来总强度可以提高至 6 倍。(*b*) 匈坂 [14] 提出，轻微的抛光可以使石英摇摆曲线的高度和宽度显著增加。在不影响聚焦尺寸的前提下，达到 2 倍或 3 倍的增益是有可能的。(*c*) 对于一些只能应用 X 射线脉冲的特殊情况，如果实验能与来自同步加速器的周期性最大辐射同步，那么同步加速器就是一个具有明显优势的源。石英的积分反射率基本上随波长呈线性递增，直到波长达到 3 Å ~ 4 Å 时（参考文献 12)。然而，在波长范围为 1.5 Å ~ 4.5 Å 时，同步辐射的强度却随波长的增加而下降，两者近似成反比。所以反射强度大致与波长无关。以前大家都会避免在长波下作实验，因为阳极材料在长波下的转换效率很低。

感谢 DESY 委员会为我们提供设备；感谢亨泽尔博士和 F41 小组所提的建议；感谢阿恩特博士和曼赫茨博士（肌肉样品是他们制备的)，以及巴林顿 · 利博士对贝雷曼单色器的计算。这台设备安置在 DESY 车间以及德国海德堡的马克斯 · 普朗克研究所中。欧洲分子生物学组织为本文的两位作者——罗森鲍姆和维茨提供了短期奖金。

（沈乃澂 翻译；尚仁成 审稿）

4. Bathow, G., Freytag, E., and Haensel, R., *J. Appl. Phys.*, **37**, 3449 (1966).

5. Witz, J., *Acta Cryst.*, **A25**, 30 (1969).

6. Berreman, D. W., *Rev. Sci. Inst.*, **26**, 1048 (1955).

7. Pringle, J. W. S., *Prog. Biophys. Mol. Biol.*, 17, 3 (edit. by Huxley, H. E., and Butler, J. A. V.) (Pergamon, Oxford, 1967).

8. *International Tables of Crystallography*, 3.

9. Morimoto, H., and Uyeda, R., *Acta Cryst.*, **16**, 1107 (1963).

10. Bearden, J. A., Marzolf, J. G., and Thomsen, J. S., *Acta Cryst.*, **A24**, 295 (1968).

11. Renninger, M., *Z. Kristallograph.*, **107**, 464 (1956).

12. Brogren, G., *Arkiv. für Fysik.*, **22**, 267 (1962).

13. Zachariasen, W. H., *Theory of X-ray Diffraction in Crystals* (Dover, New York, 1967).

14. Sakisaka, Y., *Proc. Math. Phys. Soc. Japan*, **12**, 189 (1930).

15. Evans, R. C., Hirsch, P. B., and Kellar, J. N., *Acta Cryst.*, **1**, 124 (1948); Gay, P., Hirsch, P. B., and Kellar, J. N., *Acta Cryst.*, **5**, 7 (1952).

Estimation of Nuclear Explosion Energies from Microbarograph Records

J. W. Posey and A. D. Pierce

Editor's Note

Physicists in the 1970s sought accurate means for estimating the yield of nuclear explosions, especially in tests then being conducted by the United States and the Soviet Union. Here Joe Posey and Allen Pierce report on the accuracy of a new formula they had derived to make such estimates from the detection of very weak air waves from the initial blast. Their formula relates the explosion energy to the peak-to-trough magnitude of the pressure variation in the wave, and also to the 3/2 power of its period. As they show using data collected for a number of weapons tests conducted in 1961 and 1962, their expression fitted all the data very well, especially for explosions weaker than about 11 megatons.

FOLLOWING the US and USSR atmospheric test series in 1954–1962, numerous microbarograph records[1-8] of air waves generated by nuclear bomb tests were published. Previous theoretical interpretations[7,9] of such waveforms have required some explicit knowledge of the average atmospheric temperature and wind profiles above the path connecting source to microbarograph. Such profiles are never sufficiently well known and vary from point to point, and as seemingly small changes in the profiles cause relatively large changes in the waveforms, it would seem to be difficult to estimate the explosion energy yield to even order of magnitude accuracy from such records. Recently, however, in a further account of this work to be published elsewhere, we have succeeded in deriving an approximate theoretical relationship between certain waveform features and energy yield which is insensitive to changes in atmospheric structure. This relationship is given by

$$E = 13 p_{\mathrm{FPT}} \left[r_e \sin (r/r_e) \right]^{\frac{1}{2}} H_s (c T_{1,2})^{3/2} \tag{1}$$

where E is energy release, p_{FPT} is the first peak to trough pressure amplitude (see Fig. 1), r_e is radius of the Earth, r is the great circle distance from burst point to observation point, H_s is a lower atmosphere scale height, c is a representative sound speed, and $T_{1,2}$ is the time interval between first and second peaks. The purpose of the present communication is to describe the extent to which the above relation agrees with the existing available data.

根据微压计记录估计出的核爆炸能量

波西，皮尔斯

编者按

20 世纪 70 年代，物理学家们曾试图找到一种能精确估计核爆炸输出能量的方法，特别是因为可以在美国和苏联即将实施的核爆炸试验中进行估算。乔·波西和艾伦·皮尔斯在本文中阐述了他们为估计核爆炸能量而推导出来的一个新公式的精确性，这个公式是根据对最初爆炸产生的极弱空气波的测量得到的。他们在公式中将核爆炸输出能量与次声波上的波峰–波谷间压力变化幅度及其周期的 3/2 次方联系起来。他们利用在 1961 年和 1962 年核武器试验之后所采集到的大量数据来说明上述公式与所有数据都符合得很好，尤其是对于输出能量小于 11 兆吨左右的核爆炸。

继美国和苏联在 1954 年至 1962 年间对大气所作的一系列测试之后，有不少人发表了由核弹试验产生的空气波的微压计记录数据 [1-8]。如果要用以前的理论 [7,9] 来解释这些波形，就需要较为确切地知道连接爆炸源和微压计之间路径上方的平均气温和风廓线。这样的风廓线是不可能精确得到的，并且在每一个点上的值都是有所不同的，而且因为风廓线中看似微小的变化就能使波形发生相当大的改变，所以即使只是想要根据这些微压计记录值估计出核爆炸产生能量输出的数量级恐怕也并非易事。然而，我们最近将在别处发表的一篇文章中对此项工作进行进一步的报道——我们成功地推导出了特定波形特征和能量输出之间存在的近似理论关系，这种方法对大气结构变化不是很敏感。其关系式如下：

$$E = 13\, p_{\mathrm{FPT}} \left[r_{\mathrm{e}} \sin(r/r_{\mathrm{e}})\right]^{\frac{1}{2}} H_{\mathrm{s}} (c T_{1,2})^{3/2} \tag{1}$$

式中：E 代表释放出的能量，p_{FPT} 是从第一个波峰到波谷的压力变化幅度（见图 1），r_{e} 表示地球半径，r 表示从爆炸点到观测点的大圆距离，H_{s} 表示低层大气标高，c 表示声速，而 $T_{1,2}$ 是第一个波峰与第二个波峰之间的时间间隔。本篇论文的目的在于阐述上述关系式与实际测量数据之间的符合程度。

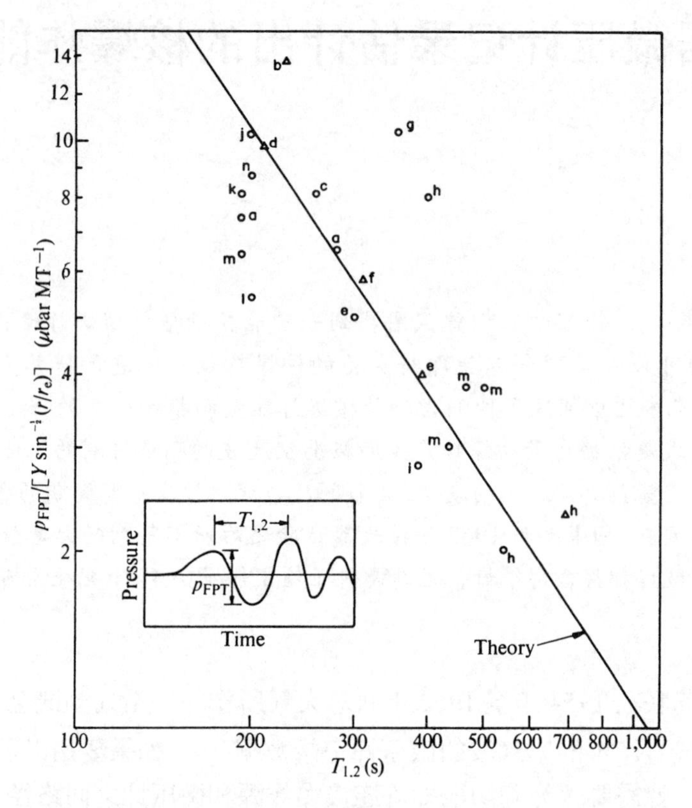

Fig. 1. Comparison of data with the theoretical relationship between amplitude and period of infrasonic waveforms generated by nuclear explosions. The data points are lettered *a* to *n* corresponding to particular events defined in the text. ○, Donn and Shaw; △, Harkrider.

The various points shown in Fig. 1 correspond to individual microbarograms recorded at Pasadena, California; Berkeley, California; Terceira, Azores; Fletcher's Ice Island; Whippany, New Jersey; Ewa Beach, Hawaii, and Palisades, New York, after the Soviet explosions of (*a*) September 10 (10 MT), (*b*) September 11 (9 MT), (*c*) September 14 (7 MT), (*d*) October 4 (8 MT), (*e*) October 6 (11 MT), (*f*) October 20 (5 MT), (*g*) October 23 (25 MT), (*h*) October 30 (58 MT), and (*i*) October 31, 1961 (8 MT) and the US explosions of (*j*) May 4 (3 MT), (*k*) June 10 (9 MT), (*l*) June 12 (6 MT), (*m*) June 27 (24 MT), and (*n*) July 11, 1962 (12 MT). Here the estimate of the yield (in equivalent megatons of TNT where one MT equals 4.2×10^{22} ergs) is taken from Båth[10]. All the records used are taken from the articles of Harkrider[7] and of Donn and Shaw[8]. Pressure amplitudes for Harkrider's records were computed using his microbarograph response data. Pressure amplitudes for the Donn and Shaw records were determined according to the premises (W. Donn, private communication) that (*a*) all records recorded by Lamont type A microbarographs are to the same scale and (*b*) the clip to clip amplitude of off scale oscillations was 350 μbars. The ordinate in Fig. 1 gives $p_{FPT}/[Y\sin^{-\frac{1}{2}}(r/r_e)]$ in μbar MT^{-1} where Y is the explosion yield in MT. The abscissa gives the period $T_{1,2}$ in s. Note that the plot is full

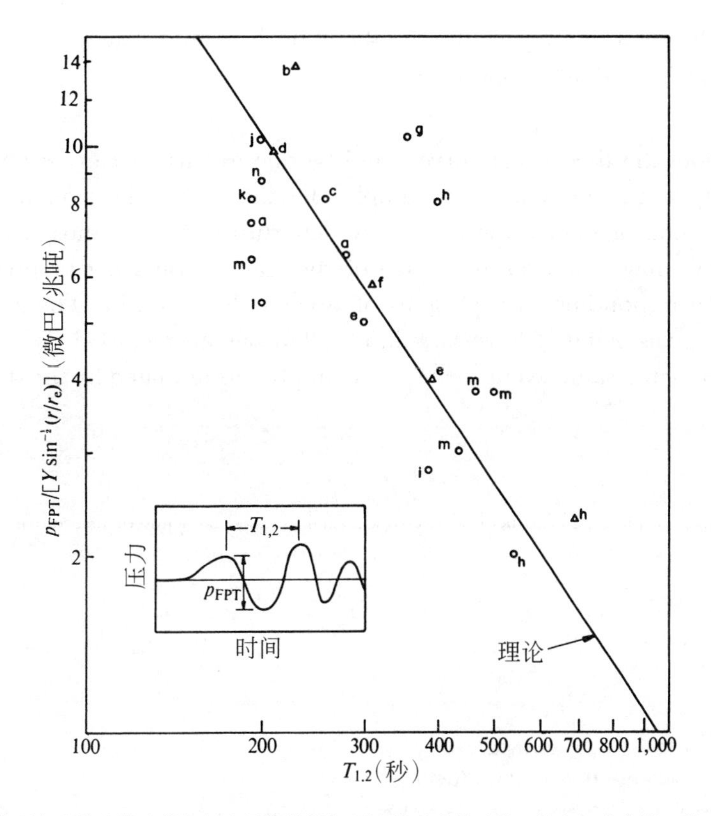

图 1. 核爆炸所产生的次声波的振幅和周期数据与理论关系式的对比。其中标有字母 *a* 到 *n* 的数据点分别对应于文章中所定义的特殊事件。○ 表示唐和肖的记录数据；△ 表示哈克赖德的记录数据。

图 1 中所示的各个点分别对应于：在苏联核爆炸之后于加州帕萨迪纳、加州伯克利、亚速尔群岛特塞拉岛、弗莱彻浮冰岛、新泽西州惠帕尼、夏威夷州埃瓦海滩和纽约州帕利塞兹测量的微压计记录结果，测量时间分别为 1961 年（*a*）9 月 10 日（10 兆吨），（*b*）9 月 11 日（9 兆吨），（*c*）9 月 14 日（7 兆吨），（*d*）10 月 4 日（8 兆吨），（*e*）10 月 6 日（11 兆吨），（*f*）10 月 20 日（5 兆吨），（*g*）10 月 23 日（25 兆吨），（*h*）10 月 30 日（58 兆吨）和（*i*）10 月 31 日（8 兆吨）；在美国核爆炸之后于上述地点测量的微压计记录结果，时间为 1962 年（*j*）5 月 4 日（3 兆吨），（*k*）6 月 10 日（9 兆吨），（*l*）6 月 12 日（6 兆吨），（*m*）6 月 27 日（24 兆吨）和（*n*）7 月 11 日（12 兆吨）。本文中对核爆炸能量输出值的估计（相当于多少兆吨三硝基甲苯爆炸所释放的能量，其中 1 兆吨 = 4.2×10^{22} 尔格）利用了贝斯的结果[10]。所有测量记录均来自哈克赖德[7]以及唐和肖[8]的文章。哈克赖德记录值中的压力变化幅度是由他的微压计响应数据计算得到的。唐和肖记录值中的压力变化幅度是根据以下假设得到的（唐，私人交流）：（*a*）所有用拉蒙特 A 型微压计所作的记录都采用了同样的标度（*b*）标度以外振动的削波幅度为 350 微巴。图 1 中纵坐标 $p_{FPT}/[Y\sin^{-\frac{1}{2}}(r/r_e)]$ 的单位为微巴 / 兆吨，其中爆炸能量输出 Y 的单位是兆吨。横坐标周期 $T_{1,2}$ 的单位为秒。请注

logarithmic. The solid line represents the theoretical relation, equation (1) with c and H_s taken as 310 m s^{-1} and 8 km, respectively.

The scatter about the theoretical curve could be due to various causes; one which seems especially likely is the undulation in amplitude due to the horizontal refraction and subsequent focusing or defocusing caused by departures of the atmosphere from perfect stratification. We may note also that much of the scatter would not be present if we had omitted data corresponding to explosions of greater than 11 MT. The general trend of longer period signals being of lower amplitudes than signals recorded elsewhere but which were generated by the same event seems to be amply substantiated by the data.

(**232**, 253; 1971)

Joe W. Posey and Allan D. Pierce: Department of Mechanical Engineering, Massachusetts Institute of Technology.

Received June 17, 1971.

References:

1. Yamamoto, R., *Bull. Amer. Meteorol. Soc.*, 37, 406 (1956).
2. Araskog, R., Ericsson, U., and Wagner, H., *Nature*, **193**, 970 (1962).
3. Carpenter, E. G., Harwood, G., and Whiteside, T., *Nature*, **192**, 857 (1961).
4. Farkas, E., *Nature*, **193**, 765 (1962).
5. Jones, R., *Nature*, **193**, 229 (1962).
6. Wexler, H., and Hass, W. A., *J. Geophys. Res.*, **67**, 3875 (1962).
7. Harkrider, D. G., *J. Geophys. Res.*, **69**, 5295 (1964).
8. Donn, W., and Shaw, D., *Rev. Geophys.*, **5**, 53 (1967).
9. MacKinnon, R., *Quart. J. Roy. Meteorol. Soc.*, **93**, 436 (1967).
10. Båth, M., Rept. A 4270-4271 (Seismological Institute, Univ. Uppsala, 1962).

意图 1 中的横纵坐标均采用对数形式。实线代表由式（1）给出的理论关系，其中 c 和 H_s 的取值分别为 310 米 / 秒和 8 千米。

理论曲线的发散可能是由多种原因造成的，其中有一个可能性非常大，即由地平大气折射以及随之而来的因大气层偏离理想层而产生的聚焦或散焦所导致的振幅波动。我们也注意到：如果忽略大于 11 兆吨的爆炸所对应的数据，那么大部分发散将不复存在。由上述数据似乎足以证明这样的趋势：对于由同样事件产生的信号，周期较长信号的幅度要低于在别处记录的信号的幅度。

（韩少卿 翻译；尚仁成 审稿）

Bomb ^{14}C in the Human Population

R. Nydal *et al.*

Editor's Note

Understanding the cycling of carbon between the atmosphere, oceans and biosphere (owing mostly to the uptake of carbon dioxide in photosynthesis) is essential for predicting how anthropogenic greenhouse gases affect climate. This paper from scientists at the Norwegian Institute of Technology shows how interest in the carbon cycle was already burgeoning in the early 1970s. It is expressed within the context of its time: it focuses on how the radioactive isotope carbon-14 (^{14}C), released into the atmosphere by nuclear weapons tests, will increase in the human body, which was considered a potential health risk. Strikingly, this increase is very short-lived, because the bomb ^{14}C is rapidly diluted by carbon dioxide released from fossil-fuel burning—today considered a much greater hazard.

IN the atmosphere ^{14}C occurs principally as $^{14}CO_2$ and is usually produced by nuclear reactions between cosmic ray neutrons and the nitrogen atoms of the air. The natural equilibrium between production and disintegration of ^{14}C determines a part of the natural background radiation to the human population. From 1955 there has been a gradual increase of ^{14}C in the atmosphere, the land biosphere and the ocean, as a result of nuclear tests. Although ^{14}C was initially not regarded as an important hazard to man[1], it was later pointed out[2-4] that ^{14}C could be a source of appreciable genetic hazard in the world's population, because of its long half life (5,700 yr).

At this laboratory we have studied the ^{14}C concentration in the human body[5]. The correspondence between ^{14}C in the atmosphere and in the human body, mediated as it is by photosynthesis, has been confirmed in 6 yr of measurements[6-8].

Since 1955, about two-thirds of the total nuclear energy liberated in nuclear tests has resulted from tests carried out in the atmosphere at high northern latitudes in 1961 and 1962 (ref. 9). The subsequent transfer of ^{14}C down to the troposphere, the biosphere and the ocean has been followed in detail[10-16]. The ^{14}C excess[16] ($\delta^{14}C$) in the northern troposphere (Fig. 1) is representative chiefly of the region between 30°N and 90°N, although the curve for the southern troposphere is representative of the region from the equator to 90°S.

进入人体的核弹^{14}C

尼达尔等

编者按

了解碳在大气、海洋和生物圈中的循环（主要归因于生物体在光合作用中对 CO_2 的吸收）对于预测人类活动所产生的温室气体将怎样影响气候相当重要。这篇由挪威技术研究所的科学家们所撰写的文章使我们了解到，在 20 世纪 70 年代早期人们是如何开始注意到碳循环的。本文反映了那个年代的焦点问题：由核武器试验释放到大气中的放射性同位素碳–14（^{14}C）在人体中的含量将会增加，这在当时被认为是一种潜在的健康威胁。显然，^{14}C 含量的增加是非常短暂的，因为核弹产生的 ^{14}C 很快就会被矿物燃料燃烧释放的 CO_2 所稀释，现在人们认为后者才是人类将要面临的更大威胁。

大气中 ^{14}C 主要以 $^{14}CO_2$ 的形式存在，通常由宇宙射线中子和大气中的氮原子通过核反应而产生。^{14}C 在产生和衰变之间的自然平衡决定了人类所受到的一部分天然本底辐射的大小。由于核试验的原因，从 1955 年起，大气层、陆地生物圈和海洋中的 ^{14}C 开始逐步增加。虽然起初人们并不认为 ^{14}C 会对人类造成重大危害 [1]，但后来有人指出 [2-4]：由于 ^{14}C 的半衰期很长（5,700 年），因而它对人类基因造成一定程度危害的可能性还是存在的。

在本实验室中，我们研究了人体内的 ^{14}C 浓度 [5]。经过 6 年时间的测试，我们已证实大气中的 ^{14}C 与人体中的 ^{14}C 是相互关联的，它们之间的媒介是光合作用 [6-8]。

自 1955 年起到现在，从核试验中释放出来的全部核能的 2/3 左右来自于 1961 年和 1962 年人们在北纬高纬度地区大气层中所进行的试验（参考文献 9）。有不少人对 ^{14}C 接下来向对流层、生物圈和海洋中的转移进行了详细的报道 [10-16]。北半球对流层中 ^{14}C 的过量值 [16]（$\delta^{14}C$）主要针对从北纬 30° 到北纬 90° 之间的区域（图 1），而南半球对流层曲线所代表的区域则是从赤道到南纬 90° 之间。

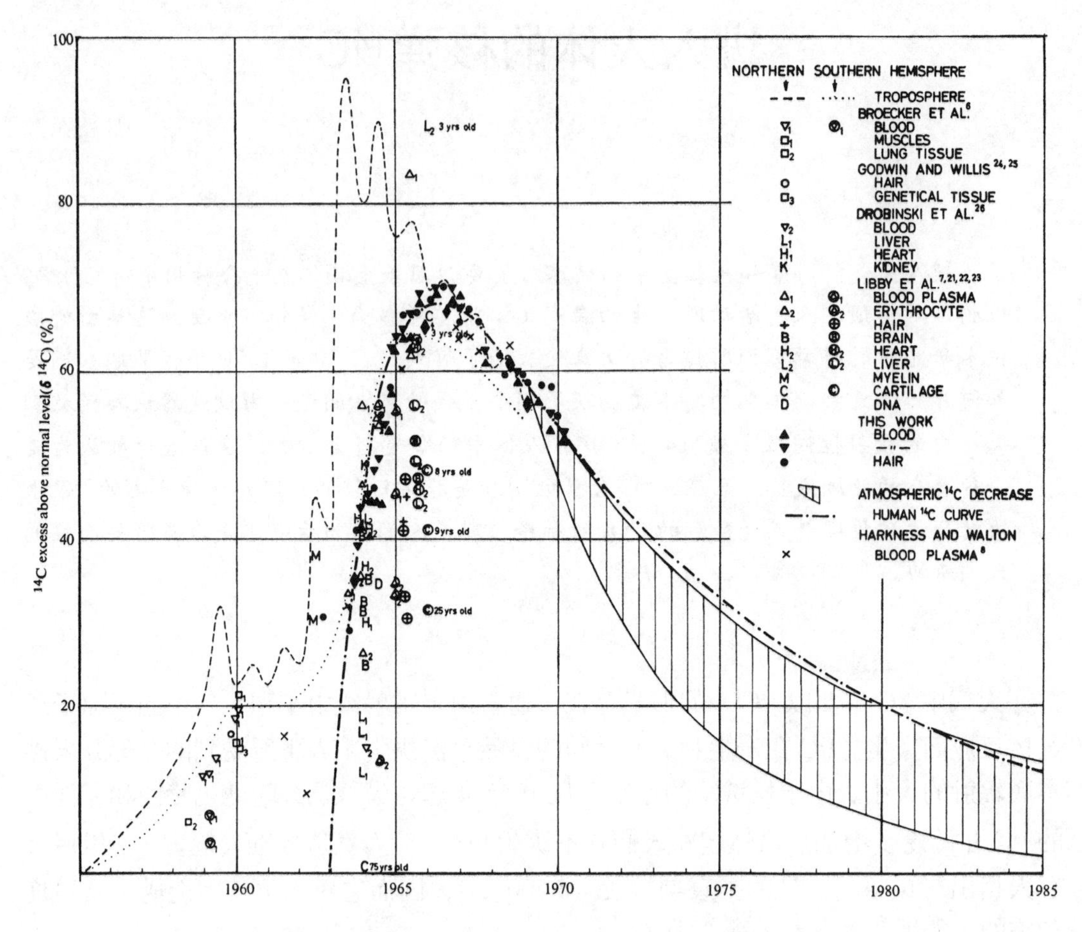

Fig. 1. Radiocarbon in the troposphere and human body.

The model for CO_2 exchange between the various reservoirs is shown in Fig. 2, in which the CO_2 in the troposphere is in exchange with CO_2 in the stratosphere, the land biosphere and the ocean. For CO_2 exchange between the troposphere and the biosphere we share the view of Münnich (for discussion, see ref. 17), who divided the biosphere into two parts. The first (b_1), which consists of leaves, grass, branches, and so on, is in rapid exchange with the troposphere and is combined with this reservoir, but the larger part of the vegetation (b_2) has a much slower exchange rate and is combined with the humus layer.

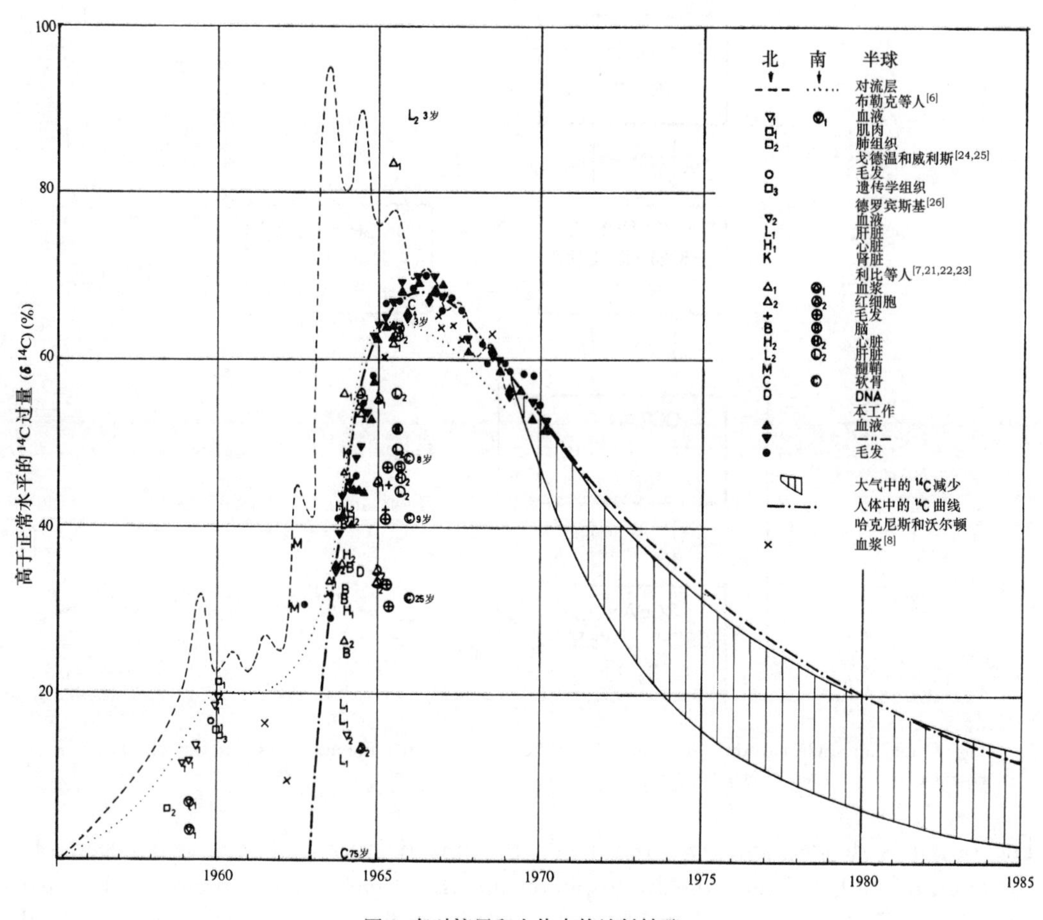

图 1. 在对流层和人体中的放射性碳

CO_2 在不同碳库之间的交换模型如图 2 所示。从图中可以看出，对流层中的 CO_2 会与平流层、陆地生物圈和海洋之中的 CO_2 发生交换。就 CO_2 在对流层和生物圈之间的交换而言，我们赞同明尼希的观点，即认为生物圈可以被分为两个部分（讨论过程见参考文献 17）。第一部分（b_1）由树叶、草、树枝等组成，这部分与对流层之间发生着快速的交换并且与之结合；但相当多的植物属于第二部分（b_2），这部分与对流层之间的交换速率很慢并且是和腐殖质层结合在一起的。

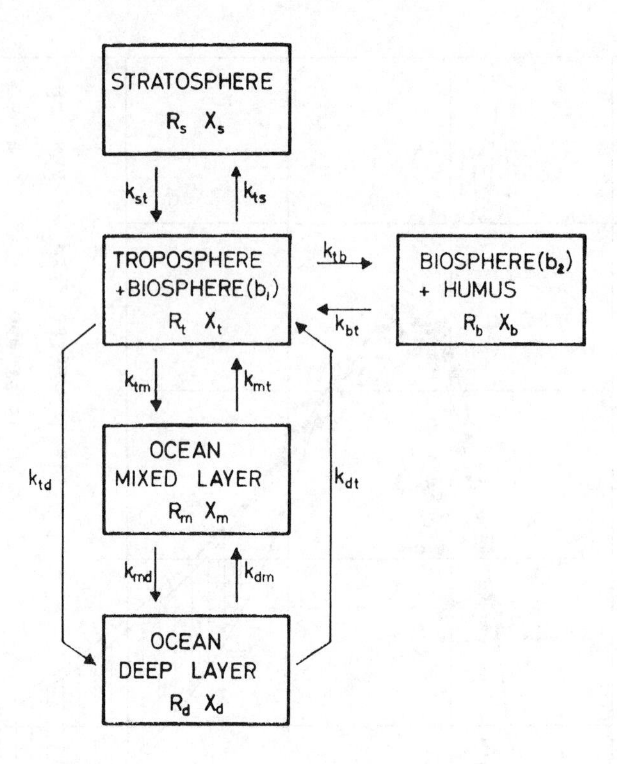

Fig. 2. Exchangeable carbon reservoirs, where: R_i, total carbon amount; X_i, ^{14}C excess; k_{ij}, exchange coefficients; i, j, t, b, m, d $(i \neq j)$.

The ocean is divided into two reservoirs, the mixed layer and the deep ocean. The exchange of CO_2 between the troposphere and the ocean occurs chiefly in the mixed layer, but according to Craig[18] there is also the possibility of a direct exchange with the deep ocean. The ^{14}C concentration in the mixed layer of the ocean is now 10 to 15% above normal[19,20].

Using the model of Fig. 2, we have treated the decrease of $\delta^{14}C$ in the troposphere in a previous article[20]. We showed that the measured variation of $\delta^{14}C$ (x_t) in the troposphere is approximately reproduced by the following two-term exponential function:

$$x_t = A_1 e^{-k_1 t} + A_2 e^{-k_2 t} \tag{1}$$

in which the parameters A_1, A_2, k_1 and k_2 depend on the various exchange coefficients. Because the errors on some of these coefficients are large, the extrapolation is uncertain and is given in the shaded area of Fig. 1.

The amount of bomb-produced ^{14}C in the atmosphere has increased the total natural amount of ^{14}C in nature by about 3%. According to Harkness *et al.*[8], the production of CO_2 from fossil fuel would lower the natural ^{14}C concentration in the atmosphere to about 16% below normal at the end of this century. It is thus reasonable that the ^{14}C excess caused by the atomic bomb will be more than compensated for by the dilution of inactive

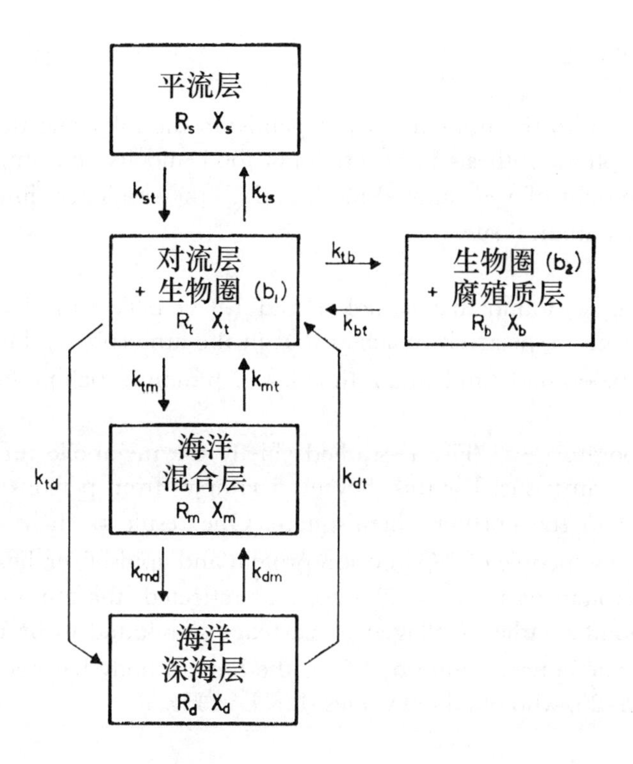

图 2. 可进行交换的碳库，其中：R_t 为总碳量；X_t 为 ^{14}C 过量；k_{ij} 为交换系数；i、j 代表 t、b、m、d ($i \neq j$)。

海洋被分成两个碳库，分别是混合层和深海层。CO_2 在对流层和海洋之间的交换主要发生在混合层，但根据克雷格的说法 [18]，直接交换也可能会发生在对流层和深海层之间。目前在海洋混合层中的 ^{14}C 浓度要高出正常值 10% ~ 15% [19,20]。

在之前发表的一篇文章中 [20]，我们利用图 2 的模型讨论了对流层中 $\delta^{14}C$ 的下降。我们认为，对流层中 $\delta^{14}C$（x_t）测量值的变化可大致由以下这个包含两项指数函数的式子来模拟：

$$x_t = A_1 e^{-k_1 t} + A_2 e^{-k_2 t} \tag{1}$$

其中，参数 A_1、A_2、k_1 和 k_2 由不同的交换系数决定。因为其中有一些系数误差很大，所以外推结果有一定的变化范围，在图 1 中用阴影部分表示。

核弹在大气中产生的 ^{14}C 已使自然界中 ^{14}C 的天然总含量增加了约 3%。根据哈克尼斯等人的说法 [8]：到本世纪末，由矿物燃料产生的 CO_2 将使大气中的天然 ^{14}C 浓度下降到比正常值低大约 16% 的水平。因此，我们可以合理地推出，非放射性碳的稀释作用完全可以补偿由核弹引起的 ^{14}C 过量（苏斯效应）。

carbon (the Suess effect).

The transfer of ^{14}C into the human body depends on the following three factors: (1) the time between the photosynthesis in vegetational food and its consumption; (2) the diet, particularly the amount of vegetational food, and (3) the residence time of the carbon in the constituents of human tissue.

Broecker *et al.*[6] (Fig. 1) found that it took 1 and 1.8 yr before the ^{14}C concentration in blood and lung tissue, respectively, reached that in the atmosphere. They also found that the $\delta^{14}C$ value of blood had a maximum time lag of 6 months behind food.

Berger and collaborators[7,21-23] (Fig. 1) studied chiefly the metabolic turnover time of the constituents of human tissue. For this they used samples from persons who had travelled from the southern to the northern hemisphere. One result of their work was that the incorporation in these people of ^{14}C in brain protein and lipids, liver, heart, plasma protein and erythrocyte protein was very similar to, and reflected, the atmospheric ^{14}C content present several months earlier. Collagen of cartilage was found to be metabolically inert in older persons. The concentration of ^{14}C in the human body has been studied by other workers (refs. 8, 24–26) who obtained values shown in Fig. 1.

At our laboratory the transfer of ^{14}C into the human body was studied by following the time-variation of $\delta^{14}C$ in blood and hair for three persons[5,27] (Tables 1–3). No separation between blood plasma and erythrocyte protein was performed, and the measured ^{14}C activity is thus a mean value for the total blood samples. Fig. 1 shows that there is excellent agreement between data obtained for the blood and for the hair samples. The values of the ^{14}C concentration in blood plasma obtained by Harkness and Walton[8] are slightly lower than ours.

Table 1. Carbon in Neck Hair, from a Boy (K.N.) Born in 1962

Time of collection		$\delta^{14}C$ %	$\delta^{13}C$/‰	$\Delta^{14}C$ %
November	1962	30.7±1.3		
June	1963	29.2±0.8	−19.7	26.7
October	1963	41.0±1.0	−19.2†	39.3
March	1964	46.0±1.0	−18.9	44.2
July	1964	54.8±1.0	−19.2†	52.9
October	1964	58.0±0.8	−18.6	55.9
February	1965	66.9±1.0	−19.2†	64.9
July	1965	67.2±1.0	−18.6	64.5
February	1966	68.7±1.0	−19.2	66.7
May	1966	70.0±0.9	−19.2†	68.0
December	1966	65.9±0.9	−19.2†	63.9
February	1967	67.5±1.1	−19.2†	65.4
July	1967	65.9±0.9	−19.2†	63.9
February	1968	61.5±0.9	−19.2†	59.6

^{14}C 向人体内的迁移过程取决于以下三个因素：(1) 从植物性食物进行光合作用到它被人类食用之间的时间；(2) 饮食结构，尤其是植物性食物所占的比重；(3) 碳在人体组织各组成成分中的滞留时间。

布勒克等人[6]（图 1）发现：要使血液和肺组织中的 ^{14}C 浓度达到大气中的浓度分别需要 1 年和 1.8 年的时间。他们还发现，从进食到血液中出现 $\delta^{14}C$ 之间的最大时间滞后可达 6 个月。

伯杰及其合作者[7,21-23]（图 1）重点研究了人体组织各组成成分的新陈代谢周转时间。他们为此选用的样本均来自于那些有过从南半球到北半球旅行经历的人。他们获得的一项研究成果显示：^{14}C 与这些人的脑蛋白以及脂类、肝、心、血浆蛋白和红细胞蛋白的结合情况非常接近于数月前大气中的 ^{14}C 含量，可以认为它能够反映数月前大气中的 ^{14}C 含量。他们发现老年人软骨组织中的胶原蛋白具有新陈代谢惰性。图 1 中还显示出了一些由其他研究者（参考文献 8 及 24~26）得到的人体中的 ^{14}C 浓度值。

在我们的实验室，对 ^{14}C 进入人体的研究是通过跟踪 3 个人血液和毛发中 $\delta^{14}C$ 随时间的变化来进行的[5,27]（表 1 ~ 表 3）。因为没有对血液中的血浆和红细胞进行分离，所以测得的 ^{14}C 放射性是整个血样的平均值。从图 1 中可以看出，由血样得到的数据和由毛发样本得到的数据吻合得非常好。哈克尼斯和沃尔顿测得的血浆中的 ^{14}C 浓度数据[8]略微低于我们测得的值。

表 1. 颈部毛发中的碳，来自于一个 1962 年出生的男孩（K.N.）

采集时间	$\delta^{14}C$ %	$\delta^{13}C$ /‰	$\Delta^{14}C$ %
1962 年 11 月	30.7 ± 1.3		
1963 年 6 月	29.2 ± 0.8	−19.7	26.7
1963 年 10 月	41.0 ± 1.0	−19.2†	39.3
1964 年 3 月	46.0 ± 1.0	−18.9	44.2
1964 年 7 月	54.8 ± 1.0	−19.2†	52.9
1964 年 10 月	58.0 ± 0.8	−18.6	55.9
1965 年 2 月	66.9 ± 1.0	−19.2†	64.9
1965 年 7 月	67.2 ± 1.0	−18.6	64.5
1966 年 2 月	68.7 ± 1.0	−19.2	66.7
1966 年 5 月	70.0 ± 0.9	−19.2†	68.0
1966 年 12 月	65.9 ± 0.9	−19.2†	63.9
1967 年 2 月	67.5 ± 1.1	−19.2†	65.4
1967 年 7 月	65.9 ± 0.9	−19.2†	63.9
1968 年 2 月	61.5 ± 0.9	−19.2†	59.6

Continued

Time of collection		$\delta^{14}C$ %	$\delta^{13}C$/‰	$\Delta^{14}C$ %
May	1968	60.2±1.1	−19.2†	58.2
August	1968	61.8±1.2	−19.2†	59.9
November	1968	59.7±1.1	−19.2†	57.7
January	1969	58.7±1.2	−19.2†	56.7
April	1969	58.2±1.2	−19.2†	56.5
May	1969	58.3±1.2	−19.2†	56.6
July	1969	58.1±1.1	−19.2†	56.4
October	1969	54.9±0.9	−19.2†	53.1
December	1969	54.7±1.1	−19.2†	52.9

† Not measured (mean value).

Table 2. Carbon in Blood Samples, from a Woman (I.N.), 26 Yr Old in 1963

Time of collection		$\delta^{14}C$ %	$\delta^{13}C$/‰	$\Delta^{14}C$ %
September	1963	34.9±1.1	−22.0	34.0±1.1
October	1963	39.3±0.8	−22.2	38.5±0.8
November	1963	43.6±0.8	−22.7	42.9±0.9
February	1964	45.6±1.1	−22.0	44.7±1.2
April	1964	48.1±1.1	−21.7	47.1±1.1
June	1964	49.6±1.1	−21.7†	48.6±1.1
July	1964	53.7±1.0	−22.1	52.8±1.1
November	1964	62.9±0.9	−19.8	61.2±0.9
January	1965	64.2±0.9	−19.7	62.4±0.9
March	1965	65.0±0.9	−20.5	63.5±1.0
May	1965	67.1±1.0	−21.1	65.6±1.0
September	1965	69.1±0.9	−22.0	67.9±1.0
November	1965	65.6±0.9	−21.4	64.2±1.0
March	1966	69.7±0.9	−22.3	68.7±0.9
July	1966	67.6±1.1	−21.2	66.2±1.1
September	1966	69.8±1.0	−21.5	68.6±1.1
December	1966	67.4±1.0	−21.9	66.3±1.0
September	1967	62.5±0.9	−21.7†	61.4±0.9
June	1968	60.6±0.9	−21.7†	59.5±1.0
August	1968	59.9±0.9	−21.7†	58.8±1.0
January	1969	56.0±0.7	−21.7†	55.0±0.7
September	1969	55.5±1.2	−21.7†	54.5±1.2
March	1970	52.6±1.2	−21.7†	51.6±1.2

† Not measured (mean value).

Table 3. Carbon in Blood Samples, from a Woman (A.L.), 26 Yr Old in 1963

Time of collection		$\delta^{14}C$ %	$\delta^{13}C$/‰	$\Delta^{14}C$ %
September	1963	35.2±0.8	−21.6	34.2±0.8
November	1963	41.5±0.6	−21.7†	40.6±0.7

续表

采集时间	$\delta^{14}C$ %	$\delta^{13}C$ /‰	$\Delta^{14}C$ %
1968 年 5 月	60.2 ± 1.1	−19.2†	58.2
1968 年 8 月	61.8 ± 1.2	−19.2†	59.9
1968 年 11 月	59.7 ± 1.1	−19.2†	57.7
1969 年 1 月	58.7 ± 1.2	−19.2†	56.7
1969 年 4 月	58.2 ± 1.2	−19.2†	56.5
1969 年 5 月	58.3 ± 1.2	−19.2†	56.6
1969 年 7 月	58.1 ± 1.1	−19.2†	56.4
1969 年 10 月	54.9 ± 0.9	−19.2†	53.1
1969 年 12 月	54.7 ± 1.1	−19.2†	52.9

† 未测量（平均值）。

表 2. 血样中的碳，来自于一位在 1963 年时年龄为 26 岁的妇女（I.N.）

采集时间	$\delta^{14}C$ %	$\delta^{13}C$/ ‰	$\Delta^{14}C$ %
1963 年 9 月	34.9 ± 1.1	−22.0	34.0 ± 1.1
1963 年 10 月	39.3 ± 0.8	−22.2	38.5 ± 0.8
1963 年 11 月	43.6 ± 0.8	−22.7	42.9 ± 0.9
1964 年 2 月	45.6 ± 1.1	−22.0	44.7 ± 1.2
1964 年 4 月	48.1 ± 1.1	−21.7	47.1 ± 1.1
1964 年 6 月	49.6 ± 1.1	−21.7†	48.6 ± 1.1
1964 年 7 月	53.7 ± 1.0	−22.1	52.8 ± 1.1
1964 年 11 月	62.9 ± 0.9	−19.8	61.2 ± 0.9
1965 年 1 月	64.2 ± 0.9	−19.7	62.4 ± 0.9
1965 年 3 月	65.0 ± 0.9	−20.5	63.5 ± 1.0
1965 年 5 月	67.1 ± 1.0	−21.1	65.6 ± 1.0
1965 年 9 月	69.1 ± 0.9	−22.0	67.9 ± 1.0
1965 年 11 月	65.6 ± 0.9	−21.4	64.2 ± 1.0
1966 年 3 月	69.7 ± 0.9	−22.3	68.7 ± 0.9
1966 年 7 月	67.6 ± 1.1	−21.2	66.2 ± 1.1
1966 年 9 月	69.8 ± 1.0	−21.5	68.6 ± 1.1
1966 年 12 月	67.4 ± 1.0	−21.9	66.3 ± 1.0
1967 年 9 月	62.5 ± 0.9	−21.7†	61.4 ± 0.9
1968 年 6 月	60.6 ± 0.9	−21.7†	59.5 ± 1.0
1968 年 8 月	59.9 ± 0.9	−21.7†	58.8 ± 1.0
1969 年 1 月	56.0 ± 0.7	−21.7†	55.0 ± 0.7
1969 年 9 月	55.5 ± 1.2	−21.7†	54.5 ± 1.2
1970 年 3 月	52.6 ± 1.2	−21.7†	51.6 ± 1.2

† 未测量（平均值）。

表 3. 血样中的碳，来自于一位在 1963 年时年龄为 26 岁的妇女（A.L.）

采集时间	$\delta^{14}C$ %	$\delta^{13}C$ /‰	$\Delta^{14}C$ %
1963 年 9 月	35.2 ± 0.8	−21.6	34.2 ± 0.8
1963 年 11 月	41.5 ± 0.6	−21.7†	40.6 ± 0.7

Continued

Time of collection		$\delta^{14}C$ %	$\delta^{13}C$/‰	$\Delta^{14}C$ %
February	1964	40.5±0.6	−20.9	39.3±0.7
March	1964	44.4±0.8	−22.5	43.5±0.9
April	1964	44.2±0.7	−23.2	43.6±0.7
June	1964	44.2±1.0	−22.1	43.3±1.0
August	1964	53.4±0.8	−21.9	52.4±0.9
November	1964	57.6±0.9	−20.7	56.2±0.9
January	1965	62.4±0.9	−19.1	60.5±0.9
March	1965	64.3±0.7	−20.1	62.6±0.7
May	1965	63.9±0.9	−21.0	62.5±1.0
September	1965	68.4±0.9	−22.0	67.3±1.0
November	1965	65.9±0.7	−21.4	64.4±0.8
March	1966	69.6±1.0	−22.0	68.5±1.1
July	1966	67.5±1.0	−21.2	66.2±1.1
September	1966	68.3±1.0	−20.9	66.9±1.0
December	1966	69.0±0.9	−20.8	67.6±1.0
September	1967	60.9±0.9	−21.7†	59.8±0.9
June	1968	61.4±0.9	−21.7†	60.3±0.9
August	1968	58.8±0.9	−21.7†	57.7±0.9
January	1969	56.6±0.7	−21.7†	55.6±0.7
May	1969	56.4±1.0	−21.7†	55.4±1.0
September	1969	53.0±1.0	−21.7†	52.0±1.0
March	1970	51.6±1.1	−21.7†	50.6±1.1

† Not measured (mean value).

The blood and hair data are almost representative for persons living in the northern hemisphere. Because $\delta^{14}C$ in the southern troposphere at present lags behind that of the northern troposphere by about 1 yr, there should also be a similar lag for $\delta^{14}C$ in the human populations of the respective hemispheres. After about 1970, the ^{14}C concentrations in people in the northern and southern hemispheres will be similar, and equal to those in the troposphere. Fig. 1 shows that $\delta^{14}C$ (x_H) in the human body appears as a pulse, delayed with respect to that of the atmosphere. The observed values for x_H can be fitted reasonably well by the following two-term exponential function:

$$x_H = 108\left(e^{-0.1t} - e^{-0.75t}\right) \tag{2}$$

The coefficients in this function were determined by a least mean squares method, using the upper limits of the measured values for ^{14}C in the human body in the period from 1963 to 1970, and extrapolated values in the troposphere in the period from 1970 to 2000. The upper limit values were chosen because there was some excess ^{14}C before 1963 which should also be considered. There is also a tendency for the blood and hair data during the last 2 yr to correspond with the upper limit of the shaded area in Fig. 1. Function (2) was simplified by assuming that all previous nuclear tests occurred within a short time interval, and that the ^{14}C increase in the human population started in about January 1963.

续表

采集时间	$\delta^{14}C$ %	$\delta^{13}C$ /‰	$\Delta^{14}C$ %
1964 年 2 月	40.5±0.6	–20.9	39.3±0.7
1964 年 3 月	44.4±0.8	–22.5	43.5±0.9
1964 年 4 月	44.2±0.7	–23.2	43.6±0.7
1964 年 6 月	44.2±1.0	–22.1	43.3±1.0
1964 年 8 月	53.4±0.8	–21.9	52.4±0.9
1964 年 11 月	57.6±0.9	–20.7	56.2±0.9
1965 年 1 月	62.4±0.9	–19.1	60.5±0.9
1965 年 3 月	64.3±0.7	–20.1	62.6±0.7
1965 年 5 月	63.9±0.9	–21.0	62.5±1.0
1965 年 9 月	68.4±0.9	–22.0	67.3±1.0
1965 年 11 月	65.9±0.7	–21.4	64.4±0.8
1966 年 3 月	69.6±1.0	–22.0	68.5±1.1
1966 年 7 月	67.5±1.0	–21.2	66.2±1.1
1966 年 9 月	68.3±1.0	–20.9	66.9±1.0
1966 年 12 月	69.0±0.9	–20.8	67.6±1.0
1967 年 9 月	60.9±0.9	–21.7†	59.8±0.9
1968 年 6 月	61.4±0.9	–21.7†	60.3±0.9
1968 年 8 月	58.8±0.9	–21.7†	57.7±0.9
1969 年 1 月	56.6±0.7	–21.7†	55.6±0.7
1969 年 5 月	56.4±1.0	–21.7†	55.4±1.0
1969 年 9 月	53.0±1.0	–21.7†	52.0±1.0
1970 年 3 月	51.6±1.1	–21.7†	50.6±1.1

† 未测量（平均值）。

来自血液和毛发的数据基本上代表了生活在北半球的所有人。因为南半球对流层的 $\delta^{14}C$ 值比北半球对流层滞后大约 1 年，所以居住在两个半球的人体内的 $\delta^{14}C$ 值也应该存在类似的滞后。大致到 1970 年之后，^{14}C 在南北半球的人体内的浓度将会趋于持平，并等于对流层中的 ^{14}C 浓度。从图 1 中可以看出：人体内部的 $\delta^{14}C$（x_H）值就像一个落后于大气中 $\delta^{14}C$ 值的脉冲。用以下这个包含两项指数函数的式子可以很好地拟合 x_H 的观测值：

$$x_H = 108\,(e^{-0.1t} - e^{-0.75t}) \tag{2}$$

函数式中的系数是通过最小均方法得到的，用到了在 1963 年～1970 年间测得的人体中 ^{14}C 的上限值，以及将对流层的数据外推到 1970 年～2000 年时的数据。选择上限值是因为还应当考虑到 1963 年以前的一些 ^{14}C 过量。在最近两年内得到的血液和毛发数据趋向于与图 1 阴影部分的上限值相符。化简式 (2) 的依据是以下两个假设：假设之前所有的核试验都集中发生在一个很短的时间间隔内，并且假设人体中 ^{14}C 含量开始增加的时间大约在 1963 年 1 月。

The first term in the brackets of equation (2) indicates that the excess ^{14}C in the human body has a mean lifetime of about 10 yr and the second term that ^{14}C enters the human body with a mean delay time of about 1.4 yr after production in the atmosphere. The latter value is probably accurate to within 30% and agrees with previous estimates[6,7].

The hazard to the human population from artificial radiocarbon arises largely from inventory radiation of the body. Natural ^{14}C contributes with a certain dose rate, r_0, as a result of its decay rate of about fourteen disintegrations per min per gram of carbon. The average value of r_0 is about 1.06 mrad/yr[9]. The dose from natural ^{14}C is distributed as follows: 1.64 mrad/yr in the bones, 1.15 mrad/yr in the cells lining bone surfaces and 0.71 mrad/yr in bone marrow and soft tissue. Applying function (2), the dose D_1 absorbed in the human body during a time t can be calculated from the formula:

$$D_1 = 1.08\, r_0 \int_0^t (e^{-0.1t} - e^{-0.75t})\, dt \tag{3}$$

For a period of about 30 yr the total radiation dose from this source will be $9r_0$. We thus obtain a total radiation dose of 16 mrad to bone, 11 mrad to cells lining bone surfaces and 7 mrad to bone marrow and soft tissue. The genetic hazard is caused by the latter. That dose constitutes about 10% of the total gonad dose from all radioactive fallout. Purdom[4] pointed out, however, that the actual gonad dose is somewhat larger because of a transmutation process in the DNA molecule, in which the decaying ^{14}C atoms are replaced by nitrogen atoms. Purdom assumed that the biological damage from the transmutation was equal to that from β-radiation.

The amount of artificial ^{14}C in the human body at about the year AD 2000 will constitute about 3% of the total amount of ^{14}C. This isotope has a half-life of 5,700 yr, and several scientists[2,3,9,28] think that it would therefore cause a most serious genetic threat. The long term radiation dose can be calculated from the formula

$$D_2 = 0.03\, r_0 \int_0^t e^{-0.000125t}\, dt \tag{4}$$

The total radiation doses (D_1+D_2) which will be received by the bone cells, the cells lining bone surfaces, and the bone marrow and soft tissue in the next 10,000 yr will be 410, 290 and 180 mrad, respectively. These doses, which are in agreement with values given in a United Nations report[9] (page 45), are more important than those from all other radioactive fallout. We question, however, the value of the long term radiation dose (D_2) because, as previously mentioned, the use of fossil fuel might reduce the ^{14}C concentration in man below normal. We are of the opinion that the only ^{14}C hazard from previous tests which should be taken into account is attributable to a total genetic dose (D) of the order of 10 mrad, received in a period of about 30 yr. This dose is, however, negligible compared with the dose received from natural sources, which constitutes about 100 mrad per yr.

式（2）括号内的第一项表明，人体内过量 ^{14}C 的平均寿命大约为 10 年；第二项表明，^{14}C 从在大气中产生到进入人体之间会有平均 1.4 年左右的时间滞后。滞后值的精确度大约可以达到 30% 以内，并与以前文献中的估计值相符 [6,7]。

人造放射性碳对人类造成的危害主要来源于人体内的辐射量。天然 ^{14}C 的剂量率 r_0 是一个固定值，由它的衰变速率——每克碳每分钟衰变约 14 次——决定。r_0 的平均值大致是 1.06 毫拉德 / 年 [9]。天然 ^{14}C 的剂量是这样分布的：在骨骼中为 1.64 毫拉德 / 年，在骨骼表层细胞中为 1.15 毫拉德 / 年，在骨髓和软组织中为 0.71 毫拉德 / 年。根据函数式（2），人体在 t 时间内吸收的剂量 D_1 可由下式计算：

$$D_1 = 1.08\, r_0 \int_0^t \left(e^{-0.1t} - e^{-0.75t}\right) dt \tag{3}$$

在大约 30 年内，来自该辐射源的总辐射剂量将为 $9r_0$。因此我们的骨骼所接受的辐射剂量为 16 毫拉德，骨骼表层细胞为 11 毫拉德，骨髓和软组织为 7 毫拉德。7 毫拉德的剂量会对基因造成危害。这一剂量大致等于生殖腺从所有放射性尘埃中所吸收的总剂量的 10%。但珀德姆 [4] 指出：实际上生殖腺所接收到的辐射剂量还要更大一些，因为 DNA 分子会发生嬗变，在嬗变过程中产生衰变的 ^{14}C 原子被替换成了 N 原子。珀德姆认为该嬗变过程对生物造成的伤害就相当于 β 辐射对生物的伤害。

到大约公元 2000 年时，人体中的人造 ^{14}C 量将占到 ^{14}C 总量的 3% 左右。这种同位素的半衰期为 5,700 年，因而有几位科学家 [2,3,9,28] 认为它将对人类基因造成严重的威胁。长期的辐射剂量可由下式计算：

$$D_2 = 0.03\, r_0 \int_0^t e^{-0.000125t}\, dt \tag{4}$$

在未来的 10,000 年里，由骨骼细胞、骨骼表层细胞以及骨髓和软组织所吸收的总辐射剂量（D_1+D_2）分别为 410、290 和 180 毫拉德。这些剂量与一份联合国报告 [9]（第 45 页）中给出的值一致，它们比所有其他的放射性尘埃都更重要。然而，我们对长期辐射剂量（D_2）有所怀疑，因为正如我们之前曾经提到的，矿物燃料的使用可能会导致人体中的 ^{14}C 浓度降低至正常水平以下。我们认为：在大约 30 年内，由以前有足够规模的核试验所造成的 ^{14}C 危害对总遗传剂量（D）的贡献只有 10 毫拉德数量级。而从天然源中吸收的剂量约为 100 毫拉德 / 年，可见，与天然源相比，人工源的剂量是不值一提的。

We thank Norges Almenvitenskapelige Forskningsråd for financial support.

(**232**, 418-421; 1971)

Reidar Nydal, Knut Lövseth and Oddveig Syrstad: Radiological Dating Laboratory, Norwegian Institute of Technology, Trondheim, Norway.

Received November 10, 1969; revised August 24, 1970.

References:

1. Libby, W. F., *Science*, **123**, 657 (1956).
2. Pauling, L., *Science*, **128**, 3333 (1958).
3. Sakharow, A. D., *Soviet Scientists on the Danger of Nuclear Tests*, 39 (Foreign Languages Publishing House, Moscow, 1960).
4. Purdom, C. E., *New Scientist*, **298**, 255 (1962).
5. Nydal, R., *Nature*, **200**, 212 (1963).
6. Broecker, W. A., Schulert, A., and Olson, E. A., *Science*, **130**, 331 (1959).
7. Libby, W. F., Berger, R., Mead, J. F., Alexander, G. V., and Ross, J. F., *Science*, **146**, 1170 (1964).
8. Harkness, D. D., and Walton, A., *Nature*, **223**, 1216 (1969).
9. *Rep. UN Sci. Comm. Effect of Atomic Radiation*, No. 14 (A/5814) (United Nations, New York, 1964).
10. Münnich, K. O., and Roether, W., *Proc. Monaco Symp.*, 93 (Vienna, 1967).
11. Bien, G., and Suess, H., *Proc. Monaco Symp.*, 105 (Vienna, 1967).
12. Rafter, T. A., *NZ J. Sci.*, **8**, 4, 472 (1965).
13. Rafter, T. A., *NZ J. Sci.*, **11**, 4, 551 (1968).
14. Young, J. A., and Fairhall, A. W., *J. Geophys. Res.*, **73**, 1185 (1968).
15. Lal, D., and Rama, *J. Geophys. Res.*, **71**, 2865 (1966).
16. Nydal, R., *J. Geophys. Res.*, **73**, 3617 (1968).
17. Nydal, R., *Symp. on Radioactive Dating and Methods of Low-Level Counting, UN Doc. SM 87/29* (International Atomic Energy Agency, Vienna, 1967).
18. Craig, H., *Second UN Intern. Conf. on the Peaceful Uses of Atomic Energy*, A/CONF. 15/P/1979 (June 1958).
19. Rafter, T. A., and O'Brien, B. J., *Proc. 12th Nobel Symp.* (Almquist and Wiksell, Uppsala, 1970).
20. Nydal, R., and Lövseth, K., *CACR Symp.* (Heidelberg, 1969); *J. Geophys. Res.*, **75**, 2271 (1970).
21. Berger, R., Fergusson, G. J., and Libby, W. F., *Amer. J. Sci., Radiocarbon Suppl.*, 7, 336 (1965).
22. Berger, R., and Libby, W. F., *Amer. J. Sci., Radiocarbon Suppl.*, **8**, 467 (1966).
23. Berger, R., and Libby, W. F., *Amer. J. Sci., Radiocarbon Suppl.*, **9**, 477 (1967).
24. Godwin, H., and Willis, E. H., *Amer. J. Sci., Radiocarbon Suppl.*, **2**, 62 (1960).
25. Godwin, H., and Willis, E. H., *Amer. J. Sci., Radiocarbon Suppl.*, **3**, 77 (1961).
26. Drobinski, jun., J. C., La Gotta, D. P., Goldin, A. S., and Terril, jun., J. G., *Health Phys.*, **11**, 385 (1965).
27. Nydal, R., and Lövseth, K., *Nature*, **206**, 1029 (1965).
28. Pauling, L., *Les Prix Nobel*, 296 (1963) (Nobelstiftelsen, Stockholm, 1964).

感谢挪威自然和人文科学研究理事会为我们提供了经费上的支持。

（邓铭瑞 翻译；刘京国 审稿）

Statistical Mechanics and Quantum Mechanics

G. E. Uhlenbeck

Editor's Note

Nature here reprints a notable address given by physicist George Uhlenbeck in a ceremony awarding him the Lorentz Medal of the Royal Netherlands Academy of Science. Uhlenbeck describes his long-standing interest in fundamental questions of statistical mechanics, and notes that while the foundational problems of quantum theory had occupied theorists for decades, phase transitions—such as the abrupt gas-liquid transition—and related phenomena continued to pose many unsolved problems. Uhlenbeck also claims that the idea of a generally apparent frontier of science—separating the known from the unknown—is mostly a romantic illusion. Instead, he suggests, there are always many frontiers, and they usually become identified whenever there is an advance, not the other way around.

ARE the problems of statistical mechanics truly fundamental? I have often changed my opinion and now I would like to elaborate on it. When I was a young student, kinetic theory of matter seemed to me an example of a theory which truly explains something. With much care and effort I worked my way through the *Lectures on Gas Theory* by Boltzmann and the *Elementary Principles of Statistical Mechanics* by Gibbs. Much escaped me and became clear only after I read the famous encyclopaedia article by the Ehrenfests. It was a revelation, not only because of its great clarity but also because it contained a careful summing up of the series of more than twelve *lacunae* in the work of the masters. These were like frontier posts and a young student could thereby learn where the real problems lay. How difficult it is to find this out nowadays. The present pollution of the scientific literature makes the finding of clear water, the fundamental concepts, an extremely time consuming occupation. This is true not only for the new student but for anyone who tries to learn something outside his own speciality, as I well know by experience.

Though I passed my examinations successfully, I knew very little of quantum theory and even less of the theory of spectra. I learned that, with Goudsmit's help, when I returned from Rome in 1925 and became Ehrenfest's assistant. I shall not elaborate on the collaboration with Goudsmit, which I shall never forget and which led to the discovery of electron spin. Both Goudsmit and I have often related our memories of this unforgettable period and I mention it only because at that time my conception of statistical mechanics changed completely. I considered it to be clearly on a secondary level. Quantum mechanics on the other hand provided a foundation from which everything should follow, including the behaviour of gases, liquids and solids. This seemed to be confirmed by the success of the electron theory of metals. Pauli and Sommerfeld (both recipients of the Lorentz Medal) showed how all difficulties disappeared as soon as one applied the true quantum statistics. A few riddles remained, such

统计力学和量子力学

乌伦贝克

编者按

《自然》杂志在这里转载了物理学家乔治·乌伦贝克在接受荷兰皇家学会授予他洛伦兹奖的仪式上发表的一篇著名演讲。乌伦贝克谈到自己长期以来一直对统计力学中的一些基本问题很感兴趣，并且提到：量子理论的基本问题在数十年内吸引了众多理论工作者，与此同时，相变（例如突发的气－液转换）以及相关的现象也不断引发许多尚待解决的问题。乌伦贝克还指出：那种认为科学存在大致明显的前沿（将未知领域与已知领域分开）的观点多半是一种浪漫的幻象。相反，他认为：科学中总有很多前沿，它们往往是在科学产生进展的时候才被意识到，而不是反过来。

统计力学的问题真是基本性的吗？在这个问题上我自己的观点也经常改变，现在我想详细讲述一下。当我是一名年轻学生的时候，关于物质的动力学理论对我来说似乎是一种真正能够解释一些事物的理论。我花了很大工夫去钻研玻尔兹曼的《气体理论讲义》以及吉布斯的《统计力学的基本原理》。很多东西我当时并没有领会，只是在读完埃伦费斯特夫妇的那篇著名的百科全书文章（译者注：是应《数学科学百科全书》主编的要求撰写的）后才豁然开朗。那篇文章极具启发性，不仅因为清楚明了，而且因为它包括了对大师工作中一系列**缺陷**（超过 12 个）的仔细总结。这些著作就像边境站一样，可以让年轻学生了解到真正的问题所在。现在要了解这些问题是非常困难的。当前科学文献的污染使人们要花费很长时间才能找到一泓清泉，也就是那些基本概念。不仅对新学生是如此，对于任何想在自己专业之外了解一些知识的人也是一样，这一点我深有体会。

尽管能成功通过多次考试，但实际上我对量子理论知之甚少，对谱理论就更缺乏了解了。我是在 1925 年从罗马回来并成为埃伦费斯特的助手时，在古德斯密特的帮助下才认识到这点的。与古德斯密特的那段合作令我终生难忘，在合作中我们发现了电子的自旋，但在这里我不打算详述那段合作。古德斯密特和我经常谈起那段难忘的时光，我在这里提起它只是因为我对统计力学的观念在那段时间发生了彻底的改变。我本来认为它显然处于第二级的地位。另一方面，量子力学则提供了一个一切事物，包括气体、液体和固体的行为，必须遵循的基本原则。这似乎已被金属电子理论的成功所证实。泡利和索末菲（都是洛伦兹奖的获得者）曾揭示过所有的困难是怎样在应用了真正的量子统计之后迎刃而解的。目前还剩下几个未解之谜，

as superconductivity, but in our optimism we felt sure that all would be straightened out eventually. My dissertation in 1927 about statistical methods in quantum theory was therefore a kind of optimistic synthesis of the encyclopaedia article of the Ehrenfests and the new quantum mechanical ideas. The number of unproved assumptions was reduced to three.

Later in Ann Arbor, the beautiful experiments of N. H. Williams on thermal noise and shot noise aroused my interest in the theory of Brownian motion, one of the nicest applications of statistical mechanics. I thought it very interesting but of course it was not fundamental. I remember very well that when I told Pauli about it he called it "desperation physics". I didn't like this but I really agreed with him. For a physicist of the quantum mechanics generation to which I also belonged, the fundamental problems of the 1930s were the theory of the positron, quantum electrodynamics and the developing theories of nuclear structure and beta radioactivity; these were the things on which to work.

Let me say at this juncture that the image of progress in science as a kind of conquest of an unknown domain with a definite "frontier" and successive "breakthroughs" seems to me more and more to be only a romantic illusion. This picture was clearly inspired by the great breakthrough of quantum mechanics and it has influenced my judgment for a long time. I might even say that I was afflicted by it.

My opinion about the fundamental character of statistical mechanics began to change in 1937 to 1938 when I, together with Boris Kahn, became involved in the so called condensation theory. The question why a gas condenses below a sharply determined critical temperature at a sharply determined density has never been called to attention since Van der Waals and the proper understanding of it seemed difficult to us. During the lively discussion about this question at the Van der Waals Congress in 1938 it was even doubted whether the basic assumptions of statistical mechanics contained the answer even in principle. This doubt was not justified, but it made a deep impression on me. As long as such common phenomena as the equilibrium between liquid and vapour and the existence of a critical temperature were not truly understood, the field was not yet conquered, not everything had been explained in principle and thus there existed fundamental aspects of statistical mechanics which I had not appreciated.

(When I showed Pauli the article by Kahn and myself on condensation theory he looked at it and said, "Yes, one should read this". He did so and ridiculed somewhat the quasi-mathematical rigour of our work but I believe that he appreciated the fundamental character of the problem.)

After the war, I continued to follow the new developments in quantum electrodynamics and the theory of beta radioactivity, and from time to time I contributed a little. But my interest moved more and more towards the fundamental questions of statistical mechanics. In the 1950s I began to write a book on statistical mechanics with my pupil and friend, the late T. H. Berlin. I hoped that it would become a modernized and expanded version of the encyclopaedia article which had made such an impression on me. I hoped that we could determine the foundations

如超导电性，但是我们从乐观的角度来看，最终所有难题必将得到解决。因此，我在 1927 年的那篇关于量子理论中统计方法的博士学位论文，便成为埃伦费斯特夫妇的百科全书文章与新的量子力学概念的一种乐观综合。未经证实的假设数目被减少到 3 个。

之后在安阿伯市，威廉姆斯关于热噪声和散粒噪声的漂亮实验引起了我对布朗运动理论的兴趣。布朗运动理论是统计力学最好的应用之一，我觉得它非常有意思，但显然布朗运动理论并非基本性的问题。我记得非常清楚，当我和泡利谈起布朗运动理论时，他称其为“绝望物理学”。虽然我不喜欢这样的表述，但很同意泡利的观点。对于一位量子力学代的物理学家（我也属于这一代）来说，20 世纪 30 年代的基本问题是正电子理论、量子电动力学和正在发展的核结构理论、β 放射性理论；这些才是当务之急。

在这个关键时刻，我想说：把科学进步看成是对有清楚“前沿”并能不断被“突破”的未知领域的一种征服，对我来说越来越像一个浪漫的幻象。这个图景显然是由量子力学的伟大突破激发产生的，它在很长一段时间里影响了我的判断。我甚至可以说它曾折磨着我。

在 1937 年到 1938 年间我对统计力学基本特征的看法开始改变，那时候我和鲍里斯 · 卡恩在合伙研究所谓的凝聚理论。为什么一种气体在严格确定的临界温度之下以严格确定的密度凝聚，这个问题自范德瓦尔斯之后从未引起过人们的注意，而我们要想正确理解它似乎并不容易。在 1938 年范德瓦尔斯会议上与会者们对这个问题的热烈讨论中，有人甚至怀疑统计力学的基本假设在原理上是否能够解答这个问题。这个怀疑没有根据，但是给我留下了深刻的印象。只要像液体和气体之间的平衡以及临界温度的存在这类常见现象尚未被真正理解，这个领域就还没有被征服，就不能说所有的事情都已经从原理上得到了解释，因而统计力学中还存在着一些基本问题是我没有意识到的。

（当我把卡恩和我所写的关于凝聚理论的文章给泡利看的时候，他看了一下说：“嗯，这个值得一读。”他确实读了，还取笑了我们文章中半吊子数学的严格性，但是我相信他理解这个问题的基本性。）

战后，我继续关注量子电动力学和 β 放射性理论的新进展，并不时发表一些自己的成果。但是我的兴趣越来越偏重于统计力学中的基本问题。在 20 世纪 50 年代，我开始和我的学生也是我的朋友——已故的伯林一起撰写一部关于统计力学的书。我希望这部书能够成为那篇曾让我留下深刻印象的百科全书文章的一个现代扩充版

of the theory in the same critical way as the Ehrenfests, and that we could make clear the nature of the still unsolved basic problems. In short, I hoped that we could discover the "structure" of statistical mechanics.

We worked hard on it but did not get very far and the sudden death of Ted Berlin in 1962 makes it doubtful that our plan will ever be realized. I have learned a lot from it, however, and I have the feeling that I more or less understand the structure of classical statistical mechanics at least. I believe that one always has to keep in mind that the task of statistical mechanics is to study the relationship between the macroscopic description of physical phenomena and the microscopic molecular description. These two pictures are in a certain sense independent; moreover, they are, so to speak, on a different level and are therefore, even qualitatively, totally different. If one sticks to this idea, one can see that Boltzmann, Gibbs, Einstein, Ehrenfest and Smoluchowski have formulated the true basis on which one has to build further. One sees also that there are still many unsolved problems, such as condensation and other so called phase transitions, on which much work is being done. All these problems are very difficult but they are, as a mathematician would say, *bien posés* and therefore one can work on them.

In my opinion the situation is somewhat different for quantum theory. The relationship between the classical and quantum mechanical descriptions of molecular phenomena is rather clear, but this is not so for quantum mechanics and the macroscopic theory. I am aware that this opinion is not shared by most of my colleagues. I believe that the most widespread opinion is that quantum mechanics can be "grafted" onto the classical statistical mechanics of Boltzmann and Gibbs and that therefore quantum mechanics does not require anything essentially new. I had thought so earlier too, but I have slowly retreated from that conviction. The recent discoveries of so called macroscopic quantization and interference phenomena in liquid helium and superconductors have had a great influence on my ideas. They seem to me to show that the existing theories of superconductivity and superfluidity do not provide a complete explanation and that the true macroscopic description of the superfluids has not yet been found. It would be no surprise to me that this is so because the foundations of quantum statistical mechanics have not yet been sufficiently clarified. One only needs to think of the persistent currents to feel doubt about the general validity of the ergodic theorems in quantum theory. Questions like these make low temperature physics so fascinating for me; they have had a rejuvenating effect on me and I am convinced that there is still much to be done on this "frontier" of physics and that profound surprises are possible.

Because I use the word "frontier" let me finally return to the romantic image of advances in science which I sketched earlier. I do not believe in it any longer. There are many frontiers and it comes down to the fact that in science one can only sometimes talk of progress. Whenever there is an advance there is a frontier, not the other way around. As to the direction of the advance, every investigator follows his own nose and does what he can. In my opinion this applies equally to space travel and to high energy physics and radio astronomy. These pursuits exist because they are possible and as long as the expense does not become too exorbitant one must, of course, continue them. But I think that one must oppose all fashion and prestige arguments. There is no natural hierarchy of problems and moreover, as Poincaré remarked

本。我希望我们能够像埃伦费斯特夫妇那样批判性地确定理论的基础，并能够弄清楚那些仍未解决的基本问题的本质。简言之，我希望我们能够发现统计力学的“架构”。

为此我们努力工作，但没有取得太大的进展，加之1962年特德·伯林突然离世，使得我们的计划能否最终实现都成了问题。但是我还是弄清了很多东西，并且感觉自己至少或多或少地理解了经典统计力学的架构。我相信一个人必须时刻记住，统计力学的任务是研究物理现象的宏观描述和微观分子描述之间的关系。这两个图景从某种意义上说是各自独立的；甚至可以说它们是在不同的层次上，因此即使定性地说也是完全不同的。如果坚持这个观点，就可以看出玻尔兹曼、吉布斯、爱因斯坦、埃伦费斯特和斯莫鲁霍夫斯基已经构造了真正的基础以供后人进一步发展。还可以看出仍然存在很多没有解决的问题，比如凝聚以及其他很多人正在研究的所谓相变过程。所有这些问题都非常艰涩，但是正如一位数学家所说的，它们是适定的，因此人们可以努力去解决它们。

按照我的观点，量子力学的情况会有所不同。分子现象的经典力学描述和量子力学描述之间的关系相当清楚，但是量子力学和宏观理论之间就另当别论了。我知道我的大部分同事都不同意这个观点。我相信最被广泛接受的观点是量子力学可以“嫁接”到玻尔兹曼和吉布斯的经典统计力学上，因此量子力学并不需要任何本质上是新的东西。以前我也这么认为，但慢慢地就不那么有信心了。最近，在液氦和超导体中发现的所谓宏观量子化以及干涉现象对我的观点影响很大。在我看来，这些发现都表明现有的超导电性和超流体性理论不能提供一个完整的解释，并且对超流体的真正宏观描述至今仍没有找到。对此我并不感到奇怪，因为量子统计力学的基础还没有被充分阐明。只要想想持续电流，就会对量子理论中遍历定理的普遍适用性产生怀疑。类似这样的问题使我感到低温物理非常有趣；它们使我精神振奋，我相信在物理学的这个“前沿”上仍有许多工作有待完成，并且还可能会有意义深刻的意外发现涌现出来。

既然使用了“前沿”这个词，让我最后回头再来看看前面简要提到的科学发展的浪漫图景。我已经不再相信它了。前沿有很多，但是归根结底的事实是，在科学中一个人只能不时地谈发展。每当有了一个进步就会出现一个前沿，而不是反过来。至于前进的方向，每位研究者都可以按自己的主意尽力而为。在我看来，这种情况对太空旅行和对高能物理学、射电天文学都同样适用。这些研究之所以存在，是因为它们有可能实现，只要代价不是太过昂贵，当然就应该继续下去。但是我觉得应该反对所有关于时髦和地位的争论。问题并无天然的等级之分，而且正如庞加莱很

long ago, a problem is never completely but always only more or less solved.

It seems better to me to view progress in science as the expansion of different circles of investigation, each autonomous and often apparently entirely independent of the others. The deep problems are to determine how these areas hang together, and how one can arrive at a larger unity. Biology and physical–chemical research, for example, form two such large circles. Their interrelationship seems to me to represent one of the deepest questions man can ask, a real *mysterium tremendum*. On a much smaller scale macroscopic and molecular physics form two such circles and statistical mechanics attempts to fathom their relationship.

(**232**, 449-450; 1971)

G. E. Uhlenbeck: Rockefeller University, New York.

久以前所指出的：一个问题永远不可能被彻底解决，只能或多或少被解决。

我更愿意把科学的发展看作是向许多不同研究领域的扩展，每个领域都是自主的，而且经常表现得完全独立于其他领域。深层次的问题是确定如何使这些领域互相结合在一起，且如何达到更广泛的统一。举例来说，生物学和物理-化学研究就是两个这样的大领域。在我看来，它们之间的相互关系是人类能够提出的最有深度的问题之一，是一个真正令人敬畏的奥秘。在小得多的尺度上，宏观物理和分子物理组成了两个这样的领域，统计力学则试图去探究它们之间的关系。

（何钧 翻译；李军刚 审稿）

The Macroscopic Level of Quantum Mechanics

C. George *et al.*

Editor's Note

Albert Einstein and other critics of quantum theory had long noted the difficulty of reconciling its description of the microscopic world—where it suggests particles often exist in a superposition of two or more states—with the existence of macroscopic objects always definite properties. Here Leon Rosenfeld and colleagues attempt to show how this inconsistency could be resolved by considering the "complementarity" of distinct modes of description at the atomic level, advocated by Niels Bohr. They argue that Bohr's view could be put into a formal framework using the density matrix of quantum statistical mechanics. This proposal, physicists later showed, did not resolve the matter, which continues to be an important foundational problem for quantum theory today.

ATOMIC theory raises the epistemological problem of harmonizing the detailed dynamical description of atomic systems given by quantum mechanics and the description of individual atomic processes and of the behaviour of matter in bulk at the level of macroscopic observation. The logical side of this problem is completely elucidated by the recognition of relationships of complementarity between the two modes of description: on the one hand, there is the complementarity expressed by the indeterminacy relations, which governs the application of the macroscopic space-time localization and momentum-energy conservation to the individual atomic processes; on the other, there is the complementarity between the account of the behaviour of a large system of atomic constituents given by quantum mechanics (and electrodynamics) and the description of the same system as a material body in terms of the concepts of macroscopic mechanics, electromagnetism and thermodynamics.

The formal side of the problem, however, which consists in establishing the consistency of the rules connecting the formalism of quantum mechanics with the concepts used in the account of macroscopic observation, still leaves scope for a presentation more in accordance with the conceptual simplicity of the actual situation. A general treatment of the quantum theory of large atomic systems which satisfies this desideratum has been given in a recent article[1]. The purpose of this note is to outline the method and the principal results.

The work pursued during the past decade by the Brussels group has clearly shown that the method best adapted to the investigation of large atomic systems is not the ergodic but the kinetic approach of statistical mechanics, applied to the limiting case of infinite systems (that is, systems of infinite degree of freedom, but such that the number of elements

宏观层次的量子力学

乔治等

编者按

量子理论对微观世界的描述认为，微观粒子常常处于两个或两个以上状态的叠加态中。阿尔伯特·爱因斯坦以及其他量子理论的批评者们很早就注意到，此一对微观世界的描述与宏观物体通常具有明确的属性这一事实之间存在着难以调和的矛盾。本文中，利昂·罗森菲尔德与其同事们试图说明这一矛盾可以通过考虑尼尔斯·玻尔所倡导的、原子层次上不同描述模式间的“互补性”来加以解决。他们认为，利用量子统计力学中的密度矩阵就可将玻尔的观点纳入一个形式的框架中。物理学家们后来指出，上述提议并没有解决这个难题，直至今日它仍然是量子理论中一个重要的基本问题。

原子理论提出了一个认识论问题——如何调和量子力学所给出的对原子系统的详细动力学描述，与在宏观观测水平上对单个原子过程以及大块物质行为的描述二者之间的关系。认识到两种描述模式之间的互补关系就可以完全阐明这一问题的逻辑性：一方面，不确定关系表达出互补性，它决定了宏观的时空局域性与动量-能量守恒可应用于单个的原子过程；另一方面，对于一个由原子组成的大系统，根据量子力学（和电动力学）计算其行为与将此系统视为材料实体按照宏观力学、电磁学以及热学的概念对其进行描述，这两者之间存在着互补性。

但是，这一问题形式上的一面，即建立能将量子力学公式与描述宏观观测所用概念联系起来的各个规律之间的一致性，仍为发展出一种与实际情况概念上的简单性更符合的表述留出了空间。我们在近期发表的一篇文章[1]中给出的用量子理论处理较大原子体系的一般性方法能满足上述急迫需求。本文旨在概述这一方法及其重要结论。

布鲁塞尔小组过去十年所从事的工作清楚地表明，最适于研究大原子系统的是曾经用于无限系统的极限情况（即具有无穷多自由度的系统，但该系统在给定相位范围内的单元是有限的）的统计力学动力学方法而不是各态遍历方法。因为这类系

within a given phase extension has a finite limit). Because such systems have essentially a continuous energy spectrum, the vexed question of "coarse graining" can be ignored and the asymptotic limit of the density operator for times very large on the atomic scale, but finite, can be directly discussed. This discussion leads to the conclusion[2] that the time evolution of the system may be split rigorously into formally independent "subdynamics", characterized, in a manner presently to be explained, by certain projection operators depending on the correlations between the constitutive elements of the system. One of these subdynamics belonging to the projector $\tilde{\Pi}$ given by equation (7) below contains all the information about equilibrium properties and linear transport properties. We may therefore define the macroscopic level of description of quantum mechanics as the reduced description in terms of the variables of the $\tilde{\Pi}$-subspace. That such a reduced description is at all possible results precisely from the fact that the $\tilde{\Pi}$-subdynamics is expressed by an independent equation of evolution. The application of this analysis to the case of a measuring apparatus, initially triggered off by an interaction of atomic duration with an atomic object, shows in a surprisingly simple way that its evolution in $\tilde{\Pi}$-space conforms to the "reduction" rule of quantum mechanics; this important conclusion follows directly from the mathematical structure of the $\tilde{\Pi}$-subspace of the apparatus, which entails the elimination of the initial phase relations between the components of the density operator of the atomic object. Our paper contains a simple and general derivation of these remarkable results, which shows more clearly how these methods not only lead to a deeper understanding of the epistemological problems of atomic theory, but even to a significant extension of the scope of quantum mechanics.

In a superspace defined as the direct product of the Hilbert space and its dual, the density operator $\rho(t)$ appears as a supervector, varying in time according to a Liouville equation of the form

$$i\dot{\rho}(t) = L\rho(t)$$

The Liouville superoperator L may be expressed in terms of the Hamiltonian H of the system. For this purpose, we may use a convenient notation for the special class of "factorizable" superoperators $O \equiv M\times N$ depending on a pair of supervectors M, N according to the definition $O\rho = M\rho N$; we may then write $L = H\times 1 - 1\times H$. Our aim being to find the long time effect of correlations, we must, to begin with, compare our system, defined by the Hamiltonian H, with a "model" system H_0, from which the interaction energy V, responsible for the correlations, is removed, in such a way that $H = H_0 + V$. The eigenstates of H_0 form a complete orthogonal basis of representation in Hilbert space, from which we construct a similar basis in superspace: the latter may be divided into two classes of supervectors, those built up of pairs of identical (or physically equivalent) eigenstates, and those built up of pairs of different eigenstates; they belong, respectively, to two orthogonal subspaces of superspace, characterized by projection superoperators P_0, P_c. Then the projections $\rho_0 = P_0\rho$ and $\rho_c = P_c\rho$ of the density supervector correspond, respectively, to the average distribution densities and the correlation amplitudes. Putting $L_{00} = P_0LP_0$, $L_{0c} = P_0LP_c$ and so on, we obtain for ρ_0 and ρ_c the coupled Liouville equations

统基本上具有连续能谱，所以可以忽略棘手的“粗粒化”问题。此外，关于时间的密度算符的渐近极限虽然在原子尺度上非常大，但仍然有限，可以直接讨论。讨论的结果[2]是：或许可以将系统的时间演化严格地分成形式上独立的“子动力学”部分，以目前的解释方式（也就是采用取决于系统组成元素之间关联性的特定投影算符）来表征。下文中式 (7) 给出的投影算符 $\tilde{\Pi}$ 的其中一个子动力学即可包含关于平衡性质和线性输运性质的所有信息。因此，我们可以将宏观层次的量子力学描述定义为基于 $\tilde{\Pi}$ 子空间变量的约化描述。这一约化描述有可能恰好是源于以下事实：$\tilde{\Pi}$ 子动力学是由一个独立的演化方程来表示的。将这种分析方法应用于一测量装置，该装置初始时通过原子寿命与原子物体之间的相互作用而触发。结果出人意料地表明：它在 $\tilde{\Pi}$ 空间中的演化遵循量子力学的“约化”规则；这一重要结论是由该装置 $\tilde{\Pi}$ 子空间的数学结构直接得出的，它消除了原子物体密度算符各分量之间的初始相位关系。在本文中我们叙述了以一种简单而通用的方式推导出上述不寻常结论的过程，由此可以更清晰地表明：这些方法如何能不仅使我们更深刻地理解原子理论中的认识论问题，甚至还能显著拓展量子力学的范围。

定义希尔伯特空间与其对偶空间的直积为超空间，其中的密度算符 $\rho(t)$ 表现为超矢量，它按照以下形式的刘维尔方程随时间变化：

$$i\dot{\rho}(t) = L\rho(t)$$

刘维尔超算符 L 可以用系统的哈密顿量 H 表示。为此，根据定义 $O\rho=M\rho N$，我们可以用简便的符号来表示取决于一对超矢量 M、N 的一类特殊的“可分解”超算符 $O\equiv M\times N$；然后，我们可以得到 $L=H\times 1-1\times H$。我们的目的是要找出关联性造成的长时效应，首先我们必须将我们的由哈密顿量 H 描述的系统与由 H_0 描述的“模型”系统进行比较。在“模型”系统中，与关联性有关的相互作用能 V 被去掉，它们的关系可以由 $H=H_0+V$ 表示。H_0 的本征态在希尔伯特空间中形成了表象的一组完备正交基，据此，我们在超空间中构造一组类似的基矢：后者可以被分为两类超矢量——由成对的全等（或物理上等效的）本征态构成的超矢量和由成对的不同本征态构成的超矢量；它们分别属于超空间中的两个正交子空间，可用投影超算符 P_0 和 P_c 来表征。因此，密度超矢量的投影算符 $\rho_0=P_0\rho$ 和 $\rho_c=P_c\rho$ 分别对应于平均分布密度和关联的振幅。取 $L_{00}=P_0LP_0$，$L_{0c}=P_0LP_c$ 等等，我们可以得到关于 ρ_0 和 ρ_c 的耦合刘维尔方程：

$$i\dot{\rho}_0 = L_{00}\rho_0 + L_{0c}\rho_c, \quad i\dot{\rho}_c = L_{cc}\rho_c + L_{c0}\rho_0 \tag{1}$$

The next step is to extract from these equations the asymptotic forms $\rho_0(t)$, $\rho_c(t)$ of ρ_0 and ρ_c for large positive values of the time variable: these are expected to express our possibilities of prediction of the future evolution of the system on the macroscopic time scale.

We must here restrict the generality of the Liouville superoperator in order to characterize the class of systems which we expect to exhibit the "normal" asymptotic behaviour, that is an approach to a state of equilibrium. To this end, we observe that the time evolution of the correlation density ρ_c is essentially governed by the superoperator $T_c = \exp(-iL_{cc}t)$, depending on the part of the Liouville superoperator which acts entirely in the correlation subspace. We assume accordingly that the asymptotic effect of this superoperator $T_c(t)$ upon any regular supervector A which is not an invariant is to reduce this supervector to zero: $\lim_{t\to\infty} T_c(t)A = 0$; we express by this assumption the fading of the system's "memory" of its correlations. This condition, which may also be expressed as an analyticity condition on the Laplace transform of $T_c(t)$, has first been verified in this form by a perturbation expansion, for infinite systems in whose description there enters a "small" physical parameter such as the coupling constant or the density[3]. More recently, it has been shown that for soluble models, such as the Friedrichs model, the analyticity assumption is satisfied rigorously (that is, independently of any perturbative approach) for a large class of interactions[4]. By means of this assumption, we readily derive from the second Liouville equation (1) the following relation between the asymptotic densities:

$$\tilde{\rho}_c(t) = \int_0^\infty d\tau\, e^{-iL_{cc}\tau}(-i\, L_{c0})\, \tilde{\rho}_0(t-\tau) \tag{2}$$

It has the form of an integral equation, showing how the asymptotic correlations build up by sequences of processes starting from the average situations through which the system passes in the course of time.

We now introduce an asymptotic time evolution operator by writing $\tilde{\rho}_0(t)$ in the form

$$\tilde{\rho}_0(t) = e^{-i\theta t}\, \tilde{\rho}_0(0) \tag{3}$$

The advantage of the representation (3) is to reduce the integral equation (2) to a simple linear relation between $\tilde{\rho}_0(t)$ and $\rho_c(t)$ taken at the same time:

$$\tilde{\rho}_c(t) = C\tilde{\rho}_0(t),\ C = C(\theta) = \int_0^\infty d\tau\, e^{-iL_{cc}\tau}(-iL_{c0})\, e^{i\theta\tau} \tag{4}$$

The first Liouville equation (1) then yields a functional equation for θ:

$$\theta = L_{00} + L_{0c}\, C(\theta) \tag{5}$$

which can be solved by iteration.

$$i\dot{\rho}_0 = L_{00}\rho_0 + L_{0c}\rho_c, \quad i\dot{\rho}_c = L_{cc}\rho_c + L_{c0}\rho_0 \tag{1}$$

下一步，从这些方程中求出当时间变量取很大正值时 ρ_0 和 ρ_c 的渐近形式 $\rho_0(t)$ 和 $\rho_c(t)$：由此我们期望能够在宏观时间尺度上预见系统未来的演变。

为了表征我们期望会出现“正常”渐近行为的那一类系统，也即趋近于平衡态的系统，我们在此必须限制刘维尔超算符的通用性。为此，我们发现关联密度 ρ_c 的时间演化实质上是由超算符 $T_c = \exp(-iL_{cc}t)$ 决定的，该超算符与刘维尔超算符中在关联子空间内完全作用的那一部分有关。因此我们假定，超算符 $T_c(t)$ 对任何正则的变化超矢量 A 的渐近作用是将这个超矢量减小到零：$\lim_{t\to\infty} T_c(t)A=0$；我们通过这个假定来表示系统对其关联性的“记忆”的衰减。这一条件也可表示为对 $T_c(t)$ 进行拉普拉斯变换的解析条件。在那些可以引入一个诸如耦合常数或密度等“小”物理参数来表示的无限系统中，利用微扰展开，人们已经首次证实该条件确实是这样的形式[3]。最近有人指出：对于可解模型，例如弗里德里克斯模型，有一大类相互作用是严格满足解析性假定的（即独立于任何微扰方法）[4]。根据上述假设，我们很容易从刘维尔方程 (1) 中的第二个式子导出渐近密度之间的下述关系：

$$\widetilde{\rho}_c(t) = \int_0^\infty d\tau\ e^{-iL_{cc}\tau}(-i\,L_{c0})\,\widetilde{\rho}_0(t-\tau) \tag{2}$$

上式具有积分方程的形式，由此可以说明系统的渐近关系是怎样通过以系统在一段时间内的平均状态为起始的一系列过程而建立的。

现在我们通过 $\widetilde{\rho}_0(t)$ 的下述表达式来引入渐近时间演化算符：

$$\widetilde{\rho}_0(t) = e^{-i\theta t}\,\widetilde{\rho}_0(0) \tag{3}$$

表达式 (3) 的优点是可将积分方程 (2) 简化为时间取相同值时 $\widetilde{\rho}_0(t)$与 $\rho_c(t)$ 之间的一种简单线性关系：

$$\widetilde{\rho}_c(t) = C\,\widetilde{\rho}_0(t),\ C = C(\theta) = \int_0^\infty d\tau\ e^{-iL_{cc}\tau}(-iL_{c0})\,e^{i\theta\tau} \tag{4}$$

因此，由刘维尔方程 (1) 中的第一个式子可以得到一个关于 θ 的函数方程：

$$\theta = L_{00} + L_{0c}\,C(\theta) \tag{5}$$

该式可用迭代法求解。

The total asymptotic density $\rho = \rho_0 + \rho_c$ thus obtained has the remarkable property of being an exact solution of the Liouville equation. According to equation (4), it may be written in the form $\rho = P_a\rho$, with $P_a = P_0+C$. It is again remarkable that this superoperator P_a has the characteristic properties of a projection operator in superspace, idempotency and "adjoint symmetry" (that is, it is such that the projection P_aA of a self-adjoint supervector A is self-adjoint). The projector P_a defines a subspace in which the asymptotic density is confined. This subspace differs from the average subspace P_0 by the adjunction of a part of the correlation subspace P_c, namely that part which is specified by the superoperator C; the latter may be interpreted as representing the building up of correlations from asymptotic average situations (we call it the superoperator of correlation creation)—it sorts out those correlation processes which have a long time effect and accordingly manifest themselves at the macroscopic level. Thus, the asymptotic density $\tilde{\rho}$ is not, as one might have expected, governed by a "kinetic equation" different from the dynamical Liouville equation: it is an exact solution of the latter, and its asymptotic character is conferred upon it by its confinement to a subspace defined by the projector P_a.

The expression for the superoperator of correlation creation C, which enters in the definition of P_a, is clearly unsymmetrical in time, and gives the projector P_a the expected bias towards a preferred direction of the time evolution. In fact, time inversion transforms P_a into a different projector $\tilde{P}_a = P_0+D$, where the superoperator

$$D = \int_0^\infty d\tau\, e^{i\eta\tau}(-iL_{0c})\, e^{-iL_{cc}\tau}$$

is the time-inverse of C; it contains the superoperator η which is the time-inverse of θ and obeys the equation $\eta = L_{00}+DL_{c0}$ derived from equation (5) by time-inversion. In contrast to C, the superoperator D describes sequences of "destructions" of correlations leading to asymptotic average situations.

An important element is still missing in the picture: we must establish a link between the asymptotic density supervector $\tilde{\rho}(t)$ and the arbitrarily chosen dynamical density supervector $\rho(0)$ from which the time-evolution is assumed to start. This is readily supplied, however, on the basis of a further remarkable property (easily derived) of the superoperator θ and its time-inverse η:

$$P_a\theta = LP_a, \quad \eta\bar{P}_a = \bar{P}_aL \tag{6}$$

With the notation $N_0=1+DC$, it follows from equations (6) and (3) that

$$N_0\,\tilde{\rho}_0(t) = N_0\, e^{-i\theta t}\tilde{\rho}_0(0) = e^{-i\eta t}\, N_0\,\tilde{\rho}_0(0)$$

and, on the other hand,

$$\bar{P}_a\rho(t)=\bar{P}_a\, e^{-iLt}\,\rho(0) = e^{-i\eta t}\bar{P}_a\rho(0)$$

由此得到的总渐近密度 $\rho = \rho_0+\rho_c$ 具有不同寻常的性质——它是刘维尔方程的一个精确解。根据式 (4)，可以将其改写成 $\rho = P_a\rho$ 的形式，其中 $P_a = P_0+C$。超算符 P_a 在超空间中具有投影算符的特性、幂等性和“伴随对称性”(即，自伴超矢量 A 的投影 P_aA 也是自伴的)，这些同样不同寻常。投影算符 P_a 定义了一个渐近密度有限的子空间。这个子空间与平均子空间 P_0 之间的差别在于叠加了关联子空间 P_c 的一部分，即由超算符 C 确定的那部分；可以把后者解释为代表了在渐近平均情况下的关联建立（我们称之为关联产生超算符）——它可以区分出那些具有长时效应并因此在宏观层次上有所显现的关联过程。因此，渐近密度 $\widetilde{\rho}$ 并非如预期的那样，是由不同于刘维尔动力学方程的“动力学方程”决定的：它是刘维尔动力学方程的一个精确解，其渐近特性是通过被限制在由投影算符 P_a 定义的子空间中而得到的。

在 P_a 的定义中引入了由关联产生的超算符 C 的表达式，该表达式在时间上显然是非对称的，它使投影算符 P_a 向时间演化的择优方向发生了预期的偏移。实际上，时间反演将 P_a 变换成了另一个投影算符 $\widetilde{P}_a = P_0+D$，其中超算符

$$D = \int_0^\infty d\tau\ e^{i\eta\tau}(-iL_{0c})\ e^{-iL_{cc}\tau}$$

是 C 的时间逆；它含有 θ 的时间逆——超算符 η，并且满足将式 (5) 进行时间反演而推导出的表达式 $\eta = L_{00}+DL_{c0}$。与 C 不同，超算符 D 描述的是导致出现渐近平均状态的关联“相消”序列。

在此图景中仍然缺少了一个重要的元素：我们必须在渐近的密度超矢量 $\widetilde{\rho}(t)$ 与任意选取的动态密度超矢量 $\rho(0)$ 之间建立起联系，其中 $\rho(0)$ 被假定为时间演化的起始点。这一点很容易实现，但要以超算符 θ 及其时间逆 η 的一个更为不同寻常的性质（容易推导出）为基础：

$$P_a\theta = LP_a, \quad \eta\overline{P}_a = \overline{P}_aL \tag{6}$$

记 $N_0 = 1+DC$，由式 (6) 和式 (3) 可得：

$$N_0\widetilde{\rho}_0(t) = N_0\ e^{-i\theta t}\widetilde{\rho}_0(0) = e^{-i\eta t}N_0\widetilde{\rho}_0(0)$$

而另一方面：

$$\overline{P}_a\rho(t) = \overline{P}_a e^{-iLt}\rho(0) = e^{-i\eta t}\overline{P}_a\rho(0)$$

This shows that $N_0\tilde{\rho}_0(t)$ and $P_a\tilde{\rho}(t)$ have the same time-evolution, governed by the superoperator $\exp(-i\eta t)$: we may therefore equate them at any instant and in this way fix the correspondence between the dynamical and the asymptotic density. This gives the quite fundamental relation, valid at any time,

$$\tilde{\rho}(t) = \tilde{\Pi}\rho(t) \text{ with } \tilde{\Pi} = P_a N_0^{-1}\bar{P}_a \tag{7}$$

from which follows

$$\tilde{\rho}(t) = \tilde{\Sigma}(t)\ \rho(0) \text{ with } \tilde{\Sigma}(t) = \tilde{\Pi}e^{-iLt}$$

the answer to our last question, completing the theory.

The striking feature about the superoperator $\tilde{\Pi}$ occurring in equation (7) is that it is also a projector in the extended sense defined above (which does not include the property of self-adjointness); moreover, it is time-reversal invariant: it defines a time-symmetrical subspace of the superspace in which the asymptotic part of the time-evolution, starting from any given situation, remains confined, exhibiting the features observed at the macroscopic level; whereas the irregular fluctuations occurring on the atomic time scale are contained in the complementary subspace orthogonal to the asymptotic one. That such a clean separation between the two aspects of the atomic system could be effected is an entirely unexpected property specific to the density supervector representation: it could never have been found by a study of the evolution of the system in Hilbert space, for it can only be formulated in terms of the superspace formalism.

The superoperator $\tilde{\Pi}$ is not factorizable: one cannot ascribe any state vector to an asymptotic situation as we have defined it, but only a density supervector $\tilde{\rho}$. In fact, as appears from equation (4), the correlation part $\tilde{\rho}_c$ of the density ρ is directly derived from the part $\tilde{\rho}_0$ expressing the average probability distributions of the system; owing to this remarkable structure of the projector $\tilde{\Pi}$, the evolution in $\tilde{\Pi}$-space may be entirely described in terms of probabilities only. In particular, the "reduction of the wave-packet" of an atomic system after its interaction with a measuring apparatus is a direct consequence of this property of the $\tilde{\Pi}$-space description: the essential point being that the apparatus must necessarily belong to the macroscopic level of quantum mechanics, and that its behaviour must accordingly be described in terms of its $\tilde{\Pi}$-space variables (loosely speaking, the behaviour of the apparatus has "thermodynamical" character, inasmuch as variables pertaining to thermodynamic equilibrium or near-equilibrium states all belong to $\tilde{\Pi}$-space). Any phase relations in the initial state of the atomic system are therefore wiped out (that is, they are rejected into the orthogonal subspace): this is the only meaning of the "reduction" of the initial state of the atomic system resulting from the measurement. As to the human observer, his interaction with the apparatus is also entirely described in the $\tilde{\Pi}$-subspace, and therefore without any influence whatsoever on whatever goes on in the orthogonal subspace.

这表明 $N_0\tilde{\rho}_0(t)$ 和 $P_a\tilde{\rho}(t)$ 的时间演化相同，均由超算符 $\exp(-i\eta t)$ 决定：因此，我们可在任意时刻令二者相等，由此得到动态密度与渐近密度之间的对应关系。这样就给出了一个非常基本的关系，并且在任意时刻均成立：

$$\tilde{\rho}(t) = \tilde{\Pi}\rho(t) \text{ 式中 } \tilde{\Pi} = P_a N_0^{-1}\bar{P}_a \tag{7}$$

由此可得：

$$\tilde{\rho}(t) = \tilde{\sum}(t)\ \rho(0) \text{ 式中 } \tilde{\sum}(t) = \tilde{\Pi}e^{-iLt}$$

这就是最后一个问题的答案，整个理论完成。

式 (7) 中超算符 $\tilde{\Pi}$ 的显著特征是：根据上文中的定义，它也是一个意义扩展了（不包含自伴性质）的投影算符；此外，它在时间反演后保持不变：它定义了超空间中的一个时间对称的子空间，从任意给定的状态开始，该子空间中时间演化的渐近部分保持有限，表现出了在宏观尺度上所观测到的性质；而在原子时间尺度上出现的不规则涨落则被纳入与渐近子空间正交的互补子空间中。对于密度超矢量表象而言，能够实现原子系统两个方面之间的明确分离完全是一个出乎意料的特性：在希尔伯特空间中对系统演化进行研究是不可能得到这一结果的，因为只有通过超空间的形式才能阐释它。

超算符 $\tilde{\Pi}$ 是不可分解的：除密度超矢量 $\tilde{\rho}$ 外，不能将其他任何态矢量归于我们所定义的渐近状态。事实上，由式 (4) 可以看出：密度 ρ 的关联部分 $\tilde{\rho}_c$ 是从表征系统平均概率分布的 $\tilde{\rho}_0$ 部分直接推导出来的；由于投影算符 $\tilde{\Pi}$ 的这种特殊结构，有可能仅用概率就能完全描述 $\tilde{\Pi}$ 空间中的演化。特别地，原子系统与测量装置发生相互作用之后的“波包坍缩”正是 $\tilde{\Pi}$ 空间描述的这一性质的直接结果：最关键的一点是，测量装置必须属于量子力学的宏观层次，于是它的行为必须通过它在 $\tilde{\Pi}$ 空间中的变量来描述（大体来说，因为与热力学平衡态或近平衡态有关的变量都属于 $\tilde{\Pi}$ 空间，所以测量装置的行为具有“热力学”的特征）。因此，我们可以去掉原子系统初态中的所有相位关系（即，使它们退化到正交子空间）：这就是原子系统的初态经测量后“坍缩”的唯一内涵。至于人类观测者，他与装置之间的相互作用也在 $\tilde{\Pi}$ 子空间中得到了完整的描述，因而他对正交子空间中发生的一切不会造成任何影响。

One further point should be mentioned. The theory gives us a simple criterion to decide whether the system shows the normal macroscopic behaviour described by thermodynamics. The superoperator of asymptotic time-evolution θ is closely related to a "collision superoperator" defined as the Laplace transform of the superoperator $L_{0c}T_c(\tau)\, L_{c0}$:

$$\Psi(z) = \int_0^\infty d\tau\, L_{0c}\, T_c(\tau) L_{c0} e^{-z\tau}$$

Indeed, equation (5) may be written as a functional relation in terms of $\Psi(z)$:

$$i\theta = iL_{00} + \Psi(-i\theta) \qquad (8)$$

Now, an homogeneous system (a system for which $L_{00} = 0$) will obviously not exhibit any tendency towards equilibrium if the collision operator vanishes identically. Equation (8) then shows that it will exhibit an irreversible tendency towards equilibrium provided that θ itself does not vanish identically. This "condition of dissipativity" is a practical one: it can be tested in concrete cases[4] by actual computation of θ.

It is thus clear that the epistemological consistency problem raised at the beginning of this note is completely answered by the neat, clear-cut representation we obtain for the complementarity between the two levels of description of atomic phenomena. It need hardly be pointed out that the problem here discussed of the macroscopic level of quantum mechanics is just a simple illustration of a general method of representation in superspace, which actually amounts to an extension of the scope of quantum mechanics. Perhaps the most significant feature of the method is the introduction of generalized projectors, involving the replacement of self-adjointness by time-inversion invariance (which reduces to the former in the absence of dissipation, that is, for systems for which the collision operator vanishes)—a generalization which may be expected to find application to a large variety of problems.

(**240**, 25-27; 1972)

C. George and I. Prigogine: Université Libre, Brussels, and Center for Statistical Mechanics and Thermodynamics, Austin, Texas.
L. Rosenfeld: Nordita, Copenhagen.

Received January 28; revised March 20, 1972.

References:

1. George, C., Prigogine, I., and Rosenfeld, L., *Det kgl. Danske Videnskabernes Selskab, mat.-fys. Meddelelser*, **38**, 12 (1972).
2. Prigogine, I., George, C., and Henin, F., *Physica*, **45**, 418 (1969).
3. Balescu, R., and Brenig, L., *Physica*, **54**, 504 (1971).
4. Prigogine, I., *Non-Equilibrium Statistical Mechanics* (Interscience, 1962).
5. Grecos, A., and Prigogine, I., *Physica*, **59**, 77 (1972).
6. Grews, A., and Prigogine, I., *Proc. US Nat. Acad. Sci.*, **69**, 1629 (1972).

另有一点值得一提。上述理论给我们提供了一条简单的标准，用以判断系统是否表现出由热力学所描述的标准宏观行为。渐近时间演化超算符 θ 与由超算符 $L_{0c}T_c(\tau)\,L_{c0}$ 的拉普拉斯变换而定义的“碰撞超算符”之间有密切的关系：

$$\Psi(z) = \int_0^\infty d\tau\, L_{0c}\, T_c(\tau) L_{c0} e^{-z\tau}$$

实际上，式 (5) 可改写成含有 Ψ(*z*) 的泛函关系式：

$$i\theta = iL_{00} + \Psi(-i\theta) \tag{8}$$

于是，当碰撞算符为零时，一个均匀系统（$L_{00} = 0$ 的系统）显然不会有任何趋于平衡的表现。由式 (8) 可知：如果 θ 本身没有同时趋于零，则系统将不可逆地趋向平衡。这种“耗散条件”是实际存在的：在具体情况下[4]，通过实际计算 θ 值可验证这一点。

由此可见，我们为得到对原子现象两种层次描述之间的互补性而采用的这种简洁、明晰的表象彻底解决了本文开篇处提出的认识论一致性的问题。不言自明的是，此处讨论的宏观层次量子力学问题仅仅是超空间表象这一通用方法的一个简单演示，它实际上拓展了量子力学的范畴。上述方法最为重要的特征或许是引入了广义投影算符，包含用时间反演不变性替换自伴性（对于碰撞算符为零的系统，在无耗散时前者可约化为后者）——可以预期这种推广将在大量问题中得到应用。

（沈乃澂 黄娆 翻译；李军刚 审稿）

Computer Analyses of Gravitational Radiation Detector Coincidences

J. Weber

Editor's Note

Within the context of general relativity, gravitational radiation is expected to come from massive and compact objects such as neutron stars and black holes. Joseph Weber designed and built two antenna systems for detecting gravitational radiation, separated by about 1,000 km. Here he analyses signals obtained from them, and finds coincidences in detection that he interprets as evidence for gravitational waves passing through the instruments. The claim is now known to be wrong, but the paper helped to launch the field of gravitational-wave detection. Indirect evidence for gravitational radiation was later seen in observations of the orbit of a millisecond pulsar, but a direct detection remains to be achieved.

MY earlier publications[1] have reported the concept, theory, and development of an antenna to detect gravitational radiation. These antennae are well isolated from the local environment by acoustic and electromagnetic shielding but they do respond to sufficiently large local disturbances. Effects of the local environments can, however, be minimized when two detectors are used which are a considerable distance apart. For this reason coincidence experiments at 1,661 Hz were carried out with two antennae, one situated at the University of Maryland and the other 1,000 km away at the Argonne National Laboratory.

For all experiments reported here the only coincidences which were recorded were those which occurred within a predetermined time Δt.

Coincidences may be due to excitation of both detectors by a common source or to chance. It is customary to compute the chance (accidental) coincidence rate by formulating a classification scheme for the coincidences and computing the number of chance coincidences in each class. A significant excess of detected event over the chance coincidence rate for a given class establishes with a certain level of confidence that not all coincidences are due to chance.

For two years the experiments were done as follows. The outputs of the Maryland and Argonne detectors are obtained from synchronous detectors which have free running crystal reference oscillators and twin channels with the reference shifted by $\pi/2$ in one of them. The two channel outputs are squared and summed to recover the power, with a time constant of 0.5 s. The envelope of the total power output of the Argonne detector was transmitted over a telephone line to Maryland, in coded digital form, and reconverted

引力辐射探测器符合计数的计算机分析

韦伯

编者按

广义相对论认为引力辐射来自大而重的致密物体，如中子星和黑洞。约瑟夫·韦伯设计并搭建了相距约 1,000 千米的两个天线系统来探测引力辐射。在这篇文章中，他分析了由这些天线系统得到的信号，并在探测中发现了符合事件，他认为这些符合事件是引力波穿过实验装置的证据。现在我们知道他的这种论断是错误的，但是这篇论文帮助人们开创了引力波探测的领域。之后，人们在对一个毫秒脉冲星轨道的观测中发现了引力辐射的间接证据，但直接探测还有待实现。

我已经在之前发表的文章 [1] 中陈述过探测引力辐射的天线的概念、理论和研制。这些天线通过声屏蔽和电磁屏蔽与局部环境充分隔离，但仍然会对足够大的局部扰动产生响应。不过，使用相距很远的两个探测器可以将局部环境的影响降到最低。因此我们使用两个天线在 1,661 赫兹频率上进行了符合计数实验，其中一个天线位于马里兰大学，另一个则位于 1,000 千米之外的阿贡国家实验室。

在本文报告的所有实验中，被记录的符合事件都发生在事先确定的时间间隔 Δt 中。

符合事件可能是因为一个共同的源激发两个探测器而引起的，也可能是偶然产生的。制定一个符合事件的分级方案并计算每个等级中的偶然符合数目，是计算偶然符合计数率的通常做法。对一个给定等级来说，如果探测到的符合事件明显超出该等级的偶然符合计数率，就表明在一定置信水平上，并非所有的符合计数都是偶然产生的。

两年时间的实验开展情况如下。马里兰和阿贡探测器的输出由同步探测器获得，该同步探测器装备了自由运转的晶体参考振荡器，并具有双通道，其中一个的参考相位移动了 $\pi/2$。设定时间常数为 0.5 秒，可将两个通道的输出求平方和来得到原先的功率。通过电话线将阿贡探测器的总功率输出包络以数字编码形式传递到马里兰，并在马里兰将其转换回模拟直流信号。输出信号由一个符合探测器接收。在时间间

to analogue d.c. at Maryland. The outputs are fed to a coincidence detector. If both channels cross some preset threshold from below, within the time interval Δt, a pulse is emitted to drive a recorder marker pen.

The amplitudes of the noise pulses have a Rayleigh distribution, and their shapes vary widely. Analysis is simplified if all pulses have roughly the same shape, because then one parameter, the pulse height, will suffice for their classification. To shape the pulses properly a second stage of filtering is carried out after coincidence detection. There is a small amount of additional filtering associated with the mass of the pen which records the output on a chart.

A mean solar day is chosen as the unit of time. Consider a given coincidence with power amplitudes P_A and P_B for the two channels. Let N_A and N_B represent the number of times per day that the powers P_A and P_B are equalled or exceeded, on the average. Let Δt be the maximum time between threshold crossings for the two channels in order to record a coincidence. The expected number of chance coincides with amplitudes equal to or exceeding those observed for an experiment with effective duration M days is η_{AB} with

$$\eta_{AB} = 2N_A N_B \Delta t M \tag{1}$$

To employ equation (1) we select some arbitrary numbers which may or may not be the same for channels A and B. A given class consists of all those coincidences for which N_A and N_B each are equal to or less than the arbitrarily selected numbers for them.

A second classification scheme was employed in order to determine if the difference between the observed and chance coincidences is sensitive to details of the statistical methods. Instead of requiring N_A and N_B each to remain within certain bounds for a given class, we require the product $N_A N_B$ to be equal to or less than some arbitrary number for a given class. The total number of such chance coincidences will be finite because neither N_A nor N_B can ever be less than one or greater than the total number of pulses observed for each channel. The most useful classifications are those for which N_A and N_B are less than some number N_S which is about an order greater than the numbers which characterize most of the real coincidences classified by equation (1).

A two dimensional space with N_A and N_B as coordinates is useful to calculate the number of chance coincidences $\tilde{\eta}_{AB}$ for which the product $N_A N_B$ is equal to or smaller than some constant and for which neither N_A nor N_B exceeds some maximum value N_S. $\tilde{\eta}_{AB}$ is expression (1) plus the integral under the hyperbola $N_A N_B$ = constant, that is

$$\tilde{\eta}_{AB} = 2N_A N_B \Delta t M \,[1+\ln(N_S^2/N_A N_B)] \tag{2}$$

In equations (1) and (2) M is smaller than the actual number of days which the experiments run, because allowance must be made for the detector relaxation time after crossing the threshold. A new coincidence cannot occur during this time.

隔 Δt 内，如果两个通道的信号都上升至超过事先设定的阈值，就会发出一个脉冲来驱动记录仪的记号笔。

噪声脉冲的幅值符合瑞利分布，且其形状变化很大。如果所有脉冲的形状都大致相同，分析将得以简化，因为这时只需使用脉冲高度这一个参数就能够对它们进行分级。对脉冲的适当整形需要在符合计数检测后再进行第二级过滤。此外还存在少量附加滤波，它们与在图纸上记录输出的记号笔的质量有关。

选择平均太阳日作为时间单位。考虑功率幅值分别为 P_A 和 P_B 的两个通道的给定符合计数。令 N_A 和 N_B 代表每天功率值达到或者超过 P_A 和 P_B 的平均次数，Δt 代表：为了记录到一次符合计数，两个通道越过功率阈值的最大时间间隔。在有效实验时间 M 天的观测中，功率幅值达到或者超过规定值的偶然符合的预期次数 η_{AB} 为：

$$\eta_{AB} = 2N_A N_B \Delta t M \tag{1}$$

为了利用公式（1），我们任意选择了一些数值，这些数值对通道 A 和 B 来说可能相同，也可能不相同。一个给定的等级包含所有满足下述条件的符合计数，即 N_A 和 N_B 都等于或者小于那些针对它们任意选择的数值。

为了确定观测到的符合计数和偶然符合计数的差值是否对统计方法的细节敏感，我们采用了另一种分级方案。我们不限定 N_A 和 N_B 每一个都必须处在给定的等级区间中，而是要求它们的乘积 $N_A N_B$ 等于或者小于为某一给定等级设定的任意值。因为不管是 N_A 还是 N_B 都不能小于 1，也不能大于在每个通道观测到的总脉冲数量，所以这样的偶然符合的总数目是有限的。最有用的等级是那些 N_A 和 N_B 小于某个数值 N_S 的，这个 N_S 比按照公式（1）进行分级的大多数代表真正符合计数的数目约大一个数量级。

考虑一个以 N_A 和 N_B 为坐标的二维空间，它有助于计算乘积 $N_A N_B$ 等于或者小于某个常数并且 N_A 和 N_B 都不超过某个最大值 N_S 的偶然符合的数目 $\tilde{\eta}_{AB}$。$\tilde{\eta}_{AB}$ 等于公式（1）加上双曲线 $N_A N_B$ = 常数下的积分，即

$$\tilde{\eta}_{AB} = 2N_A N_B \Delta t M\,[1+\ln(N_S^2/N_A N_B)] \tag{2}$$

公式（1）和（2）中的 M 小于实验运行的实际天数，这是因为必须考虑越过阈值后探测器的弛豫时间，在这段时间内不会出现新的符合计数。

Early experiments[5,6] observed that the number of coincidences for certain values of N_A, N_B, N_AN_B substantially exceeded the accidental rate computed from equations (1) and (2). These gave about the same numbers for the difference of the observed and chance coincidences. The conclusion that positive results are obtained is based on certain assumptions—for example that the classification scheme based on amplitude alone with the successive filtering is sound. To test this, a parallel experiment was done to measure the accidental rate by inserting time delays of varying length in one channel or the other.

This time delay experiment showed a large decrease in the number of coincidences for which N_A, $N_B \leq 100$; and also those coincidences for which $N_AN_B \leq 6{,}000$ with $N_S \leq 1{,}000$ for all time delays significantly larger than Δt. There is a subjective element in the data processing, in that for each coincidence mark the leading edge of the pulse must be examined. If it increases smoothly to a peak then the peak value is the required amplitude. If there is a discontinuity in slope before reaching the peak it is concluded that the excitation which results in a threshold crossing was followed at the discontinuity by a heat bath noise excitation. The required pulse height is then only the value measured to the point of discontinuity. Sometimes the discontinuity is not well defined and a human decision is needed.

Using a magnetic tape and a computer provides an independent procedure for data processing. The human element is only involved in the program preparation. A variety of thresholds and time delays can be applied to any stretch of data. The synchronous detector output with 0.5 s filter time constant was recorded in digital form every 0.1 s with both channels on one tape. A computer program was prepared by Mr. Brian K. Reid in the following way: thresholds were set such that if both channels crossed from below within 0.5 s the computer measured the pulse heights after a second stage of 5 s time constant filtering and a third stage of filtering with 0.5 s time constant to simulate the ink recorder pen. Then the computer counted the number of times the filtered data amplitude was equalled or exceeded for the previous hour, the previous 6 h, and the given day following the schematic diagram in Fig. 2. The same procedure was repeated for time delays of 1 s, 2 s, 5 s, 10 s, 20 s and 40 s.

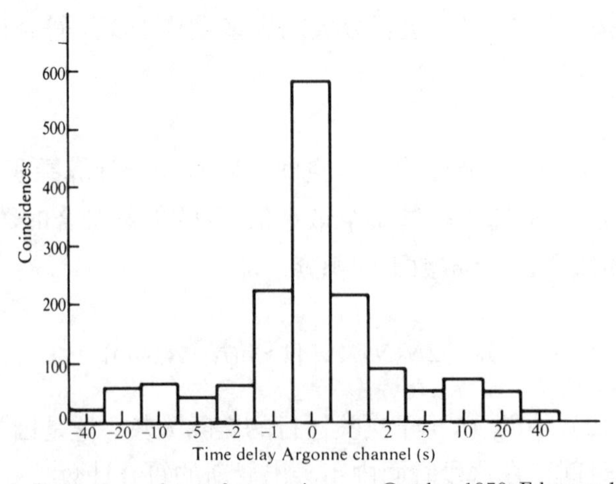

Fig. 1. Computer analyses of magnetic tapes, October 1970–February 1971.

由早期的一些实验[5,6]发现：对于特定的 N_A、N_B、N_AN_B 值，符合计数的数目大大超过了由公式（1）和（2）计算得到的偶然符合计数率。这些实验给出的符合计数观测值和偶然符合计数的差值大致相同。认为能够获得肯定性结果的结论建立在某些假设的基础之上——例如，假设单纯基于幅值然后再过滤的分级方案是合理的。为了检验这个假设，我们进行了一项对照实验，通过在一个或者另一个通道中加入不同长度的时间延迟来测量偶然符合计数率。

这个时间延迟实验显示，在时间延迟显著大于 Δt 的情况下：当 N_A、$N_B \leq 100$ 时，符合计数的数目大大降低；当 $N_AN_B \leq 6,000$ 且 $N_S \leq 1,000$ 时，符合计数的数目也大大降低。数据处理中的一个主观因素在于，标记每个符合计数时都必须检查脉冲的前沿。如果它平滑地上升到峰顶，那么这个峰值就是所需的幅值。如果在到达峰顶前的上坡中间有一个间断点，我们就认为：导致超越阈值的激发之后，在间断点处紧跟着一个热库噪声激发。这样所需的脉冲高度就只能是间断点处的值。有时候这个间断不是很明显，需要人为主观判断。

磁带和计算机的应用提供了一个独立的数据处理程序。只有在程序准备的过程中会牵涉到人为因素。可以对任何一部分数据应用不同的阈值和时间延迟。设过滤时间常数为 0.5 秒，同步探测器两个通道的输出都以数字形式每隔 0.1 秒记录到同一个磁带上。布赖恩·里德先生准备了一个计算机程序，情况如下：设定好阈值，一旦两个通道在 0.5 秒的时间范围内都上升超过阈值，计算机就会在第二级时间常数为 5 秒的过滤以及第三级时间常数为 0.5 秒的过滤之后模仿记录笔来测量脉冲高度。接下来计算机按照图 2 所示的流程计算出前 1 小时、前 6 小时以及某一天之内过滤后的数据幅值达到或者超过阈值的次数。对 1 秒、2 秒、5 秒、10 秒、20 秒和 40 秒的时间延迟，也重复运行同样的程序。

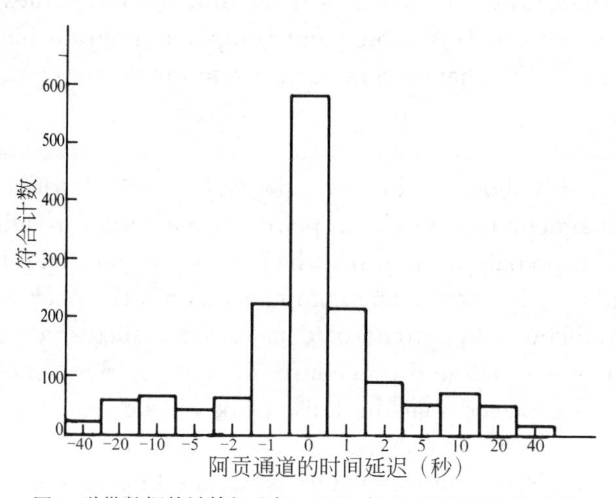

图 1. 磁带数据的计算机分析，1970 年 10 月至 1971 年 2 月

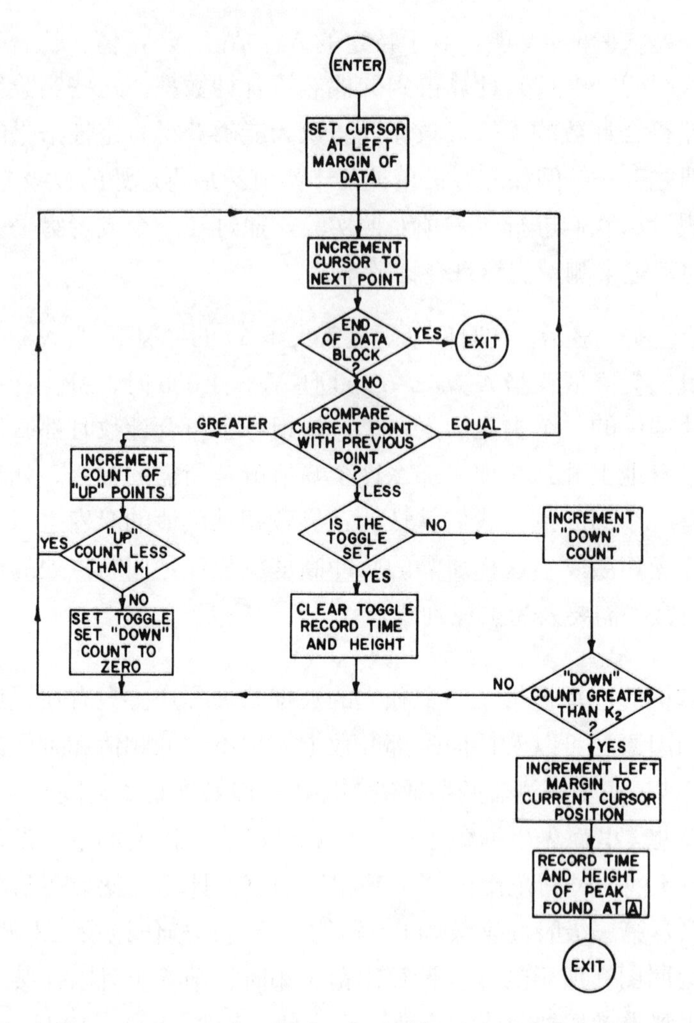

Fig. 2. Computer program for statistics of coincidences recorded on magnetic tape.

The program was inaccurate beyond 20 s delay and the bin values given in Fig. 1 for 40 s delay are much too small. A subsequent computer program has verified that there is no further decrease in the chance coincidence rate for delays exceeding 100 times the coincidence window.

The pen and ink records show pulses with relatively smooth leading edges. The greater time resolution of magnetic tape results in pulses with a certain roughness everywhere. It is therefore difficult to prepare a program which measures the pulse height to a point of discontinuity in leading edge slope. The computer defines the peak as the point where a trend of increases is followed by a trend of decreases. Detailed study of greatly expanded pulses indicates that this is a sound procedure for most of the coincidences. Because of the 5 s averaging it is expected that the pulse peak will be reached between 2 and 15 s after the threshold is crossed and only those coincidences are recorded which reach peaks within such an interval. This range of values is required because of noise and the initial

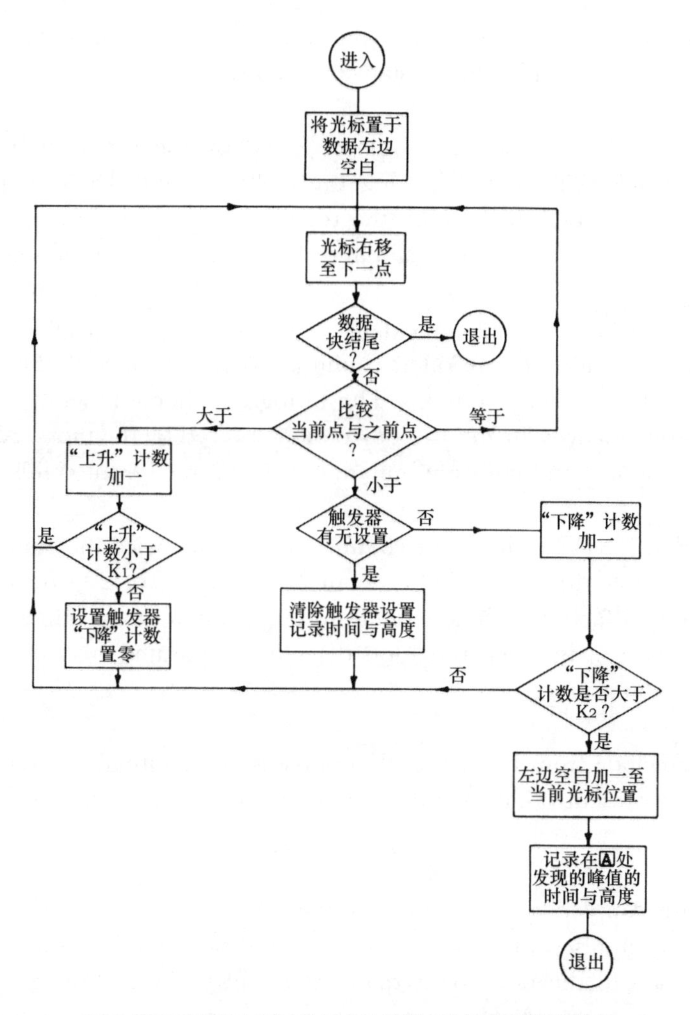

图 2. 用于对磁带上记录的符合计数进行统计的计算机程序

这个程序对20秒以上的延迟不是很精确。图1中40秒延迟的柱高的数值太小了。随后的一个计算机程序证实：对于超过符合时间窗100倍以上的延迟，偶然符合计数率并没有进一步的降低。

笔墨记录的脉冲前沿比较平滑。磁带记录的时间分辨率较高，导致脉冲在各点处都有点粗糙。因此很难用一个程序去测量到前沿上坡间断点处的脉冲高度。计算机对峰顶的定义是从上升走向变为下降走向的转折点。对大大扩展的脉冲进行的详细研究显示：对绝大多数符合计数，这一定义是合理的。由于平均时间为5秒，一般预期在超越阈值后2秒到15秒之间达到脉冲峰顶，也只有当峰顶出现在这一区间时，符合才会被记录。规定这一数值范围是因为考虑到了噪声和滤波电路的电容器的初始充电情况。计算机也测量脉冲的面积，不过没有包括在统计分析中。

conditions of charge on the capacitors of the filter. The computer measures the area of the pulse, but this is not included in the statistical analysis.

A small fraction of the coincidences are not found by the program because no peak is found in the required interval following threshold crossing. Detailed plotting usually permits the peak to be identified within the required time after threshold crossing. Such coincidences are not included in these data in order to conserve computing time and costs.

Results are shown in Fig. 1 for all coincidences for which $N_A N_B \leq 10{,}000$ and $N_S \leq 1{,}000$. The low intensity coincidences are often counted more than once if threshold is crossed and re-crossed in the presence of noise. The histogram therefore shows about twice the real number of coincidences for the 120 day period covered by the tapes. All data of Fig. 1 came from the computer and not from human examination of printed lists.

It is significant that for delays in either channel as small as twice the coincidence interval Δt, the coincidence rate drops by a factor about 2.5. At delays of 20 s the coincidence rate is down by roughly a factor 10. These data, untouched by human hands, leave no doubt whatsoever that the gravitational radiation detectors, separated by 1,000 km, are being excited by a common source.

Before we can conclude from Fig. 1 that the source is gravitational radiation, it is essential to rule out other presently known interactions including seismic, electromagnetic, and cosmic ray effects.

The detectors are coupled to each other by the Earth and respond to sufficiently large seismic events. A number of seismometers were developed including a vertical axis accelerometer tuned to the detector frequency of 1,661 Hz, a three axis accelerometer covering frequencies near 100 Hz, and a two axis tilt meter. No significant correlations were observed between seismic activity and the coincidences.

Electromagnetic signals may enter the instrumentation through the vacuum chambers and through the cables and shields of the electronic equipment. Experiments were carried out to measure the susceptibility to electromagnetic excitation. It was found that 1,661 Hz magnetic fields of amplitude about 0.1 Gauss will excite the detector, and much larger fields at 830.5 Hz will also excite the detector. Non-linear effects are expected because the electromagnetic stress tensor is quadratic in the fields, and electromagnetic stresses can cause mechanical forces on the detector cylinders. The response is large at 1,661 Hz because the electronics has its acceptance band there and because 1,661 Hz currents in the aluminium cylinder give forces by interacting with the Earth's magnetic field.

Both laboratory sites are to some degree shielded as a result of grounded metal ceilings and structures around the apparatus. A radio receiver was employed at the Maryland site, with non-linear pre-amplifier and post-amplifier tuned to a narrow band of frequencies near 1,661 Hz. The receiver sensitivity could be adjusted to respond to fields much

由于在超越阈值后所要求的时间区间内未能发现峰顶，有一小部分符合计数没有被程序捕捉到。一般来说，通过详细的测绘可在超越阈值后所要求的时间内识别出这个峰顶。为了节约计算时间和成本，这样的符合计数没有被包括在这些数据中。

图 1 是 $N_AN_B \le 10{,}000$ 且 $N_S \le 1{,}000$ 的所有符合计数结果。当阈值被超越且由于噪声的存在而被再次超越时，低强度的符合经常被多次计数。因此，直方图给出的符合计数是磁带记录的 120 天内真实符合计数的两倍左右。图 1 中所有数据都来源于计算机，而不是对打印列表人工检验的结果。

值得注意的是：任一通道的时间延迟，尽管小到只有符合计数时间间隔 Δt 的两倍，也使得符合计数率下降到原来的 2/5。对 20 秒的延迟，符合计数率下降到原来的 1/10 左右。这些未经人工处理的数据毫无疑问地表明，相隔 1,000 千米的引力辐射探测器受到了同一个源的激发。

在我们根据图 1 确定这个源就是引力辐射之前，必须排除其他目前已知的干扰。这些干扰包括地震、电磁以及宇宙线的影响。

这些探测器通过地球互相耦合，会对足够大的地震事件产生响应。我们研制了许多台地震计，包括一台探测器频率被调到 1,661 赫兹的垂直轴加速度计、一台覆盖 100 赫兹左右频率的三轴加速度计以及一台二轴倾斜仪。在地震活动与符合计数之间没有发现显著的关联。

电磁信号可能会通过真空室或者电子设备的电缆及屏蔽物进入仪器。我们进行了一些实验来测量系统对电磁干扰的敏感度。结果发现强度约为 0.1 高斯的 1,661 赫兹磁场能够激发探测器。若为 830.5 赫兹，则要激发探测器所需的磁感应强度就要大得多。因为电磁应力张量是电磁场的二次型，并且电磁应力可以在探测器圆筒上产生机械力，所以这种非线性效应是可以预期的。1,661 赫兹处的响应很强是因为电子在那里有一个接收带，并且因为在铝制圆筒中 1,661 赫兹的电流通过与地磁场的相互作用而产生了力。

由于金属天花板和仪器周围的金属结构接地，两个实验站在某种程度上都是被屏蔽的。我们在马里兰站使用了一台无线电接收器，将其非线性前置放大器和后置放大器调频到 1,661 赫兹附近的窄波段。接收器的灵敏度可调，哪怕场强小到远远

smaller than those which excite the gravitational radiation detectors. Magnetic and electric dipole antennae were employed at different locations and with different orientations at various times. Local electromagnetic fields associated with air conditioning, magnetic tape recorders, and power line fluctuations limited the radio receiver sensitivity to values about three orders smaller than those to which the gravitational radiation detector responds. No significant correlations were observed with the gravitational radiation detector coincidences.

There is additional evidence that the coincidences are not due to electromagnetic or seismic interactions. The gravitational radiation detector system input temperature has been measured by a noise generator. This resulted in an accurate accounting of all significant signal sources and ruled out the possibility that any significant fraction of the input temperature is due to some background other than internal noise. If the coincidences are caused by terrestrial effects such as seismic and electromagnetic activity, these must give signals arriving simultaneously or nearly simultaneously at both sites from a region on or within the Earth. For each coincident arrival at the two locations, however, there must be many individual arrivals, and these would result in a higher system noise level than is observed. Measurements were also carried out of coincidence rates for a pair of gravitational radiation detectors at the Maryland site over a period of six months. Within limits of experimental error the one-site, two-detector coincidence rate was identical with the separated-site two-detector coincidence rate. This indicates that for the sensitivities reported here, the isolation from the local environments was adequate and that electromagnetic and seismic interactions do not cause the observed coincidences.

The separation of 1,000 km makes a cosmic ray shower explanation very unlikely. Nonetheless it was important to investigate the effect of cosmic ray charged particles. Professors N. Sanders Wall and Gaurang B. Yodh and Dr. David Ezrow instrumented one of the Maryland site gravitational radiation detectors with Čerenkov radiation counters and later with meter square plastic scintillators. They observed no significant correlations with the gravitational radiation detector coincidences[7].

A search was made for coincidences between a 1,661 Hz gravitational radiation detector and a 5,000 Hz mode of a second gravitational radiation detector, both at the Maryland site. No excess of coincidences above the chance rate was observed. Subsequent investigation showed that the 5,000 Hz mode was a bending mode and not a compressional mode. General relativity theory predicts that this bending mode should not interact with gravitational radiation. The lack of coincidences is thus evidence against a non-gravitational radiation origin for the 1,661 Hz coincidences.

Other papers have presented evidence that there is anisotropy of the observed coincidences, the maxima occurring in the direction of the galactic centre[8]. Experiments with a disk were consistent with predictions of Einstein's pure tensor theory. Small effects of the local environment can affect the anisotropy in a significant way. To average these properly at least six months of data are required. The magnetic tapes considered here

不够激发引力辐射探测器，也可以产生响应。我们在不同的位置使用磁偶极和电偶极天线，在不同的时间采用了不同的方向。与空调、磁带记录仪和电源线波动相关的局部电磁场限制了无线电接收器的灵敏度，其数值大概比引力辐射探测器响应的小 3 个数量级。在电磁干扰与引力辐射探测器符合计数之间没有观测到显著关联。

还有其他证据可以证明符合计数不是电磁或地震干扰引起的。我们曾用噪声发生器测量引力辐射探测器系统的输入温度。这样可以准确说明所有显著信号源的情况，并排除输入温度的任何主要部分来源于内噪声以外的其他背景的可能性。如果符合计数是由像地震或者电磁活动这样的地球效应引起的，那么这样的信号会同时或者几乎同时到达地表或地球内部同一区域的两个实验站。然而，对于每一个同步到达两个地点的符合事件，都应该伴随有很多单个的到达，这样就会产生高于实际观测值的系统噪声水平。我们还测量了 6 个月内马里兰站一对引力辐射探测器的符合计数率。在实验误差范围内，同一站中两个探测器的符合计数率与另一站中两个探测器的符合计数率是完全相同的。这个结果表明：在本文涉及的灵敏度范围内，与局部环境的隔离是足够好的，并且观测到的符合计数不是由电磁和地震干扰引起的。

1,000 千米的距离太远，因此不太可能用宇宙线簇射来解释符合事件。尽管如此，研究宇宙线带电粒子的影响还是很重要的。桑德斯·沃尔教授、高朗·尤德教授和戴维·埃兹罗博士在马里兰站的一个引力辐射探测器上装备了切伦科夫辐射计数器，后来又装备了米方塑料闪烁体。他们在宇宙线簇射与引力辐射探测器符合计数之间没有观测到显著的关联[7]。

我们对同样位于马里兰站的一个 1,661 赫兹引力辐射探测器和另一个 5,000 赫兹模式引力辐射探测器之间的符合计数进行了研究。没有观测到超过偶然符合计数率的符合计数。接下来的研究表明：5,000 赫兹模式是一个弯曲模式，而不是一个压缩模式。广义相对论预言这种弯曲模式不会和引力辐射发生相互作用。两个探测器之间缺乏符合计数的事实就成为否定 1,661 赫兹下符合计数来自非引力辐射源的证据。

另有一些文章提出证据指出：观测到的符合计数具有各向异性，最大值出现在银河系中心方向[8]。圆盘实验的结果与爱因斯坦纯张量理论的预言一致。局部环境的小作用会显著影响各向异性。为了适当地平均这些影响，最少需要 6 个月的数据。本文采用的磁带记录在时间上不连续。空白以及记录器故障使得数据不足以用来进

did not have times continuously written. Gaps and recorder failures left insufficient data for a study of the anisotropy. Visual examination of the lists of coincidences found by the computer implied that the data are consistent with the earlier anisotropy experiments based on human observer study of pen and ink recorder charts.

The present data are free of human observer bias. No assumptions are made concerning the duration of expected pulses or their shapes beyond the requirement of a fairly smooth leading edge. This is important because addition of noise will change the shape of received pulses. It is considered established beyond reasonable doubt that the gravitational radiation detectors at ends of a 1,000 km baseline are being excited by a common source as a result of interactions which are neither seismic, electromagnetic nor those of charged particles of cosmic rays.

I thank F. J. Dyson, V. Trimble, and G. R. Ringo for valuable discussions and D. J. Gretz and J. Peregrin for maintaining and operating the gravitational radiation detectors. This work was supported in part by NSF and in part by the Computer Science Center of the University of Maryland.

(**240**, 28-30; 1972)

J. Weber: Department of Physics and Astronomy, University of Maryland, College Park, Maryland.

Received April 19; revised August 4, 1972.

References:

1. Weber, J., *Phys. Rev.*, 117, 306 (1960).
2. Weber, J., in *General Relativity and Gravitational Waves*, chap. 8 (Interscience, New York, London, 1961).
3. Weber, J., in *Relativity Groups and Topology*, 875 (Gordon and Breach, New York, 1964).
4. Weber, J., *Nuovo Cim. Lett.*, Series I, 4, 653 (1971).
5. Weber, J., *Phys. Rev. Lett.*, **22**, 1320 (1969).
6. Weber, J., *Phys. Rev. Lett.*, **24**, 276 (1970).
7. Ezrow, D., Wall, N. S., Weber, J., and Yodh, G. B., *Phys. Rev. Lett.*, **24**, 17 (1970).
8. Weber, J., *Phys. Rev. Lett.*, **25**, 180 (1970).
9. Weber, J., *Nuovo Cim.*, **4B**, 197 (1971).

行各向异性的研究。对计算机发现的符合事件列表进行的表观检查显示，这些数据与之前基于对笔墨记录表格进行人工观测研究而得到的各向异性实验结果一致。

现有数据没有人为观测偏差。除了要求有相当平滑的前沿之外，对预期脉冲的持续时间或其形状都没有作任何别的假设。这一点的重要性在于，噪声的加入会改变接收到的脉冲的形状。我们认为已经可以毫无疑义地接受，在 1,000 千米基线两端的引力辐射探测器是由一个共同的源激发的。这是相互作用的结果，这个作用既不是地震或电磁，也不是来自宇宙线带电粒子。

感谢戴森、特林布尔和林戈与我进行了有价值的讨论，还要感谢格雷茨和佩雷格林对引力辐射探测器的维护和操作。美国国家科学基金会和马里兰大学计算机科学中心都为这项工作提供了部分支持。

（何钧 翻译；张元仲 审稿）

Image Formation by Induced Local Interactions: Examples Employing Nuclear Magnetic Resonance

P. C. Lauterbur

Editor's Note

Physicists in the 1940s learned to exploit the response of nuclear spins to applied magnetic fields to probe the structure of solids and liquids. A magnetic field splits the energies of spin states of a nucleus, and radio-frequency radiation can induce transitions between them. Because different nuclei absorb energy at different frequencies, and because the chemical environment also influences this frequency, the technique of nuclear magnetic resonance (NMR) can be used to probe the chemical structure of a sample. Here chemist Paul Lauterbur shows how to adapt this technique to produce detailed spatial images. The technique, known now as magnetic resonance imaging (MRI), is used ubiquitously in basic and applied science, especially medicine. In 2003 Lauterbur shared a Nobel Prize with Peter Mansfield, who developed methods for analysing MRI signals.

AN image of an object may be defined as a graphical representation of the spatial distribution of one or more of its properties. Image formation usually requires that the object interact with a matter or radiation field characterized by a wavelength comparable to or smaller than the smallest features to be distinguished, so that the region of interaction may be restricted and a resolved image generated.

This limitation on the wavelength of the field may be removed, and a new class of image generated, by taking advantage of induced local interactions. In the presence of a second field that restricts the interaction of the object with the first field to a limited region, the resolution becomes independent of wavelength, and is instead a function of the ratio of the normal width of the interaction to the shift produced by a gradient in the second field. Because the interaction may be regarded as a coupling of the two fields by the object, I propose that image formation by this technique be known as zeugmatography, from the Greek ζευγμα, "that which is used for joining".

The nature of the technique may be clarified by describing two simple examples. Nuclear magnetic resonance (NMR) zeugmatography was performed with 60 MHz (5 m) radiation and a static magnetic field gradient corresponding, for proton resonance, to about 700 Hz cm^{-1}. The test object consisted of two 1 mm inside diameter thin-walled glass capillaries of H_2O attached to the inside wall of a 4.2 mm inside diameter glass tube of D_2O. In the first experiment, both capillaries contained pure water. The proton resonance

诱导局域相互作用成像：核磁共振应用实例

劳特布尔

编者按

20世纪40年代，物理学家们掌握了利用核自旋对外加磁场的响应来探索固体和液体的结构的方法。磁场使核自旋态发生能级分裂，而射频辐射可使原子核发生能级跃迁。由于不同的核吸收不同频率的能量，并且化学环境也会影响到这个频率，因此核磁共振技术就可以用来探索样品的化学结构。本文中，化学家保罗·劳特布尔展示了如何利用这一技术获得清晰的空间影像。这一技术——现在被称为磁共振成像，已在基础和应用科学，尤其是在医学中都得到了广泛应用。2003年，劳特布尔和彼得·曼斯菲尔德（后者发展了磁共振成像信号分析方法）共同获得了诺贝尔奖。

一个物体的像可以定义为它的一个或多个特征量的空间分布的图像化表示。成像通常需要该物体与某种物质或辐射场相互作用，且与之作用的物质或辐射场的波长应与期望分辨的物体特征量的最小尺寸相当或更小，从而可能限定相互作用的区域，并产生足够分辨率的图像。

利用诱导局域相互作用可以克服上述波长对成像分辨率的限制，并由此可发展出一类新成像方法：通过施加一个额外场可以将物体与辐射场的相互作用限制在一个有限区域。由此而获得的像的分辨率只是相互作用的标准宽度与由额外场梯度产生的变化量的比值相关的函数，而与波长无关，因为这个相互作用可以看作是两个场借助物体而产生耦合，故建议将基于此技术的成像命名为“zeugmatography”（常被翻译为“共轭成像法”，“结合成像法”或“组合层析成像法”），其来源于希腊语“ζευγμα”，是“结合”的意思。

下面举两个简单的例子来说明该技术的本质。核磁共振共轭成像（现在一般称为“核磁共振成像”或“磁共振成像”）实验采用60 MHz（波长为5m）的射频场和一个带梯度的静磁场，对质子共振而言，该静磁场的梯度场强为700 Hz · cm^{-1}。实验样品是两根装有水（H_2O）的内径为1 mm的薄壁玻璃毛细管，它们被置于一根装有重水（D_2O）的内径为4.2 mm的玻璃管中。在第一个实验中，两根毛细管中均装

line width, in the absence of the transverse field gradient, was about 5 Hz. Assuming uniform signal strength across the region within the transmitter-receiver coil, the signal in the presence of a field gradient represents a one-dimensional projection of the H_2O content of the object, integrated over planes perpendicular to the gradient direction, as a function of the gradient coordinate (Fig. 1). One method of constructing a two-dimensional projected image of the object, as represented by its H_2O content, is to combine several projections, obtained by rotating the object about an axis perpendicular to the gradient direction (or, as in Fig. 1, rotating the gradient about the object), using one of the available methods for reconstruction of objects from their projections[1–5]. Fig. 2 was generated by an algorithm, similar to that of Gordon and Herman[4], applied to four projections, spaced as in Fig. 1, so as to construct a 20×20 image matrix. The representation shown was produced by shading within contours interpolated between the matrix points, and clearly reveals the locations and dimensions of the two columns of H_2O. In the second experiment, one capillary contained pure H_2O, and the other contained a 0.19 mM solution of $MnSO_4$ in H_2O. At low radio-frequency power (about 0.2 mgauss) the two capillaries gave nearly identical images in the zeugmatogram (Fig. 3*a*). At a higher power level (about 1.6 mgauss), the pure water sample gave much more saturated signals than the sample whose spin-lattice relaxation time T_1 had been shortened by the addition of the paramagnetic Mn^{2+} ions, and its zeugmatographic image vanished at the contour level used in Fig. 3*b*. The sample region with long T_1 may be selectively emphasized (Fig. 3*c*) by constructing a difference zeugmatogram from those taken at different radio-frequency powers.

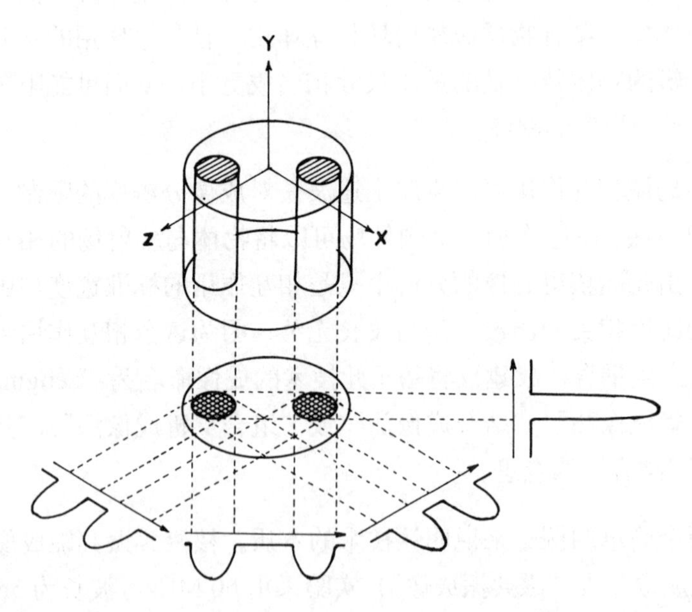

Fig. 1. Relationship between a three-dimensional object, its two-dimensional projection along the Y-axis, and four one-dimensional projections at 45° intervals in the XZ-plane. The arrows indicate the gradient directions.

的是纯水，在未施加横向梯度场时，质子共振线宽约为 5 Hz。假定发射—接收线圈区域内的信号强度均匀，在施加梯度场后获得的信号代表了实验样品中水含量分布沿梯度垂直方向的一维投影，它是与梯度坐标相关的函数（如图 1 所示）。绕垂直于梯度场方向的轴转动实验样品（或如图 1 所示，以样品为轴转动梯度场），获得实验样品的一系列一维投影。以此为基础，采用投影重建的方法 [1-5] 可构建出表示实验样品水含量分布的二维投影像。利用如图 1 所示的四个不同方向的一维投影并采用类似于戈登和赫尔曼的投影重建算法 [4]，可以构建出如图 2 所示的 20 × 20 的图像矩阵，此图像是由矩阵点间轮廓线插值的描影法来显示，它清楚地表征了这两组圆柱体水的位置和大小。在第二个实验中，一根毛细管内装的是纯水，另一根毛细管中装的是 0.19 mM 浓度的 $MnSO_4$ 水溶液。在低射频功率（功率约为 0.2 mGs）下，这两组毛细管水呈现出几乎相同的核磁共振像（如图 3*a* 所示）。但在高功率射频场（功率约为 1.6 mGs）作用下，纯水样品的信号被饱和，导致其在同标准等高线图显示的核磁共振像消失；而对于加入顺磁性 Mn^{2+} 离子的水溶液样品，由于其自旋 – 晶格弛豫时间 T_1 变短，故信号不容易饱和，其核磁共振像仍然出现，如图 3*b* 所示。比较不同射频功率作用下生成的核磁共振像的差分像，即可发现样品中 T_1 时间长的区域得到选择性的突出显示（如图 3*c* 所示）。

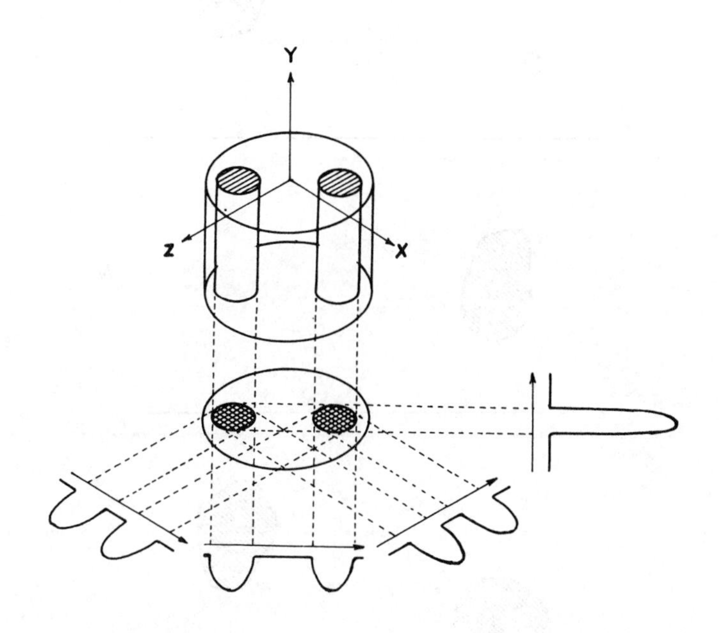

图 1. 核磁共振成像实验的三维示意图：其二维投影方向沿着 Y 轴，四个 45° 间隔的一维投影位于 XZ 平面内，箭头所示方向为梯度场方向。

Fig. 2. Proton nuclear magnetic resonance zeugmatogram of the object described in the text, using four relative orientations of object and gradients as diagrammed in Fig. 1.

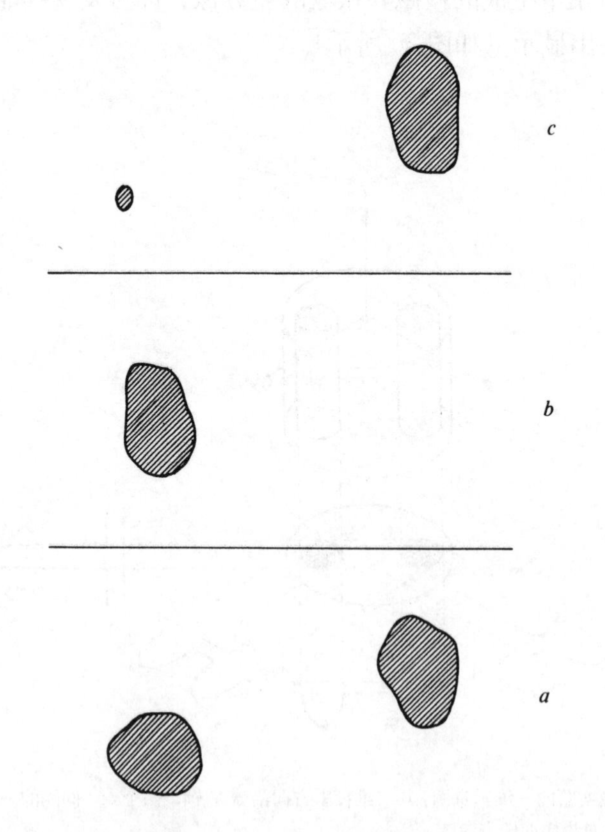

Fig. 3. Proton nuclear magnetic resonance zeugmatograms of an object containing regions with different relaxation times. *a*, Low power; *b*, high power; *c*, difference between *a* and *b*.

图 2. 样品的质子核磁共振像：利用如图 1 所示的四个不同方向的一维投影按文中所述的投影重建方法构建而成

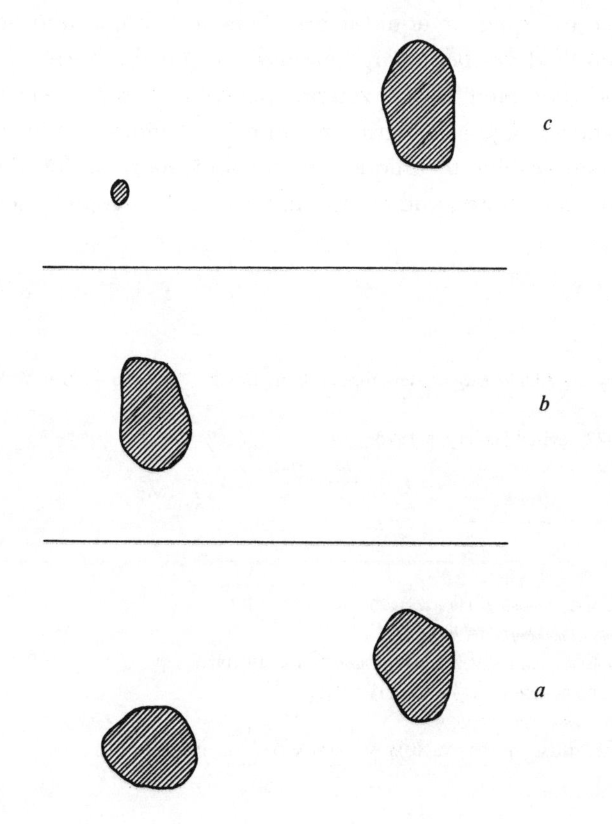

图 3. 样品中弛豫时间不同的区域的质子核磁共振像：*a*, 低强度射频；*b*, 高强度射频；*c*, 图 *a* 和图 *b* 的差分像

Applications of this technique to the study of various inhomogeneous objects, not necessarily restricted in size to those commonly studied by magnetic resonance spectroscopy, may be anticipated. The experiments outlined above demonstrate the ability of the technique to generate pictures of the distributions of stable isotopes, such as H and D, within an object. In the second experiment, relative intensities in an image were made to depend upon relative nuclear relaxation times. The variations in water contents and proton relaxation times among biological tissues should permit the generation, with field gradients large compared to internal magnetic inhomogeneities, of useful zeugmatographic images from the rather sharp water resonances of organisms, selectively picturing the various soft structures and tissues. A possible application of considerable interest at this time would be to the *in vivo* study of malignant tumours, which have been shown to give proton nuclear magnetic resonance signals with much longer water spin-lattice relaxation times than those in the corresponding normal tissues[6].

The basic zeugmatographic principle may be employed in many different ways, using a scanning technique, as described above, or transient methods. Variations on the experiment, to be described later, permit the generation of two- or three-dimensional images displaying chemical compositions, diffusion coefficients and other properties of objects measurable by spectroscopic techniques. Although applications employing nuclear magnetic resonance in liquid or liquid-like systems are simple and attractive because of the ease with which field gradients large enough to shift the narrow resonances by many line widths may be generated, NMR zeugmatography of solids, electron spin resonance zeugmatography, and analogous experiments in other regions of the spectrum should also be possible. Zeugmatographic techniques should find many useful applications in studies of the internal structures, states, and compositions of microscopic objects.

(**242**, 190-191; 1973)

P. C. Lauterbur
Department of Chemistry, State University of New York at Stony Brook, Stony Brook, New York 11790

Received October 30, 1972; revised January 8, 1973.

References:

1. Bracewell, R. N., and Riddle, A. C., *Astrophys. J.*, **150**, 427 (1967).
2. Vainshtein, B. K., *Soviet Physics-Crystallography*, **15**, 781 (1971).
3. Ramachandran, G. N., and Lakshminarayan, A. V., *Proc. US Nat. Acad. Sci.*, **68**, 2236 (1971).
4. Gordon, R., and Herman, G. T., *Comm. Assoc. Comput. Mach.*, **14**, 759 (1971).
5. Klug, A., and Crowther, R. A., *Nature*, **238**, 435 (1972).
6. Weisman, I. D., Bennett, L. H., Maxwell, Sr., L. R., Woods, M. W., and Burk, D., *Science*, **178**, 1288 (1972).

核磁共振成像是不同于那些传统的核磁共振谱技术，不太受样品大小（以及均匀性）的限制，可以预期，此技术将会被应用于诸多非均匀样品的研究。上述实验证明，这项技术能够用于获得样品中稳定同位素（如H和D）含量的分布图像。在第二个实验中，图像的相对强度取决于对应核的弛豫时间的长短。根据生物组织中水的含量和质子弛豫时间的不同，当外加的梯度场远大于生物组织的内部梯度场时，可以利用有机体中相对窄的水共振信号获得有用的核磁共振像，实现各种生物体软组织的选择性成像。当前，一个非常诱人的可能应用是恶性肿瘤的活体研究。因为恶性肿瘤组织中水的质子自旋–晶格弛豫时间比相应正常组织中的要长得多[6]。

基于上文提到的扫描技术或瞬态方法的共轭成像的基本原理有着诸多不同的潜在应用。将上述实验进行拓展延伸，能够产生显示样品的化学成分、扩散系数以及依靠谱学技术测量的其他特征量的二维或三维图像，这些在以后描述。核磁共振成像技术在液体和液体状的体系中的应用简单且引人注目，这是因为此时梯度场足够大且足以移动较窄的共振信号使其空间位置得以区分。而类似的固体核磁共振共轭成像、电子自旋共振共轭成像以及其他频谱学的共轭成像同样有实现的可能。可以预见，共轭成像技术未来将在微观物质的内部结构、状态和构成的研究中也会被广泛应用。

（王耀杨 翻译；刘朝阳 陈方 审稿）

Experiments on Polishing of Diamond

J. Wilks

Editor's Note

The interest of this contribution from John Wilks at Oxford is not so much in the specific findings—that polished diamond is microscopically rough—but in the way it touches on several themes in materials science that have become increasingly relevant in recent decades: the microscopic mechanisms of lubrication and friction, the microtopography of surfaces, and the quest for superhard materials. Wilks highlights the fact that polishing a solid surface is usually different to abrading it, involving local melting because of intense frictional heating. But as only diamond powder itself can polish diamond, abrasion must here be the mechanism: the powder chips the surface and leaves it rough in a way that depends on the orientation of the crystal planes.

Polished diamond surfaces have unusual properties because polishing of the brittle material generally proceeds by mechanical chipping.

DIAMONDS are shaped and polished to produce gem stones by methods which have remained essentially unchanged for several hundred years. Yet it is only recently that the physical basis of these operations has been studied, and their unusual features fully appreciated. This article outlines the principal features of the polishing process and describes some recent experiments which illustrate the very characteristic nature both of the process itself and the polished surfaces produced. We shall show that these perhaps unique properties result from the very hard and brittle nature of the diamond.

The usual method of polishing a diamond is to hold it either against the face of a rotating cast-iron wheel (or scaife) charged with a suspension of fine diamond powder in a light oil, or against the face of a wheel in which diamond powder is bonded into a metal matrix. One of the most striking features of this method is that if a cube face is prepared on the diamond, it turns out that the rate of removal of material during polishing with a bonded wheel in a direction parallel to a cube axis is about 100 times greater than if the same face is polished in a direction at 45° to a cube axis. Similar effects are observed on other faces, and even greater differences are evident when polishing on a cast-iron scaife, as the more resistant directions of the diamond tend to knock the powder out of the scaife.

In preparing a gem stone from a rough diamond it is necessary to remove a certain amount of material to position the facets before polishing them. Another unusual feature is that this removal of material is effected in almost the same way as polishing, except that larger sized diamond powder is used to remove material quickly. For most materials,

金刚石抛光实验

威尔克斯

编者按

来自牛津的约翰·威尔克斯所做的贡献中令人感兴趣的不仅仅在于抛光的金刚石在显微镜下是粗糙的这一特定研究结果，还在于近几十年来它越来越多地涉及材料科学的若干主题：润滑和摩擦的微观机制、表面的微观形貌和对超硬物质的探求。威尔克斯强调固体表面的抛光与磨削是有区别的，抛光包括由强烈的摩擦热产生的局部熔化。但是只有金刚石粉本身能抛光金刚石，磨削一定是这种机制：粉末切削表面后，表面的粗糙情况在某种程度上取决于晶面的取向。

抛光的金刚石表面具有不同寻常的性质，因为这种脆性物质的抛光通常是通过机械切削来进行的。

在过去的几百年中，将金刚石加工成形和抛光以生产宝石所用的方法在本质上保持不变。一直到最近，这些操作的物理学基础才得到了研究，它们那不同寻常的特征也得到了充分的理解。本文概述了抛光过程的主要特征，并且描述了一些最近的实验，它们阐明了对于抛光过程本身和所产生的抛光表面都极有特征性的本性。我们将要说明，这些独特性质可能是由金刚石极硬且脆的本性所导致的。

抛光金刚石的常规方法是将金刚石抵在覆盖着细小金刚石粉末的轻油悬浊液的旋转铸铁轮（或磨光盘）表面上，或者是抵在表面嵌入金刚石粉末的金属基体的轮表面上。这种方法一个最令人吃惊的特征是，如果要在金刚石上磨制一个立方体表面，就会发现在用黏合了粉末的轮子进行的抛光过程中，平行于立方轴方向上的磨除速率约是同一表面在与立方轴呈 45° 角方向上的速率的 100 倍。在其他面上也观测到了类似的效应，在用铸铁磨光盘进行抛光时甚至发现了更为显著的速率差异，因为在金刚石更具刚性的方向上更容易磨掉磨光盘上的粉末。

在把金刚石原料加工成宝石时，必须要在抛光定位刻面前切除一定量的余料。另一个不同寻常的特征就是这种材料的切除会产生与抛光几乎一样的效应，只不过为了快速切除会使用较大尺寸的金刚石粉末。对于大多数材料来说，抛光和通过磨

polishing and the removal of material by abrasion are quite different processes. Removal of material is generally achieved by the use of an abrasive powder in what is essentially a cutting process; the powder is harder than the specimen, and it either gouges out material, or sets up stresses which lead to cracking after it has moved on. By contrast, polishing is generally a smearing process, brought about by rubbing with a second material to produce high temperatures in the small regions of true contact between the asperities on the two surfaces; the high temperatures result in local melting of the asperities, which are then smoothed out. It follows that to polish a specimen, we must use polishing material with a higher melting point, so that on rubbing, the specimen melts and flows first. For further details of the general features of abrasion and polishing, see for example, Bowden and Tabor[1] and Cottrell[2].

In view of all this, one would expect to encounter difficulties both in removing material from diamonds and in polishing them. As far as the removal of material is concerned, there is no harder substance than diamond which can be used to produce abrasive cutting. In rubbing operations, the high thermal conductivity of diamond will tend to reduce the temperature of the hot spots at the areas of local contact, whereas high temperatures are required to effect even the conversion of diamond to graphite. Thus, conditions seem unfavorable for both abrasion and polishing, but in fact the traditional methods used by the diamond industry abrade and polish diamond quite successfully. The rate of abrasion is relatively low, but a high polish can readily be obtained, and a diamond surface may be worked to the standard of a good optical flat[3].

The first scientific experiments on the polishing process were made more than 50 years ago by Tolkowsky[4], who proposed that both abrasion and polishing proceed by mechanical chipping on a microscopic scale. This chipping will be controlled by the lie of the cleavage planes relative to the direction of abrasion, and will therefore occur more readily in some directions than others, thus giving rise to very considerable differences in rates of abrasion. To indicate the geometrical arrangement of the cleavage planes, Tolkowsky made use of a model of diamond built up from small identical octahedral and tetrahedral shaped blocks, and thus explained the variations of the hardness on both the cube and the dodecahedron planes[4,5].

Tolkowsky was principally interested in practical applications and his conclusions about how to polish and abrade diamonds soon found their way into the diamond trade, as did his other research on the best way to lay out the facets of a diamond so as to produce the well known brilliant cut. On the other hand, his account of the underlying mechanism was not widely published, and attracted little attention. His treatment was criticized on the grounds that diamond is certainly not built up of equally sized elementary blocks as in his model, but the essential function of the model is to indicate the positions of the cleavage planes—and this it does correctly. It now turns out, as the result of various experiments on diamond over the past 15 years, together with increasing knowledge of the nature of brittle materials, that Tolkowsky's views are essentially correct.

削来切除材料是相当不同的过程。原料的切除一般是通过使用磨料粉末以本质上是切削的过程来实现的；粉末比样品更坚硬，它们在移动中或者挖去原料，或者施加应力而使样品产生破裂。与此相反，抛光一般是一个涂抹的过程，通过用另一种材料进行摩擦，在两个表面的粗糙部位之间真正接触的小区域中产生高温，高温引起粗糙部位的局部熔融，继而使其变平滑。由此可知，为了将样品抛光，我们必须使用具有较高熔点的抛光材料，以便使样品在摩擦时首先熔融和流动。关于磨削与抛光过程一般特征的更详细资料可参见鲍登和泰伯[1]与科特雷尔[2]的文章。

考虑到上述这一切，可以预料人们在从金刚石上切除余料以及对其进行抛光时都会遇到困难。就原料的切除而言，不存在比金刚石更硬的物质可以用来进行磨削切削。在摩擦操作中，金刚石的热传导性高，将会倾向于降低局部接触区域中热位点的温度，而高温甚至是从金刚石到石墨的转化过程所必需的。因此，情况对于磨削和抛光似乎都是不利的，但事实上，金刚石工业所用的磨削和抛光金刚石的传统方法却是相当成功的。磨削速率相对来说较低，但是很容易获得高度抛光，而且金刚石表面可以作为一个良好的光学标准平面[3]。

关于抛光过程的第一个科学实验是在50多年前由托尔考斯基[4]进行的，他提出，磨削和抛光过程都是通过微观尺度上的机械切削实现的。这一切削过程将会受到解理面的形貌相对于磨削方向的位置的控制，并且会因此在某些方向上比在另一些方向上更容易发生切削过程，从而导致颇为可观的磨削速率差异。为了说明解理面的几何排列方式，托尔考斯基使用了以同样的八面体和四面体形小块搭建的金刚石模型，并由此解释了立方体和十二面体平面上的硬度变化[4,5]。

托尔考斯基主要关心实践应用，因而他关于如何抛光和磨削金刚石的结论很快就在金刚石行业得到应用，就像他另一个关于展示金刚石刻面的最佳方式的研究导致了广为人知的明亮琢型的产生一样。另一方面，他关于潜在机制的说明却没有得到广泛发表，因此没有引起多少注意。他的处理方法遭到了批评，根据是金刚石肯定不是像他的模型中那样由同样大小的基本模块搭建而成的，但是该模型的实质功能在于说明解理面的位置——而它正确地做到了。过去15年来关于金刚石的若干实验的结果，再加上对于脆性材料本性越来越多的了解，可以证实托尔考斯基的观点实质上是正确的。

One of the crucial features of the above experiment observed but not stressed by Tolkowsky[4], is that the rate of removal of material is proportional to the total number of revolutions of the polishing scaife but does not depend on its speed. This result, confirmed by Wilks and Wilks[5,6], shows that thermally activated processes play no significant part in the removal of material. The rubbing together of two materials produces heating at the areas of true local contact, and this excess temperature increases rapidly with the speed of rubbing[1]. The fact that the abrasion per revolution does not increase with rising speed of rubbing clearly demonstrates that local hot spots play no significant part in the wear process. In particular, it rules out the possibility that the diamond is worn away by either burning or conversion to graphite, as suggested by Seal[7]. On the other hand, one would expect the same amount of material to be removed by each revolution of the wheel if the abrasion proceeds by a mechanical chipping or cleavage process.

The conditions necessary for materials to fail by chipping or brittle fracture have been discussed by several authors, for example Kelly[8]. Kelly, Tyson and Cottrell[9] have considered whether an ideal crystal free of imperfections will fail plastically by shear or by brittle fracture under tension. They show that for face centred metals such as copper, gold and silver, the stress required to produce fracture is about 30 times greater than that to produce shear, and that failure always occurs by plastic deformation rather than cleavage. On the other hand, the calculations show that, of the substances considered, sodium chloride and diamond are most likely to fail by brittle fracture. The critical stresses for shear and cleavage in sodium chloride and diamond are approximately equal, but the calculations are not sufficiently refined to indicate which process is preferred. These calculations are, however, for ideal crystals and the essential question for real materials is whether the presence and motion of dislocations will permit plastic flow before fracture.

Dislocations in diamond have been observed by Evans[10] who prepared specimens of diamond by etching, and viewed the dislocations directly in the transmission electron microscope. They are also clearly delineated by the X-ray topograph techniques of Frank and Lang[11]. Calculations mentioned by Evans[12] suggest that the movement of the dislocations will be inhibited by the necessity of breaking carbon-carbon bonds. Their motion has been studied experimentally by Evans and Wild[12,13] who took specimens of diamond in the form of thin slabs, loaded them in the middle, and observed whether they failed by fracture or plastic deformation. They found that at low temperatures the beam always failed by fracture, but that an increasing amount of plastic flow was observed as the temperature exceeded 1,500 °C. Hence dislocations in diamond may be quite mobile, but at room temperature are held back by obstacles which can only be overcome by considerable thermal activation. As the rates of abrasion and polish in the experiments referred to above were independent of the temperature of local hot spots, one concludes that the abrasion was not accompanied by any thermally activated motion of dislocations.

It is sometimes suggested that plastic flow in diamond may be produced by conditions of intense local pressure. The effect of pressure on brittle materials is sometimes surprising, for example it is possible to compress a slab of brittle material, such as rock salt, to half its

上述实验中托尔考斯基观察到了但并未强调的一个关键特征[4]，就是切除材料的速率与抛光磨盘总转数成比例，而与其速率无关。这个结果得到了威尔克斯和威尔克斯[5,6]的确认，它表明热激发过程在材料切除中的作用并不重要。两种材料的相互摩擦使真正接触的局部区域产生了热，而且过剩温度随着摩擦速率的上升而快速增加[1]。每转的磨削不随摩擦速率提高而增加这一事实清晰地证明局部热位点在磨削过程中不起重要作用。尤其是，它排除了西尔[7]的观点，即金刚石通过燃烧或转化为石墨而磨削的可能性。另一方面，人们可以预想，如果磨削是通过机械切削或解理过程而进行的，那么轮的每一转会切除相同量的材料。

若干位作者，比如凯利[8]，已经讨论了关于材料通过切削或脆性断裂而破坏的必需条件。凯利、泰森和科特雷尔[9]曾考虑，毫无缺陷的理想晶体是否会通过切变或在张力下脆性断裂发生塑性破坏。他们指出，对于诸如铜、金和银等面心金属来说，产生断裂所需的应力是产生切变所需应力的约 30 倍，而且破坏的产生经常是由于塑性形变而不是解理。另一方面，计算表明，在所考虑的各种物质中，氯化钠和金刚石是最容易由于脆性断裂而破坏的。氯化钠与金刚石中切变和解理的临界应力是近似相等的，但是计算还没有精细到足以指出哪一过程是更为有利的。不过，这些计算都是针对理想晶体的，而对于真实材料的实质问题是，位错的存在和运动是否会允许先塑性流动后再断裂。

金刚石中的位错已被埃文斯[10]所观测到，他用蚀刻法制备金刚石样品，然后在透射电子显微镜下直接观察到了位错。弗兰克和兰[11]也用 X 射线照相技术对其进行了清晰的描述。埃文斯[12]所提到的计算指出位错的运动会因为必须要打破碳—碳键而被抑制。位错的运动已由埃文斯和维尔德[12,13]进行了实验研究，他们使用具有薄板形式的金刚石样品，在其中部加置负荷，然后观察它们是因断裂还是塑性形变而破坏。他们发现，低温时样品板总是会因断裂而破坏，但是当温度超过 1,500 ℃时，可以观测到塑性流动量的增加。因此金刚石中的位错可能具有相当的流动性，不过在室温时被某种阻碍所抑制，只有通过相当大的热激发才能克服这种阻碍。由于上面所提到的实验中的磨削和抛光速率与局部热位点的温度无关，有人断言，磨削并不伴随着任何位错的热激发运动。

不时有人提出，金刚石中的塑性流动可能是由强烈的局部压力而产生的。压力对于脆性材料的影响有时会是令人惊讶的，例如，将诸如岩盐等脆性材料的平板压

initial thickness without producing cracking[14], and Howes and Tolansky[15] have argued that the ring cracks produced by indenting diamond with steel spheres show evidence of plastic flow. Later studies by Lawn and Komatsu[16] of the region round such a crack, using X-ray topography, show no sign of plastic deformation, and Frank and Lawn[17] have since shown that the cracks may be described in terms of fracture processes.

We conclude that the mechanism of abrasion is one of mechanical chipping. This conclusion is supported by a study of the area around an abrasion mark, by Frank, Lawn, Lang and Wilks[18] using X-ray topography, which shows that the state of strain in the surface is consistent with stresses set up by fracture processes and not by plastic flow. We believe that there is no convincing evidence for the motion of dislocations in any of the relevant abrasion and polishing experiments, although claims to have observed plastic flow in diamond at room temperature have been made by Brookes[19] and by Gane and Cox[20]. Brooks indented a diamond surface with a Knoop indenter made of diamond and claimed that the form of the indentation indicated that plastic flow had occurred. A careful study of Brookes's indentations, however, show clear indications of brittle fracture (Fig. 1), which appear to have been obscured in the earlier micrographs. The observations of Gane and Cox were based on the study of thin diamond wedges pressed against each other, but their evidence is at best inconclusive. Even under these rather specialized conditions, there is no clear evidence that plastic flow and motion of dislocations took place.

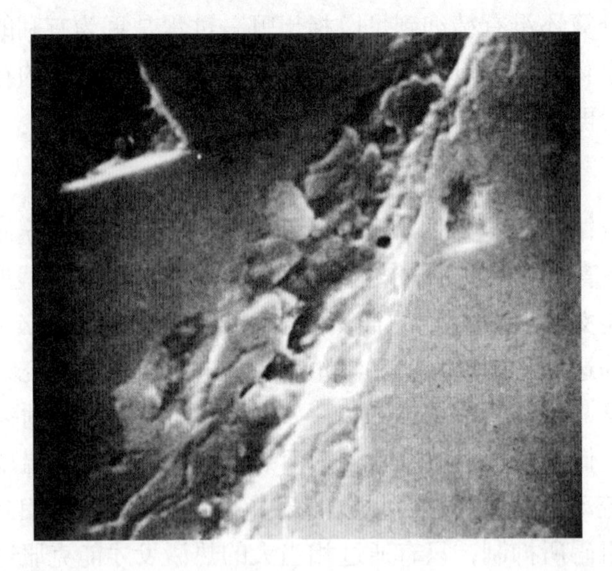

Fig. 1. Scanning electron microscope micrograph of a diamond surface indented with a Knoop indenter (×8,400).

We shall now describe three sets of experiments which confirm the fracture process and illustrates the unusual properties of polished diamond surfaces. In the first set a simple micro-abrasion tester[21] was used to study the variation of the rate of removal of material—in short the hardness of the diamond—with the orientation of the surface and the direction of polish. The hardness is often extremely sensitive to any change in the orientation of the facet being abraded. For example, Fig. 2 shows the relative rates

缩到其初始厚度的一半而不产生裂痕[14]是有可能的，而且豪斯和托兰斯基[15]认为，通过用钢球挤压金刚石而产生的环形裂纹可以作为塑性流动的证据。后来朗和小松[16]利用X射线照相法对这类裂纹周围的区域的研究指出没有这种塑性形变的迹象，弗兰克和朗[17]由此指出，可以按照断裂过程对裂纹加以描述。

我们断言磨削的机制是一种机械切削。这一结论为弗兰克、朗、兰和威尔克斯[18]用X射线照相法对磨削痕迹周围区域的研究所证实，该研究表明，表面的应变状态与断裂过程而非塑性流动所导致的应力相一致。我们相信，在任何有关磨削和抛光的实验中都不存在位错运动的可靠证据，尽管布鲁克斯[19]以及甘恩和考克斯[20]曾宣称在室温下观测到金刚石中的塑性流动。布鲁克斯用金刚石制成的努普压头挤压金刚石表面，并且宣称凹痕的形态表明曾发生过塑性流动。但是，对布鲁克斯凹痕的仔细研究可以看出清晰的脆性断裂迹象（图1），它在以前的显微图像中可能一直都是模糊的。甘恩和考克斯的观测是基于对彼此挤压的小块楔形金刚石的研究，但是他们的证据充其量是不确定的。即便是在这些相当特殊的条件下，也不存在发生塑性流动和位错运动的明确证据。

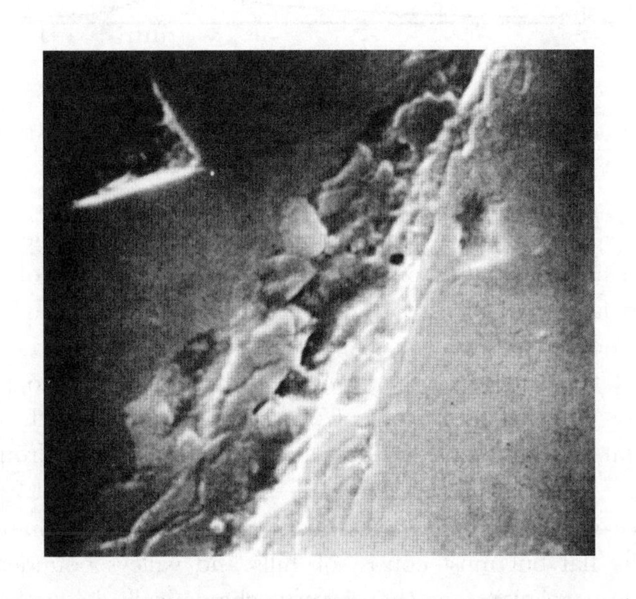

图1. 用努普压头挤压金刚石表面得到的扫描电子显微图像（×8,400）

现在我们要介绍三组实验，它们确证了断裂过程并且描绘出抛光后金刚石表面不同寻常的性质。第一组实验是用一个简单的微型磨削检测器来研究材料切除速率[21]——简而言之就是金刚石的硬度——随表面指向和抛光方向的变化。硬度通常对于磨削刻面的任何指向变化都极为敏感。例如，图2所示为通过使八面体平面偏

of removal of material on surfaces obtained by tilting an octahedron plane about the (011) axis indicated in the sketch plan. The crosses show the rate of removal of material for abrasion in the direction A_x, and the circles the rate in the direction A_o. We see that on a true octahedron plane the rate of removal in the direction A_o is more than twice that in the direction A_x, but that a tilt of only about 1° is sufficient to reverse the relative hardness.

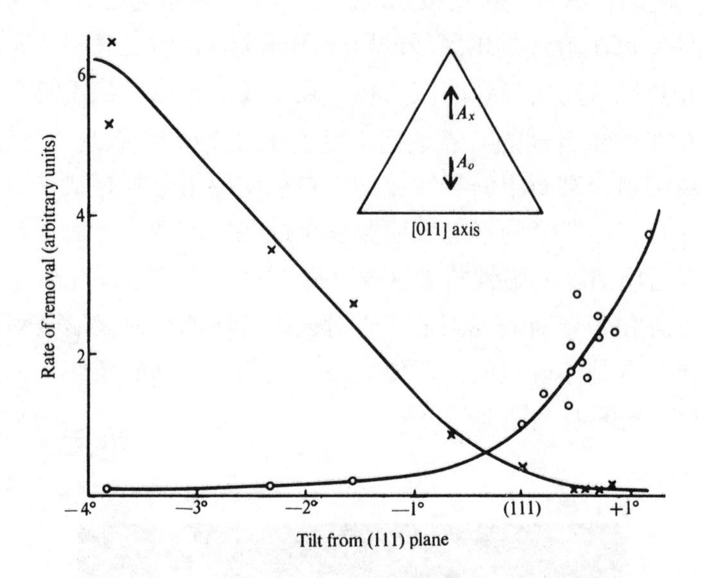

Fig. 2. The relative rates of removal of material by grinding on surfaces near to an octahedron plane. The symbols ○ and × correspond to the directions shown in the inset.

These orientation effects are not explained by Tolkowsky's treatment, which also has difficulty in discussing the octahedron face. The variation in rates of abrasion on cube and dodecahedron faces is explained by the lie of the cleavage planes which results in the production of a different structure by abrasion in different directions. As the cleavage planes run parallel to an octahedron face, the treatment seems to predict a uniform surface with no easy and hard directions of abrasion, in contrast with the measurements. To explain this point, we must consider the behavior of tilted octahedron facets.

A diamond surface which has been polished by a process of mechanical chipping will not be atomatically flat but must consist of hills and valleys bounded by an irregular arrangement of cleavage planes as shown very schematically in Fig. 3*A*. It follows that abrasion proceeds by the removal of material from the tops or sides of the microhills. If Fig. 3*A* represents a section through an octahedron face, parallel to the directions A_x and A_o, the lie of the cleavage planes is such that the most likely sites for the removal of material are the edges *a* and *b* (ref. 5). It also follows that the easiest direction of abrasion is A_o, and that most of the material is then removed from the edges *b*. If, on the other hand, a face is prepared inclined to the octahedron face by a small angle, the lie of the cleavage planes remains the same, so the structure of the microhills is as shown schematically in

离示图中所标出的（011）轴而得到的表面的材料切除的相对速率。叉形符号表示 A_x 方向上由磨削所导致的材料切除速率，圆形符号则表示 A_o 方向上的速率。我们看到在一个真正的八面体平面上，A_o 方向上的切除速率比 A_x 方向上的 2 倍还要大，但是只偏离 1° 就足以使相对硬度关系逆转。

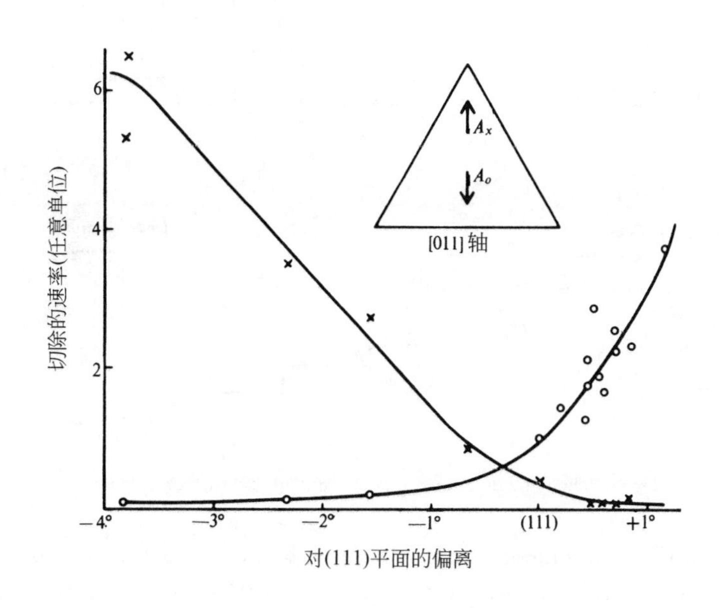

图 2. 在八面体平面附近的表面上碾磨所产生的材料切除的相对速率。符号 ○ 和 × 对应于插图中所示的方向。

托尔考斯基的处理方法并未解释这种指向效应，他的方法在讨论八面体的面时也是有困难的。在立方体表面和十二面体面上的磨削速率差异可以利用通过在不同方向上磨削而导致不同结构产生的解理面的形貌来解释。由于解理面趋向于与一个八面体的面平行，这种处理方法似乎预言了一个不存在磨削难易方向的均一表面，而这与测量结果是违背的。为了解释这一点，我们必须考虑偏离的八面体刻面的行为。

如同图 3*A* 中颇为概略地显示的那样，通过机械切削过程抛光后得到的一个金刚石表面不是平坦的，而是包含着由解理面无规则排布所围成的峰和谷。由此可知，磨削是通过从这些微小山峰的顶部和侧部切除材料而实现的。如果图 3*A* 表示通过八面体的一个面的截面，平行于 A_x 和 A_o 方向，那么解理面就会具有这样的形貌：最有可能发生材料切除的位置是边 *a* 和 *b*（参考文献 5）。还可以知道最容易磨削的方向是 A_o，以及大部分材料因此是从 *b* 边切除的。另一方面，如果令一个面与八面体的面倾斜成一个小角度，那么解理面的形貌仍保持原样，所以微小山峰的结构就如图 3*B* 中概略地显示的那样。从形貌的角度看，磨削轮一定会得到数量增加的 *b* 类

Fig. 3*B*. Topographically, the abrading wheel must be presented with an increased number of *b* type edges, and the hardness correspondingly reduced. Similarly a tilt in the opposite direction increases the hardness.

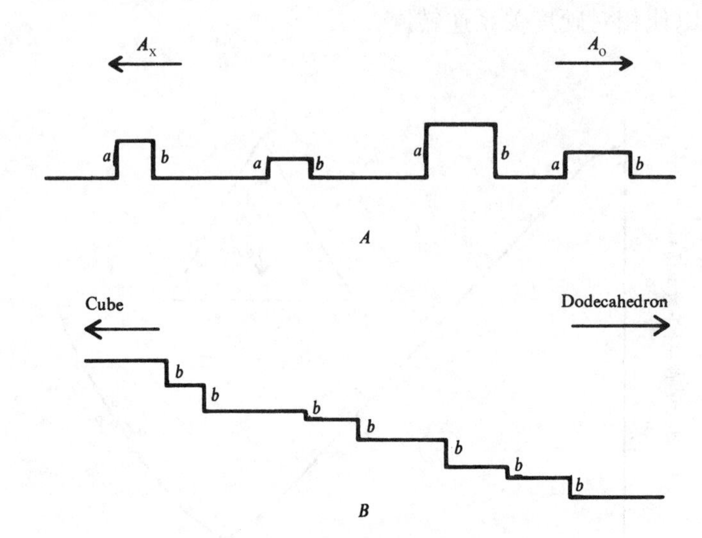

Fig. 3. Schematic diagrams of polished diamond surfaces (see text).

One also expects that the change in hardness with tilt may be very rapid. If the linear dimensions of the microhills are all of the same order of magnitude h, the average distance between hills in contact with the abrading wheel will be much larger because the areas of real contact are less than the nominal areas by factors of order 10^4. Thus the distance between the hills may be of the order of 100 h, so that a tilt of 1° would be sufficient to change the topography of Fig. 3*A* to that of Fig. 3*B*, and thus approximately double the number of sites of easy abrasion. Similar arguments may be applied to account for the hardness of both true and tilted cube and dodecahedron faces. All the predictions that can be made from the model are in accord with experiment[5].

We have also made studies of the jagged nature of polished diamond surfaces, implied by our abrasion experiments, by observing the friction of diamond sliding on diamond in air. Some time ago Seal[7] showed that the friction of a polished cube surface varies with the azimuthal angle of sliding, with a four-fold symmetry. This result is at first surprising, as Bowden and Hanwell[22] have shown that diamond surfaces in air are covered by a strongly adsorbed film which reduces the friction by an order of magnitude below its value for clean surfaces in high vacuum. Thus even though the diamond surface is covered by a tenacious film of adsorbed gases, the symmetry of the diamond is projected through the film.

The friction between diamonds in air was first measured by Bowden and Young[23], and Bowden and Tabor[24] later pointed out that their results showed an approximate relation between the friction μ and the load W of the form $\mu \propto W^{-1/3}$. They then suggested that

型的边，而硬度则相应地下降。类似地，相反方向上的偏离则会增加硬度。

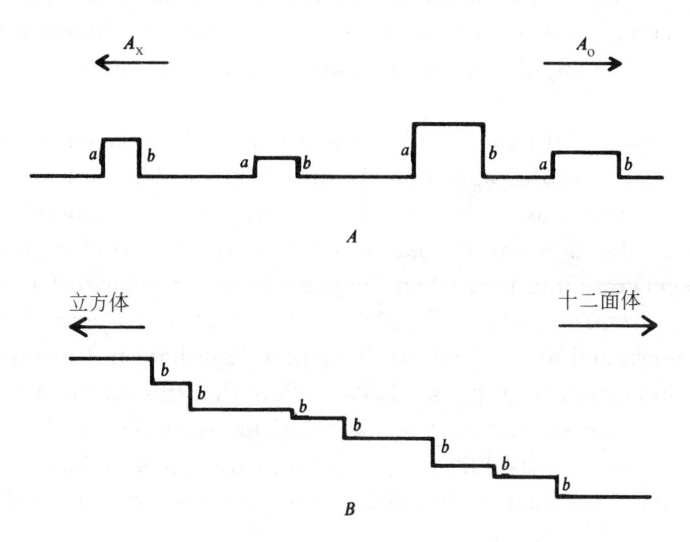

图 3. 抛光后金刚石表面的示意图（见正文）

我们还可以预期硬度随偏离的变化会非常快。如果微小山峰的线性尺度全都具有同样的数量级 h，与磨削轮接触的山峰之间的平均距离就会大很多，因为真正的接触面积不到名义上的面积的 10^{-4}。因此山峰之间的距离可能具有 $100h$ 的数量级，才能使 1° 的偏离就足以使图 3*A* 中的形貌改变成图 3*B* 中那样，并由此而使易磨削位置的数目近似翻了一倍。类似的论证可以用来解释正的与倾斜的立方体面和十二面体面的硬度。从这个模型所能得到的所有预言都与实验结果吻合 [5]。

我们还对磨削实验显示的抛光后金刚石表面的粗糙性质进行了研究，方法是观测在空气中金刚石在金刚石表面滑行时的摩擦力。一段时间之前，西尔 [7] 曾指出抛光立方体表面的摩擦力随着滑行方位角的变化而变化，并具有四重对称性。最初看来这个结果是令人惊讶的，因为鲍登和汉威尔 [22] 曾指出，空气中的金刚石表面为一层强烈吸附的薄膜所覆盖，它使摩擦力比起在高真空中清洁表面降了一个数量级。因此，即使金刚石表面被黏着性的吸附气体薄膜所覆盖，金刚石的对称性也会透过薄膜表现出来。

空气中的金刚石间摩擦力是由鲍登和扬 [23] 首先测得的，后来鲍登和泰伯指出，他们的结果表明摩擦力 μ 与负载 W 之间具有形如 $\mu \propto W^{-1/3}$ 的近似关联。接着，他们提出，这种关联性会由于实际上的真实接触区域之间的附着过程而增加，而且这

this relation arose because of adhesion processes between the actual areas of true contact, and that this adhesion was responsible for the friction. Recent measurements by Casey and Wilks[25] show, however, that the friction is independent of the load over a wide range of loads, including the range of the earlier experiments. It therefore seems that the mechanism responsible for friction is not one of adhesion, a result which is confirmed by the fact that the friction remains unchanged when the diamond surface is lubricated with light oil.

The observed behavior of the friction is readily explained by a roughness (or ratchet) type mechanism associated with the jagged nature of a polished diamond surface. That is, the friction force arises from the work required to move the two diamond surfaces against the normal load as they are forced apart when asperities ride over each other. Some of this work will degenerate into heat when the jagged and irregular surfaces come together again, because the return motion will tend to be abrupt and irreversible. The work of separation is proportional to the load, so this type of mechanism leads to a value of the friction which is independent of the load. We believe that this mechanism accounts quite generally for the friction of diamond on diamond in air as discussed in more detail by Casey and Wilks[25]. The adsorbed film acts as a lubricant, which reduces adhesion and the high friction observed in a vacuum, but does not obscure the structure of the surface.

Fig. 4*a* shows the results of recent measurements on the friction of a diamond sliding over a polished cube surface which are similar to those of Seal[7]. The diamond surface had been polished in the usual way and was therefore covered by an array of fine polishing lines or grooves running parallel to a cube axis. Nevertheless, the friction shows a full four-fold symmetry, having the same value in directions parallel and perpendicular to the grooves. We then repolished a surface by abrasion in a hard direction at 45° to a cube axis, until the original polishing lines were replaced by a new set at 45° to the original direction. The friction on this new surface is shown in curve *b*; the friction has quite different values and exhibits only a two-fold symmetry, the lowest values being for directions parallel to the direction of polish.

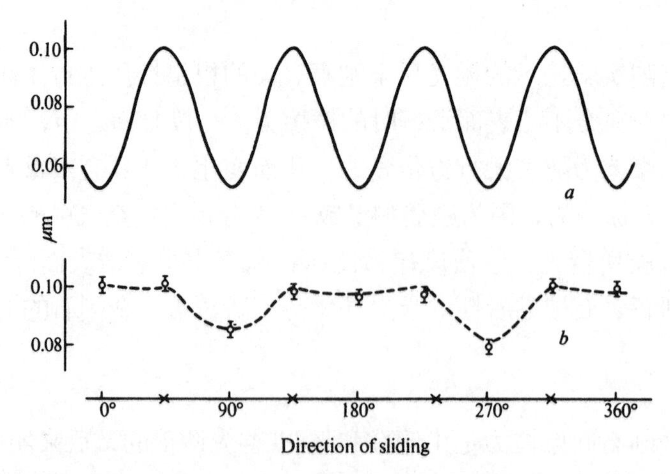

Fig. 4. The friction of a diamond sliding over a polished cube surface as a function of azimuthal angle. *a*, After polishing in the normal soft direction; *b*, after polishing in a hard direction.
The symbols × indicate the positions of the cube axes.

种附着就是摩擦力产生的原因。不过，凯西和威尔克斯最近的测量结果[25]表明，摩擦力在一个宽广的负载范围内是与负载无关的，其中包含着早期实验中的负载范围。于是，看起来附着机制并不是导致摩擦力产生的机制，这个结果被下面的事实所证实，即用轻油将金刚石表面润滑后摩擦力保持不变。

利用与抛光的金刚石表面粗糙性质有关的粗糙度（或棘齿）型机制，很容易解释已观测到的摩擦力行为。也就是说，摩擦力缘起于承担着正常负载的两个金刚石表面在其凹凸部分越过彼此被迫分离时所需要的功。这些功的一部分会在粗糙且不规则的表面再次靠拢时转化为热，因为返回运动会倾向于不连续和不可逆。这种分离功与负载成比例，因此这种机制导致摩擦力数值与负载无关。我们相信这种机制更普遍地解释了空气中金刚石间的摩擦力，如同凯西和威尔克斯[25]更为详细地讨论的那样。吸附膜承担润滑剂的作用，它减少了真空中所观测到的附着和强摩擦，但是并不掩盖表面的结构。

图 4*a* 显示了最近关于金刚石在抛光立方表面上滑行的摩擦力的测量结果，它与西尔的结果[7]类似。金刚石的表面已经用常规方式进行了抛光，从而为平行于立方轴的一排精细的抛光线或者说凹槽所覆盖。尽管如此，摩擦还呈现出完全的四重对称性，在平行于和垂直于凹槽的方向上具有相同的数值。接着，我们在与立方轴成 45° 角的硬方向用磨削方法对一个表面进行重新抛光，直到原来的抛光线被一组与原方向成 45° 角的新抛光线所取代。曲线 *b* 显示了这个新表面上的摩擦力；摩擦力具有不同的数值，而且只呈现出二重对称性，最低数值出现在与抛光方向平行的方向上。

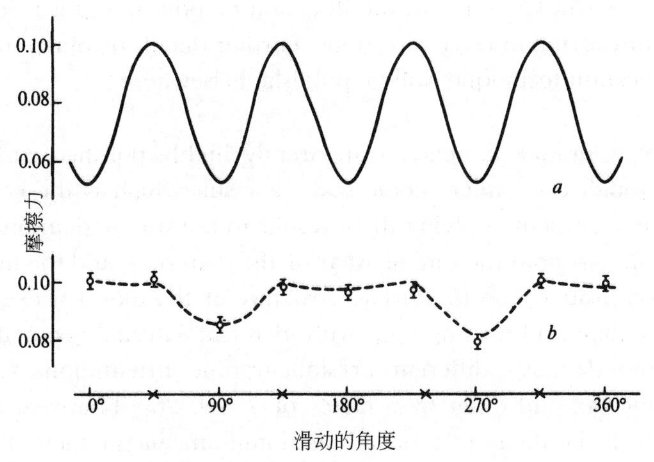

图 4. 金刚石在抛光的立方体表面滑动时的摩擦力作为方位角的函数。*a*，正常的软方向上的抛光后；*b*，硬方向上的抛光后。符号 × 表示立方轴的位置。

These results confirm that the topography of a polished surface is determined by mechanical chipping. Just as abrasion proceeds at different rates in different directions, because of the orientation of the cleavage planes, so surfaces prepared by polishing in different directions will have different structures. As discussed by Tolkowsky, a surface polished in the usual direction parallel to a cube axis will present a jagged surface with a four-fold symmetry. If, however, the diamond is polished in one of the hard directions, the surface structure will tend to take the form of grooving parallel to the direction of polish, with the two-fold symmetry shown in the friction measurements. We are at present extending our studies to observe the wear effects associated with repeated passes of the stylus over the same area of surface.

The third set of experiments are concerned with observing the surface structure of polished diamonds by using high resolution electron microscopy and carbon replica techniques. The structure turns out to be irregular and on a very small scale; therefore, particular care must be taken to ensure that the micrograph is in focus and free of astigmatism. Fig. 5 shows micrographs prepared under carefully controlled and similar conditions, except that the replica of Fig. 5*a* was taken from a freshly cleaved surface of mica, and those of Fig. 5*b* and *c* from polished surfaces of diamond. The greater contrast in the diamond micrographs indicates that the surface is much rougher than the mica and that the scale of the irregularities is of the order of 50 Å.

The micrographs in Fig. 5*b* and *c* were prepared from the same stone, using the same technique, but Fig. 5*b* was taken from a cube surface polished in the usual way, and Fig. 5*c* from a polished octahedron face. We have also observed similar structures on the cube and octahedron faces of a second diamond, and also a different characteristic structure on the dodecahedron faces. These results confirm the fact, well known to diamond polishers, that the quality of polish depends on the particular face being polished, as follows from the mechanical nature of the polishing process. One expects the surface structure to be delineated, at least to some extent, by cleavage planes. On a cube surface, the traces of these cleavage planes will lie at 45° to the direction of polish, and it is interesting to note that there are features lying in these directions. Further details of our results, together with details of the replication technique, will be published elsewhere.

Diamond is perhaps unique in that an apparently highly polished surface may in fact be mechanically rough on a microscopic scale, a result which is the consequence of its brittleness. We are at present applying these results to the use of diamond as cutting tools for machining metals, as both the rate of wear of the diamond, and the finish produced on the metal, depend sensitively on the surface structure of the tool. For example, the rate of wear of two single-diamond turning tools with identical external geometry, but fabricated so that the diamonds have different crystallographic orientations, show wear rates which are reproducible, and differ by a factor of 7 (ref. 26). The wear in this particular experiment, in which the diamonds turned an aluminum-silicon alloy of the type used to manufacture motor car pistons, was almost certainly controlled by the lie of the fracture planes as discussed above, and these are quite different in the two crystallographic

上述结果证实，抛光表面的形貌是由机械切削所决定的。正如磨削过程在不同方向上以不同速率进行一样，由于解理面的取向，通过不同方向上的抛光所制得的表面就会有不同的结构。根据托尔考斯基所讨论的，在平行于立方轴的常规方向上抛光的表面会呈现为具有四重对称性的粗糙表面。但是，如果将金刚石在某一个硬方向上抛光，表面结构将会倾向于形成平行于抛光方向的凹槽，并具有摩擦力测量中所显示的二重对称性。目前我们还将研究扩展到观测用铁针反复划过同一表面区域时所涉及的磨损效应。

第三组实验是利用高分辨电子显微技术和碳复型技术对抛光金刚石表面结构进行观测。研究表明该结构是不规则的并且具有很小的尺度；因此必须要特别小心以保证显微图像是清晰的而且没有受到像散的影响。图 5 显示了在类似条件下经谨慎控制所制备的显微图像，仅有的区别在于图 *5a* 的样品是云母的新鲜解理面，而图 *5b* 和 *5c* 则是抛光的金刚石表面。金刚石显微图像中更为鲜明的对比说明其表面比云母要粗糙很多，而且其不规则性的尺度具有 50 Å 的数量级。

图 *5b* 和 *5c* 中的显微图像是从同一块宝石以同样的技术得到的，不过图 *5b* 是来自以常规方式抛光的立方体表面，而图 *5c* 则是来自抛光的八面体表面。我们还曾在另一块金刚石立方体表面和八面体表面上观察到类似的结构，而且在十二面体表面上发现了不同的特征结构。这些结果确认了一个为金刚石抛光者所熟悉的事实，即抛光的质量取决于所抛光的特定的面，如同从抛光过程的机械性质中所得知的那样。人们期望至少在某种程度上表面结构可以用解理面来加以描述。在一个立方体表面上，这些解理面的交线会位于与抛光方向成 45° 角的方向，而且值得注意的是，在这些方向上存在着形貌特征。关于我们的结果的更多具体内容以及制样技术的细节，将会在其他文章中发表。

金刚石的独特在于它外观上高度抛光的表面可能实际上在显微镜尺度下却是机械粗糙的，这是其脆性所导致的后果。由于金刚石的磨损速率和金属表面的光亮效果都高度依赖于工具的表面结构，因此目前我们正在将这些结果运用在将金刚石制成金属加工的切割工具上。例如，两个具有相同外部几何形状但是制作成具有不同结晶学取向的单晶金刚石车刀，其磨损速率表现为可重现的，而且相差 7 倍（参考文献 26）。在这个特别的实验中，用两个结晶学取向完全不同的金刚石车刀车削一块用于制造汽车活塞的铝硅合金，其磨损几乎可以肯定是由上面所讨论的断裂面形

orientations. Further experiments are being continued on both the wear of diamonds and the finish they produce.

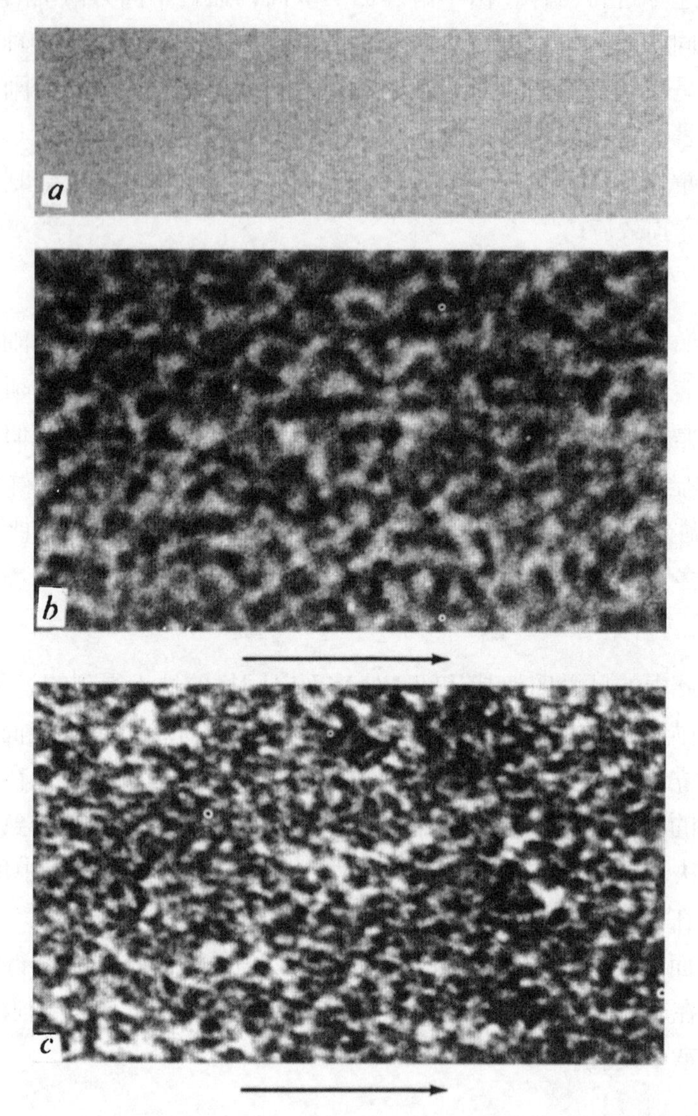

Fig. 5. Micrographs of replicas taken from *a*, cleaved mica; *b*, a polished cube face of diamond; *c*, a polished octahedron face of diamond. The arrows indicate the directions of polish. (× 1,000,000).

I thank Mr M. Casey for Figs. 1 and 4, Dr E. M. Wilks for Fig. 2, Dr D. Driver and Mr A. G. Thornton for the micrographs in Fig. 5, and the Science Research Council and De Beers Industrial Diamond Division for their support of this work.

(**243**, 15-18; 1973)

J. Wilks
Clarendon Laboratory, University of Oxford

貌所控制的。对于金刚石的磨损和它们所产生的光亮效果正进行进一步实验。

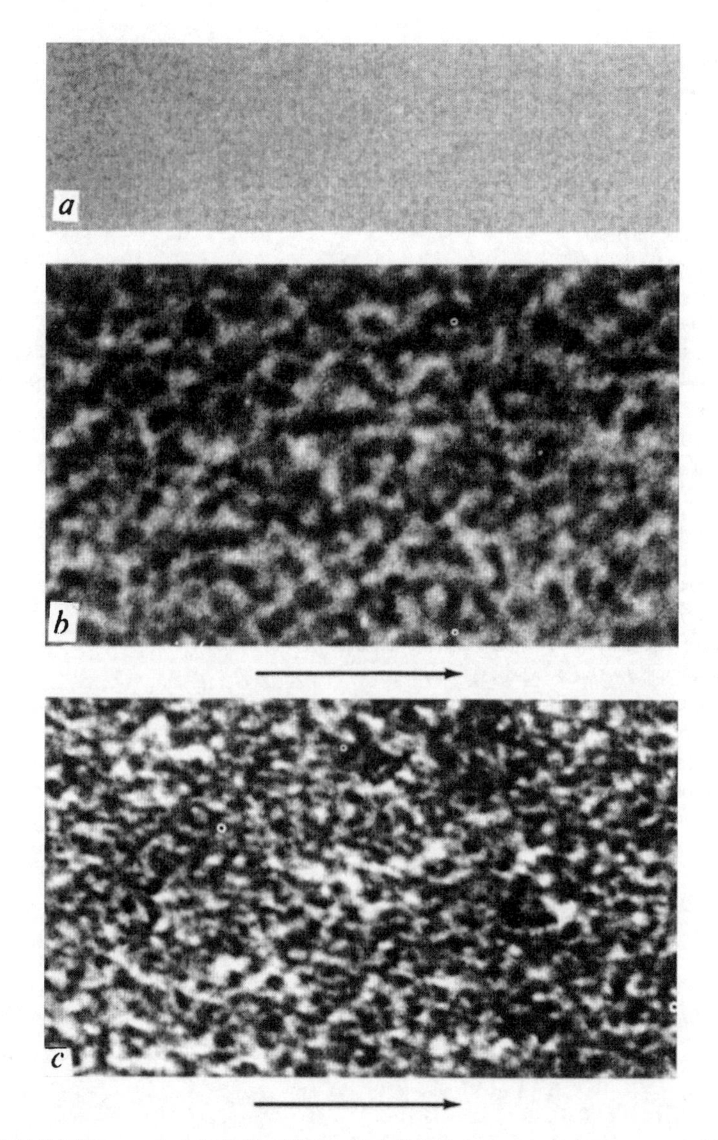

图 5. 样品的显微图像，*a*，取自解理的云母；*b*，取自抛光的金刚石立方表面；*c*，取自抛光的金刚石十二面体表面。箭头表示抛光方向。(×1,000,000)。

在此我要感谢凯西先生提供图 1 和图 4，感谢威尔克斯博士提供图 2，感谢德赖弗博士和桑顿先生提供图 5 中的显微图像，并且感谢科学研究理事会与戴比尔斯工业钻石部对这项研究的支持。

（王耀杨 翻译；郝伟 审稿）

References:

1. Bowden, F. P., and Tabor, D., *The Friction and Lubrication of Solids*, **1** (Clarendon, 1950); **2** (Clarendon, 1964).
2. Cottrell, A. H., *The Mechanical Properties of Matter* (Wiley, 1964).
3. Wilks, E. M., *J. Opt. Soc. Amer.*, **2**, 84 (1953).
4. Tolkowsky, M., thesis, Univ. London (1920).
5. Wilks, E. M., and Wilks, J., *J. Phys. D.*, **5**, 1902 (1972).
6. Wilks, E. M., and Wilks, J., *Phil. Mag.*, **38**, 158 (1959).
7. Seal, M., *Proc. Roy. Soc.*, **A248**, 379 (1958).
8. Kelly, A., *Strong Solids* (Clarendon, 1966).
9. Kelly, A., Tyson, W. R., and Cottrell, A. H., *Phil. Mag.*, **15**, 567 (1967).
10. Evans, T., *Physical Properties of Diamond* (edit. by Berman, R.), 116 (Clarendon, 1965).
11. Frank, F. C., and Lang, A. R., *Physical Properties of Diamond* (edit. by Berman, R.), 69 (Clarendon, 1965).
12. Evans, T., *Science and Technology of Industrial Diamonds*, **1** (edit. by Burls, J.), 105 (Industrial Diamond Information Bureau, London, 1966).
13. Evans, T., and Wild, R. K., *Phil. Mag.*, **12**, 479 (1965).
14. Bowden, F. P., and Tabor, D., *The Friction and Lubrication of Solids*, **2**, 118 (Clarendon, 1964).
15. Howes, V. R., and Tolansky, S., *Proc. Roy. Soc.*, **A230**, 287, 294 (1955).
16. Lawn, B. R., and Komatsu, H., *Phil. Mag.*, **14**. 689 (1966).
17. Frank, F. C., and Lawn, B. R., *Proc, Roy. Soc.* **A299**, 291 (1967).
18. Frank, F. C., Lawn, B. R., Lang, A. R., and Wilks, E. M., *Proc. Roy. Soc.*, **A301**, 239 (1967).
19. Brookes, C. A., *Nature*, **228**, 660 (1970).
20. Gane, N., and Cox, J. M., *J. Phys. D.*, **3**, 121 (1970).
21. Wilks, E. M., and Wilks, J., *The Physical Properties of Diamond* (edit. by Berman, R.), 221 (Clarendon, 1965).
22. Bowden, F. P., and Hanwell, A. E., *Proc. Roy. Soc.* **A295**, 233 (1966).
23. Bowden, F. P., and Young, J. E., *Proc. Roy. Soc.* **A208**, 444 (1951).
24. Bowden, F. P., and Tabor, D., *The Friction and Lubrication of Solids*, **2**, 169 (Clarendon, 1964).
25. Casey, M., and Wilks, J., *Diamond Research 1972* (Suppl. to *Ind. Diam. Rev.*), 6 (1972).
26. Casey, M., and Wilks, J., *Diamond Research 1972* (Suppl. to *Ind. Diam. Rev.*), **11** (1972).

Black Hole Explosions?

S. W. Hawking

[illegible]

[illegible]

[illegible]

Black Hole Explosions?

S. W. Hawking

Editor's Note

It was realized more than two hundred years ago that there is a critical mass and radius beyond which light cannot escape the gravitational field of an object—such an object becomes a "black hole". This idea was rigorously validated by the theory of general relativity. Here Stephen Hawking shows that black holes have effective temperatures that are inversely related to their mass, and should therefore radiate photons and neutrinos from their event horizons—they are not fully "black". As this radiation proceeds, the black hole loses mass. Finally it emits large quantities of X-rays and gamma-rays, and disappears in an explosion. "Hawking radiation" from black holes is now widely expected, but has not yet been seen.

QUANTUM gravitational effects are usually ignored in calculations of the formation and evolution of black holes. The justification for this is that the radius of curvature of space-time outside the event horizon is very large compared to the Planck length $(G\hbar/c^3)^{1/2} \approx 10^{-33}$ cm, the length scale on which quantum fluctuations of the metric are expected to be of order unity. This means that the energy density of particles created by the gravitational field is small compared to the space-time curvature. Even though quantum effects may be small locally, they may still, however, add up to produce a significant effect over the lifetime of the Universe $\approx 10^{17}$ s which is very long compared to the Planck time $\approx 10^{-43}$ s. The purpose of this letter is to show that this indeed may be the case: it seems that any black hole will create and emit particles such as neutrinos or photons at just the rate that one would expect if the black hole was a body with a temperature of $(\kappa/2\pi)$ $(\hbar/2k) \approx 10^{-6}$ $(M_\odot/M)K$ where κ is the surface gravity of the black hole[1]. As a black hole emits this thermal radiation one would expect it to lose mass. This in turn would increase the surface gravity and so increase the rate of emission. The black hole would therefore have a finite life of the order of 10^{71} $(M_\odot/M)^{-3}$ s. For a black hole of solar mass this is much longer than the age of the Universe. There might, however, be much smaller black holes which were formed by fluctuations in the early Universe[2]. Any such black hole of mass less than 10^{15} g would have evaporated by now. Near the end of its life the rate of emission would be very high and about 10^{30} erg would be released in the last 0.1 s. This is a fairly small explosion by astronomical standards but it is equivalent to about 1 million 1 Mton hydrogen bombs.

To see how this thermal emission arises, consider (for simplicity) a massless Hermitean scalar field ϕ which obeys the covariant wave equation $\phi_{;ab}g^{ab} = 0$ in an asymptotically flat space time containing a star which collapses to produce a black hole. The Heisenberg operator ϕ can be expressed as

黑洞爆炸?

霍金

编者按

人们在两百多年前就已经意识到，对于给定的半径有一个临界质量，当物体的质量超出临界质量时，其引力场就强到甚至连光都不能逃脱——这样的物体变成了“黑洞”。这一想法在广义相对论中得到了严格验证。本文中，斯蒂芬·霍金向我们表明黑洞有跟其质量成反比的等效温度，因此它必须从其事件视界向外辐射光子和中微子——即它们并不完全是“黑”的。黑洞在辐射过程中会损失质量，最终释放出大量的X射线和伽马射线，并在一次爆炸之后消失。虽然目前还没有直接观测到，但人们普遍相信黑洞存在“霍金辐射”。

在计算黑洞的形成和演化时，一般可忽略量子引力效应。这一点的合理性在于，在事件视界外的时空曲率半径远大于普朗克长度 $(G\hbar/c^3)^{1/2} \approx 10^{-33}$ cm，而在此尺度上预期度规的量子涨落是 1 的量级。这意味着由引力场产生的粒子的能量密度和时空曲率相比要小。虽然量子效应在局部很小，然而它们仍然可能在宇宙的寿命 $\approx 10^{17}$ s 内积累产生重大的影响，这个时间远长于普朗克时间 $\approx 10^{-43}$ s。这篇快报的目的是说明，似乎任何黑洞都将以预期的速率产生和发射粒子，如中微子或光子，正如同黑洞是一个温度为 $(\kappa/2\pi)(\hbar/2k) \approx 10^{-6}\,(M_\odot/M)\,K$ 的物体所表现的那样，其中 κ 是黑洞的表面引力 [1]。当黑洞发射这类热辐射时，我们预期它将损失质量。这本身将增大它的表面引力，因而增大其发射速率。从此，黑洞将具有 $10^{71}(M_\odot/M)^{-3}$ s 量级的有限寿命。对于太阳质量的黑洞，这将比宇宙年龄更长。然而，可能存在许多较小的黑洞，它们是由早期宇宙中的涨落形成的 [2]。任何这类质量小于 10^{15} g 的黑洞到现在都应该蒸发殆尽了。在接近它生命终了时，其粒子发射速率将非常高，在最后 0.1 s 将释放约 10^{30} erg 能量。以天文学的标准来看，这是一个相当小的爆炸，但它相当于大约一百万个 100 万吨量级的氢弹爆炸。

为了解释黑洞热辐射是如何产生的，为简单起见，在一个包含由一颗恒星塌缩形成的一个黑洞的渐近平直时空中，考虑一个无质量的厄米标量场 ϕ，且它遵守协变波动方程 $\phi_{;ab}g^{ab}=0$。海森堡算符 ϕ 可表示为：

$$\phi = \sum_i \{f_i a_i + \bar{f}_i a_i^+\}$$

where the f_i are a complete orthonormal family of complex valued solutions of the wave equation $f_{i;ab}g^{ab} = 0$ which are asymptotically ingoing and positive frequency—they contain only positive frequencies on past null infinity I^-[3,4,5]. The position-independent operators a_i and a_i^+ are interpreted as annihilation and creation operators respectively for incoming scalar particles. Thus the initial vacuum state, the state containing no incoming scalar particles, is defined by $a_i|0_-\rangle = 0$ for all i. The operator ϕ can also be expressed in terms of solutions which represent outgoing waves and waves crossing the event horizon:

$$\phi = \sum_i \{p_i b_i + \bar{p}_i b_i^+ + q_i c_i + \bar{q}_i c_i^+\}$$

where the p_i are solutions of the wave equation which are zero on the event horizon and are asymptotically outgoing, positive frequency waves (positive frequency on future null infinity I^+) and the q_i are solutions which contain no outgoing component (they are zero on I^+). For the present purposes it is not necessary that the q_i are positive frequency on the horizon even if that could be defined. Because fields of zero rest mass are completely determined by their values on I^-, the p_i and the q_i can be expressed as linear combinations of the f_i and the $\bar{f}_i$:

$$p_i = \sum_j \{\alpha_{ij} f_j + \beta_{ij} \bar{f}_j\} \text{ and so on}$$

The β_{ij} will not be zero because the time dependence of the metric during the collapse will cause a certain amount of mixing of positive and negative frequencies. Equating the two expressions for ϕ, one finds that the b_i, which are the annihilation operators for outgoing scalar particles, can be expressed as a linear combination of the ingoing annihilation and creation operators a_i and a_i^+

$$b_i = \sum_j \{\bar{\alpha}_{ij} a_j - \bar{\beta}_{ij} a_j^+\}$$

Thus when there are no incoming particles the expectation value of the number operator $b_i^+ b_i$ of the ith outgoing state is

$$\langle 0_- | b_i^+ b_i | 0_- \rangle = \sum_j |\beta_{ij}|^2$$

The number of particles created and emitted to infinity in a gravitational collapse can therefore be determined by calculating the coefficients β_{ij}. Consider a simple example in which the collapse is spherically symmetric. The angular dependence of the solution of the wave equation can then be expressed in terms of the spherical harmonics Y_{lm} and the dependence on retarded or advanced time u, v can be taken to have the form $\omega^{-1/2} \exp(i\omega u)$ (here the continuum normalisation is used). Outgoing solutions $p_{lm\omega}$ will now be expressed as an integral over incoming fields with the same l and m:

$$p_\omega = \int \{\alpha_{\omega\omega'} f_{\omega'} + \beta_{\omega\omega'} \bar{f}_{\omega'}\} d\omega'$$

$$\phi = \sum_i \{f_i a_i + \bar{f}_i a_i^+\}$$

其中，f_i 是波动方程 $f_{i;ab}g^{ab} = 0$ 的一族渐近向内、频率为正且完备正交归一复数解，它们在过去类光无穷远 I^-[3,4,5] 只含有正频率。对于入射的标量粒子，位置无关的算符 a_i 和 a_i^+ 分别解释为湮灭和产生算符。因此对于所有的 i，初始真空态，即不含有向内传播的标量粒子的态，可定义为 $a_i\,|0_-\rangle = 0$。算符 ϕ 也可以用代表向外的波和穿过事件视界的波的解表示：

$$\phi = \sum_i \{p_i b_i + \bar{p}_i b_i^+ + q_i c_i + \bar{q}_i c_i^+\}$$

其中，p_i 是波动方程的解，它们在视界上为零且是渐近向外的正频波（在未来类光无穷远 I^+ 为正频率），而 q_i 是不含向外成分的解（它们在 I^+ 为零）。就现在的目的而言，即使可以被定义，q_i 在视界处也不一定是正频的。因为零静止质量的场完全被它们在 I^- 的值确定，p_i 和 q_i 可以表示为 f_i 和 $\bar{f}_i$ 的线性组合：

$$p_i = \sum_j \{\alpha_{ij} f_j + \beta_{ij}\,\bar{f}_j\} \text{ 等}$$

因为在塌缩期间度规的时间依赖性将导致一定量的正频和负频的混合，所以 β_{ij} 将不为零。令 ϕ 的两个表达式相等，可以发现向外传播标量粒子的湮灭算符 b_i 可以表示为向内湮灭和产生算符 a_i 和 a_i^+ 的线性叠加，即：

$$b_i = \sum_j \{\bar{\alpha}_{ij} a_j - \bar{\beta}_{ij} a_j^+\}$$

于是在没有向内态的粒子时，第 i 个向外态的粒子数算符 $b_i^+ b_i$ 的期望值为：

$$\langle 0_- | b_i^+ b_i | 0_- \rangle = \sum_j |\beta_{ij}|^2$$

因此，在一次引力塌缩中产生并发射到无穷远的粒子的数目可以通过计算系数 β_{ij} 确定。考虑一个简单的例子，塌缩是球对称的。波动方程的解对角度的依赖可以用球谐函数 Y_{lm} 表示，对推迟或超前时间 u、v 的依赖可以取为 $\omega^{-1/2}\exp(i\omega u)$（这里使用了连续归一化）。向外的解 $p_{lm\omega}$ 现在可以表示为对相同 l 和 m 的向内的场的积分：

$$p_\omega = \int \{\alpha_{\omega\omega'} f_{\omega'} + \beta_{\omega\omega'} \bar{f}_{\omega'}\} \mathrm{d}\omega'$$

(The lm suffixes have been dropped.) To calculate $\alpha_{\omega\omega'}$ and $\beta_{\omega\omega'}$ consider a wave which has a positive frequency ω on I^+ propagating backwards through spacetime with nothing crossing the event horizon. Part of this wave will be scattered by the curvature of the static Schwarzschild solution outside the black hole and will end up on I^- with the same frequency ω. This will give a $\delta(\omega-\omega')$ behaviour in $\alpha_{\omega\omega'}$. Another part of the wave will propagate backwards into the star, through the origin and out again onto I^-. These waves will have a very large blue shift and will reach I^- with asymptotic form

$$C\omega^{-1/2} \exp \{-i\omega\kappa^{-1} \log (v_0-v) + i\omega v\} \textit{ for } v < v_0$$

and zero for $v \geq v_0$, where v_0 is the last advanced time at which a particle can leave I^-, pass through the origin and escape to I^+. Taking Fourier transforms, one finds that for large ω', $\alpha_{\omega\omega'}$ and $\beta_{\omega\omega'}$ have the form:

$$\alpha_{\omega\omega'} \approx C \exp [i(\omega-\omega')v_0](\omega'/\omega)^{1/2} \cdot \Gamma(1-i\omega/\kappa) [-i(\omega-\omega')]^{-1+i\omega/\kappa}$$

$$\beta_{\omega\omega'} \approx C \exp [i(\omega+\omega')v_0](\omega'/\omega)^{1/2} \cdot \Gamma(1-i\omega/\kappa) [-i(\omega+\omega')]^{-1+i\omega/\kappa}$$

The total number of outgoing particles created in the frequency range $\omega \rightarrow \omega+d\omega$ is $d\omega\int_0^\infty |\beta_{\omega\omega'}|^2 d\omega'$. From the above expression it can be seen that this is infinite. By considering outgoing wave packets which are peaked at a frequency ω and at late retarded times one can see that this infinite number of particles corresponds to a steady rate of emission at late retarded times. One can estimate this rate in the following way. The part of the wave from I^+ which enters the star at late retarded times is almost the same as the part that would have crossed the past event horizon of the Schwarzschild solution had it existed. The probability flux in a wave packet peaked at ω is roughly proportional to $\int_{\omega_1'}^{\omega_2'} \{|\alpha_{\omega\omega'}|^2 - |\beta_{\omega\omega'}|^2\} d\omega$ where $\omega_2' \gg \omega_1' \gg 0$. In the expressions given above for $\alpha_{\omega\omega'}$ and $\beta_{\omega\omega'}$ there is a logarithmic singularity in the factors $[-i(\omega-\omega')]^{-1+i\omega/\kappa}$ and $[-i(\omega+\omega')]^{-1+i\omega/\kappa}$. Value of the expressions on different sheets differ by factors of $\exp(2\pi n\omega\kappa^{-1})$. To obtain the correct ratio of $\alpha_{\omega\omega'}$ to $\beta_{\omega\omega'}$ one has to continue $[-i(\omega+\omega')]^{-1+i\omega/\kappa}$ in the upper half ω' plane round the singularity and then replace ω' by $-\omega'$. This means that, for large ω',

$$|\alpha_{\omega\omega'}| = \exp (\pi\omega/\kappa) |\beta_{\omega\omega'}|$$

From this it follows that the number of particles emitted in this wave packet mode is $(\exp(2\pi\omega/\kappa) -1)^{-1}$ times the number of particles that would have been absorbed from a similar wave packet incident on the black hole from I^-. But this is just the relation between absorption and emission cross sections that one would expect from a body with a temperature in geometric units of $\kappa/2\pi$. Similar results hold for massless fields of any integer spin. For half integer spin one again gets a similar result except that the emission cross section is $(\exp(2\pi\omega/\kappa)+1)^{-1}$ times the absorption cross section as one would expect for thermal emission of fermions. These results do not seem to depend on the assumption of exact spherical symmetry which merely simplifies the calculation.

这里略去了 lm 下标。为计算 $\alpha_{\omega\omega'}$ 和 $\beta_{\omega\omega'}$，考虑一个在 I^+ 为正频 ω，在时空中反向传播的不穿过视界的波。这个波的一部分将被黑洞外的静态史瓦西解的曲率散射并将在 I^- 上以相同的频率 ω 终止。这将导致 $\alpha_{\omega\omega'}$ 的 $\delta(\omega-\omega')$ 行为。这个波的另外一部分将向后传播到恒星中，通过原点然后向外再到 I^-。这些波将有非常大的蓝移，并将以渐近形式接近 I^-：

$$C\omega^{-1/2}\exp\{-i\omega\kappa^{-1}\log(v_0-v)+i\omega v\}，\text{当 } v<v_0 \text{ 时}$$

当 $v\geqslant v_0$ 时，这个蓝移值为 0。其中，v_0 是最后的超前时间，此时的粒子尚可脱离 I^-，通过原点并逃向 I^+。作傅里叶变换可以发现对于大的 ω'，$\alpha_{\omega\omega}$ 和 $\beta_{\omega\omega'}$ 有如下形式：

$$\alpha_{\omega\omega'}\approx C\exp[i(\omega-\omega')v_0](\omega'/\omega)^{1/2}\cdot\Gamma(1-i\omega/\kappa)[-i(\omega-\omega')]^{-1+i\omega/\kappa}$$

$$\beta_{\omega\omega'}\approx C\exp[i(\omega+\omega')v_0](\omega'/\omega)^{1/2}\cdot\Gamma(1-i\omega/\kappa)[-i(\omega+\omega')]^{-1+i\omega/\kappa}$$

在频率范围 $\omega\to\omega+d\omega$ 产生的向外粒子的总数为 $d\omega\int_0^\infty|\beta_{\omega\omega'}|^2d\omega'$，并由上面的表达式可以看出这个量是无限的。通过考虑在较晚的推迟时间的、峰值频率 ω 的向外波包，可以看到这个无限的粒子数对应于较晚推迟时间的一个稳态发射率。可以通过以下方法估计这个发射率。波中来自 I^+ 的在较晚推迟时间进入恒星的部分（如果这部分存在的话）和穿过史瓦西解的过去视界的部分几乎相同，峰值在 ω 处的波包中的概率流大致正比于 $\int_{\omega_1'}^{\omega_2'}\{|\alpha_{\omega\omega'}|^2-|\beta_{\omega\omega'}|^2\}d\omega$，其中，$\omega_2'\gg\omega_1'\gg 0$。在上面给出的 $\alpha_{\omega\omega'}$ 和 $\beta_{\omega\omega'}$ 的表达式中，$[-i(\omega-\omega')]^{-1+i\omega/\kappa}$ 和 $[-i(\omega+\omega')]^{-1+i\omega/\kappa}$ 因子中有一个对数奇点。这个表达式在不同面上的值相差一个 $\exp(2\pi n\omega\kappa^{-1})$ 因子。为得到正确的 $\alpha_{\omega\omega'}$ 和 $\beta_{\omega\omega'}$ 的比值，必须在上半 ω' 平面围绕奇点对 $[-i(\omega+\omega')]^{-1+i\omega/\kappa}$ 进行延拓并将 ω' 换为 $-\omega'$。这意味着对于大的 ω'：

$$|\alpha_{\omega\omega'}|=\exp(\pi\omega/\kappa)|\beta_{\omega\omega'}|$$

由此可得，这个波包模式中发射的粒子数是从 I^- 入射到黑洞上的类似波包中已被吸收的粒子数的 $[\exp(2\pi\omega/\kappa)-1]^{-1}$ 倍，但这正是根据几何单位下温度为 $\kappa/2\pi$ 的物体所预期的吸收和发射截面之间的关系。类似结果对任何整数自旋的无质量场也同样成立。对半整数自旋，正如我们对费米子的热辐射所预期的那样，也能得到类似结果，只是发射截面是吸收截面的 $[\exp(2\pi\omega/\kappa)+1]^{-1}$ 倍，这些结果似乎并不依赖于只是为了简化计算而采取的精确球对称性假设。

Beckenstein[6] suggested on thermodynamic grounds that some multiple of κ should be regarded as the temperature of a black hole. He did not, however, suggest that a black hole could emit particles as well as absorb them. For this reason Bardeen, Carter and I considered that the thermodynamical similarity between κ and temperature was only an analogy. The present result seems to indicate, however, that there may be more to it than this. Of course this calculation ignores the back reaction of the particles on the metric, and quantum fluctuations on the metric. These might alter the picture.

Further details of this work will be published elsewhere. The author is very grateful to G. W. Gibbons for discussions and help.

(**248**, 30-31; 1974)

S. W. Hawking
Department of Applied Mathematics and Theoretical Physics and Institute of Astronomy, University of Cambridge

Received January 17, 1974.

References:

1. Bardeen, J. M., Carter, B., and Hawking, S. W., *Commun. math. Phys*., **31**, 161–170 (1973).
2. Hawking, S. W., *Mon. Not. R. astr. Soc.*, **152**, 75-78 (1971).
3. Penrose, R., in *Relativity, Groups and Topology* (edit. by de Witt, C. M., and de Witt, B. S). Les Houches Summer School, 1963 (Gordon and Breach, New York, 1964).
4. Hawking, S. W., and Ellis, G. F. R., *The Large-Scale Structure of Space-Time* (Cambridge University Press, London 1973).
5. Hawking, S. W., in *Black Holes* (edit. by de Witt, C. M., and de Witt, B. S), Les Houches Summer School, 1972 (Gordon and Breach, New York, 1973).
6. Beckenstein, J. D., *Phys. Rev.*, D7, 2333–2346 (1973).

贝肯斯坦[6]在热力学基础上提出，κ的某个倍数应该被看作黑洞的温度。然而，他并未提出，黑洞可以像吸收粒子一样发射粒子。因此，巴丁、卡特和我曾经认为，κ和温度之间的热力学的相似性只是一种类比。然而，目前的结果似乎表明，可能存在比这更多的内容。当然，这个计算忽略了粒子对度规的反作用以及度规本身的量子涨落，这些不排除会改变这个物理图像。

这项工作的进一步细节将在其他地方发表。作者非常感谢吉本斯给予的建议和帮助。

（沈乃澂 翻译；肖伟科 审稿）

Observation of Cold Nuclear Fusion in Condensed Matter

S. E. Jones *et al.*

Editor's Note

In 1989, two groups of researchers working in Utah believed they had found evidence that electrolysis of heavy water using palladium electrodes could trigger nuclear fusion of the deuterium atoms, offering vast amounts of energy in a simple benchtop process. Both groups submitted their results to *Nature*. After questions raised by referees, one team—Martin Fleischmann, Stanley Pons and their student Marvin Hawkins at the University of Utah—withdrew their paper and published it elsewhere. This is the other paper, by Steven Jones and colleagues at Brigham Young University. The controversial claim of cold fusion spurred worldwide activity to replicate the results, but no such vindication was ever found, and electrolytic "cold fusion" is now almost universally discredited.

When a current is passed through palladium or titanium electrodes immersed in an electrolyte of deuterated water and various metal salts, a small but significant flux of neutrons is detected. Fusion of deuterons within the metal lattice may be the explanation.

FUSION of the nuclei of isotopes of hydrogen is the principal means of energy production in the high-temperature interiors of stars. In relatively cold terrestrial conditions, the nuclei are surrounded by electrons and can approach one another no more closely than is allowed by the molecular Coulomb barrier. The rate of nuclear fusion in molecular hydrogen is then governed by quantum-mechanical tunnelling through that barrier, or equivalently, the probability of finding the two nuclei at zero separation. In a deuterium molecule, where the equilibrium separation between deuterons (d) is 0.74 Å, the d–d fusion rate is exceedingly slow, about 10^{-74} per D_2 molecule per second[1].

By replacing the electron in a hydrogen molecular ion with a more massive charged particle, the fusion rate is greatly increased. In muon-catalysed fusion, the internuclear separation is reduced by a factor of ~200 (the ratio of the muon to electron mass), and the nuclear fusion rate correspondingly increases by about eighty orders of magnitude. Muon-catalysed fusion has been shown to be an effective means of rapidly inducing fusion reactions in low-temperature mixtures of hydrogen isotopes[2,3].

A hypothetical quasi-particle a few times as massive as the electron would increase the cold fusion rate to readily measurable levels of ~10^{-20} fusions per d–d molecule per

凝聚态物质中观察到的冷核聚变

琼斯等

编者按

1989 年，犹他州的两个研究小组认为他们已经找到了证据证明，在一个简单的桌面装置中，用钯电极电解重水会引发氘原子核聚变，释放巨大的能量。这两组都把他们的结果投到了《自然》。在审稿人提出问题之后，其中的一组——犹他大学的马丁·弗莱施曼，斯坦利·庞斯和他们的学生马尔温·霍金斯——撤回了他们的论文并在别处发表。本文则是另一篇论文，由杨百翰大学的史蒂文·琼斯及其同事们所作。冷核聚变的提出颇具争议，激励着世界范围内的研究行动来重复该结果，但是一直没有得到证实，如今，由电解产生的“冷核聚变”受到人们的普遍质疑。

当电流通过浸没在由氘化水和各种金属盐组成的电解质中的钯或钛电极时，我们探测到了一个微弱却意义重大的中子流。对于该现象，金属晶格中的氘核聚变或许可以做出解释。

在高温的恒星内部，氢同位素的核聚变是能量产生的主要途径。而在相对寒冷的陆地环境中，原子核被电子包围着，并且彼此间的距离不会比分子库仑势垒所允许的距离更小。因此，氢气分子的核聚变率受穿过势垒的量子力学隧道效应的控制，相当于受两核间距为零的现象发生概率的控制。在氘分子中，氘核（即 d）间的平衡距离为 0.74 Å，d–d 聚变率非常小，大约每秒钟每个氘气分子中发生 10^{-74} 次聚变[1]。

用一个更大的带电粒子来替代氢分子离子中的电子，聚变率会大大提高。在 μ 子催化聚变中，核间距会缩小约 200 倍（即 μ 子与电子质量之比），相应地核聚变率也会提升约 80 个数量级。μ 子催化聚变已经被证实为能够在氢同位素低温混合物中迅速引起聚变反应的一个有效途径[2,3]。

一个假定的比电子大几倍的准粒子会将冷核聚变率增加到易于测量到的水平，即每秒钟每个氘核对中约 10^{-20} 次聚变[1]。本文报告的结果表明，在特定条件下将氢

second[1]. The results reported here imply that a comparable distortion of the internuclear wavefunction can be realized when hydrogen isotope nuclei are loaded into metals under certain conditions. We have discovered a means of inducing nuclear fusion without the use of either high temperatures or radioactive muons.

Indirect Evidence

Observations of naturally occurring ^3He in the Earth suggested to us new directions for laboratory investigations of nuclear fusion in condensed matter. ^3He is produced by the following fusion reactions:

$$p + d \rightarrow {}^3He + \gamma(5.4\ MeV) \quad (1)$$

$$d + d \rightarrow {}^3He(0.82\ MeV) + n(2.45\ MeV) \quad (2a)$$

$$\rightarrow t(1.01\ MeV) + p(3.02\ MeV) \quad (2b)$$

Tritium (t) decays with a 12.4-yr half-life to produce ^3He. The well established high $^3He/^4He$ ratio in solids, liquids and gases associated with volcanoes and other areas of high heat flow[4-6] suggests fusion as a possible source for the ^3He.

To estimate a possible rate of fusion in the Earth, we assume a simple, steady-state model in which the known flux of ^3He out of the mantle, 2×10^{19} ^3He atoms per second[7], arises from p–d fusion occurring uniformly in the mantle water reservoir, taken as $\sim1.4\times10^{24}$ g (R. Poreda, personal communication). Note that if the Earth contains "primordial" ^3He, our calculated rate will be an upper limit; on the other hand, if fusion-produced ^3He is stored in the mantle (so that the outward flux does not equal the production rate), our value will be a lower limit. As each p–d fusion produces one ^3He atom, and as the isotopic abundance of deuterium in water is $\sim1.5\times10^{-4}$ deuterons per proton, we infer a geological fusion rate constant, λ_f, of

$$\lambda_f \approx \frac{2\times10^{19}\ {}^3He\ atoms\ s^{-1}}{1.4\times10^{43}\ deuterons} \approx 10^{-24}\ fusions\ d^{-1}\ s^{-1} \quad (3)$$

This rate is fifty orders of magnitude larger than that expected in an isolated HD molecule, and fusion at this rate could be detected if reproduced in the laboratory.

Cold nuclear fusion may be important in celestial bodies other than the Earth. Jupiter, for example, radiates about twice as much heat as it receives from the Sun. It is interesting to consider whether cold nuclear fusion in the core of Jupiter, which is probably metallic hydrogen plus iron silicate, could account for its excess heat. Heat is radiated at an approximate rate of 10^{18} watts, which could be produced by p–d fusions occurring at a rate of 10^{30} s^{-1}[1]. Assuming a core of radius 4.6×10^9 cm, containing mostly hydrogen, with density ~10 g cm^{-3} and a deuteron/proton ratio of $\sim10^{-4}$, we deduce a required p–d fusion rate of $\lambda_f \approx 10^{-19}$ fusions d^{-1} s^{-1} if all the heat derives from fusion. Catalysed nuclear

同位素原子核注入金属中，可以实现原子核间波函数相应的变化。我们已经发现了一种在既不要求高温也不使用放射性 μ 子的情况下实现核聚变的方法。

间接证据

对地球上自然产生的 ^{3}He 的观测，为我们在实验室条件下研究凝聚态物质中的核聚变指明了新的方向。^{3}He 在下列聚变反应中产生：

$$p + d \rightarrow {}^3He + \gamma(5.4\ MeV) \tag{1}$$

$$d + d \rightarrow {}^3He(0.82\ MeV) + n(2.45\ MeV) \tag{2a}$$

$$\rightarrow t(1.01\ MeV) + p(3.02\ MeV) \tag{2b}$$

氚（即 t）以 12.4 年的半衰期衰变产生 ^{3}He。与火山爆发和其他高温热流相关的固体、液体、气体中存在的已经确定的 $^3He/^4He$ 的高比值[4-6] 表明，聚变是 ^{3}He 的一个可能来源。

为估算地球中的可能聚变率，我们假设一个简单的稳态模型，在该模型中，已知从地幔中流出的 ^{3}He 流为每秒钟 2×10^{19} 个 ^{3}He 原子[7]，是由在地幔储水层（约 1.4×10^{24} g）中均匀发生的 p–d 聚变引起的（波雷达，个人交流）。需要注意的是，一方面，如果地球含有“原始”^{3}He，那么我们计算出的上述聚变率就是上限；另一方面，如果核聚变产生的 ^{3}He 是储藏在地幔中的（这样一来，向外流出的量就与产生率不相等），那么我们的计算值就是下限。由于每次 p–d 聚变产生一个 ^{3}He 原子且水中氘的同位素丰度约为每个质子对应 1.5×10^{-4} 个氘核，我们推断出一个地球的聚变率常数 λ_f：

$$\lambda_f \approx \frac{2\times10^{19}\ {}^3He\ 原子/秒}{1.4\times10^{43}\ 氘核} \approx 10^{-24}\ 聚变/（氘核 \cdot 秒） \tag{3}$$

这个聚变率要比孤立 HD 分子中的预期聚变率高出 50 个数量级，如果在实验室条件下重现该过程，处于这种聚变率的聚变是可以被探测到的。

与地球不同，天体中的冷核聚变或许是很重要的。例如木星，它辐射出的热量大约是其从太阳那里吸收的热量的两倍。那么，发生在可能由金属氢加铁硅酸盐组成的木星核心处的冷核聚变能否为多余的热量做出解释呢？这是一个很有趣的问题。热量辐射率大约为 10^{18} W，而该热量可以由聚变率为 $10^{30}\ s^{-1}$ 的 p–d 聚变产生[1]。假设有这样一个核心，其半径为 4.6×10^9 cm，主要由氢组成，密度约为 $10\ g \cdot cm^{-3}$，氘核与质子之比约为 10^{-4}，且如果所有的热量全部来自聚变，那么我们推断出 p–d 聚变率 λ_f 必须为每秒钟每个氘核中约 10^{-19} 次聚变。处于这个聚变率的催化核聚变在

fusion at this rate could be readily measured in the laboratory.

Further evidence for cold nuclear fusion in condensed matter comes from studies of ^{3}He and ^{4}He in metals. There have been several reports of high ^{3}He concentrations in metal crucibles and foils (H. Craig, R. Poreda, A. Nier, personal communications), consistent with *in situ* formation by cold fusion. In particular, Mamyrin *et al.*[8] report the occurrence of patchy, high concentrations of ^{3}He in a number of metal foils. Electrolytic refining of the metals could have provided the appropriate conditions for the cold nuclear fusion reactions (1) and possibly (2). Among several possible explanations for the observations, the authors suggest an analogue of muon catalysis[8].

Detection of Cold-fusion Neutrons

The considerations outlined above led to laboratory experiments performed at Brigham Young University to determine whether cold nuclear fusion can actually occur in condensed matter. We now report the observation of deuteron–deuteron fusion at room temperature during low-voltage electrolytic infusion of deuterons into metallic titanium or palladium electrodes. The fusion reaction (2*a*) is apparently catalysed by the deposition of d^{+} and metal ions from the electrolyte at (and into) the negative electrode. Neutrons with an energy of ~2.5 MeV are clearly detected with a sensitive neutron spectrometer. The experimental layout is shown in Fig. 1.

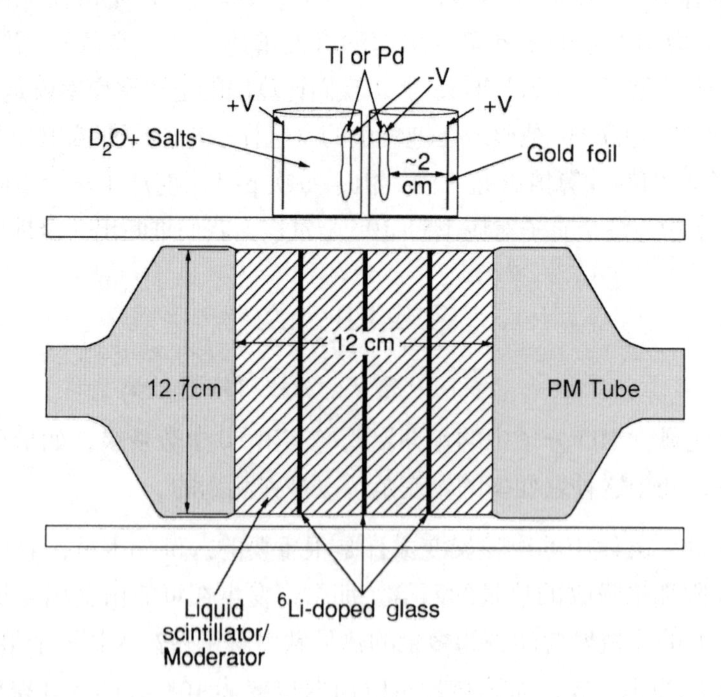

Fig. 1. Schematic diagram of the experiment. Electrolytic cells are shown on top of the neutron spectrometer.

实验室条件下可以很容易地被探测到。

关于凝聚态物质中冷核聚变的进一步证据来源于对金属中的 ^{3}He 和 ^{4}He 的研究。目前已经有一些关于金属坩埚和金属箔中的高浓度 ^{3}He 的报道（克雷格，波雷达，尼尔，个人交流），且与冷核聚变在原位形成相一致。特别是，马梅林等人声称在许多金属箔中观察到了分布不均匀的高浓度 ^{3}He[8]。金属的电解精炼可能为冷核聚变反应（1），甚至可能包括反应（2）提供了适宜的条件。在对该现象的一些可能的解释中，作者比较赞同类似于 μ 子的催化作用的解释[8]。

冷核聚变中子的探测

鉴于上面概述的种种考虑，在杨百翰大学进行了实验以验证冷核聚变能否在凝聚态物质中发生。我们在这里报道了室温下将氘核向金属钛或钯电极低压电解注入期间的 d–d 聚变。聚变反应（2*a*）显然受到了负电极处（及其内部）电解质中的 d^+ 和金属离子沉积的催化。灵敏的中子谱仪也明显探测到能量约为 2.5 MeV 的中子。实验装置如图 1 所示。

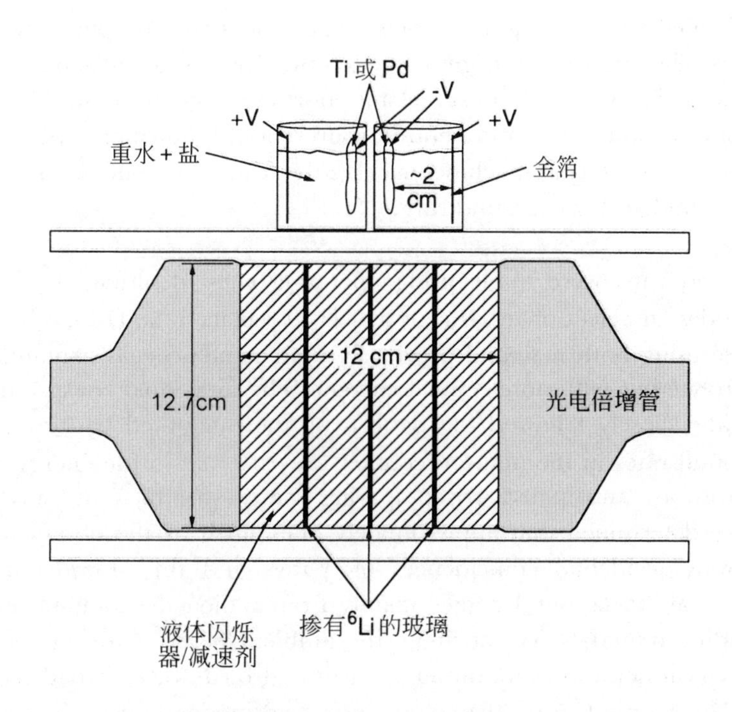

图 1. 实验装置简图。在中子谱仪上放置的是电解池。

The neutron spectrometer, developed at Brigham Young University over the past few years (ref. 9 and manuscript in preparation) has been crucial to the identification of this cold fusion process. The detector consists of a liquid organic scintillator (BC-505) contained in a glass cylinder 12.5 cm in diameter, in which three glass scintillator plates doped with lithium-6 are embedded. Neutrons deposit energy in the liquid scintillator through multiple collisions, and the resulting light output yields energy information. As their energy decreases, the neutrons are scavenged by ^6Li nuclei, and the reaction $n + {}^6Li \rightarrow t + {}^4He$ results in scintillations in the glass. Pulse shapes and amplitudes from the two scintillators differ; the two distinct signals are registered by two photomultiplier tubes, whose signals are summed. A coincidence of identified signals from the two media within 20 μs identifies an incoming neutron that has stopped in the detector.

The spectrometer was calibrated using 2.9- and 5.2-MeV neutrons generated by deuteron–deuteron interactions at 90° and 0°, respectively, with respect to a deuteron beam from a Van de Graaff accelerator. The observed energy spectra show broad structures which imply that 2.45-MeV neutrons should appear in the multichannel analyser spectrum in channels 45-150. The stability of the detector system was checked between data runs by measuring the counting rate for fission neutrons from a broad-spectrum californium-252 source.

We have performed extensive tests to verify that the neutron spectrometer does not respond preferentially in this pulse height range to other sources of radiation such as thermal neutrons. In particular, we made unsuccessful efforts to generate false 2.5-MeV neutron "signals" by using various γ-ray and neutron sources and by turning auxiliary equipment on and off. Neutron-producing machines such as the Van de Graaff accelerators were off during all foreground and background runs.

Many background runs were made using operating cells (described below) containing standard electrodes and electrolytes, except that H_2O replaced the D_2O; other background runs were made using both new and previously used standard cells containing D_2O plus the usual electrolyte but with no electrical current. The individual background runs were all featureless and closely followed the pattern of the integrated background shown in Fig. 2. Background rates in the neutron counter are $\sim 10^{-3}\ s^{-1}$ in the energy region where 2.5-MeV neutrons are anticipated. By comparing energy spectra from γ-ray and neutron sources we have determined that approximately one-fourth of the observed background events arise from accidental coincidences of γ-rays and three-fourths from ambient neutrons. The γ-ray background comes mainly from radioactive radium and potassium in the surrounding materials. We attribute the ambient neutrons to cosmic-ray sources. Although the typical neutron evaporation spectrum (at birth) has a broad maximum near 2.5 MeV (ref. 10), Monte Carlo calculations show that moderation in the source medium (predominantly the shielding surrounding the detector) will wash out this structure and produce a smoothly decreasing background spectrum above 0.5 MeV, as observed.

杨百翰大学在过去几年中发展的中子谱仪（参考文献 9 及处于准备阶段的稿件）对于识别这种冷核聚变过程是非常重要的。探测器包括一个放置在直径为 12.5 cm 的玻璃圆筒中的液体有机闪烁器（BC–505），该圆筒内嵌着三个掺杂有 Li–6 的玻璃闪烁器平板。中子通过多次碰撞将能量存积在液体闪烁器中，然后生成的光输出产生能量信息。由于能量减少了，中子就会被 ^{6}Li 核全部吸收，并且该反应 n + ^{6}Li → t + ^{4}He 会在玻璃中引起闪烁。来自两个闪烁器的脉冲波形和振幅不相同；这两个不同的信号会被两个光电倍增管记录下来，并把倍增管的信号相加。在 20 μs 内来自两个倍增管的可分辨的一致信号就能够识别出一个停在探测器内的入射中子。

用 d–d 反应产生的，能量分别为 2.9 MeV 和 5.2 MeV 的中子来校准谱仪，这些中子相对于范德格拉夫加速器氘核束流的方向在 90° 和 0° 方向射出。观测到的能谱比较宽，这意味着 2.45 MeV 中子应该出现在多道分析器的第 45 到第 150 道之间。通过对锎–252 广谱源裂变中子计数率的多次测量，检验了探测系统的稳定性。

我们已做了大量的测试证实，在这种脉冲高度范围内，中子谱仪不会对像热中子这样的其他辐射源优先响应。特别是，我们企图通过使用多种 γ 射线、中子源和开 / 关辅助设备来产生赝 2.5 MeV 中子“信号”的努力都失败了。在所有的信号和本底运行中，我们关闭了像范德格拉夫加速器这样的中子产生装置。

许多本底运行是用含标准电极和电解质的正在运行的电解池（下文将介绍）完成的，只是将其中的 D_2O 换成了 H_2O；而其他的本底运行是用新的和之前用过的包含 D_2O 和平常电解质的无电流标准电解池完成。单个的本底运行是毫无特征的，且与图 2 所示的整体本底模式高度一致。在预期的 2.5 MeV 中子出现的能量范围内，中子计数器的本底率约为 10^{-3} s^{-1}。通过对比 γ 射线和中子源的能谱，我们已经确定，在观测到的本底事件中，大约 1/4 是由于 γ 射线的偶然重叠，3/4 由周边环境中的中子引起。γ 射线本底主要来源于周围材料中的放射性镭和钾。我们将环境中的中子归因于宇宙射线源。尽管典型的中子的发射谱（初生的）在接近 2.5 MeV 处有一个很宽的峰（参考文献 10），但蒙特卡罗计算表明，由于源介质的小曼化（主要是对探测器周围的屏蔽层引起）将消除这种结构，并产生一个在 0.5 MeV 以上的平滑递减的本底谱，这和观测结果一致。

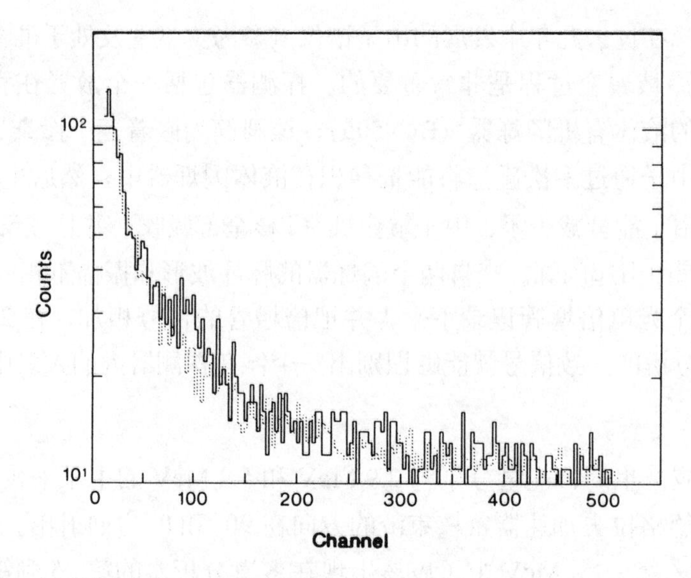

Fig. 2. Foreground (solid) and background (dashed) counts as a function of pulse height (corresponding to neutron energy) in the neutron spectrometer. Ten counts have been added to each three-channel bin for clarity of presentation.

The predicted and measured absence of structure in the spectrum of cosmic-ray-produced neutrons will not be influenced by the relatively small temporal variations that may occur in the cosmic-ray flux, such as the observed decreases that may accompany solar flares. This means that the observed peak at 2.5 MeV cannot be accounted for by ambient-neutron background variations, because, as explained below, the analysis is based on the shape of the spectra and not simply on rates. Low-energy cosmic-ray muons would be rapidly scavenged by nuclei with high atomic number, so as to reduce muon-catalysed d–d fusion to a negligible level[2,3]. Considering volume and solid angle, the rate of production of neutrons by muons absorbed by carbon nuclei in the detector exceeds that from muons absorbed by oxygen nuclei in the electrolytic cells by a factor of ~60. Thus, the presence or absence of electrolytic cells is an unimportant perturbation in the background.

During the search for suitable catalytic materials, the following (unoptimized) prescription for the electrolytic cells evolved. It began with salts typical of volcanic hot springs and included electrode-metal ions. The electrolyte is typically a mixture of ~160 g D_2O plus various metal salts in ~0.1 g amounts each: $FeSO_4 \cdot 7H_2O$, $NiCl_2 \cdot 6H_2O$, $PdCl_2$, $CaCO_3$, $Li_2SO_4 \cdot H_2O$, $Na_2SO_4 \cdot 10H_2O$, $CaH_4(PO_4)_2 \cdot H_2O$, $TiOSO_4 \cdot H_2SO_4 \cdot 8H_2O$, and a very small amount of AuCN. The pH is adjusted to $\leqslant 3$ with HNO_3. All 14 runs reported here began with this basic electrolyte.

Titanium and palladium, initially selected because of their large capacities for holding hydrogen and forming hydrides, were found to be effective negative electrodes. Individual electrodes consisted of ~1 g purified "fused" titanium in pellet form, or 0.05 g of 0.025-mm-thick palladium foils, or 5 g of mossy palladium. Typically 4–8 cells were used simultaneously. The palladium pieces were sometimes re-used after cleaning and

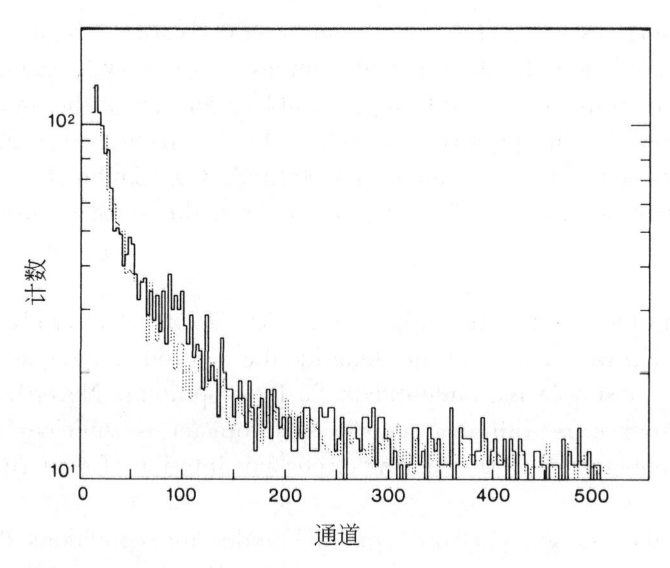

图 2. 信号(实线)和本底(虚线)计数关于中子谱仪中脉冲高度(对应于中子能量)的函数。为了表述清楚,我们在每 3 通道中加入 10 个计数。

预期和测到的宇宙射线产生的中子谱中的这种结构的缺失，不会受宇宙射线流中可能发生的相对微小而短暂的变化（如可观测到的与太阳耀斑相伴出现的减小）的影响。这就意味着，周边环境中子本底的变化不能对 2.5 MeV 处所观测到的峰作出解释，因为，正像下文解释的那样，分析是基于谱的形状而不是简单地依据比率得出的。低能宇宙射线 μ 子应该被具有较高原子序数的原子核迅速清除掉，以便将 μ 子催化的 d–d 聚变降到可以忽略的水平上[2,3]。考虑到体积和立体角，由检测器中碳核吸收的 μ 子产生中子的速率比在电解池中氧核吸收的 μ 子产生中子的速率要快近 60 倍。因此，电解池存在与否对本底的影响并不大。

在寻找合适的催化剂材料的过程中，逐渐形成了下面对电解池的配方（非最优化的）。它从火山温泉中典型的盐类化合物开始，并要包含电极 – 金属离子。典型的电解质是一种混合物，包含约 160 g D_2O 与多种金属盐，每种金属盐的总量约为 0.1 g，其中包括：$FeSO_4 \cdot 7H_2O$、$NiCl_2 \cdot 6H_2O$、$PdCl_2$、$CaCO_3$、$Li_2SO_4 \cdot H_2O$、$Na_2SO_4 \cdot 10H_2O$、$CaH_4(PO_4)_2 \cdot H_2O$、$TiOSO_4 \cdot H_2SO_4 \cdot 8H_2O$ 以及极少量的 AuCN。用 HNO_3 将 pH 值调整到 ≤3。本文中所报道的这 14 次运行均从这一基础的电解质开始。

最初选择钛和钯是由于它们能够大量储存氢并形成氢化物，结果发现它们还是有效的负电极。单个电极是由约 1 g 纯化的小球形状的“熔融”钛组成，或者由 0.05 g 厚度为 0.025 mm 的钯箔组成，也可以由 5 g 表面粗糙的钯组成。通常会同时使用 4~8 个电解池。有时我们用稀释过的酸或磨蚀材料对钯箔的表面进行清洗、粗化，

roughening the surfaces with dilute acid or abrasives. Hydrogen bubbles were observed to form on the Pd foils only after several minutes of electrolysis, suggesting the rapid absorption of deuterons into the foil; oxygen bubbles formed at the anode immediately. Gold foil was used for the positive electrodes. Direct-current power supplies provided 3–25 volts across each cell at currents of 10–500 mA. Correlations between fusion yield and voltage, current density, or surface characteristics of the metallic cathode have not yet been established.

Small jars, ~4 cm high and 4 cm in diameter, held ~20 ml of electrolyte solution each. The electrolytic cells were placed on or alongside the neutron counter, as shown in Fig. 1. The present cells are simple and undoubtedly far from optimum. Nevertheless, the present combination of our cells with the neutron spectrometer is sufficient to establish the phenomenon of cold nuclear fusion during electrolytic infusion of deuterium into metals.

Figure 2 shows the energy spectrum obtained under the conditions described above, juxtaposed with the (scaled) background spectrum. We acquired about twice as much background data as foreground data. Assuming conservatively that all deviations from background are statistical fluctuations, we scale the background counts by a factor of 0.46 to match the total number of foreground counts over the entire energy range shown in Fig. 2. A feature in channels 45–150 rises above background by nearly four standard deviations. This implies that our assumption is too conservative and that this structure represents a real physical effect. After re-scaling the background by a factor of 0.44 to match the foreground levels in regions just below and just above this feature, the difference plot (Fig. 3) is obtained. It shows a robust signal centred near channel 100, with a statistical significance of almost five standard deviations. A gaussian fit to this peak yields a centroid at channel 101 with a standard deviation of 28 channels, and an amplitude of 23.2±4.5 counts. Both the position and width of this feature correspond to those expected for 2.5–MeV neutrons, according to the spectrometer calibration. The fact that a significant signal appears above background with the correct energy for d–d fusion neutrons (~2.5 MeV) provides strong evidence that room-temperature nuclear fusion is occurring at a low rate in the electrolytic catalysis cells.

之后将其回收利用。仅仅在电解发生的几分钟之后，我们就观察到钯箔上形成了氢气泡，这说明氘核已经被迅速吸收进箔里；氧气泡也在阳极迅速形成。阳极材料为金箔。直流电源对每个电解池提供了 3~25 V 电压和 10~500 mA 电流。但是，聚变产额与电压、电流密度或金属阴极的表面特征间的关联还不确定。

在每个高和直径均约为 4 cm 的小广口瓶中盛有约 20 ml 的电解质溶液。如图 1 所示，我们将电解池放在中子计数器的上面或旁边。无疑，目前的电解池非常简单，而且远远未达到最佳效果。但对于证实在氘电解注入金属期间的冷核聚变现象而言，这种带有中子谱仪的电解池组合已经足够了。

图 2 给出了在上述条件下得到的能谱，且并列给出本底谱（已定标）。我们取得的本底数据大约是信号数据的两倍。保守地假设来自本底的所有偏差都是统计涨落，我们用大小为 0.46 的因子来定标本底计数，使它与图 2 所示的整个能量范围内的信号计数相匹配。在 45~150 道之间的一个特征峰高出本底将近 4 个标准偏差。这意味着，我们的假设太过保守，且这个结构显示了真实的物理效应。在以大小为 0.44 的因子对本底进行重新定标之后，我们得到了二者的差分图（图 3），重新定标旨在使本底计数在稍低于和稍高于该特征峰的区域内与信号水平相匹配。该图展示了一个集中于 100 道附近的强烈信号，其统计显著性几乎是 5 个标准偏差。高斯拟合后，有一个峰，其中心在 101 道附近，其标准偏差是 28 个通道，幅度为 23.2 ± 4.5 个计数。根据光谱仪校准，该特征峰的位置和宽度对应于 2.5 MeV 中子的期望值。在本底之上出现的显著信号对应的能量恰为 d–d 聚变的中子能量（约 2.5 MeV），这一事实为电解质催化电解池中存在反应率很低的核聚变提供了强有力的证据。

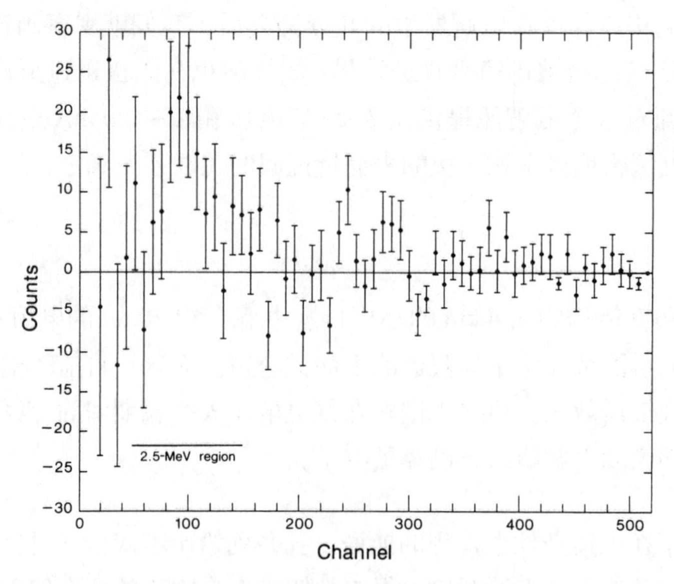

Fig. 3. Difference spectrum obtained by substracting scaled background from the foreground. Statistical errors ($\pm 1\sigma$) are shown for each eight-channel bin.

Fusion Rate Determination

It is instructive to examine the fourteen individual runs which enter into the combined data discussed above. These runs were performed over the period 31 December 1988 to 6 March 1989. Figure 4 displays, for each run, the ratio of foreground count rate in the 2.5-MeV energy region to the background rate obtained for each run. Electronic changes were made in the apparatus during the course of the experiment which altered the observed background rates, so we plot the data in terms of foreground-to-background ratios rather than absolute rates. In one set of data (runs 1 to 8) for which the system was kept as untouched as possible to avoid changes in background rates, the measured rate of detection of 2.5-MeV neutrons was $(6.2\pm1.3)\times10^{-4}$ s^{-1} above background. For this set of data, the background and foreground rates for all energies above ~3 MeV (that is, for all channels from 190 to 512) are equal, at $(1.4\pm0.1)\times10^{-3}$ s^{-1}.

Run 6 is particularly noteworthy, with a statistical significance of approximately five standard deviations above background. Fused titanium pellets were used as the negative electrode, with a total mass of ~3 g. The neutron production rate increased after about one hour of electrolysis. After about eight hours, the rate dropped dramatically, as shown in the follow-on run 7. At this time, the surfaces of the titanium electrodes showed a dark grey coating. An analysis using electron microscopy with a microprobe showed that the surface coating was mostly iron, deposited with deuterons at the cathode. The same phenomenon of a decrease in the neutron signal after about eight hours of operation appears in run 13 followed by run 14. Runs 13 and 14 use the same eight electrochemical cells, and again the negative electrodes developed coatings after a few hours of electrolysis. These observations suggest the importance of surface conditions for the cold fusion process. Variations in surface conditions and electrolyte composition are anticipated

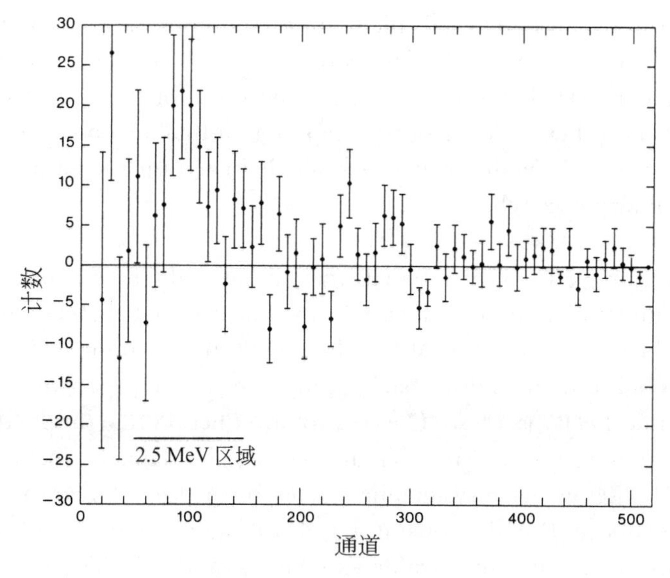

图 3. 通过从信号中减去定标的本底得到的差分谱图。图中给出每 8 个通道的统计误差（±1σ）。

测定聚变率

对列入上文讨论的综合数据中的那 14 次运行分别进行考查是有益的。这些运行是在 1988 年 12 月 31 日至 1989 年 3 月 6 日期间完成的。图 4 给出了每一次运行在 2.5 MeV 的能量区域内的信号计数率与本底计数率之比。在改变观察到的本底率的实验过程中，仪器中存在电子学的变化，所以我们根据信号 / 本底计数率的比值而非绝对计数率来绘制数据。在一组数据中（运行 1 到 8），我们尽量使系统不受影响以避免本底率的改变，对 2.5 MeV 中子探测的测量率高出本底 $(6.2\pm1.3)\times10^{-4}\ s^{-1}$。在这组数据中，本底率和信号率对于所有约 3 MeV 以上（即 190 道到 512 道之间的所有道）的能量来说都是相等的，为 $(1.4\pm0.1)\times10^{-3}\ s^{-1}$。

特别值得注意的是运行 6，其在本底之上的统计显著性为将近 5 个标准偏差。我们将质量约为 3 g 的熔融钛小球作为负极。在电解开始一个小时之后，中子产率增加了。约 8 个小时之后，中子产率却急剧地降低了，这点可以从后续的运行 7 看出。此时，在钛电极表面出现了一个深灰色的涂层。使用电子显微镜上的微探针进行分析后发现，该表面涂层主要是铁，而氘核在阴极沉积。同样的现象，即电解进行约 8 个小时之后中子信号减弱，也发生在运行 13 及其后的运行 14 中。运行 13 和运行 14 用的是相同的 8 个电解池，并且在电解开始的几个小时之后涂层又出现在了阴极上。这些观察结果揭示了表面状况对于冷核聚变过程的重要性。由于物质会从溶液中析出，所以每次测试运行中表面状况及电解质成分的变化是可以预料到的；

during each test run because materials plate out of solution; the solution pH also changes significantly during a run. These 14 runs represent two choices of electrode material plus various operating currents. These variations may account for the fluctuations in the signal level that are evident in Fig. 4. As these runs represent a total of only ~200 signal neutrons at an average rate of ~2 per hour, it was difficult to optimize experimental conditions. This is a task for future research.

The observed "turning off" of the signal after about eight hours may account for low signal-to-background ratios in runs 1 and 3, in that a signal that lasted for only a few hours may have been overwhelmed after a long (~20-hour) running time. When run 10 started with rates substantially above background, we stopped the run and removed half of the electrochemical cells as a test. The neutron production rate dropped off as expected (run 11). In determining the statistical significance of the data, we included runs 1, 3, 7, 11 and 14, even though we see a systematic reason for their low foreground-to-background ratios as explained above. Run 8, shown in Fig. 4, was inadvertently lost from the magnetic storage device and could not be included in Figs 2 and 3. This does not change our conclusions.

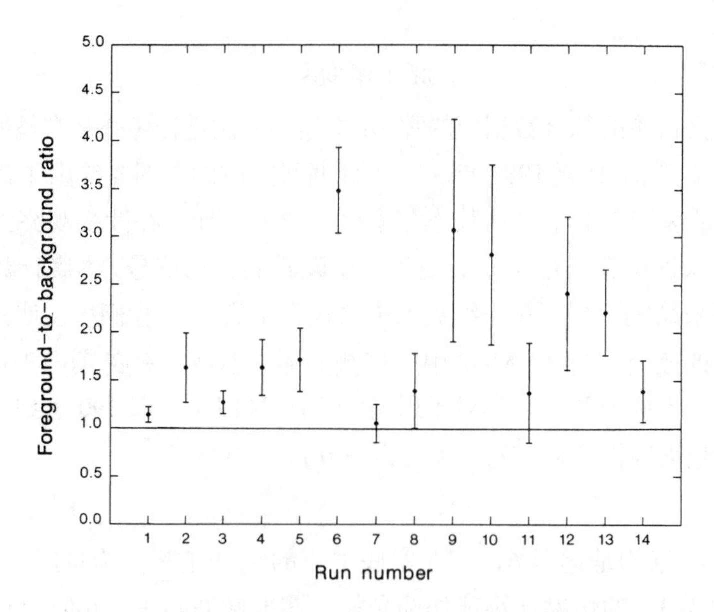

Fig. 4. Ratio of foreground rate to background rate for each run, in the 2.5-MeV energy region of the pulse-height spectrum. Statistical errors ($\pm 1\sigma$) are shown.

We can estimate the rate for the neutron-production branch of d–d fusion during electrolysis, specifically for run 6, as follows:

$$\text{Fusions per deuteron pair per second} = \frac{R/\varepsilon}{M \times \frac{\mathrm{d}}{2M}} \tag{4}$$

where the observed rate of neutron detection, $R = (4.1 \pm 0.8) \times 10^{-3}\ \mathrm{s}^{-1}$, is based on

并且溶液的 pH 值在一个运行过程中也会显著地变化。这 14 次运行展示了对于电极材料和多种工作电流的两个选择。这些变化或许可以解释图 4 中信号电平的明显波动。鉴于这些运行展示了总量只有约 200 个的信号中子，其平均产生率约为每小时 2 个，优化该实验条件是非常困难的。这是未来研究的一项任务。

实验观察到的约 8 小时之后出现的信号“中断”或许可以解释运行 1 和 3 中的信号 / 本底的低比值，因为信号只能持续几个小时，在一个较长的运行时间（约 20 小时）之后，该信号可能早就被覆盖掉了。当运行 10 以明显高于本底的速率开始后，作为测试，我们停止该运行，并撤去一半的电解池。正如我们所料，中子产生率逐渐减少（运行 11）。在检测数据的统计显著性时，我们将运行 1、3、7、11 和 14 包括进来，尽管我们看到了造成上文提到的它们较低的信号 / 本底之比的系统性原因。图 4 中所示的运行 8 从磁存储器中不幸丢失，因而在图 2 和图 3 中未显示出来，但这并不影响我们的结论。

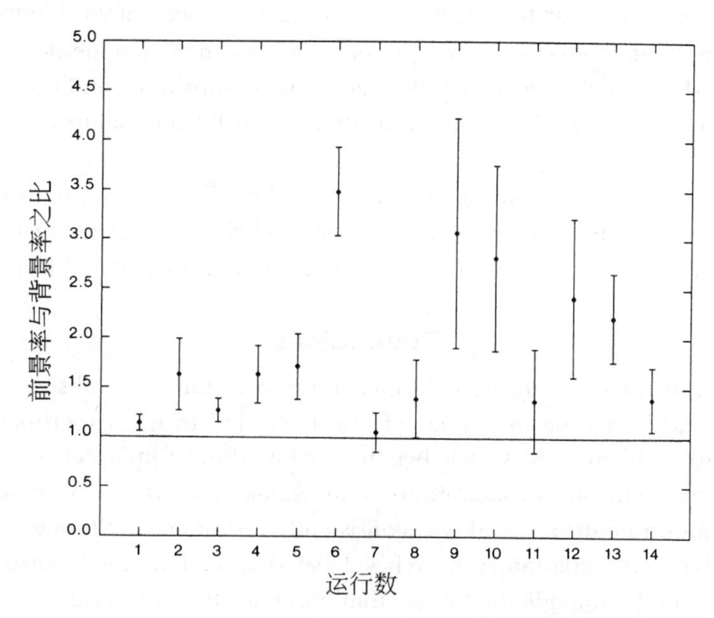

图 4. 在脉冲高度谱的 2.5 MeV 能区内，每个运行的信号率与本底率之比。图中给出统计误差（$\pm 1\sigma$）。

特别是对于运行 6，我们可以用以下公式估计电解过程中 d–d 聚变产生中子分支的产率：

$$\text{每秒钟每个氘核对的聚变次数} = \frac{R / \varepsilon}{M \times \frac{\mathrm{d}}{2M}} \tag{4}$$

其中，观测到的中子探测率 $R = (4.1 \pm 0.8) \times 10^{-3}\ \mathrm{s}^{-1}$ 是基于 45~150 道间的信号计数

foreground minus corresponding background counts in channels 45–150; the neutron detection efficiency, including geometrical acceptance, is calculated using a Monte Carlo neutron–photon transport code[11] to be $\varepsilon = (1.0\pm0.3)\%$; $M \approx 4\times10^{22}$ titanium atoms for 3 g of titanium; and the ratio of deuteron pairs to metal ions, $d/2M \approx 1$, is based on the assumption that nearly all tetrahedral sites in the titanium lattice are occupied, forming the γ-TiD_2 hydride. Then the estimated cold nuclear fusion rate for the neutron-production branch, by equation (4), is $\lambda_f \approx 10^{-23}$ fusions per deuteron pair per second. If most fusions take place near the surface, or if the titanium lattice is far from saturated with deuterons, or if conditions favouring fusion occur intermittently, then the inferred fusion rate must be much larger, perhaps 10^{-20} fusions per deuteron pair per second.

We note that such a fusion rate could be achieved by "squeezing" the deuterons to about half their normal (0.74-Å) separation in molecules. That such rates are now observed in condensed matter suggests catalysed "piezonuclear" fusion as the explanation[1]. A possible cause is that quasi-electrons form in the deuterated metal lattice, with an effective mass a few times that of a free electron. Isotopes of hydrogen are known to accumulate at imperfections in metal lattices[12], and a local high concentration of hydrogen ions might be conducive to piezonuclear fusion. Because we have not seen any evidence for fusion in equilibrated, deuterated metals or compounds such as methylamine-d_2 deuteriochloride or ammonium-d_4 chloride, we conclude that non-equilibrium conditions are essential. Electrolysis is one way to produce conditions that are far from equilibrium.

It may seem remarkable that one might influence the effective rate of fusion by varying external parameters such as pressure, temperature and electromagnetic fields, but just such effects are seen in another form of cold nuclear fusion, muon-catalysed fusion[13].

Conclusions

The correlation of ideas regarding cold piezonuclear fusion[1] with observations of excess ^3He in metals and in geothermal areas of the Earth led to our experimental studies of fusion in electrochemical cells, which began in May 1986. Our electrolyte compositions evolved from geochemical considerations, and changed as results were observed. The presence of a fusion neutron signal was consistently reproduced, although the rate varied widely. Now that our exploratory searches have disclosed a small piezonuclear fusion effect, it remains to disentangle the factors that influence the fusion rate.

The need for off-equilibrium conditions is clearly implied by our data, and suggests that techniques other than electrochemistry may also be successful. We have begun to explore the use of ion implantation and of elevated pressures and temperatures, mimicking geological conditions. Cold nuclear fusion in condensed matter may be of interest as a novel probe of metal-hydrogen systems, including geological ones, and as a source of monoenergetic neutrons. If deuteron–deuteron fusion can be catalysed, then the d–t fusion reaction is possibly favoured because of its much larger nuclear cross-section. Although the fusion rates observed so far are small, the discovery of cold nuclear fusion in condensed

减去相应的本底计数得出的；用蒙特卡罗中子–光子传输编码[11]计算出的中子探测效率（包括几何接受角）是 $\varepsilon = (1.0 \pm 0.3)\%$；而 3 g 钛中含 $M \approx 4 \times 10^{22}$ 个钛原子；基于钛晶格中几乎所有的四面体位点均被占据的假设，氘核对与金属离子之比 $d/2M \approx 1$，形成 γ–TiD_2 氢化物。利用方程（4），对产生中子分支的冷核聚变率的估算值 λ_f 为每秒钟每个氘核对内约 10^{-23} 次聚变。如果大多数聚变发生在表面附近，或者钛晶格远远没有被氘核占据，又或者利于聚变产生的条件间歇性出现，那么推断出的聚变率一定会大很多，大概是每秒钟每个氘核对 10^{-20} 次聚变。

我们注意到，这样的聚变率可以通过将分子中的氘核间距“挤压”至其正常值（0.74 Å）的一半来得到。这样的聚变率如今可以在凝聚态物质中观察到，表明可以用催化的“压核”聚变来解释[1]。一个可能的原因是，在氘化金属晶格中形成了准电子，其有效质量是一个自由电子的几倍。众所周知，氢的同位素一般积聚在金属晶格的缺陷中[12]，而局部的高浓度氢离子可能有利于压核聚变。因为在平衡氘化金属或混合物（比如甲胺–d_2 氘化氯化物或铵–d_4 氯化物）中，我们没有发现任何聚变发生的证据，所以我们推断非平衡条件是必需的。电解就是一种产生远离平衡态的条件的方法。

通过改变诸如压强、温度和电磁场等外部参数，或许可以影响聚变的有效率，但是这种影响被看作是冷核聚变的另一种形式，即 μ 子催化聚变[13]。

结　论

关于冷压核聚变[1]的有关想法以及对存在于金属和地球地热区中的过量 ^3He 的观测，促使我们对电解池中的聚变进行了实验研究，该研究始于 1986 年 5 月。我们的电解质成分演变发端于地球化学的考虑，并且随着观测到结果而逐渐加以改变。虽然产生率波动很大，但聚变中子信号的出现一再被重现出来。既然我们的探索性研究已经揭露了一个小的压核聚变效应，接下来就要弄清影响聚变率的各种因素。

我们的数据清楚地表明了需要远离平衡态条件的要求，并且该要求提示那些不同于电化学的方法也许同样可行。我们已经开始探索离子注入、增大压强和提高温度的作用，这些操作是为了模拟相应的地质条件。将凝聚态物质中的冷核聚变作为一种包括地质系统在内的金属氢系统的新型探测手段，以及作为一种单能中子源，或许是非常有用的。如果 d–d 聚变能够被催化，那么 d–t 聚变反应可能会更受青睐，因其核截面要大得多。尽管目前观测到的聚变率很低，但凝聚态物质中冷核聚变的

matter opens the possibility, at least, of a new path to fusion energy.

(**338**, 737-740; 1989)

S. E. Jones*, E. P. Palmer*, J. B. Czirr*, D. L. Decker*, G. L. Jensen*, J. M. Thorne*, S. F. Taylor* & J. Rafelski†
* Departments of Physics and Chemistry, Brigham Young University Provo, Utah 84602, USA
† Department of Physics, University of Arizona, Tucson, Arizona 85721, USA

Received 24 March; accepted 14 April 1989.

References:

1. Van Siclen, C. D. & Jones, S. E. *J. Phys.* G. **12**, 213-221 (1986).
2. Jones, S. E. *Nature* **321**, 127-133 (1986).
3. Rafelski, J. & Jones, S. E. *Scient. Am.* **257**, 84-89 (July 1987).
4. Craig, H., Lupton, J. E., Welhan, J. A. & Poreda, R. *Geophys. Res. Lett.* **5**, 897-900 (1978).
5. Lupton, J. E. & Craig, H. *Science* **214**, 13-18 (1981).
6. Mamyrin, B. A. & Tolstikhin, L. N. *Helium Isotopes in Nature* (Elsevier, Amsterdam, 1984).
7. Craig, H. & Lupton, J. E. in *The Sea* Vol. 7 (ed. Emiliani, C.) Ch. 11 (Wiley, New York, 1981).
8. Mamyrin, B. A., Khabarin, L. V. & Yudenich, V. S. *Soviet Phys. Dokl.* **23**, 581-583 (1978).
9. Jensen, G. L., Dixon, D. R., Bruening, K. & Czirr, J. B. *Nucl. Instrum. Meth.* **220**, 406-408 (1984).
10. Hess, W. N., Patterson, H. W. & Wallace, R. *Phys. Rev.* **116**, 445-457 (1959).
11. *MCNP: Monte Carlo Neutron and Photon Transport Code*, CCC-200 (Version 3) (Radiation Shielding Information Center, Oak Ridge Natn. Lab., 1983).
12. Bowman, R. C. Jr in *Metal Hydrides* (ed. Bambakides, G.) 109-144 (Plenum, New York, 1981).
13. Jones, S. E. *et al. Phys. Rev. Lett.* **51**, 1757-1760 (1983).

Acknowledgements. We acknowledge valuable contributions of James Baer, David Mince, Rodney Price, Lawrence Rees, Eugene Sheely and J. C. Wang of Brigham Young University, and of Mike Danos, Fraser Goff, Berndt Müller, Albert Nier, Göte Ostlund and Clinton Van Siclen. We especially thank Alan Anderson for advice on the data analysis and Harmon Craig for continuing encouragement. This research is supported by the Advanced Energy Projects Division of the US Department of Energy.

发现至少为一条通往聚变能的新途径提供了可能性。

（牛慧冲 翻译；李兴中 审稿）

Problems with the γ-ray Spectrum in the Fleischmann *et al.* Experiments

R. D. Petrasso *et al.*

Editor's Note

The claim by electrochemists Martin Fleischmann and Stanley Pons to have conducted "cold" nuclear fusion by electrolysis led to attempts worldwide to replicate the findings. Others focused on whether Fleischmann and Pons' evidence supported their claims. A key argument was that they detected the neutrons and gamma-rays expected to be emitted, at well-defined energies, from the fusion reaction and its by-products. Here fusion expert Richard Petrasso of MIT and his colleagues investigate a gamma-ray spectrum shown on US television by Fleischmann and Pons (the only time they revealed it), and show that the alleged gamma-ray peak is in the wrong place. This challenge led to the first suspicions of fraud, rather than poor experimentation, in the cold-fusion claims.

SIR—Fleischmann, Pons and Hawkins[1] recently announced the observation of significant heating in their cold-fusion experiments, a result that they attribute to copious fusion reactions. As compelling evidence that fusion had occurred, they reported the observation of the 2.22-MeV γ-ray line that originates from neutron capture by hydrogen nuclei[2,3]

$$n + p \rightarrow d + \gamma \text{ (2.22 MeV)} \quad (1)$$

(Here d represents a deuteron.) They contend that the neutron in reaction (1) is generated by the reaction

$$d + d \rightarrow n + {}^{3}He \quad (2)$$

and conclude, therefore, first that the 2.22-MeV γ-ray confirms that the fusion process (2) is occurring, and second that a neutron production rate of the order of 4×10^4 neutrons s^{-1} is derivable from their γ-ray signal rate. They further state that most of the heat generation occurs not through process (2), but through a hitherto unknown nuclear-fusion process.

Here we focus solely on the identity of the reported γ-ray line, which we shall henceforth call the signal line. We argue that the claim of Fleischmann *et al.* to have observed the 2.22-MeV line characteristic of reaction (1) is unfounded. We do so on the basis of three quantitative considerations: (1) that the linewidth is a factor of two smaller than their instrumental resolution would allow; (2) that a clearly defined Compton edge[4], which should be evident in their published data at 1.99 MeV, is not in fact present; and (3) that

弗莱施曼等人实验中 γ 射线谱的问题

佩特拉索等

编者按

电化学家马丁·弗莱施曼和斯坦利·庞斯宣称通过电解实现了“冷”核聚变，这引发了世界范围内去重复这一发现的尝试。而另一些人则关注弗莱施曼和庞斯的证据是否能够支持他们的结论。争论的关键在于他们是否在聚变反应和它的副产物中观测到了预期中的具有特定能量的中子和γ射线发射。麻省理工学院的核聚变专家理查德·佩特拉索和他的合作者研究了弗莱施曼和庞斯在美国的电视节目上演示的γ射线谱（这是他们唯一一次展示该图谱），并指出他们所声称的γ射线峰处于错误的位置。这一质疑引发了人们对冷核聚变的报告中存在欺诈的首次怀疑，而不仅仅只是对简陋实验的质疑。

弗莱施曼、庞斯和霍金斯[1]最近宣称在他们的冷核聚变实验中观察到了明显的放热现象，他们将这一结果归因于大量的聚变反应。作为核聚变发生的强有力的证据，他们在报道中说发现了 2.22 MeV 的 γ 射线，该射线源于氢原子核对中子的捕获[2,3]

$$n + p \rightarrow d + \gamma \ (2.22\ \text{MeV}) \tag{1}$$

（其中 d 代表一个氘核。）他们认为反应 (1) 中的中子产生于下述反应

$$d + d \rightarrow n + {}^{3}\text{He} \tag{2}$$

由此得出如下结论：首先，2.22 MeV 的 γ 射线证实了核聚变过程 (2) 的发生；其次，量级为每秒 4×10^4 个中子的中子产生速率可以由他们的 γ 射线信号的产生速率推导而来；他们进一步声明大多数热量的产生并非通过过程 (2)，而是通过一个迄今为止并不清楚的核聚变过程。

我们在本文中主要关注如何识别报道的那条 γ 射线，本文中我们将称之为信号线。我们认为弗莱施曼等人所声称的反应 (1) 中观察到的 2.22 MeV 的特征谱线其实并未被找到。我们这么认为主要基于以下三个定量的考虑：(1) 谱线宽度是他们的仪器所允许的分辨率的一半；(2) 在他们发表的数据中本应位于 1.99 MeV 的清晰可辨

their estimated neutron production rate is too large by a factor of 50. In addition, from a consideration of the terrestrial γ-ray background, we argue that their purported γ-ray line actually resides at 2.5 MeV rather than 2.22 MeV. These conclusions are, in part, based on our studies of neutron capture by hydrogen, using a neutron source submerged in water. These measurements allow us to compare the results of Fleischmann *et al.* directly with a controlled experiment.

We measured terrestrial γ-ray background spectra in order to compare our detector characteristics with those of Fleischmann *et al.* Figure 1*a* shows a typical terrestrial γ-ray background spectrum obtained with a 3 in. × 3 in. NaI(Tl) crystal spectrometer system (see ref. 5 for details). The main features of the background spectrum are quite similar throughout the terrestrial environment[6,7]. Fleischmann *et al.* showed a similar γ-ray spectrum on television (Fig. 1*b*). (We believe that we have viewed all the cold-fusion γ-ray spectra that have been shown on KSL-TV (Utah) up to 19 April. This information was obtained from Utah News Clips, Inc., Utah. As far as we can tell, all spectra are identical to that of Fig. 1*b*.) This spectrum was obtained in the course of the Fleischmann *et al.* experiments (M. Hawkins, personal communication). Their spectrometer system consisted of a Nuclear Data ND-6 portable analyser with a 3 in. × 3 in. NaI (Tl) crystal (ref. 1 and M. Hawkins and R. Hoffmann, personal communications). A $\frac{3}{8}$-in.-thick Pb annulus encompassed the scintillator. It is clear from Fig. 1*a* and *b*, and particularly from the ^{40}K (1.46 MeV) and ^{208}Tl (2.61 MeV) lines, that our resolution is comparable to or better than that of the spectrometer used by Fleischmann *et al.*, a point to which we return later.

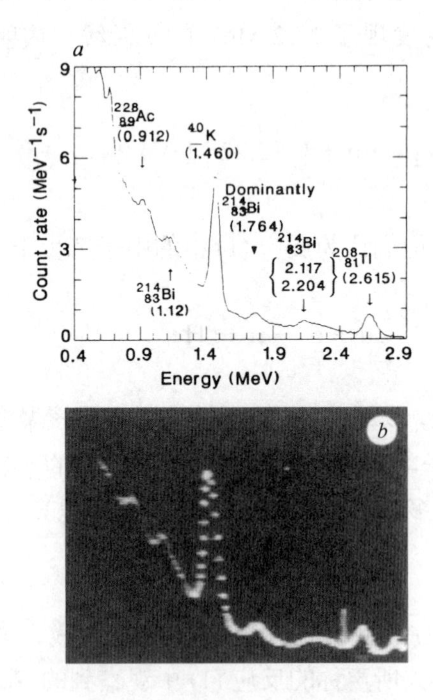

Fig. 1. *a*, The γ-ray background spectrum measured with a 3 in. × 3 in. NaI(Tl) detector at MIT. Some important terrestrial γ-ray lines have been identified in this figure[6,7,12]. (As explained in ref. 12, the

的康普顿边 [4]，却没有出现；(3) 他们估算的中子产率大了 50 倍。更进一步地说，考虑到地面的 γ 射线本底，我们怀疑他们所声称的 γ 射线实际上位于 2.5 MeV 处，而非 2.22 MeV 处。这些结论部分基于我们对氢捕获中子的研究，该研究通过将中子源浸没在水中进行。这些测量使得我们可以将弗莱施曼等人的结果与对照实验进行直接对比。

为了比较我们与弗莱施曼等人的探测器的特性，我们测量了地面的本底 γ 射线谱。图 1*a* 显示了一个典型的地面本底 γ 射线谱，该光谱由 3 in × 3 in 碘化钠（铊）晶体光谱仪系统（详见参考文献 5）测得。在整个地面环境中 γ 射线本底谱的主要特征都非常相似 [6,7]。弗莱施曼等人在电视上展示了一个类似的 γ 射线谱(图 1*b*)。(我们相信我们看到了截止到 4 月 19 日在 KSL 电视台（犹他州）播出的所有的冷核聚变的 γ 射线谱。此信息从犹他州的犹他新闻剪影公司获得。在我们所能分辨的范围内，所有的光谱都与图 1*b* 所示的相同。）这个 γ 谱是从弗莱施曼等人的实验中获得的（霍金斯，个人交流）。他们的光谱仪系统由一个装有 3 in × 3 in 碘化钠（铊）晶体的 ND-6 型便携式核数据分析仪组成（参考文献 1 以及与霍金斯和霍夫曼的个人交流）。一个 3/8 in 厚的铅圆筒包裹着闪烁体。从图 1*a* 和 1*b* 中，尤其是 ^{40}K（1.46 MeV）和 ^{208}Tl（2.61 MeV）的谱线中看得非常清楚，我们的分辨率与之相当或者优于弗莱施曼等人的 γ 谱仪，这一点我们后面还会谈到。

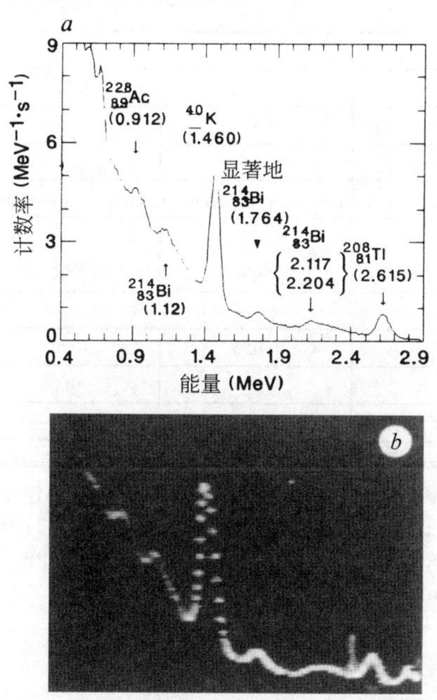

图 1. *a*，用一个 3 in × 3 in 碘化钠（铊）探测器在麻省理工学院测得的 γ 射线本底谱。一些重要的地面 γ 射线谱线在图中被标示出来 [6,7,12]。(正如参考文献 12 中所解释的那样，最终衰变产物的瞬态母核是可

immediate parent of the final decay product is identified. For example, ^{40}K β^+ decays into an excited nuclear state of ^{40}Ar, which actually then emits the 1.460-MeV photon discussed in the text.) The spectrum is averaged over an 84-hour run. *b*, the γ-ray spectrum shown on television by Fleischmann *et al.* The main characteristics of the two spectra are similar; one can also tell that the two detectors have comparable spectral resolution. In *b*, note the curious structure at about 2.5 MeV and that beyond the ^{208}Tl peak (2.61 MeV), which appear to be artefacts. (The spectrum can also be obtained from KSL-TV in Utah (M. Hawkins, personal communication).)

In the interval 1.46–2.61 MeV, the energy resolution of a NaI(Tl) spectrometer, which determines the γ-ray linewidth, can be well described by the formula[8,9]

$$R(E) = \frac{\Delta E}{E} \approx R(E_n)\sqrt{E_n/E} \tag{3}$$

Here ΔE is the full width at half maximum (FWHM) of the line, E is the energy of the photon and $R(E_n)$ is the measured "reference" resolution at energy E_n. $R(E_n)$ can be accurately determined using a ^{60}Co source (that is, the ^{60}Co line at 1.33 MeV), or it can be fairly well approximated by the ^{40}K decay line at 1.46 MeV. (From Fig. 1*b*, the ^{40}K decay line allows one to estimate Fleischmann *et al.*'s resolution as ~8%.) Table 1 lists the resolution data for our detectors and for that of Fleischmann *et al.*.

Table 1. Comparison of energy resolutions of the γ-ray spectrometers

(*a*) Resolution of MIT spectrometers					
Energy (MeV)	1.17 ^{60}Co	1.33 ^{60}Co	1.46 ^{40}K	2.22 n(p, γ)d	2.61 ^{208}Tl
origin					
Natural background			0.055		0.043 (0.041)
^{60}Co	0.056	0.051			
Pu/Be neutron source				0.05 (0.045)	
(*b*) Resolution of the Fleischmann *et al.* spectrometer[1]					
Energy (MeV)		1.33 ^{60}Co	1.46 ^{40}K	2.22 n(p, γ)d	2.61 ^{208}Tl
Reference					
Hoffman*		0.056	0.065		
TV news†			~0.08		~0.05 (0.049)
Ref. 1 (errata)				0.025 (0.053)	

The resolution is defined as the full width at half maximum (FWHM) divided by the peak energy. Numbers in parentheses are predicted values based on the detector resolution at 1.46 MeV (see text). In *b*, the prediction is based on the resolution value (0.065 at 1.46 MeV) provided by R. Hoffman (personal communication)

*R. Hoffman (personal communication).

† Derived from images of the televised news broadcasts.

We now compare the signal line of Fleischmann *et al.* (Fig. 1*a* of ref. 1 (errata), shown as Fig. 2 here) with our measured spectrum obtained from the experiments on neutron capture by hydrogen (Fig. 3 here, and Fig. 4 of ref. 5). In these experiments, a Pu/Be eutron

以辨认的。例如，^{40}K 经 β^+ 衰变后转变为处于核激发态的 ^{40}Ar，其实这个核然后会放射出文章中讨论的 1.460 MeV 的光子。）γ 谱是经过 84 小时测量后所取的平均值。*b*，弗莱施曼等人在电视上展示的 γ 射线谱。两张 γ 谱的主要特征是相似的；两个探测器具有相当的 γ 谱分辨率。*b* 中，注意在大约 2.5 MeV 处和超出 ^{208}Tl 峰（2.61 MeV）处的奇异结构，看上去像是伪造的。（γ 谱也可从犹他州的 KSL 电视台获得（霍金斯，个人交流）。）

在 1.46~2.61 MeV 区间内，决定 γ 射线线宽的碘化钠（铊）γ 谱仪能量分辨率可以由下式描述[8,9]

$$R(E) = \frac{\Delta E}{E} \approx R(E_n)\sqrt{E_n/E} \tag{3}$$

其中 ΔE 是 γ 谱线的半高宽（FWHM），E 是光子的能量，$R(E_n)$ 为能量 E_n 处测量的“参考”分辨率。$R(E_n)$ 可以使用 ^{60}Co 源（即在 1.33 MeV 处 ^{60}Co 的谱线）进行准确的测定，或者通过 ^{40}K 在 1.46 MeV 处的衰变谱线进行相对准确的估算。（通过图 1*b* 中 ^{40}K 的衰变谱线可以估算出弗莱施曼等人的 γ 谱仪分辨率约为 8%。）表 1 列出了我们和弗莱施曼等人的探测器的分辨率数据。

表 1 . 两组 γ 射线谱仪能量分辨率的对比

(*a*) 麻省理工学院 γ 谱仪的分辨率					
能量(MeV)	1.17 ^{60}Co	1.33 ^{60}Co	1.46 ^{40}K	2.22 n(p,γ)d	2.61 ^{208}Tl
源					
自然本底			0.055		0.043 (0.041)
^{60}Co	0.056	0.051			
钚/铍中子源				0.05 (0.045)	
(*b*) 弗莱施曼等人的 γ 谱仪的分辨率[1]					
能量(MeV)		1.33 ^{60}Co	1.46 ^{40}K	2.22 n(p,γ)d	2.61 ^{208}Tl
参考					
霍夫曼*		0.056	0.065		
电视新闻†			~0.08		~0.05 (0.049)
参考文献 1 (勘误)				0.025 (0.053)	

本分辨率定义为半高宽（FWHM）除以峰值的能量。括号中的数字为基于探测器在1.46 MeV 处的分辨率预测的数值(见文中)。在 *b* 中，预测值是基于霍夫曼（个人交流）提供的分辨率数值（1.46 MeV 处为 0.065)得出的。
* 霍夫曼（个人交流）。
† 从电视新闻节目的图像获得。

我们现在将弗莱施曼等人的信号线（参考文献 1（勘误）中的图 1*a*，即本文所示的图 2）与我们在氢捕获中子的实验中测量的 γ 谱（本文的图 3 和参考文献 5 的图 4）来进行对比。在这些实验中，一个钚 / 铍的中子源被放置在水槽中。$^{239}_{94}Pu$

source was placed in a water tank. $^{239}_{94}Pu$ emits energetic α-particles, which produce neutrons through (α, n) reactions with Be (refs 4,9). The neutrons are thermalized in water, and we observe the emitted neutron-capture γ-rays with our spectrometers. The measured resolution at 2.22 MeV is ~5% (Table 1*a*), and is reasonably well predicted by equation (3). As a consequence, this calls into immediate question the identity of Fleischmann *et al.*'s signal line as a γ-ray line. Specifically, Fig. 2 shows the signal line to have a resolution of 2.5%. This is about a factor of two smaller than that predicted by equation (3) on the basis of the known resolution (Table 1*b*) from either the ^{40}K decay line (1.46 MeV) or from the ^{60}Co source (1.33 MeV) (R. Hoffman, personal communication). But we know from Table 1 that the spectrometer used by Fleischmann *et al.* has a resolution that is at best comparable to our own for the entire region from 1.46 to 2.61 MeV (see also Fig. 1), so it is inconsistent that their linewidth at 2.22 MeV is a factor of two below the predicted value.

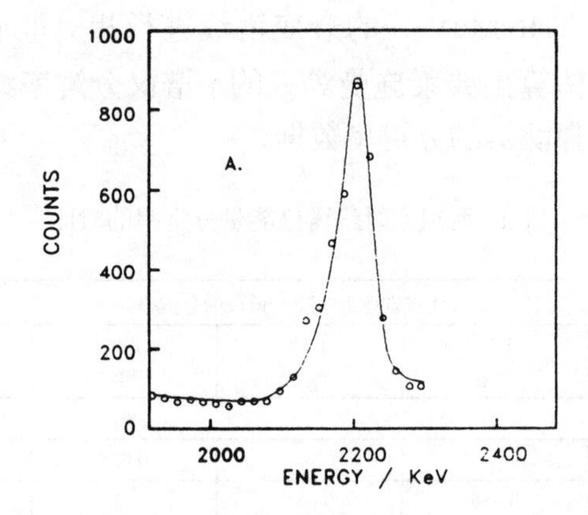

Fig. 2. A reproduction of the purported 2.22-MeV γ-ray signal line of Fleischmann *et al.* (Fig. 1*a* of errata to ref. 1). The resolution, based on the linewidth, is about 2.5%. With such resolution, one would expect to see a clearly defined Compton edge at 1.99 MeV. No edge is evident. Also, a resolution of 2.5% is inconsistent with their spectral resolution. Furthermore, we argue that the signal line may reside at 2.5 MeV, not at 2.22 MeV as is claimed by Fleischmann *et al.* and depicted here.

There is a second crucial inconsistency with the published signal line (Fig. 2). If we assume a resolution of 2.5% at 2.22 MeV, then there should be a clearly defined Compton edge[4] at 1.99 MeV. For example, in Fig. 3 the Compton edge is evident even for our measured resolution of only 5%. For a resolution of 2.5%, the definition of the Compton edge would be distinctly sharper. The lack of a Compton edge at 1.99 MeV for the signal line therefore negates the conclusion of Fleischmann *et al.* that they have observed the 2.22-MeV γ-rays from neutron capture by hydrogen.

放射出具有一定能量的 α 粒子，α 粒子与铍通过（α，n）反应产生中子（参考文献 4、9）。中子在水中达到热平衡，我们通过我们的 γ 谱仪对产生的中子捕获 γ 射线进行观测。测量的分辨率在 2.22 MeV 处约为 5%(表 1*a*)，这一分辨率可以通过方程(3)合理地预测出来。因而，这立即引出了弗莱施曼等人的信号线作为 γ 射线峰的验证问题。具体来说，图 2 表明信号线具有 2.5% 的分辨率。这是在从 ^{40}K 的衰变谱峰（1.46 MeV）或者 ^{60}Co 源（1.33 MeV）获知的分辨率（表 1*b*）的基础上根据方程（3）计算出的分辨率的二分之一（霍夫曼，个人交流）。但是我们从表 1 知道，弗莱施曼等人使用的 γ 谱仪分辨率在 1.46 MeV 到 2.61 MeV 的整个范围内至多与我们的仪器相当（见图 1），而这与他们在 2.22 MeV 处的线宽只有推算值的二分之一不相符。

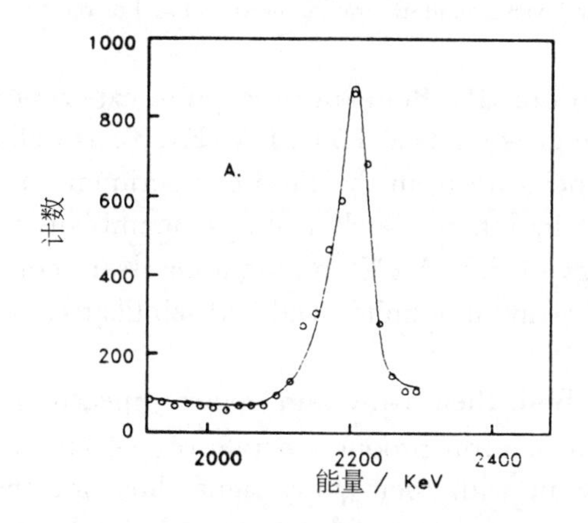

图 2. 弗莱施曼等人所声称的 2.22 MeV 的 γ 射线信号线的重复实验（参考文献 1 的勘误中的图 1*a*）。基于线宽，分辨率约为 2.5%。在这样的分辨率下，预期可以在 1.99 MeV 处看到清晰可辨的康普顿边。可是显然没有康普顿边。另外，2.5% 的分辨率与他们的 γ 谱分辨率不符。进一步说，我们怀疑那个信号线可能位于 2.5 MeV 处，而不是弗莱施曼等人所声称的以及这里所展示的 2.22 MeV 处。

发表出来的信号线（图 2）还有第二个严重的矛盾。如果我们假设 2.22 MeV 处的分辨率为 2.5%，那么在 1.99 MeV 处应该有一个清晰可辨的康普顿边 [4]。比如，即使对于我们的测量分辨率仅有 5% 的结果而言，图 3 中的康普顿边也非常明显。对于 2.5% 的分辨率而言，康普顿边的轮廓应该更锐利。信号线在 1.99 MeV 处康普顿边的缺失否定了弗莱施曼等人观察到的氢捕获中子产生 2.22 MeV 的 γ 射线的结论。

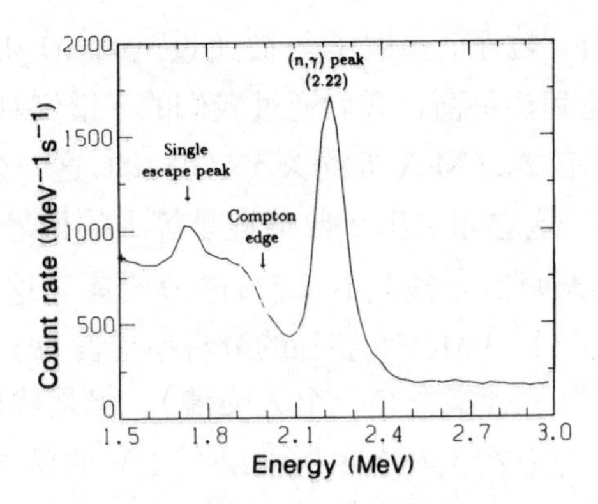

Fig. 3. The γ-ray spectrum measured by a 3 in. × 3 in. NaI(Tl) spectrometer during a neutron-capture-by-hydrogen experiment using a (Pu/Be) neutron source submerged in water. Because of the finite size of the crystal (which is identical to that of Fleischmann *et al.*[1]), we also see an escape peak[2-4] and, of particular importance here, the Compton edge[4]. In this figure, the digitization energy width is 0.024 MeV per channel. The full Pu/Be and background spectra are shown in Fig. 4 of ref. 5.

We also point out that in our (Pu/Be) neutron-capture experiments, a conspicuous e^+–e^- annihilation single-escape peak exists at 1.71 MeV (Fig. 3), as well as a double-escape peak at 1.20 MeV. (The full spectrum from the Pu/Be experiment, as well as the background spectrum, can be found in ref. 5.) Such features unambiguously identify the primary γ-rays as having an energy of 2.22 MeV, and are a necessary consequence of the physical processes of detection of γ-rays in a finite-sized NaI scintillator.

Based independently on both their γ-ray and neutron measurements, Fleischmann *et al.* claim to have observed a neutron production rate of $\sim 4\times10^4$ neutrons s^{-1} (ref. 1). This claim is clearly inconsistent with their γ-ray signal line, for the following quantitative reasons. The Pu/Be neutron source used in our experiment is absolutely calibrated to within 10% of 1.5×10^6 neutrons s^{-1} (ref. 10 and MIT Reactor Radiation Protection Office). In obtaining the data in Fig. 3, we used an experimental setup similar to that of Fleischmann *et al.* (ref. 1; televised broadcasts; and M. Hawkins and R. Hoffman, personal communications). Our Pu/Be source was submerged 6 in. into a large water tank. The rate at the 2.22-MeV peak, after subtracting the background continuum, is about 1.4×10^3 MeV^{-1} s^{-1} (see Fig. 3). Scaling this rate to a neutron source of 4×10^4 neutrons s^{-1} (the level given by Fleischmann *et al.*), and integrating over the linewidth, gives a total 2.22-MeV γ-ray rate of about 4.5 counts per second. This value is a factor of 50 times higher than the rate that would be calculated on the basis of the results in Fig. 1*a* (that is, 0.081 counts per second). (Fleischmann *et al.* state that their neutron count rate is measured with a BF_3 neutron counter over a 0.4 mm × 10 cm Pd cell, and that the γ-ray measurement is over a 0.8 mm × 10 cm Pd cell[1]. If the total reaction rate is proportional to the volume of Pd rod, as they state, the inconsistency in the reported neutron rate is by a factor of 200 rather than 50.) While differences in rates of a factor of two might possibly be explained by geometrical considerations, a factor of 50 is inexplicable.

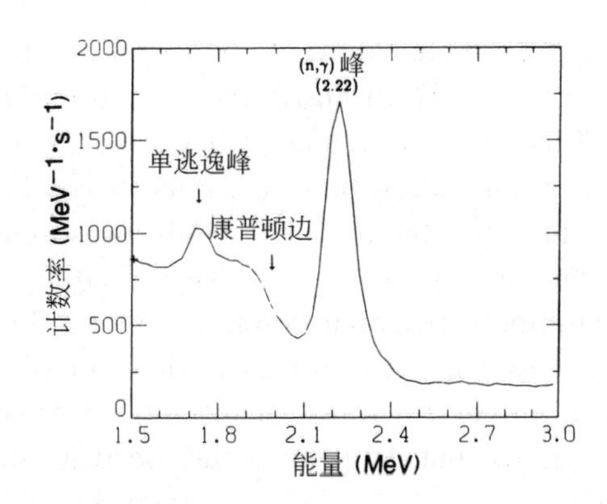

图 3. 在使用浸入水中的（钚 / 铍）中子源进行的氢捕获中子的实验中，通过一个 3 in × 3 in 碘化钠（铊）γ 谱仪测得的 γ 射线谱。因为晶体的有限尺寸（与弗莱施曼等人 [1] 使用的完全相同），我们也看到了逃逸峰 [2-4]，尤其重要的是，我们看到了康普顿边 [4]。本图中，数字化能量的宽度为每通道 0.024 MeV。钚 / 铍以及本底的全谱示于参考文献 5 中的图 4。

我们也要指出，在我们的（钚 / 铍）中子捕获实验中，一个明显的正负电子湮灭的单逃逸峰出现在 1.71 MeV 处（图 3），在 1.20 MeV 处也存在一个双逃逸峰。（钚 / 铍实验中的全谱以及本底光谱都可以在参考文献 5 中找到。）这些特征毫无疑问地确认了初始的 γ 射线具有 2.22 MeV 的能量，这也是用一个有限尺寸的碘化钠闪烁体探测 γ 射线的物理过程的必然结果。

分别基于他们的 γ 射线和中子测量的结果，弗莱施曼等人声称观察到了约为每秒 4×10^4 个中子的中子产生速率（参考文献 1）。基于下述的几个定量分析，这一结论与他们的 γ 射线信号线明显不符。我们实验中使用的钚 / 铍中子源经过了校准，中子产率为每秒 1.5×10^6 个中子，该值的校准精度为 10%（参考文献 10 和麻省理工学院反应堆辐射防护办公室）。在获得图 3 的数据的过程中，我们使用的实验装置与弗莱施曼等人所使用的装置类似（参考文献 1；电视广播；与霍金斯和霍夫曼的个人交流）。我们的钚 / 铍中子源被浸入水槽中的水下 6 in 处。对应于 2.22 MeV 的峰的中子产生速率在扣除连续的本底之后大约为每 MeV 每秒 1.4×10^3 个中子（见图 3）。将这个速率放大到每秒 4×10^4 个中子（弗莱施曼等人所给出的水平），对整个线宽进行积分，得出整个 2.22 MeV 的 γ 射线的产生速率为每秒 4.5 个中子。这一数值是根据图 1*a* 中的结果计算出来的速率（即每秒 0.081 个中子）的 50 倍。（弗莱施曼等人指出，他们的中子计数速率是用一个三氟化硼中子计数器在一个 0.4 mm × 10 cm 的钯电解池上测得的，而 γ 射线的测量是在一个 0.8 mm × 10 cm 的钯电解池上进行的 [1]。如果整个反应的速率如他们所说的正比于钯棒的体积，那么报道的中子产生速率的差别将达到 200 倍而不是 50 倍。）反应速率存在两倍的差异也许可以从几何尺寸方面给予解释，但相差 50 倍的原因就无法解释了。

A further point concerning the identification of the signal line is the precise value of the energy at which the peak occurred. From Fig. 2, the background in the neighbourhood of the peak is seen to be ~80 counts per channel, a level that corresponds to ~400 counts per channel for a 48-hour accumulation time (the data in Fig. 2 were accumulated for a period of 10 hours[1]). On the other hand, in the Utah measurements of terrestrial γ-ray background, the level in the vicinity of the 2.22-MeV feature was found to be ~4,000 counts per channel (R. Hoffman, personal communication). The only relevant part of the entire γ-ray spectrum (between 1.46 and 2.61 MeV) in which the background was as low as 400 counts was at an energy in the vicinity of 2.5 MeV (R. Hoffman, personal communication). Thus, we argue that the peak in the spectrum shown in Fig. 2 may be at 2.5 MeV, not at 2.22 MeV.

The importance of properly identifying the energy of the feature claimed by Fleischmann *et al.* can hardly be overemphasized. Thus, it is extremely unfortunate that they chose to display only the energy range 1.9–2.3 MeV in their published Fig. 1*a*, thereby not providing the supporting evidence of the ^{40}K (1.46-MeV) and ^{208}Tl (2.61-MeV) features which must be present in their spectra in order for their identification to be correct.

Therefore, although Fleischmann *et al.* may have observed a change in their γ-ray spectra that bears some relation to detector location, we conclude that it is unrelated to the 2.22-MeV neutron-capture γ-rays, and that it is also unrelated to the background $^{214}_{83}$Bi line (2.20 MeV; Fig. 1*a*), as has been suggested elsewhere[11]. We can offer no plausible explanation for the feature other than it is possibly an instrumental artefact, with no relation to a γ-ray interaction.

(**339**, 183-185; 1989)

R. D. Petrasso, X. Chen, K. W. Wenzel, R. R. Parker, C. K. Li and C. Fiore
Plasma Fusion Center, Massachusetts Institute of Technology, Cambridge, Massachusetts 02139, USA

References:

1. Fleischmann, M., Pons, S. & Hawkins, M. *J. electroanalyt. Chem.* **261**, 301-308 (1989); and errata.
2. Hamermesh, B. & Culp, R. J. *Phys. Rev.* **92**, 211 (1953).
3. Greenwood, R. C. & Black, W. W. *Phys. Lett.* **6**, 702 (1966).
4. Knoll, G. F. *Radiation Detection and Measurements* (Wiley, New York, 1979).
5. Petrasso, R. D. *et al.* MIT Plasma Fusion Center Report PFC/JA-89-24 Rev. (1989).
6. Eisenbud, M. *Envir. Radioactivity* (Academic, New York, 1973).
7. Adams, J. A. S. & Lowder, W. M. *The Natural Radiation Environment* (Univ. of Chicago Press, 1964).
8. *Harshaw Radiation Detectors Scintillation Counting Principles*, Solon, Ohio, 44139 (1984).
9. Crouthamel, C. E. *Applied Gamma-Ray Spectrometry* 2nd edn (eds Adam, F. & Dams, R.) (Pergamon, Oxford, 1970).
10. Reilly, W. F. thesis, Massachusetts Institute of Technology (1959).
11. Koonin, S. E., Bailey, D. C. *Am. phys. Soc. Meet.* Special Session on Cold Fusion, Baltimore, Maryland, 1-2 May (1989).
12. Lederer, M. C., Hollander, J. M. & Perlman, I. *Table of Isotopes* 6th edn (Wiley, New York, 1967).

Acknowledgements. We thank V. Kurz, J. S. Machuzak, F. F. McWilliams and Dr S. C. Luckhardt. For the use of a spectrometer system, we thank Professor G. W. Clark. For discussions, we thank M. Hawkins and R. Hoffman of the University of Utah. For suggestions and criticisms, we are grateful to Dr G. R. Ricker Jr and Professor D. J. Sigmar. We are indebted to J. K. Anderson for assembling this document. For locating important references, we thank K. A. Powers. Supported in part by the US Department of Energy.

关于信号线的确认更进一步的问题是谱峰出现位置的能量的准确值。从图 2 中，临近谱峰的本底看上去约为 80 计数 / 通道，这个量级与累计 48 小时约 400 计数 / 通道的值是相对应的（图 2 中的数据累积了 10 小时 [1]）。另一方面，在犹他州测量的地面 γ 射线本底中，在 2.22 MeV 特征峰附近的本底数量级约为 4,000 计数 / 通道（霍夫曼，个人交流）。而整个 γ 射线谱（1.46 ~ 2.61 MeV）的本底都低至 400 计数，其中唯一与之相当的部分就是在能量 2.5 MeV 附近（霍夫曼，个人交流）。因此，我们质疑图 2 中 γ 谱的谱峰可能位于 2.5 MeV 处而不是在 2.22 MeV 处。

对弗莱施曼等人声称的特征峰进行正确分辨的重要性再怎么强调都不为过。然而，非常遗憾的是在他们发表的图 1*a* 中仅展示了 1.9 ~ 2.3 MeV 的能量范围，却没有提供 ^{40}K（1.46 MeV）和 ^{208}Tl（2.61 MeV）的特征峰作为支持的证据，而为了证实他们的观点是正确的，这些峰本应出现在他们的谱图中。

因此，尽管弗莱施曼等人可能在他们的 γ 射线光谱中观察到了某种由探测器位置改变所造成的变化，但正如之前人们所提出的那样 [11]，我们认为那与 2.22 MeV 的中子捕获 γ 射线并无关联，也与本底的 $^{214}_{83}$Bi 谱线（2.20 MeV；图 1*a*）无关。除了认为可能是仪器出错，而与 γ 射线的作用无关以外，我们不能够为这个现象提供其他可信的解释。

（李琦 翻译；李兴中 审稿）

Upper Bounds on "Cold Fusion" in Electrolytic Cells

D. E. Williams *et al.*

Editor's note

When electrochemists Martin Fleischmann and Stanley Pons claimed to have found evidence of "cold" nuclear fusion happening in a small flask of salty heavy water undergoing electrolysis, conflicting reports of verification or failure quickly came from all over the world. This paper by chemist David Williams and his co-workers in England was one of the first to turn the tide of opinion towards the idea that cold fusion is an illusion. It reports a very comprehensive attempt to replicate the results of Fleischmann and Pons, and finds no evidence of excess heat, neutron emission or the formation of tritium—the last two being diagnostic of deuterium fusion.

Experiments using three different calorimeter designs and high-efficiency neutron and γ-ray detection on a wide range of materials fail to sustain the recent claims of cold fusion made by Fleischmann *et al.*[1] and Jones *et al.*[2]. Spurious effects which, undetected, could have led to claims of cold fusion, include noise from neutron counters, cosmic-ray background variations, calibration errors in simple calorimeters and variable electrolytic enrichment of tritium.

RECENT publications[1,2] reporting electrochemically induced nuclear fusion, at room temperature, have aroused great interest. The signatures reported are excess heat output, neutron emission and tritium generation from cells with palladium cathodes[1], and neutron emission alone at a much lower level from cells with titanium cathodes[2]. Conventional nuclear physics predicts that fusion between light nuclei requires either very high temperature (as in a tokamak) or unusually close proximity of the two nuclei (as in muon-catalysed fusion). The calculated fusion rate at the internuclear separation in the deuterium molecule (0.74 Å) is $\sim 3\times 10^{-64}\ s^{-1}$ (ref. 3), so the rates reported in ref. 2 ($\sim 10^{-23}$ d-d-pair^{-1} s^{-1}; d is deuteron) are not easy to understand in the context of the known interstitial-site separations (~3 Å), even more so when it is realized that this rate is a severe underestimate because of incomplete deuterium loading in titanium (see Materials characterization section). The reaction rates reported in ref. 1 are even more difficult to understand[4], and the existence, despite strong arguments to the contrary (ref. 5, for example), of an unknown mechanism, which results both in an extraordinary enhancement of the reaction rate and a suppression of the normal nuclear-reaction channels, has been postulated.

电解池中“冷核聚变”的上限

威廉斯等

编者按

当电化学家马丁·弗莱施曼和斯坦利·庞斯宣称他们在小瓶的电解含盐重水中发现了“冷”核聚变的证据时，世界各地很快出现了一些证实其成功或失败的相矛盾的报告。这篇来自英国化学家戴维·威廉斯和他的合作者的文章是首篇将流行观点转向认为冷核聚变只是错觉的文章之一。文章报告了对弗莱施曼和庞斯实验结果的一次非常全面的重复尝试，结果发现没有“过热”、中子发射或者氚生成的证据，而后两者被认为是氘发生聚变的判据。

使用三种不同设计的量热计、高效率的中子和γ射线探测装置对多种材料进行研究的实验未能支持最近弗莱施曼等人[1]和琼斯等人[2]关于冷核聚变的声明。可能是一些未探测到的虚假效应导致了这一关于冷核聚变的断言，其中包括中子计数器的噪声、宇宙射线本底的变化、简易量热计的标定误差和可变的氚的电解富集。

最近发表的论文[1,2]报道了室温下电化学引发的核聚变，该结果引起了极大的关注。报道中发生核聚变的鲜明特征是配有钯阴极的电解池中过剩的热量输出、中子（即n）发射和氚（即t）的产生[1]，以及配有钛阴极的电解池中仅有很低水平的中子发射[2]。传统核物理预测轻核之间的聚变需要非常高的温度（如在托卡马克装置中那样）或者两个核异常接近（如在μ子催化核聚变中那样）。根据氘（即d）分子的核间距（0.74 Å）计算出的核聚变速率约为 $3\times10^{-64}\ s^{-1}$（参考文献3），所以在已知间隙位间距（约为3 Å）的情况下，很难理解参考文献2报道的反应速率（每对氘核每秒约产生 10^{-23} 次聚变效应），如果考虑到氘在钛中并未完全充满（参见材料表征部分），这一速率（约为 $3\times10^{-64}\ s^{-1}$）还是被大大低估了，参考文献2中的速率就更不易理解了。而参考文献1中报道的反应速率甚至更难理解[4]，尽管有强大的反对意见（如参考文献5），他们还是假定存在一种未知的机理，可以使反应速率异常增加，并抑制正常的核反应通道。

In the first reports of the electrolytic cold fusion effect it was stated that the effect is not consistently reproducible, and that it both takes some time to appear and that it may subsequently disappear. Because electrochemical phenomena can be sensitively affected by the state of the surface, some irreproducibility is not, in itself, surprising, and other recent reports (refs 6–8, for example) give well documented accounts of failures as well as successes (R. A. Huggins and A. J. Appleby, Workshop on Cold Fusion, Santa Fe, May 1989) in observing the claimed effects. The clear lack of reproducibility necessitates significant replication, with controls at least equal in number to the number of tests, if positive results are to be viewed with confidence, and exploration of many different, well characterized, material and electrolyte combinations. Particularly, the timescales and achievable concentrations for electrolytic loading of deuterium into palladium and titanium, the quantities of hydrogen "impurity", and the species detectable by surface analysis of used cathodes ought to be determined.

Calorimetry

We used three types of calorimeter. First, we built calorimeters of size and design similar to those used by Fleischmann *et al.* (ref. 1 and M. Fleischmann, personal communication) (Fig. 1). We found these to be inaccurate instruments with some very subtle sources of error which it is necessary to appreciate and analyse in detail. We used sixteen such cells, containing different-size cathodes (1-, 2-, 4- or 6-mm Pd rods) and different electrolytes (0.1 M LiOD, 0.1 M LiOH, 0.1 M NaOD or 0.1 M NaOH). Figure 2*a*, *b* shows the results obtained for a typical cell. An immediately evident characteristic is the sloping baseline, with the sawtooth pattern a consequence of the regular refilling of the cell. The baseline slope can be quantitatively accounted for by a variation of the calibration constant, *k*, with the level of liquid in the cell. This is a result of radiative losses through the vacuum jacket[9] and conduction of the glass inner wall of the cell[10] (see Fig. 2 legend). A consequence of the sloping baseline is that any calibration is only valid for a particular liquid level. We regularly refilled these cells to a reference level at which the calibration had been performed. We calibrated the cells with the heaters before the electrolysis was started and again much later in the run, during the electrolysis, as is evident in Fig. 2*a* (see Fig. 1 legend for details of the calibration procedure). There was no statistically significant difference between the different calibration sets, so all data were combined.

第一批电解冷核聚变效应的报道指出，该效应并非自始至终都是可重复的，它耗时少许才出现并可能随后消失。因为电化学现象对表面状态非常敏感，一些不可重复性本身并不令人惊奇，最近其他的报道（如参考文献 6～8）在观察所谓的冷核聚变效应时给出了失败和成功的完备记录（哈金斯和阿普尔比，冷核聚变专题讨论会，圣菲，1989 年 5 月）。重复性的明显缺乏使得效果显著的重复实验变得很有必要，要想使结果具有较高的置信水平，对照实验至少在数目上应该与之前的实验相当，同时要探究许多不同的、特性明确的材料和电解液组合。尤其是钯电极和钛电极电解法充氘的时间尺度、可达到的浓度，氢“杂质”的量化，以及在用过的阴极上通过表面分析可以检测到的各种物质都应该测定。

量 热 法

我们使用了三种量热计。首先，我们构建了与弗莱施曼等人所用的（参考文献 1；弗莱施曼，个人交流）尺寸和设计相似的量热计（图 1）。我们发现这种量热计是不精准的装置，具有一些不太容易被发现的误差来源，很有必要对其进行仔细研究和分析。我们使用了 16 个这样的电解池，包括不同尺寸的阴极（1 mm、2 mm、4 mm 和 6 mm 的钯棒）和不同的电解液（0.1 M LiOD，0.1 M LiOH，0.1 M NaOD 或者 0.1 M NaOH）。图 2*a* 和 2*b* 显示了其中一个典型的电解池获得的结果。一个直接的明显特征是具有锯齿状形式的倾斜基线，这是电解池定期地添加电解液的结果。基线的斜率可以通过仪器常数 k 随电解池中液面水平的变化定量地求出。这是由杜瓦瓶内的真空夹层的热辐射损失 [9] 和电解池玻璃内壁的热传导 [10] 产生的结果（见图 2 的图注）。基线倾斜的结果就是任何标定只有针对特定的液面水平才是有效的。我们定期将这些电解池重新灌注到进行标定时的参考水平。在电解开始前，我们用加热器对电解池进行标定，电解期间，运行了相当一段时间后再标定一次，如图 2*a* 所示（标定过程的细节见图 1 的图注）。两种标定设置没有统计学上的显著差异，所以我们将所有的数据都结合起来了。

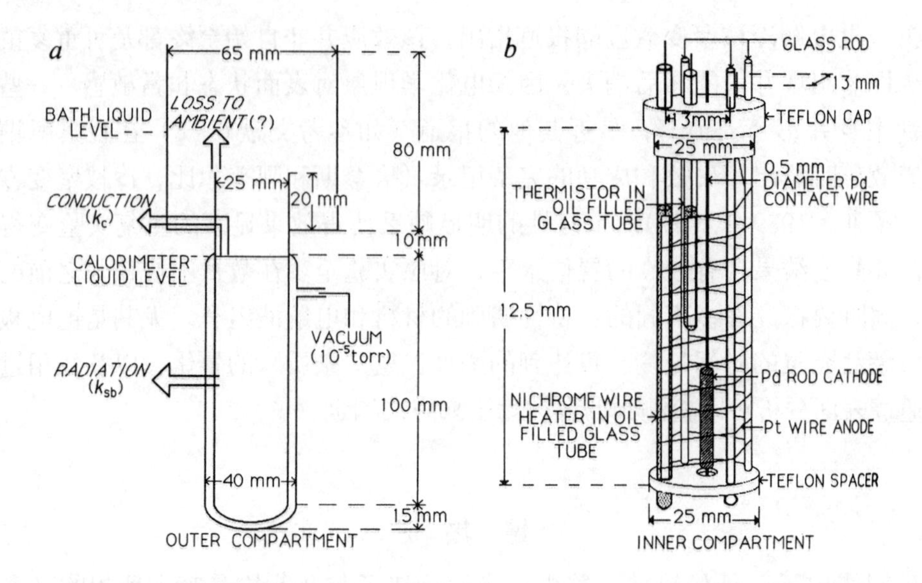

Fig. 1. Schematic diagram of the FPH-type heat-flow calorimeter used here. Heat flow paths are indicated here and discussed further in Fig. 2 legend.

METHODS. As well as the Fleischmann, Pons and Hawkins (FPH) type, we used[10] two other calorimeter types—an improved heat-flow calorimeter (IHF) and an isothermal calorimeter. The IHF calorimeters differed from the FPH type in three important ways. First, they were larger: a 500-ml-capacity cylindrical vessel constituted the cell, with anode-cathode spacing ~2 cm. Second, the electrolysis vessel was inserted, using a film of oil for thermal contact, into a tightly fitting aluminium can which was itself packed around with insulating material and placed in a Dewar flask. Whereas the temperature of the cell contents in the FPH calorimeter was determined using a glass-clad thermistor immersed in the cell itself, in the IHF design the temperature of the aluminium can was measured: the can defined an isothermal surface for conduction of heat away from the cell and its use as the temperature measurement surface eliminated the sloping-baseline problem of the FPH design. Third, the space within the Dewar flask above the electrolysis vessel was filled with a polystyrene cap extending well above the Dewar flask, with the aim of significantly reducing unquantified heat losses to the atmosphere. The two types of heat-flow calorimeter were operated in water baths held at a constant temperature of 20 °C(±0.08 °C). The tops of the water baths were covered with a polystyrene lid. In both cases the cells comprised a spirally wound Pt wire anode (0.25 mm) and a central Pd cathode (Johnson Matthey). For the FPH cells we prepared the different electrolytes using either conductivity water (H_2O cells) or slightly tritiated D_2O (specific activity 13 kBq ml^{-1}) of isotopic purity initially >99.9%. We used 0.1M LiOD (D_2O from Aldrich, measured >99.9% initial isotopic purity) as the electrolyte in the IHF calorimeters. The Pt and Pd contact wire was shrouded in glass tubing in the IHF cells to prevent any possible catalytic recombination of the electrolysis products. In later experiments with the FPH cells we also used screened electrode contacts, but this had no effect on the results obtained. The IHF cells contained a Pd wire electrode insulated to the very tip, which was positioned about three-quarters of the way up the cell. This was used to define the internal liquid level during filling, or refilling, of the cell. The larger volume of electrolyte in the IHF cells (300 ml) meant that refilling was required only occasionally, but the large thermal mass meant that the response was slow (time constant ~12 h): thus they were not sensitive to small bursts of heat. We calibrated the heat-flow calorimeters using a nichrome wire heater placed in an oil-filled glass tube which was in contact with the cell contents. The calibration procedure involved operating the heater, either without electrolytic current, or with a current significantly less (0.2–0.4 times) than the normal electrolysis current, until a steady-state temperature was attained. With the FPH cells, use of a lower electrolysis current ensured continued stirring while diminishing the errors arising from the baseline drift; before each individual calibration the cells were refilled. An empirical calibration curve was thus obtained by fitting the observed thermistor resistance, R, to a range of applied powers, P, using the equation $P = a-b(\log R)+c(\log R)^2$, where a, b and c are the fitted parameters. The

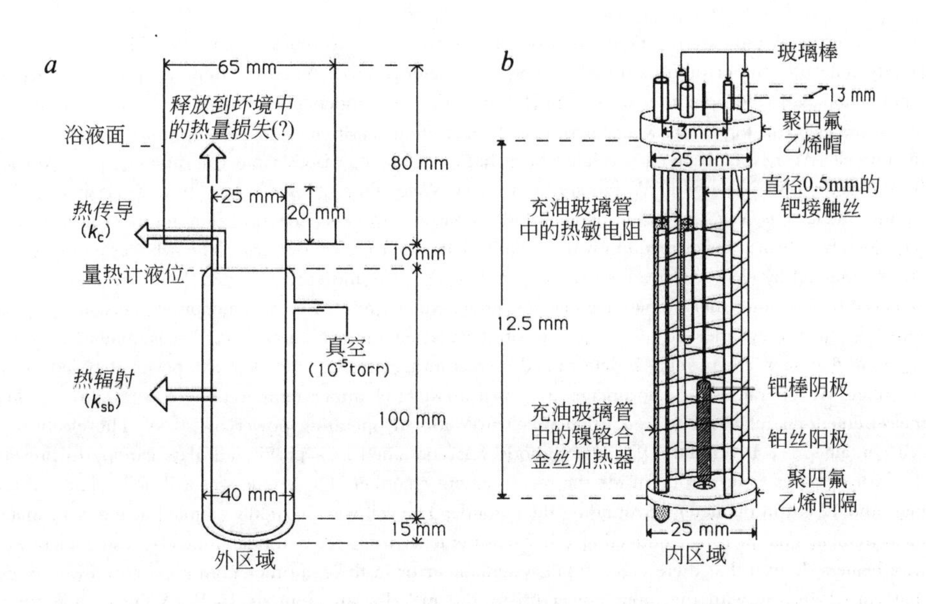

图1. 本文使用的FPH型热流量热计示意图。在图中标出了热流的路径，更进一步的讨论如图2的图注所示。

方法：除了弗莱施曼、庞斯和霍金斯（FPH）型的量热计之外，我们还使用了[10]两种其他类型的量热计——改进的热流量热计（IHF）和等温式量热计。IHF 量热计与 FPH 量热计的不同主要表现在三个方面。首先，IHF 量热计更大：500 ml 容积的柱状容器构成了电解池，其中阳极–阴极间距约为 2 cm。其次，电解容器被塞进一个紧密贴合的铝罐中，用一层油膜保证热接触，铝罐被隔热材料包裹起来并放入一个杜瓦瓶中。FPH 量热计中电解池的温度是通过浸入电解池内的、由玻璃保护的热敏电阻进行测量的，而在 IHF 结构中，测量的是铝罐的温度：对于将电解池中的热量传导出去的过程而言，罐体可以被认为是一个恒温的表面，将它作为温度测量表面解决了 FPH 结构中基线倾斜的问题。第三，将杜瓦瓶中电解容器上方的空间用一个高高地伸出到杜瓦瓶上方的聚苯乙烯盖子填满，这样做的目的是显著减小未知的释放到大气中的热量损失。两种类型的热流量热计都在水浴中运行，水浴控制在恒定的温度 20 ℃（±0.08 ℃）。水浴的顶部用一个聚苯乙烯的盖子盖住。在两种量热计中，电解池均包括一个弯成螺旋形的铂丝阳极（0.25 mm）和一个中央的钯阴极（庄信万丰公司）。对 FPH 电解池我们制备了不同的电解液，它们由电导水（H_2O 电解池）或者初始同位素纯度大于 99.9% 轻微氚化了的 D_2O（比活性为 13 kBq · ml^{-1}）制成。我们用 0.1 M LiOD（D_2O 购自奥德里奇公司，测得初始同位素纯度大于 99.9%）作为 IHF 量热计中的电解液。在 IHF 电解池中，铂和钯的接触丝被封装在玻璃管中，以避免电解产物任何可能的催化复合过程。在后面使用 FPH 电解池的实验中我们也使用了屏蔽的电极接触丝，但是这对获得的结果没有任何影响。IHF 电解池包含一根用绝缘套管覆盖到顶端的钯丝电极，它被放置于距电解池底端四分之三处。这是在向电解池添加或者重复添加液体时用来规定内部液面水平的参照。IHF 电解池中更大的电解液体积（300 ml）意味着重复添加仅需要偶尔进行，但是较大的热质则意味着响应会比较慢（时间常数约为 12 小时）：因此他们对热量的少量释放并不敏感。我们用置于填满油的玻璃管中的镍铬合金丝加热器对热流量热计进行了标定，该玻璃管与电解池的内容物相接触。标定的过程包括运行加热器，不施加电解电流或者施加显著小于（0.2~0.4 倍）通常电解所需的电流，直到体系获得一个稳态的温度。对于 FPH 电解池，使用较小的电解电流旨在确保持续搅拌的同时减小由于基线漂移带来的误差；在每次单独的标定之前电解池都要再添加一次电解液。利用方程 $P = a-b(\log R) + c(\log R)^2$ 将得到的热敏电阻的阻值 R 在应用的功率 P 的范围内对 P 进行拟合，其中 a，b 和 c 是拟合参数，这样一条经

standard error of estimate on the fitted curve varied from 10 to as much as 70 mW with the FPH cells, largely reflecting the error in extrapolation to the reference level. Neutron counting was performed in conjunction with the operation of the FPH cells, using two independent banks of counters mounted above the cells, on top of the water-bath cover[10]. Detection sensitivity above background was 3 event s^{-1} in the cell. We never observed a signal above background. The isothermal calorimeter (J. A. Mason, R. W. Wilde, J. C. Vickery, B. W. Hooton and G. M. Wells, Proc. 29th A. Meet. Inst. Nuclear Materials Management, Las Vegas, Nevada, June 1988) comprises three concentric aluminium cylinders each separated by a heat transfer medium with a relatively low thermal conductivity. The cylinder temperatures are maintained by electrical heaters wound as helical coils around each cylinder. Temperature control is achieved by resistance thermometers on each cylinder which are used in conjunction with classical control software and the cylinder heaters. The rate of thermal energy evolution in the measurement chamber (12.5-cm diameter, 26-cm high) is determined by measuring precisely the electrical power supplied to the chamber. We operated the calorimeter at a measurement-chamber temperature of 42±0.001 °C. The measurement-chamber power resolution is <5 mW for an operating power of 20 W. The electrolytic cell contained ~1 L 0.1M LiOD. The cathode was contained in a perforated glass canopy to prevent the evolved gases from mixing inside the measurement chamber. The anode was a Pt foil cylinder 3 cm high and 12 cm in diameter surrounding the cathode. The cell was thermally coupled to the calorimeter measurement chamber by conducting oil. Measurements using Pt cathodes and also using nichrome wire heaters showed that there was a small systematic error in the calorimeter, an apparent power excess that varied linearly with the input power up to 100 mW for an input of 15 W. A linear fit to these measurements (12 points, standard error of estimate 8.8 mW) was therefore used to apply a correction to the apparent excess measured for the Pd cathodes. The measured output power of both the isothermal and IHF cells was corrected for the power loss that is due to evaporation of the electrolyte, assuming that the electrolysis gases were saturated in water vapour as they passed out of the cell (25.0 mW A^{-1} at 42 °C): $q_V = (p_V/p_0)(1.5)\Delta H_V I/(2F)$ where q_V is the power loss, p_V denotes the saturation vapour pressure and ΔH_V the latent heat of evaporation of the electrolyte, p_0 is the atmospheric pressure, F is the Faraday constant and I is the current (assumes ideal gases).

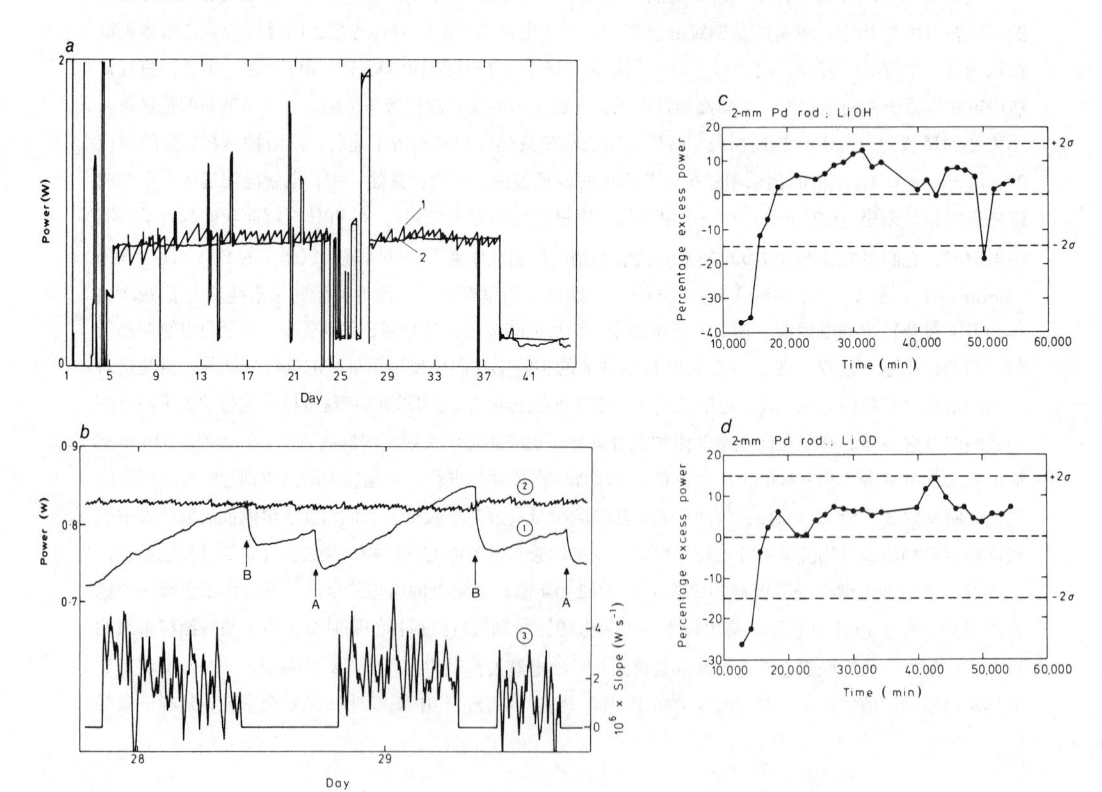

验的标定曲线就得到了。对拟合曲线估算出的 FPH 电解池的标准误差在 10 mW 到高达 70 mW 之间变化，这主要反映了外推到参考水平时的误差。中子计数使用了安装在电解池上方、水浴盖顶部[10]的两个独立的计数管组合体，是与 FPH 电解池的运行同时进行的。高于本底的探测灵敏度在电解池中为每秒 3 个计数。我们没有观察到任何高于本底的信号。等温式量热计（梅森、怀尔德、维克里、胡顿和韦尔斯，第 29 届核材料管理协会年度会议录，拉斯维加斯，内华达州，1988 年 6 月）包括三个同轴的铝筒，相互之间用一种较低热导率的导热介质隔开。圆筒的温度通过螺旋缠绕在各圆筒上的电加热器来维持。温度通过每个圆筒上的电阻式温度计来控制，该温度计与经典控制软件和圆筒加热器连接在一起。测量室（直径 12.5 cm，高 26 cm）中的热能产生速率可以通过精确测量供给测量室的电能来确定。我们在测量室温度为 42±0.001℃的条件下运行量热计。运行功率为 20 W 时，测量室的功率分辨率小于 5mW。电解池中含有约 1 L 0.1 M 的 LiOD。阴极置于一个打孔的盖状玻璃中，以阻止产生的气体在测量室中混合。阳极是一个绕着阴极的高 3 cm、直径 12 cm 的铂箔圆筒。电解池与量热计的测量室通过导热油进行热耦合。使用铂阴极以及镍铬铁合金丝加热器测量，测量结果显示量热计中存在一个小的系统误差，表观的功率过剩随输入功率呈线性变化，输入功率为 15 W 时，功率过剩高达 100 mW。将这些测量值（12 个点，估计值的标准误差为 8.8 mW）的线性拟合用于对钯阴极所测表观过剩的校正。再对等温式和 IHF 电解池所测的输出功率进行功率损失校正，功率损失是由于电解液的挥发造成的，假设当电解气体流出电解池（42℃时为 25.0 mW · A^{-1}）时电解气体在水蒸气中已达到饱和：$q_v = (p_v/p_0)(1.5)\Delta H_v I/(2F)$，其中 q_v 为功率损失，p_v 代表饱和蒸气压，ΔH_v 为电解液挥发的相变潜热，p_0 为大气压，F 为法拉第常数，I 为电流（假设为理想气体）。

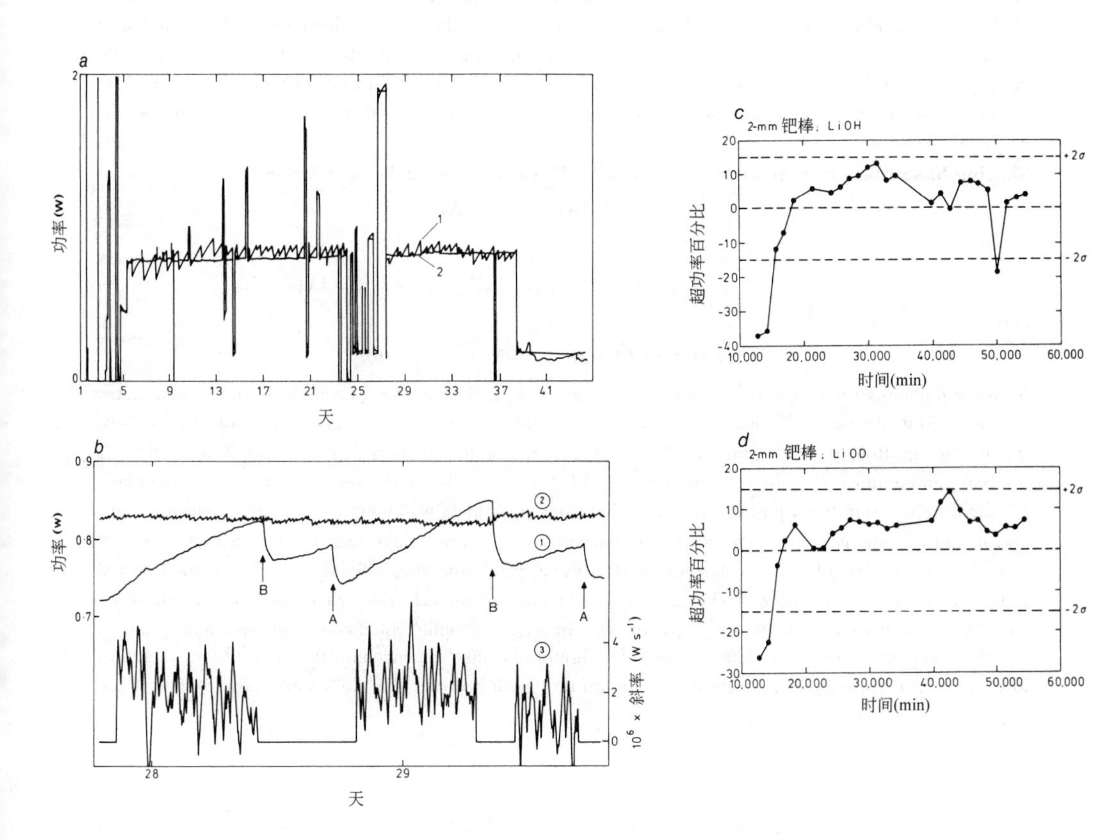

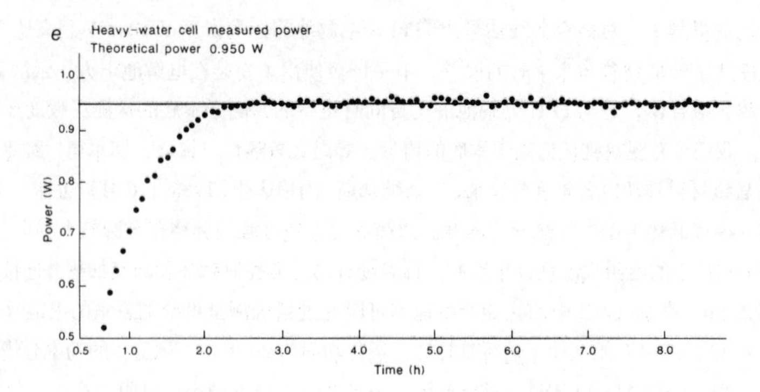

Fig. 2. *a*, Raw data from FPH-type calorimeter containing a 4-mm Pd rod (1.5 cm long, Johnson Matthey "specpure", drawn from sintered stock) in LiOH elctrolyte. Line 1 represents the output power calculated from the thermistor reading and line 2 represents the Joule input power to the cell, $P_{in} = I(V - V_0)$ where $V_0 = \Delta H_d/2F$ (1.527 V for D_2O and 1.481 V for H_2O, ΔH_d being the enthalpy of dissociation, for example, $D_2O(l) \rightarrow D_2(g) + 1/2O_2(g)$). The large step variations are calibrations. *b*, An expanded region of *a*, which emphasizes the sloping baseline. Lines 1 and 2 are as in *a*. Line 3 is the gradient of the apparent output power calculated by differentiation of the data using a seven-point Savitzky–Golay routine[23], stepping one point at a time. At points A the calorimeter was topped up to the reference mark with H_2O pre-warmed to the cell temperature. At points B a volume of liquid estimated from the electrolysis rate was added. *c* and *d*. Results for the *c* and *d*, Results for the two most consistently exothermic FPH calorimeters—0.2×3-cm Pd rod in (1) LiOD and (2) LiOH—in the form of percentage excess of apparent output power over the Joule input power. The error bars represent the control limits $\pm 2\sigma$ calculated from the results obtained for all the different cells (see text). *e*, Raw data obtained using the isothermal calorimeter (20×2-mm diameter Pd cathode), immediately following the application of the input power, with the line representing the Joule input power and the dots the observed cell power. The response time of this calorimeter is governed by the time to obtain thermal mixing of the cell contents, and is faster than that of the simple calorimeters, despite the large solution volume.

Sloping baseline. The response of the "simple" FPH calorimeter can be modelled by[10]

$$P_{in} = k\Delta T = (k_{sb} + k_c)\Delta T \quad (3)$$

where[9],

$$k_{sb} = \sigma A_1[T_i^4 - T_0^4]/\Delta T \approx 4\sigma A_1 T_0^3[1 + 3\Delta T/(2T_0)] = k_{sb,0}(1 + 3\Delta T/(2T_0)) \quad (4)$$

and[10]

$$k_c \approx \kappa A_2/l = \kappa A_2/(l_0 + IV_m\delta t/(2F\pi r_i^2)) \approx k_{c,0}(1 - \alpha\delta t)$$

(P_{in} is the Joule input power, k is the calorimeter constant, k_{sb} and k_c are the contributions to the calorimeter constant of radiative losses through the vacuum jacket and conduction up the glass inner wall respectively, σ is the Stefan–Boltzmann constant, A_1 is the contact area of the solution with the wall of the cell, T_0 is the bath temperature, T_i is the cell temperature, $\Delta T = T_i - T_0$, κ is the thermal conductivity of the glass, l is the distance from the liquid to the point where the inner glass wall comes into contact with the bath and its initial value is l_0, A_2 is the cross-sectional area of the inner glass wall, I is the cell current (300 mA), V_m is the molar volume of cell solution, δt is the elapsed time after refilling the cell, r_i is the internal radius of the glass vessel and $\alpha = IV_m/(2F\pi r_i^2 l_0)$). We measured the calorimeter constant for the cells to be ~0.1 W K^{-1}, with the calculation using equation (4) indicating roughly equal contributions from k_{sb} and k_c. A more complete description of the calorimeter should also take into account the effects of the solution loss on the heat capacity of the calorimeter. It can be shown however that the contribution of this to the sloping baseline is insignificant[10].

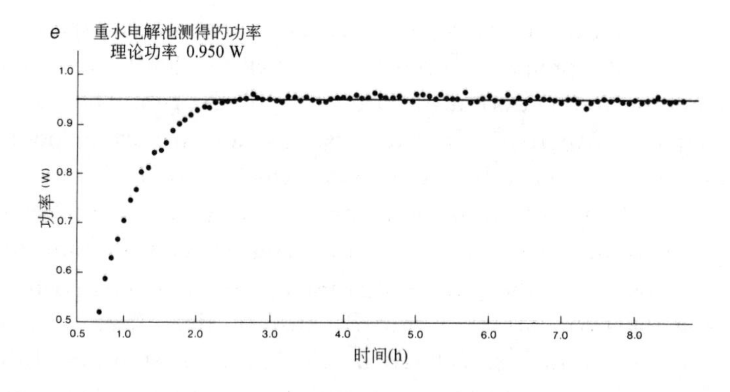

图 2. a，从 LiOH 电解液中带有一个 4 mm 钯棒（1.5 cm 长，庄信万丰公司的“光谱纯”样品，从熔融态得到）的 FPH 型量热计得到的原始数据。线 1 表示根据温度计读数计算出的输出功率，线 2 表示电解池的焦耳输入功率，$P_{in}=I(V-V_0)$，其中 $V_0=\Delta H_d/2F$（对于 D_2O 为 1.527 V，H_2O 为 1.481 V，ΔH_d 为解离焓，例如，$D_2O(l)\rightarrow D_2(g)+1/2O_2(g)$）。大的台阶式变化为标定值。$b$，$a$ 的一个放大区域，突出了倾斜的基线。线 1 和线 2 与 a 中相同。线 3 是使用七点赛威特斯基－高勒程序[23]对数据取微分计算出的表观输出功率的斜率，一步一个点。在点 A，将预热到电解池温度的水添加到量热计的参考标记处。在点 B，添加进根据电解速率估算的一定体积的液体。c 和 d，两个最为一致的 FPH 型量热计的结果——0.2 cm × 3 cm 的钯棒在（1）LiOD 和（2）LiOH 中——以表观输出功率超出焦耳输入功率的百分比形式表示。误差棒代表从所有不同的电解池中所得的结果计算出的控制极限 $\pm 2\sigma$（见正文）。e，使用等温式量热计（20 mm × 2 mm 直径的钯阴极）得到原始数据，应用输入功率后立即开始测量，曲线代表焦耳输入功率，点代表观察到的电解池功率。这个量热计的响应时间由电解池内容物取得热量混合的时间决定，尽管具有较大的溶液体积，它仍快于简单量热计的响应时间。

倾斜的基线“简易的”FPH 型量热计的响应可以模型化为[10]

$$P_{in}=k\Delta T=(k_{sb}+k_c)\Delta T \tag{3}$$

其中[9]

$$k_{sb}=\sigma A_1[T_i^4-T_0^4]/\Delta T\approx 4\sigma A_1T_0^3[1+3\Delta T/(2T_0)]=k_{sb,0}(1+3\Delta T/(2T_0)) \tag{4}$$

同时[10]

$$k_c\approx \kappa A_2/l=\kappa A_2/(l_0+IV_m\delta t/(2F\pi r_i^2))\approx k_{c,0}(1-\alpha\delta t)$$

（其中 P_{in} 是焦耳输入功率，k 是量热计常数，k_{sb} 和 k_c 分别为通过杜瓦瓶内的真空夹层的辐射损失和沿玻璃内壁的传导对量热计常数的贡献，σ 为斯特凡—波尔兹曼常数，A_1 为溶液与电解池壁的接触面积，T_0 为水浴温度，T_i 为电解池的温度，$\Delta T=T_i-T_0$，κ 为玻璃的热导率，l 为液体到玻璃内壁与水浴的接触点间的距离，它的初始值为 l_0，A_2 为玻璃内壁的横截面积，I 为电解池电流（300 mA），V_m 为电解池溶液的摩尔体积，δt 是从添加电解液后算起的时间，r_i 为玻璃容器的内半径，$\alpha=IV_m/(2F\pi r_i^2l_0)$）。我们测量的电解池的量热计常数约为 0.1 W · K^{-1}，是通过方程（4）计算的，这表明 k_{sb} 和 k_c 的贡献大致相当。要对量热计进行更完整的描述就必须考虑溶液损失对量热计热容的影响。然而，可以证明这对倾斜基线的影响是无足轻重的[10]。

Fig. 2*c* shows results for two of the calorimeters at the calibrated liquid level, in the form of percentage excess of apparent output power over the Joule input power. There was apparently an endothermic period at the beginning of the run, whose duration increased with cathode diameter. We speculate that this was due in part to poor stirring of the solution during hydrogen uptake by the cathode which caused a temperature gradient in the cell (gas is not at first evolved at the cathode and the larger diameter cathodes were shorter, to keep a constant surface area). An analogous effect at the start of the electrolysis was reported by Lewis *et al.*[8], who showed that the apparent heating coefficient of a similar cell varied during the first part of a run. Because of this effect, we excluded the first 10,000 minutes of data from each cell when calculating the statistics. Table 1*a* shows the mean absolute power deviation for each cell and the standard deviation in this value. The standard deviation for all of the H_2O cells was not significantly different from that of the D_2O cells (F test, 1% level) so that the data from all of the cells can be used to estimate the error: $\sigma = 0.048$ W (that is ±5–10%). A more detailed analysis of this error[10] showed it to be largely determined by the variability in the liquid level after refilling the cell, but also with a contribution from unquantified variations in the heat loss by conduction up the calorimeter wall to the air above the water bath. Compared with these errors, effects that resulted from temperature gradients inside the cells were minor. The design used here varied from that in ref. 1 in that the glass sleeve that supported the calorimeter and which was in direct contact with the inner wall, provided a large area of thermal contact with the water bath, and thus reduced the effect of the ambient temperature variations.

Table 1. Calorimetry results

(*a*) FPH Calorimeters: current 300 mA, current density 80–110 mA cm^{-2}, 0.1 M LiOH, NaOH, LiOD, NaOD

Pd[a]			H_2O Cells		D_2O Cells[b]	
Diameter / Surface area	Volume / Charging time	Cation	Mean Joule input[c] (mW)	Excess[d,e] (mW)	Mean Joule input[c] (mW)	Excess[d,e] (mW)
6 mm	0.28 cm^3	Li	820	−45±50	960	12±57
3.9 cm^2	870 h	Na	590	14±71	980	57±44
4 mm	0.19 cm^3	Li	800	−19±28	960	−20±29
3.5 cm^2	810 h	Na	610	−55±53	960	−11±49
2 mm	0.094 cm^3	Li	780	40±53	930	61±26
3.1 cm^2	740 h	Na	620	−12±38	940	36±40
1 mm	0.071 cm^3	Li	830	−46±64	980	43±64
2.8 cm^2	670 h	Na	620	−18±60	980	−4±35
Mean excess	$\pm(\sigma/\sqrt{n})$ (mW)[f]			−18±4 (n = 143)		+21±4 (n = 146)

图 2c 显示了两个量热计的液面水平都在同一个标定面时的结果，以表观输出功率超过焦耳输入功率的百分比表示。实验开始时有一个明显的吸热反应阶段，这个阶段随阴极直径变大而延长。我们推测在阴极吸氢过程中搅拌不充分是原因之一，因为它在电解池中造成了一个温度梯度（电解开始时并无气体从阴极放出，为了保持表面积不变，直径较大的阴极更短一些）。电解开始时一个类似的效应由刘易斯等人 [8] 报道过，他们指出类似的电解池的表观传热系数在实验的初始阶段会变化。由于这个效应，我们在进行统计计算时排除了每个电解池开始运行的前 10,000 分钟的数据。表 1a 显示了每个电解池的平均绝对功率偏差值和这个值的标准偏差。所有 H_2O 电解池的标准偏差同 D_2O 电解池的没有明显差异（F 检验，1% 的水平），所以所有电解池的数据都可以用于估算误差：$\sigma = 0.048$ W（即 ±5%~10%）。对这个误差进行更加详细的分析 [10]，揭示了它主要由重新灌注电解池后液面水平的变化决定，也有由量热计壁向水浴上方空气传导造成的热量损失这一未经量化的变化量的贡献。与这些误差相比，电解池内的温度梯度产生的影响很小。本文中使用的设计不同于参考文献 1，我们用玻璃套筒来支撑量热计，而且玻璃套筒与内壁直接接触，由此玻璃套筒就可与水浴有较大面积的热接触，进而减小了室温变化的影响。

表 1. 量热法结果

(a) FPH 量热计：电流 300 mA，电流密度 80~110 mA · cm^{-2}，0.1 M LiOH、NaOH、LiOD、NaOD						
Pd^a		H_2O电解池			D_2O电解池^b	
直径 / 表面积	体积 / 运行时间	阳离子	平均焦耳输入^c (mW)	过剩^{d,e} (mW)	平均焦耳输入^c (mW)	过剩^{d,e} (mW)
6 mm	0.28 cm^3	Li	820	-45 ± 50	960	12 ± 57
3.9 cm^2	870 h	Na	590	14 ± 71	980	57 ± 44
4 mm	0.19 cm^3	Li	800	-19 ± 28	960	-20 ± 29
3.5 cm^2	810 h	Na	610	-55 ± 53	960	-11 ± 49
2 mm	0.094 cm^3	Li	780	40 ± 53	930	61 ± 26
3.1 cm^2	740 h	Na	620	-12 ± 38	940	36 ± 40
1 mm	0.071 cm^3	Li	830	-46 ± 64	980	43 ± 64
2.8 cm^2	670 h	Na	620	-18 ± 60	980	-4 ± 35
平均过剩	$\pm(\sigma/\sqrt{n})$(mW)^f			-18 ± 4 ($n = 143$)		$+21 \pm 4$ ($n = 146$)

Continued

(*b*) IHF and isothermal calorimeters: 0.1 M LiOD

Calorimeter type	Pd: Type / Surface area	Pd: Total charging time / Volume	Current (mA cm^{-2})	Time (h)	Joule input[c] (W)	Excess[d] (mW)
IHF	2-mm rod[a]	797 h[l]	156	231	2.341	78±77[m]
			156	206	2.380	27±48
	3.1 cm^2	0.16 cm^3	219	187	3.842	−1±58
			279	173	5.188	−55±101
	Cast[g]	797 h[l]	35	236	2.172	100±40
			36	201	2.219	47±36
	14 cm^2	0.88 cm^3	49	187	3.489	−31±51
			62	173	4.863	−120±86
Isothermal	2-mm rod[a] 1.3 cm^2	284 h[b] 0.063 cm^3	159	284	0.925[o]	−10±12[n]
	Cast beads[h]	355 h[b]	50	7	0.985	33±12
			30	15	0.415	11±11
			70	7.8	1.796	5±12
			20	15	0.214	10±11
			100	8.6	3.405	−5±12
	5 cm^2	0.5 cm^3	120	270	4.825	2±13
			40	8.1	0.675	7±12
			80	5.7	2.311	9±12
			200	4.8	12.314	−10±12
			160	4.1	8.140	−17±12
			220	4.8	14.675	−9±12
	Melt-spun-ribbon[i] 74 cm^2	74 h[l] 0.35 cm^3	14	74	8.991	64±17
	2-mm rod[j]	323 h[l]	152	70	1.091	7±11
	1.3 cm^2	0.066 cm^3	530	253	10.358	15±14
	8-mm bar[k] 28 cm^2	520 h[l] 1.5 cm^3	30	520	8.974	36±15

[a] Johnson Matthey (JM) "Specpure"; drawn from sintered stock prepared from high-purity powder.

[b] D_2O from Harwell reference stock, contains 13 kBq ml^{-1} tritium.

[c] Mean Joule input power supplied to cell (see Fig. 2 legend). Values for IHF and isothermal calorimeters have been corrected for heat loss that is due to evaporation (see Fig. 1 legend).

[d] Excess power = measured cell output power − calculated Joule input power.

[e] (Mean±1σ) of values calculated after each refilling to the reference level, excluding first 10,000 minutes of polarization (see text).

[f] Mean and standard deviation of the mean calculated for all H_2O data points and all D_2O data points.

[g] Specially produced material supplied by JM—prepared from cast Pd stock that was argon-arc melted into rod form using a gravity casting process. The rods were subsequently sliced and bent to decrease the loading time required (maximum distance from bulk to surface ~1 mm). The sample was cleaned using acetone, 10% HCl and distilled water.

续表

(*b*) IHF 型和等温式量热计：0.1 M LiOD						
量热计类型	Pd		电流 ($mA \cdot cm^{-2}$)	时间 (h)	焦耳输入[c] (W)	过剩[d] (mW)
	种类 表面积	总运行时间 体积				
IHF	直径 2 mm 的钯棒[a]	797 h[l]	156	231	2.341	78 ± 77[m]
			156	206	2.380	27 ± 48
	3.1 cm^2	0.16 cm^3	219	187	3.842	−1 ± 58
			279	173	5.188	−55 ± 101
	铸造[g]	797 h[l]	35	236	2.172	100 ± 40
			36	201	2.219	47 ± 36
	14 cm^2	0.88 cm^3	49	187	3.489	−31 ± 51
			62	173	4.863	−120 ± 86
等温式	直径 2 mm 的钯棒[a] 1.3 cm^2	284 h[b] 0.063 cm^3	159	284	0.925[o]	−10 ± 12[n]
	铸造的颗粒[h]	355 h[b]	50	7	0.985	33 ± 12
			30	15	0.415	11 ± 11
			70	7.8	1.796	5 ± 12
			20	15	0.214	10 ± 11
			100	8.6	3.405	−5 ± 12
	5 cm^2	0.5 cm^3	120	270	4.825	2 ± 13
			40	8.1	0.675	7 ± 12
			80	5.7	2.311	9 ± 12
			200	4.8	12.314	−10 ± 12
			160	4.1	8.140	−17 ± 12
			220	4.8	14.675	−9 ± 12
	熔体快淬带[i] 74 cm^2	74 h[l] 0.35 cm^3	14	74	8.991	64 ± 17
	直径2 mm的钯棒[j]	323 h[l]	152	70	1.091	7 ± 11
	1.3 cm^2	0.066 cm^3	530	253	10.358	15 ± 14
	直径8 mm的钯棒[k] 28 cm^2	520 h[l] 1.5 cm^3	30	520	8.974	36 ± 15

[a] 庄信万丰公司（JM）的“光谱纯”；由高纯粉末制备的烧结块料拉制而成。

[b] 购自哈韦尔公司的 D_2O，含 13 $kBq \cdot ml^{-1}$ 的氚。

[c] 提供给电解池的平均焦耳功率（参见图 2 的图注）。IHF 和等温式量热计的数值都就蒸发产生的热量损失进行了校正（参见图 1 的图注）。

[d] 过剩功率 = 测量的电解池输出功率 − 计算的焦耳输入功率。

[e] 在每一次将液面加至参考液面时算出的数值（平均值 $\pm 1\sigma$），排除了开始 10,000 分钟极化过程的数据（见正文）。

[f] 所有 H_2O 和 D_2O 数据点的平均值和平均值的标准偏差。

[g] 由 JM 提供的特制材料——从库存的铸造钯制备得到，使用一种重力铸造工艺通过氩弧将之熔化成棒状。棒随后被切成片，并弯曲以减少所需的吸附时间（从体相到表面的最大距离约 1 mm）。样品用丙酮、10% 的盐酸和蒸馏水清洗。

^{h} "Specpure" Pd arc melted three times under argon on a water-cooled copper hearth.

^{i} A variety of ribbons prepared (JM) by melt spinning of cast or sintered Pd. A proportion of the ribbons were heat treated (JM) for 20 min at 100 °C under 10% H_2/N_2. Ribbon thickness, 125 μm.

^{j} Type ^{a} that was subsequently vacuum degassed at 1,200 °C, and loaded with deuterium at a pressure of 40 bar. The sample was cooled to liquid-nitrogen temperature before transferring to the calorimeter to minimize loss of D_2.

^{k} Sintered high-purity bar, sliced and bent to decrease the loading time required (maximum distance from surface to bulk ~1 mm).

^{l} D_2O from Aldrich Chemical Co., contains ~15 kBq ml^{-1} tritium.

^{m} Mean and standard deviation of all data points after temperature stabilization: data point every 3 min.

^{n} Standard deviation given by $\sigma = \sqrt{\sigma_b^2 + \sigma_y^2 + \sigma_c^2}$ where σ_b is the standard deviation of the baseline measurement, σ_y that of the power measurement with the cell running and σ_c is the standard error of estimate of the correction line (see Fig. 1 legend). σ_b and σ_y were typically 6 mW, σ_c was 8.8 mW.

^{o} Small error in this particular measurement gave rise to the apparent small endotherm.

As we expected, occasional points from both H_2O and D_2O cells lay outside the "control limits" of $\pm 2\sigma$ (Fig. 2*c*). No points lay outside $\pm 3\sigma$. No cell showed two or more consecutive points outside $\pm 2\sigma$, and the number of points lying above the control limit in the D_2O-cell experiments was no different from that in the H_2O-cell experiments. Therefore, we conclude that, within the experimental error, there was no significant anomaly in the behaviour of the D_2O cells compared with that of the H_2O cells. If, however, the mean power deviation (Table 1*a*) for all eight of the D_2O cells (+9.91±0.47% or 0.021±0.004 W; mean Joule input 0.96 W) is tested against that for all eight of the H_2O cells (−2.46±0.72% or −0.018±0.004 W, mean Joule input 0.81 W for LiOH cells and 0.61 W for NaOH cells), it is clearly highly significant. Because no sequence of individual points lay outside the control limits established above, we suspect that there is another, unknown source of error that scales with the input power. This, of course, could be the postulated "fusion effect", but the magnitude of the effect is commensurate with the errors. The only reliable way of checking this is to construct calorimeters that are free of major sources of errors, and in particular, do not have a sloping baseline. We therefore also used both isothermal and steady state (heat flow) calorimeters which satisfy this criterion.

We have also analysed in detail the slope of the output curve[10] to look for momentary power pulses on timescales shorter than the interval between calibrated points. If the Joule input remains constant then this slope should not vary. Any sudden or momentary extra power input, q_c, would change the slope by approximately $(k_{sb} + k_{c,0})\, q_c/M_0$ (M_0 is the water equivalent of the calorimeter, other symbols defined in Fig. 2 legend). The number of significant deviations from the mean slope was found to be roughly the same for both the H_2O and the D_2O cells. For the whole data set, the largest power excursion for a D_2O cell (6-mm rod) was ~45 mW and for an H_2O cell (6-mm rod) ~40 mW. These are small compared with the input power. It is certainly clear that no unusually large power pulses occurred. Given the difficulties in operating these calorimeters, very occasional occurrences of small fluctuations cannot be considered as support for a "fusion" hypothesis.

The other two types of calorimeters used in this study are described in Fig. 1 legend and results are given in Table 1*b*. We explored a range of different preparations of palladium and of current density, up to nearly 600 mA cm^{-2}. With the improved heat-

[h] “光谱纯”的钯在一个水冷的铜炉床上，在氩气保护下用电弧熔融三次。
[i] 由铸造或者烧结的钯通过熔体快淬制备得到（JM）的各种快淬带。一部分带经过了在含氢 10% 的氮气下 20 分钟 100 ℃的热处理。带的厚度为 125 μm。
[j] a 型随后在 1,200 ℃下进行真空除气处理，在 40 巴的气压下吸附氘。转移到量热计之前样品被冷却至液氮的温度，以减少 D_2 的流失。
[k] 烧结的高纯度钯棒被切成片并弯曲，以减小所需的吸附时间（表面到体相的最大距离为约 1 mm）。
[l] 购自奥尔德里奇化学品公司的 D_2O，约含 15 kBq · ml^{-1} 的氚。
[m] 所有数据点的平均值和标准偏差都在温度稳定后采集：每 3 分钟采集一个数据。
[n] 标准偏差由公式 $\sigma=\sqrt{\sigma_b^2+\sigma_y^2+\sigma_c^2}$ 计算，其中 σ_b 为基线测量的标准偏差，σ_y 为电解池运行时功率测量值的标准偏差，σ_c 为根据校正曲线估算的标准误差（见图 1 的图注）。σ_b 和 σ_y 的典型值为 6 mW，σ_c 为 8.8 mW。
[o] 这次测量中产生表观少量吸热的小的误差。

正如我们所预期，H_2O 和 D_2O 电解池给出的数据点都有少量处于 $\pm 2\sigma$ 的“控制界限”之外（图 2*c*）。没有数据点是在 $\pm 3\sigma$ 之外的。所有电解池都没有两个或者更多的连续数据点处于 $\pm 2\sigma$ 之外，D_2O 电解池实验中超出控制界限的数据点的数目与 H_2O 电解池实验并无不同。因此，我们得出结论，在实验误差范围内，D_2O 电解池的特性与 H_2O 电解池相比没有明显的异常。然而，如果根据表 1*a*，将所有 8 个 D_2O 电解池（+9.91%±0.47% 或者 0.021±0.004 W；平均焦耳输入 0.96 W）与所有 8 个 H_2O 电解池的平均功率偏差（–2.46%±0.72% 或者 –0.018±0.004 W，LiOH 电解池的平均焦耳输入为 0.81 W 而 NaOH 电解池的为 0.61 W）进行对比的话，其差别是十分明显的。因为没有一系列的独立数据点处于上述的控制界限之外，我们认为有另外一个未知的误差来源，与输入功率成比例。当然，这可以是假设的“核聚变效应”，但是这个效应的数量级与误差相同。检验这个假设的唯一可信赖的方法就是构建不受主要误差来源影响的量热计，尤其是不倾斜的基线。我们因此也使用了满足这一准则的等温式量热计和稳态（热流）量热计。

我们也仔细分析了输出曲线的斜率 [10]，以寻找时间尺度上短于校准点间隔的瞬时功率脉冲。如果焦耳输入保持恒定，则斜率应该不变。任何突然或瞬间额外功率 q_c 的输入，都会将斜率改变大约 $(k_{sb}+k_{c,0})q_c/M_0$（M_0 为与量热计等效的水的量，其他符号沿用图 2 的图注中的定义）。H_2O 和 D_2O 电解池中与平均斜率明显偏离的数目大体相同。就所有数据而言，D_2O 电解池（6 mm 的棒）最大的功率偏移约为 45 mW，而 H_2O 电解池（6 mm 的棒）约为 40 mW。这些值与输入功率相比较小。很显然没有异常大的功率脉冲出现。鉴于控制量热计的难度，不能将非常偶然的小波动的出现视为是支持“核聚变”假说的证据。

本研究中使用的另外两种量热计在图 1 的图注中进行了描述，结果列于表 1*b* 中。我们尝试了一系列不同的钯制备方法和最大值接近 600 mA · cm^{-2} 的一系列电流密度。使用改进的热流量热计（IHF），功率过剩的符号和数量级随焦耳输入功率而

flow calorimeters (IHF), the sign and magnitude of the power excess varied with Joule input power, but was always < 5%. Expressed in terms of the volume of the palladium cathode, this sets a limit of 100–500 mW cm^{-3}. The most accurate calorimeter used was the isothermal calorimeter (minimum-detectable power change ~10 mW; minimum-detectable energy in any brief burst ~40 J). We analysed these results at four-hour intervals by averaging both the Joule input power and the measured output power over a period of ~20 min. Inspection of the data collected every four minutes between the regions of analysis showed no obvious signs of any short heat "bursts" and we found no trend with time of the measured output power under any of the conditions used. Table 1*b* shows that we obtained thermal balance to better than 20 mW (24–240 mW cm^{-3} Pd). We observed slight thermal excesses (30–60 mW) during the initial charging period of the palladium beads, and during runs with high-surface-area cathodes at high current: it can reasonably be assumed that a small amount of recombination (4% at most) was responsible for this effect.

Neutron Counting

We investigated the emission of neutrons from a wide range of cells using three different detector systems (Table 2). The large, high-efficiency detector with which most of the neutron measurements were made is an oil-moderated assembly of 56 $^{10}BF_3$ proportional counters configured as 5 concentric rings[11,12]. The total efficiency for 2.45-MeV d–d neutrons is 44%. We built an automatic cell shuttle mechanism to exchange regularly two nominally identical cells, only one of which was powered. This enabled the background (which is due mostly to cosmic rays, there being no anti-coincidence counter arrangements) and any signal to be counted virtually simultaneously. In operation the cells were exchanged every 5 min, and the data from the 5 rings were recorded separately. Data from a typical run are shown in Fig. 3 as differences in the count rate between the powered and unpowered cell. Although in the particular example shown, two spikes can be seen in the count rate differences from the detector as a whole, it is clear that these spikes are due entirely to the misbehaviour of ring 4, and are therefore spurious.

Table 2. Measured neutron and γ-ray emission rates for cold fusion electrolytic cells

(*a*) Neutron emission rates								
Cathode			Electrolyte	Anode	Details of electrolysis			Measured neutron yield[w] ($n\ s^{-1}$)
Material	Mass (g)	Surface area (cm^2)			Current (mA)	Typical voltage	Duration (h)	
Pd rod 2mm[a]	3.4	5.7	0.1 M LiOD	Pt wire	360	5	914	These cells monitored initially by low-efficiency n-detectors with lower detection limit of ~100 s^{-1}. Later moved to a dual-cavity neutron detector[10] with lower detection limit ~2 s^{-1}.
Pd wire 1 mm[a]	0.94	3.1	0.1 M LiOD	Pt wire	200[r]	4	916	
Pd wire 1 mm[b]	0.94	3.1	0.1 M LiOD	Pt wire	200	4	916	
Pd wire 1 mm[b]	1.4	4.7	0.1 M LiOD	Pt wire	300[s]	5	917	
Pd wire 1 mm[c]	0.47	1.6	0.1 M LiOD	Pt wire	750	11	856	
Pd plate 1 mm[d]	7.5	13.5	1 M LiOD	Pt sheet	2,000	10	307	
Pd foil	0.5	8.0	0.1 M LiOD	Vitreous C	400–120[t]	4–20	142	
Pd/22Ag[e]	1.6	3.8	0.1 M LiOD	Pt wire	380	5	859	No detected n-emission.
Pd/22Ag[e]	1.6	3.8	0.1 M LiOD	Pt wire	380[u]	5	547	

变化，但总是小于 5%。用钯阴极的体积表示，这个值的范围在 100~500 $mW \cdot cm^{-3}$ 之间。最精确的量热计是等温式量热计（可探测的最小功率变化约为 10 mW；可探测的任何最小猝发能量约为 40 J）。我们分析这些以 4 小时为间隔的结果，每约 20 分钟将焦耳输入功率和测量的输出功率作平均。对分析区域每 4 分钟获得的数据进行的研究显示，没有任何短时热量“猝发”的明显迹象，而且我们发现在任何我们使用的条件下，测得的输出功率没有随着时间变化的趋势。表 1*b* 显示我们获得的热平衡优于 20 mW（对钯而言为 24~240 $mW \cdot cm^{-3}$）。在最初的钯电极充电期间以及高表面积阴极在大电流条件下运行的过程中，我们观察到了少量的热量过剩（30~60 mW）：可以合理地假设是少量的复合过程（至多 4%）产生了此效应。

中 子 计 数

我们使用三种不同的检测系统研究了来自多种电解池的中子发射（表 2）。我们用大而高效的探测器来进行大部分的中子测量，它是一个由 56 个 $^{10}BF_3$ 正比计数器排布成的、由 5 个同心圆环阵列 [11,12] 所组成的装置。2.45 MeV的 d–d 中子总效率为 44%。我们构造了一个自动电解池滑梭装置来定期调换两个形式上完全相同的电解池，只有其中一个是通电的。这使得本底（本底主要来自宇宙射线，没有布置反符合计数管）和任何信号可以实质上同时计数。操作时，电解池每隔 5 分钟调换一次，5 个圆环上的数据分别被记录下来。从一次典型实验中得到的数据如图 3 所示，它们是通电和不通电的电解池的计数率之差。尽管在所展示的特例中，从探测器整体的计数率之差可以看到两个尖峰信号，但是这些尖峰信号显然完全是由于环 4 的错误响应引起的，因而是假象。

表 2. 所测的各冷核聚变电解池的中子和 γ 射线发射速率

(*a*)中子发射速率

阴极			电解液	阳极	电解细节			测得的中子产率[w]（个/秒）
材料	质量(g)	表面积(cm^2)			电流(mA)	典型电压	时长(h)	
2 mm 的钯棒[a]	3.4	5.7	0.1 M LiOD	铂丝	360	5	914	这些电解池最初使用具有较低检测限（约100 s^{-1}）的低效率中子探测器进行检测。后来又使用一个具有更低检测限（约 2 s^{-1}）的双腔中子探测器[10]。
1 mm 的钯丝[a]	0.94	3.1	0.1 M LiOD	铂丝	200[r]	4	916	
1 mm 的钯丝[b]	0.94	3.1	0.1 M LiOD	铂丝	200	4	916	
1 mm 的钯丝[b]	1.4	4.7	0.1 M LiOD	铂丝	300[s]	5	917	
1 mm 的钯丝[c]	0.47	1.6	0.1 M LiOD	铂丝	750	11	856	
1 mm 的钯盘[d]	7.5	13.5	1 M LiOD	铂片	2,000	10	307	
钯箔	0.5	8.0	0.1 M LiOD	玻碳	400~120[t]	4~20	142	
钯银 22 合金[e]	1.6	3.8	0.1 M LiOD	铂丝	380	5	859	未检测到中子发射
钯银 22 合金[e]	1.6	3.8	0.1 M LiOD	铂丝	380[u]	5	547	

Continued

(*a*) Neutron emission rates

Cathode			Electrolyte	Anode	Details of electrolysis			Measured neutron yield[w] (n s^{-1})
Material	Mass (g)	Surface area (cm^2)			Current (mA)	Typical voltage	Duration (h)	
Pd foil[f]	0.075	1.0	0.1 M LiOD	Pt wire	30	3	87	−0.019±0.042
Pd foil[g]	0.075	1.0	0.1 M LiOD	Pt wire	250	7	6	0.00±0.08
Pd foil[h]	0.075	1.0	0.1 M LiOD +17 mM Na_2S	Pt wire	250	7	16	0.04±0.11
Pd foil[i]	0.075	1.0	0.1M LiOD	Pt wire	250	8	16	0.005±0.063
Pd foil[j]	0.075	1.0	0.1 M LiOD	Pt wire	250	8	16	0.051±0.061
Pd foil[k]	0.075	1.0	0.1 M LiOD	Pt wire	1,000	15	56	0.068±0.061
Pd foil	0.63	8.0	0.1 M LiOD	Pt foil	1,000	14	17	−0.110±0.082
Pd foil	0.63	8.0	0.1 M LiOD	Pd foil	1,000	18	22	0.012±0.079
Pd ribbon[l]	0.45	0.92	0.1 M LiOD	Pd foil	600[v]	15	110	−0.063±0.096
Pd foil	0.60	8.0	0.1 M LiOD	Au wire	1,000	12	43	0.068±0.058
Pd pellet[m]	3.0	2	0.1 M LiOD	Pt foil	650	18	68	0.007±0.042
Pd pieces[n]	4.4		0.1 M LiOD	Pd foil	500	8	88	0.012±0.033
Pd pieces	4.4		0.1 M LiOD	Pd foil	500	15	88	0.003±0.046[x]
Ti foil	0.038	1.0	0.1 M D_2SO_4	Au wire	250	5	34	−0.091±0.054
Ti foil	0.038	1.0	0.1 M D_2SO_4	Au foil	1,000	5	3	−0.22±0.12[y]
(continued)			0.1 M D_2SO_4 +0.02M $PdCl_2$		1,000	5	4	−0.22±0.12[y]
Ti rod	2.2	4.3	Jones[q]	Au foil	100	4	3	−0.06±0.15[y]
(continued)					500	6	2	−0.42±0.21[y]
Ti granules[o]	0.5		0.1 M D_2SO_4	Pt wire	250	7	22	0.00±0.072
Ti granules	0.5		0.1 M D_2SO_4 +3 mM $Na_4P_2O_7$	Pt wire	250	7	66	0.044±0.054
Ti granules	40		0.1 M D_2SO_4	Pt wire	250	25	22	−0.009±0.058
Ti granules	40		0.1 M D_2SO_4 +3 mM $Na_4P_2O_7$	Pt foil	250	25	22	−0.010±0.060
Ti granules	17		Jones	Au foil	500	26	24	0.021±0.054
Ti granules								
Ti/6Al/4V	2.8	15.0	0.1 M D_2SO_4	Pt wire	670	5	87	−0.003±0.031[x]
Ti/6Al/4V[p]	2.7	14.3	0.1 M D_2SO_4	Pt wire	660	5	19	0.031±0.044[x]
TiFe granules	1		0.1 M D_2SO_4	Pt foil	250	8	24	0.12±0.06
$CeAl_2$ granules	11		0.1 M LiOD	Pt foil	500	22	22	−0.051±0.056
UPt_3 granules	57		0.1 M LiOD	Pt foil	500	20	94	0.10±0.06

(*b*) γ-ray emission rates

Cathode			Electrolyte[z]	Anode	Details of electrolysis			Measured 5,488-keV γ-ray yield[aa] (γ s^{-1})
Material	Mass (g)	Surface area (cm^2)			Current (mA)	Typical voltage	Duration (h)	
Pd foil	0.58	7.5	0.1 M LiOH	Pt foil	450	9	45	$(-4.9\pm7.43)\times10^{-3}$
Pd foil	0.74	9.5	0.1 M LiOH	Pt foil	510	9	164	−0.011±0.004
Pd sheet	7.7	12.9	0.1 M LiOH*	Pt foil	200	4.6	50	−0.017±0.012
Pd rod	3.1	2.3	0.1 M LiOH	Au foil	1,000	26	17	0.008±0.011

续表

(*a*)中子发射速率

阴极			电解液	阳极	电解细节			测得的中子产率[w]（个/秒）
材料	质量(g)	表面积(cm^2)			电流(mA)	典型电压	时长(h)	
钯箔[f]	0.075	1.0	0.1 M LiOD	铂丝	30	3	87	–0.019 ± 0.042
钯箔[g]	0.075	1.0	0.1 M LiOD	铂丝	250	7	6	0.00 ± 0.08
钯箔[h]	0.075	1.0	0.1 M LiOD +17 mM Na_2S	铂丝	250	7	16	0.04 ± 0.11
钯箔[i]	0.075	1.0	0.1 M LiOD	铂丝	250	8	16	0.005 ± 0.063
钯箔[j]	0.075	1.0	0.1 M LiOD	铂丝	250	8	16	0.051 ± 0.061
钯箔[k]	0.075	1.0	0.1 M LiOD	铂丝	1,000	15	56	0.068 ± 0.061
钯箔	0.63	8.0	0.1 M LiOD	钯箔	1,000	14	17	–0.110 ± 0.082
钯箔	0.63	8.0	0.1 M LiOD	钯箔	1,000	18	22	0.012 ± 0.079
钯带[l]	0.45	0.92	0.1 M LiOD	钯箔	600[v]	15	110	–0.063 ± 0.096
钯箔	0.60	8.0	0.1 M LiOD	金丝	1,000	12	43	0.068 ± 0.058
钯颗粒[m]	3.0	2	0.1 M LiOD	铂箔	650	18	68	0.007 ± 0.042
钯片[n]	4.4		0.1 M LiOD	钯箔	500	8	88	0.012 ± 0.033
钯片	4.4		0.1 M LiOD	钯箔	500	15	88	0.003 ± 0.046[x]
钛箔	0.038	1.0	0.1 M D_2SO_4	金丝	250	5	34	–0.091 ± 0.054
钛箔	0.038	1.0	0.1 M D_2SO_4	金箔	1,000	5	3	–0.22 ± 0.12[y]
（续）			0.1 M D_2SO_4 +0.02 M $PdCl_2$		1,000	5	4	–0.22 ± 0.12[y]
钛棒	2.2	4.3	琼斯[q]	金箔	100	4	3	–0.06 ± 0.15[y]
（续）					500	6	2	–0.42 ± 0.21[y]
钛颗粒[o]	0.5		0.1 M D_2SO_4	铂丝	250	7	22	0.00 ± 0.072
钛颗粒	0.5		0.1 M D_2SO_4 +3 mM $Na_4P_2O_7$	铂丝	250	7	66	0.044 ± 0.054
钛颗粒	40		0.1 M D_2SO_4	铂丝	250	25	22	–0.009 ± 0.058
钛颗粒	40		0.1 M D_2SO_4 +3 mM $Na_4P_2O_7$	铂箔	250	25	22	–0.010 ± 0.060
钛颗粒	17		琼斯	金箔	500	26	24	0.021 ± 0.054
钛颗粒								
Ti/6Al/4V合金	2.8	15.0	0.1 M D_2SO_4	铂丝	670	5	87	–0.003 ± 0.031[x]
Ti/6Al/4V合金[p]	2.7	14.3	0.1 M D_2SO_4	铂丝	660	5	19	0.031 ± 0.044[x]
TiFe 颗粒	1		0.1 M D_2SO_4	铂箔	250	8	24	0.12 ± 0.06
$CeAl_2$ 合金颗粒	11		0.1 M LiOD	铂箔	500	22	22	–0.051 ± 0.056
UPt_3 颗粒	57		0.1 M LiOD	铂箔	500	20	94	0.10 ± 0.06

(*b*) γ 射线发射速率

阴极			电解液[z]	阳极	电解细节			测得的5,488 keV 的γ射线的产率[aa]（γ射线/秒）
材料	质量(g)	表面积(cm^2)			电流(mA)	典型电压	时长(h)	
钯箔	0.58	7.5	0.1 M LiOH	铂箔	450	9	45	$(-4.9 \pm 7.43) \times 10^{-3}$
钯箔	0.74	9.5	0.1 M LiOH	铂箔	510	9	164	–0.011 ± 0.004
钯片	7.7	12.9	0.1 M LiOH*	铂箔	200	4.6	50	–0.017 ± 0.012
钯棒	3.1	2.3	0.1 M LiOH	金箔	1,000	26	17	0.008 ± 0.011

Continued

(*b*) γ-ray emission rates								
Cathode			Electrolyte[z]	Anode	Details of electrolysis			Measured 5,488-keV γ-ray yield[aa] ($\gamma\ s^{-1}$)
Material	Mass (g)	Surface area (cm^2)			Current (mA)	Typical voltage	Duration (h)	
Ti foil	0.089	2.0	Jones*	Au foil	320	5	42	−0.022±0.011
Ti foil	2.2	32	Jones	Au foil	1,000	4	12	−0.015±0.034
Ti foil	2.2	32	Jones	Au foil	1,000	4	6	0.012±0.007
Ti granules	25		Jones	Au foil	710	13	8	0.032±0.036
Ti granules	25		0.1 M H_2SO_4	Au foil	400	20	25	−0.019±0.018
Ti granules	25		0.1 M H_2SO_4	Au foil	765	16	16	0.017±0.022
UPt_3 granules	42		0.1 M LiOH	Pt foil	300	31	166	$(3.1\pm6.0)\times10^{-3}$

[a] Cells provided by M. Fleischmann. Cathodes analysed for H, D after use—results H/Pd: 0.01, 0.02 D/Pd: 0.84, 0.72.

[b] Cathode wound into a tight spiral. Examined initially in the large n-detector, without shuttle, for 147 h with estimated detection limit 1 n s^{-1} and a further 5 h with detection limit 0.2 n s^{-1}. Surface of cathode then rubbed with S before further use.

[c] Cathode cleaned with emery paper after 300 h.

[d] Cathode examined initially in the large n-detector, without shuttle, in 0.1 M LiOD for 183 h; estimated detection limit 1 n s^{-1}. Abraded with 400-grit SiC paper before use.

[e] Tubes from D storage system for Tokamak.

[f] Cathode analysed for H, D after use—results: H/Pd: 0.03 D/Pd: 0.80.

[g] Electrolyte of the above run reused.

[h] Cathode and electrolyte of the above run reused; Na_2S added as concentrated aqueous solution.

[i] Foil vacuum degassed 1,000 °C before use. Analysed for H, D after use—results: H/Pd: 0.02 D/Pd: 0.83.

[j] Foil dipped in Na_2S (concentrated solution) before use. Analysed for H, D after use—results: H/Pd: 0.05 D/Pd: 0.80.

[k] Cathode cut in half and analysed for H, D after use—results: H/Pd: 0.02, 0.02 D/Pd: 0.85, 0.78.

[l] Melt-spun ribbon provided by Johnson-Matthey Technology Centre.

[m] Arc remelted twice; electrolytically charged for 1 month in 0.1 M LiOD then frozen in liquid nitrogen, dipped in concentrated Na_2S solution and transferred to n-counting cell.

[n] Four 1–2-mm-thick discs of different types of Pd (Johnson-Matthey) spot welded to Pd wires plus strained Pd Wire.

[o] Electrolyte reused after a previous run with a Ti cathode and Pt anode.

[p] Material heated to 900 °C then quenched in water before use.

[q] AuCN in Jones electrolyte replaced with $NaAuCl_4$.

[r] Current on for 36 h then changed between 200 mA and 20 mA every hour.

[s] Current on for 36 h then changed between 300 mA and 30 mA every 5 min.

[t] Current changed slowly over range shown during electrolysis period as anode disintegrated.

[u] Current on for 36 h then cycled off for 10 h and on for 2 h for period of 270 h. Then 5,000 s at 380 mA cathodic and 4,000 s at 10 mA anodic for rest of period.

[v] Current cycled on/off every 2 h. Neutron yield is for "on" cycle.

[w] The errors assigned (1σ, calculated over full run duration) vary somewhat because in an attempt (mostly in the earlier stages of the programme) to cover as wide a range of cell configurations as possible, pairs of unpowered + unpowered and powered + unpowered runs were not always carried out, and consequently allowance has to made for slight differences in overall cosmic-ray neutron detection efficiencies caused by slight differences between the two nominally identical cells.

[x] These data were obtained using a different data acquisition system to drive the shuttle and accumulate data, set up to look for neutron bursts[10].

[y] Cells exchanged every 5 min by hand.

[z] All electrolytes used for γ-ray work were 50:50 H_2O:D_2O except * which were 41:59 H_2O:D_2O.

[aa] For the present measurements, the system was calibrated with a set of standard γ-ray sources and a $^{238}Pu/^{13}C$ (α, nγ) source emitting 6,129-keV γ-rays. The cells were positioned such that their cathodes were as close as possible to the detector crystal, and the detection efficiencies were calculated by integrating previously measured[21] point efficiency functions over the cathode volume. Peaks were searched for at 5,488, 4,977 and 4,466 keV corresponding to the expected location of the full energy, single-escape and double-escape peaks using the method recommended in ref. 22.

续表

(b) γ射线发射速率								
阴极			电解液[z]	阳极	电解细节			测得的5,488 keV的γ射线的产率[aa](γ射线/秒)
材料	质量(g)	表面积(cm^2)			电流(mA)	典型电压	时长(h)	
钛箔	0.089	2.0	琼斯*	金箔	320	5	42	–0.022 ± 0.011
钛箔	2.2	32	琼斯	金箔	1,000	4	12	–0.015 ± 0.034
钛箔	2.2	32	琼斯	金箔	1,000	4	6	0.012 ± 0.007
钛颗粒	25		琼斯	金箔	710	13	8	0.032 ± 0.036
钛颗粒	25		0.1 M H_2SO_4	金箔	400	20	25	–0.019 ± 0.018
钛颗粒	25		0.1 M H_2SO_4	金箔	765	16	16	0.017 ± 0.022
UPt_3颗粒	42		0.1 M LiOH	铂箔	300	31	166	$(3.1 \pm 6.0) \times 10^{-3}$

a 弗莱施曼提供的电解池。使用之后分析了阴极的氢和氘——结果 H/Pd 为 0.01、0.02，D/Pd 为 0.84、0.72。

b 阴极绕成一个紧密的螺旋状。初始阶段在大型中子检测器中进行监测，没有滑梭，持续了 147 小时，估算的检测限为 1 n · s^{-1}，之后的 5 小时检测限为 0.2 n · s^{-1}。再次使用之前阴极的表面用硫擦过。

c 300 小时后用砂纸清洁阴极。

d 初始阶段阴极在大的中子检测器中进行监测，没有滑梭，在 0.1 M LiOD 中进行了 183 小时；估算的检测限为 1 n · s^{-1}。使用之前用 400 目的金刚砂纸打磨。

e 管材来自托卡马克装置中的氘存储系统。

f 使用后分析了阴极中的氢和氘——结果：H/Pd 为 0.03，D/Pd 为 0.80。

g 上面实验中的电解液重复使用。

h 上面实验中的电解液和阴极重复使用；Na_2S 以高浓度水溶液的形式加入。

i 使用前在 1,000 ℃真空下对箔进行了除气。使用后分析了氢和氘——结果：H/Pd 为 0.02，D/Pd 为 0.83。

j 使用之前将箔浸入 Na_2S（浓溶液）中。使用后分析了氢和氘——结果：H/Pd 为 0.05，D/Pd 为 0.80。

k 将阴极切掉一半，使用后分析了氢和氘——结果：H/Pd 为 0.02、0.02，D/Pd 为 0.85、0.78。

l 熔体快淬带，由庄信万丰工程中心提供。

m 电弧熔化两次；在 0.1 M LiOD 溶液中电化学充氘一个月，然后在液氮中冷冻，浸入浓的 Na_2S 溶液并转移到中子计数电解池。

n 包括点焊在钯丝上的四个厚度为 1~2mm、不同型号的圆形钯片（庄信万丰公司）以及受到应力的钯丝。

o 之前实验的电解液重复使用，使用钛阴极和铂阳极。

p 使用之前将材料加热到 900 ℃然后在水中淬火。

q 将琼斯使用的电解液中的 AuCN 用 $NaAuCl_4$ 替代。

r 电流开启 36 小时，然后每小时在 200 mA 和 20 mA 之间变换一次。

s 电流开启 36 小时，然后每 5 分钟在 300 mA 和 30 mA 之间变换一次。

t 由于电解期间阳极的分解，电流在所示范围内缓慢变化。

u 电流开启 36 小时，然后关闭 10 小时、开启 2 小时并循环这一过程 270 小时。然后余下的时间中 5,000 秒为 380 mA 的阴极电流，4,000 秒为 10 mA 的阳极电流。

v 电流每 2 小时开启或者关闭，依次循环。中子产率指的是“开启”期间的。

w 给定的误差（1σ，由整个实验期间计算而来）多少会有些变化，因为为了涵盖尽可能多的电解池的设置（主要是在规划的早期阶段），并不总是能够成对地开展未通电 + 未通电和通电 + 未通电的实验，从而使得整个宇宙射线中子的探测效率存在少许差异，这是由于两个形式上相同的电解池存在少许差异造成的。

x 采集这些数据时使用了一个不同的数据采集系统驱动滑梭和累计数据，旨在探寻猝发中子 [10]。

y 电解池每过 5 分钟进行手动交换。

z 进行 γ 射线研究的所有电解液中 H_2O : D_2O 都为 50 : 50，标 * 的 H_2O : D_2O 为 41 : 59。

aa 当前测量中，用一系列标准的 γ 射线源和一个放射出 6,129 keV γ 射线的 $^{238}Pu/^{13}C$（α，nγ）源对系统进行了校准。放置各个电解池时它们的阴极应尽可能地靠近探测晶体，探测效率通过将之前测量 [21] 的点效率函数对阴极体积进行积分计算出来。在预期的全能峰、单逃逸峰和双逃逸峰的位置 (5,488 keV、4,977 keV 和 4,466 keV) 处寻找峰，这些位置是用参考文献 22 中推荐的方法预测的。

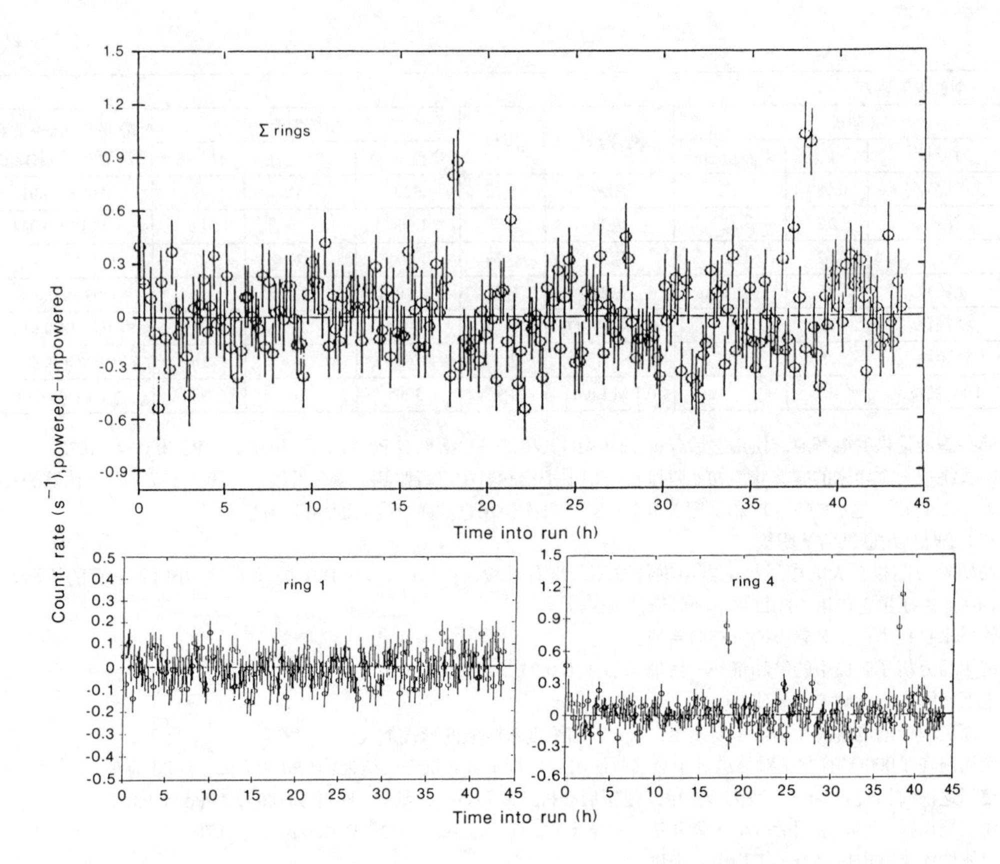

Fig. 3. Data from a typical run on the large high-efficiency neutron detector. These show the differences between count rates for powered and unpowered cells resulting from successive alternate shuttle positions for 40 g of titanium granules in 0.1 M D_2SO_4. The errors shown are 1σ errors for each five-minute counting period. In addition to results from the detector as a whole, results from two of the five rings of 2-inch diameter 107-cm active length $^{10}BF_3$ proportional counters are also shown. The counts in rings 2, 3 and 5 (not shown) were very similar to those in ring 1. The apparent bursts were seen only in ring 4 and are therefore a spurious effect. A tube (90-mm inner diameter) passes through the centre of the detector, and the shielding consists of 6 inches of borated resin, 1 mm of cadmium and 2 inches of lead. The neutron detection efficiency is high and is largely independent of neutron energy, varying from 48% for 0.5-MeV (Am/Li) neutrons to 40% for 4.2-MeV (Am/Be) neutrons. The background count rate is 4–5 count s^{-1}, and most neutrons are from cosmic rays. Each of the five rings of counters has its own independent pre-amplifier, pulse-shaping amplifier and discriminator, and the mean energy of neutrons counted can be obtained from the ratio of counts in the outermost ring of counters (ring 5) to counts in the innermost (ring 1). Because neutron counts are distributed over five rings, and because of the 135-μs-mean time to capture, neutrons emitted simultaneously from a source as a burst are counted separately (as seen with the ^{252}Cf source, for example). The pre-amplifiers, high voltage components and insulators are all contained within a desiccated electrically screened box, and we eliminate earth loops and induced electromagnetic pick-up ("aerial") effects from external coaxialcable runs to the PDP-11/45 data-acquisition computer by using isolating high-frequency pulse transformers and by winding the coaxial cables many times around ferrite rings. The neutron detector and its immediately associated electronics are located in a temperature-controlled air-conditioned blockhouse with two-foot-thick concrete walls and roof.

The details of the cells and results of the measurements are given in Table 2. The lowest limits (2σ) on neutron emission derivable from Table 2 for palladium are 1.5×10^{-2} n $s^{-1}g^{-1}$

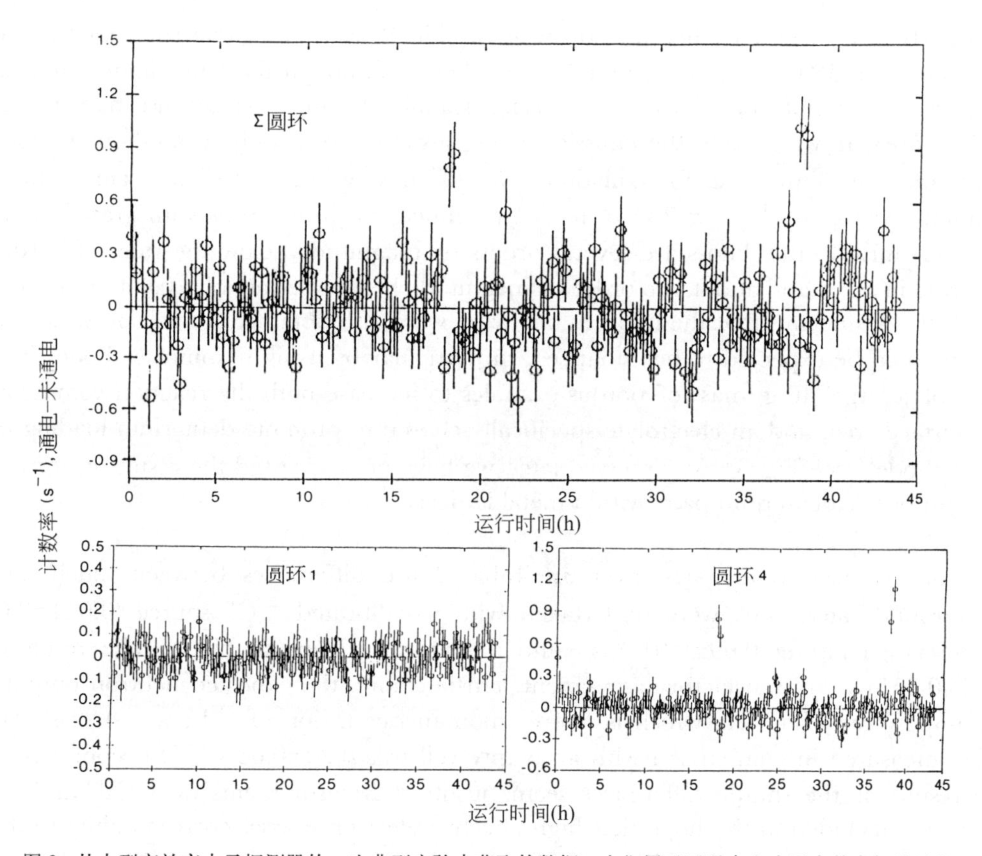

图 3. 从大型高效率中子探测器的一次典型实验中获取的数据。它们展示了通电和未通电的电解池计数率的差异，这是通过持续地变换滑梭位置实现的，具有 0.1 M D_2SO_4 的电解池中含有 40 g 钛颗粒。显示的误差为每 5 分钟计数区间的 1σ 误差。除了将探测器作为一个整体得到的结果之外，还显示了五个圆环中的两个获得的结果，组成圆环的 $^{10}BF_3$ 正比计数器直径为 2 in，活性长度为 107 cm。圆环 2、3 和 5（未显示）的计数与圆环 1 的非常相似。只在圆环 4 中观察到了明显的猝发，因此这是一个假信号。一只管（内直径 90 mm）穿过探测器的中心，屏蔽设备包括 6 in 的硼酸树脂、1 mm 的镉和 2 in 的铅。中子探测效率比较高，并且与中子能量无关，在 0.5 MeV 中子（Am/Li）的 48% 到 4.2 MeV 中子 (Am/Be) 的 40% 之间变化。本底计数率为每秒 4~5 个计数，大部分中子来自宇宙射线。五个圆环计数器中的每一个都有自己独立的前置放大器、脉冲形状放大器和甄别器，计数的中子的平均能量通过最外侧的环形计数器（环 5）的计数与最内侧的（环 1）计数之比得到。因为中子计数分布于五个环上，也因为平均捕获时间为 135 微秒，从一个源中同时猝发的多个中子可以被分别计数（例如，就像 ^{252}Cf 源一样）。前置放大器、高压部件和绝缘体都被囊括在一个干燥的电屏蔽箱中，我们消除了运行于 PDP-11/45 型数据采集计算机上同轴电缆的接地回路和诱发的电磁捕获（“天线”）效应，这是通过使用隔离的高频脉冲转换器并将同轴电缆在铁氧体环上缠绕多次实现的。中子探测器和与之相连的电子器件被置于一个控温的空调房中，这个空调房有两英尺厚的混凝土墙壁和屋顶。

电解池细节和测量结果列在表 2 中。如果在整个运行过程中中子持续发射，则从表 2 中推导出钯的中子发射的最低限（2σ）为 $1.5\times10^{-2}\,n\cdot s^{-1}\cdot g^{-1}$ 或

or 1.5×10^{-2} n s^{-1} cm^{-2}, if the emission were sustained over the whole run. For titanium, the values are 3×10^{-3} n s^{-1} g^{-1} and 4×10^{-3} n s^{-1} cm^{-2}, although the latter limit is too high because it does not apply to the runs using granules having a large and indeterminate surface area. If we assume the emission to be sustained over only a one-hour period at most, then the limits are, for palladium: 7×10^{-2} n s^{-1} g^{-1} or 4×10^{-2} n s^{-1} cm^{-2}, and for titanium: 8×10^{-3} n s^{-1} g^{-1} or 2×10^{-2} n s^{-1} cm^{-2} (much less in the runs with granules). The neutron-emission rate limits are several orders of magnitude below the rates of $\sim10^4$ s^{-1} reported in ref. 1 and about one order of magnitude below the rate of ~0.4 s^{-1} reported in ref. 2. It is significant that the limits in our work were also obtained for cells in which cold fusion could be expected to be enhanced, in particular by using titanium cathodes in the form of a large (40 g) mass of porous granules to increase both the reaction volume and the surface area, and an electrolyte specifically chosen to promote deuterium loading into the cathode[13] (see Fig. 3). As discussed later, we have not expressed the results in terms of the number of deuterium pairs within metal lattices.

Because the net count rates given in Table 2 are differences between much larger background rates, runs were undertaken with a calibrated ^{252}Cf source (0.024±0.003 fissions s^{-1}) emitting 0.09±0.01 n s^{-1} and a physically identical blank. The result was 0.14±0.05 n s^{-1}, in satisfactory agreement. Furthermore, the expected neutron output of the unpowered UPt_3 cell, which is due to spontaneous fission of ^{238}U, was 0.22 s^{-1}. The value measured in comparison with an empty cell was 0.20±0.06 s^{-1}. Our confidence in the results of the shuttle differences seems justified. Measurements on $CeAl_2$ and UPt_3 cells were included in the hope that high effective electron masses corresponding to these metal crystal lattices would mimic in some way the fusion enhancement effect of "heavy electrons" such as muons in binding the deuterium nuclei closer together. The palladium ribbon run with the cell power switched on and off every two hours (which was about eight times the characteristic diffusion time in the ribbon) was undertaken to enhance the appearance of non-equilibrium effects, and results (negative) are given in Fig. 4.

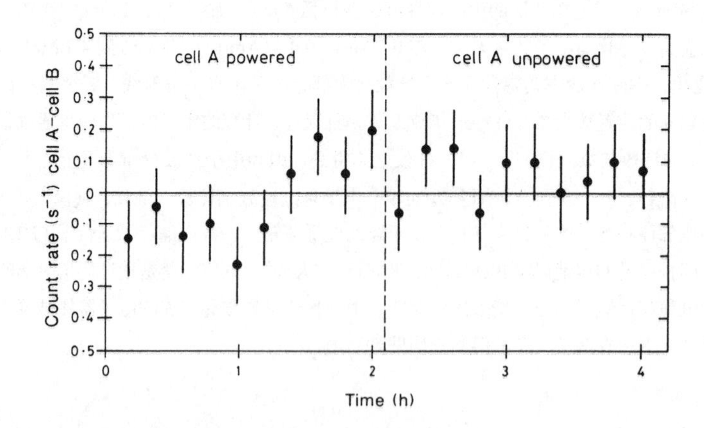

Fig. 4. Time dependence of neutron emission from an electrolytic cell with melt-spun Pd ribbon cathode and Pd foil anode, with power alternately turned on for deuterium loading and off for relaxation every 2 h by the data-acquisition computer: see Table 2 footnote *l*. The total length of the run was 110 h, and the sums of the "on" and "off" data are shown. After 90 h of running, we added 1 μM Pb^{2+} (to poison

$1.5 \times 10^{-2}\ n \cdot s^{-1} \cdot cm^{-2}$。就钛而言，这个数值为 $3 \times 10^{-3}\ n \cdot s^{-1} \cdot g^{-1}$ 或 $4 \times 10^{-3}\ n \cdot s^{-1} \cdot cm^{-2}$，然而后面的限值太高了，因为它并不适用于所用颗粒表面积大而不确定的实验。如果我们假设发射至多持续了一个小时的时间，那么限值为，钯：$7 \times 10^{-2}\ n \cdot s^{-1} \cdot g^{-1}$ 或 $4 \times 10^{-2}\ n \cdot s^{-1} \cdot cm^{-2}$，钛：$8 \times 10^{-3}\ n \cdot s^{-1} \cdot g^{-1}$ 或 $2 \times 10^{-2}\ n \cdot s^{-1} \cdot cm^{-2}$（比用小颗粒的实验小得多）。中子发射速率限值比参考文献 1 报告的约 $10^4\ s^{-1}$ 低数个数量级，比参考文献 2 报告的约 $0.4\ s^{-1}$ 低一个数量级。值得注意的是，我们研究的限值也可以在那些冷核聚变可能会增强的电解池中得到，尤其是使用钛阴极，以大质量（40 g）的多孔颗粒来提高反应体积和表面积，而且特意选择能够促使氘吸附到阴极的电解液[13]（见图 3）。正如后面讨论的，我们没有展示氘在金属晶格中数目的结果。

由于表 2 给出的纯计数率是大得多的本底速率之间的差值，我们使用发射速率为 $0.09 \pm 0.01\ n \cdot s^{-1}$ 的校准过的 ^{252}Cf 源（每秒 0.024 ± 0.003 次裂变）和一个物理上一样的空白进行了实验。结果为 $0.14 \pm 0.05\ n \cdot s^{-1}$，其一致性令人满意。而且，由于 ^{238}U 的自发裂变，不通电的 UPt_3 电解池预期的中子输出为 $0.22\ s^{-1}$。与一个空电解池对比后测量值为 $0.20 \pm 0.06\ s^{-1}$。我们对滑梭差异这一结果的信心似乎得到了验证。在 $CeAl_2$ 和 UPt_3 电解池中进行了测试，以期源自那些金属晶格的大的有效电子质量可以以某种方式模拟“重电子”对核聚变的增强效应，就像 μ 子将两个氘核束缚得非常接近时一样。将钯带实验电解池的电源开关每隔两个小时（约八倍于带中的特征扩散时间）转换一次，以增强非平衡效应的表现，结果（负结果）如图 4 所示。

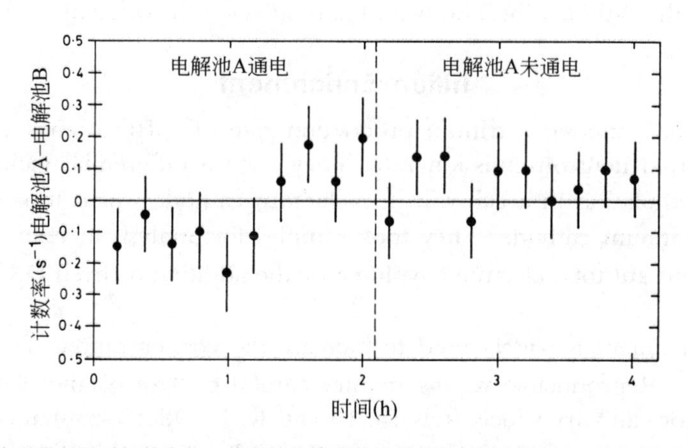

图 4. 电解池中子发射随时间的变化，使用熔体快淬的钯带状阴极和钯箔阳极，以两个小时为周期，轮换着打开电源使氘吸附、关上电源使之弛豫，这通过数据采集的计算机来实现：见表 2 的注脚 *l*。总的运行时长为 110 小时，“打开”和“关闭”数据的总和如图所示。运行 90 小时以后，我们向溶液中添加了 1 μM 的 Pb^{2+} 离子（通过铅的沉积使表面中毒）。有或者没有 Pb^{2+} 离子的数据之间没有差异，所

the surface by Pb deposition) to the solution. There was no difference between the data in the presence and absence of Pb^{2+}, and so data for the whole run was combined. This run was undertaken to enhance the appearance of the non-equilibrium effects discussed in the text. It is assumed that the ribbon would have had a high density of grain boundaries, dislocations and other lattice defects. Because the "on" and "off" periods were several times the characteristic diffusion time for the ribbon, it is assumed that the composition was cycling in the β-phase region between the fully loaded condition (D/Pd ≈ 0.83) and the limit of the (α + β) phase field (D/Pd ≈ 0.65).

γ-ray Counting

As it has been shown that cold proton-deuteron (p–d) fusion is expected to proceed at rates greater by ~8.5 orders of magnitude than d–d fusion[3], we carried out an alternative investigation of electrolytic enhancement of hydrogen-isotope fusion by looking for the $D(p, \gamma)^3He$ 5,488-keV γ-rays from p–d fusion. We used a lead-shielded 113-cm^3 n-type high-purity germanium (HPGe) crystal γ-ray spectrometer to search for any γ-rays in the energy range 0.1–7 MeV emitted by a variety of cold-fusion cells operating with a mixture of light and heavy water. The results, also given in Table 2, are consistent with no γ-ray emission. The lowest limits (2σ) on γ-emission derivable from Table 2 are, for palladium, $7\times10^{-3}\ \gamma\ s^{-1}\ g^{-1}$ or $1\times10^{-3}\ \gamma\ s^{-1}\ cm^{-2}$, and for titanium $1.4\times10^{-3}\ \gamma\ s^{-1}\ g^{-1}$ or $4\times10^{-4}\ \gamma\ s^{-1}\ cm^{-2}$, if we assume that the emission is sustained over the whole run. Jones *et al.*[2] report that the fusion activity may last for only 4–8 h and begin ~1 h after the cell is powered, and in these circumstances typical values for the standard deviation in the γ-ray emission rates are ~0.05 s^{-1} and the 2σ limits for titanium are $4\times10^{-3}\ \gamma\ s^{-1}\ g^{-1}$ or $3\times10^{-3}\ \gamma\ s^{-1}\ cm^{-2}$. Post-run analysis of the hydrogen isotopes taken up by the palladium cathodes with the mixed light and heavy-water electrolytes used gave very satisfactory D:H ratios, lying in the range 1.5–2.5. We note that if the proposed enhancement of the fusion process is as valid for p–d as for d–d fusion then, because of the enhanced tunnelling in the lighter d–p system[3], the γ-ray measurements actually provide a much more stringent limit on the cold fusion process than do the neutron measurements. The interpretation of the results from some of these cells was complicated by the dissolution of gold anodes and the consequent gold deposition onto the cathode: this also would have affected the original work[2].

Tritium Enrichment

Fleischmann *et al.*[1] claimed a tritium production rate of ~10^4 atom s^{-1}, commensurate with their reported neutron emission rate. They used a differential technique in which the tritium accumulation in a cell with a palladium cathode was compared with that in a cell with a platinum cathode. They took samples for analysis at regular intervals and maintained a constant total electrolyte volume by the addition of fresh D_2O.

Electrolytic enrichment is widely used to increase the concentration of tritium in water before analysis[14]. Reproducible results require careful control of the electrolysis, as the enrichment factor can vary widely (refs 15, 16 and R. L. Otlet, personal communication): important effects are seen with change of electrode materials, with variation in the condition (activity) of the electrode surface, with the current density (overvoltage) and with the temperature. Without precautions, the variations in enrichment factor can be more than a factor of two[16]. The claims therefore need to be assessed against this known

以将整个运行期间的数据都结合起来了。这次运行目的在于增强文中讨论的非平衡效应的表现。假设钯带具有高密度的晶界、位错和其他的晶格缺陷。因为“打开”和“关闭”的区间是钯带特征扩散时间的几倍，所以假设钯带的组成在完全充氘的条件（$D/Pd \approx 0.83$）和（$\alpha + \beta$）相区域的限制值（$D/Pd \approx 0.65$）之间的 β 相区域循环。

γ 射线计数

由于已经表明质子－氘核冷核聚变预计进行的速率比氘核－氘核核聚变要快约 8.5 个数量级[3]，我们通过寻找质子－氘核核聚变产生的 $D(p, \gamma)^3He$ 5,488 keV 的 γ 射线对氢－同位素核聚变的电解增强进行了研究。我们使用了一个 113 cm^3 的铅防护 n 型高纯锗晶体 γ 射线谱仪，在内有轻水和重水混合物的多种冷核聚变电解池中探寻一切能量范围在 0.1~7 MeV 的 γ 射线。结果也在表 2 中列出，它们一致表明没有 γ 射线放出。如果我们假设放射在整个运行过程中得以维持的话，那么由表 2 中可获得的 γ 放射值的最低限（2σ），钯为 $7 \times 10^{-3}\ \gamma \cdot s^{-1} \cdot g^{-1}$ 或 $1 \times 10^{-3}\ \gamma \cdot s^{-1} \cdot cm^{-2}$，钛为 $1.4 \times 10^{-3}\ \gamma \cdot s^{-1} \cdot g^{-1}$ 或 $4 \times 10^{-4}\ \gamma \cdot s^{-1} \cdot cm^{-2}$。琼斯等人[2]报道核聚变的活性仅可保持 4 ~ 8 小时，而且始于电解开始后约 1 小时，在这样的条件下 γ 射线放射速率的标准偏差典型值约为 0.05 s^{-1} 而 2σ 值对钛而言为 $4 \times 10^{-3}\ \gamma \cdot s^{-1} \cdot g^{-1}$ 或 $3 \times 10^{-3}\ \gamma \cdot s^{-1} \cdot cm^{-2}$。运行后对钯阴极的吸附物及所用的轻水和重水混合物电解液的氢同位素进行分析，给出了令人满意的氘和氢的比值，在 1.5~2.5 的范围内。我们注意到，如果假定的核聚变过程的增强对质子－氘核核聚变像对氘核－氘核核聚变一样有效的话，那么，因为在更轻的氘核－质子核系统中的隧道效应增强[3]，γ 射线测量对冷核聚变过程实际能提供比中子测量更加严格的限值。由于金阳极上有金的溶解且在阴极上有相应的金的沉积，对从一些电解池获得的结果的解释变得较为复杂：这也影响了最初的工作[2]。

氚 的 富 集

弗莱施曼等人[1]宣称氚产生速率约为每秒 10^4 个原子，与他们报道的中子发射速率相等。他们使用了一种差分技术，这种技术是将钯阴极电解池中氚的累积与另一个铂阴极电解池中氚的累积进行比较。他们以规定的时间间隔取样分析，同时添加新的重水以保持总电解液体积不变。

电解富集被广泛用于分析前增大水中氚的浓度[14]。要实现可重复的结果需要仔细控制电解，因为富集因子变化很大（参考文献 15、16；奥特莱，个人交流）：改变电极材料、电极表面状态（活性）、电流密度（过电压）和温度的改变都会造成巨大的影响。如果不加注意，富集因子的变化可以超过两倍[16]。因此只有考虑这个已知的可变电解富集因素才能对弗莱施曼等人的宣称作出评估。令人惊异的是，弗莱

variable electrolytic enrichment. Surprisingly, Fleischmann *et al.*[1] claimed that there was no electrolytic enrichment in their platinum cell.

If the rates of electrolytic evolution of hydrogen isotopes are written as follows

$$D_2O \rightarrow D_2 + \tfrac{1}{2}O_2 \qquad \text{rate} = r$$

$$DTO \rightarrow DT + \tfrac{1}{2}O_2 \qquad \text{rate} = XSr$$

where r denotes the electrolysis rate, S the tritium-deuterium separation factor and X the mole fraction of tritium, T, in the solution, then the tritium accumulation is given by[10]

$$\frac{X}{X_0} = \frac{1}{(S+\alpha+\gamma)}\left\{1+\alpha+\gamma-(1-S)\exp\left[-\frac{(S+\alpha+\gamma)rt}{N_0}\right]\right\} \tag{1}$$

where the total solution volume is maintained constant by the addition of fresh D_2O (containing a mole fraction X_0 of tritium, directly proportional to the disintegration rate), the ratio of the sampling rate of the solution to the electrolysis rate is α and γ denotes the ratio of the rate of evaporation of the solution to the electrolysis rate—a small correction that may be calculated assuming that the electrolysis gases passing out of the cell are saturated with water vapour. On the timescales of interest, variability in S would give a variable enrichment

$$\delta\left(\frac{X}{X_0}\right) = -\frac{rt}{N_0}\delta S \tag{2}$$

If the solution becomes contaminated by hydrogen absorption from the atmosphere, then if S_{obs} denotes the apparent DT enrichment factor derived from the application of equation (1), $S_{obs} = S/[0.5(1+f)]$, where $f = (1-X_H)/(1+X_H(S_{HD}-1))$ with X_H denoting the mole fraction of hydrogen and S_{HD} the HD separation factor (relative rate of reaction of HDO and D_2O). Therefore, as well as the inherent variability from one electrode to another, it is evident that any variability in the amount of hydrogen pickup will give a variation in the apparent enrichment factor:

$$\frac{dS_{obs}}{dX_H} = \frac{S_{HD}}{2[1+X_H(S_{HD}-1)^2]}$$

We found that, unless exceptional precautions were taken (and we believe that we used experimental procedures very similar to those used in ref. 1 (M. Fleischmann, personal communication)), values of $X_H \approx 0.07$ were common. Under these conditions, a change of X_H of only 0.01 would give a change in S_{obs} of 0.02, which could be significant given the smallness of the claimed effect.

Any assessment of whether differential enrichment can be considered to account for the results in ref. 1 depends critically on the value of X_0, which was not reported. Using values[1] of r (1.24×10^{18} atom s^{-1}) and N_0 (14.6×10^{23} atom) we calculate from equation (2)

施曼等人[1]宣称他们的铂电解池中没有电解富集。

如果氢同位素的电解过程速率表述如下

$$D_2O \rightarrow D_2 + \tfrac{1}{2}O_2 \qquad \text{反应速率为 } r$$
$$DTO \rightarrow DT + \tfrac{1}{2}O_2 \qquad \text{反应速率为 } XSr$$

其中 r 代表电解反应速率，S 为氚 – 氘分配系数，X 为氚（T）在溶液中的摩尔分数，那么氚的累积可以表示为[10]

$$\frac{X}{X_0}=\frac{1}{(S+\alpha+\gamma)}\left\{1+\alpha+\gamma-(1-S)\exp\left[-\frac{(S+\alpha+\gamma)rt}{N_0}\right]\right\} \tag{1}$$

其中通过添加新的重水（包含一定量摩尔分数为 X_0 的氚，与衰变速率成正比）保持溶液总体积不变，溶液取样速率与电解速率之比为 α，γ 代表溶液蒸发的速率与电解速率之比——需要进行一个小的纠正，可以假设从电解池中溢出的电解气体在水蒸气中是饱和的。在目标时间尺度上，S 的变化可以给出富集情况的变化

$$\delta\left(\frac{X}{X_0}\right)=-\frac{rt}{N_0}\delta S \tag{2}$$

如果溶液被大气中吸附而来的氢污染了，那么如果 S_{obs} 代表由方程（1）推导的表观 DT 富集系数，则 $S_{obs}=S/[0.5(1+f)]$，其中 $f=(1-X_H)/(1+X_H(S_{HD}-1))$，$X_H$ 代表氢的摩尔分数，S_{HD} 代表 HD 的分配系数（HDO 和 D_2O 反应的相对速率）。因此，除了一个电极到另一个电极的固有变化之外，很显然任何吸附氢的量变化都将会表现出表观富集系数的变化：

$$\frac{dS_{obs}}{dX_H}=\frac{S_{HD}}{2[1+X_H(S_{HD}-1)^2]}$$

我们发现除非采取非常严格的预防措施（而且我们相信我们采用的实验步骤与参考文献 1 中所用的是非常相似的（弗莱施曼，个人交流）），$X_H\approx 0.07$ 这样的值是常见的。在这些条件下，X_H 仅 0.01 的变化将会使 S_{obs} 变化 0.02，鉴于所宣称的效应很小，这个变化是非常显著的。

关于差分富集是否能够用来说明参考文献 1 的结果，任何这样的评价关键取决于 X_0 的值，这个值未曾被报告过。使用 r（1.24×10^{18} 原子 / 秒）和 N_0（14.6×10^{23} 个原子）的值[1]，我们从方程（2）计算出，如果 X_0 处于商用 D_2O 范围的极低端

that if X_0 were at the extreme low end of the range for commercial D_2O (3 Bq ml^{-1}), a value of $\delta S = 0.46$ would be required; if it were moderate (10–15 Bq ml^{-1}) a value $\delta S \approx 0.1$ would be needed; if it were high (80 Bq ml^{-1}) $\delta S \approx 0.02$ would suffice.

The applicability of equation (1) was confirmed experimentally, on both platinum and palladium cathodes, in conjunction with calorimetric and neutron-counting experiments (see Fig. 5 legend) using D_2O with an initial tritium content of 13 kBq ml^{-1} (efficiency corrected). The fit of this equation to all the experimental data (Fig. 5) gave, for palladium, $S_{obs} = 0.59$, and for platinum, $S_{obs} = 0.61$. Correction for the uptake of hydrogen, using $S_{HD} = 6$ (ref. 16), gave $S \approx 0.48$, in agreement with previous work and theoretical expectations ($S = 0.46 \pm 0.02$, (refs 14, 15 and D. S. Rawson and R. L. Otlet, personal communication)). For individual electrodes, S_{obs} for palladium varied from 0.46 to 0.65 (corrected S from 0.42 to 0.58), whereas S_{obs} for platinum varied from 0.56 to 0.85 (corrected S from 0.43 to 0.69)—a total range for individual electrodes of $\delta S_{obs} = 0.4$, which is enough to account for the results of Fleischmann *et al.*[1].

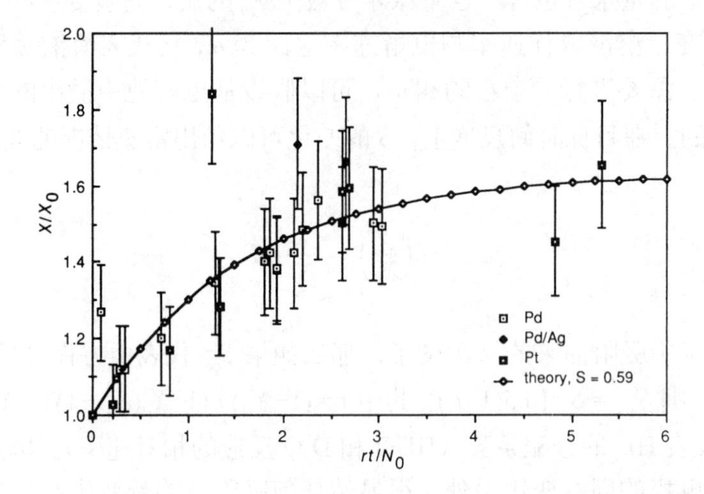

Fig. 5. Tritium enrichment by electrolysis in open cells at constant volume, with both Pd and Pt cathodes: relative count rate X/X_0 against amount of electrolysis rt/N_0 (symbols defined in the text). Initial count rate (efficiency corrected) was 13.0 ± 0.6 kBq ml^{-1}. Errors are 1σ. The Pd cells were the FPH calorimeters, sampled after 490–660 h, the cells in the first section of Table 2 (footnotes *a–e*) sampled at the end of the run and one cell from the high-efficiency neutron detector—footnote *k* in Table 2. The line is the fit to equation (1).

It is clear from these results and discussion that more evidence needs to be presented before the tritium accumulation reported[1] can be considered as experimentally reliable evidence for the occurrence of a fusion process.

Materials Characterization

With 125-μm palladium foils, hydrogen loadings (determined[10] by hot-extraction mass spectrometry of specimens frozen in liquid nitrogen, and independently by electrochemical extraction) of $H/Pd = 0.95 \pm 0.05$ were achieved in ~1 h at 100 mA cm^{-2} in 0.1 M

(3 Bq · ml^{-1})，需要 $\delta S = 0.46$；如果它属于中等（10~15 Bq · ml^{-1}），需要 $\delta S \approx 0.1$；如果它比较高（80 Bq · ml^{-1}），$\delta S \approx 0.02$ 就足够了。

方程（1）在铂和钯阴极上的适用性都经过了实验的验证，是同量热和中子计数实验联合进行的（见图 5 的图注），使用氚初始含量为 13 kBq · ml^{-1} 的 D_2O（效率经过了校正）。用这个方程拟合所有给出的实验数据（图 5），对钯而言，$S_{obs} = 0.59$，对于铂，$S_{obs} = 0.61$。对氢吸附量做校正，采用 $S_{HD} = 6$（参考文献 16），得出 $S \approx 0.48$，与之前的工作和理论预期一致（$S = 0.46 \pm 0.02$，（参考文献 14、15；罗森和奥特莱，个人交流））。对于单独的电极，钯的 S_{obs} 在 0.46 到 0.65 之间变化（校正后在 0.42 到 0.58 之间），而铂的 S_{obs} 在 0.56 到 0.85 之间变化（校正后在 0.43 到 0.69 之间）——对于单独电极的总范围 $\delta S_{obs} = 0.4$，这足以对弗莱施曼等人的结果[1]给予解释。

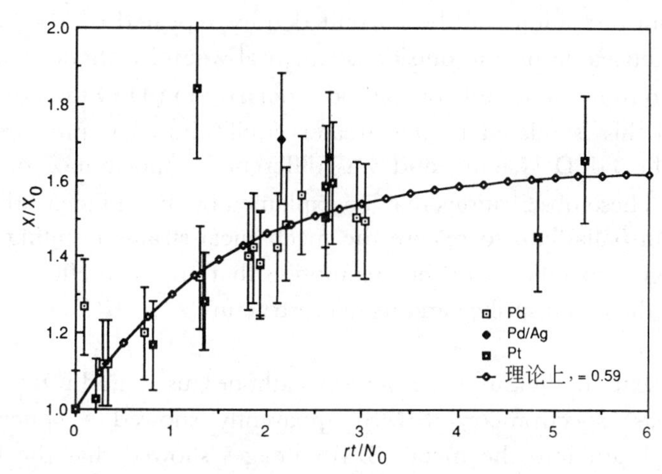

图 5. 一定体积的开放电解池中氚的电解富集，使用钯和铂阳极：相对计数率 X/X_0 相对于电解量 rt/N_0 的关系（符号在文中进行了定义）。初始的计数率（经过了效率校正）为 13.0 ± 0.6 kBq · ml^{-1}。误差为 1σ。钯电解池是 FPH 量热计，490~660 小时后取样，表 2 中第一部分的电解池（脚注 *a~e*）在运行结束的时候取样，一个电解池通过高效中子探测器取样——表 2 的脚注 *k*。曲线是对方程（1）的拟合。

从这些结果和讨论中可以很清楚地看到，要证明所报告的氚的累积[1]是发生核聚变的可靠的实验证据，还需要更多的证据。

材料表征

使用 125 μm 厚的钯箔，在 0.1 M 的 LiOH 溶液中以铂为阳极、电流密度为 100 mA · cm^{-2} 的条件下，充氢率（通过冷冻在液氮中样品的热萃取质谱和独立的

LiOH solution with a platinum anode. The limit of deuterium loading in 0.1 M LiOD was lower (0.84±0.03). Rods (1- and 2-mm diameter) polarized for extended periods in neutron counting and calorimetry experiments showed D/Pd = 0.76±0.06. It is well known[17] that the equilibrium pressure for a given deuterium loading is higher than that for the same hydrogen loading. Current-interruption methods confirmed that overpotentials in the range 0.8–1 V (ref. 1) were being obtained. From an initial composition >99.9% D_2O, the solutions degraded to ~98.5% in 24 h and to 88–98% (analysis by infrared spectrometry) following electrolysis for many weeks. The resulting ratio H/D in the palladium was 0.02–0.04.

During electrolysis, all of the palladium cathodes became covered with a layer that varied in appearance from a dull tarnish to a dense jet black. In the latter case, loose black material was also formed, which in extreme cases came off during electrolysis, resulting in quite heavy erosion of the cathode. The layer itself evidently represented a modification of the morphology of the cathode at the surface. The formation of a thick black layer was enhanced at high current density, at high temperature, on smaller diameter wire and by frequent abrupt alterations of the current density repeated over a long period. The layer was more noticeable on the outside of a spiral-wound cathode than on the inside. The layer was also more noticeable on cathodes polarized in D_2O than on those polarized in H_2O (perhaps this is related to the greater equilibrium gas pressure for equivalent composition in the Pd–D system), and was different in appearance on materials from different sources. These observations may be explained by the old idea[17] that microfissures, or rifts, develop in palladium to release the mechanical strains resulting from the heavy loading of hydrogen, together with the assumption that any such effect would depend on the stress state of the metal surface and its microstructure.

Lithium was present in the surface layer on cathodes used in LiOD, and analysis by secondary-ion mass spectrometry (SIMS) apparently showed a concentration profile extending about 1 μm into the metal. SIMS images showed that the lithium was not uniformly distributed, however, and it seemed likely that it was trapped in microfissures in the surface layer.

Surface analysis showed a number of other species on and in the surface layer, notably small quantities of platinum and traces of copper, zinc, iron, lead and silicon: platinum would have originated from the anode and silicon from the glass container. No doubt the majority of the other contamination would have come from the solution: the levels found (a few atom percent, confined to the surface layers) were consistent with deposition by extended electrolysis from a solution of concentration around 10^{-9}–10^{-10} M. One possible criticism is that low levels of such deposition could poison any essential catalytic activity: we therefore used pre-electrolysis in several experiments, in an attempt to lower the surface contamination. We either made repeat experiments on the same solution, simply changing the cathode, or treated the solution beforehand in a separate cell.

Because of claims that an unusual mechanism might lead to nuclear reactions proceeding predominantly non-radiatively to ^4He, four cathodes (Table 2, first four entries) were

电化学萃取获得[10]）达到 H/Pd = 0.95±0.05 约耗时 1 小时。在 0.1 M LiOD 中充氘率的上限更低（0.84±0.03）。在中子计数和量热实验中极化时间更长的钯棒（直径 1 mm 和 2 mm）显示 D/Pd = 0.76±0.06。对于给定的充氘率，其平衡压力要高于同样充氢率的平衡压力，这是众所周知的[17]。用中断电流的方法确定了所得到的过电压在 0.8~1 V 的范围内（参考文献 1）。D_2O 从初始的 99.9%，24 小时内降为约 98.5%，继续电解数个星期降为 88%~98%（通过红外光谱进行分析）。导致钯中 H/D 的比值变为 0.02~0.04。

电解期间，所有的钯阴极都会覆盖上一层从暗无光泽到致密黑色这样外观不同的物质。在后面的实例中，也发现过疏松的黑色物质，在一些极端的情况下它会在电解进行期间掉落，结果导致阴极的严重腐蚀。该层本身清楚反映了阴极表面形态的改变。在高电流密度、高温、较小直径的棒上以及在长期频繁的阶跃式改变电流密度的条件下，容易生成厚的黑色层。该层在螺旋状阴极的外侧要比内侧更容易观察到，而且在 D_2O 中极化过的阴极上也比在 H_2O 中极化过的阴极上更容易观察到（这可能与相同含量情况下 Pd–D 体系有更高的平衡气压有关），而且在不同来源的材料上其外观也不相同。这一发现可以用已有的理论[17]给予解释：由于充氢率很高，钯中将出现微裂缝和裂缝以释放掉由此而来的机械应力，同时也有假说认为这种效应依赖于金属表面的应力状态和它的微结构。

在 LiOD 中使用的阴极表面层含有锂，通过二次离子质谱（SIMS）分析表明锂的浓度分布已经深入金属内 1 μm。SIMS 图像显示锂并不是规则分布的，而是看起来像被捕获在表面层中的微裂缝里。

表面分析表明有一定数目的其他元素位于表面层之上或者其中，有极少量的铂，痕量的铜、锌、铁、铅和硅：铂可能来源于阳极，而硅来自玻璃容器。毫无疑问，主要的其他污染物应该来自溶液：我们发现的数量级（按原子数计算是百分之几，限于表面层）与在浓度为 10^{-9}~10^{-10} M 的溶液中长时间电解后的沉积物是一致的。可能有人批评，低水平的沉积物会使潜在的催化活性中毒：因此我们在一些实验中使用了电解预处理，以期减少表面的污染物。我们通过在同一溶液中重复实验，简单地改变阴极，或者事先将溶液在一个分开的电解池中进行处理。

因为有报告声称有一种非同寻常的机制导致核反应进行过程中优先生成非辐射性的 ^4He，我们用真空熔融 / 质谱仪对四个阴极（表 2 中的前 4 组）进行了分析。由

analysed by vacuum fusion/mass spectrometry. The high vapour pressure of palladium at the melting point caused difficulties, so internal standards were prepared by ion implantation of 10^{13} and 10^{15} atoms of ^{4}He into samples cut from the cathodes. Detection limits for ^{3}He and ^{4}He determined in this way were ~8×10^{10} atoms per sample—$(1–10)\times10^{11}$ atom g^{-1}. We found no ^{3}He or ^{4}He. The expected level, if fusion had been occurring at the rate reported in ref. 1, was ~10^{16} atom g^{-1}.

Evolution of hydrogen at a titanium cathode resulted, as is well known[13], in a dense network of hydride precipitates, observable by standard metallographic methods, penetrating below the surface. In 0.1 M D_2SO_4 electrolyte at 100 mA cm^{-2}, the network penetrated ~30 μm in 1 h. Precious-metal deposition inhibited the electrolytic uptake, presumably by lowering the overvoltage and promoting gas evolution. Electrolysis in the "brew" used by Jones *et al.*[2] resulted in hydride precipitates confined to the grain boundaries; other experiments showed that the presence of $PdCl_2$ in the electrolyte caused this effect, presumably as a consequence of palladium plating on the cathode. It is clear that expressions of the fusion rate that assume that the cathode has composition TiD_2 are completely misleading: metallography shows that the number of deuterium pairs is far fewer, hence the claimed fusion rate per deuterium pair is far higher than the figure given, and so the results are even more difficult to reconcile with expectations than had been implied[2].

Discussion

The interest in cold fusion has generated a large number of neutron counting experiments. It is well known that, in general, it is inadvisable to measure the signal + background and background of counting measurements at different times and in different physical locations (as in ref. 1), because of unexpected systematic variations. This is especially true for low-count-rate experiments. Compensating for variation of the background rate and assessing appropriate errors for the procedure chosen are particular problems. The work of Jones *et al.*[2] can be criticized for such errors[18]. It is notable that in ref. 2, only one run in fourteen showed a significant effect and then only because this particular run was assigned a smaller counting error than the others. Here we have attempted to minimize, by the shuttle procedure, uncertainty about background and counter variability and about error calculation. Given some of the more spectacular claims that have been made, we note that further caution is advisable because of the notorious sensitivity of $^{10}BF_3$ and ^{3}He proportional counters to humidity and of counter-amplifier systems to earth loops. In our work the neutron detectors were segmented, and the relationship between signals from the segments was well known, so spurious effects giving inconsistent signals from the segments could be identified.

Failure to reproduce the effects has been attributed by some to the need for rigorous exclusion of hydrogen and claims have been made that palladium electrodes must be cast and carefully degassed before use, to remove all traces of carbon and hydrogen impurity which might decorate dislocations or other high-energy sites (ref. 19 and R. A.

于钯在熔点处的高蒸气压导致实验发生困难，所以内部标样是用离子注入法向阴极上切下的样品注入 10^{13} 和 10^{15} 个 ^4He 原子而得到的。这种确定 ^3He 和 ^4He 的方法的检测限约为每份样品 8×10^{10} 个原子，即每克 $(1\sim10)\times10^{11}$ 个原子。我们没有发现 ^3He 和 ^4He。如果核聚变以参考文献 1 中所报告的速率发生，预期的测量值应约为每克 10^{16} 个原子。

众所周知[13]，氢在钛阴极上析出得到的是一个致密的网状氢化物沉淀，该氢化物可以采用标准的金相学方法进行观察，它会渗入到表面以下。在 0.1 M D_2SO_4 电解液中，100 mA · cm^{-2} 的电流密度下，网状物 1 小时渗入约 30 μm。贵金属的沉积会阻止电解液的吸附，这可能是通过降低过电压和促进气体产生造成的。琼斯等人[2]采用的在“啤酒”（译者注：此处有调侃琼斯等人使用的多种盐混合物溶液的意思）中电解得到的氢化物沉淀局限于颗粒边界；其他实验表明电解液中 $PdCl_2$ 的存在会导致这一效应，可能是钯在阴极电镀的结果。很清楚的是，假设含有化合物 TiD_2 阴极的核聚变速率表达式完全是误导人的：金相学表明成对氘的数目非常少，因此，所声称的每个氘对的核聚变速率远高于给出的数值，所以相比之前的解释[2]，这一结果更难与预期的相符。

讨　论

对冷核聚变的兴趣催生了大量中子计数实验。众所周知，通常情况下，由于不可预期的系统变化，在不同的时间和位置对信号 + 本底和本底的计数测量（如参考文献 1 所为）是不可取的。对于低计数率实验尤为如此。对本底速率变化的补偿和选定过程的误差评估都很成问题。琼斯等人的工作[2]存在这些错误[18]，是应该被批评的。在参考文献 2 中需要注意的是，十四次实验只有一次显示出明显的效应，而且仅仅是因为给这一次特殊的实验赋予了比其他实验小的计数误差。这篇文章中我们试图用滑梭的办法来减小本底和计数器的变化以及误差计算的不确定性。对一些更加惊人的声称进行思考时，我们注意到，因为人所共知的 $^{10}BF_3$ 和 ^3He 正比计数器对湿度的敏感性，以及计数器放大器系统对接地回路的敏感性，因此更加的谨慎是明智的。在我们的研究中，中子计数器被分成几部分，而且从不同部分得来的信号之间的关系是已知的，所以不同部分给出不一致的信号而产生的假信息是可以分辨出来的。

重现这些效应之所以失败，是由于严格除氢的需要造成的，已经有结论指出钯电极在使用前必须经过铸造并小心地除气，以除去所有痕量的碳和杂质氢（译者注：氢在此被看作杂质是相对于氘而言），它们可能存在于位错点或者其他的一些高能位

Huggins, Workshop on Cold Fusion, Santa Fe, May 1989): this is however inconsistent with the postulate of a "fusion" origin for the effect because, at low energies, p–d fusion is expected to be significantly faster than d–d fusion[3]. Furthermore, the original reports[1,2] did not mention any special precautions to exclude atmospheric water vapour apart from careful covering of the electrolytic cells. We followed similar procedures, and found that degradation of the heavy water by exchange with atmospheric moisture occurred quite rapidly. Alternatively, it is claimed that it is essential to maintain a high current density for a considerable period (A. J. Appleby, Workshop on Cold Fusion, Santa Fe, May 1989) although it is not completely clear whether these latter claims are in fact reproductions of the effect reported in ref. 1 or are something different. In our neutron counting experiments, fresh dislocations were introduced by plastic deformation, some counting experiments were carried out at current density as high as 1 A cm^{-2} (Table 2) and in one experiment in the isothermal calorimeter a high current density was maintained for a considerable period (Table 1). Trace deposition of platinum on the cathode, supposedly causing a lowering of the overpotential for deuterium evolution and hence a lowering of the attainable deuterium level in the cathode, has also been suggested as an explanation for irreproducibility (M. Fleischmann, personal communication). We observed no effect when we used palladium anodes in our neutron counting experiments.

Timescales for hydrogen and deuterium loading consistent with the expected diffusion time (x^2/D where the diffusion coefficient, $D = 10^{-7}$ cm^2 s^{-1} and x is the radius or half-thickness of the specimen) were measured[10] and nuclear counting and calorimetric experiments were always conducted over periods much longer than this. Furthermore, in the process of electrolytic loading of palladium, there will clearly be a concentration gradient, a moving phase boundary and the outer atomic layers of the metal will be saturated (possibly supersaturated) with deuterium[17]. It might be expected, therefore, that sufficiently sensitive equipment would detect any fusion process well before the material is completely loaded. Because our neutron detection sensitivity was $\sim 10^5$–10^6 times greater than that of Fleischmann *et al.*[1], it seems unlikely that any greatly enhanced fusion process associated with the absorption of deuterium into palladium, giving rise to neutron emission, is occurring. We are, of course, aware that it is always possible to construct essentially untestable theories involving hypothetical special conditions of the metal or of its surface. Careful characterization of materials for which positive results are claimed is therefore of great importance.

It has been argued that the neutron branch of the d–d reaction might be completely suppressed in favour of the (t + p) branch and it has been further argued (S. Pons, personal communication) that the tritium produced as a result of a nuclear process inside the electrode need not necessarily exchange with the electrolyte and might not therefore be detected. However, the other product of such a nuclear process, a high-energy proton, should be detectable by its interaction with the lattice, including neutron emission: we estimate, knowing the rate of energy loss of the protons and by comparison with (p, n) reaction cross-sections for neighbouring elements, a yield of $\sim 10^{-6}$ neutron per proton, implying a neutron yield in our counting experiments of as much as 10^4 s^{-1} if fusion at the

点上（参考文献 19；哈金斯在 1989 年 5 月圣菲举行的冷核聚变专题讨论会上的报告）：然而这与该效应“核聚变”起源的假定不符，因为在低能量范围，质子－氘核核聚变明显快于氘核－氘核核聚变[3]。而且，最初的报告[1,2]除了小心盖好电解池以外，并没有指出任何特殊预防措施以排除空气中的水蒸气。我们沿用类似的步骤，发现重水与空气中的水汽交换而退化的现象迅速发生。另外，有人指出在相当的一段时期内维持一个高的电流密度是必需的（见阿普尔比在 1989 年 5 月圣菲举行的冷核聚变专题研讨会上的报告），尽管我们并不完全清楚后面的这个结论是对参考文献 1 所报道现象的重现还是有所不同。在我们的中子计数实验中，新形成的位错是通过塑性形变产生的，一些计数实验是在高达 $1\ A\cdot cm^{-2}$ 的电流密度下进行的（表 2），在使用等温量热计的一个实验中，其高电流密度保持了相当长的一段时间（表 1）。痕量的铂在阴极上沉积，假设这会导致氘析出的过电压降低，从而降低阴极中氘可达到的浓度水平，也被认为是对不可重复性的一种解释（弗莱施曼，个人交流）。当我们在我们的中子计数实验中使用钯阳极时，并没有观测到任何效应。

充氢和充氘的时间与预期的扩散时间一致 (x^2/D，其中扩散系数 $D = 10^{-7}\ cm^2\cdot s^{-1}$，$x$ 为样品的半径或其厚度的一半)[10]，并且核计数和量热实验进行的时间也总是远大于这一时间尺度。此外，在钯的电解充氘（氢）过程中，显然会有一个浓度梯度，一个移动的相边界，并且金属外表面的原子层会被氘所饱和[17]（也可能过饱和）。因此，可以预期，足够灵敏的设备可以在这种材料完全充氘以前检测到核聚变过程。因为我们的中子探测实验的灵敏度是弗莱施曼等人[1]的约 10^5~10^6 倍，看上去似乎不太可能发生任何伴随着钯充氘的大大增强的核聚变并产生中子发射。当然，我们注意到总是有可能构建一个实质上尚难检验的理论，该理论涉及金属或其表面假定的特殊情况。因此，仔细表征具有正结果的材料是非常重要的。

氘核－氘核反应的中子分支会被完全抑制掉而代之以（t+p）分支，这一观点已得到一些论证，进一步的论证指出（庞斯，个人交流）在电极内部作为核过程产物的氚并不一定与电解液发生交换，因而也就可能检测不到。然而，这样一个核过程的其他的产物——高能质子在与晶格相互作用的时候应该能被检测到，产物中还应包括中子发射：我们估计，已知质子损失能量的速率，通过与相邻元素 (p，n) 反应截面的比较，产率约为每质子 10^{-6} 个中子，如果核聚变以参考文献 1 中所报道的速

rate reported in ref. 1 were to proceed entirely through the (t + p) branch.

In view of the rather large d–d separations in both PdD and TiD_2, it might be argued that fusion requires some non-equilibrium state in the lattice—perhaps at the α/β phase boundary in palladium or at the tips of the growing TiD_2 needles or at a lattice defect—where the d–d or p–d separation could be greatly reduced. Here we created non-equilibrium situations by pulsing the current but we detected no neutron emission (Fig. 4).

Some explanations of the apparent excess heat production have emphasized recombination processes at catalytic metal surfaces[20]: apart from occasional explosions, however, which did not result in any detectable neutron emission, we found, by comparison of the volume of water added to maintain the cell volume with the electrolysis charge passed, that this was not a significant effect (in agreement with others[9]). Recombination in the gas space above the liquid, either on exposed cathode surface or catalysed by colloidal metal particles eroded from the cathode, might however account for some of the observations of bursts of heat recently reported (M. Fleischmann and S. Pons, Electrochemical Society Meeting, Los Angeles, May 1989). In discussing the claims in ref. 1, we prefer to focus on characteristics of the "simple" Fleischman, Pons and Hawkins (FPH) calorimeters, because we have only observed small effects (at the level of the inherent uncertainties), which might mistakenly be claimed as arising from cold fusion, in the one type of calorimeter (FPH type) that has major calibration difficulties. Cells using the same electrode and electrolyte materials operated in calorimeters that did not have these problems exhibited none of these effects.

There are two points regarding the calibration that could have a profound effect on the apparent results obtained with FPH-type calorimeters: the first concerns when the calibration is performed and the second how it is performed. Concerning the first point, it seems from our work and that of Lewis *et al.*[8] that a calibration performed during the first 10,000 min of electrolysis could be seriously in error and lead to an erroneous conclusion that subsequently, rather than being in balance, the cells were exothermic. Concerning the second point, Fleischmann *et al.*[1] describe calibration using the internal-resistance heater, by measurement of Newton's-law-of-cooling losses. Typically, this procedure might involve the application of power to the heater while electrolysis was occurring, following the temperature-time trace until a steady state was obtained, then switching the heater off and following the cooling curve. This procedure gives an approximation to the differential calorimeter constant, $k_d = d(\Delta P)/d(\Delta T)$ (where P is power and T is temperature) at the operating temperature of the cell and can give rise to errors in two ways. First, because any calibration sequence would require 5–10 h, extrapolation would be required to obtain the correct value at the reference liquid level: an estimated error of 20% or more could result. Furthermore, because the evaporation of the cell contents increases markedly with increasing temperature, the baseline slope would increase with increasing temperature and the effects of this sort of error would become correspondingly more marked: the claimed effects were indeed greatest in the cells run at the highest input power. Second, these calorimeters are nonlinear (equation (3), Fig. 2 legend), with the effect being significant

率完全通过（t+p）分支进行的话，我们的计数实验可以产生的中子产率为 $10^4\ s^{-1}$。

由于在 PdD 和 TiD_2 中都存在相对较大的氘核 – 氘核间距，可能有争论说核聚变需要一些晶格中的非平衡状态——也许在钯的 α/β 相边界，或者在生长中的 TiD_2 针状物的尖端，又或者在一个晶格缺陷上，在这些地方，也许氘核 – 氘核或者质子 – 氘核的间距会大大减小。本文中，我们通过脉冲电流创造了一个非平衡的状态，但是仍没有检测到中子发射（图 4）。

一些对产生表观“过热”现象的解释强调的是在具有催化性的金属表面的复合[20]：然而除了偶尔的突然爆发，它不会导致任何可观测的中子发射，通过比较为了维持电解池的体积而加入的水的体积与通过电解池的电荷数量，我们发现，这并不是一个重要的效应（与其他研究者的结论一致[9]）。在液体上方的气体空间复合，不是发生在暴露的阴极表面，就是被从阴极上腐蚀下来的胶体金属颗粒所催化，但是，这仅能解释最近报道的所观察到放热实验中的一些现象（见弗莱施曼和庞斯在 1989 年 5 月于洛杉矶举行的电化学会议上的报告）。在讨论参考文献 1 的结论时，我们倾向于关注“简易的”弗莱施曼、庞斯和霍金斯（FPH）量热计的特征，因为我们只观测到了一些小的效应（在固有的不确定性的水平上），在一种标定很困难的量热计（FPH 型）上，这些效应可能被错误地认为是来源于冷核聚变。使用同样的电极和电解液材料并利用没有这些问题的量热计，发现这样的电解池并未显示出这些效应。

有两点关于标定的问题可能对于 FPH 型量热计所获得的表观结果产生重大的影响：首先是标定何时进行，其次是标定如何进行。关于第一点，从我们和刘易斯等人[8]的工作来看，似乎在电解进行的前 10,000 分钟进行的标定是非常错误的，这将会导致错误的结论，即把处在平衡态的电解池误认作处在放热状态。关于第二点，弗莱施曼等人[1]对使用内部的电阻加热器、通过测量牛顿冷却定律的损失来进行标定的方法进行了描述。通常情况下，这一过程包括了在电解出现时给加热器施加功率，紧接着跟踪温度 – 时间轨迹，直到达到稳定状态，然后把加热器关掉，记录冷却曲线。这个过程可以算出工作温度下一个电解池的微分量热计常数的估计值，$k_d = d(\Delta P) / d(\Delta T)$（其中 P 为功率，T 为温度），有两方面的因素会造成这个值的误差变大。首先，因为任何的标定程序都需要 5~10 小时，要得到参考液面处的正确数值，需要进行外推：这估计会对结果造成 20% 或更大的误差。此外，因为电解池的蒸发量会随着温度的升高而明显增大，随着温度的升高，基线斜率也会增大，这种错误所产生的效应相应的会更加明显：声称的效应实际上在电解池以最高的输入功率运行时是最大的。第二，这种量热计是非线性的（方程（3），图 2 的图注），当温

when the temperature gradient is large: they cannot be described by a simple Newton's law-of-cooling constant. If a differential calorimeter constant is used to derive the input power, the calculated output would be (from Fig. 2 legend, equation (3))

$$P_{app} = k_d\Delta T = (k_{sb,0} + k_c)\,\Delta T + 3(k_{sb,0}/T_0)\,(\Delta T)^2$$

so that, in comparison with the correct output (equation (3))

$$P_{app} - P = 3k_{sb,0}(\Delta T)^2/2T_0$$

If the power applied to the heater is significant compared with the electrolysis power then this error will be even greater. Back calculation from the data given in ref. 1 shows that the claimed effects were largest for Joule input powers on the order of 6 W. For our FPH-type calorimeters, this would have given $\Delta T \approx$ 50–60°C. Therefore, had these cells been calibrated in this way, an apparent heat excess on the order of 0.8–1 W would have been observed. Our method of calibration (see Fig. 1 caption) considerably reduced the effect of these sources of error. The original report[1] was clearly preliminary in nature, and it is evident from the above that the claims made therein cannot be assessed in the absence of a detailed description of the experimental procedure used and of the methods used to compensate for the systematic errors inherent in the use of a simple calorimeter.

We feel that our work has served to establish clear bounds for the non-observance of cold fusion in electrolysis cells, under carefully controlled and well understood experimental conditions and using well characterized materials. Further details are given in ref. 10. Claims of observations of cold fusion ought now to meet similar standards of data analysis and materials characterization so that a proper assessment can be made.

(**342**, 375-384; 1989)

D. E. Williams*, D. J. S. Findlay†, D. H. Craston*, M. R. Sené†, M. Bailey†, S. Croft†, B. W. Hooton‡, C. P. Jones*, A. R. J. Kucernak*, J. A. Mason§ & R. I. Taylor*

* Materials Development Division, † Nuclear Physics and Instrumentation Division and ‡ Nuclear Materials Control Office, Harwell Laboratory, UK Atomic Energy Authority, Didcot, Oxfordshire, OX11 ORA, UK

§ Reactor Centre and Centre for Fusion Studies, Imperial College, Silwood Park, Ascot SL5 7PY, UK

Received 8 August; accepted 16 October 1989.

References:

1. Fleischmann, M., Pons, S. & Hawkins, M. *J. electroanal. Chem.* **261**, 301-308 (1989); erratum **263**, 187 (1989).
2. Jones, S. E. *et al. Nature* **338**, 737-740 (1989).
3. Koonin, S. E. & Nauenberg, M. *Nature* **339**, 690-691 (1989).
4. Sun, Z. & Tomanek, D. *Phys. Rev. Lett.* **63**, 59-61 (1989).
5. Leggett, A. J. & Baym, G. *Nature* **340**, 45-46 (1989).
6. Gai, M. *et al. Nature* **340**, 29-34 (1989).
7. Zeigler, J. F. *et al. Phys. Rev. Lett.* **62**, 2929-2932 (1989).

度梯度较大时，非线性效应比较明显：它们不能用简单的牛顿冷却定律来描述。如果一个微分量热计常数通过输入功率来导出，计算出的输出功率应该为（来自图2的图注中的，方程（3））

$$P_{\mathrm{app}} = k_{\mathrm{d}}\Delta T = (k_{\mathrm{sb,0}} + k_{\mathrm{c}})\,\Delta T + 3(k_{\mathrm{sb,0}}/T_0)\,(\Delta T)^2$$

与正确的输出功率（方程（3））进行比较，则有

$$P_{\mathrm{app}} - P = 3k_{\mathrm{sb,0}}(\Delta T)^2/2T_0$$

如果与电解功率相比，施加在加热器上的功率很大，那么这个误差会更大。从参考文献1给出的数据进行反演计算表明，声称的效应在焦耳输入功率为6 W的数量级时最大。对我们的FPH型量热计而言，这将得出$\Delta T \approx 50$~60 ℃。因此，如果这些电解池经过这种方法进行标定，将能观察到0.8~1 W量级的表观过热。我们的标定方法（见图1的图注）显著地减小了这种来源的误差。最原始的报告[1]实质上很明显是初步的，而且从上面的证据来看，在没有详细描述所采用的实验步骤和用于抵消使用一个简单量热计带来的系统误差的方法之前，最初的报告[1]中的宣称是难以评估的。

在仔细控制、很好理解实验条件并使用完备表征材料的条件下，我们认为我们的工作为电解池中尚未观察到冷核聚变建立了清晰的界限。更进一步的细节参见参考文献10。现在对观察到冷核聚变的声明应该符合类似的数据分析和材料表征标准，这样才能对其作出适当的评估。

（李琦 翻译；李兴中 审稿）

8. Lewis, N. *et al. Nature* **340**, 525-530 (1989).
9. Cunnane, V. J., Scannell, R. A. & Schriffrin, D. J. *J. electroanal. Chem.* **269**, 163-174 (1989).
10. Williams, D. E. *et al. Harwell Report AERE R*-13606 (HMSO, London, 1989).
11. Edwards, G., Findlay, D. J. S. & Lees, E. W. *Ann. Nucl. Energy* **9**, 127-135 (1989).
12. Lees, E. W., Patrick, B. H. & Bowey, E. M. *Nucl. Instrum. Meth.* **171**, 29-41 (1980).
13. Fouroulis, Z. A. *J. electrochem. Soc.* **128**, 219-221 (1981).
14. Östlund, H. G. & Werner, E. in *Tritium in the Physical and Biological Sciences* Vol. 1, 95-104 (Int. Atomic Energy Agency, Vienna, 1962).
15. Kaufmann, S. & Libby, W. F. *Phys. Rev.* **93**, 1334-1337 (1954).
16. von Buttlar, H., Vielstich, W. & Barth, H. *Ber. Bunsenges phys. Chem.* **67**, 650-657 (1963).
17. Smith, D. P. *Hydrogen in Metals* (University of Chicago Press, 1948).
18. Carpenter, J. M. *Nature* **338**, 711 (1989).
19. Gittus, J. & Bockris, J. *Nature* **339**, 105 (1989).
20. Kreysa, G., Marx, G. & Plieth, W. *J. electroanal. Chem.* **266**, 437-441 (1989).
21. Croft, S. *Nucl. Instrum. Meth. Phys. Res.* **A281**, 103 (1989).
22. Anicin, I. V. & Yap, C. T. *Nucl. Instrum. Meth. Phys. Res.* **A259**, 525-528 (1987).
23. Savitzky, A. & Golay, M. J. E. *Anal. Chem.* **36**, 1627-1639 (1964).

Acknowledgements. We thank Johnson-Matthey (Dr I. McGill and Dr M. Doyle) for assistance and for the loan of special samples of palladium. IMI Titanium (Mr J. R. B. Gilbert) provided titanium and information. We thank the following for assistance: P. Fozard, M. Newan, J. Monahan, R. Morrison, H. Bishop, V. Moore and colleagues, J. Asher, P. W. Swinden, A. M. Leatham, R. A. P. Wiltshire, D. A. Webb, J. O. W. Norris, R. Roberts, R. Crispin, C. Westcott, R. Neat, D. Robinson, H. Watson and G. McCracken. We also thank Dr D. Schriffrin, Dr P. T. Greenland and Dr D. Morrison for discussions and Dr R. Bullough for support. This work was financed by the Underlying Science programme of the UKAEA, and UK department of Energy and the Commission of the European Communities.

Limits on the Emission of Neutrons, γ-rays, Electrons and Protons from Pons/Fleischmann Electrolytic Cells

M. H. Salamon *et al.*

Editor's Note

Of the many experiments conducted worldwide to investigate the claim of Martin Fleischmann and Stanley Pons to have carried out nuclear fusion by benchtop electrolysis in 1989, those reported here provided some of the most compelling contrary results. Michael Salamon was a physicist at the same university (Utah) as Pons, and he obtained Pons' agreement to re-run the experiments using the same apparatus. After exhaustive trials, Salamon's team found no evidence of "cold fusion". Pons' claim that excess heat had been generated in one event after a power failure had prevented the computer from collecting data only added to the growing scepticism and suspicion about the original claim. At one point, these experiments provoked threats of legal action against Salamon.

Emissions of γ-rays from the cold-fusion cells used by Pons and Fleischmann were monitored in Pons' laboratory at the University of Utah by NaI detectors nearly continuously over a five-week period. No evidence of fusion activity was observed above power limits varying between 10^{-12} and 10^{-6} W for the known fusion reactions. In addition, neutron-track detectors indicated an integrated upper limit of approximately 1 emitted neutron per second from any of the cold-fusion cells over a period of 67 hours.

PONS and Fleischmann[1] claim to have achieved "cold fusion" of deuterium nuclei in electrolytic cells containing palladium cathodes and D_2O (with 0.1 M LiOD) electrolyte, which produced excess heat of the order of a few watts. Their claim was based on the lack of a known electrochemical mechanism for the observed heat excess and the emission of γ-rays and neutrons at levels slightly above background. These latter data have been criticized[2-4], but reports of tritium production in similar cells[5] and the discrepancy between cell power inputs and outputs in several laboratories[6] have sustained the controversy despite a large number of negative results[4].

The known d + d and d + p reactions and their energy yields (Q) are[7]: d(d, p)t (Q = 4.03 MeV); d(d, n)^{3}He (Q = 3.27 MeV); d(d, γ)^{4}He (Q = 23.85 MeV); d(d, e^-)^{4}He (internal conversion) (Q = 23.85 MeV); d(p, γ)^{3}He (Q = 5.49 MeV).

庞斯/弗莱施曼电解池的中子、γ射线、电子和质子发射的上限

萨拉蒙等

编者按

全世界进行了许多实验来研究马丁·弗莱施曼和斯坦利·庞斯所声称的已在 1989 年利用台式电解实现了核聚变的断言，这篇文章中所报道的内容则提供了一些最具说服力的反面结果。迈克尔·萨拉蒙是与庞斯同一所学校（犹他大学）的物理学家，他得到了庞斯的允许，使用同一装置重复了那些实验。在彻底的尝试之后，萨拉蒙的团队没有找到“冷核聚变”的证据。庞斯声称在断电妨碍了计算机收集数据之后，确实有过剩热量产生。这一说法只会增加大家对最初断言的怀疑。这些实验曾一度为萨拉蒙招致诉讼威胁。

犹他大学庞斯的实验室通过 NaI 探测器对庞斯和弗莱施曼使用的冷核聚变电解池放射出的 γ 射线进行了五个星期近乎连续的记录。已知的聚变反应的功率上限是 10^{-12}~10^{-6} W，在这个功率限制之上没有观察到任何聚变活动的证据。此外，中子径迹探测器表明，在 67 小时内，从其中任何一个冷核聚变电解池中发射出来的中子总数上限约为每秒一个中子。

庞斯和弗莱施曼 [1] 宣称在含有钯阴极和 D_2O 电解质（含 0.1 M LiOD）的电解池中实现了氘核的“冷核聚变”，该反应产生了数量级为几瓦的过剩热量。他们的结论是基于以下理由得出的：缺乏已知的电化学机制来解释热量过剩以及略微高于本底的 γ 射线和中子的放射。上述数据的后者已经遭到批判 [2-4]，尽管存在大量的负面结果 [4]，但是在类似的电解池中产生氚的报告 [5] 和一些实验室 [6] 中电解池输入功率和输出功率之间的差异仍然使得这场论战得以持续。

已知的 d+d 和 d+p 反应和它们的能量产率(Q)为 [7]：d(d, p)t (Q = 4.03 MeV)；d(d, n)^{3}He (Q = 3.27 MeV)；d(d, γ)^{4}He (Q = 23.85 MeV)；d(d, e$^-$)^{4}He (内转换)(Q = 23.85 MeV)；d(p, γ)^{3}He (Q = 5.49 MeV)。

Several weeks after his initial press announcement, Pons allowed us into his laboratory to make independent measurements of any radiation emanating from his operating electrolytic cells. Using equipment that was immediately available, a lead-shielded sodium iodide (NaI) detector (8 × 4 in.) was installed below his cells (Fig. 1) and collected data nearly continuously for over five weeks in a γ-ray energy range of 0.1–25.5 MeV. In addition, several neutron detectors, which integrated the neutron flux over a period of approximately three days, were placed within the water tank adjacent to the cells; these were made of ^{235}U foils sandwiched between nuclear-track-detecting plastic film.

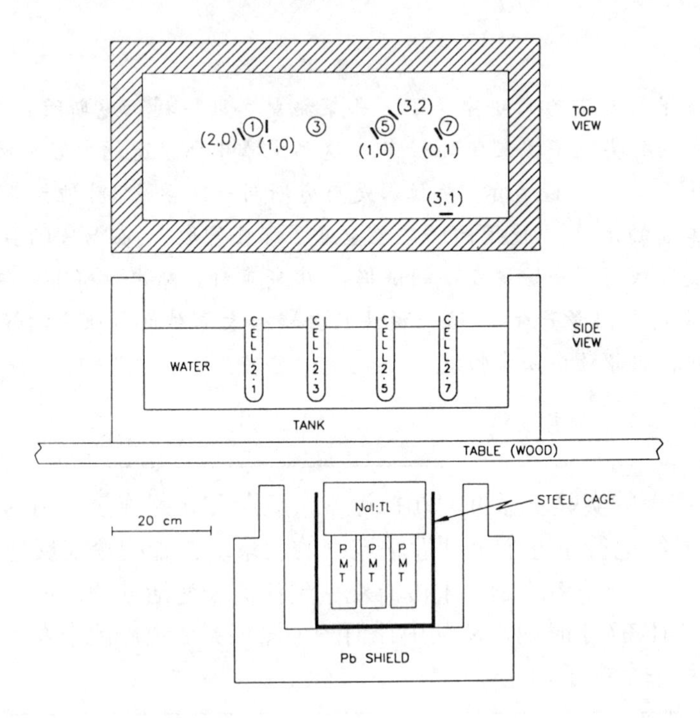

Fig. 1. Side view of the geometry of the electrolytic cells and NaI detector, and top view of the cells with neutron-detecting sandwiches in place (thickline segments). Cells 2-1, 2-5 and 2-7 have palladium cathodes with diameters/lengths (in cm) respectively of 0.4/1.25, 0.4/10.0 and 0.1/10.0; cell 2-3 has a platinum cathode of dimensions 0.1/10.0. The two numbers (*n*, *m*) shown for each sandwich in the top view are the number of fission fragments counted in the plastic film adjacent to the uranium foil (without, with) Cd covers. Only one plastic film per foil was analysed.

NaI Detection of γ-rays

Four open cells (no gas collection) underwent electrolysis nearly continuously while the NaI detector was collecting data (9 May to 16 June). These cells consisted of D_2O (0.1 M LiOD) electrolyte with platinum anodes; the cathodes were as described in Fig. 1. The cells were run in constant-current mode, and current settings were varied over the five-week interval.

The NaI detector system was placed under the table supporting the water tank and cells, so as not to interfere with other research activities in Pons' laboratory. Any γ-rays produced in the cells would thus pass through water, water tank and table before being detected by the

在最初的声明发表之后数周，庞斯允许我们进入他的实验室，对从他运行的电解池中放射的所有辐射进行独立的测量。我们使用的是即时可用装置，将一个铅屏蔽的碘化钠（NaI）探测器（8 in × 4 in）安装在他的电解池下方（图 1），并在五周内几乎连续地对能量范围在 0.1~25.5 MeV 之内的 γ 射线进行了数据收集。此外，还将几个中子探测器放置在水槽中，与电解池相邻，它们对大约 3 天的时间段内的中子流进行了累计；这些中子探测器由夹在核径迹探测塑料薄膜之间的 ^{235}U 箔组成。

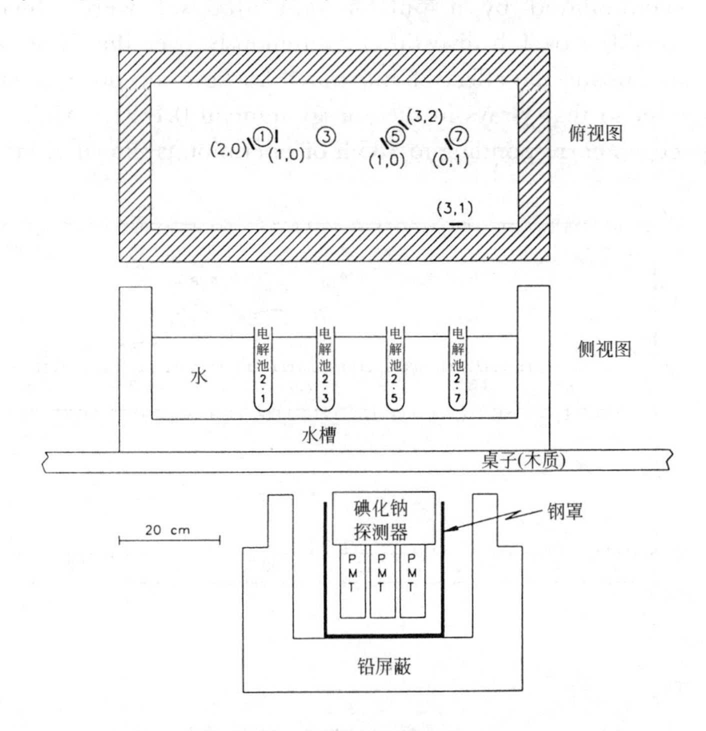

图 1. 电解池和 NaI 探测器布局的侧视图，以及相应位置（粗线部分）的可探测中子的三明治结构的电解池俯视图。电解池 2-1，2-5 和 2-7 的钯阴极直径 / 长度（以厘米为单位）分别为 0.4/1.25、0.4/10.0 和 0.1/10.0；电解池 2-3 有尺寸为 0.1/10.0 的铂阴极。俯视图中每一个三明治结构显示的两个数字 (*n*，*m*) 分别是没有镉覆盖层和有镉覆盖层的铀箔塑料薄膜中记录的裂变碎片数目。只分析每个箔中的一个塑料薄膜。

γ 射线的 NaI 探测

让四个开放的电解池（不进行气体收集）近乎连续地进行电解，同时 NaI 探测器在收集数据（5 月 9 日至 6 月 16 日）。这些电解池由 D_2O（0.1 M LiOD）电解质和铂阳极构成；阴极如图 1 所示。电解池在恒电流模式下运行，电流设定值在五周内是变动的。

NaI 探测器系统被放置于支撑水槽和电解池的桌子下方，这样就不会对庞斯实验室里的其他研究活动形成干扰。电解池中产生的任何 γ 射线在被 8 in × 4 in 的 NaI

8 × 4 in. NaI detector. The energy-dependent, absolute efficiency for γ-ray detection in the photo-peak (corresponding to complete γ-ray energy conversion within the scintillator) was determined with standard γ-ray sources in a separate laboratory, where an identical configuration of water, tank, table and detector could be installed or removed at will. Radiation limits given below are based on the absolute efficiency for a source at the location of cell 2-1.

Spectral data accumulated by a pulse-height analyser were downloaded to a microcomputer every 0.5 or 1 h (live time) continuously over the five-week observation period, thereby minimizing the effect of integrated background on transient signals. The system's gain was set so that γ-rays in the energy interval 0.1–25.5 MeV were recorded. An aggregate spectrum corresponding to 785 h of operation is shown in Fig. 2*a, b*.

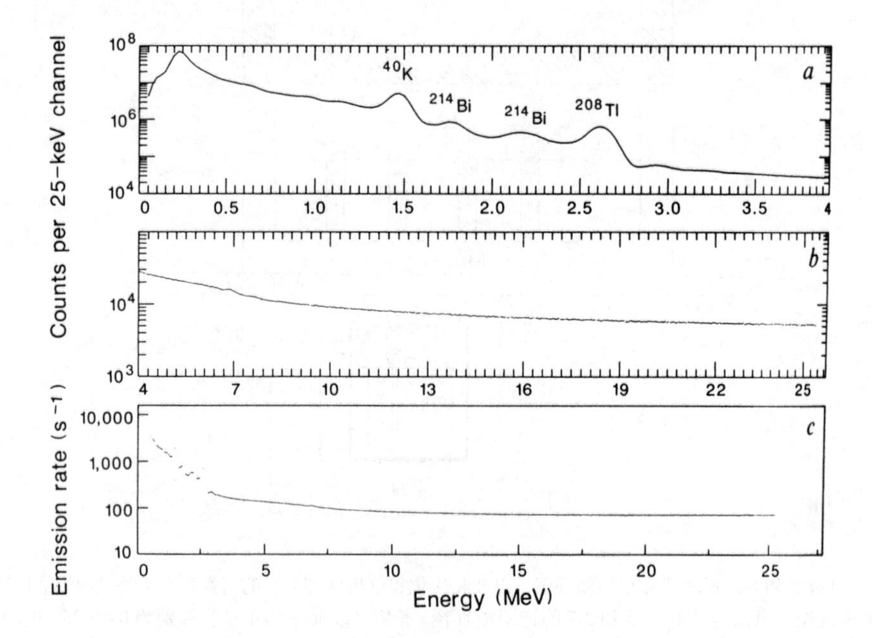

Fig. 2. *a, b*, Aggregate γ-ray spectrum for 785 h of live collection time. The dominant background lines are ^{40}K (1.4608 MeV), ^{214}Bi (1.7645 Mev), ^{208}Tl (2.6146 MeV), a backscatter peak at ~0.23 MeV, and a broad peak at ~2.17 MeV due to two lines, ^{214}Bi (2.1186, 2.2042 MeV). The detector energy resolution is given by $R = 0.07E^{-1/2}(1+1.3/E)^{1/2}$, with E in MeV, where the leading factor is the detector's optimal resolution and the additional factor is caused by the presence of noise at the charge-integrating input of the pulse-height analyser's analog-to-digital converter. The integral nonlinearity of the system electronics was found to be <0.1% over the interval 0.1–3.5 MeV, and 0.2% over the full interval 0.1–25.5 MeV. System gain variations (owing to phototube drift, for example) were monitored and found to be <3.5% over the full five-week interval and <0.5% over any 24-h interval; *c*, γ-ray source emission-rate limit against energy for the general γ-ray search over the range 0.1–25.5 MeV (see text). The probability of falsely rejecting a signal at these limits is $<10^{-3}$.

Individual spectra (1,116), each of 0.5 or 1.0 h live integration time (for a total of 831.0 h of live time), were searched for transient γ-ray signals above background. Background for each individual spectrum was obtained by averaging several (10–50) successive spectra to

探测器探测到以前，都会先通过水、水槽和桌子。在光电峰中对γ射线进行探测的与能量有关的绝对效率（与在闪烁体中γ射线完全的能量转换相对应）可在一个独立实验室中用标准γ射线源进行确认，该独立实验室中可对水、水槽、桌子和探测器进行同样的配置和任意的去除。下面给出的辐射限是基于放置在电解池 2-1 处的源的绝对效率得到的。

将通过脉冲高度分析仪累积的γ谱数据在五周的观察时间内每隔半小时或者 1 小时（活时间）连续地下载到一台微机上，从而将瞬态信号的本底累积效应降至最低。设定了系统的增益以使 0.1~25.5 MeV 能量区间内的γ射线可以被记录下来。仪器运行 785 小时所得到的累积γ谱如图 2*a* 和 2*b* 所示。

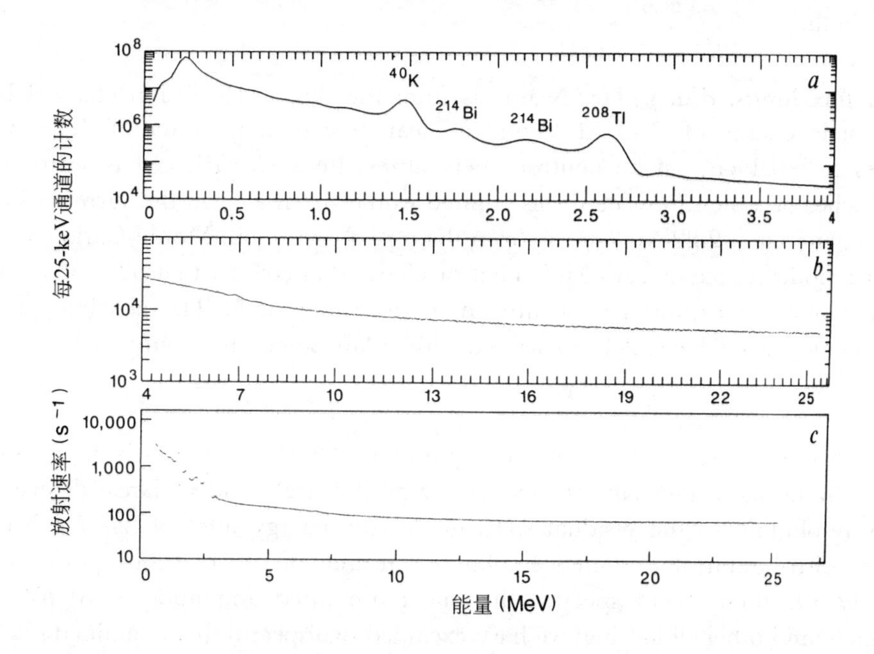

图 2. *a* 和 *b*, 785 小时的活时间收集的累积γ射线谱。主要的本底峰为 ^{40}K(1.4608 MeV)、^{214}Bi(1.7645 MeV) 和 ^{208}Tl（2.6146 MeV），而一个约 0.23 MeV 处的背散射峰和一个约 2.17 MeV 处的宽峰源自 ^{214}Bi 的两个峰（2.1186 MeV 和 2.2042 MeV）。探测器能量分辨率由 $R = 0.07E^{-1/2}(1+1.3/E)^{1/2}$ 给出，E 的单位为 MeV，式中的主导因素是探测器的最优分辨率，附加的因素是由于脉冲高度分析仪的模数转换器的电荷积分输入存在噪音而导致的。系统电子部件的积分非线性在 0.1~3.5 MeV 的区间内小于 0.1%，在整个 0.1~25.5 MeV 区间为 0.2%。对系统增益变化（比如由于光电流漂移）的监测表明，在整个五周的时间内，变化小于 3.5%，而在任何 24 小时区间内，变化小于 0.5%；*c*，在 0.1~25.5 MeV 的范围内，对全体γ射线进行搜寻得到的γ射线源的放射速率限值与能量的关系（见正文部分）。在这些限值以内的信号被漏记的概率小于 10^{-3}。

逐条γ谱（1,116 条）、逐段时间（每一段活时间积分的时长为 0.5 或 1 小时，活时间共计 831.0 小时）地寻找瞬时高于本底的γ射线信号。每条γ谱的本底要用若干（10~50）条连续的本底γ谱取平均而得到一个合成γ谱来求得，然后，从原始的

form an aggregate spectrum, which was then subtracted from the original spectrum to form a "residual spectrum". This was performed for each individual spectrum, yielding 1,116 residual spectra. Any transient anomalous signals of duration <10–50 h would then appear in the residual spectra as positive excesses amidst Poisson fluctuations about zero. Signals of longer duration were searched for with even greater sensitivity by performing a similar operation on the collection of aggregate spectra.

To correct for temporal variations in the detector system's gain and pedestal (the pulse-height-analyser channel corresponding to zero energy), the gain and pedestal were determined for each individual spectrum by fitting to the background γ-ray lines. All spectra were then rescaled to a common gain with zero offset before background subtraction, thus minimizing artefacts in the residual spectra arising from gain or zero shifts, or both.

Neutron flux limits: d(d, n)^{3}He. Neutrons from the d(d, n)^{3}He fusion channel have an initial kinetic energy of 2.45 MeV and a mean free path in water of 4.9 cm, which decreases to ~0.4 cm as the neutron thermalizes. Because each cell is surrounded by several inches of water, most neutrons emitted would thermalize in the surrounding water bath and generate a 2.22-MeV γ-ray from the n(p, d)γ reaction; Monte Carlo calculations show for a point source of 2.45-MeV neutrons located at cell 2-1 that 62% of the emitted neutrons would be captured by hydrogen in the water bath. The absolute photopeak detection efficiency (for cell 2-1), measured with a Am-Be neutron source, was found to be 3.0×10^{-3}.

The presence of a γ-ray signal above background at 2.2 MeV was sought in each residual spectrum by fitting a gaussian (of central energy 2.2 MeV and variance determined by detector resolution) to the residual spectrum in the energy interval 2.0–2.4 MeV. The frequency distribution of the fitted amplitudes, in units of pW of fusion power, is shown in Fig. 3*a*. Of these 1,114 spectra, the maximum fitted amplitude is 10 pW. In this distribution (and others following) we have excluded two spectra that contained photopeaks due to ^{22}Na and ^{137}Cs sources brought into the laboratory by other personnel.

γ 谱中扣去此本底而得到“剩余谱”。对于每条 γ 谱进行这一扣除后就得到 1,116 条剩余谱。于是，在小于 10~50 小时内任何瞬时的异常信号都会在剩余谱中以零点附近一片泊松涨落之上的正剩余的形式显示出来。持续时间更长的信号还可以通过对合成 γ 谱的集合进行相似的操作来寻找，而且其灵敏度还更高些。

为了能够对探测器系统的增益和消隐脉冲电平（对应于能量为零的脉冲高度分析仪的通道）的短暂变化进行校正，通过把每一个单独的 γ 谱对 γ 射线本底进行拟合的方法来对增益和消隐脉冲电平进行确定。然后在扣除本底之前，所有的 γ 谱被重新调节到通常具有零偏置的增益水平，这样可以减小剩余 γ 谱中由增益或零点漂移或者两者兼而有之所带来的伪差。

中子流的限值：$d(d, n)^3He$ 来自核聚变通道 $d(d, n)^3He$ 的中子具有 2.45 MeV 的初始动能和在水中 4.9 cm 的平均自由程，由于中子热化，它的自由程会缩减为 0.4 cm。由于每一个电解池都被数英寸的水浴包围，大多数放射出的中子会在周围的水浴中热化并通过 n(p, d)γ 反应产生 2.22 MeV 的 γ 射线；蒙特卡洛计算表明，对位于 2-1 电解池处的 2.45 MeV 的点中子源而言，62% 放射出的中子会被水浴中的氢捕获。通过一个 Am-Be 中子源测得，绝对的光电峰探测效率（对 2-1 电解池）为 3.0×10^{-3}。

通过对 2.0~2.4 MeV 能量区间的剩余 γ 谱进行高斯拟合(以 2.2 MeV 为中心能量，方差由探测器分辨率确定)，每一个剩余 γ 谱中都可以观察到一个位于 2.2 MeV 处的高于本底的 γ 射线信号。拟合幅值的频数分布以 pW 为核聚变功率的单位，如图 3*a* 所示。在这 1,114 个 γ 谱中，拟合出的最大幅值为 10 pW。在这个分布（以及后面其他的分布）中我们排除了两个图谱，因为它们含有来自 ^{22}Na 和 ^{137}Cs 源的光电峰，这是由于其他人将这两种放射源带入实验室造成的。

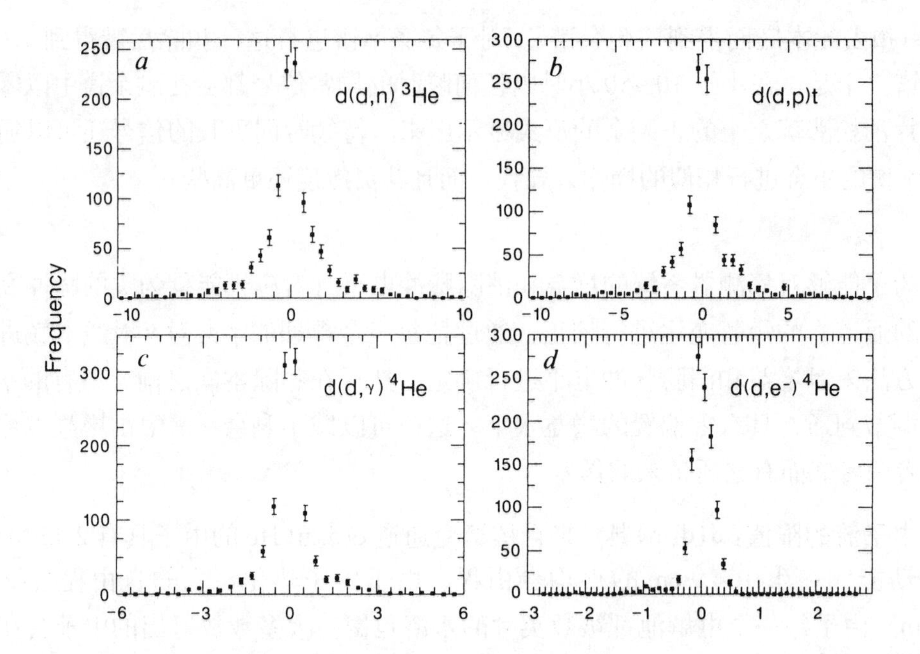

Fig. 3. Estimated rates for the four d + d fusion reactions. For each of the 1,114 γ-ray spectra obtained from data-collection episodes lasting 0.5–1 h, a residual spectrum was obtained by subtracting an averaged background spectrum from the signal. A fit was then made to each residual by scaling the expected form of the γ-ray spectrum for each reaction; the fit yields a magnitude that can be positive or negative. The figures here are histograms of those 1,114 magnitudes. In *a* and *b*, the rate is expressed in units of fusion power (*a*: pW; *b*: mW); for *c* and *d*, reaction rates are used (*c*: s^{-1}; *d*: $10^{3}\ s^{-1}$). In each case the histogram is approximately a gaussian distribution of amplitudes centred around zero, indicating no measurable rates for any of the fusion reactions.

Proton flux limits: d(d, p)t. The fusion channel d(d, p)t (Q = 4.03 MeV) has been a leading candidate for the cold-fusion process, both because of reports of excess tritium production[5] and because its reaction products come to rest within the palladium electrode (the range of a 3-MeV proton in palladium is ~ 30 μm; ref. 8), thereby presumably avoiding the paradox of watts of fusion power being generated without observed particle emissions. In fact, a strong and distinct γ-ray signature exists for this reaction: the 3.02-MeV protons cause Coulomb excitation of the even–even isotopes of Pd, whose radiative de-excitations, between 0.37 and 0.56 MeV, are detectable by the NaI detector with efficiencies η given in Table 1. This table, adapted from ref. 9, lists for each Pd isotope its E2 (electric quadrupole) γ-ray energy and thick-target radiation yield (excitations per microcoulomb of protons) for 100% isotopically enriched samples, along with photopeak detection efficiency assuming cell 2-1 as the source, plus a factor accounting for absorption of the E2 γ-ray within the Pd cathode. Figure 4 shows a NaI γ-ray spectrum from a 3-mm target of natural Pd exposed to a beam of 3.02-MeV protons from a Van de Graaf accelerator (W. Schier, personal communication). Peaks at 0.37, 0.43 and 0.51 MeV are observed with their expected strengths.

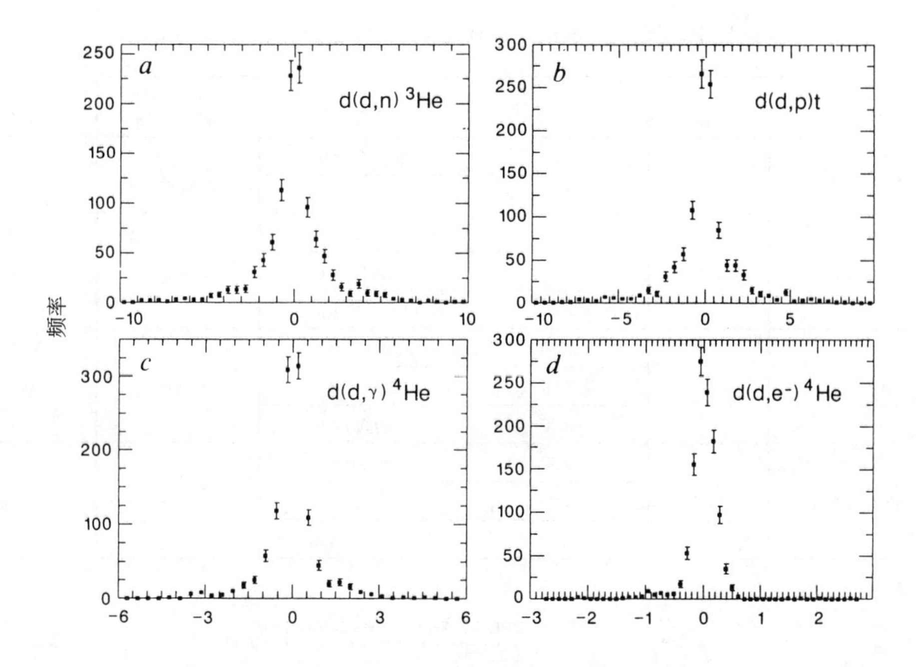

图 3. 对四个 d + d 核聚变反应估算的速率。对于数据采集区间持续 0.5~1 小时的 1,114 条 γ 射线谱中的每一条，从信号中减去平均化的本底 γ 谱可以得到一条剩余 γ 谱。通过调节每一个反应的已知形式的 γ 射线谱的比例，对每一个剩余光谱进行拟合；拟合得到的数量级可以为正也可以为负。这里的图是所有 1,114 条 γ 谱的柱状图。在 *a* 和 *b* 中，速率以聚变的功率为单位（*a*: pW；*b*: mW）；*c* 和 *d* 中，反应速率为通常使用的单位（*c*: s^{-1}；*d*: $10^3\ s^{-1}$）。每一个图中，柱状图大致呈现为零点附近的高斯分布，表明对于任何一个核聚变反应没有可测量的速率。

质子流的限值：d(d, p)t 核聚变通道 d(d, p)t (Q = 4.03 MeV) 是冷核聚变过程的一个最有可能的反应，这既是因为有过量氚产生的报告[5]，也是因为它的反应产物被固定在了钯电极中（3 MeV 的质子在钯中的范围约为 30 μm；参考文献 8），因此有可能避免“产生了瓦特量级的聚变功率却没有观察到粒子发射”的佯谬。事实上，一个强而明显的 γ 射线特征存在于这一反应中：3.02 MeV 的质子导致钯同位素偶 – 偶核的库仑激发，退激发时它辐射出的 0.37 MeV 到 0.56 MeV 之间的辐射可以被 NaI 探测器以表 1 中给出的效率 η 探测到。这个表由参考文献 9 改编而来，列出了每一种钯同位素的 E2（电四极）γ 射线的能量和 100% 同位素富集的样品的厚靶辐射产额(每微库仑质子的激发量)，同时还假设了电解池 2-1 为源的光电峰探测效率，并考虑到在钯阴极内部 E2 γ 射线吸收的修正因子。图 4 显示了一个从暴露在源自范德格拉夫加速器的 3.02 MeV 质子束中的 3 mm 的天然钯靶所放射出的 NaI γ 射线谱（希尔，个人交流）。在 0.37 MeV、0.43 MeV 和 0.51 MeV 处观察到了峰值，并且与预期的强度一致。

Table 1. Radiation yields of 3-MeV protons in palladium electrodes

Isotope (E2 energy) (MeV)	Isotopic fraction	Proton energy (MeV)	Excitation/μC	NaI detector efficiency η	Pd escape probability
^{104}Pd(0.555)	11.0	3.30	5.08(0.16)×10^5	2.3×10^{-3}	0.81
		2.40	4.35(0.17) ×10^4		
^{106}Pd(0.513)	27.3	3.30	7.47(0.18) ×10^5	2.3×10^{-3}	0.80
		3.00	4.00(0.12) ×10^5		
		2.70	1.98(0.06) ×10^5		
^{108}Pd(0.433)	26.7	3.30	1.19(0.03) ×10^6	2.2×10^{-3}	0.77
		3.00	7.02(0.19) ×10^5		
		2.70	3.76(0.08) ×10^5		
^{110}Pd(0.374)	11.8	3.30	1.82(0.04) ×10^6	2.1×10^{-3}	0.73
		3.00	1.11(0.02) ×10^6		
		2.70	6.15(0.13) ×10^5		

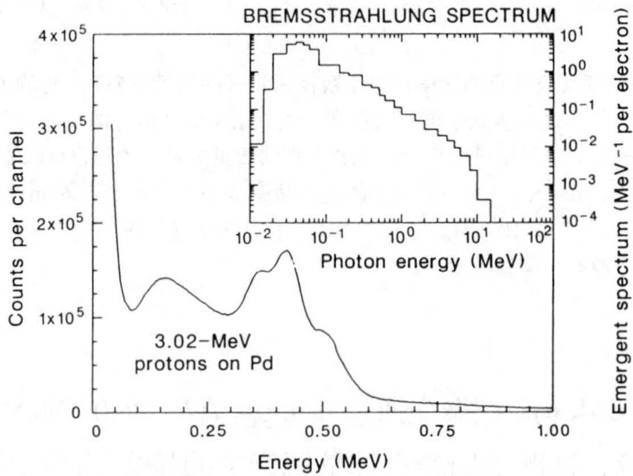

Fig. 4. Electric-quadrupole (E2) γ-ray lines from the isotopes of Pd, excited by 3.02-MeV protons, as measured by a 3 × 3 in. NaI detector. Three lines at 0.37, 0.43 and 0.51 MeV constitute a clear γ-ray signature for the reaction d(d, p)t (a fourth line at 0.56 MeV from ^{104}Pd is too weak to be seen here). Their collective backscatter peak is at ~0.16 MeV. Inset: The emergent bremsstrahlung spectrum for an electron emitted at 20 MeV at the centre of a cylinder of H_2O, of height 36 cm and diameter 36 cm. The normalization is for a single electron, with the γ-ray flux integrated over the cylinder surface; the area under the spectrum corresponds to 1.1 photons per electron, including a small δ-function contribution of 0.511-MeV annihilation quanta not shown in the figure. This spectrum was obtained using the Monte Carlo code ETRAN[15], which treats the coupled electron–photon transport using a recently developed set of bremsstrahlung cross-sections[16].

Convolving these line strengths with the resolution of the NaI detector, and using the efficiency factors from Table 1, a theoretically generated spectrum was scaled to optimally fit the 0.362–0.613-MeV region of each residual spectrum. The frequency distribution of the 1,114 fitted scaling factors (in units of mW of fusion power) is shown in Fig. 3*b*. None of the scaling factors exceeds 10 mW. This limit is comparable to the sensitivity of the calorimetric measurements made on these cells.

表 1. 3 MeV 的质子在钯电极中的辐射产额

同位素（E2 能量）(MeV)	同位素丰度	质子能量（MeV）	激发/μC	NaI 探测器效率 η	钯的逃逸概率
^{104}Pd(0.555)	11.0	3.30	$5.08(0.16)\times10^5$	2.3×10^{-3}	0.81
		2.40	$4.35(0.17)\times10^4$		
^{106}Pd(0.513)	27.3	3.30	$7.47(0.18)\times10^5$	2.3×10^{-3}	0.80
		3.00	$4.00(0.12)\times10^5$		
		2.70	$1.98(0.06)\times10^5$		
^{108}Pd(0.433)	26.7	3.30	$1.19(0.03)\times10^6$	2.2×10^{-3}	0.77
		3.00	$7.02(0.19)\times10^5$		
		2.70	$3.76(0.08)\times10^5$		
^{110}Pd(0.374)	11.8	3.30	$1.82(0.04)\times10^6$	2.1×10^{-3}	0.73
		3.00	$1.11(0.02)\times10^6$		
		2.70	$6.15(0.13)\times10^5$		

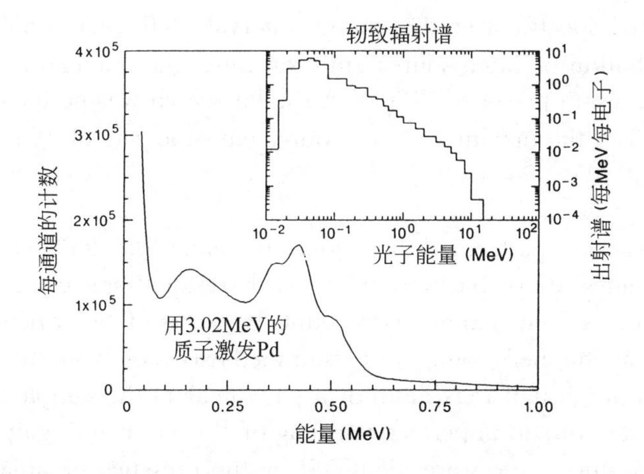

图 4. 来自钯同位素的电四极（E2）γ 射线峰，通过 3.02 MeV 的质子激发，用 3 in × 3 in 的 NaI 探测器测量。0.37 MeV、0.43 MeV 和 0.51 MeV 处的三个峰构成了 d(d, p)t 反应清晰的 γ 射线特征（0.56 MeV 处 ^{104}Pd 的第四个峰很弱，在这里难以看清）。它们共有的背散射峰在约 0.16 MeV 处。插图：从一个高 36 cm、直径 36 cm 的水柱中心出射的在 20 MeV 处的电子的出射韧致辐射谱。对单个电子进行归一化，将 γ 射线流沿柱状体的表面积分；光谱下的面积对应于每电子 1.1 个光子，包括 0.511 MeV 的湮灭光子的 δ 函数导致的一小部分（没有在图中显示）。这个 γ 谱使用蒙特卡洛代码 ETRAN[15] 得到，该代码使用最近发展的韧致辐射截面方法 [16] 对耦合的电子 – 光子输运进行了处理。

以 NaI 探测器的分辨率对这些峰强度进行卷积，并使用表 1 中的有效因子，对理论产生的 γ 谱按比例进行缩放以使其与每个剩余 γ 谱的 0.362~0.613 MeV 区域拟合最佳。1,114 个拟合比例因子(以聚变功率的 mW 为单位)的频数分布示于图 3*b* 中。没有一个比例因子超过 10 mW。这个限值与在这些电解池上进行的量热测量的灵敏度相似。

More stringent limits can be placed on this channel by recognizing that the tritium, emitted with kinetic energy of 1.0 MeV, can initiate the t(d, n)^{4}He reaction. By integrating the energy-dependent reaction cross-section[10] over the range of the tritium within the Pd, we obtain a d + t fusion probability of 4×10^{-5} per emitted tritium, assuming a 1:1 d:Pd ratio in the lattice. (We note that the probability p that a 1.0-MeV triton will produce a neutron within a deuterated palladium lattice via the reaction t(d, n)^{4}He can be expressed as a function of r, the deuterium-to-palladium ratio (by number): $p = a_0+a_1r+a_2r^2$, where $a_0 = 9.63\times10^{-8}$, $a_1 = 4.48\times10^{-5}$ and $a_2 = -5.40\times10^{-6}$. This expression, based on nuclear cross-sections from ref. 10 and hydrogen stopping power from ref. 17, is accurate to $\lesssim$10% and is valid for $0.4\leqslant r \leqslant 1.5$.) Monte Carlo calculations show that 39% of the resulting ~14-MeV (centre-of-mass) neutrons will thermalize and be captured by protons within the water tank, yielding a 2.2-MeV γ-ray signal. From the absence of this signal, we obtain an upper limit of the fusion power amplitude of 0.4 μW.

γ-ray flux limits: d(d, γ)^{4}He, d(p, γ)^{3}He, monoenergetic γ. A search was performed for 23.85-MeV γ-ray from d(d, γ)^{4}He by fitting the expected line profile of a 23.85-MeV γ-ray to the residual spectra over the energy interval 23.6–24.1 MeV. Figure 3*c* shows the resulting distribution of fitted source emission rates; the maximum fitted rate is 5 s^{-1}, corresponding to a fusion power of 20 pW. A similar search was performed for 5.49-MeV γ-rays from d(p, γ)^{3}He; the maximum fitted source emission rate of 10 s^{-1} corresponds to a fusion power of ~10 pW.

A general search was also performed throughout the entire 0.1–25.5-MeV energy interval for possible γ-ray lines above background in each residual spectrum. Candidates were required to have at least one channel with counts in excess of 3σ (σ being the propagated Poisson error for that channel's count) and a summed excess of 9σ in the adjacent channels spanning the full width at half maximum of a γ-ray peak at the sampled energy. The only candidates found were due to imperfect rescaling of the spectrum's gain and zero relative to averaged background; these were identified by the presence of adjacent positive and negative excursions of equal magnitude, with a zero-crossing at a known background peak position. No other candidate γ-ray lines were found. Allowing for detector efficiency, this yields the relationship between γ-ray emission-rate limit and energy (for cell 2-1) shown in Fig. 2*c*.

Internal-conversion electron flux limits: d(d, e$^-$)^{4}He. Even though nuclear de-excitation by internal conversion is greatly suppressed relative to radiative de-excitation for low-atomic-mass nuclei and for photon energies much greater than the electron mass, it has been suggested that cold fusion may in fact be occurring via internal conversion[11], which would produce an electron that carries off 23.8 MeV in kinetic energy.

Figure 4 shows the calculated bremsstrahlung spectrum emerging from a point-isotropic source of 20-MeV electrons located at the centre of a cylinder of water 36 cm in diameter (~4 MeV are assumed to have been lost within the Pd cathode[12]). The spectrum, normalized to one source electron, has an integrated area of 1.1 photons per electron.

通过认定以动能 1.0 MeV 放射出来的氚能够引发 $t(d, n)^4He$ 反应，可以确定这个通道更为严谨的限值。通过对钯中氚的能量范围内能量依赖反应的截面 [10] 进行积分，假设晶格中 d∶Pd 的比例为 1∶1，每发射一个氚引发 d + t 聚变的概率为 4×10^{-5}。（我们注意到 1.0 MeV 的氚核在氘化钯的晶格中通过反应 $t(d, n)^4He$ 产生一个中子的概率 p 可以表达为氘与钯之比（原子数之比）r 的函数：$p = a_0 + a_1 r + a_2 r^2$，其中 $a_0 = 9.63 \times 10^{-8}$，$a_1 = 4.48 \times 10^{-5}$，$a_2 = -5.40 \times 10^{-6}$。这个表达式是基于参考文献 10 中的核反应截面和文献 17 中的氢的遏止率得到的，其精度 ≤10%，在 $0.4 \leqslant r \leqslant 1.5$ 范围内有效。）蒙特卡罗计算显示，得到的约 14 MeV（质心）的中子中，39% 会热化并被水槽中的质子捕获，得到一个 2.2 MeV 的 γ 射线信号。从这一信号的缺失情况来看，我们得到核聚变功率幅值的上限为 0.4 μW。

γ 射线流的限值：$d(d, \gamma)^4$**He**，$d(p, \gamma)^3$**He**，**单能 γ 射线** 通过将预期的 23.85 MeV 的 γ 射线谱线轮廓对剩余 γ 谱在 23.6~24.1 MeV 能量区间内进行拟合，来搜索来自反应 $d(d, \gamma)^4He$ 的 23.85 MeV γ 射线。图 3*c* 显示了源发射速率分布的拟合结果；最大拟合速率为 5 s^{-1}，对应的核聚变功率为 20 pW。对 $d(p, \gamma)^3He$ 反应放出的 5.49 MeV 的 γ 射线也进行了类似的搜索；最大的源放射拟合速率 10 s^{-1} 对应于约 10 pW 的核聚变功率。

在每个剩余 γ 谱中对整个 0.1~25.5 MeV 能量区间内高于本底的可能 γ 射线峰进行了总体搜寻。符合条件的可能情况需要有至少一个通道计数高于 3σ（σ 为那个通道计数的传递泊松误差），并且延展到取样能量处 γ 射线峰的半高宽的临近通道计数和要大于 9σ。找到的唯一可能的 γ 射线峰是由于 γ 谱增益不完善的缩放和平均本底零点不准而造成的；这些可以通过其邻近的正漂移和负漂移的数量级相同，以及在一个已知的本底峰的位置过零点，来进行识别。没有发现其他的可能的 γ 射线峰。结合探测器效率，得到了 γ 射线放射速率限值和能量的相互关系（对电解池 2-1 而言），如图 2*c* 所示。

内转换电子流的限值：$\mathbf{d(d, e^-)^4He}$ 即使对于低原子量的核以及光子能量远高于电子质量的情形，内转换引起的核退激相对于辐射退激是被大大地抑制了的，仍有人认为冷核聚变实际上可以通过内转换发生 [11]，这一过程会产生携带有 23.8 MeV 动能的电子。

图 4 显示了计算出的一个位于 36 cm 直径的水圆柱体中心各向同性的 20 MeV 电子点源所产生的韧致辐射光谱（假定在钯阴极内有约 4 MeV 的损失 [12]）。归一化到一个源电子，光谱拥有的积分面积为每电子 1.1 个光子。这个韧致辐射光谱，经

This bremsstrahlung spectrum, modified by detector efficiency, was scaled to obtain an optimal fit to each residual spectrum, giving a best-fit number of bremsstrahlung photons for each spectrum. Figure 3*d* shows the frequency distribution of the electron emission rate corresponding to the fitted bremsstrahlung photon number. The maximum fitted election emission rate, 2.6×10^3 electrons per second, corresponds to a fusion power level of 10 nW.

Neutron Detection with Nuclear Track Detectors

A significantly lower limit than the 10 pW discussed above on the mean neutron production rate during a 67-h interval (16–19 May) was obtained with neutron-detecting sandwiches made of ^{235}U-enriched (80%) uranium foils and nuclear-track-detecting plastic, Lexan polycarbonate. Six sandwiches were installed in the water tank containing the cells (Fig. 1); each consisted of two uranium foils, each with an area of 1.4 cm^2 and mass ~0.078 g, held between two pieces of 0.01-in.-thick Lexan polycarbonate. One of the two foils was shielded against slow neutrons (<0.5 eV) by two cadmium covers.

The fission capture cross-section for ^{235}U is 1.4 barn at 2.5 MeV, increasing to 580 barn at thermal energies. Some of the neutrons emitted at 2.45 MeV from the cells will be captured by ^{235}U nuclei during their diffusion within the water bath. The fission fragments have a typical range of a few micrometres in U; those that escape the foil enter the plastic film and rupture polymer bonds, thereby creating a nuclear track that can be viewed microscopically after chemical etching with NaOH (ref. 13). (The track registration threshold of Lexan is such that the numerous α particles produced by the ^{238}U component of the U foils do not produce visible tracks.)

The absolute neutron detection efficiency was measured with a ^{252}Cf source in a large water tank, with foil sandwiches placed relative to the source identically to those in Pons' laboratory, and was found to be 1.3×10^{-5} fission tracks per emitted neutron for a foil with no Cd cover; for those foils with Cd covers, the efficiency was a factor of about three lower.

Fission track counts for each sandwich's pair of foils are shown in Fig. 1, the first number being the count for the foil without a Cd cover and the second being that with. There was no indication of an excess neutron signal from any of the cells; in fact, the control sandwich, adjacent to the water-tank wall, registered one of the highest fission-track counts, these being a measure of the dominant background source, cosmic-ray neutrons[14]. If we assume zero background, however, the fission track counts at cell 2-5 correspond to a mean neutron emission rate of 0.8 s^{-1} over the 67-h integration period, with a 99% confidence-level upper limit on neutron emission rate from cell 2-5 (based on counts in foils without Cd covers) of 1.8 s^{-1}, corresponding to a fusion power level of 0.9 pW.

Pons has informed us that "no neutron detectors were ever placed in a tank when a cell in that tank was generating excess enthalpy".

过探测器效率修正后，调整其比例以得到对每一剩余光谱的最佳拟合，从而给出每一个光谱最佳拟合的轫致辐射光子的数目。图 3*d* 显示对应于拟合的轫致辐射光子数目的电子发射速率频数分布。电子发射速率的最大拟合值为每秒 2.6×10^3 个电子，对应于核聚变功率水平为 10 nW。

使用核径迹探测器的中子探测

通过由富集 ^{235}U（80%）铀箔和核径迹探测塑料（即莱克森聚碳酸酯）组成的中子探测三明治结构，得到了一个明显小于 10 pW 的上限，这个值是根据前面讨论的基于 67 小时区间（5 月 16 日至 19 日）的平均中子产率得到的。六个三明治结构被安装在了含有电解池的水槽中（图 1）；每一个含有两层铀箔，每一层铀箔面积约为 1.4 cm^2，质量约为 0.078 g，夹在两层 0.01 in 厚的莱克森聚碳酸酯之间。两片铀箔之一用两层镉屏蔽掉慢中子（<0.5 eV）。

^{235}U 的裂变捕获截面在 2.5 MeV 处为 1.4 靶恩，在热能下增大到 580 靶恩。一些从电解池中放射的 2.45 MeV 的中子在水浴里扩散期间会被 ^{235}U 捕获。裂变碎片在铀中有一个数微米的长度范围；那些逃离铀箔的裂变碎片会进入塑料薄膜并破坏高分子化学键，从而产生一个核径迹，该径迹经 NaOH 腐蚀后可以通过显微镜观察到(参考文献 13)。(莱克森聚碳酸酯的径迹记录阈值使得铀箔中的 ^{238}U 组分产生的为数众多的 α 粒子不会产生可见的径迹。)

通过一个位于大水槽中的 ^{252}Cf 源对绝对的中子探测效率进行了测量，箔的三明治结构相对于源的放置位置与庞斯实验室的装置位置相同，结果在没有镉覆盖层的箔，每个放射的中子有 1.3×10^{-5} 个裂变径迹；对那些有镉覆盖层的箔，效率变为原来的$\frac{1}{3}$。

每一对由薄膜组成的三明治结构的裂变径迹计数如图 1 所示，第一个数字是没有镉覆盖层的箔，第二个对应有镉覆盖层的箔。没有迹象表明从任何一个电解池中曾放射出电中子本底高出 1 个中子的信号；实际上在临近水槽壁、用于对照的三明治结构中，记录到一个最高的裂变径迹计数，这些是对主要的本底源——宇宙射线中子[14]的测量。然而，如果我们假定本底为零，电解池 2-5 的裂变径迹计数在 67 小时积分区间内对应的平均中子发射速率就为 0.8 s^{-1}，来自电解池 2-5 的中子放射速率的上限置信水平为 99%,(基于没有镉覆盖层的箔所得的计数) 为 1.8 s^{-1},对应的核聚变功率水平为 0.9 pW。

庞斯已经告知我们“当水槽中的电解池产生过剩焓时，水槽中并没有放置过中子探测器”。

Fusion Limits and Excess Heat Production

During the 831 h (live time) of monitoring γ-ray emissions from electrolytic cells in Pons' laboratory, no evidence was seen of radiation from any known d + d (or p + d) fusion reaction. The upper limits placed on power from these reactions range between 10^{-12} and 10^{-6} W, which are many orders of magnitude lower than the sensitivity of the calorimetric measurements made by Pons' group; therefore, if a heat excess were to have occurred during our period of observation, one could conclude that no known fusion process contributed significantly to that excess.

At one point, the D_2O in cell 2-1 was observed to boil for ~2 h. Figure 5*a* shows an aggregate γ-ray spectrum for the 2.5 h that included this boiling episode, and Fig. 5*b* is the residual spectrum after subtraction of the previous 2.5 h of data. No spectral features of fusion are present in this residual spectrum. After completing our analysis, we were informed that we should not "reference these events as being due to release of excess thermal energy" (S. Pons, personal communication), because this boiling event may very well have a conventional explanation. Unfortunately we have not received any numerical data on excess heat production during the 831 h of our monitoring, so we are not able to correlate the absence of nuclear signatures with the presence of anomalous heat.

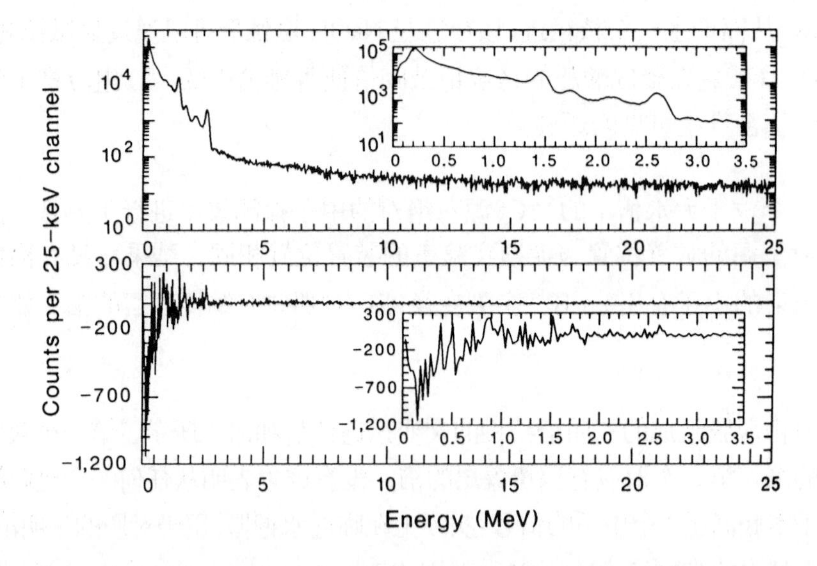

Fig. 5. *a*, γ-ray spectrum accumulated over a 2.5-h interval that included a ~2-h period during which cell 2-1 was observed to boil the D_2O electrolyte. The inset is an expansion of the low-energy end of the spectrum. *b*, Residual spectrum after subtraction of the spectrum accumulated for the preceding 2.5 h. The negative excess at low energy is due to a slight (0.1%) gain shift that occurred during this 5-h period (these spectra were not rescaled).

We were told, however, that "there was a two-hour segment in which there was excessive thermal release from cell 2-1... Unfortunately, your computer and detector were not under power at that time since they had not been reset from a power failure which had occurred in the lab" (S. Pons, personal communication). Although 48 h of data were

核聚变限值和过剩热量的产生

在庞斯实验室中监测电解池γ射线放射的831小时（活时间）期间，没有发现任何已知的d + d（或p + d）核聚变反应辐射的证据。这些反应的功率上限在10^{-12} W到10^{-6} W之间，这比庞斯研究组进行的量热测量的灵敏度低很多个数量级；因此，如果在我们的观察期间发生热量过剩，我们能够得出的结论是，没有已知的核聚变过程对这一过剩有明显贡献。

我们曾一度观察到电解池2-1中的D_2O沸腾了约2小时。图5*a*显示了包括这个沸腾过程的2.5小时的累积γ射线谱，图*b*是减去沸腾开始之前2.5小时的数据后的剩余γ谱，剩余γ谱没有出现核聚变的γ谱特征。在完成我们的分析之后，我们被告知我们不应该“把这些事件归因于有过量热能释放”（庞斯，个人交流），因为对这个沸腾的事件可以有一个很好的传统解释。不幸的是，在我们监测的831小时期间，我们并没有获得任何关于过剩热量产生的计算数据，所以我们不能够将核信号的缺失与异常热量的存在关联起来。

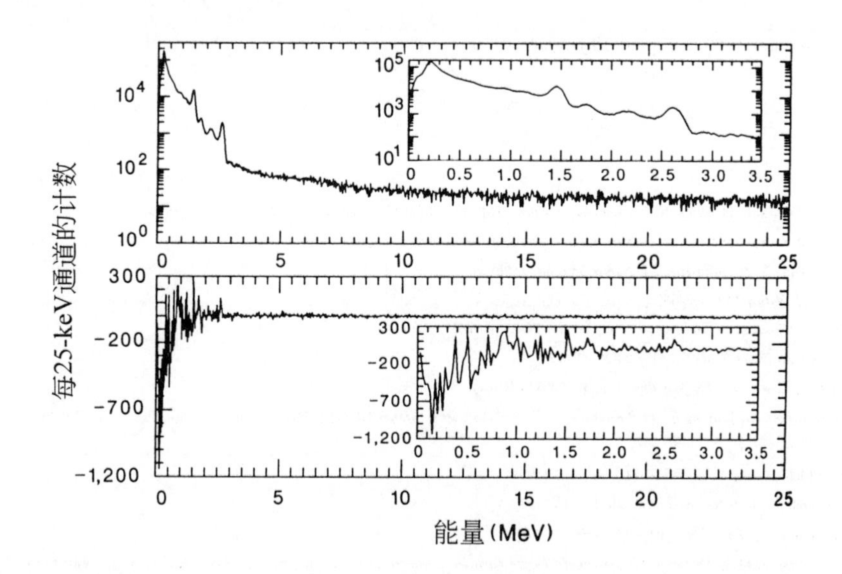

图5. *a*，累积2.5小时的γ射线谱。其中包括一个约2小时的区间，在此期间在电解池2-1中观察到了D_2O电解质沸腾。插图是对γ谱低能端的放大。*b*，扣除沸腾开始之前2.5小时累积的γ谱之后的剩余γ谱。低能量处的负盈余是源于发生在整个5小时期间的一个微小的（0.1%）增益漂移（这些γ谱没有经过再标定）。

然而，我们被告知“曾经有两个小时的时间段内电解池2-1有过剩的热量释放……不幸的是，当时你们的计算机和探测器并没有在工作，因为在实验室发生断电之后，它们没有得到重启”（庞斯，个人交流）。尽管因为雷击，48小时的数据确实丢失了，然而我们能估算出在这2小时的时段内，核聚变能量的平均上限对于

indeed lost because of a lightning strike, we can nevertheless estimate mean upper limits for fusion power of $\sim 10^{-2}$ W for d(d, p)t and 10^{-6} W for d(d, n)^{3}He during this 2-h episode, because a fraction of the neutrons produced from these reactions would activate the ^{23}Na in the NaI detector, producing ^{24}Na. As ^{24}Na decays with a 15.0-h half-life, a spectral signature of a neutron burst would be present even several days after the burst. None was observed, leading to the conclusion that neither the d(d, p)t nor the d(d, n)^{3}He reaction was responsible for this anomalous heat burst.

In addition, we later learned that a low-level, d.c. heat excess was observed during our monitoring period (S. Pons, EPRI Conference, University of Utah, 16 August 1989); if this is the case, this excess did not originate from known nuclear processes.

(**344**, 401-405; 1990)

M. H. Salamon*, M. E. Wrenn†, H. E. Bergeson*, K. C. Crawford‡, W. H. Delaney*†, C. L. Henderson†‡, Y. Q. Li*, J. A. Rusho*, G. M. Sandquist‡ & S. M. Seltzer§

* Department of Physics, † Environmental Radiation and Toxicology Laboratory and ‡ Nuclear Engineering Department, University of Utah, Salt Lake City, Utah 84112, USA

§ National Institute of Standards and Technology, Gaithersburg, Maryland 20899, USA

Received 26 September 1989; accepted 30 January 1990.

References:

1. Fleischmann, M., Pons, S. & Hawkins, M. *J. electroanalyt. Chem.* **261**, 301-308(1989); and erratum, **263**, 187-188 (1989).
2. Petrasso, R. D. *et al. Nature* **339**, 183-185(1989); and erratum, **339**, 264 (1989).
3. Fleischmann, M., Pons, S. & Hoffman, R. J. *Nature* **339**, 667 (1989).
4. Petrasso, R. D. *et al. Nature* **339**, 667-669 (1989).
5. Wolf, K. *et al. Proc. Workshop on Cold Fusion* Santa Fe, May 23-25, (1989).
6. *Proc. Workshop on Cold Fusion* Santa Fe, May 23-25, (1989).
7. Zel'dovich, Ya. B. & Gershtein, S. S. *Sov. Phys. Usp.* **3**, 593-623 (1961).
8. Barkas, W. H. & Berger, M. J. in *Studies in Penetration of Charged Particles in Matter* 103-172 (Natn. Acad. Sci. Natn. Res. Council Publn 1133, Washington, DC, 1964).
9. Stelson, P. H. & McGowan, F. K. *Phys. Rev.* **110**, 489-506 (1958).
10. Fowler, J. L. & Brolley, J. E. Jr *Rev. mod. Phys.* **28**, 103-134 (1956).
11. Walling, C. & Simons, J. *J. Phys. Chem.* **93**, 4693-4696 (1989).
12. Berger, M. J. & Seltzer, S. M. in *Studies in Penetration of Charged Particles in Matter* 69-98 (Natn. Acad. Sci. Natn. Res. Council Publn 1133, Washington, DC, 1964).
13. Fleischer, R. L., Price, P. B. & Walker, R. M. *Nuclear Tracks in Solids* (University of California Press, Berkeley, 1975).
14. Hess, W. N., Canfield, E. H. & Lingenfelter, R. E. *J. Geophys. Res.* **66**, 665-677 (1961).
15. Seltzer, S. M. in *Monte Carlo Transport of Electrons and Photons* (eds Jenkins, T. M., Nelson, W. R. & Rindi, A.) (Plenum, New York, 1988).
16. Seltzer, S. M. & Berger, M. J. *Nucl. Instrum. Meth.* **B12**, 95-134 (1985).
17. Andersen, H. H. & Ziegler, J. F. *Hydrogen Stopping Powers and Ranges in all Elements* (Pergamon, New York, 1977).

Acknowledgements. We thank S. Pons, M. Anderson and M. Hawkins for their hospitality during our work in their laboratory. We thank K. Wolf for alerting us to the t(d, n)^{4}He reaction that follows d(d, p)t, and K. Drexler for suggesting neutron detection via activation of ^{23}Na within our detector. We also thank R. Lloyd, W. Schier, D. Leavitt, F. Steinhausler and P. Bergstrom for valuable assistance, R. Petrasso for a careful review of an earlier manuscript, P. B. Price, M. Solarz, S. Barwick, R. Huber, R. Price and C. DeTar for helpful conversations, and R. Cooper for technical assistance. This work was supported by the State of Utah. S.M.S. was supported by the Office of Health and Environmental Research, US Department of Energy.

d(d, p)t 反应约为 10^{-2} W，对于 d(d, n)^{3}He 反应为 10^{-6} W，因为这些反应产生的中子的一部分会活化 NaI 探测器中的 ^{23}Na，从而产生 ^{24}Na。由于 ^{24}Na 以 15 小时的半衰期进行衰变，一个猝发中子的光谱信号在猝发之后的数天还可以观察到。由于什么也没有观察到，我们得出了以下结论，即不论是 d(d, p)t 还是 d(d, n)^{3}He 反应都不是这一异常的猝发热量的原因。

另外，我们后来知道，在我们的监测期间观察到了一个低水平的热量过剩（庞斯，美国电力科学研究院（EPRI）会议，犹他大学，1989 年 8 月 16 日）；如果这是事实，这一过剩并非源自已知的核过程。

（李琦 翻译；李兴中 审稿）

Positioning Single Atoms with a Scanning Tunnelling Microscope

D. M. Eigler and E. K. Schweizer

Editor's Note

The scanning tunnelling microscope was invented in the early 1980s as a tool for imaging the surfaces of materials at atomic resolution. As its needle-like metal tip moves just above the surface, electrons can jump the gap to produce an electrical current that changes as the gap size varies, offering a topographic map. In the late 1980s it became clear that electrical pulses applied to the tip could remove or alter the surface material. Here Don Eigler and Eric Schweizer at IBM's research laboratories in California show that the tip can be used to drag individual xenon atoms over a surface and position them at will. Their spelling of the company name in 35 atoms became an icon of nanotechnology.

Since its invention in the early 1980s by Binnig and Rohrer[1,2], the scanning tunnelling microscope (STM) has provided images of surfaces and adsorbed atoms and molecules with unprecedented resolution. The STM has also been used to modify surfaces, for example by locally pinning molecules to a surface[3] and by transfer of an atom from the STM tip to the surface[4]. Here we report the use of the STM at low temperatures (4 K) to position individual xenon atoms on a single-crystal nickel surface with atomic precision. This capacity has allowed us to fabricate rudimentary structures of our own design, atom by atom. The processes we describe are in principle applicable to molecules also. In view of the device-like characteristics reported for single atoms on surfaces[5,6], the possibilities for perhaps the ultimate in device miniaturization are evident.

THE tip of an STM always exerts a finite force on an adsorbate atom. This force contains both Van der Waals and electrostatic contributions. By adjusting the position and the voltage of the tip we may tune both the magnitude and direction of this force. This, taken together with the fact that it generally requires less force to move an atom along a surface than to pull it away from the surface, makes it possible to set these parameters such that the STM tip can pull an atom across a surface while the atom remains bound to the surface. Our decision to study xenon on nickel (110) was dictated by the requirement that the corrugations in the surface potential be sufficiently large for the xenon atoms to be imaged without inadvertently moving them, yet sufficiently small that, when desired, enough lateral force could be exerted to move xenon atoms across the surface.

用扫描隧道显微镜定位单个原子

艾格勒，施魏策尔

编者按

扫描隧道显微镜发明于 20 世纪 80 年代初，是在原子级分辨率上对材料表面成像的一种工具。随着它的针状金属尖端在材料表面上移动，电子通过隧道效应穿过针尖与样品表面的间隙从而产生电流，电流强度随间隙大小不同而发生改变，从而描绘出材料表面的形貌。20 世纪 80 年代后期，通过在探针尖端应用电脉冲可以移动或改变材料表面。本文中加利福尼亚 IBM 研究实验室的唐 · 艾格勒和埃里克 · 施魏策尔描述了利用针尖在材料表面拖动单个氙原子并进行任意定位。他们用 35 个原子拼写出的公司名称成为纳米技术的一个标志。

自从宾尼希和罗雷尔在 20 世纪 80 年代初发明扫描隧道显微镜（STM）[1,2] 以来，它就以空前的分辨率为我们提供了关于材料表面和被吸附原子及分子的图像。STM 还被用来修饰表面，例如将分子局部地固定在表面上[3]，以及将一个原子从 STM 的探针尖端转移到材料表面[4]。本文中我们报道了在低温下（温度为 4 K）利用 STM 以原子级精度把单个氙原子定位在单晶镍表面上。这种能力使我们可以逐个原子地制作我们设计的基本结构。我们所描述的这个过程原则上同样适用于分子。从已报道的表面上单个原子的独特性质[5,6] 来看，最终实现功能器件最小化的可能性是显而易见的。

STM 的探针尖端总是对被吸附原子施加一个有限的力。这个力中同时包含着范德华力和静电力的贡献。通过调整针尖的位置和电压，我们可以调节此力的大小和方向。此外，考虑到沿着表面移动原子比将其拖离表面更省力这一事实，我们可以通过设置这些参数，使 STM 针尖拉动一个原子穿过表面的同时，仍保持该原子结合在表面上。出于以下原因我们决定研究镍（110）面上的氙：表面电势起伏可以足够大，使得我们可以对氙原子成像，而又不会不经意地移动它们；另外在需要时，表面电势的起伏还可以足够小，使得我们可以施加足够的侧向力在表面上移动氙原子。

The experiments were performed using an STM contained in an ultra-high-vacuum system and cooled to 4 K. The entire chamber housing the microscope was cooled to 4 K, which so reduced the contamination rate of the sample surface through adsorption of residual gases that no measurable contamination occurred over weeks. The stability of the sample and of the microscope due to the low temperature are such that one may perform experiments on a single atom for days at a time. This stability proved to be an operational necessity for performing these experiments. This notwithstanding, it is important to realize that the process that we describe for sliding atoms on a surface is fundamentally temperature independent.

The nickel sample was processed by cycles of argon-ion sputtering, annealing in a partial pressure of oxygen to remove surface carbon, and flash annealing. After cooling to 4 K the sample was imaged with the STM and found to be of acceptable quality. Figure 1*a* is an image of the surface taken under constant-current scanning conditions after dosing with xenon to the desired coverage. This image was obtained with a tip bias voltage of 0.010 V relative to the sample, and a tunnel current of 10^{-9} A. Each xenon atom appears as a 1.6-Å-high bump on the surface, apparently at random locations. At this gap impedance the interaction of the xenon with the tip is sufficiently weak to leave the xenon essentially unperturbed during the imaging process.

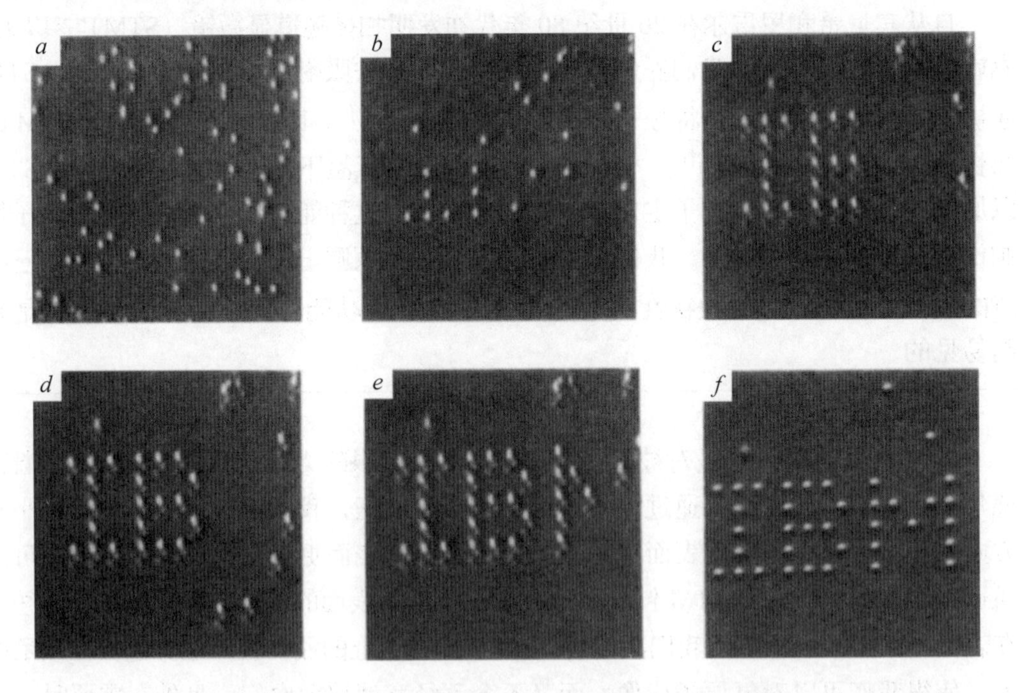

Fig. 1. A sequence of STM images taken during the construction of a patterned array of xenon atoms on a nickel (110) surface. Grey scale is assigned according to the slope of the surface. The atomic structure of the nickel surface is not resolved. The $\langle 1\bar{1}0 \rangle$ direction runs vertically. *a*, The surface after xenon dosing. *b–f*, Various stages during the construction. Each letter is 50 Å from top to bottom.

实验在一台含有超高真空系统的STM上进行，同时系统被冷却至4 K。将容纳显微镜的整个样品室冷却至4 K，通过吸收残余气体可以降低样品表面的污染速率，使其在几星期内没有可测量到的污染发生。低温保证了样品和显微镜的稳定性，这使得人们可以对单个原子进行一次长达数天的实验。我们已证明这一稳定性是进行实验的必要操作条件。尽管如此，还要注意到，我们所描述的在表面上移动原子的过程本质上是不依赖于温度的。

我们利用氩离子对镍样品进行反复喷溅，在氧分压中退火以除去镍表面的碳，且经快速退火完成样品处理。在冷却至4 K后，用STM对样品成像，确定其表面足够清洁。图1*a*是将氙添加到预期覆盖度后在恒定电流扫描条件下获得的表面图像。该图像是在相对于样品0.010 V的针尖偏压和10^{-9} A的隧道电流条件下获得的。每个氙原子看起来就像表面上一个1.6 Å高的团块，并且明显地处于随机位置。在隧道结阻抗的条件下，氙原子与针尖的相互作用非常弱，因此在成像过程中，氙原子基本不受到扰动。

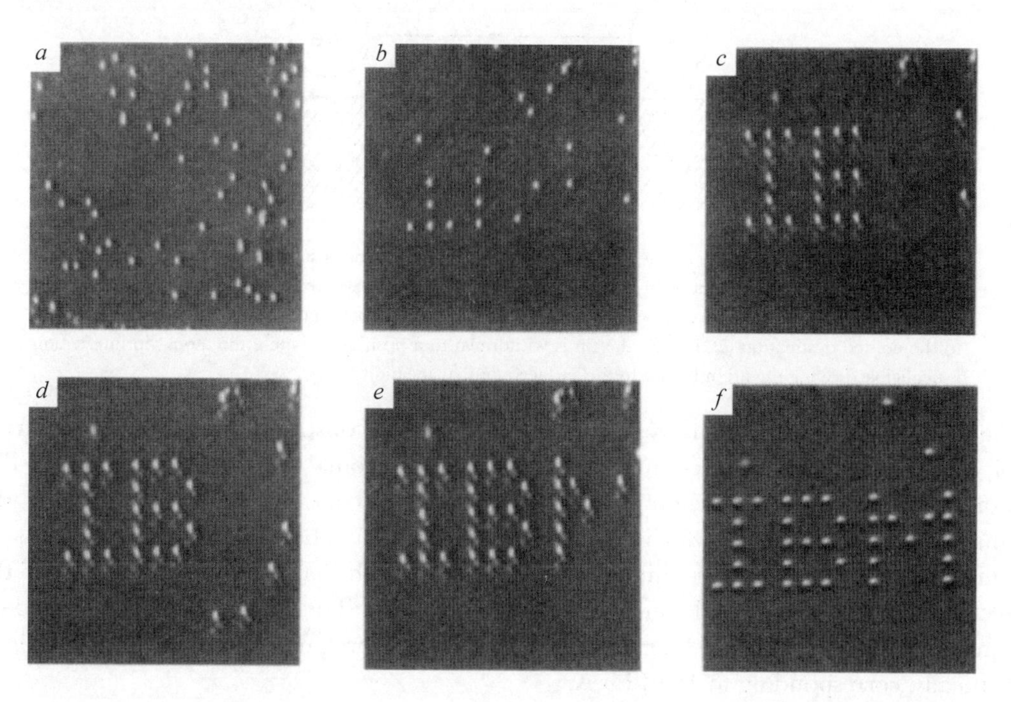

图1. 在镍（110）表面上构建排列成图案的氙原子阵列过程中拍摄的一系列STM图像。根据表面的斜率确定灰度。对镍表面的原子结构未进行分辨。$\langle 1\bar{1}0 \rangle$方向为垂直指向。*a*，加入氙之后的表面。*b*~*f*，构建过程中的各个阶段。每个字母从顶部到底部是50 Å。

To move an atom we follow the sequence of steps depicted in Fig. 2. We begin by operating the microscope in the nonperturbative imaging mode described above to locate the atom to be moved and to target its destination. We stop scanning and place the tip directly above the atom to be moved (*a*). We then increase the tip–atom interaction by lowering the tip toward the atom (*b*); this is achieved by changing the required tunnel current to a higher value, typically in the range $1–6\times10^{-8}$ A, which causes the tip to move towards the atom until the new tunnel current is reached. We then move the tip under closed-loop conditions across the surface, (*c*), at a speed of 4 Å per second to the desired destination (*d*), dragging the xenon atom with it. The tip is then withdrawn (*e*) by reducing the tunnel current to the value used for imaging. This effectively terminates the attraction between the xenon and the tip, leaving the xenon bound to the surface at the desired location.

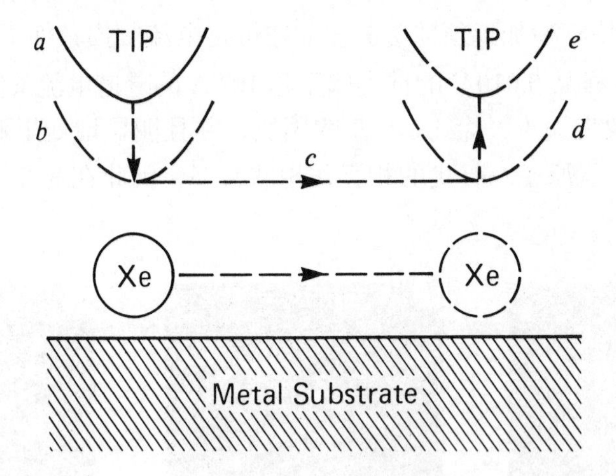

Fig. 2. A schematic illustration of the process for sliding an atom across a surface. The atom is located and the tip is placed directly over it (*a*). The tip is lowered to position (*b*), where the atom–tip attractive force is sufficient to keep the atom located beneath the tip when the tip is subsequently moved across the surface (*c*) to the desired destination (*d*). Finally, the tip is withdrawn to a position (*e*) where the atom–tip interaction is negligible, leaving the atom bound to the surface at a new location.

Figure 1 is a sequence of images taken during our first construction of a patterned array of atoms, and demonstrates our ability to position atoms with atomic precision. The exact periodicity of the xenon spacing is derived from the crystalline structure of the underlying nickel surface (which is not resolved in Fig. 1). The nickel (110) surface has an unreconstructed rectangular unit cell and is oriented such that the short dimension of the rectangular surface unit cell runs vertically in the image. The xenon atoms are spaced on a rectangular grid which is four nickel unit cells long horizontally and five unit cells long vertically, corresponding to 14×12.5 Å.

Although the STM tip is made from tungsten wire, the true chemical identity and the structure of the outermost atoms of the tip are not known to us. We find that for any given tip and bias voltage, there is a threshold height below which the tip must be located to be able to move xenon atoms parallel to the rows of nickel atoms, and a lower threshold

按照图 2 所示的一系列步骤来移动单个原子。用上文所描述的无干扰成像模式操作显微镜，找到待移动原子，瞄准其目的地。然后停止扫描，将针尖置于待移动原子的正上方（*a*)。接着，我们通过将针尖向着原子下降（*b*）来增加针尖–原子之间的相互作用；通过增大所需隧道电流值，一般是达到 1×10^{-8} A 至 6×10^{-8} A 的范围时，即可实现这一点，这一电流范围使得针尖移向原子，直至达到新的隧道电流值。接着，我们在闭合回路的条件下移动针尖穿过表面（*c*)，以 4 Å/s 的速度移向目的地(*d*)，同时始终拖带着氙原子。然后，通过将隧道电流减少到成像所需的值使针尖后撤(*e*)。这就有效地终止了氙与针尖之间的吸引力，将氙留在了表面上所预期的位置。

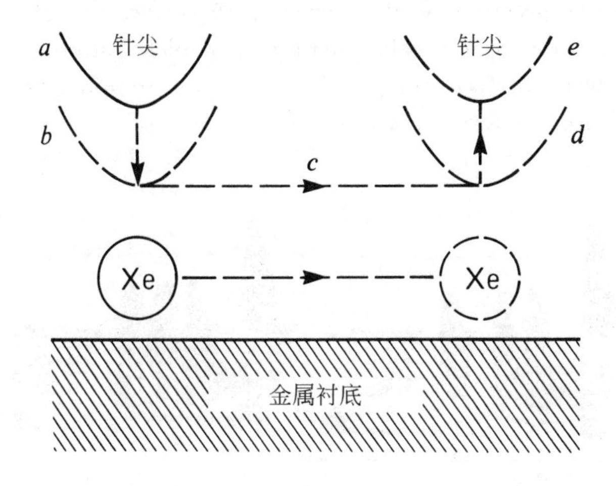

图2. 在表面移动原子过程的示意图。定位原子，针尖位于其正上方（*a*)。将针尖降低到一定位置（*b*)，在针尖随后移过表面 (*c*) 到达目的地 (*d*) 期间，原子–针尖之间的吸引力足以维持原子固定在针尖下方。最后，将针尖后撤到一个位置 (*e*)，此时原子–针尖相互作用小到可忽略，把原子留在表面上的一个新位置。

图 1 是在我们首次构建排列成图案的氙原子阵列过程中所拍摄的一系列图像，证明了我们能够以原子级精度放置原子。氙原子间距的精确周期可以利用背景镍表面的晶体结构（在图 1 中没有进行分辨）得到。镍（110）表面有一个未重现出来的矩形单胞，其方向为矩形表面单胞的短边处于图像中的垂直方向。这些氙原子间隔排列在一个矩形网格中，水平方向上为四个镍单胞长，垂直方向上为五个晶胞长，面积相当于 14 Å × 12.5 Å。

尽管 STM 针尖是用钨丝制成的，但是我们并不知道针尖最外围原子的真实化学特性与结构。我们发现对于任何给定的针尖和偏压，都存在着一个临界高度，针尖必须位于这个高度以下才能沿着平行于镍原子阵列的方向移动氙原子，还存在着一

height for movement perpendicular to the rows of nickel atoms. This is consistent with a simple model wherein the xenon interaction with the metal surface is approximated by pairwise interactions with the individual nickel atoms. Simple investigations showed that the magnitude or sign of the applied voltage had no significant effect on the threshold tip height. This suggests that the dominant force between the tip and the xenon atom is due to the Van der Waals interaction, but this tentative conclusion requires further investigation.

In Fig. 3 we show a sequence of images demonstrating how a simple structure, the linear multimer, may be fabricated using this process. First we slide xenon atoms into a chosen row of nickel atoms. We next slide a xenon atom along the row to a position where it will bind with a neighbouring xenon to form a dimer. We repeat the process, forming a linear trimer, then a linear tetramer, and so on. From these images we find that the linear xenon chain along the $\langle 1\bar{1}0 \rangle$ direction of the nickel (110) surface is stable, the xenon atoms occupying every other surface unit cell along a row of nickel atoms. Attempts to pack the xenon atoms closer were unsuccessful. We find the xenon–xenon spacing along the row to be uniform (excluding end effects) to within 0.2 Å.

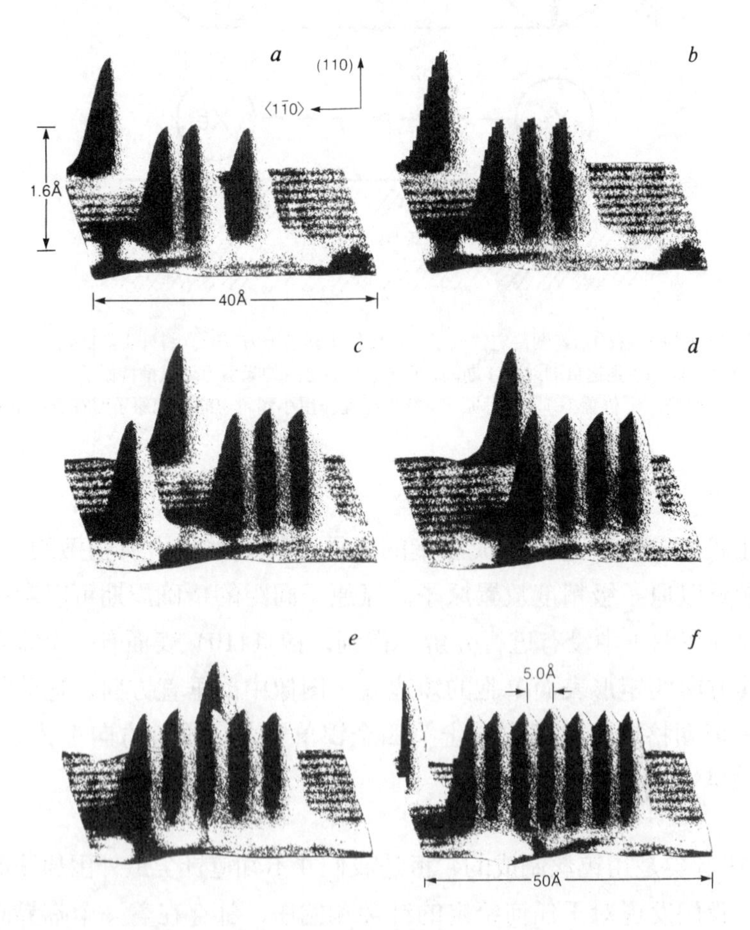

Fig. 3. Various stages in the construction of a linear chain of xenon atoms on the nickel (110) surface. The individual xenon atoms appear as 1.6-Å-high protrusions in these images. The rows of nickel atoms

个垂直于镍原子阵列移动时的更低临界高度。这与如下的简单模型是相一致的，其中氙与金属表面的相互作用近似于单个镍原子的成对相互作用。简单的研究表明外加电压的大小和方向对于临界针尖高度没有显著影响。这意味着针尖与氙原子之间的主要作用力应归结为范德华相互作用，但是这个初步结论还需要进一步的研究。

在图 3 中，我们给出了一系列图像来说明怎样利用这个过程来搭建一个简单结构，即线型多聚体。首先，将氙原子滑入一个选定的镍原子行中。接下来，沿着该行滑动一个氙原子使其到达一个能与邻近氙原子连接形成二聚体的位置。我们重复这一过程，形成一个线型三聚体，接着是线型四聚体，依此类推。从这些图像中，我们发现沿着镍原子（110）表面上〈$1\bar{1}0$〉方向的线型氙原子链是稳定的，氙原子沿着一行镍原子间隔占据表面单胞。使氙原子堆积得更近的尝试是不成功的。我们发现沿着行方向的氙－氙间距是相同的（除了末端效应以外），误差在 0.2 Å 以内。

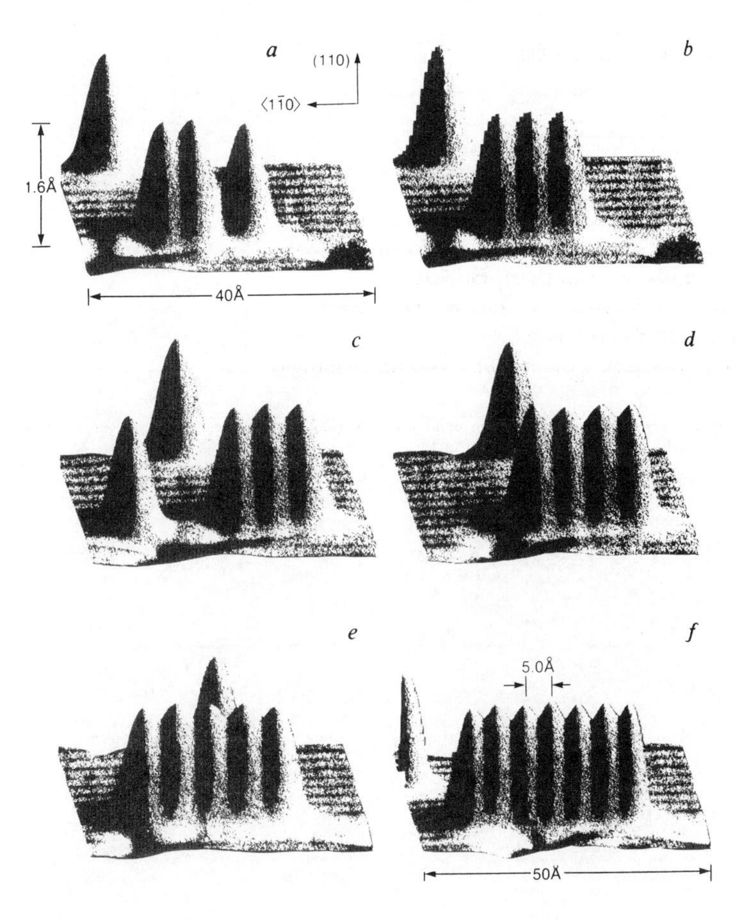

图 3. 在镍（110）表面上构建线型氙原子链的各个阶段。在这些图像中单个氙原子显示为 1.6 Å 高的突出物。镍原子行看起来犹如明暗交替的条纹。可以看到镍表面有若干处点缺陷。*a*，已装配好的氙原子

are visible as alternating light and dark stripes. Several point defects are visible in the nickel surface. *a*, The assembled xenon dimer. To the right of the dimer, a xenon atom has been moved into position for forming a xenon trimer. *b*, Formation of the xenon linear trimer. *c–f*, Various stages in construction of the linear heptamer, a process that can be completed in an hour. The xenon atoms are 5 Å apart, occupying every other unit cell of the nickel surface.

We anticipate that there will be a limiting class of adsorbed atoms and molecules that may be positioned by this method. Many new avenues of investigation are open to us. It should be possible to assemble or modify certain molecules in this way. We can build novel structures that would otherwise be unobtainable. This will allow a new class of surface studies that use the STM both to fabricate overlayer structures and to probe their properties. The prospect of atomic-scale logic circuits and other devices is a little less remote.

(**344**, 524-526; 1990)

D. M. Eigler & E. K. Schweizer*
IBM Research Division, Almaden Research Center, 650 Harry Rd, San Jose, California 95120, USA
*Permanent address: Fritz-Haber-Institut, Faradayweg 4-6, D-7000 Berlin 33, FRG.

Received 17 January; accepted 9 March 1990.

References:

1. Binnig, G., Rohrer, H., Gerber, Ch. & Weibel, E. *Appl. Phys. Lett.* **40**, 178-180 (1982).
2. Binnig, G., Rohrer, H., Gerber, Ch. & Weibel, E. *Phys. Rev. Lett.* **49**, 57-61 (1982).
3. Foster, J. S., Frommer, J. E. & Arnett, P. C. *Nature* **331**, 324-326 (1988).
4. Becker, R. S., Golovchenko, J. A. & Swartzentruber, B. S. *Nature* **325**, 419-421 (1987).
5. Lyo, I.-W. & Avouris, P. *Science* **245**, 1369-1371 (1989).
6. Bedrossion, P., Chen, D. M., Mortensen, K. & Golovchenko, J. A. *Nature* **342**, 258-260 (1989).

Acknowledgements. This work would not have occurred were it not for the patient and visionary management of the IBM Research Division.

二聚体。在二聚体的右侧，一个氙原子已进入恰当的位置，准备形成一个三聚体。*b*，氙原子线型三聚体的形成。*c~f*，构建线形七聚体的各个阶段，这个过程能够在一小时内完成。氙原子间距为 5 Å，各自占据相邻的镍表面单胞。

我们预期将会有一类特定的被吸附的原子和分子可以用这种方法来定位。很多新的研究途径展现在我们面前。应该有可能以这种方式来组装或修饰某些分子。我们能够构建出用其他办法无法得到的新型结构。这将引出一系列新的表面研究工作，即利用 STM 构建吸附层结构并探查其性质。这样，离实现原子尺度的逻辑电路和其他器件的梦想又近了一步。

（王耀杨 翻译；王琛 审稿）

Light-emitting Diodes Based on Conjugated Polymers

J. H. Burroughes *et al.*

Editor's Note

After the discovery of electrically conducting polymers in the 1970s, subsequent work identified several organic polymers with semiconducting electrical behaviour. One of them—poly (*p*-phenylene vinylene) or PPV—was prepared by Richard Friend and his co-workers at Cambridge. Here they show that PPV can be used in polymer light-emitting diodes (LEDs), which emit a yellowish light when an electrical current is passed through them. In conventional inorganic LEDs, new colours demand new materials. But the wavelength of light from polymer LEDs can be tuned by chemically modifying the polymer backbone. Such devices, spanning the entire visible spectrum, are now used in prototype commercial light-emitting displays. Being all-plastic, they can be rolled up like paper and fabricated using cheap printing technologies.

Conjugated polymers are organic semiconductors, the semiconducting behaviour being associated with the π molecular orbitals delocalized along the polymer chain. Their main advantage over non-polymeric organic semiconductors is the possibility of processing the polymer to form useful and robust structures. The response of the system to electronic excitation is nonlinear—the injection of an electron and a hole on the conjugated chain can lead to a self-localized excited state which can then decay radiatively, suggesting the possibility of using these materials in electroluminescent devices. We demonstrate here that poly (*p*-phenylene vinylene), prepared by way of a solution-processable precursor, can be used as the active element in a large-area light-emitting diode. The combination of good structural properties of this polymer, its ease of fabrication, and light emission in the green–yellow part of the spectrum with reasonably high efficiency, suggests that the polymer can be used for the development of large-area light-emitting displays.

THERE has been long-standing interest in the development of solid-state light-emitting devices. Efficient light generation is achieved in inorganic semiconductors with direct band gaps, such as GaAs, but these are not easily or economically used in large-area displays. For this, systems based on polycrystalline ZnS have been developed, although low efficiencies and poor reliability have prevented large-scale production. Because of the high photoluminescence quantum yields common in organic molecular semiconductors, there has long been interest in the possibility of light emission by these organic semiconductors through charge injection under a high applied field (electroluminescence)[1-7]. Light-emitting devices are fabricated by vacuum sublimation of the organic layers, and although the

基于共轭聚合物的发光二极管

伯勒斯等

编者按

20 世纪 70 年代发现了导电聚合物，几种具有半导体电性能的有机聚合物被随后的研究所证实。其中之一的聚对苯撑乙烯（PPV）是由剑桥大学的理查德·弗兰德及其同事们制备的，它可被用在聚合物发光二极管（LED）中，当有电流通过时该器件发出淡黄色的光。对于传统的无机 LED 来说，若想得到新的发光颜色需要新的材料。但是，聚合物 LED 的发光波长可以通过对聚合物骨架进行化学修饰来调节，器件的发射光谱覆盖整个可见光谱区，现在它们被用作商业发光显示器的原型器件。可以采用低成本的印刷工艺制作全塑料的聚合物 LED，它能像纸一样卷起。

共轭聚合物是有机半导体，其半导体性能与沿聚合链离域的 π 分子轨道有关。与非聚合物的有机半导体相比，共轭聚合物的主要优点是它的可加工性，能形成有用而又耐用的结构。聚合物对电激发的响应是非线性的，在共轭链上注入一个电子和一个空穴能够形成一个自局域激发态，随后它能够以辐射方式衰减，这就表明可以在电致发光器件中应用这些材料。这里我们要论述的聚对苯撑乙烯，是通过它的前驱体溶液进行加工而获得的，它被用作大面积发光二极管中的活性成分。该聚合物具有良好的结构性质，制备简单，在光谱中的绿–黄光波段有相当高的发光效率，这些优点使该聚合物能够用于发展大面积发光显示器。

人们对于固态发光器件的发展保持着长久的兴趣。诸如砷化镓等直接带隙无机半导体能实现有效发光，但是将它们用于大面积显示器既不容易也不经济。基于此，已发展出基于多晶硫化锌的体系，但效率低和可靠性较差等缺点使其不宜进行大规模生产。由于有机分子半导体普遍具有高的光致发光量子效率，长期以来，人们一直在关注利用这些有机半导体在外加强场条件下通过电荷注入而发光（电致发光）的可能性[1-7]。发光器件是通过有机层的真空升华而制成的，尽管它的发光效率很高且发光颜色选择范围很大，但是通常存在着这样的问题：升华而得到的有机薄

efficiencies and selection of colour of the emission are very good, there are in general problems associated with the long-term stability of the sublimed organic film against recrystallization and other structural changes.

One way to improve the structural stability of these organic layers is to move from molecular to macromolecular materials, and conjugated polymers are a good choice in that they can, in principle, provide both good charge transport and also high quantum efficiency for the luminescence. Much of the interest in conjugated polymers has been in their properties as conducting materials, usually achieved at high levels of chemical doping[8], and there has been comparatively little interest in their luminescence. One reason for this is that polyacetylene, the most widely studied of these materials, shows only very weak photoluminescence. But conjugated polymers that have larger semiconductor gaps, and that can be prepared in a sufficiently pure form to control non-radiative decay of excited states at defect sites, can show high quantum yields for photoluminescence. Among these, poly(*p*-phenylene vinylene) or PPV can be conveniently made into high-quality films and shows strong photoluminescence in a band centred near 2.2 eV, just below the threshold for π to π^* interband transitions[9,10].

We synthesized PPV (I) using a solution-processable precursor polymer (II), as shown in Fig. 1. This precursor polymer is conveniently prepared from α,α'-dichloro-*p*-xylene (III), through polymerization of the sulphonium salt intermediate (IV)[11-13]. We carried out the polymerization in a water/methanol mixture in the presence of base and, after termination, dialysed the reaction mixture against distilled water. The solvent was removed and the precursor polymer redissolved in methanol. We find that this is a good solvent for spin-coating thin films of the precursor polymer on suitable substrates. After thermal conversion (typically $\geqslant$250 °C, *in vacuo*, for 10 h), the films of PPV (typical thickness 100 nm) are homogeneous, dense and uniform. Furthermore, they are robust and intractable, stable in air at room temperature, and at temperatures >300 °C in a vacuum[11].

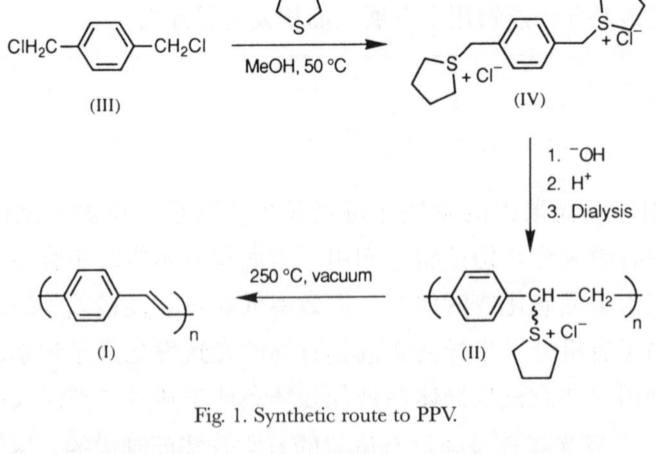

Fig. 1. Synthetic route to PPV.

Structures for electroluminescence studies were fabricated with the PPV film formed on a bottom electrode deposited on a suitable substrate (such as glass), and with the top electrode

膜的长期稳定性不够强，导致其易于发生重结晶和其他结构性变化。

改善这些有机层的结构稳定性的一种方法是从分子材料转向大分子材料，而共轭聚合物就是一个好的选择，因为从原理上讲，它们不仅能提供良好的电荷传输并能为发光提供高的量子效率。对于共轭聚合物的极大兴趣一直集中于它们作为导体材料的性质上，这些性质通常是通过大量的化学掺杂来实现的[8]，而它们的发光现象极少引起人们的注意。出现这种情况的一个原因是它们中被广泛研究的聚乙炔只呈现出很弱的光致发光。但是，那些具有较宽半导体能隙并且可以制备成十分纯的形式来控制其在缺陷位处的激发态非辐射衰减的共轭聚合物，能够表现出高光致发光的量子效率。其中，聚对苯撑乙烯（PPV）能很方便地制成高质量的薄膜，并且在一个以 2.2 eV 为中心的带显示出很强的光致发光，刚好低于从 π到 π^* 的带间跃迁的阈值[9,10]。

我们利用一种可溶液加工的前驱体聚合物（II）合成了 PPV（I），如图 1 所示。该前驱体聚合物制备起来很方便，是由 α,α'-二氯对二甲苯（III）通过巯盐中间产物（IV）的聚合[11-13]得到的。在碱性条件下的水/甲醇混合物中进行聚合。反应终止后，对反应混合物进行蒸馏水透析。除去溶剂，将前驱体聚合物重新溶于甲醇中。我们发现对于在适当基质上旋涂制备前驱体聚合物薄膜来说，甲醇是一种好的溶剂。经过热转化之后（典型条件是 ≥250 ℃，真空中持续 10 小时），形成的 PPV 薄膜（典型厚度为100 nm）是同质、致密而均匀的。此外，该薄膜坚固而无延展性，在室温下时空气中以及温度＞300 ℃时在真空中都是稳定的[11]。

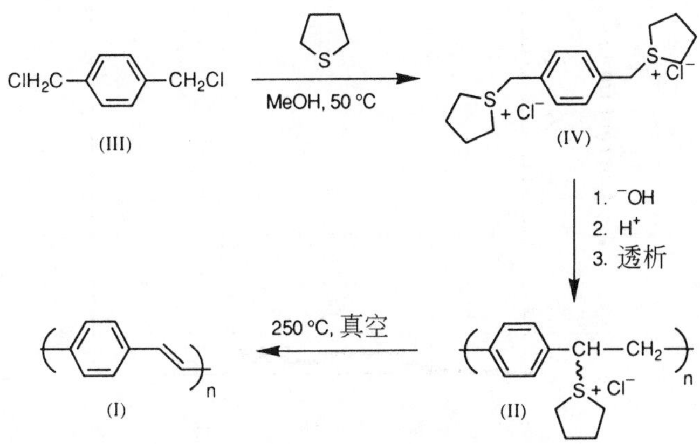

图 1. PPV的合成路线

用于进行电致发光研究的器件的制备过程如下：先在适当基底（例如玻璃）上沉积一个底部电极，之后在该电极上形成 PPV 薄膜，最后在全部转换的 PPV 薄膜

formed onto the fully converted PPV film. For the negative, electron-injecting contact we use materials with a low work function, and for the positive, hole-injecting contact, we use materials with a high work function. At least one of these layers must be semi-transparent for light emission normal to the plane of the device, and for this we have used both indium oxide, deposited by ion-beam sputtering[14] and thin aluminium (typically 7–15 nm). We found that aluminium exposed to air to allow formation of a thin oxide coating, gold and indium oxide can all be used as the positive electrode material, and that aluminium, magnesium silver alloy and amorphous silicon hydrogen alloys prepared by radiofrequency sputtering are suitable as the negative electrode materials. The high stability of the PPV film allows easy deposition of the top contact layer, and we were able to form this contact using thermal evaporation for metals and ion-beam sputtering for indium oxide.

Figures 2 and 3 show typical characteristics for devices having indium oxide as the bottom contact and aluminium as the top contact. The threshold for substantial charge injection is just below 14 V, at a field of 2×10^{6} V cm^{-1}, and the integrated light output is approximately linear with current. Figure 4 shows the spectrally resolved output for a device at various temperatures. The spectrum is very similar to that measured in photoluminescence, with a peak near 2.2 eV and well resolved phonon structure[9,10]. These devices therefore emit in the green–yellow part of the spectrum, and can be easily seen under normal laboratory lighting. The quantum efficiency (photons emitted per electron injected) is moderate, but not as high as reported for some of the structures made with molecular materials[2-7]. The quantum efficiencies for our PPV devices were up to 0.05%. We found that the failure mode of these devices is usually associated with failure at the polymer/thin metal interface and is probably due to local Joule heating there.

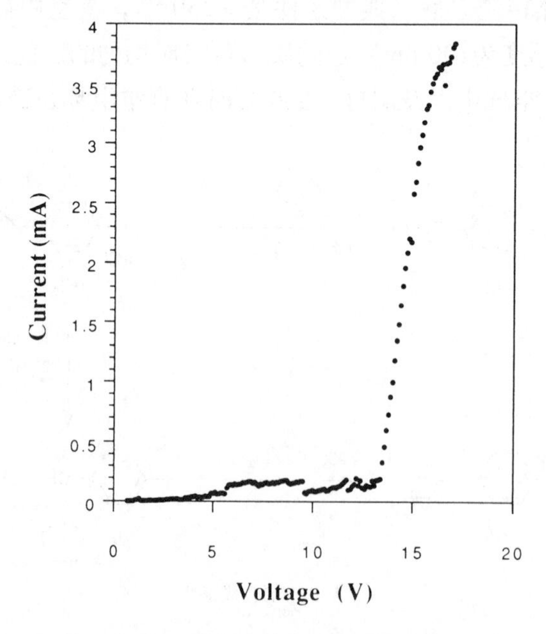

Fig. 2. Current–voltage characteristic for an electroluminescent device having a PPV film 70 nm thick and active area of 2 mm^{2}, a bottom contact of indium oxide, and a top contact of aluminium. The forward-bias regime is shown (indium oxide positive with respect to the aluminium electrode).

上形成顶部电极。对于负的电子注入接触，我们使用低功函材料；而对于正的空穴注入接触，我们使用高功函材料，其中至少有一层对于与器件平面正交的发光必须是半透明的。为此，我们同时使用了通过离子束溅射法沉积的氧化铟[14]和薄层的铝（典型厚度为 7~15 nm）。我们发现暴露在空气中的铝能够形成薄的氧化物覆盖层，金和氧化铟都能被用作正电极材料，而铝、镁银合金和通过射频溅射所制得的非晶硅氢合金都适合作负电极材料。PPV 薄膜的高度稳定性使顶部接触层的沉积很容易实现，因而我们可以利用金属的热蒸镀法和氧化铟的离子束溅射法形成这种接触。

图 2 和图 3 体现出以氧化铟作为底部接触且以铝作为顶部接触的器件的典型特性。大量电荷注入的阈值正好低于 14 V，对应的电场强度为 2×10^{6} V·cm^{-1}，积分光输出随电流的变化近似为线性的。图 4 给出了电致发光器件在不同温度下的光谱分辨输出，其光谱与光致发光谱非常类似，在 2.2 eV 附近存在峰值，且清晰的声子结构[9,10]。这些器件发出光谱中的绿 – 黄光波段的光，而且在正常的实验室光照条件下很容易被看到。量子效率（每注入一个电子所发出的光子类）适中，但是没有某些利用分子材料制成的结构所报道的那样高[2-7]。我们的 PPV 器件的量子效率达到 0.05%。我们发现这些器件中的失效模式通常都与聚合物/薄金属界面处的失效有关，这可能是因为那里的局部焦耳热引起的。

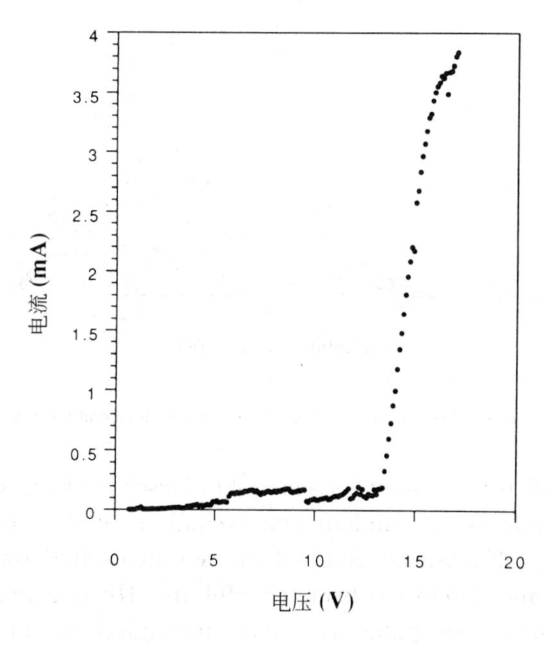

图 2. PPV 薄膜电致发光器件的电流 – 电压特性图。PPV 薄膜厚度为 70 nm，发光面积为 2 mm²，底部接触为氧化铟，顶部接触为铝。图中显示的是正向偏置状态的曲线（氧化铟相对于铝电极为正）。

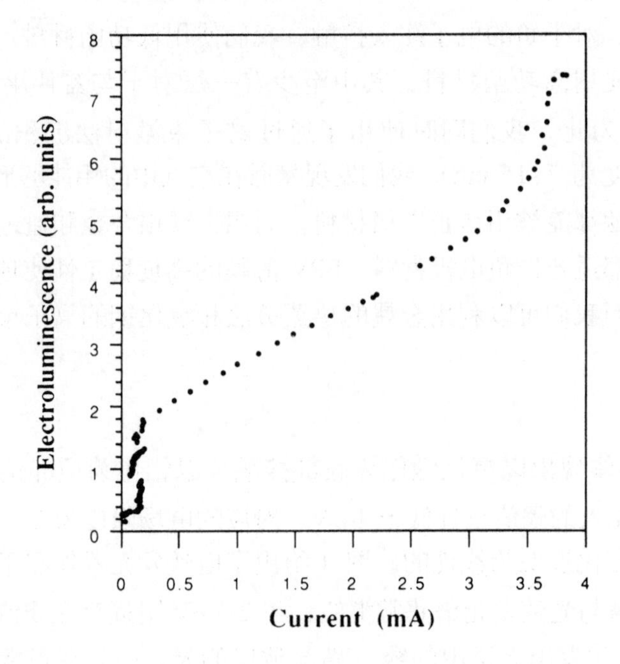

Fig. 3. Integrated light output plotted against current for the electroluminescent device giving the current–voltage characteristic in Fig. 2.

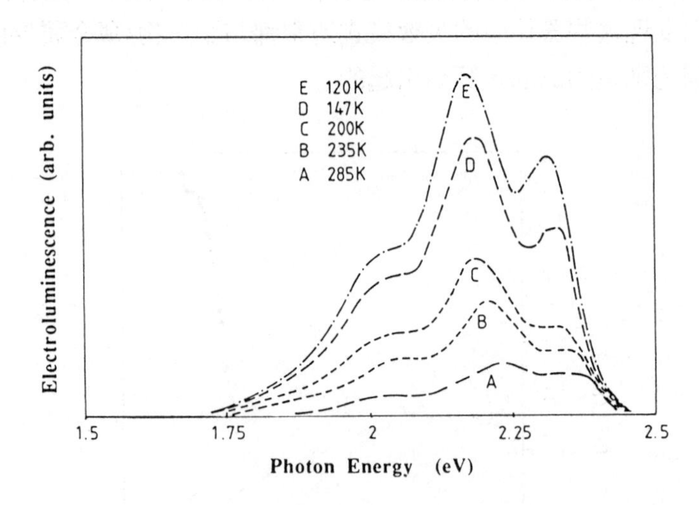

Fig. 4. Spectrally resolved output for an electroluminescent device at various temperatures.

The observation and characterization of electroluminescence in this conjugated polymer is of interest in the study of the fundamental excitations of this class of semiconductor. Here, the concept of self-localized charged or neutral excited states in the nonlinear response of the electronic system has been a useful one. For polymers with the symmetry of PPV , these excitations are polarons, either uncharged (as the polaron exciton) or charged (singly charged as the polaron, and doubly charged as the bipolaron)[15,16]. We have previously assigned the photoluminescence in this polymer to radiative recombination

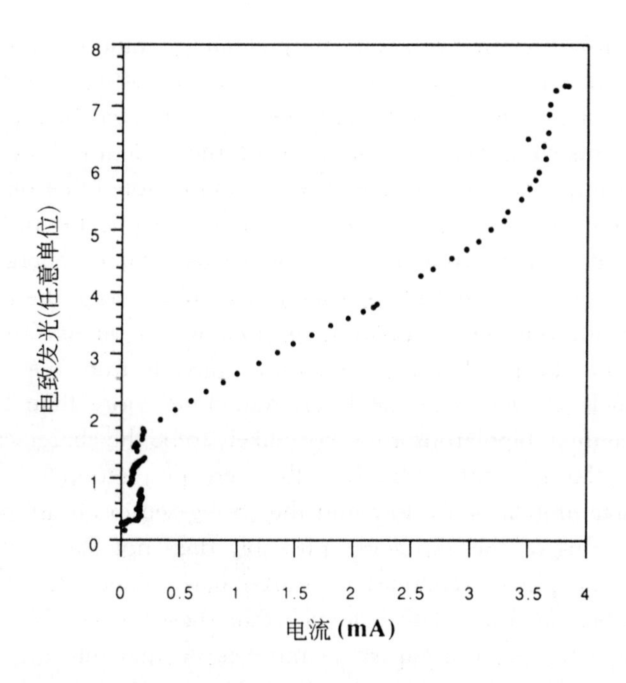

图 3. 电致发光器件的积分光输出与电流的关系图（该器件的电流–电压特性曲线如图 2 所示）

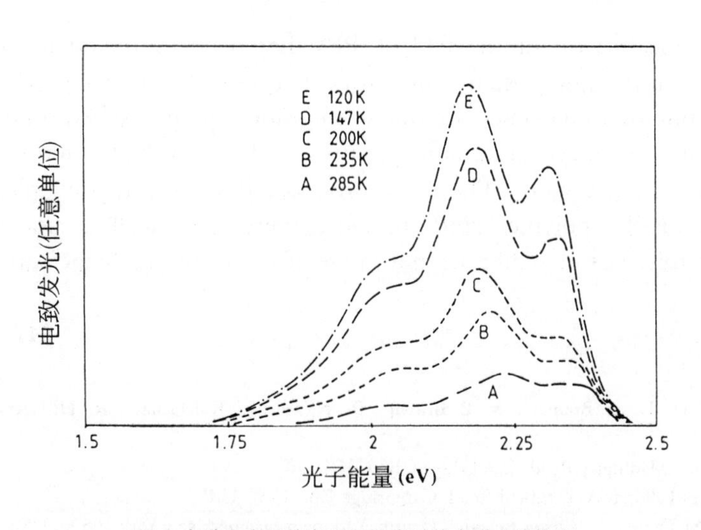

图 4. 电致发光器件在不同温度时的光谱分辨输出

观测和描述该共轭聚合物中电致发光现象是对这类半导体基本的激发过程进行研究的兴趣所在。这里，电子体系中非线性响应的自局域带电或中性激发态概念已成为一个有用的概念。对于具有像 PPV 一样对称性的聚合物，这些激发是极化子，或者是不带电荷的（如极化激子）或者是带电荷的（带单个电荷的极化子，以及带两个电荷的双极化子）[15,16]。之前我们将这种聚合物中的光致发光归因于由链间激发

of the singlet polaron exciton formed by intrachain excitation[9,10] and, in view of the identical spectral emission here, we assign the electroluminescence to the radiative decay of the same excited state. The electroluminescence is generated by recombination of the electrons and holes injected from opposite sides of the structure, however, and we must consider what the charge carriers are. We have previously noted that bipolarons, the more stable of the charged excitations in photoexcitation and chemical doping studies, are very strongly self-localized, with movement of the associated pair of energy levels deep into the semiconductor gap, to within 1 eV of each other[9]. In contrast, the movement of these levels into the gap for the neutral polaron exciton, which one-electron models predict to be the same as for the bipolaron[15], is measured directly from the photoluminescence emission to be much smaller, with the levels remaining more than 2.2 eV apart. For electroluminescence then, bipolarons are very unlikely to be the charge carriers responsible for formation of polaron excitons, because their creation requires coalescence of two charge carriers, their mobilities are low and the strong self-localization of the bipolaron evident in the positions of the gap states probably does not leave sufficient energy for radiative decay at the photon energies measured here. Therefore, the charge carriers involved are probably polarons. The evidence that they can combine to form polaron excitons requires that the polaron gap states move no further into the gap than those of the polaron exciton and may account for the failure to observe the optical transitions associated with the polaron.

The photoluminescence quantum yield of PPV has been estimated to be ~8%. It has been shown[10,17] that the non-radiative processes that limit the efficiency of radiative decay as measured in photoluminescence are due to migration of the excited states to defect sites which act as non-radiative recombination centres, and also, at high intensities, to collisions between pairs of excited states. These are processes that can, in principle, be controlled through design of the polymer, and therefore there are excellent possibilities for the development of this class of materials in a range of electroluminescence applications.

(**347**, 539-541; 1990)

J. H. Burroughes[*‡], **D. D. C. Bradley**[*], **A. R. Brown**[*], **R. N. Marks**[*], **K. Mackay**[*], **R. H. Friend**[*], **P. L. Burns**[†] & **A. B. Holmes**[†]

[*] Cavendish Laboratory, Madingley Road, Cambridge CB3 OHE, UK

[†] University Chemistry Laboratory, Lensfield Road, Cambridge CB2 1EW, UK

[‡] Present address: IBM Thomas J. Watson Research Centre, Yorktown Heights, New York 10598, USA

Received 21 August; accepted 18 September 1990.

References:

1. Vincent, P. S., Barlow, W. A., Hann, R. A. & Roberts, G. G. *Thin Solid Films* **94**, 476-488 (1982).
2. Tang, C. W. & VanSlyke, S. A. *Appl. Phys. Lett.* **51**, 913-915 (1987).
3. Tang, C. W., VanSlyke, S. A. & Chen, C. H. *J. appl. Phys.* **65**, 3610-3616 (1989).
4. Adachi, C., Tokito, S., Tsutsui, T. & Saito, S. *Jap. J. appl. Phys.* **27**, 59-61 (1988).

形成的单重态极化激子的辐射复合[9,10]，而考虑到这里有同样的辐射光谱，我们将电致发光归结为相同激发态的辐射衰减。但是电致发光是由器件的相反两边注入的电子和空穴复合而产生，我们必须考虑电荷载流子是什么。之前我们已经注意到双极化子是指在光致激发与化学掺杂研究中更为稳定的带电激发，它们是极强的自局域的，伴随着双极化子能级向半导体带隙深入运动，彼此间在 1 eV 之内[9]。相反，对于中性极化子激子来说，根据单电子模型这些能级向能隙的移动程度与双极化子的相同[15]，由光致发光直接测得的能级移动是很小的，并且保持着超过 2.2 eV 的间隔。那么对于电致发光来说，双极化子不可能是使极化激子得以形成的电荷载流子，因为它们的产生需要两个电荷载流子的结合，它们的迁移率很低，而且明显出现在能隙状态所处位置的双极化子的强自局域，可能没有给在这里测得的光子能量留下足够的能量进行辐射衰减。因此，参与其中的载流子可能是极化子。关于它们能够结合而形成极化激子的这一证据，要求极化子能隙状态向能隙的运动不超过极化激子的，从而可以解释为什么无法观测到与极化子有关的光跃迁。

PPV 的光致发光量子效率估算约为 8%。研究表明[10,17]：无辐射跃迁过程限制光致发光中测得的辐射衰减效率，该无辐射过程是来自于激发态向作为无辐射复合中心的缺陷位点的迁移；而在高强度时，该无辐射过程还可归结为激发态对之间的碰撞。这些过程原则上是能够通过聚合物的设计来控制的，因此在一系列电致发光应用中发展这类材料具有极大的可能性。

（王耀杨 翻译；于贵 朱道本 审稿）

5. Adachi, C., Tsutsui, T. & Saito, S. *Appl. Phys. Lett.* **55**, 1489-1491 (1989).
6. Adachi, C., Tsutsui, T. & Saito, S. *Appl. Phys. Lett.* **56**, 799-801 (1989).
7. Nohara, M., Hasegawa, M., Hosohawa, C., Tokailin, H. & Kusomoto, T. *Chem. Lett.* 189-190 (1990).
8. Basescu, N. *et al. Nature* **327**, 403-405 (1987).
9. Friend, R. H., Bradley, D. D. C. & Townsend, P. D. *J. Phys.* **D20**, 1367-1384 (1987).
10. Bradley, D. D. C. & Friend, R. H. *J. Phys.: Condensed Matter* **1**, 3671-3678 (1989).
11. Bradley, D. D. C. *J. Phys.* **D20**, 1389-1410 (1987).
12. Murase, I., Ohnishi, T., Noguchi, T. & Hirooka, M. *Synthetic Metals* **17**, 639-644 (1987).
13. Stenger-Smith, J. D., Lenz, R. W. & Wegner, G. *Polymer* **30**, 1048-1053 (1989).
14. Bellingham, J. R., Phillips, W. A. & Adkins, C. J. *J. Phys.: Condensed Matter* **2**, 6207-6221 (1990).
15. Fesser, K., Bishop, A. R. & Campbell, D. K. *Phys. Rev.* **B27**, 4804-4825 (1983).
16. Brazovskii, S. A. & Kirova, N. N. *JEPT Lett.* **33**, 4-8 (1981).
17. Bradley, D. D. C. *et al. Springer Ser. Solid St. Sci.* **76**, 107-112 (1987).

Acknowledgements. We thank J. R. Gellingham, C. J. Adkins and W. A. Phillips for their help in preparing the indium oxide films. We thank SERC and Cambridge Research and Innovation Ltd for support.

Superconductivity at 18 K in Potassium-doped C_{60}

A. F. Hebard *et al.*

Editor's Note

The discovery in 1990 of a method for making large quantities of the cage-like carbon molecule C_{60} led to intensive research into its physical and chemical behaviour. Here Art Hebard and colleagues at AT&T Bell Laboratories in New Jersey report one of the most striking outcomes of that work: when doped with alkali metals, C_{60} becomes superconducting at low temperatures, conducting electricity without resistance. The discovery followed soon after the Bell Labs team found that the doped solid had metallic conductivity. The onset temperature of superconductivity is relatively high compared to normal metals, and inspired hopes that it might be made higher still—as indeed it soon was. The finding never spawned practical applications, but was of immense theoretical interest.

The synthesis of macroscopic amounts of C_{60} and C_{70} (fullerenes)[1] has stimulated a variety of studies on their chemical and physical properties[2,3]. We recently demonstrated that C_{60} and C_{70} become conductive when doped with alkali metals[4]. Here we describe low-temperature studies of potassium-doped C_{60} both as films and bulk samples, and demonstrate that this material becomes superconducting. Superconductivity is demonstrated by microwave, resistivity and Meissner-effect measurements. Both polycrystalline powders and thin-film samples were studied. A thin film showed a resistance transition with an onset temperature of 16 K and essentially zero resistance near 5 K. Bulk samples showed a well-defined Meissner effect and magnetic-field-dependent microwave absorption beginning at 18 K. The onset of superconductivity at 18 K is the highest yet observed for a molecular superconductor.

THE sensitivity to air of alkali-metal-doped fullerenes (A_xC_n) limits the choice of sample preparation and characterization techniques. To avoid sample degradation, we carried out reactions with the alkali metal vapour and C_{60} in sealed tubes either in high vacuum or under a partial pressure of helium. The C_{60} was purified by chromatography[1] of fullerite[2] and was heated at 160 °C under vacuum to remove solvents.

Small amounts of the individual fullerenes (~0.5 mg) were placed in quartz tubes with alkali metals and sealed under vacuum. These samples were subjected to a series of heat treatments and tests for superconductivity by 9-GHz microwave-loss experiments[5]. Preliminary tests indicated that only the K-doped C_{60} showed a response consistent with a superconducting transition (Fig. 1). For this reason, together with the fact that K_xC_{60} showed the highest film conductivity[1], we focused our studies on the K-doped compound.

掺钾的 C_{60} 在 18 K 时的超导性

赫巴德等

编者按

1990 年制备大量笼状碳 C_{60} 分子方法的发现引起了人们对其物理和化学性质的集中研究。本文中，新泽西州的美国电话电报公司贝尔实验室的阿特·赫巴德和同事们报道了这项工作中的一个最显著的成果：当掺入碱金属时，C_{60} 在低温下变成超导体，能够无阻导电。这是贝尔实验室继发现掺杂入金属的 C_{60} 具有金属导电性后的又一个发现。与一般的金属相比，它的超导起始温度相对较高，从而激发了人们使之更高的期望，不久，这就成为了事实。尽管此发现还未推广至实际应用中，但其引起了巨大的理论研究兴趣。

宏观量的 C_{60} 和 C_{70}（富勒烯）的合成[1] 已经激发了很多关于其化学和物理性质的研究[2,3]。最近我们的实验表明：掺碱金属后的 C_{60} 和 C_{70} 具有导电性[4]。本文中我们论述了针对掺钾 C_{60} 的薄膜样品和块状样品的低温研究，并且说明了这种材料能变得具有超导性。超导性是通过微波、电阻率和迈斯纳效应的测试来说明的。我们对多晶粉末和薄膜样品分别进行了研究。结果表明，薄膜呈现出以 16 K 为起始温度的电阻转变，而且在温度接近 5 K 时就具有实质上的零电阻。块状样品呈现出明显的迈斯纳效应，且温度达到 18 K 时出现了随磁场变化的微波吸收现象。到目前为止，18 K 是人们所观测到的分子超导体最高的超导性起始温度值。

掺碱金属的富勒烯（A_xC_n）对于空气的敏感性限制了样品的制备和表征技术的选择。为了避免样品变质，我们让碱金属蒸气和 C_{60} 在高真空或者是具有氦分压的密封管中进行反应。C_{60} 是利用色谱法[1]对富勒烯固体[2]进行提纯且于 160 ℃在真空中加热除去溶剂而得到的。

将少量的单体富勒烯（约为 0.5 mg）与碱金属同置于石英管中，并在真空条件下密封。将这些样品进行一系列热处理后，通过 9 GHz 微波损失实验进行超导性检测[5]。初步的检测表明，只有掺钾的 C_{60} 表现出与超导转变相一致的反应（如图1所示）。由于这个原因，再加上 K_xC_{60} 显示出最高的薄膜导电性的事实[1]，我们集中研究了掺钾的化合物。

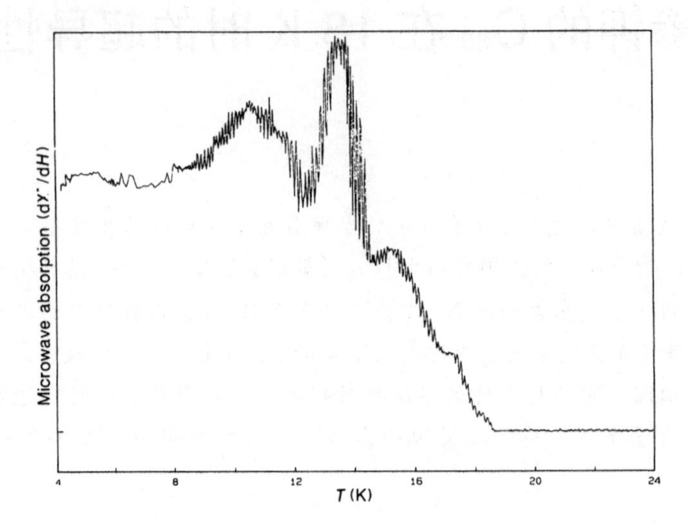

Fig. 1. Microwave loss as a function of temperature for K_xC_{60} in a static field of 20 Oe.

The conductivity measurements were performed on potassium-doped films of C_{60} that were prepared in a one-piece all-glass version of the apparatus described previously[4]. This reaction vessel was sealed under a partial pressure of helium before reaction. This configuration allowed both *in situ* doping and low-temperature studies of thin films. All measurements were made in a four-terminal Van der Pauw configuration using a 3-μA a.c. current at 17 Hz. Figure 2 shows the temperature dependence of the resistivity of a 960-Å-thick K_xC_{60} film. The film was doped with potassium until the resistivity had fallen to 5×10^{-3} Ω cm. The resistivity increases by a factor of two on cooling the sample to near 20 K. Below 16 K, the resistivity starts to decrease; zero resistivity (<10^{-4} of the normal state) is obtained below 5 K. The 10–90% width of the transition is 4.6 K. At 4 K we measured the lower bound to the critical current to be 40 A cm^{-2}.

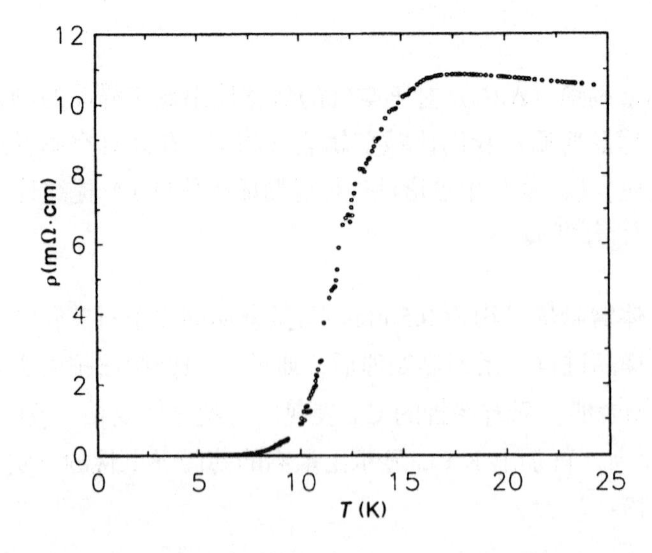

Fig. 2. Temperature dependence of the electrical resistivity of a 960-Å-thick film of K_xC_{60}.

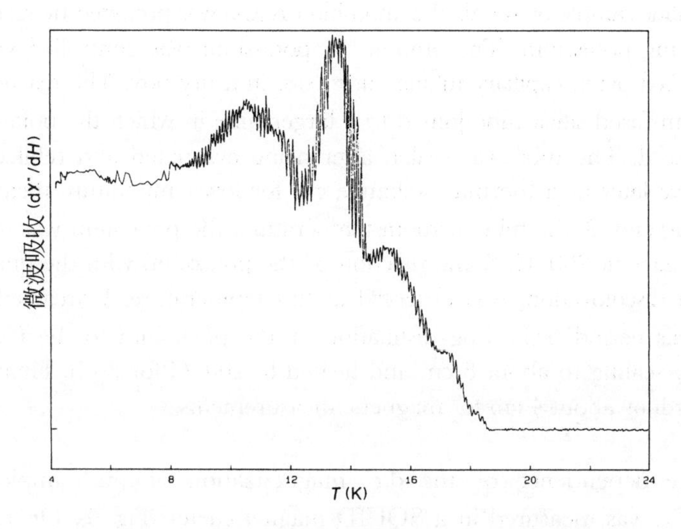

图 1. K_xC_{60} 在 20 Oe 的静场中的微波损失关于温度的函数

对掺钾的 C_{60} 薄膜进行导电性测试，薄膜于先前曾描述过的[4]整体型全玻璃装置中制备。这种反应容器是在反应前具有氦分压的条件下密封的。这种构造可以同时允许原位掺杂和薄膜的低温研究。所有的测量均在四端范德堡配置中进行，条件为 17 Hz 频率的 3 μA 交流电。图 2 显示出厚度为 960 Å 的 K_xC_{60} 薄膜的电阻率随温度的变化关系。用钾掺杂薄膜，直到电阻率下降至 5×10^{-3} Ω·cm 为止。在将样品冷却至 20 K 左右时，电阻率增大了 2 倍。在温度低于 16 K 时，电阻率开始减小；在温度为 5 K 以下时得到零阻抗（<正常态时的 10^{-4}）。电阻变化 10%~90% 所对应的温度间隔为4.6 K。温度为 4 K 时我们测得临界电流的下限值为 40 A·cm^{-2}。

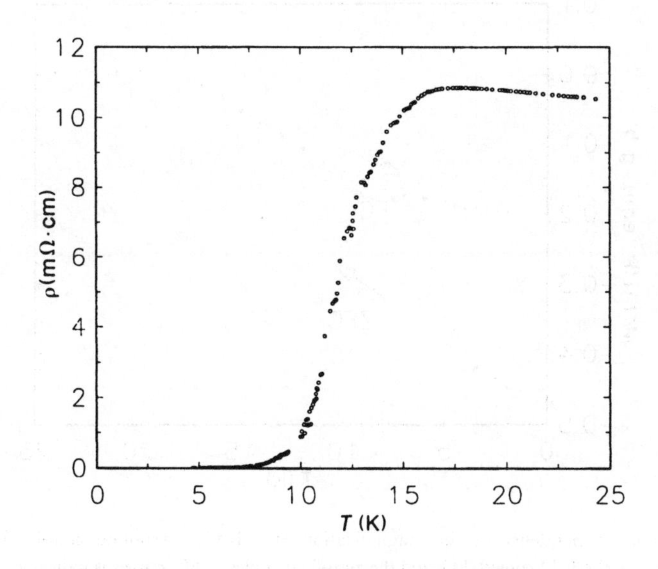

图 2. 厚度为 960 Å 的 K_xC_{60} 薄膜电阻率随温度的变化关系

A bulk polycrystalline sample of nominal composition K_3C_{60} was prepared by reaction of 29.5 mg of C_{60} with 4.8 mg potassium. The amount of potassium was controlled volumetrically by using potassium-filled pyrex capillary tubing cut to size in a dry box. The reaction was run with the C_{60} in a 5-mm fused silica tube joined to a larger tube in which the potassium-containing capillary was placed. The tube was sealed after being evacuated and refilled with 10^{-2} torr of helium to serve later as a thermal-exchange gas for low-temperature measurements. With the C_{60}-containing end of the tube at room temperature, the potassium was distilled from the capillary in a furnace at 200 °C. Some reaction of the potassium with the quartz tube, visible as a dark brown discoloration, was observed at this temperature. Unreacted potassium was observed after this period. Following distillation of the potassium to the C_{60} end, the tube was shortened by sealing to about 8 cm and heated to 200 °C for 36 h. Finally, the tube was resealed to a length of about 4 cm for magnetic measurements.

The temperature dependence of the d.c. magnetization of the sample with nominal composition K_3C_{60} was measured in a SQUID magnetometer (Fig. 3). On zero-field cooling the sample to 2 K, a magnetic field of 50 Oe was applied. On warming, this field is excluded by the sample to 18 K; this verifies the presence of a superconducting phase. The bulk nature of superconductivity in the sample is demonstrated unambiguously by cooling in a field of 50 Oe. A well defined Meissner effect (flux expulsion) develops below 18 K. The shape of the magnetization curve, in particular the temperature-independent signal at low temperature, indicates good superconducting properties for this sample. Also noteworthy is the relatively narrow transition width. The magnitude of the flux exclusion for the zero-field-cooled curve corresponds to 1% volume fraction. This small fraction is possibly due to non-optimal doping or the granular nature of the sample. The large value of the Meissner effect for the field-cooled curve relative to the total exclusion, however, indicates bulk superconductivity in the electrically connected regions.

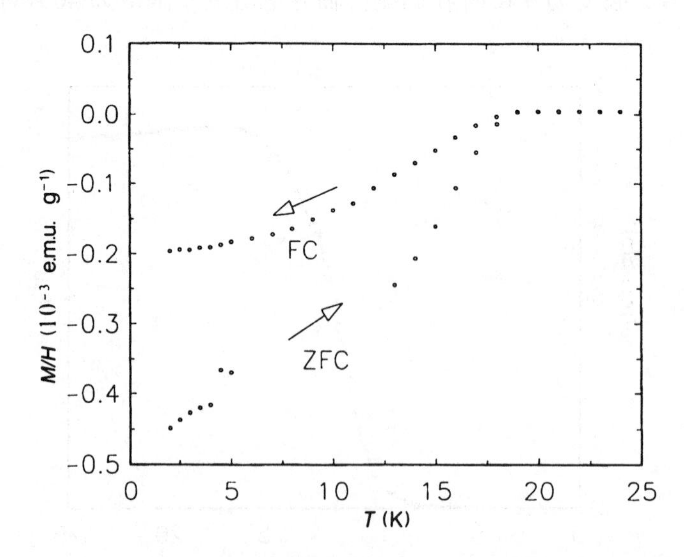

Fig. 3. Temperature dependence of the magnetization of a K_xC_{60} crystalline sample. The direction of temperature sweep in the field-cooled (FC) and the zero-field-cooled (ZFC) curves is indicated by the arrows.

标称成分为K_3C_{60}的块状多晶样品是通过将29.5 mg C_{60}与4.8 mg钾反应来制备的。通过控制体积调节钾的含量，即在干燥器中把充满钾的耐热玻璃毛细管裁割至适当体积。将一个装有C_{60}的5 mm熔封石英管与一个放有含钾的毛细管的较大管子联结在一起，使反应得以进行。在抽真空并重新填入10^{-2} torr的氦作为随后的低温测量中的热交换气体后，将管密封。保持含C_{60}的管末端处于室温状态，在温度为200 ℃的炉子中将钾从毛细管中蒸馏出来。在这个温度下观察到钾与石英管的某些反应，可以看到有深褐色的褪色斑点。之后能够观察到未反应的钾。接着将钾蒸馏到C_{60}末端，用密封方法将管子截短至约8 cm，并将其加热到温度为200 ℃，保持36小时。最后，将管子再次密封至约4 cm的长度以备磁学检测。

用超导量子干涉磁强计测试了标称成分为K_3C_{60}的样品的直流磁化随温度变化的关系（如图3所示）。在零场中，冷却样品到2 K，再对样品加50 Oe的磁场。加热温度直到18 K时，样品都可将该场排出；这就证明了一个超导相的存在。通过在50 Oe的场冷却，已明确地说明了样品中超导性的体特征。温度低于18 K时出现了明显的迈斯纳效应（磁通排出）。磁化曲线的形状，特别是低温下与温度无关的现象，表明该样品具有良好的超导性。同样值得我们注意的是相当狭窄的转变温度范围。零场冷却曲线对应的磁通排出的大小表明超导体积比为1%。如此小的超导体积比可能是由于非理想掺杂或者是样品的粒状性质。但是，场冷却曲线的迈斯纳效应相对于总排出量的大的数值表明样品电连通区域中显著的体超导性。

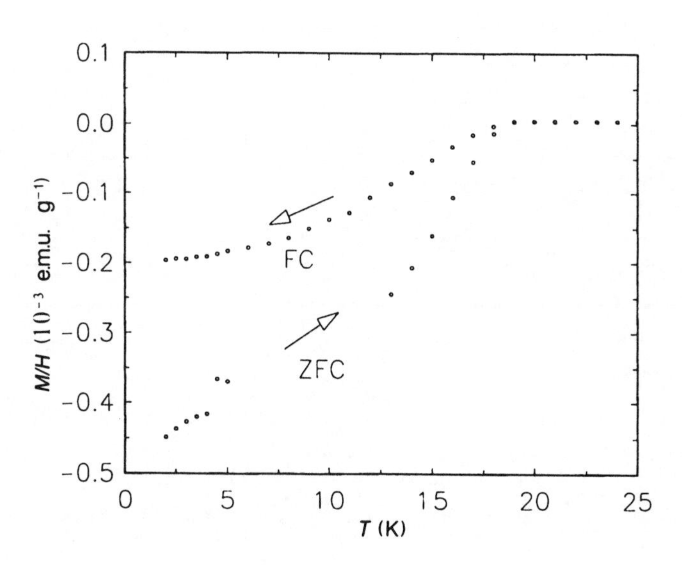

图3. K_xC_{60}晶体样品的磁化随温度的变化关系。箭头指出了场冷却（FC）与零场冷却（ZFC）曲线中的温度扫描方向。

The universally accepted tests for superconductivity, namely a transition to zero resistance and a Meissner effect showing the expulsion of magnetic field, demonstrate unequivocally the existence of superconductivity in K_xC_{60}. The 18-K transition temperature is the highest yet reported for a molecular superconductor. This may be compared with the previously reported occurrence of superconductivity at 0.55 K in potassium-intercalated graphite[6]. We expect that optimization of composition and crystallinity will lead to further improvement in the superconducting properties.

(**350**, 600-601; 1991)

A. F. Hebard, M. J. Rosseinsky, R. C. Haddon, D. W. Murphy, S. H. Glarum, T. T. M. Palstra, A. P. Ramirez & A. R. Kortan
AT&T Bell Laboratories, Murray Hill, New Jersey 07974-2070, USA

Received 26 March; accepted 4 April 1991.

References:

1. Kroto, H. W., Heath, J. R., O'Brien, S. C., Curl, R. F. & Smalley, R. E. *Nature* **318**, 162-164 (1985).
2. Kratschmer, W., Lamb, L. D., Fostiropoulos, K. & Huffman, D. R. *Nature* 347, 354-358 (1990).
3. Meijer, G. & Bethune, D. S. *J. Chem. Phys.* **93**, 7800-7802 (1990).
4. Haddon, R. C. *et al. Nature* **350**, 320-322 (1991).
5. Haddon, R. C., Glarum, S. H., Chichester, S. V., Ramirez, A. P. & Zimmerman, N. M. *Phys. Rev.* B **43**, 2642- 2647 (1991).
6. Hannay, N. B. *et al. Phys. Rev. Lett.* **14**, 225-226 (1965).

Acknowledgements. We thank G. Dabbagh, S. J. Duclos, R. H. Eick, A. T. Fiory, R. M. Fleming, M. L. Kaplan, K. B. Lyons, A. V. Makhija, B. Miller, A. J. Muller, K. Raghavachari, J. M. Rosamilia, L. F. Schneemeyer, F. A. Thiel, J. C. Tully, R. Tycko, R. B. Van Dover, J. V. Waszczak, W. L. Wilson and S. M. Zahurak for valuable contributions to this work.

得到普遍认可的超导性检测，即零电阻的转变与表现为磁场排出的迈斯纳效应，明确地证明了 K_xC_{60} 中超导性的存在。18 K 的转变温度是到目前为止关于分子超导体报道中最高的温度值。这可以与以前报道的插钾层的石墨在 0.55 K 时超导性的出现[6] 相比较。我们期望对成分与结晶度的优化能够进一步改善超导性质。

（王耀杨 翻译；韩汝珊 郭建栋 审稿）

Curling and Closure of Graphitic Networks under Electron-beam Irradiation

D. Ugarte

Editor's Note

After the discovery of cage-like carbon molecules called fullerenes and tubular structures called carbon nanotubes, it was suspected that sheets of graphite-like carbon might adopt a range of nanoscale structures in a kind of molecular origami. Here Daniel Ugarte at EPFL in Lausanne, Switzerland, reports another such variant: onion-like structures comprised of nested spherical shells of carbon. Some suggested that these were like giant fullerenes, although it transpired that the shells contain many crystal defects and breaks. Carbon "onions" with hollow centres have since been used to study encapsulated materials, and intense pressures at their centres may transform graphite-like carbon into diamond.

The discovery[1] of buckminsterfullerene (C_{60}) and its production in macroscopic quantities[2] has stimulated a great deal of research. More recently, attention has turned towards other curved graphitic networks, such as the giant fullerenes (C_n, $n>100$)[3,4] and carbon nanotubes[5-8]. A general mechanism has been proposed[9] in which the graphitic sheets bend in an attempt to eliminate the highly energetic dangling bonds present at the edge of the growing structure. Here, I report the response of carbon soot particles and tubular graphitic structures to intense electron-beam irradiation in a high-resolution electron microscope; such conditions resemble a high-temperature regime, permitting a degree of structural fluidity. With increased irradiation, there is a gradual reorganization of the initial material into quasi-spherical particles composed of concentric graphitic shells. This lends weight to the nucleation scheme proposed[9] for fullerenes, and moreover, suggests that planar graphite may not be the most stable allotrope of carbon in systems of limited size.

THE remarkable stability of the C_{60} molecule has been attributed to its highly symmetrical structure where the carbon atoms are arranged at the vertices of a truncated icosahedron[1]. The C_{60} molecule may be viewed as a hexagonal graphitic sheet which, by incorporating pentagons, has eliminated all dangling bonds, curling to form a hollow ball (7.1 Å in diameter). For all fullerenes, the strain due to the bending of the sp^2 orbitals tends to concentrate at the vertices of the pentagons. The outstanding stability of the C_{60} molecule is due both to the fact that this is the smallest carbon cage where there are no adjacent pentagons, and to its spherical form which allows the strain to be symmetrically distributed over all atoms[10].

石墨网络在电子束照射下的卷曲与闭合

丹尼尔·乌加特

编者按

在发现称为富勒烯的笼状碳分子和称为碳纳米管的微管结构分子后，有人估计层状石墨碳可能产生一系列折纸状纳米尺度分子。本文中，瑞士洛桑联邦理工学院的丹尼尔·乌加特报道了另一个这样的变体：由碳组成的球壳嵌套构成的洋葱状结构。一些人认为它们像巨型富勒烯，尽管人们知道其外壳中含有许多晶体缺陷和断裂。具有空心中心的“洋葱”碳已被用于封装材料的研究，在其中心的巨大压强可以把石墨状碳变成金刚石。

巴克敏斯特富勒烯（C_{60}）的发现[1]和它的批量制备[2]已经激发了大量研究。最近，人们已将注意力转向其他弯曲的石墨网络，例如巨型富勒烯（$C_n, n > 100$）[3,4]和碳纳米管[5-8]。通常理论认为[9]，石墨层之所以发生弯曲，其目的在于消除出现于生长中的结构边缘的高能悬挂键。这里我要报道在高分辨率电子显微镜中的炭黑粒子和微管形石墨结构对于强电子束照射的反应；这样的条件类似于高温环境，能使结构获得一定程度的流动性。随着照射增强，存在着从初始物质向由同心石墨层组成的准球形颗粒的渐变式重组织化。这为提出[9]的富勒烯成核理论增加了支持，而且进一步指出，平面型石墨可能不是碳在有限尺寸体系中最稳定的同素异形体。

一直以来人们都将 C_{60} 分子的显著稳定性归因于其高度对称性结构，其中碳原子位于截顶二十面体的顶点处[1]。C_{60} 分子可以被看作是一个六角石墨层通过引入正五边形以消除所有悬挂键而卷曲成的一个空心球(直径为 7.1 Å)。对于所有的富勒烯，由 sp^2 轨道弯曲而产生的应力倾向于集中在五边形的顶点处。C_{60} 分子出色的稳定性归因于以下两个事实：它是不包含相邻五边形的最小碳笼，并且它的球状形态使应力得以对称地分布在所有原子上[10]。

In addition to C_{60}, cylindrical carbon structures have been observed[5] in which the graphitic sheet has a helical arrangement. The natural question arises: how is it possible to generate such symmetrical, low-entropy forms from the random condensation of carbon vapour? Perhaps it is worthwhile to contemplate the ease with which these carbon hexagonal networks grow as curved or closed sheets rather than the traditionally planar ones. Chemists are conditioned to think of graphitic sheet structures as flat, where carbon atoms are bound in infinite hexagonal sheets, like chicken wire. In carbon vapour, pieces of graphite sheet would have many dangling bonds; they would have little reason to remain flat, and the physical tendency to reach the lowest energy level available would induce the sheets to eliminate their dangling bonds by curling up[12]. By heating and properly annealing pure carbon in the absence of other chemically active elements, and under conditions that favour sp^2 carbon network formation, it should be possible to synthesize curled nets and closed cages. In fact, this is the situation in arc discharges, which is the technique used at present to produce macroscopic quantities of fullerenes[2], metallofullerenes[13] and graphitic tubules[14].

Under the conditions of observation in an electron microscope, strong irradiation in some respects resembles a high-temperature regime, allowing structural fluidity; for example, amorphous carbon films usually develop slight graphitization under the electron beam. In particular, irradiation usually heats the sample by energy absorption and ruptures bonds through electron excitations. Furthermore, high-energy particles can transfer momentum to the nuclei, displacing atoms to interstitial lattice sites ("knock on"). Such conditions may be realized by electron bombardment in a high-resolution electron microscope (HREM), and consequently the evolution of a sample may be observed, even up to atomic details under favourable conditions. We must note, however, that electron-beam heating may not lead to the same result as thermal heating, because of the contribution of the excitation processes.

We have irradiated carbon soot in a 300-kV HREM microscope (Philips EM430 ST), using an electron dose up to 10–20 times higher than under normal operating conditions (the usual dose is 10–20 A cm^{-2}). Figure 1 shows a sequence of images of a group of irradiated graphitic particles, taken at 10-minute intervals. The original soot, collected in an arc-discharge apparatus, contains mostly nanometric needles formed by coaxial graphitic tubes, with some polyhedral graphitic particles formed by the junction of small flakes of planar graphite (Fig. 1*a*). All the particles are covered with a thin amorphous carbon layer. In the intermediate image of the sequence (Fig. 1*b*), particles now show more marked curvature and, in particular, the tubular structures are collapsing. At this stage, the amorphous carbon layer has graphitized epitaxially onto the particles. Finally (Fig. 1*c*), the electron annealing leads to a sample composed almost entirely of spherical particles. Detailed examination of the particles shows that they consist of an assembly of concentric spherical graphitic cages (see Fig. 2), the distance between layers agreeing with that for bulk graphite (d_{002} = 3.34 Å). The apparent disorder in the spherical shells arises as a consequence of the low electron dose necessary for taking the micrographs; under such conditions, the heating effect of the electron beam is insufficient to permit structural

除 C_{60} 之外，还观测到了圆柱形碳结构[5]，其中石墨层具有螺旋形排布方式。由此自然产生一个问题：碳蒸气的随机凝聚是以何种方式产生出这种对称的低熵形态的呢？也许以下问题是值得思考的，即碳的六边形网络作为弯曲或闭合层的生长比起传统的平面层生长更具有便易性。化学家习惯于认为石墨层结构是平面的，碳原子被限定于无限的六边形层中，就像铁丝网一样。在碳蒸气中，石墨层碎片可以含有很多悬挂键；它们很难保持平面形，而趋向于达到可能的最低能量的物理性质将会促使层通过卷曲消除其悬挂键[12]。在没有其他化学活性元素存在，以及有利于 sp^2 碳网络形成的条件下，通过加热纯碳并适当地退火，应该有可能合成卷曲网状和闭合笼状的物质。实际上，这就是电弧放电的情况，这是一种目前用来大量制备富勒烯[2]、金属富勒烯[13] 和石墨微管[14] 的技术。

在电子显微镜观测条件下，强烈的照射在某些方面类似于高温环境，使结构具有了流动性；例如，无定形碳薄膜在电子束照射下通常会产生轻微的石墨化。特别是，照射通常以能量吸收的方式加热样品并以电子激发的方式打断样品化学键。此外，高能粒子能将动量传递给核，使原子移动到晶格间隙位点（“撞出”）。利用高分辨率电子显微镜（HREM）中的电子轰击可以实现上述条件，同时可以观测到样品的变化过程，在有利条件下甚至可以观测到原子水平的细节。但是我们必须注意，由于激发过程的贡献，电子束加热可能不会导致与传统加热方式同样的结果。

我们在一台 300 kV HREM 显微镜（飞利浦 EM430 ST）中照射炭黑，使用的电子束剂量超过正常操作条件下的 10~20 倍（通常剂量为 $10\text{~}20\ \mathrm{A\cdot cm^{-2}}$）。图 1 显示了一组经照射的石墨颗粒的图像序列，拍摄间隔为 10 分钟。初始炭黑是从电弧放电装置中收集到的，其中主要包含由共轴石墨管形成的纳米尺寸针状物，以及由小的平面型石墨薄片接合而形成的一些多面体石墨颗粒（图 1*a*）。所有颗粒上都覆盖着一层薄的无定形碳。在图像序列的中间部分（图 1*b*）里，颗粒呈现出更为明显的曲率，特别是，微管型结构正在瓦解。在这个阶段中，无定形碳层石墨化并外延到颗粒上。最后（图 1*c*），电子退火产生出一种几乎完全由球状颗粒组成的样品。对颗粒的详细检测表明，它们是由同心球形石墨笼的聚集体构成的（参见图 2），层与层之间的距离和块状石墨中的相吻合（$d_{002} = 3.34$ Å）。球层中明显的无序性是由拍摄显微图像所必需的低电子剂量产生的；在这样的条件下，电子束的加热效应不足以使结构重

reorganization back to the closed-shell form. Imaging on a much shorter timescale, however, permits the observation of more complete structures. The existence of these structures has already been proposed[15-17].

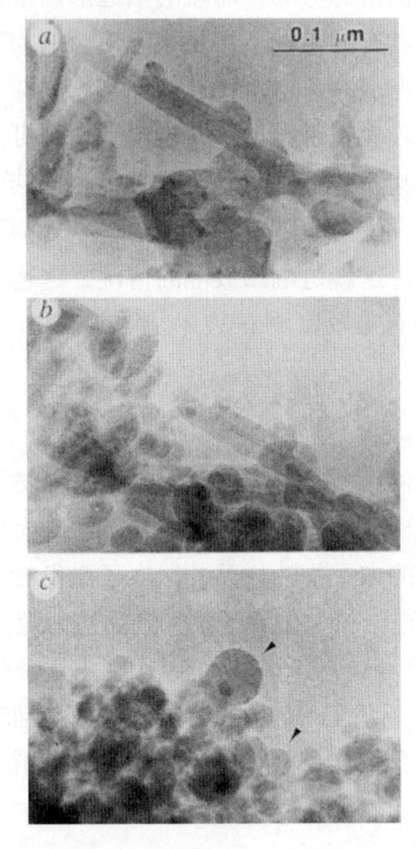

Fig. 1. Sequence of transmission electron micrographs of carbon soot subjected to strong electron beam irradiation. *a*, Original soot containing tubular or polyhedral graphitic particles; *b*, after 10 minutes of strong irradiation, there is a noticeable tendency for the particles forming the soot to become more spherical, especially the graphitic needles; *c*, after 20 minutes, the soot is nearly exclusively composed of quasi-spherical graphitic particles. Further irradiation does not produce significant observable changes in structure.

The final spherical structures do not correspond to the deformation of a tube into a sphere with the same number of atoms (see scheme for single-shell particles in Fig. 3*a*, *b*), but rather to the formation of a multiple-shell sphere with a very small central cage (Fig. 3*c*). The reduced dimension of the final inner shell (0.6–1 nm) allows an increase in the number of shells. Following Euler's theorem, any closed hexagonal network contains exactly 12 pentagonal rings; this is the case in the tube (Fig. 3*a*, the pentagons being situated at the hemispherical domes at the extremities of the cylinder) or in the spherical cage (Fig. 3*b*). The onion-like particle of Fig. 3*c* is formed by four closed shells; in consequence, it contains four times as many pentagons. Our results present clear experimental evidence of the spontaneous tendency of graphite to include pentagons in its hexagonal network and form curved structures. Hence, they support the dangling bond minimization scheme, proposed to explain the growth of fullerenes from carbon vapour[9].

组回闭合壳层结构。不过，在短得多的时间尺度上的成像使我们得以观测更完整的结构。已有人提出这些结构是存在的[15-17]。

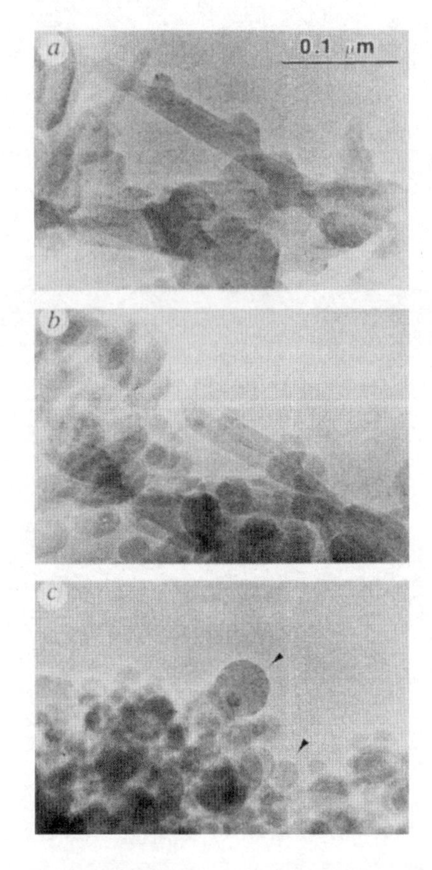

图 1. 炭黑遭到强电子束照射的透射电子显微图序列。*a*，含微管或多面体石墨颗粒的初始炭黑；*b*，强照射 10 分钟后，炭黑中的颗粒，尤其是石墨针状物中，出现了明显向球形转变的趋势；*c*，20 分钟后，炭黑基本上毫无例外地由准球形石墨颗粒组成。继续照射不再产生结构上可观测的显著变化。

最终的球形结构并不对应于具有相同原子数目的管到球的变形（参见图 3*a*、*b* 中单壳层颗粒的图解），而是形成包含一个很小的中心笼的多层球体（图 3*c*）。最内一层尺寸的减少（0.6~1 nm）使得壳层数目增加。根据欧拉定理，任一闭合六方网络恰好包含 12 个五边形环；这就是管（图 3*a*，五边形位于圆柱体末端的半球拱处）或者球形笼(图 3*b*)的情况。图 3*c* 中的洋葱状颗粒是由四个闭合壳层形成的。因此，它包含四倍的五边形数。我们的结果为石墨在其六边形网络中包含五边形并且形成弯曲结构的自发倾向提供了清晰的实验证据。因此，它们支持了为解释富勒烯能从碳蒸气中生长出来而提出的悬挂键数目最小化的理论[9]。

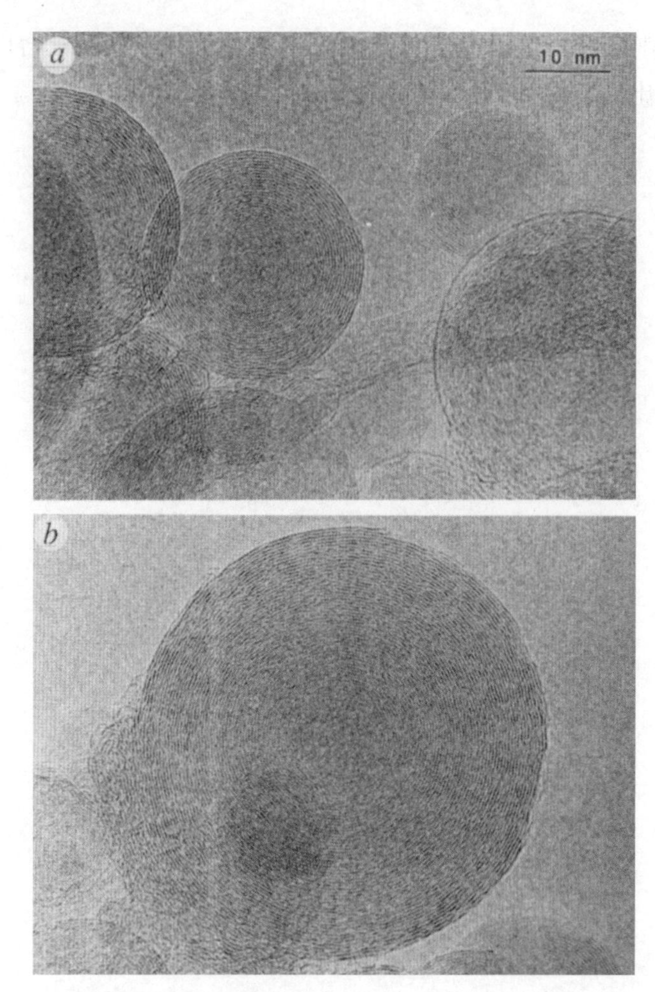

Fig. 2. Detailed structure of the graphitic particles marked with an arrow in Fig. 1*c*. Dark contrast rings correspond roughly to atomic positions, and the distance between rings corresponds to the (002) lattice parameter of bulk graphite. Note the remarkable sphericity of the particles.

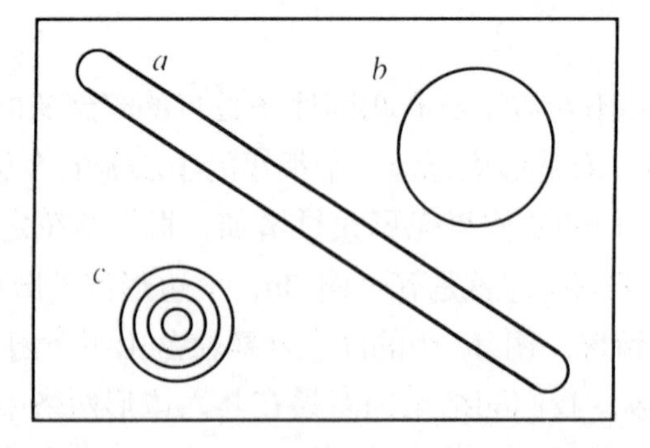

Fig. 3. Schematic representation of three-dimensional particles formed by an equal graphitic surface: *a*, cylindrical structure closed by hemispherical domes (diameter Ø≈1 nm and 13.7 nm long); *b*, spherical cage (Ø≈3.74 nm); *c*, onion-like structure formed by 4 shells (Ø≈2.72 nm), the central one being a C_{60} molecule.

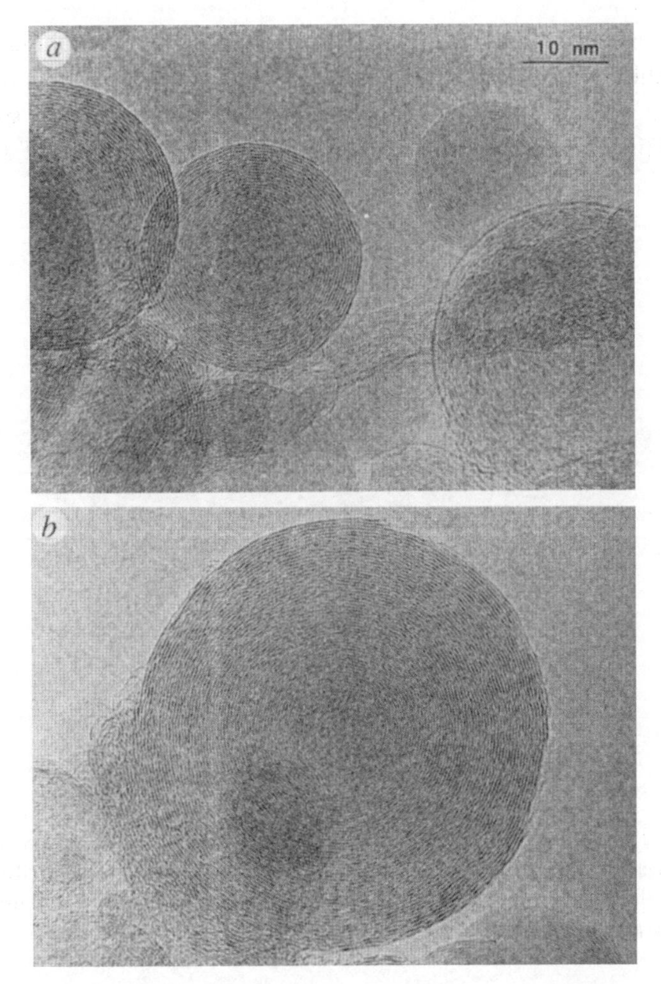

图 2. 图 1c 中箭头所指示的石墨颗粒的详细结构。暗衬度环大致上对应于原子的位置，而环之间的距离则对应于块状石墨的 (002) 晶格参数。注意颗粒明显为球形。

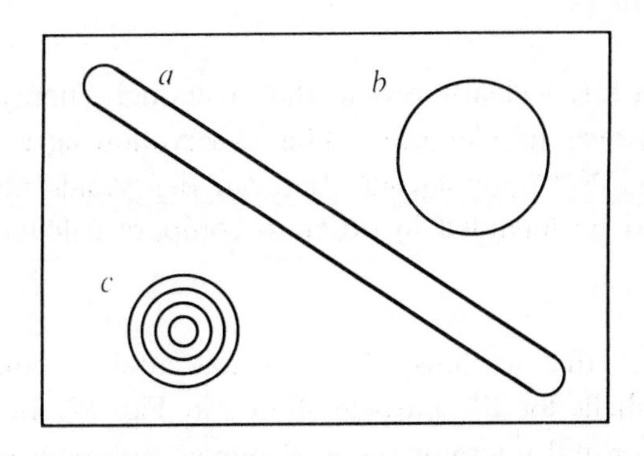

图 3. 由相等的石墨表面形成的三维结构示意图：a，用半球圆拱封口的圆柱形结构（直径 Ø ≈ 1 nm，长 13.7 nm）；b，球形笼（Ø ≈ 3.74 nm）；c，由四个壳层形成的洋葱状结构（Ø ≈ 2.72 nm），中心的那个是 C_{60} 分子。

Curved graphitic sheets can also be formed by irradiation of amorphous carbon particles (Fig. 4). The graphitic structures generated are naturally curled, and two nucleation centres are easily recognizable (marked with arrows in Fig. 4*b*), from which spherical particles will be formed. Further irradiation annealing leads to the separation of the two graphitic spheres.

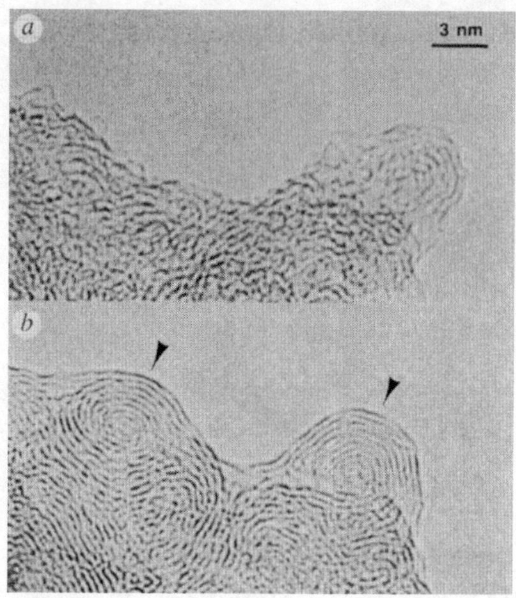

Fig. 4. Transmission electron micrographs of amorphous carbon subjected to electron irradiation. *a*, Original particle; *b*, after 10 minutes of strong irradiation, graphitization (marked with arrows) is present, and the sheets show a clear tendency to form closed cages. Further irradiation would lead to the separation of two graphitic spheres.

The formation mechanism for these multiple-shell spheres is based on irradiation-stimulated graphitization, and is rather different from the accretion mechanism originally considered in laser vaporization and arc-discharge experiments, which would produce spiral multiple-shell particles[9].

The sequence shown in Fig. 1 clearly reveals that if enough energy is provided, spherical structures are favoured over tubular ones. This observation agrees with the predictions made for giant fullerenes[18,19] (monolayers), but van der Waals interaction between the concentric layers should be included in order to compare calculations with the present experiments.

The "spherical graphite" that we have observed may attain a considerable size (47 nm in diameter and ~70 shells for the particle shown in Fig. 2*b*). In a few cases, we have even observed spheres several micrometres in diameter, although in this range of sizes a prolonged irradiation period is required before the particles become spherical. We should not rule out the possibility that even larger (possibly macroscopic) graphitic spheres could be generated by adequate annealing of carbon; the maximum size attainable will give

弯曲石墨层还能通过对无定形碳颗粒的照射而形成（图 4)。所生成的石墨结构是天然弯曲的，并且很容易识别出两个成核中心（图 4*b* 中以箭头标出)，球形颗粒就从这里开始形成。继续照射退火导致两个石墨球形颗粒的分离。

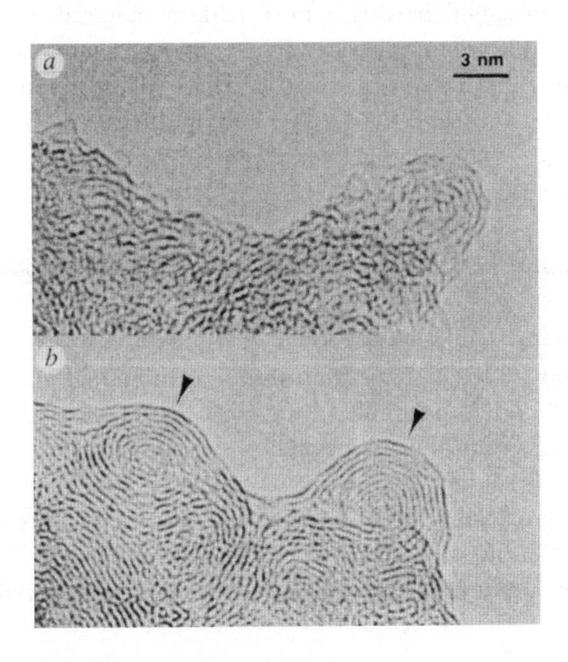

图 4. 接受电子束照射的无定形碳的透射电子显微图像。*a*，初始颗粒；*b*，经过 10 分钟强照射后，出现石墨化（用箭头标出)，层呈现出清晰的形成闭合笼的倾向。继续照射会导致两个石墨球体的分离。

这些多壳层球体的形成机制是基于照射激发石墨化的原理，并且与最初在激光气化和电弧放电实验中所考虑的堆积机制有很大不同，后者会产生出螺旋形多壳层颗粒 [9]。

图 1 中显示的图像序列清楚地表明，如果提供足够的能量，球形结构比微管结构更容易形成。这一观测结果与对巨型富勒烯（单层）的预测 [18,19] 是一致的，但是要将计算结果与当前实验进行比较，则同心层之间的范德华相互作用也应包含在内。

我们所观测到的“球形石墨”可能具有相当大的尺寸（图 2*b* 中显示的颗粒，直径达到 47 nm，约 70 个壳层)。在少数情况下，我们甚至曾观测到直径达几个微米的球体，不过要达到这个范围的尺寸需要在颗粒变成球形之前延长照射时间。我们不应该排除下列可能性：更大的（可能是宏观水平的）石墨球体可以利用碳的充分退火而生产出来；所得到的最大尺寸将为我们提供关于闭合石墨层中存在的变形的

us information about the distortion present in the closed graphite sheets. The traditional idea that planar graphite is the most stable form of pure carbon would then have to be seriously reviewed: a flat graphite flake cannot be perfect, and includes many dangling bonds which are usually eliminated by attaching impurities (for example, hydrogen). The "spherical graphite" presented here is a pure carbon material, which has no dangling bonds, and moreover, having a spherical shape, allows a uniform distribution of the strain because of the out-of-plane geometry. Those of us accustomed to traditional planar graphite, initially surprised by the fascinating fullerenes, are now confronted with supplementary evidence that spherical carbon networks can be favoured under high temperature or strong irradiation regimes.

This notion also raises a point concerning the solid allotropes of carbon. When Krätschmer *et al.*[2] synthesized large amounts of the C_{60} molecule for the first time, they prepared a new, third form, of solid carbon (called fullerite), which is a three-dimensional packing of C_{60} spheres and is distinct from the two traditional crystalline carbon forms, graphite and diamond. Considering the observed tendency of graphite to form multiple-shell spheres ("onions"), of which single-shell fullerenes are only the first member, we speculate that fullerite is the first member of a family of new solid forms of carbon that could be formed by the packing of these onion-like graphitic spheres, interacting through van der Waals forces. Further experimental work will be needed to produce and isolate multishelled graphitic spheres, but a huge family of carbon materials awaits the skill of experimentalists.

(**359**, 707-709; 1992)

Daniel Ugarte

Institut de Physique Expérimentale, Ecole Polytechnique Fédérale de Lausanne, 1015 Lausanne, Switzerland

Received 7 September; accepted 2 October 1992.

References:

1. Kroto, H. W., Heath, J. R., O'Brien, S. C., Curl, R. F. & Smalley, R. E. *Nature* **318**, 162-163 (1985).
2. Krätschmer, W., Lamb, L. D., Fostiropoulos, K. & Huffman, D. *Nature* **347**, 354-358 (1990).
3. Kroto, H. W. & McKay, K. *Nature* **331**, 328-331 (1988).
4. Lamb, L. D. *et al. Science* **255**, 1413-1416 (1992).
5. Iijima, S. *Nature* **354**, 56-58 (1991).
6. Mintimire, J. W., Dunlap, B. I. & White, C. T. *Phys. Rev. Lett.* **68**, 631-634 (1992).
7. Hamada, N., Sawada, S. & Oshiyama, A. *Phys. Rev. Lett.* **68**, 1579-1581 (1992).
8. Tanaka, K., Okahara, K., Okada, M. & Yamade, T. *Chem. Phys. Lett.* **191**, 469-472 (1992).
9. Zhang, Q. L. *et al. J. Phys. Chem.* **90**, 525-528 (1986).
10. Kroto, H. W. *Nature* **329**, 529-531 (1987).
11. Bundy, F. P. *Physica* **A156**, 169-178 (1989).
12. Robertson, D. H., Brenner, D. W. & White, C. T. *J. Phys. Chem.* **96**, 6133-6135 (1992).
13. Chai, Y. *et al. J. Phys. Chem.* **95**, 7564-7568 (1991).
14. Ebbesen, T. W. & Ajayan, P. M. *Nature* **358**, 220-222 (1992).

信息。平面型石墨是纯碳的最稳定形态这一传统观念就不得不重新接受严肃的审视：平坦的石墨薄片不可能是完美的，其中包含很多悬挂键，它们通常是通过吸附杂质（例如氢）而得以消除的。这里所展示的“球形石墨”是一种纯净的碳物质，其中不含悬挂键，而且凭借其球形形状的非平面几何性质而使应力得到均匀分布。那些习惯于传统平面型石墨且最初为神奇的富勒烯而惊奇的研究者，现在必须面对足以证明在高温或强照射环境中更有利于球型碳网络形成的证据。

这一观念还引出了一个关于碳的固态同素异形体的问题。当克雷奇默等人[2]第一次合成大量 C_{60} 分子时，他们制备出了一种新型——第三种形式——的固态碳（称为富勒烯），它是 C_{60} 球体的三维堆积，有别于两种传统晶体形式的碳：石墨和金刚石。考虑到我们观测到的石墨形成多壳层球体（“洋葱”）的趋势——单壳层富勒烯只是第一位成员而已，我们猜测富勒烯固体是一族新的固态碳物质的第一位成员，这族固态碳物质可以通过范德华相互作用由这些洋葱状石墨球体堆积形成。要制备和分离多壳层石墨球体还需要进一步的实验研究，但是一大族碳材料正等待着实验科学家技能的发掘。

（王耀杨 翻译；顾镇南 审稿）

15. Curl, R. F. & Smalley, R. E. *Scient. Am.* **265**, 32-41 (October, 1991).

16. Iijima, S. *J. Cryst. Growth* **50**, 657-683 (1980).

17. Iijima, S. *J. Phys. Chem.* **91**, 3466-3467 (1987).

18. Adams, G. B., Sankey, O. F., Page, J. B., O'Keeffe, M. & Drabold, D. A. *Science* **256**, 1792-1795 (1992).

19. Scuseria, G. E. in *Buckminsterfullerene* (eds Billups, W. E. & Ciufolini, M. A.) (VCH, New York, in the press).

Acknowledgements. We thank H. W. Kroto and W. de Heer for discussions and comments, and D. Reinhard and B. D. Hall for critical reading of the manuscript.

Superconductivity above 130 K in the Hg–Ba–Ca–Cu–O System

A. Schilling *et al.*

Editor's Note

In 1986, physicists in Switzerland discovered a ceramic material that became superconducting at temperatures below about 30 K when doped with impurities. Hitherto, many physicists suspected that superconductivity would be impossible above about 20 K, and the discovery of so-called high-temperature superconductivity kicked off a race both to explain the effect and to design new materials with even higher transition temperatures (T_c). By 1993, the record had reached 125 K; here Andreas Schilling and colleagues improve on that with a material having a T_c of 133 K. Since then the record for copper-oxide materials has been raised to 138 K. But a definitive theory to explain the phenomenon remains an outstanding problem in physics.

The recent discovery[1] of superconductivity below a transition temperature (T_c) of 94 K in $HgBa_2CuO_{4+\delta}$ has extended the repertoire of high-T_c superconductors containing copper oxide planes embedded in suitably structured (layered) materials. Previous experience with similar compounds containing bismuth and thallium instead of mercury suggested that even higher transition temperatures might be achieved in mercury-based compounds with more than one CuO_2 layer per unit cell. Here we provide support for this conjecture, with the discovery of superconductivity above 130 K in a material containing $HgBa_2Ca_2Cu_3O_{1+x}$ (with three CuO_2 layers per unit cell), $HgBa_2CaCu_2O_{6+x}$ (with two CuO_2 layers) and an ordered superstructure comprising a defined sequence of the unit cells of these phases. Both magnetic and resistivity measurements confirm a maximum transition temperature of ~133 K, distinctly higher than the previous established record value of 125–127 K observed in $Tl_2Ba_2Ca_2Cu_3O_{10}$ (refs 2, 3).

THE structural similarity of $HgBa_2CuO_{4+\delta}$ (Hg-1201, ref. 1) to a member of the thallium-containing family of copper oxides, $TlBa_2CuO_5$ (Tl-1201), suggests the existence of compounds with the general composition $HgBa_2Ca_{n-1}Cu_nO_{2n+2+\delta}$. The transition temperatures of the thallium-containing analogues, $TlBa_2Ca_{n-1}Cu_nO_{2n+3}$, range from < 10 K ($n = 1$, ref. 4) to ~110 K ($n = 3$, ref. 5). In this sense, transition temperatures exceeding 100 K may be expected also in the Hg–Ba–Ca–Cu–O (HBCCO) system. Although the successful synthesis of $HgBa_2RCu_2O_{6+x}$ (Hg-1212) with R being (Eu, Ca) has been reported, no superconductivity was found in that system[6].

Hg-Ba-Ca-Cu-O 体系在 130 K 以上时的超导性

席林等

编者按

1986 年，两位瑞士物理学家发现了一种氧化物陶瓷材料，在温度低于约 30 K 时具有超导性。此前，许多物理学家们推测超导温度不可能高于 20 K，而高温超导的发现开启了理论解释和实验探索具有更高转变温度 (T_c) 的新超导材料的竞赛。在 1993 年以前，T_c 记录值已达到 125 K。本文中的安德烈亚斯·席林及其同事们发现一种 T_c 为 133 K 的新型超导材料。从此，铜氧化物材料的 T_c 温度记录提升至 138 K。但是目前在物理上并没有一种明确的理论解释。

最近对于 $HgBa_2CuO_{4+\delta}$ 在低于 94 K 转变温度 (T_c) 的超导性的发现[1]，进而扩大到对适当 (层状) 结构中含有铜氧面的高 T_c 超导体的研究。以前用含铋和铊而不是汞的类似复合物所进行的实验表明，每个单胞中 CuO_2 层多于一个的汞基复合物有可能达到更高的转变温度。本文对这一猜测提供了支持：因为发现高于 130 K 的超导性存在于含有 $HgBa_2Ca_2Cu_3O_{1+x}$(每个单胞中有三个 CuO_2 层)、$HgBa_2CaCu_2O_{6+x}$(含两个 CuO_2 层) 的材料中，而且这两相的单胞以确定的序列形成有序的超结构。磁性和电阻测量都确认最大转变温度约为 133 K，明显高于过去在 $Tl_2Ba_2Ca_2Cu_3O_{10}$ (参考文献 2，3) 中观测到的记录值 125~127 K。

$HgBa_2CuO_{4+\delta}$(Hg–1201，参考文献 1) 与铊系铜氧化物，如 $TlBa_2CuO_5$ (Tl–1201)，在结构上的相似性暗示着具有组成通式为 $HgBa_2Ca_{n-1}Cu_nO_{2n+2+\delta}$ 的化合物的存在。含铊化合物 $TlBa_2Ca_{n-1}Cu_nO_{2n+3}$ 的转变温度范围在小于 10 K($n = 1$，参考文献 4) 至约 110 K($n = 3$，参考文献 5) 之间。从这种意义上来看，我们可以预期 Hg–Ba–Ca–Cu–O (HBCCO) 体系也会有超过 100 K 的转变温度。尽管已有 $HgBa_2RCu_2O_{6+x}$(Hg–1212) (R 为 Eu，Ca) 的成功合成的报道，但却没有在该体系中发现超导性[6]。

We prepared the samples following the procedure described in ref. 1 for Hg-1201. A precursor material with the nominal composition $Ba_2CaCu_2O_5$ was obtained from a well ground mixture of the respective metal nitrates, sintered at 900 °C in O_2. After regrinding and mixing with powdered HgO, the pressed pellets were sealed in evacuated quartz tubes. These tubes were placed horizontally in tight steel containers and held at 800 °C for 5 hours. On opening the containers, we found that the quartz tubes were broken. It was not possible to reconstruct at which stage of the heating, cooling or opening procedure this happened. Some of the pellets were finally annealed for 5 hours at 300 °C in flowing oxygen. During the preparation and the characterization of the samples, all possible measures were taken to avoid any contamination with toxic mercury or mercury-containing compounds.

After annealing, the resulting black material was characterized by X-ray diffraction using the Guinier technique, by energy-dispersive X-ray spectrometry (EDS), and by selected-area electron-diffraction techniques (SAED) and high-resolution transmission electron microscopy (HRTEM). The EDX analysis showed that the samples are composites of isolated grains of $BaCuO_2$ (~30%), CuO (~30%), an unidentified oxide containing Hg, Ca and Cu (~15%), an oxide with Ca and Cu (~5%), and ~5% impurities with unspecified composition. About 15% of the total sample volume consisted of plate-like grains containing Hg, Ba, Ca and Cu. Some of these were investigated in detail by SAED and HRTEM techniques on a Phillips CM 30-ST transmission electron microscope. Both techniques showed clearly that these identified grains consist mostly of pure $HgBa_2Ca_2Cu_3O_{8+x}$ (Hg-1223), disordered mixtures of Hg-1223 and Hg-1212, and a periodic stacking sequence of the latter unit cells. We found no grains or intergrowths associated with the Hg-1201 structure. As the volume fraction of the phases of interest is fairly small, we could not measure the lattice parameters precisely with the X-ray Guinier technique. Nevertheless, from the SAED patterns, we deduce the lattice constants $c = 12.7(2)$ Å and $c = 16.1(3)$ Å for the tetragonal Hg-1212 and Hg-1223 units, respectively, and $a = 3.93(7)$ Å, valid for both types of compounds. The results for Hg-1212 are in good agreement with the values obtained in ref. 6. Figure 1 is a representative HRTEM image showing, as an example, a stacking containing both Hg-1212 and Hg-1223 layers. The stacking sequence 1223/1223/1212/1212/1223/1212 with a supercell c-axis $c \approx 86.4$ Å extends beyond 2,000 Å, thus qualifying this superstructure as a proper phase. HRTEM images as well as SAED patterns gave no evidence for the presence of HgO-double layers.

我们根据参考文献 1 中所描述的制备 Hg–1201 的过程制备了样品。将每种金属的硝酸盐的混合物充分研磨，在 900 ℃于氧气中烧结，得到一种具有标称成分为 $Ba_2CaCu_2O_5$ 的前驱物。烧结物与 HgO 粉末混合并重新研磨后，压制成片并密封于抽成真空的石英管中。再将石英管水平地放置于钢质密闭容器中，于温度为 800 ℃下保持 5 小时。打开容器时，我们发现石英管已经破裂。我们无法推想破裂发生在加热、冷却或开启过程中的哪个阶段。最后将部分的样片在流动的氧气中于 300 ℃退火 5 小时。在样品的制备和表征过程中，采用了各种可能的措施以避免有毒的汞或含汞化合物的污染。

在退火之后，将所得黑色物质采用吉尼耶技术的 X 射线衍射、能量色散 X 射线光谱（EDS）、选区电子衍射技术（SAED）和高分辨透射电子显微镜（HRTEM）进行表征。EDX 分析表明，样品是由下列物质的分离的颗粒所组成：包括 $BaCuO_2$（约 30%）、CuO（约 30%）、一种含 Hg、Ca 和 Cu 组成不明的氧化物（约 15%）、一种含 Ca 和 Cu 的氧化物（约 5%）和约 5% 含未确定成分的杂质。样品总体积中约 15% 的部分是由含 Hg、Ba、Ca 和 Cu 的片状颗粒组成。我们对其中某些区域用菲利普斯 CM 30–ST 透射电子显微镜进行了 SAED 和 HRTEM 技术的详细研究。两种技术都清楚地表明，上述所检验的颗粒主要包含纯的 $HgBa_2Ca_2Cu_3O_{8+x}$（Hg–1223）、无序的 Hg–1223 与 Hg–1212 的混合物，以及这两种单胞的周期性堆垛序列。我们没有发现与 Hg–1201 结构有关的颗粒或共生物。由于有意义的相只占相当小的体积分数，我们未能用 X 射线吉尼耶技术精确地测量其晶格参数。尽管如此，根据 SAED 图像，我们推测四方晶系 Hg–1212 和 Hg–1223 的晶格常数分别为 $c = 12.7\ (2)$ Å 和 $c = 16.1\ (3)$ Å，以及对两种化合物都有效的 $a = 3.93\ (7)$ Å。对 Hg–1212 的观测结果与参考文献 6 所得到的数值非常吻合。图 1 是典型的 HRTEM 图像，例如，它显示的堆垛同时含有 Hg–1212 和 Hg–1223 层。具有超晶格 c 轴 $c \approx 86.4$ Å 按照 1223/1223/1212/1212/1223/1212 堆垛序列延展到超过 2,000 Å，因而可将这种超结构视为本征相。HRTEM 和 SAED 图像都未能证明有 HgO 双层的存在。

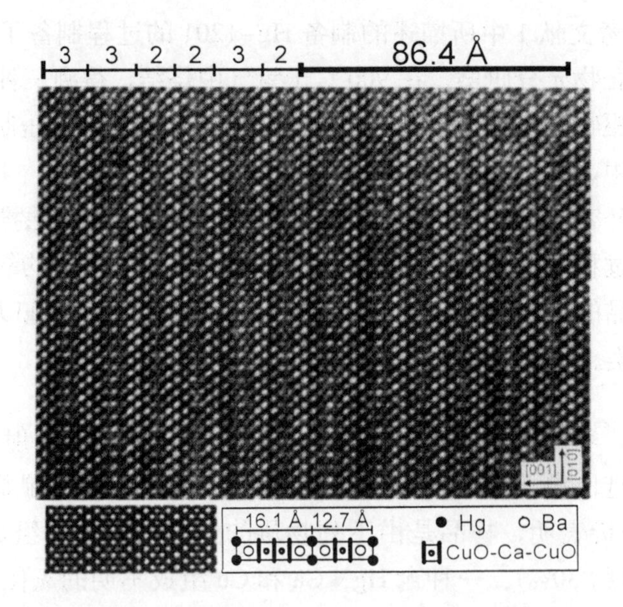

Fig. 1. HRTEM image of a grain in [100] orientation, containing layers of Hg-1212 and Hg-1223. Here, they are stacked in a periodic sequence forming a supercell with $c \approx 86.4$ Å (see text). A contrast simulation ($c_s = 1.1$ mm, $E = 300$ keV, defocus -870 Å, specimen thickness 23 Å) is inserted. The stacking sequence in terms of the number of Cu–O planes and an enlarged schematic drawing of the involved unit cells are included.

We measured the magnetic susceptibility of the specimens using a SQUID-magnetometer (Quantum Design). Figure 2 shows the result obtained for an oxygen-annealed sample. In an external field $H = 27$ Oe, the zero-field cooling susceptibility (ZFC) amounts to ~100% of $1/4\pi$ at temperature $T = 6$ K, indicating complete magnetic screening. For this estimate, we assumed an average density $\rho \approx 6$ g cm^{-3}. The field-cooling (FC) susceptibility reaches ~10% of the maximum possible value. This value represents a lower-bound value for the true superconducting volume fraction in the sample, indicating the bulk nature of superconductivity. The onset temperature of diamagnetism is $T_c \approx 133.5$ K, seen both in FC and ZFC experiments (see Fig. 2, inset). The FC susceptibility reaches ~60% of its full low-temperature value at 125 K, strongly indicating that the phase with $T_c \approx 133.5$ K dominates all other superconducting phases. In the ZFC curves, additional features are seen at 126 K and 112 K, which we ascribe to different superconducting phases with lower transition temperatures.

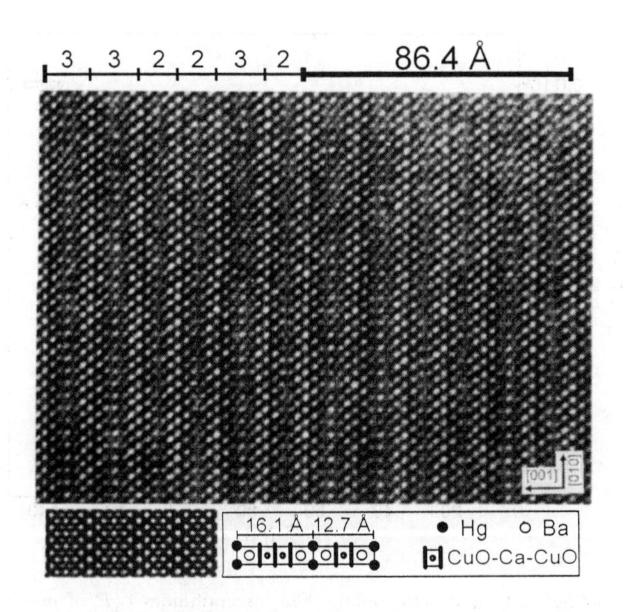

图 1. 晶粒在 [100] 取向上的 HRTEM 图像，包含 Hg–1212 和 Hg–1223 层。按照周期顺序，它们堆垛形成 $c \approx 86.4$ Å（见正文）的超晶格。插图是一个衬度模拟（$c_s = 1.1$ mm，$E = 300$ keV，散焦 −870 Å，样品厚度 23 Å）。右下方表示的是以 Cu–O 平面层的数目表示的堆垛序列和有关晶胞的放大示意图。

我们使用一台超导量子干涉磁强计（量子设计）测量了样品的磁化率。图 2 显示的是有氧退火样品所测得的结果。在 $H = 27$ Oe 的外场下，零场冷却磁化率（ZFC）相当于温度 $T = 6$ K 时达到 $1/4\pi$ 的 100% 左右，显示出完全的磁屏蔽。基于这一点估计，我们假定其平均密度 ρ 约为 6 g · cm^{-3}。场冷却（FC）磁化率达到最大可能值的 10% 左右。这个数值相当于样品中真正超导部分的体积分数的下限值，表明了体超导特性。抗磁性的起始温度 T_c 约为 133.5 K，在 FC 和 ZFC 实验中均是如此（见图 2 中插图）。FC 磁化率在温度为 125 K 时达到其完全低温数值的 60% 左右，强烈地表明 $T_c \approx 133.5$ K 的相比其他所有超导相都有优势。在 ZFC 曲线中，在温度为 126 K 和 112 K 处可以看到其他的特征，我们将其归结为另外的具有较低转变温度的不同超导相所引起的。

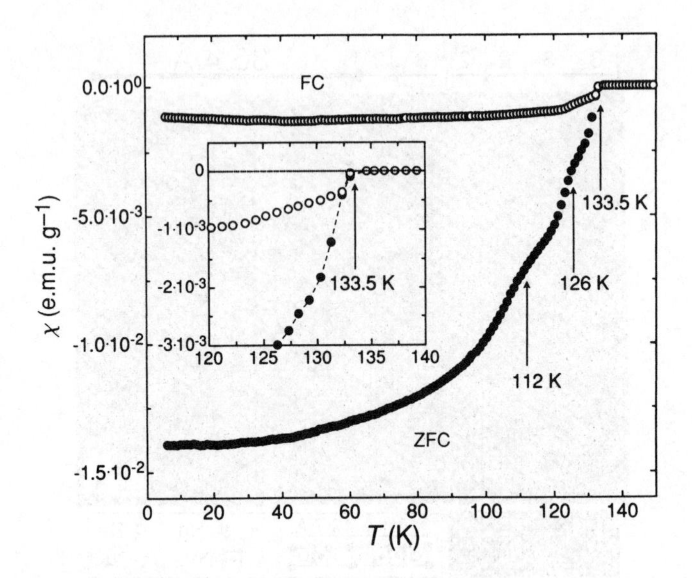

Fig. 2. Zero-field cooling (ZFC) and field cooling (FC) susceptibilities $\chi(T)$ of one of the investigated oxygen-annealed HBCCO samples, measured in $H = 27$ Oe. The ZFC curve indicates the presence of several different superconducting phases.

The resistivity R as a function of temperature T of an annealed sample is shown in Fig. 3. At $T \approx 132.5$ K, $R(T)$ drops sharply with a maximum in the differential $\mathrm{d}R/\mathrm{d}T$, and reaches zero at $T = 95$ K within the resolution of the four-probe a.c.-resistance bridge used. This temperature is still considerably higher than the zero-resistance temperature $T \approx 35$ K, reported for Hg-1201 (ref. 1) The final oxygen treatment was very effective in increasing the critical temperature; the as-sintered samples showed a maximum T_c of only ~117 K.

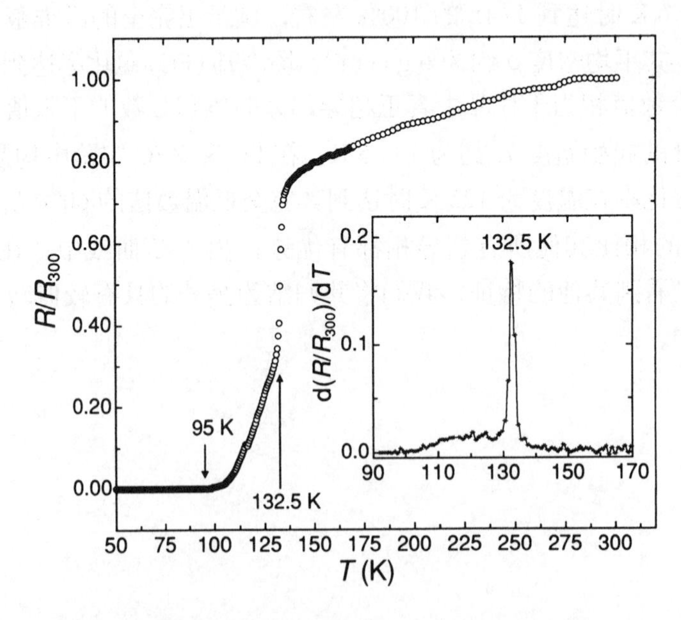

Fig. 3. Resistivity $R(T)$ of an annealed HBCCO specimen, normalized with respect to the resistance value $R(300) \approx 0.10\ \Omega$. The inset displays the temperature derivative $\mathrm{d}R/\mathrm{d}T$ to show the maximum resistivity drop at $T \approx 132.5$ K. Zero resistance is attained at $T = 95$ K.

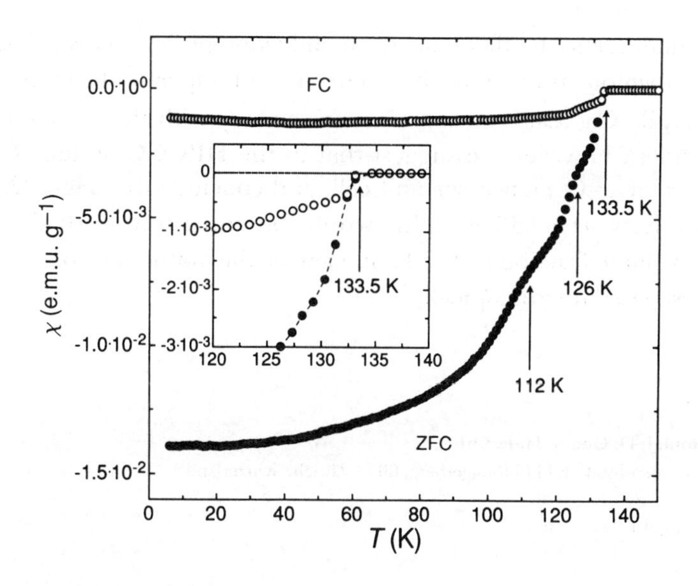

图 2. 一份有氧退火的 HBCCO 研究样品的零场冷却（ZFC）和场冷却（FC）磁化率 $\chi(T)$，在 $H=27$ Oe 中测得。ZFC 曲线指出若干个不同超导相的存在。

图 3 中显示了经退火处理样品的电阻 R 作为温度 T 的函数的变化。在温度 T 约为 132.5 K 处，$R(T)$ 急剧下降，其微分值 dR/dT 达到最大值，并且在四电极交流电阻桥的分辨率下观测到在温度 T 为 95 K 处达到零电阻。这个温度比起对于 Hg–1201 所报道的（参考文献 1）零电阻温度 T 约 35 K 还是高出很多。最后的氧处理对于提高临界温度非常有效；那些原始烧结的样品具有的最高温度 T_c 只有约 117 K。

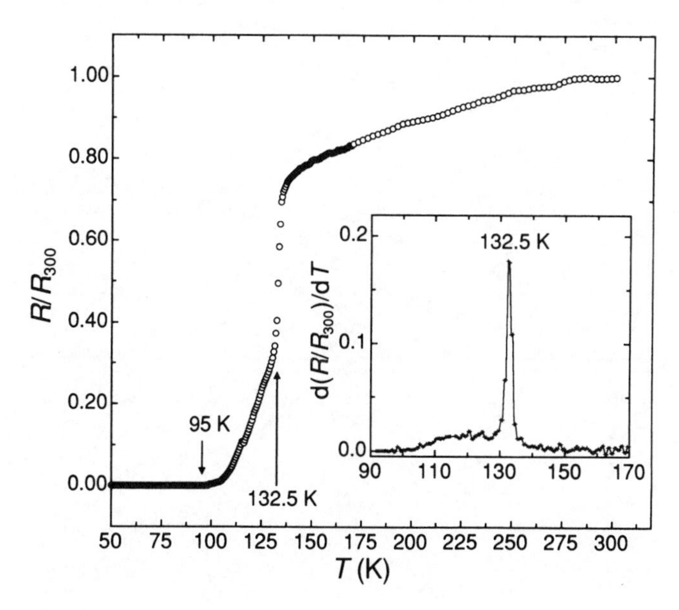

图 3. 一份经退火处理的 HBCCO 样品的电阻 $R(T)$，以阻值 $R(300)\approx 0.10\ \Omega$ 为标准。插图显示了温度的导数 dR/dT 以表明 $T\approx 132.5$ K 处的最大电阻降。在 $T=95$ K 时得到零电阻。

At present we cannot relate the different superconducting phases to crystallographic phases. There is no unambiguous proof that the occurrence of superconductivity in our samples stems from the $HgBa_2Ca_{n-1}Cu_nO_{2n+2+\delta}$ phases. In analogy with the thallium- and bismuth-based copper oxides[5], however, we suggest that in the HBCCO system T_c also increases with the number of Cu–O planes per unit cell, and conclude that Hg-1223 is responsible for superconductivity at ~133 K. This would be consistent with the large relative superconducting volume fraction at 125 K, in view of the dominance of Hg-1223 observed in the grains investigated microscopically.

(**363**, 56-58; 1993)

A. Schilling, M. Cantoni, J. D. Guo & H. R. Ott
Laboratorium für Festkörperphysik, ETH Hönggerberg, 8093 Zürich, Switzerland

Received 14 April; accepted 15 April 1993.

References:

1. Putilin, S. N., Antipov, E. V., Chmaissem, O. & Marezio, M. *Nature* **362**, 226-228 (1993).
2. Kaneko, T., Yamauchi, H. & Tanaka, S. *Physica* **C178**, 377-382 (1991).
3. Parkin, S. S. P. *et al. Phys. Rev. Lett.* **60**, 2539-2542 (1988).
4. Gopalakrishnan, I. K., Yakhmi, J. V. & Iyer, R. M. *Physica* **C175**, 183-186 (1991).
5. Parkin, S. S. P. *et al. Phys. Rev. Lett.* **61**, 750-753 (1988).
6. Putilin, S. N., Bryntse, I. & Antipov, E. V. *Mat. Res. Bull.* **26**, 1299-1307 (1991).

Acknowledgements. We thank S. Ritsch for his help in the structural characterization. This work was supported in part by the Schweizerische Nationalfonds zur Förderung der wissenschaftlichen Forschung.

目前我们还无法将不同的超导相与各个晶相一一对应起来。还没有明确的证据表明我们的样品的超导性是源于 $HgBa_2Ca_{n-1}Cu_nO_{2n+2+\delta}$ 相。不过，与铊基和铋基铜的氧化物[5] 类比，我们认为，HBCCO 体系的 T_c 也会随着每个单胞中 Cu–O 平面数目的增加而升高，并断言 Hg–1223 是温度约为 133 K 时超导性产生的原因。考虑到用显微镜研究颗粒时所观测到的 Hg–1223 占优势，这与 125 K 时大的相对超导体积分数是一致的。

（王耀杨 翻译；韩汝珊 郭建栋 审稿）

Experimental Quantum Teleportation

D. Bouwmeester *et al.*

Editor's Note

Exploitation of the principles of quantum physics to manipulate information in new ways is giving rise to the discipline of quantum information technology, which promises superfast quantum computers and secure quantum cryptography. One exotic possibility allowed by quantum physics is the instantaneous transfer of a quantum state over large distances: quantum teleportation. This is enabled by using the phenomenon of entanglement, in which the states of two or more quantum particles are interdependent. Here Anton Zeilinger and colleagues at the University of Innsbruck report the first experimental demonstration of quantum teleportation, in which the polarization state of one photon is transferred to another entangled photon some distance away. Realized within the laboratory here, quantum teleportation has now been achieved over more than a hundred kilometres.

Quantum teleportation—the transmission and reconstruction over arbitrary distances of the state of a quantum system—is demonstrated experimentally. During teleportation, an initial photon which carries the polarization that is to be transferred and one of a pair of entangled photons are subjected to a measurement such that the second photon of the entangled pair acquires the polarization of the initial photon. This latter photon can be arbitrarily far away from the initial one. Quantum teleportation will be a critical ingredient for quantum computation networks.

THE dream of teleportation is to be able to travel by simply reappearing at some distant location. An object to be teleported can be fully characterized by its properties, which in classical physics can be determined by measurement. To make a copy of that object at a distant location one does not need the original parts and pieces—all that is needed is to send the scanned information so that it can be used for reconstructing the object. But how precisely can this be a true copy of the original? What if these parts and pieces are electrons, atoms and molecules? What happens to their individual quantum properties, which according to the Heisenberg's uncertainty principle cannot be measured with arbitrary precision?

Bennett *et al.*[1] have suggested that it is possible to transfer the quantum state of a particle onto another particle—the process of quantum teleportation—provided one does not get any information about the state in the course of this transformation. This requirement can be fulfilled by using entanglement, the essential feature of quantum mechanics[2]. It describes correlations between quantum systems much stronger than any classical correlation could be.

量子隐形传态实验

鲍夫梅斯特等

编者按

基于量子物理的基本原理对信息以全新的方式进行操控，催生了量子信息技术，使我们有望实现超快速的量子计算和安全的量子密码。而由量子物理带来的其中一个神奇的现象是在远距离下量子态的瞬时转移：量子隐形传态。这项技术是利用纠缠态（在两个或多个粒子的纠缠态中，这些粒子的状态都是相互依赖的）的特性来实现的。在这篇文章中，来自因斯布鲁克大学的安东·蔡林格和他的同事们报道了第一个量子隐形传态的实验演示。这个实验是将一个光子的偏振态转移给远处另一个处在纠缠态的光子。自这个实验以后，如今已经可以在一百千米以上的距离实现量子隐形传态。

我们实验演示了量子隐形传态，也就是将量子系统的状态传递到任意距离之外并加以重建。本次实验过程涉及三个光子，第一个光子携带将被转移的偏振信息。第二个和第三个光子处于纠缠态，对第一个和第二个光子进行联合测量，第三个光子就可以获得第一个光子的偏振信息。第三个光子与第一个光子之间的距离可以任意远。量子隐形传态将成为量子计算网络的关键组成。

隐形传态是想通过在远处重现的方法来实现旅行的梦想。远距传送的对象可以完全由其性质来表示，在经典物理中这些性质可以由测量决定。要在一定距离之外造出对象的一个副本，我们并不需要原来的部分和零件，只要把扫描得到的信息传送过去用来重建对象就可以了。但是这样得到的原件的副本有多精确？如果那些部分和零件是电子，原子和分子呢？它们的个体量子性质会是什么情况？根据海森堡的测不准原理，这些个体量子性质是无法以任意精度测量的。

本内特等[1]提出，一个粒子的量子态可以被转移到另一个粒子上，前提是在转移过程中关于这个态的信息不可知，这就是量子隐形传态过程。其前提条件可以由作为量子力学本质特征的纠缠[2]来实现。它描述了比任何经典的相关都要强的量子系统之间的关联。

The possibility of transferring quantum information is one of the cornerstones of the emerging field of quantum communication and quantum computation[3]. Although there is fast progress in the theoretical description of quantum information processing, the difficulties in handling quantum systems have not allowed an equal advance in the experimental realization of the new proposals. Besides the promising developments of quantum cryptography[4] (the first provably secure way to send secret messages), we have only recently succeeded in demonstrating the possibility of quantum dense coding[5], a way to quantum mechanically enhance data compression. The main reason for this slow experimental progress is that, although there exist methods to produce pairs of entangled photons[6], entanglement has been demonstrated for atoms only very recently[7] and it has not been possible thus far to produce entangled states of more than two quanta.

Here we report the first experimental verification of quantum teleportation. By producing pairs of entangled photons by the process of parametric down-conversion and using two-photon interferometry for analysing entanglement, we could transfer a quantum property (in our case the polarization state) from one photon to another. The methods developed for this experiment will be of great importance both for exploring the field of quantum communication and for future experiments on the foundations of quantum mechanics.

The Problem

To make the problem of transferring quantum information clearer, suppose that Alice has some particle in a certain quantum state $|\psi\rangle$ and she wants Bob, at a distant location, to have a particle in that state. There is certainly the possibility of sending Bob the particle directly. But suppose that the communication channel between Alice and Bob is not good enough to preserve the necessary quantum coherence or suppose that this would take too much time, which could easily be the case if $|\psi\rangle$ is the state of a more complicated or massive object. Then, what strategy can Alice and Bob pursue?

As mentioned above, no measurement that Alice can perform on $|\psi\rangle$ will be sufficient for Bob to reconstruct the state because the state of a quantum system cannot be fully determined by measurements. Quantum systems are so evasive because they can be in a superposition of several states at the same time. A measurement on the quantum system will force it into only one of these states—this is often referred to as the projection postulate. We can illustrate this important quantum feature by taking a single photon, which can be horizontally or vertically polarized, indicated by the states $|\leftrightarrow\rangle$ and $|\updownarrow\rangle$. It can even be polarized in the general superposition of these two states

$$|\psi\rangle = \alpha\,|\leftrightarrow\rangle + \beta\,|\updownarrow\rangle \tag{1}$$

where α and β are two complex numbers satisfying $|\alpha|^2+|\beta|^2 = 1$. To place this example in a more general setting we can replace the states $|\leftrightarrow\rangle$ and $|\updownarrow\rangle$ in equation (1) by $|0\rangle$ and

量子信息传送的可能性是新兴的量子通信和量子计算领域的基石之一[3]。虽然量子信息处理的理论描述发展很快，但是由于操纵量子系统的困难性，新方案的实验演示相对滞后。除了已经演示了充满希望的量子密码[4](第一个被证实的保密通信方式)之外，我们还只是在前不久又成功演示了量子密集编码这种基于量子力学实现增强数据压缩的可能性[5]。实验进展如此之慢的主要原因在于，虽然存在产生纠缠光子对的方法[6]，但原子的纠缠到最近才被证实[7]，并且目前为止无法产生两个量子以上的纠缠态。

本文中我们报道了量子隐形传态的首次实验证明。通过参量下转换产生纠缠光子对，并利用双光子干涉测量来分析纠缠，我们可以做到将量子特性(在我们的实验中是偏振态)从一个光子转移到另一个光子。本实验发展的方法将在量子通信领域以及未来量子力学基础检验中发挥重要作用。

问 题 阐 述

为了阐明量子信息转移的问题，我们假设艾丽斯有一个粒子处于某个量子态 $|\psi\rangle$ 并且她希望远处的鲍勃也有一个粒子处于这个态。当然，一种可能的方式是直接把自己的粒子送到鲍勃那里。但是如果艾丽斯和鲍勃之间的通信通道不够好不足以保持必要的量子相干性，或者传送过程需要的时间太长(若 $|\psi\rangle$ 是更复杂的或者更大的对象的态函数，就会很容易出现这些情况)，艾丽斯和鲍勃能采取什么策略呢?

正如前面提到的一样，因为量子系统的态是不能完全由测量确定的，艾丽斯对 $|\psi\rangle$ 所能进行的测量，都不能精确到让鲍伯能够重建这个态。量子叠加性原理可以使量子系统同时处于几个态的叠加。对量子系统进行测量会迫使它处于其中的一个态，这经常被称为投影假设。我们可以通过一个光子的例子来说明这个重要的量子特征。这个光子可以处于水平偏振或者垂直偏振，分别以 $|\leftrightarrow\rangle$ 和 $|\updownarrow\rangle$ 表示。其偏振态甚至可以是这两个态的普遍叠加

$$|\psi\rangle = \alpha\,|\leftrightarrow\rangle + \beta\,|\updownarrow\rangle \tag{1}$$

其中 α 和 β 是两个复数满足 $|\alpha|^2+|\beta|^2=1$。为了让这个例子更具普遍性，我们可以把

$|1\rangle$, which refer to the states of any two-state quantum system. Superpositions of $|0\rangle$ and $|1\rangle$ are called qubits to signify the new possibilities introduced by quantum physics into information science[8].

If a photon in state $|\psi\rangle$ passes through a polarizing beam splitter—a device that reflects (transmits) horizontally (vertically) polarized photons—it will be found in the reflected (transmitted) beam with probability $|\alpha|^2$ ($|\beta|^2$). Then the general state $|\psi\rangle$ has been projected either onto $|\leftrightarrow\rangle$ or onto $|\updownarrow\rangle$ by the action of the measurement. We conclude that the rules of quantum mechanics, in particular the projection postulate, make it impossible for Alice to perform a measurement on $|\psi\rangle$ by which she would obtain all the information necessary to reconstruct the state.

The Concept of Quantum Teleportation

Although the projection postulate in quantum mechanics seems to bring Alice's attempts to provide Bob with the state $|\psi\rangle$ to a halt, it was realised by Bennett *et al.*[1] that precisely this projection postulate enables teleportation of $|\psi\rangle$ from Alice to Bob. During teleportation Alice will destroy the quantum state at hand while Bob receives the quantum state, with neither Alice nor Bob obtaining information about the state $|\psi\rangle$. A key role in the teleportation scheme is played by an entangled ancillary pair of particles which will be initially shared by Alice and Bob.

Suppose particle 1 which Alice wants to teleport is in the initial state $|\psi\rangle_1 = \alpha\,|\leftrightarrow\rangle_1 + \beta\,|\updownarrow\rangle_1$ (Fig. 1a), and the entangled pair of particles 2 and 3 shared by Alice and Bob is in the state:

$$|\psi^-\rangle_{23} = \frac{1}{\sqrt{2}}\left(|\leftrightarrow\rangle_2|\updownarrow\rangle_3 - |\updownarrow\rangle_2|\leftrightarrow\rangle_3\right) \tag{2}$$

That entangled pair is a single quantum system in an equal superposition of the states $|\leftrightarrow\rangle_2|\updownarrow\rangle_3$ and $|\updownarrow\rangle_2|\leftrightarrow\rangle_3$. The entangled state contains no information on the individual particles; it only indicates that the two particles will be in opposite states. The important property of an entangled pair is that as soon as a measurement on one of the particles projects it, say, onto $|\leftrightarrow\rangle$ the state of the other one is determined to be $|\updownarrow\rangle$, and vice versa. How could a measurement on one of the particles instantaneously influence the state of the other particle, which can be arbitrarily far away? Einstein, among many other distinguished physicists, could simply not accept this "spooky action at a distance". But this property of entangled states has now been demonstrated by numerous experiments (for reviews, see refs 9, 10).

公式 (1) 中的态 $|\leftrightarrow\rangle$ 和 $|\updownarrow\rangle$ 换作 $|0\rangle$ 和 $|1\rangle$，来表示任何双态量子系统的两个态。$|0\rangle$ 和 $|1\rangle$ 的叠加被称为量子比特，以显示量子物理为信息科学引入新的可能性 [8]。

如果一个处于 $|\psi\rangle$ 态的光子通过一个偏振分束器 (这种器件反射水平偏振光而透射垂直偏振光)，则在反射光束中找到该光子的概率是 $|\alpha|^2$，在透射光束中找到该光子的概率是 $|\beta|^2$。测量的行为使得普通态函数 $|\psi\rangle$ 被投影到 $|\leftrightarrow\rangle$ 或者 $|\updownarrow\rangle$ 上。我们的结论是：量子力学的规则，特别是投影假设，使得艾丽斯不可能对 $|\psi\rangle$ 进行一次测量以得到所有必要的信息来重建这个态。

量子隐形传态的概念

尽管量子力学的投影假设看上去阻止了艾丽斯将 $|\psi\rangle$ 态提供给鲍勃的意图，本内特等 [1] 却意识到就是这个投影假设使得 $|\psi\rangle$ 从艾丽斯到鲍勃的隐形传态成为可能。在隐形传态过程中，艾丽斯将对自己这边的量子态进行测量破坏，这时，鲍勃那边才接收到这个量子态，这个过程中，两个人都没有获知 $|\psi\rangle$ 态的信息。最初由艾丽斯和鲍勃共享的处于纠缠态的辅助粒子对在这个隐形传态方案中起到关键作用。

假定艾丽斯想要传送的粒子 1 处于初始态 $|\psi\rangle_1 = \alpha\,|\leftrightarrow\rangle_1 + \beta\,|\updownarrow\rangle_1$ (图 1a)，并且艾丽斯和鲍勃共享的纠缠粒子对，即粒子 2 和粒子 3 处于下列态中：

$$|\psi^-\rangle_{23} = \frac{1}{\sqrt{2}}\left(|\leftrightarrow\rangle_2|\updownarrow\rangle_3 - |\updownarrow\rangle_2|\leftrightarrow\rangle_3\right) \tag{2}$$

这个纠缠粒子对是 $|\leftrightarrow\rangle_2|\updownarrow\rangle_3$ 态和 $|\updownarrow\rangle_2|\leftrightarrow\rangle_3$ 态等量叠加的一个简单量子系统。纠缠态不包含任何个体粒子的信息，它仅仅表明两个粒子是处于正交的态。纠缠对的一个重要特性就是当对其中一个粒子进行测量而将其投影到比如说 $|\leftrightarrow\rangle$ 上，另一个粒子的态就确定为 $|\updownarrow\rangle$，反过来也是一样。对一个粒子的测量如何能够瞬间影响另一个可以是任意远的粒子的态？爱因斯坦还有其他很多杰出物理学家干脆不能接受这种“幽灵般的超距作用”。但是纠缠态的这个性质已经被很多实验所证实 (请看参考文献 9 和 10)。

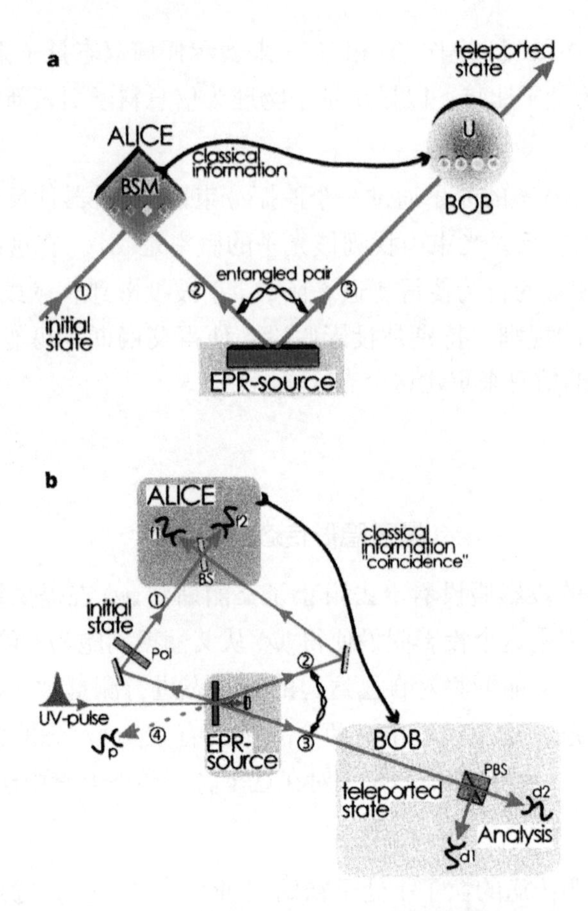

Fig. 1. Scheme showing principles involved in quantum teleportation (**a**) and the experimental set-up (**b**). **a**, Alice has a quantum system, particle 1, in an initial state which she wants to teleport to Bob. Alice and Bob also share an ancillary entangled pair of particles 2 and 3 emitted by an Einstein–Podolsky–Rosen (EPR) source. Alice then performs a joint Bell-state measurement (BSM) on the initial particle and one of the ancillaries, projecting them also onto an entangled state. After she has sent the result of her measurement as classical information to Bob, he can perform a unitary transformation (U) on the other ancillary particle resulting in it being in the state of the original particle. **b**, A pulse of ultraviolet radiation passing through a nonlinear crystal creates the ancillary pair of photons 2 and 3. After retroflection during its second passage through the crystal the ultraviolet pulse creates another pair of photons, one of which will be prepared in the initial state of photon 1 to be teleported, the other one serving as a trigger indicating that a photon to be teleported is under way. Alice then looks for coincidences after a beam splitter BS where the initial photon and one of the ancillaries are superposed. Bob, after receiving the classical information that Alice obtained a coincidence count in detectors f1 and f2 identifying the $|\psi^{-}\rangle_{12}$ Bell state, knows that his photon 3 is in the initial state of photon 1 which he then can check using polarization analysis with the polarizing beam splitter PBS and the detectors d1 and d2. The detector p provides the information that photon 1 is under way.

The teleportation scheme works as follows. Alice has the particle 1 in the initial state $|\psi\rangle_1$ and particle 2. Particle 2 is entangled with particle 3 in the hands of Bob. The essential point is to perform a specific measurement on particles 1 and 2 which projects them onto the entangled state:

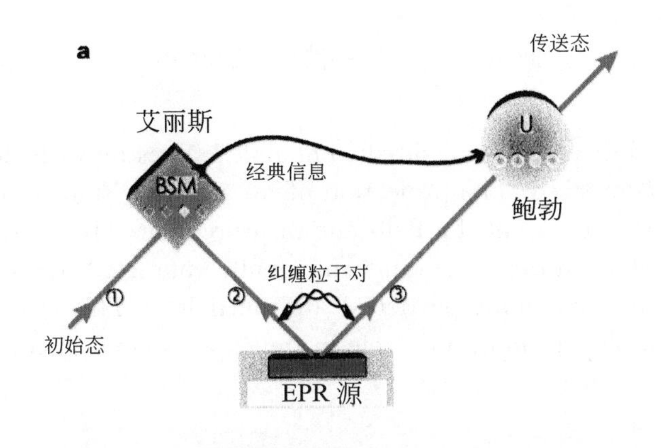

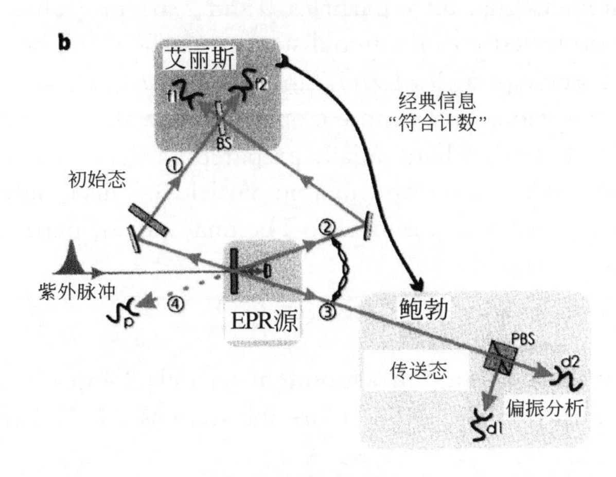

图 1. 表现量子隐形传态相关原理(**a**)和实验装置(**b**)的示意图。**a**，艾丽斯想把自己的量子系统也就是粒子 1 所处的初始态传送给鲍勃。艾丽斯和鲍勃共享一个从 EPR 源发出的辅助纠缠粒子对 2 和 3。艾丽斯接着对初始粒子 1 和辅助对中的一个粒子进行一个联合贝尔态测量，将其投影到一个纠缠态上。在她将测量结果以经典信息形式传送给鲍勃后，他就可以对另一个辅助粒子进行幺正变换，使之处于初始粒子 1 的状态中。**b**，一束紫外线脉冲通过非线性晶体时产生了辅助光子对 2 和 3。逆反射后第二次穿过晶体时这个紫外脉冲又产生一对光子，其中一个将被制备到待传递的光子 1 的初始态，另一个用作触发器，表示一个待传态的光子已经上路。于是，艾丽斯查看经过分束器 BS 后初始光子和一个辅助光子叠加时的符合计数。鲍勃在收到艾丽斯已经在探测器 f1 和 f2 处获得验证贝尔态 $|\psi^-\rangle_{12}$ 的符合计数的经典信息后，就知道他的光子 3 已处于光子 1 的初始态中，并可用偏振分光器 PBS 和探测器 d1 和 d2 进行偏振分析来验证。探测器 p 提供光子正在路上的信息。

隐形传态方案如下面所述。艾丽斯现有粒子 2 和处于初始态 $|\psi\rangle_1$ 的粒子 1，粒子 2 与鲍勃处的粒子 3 处于纠缠态。关键之处是对粒子 1 和粒子 2 进行一个特定的测量，将它们投影到纠缠态上：

$$|\psi^-\rangle_{12} = \frac{1}{\sqrt{2}} (|\leftrightarrow\rangle_1 |\updownarrow\rangle_2 - |\updownarrow\rangle_1 |\leftrightarrow\rangle_2) \tag{3}$$

This is only one of four possible maximally entangled states into which any state of two particles can be decomposed. The projection of an arbitrary state of two particles onto the basis of the four states is called a Bell-state measurement. The state given in equation (3) distinguishes itself from the three other maximally entangled states by the fact that it changes sign upon interchanging particle 1 and particle 2. This unique antisymmetric feature of $|\psi^-\rangle_{12}$ will play an important role in the experimental identification, that is, in measurements of this state.

Quantum physics predicts[1] that once particles 1 and 2 are projected into $|\psi^-\rangle_{12}$, particle 3 is instantaneously projected into the initial state of particle 1. The reason for this is as follows. Because we observe particles 1 and 2 in the state $|\psi^-\rangle_{12}$ we know that whatever the state of particle 1 is, particle 2 must be in the opposite state, that is, in the state orthogonal to the state of particle 1. But we had initially prepared particle 2 and 3 in the state $|\psi^-\rangle_{23}$, which means that particle 2 is also orthogonal to particle 3. This is only possible if particle 3 is in the same state as particle 1 was initially. The final state of particle 3 is therefore:

$$|\psi\rangle_3 = \alpha|\leftrightarrow\rangle_3 + \beta|\updownarrow\rangle_3 \tag{4}$$

We note that during the Bell-state measurement particle 1 loses its identity because it becomes entangled with particle 2. Therefore the state $|\psi\rangle_1$ is destroyed on Alice's side during teleportation.

This result (equation (4)) deserves some further comments. The transfer of quantum information from particle 1 to particle 3 can happen over arbitrary distances, hence the name teleportation. Experimentally, quantum entanglement has been shown[11] to survive over distances of the order of 10 km. We note that in the teleportation scheme it is not necessary for Alice to know where Bob is. Furthermore, the initial state of particle 1 can be completely unknown not only to Alice but to anyone. It could even be quantum mechanically completely undefined at the time the Bell-state measurement takes place. This is the case when, as already remarked by Bennett *et al.*[1], particle 1 itself is a member of an entangled pair and therefore has no well-defined properties on its own. This ultimately leads to entanglement swapping[12,13].

It is also important to notice that the Bell-state measurement does not reveal any information on the properties of any of the particles. This is the very reason why quantum teleportation using coherent two-particle superpositions works, while any measurement on one-particle superpositions would fail. The fact that no information whatsoever is gained on either particle is also the reason why quantum teleportation escapes the verdict of the no-cloning theorem[14]. After successful teleportation particle 1 is not available in its original state any more, and therefore particle 3 is not a clone but is really the result of teleportation.

$$|\psi^{-}\rangle_{12}=\frac{1}{\sqrt{2}}\left(|\leftrightarrow\rangle_{1}|\updownarrow\rangle_{2}-|\updownarrow\rangle_{1}|\leftrightarrow\rangle_{2}\right) \tag{3}$$

这仅仅是任意双粒子态可以分解成的四个可能的最大纠缠态之一。双粒子任意态在四个最大纠缠态组成的一组正交完备基中任何一个态上的投影测量叫贝尔态测量。公式 3 中的态和其他三个最大纠缠态的不同之处在于交换粒子 1 和粒子 2 将导致其符号改变。$|\psi^{-}\rangle_{12}$ 态的这个独特的反对称特征在实验确认也就是在这个态的测量中将起到重要作用。

量子物理预言 [1] 当粒子 1 和 2 被投影到 $|\psi^{-}\rangle_{12}$ 态时，粒子 3 将被即时投影到粒子 1 的初始态。原因如下：因为我们测量得知粒子 1 和 2 处于 $|\psi^{-}\rangle_{12}$ 态中，所以我们知道无论粒子 1 处于任何态，粒子 2 一定处于正交态，也就是和粒子 1 的态正交的态中。但是我们一开始将粒子 2 和 3 置于 $|\psi^{-}\rangle_{23}$ 态中，也就是说粒子 2 也和粒子 3 正交。要实现这一点唯一的可能就是粒子 3 处于和粒子 1 的初始态相同的态中。由此粒子 3 的终态就是

$$|\psi\rangle_{3}=\alpha|\leftrightarrow\rangle_{3}+\beta|\updownarrow\rangle_{3} \tag{4}$$

注意在贝尔态测量过程中粒子 1 因为与粒子 2 纠缠而失去了自身特征。因此艾丽斯这边的 $|\psi\rangle_{1}$ 态在隐形传态过程中被破坏。

这个结果（公式 4）值得进一步讨论。从粒子 1 到粒子 3 的量子信息的转移可以是通过任意距离，所以才有隐形传态的名字。实验证明量子纠缠能够建立在 10 千米量级的距离上 [11]。注意隐形传态方案中艾丽斯不需要知道鲍勃在哪里。另外不光是艾丽斯，其他任何人都不用知道粒子 1 的初始态。在进行贝尔态测量时，它甚至可以是在量子力学意义上完全未定义的。正像本内特等 [1] 已经指出过的，当粒子 1 自己是一个纠缠对中的一员因而没有自己的明确定义的性质时，就是这种情况。这最终导致纠缠交换 [12,13]。

另外重要的一点是贝尔态测量不揭示任何一个粒子性质的任何信息。这是为什么利用相干双粒子叠加的量子隐形传态能够成功，而任何单粒子叠加的测量都会失败的原因。未能获取任一粒子的任何信息的事实，也是量子隐形传态并不违反量子不可克隆定理 [14] 的原因。在成功的隐形传态之后，粒子 1 不再处于初始态，因而粒子 3 确实是隐形传态的结果而不是一个克隆。

A complete Bell-state measurement can not only give the result that the two particles 1 and 2 are in the antisymmetric state, but with equal probabilities of 25% we could find them in any one of the three other entangled states. When this happens, particle 3 is left in one of three different states. It can then be brought by Bob into the original state of particle 1 by an accordingly chosen transformation, independent of the state of particle 1, after receiving via a classical communication channel the information on which of the Bell-state results was obtained by Alice. Yet we note, with emphasis, that even if we chose to identify only one of the four Bell states as discussed above, teleportation is successfully achieved, albeit only in a quarter of the cases.

Experimental Realization

Teleportation necessitates both production and measurement of entangled states; these are the two most challenging tasks for any experimental realization. Thus far there are only a few experimental techniques by which one can prepare entangled states, and there exist no experimentally realized procedures to identify all four Bell states for any kind of quantum system. However, entangled pairs of photons can readily be generated and they can be projected onto at least two of the four Bell states.

We produced the entangled photons 2 and 3 by parametric down-conversion. In this technique, inside a nonlinear crystal, an incoming pump photon can decay spontaneously into two photons which, in the case of type II parametric down-conversion, are in the state given by equation (2) (Fig. 2)[6].

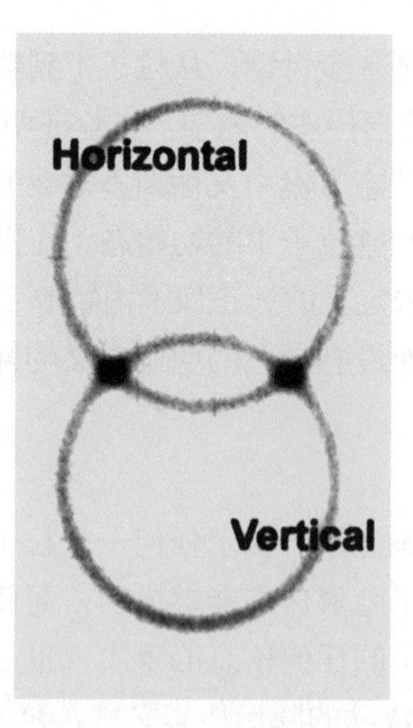

Fig. 2. Photons emerging from type II down-conversion (see text). Photograph taken perpendicular to the propagation direction. Photons are produced in pairs. A photon on the top circle is horizontally polarized while its exactly opposite partner in the bottom circle is vertically polarized. At the intersection points their polarizations are undefined; all that is known is that they have to be different, which results in entanglement.

一个完整的贝尔态测量不仅能给出两个粒子 1 和 2 处于反对称态的结果，我们还可能发现它们处于其他三个纠缠态中的任何一个的概率都是相等的 25%。如果发生这种情况，粒子 3 就处于三个不同的态中的一个。接着鲍勃可以在通过经典通信通道收到艾丽斯获取的是哪一个贝尔态的信息后，通过选择相应的变换，在与粒子 1 所处态无关的情况下，将粒子 3 置于粒子 1 的初始态中。我们要着重强调，即使像前面所讨论的那样我们选择只确认四个贝尔态中的一个，隐形传态仍然可以成功实现，尽管只是在四分之一的事例中。

实验的实现

实现隐形传态需要生成和测量纠缠态。任何实验实现都要面临这两个难题。到目前为止，能够作出纠缠态的实验技术寥寥无几，而且没有成功的实验程序来确认任何一种量子系统的所有四个贝尔态。虽然如此，纠缠光子对可以由简易方法产生并被投影到至少两个贝尔态上。

我们通过参量下转换产生纠缠光子对 2 和 3。在这项技术中，非线性晶体中进来一个泵浦光子可以即时衰变为两个光子。在二型参量下转换情形中，这两个光子处于公式 2 给出的态（图 2）[6]。

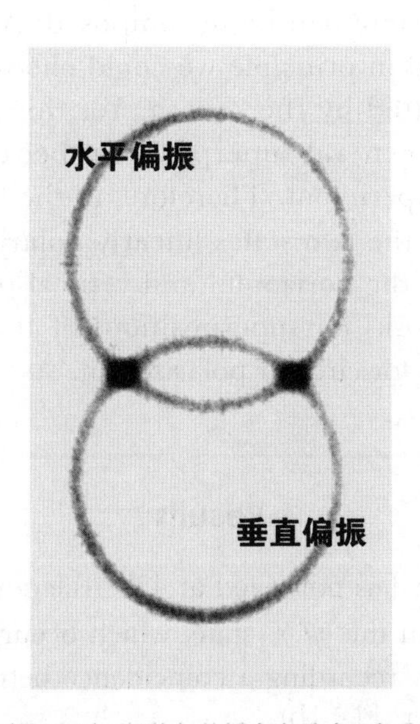

图 2. 二型参量下转换产生的光子（见正文）。相片与传播方向垂直。光子成对产生。顶圈的光子水平偏振，其严格相反的底圈同伴为垂直偏振。在相交的两点它们的偏振态是不明确的，我们只知道它们必须是不同的，这就导致纠缠。

To achieve projection of photons 1 and 2 into a Bell state we have to make them indistinguishable. To achieve this indistinguishability we superpose the two photons at a beam splitter (Fig. 1b). Then if they are incident one from each side, how can it happen that they emerge still one on each side? Clearly this can happen if they are either both reflected or both transmitted. In quantum physics we have to superimpose the amplitudes for these two possibilities. Unitarity implies that the amplitude for both photons being reflected obtains an additional minus sign. Therefore, it seems that the two processes cancel each other. This is, however, only true for a symmetric input state. For an antisymmetric state, the two possibilities obtain another relative minus sign, and therefore they constructively interfere[15,16]. It is thus sufficient for projecting photons 1 and 2 onto the antisymmetric state $|\psi^{-}\rangle_{12}$ to place detectors in each of the outputs of the beam splitter and to register simultaneous detections (coincidence)[17-19].

To make sure that photons 1 and 2 cannot be distinguished by their arrival times, they were generated using a pulsed pump beam and sent through narrow-bandwidth filters producing a coherence time much longer than the pump pulse length[20]. In the experiment, the pump pulses had a duration of 200 fs at a repetition rate of 76 MHz. Observing the down-converted photons at a wavelength of 788 nm and a bandwidth of 4 nm results in a coherence time of 520 fs. It should be mentioned that, because photon 1 is also produced as part of an entangled pair, its partner can serve to indicate that it was emitted.

How can one experimentally prove that an unknown quantum state can be teleported? First, one has to show that teleportation works for a (complete) basis, a set of known states into which any other state can be decomposed. A basis for polarization states has just two components, and in principle we could choose as the basis horizontal and vertical polarization as emitted by the source. Yet this would not demonstrate that teleportation works for any general superposition, because these two directions are preferred directions in our experiment. Therefore, in the first demonstration we choose as the basis for teleportation the two states linearly polarized at −45° and +45° which are already superpositions of the horizontal and vertical polarizations. Second, one has to show that teleportation works for superpositions of these base states. Therefore we also demonstrate teleportation for circular polarization.

Results

In the first experiment photon 1 is polarized at 45°. Teleportation should work as soon as photon 1 and 2 are detected in the $|\psi^{-}\rangle_{12}$ state, which occurs in 25% of all possible cases. The $|\psi^{-}\rangle_{12}$ state is identified by recording a coincidence between two detectors, f1 and f2, placed behind the beam splitter (Fig. 1b).

If we detect a f1f2 coincidence (between detectors f1 and f2), then photon 3 should also be polarized at 45°. The polarization of photon 3 is analysed by passing it through

要将光子 1 和 2 投影到一个贝尔态上我们必须让它们变得不可区分。为实现这个不可区分性我们将这两个光子在分束器处叠加(图 1b)。这样的话，如果它们从两边一边一个地入射，能不能做到出射时也是一边一个呢？显然这是可能的，只要它们都被反射或者都被透射就行。在量子物理中我们必须将这两个可能的振幅叠加。幺正性意味着两个光子都被反射的那个可能情形的振幅需要多加一个负号。这样的话这两个过程看上去互相抵消。但是只有对称入射态才是这种情况。对于反对称入射态，这两个可能情形要再加一个相对的负号，这样它们之间的干涉是相长的[15,16]。由此，只要将探测器放在分束器的每个输出端记录同时发生的响应(符合计数)，就可以将光子 1 和 2 投影到反对称态 $|\psi^{-}\rangle_{12}$[17-19]。

为了确保不能通过到达时间来区分光子 1 和 2，我们通过脉冲泵浦光束产生光子 1 和 2 并使其通过窄带宽滤波器来产生一个比泵浦脉冲长度长得多的相干时间[20]。在实验中泵浦脉冲持续时间为 200 fs，重复率为 76 MHz。观测下转换产生的光子(该光子波长为 788 nm，带宽为 4 nm)得到相干时间为 520 fs。这里应当提到的是，由于光子 1 也是作为纠缠对的一部分产生的，可以用其同伴来显示它已被发出。

如何用一个实验证明未知量子态被隐形传送？首先要证明隐形传态能够在一组(完备)基矢上成功实现，任何其他态都能分解到这组已知基矢上。偏振态的基矢只有两个成分，原理上我们可以把源发射的水平和垂直偏振作为基矢。但是因为这两个方向在我们实验中是优先方向，这种演示不能证明隐形传送对任何普遍叠加态都能实现。出于这个原因，在第一个实验中我们选择 -45° 和 $+45^{\circ}$ 两个线性偏振态作为拟传送态的基矢。它们都已经是水平和垂直偏振的叠加。其次，我们必须证明对这些基矢的叠加也能实现隐形传态。为此我们还证明了圆偏振的隐形传态。

实验结果

在第一个实验中光子 1 处于 45° 偏振。在所有可能情形中，光子 1 和 2 被检测到处于 $|\psi^{-}\rangle_{12}$ 态的概率是 25%。一旦出现这种情形，隐形传态就应当实现。$|\psi^{-}\rangle_{12}$ 态的确认是通过置于分束器后面两个探测器 f1 和 f2 的符合计数来实现的(图 1b)。

如果我们探测到一个 f1f2(探测器 f1 和 f2 之间的)符合事例，那么光子 3 应当也是 45° 偏振。光子 3 的偏振态的分析方法是让它通过一个选择 $+45^{\circ}$ 和 -45° 偏振

a polarizing beam splitter selecting +45° and −45° polarization. To demonstrate teleportation, only detector d2 at the +45° output of the polarizing beam splitter should click (that is, register a detection) once detectors f1 and f2 click. Detector d1 at the −45° output of the polarizing beam splitter should not detect a photon. Therefore, recording a three-fold coincidence d2f1f2 (+45° analysis) together with the absence of a three-fold coincidence d1f1f2 (−45° analysis) is a proof that the polarization of photon 1 has been teleported to photon 3.

To meet the condition of temporal overlap, we change in small steps the arrival time of photon 2 by changing the delay between the first and second down-conversion by translating the retroflection mirror (Fig. 1b). In this way we scan into the region of temporal overlap at the beam splitter so that teleportation should occur.

Outside the region of teleportation, photon 1 and 2 each will go either to f1 or to f2 independent of one another. The probability of having a coincidence between f1 and f2 is therefore 50%, which is twice as high as inside the region of teleportation. Photon 3 should not have a well-defined polarization because it is part of an entangled pair. Therefore, d1 and d2 have both a 50% chance of receiving photon 3. This simple argument yields a 25% probability both for the −45° analysis (d1f1f2 coincidences) and for the +45° analysis (d2f1f2 coincidences) outside the region of teleportation. Figure 3 summarizes the predictions as a function of the delay. Successful teleportation of the +45° polarization state is then characterized by a decrease to zero in the −45° analysis (Fig. 3a), and by a constant value for the +45° analysis (Fig. 3b).

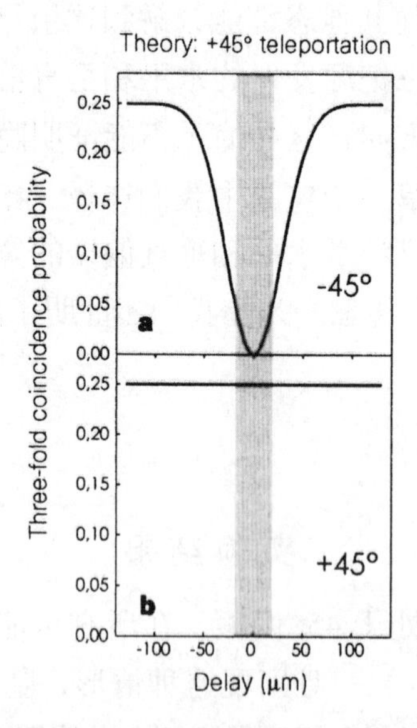

Fig. 3. Theoretical prediction for the three-fold coincidence probability between the two Bell-state detectors (f1, f2) and one of the detectors analysing the teleported state. The signature of teleportation of a photon polarization state at +45° is a dip to zero at zero delay in the three-fold coincidence rate with the detector analysing −45° (d1f1f2) (**a**) and a constant value for the detector analysis +45° (d2f1f2) (**b**). The shaded area indicates the region of teleportation.

的偏振分束器。若要证明隐形传态的实现，当探测器 f1 和 f2 响应时，必须只有偏振分束器 +45°输出端的探测器 d2 有响应，也就是记录下了一次探测，而偏振分束器 −45°输出端的探测器 d1 不应探测到光子。因此，一个三重符合计数 d2f1f2（+45°偏振分析）的记录，加上三重符合计数 d1f1f2（−45°偏振分析）的缺失，就成为光子 1 的偏振态已经被隐形传态到光子 3 的证据。

为满足时间重叠条件，我们通过平移逆反射镜改变第一次和第二次下转换之间的延迟来微调光子 2 的到达时间（图 1b）。用这种办法，我们逐渐进入分束器处时间重叠的区间，保证隐形传态发生。

在隐形传态区间之外，光子 1 和 2 各自可能独立奔向 f1 或者 f2。因此在 f1 和 f2 得到一个符合计数的概率是 50%。这个概率是在隐形传态区间中的两倍。光子 3 因为是纠缠对中一员，所以不应有明确定义的偏振。这样的话，d1 和 d2 都有 50%的可能接收到光子 3。这个简单的论证得到的结果是，在隐形传态区间之外，−45°偏振分析（d1f1f2 符合计数）和 +45°偏振分析（d2f1f2 符合计数）都有 25%的概率。图 3 总结了作为时间延迟函数的预测概率值。−45°分析下降到零（图 3a）和 +45°分析保持为一个常数（图 3b），表征 +45°偏振态的隐形传态得到成功实现。

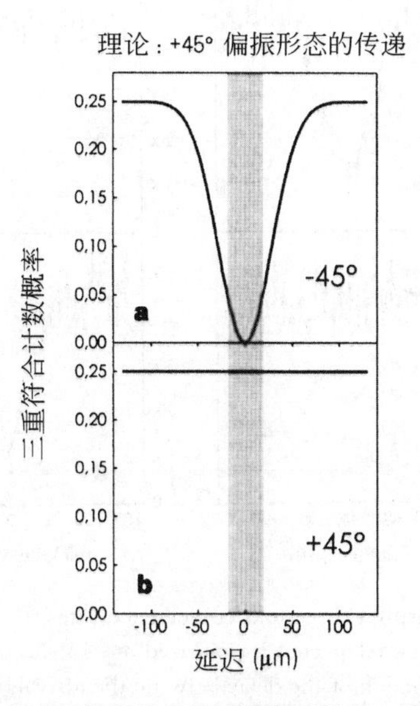

图 3. 两个贝尔态探测器（f1,f2）和分析被传递态的两个探测器之一的三重符合计数概率的理论预期。+45°光子偏振态传递的特征是：在零延迟处，−45°偏振分析探测器三重符合计数率（d1f1f2）有一个到零概率值的下降（**a**），而 +45°分析探测器三重符合计数率（d2f1f2）保持为常数（**b**）。阴影区为隐形传态区间。

The theoretical prediction of Fig. 3 may easily be understood by realizing that at zero delay there is a decrease to half in the coincidence rate for the two detectors of the Bell-state analyser, f1 and f2, compared with outside the region of teleportation. Therefore, if the polarization of photon 3 were completely uncorrelated to the others the three-fold coincidence should also show this dip to half. That the right state is teleported is indicated by the fact that the dip goes to zero in Fig. 3a and that it is filled to a flat curve in Fig. 3b.

We note that equally as likely as the production of photons 1, 2 and 3 is the emission of two pairs of down-converted photons by a single source. Although there is no photon coming from the first source (photon 1 is absent), there will still be a significant contribution to the three-fold coincidence rates. These coincidences have nothing to do with teleportation and can be identified by blocking the path of photon 1.

The probability for this process to yield spurious two- and three-fold coincidences can be estimated by taking into account the experimental parameters. The experimentally determined value for the percentage of spurious three-fold coincidences is $68\% \pm 1\%$. In the experimental graphs of Fig. 4 we have subtracted the experimentally determined spurious coincidences.

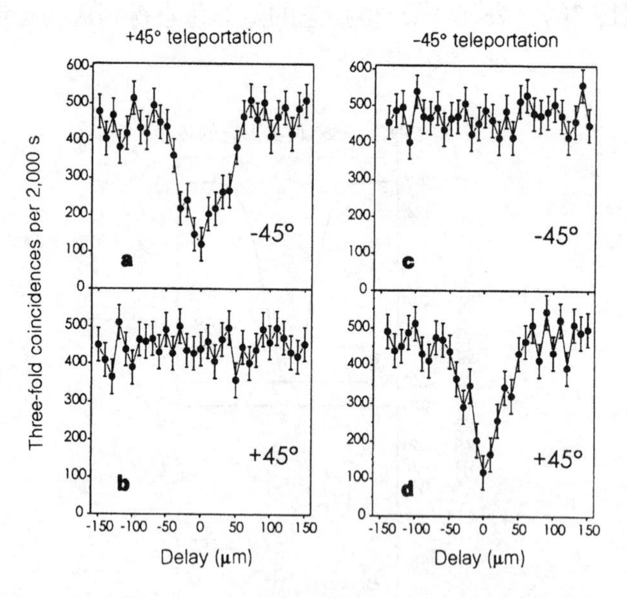

Fig. 4. Experimental results. Measured three-fold coincidence rates d1f1f2 (−45°) and d2f1f2 (+45°) in the case that the photon state to be teleported is polarized at +45° (**a** and **b**) or at −45° (**c** and **d**). The coincidence rates are plotted as function of the delay between the arrival of photon 1 and 2 at Alice's beam splitter (see Fig. 1b). The three-fold coincidence rates are plotted after subtracting the spurious three-fold contribution (see text). These data, compared with Fig. 3, together with similar ones for other polarizations (Table 1) confirm teleportation for an arbitrary state.

图 3 中的理论预计图可以这样来简单理解：与隐形传态区间之外相比较，在延迟为零时，两个贝尔态探测器 f1 和 f2 的符合计数率下降一半；如此一来，如果光子 3 的偏振与其他光子完全无关，三重符合计数也应该降到一半。在图 3a 中曲线下降一直到零而在图 3b 中被填平成为平坦曲线，表明设计的隐形传态得到实现。

需要指出的是，与光子 1、2 和 3 的产生同样可能发生的是单一源发出两对下转换光子。虽然这种情况中没有光子来自第一个源(光子 1 未出现)，但是它对三重符合计数率仍有显著贡献。这些符合计数与隐形传态无关，可以用阻断光子 1 路径的办法确认。

这一产生乱真二重和三重符合计数的过程发生的概率可以从实验参数估计。由实验确定的乱真三重符合计数的百分比是 68% ± 1%。图 4 中的实验图形中我们已经扣除这个实验确定的乱真符合计数。

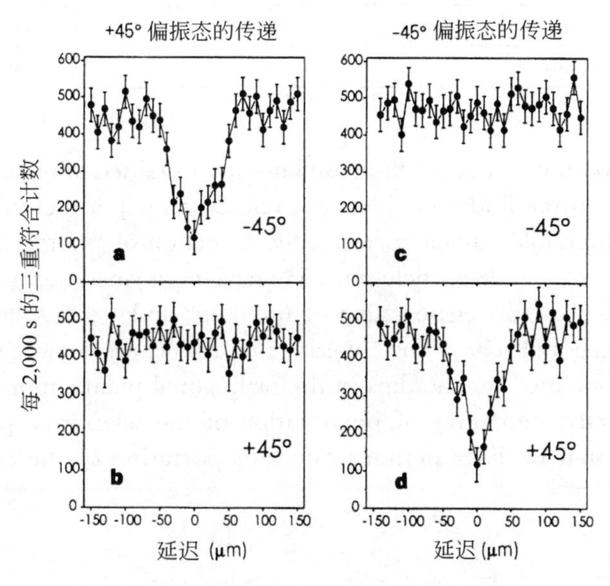

图 4. 实验结果。在待传光子为 +45° 偏振(**a** 和 **b**)或者 −45° 偏振(**c** 和 **d**)的情况下，测量 d1f1f2 (−45°) 和 d2f1f2 (+45°) 三重符合计数率。画出的符合计数率是光子 1 和 2 分别到达艾丽斯的分束器的延迟的函数(见图 1b)，并扣除了乱真三重计数的贡献(见正文)。这些数据与图 3 对比，再加上其他偏振的类似结果(表 1)，证实了任意一个态都可被传递。

The experimental results for teleportation of photons polarized under +45° are shown in the left-hand column of Fig. 4; Fig. 4a and b should be compared with the theoretical predictions shown in Fig. 3. The strong decrease in the −45° analysis, and the constant signal for the +45° analysis, indicate that photon 3 is polarized along the direction of photon 1, confirming teleportation.

The results for photon 1 polarized at −45° demonstrate that teleportation works for a complete basis for polarization states (right-hand column of Fig. 4). To rule out any classical explanation for the experimental results, we have produced further confirmation that our procedure works by additional experiments. In these experiments we teleported photons linearly polarized at 0° and at 90°, and also teleported circularly polarized photons. The experimental results are summarized in Table 1, where we list the visibility of the dip in three-fold coincidences, which occurs for analysis orthogonal to the input polarization.

Table 1. Visibility of teleportation in three-fold coincidences

Polarization	Visibility
+45°	0.63 ± 0.02
-45°	0.64 ± 0.02
0°	0.66 ± 0.02
90°	0.61 ± 0.02
Circular	0.57 ± 0.02

As mentioned above, the values for the visibilities are obtained after subtracting the offset caused by spurious three-fold coincidences. These can experimentally be excluded by conditioning the three-fold coincidences on the detection of photon 4, which effectively projects photon 1 into a single-particle state. We have performed this four-fold coincidence measurement for the case of teleportation of the +45° and +90° polarization states, that is, for two non-orthogonal states. The experimental results are shown in Fig. 5. Visibilities of 70% ± 3% are obtained for the dips in the orthogonal polarization states. Here, these visibilities are directly the degree of polarization of the teleported photon in the right state. This proves that we have demonstrated teleportation of the quantum state of a single photon.

图 4 左栏是 +45°偏振光子的隐形传态的实验结果；我们应当把图 4a 和 b 与图 3 中的理论预计曲线相比较来看。−45°分析中的猛烈下降和 +45°分析中的稳定信号表明光子 3 与光子 1 的偏振方向相同，证实了隐形传态的实现。

−45°偏振的光子 1 的结果证明能够对一组完备的偏振基态实现隐形传态（图 4 右栏）。为了排除对实验结果的任何经典解释，我们用另外的实验进一步证实我们的方法是成功的。在这些附加实验中我们隐形传送了 0°和 90°的线偏振光子以及圆偏振光子，结果总结在附表 1 中。实验中与入射偏振正交方向的三重符合计数出现下降，表中列出了这个下降的可见度。

表 1. 三重符合计数下隐形传态的可见度

偏振方向	可见度
+45°	0.63 ± 0.02
−45°	0.64 ± 0.02
0°	0.66 ± 0.02
90°	0.61 ± 0.02
圆偏振	0.57 ± 0.02

如前所述，这些可见度是扣除乱真三重符合计数引起的偏差之后得到的数值。实验上排除这些计数的办法是对三重计数再附加一个光子 4 计数的条件。这使得光子 1 实际上被投影到一个单粒子态上。我们将这个四重符合计数测量方法用到 +45°和 +90°偏振态这两个非正交态的隐形传态中。实验结果在图 5 中给出。得到的正交偏振态下降的可见度在 70% ± 3%。这些可见度直接就是被传态光子在其本态的偏振度。这表示我们已经证明了单个光子的量子态的隐形传递。

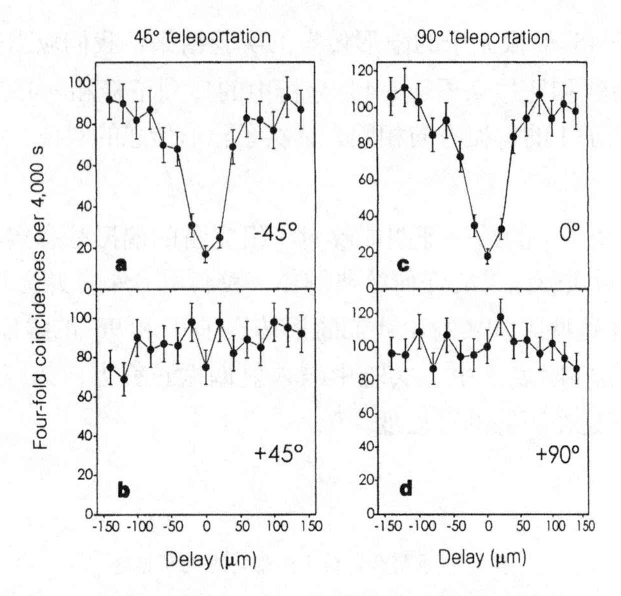

Fig. 5. Four-fold coincidence rates (without background subtraction). Conditioning the three-fold coincidences as shown in Fig. 4 on the registration of photon 4 (see Fig. 1b) eliminates the spurious three-fold background. **a** and **b** show the four-fold coincidence measurements for the case of teleportation of the +45° polarization state; **c** and **d** show the results for the +90° polarization state. The visibilities, and thus the polarizations of the teleported photons, obtained without any background subtraction are 70% ± 3%. These results for teleportation of two non-orthogonal states prove that we have demonstrated teleportation of the quantum state of a single photon.

The Next Steps

In our experiment, we used pairs of polarization entangled photons as produced by pulsed down-conversion and two-photon interferometric methods to transfer the polarization state of one photon onto another one. But teleportation is by no means restricted to this system. In addition to pairs of entangled photons or entangled atoms[7,21], one could imagine entangling photons with atoms, or phonons with ions, and so on. Then teleportation would allow us to transfer the state of, for example, fast-decohering, short-lived particles, onto some more stable systems. This opens the possibility of quantum memories, where the information of incoming photons is stored on trapped ions, carefully shielded from the environment.

Furthermore, by using entanglement purification[22]—a scheme of improving the quality of entanglement if it was degraded by decoherence during storage or transmission of the particles over noisy channels—it becomes possible to teleport the quantum state of a particle to some place, even if the available quantum channels are of very poor quality and thus sending the particle itself would very probably destroy the fragile quantum state. The feasibility of preserving quantum states in a hostile environment will have great advantages in the realm of quantum computation. The teleportation scheme could also be used to provide links between quantum computers.

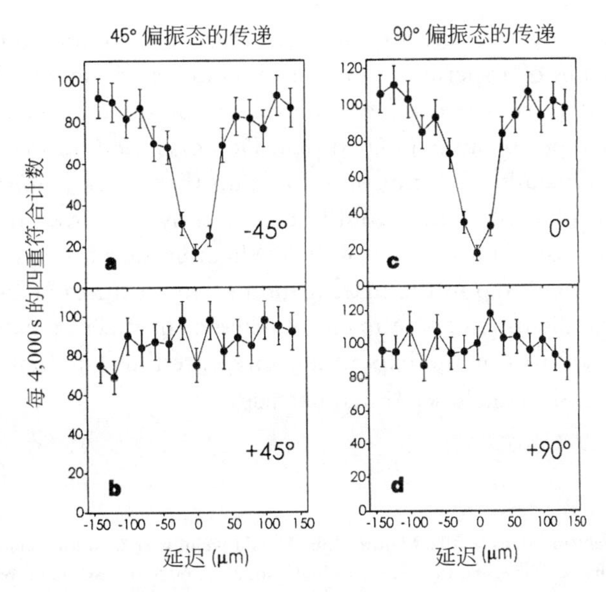

图 5. 四重符合计数率(未扣除背景)。对图 4 的三重符合计数附加一个光子 4 计数的条件(见图 1b)消除了乱真三重计数的背景。**a** 和 **b** 显示传递 +45° 偏振态的四重符合计数测量;**c** 和 **d** 是 +90° 偏振态的结果。在没有去除背景的情况下,得到的可见度,也就是被传态光子的偏振度,是 70%±3%。这两个非正交态的传递结果表明了我们已经证实单个光子的量子态的隐形传态。

展　望

我们的实验使用脉冲下转换产生的偏振纠缠光子对,并利用双光子干涉测量法来将一个光子的偏振态传递给另一个光子。但是隐形传态不只局限于这个系统。除了纠缠光子对或者纠缠原子对 [7,21],还可以设想光子和原子之间纠缠,声子和离子之间纠缠等等。隐形传态使得我们能够把比如快速退相干的短寿命粒子的态转移到其他更稳定的系统中,这使得量子存储成为可能。在量子存储中,进来的光子的信息被储存到被俘获的离子上,外界的影响被很好地屏蔽掉。

纠缠纯化 [22] 是一种在存储粒子或在噪声信道传送粒子的过程中因退相干导致纠缠退化时改进纠缠质量的方案。利用纠缠纯化可以更进一步将一个粒子的量子态传至异地,哪怕当前量子通道的质量很差并且使得传递粒子本身很可能会把脆弱的量子态破坏。在不利环境中保存量子状态的可行性在量子计算领域中会带来很大优势。隐形传态的方案还可以用于两台量子计算机之间的通信。

Quantum teleportation is not only an important ingredient in quantum information tasks; it also allows new types of experiments and investigations of the foundations of quantum mechanics. As any arbitrary state can be teleported, so can the fully undetermined state of a particle which is member of an entangled pair. Doing so, one transfers the entanglement between particles. This allows us not only to chain the transmission of quantum states over distances, where decoherence would have already destroyed the state completely, but it also enables us to perform a test of Bell's theorem on particles which do not share any common past, a new step in the investigation of the features of quantum mechanics. Last but not least, the discussion about the local realistic character of nature could be settled firmly if one used features of the experiment presented here to generate entanglement between more than two spatially separated particles[23,24].

(**390**, 575-579; 1997)

Dik Bouwmeester, Jian-Wei Pan, Klaus Mattle, Manfred Eibl, Harald Weinfurter & Anton Zeilinger
Institut für Experimentalphysik, Universität Innsbruck, Technikerstr. 25, A-6020 Innsbruck, Austria

Received 16 October; accepted 18 November 1997.

References:

1. Bennett, C. H. *et al.* Teleporting an unknown quantum state via dual classic and Einstein-Podolsky-Rosen channels. *Phys. Rev. Lett.* **70**, 1895-1899 (1993).
2. Schrödinger, E. Die gegenwärtige Situation in der Quantenmechanik. *Naturwissenschaften* **23**, 807-812; 823-828; 844-849 (1935).
3. Bennett, C. H. Quantum information and computation. *Phys. Today* **48(10)**, 24-30, October (1995).
4. Bennett, C. H., Brassard, G. & Ekert, A. K. Quantum Cryptography. *Sci. Am.* **267(4)**, 50-57, October (1992).
5. Mattle, K., Weinfurter, H., Kwiat, P. G. & Zeilinger, A. Dense coding in experimental quantum communication. *Phys. Rev. Lett.* **76**, 4656-4659 (1996).
6. Kwiat, P. G. *et al.* New high intensity source of polarization-entangled photon pairs. *Phys. Rev. Lett.* **75**, 4337-4341 (1995).
7. Hagley, E. *et al.* Generation of Einstein-Podolsky-Rosen pairs of atoms. *Phys. Rev. Lett.* **79**, 1-5 (1997).
8. Schumacher, B. Quantum coding. *Phys. Rev. A* **51**, 2738-2747 (1995).
9. Clauser, J. F. & Shimony, A. Bell's theorem: experimental tests and implications. *Rep. Prog. Phys.* **41**, 1881-1927 (1978).
10. Greenberger, D. M., Horne, M. A. & Zeilinger, A. Multiparticle interferometry and the superposition principle. *Phys. Today* August, 22-29 (1993).
11. Tittel, W. *et al.* Experimental demonstration of quantum-correlations over more than 10 kilometers. *Phys. Rev. Lett.* (submitted).
12. Zukowski, M., Zeilinger, A., Horne, M. A. & Ekert, A. "Event-ready-detectors" Bell experiment via entanglement swapping. *Phys. Rev. Lett.* **71**, 4287-4290 (1993).
13. Bose, S., Vedral, V. & Knight, P. L. A multiparticle generalization of entanglement swapping. (preprint).
14. Wootters, W. K. & Zurek, W. H. A single quantum cannot be cloned. *Nature* **299**, 802-803 (1982).
15. Loudon, R. *Coherence and Quantum Optics VI* (eds Everly, J. H. & Mandel, L.) 703-708 (Plenum, New York, 1990).
16. Zeilinger, A., Bernstein, H. J. & Horne, M. A. Information transfer with two-state two-particle quantum systems. *J. Mod. Optics* **41**, 2375-2384 (1994).
17. Weinfurter, H. Experimental Bell-state analysis. *Europhys. Lett.* **25**, 559-564 (1994).
18. Braunstein, S. L. & Mann, A. Measurement of the Bell operator and quantum teleportation. *Phys. Rev. A* **51**, R1727-R1730 (1995).
19. Michler, M., Mattle, K., Weinfurter, H. & Zeilinger, A. Interferometric Bell-state analysis. *Phys. Rev. A* **53**, R1209-R1212 (1996).
20. Zukowski, M., Zeilinger, A. & Weinfurter, H. Entangling photons radiated by independent pulsed sources. *Ann. NY Acad. Sci.* **755**, 91-102 (1995).
21. Fry, E. S., Walther, T. & Li, S. Proposal for a loophole-free test of the Bell inequalities. *Phys. Rev. A* **52**, 4381-4395 (1995).
22. Bennett, C. H. *et al.* Purification of noisy entanglement and faithful teleportation via noisy channels. *Phys. Rev. Lett.* **76**, 722-725 (1996).
23. Greenberger, D. M., Horne, M. A., Shimony, A. & Zeilinger, A. Bell's theorem without inequalities. *Am. J. Phys.* **58**, 1131-1143 (1990).
24. Zeilinger, A., Horne, M. A., Weinfurter, H. & Zukowski, M. Three particle entanglements from two entangled pairs. *Phys. Rev. Lett.* **78**, 3031-3034 (1997).

Acknowledgements. We thank C. Bennett, I. Cirac, J. Rarity, W. Wootters and P. Zoller for discussions, and M. Zukowski for suggestions about various aspects of the experiments. This work was supported by the Austrian Science Foundation FWF, the Austrian Academy of Sciences, the TMR program of the European Union and the US NSF.

Correspondence and requests for materials should be addressed to D.B. (e-mail: Dik.Bouwmeester@uibk.ac.at).

量子隐形传态不仅仅是量子信息处理任务的一个重要组成，它还为量子力学基础提供新的实验和研究方法。既然任意一个态都可以被传递，作为一个纠缠对中一员的粒子的完全未确定的量子态也可以被传递。这种做法等于在粒子间传递纠缠。这样我们不但能够用长距离链式传递量子态解决退相干在途中破坏量子态的问题，而且能够对没有共同历史的粒子进行贝尔定律检测（量子力学特征研究的新步骤）。最后同样重要的一点是，如果利用本文描述的实验要素来产生两个以上的空间分离粒子的纠缠，就可能确实解决自然世界定域实在特征的争论[23,24]。

（何钧 沈乃澂 翻译；陆朝阳 李军刚 审稿）

Localization of Light in a Disordered Medium

D. Wiersma *et al.*

Editor's Note

Quantum physics implies that particle diffusion inside disordered materials may be inhibited by destructive wave interference, if the disorder is sufficiently pronounced. This phenomenon of strong localization—or "Anderson localization", as it was originally described by physicist Philip Anderson—affects not only the behaviour of electrons in many solids, but that of photons as well. Here Diederik Wiersma and colleagues provided experimental confirmation of Anderson localization for light in semiconductor powders. They studied light transmission through gallium arsenide in a crystal and in powders with various particle sizes. They found a sharp threshold of particle size at which disorder-induced localization ultimately blocks light transmission, causing the transmission coefficient through a sample to decrease exponentially with sample thickness rather than linearly.

Among the unusual transport properties predicted for disordered materials is the Anderson localization[1] phenomenon. This is a disorder-induced phase transition in the electron-transport behaviour from the classical diffusion regime, in which the well-known Ohm's law holds, to a localized state in which the material behaves as an insulator. The effect finds its origin in the interference of electrons that have undergone multiple scattering by defects in the solid[2-10]. A similar phenomenon is anticipated for multiple scattering of electromagnetic waves, but with one important simplification: unlike electrons, photons do not interact with one another. This makes transport of photons in disordered materials an ideal model system in which to study Anderson localization[10-17]. Here we report direct experimental evidence for Anderson localization of light in optical experiments performed on very strongly scattering semiconductor powders.

MULTIPLE scattering of light is a common phenomenon in daily life, occurring for example in sugar, fog, white paint and clouds. The propagation of light in these media can in general be described by a normal diffusion process. For diffusion of light through a disordered material the same Ohm's law holds as for diffusion of electrons through any common resistor: the transmission, or conductance, decreases linearly with the system length (thickness).

Anderson localization brings classical diffusion to a complete halt. That is, on increasing the amount of scattering beyond a critical value, the material makes a transition into a localized state (see Fig. 1). This transition can best be observed in the transmission properties of the system. In the localized state, the transmission coefficient decreases exponentially instead of linearly with the thickness of a sample. At the transition, the

无序介质中光的局域化

维尔斯马等

编者按

量子物理学指出，在材料的无序程度足够显著的情况下，粒子在该材料中的扩散会被波的相消干涉所抑制。这种强烈的局域化现象，或者叫“安德森局域化”(因为该现象最早是由物理学家菲利普·安德森描述出来的)，影响了电子在多种固体中的行为，而且也会影响光子的行为。在本文中，迪德里克·维尔斯马和他的同事们用实验证实了光在半导体粉末中的安德森局域化。他们研究了光在砷化镓晶体中和在具有不同粒度的几种砷化镓粉末中的透射。他们发现，光透射率在某一粒度值处出现一个显著的阈值，在该处，无序诱导的局域化完全阻挡了光的透射，导致样品的透射系数随样品厚度呈指数下降而非线性下降。

安德森局域化[1]现象是科学家预测到的无序材料不同寻常的传输性质之一。这是一种由无序环境诱导产生的电子传输行为上的相变，表现为由经典的扩散向局域态的转变，而经典扩散是众所周知的欧姆定律适用的范围，局域态则对应绝缘体材料。这个效应来源于被固体中的缺陷多次散射的电子之间的干涉[2-10]。可以预料的是，电磁波的多次散射会产生相似的现象，但是有一个重要的简化：和电子不同，光子之间不会有相互作用。这使光子在无序材料中的传输成为研究安德森局域化的理想模型[10-17]。在本文中，利用基于强散射性半导体粉末的光学实验，我们报道了光的安德森局域化的直接实验证据。

光的多次散射是生活中常见的现象，会出现在例如糖、雾、白色颜料、云中。光在这些介质中的传播通常可以用标准扩散过程来描述。对于光在无序材料中的扩散，欧姆定律依然适用，如同电子在电阻器中的扩散一样，即传导性随系统长度(厚度)线性下降。

安德森局域化使经典扩散不再适用。也就是说，一旦散射量高于某一临界值，材料就出现局域态(见图 1)。观察这种变化的最佳方式是测试系统的透射。在局域态中，透射系数随样品厚度呈指数减小，而非呈线性减小。转变过程中，透射系数

transmission coefficient is expected to have a power-law dependence on the inverse thickness, which is probably quadratic[12,18].

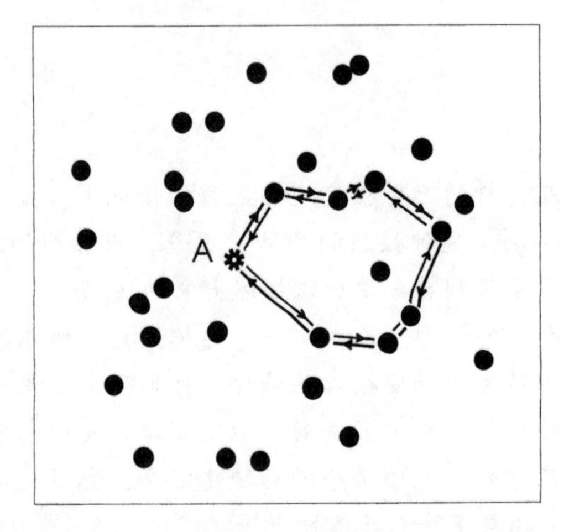

Fig. 1. Anderson localization of waves in disordered systems originates from interference in multiple elastic scattering. Here we consider a light source (like an excited atom emitting a photon) in a disordered medium at position A. The light source is denoted by a star symbol and the spheres denote the scattering elements. A random light path that returns to the light source can be followed in two opposite directions. The two waves which propagate in opposite directions along this loop will acquire the same phase and therefore interfere constructively in A. This leads to a higher probability of the wave coming back to A and consequently a lower probability of propagating away from A. On decreasing the mean free path l, the probability for such looped paths increases and at strong enough scattering the system makes a phase transition from the normal conducting state into a localized state, due to interference. In the localized regime, the system behaves as a non-absorbing insulator. Light which is incident on, for example, a slab would be almost completely reflected and the remaining transmission would decrease exponentially with the slab thickness.

The main difficulty in the search for localization of light has been the realization of strong enough scattering. The appropriate measure for the amount of scattering is the mean free path l for the light in the medium, times the magnitude of the wavevector k. Localization is expected for $kl \leqslant 1$ (ref. 19), which is known as the (modified) Ioffe–Regel criterion. This criterion can be understood intuitively if one realizes that below $kl = 1$, the electric field can not even perform one oscillation before the wave is scattered again. So far, localization effects have been reported only for microwaves in a two-dimensional system of rods[20] and for microwaves in a confined geometry (a copper tube filled with metallic and dielectric spheres)[21,22]. In the latter experiment the absorption was very large, which makes the interpretation of the data complicated. The disadvantage of experiments with microwaves compared to light waves is that it is difficult to avoid absorption.

We have been able to realize very strongly scattering samples for light waves, using semiconductor powders. Semiconductors can have a very large refractive index, while, for wavelengths in the bandgap, the absorption is extremely small. We used light at wavelength $\lambda = 1{,}064$ nm, at which the absorption coefficient κ of pure GaAs is $\kappa \ll 1\ \mathrm{cm}^{-1}$ and

预计会与厚度的倒数呈幂律相关，且很有可能是二次幂律[12,18]。

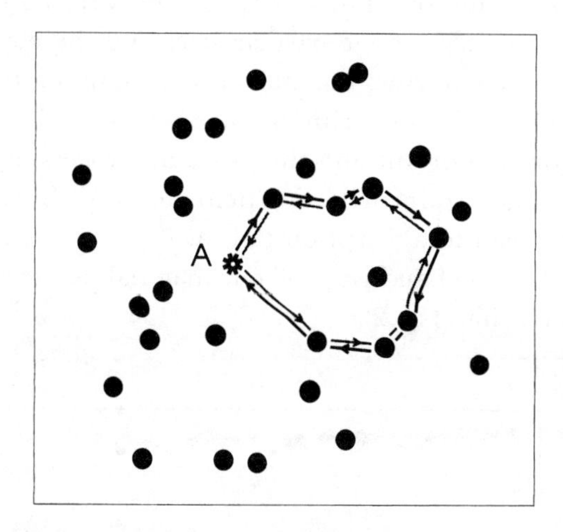

图 1. 无序系统中波的安德森局域化来源于多次弹性散射的干涉。在这里我们把无序介质中的 A 点看作是一个光源（好比一个受激原子发射出一个光子）。这个光源用星号标记，而圆点表示散射点。一条能返回到光源的不规则光路可以从相反的两个方向去实现。这两支沿着这条回路反向传播的波会具有相同的相位，从而在 A 点形成相长干涉。这种情况导致波回到 A 点的概率增加而向远离 A 点方向传播的概率减小。一旦平均自由程 l 减小，出现这种回路的概率就增加了，并且在散射足够强的情况下，干涉会导致系统发生由普通传导态到局域态的相变。在局域区，系统表现为无吸收的光绝缘体。例如，照在一块平板上的光几乎都会被反射，仅剩的一点透射光会随板的厚度呈指数衰减。

探寻光的局域化的最大难点在于实现足够强的散射。衡量散射量最恰当的方法是用光在介质中的平均自由程 l 乘以波矢量 k 的大小。局域化发生的条件为 $kl \leqslant 1$（参考文献 19），即（修正的）Ioffe–Regel 判据。如果能意识到，低于 $kl = 1$ 时，在波被再次散射前电场不能完成一次振荡，我们便能直观地理解这条判据。迄今为止，对局域化效应的报告仅限于二维棒状系统中的微波[20]和封闭空间（充满金属和电介质球的铜管）里的微波[21,22]。在后者中，微波被大量吸收，导致数据解释变得复杂。和光波实验比起来，微波实验的劣势在于吸收难以避免。

我们已利用半导体粉末实现光波的强散射。半导体可以具有很大的折射率，而对于带隙中的波长来说，吸收是极小的。我们使用波长 $\lambda = 1{,}064$ nm 的光，在这个波长下纯砷化镓的吸收系数 $\kappa \ll 1\ \text{cm}^{-1}$，折射率为 3.48。我们的样品由纯的（99.999%）

the refractive index is 3.48. Our samples consist of pure (99.999%) gallium arsenide (GaAs) crystals which were ground (as a suspension in methanol) by hand in a ceramic mortar and in a planetary micromill at low speed. By varying the grinding time, we obtained samples with different average particle sizes and thereby different amounts of scattering. In grinding semiconductors one has to be careful not to introduce absorption at the wavelength at which the experiments are performed. For a powder, surface states could become more important and the grinding process (even if performed with little force) could introduce strain or lattice deformations. Surface states, strain and lattice deformations can lead to absorption tails at the edge of the bandgap. We have characterized the change of the band edge of our material by measuring the temperature dependence of the transmission (Fig. 2).

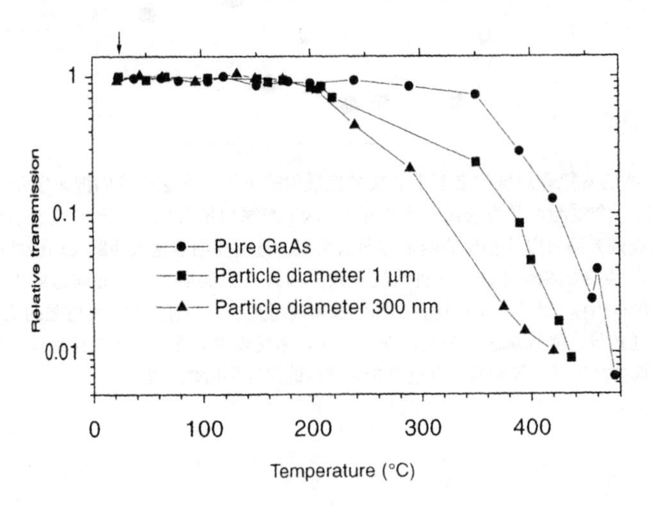

Fig. 2. Comparison of the temperature dependence of the transmission of pure GaAs crystals and powders. Data obtained using particles of size of 1 μm and 300 nm are shown. The transmission of the powder samples is measured at a thickness where the total transmission equals 1% at room temperature. All curves are scaled to have a maximum of 1. On increasing the temperature T, the bandgap shifts to lower energies $E_g(T)$ corresponding to higher wavelengths $\lambda_g = 1.24/E_g$ (in μm). The temperature dependence is given by the empirical relation: $E_g(T) = E_g(0) - \alpha T^2/(T+\beta)$, where (for GaAs) $E_g(0) = 1.522$, $\alpha = 8.871 \times 10^{-4}$ and $\beta = 572$, with T in K. This temperature-dependent shift enables the scanning of the region of the bandgap around the laser wavelength, without changing the laser wavelength itself. For example, a temperature of 200 °C corresponds to a wavelength shift of 65 nm. The arrow at top left shows the temperature where we performed all other experiments. We note that the band edge becomes less steep upon grinding the GaAs crystal; this is probably due to the increased importance of surface states and, for example, lattice deformations. But the region close to the laser wavelength remains unabsorbing.

We observe (Fig. 2) that the band edge indeed becomes less steep on grinding, and that the onset of absorption shifts into the bandgap. It is also clear, however, that the region of the absorption tails is still far enough (65 nm) below our laser wavelength as to ensure that these tails will not introduce absorption in our experiments.

To characterize the mean free path l of our samples, we used coherent backscattering. This phenomenon is a general interference effect between counter-propagating waves, which leads to a narrow cone in exact backscattering[23]. It is seen as the precursor to

砷化镓(GaAs)晶体构成，是在陶瓷研钵里手动碾磨(以悬浮于甲醇的形式)以及在微型行星式高能球磨机中低速碾磨而成的。通过改变碾磨时间，我们得到了具有不同平均粒度从而具有不同散射量的样品。碾磨半导体时，操作人员必须非常小心，以确保实验中所用波长的光不会被吸收。对于粉末来说，表面态更为重要，并且碾磨过程(即使只用了很小的力)会引起应变或晶格形变。表面态、应变和晶格形变都会导致带边出现吸收带尾。通过测量透射率与温度的相关性，我们描绘了材料的带边变化(图 2)。

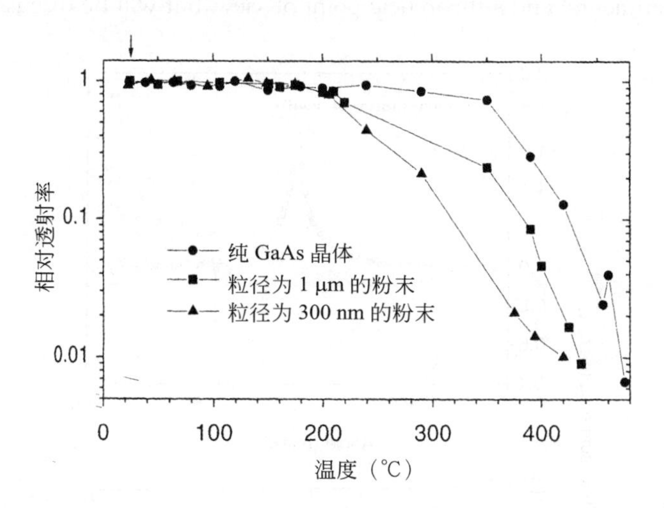

图 2. 纯 GaAs 晶体与粉末之间透射率与温度相关性的对比。图中展示了粒径为 1 μm 和 300 nm 的数据。粉末样品的透射率是在一个特定的厚度下测得的，即室温下总透射率为 1% 时的厚度。所有曲线都被按比例缩放，使得最高点是 1。一旦温度 T 升高，带隙就移向低能量 $E_g(T)$，对应于高波长 $\lambda_g = 1.24/E_g$(单位是 μm)。和温度的相关性满足经验关系：$E_g(T) = E_g(0) - \alpha T^2/(T+\beta)$，其中(GaAs 的)$E_g(0) = 1.522$，$\alpha = 8.871 \times 10^{-4}$，$\beta = 572$，$T$ 的单位是 K。这种依赖于温度的移动，使在不改变激光波长的情况下，遍历带隙中激光波长附近的能量成为可能。例如，200 ℃的温度对应于 65 nm 的波长改变量。左上角的箭头指出了我们其余全部实验的操作温度。我们注意到，在 GaAs 晶体被碾磨得更细的时候，吸收带边缘变得没那么陡直了；这很可能是因为表面态和晶格形变的影响加重了。但是激光波长附近的波长区域仍然没有吸收。

从图 2 中我们发现，在碾磨得更细的时候，吸收带边缘变得没那么陡直了，而且吸收的起始点移入带隙中了。然而另一个明确的方面是，吸收带尾所处的波长区域远比我们使用的激光的波长要小(65 nm)，这样就保证了这种带尾不会使我们的实验里出现激光的吸收。

为确定此材料中的平均自由程 l，我们应用了相干背散射。该现象是反向传播的波之间的普通干涉效应，使背散射谱图上出现一个窄的锥形[23]。这被看作是安德森

Anderson localization and is therefore also called "weak localization of light"[24,25]. The width of the backscattering cone and the angle of its cusp are inversely proportional to l (refs 26, 27) and therefore enable the determination of the amount of scattering inside the samples. Furthermore, the cusp is due to a summation up to (in principle) infinite path length, and is therefore only present in the absence of absorption[26-28]. In Figure 3, we show backscattering cones for a coarse-grained (Fig. 3a) and a fine-grained (Fig. 3b) powder with an average particle diameter of 10 μm and 1 μm, respectively. From the angle of the cusp we find for the coarse-grained powder $l = 8.5$ μm and hence $kl = 76$. For the fine-grained powder we find $l \approx 0.17$ μm which corresponds to $kl \approx 1.5$. The observed value of $kl = 1.5$ is four times smaller than the smallest kl values that have been reported so far[30] for light waves. The shape of the coherent backscattering cone at the localization transition is interesting from both an experimental and a theoretical point of view, but will be discussed elsewhere.

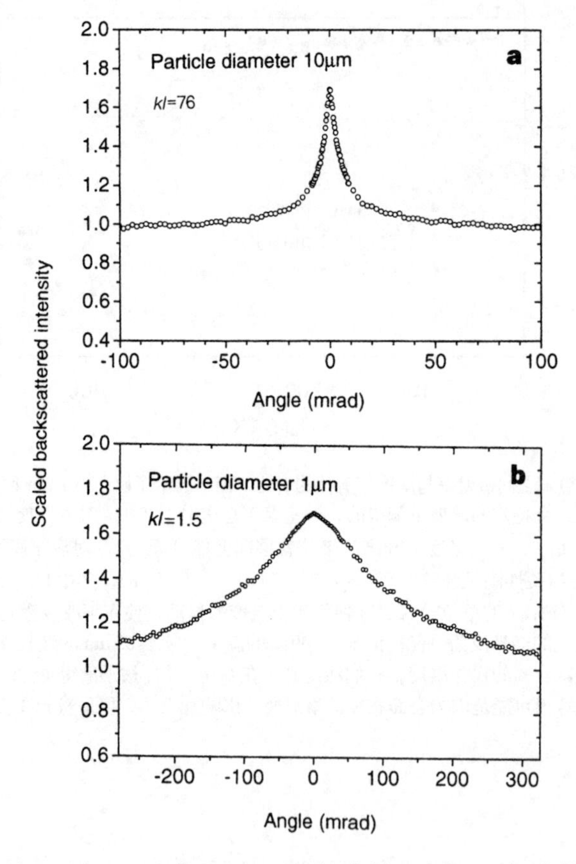

Fig. 3. Coherent backscattering cones from coarse-grained (**a**) and fine-grained (**b**) GaAs powder. The samples are strongly polydisperse with an average particle diameter of 10 μm and 1 μm, respectively. The sample is illuminated with a mode-locked Nd:YAG laser operating at 76 MHz, with a wavelength of 1,064 nm, pulse duration 100 ps, beam diameter 6 mm and incident power 100 mW. In all experiments we have checked that nonlinear absorption processes like two-photon absorption do not play a role; this was done by lowering the incident power to 20 mW, and finding that this reduction did not influence the results. Detection is performed in the polarization-conserving channel. The mean free path is calculated from the angle of the cusp[27], taking into account internal reflection[29]. This yields $l = 8.5$ μm and $kl = 76$ for the data in **a**, and $l = 0.17$ μm resulting in $kl \approx 1.5$ for the data in **b**.

局域化的前身，也因此被称为“光的弱局域化”[24,25]。背散射的锥形的宽度和尖端的角度都与 l 成反比(参考文献 26 和 27)，从而使确定样品中的散射量成为可能。而且，尖端是由(理论上)无限长的光程得到的，所以只在没有吸收的情况下出现 [26-28]。图 3 中，我们展示了粗粒粉末(图 3a)和细粒粉末(图 3b)的背散射锥形，粉末的平均粒径分别为 10 μm 和 1 μm。从尖端的角度中我们发现，粗粒粉末中 $l = 8.5$ μm，所以 $kl = 76$，细粒粉末中 $l \approx 0.17$ μm，所以 $kl \approx 1.5$。观测值 $kl = 1.5$ 是迄今报道过的光波的最小 kl 值的四分之一 [30]。无论从实验角度还是理论角度来看，向局域化转化的过程中，相干背散射锥的形状都很有趣，这一点会在别处讨论。

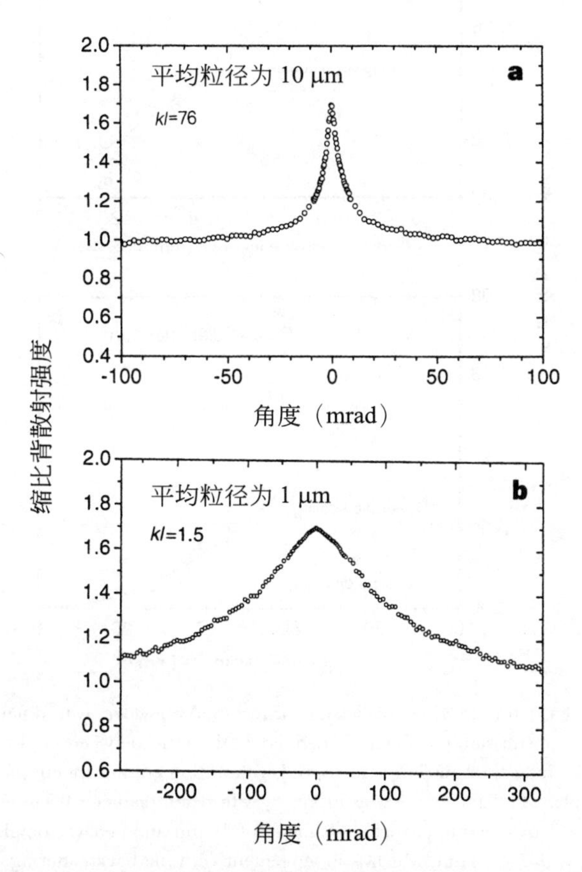

图 3. 从粗粒(**a**)和细粒(**b**)GaAs 粉末中得到的相干背散射锥。两种样品都有很大的多分散性，平均粒径分别为 10 μm 和 1 μm。用来照射样品的是锁模 Nd : YAG 激光，频率为 76 MHz，波长为 1,064 nm，脉冲宽度为 100 ps，光束直径为 6 mm，入射功率为 100 mW。在全部实验中，我们都避免了非线性吸收(如双光子吸收)。方法是把入射功率降低到 20 mW 后发现功率的降低并没有影响实验结果。检测是在保偏通道中进行的。平均自由程由锥形尖端的角度计算得出 [27]，内反射也被考虑在内 [29]。图 **a** 中数据得出的结果是 $l = 8.5$ μm，$kl = 76$，图 **b** 则是 $l = 0.17$ μm，$kl \approx 1.5$。

To observe a possible localization transition, we have measured the transmission coefficient as a function of sample thickness. In Fig. 4a, the transmission coefficient is shown for the coarse-grained powder. We see that the transmission behaves completely classically with $l = 9.8$ μm, which is consistent with the coherent backscattering data. At very large sample thicknesses ($L > 500$ μm), the transmission decreased more rapidly than linear due to the onset of absorption. This gives an upper limit for κ (defined as $\kappa = l_{in}^{-1}$) for the coarse-grained powder, of ~0.13 cm^{-1}, which shows again that the grinding process did not introduce any absorption.

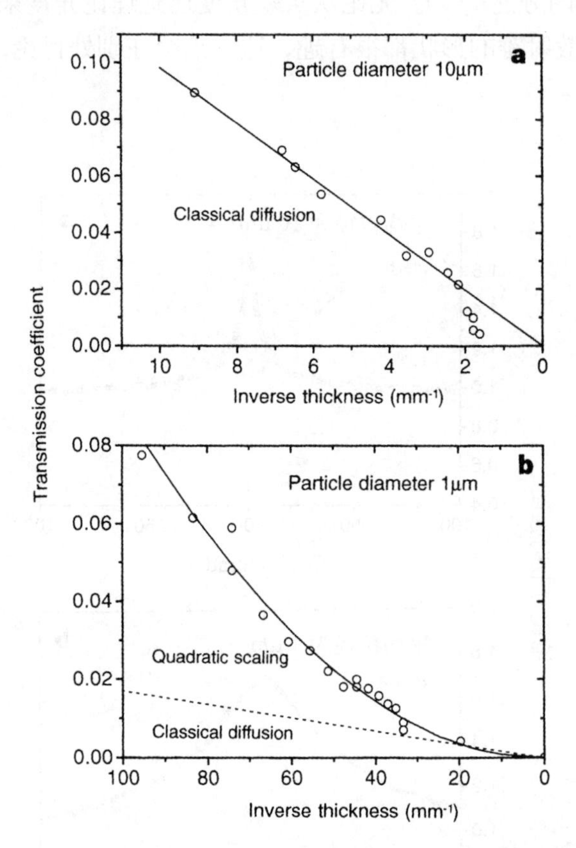

Fig. 4. Comparison of the transmission coefficients for two GaAs powders with different average particle diameters. Here the transmission coefficient is defined as the ratio between total transmitted flux and the incident flux. The total transmission is measured with an integrating sphere placed in contact with the back of the sample. The light source is as in Fig. 3, with beam diameter 0.5 mm and incident power 20 μW. The data in **a** correspond to particle a diameter of 10 μm and behave completely classically. The solid line is $T = l/L$, with $l = 9.8$ μm, which is in agreement with the backscattering data. The data in **b** correspond to a particle diameter of ~1 μm. The dashed line is the theoretical curve for classical diffusion with $l = 0.17$ μm, as obtained from the backscattering data. The solid line is a quadratic fit to the data. This quadratic dependence of the transmission coefficient on the inverse thickness is the expected behaviour at an Anderson localization transition.

In Fig. 4b, the transmission coefficient T is shown for the fine-grained powder. The dashed line is the theoretical curve for classical diffusion assuming $l = 0.17$ μm as obtained from the backscattering data. Whereas for classical diffusion T would decrease linearly with the sample

为了观测到一个可能的向局域化转化的过程，我们测量了透射系数随样品厚度的变化。图 4a 展示了粗粒粉末的透射系数。可以看出，这种透射完全表现为经典扩散，其中 $l = 9.8\ \mu m$ ，与相干背散射得到的数据一致。在样品厚度特别大的时候 ($L > 500\ \mu m$)，由于吸收的出现，透射率的降低速度会超线性。因此粗粒粉末的 κ 值（定义为 $\kappa = l_{in}^{-1}$）存在上限，约为 0.13 cm^{-1}，再次说明了碾磨过程不会引入吸收。

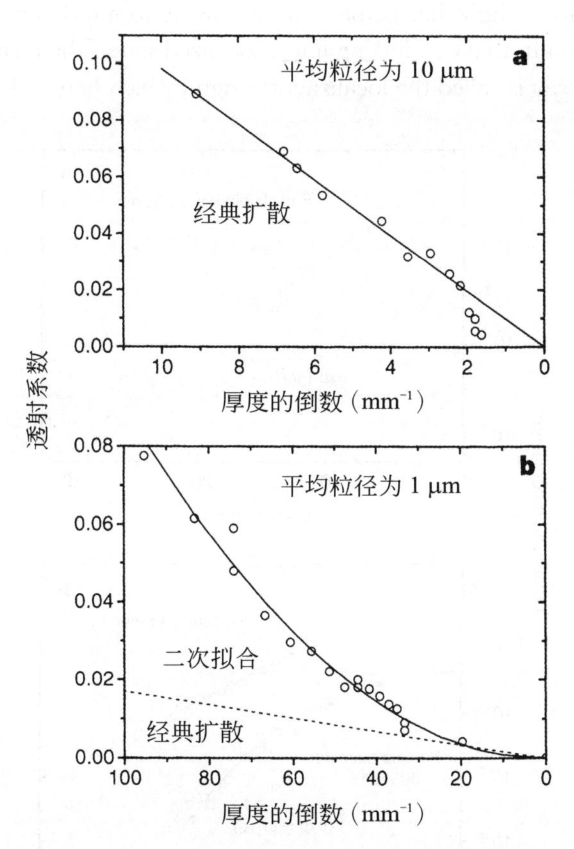

图 4. 平均粒径不同的两种 GaAs 粉末透射系数的对比。此处，透射系数被定义为总透射通量与入射通量之比。总透射量是由紧贴样品背面的积分球测出的。光源和图 3 中的一样，光束直径为 0.5 mm，发射功率为 20 μW。图 **a** 中的数据来自直径为 10 μm 的颗粒，完全表现为经典扩散。实线为 $T = l / L$，其中 $l = 9.8\ \mu m$，这和背散射数据相符。图 **b** 中的数据对应粒径约 1 μm 的情况。虚线是按经典扩散作出的理论曲线，其中 $l = 0.17\ \mu m$，由背散射数据所得。实线是实验数据的二次拟合。在转变为安德森局域化的过程中，透射系数与厚度的倒数的二次相关符合预期。

图 4b 展示了细粒粉末的透射系数 T。虚线是根据经典扩散画出的理论曲线，其中假设 $l = 0.17\ \mu m$，即由背散射数据算出的值。虽然经典扩散中的 T 值会随样品厚

thickness L, we find for these samples a quadratic dependence ($T \propto L^{-2}$). This is exactly the behaviour predicted by the scaling theory of localization at the localization transition[12,18]. We note that no classical diffusion process can show a quadratic system-size dependence.

If we decrease the particle size even further we expect (from Mie-scattering theory) to obtain even stronger scattering. In Fig. 5, the transmission coefficient is shown for a powder with an average particle diameter of 300 nm. In Fig. 5b we see that the quadratic behaviour has changed into an exponential decay in this case, as expected in the localized regime. The transport of light has come to a halt owing to interference, and the system has made a phase transition from a conducting into a localized state. The characteristic length scale for the exponential decay is called the localization length, which here is 4.3 μm.

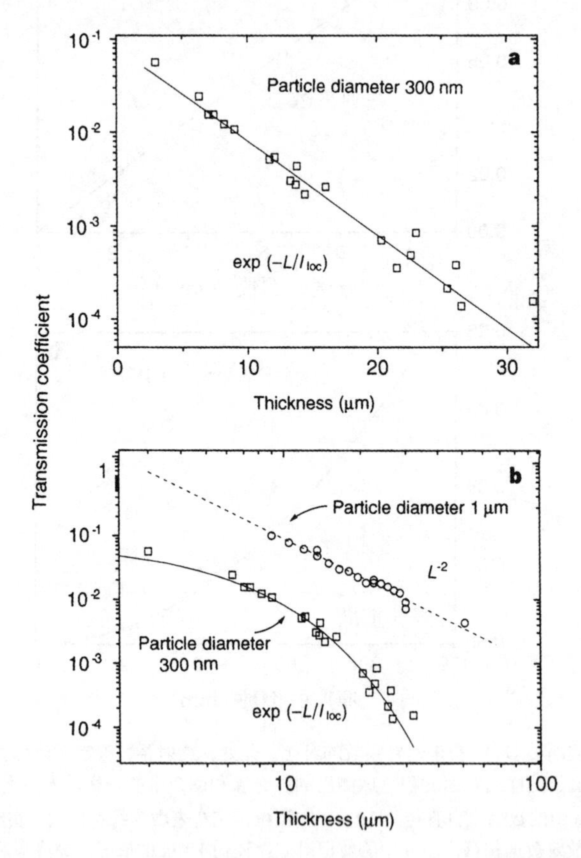

Fig. 5. The transmission coefficient of very fine GaAs powder as a function of thickness. The average particle diameter is 300 nm. In **a**, the data are plotted on a semi-logarithmic scale. The solid line is an exponential fit $\exp(-L/l_{loc})$ with a localization length of $l_{loc} = 4.3$ μm. In **b**, the same data are plotted on a double-logarithmic scale and the data of Fig. 4b are imported for comparison. We see that for the very fine powder, the quadratic behaviour changes to an exponential decay. The system goes into a localized state where it behaves as a (non-absorbing) insulator.

Different techniques could be used to map out the complete localization transition. The amount of scattering could be varied by changing the average particle diameter, the refractive index contrast, the particle density and the wavelength of the light. An

度 L 的增大呈线性下降，但我们发现这些样品中出现了二次相关（$T \propto L^{-2}$）。这和以局域化标度理论预测出的向局域化转化的表现完全一致[12,18]。我们在此特别指出，任何经典扩散过程都不会呈现与系统尺度的二次相关。

如果我们继续降低粒度，我们会（根据米氏散射理论）得到更强的散射。图 5 给出了一种平均粒径为 300 nm 的粉末的透射系数。从图 5b 中我们可以看出，在这种情况下二次相关已经变为指数衰减，符合局域区的预期。由于干涉，光不再透射，系统实现了由传导态到局域态的相变。指数衰减的特征长度尺度被称为局域化长度，此处为 4.3 μm。

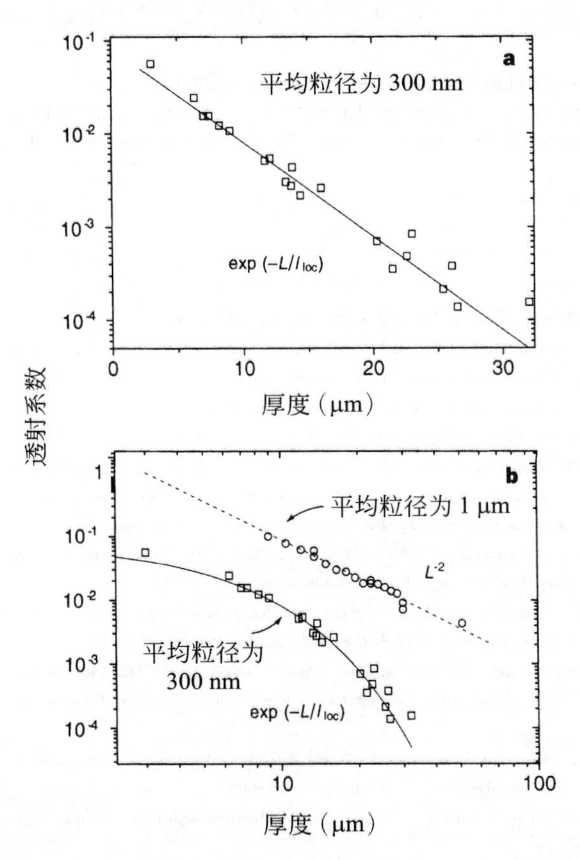

图 5. 极细 GaAs 粉末的透射系数与厚度的关系。平均粒径为 300 nm。图 **a** 中，数据被标在半对数坐标中。实线是指数化拟合曲线 $\exp(-L/l_{loc})$，其中局域化长度 $l_{loc} = 4.3$ μm。图 **b** 中，同样的数据被标在双对数坐标中，与图 4b 中的数做比较。我们可以看出，在极细的粉末中，二次相关变成了指数衰减。系统变为局域态，即表现为（无吸收的）光绝缘体。

多种方法可以用来描绘完整的向局域化转化的过程。通过改变平均粒径、折射率对比度、颗粒密度以及光的波长，可以改变散射量。时间分辨透射实验是一种重

important experiment would be a time-resolved transmission experiment in which the reduction of the diffusion constant at the Anderson localization transition is observed. Localization of classical waves is, in many ways, similar to Anderson localization of electrons. There are, however, also interesting differences between light and electrons. For experiments with light waves, coherent (laser) sources are available; the wavelength of such sources may be easily adjusted. Furthermore, for electromagnetic waves photon–photon interactions can be neglected, whereas in the case of electrons, electron–electron interactions always play a role. This latter property in particular makes light in strongly disordered media an interesting system in which to study the Anderson localization transition.

(**390**, 671-673; 1997)

Diederik S. Wiersma*, Paolo Bartolini*, Ad Lagendijk† & Roberto Righini*

* European Laboratory for Non-Linear Spectroscopy, Largo E. Fermi 2, 50125 Florence, Italy

† Van der Waals-Zeeman Laboratory, Valckenierstraat 65-67, 1018 XE Amsterdam, The Netherlands

Received 16 June; accepted 15 October 1997.

References:

1. Anderson, P. W. Absence of diffusion in certain random lattices. *Phys. Rev.* **109**, 1492-1505 (1958).
2. Bergmann, G. Quantitative analysis of weak localization in thin Mg films by electroresistance measurements. *Phys. Rev. B* **25**, 2937-2939 (1982).
3. Altshuler, B. L., Aronov, A. G., Khmel'nitskii, D. E. & Larkin, A. I. in *Quantum Theory of Solids* (ed. Lifshits, I. M.) 130-237 (MIR, Moskow, 1983).
4. Khmel'nitskii, D. E. Localization and coherent scattering of electrons. *Physica B* **126**, 235-241 (1984).
5. Lee, P. A. & Ramakrishnan, T. V. Disordered electronic systems. *Rev. Mod. Phys.* 57, 287-337 (1985).
6. Condat, C. A., Kirkpatrick, T. R. & Cohen, S. M. Acoustic localization in one dimension in the presence of a flow field. *Phys. Rev. B* **35**, 4653-4661 (1987).
7. Souillard, B. in *Chance and Matter* (eds Souletie, J., Vannimenus, J. & Stora, R.) Ch. 5 (North-Holland, Amsterdam, 1987).
8. Ando, T. & Fukuyama, H. (eds) *Anderson Localization, Springer Proceedings in Physics* Vol. 28 (Springer, Berlin, 1988).
9. Vollhardt, D. & Wölfle, P. in *Electronic Phase Transitions* (eds Hanke, W. & Kopaev, Yu. V.) 1-78 (Elsevier, Amsterdam, 1992).
10. Sheng, P. *Introduction to Wave Scattering, Localization, and Mesoscopic Phenomena* (Academic, San Diego, 1995).
11. John, S. Electromagnetic absorption in a disordered medium near a photon mobility edge. *Phys. Rev. Lett.* **53**, 2169-2172 (1984).
12. Anderson, P. W. The question of classical localization: a theory of white paint? *Phil. Mag. B* **52**, 505-509 (1985).
13. Sheng, P. & Zhang, Z.-Q. Scalar-wave localization in a two-component composite. *Phys. Rev. Lett.* 57, 1879-1882 (1986).
14. Arya, K., Su, Z. B. & Birman, J. L. Anderson localization of electromagnetic waves in a dielectric medium of randomly distributed metal particles. *Phys. Rev. Lett.* 57, 2725-2728 (1986).
15. Lagendijk, A., van Albada, M. P. & van der Mark, M. B. Localization of light: the quest for the white hole. *Physica A* **140**, 183-190 (1986).
16. Kaveh, M. Localization of photons indisordered systems. *Phil. Mag. B* **56**, 693-703 (1987).
17. Soukoulis, C. M., Economou, E. N., Grest, G. S. & Cohen, M. H. Existence of Anderson localization of classical waves in a random two-component medium. *Phys. Rev. Lett.* **62**, 575-578 (1989).
18. Abrahams, E., Anderson, P. W., Licciardello, D. C. & Ramakrishnan, T. V. Scaling theory of localization: absence of quantum diffusion in two dimensions. *Phys. Rev. Lett.* **42**, 673-676 (1979).
19. Mott, N. F. *Metal-Insulator Transitions* (Taylor & Francis, London, 1974).
20. Dalichaouch, R., Armstrong, J. P., Schultz, S., Platzman, P. M. & McCall, S. L. Microwave localization by two-dimensional random scattering. *Nature* **354**, 53-55 (1991).
21. Garcia, N. & Genack, A. Z. Anomalous photon diffusion at the threshold of the Anderson localization transition. *Phys. Rev. Lett.* **66**, 1850-1853 (1991).
22. Genack, A. Z. & Garcia, N. Observation of photon localization in a three-dimensional disordered system. *Phys. Rev. Lett.* **66**, 2064-2067 (1991).
23. Kuga, Y. & Ishimaru, J. Retroflection of a dense distribution of spherical particles. *J. Opt. Soc. Am. A* **1**, 831-835 (1984).
24. Albada, M. P. & Lagendijk, A. Observation of weak localization of light in a random medium. *Phys. Rev. Lett.* **55**, 2692-2695 (1985).
25. Wolf, P. E. & Maret, G. Weak localization and coherent backscattering of photons in disordered media. *Phys. Rev. Lett.* **55**, 2696-2699 (1985).
26. Mark, M. B., Albada, M. P. & Lagendijk, A. Light scattering in strongly scattering media: multiple scattering and weak localization. *Phys. Rev. B* **37**, 3575-3592 (1988).

要的实验，在该种实验中，向安德森局域化转化的过程中，扩散常数的减小是可观测到的。经典波的局域化在很多方面都类似于电子的安德森局域化。然而，光和电子有一些有趣的不同点。在光波实验中，相干（激光）光源是可实现的；这种光源的波长是很容易调节的。而且，电磁波中的光子－光子相互作用可以忽略不计，但是电子体系中的电子－电子相互作用总是对实验有影响。后者这种特性使在强无序介质中的光成为一个研究安德森局域化转变的有趣系统。

（葛聆沨 翻译；翟天瑞 审稿）

27. Akkermans, E., Wolf, P. E., Maynard, R. & Maret, G. Theoretical study of the coherent backscattering of light by disordered media. *J. Phys.* (*Paris*) **49**, 77-98 (1988).
28. Edrei, I. & Stephen, M. J. Optical coherent backscattering and transmission in a disordered media near the mobility edge. *Phys. Rev. B* **42**, 110-117 (1990).
29. Zhu, J. X., Pine, D. J. & Weitz, D. A. Internal reflection of diffusive light in random media. *Phys. Rev. A* **44**, 3948-3957 (1991).
30. Wiersma, D. S., Albada, M. P., van Tiggelen, B. A. & Lagendijk, A. Experimental evidence for recurrent multiple scattering events of light in disordered media. *Phys. Rev. Lett.* **74**, 4193-4196 (1995).

Acknowledgements. We thank F. Bogani, M. Colocci, R. Torre and M. Gurioli for discussions, and M. Colocci also for supplying GaAs crystals. D.S.W. thanks M. van Albada for advice and continuous support during the experiments, and M. Brugmans for reading of the manuscript. A.L. was supported by the "Stichting voor Fundamenteel Onderzoek der Materie" (FOM). This work was supported by the Commission of the European Community.

Correspondence and requests for materials should be addressed to D.S.W. (e-mail: wiersma@lens.unifi.it).

Extraordinary Optical Transmission through Sub-wavelength Hole Arrays

T. W. Ebbesen *et al.*

Editor's Note

Light rays are generally considered to be blocked by apertures smaller than the light's wavelength. This limits the resolution of optical microscopes. But here Thomas Ebbesen of the NEC Corporation in Princeton and colleagues report the surprising ability of light to "squeeze through" an array of sub-wavelength-sized holes in a thin metal sheet. The effect stems from the interaction of light with mobile electrons on the surface of the metal, which can be excited into wavelike motions called plasmons. A plasmon wave can pass through the hole and then re-radiate light on the far side. This paper stimulated interest in the topic now called plasmonics, which takes advantage of light-plasmon applications to achieve a range of unusual effects, including "invisibility shields".

The desire to use and control photons in a manner analogous to the control of electrons in solids has inspired great interest in such topics as the localization of light, microcavity quantum electrodynamics and near-field optics[1-6]. A fundamental constraint in manipulating light is the extremely low transmittivity of apertures smaller than the wavelength of the incident photon. While exploring the optical properties of submicrometre cylindrical cavities in metallic films, we have found that arrays of such holes display highly unusual zero-order transmission spectra (where the incident and detected light are collinear) at wavelengths larger than the array period, beyond which no diffraction occurs. In particular, sharp peaks in transmission are observed at wavelengths as large as ten times the diameter of the cylinders. At these maxima the transmission efficiency can exceed unity (when normalized to the area of the holes), which is orders of magnitude greater than predicted by standard aperture theory. Our experiments provide evidence that these unusual optical properties are due to the coupling of light with plasmons—electronic excitations—on the surface of the periodically patterned metal film. Measurements of transmission as a function of the incident light angle result in a photonic band diagram. These findings may find application in novel photonic devices.

A variety of two-dimensional arrays of cylindrical cavities in metallic films were prepared and analysed for this study. Typically, a silver film of thickness $t = 0.2$ μm was first deposited by evaporation on a quartz substrate. Arrays of cylindrical holes were fabricated through the film by sputtering using a Micrion focused-ion-beam (FIB) System 9500 (50 keV Ga ions, 5 nm nominal spot diameter). The individual hole diameter d was varied between 150 nm and 1 μm and the spacing between the holes (that is, the periodicity) a_0, was between 0.6 and 1.8 μm. The zero-order transmission spectra, where the incident

通过亚波长孔洞阵列的超常光透射现象

埃贝森等

编者按

一般认为，当孔径小于光束波长时，光线将被阻碍，从而使光学显微镜的分辨率受到限制。但在本文中，普林斯顿日本电气公司研究所的托马斯·埃贝森及其同事报道了光具有“挤过”金属薄膜中亚波长尺寸孔洞阵列的惊人能力。该效应源于光与金属表面可迁移电子间的相互作用，使其激发成为与等离子体激元类似的波的运动。等离子体激元波可以通过孔洞，进而在远端发生光的再辐射。本文诱发了对目前名为等离子体光子学领域的兴趣，该领域旨在利用光–等离子体激元来实现包括“隐形盾”在内的一系列不寻常的效应。

通过与控制固体中电子类似的方式来应用和控制光子的愿望，激起了人们对于光定域化、微腔量子电动力学和近场光学等主题的强烈兴趣[1-6]。调制光时遇到的基本限制就是当孔隙小于入射光子波长时透射率将非常低。但是，在探索金属薄膜中亚微米级圆柱形谐振腔的光学性质时，我们发现，当波长大于阵列周期时，这些孔洞阵列呈现出极不寻常的零级透射光谱(入射光与所探测的光二者共线)，并且没有发生衍射。更为特别的是，在波长为圆柱直径的 10 倍处观测到尖锐的透射峰。而且在极大值处透射率可以超过 1(按孔面积归一化后)，这比标准孔隙理论的预期值高出几个数量级。我们的实验为下列说法提供了证据：这些不寻常的光学性质源自有周期性排列孔洞的金属薄膜表面上光与等离子体激元的耦合——电子激发作用。以透射作为光线入射角函数的观测获得了光子带图。这些发现可能会在新型光子器件中获得应用。

在这项研究中，我们制备了具有多种二维阵列的圆柱形谐振腔的金属薄膜，并对实验结果进行了分析。典型的一例是，首先通过蒸镀在石英基质上沉积一层 $t = 0.2$ μm 厚的银膜。继而用 Micrion 聚焦离子束系统 9500(50 keV 镓离子，5 nm 标称点直径) 进行溅镀，并在薄膜上加工出圆柱形孔洞阵列。单个孔直径 d 在 150 nm 到 1 μm 之间，孔洞之间的距离 a_0(亦即周期) 则在 0.6 μm 和 1.8 μm 之间。零级透射

and detected light are collinear, were recorded with a Cary 5 ultraviolet–near infrared spectrophotometer with an incoherent light source, but the arrays were also studied on an optical bench for transmission, diffraction and reflection properties using coherent sources.

Figure 1 shows a typical zero-order transmission spectrum for a square array of 150 nm holes with a period a_0 of 0.9 μm in a 200 nm thick Ag film. The spectrum shows a number of distinct features. At wavelength $\lambda = 326$ nm the narrow bulk silver plasmon peak is observed which disappears as the film becomes thicker. The most remarkable part is the set of peaks which become gradually stronger at longer wavelengths, increasingly so even beyond the minimum at the periodicity a_0. There is an additional minimum at $\lambda = a_0\sqrt{\varepsilon}$ corresponding to the metal–quartz interface (where ε is the dielectric constant of the substrate). For $\lambda > a_0\sqrt{\varepsilon}$, there is no diffraction from the array nor from the individual holes. As expected, the first-order diffraction spots can be seen to be grazing the surface (that is, the diffraction angle approaches 90°) as the wavelength approaches the period from below (this might, in fact, enhance the coupling to be discussed in the next paragraphs). The maximum transmitted intensity occurs at 1,370 nm, nearly ten times the diameter of an individual hole in the array. Even more surprising is that the absolute transmission efficiency, calculated by dividing the fraction of light transmitted by the fraction of surface area occupied by the holes, is $\geqslant 2$ at the maxima. In other words, more than twice as much light is transmitted as impinges directly on the holes. Furthermore, the transmittivity of the array scales linearly with the surface area of the holes. This is all the more remarkable considering that the transmission efficiency of a single sub-wavelength aperture is predicted by Bethe[7] to scale as $(r/\lambda)^4$ where r is the hole radius; accordingly for a hole of 150 nm diameter one expects a transmission efficiency on the order of 10^{-3}. In addition, the intensity (I) of the zero-order transmission from a grating is expected to decrease monotonically at larger wavelengths ($I \propto \lambda^{-1}$) (ref. 8). Therefore our results must imply that the array itself is an active element, not just a passive geometrical object in the path of the incident beam.

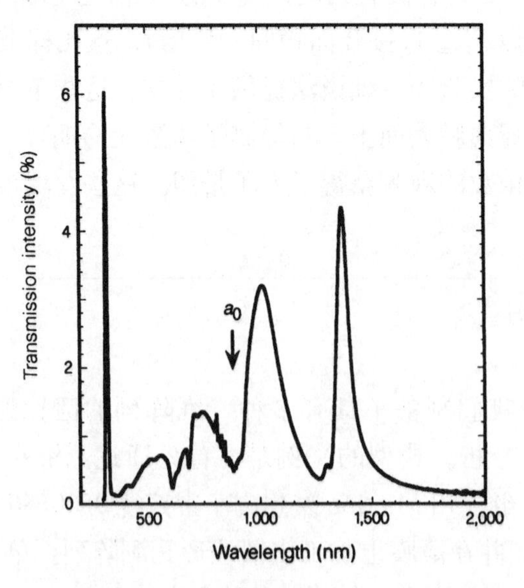

Fig. 1. Zero-order transmission spectrum of an Ag array ($a_0 = 0.9$ μm, $d = 150$ nm, $t = 200$ nm).

光谱——其中入射光和所探测的光共线——是用一台具有非相干光源的 Cary 5 型紫外–近红外分光光度计记录的，也用相干光源对光具座上的阵列就透射、衍射和反射性质进行了研究。

图 1 显示了在 200 nm 厚的银膜中，周期 a_0 为 0.9 μm 的 150 nm 孔洞方形阵列典型零级透射光谱。光谱呈现出很多不同的特征。在波长 $\lambda = 326$ nm 处可以看到本底银狭窄的等离子体峰，当薄膜厚度增加时它会消失。最值得关注的部分是在波长较长处逐渐变强的一组峰，甚至增加到超过位于周期 a_0 处的最小值。在 $\lambda = a_0\sqrt{\varepsilon}$ 处有一个额外的最小值，它对应于金属–石英界面(其中 ε 是基质的介电常数)。在 $\lambda > a_0\sqrt{\varepsilon}$ 处，阵列和单个孔洞都没有产生衍射。如同预期的那样，当波长由低处趋近周期值时，可以看到一级衍射点随之掠向表面(即衍射角接近 90°)的现象(实际上，它可能增强我们在下面段落中将要讨论到的耦合作用)。最大透射强度出现在 1,370 nm 处，这一波长大约是阵列中单独一个孔洞的直径的 10 倍。更令人吃惊的是绝对透射率——由光的透射部分除以孔洞占据的表面积即可算出——在最大处 ≥2。换句话说，透射出的光强比直接撞击孔洞的光强要高出一倍以上。此外，阵列的透射率随着孔洞表面积呈线性变化。如果考虑到贝特[7]关于单个亚波长孔道透射率与 $(r/\lambda)^4$ 成比例的推断——其中 r 是孔洞半径，这一点就更加值得注意；据此推断，对于直径 150 nm 的孔洞来说，可以预期具有 10^{-3} 数量级的透射率。另外，预期光栅产生的零级透射强度(I)会在较大波长($I \propto \lambda^{-1}$)处单调地下降(参考文献 8)。因此我们的结果必然意味着阵列本身就是一个有源元件，而不只是入射光束路径上的一个无源几何体。

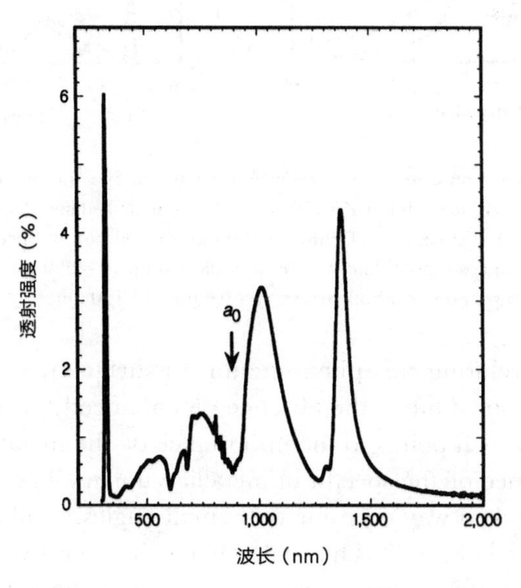

图 1. 一个银阵列($a_0 = 0.9$ μm，$d = 150$ nm，$t = 200$ nm)的零级透射光谱

To understand the origin of this phenomenon, we tested the dependence on all the possible variables such as hole diameter, periodicity, thickness and type of metal. It is beyond the scope of this Letter to describe the details of all these experiments. Instead we summarize our observations and illustrate some of the key factors that determine the shape of these spectra. To begin with, the periodicity of the array determines the position of the peaks. The positions of the maxima scale exactly with the periodicity, as shown in Fig. 2a, independent of metal (Ag, Cr, Au), hole diameter and film thickness. The width of the peaks appears to be strongly dependent on the aspect ratio (t/d or depth divided by diameter) of the cylindrical holes (Fig. 2a). For $t/d = 0.2$, the peaks are very broad and just discernible and when the ratio reaches ~1, the maximum sharpness is obtained. Further narrowing might depend on the quality of the individual holes. The thickness dependence of the spectra is displayed in Fig. 2b for 0.2 and 0.5 μm Ag films. While the intensity of the bulk plasmon peak decreases rapidly in this range, that of the longer-wavelength peaks decreases approximately linearly with thickness. The spectra change significantly with the type of lattice, for example whether the array is a square or a triangular lattice.

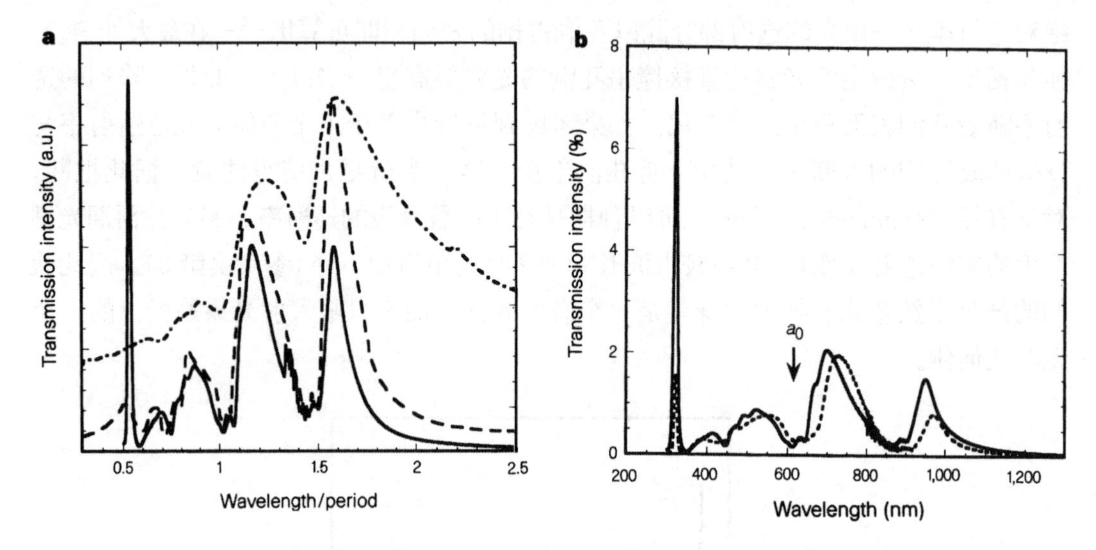

Fig. 2. Effects of parameters on zero-order transmission spectra. **a**, Spectra for various square arrays as a function of λ/a_0. Solid line: Ag, $a_0 = 0.6$ μm, $d = 150$ nm, $t = 200$ nm; dashed line: Au, $a_0 = 1.0$ μm, $d = 350$ nm, $t = 300$ nm; dashed-dotted line: Cr, $a_0 = 1.0$ μm, $d = 500$ nm, $t = 100$ nm. **b**, Spectra for two identical Ag arrays with different thicknesses. Solid line: $t = 200$ nm; dashed line: $t = 500$ nm (this spectrum has been multiplied by 1.75 for comparison). For both arrays: $a_0 = 0.6$ μm; $d = 150$ nm.

Two important clues relating this phenomenon to surface plasmons (SPs) come from the following observations. One is the absence of enhanced transmission in hole arrays fabricated in Ge films which points to the importance of the metallic film. The other clue is the angular dependence of the spectra in metallic samples. The zero-order transmission spectra change in a marked way even for very small angles, as illustrated in Fig. 3 where the spectra were recorded every 2°. The peaks change in intensity and split into new peaks which move in opposite directions. This is exactly the behaviour observed when light couples with SPs in reflection gratings[9-14]. SPs are oscillations of surface charges at the metal

为理解这种现象的来由，我们检测了它对于诸如孔洞直径、周期、金属膜厚度和类型等所有可能变量的依赖性。由于本快报篇幅有限，无法描述所有实验的细节，因此我们总结了观察结果并说明了决定这些光谱的形状的一些关键因素。首先，阵列的周期决定了峰的位置。如图 2a 所示，最大值的位置严格地与周期成比例，而不依赖于所用金属（银、铬、金）、孔洞直径和薄膜厚度。峰的宽度显然强烈地依赖于圆柱形谐振腔的纵横比（t/d，或厚度除以直径）（图 2a）。当 $t/d = 0.2$ 时，峰很宽而且刚刚可以分辨出来，而当该比值达到约 1 时具有最高清晰度。进一步的变窄可能和单个孔洞的性质有关。图 2b 中显示出光谱对于薄膜厚度的依赖性，其中银膜厚度分别为 0.2 μm 和 0.5 μm。本底等离子体激元峰的强度在这个范围内快速下降，而较长波长处的峰强度随着厚度变化大致呈线性下降。光谱随着栅格类型——例如阵列究竟是方形还是三角形栅格——发生显著变化。

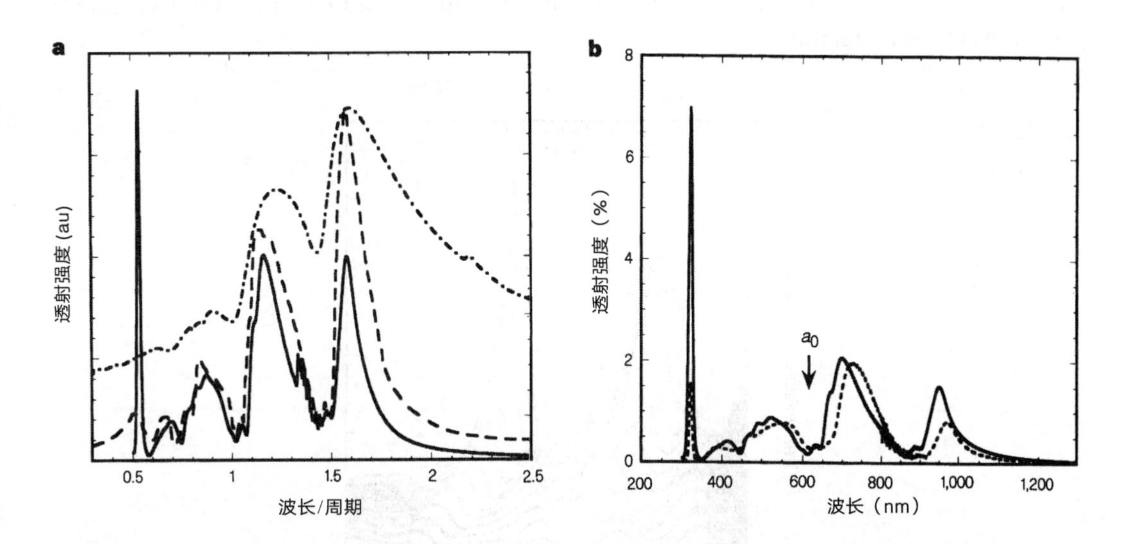

图 2. 参数对于零级透射光谱的影响。**a**，作为 λ/a_0 的函数的各种方形阵列的光谱。实线：银，$a_0 = 0.6$ μm，$d = 150$ nm，$t = 200$ nm；虚线：金，$a_0 = 1.0$ μm，$d = 350$ nm，$t = 300$ nm；点划线：铬，$a_0 = 1.0$ μm，$d = 500$ nm，$t = 100$ nm。**b**，只有厚度不同的两个相同银阵列的光谱。实线：$t = 200$ nm；虚线：$t = 500$ nm（为便于比较，此光谱已乘以倍数 1.75）。对两个阵列同时有：$a_0 = 0.6$ μm；$d = 150$ nm。

将这种现象与表面等离子体激元（SP）联系起来的两条重要线索来自下面的观测结果。其一是，由锗薄膜加工制成的孔洞阵列中不存在透射增强现象，表明了薄膜金属的重要性。另一条线索是在金属样品中光谱的角度依赖性。如图 3 所示，即使角度变化很小时，零级透射光谱也会发生明显的变化。图 3 是角度间隔为 2°时记录下来的光谱。谱峰强度发生变化，并且分裂成向相反方向移动的新谱峰。这正是光与反射光栅中表面等离子体激元耦合时观测到的行为 [9-14]。表面等离子体激元是金属

interface and are excited when their momentum matches the momentum of the incident photon and the grating as follows:

$$k_{sp} = k_x \pm nG_x \pm mG_y$$

where k_{sp} is the surface plasmon wavevector, $k_x = (2\pi/\lambda)\sin\theta$ is the component of the incident photon's wavevector in the plane of the grating and $G_x = G_y = 2\pi/a_0$ are the grating momentum wavevectors for a square array. Therefore if the angle of incidence θ is varied, the incident radiation excites different SP modes and by recording the peak energies as a function of k_x we obtain the dispersion relation shown in Fig. 4. This figure reveals the band structure of SP in the two-dimensional array and it clearly demonstrates the presence of gaps with energies around 30 to 50 meV; these are due to the lifting of the degeneracy by SPs interacting with the lattice. An extrapolation of the dispersion curves yields an intercept with the k-axis at a value which is within the experimental error (~10%) of that expected from the periodicity ($2\pi/a_0$). The results of Figs 3 and 4 are also sensitive to polarization as expected, but a clear assignment has not yet been possible because of inherent structure of the grains in the metal film.

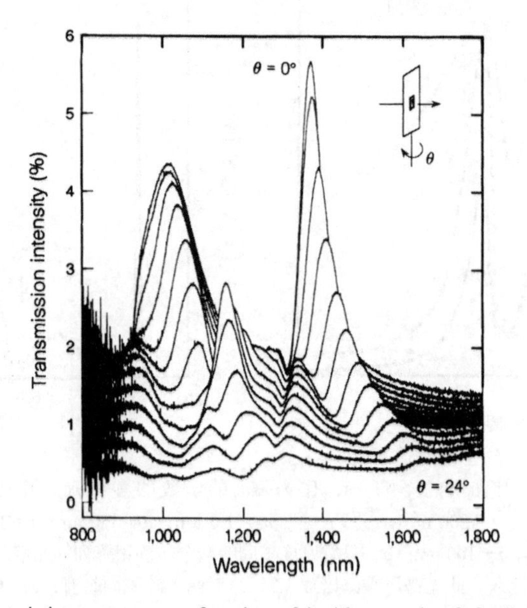

Fig. 3. Zero-order transmission spectra as a function of incident angle of the light. Spectra were taken every 2° up to 24° for a square Ag array ($a_0 = 0.9$ μm, $d = 150$ nm, $t = 200$ nm). The individual spectra are offset vertically by 0.1% from one another for clarity.

界面上表面电荷的振荡，当它们的动量与入射光子以及光栅的动量满足如下表达式时将受到激发：

$$k_{sp} = k_x \pm nG_x \pm mG_y$$

其中 k_{sp} 是表面等离子体激元波矢，$k_x = (2\pi/\lambda)\sin\theta$ 是入射光子波矢在光栅平面内的分量，而 $G_x = G_y = 2\pi/a_0$ 是方形阵列的光栅动量波矢。因此，如果入射角 θ 变化，入射辐射将激发不同的表面等离子体激元模式；通过记录作为 k_x 的函数的峰值能量，我们得到了图 4 中所显示的色散关系。这幅图揭示了二维阵列中表面等离子体激元的带结构，而且它清楚地阐明了具有约 30 meV 到 50 meV 能量的带隙的存在；它们来源于表面等离子体激元与栅格相互作用导致的简并提升。色散曲线的外推得到在 k 轴上的截距，其数值在依据周期 $(2\pi/a_0)$ 预期的实验误差（约 10%）之内。如同预期，图 3 和图 4 中的结果也对极化很敏感，但是由于金属薄膜中晶粒内在结构间的差异，目前还不能做出明确的归属。

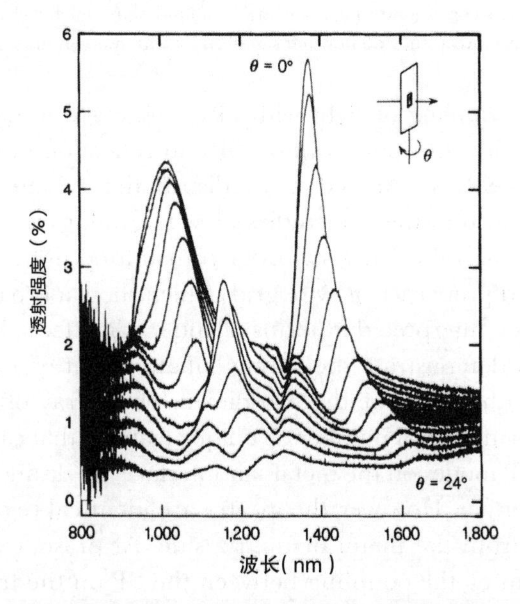

图 3. 作为光线入射角函数的零级透射光谱。对一个方形银阵列每隔 2° 测一次谱图，一直测到 24°（$a_0 = 0.9$ μm，$d = 150$ nm，$t = 200$ nm）。为清晰起见，各个谱之间在垂直方向上错开 0.1%。

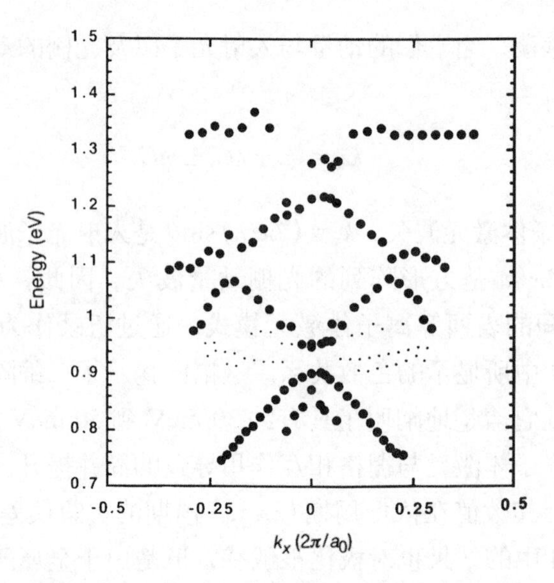

Fig. 4. Dispersion curves (solid circles) along the [10] direction of the array. These curves are extracted from the energy of the transmission peaks of Fig. 3. The momentum k_x is in the plane of the array and is given by $k_x = (2\pi/\lambda)\sin\theta$ where θ is the incident angle of the beam (the units are normalized to $2\pi/a_0$). The curves with the smaller dots correspond to peaks whose amplitudes are much weaker and may or may not be related to the band structure as they do not shift significantly with momentum.

In our experiments the coupling of light with SPs is observed in transmission rather than reflection, in contrast with previous work on SPs on reflection gratings. In those studies the coupling of light to SPs is observed as a redistribution of intensity between different diffracted orders. Even in transmission studies of wire gratings by Lochbihler[15], the effect of SPs is observed through dips in zero-order transmitted light. There is an extensive literature on the infrared properties of wire grids which show a broad transmission centred at $\lambda \approx 1.2a_0$; this has been interpreted in terms of induction effects, in analogy with electric circuits[16,17]. Our results demonstrate the strong enhancement of transmitted light due to coupling of the light with the SP of the two-dimensional array of sub-wavelength holes. Furthermore our results indicate a number of unique features that cannot be explained with existing theories. The SP modes on the metal–air interface are distinctly different from those at the metal–quartz interface. However, the spectra are identical regardless of whether the sample is illuminated from the metal or quartz side. At present we do not understand the detailed mechanism of the coupling between the SP on the front and back surfaces which results in larger than unity transmission efficiency of the holes. The thickness dependence (see Fig. 2b) suggests that the holes play an important part in mediating this coupling and that nonradiative SP modes are transferred to radiative modes by strong scattering in the holes.

In photonic bandgap arrays[2,4], the material is passive and translucent at all wavelengths except at the energies within the gap. In the present arrays, the material plays an active role (through the plasmons) and it is opaque at all wavelengths except those for which coupling occurs. The combination, or integration, of these two types of phenomena might lead to

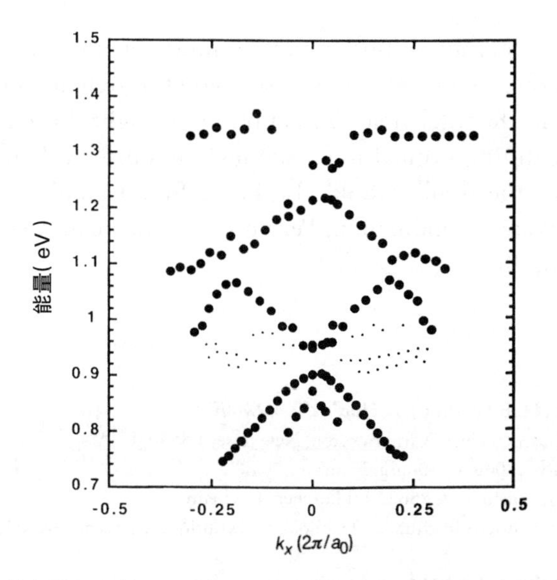

图 4. 沿阵列 [10] 方向的色散曲线(实心圆点)。这些曲线是利用图 3 中透射峰的能量得到的。动量 k_x 位于阵列平面内，由 $k_x = (2\pi/\lambda)\sin\theta$ 给出，其中 θ 是光束的入射角(单位归一化为 $2\pi/a_0$)。由较小的点组成的曲线所对应的峰，其幅度要弱很多，由于随动量变化发生的偏移并不明显，因此与带结构的关联不能确定。

在我们的实验中，光与表面等离子体激元的耦合是在透射中而不是在反射中观测到的，这与此前对反射光栅上的表面等离子体激元的研究相反。在那些研究中，光与表面等离子体激元的耦合是作为强度在不同衍射级之间的重新分配观测到的。即使是在洛赫比勒对线光栅的透射研究中 [15]，表面等离子体激元的影响也是通过零级透射光中的凹陷观测到的。关于线栅的红外性质，已经有大量文献表明它有一个以 $\lambda \approx 1.2a_0$ 为中心的宽透射；可以用类似于电路的感应效应对此进行解释 [16,17]。我们的结果表明，由于光与亚波长孔洞二维阵列的表面等离子体激元耦合，透射光将明显增强。此外，我们的结果还指出了很多用现有理论无法解释的独特性状。位于金属–空气界面上的表面等离子体激元模式明显不同于位于金属–石英界面上的。但是，无论是从金属一侧还是石英一侧对样品进行光照，所得光谱都是一样的。目前，我们还不了解导致孔洞透射效率大于 1 的前后表面上等离子体激元之间耦合的详细机制。厚度依赖性(参见图 2b)意味着孔洞对于调制这类耦合起着重要的作用，也意味着通过孔洞中的强烈散射，非辐射表面等离子体激元模式将转化为辐射模式。

在光子带隙阵列中 [2,4]，材料在除去能隙中的能量之外的所有波长处都是无源、半透明的。在当前的阵列中，材料起到有源的作用(通过等离子体激元)，并且在除去发生耦合的那些波长之外的所有波长处都是不透明的。这两种现象的结合或整合

optical features that are very interesting from both fundamental and technological points of view. The demonstration of efficient light transmission through holes much smaller than the wavelength and beyond the inter-hole diffraction limit might, for example, inspire designs for novel nearfield scanning optical microscopes[6], or sub-wavelength photolithography. Theoretical analysis of the results would also be useful for gaining better insight into this extraordinary transmission phenomenon. Perhaps only then can we expect to grasp the full implications of these findings.

(**391**, 667-669; 1998)

T. W. Ebbesen*†, H. J. Lezec‡, H. F. Ghaemi*, T. Thio* & P. A. Wolff*§

* NEC Research Institute, 4 Independence Way, Princeton, New Jersey 08540, USA

† ISIS, Louis Pasteur University, 67000 Strasbourg, France

‡ Micrion Europe GmbH, Kirchenstraße 2, 85622 Feldkirchen, Germany

§ Department of Physics, Massachusetts Institute of Technology, Cambridge, Massachusetts 02139, USA

Received 15 July; accepted 24 November 1997.

References:

1. John, S. Localization of light. *Phys. Today* 32 (May 1991).
2. Yablonovitch, E. & Leung, K. M. Hope for photonic bandgaps. *Nature* **351,** 278 (1991).
3. Dalichaouch, R., Armstrong, J. P., Schultz, S., Platzman, P. M. & McCall, S. L. Microwave localization by two-dimensional random scattering. *Nature* **354,** 53-55 (1991).
4. J. D. Joannopoulos, Meade R. D. & Winn, J. N. *Photonic Crystals* (Princeton Univ. Press, Princeton, 1995).
5. Haroche, S. & Kleppner, D. Cavity quantum electrodynamics. *Phys. Today* **24** (January 1989).
6. Betzig, E. & Trautman, J. K. Near-field optics: Microscopy, spectroscopy, and surface modification beyond the diffraction limit. *Science* **257,** 189-194 (1992).
7. Bethe, H. A. Theory of diffraction by small holes. *Phys. Rev.* **66,** 163-182 (1944).
8. Born, M. & Wolf, E. *Principles of Optics* (Pergamon, Oxford, 1980).
9. Ritchie, R. H., Arakawa, E. T., Cowan, J. J. & Hamm, R. N. Surface-plasmon resonance effect in grating diffraction. *Phys. Rev. Lett.* **21,** 1530-1533 (1968).
10. Raether, H. *Surface Plasmons* (Springer, Berlin, 1988).
11. Chen, Y. J., Koteles, E. S., Seymour, R. J., Sonek, G. J. & Ballantyne, J. M. Surface plasmons on gratings: coupling in the minigap regions. *Solid State Commun.* **46,** 95-99 (1983).
12. Kitson, S. C., Barnes, W. L. & Sambles, J. R. Full photonic band gap for surface modes in the visible. *Phys. Rev. Lett.* **77,** 2670-2673 (1996).
13. Watts, R. A., Harris, J. B., Hibbins, A. P., Preist, T. W. & Sambles, J. R. Optical excitations of surface plasmon polaritons on 90 and 60 bi-gratings. *J. Mod. Opt.* **43,** 1351-1360 (1996).
14. Derrick, G. H., McPhedran, R. C., Maystre, D. & Neviere, M. Crossed gratings: a theory and its applications. *Appl. Phys.* **18,** 39-52 (1979).
15. Lochbihler, H. Surface polaritons on gold-wire gratings. *Phys. Rev. B* **50,** 4795-4801 (1994).
16. Ulrich, R. Far-infrared properties of metallic mesh and its complimentary structure. *Infrared Phys.* **7,** 37-55 (1967).
17. Larsen, T. A survey of the theory of wire grids. *I.R.E. Trans. Microwave Theory Techniques* **10,** 191-201 (1962).

Acknowledgements. We thank S. Kishida, G. Bugmann and J. Giordmaine for their encouragement, and R. Linke, R. McDonald, M. Treacy, J. Chadi and C. Tsai for discussions. We also thank G. Lewen, G. Seidler, A. Krishnan, A. Schertel, A. Dziesiaty and H. Zimmermann for assistance.

Correspondence should be addressed to T.W.E.

可能产生无论从基础原理还是技术性角度来讲都极为有趣的光学特性。光能有效地透过比波长小很多的孔洞并突破孔洞间衍射的限制，对这一点的证明可以对诸如新型近场扫描光学显微镜[6]或者亚波长光刻法的设计等有所启发。对这些结果的理论分析可能会有助于对这种超强透射现象有更深入的理解。也许只有到那时我们才有望掌握这些发现的全部意义。

（王耀杨 翻译；宋心琦 审稿）

A Silicon-based Nuclear Spin Quantum Computer

B. E. Kane

Editor's Note

In the 1980s, physicists began trying to build computing devices that exploit quantum effects, which could be much more powerful than classical devices, at least for certain computational tasks. The practical development of such machines was immensely challenging because of their inherent sensitivity to disruption from the environment. Here Australian physicist Bruce Kane proposes a scheme for implementing a quantum computer by storing its information in the relatively isolated spins of nuclei of dopant atoms inserted into silicon electronic devices. Logical operations, he suggests, could be performed on these spins with external electric fields, and measurements made with currents of spin-polarized electrons. Practical quantum computers are now being developed, and this strategy remains one of many being explored.

Quantum computers promise to exceed the computational efficiency of ordinary classical machines because quantum algorithms allow the execution of certain tasks in fewer steps. But practical implementation of these machines poses a formidable challenge. Here I present a scheme for implementing a quantum-mechanical computer. Information is encoded onto the nuclear spins of donor atoms in doped silicon electronic devices. Logical operations on individual spins are performed using externally applied electric fields, and spin measurements are made using currents of spin-polarized electrons. The realization of such a computer is dependent on future refinements of conventional silicon electronics.

ALTHOUGH the concept of information underlying all modern computer technology is essentially classical, physicists know that nature obeys the laws of quantum mechanics. The idea of a quantum computer has been developed theoretically over several decades to elucidate fundamental questions concerning the capabilities and limitations of machines in which information is treated quantum mechanically[1,2]. Specifically, in quantum computers the ones and zeros of classical digital computers are replaced by the quantum state of a two-level system (a qubit). Logical operations carried out on the qubits and their measurement to determine the result of the computation must obey quantum-mechanical laws. Quantum computation can in principle only occur in systems that are almost completely isolated from their environment and which consequently must dissipate no energy during the process of computation, conditions that are extraordinarily difficult to fulfil in practice.

Interest in quantum computation has increased dramatically in the past four years because of two important insights: first, quantum algorithms (most notably for prime factorization[3,4]

硅基核自旋量子计算机

凯恩

编者按

20世纪80年代，物理学家开始尝试构建利用量子效应的计算设备，至少对于某些计算任务而言，它可能比传统设备性能强大得多。由于其对来自环境的干扰具有内在的敏感性，这类机器的实际研发极具挑战性。本文中，澳大利亚物理学家布鲁斯·凯恩提出了一种实现量子计算机的方案，将其信息存储在硅电子器件中内嵌掺杂原子的相对孤立的核自旋中。他建议，逻辑运算可以在外部电场辅助下用这些自旋进行，并用自旋极化电子的电流进行测量。可行的量子计算机正在开发中，该方案仍然是许多正在被探索的策略之一。

量子计算机的计算效率可能会超过传统经典计算机，因为量子算法允许用更少的步骤来执行某些任务。但是这些机器的实际研发仍然是一个艰巨的挑战。这里我将提出一种构建基于量子力学的计算机的方法。计算信息被编码在掺杂硅的电子器件中施主原子的核自旋上。单个自旋的逻辑运算用外部施加电场来实现，而自旋测量基于自旋极化电子的电流。这样一个计算机的实现依赖于传统硅器件未来的精密程度。

尽管所有现代计算机所依赖的信息概念本质上都是基于经典理论的，然而物理学家知道自然界遵从量子力学定律。量子计算机的概念在理论上已经发展了几十年，旨在阐明使用量子力学来处理信息的计算机的能力和极限等基本问题[1,2]。具体而言，在量子计算机中，经典计算机中的0和1被一个二能级系统的量子态（量子比特）所取代。在量子比特上进行的逻辑运算及用来决定计算结果的测量必须遵守量子力学定律。量子计算理论上只能发生在与外界环境几乎完全隔离的系统中，这样在计算过程中不会存在能量的耗散，这样的苛刻条件在现实中很难满足。

在过去四年中，两个重大的突破使人们对量子计算的兴趣与日俱增。第一个是量子算法（尤其值得一提的是质因子分解[3,4]和穷举搜索[5]）的开发，而且这些算法

and for exhaustive search[5]) have been developed that outperform the best known algorithms doing the same tasks on a classical computer. These algorithms require that the internal state of the quantum computer be controlled with extraordinary precision, so that the coherent quantum state upon which the quantum algorithms rely is not destroyed. Because completely preventing decoherence (uncontrolled interaction of a quantum system with its surrounding environment) is impossible, the existence of quantum algorithms does not prove that they can ever be implemented in a real machine.

The second critical insight has been the discovery of quantum error-correcting codes that enable quantum computers to operate despite some degree of decoherence and which may make quantum computers experimentally realizable[6,7]. The tasks that lie ahead to create an actual quantum computer are formidable: Preskill[8] has estimated that a quantum computer operating on 10^6 qubits with a 10^{-6} probability of error in each operation would exceed the capabilities of contemporary conventional computers on the prime factorization problem. To make use of error-correcting codes, logical operations and measurement must be able to proceed in parallel on qubits throughout the computer.

The states of spin 1/2 particles are two-level systems that can potentially be used for quantum computation. Nuclear spins have been incorporated into several quantum computer proposals[9-12] because they are extremely well isolated from their environment and so operations on nuclear spin qubits could have low error rates. The primary challenge in using nuclear spins in quantum computers lies in measuring the spins. The bulk spin resonance approach to quantum computation[11,12] circumvents the single-spin detection problem essentially by performing quantum calculations in parallel in a large number of molecules and determining the result from macroscopic magnetization measurements. The measurable signal decreases with the number of qubits, however, and scaling this approach above about ten qubits will be technically demanding[37].

To attain the goal of a 10^6 qubit quantum computer, it has been suggested that a "solid state" approach[13] might eventually replicate the enormous success of modern electronics fabrication technology. An attractive alternative approach to nuclear spin quantum computation is to incorporate nuclear spins into an electronic device and to detect the spins and control their interactions electronically[14]. Electron and nuclear spins are coupled by the hyperfine interaction[15]. Under appropriate circumstances, polarization is transferred between the two spin systems and nuclear spin polarization is detectable by its effect on the electronic properties of a sample[16,17]. Electronic devices for both generating and detecting nuclear spin polarization, implemented at low temperatures in $GaAs/Al_xGa_{1-x}As$ heterostructures, have been developed[18], and similar devices have been incorporated into nanostructures[19,20]. Although the number of spins probed in the nanostructure experiments is still large ($\sim 10^{11}$; ref. 19), sensitivity will improve in optimized devices and in systems with larger hyperfine interactions.

在执行相同任务时优于经典计算机上最著名的算法。这些算法要求量子计算机的内部状态被准确无误地控制，这样量子算法所依赖的相干量子态就不会被破坏。因为完全阻止量子退相干(量子系统和外界环境之间非受控的相互作用)是不可能的，所以量子算法的存在并不能保证它们可以在真实的计算机上面得以实现。

第二个重大的突破是量子纠错码的发现，它可以使量子计算机运行时容忍一定程度的量子退相干，这使得量子计算机在实验上的实现成为可能[6,7]。眼下制造一台真正的量子计算机的任务是艰巨的：据普雷斯基尔[8]估计，一台使用 10^6 个量子比特且每次运算误码率为 10^{-6} 的量子计算机可以在质因子分解问题上超过同时代的传统计算机。为了利用纠错码，逻辑运算和测量必须能够在整个计算机的量子比特上并行进行。

自旋为 1/2 的粒子的状态是二能级系统，具备应用于量子计算的潜质。核自旋已经被纳入数个量子计算机提案[9-12]中，因为它们与外界环境间被很好地隔离，因此利用核自旋比特来运算，误码率可能很低。在量子计算机中使用核自旋的主要挑战在于测量自旋。量子计算的系综自旋共振方法[11,12]基本上是通过在大量分子上的并行量子计算并由宏观磁化测量确定结果来绕开单个自旋的测量问题。然而可测量信号随着量子比特数目的增多而减少，因此将这种方法扩展到 10 个量子比特以上将会对技术要求极为苛刻[37]。

为了实现 10^6 量子比特计算机，有人建议使用“固态”方法[13]，这种方法也许可以最终复制现代电子器件微加工技术的巨大成功。核自旋量子计算的一种有吸引力的替代方法是将核自旋融合到电子器件中，并且通过电学方式来测量自旋以及控制其之间的相互作用[14]。电子和核自旋被超精细相互作用耦合在一起[15]。在适当情况下，极化可以在两个自旋系统中转移，核自旋极化可以通过其对样品的电子性质的影响来测量[16,17]。用来产生和探测核自旋极化的电子器件也已经在低温砷化镓(GaAs)/铝镓砷($Al_xGa_{1-x}As$)异质结构中实现[18]，类似的器件也已经被整合到纳米结构中[19,20]。尽管在纳米结构试验中研究的自旋数目仍然很大(约为 10^{11}；文献 19)，但在经过优化的器件中以及更强的超精细相互作用系统中，灵敏度会得到进一步的改善。

Here I present a scheme for implementing a quantum computer on an array of nuclear spins located on donors in silicon, the semiconductor used in most conventional computer electronics. Logical operations and measurements can in principle be performed independently and in parallel on each spin in the array. I describe specific electronic devices for the manipulation and measurement of nuclear spins, fabrication of which will require significant advances in the rapidly moving field of nanotechnology. Although it is likely that scaling the devices proposed here into a computer of the size envisaged by Preskill[8] will be an extraordinary challenge, a silicon-based quantum computer is in a unique position to benefit from the resources and ingenuity being directed towards making conventional electronics of ever smaller size and greater complexity.

Quantum Computation with a ^{31}P Array in Silicon

The strength of the hyperfine interaction is proportional to the probability density of the electron wavefunction at the nucleus. In semiconductors, the electron wavefunction extends over large distances through the crystal lattice. Two nuclear spins can consequently interact with the same electron, leading to electron-mediated or indirect nuclear spin coupling[15]. Because the electron is sensitive to externally applied electric fields, the hyperfine interaction and electron-mediated nuclear spin interaction can be controlled by voltages applied to metallic gates in a semiconductor device, enabling the external manipulation of nuclear spin dynamics that is necessary for quantum computation.

The conditions required for electron-coupled nuclear spin computation and single nuclear spin detection can arise if the nuclear spin is located on a positively charged donor in a semiconductor host. The electron wavefunction is then concentrated at the donor nucleus (for *s* orbitals and energy bands composed primarily of them), yielding a large hyperfine interaction energy. For shallow-level donors, however, the electron wavefunction extends tens or hundreds of ångströms away from the donor nucleus, allowing electron-mediated nuclear spin coupling to occur over comparable distances. The quantum computer proposed here comprises an array of such donors positioned beneath the surface of a semiconductor host (Fig. 1). A quantum mechanical calculation proceeds by the precise control of three external parameters: (1) gates above the donors control the strength of the hyperfine interactions and hence the resonance frequency of the nuclear spins beneath them; (2) gates between the donors turn on and off electron-mediated coupling between the nuclear spins[13]; (3) a globally applied a.c. magnetic field B_{ac} flips nuclear spins at resonance. Custom adjustment of the coupling of each spin to its neighbours and to B_{ac} enables different operations to be performed on each of the spins simultaneously. Finally, measurements are performed by transferring nuclear spin polarization to the electrons and determining the electron spin state by its effect on the orbital wavefunction of the electrons, which can be probed using capacitance measurements between adjacent gates.

这里我将介绍一种利用位于硅材料施主上面的核自旋阵列来实现量子计算机的方法，硅材料是大多数传统计算机电子学中使用的半导体。理论上逻辑运算和测量可以在阵列中的各个自旋上独立且并行执行。我将介绍特定的电子器件来实现核自旋的控制和测量，其制造需要迅猛发展的纳米科技领域的重大突破。把这里提出的器件缩放到普雷斯基尔[8]设想的尺寸的计算机中，这也许是一个巨大的挑战，然而，由于人们为了将传统电子器件做得更小且更加复杂的过程中积累了资源和聪明才智，硅基量子计算机仍然具有独特的地位。

使用硅中 ^{31}P 阵列的量子计算

超精细相互作用的强度正比于原子核处电子波函数的概率密度。在半导体中，电子波函数在晶格上延展很长的距离。两个核自旋因此可以和同一个电子相互作用，这就导致了电子介导的或者间接的核自旋耦合[15]。由于电子对于外加电场的敏感，超精细相互作用以及电子介导的核自旋相互作用可以通过调节施加在半导体器件中金属栅极上的电压来控制，这使量子计算所要求的核自旋动力学的外部调制成为可能。

如果核自旋位于半导体宿主中带正电的施主上面，电子耦合核自旋计算以及单核自旋检测所需要的条件会被满足。电子波函数将会集中于施主的原子核(对于 s 轨道和主要由它们组成的能带)，产生大的超精细相互作用能。然而对于浅能级施主来说，电子波函数从原子核向外扩展了几十甚至几百埃，从而允许电子介导核自旋耦合在相对远的距离发生。本文提出的量子计算机包括在半导体宿主表面下方的一系列这样的施主(如图 1)。量子力学计算通过精确控制三个外部参数来进行：(1) 施主上方的栅极控制超精细相互作用的强度，因此也就控制了栅极下方的核自旋的共振频率；(2) 施主之间的栅极可以用来开通和关闭核自旋之间的电子介导耦合[13]；(3) 一个对全体施加的交变磁场 B_{ac} 用来翻转与磁场共振的核自旋。单独调节每个自旋之间的耦合以及自旋与外加磁场 B_{ac} 的耦合，使每个自旋上的不同运算可以同时进行。最后，通过转移核自旋极化到电子身上进而影响电子轨道波函数，最终通过测试其对相邻的栅极之间的电容的改变来测得自旋态。

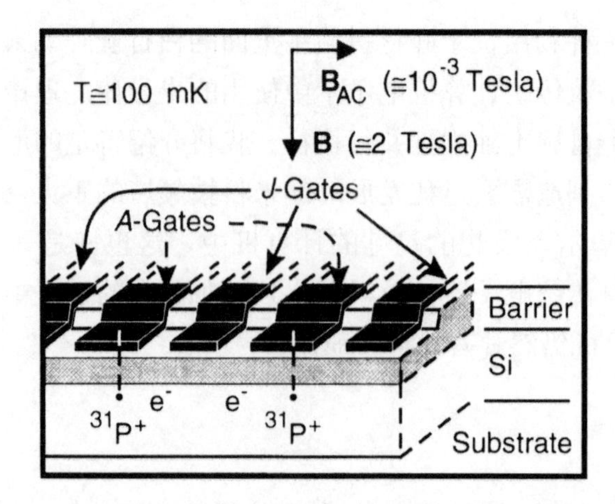

Fig. 1. Illustration of two cells in a one-dimensional array containing ^{31}P donors and electrons in a Si host, separated by a barrier from metal gates on the surface. "*A* gates"control the resonance frequency of the nuclear spin qubits; "*J* gates"control the electron-mediated coupling between adjacent nuclear spins. The ledge over which the gates cross localizes the gate electric field in the vicinity of the donors.

An important requirement for a quantum computer is to isolate the qubits from any degrees of freedom that may lead to decoherence. If the qubits are spins on a donor in a semiconductor, nuclear spins in the host are a large reservoir with which the donor spins can interact. Consequently, the host should contain only nuclei with spin $I = 0$. This simple requirement unfortunately eliminates all III–V semiconductors as host candidates, because none of their constituent elements possesses stable $I = 0$ isotopes[21]. Group IV semiconductors are composed primarily $I = 0$ isotopes and can in principle be purified to contain only $I = 0$ isotopes. Because of the advanced state of Si materials technology and the tremendous effort currently underway in Si nanofabrication, Si is the obvious choice for the semiconductor host.

The only $I = 1/2$ shallow (group V) donor in Si is ^{31}P. The Si:^{31}P system was exhaustively studied 40 years ago in the first electron–nuclear double-resonance experiments[22,23]. At sufficiently low ^{31}P concentrations at temperature $T = 1.5$ K, the electron spin relaxation time is thousands of seconds and the ^{31}P nuclear spin relaxation time exceeds 10 hours. It is likely that at millikelvin temperatures the phonon limited ^{31}P relaxation time is of the order of 10^{18} seconds (ref. 24), making this system ideal for quantum computation.

The purpose of the electrons in the computer is to mediate nuclear spin interactions and to facilitate measurement of the nuclear spins. Irreversible interactions between electron and nuclear spins must not occur as the computation proceeds: the electrons must be in a non-degenerate ground state throughout the computation. At sufficiently low temperatures, electrons only occupy the lowest energy-bound state at the donor, whose twofold spin degeneracy is broken by an applied magnetic field B. (The valley degeneracy of the Si conduction band is broken in the vicinity of the donor[25]. The lowest donor excited state is approximately 15 meV above the ground state[23].) The electrons will only occupy the lowest energy spin level when $2\mu_B B \gg kT$, where μ_B is the Bohr magneton. (In Si, the Landé

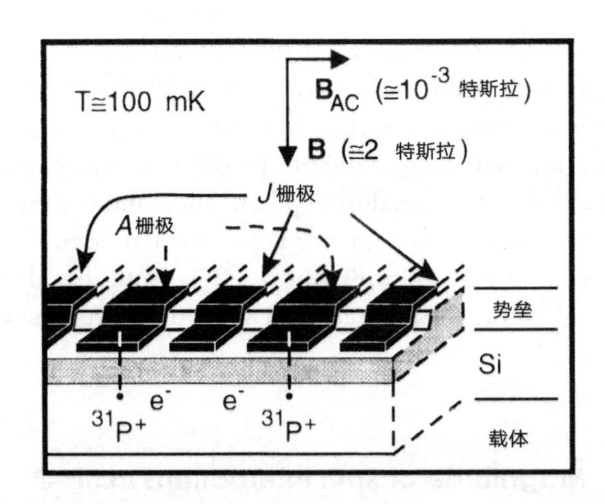

图 1. 硅宿主中 ^{31}P 施主和电子的一维阵列中的两个单元的示意图，图中的势垒层将该阵列与表面的金属栅极隔离开。“A 栅极”用来控制核自旋量子比特的共振频率。“J 栅极”用来控制相邻核自旋的电子介导耦合。栅极末端的台阶状设计使得加在施主附近的电场局域化。

对于量子计算机来说，一个重要的要求是量子比特应该与可能导致退相干的任何自由度隔离开来。如果量子比特是在半导体中的施主上的自旋，施主自旋可以和宿主材料中其他大量核自旋构成的群体相互作用。因此宿主应该只含有自旋 $I=0$ 的原子核。这个简单的要求遗憾地排除了所有 III–V 半导体作为宿主的可能性，因为他们中的任何元素都没有稳定的 $I=0$ 的同位素 [21]。而 IV 半导体主要包含 $I=0$ 同位素，而且理论上可以被提纯到只含有 $I=0$ 的同位素。由于硅材料技术的领先状态以及现阶段人们对硅纳米加工方面的巨大投入，硅理所当然地成为半导体宿主的选择。

硅中唯一的 $I=1/2$ 浅施主（第五主族）是 ^{31}P。在 40 年以前人们发现的第一个电子–原子核双共振的试验中 [22,23]Si:^{31}P 系统已经得到了深入的研究。温度 $T=1.5$ K 时，当 ^{31}P 的浓度足够低时，电子自旋的弛豫时间是几千秒，^{31}P 的核自旋弛豫时间超过 10 小时。这样的话，在毫开尔文的温度下，声子散射受限的 ^{31}P 的弛豫时间在 10^{18} 秒的量级（文献 24），使得这种系统成为量子计算的理想选择。

该计算机利用电子来促成核自旋之间的相互作用以及核自旋的测量。电子和核自旋之间不可逆的相互作用必须在计算过程中避免：电子必须在整个计算过程中处于非简并基态。在足够低的温度下，通过外加磁场 B 打破电子基态的双重自旋简并度，这样电子只占据施主的最低能量束缚态。（硅的导带的谷简并在施主附近被破坏了 [25]。施主的最低激发态大概是基态上方 15 meV[23]。）当 $2\mu_B B \gg kT$ 时，电子仅占据最低自旋能级，其中 μ_B 是波尔磁子。（在硅中，朗德 g 因子非常接近 +2，所以在我

g-factor is very close to $+2$, so $g=2$ is used throughout this discussion.) The electrons will be completely spin-polarized ($n_{\uparrow}/n_{\downarrow}<10^{-6}$) when $T\leqslant 100$ mK and $B\geqslant 2$ tesla. A quantum-mechanical computer is non-dissipative and can consequently operate at low temperatures. Dissipation will arise external to the computer from gate biasing and from eddy currents caused by B_{ac}, and during polarization and measurement of the nuclear spins. These effects will determine the minimum operable temperature of the computer. For this discussion, I will assume $T=100$ mK and $B=2$ T. Note that these conditions do not fully polarize the nuclear spins, which are instead aligned by interactions with the polarized electrons.

Magnitude of Spin Interactions in Si:^{31}P

The size of the interactions between spins determines both the time required to do elementary operations on the qubits and the separation necessary between donors in the array. The hamiltonian for a nuclear spin–electron system in Si, applicable for an $I=1/2$ donor nucleus and with $B\|z$ is $H_{en}=\mu_B B\sigma_z^e - g_n\mu_n B\sigma_z^n + A\sigma^e\cdot\sigma^n$, where σ are the Pauli spin matrices (with eigenvalues ± 1), μ_n is the nuclear magneton, g_n is the nuclear g-factor (1.13 for ^{31}P; ref. 21), and $A=\frac{8}{3}\pi\mu_B g_n\mu_n|\Psi(0)|^2$ is the contact hyperfine interaction energy, with $|\Psi(0)|^2$, the probability density of the electron wavefunction, evaluated at the nucleus. If the electron is in its ground state, the frequency separation of the nuclear levels is, to second order

$$h\nu_A = 2g_n\mu_n B + 2A + \frac{2A^2}{\mu_B B} \tag{1}$$

In Si:^{31}P, $2A/h=58$ MHz, and the second term in equation (1) exceeds the first term for $B<3.5$ T.

An electric field applied to the electron–donor system shifts the electron wavefunction envelope away from the nucleus and reduces the hyperfine interaction. The size of this shift, following estimates of Kohn[25] of shallow donor Stark shifts in Si, is shown in Fig. 2 for a donor 200 Å beneath a gate. A donor nuclear spin–electron system close to an "A gate" functions as a voltage-controlled oscillator: the precession frequency of the nuclear spin is controllable externally, and spins can be selectively brought into resonance with B_{ac}, allowing arbitrary rotations to be performed on each nuclear spin.

们的讨论中使用 $g=2$。) 当 $T \leqslant 100$ mK 且 $B \geqslant 2$ T 的时候，电子将会被完全自旋极化 ($n_{\uparrow}/n_{\downarrow} < 10^{-6}$)。量子力学计算机是无耗散的，因此可以在低温状态下运行。然而当栅极偏压出现以及 B_{ac} 引起的涡电流出现时，计算机外部的功耗会上升，这在核自旋的极化和测量时也会出现。这些效应决定了计算机的最低工作温度。为了讨论这些问题，我们假设 $T=100$ mK 且 $B=2$ T。注意这些条件并不是完全极化了核自旋，核自旋是通过和极化电子的作用来对齐的。

在 Si:^{31}P 中的自旋相互作用的量级

自旋相互作用的大小决定了量子比特基本运算所需要的时间以及阵列中施主之间需要的间隔。硅中的核自旋–电子系统的哈密顿量，在施主原子核 $I=1/2$、磁场 B 方向平行于 z 方向时，为 $H_{en}=\mu_B B\sigma e_z - g_n\mu_n B\sigma e_z + A\sigma^e \cdot \sigma^n$，这里 σ 是泡利自旋矩阵 (本征值为 ± 1)，μ_n 是核磁子，g_n 是原子核的 g 因子 (对于 ^{31}P 而言，g 因子为 1.13；文献 21)，$A=\frac{8}{3}\pi\mu_B g_n\mu_n|\Psi(0)|^2$ 是接触超精细相互作用能，其中 $|\Psi(0)|^2$ 是电子波函数在原子核的概率密度。如果电子处于基态，那么原子核能级的频率间隔 (取到二阶) 是，

$$h\nu_A = 2g_n\mu_n B + 2A + \frac{2A^2}{\mu_B B} \tag{1}$$

在 Si:^{31}P 中，$2A/h = 58$ MHz；在 $B < 3.5$ T 的时候，公式 (1) 中的第二项大于第一项。

一个施加于电子–施主系统的外加电场会使得电子波函数的包络线向远离原子核的方向平移从而减少超精细相互作用。根据科恩 [25] 对硅中浅施主斯塔克位移的估测，图 2 给出了栅极下方 200 Å 处施主上述位移的大小。靠近“A 栅极”的施主核自旋–电子系统起到电压控制振荡器的作用：这样核自旋的进动频率就可以外部控制了，自旋也可以选择性地与 B_{ac} 共振，这就使得每一个核自旋可以任意地旋转。

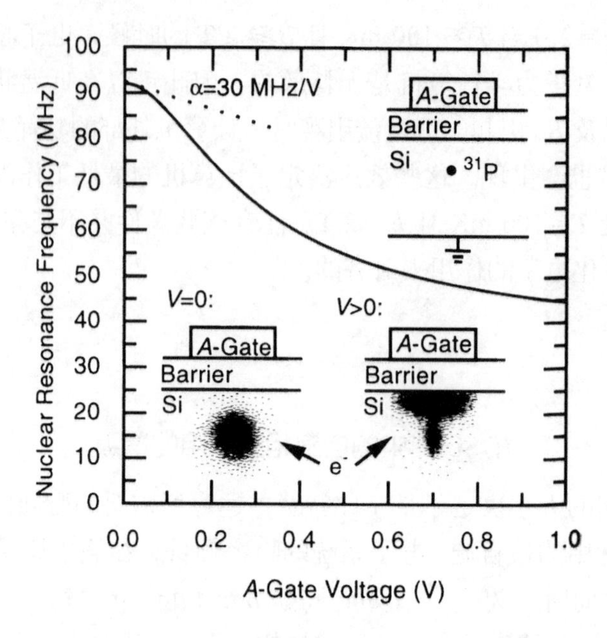

Fig. 2. An electric field applied to an A gate pulls the electron wavefunction away from the donor and towards the barrier, reducing the hyperfine interaction and the resonance frequency of the nucleus. The donor nucleus–electron system is a voltage-controlled oscillator with a tuning parameter α of the order of 30 MHz V^{-1}.

Quantum mechanical computation requires, in addition to single spin rotations, the two-qubit "controlled rotation" operation, which rotates the spin of a target qubit through a prescribed angle if, and only if, the control qubit is oriented in a specified direction, and leaves the orientation of the control qubit unchanged[26,27]. Performing the controlled rotation operation requires nuclear-spin exchange between two donor nucleus–electron spin systems[13], which will arise from electron-mediated interactions when the donors are sufficiently close to each other. The hamiltonian of two coupled donor nucleus–electron systems, valid at energy scales small compared to the donor–electron binding energy, is $H = H(B) + A_1\sigma^{1n}\cdot\sigma^{2e} + A_2\sigma^{2n}\cdot\sigma^{2e} + J\sigma^{1e}\cdot\sigma^{2e}$, where $H(B)$ are the magnetic field interaction terms for the spins. A_1 and A_2 are the hyperfine interaction energies of the respective nucleus–electron systems. $4J$, the exchange energy, depends on the overlap of the electron wavefunctions. For well separated donors[28]

$$4J(r) \cong 1.6\,\frac{e^2}{\epsilon a_B}\left(\frac{r}{a_B}\right)^{\frac{5}{2}}\exp\left(\frac{-2r}{a_B}\right) \tag{2}$$

where r is the distance between donors, ϵ is the dielectric constant of the semiconductor, and a_B is the semiconductor Bohr radius. This function, with values appropriate for Si, is plotted in Fig. 3. Equation (2), originally derived for H atoms, is complicated in Si by its valley degenerate anisotropic band structure[29]. Exchange coupling terms from each valley interfere, leading to oscillatory behaviour of $J(r)$. In this discussion, the complications introduced by Si band structure will be neglected. In determining $J(r)$ in Fig. 3, the transverse

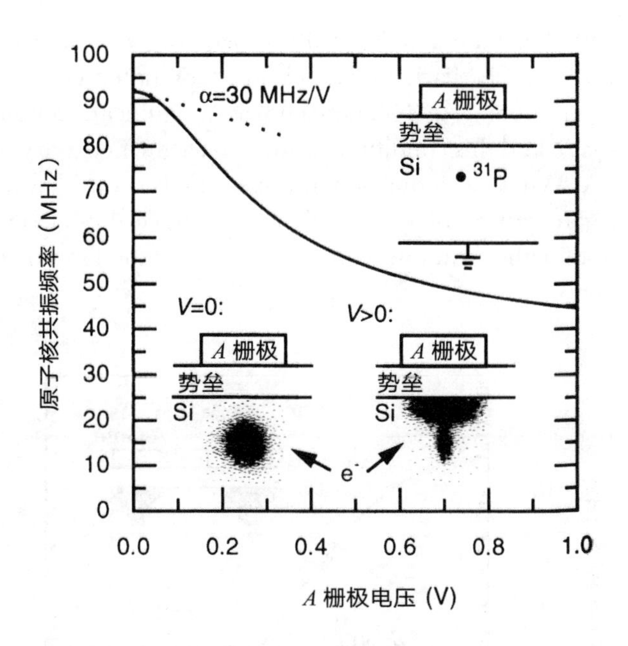

图 2. 施加于 A 栅极的电场使得电子波函数远离施主而朝向势垒，这就减少了超精细相互作用以及原子核的共振频率。施主的原子核–电子系统是一个电压调节的振荡器，调谐参数 α 量级为 30 MHz · V^{-1}。

除了单个自旋的旋转，量子力学计算需要双量子比特的“控制旋转”操作，当且仅当控制量子比特的取向在特定方向时，该操作将会使目标量子比特的自旋按照指定的角度旋转，从而使得控制量子比特的取向不变[26,27]。要实现这种控制旋转操作需要两个施主原子核–电子自旋系统中的核自旋交换[13]，当两个施主足够接近的时候，这将会由于电子介导相互作用而发生。当能量尺度与施主–电子结合能相比较小的时候，两个耦合的施主原子核–电子系统的哈密顿量可以表示为 $H = H(B) + A_1\sigma^{1n}\cdot\sigma^{2e} + A_2\sigma^{2n}\cdot\sigma^{2e} + J\sigma^{1e}\cdot\sigma^{2e}$。其中 $H(B)$ 为自旋的磁场相互作用项。A_1 和 A_2 为两个原子核–电子系统中各自的超精细相互作用能量。$4J$ 为依赖于电子波函数重叠的交换能。对于足够隔离的施主[28]而言，

$$4J(r) \cong 1.6\frac{e^2}{\epsilon a_{\mathrm{B}}}\left(\frac{r}{a_{\mathrm{B}}}\right)^{\frac{5}{2}}\exp\left(\frac{-2r}{a_{\mathrm{B}}}\right) \tag{2}$$

其中 r 是施主之间的距离，ϵ 是半导体的介电常数，a_{B} 是半导体的玻尔半径。根据硅来选取适当的参数，这个函数的图像如图 3 所示。起初由氢原子推导出来的公式 (2) 在硅中是非常复杂的，这是因为硅的谷简并各向异性能带结构[29]。每一个谷干涉的交换耦合项都会导致 $J(r)$ 的振荡行为。在这里的讨论中，我们将忽略由于硅的能带结构导致的复杂问题。为了在图 3 中确定 $J(r)$，采用硅的横质量 ($\simeq 0.2m_e$)， mass for Si ($\cong 0.2m_e$) has been used, and $a_B = 30$ Å. Because J is proportional to the electron wave function overlap, it can be varied by an electrostatic potential imposed by a "J-gate" positioned between the donors[13]. As shall be seen below, significant coupling between nuclei will occur when $4J \approx \mu_B B$, and this condition approximates the necessary separation between donors of 100–200 Å. Whereas actual separations may be considerably larger than this value because the J gate can be biased positively to reduce the barrier between donors, the gate sizes required for the quantum computer are near the limit of current electronics fabrication technology.

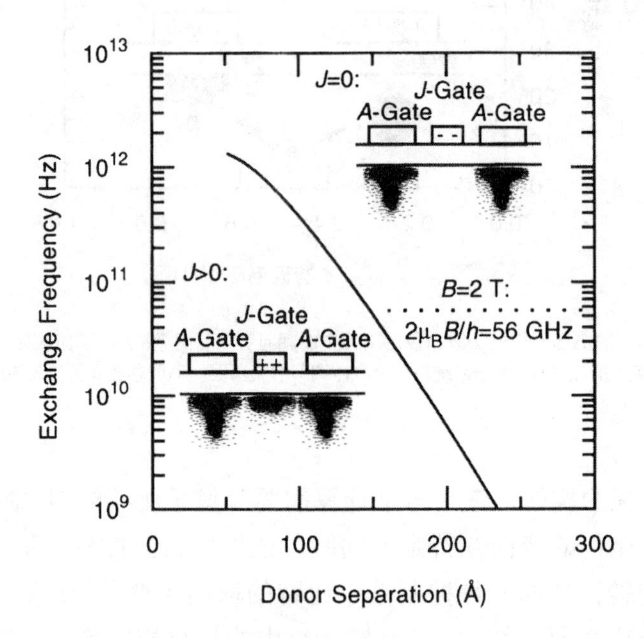

Fig. 3. J gates vary the electrostatic potential barrier V between donors to enhance or reduce exchange coupling, proportional to the electron wavefunction overlap. The exchange frequency ($4J/h$) when $V=0$ is plotted for Si.

For two-electron systems, the exchange interaction lowers the electron singlet ($|\uparrow\downarrow - \downarrow\uparrow\rangle$) energy with respect to the triplets[30]. (The $|\uparrow\downarrow\rangle$ notation is used here to represent the electron spin state, and the $|01\rangle$ notation the nuclear state; in the $|\downarrow\downarrow 11\rangle$ state, all spins point in the same direction. For simplicity, normalization constants are omitted.) In a magnetic field, however, $|\downarrow\downarrow\rangle$ will be the ground state if $J < \mu_B B/2$ (Fig. 4a). In the $|\downarrow\downarrow\rangle$ state, the energies of the nuclear states can be calculated to second order in A using perturbation theory. When $A_1 = A_2 = A$, the $|10-01\rangle$ state is lowered in energy with respect to $|10+01\rangle$ by:

$$h\nu_J = 2A^2\left(\frac{1}{\mu_B B - 2J} - \frac{1}{\mu_B B}\right) \tag{3}$$

The $|11\rangle$ state is above the $|10+01\rangle$ state and the $|00\rangle$ state below the $|10-01\rangle$ state by an energy $h\nu_A$, given in equation (1). For the Si:^{31}P system at $B = 2$ T and for $4J/h = 30$ GHz, equation (3) yields $\nu_J = 75$ kHz. This nuclear spin exchange frequency approximates the rate at which binary operations can be performed on the computer (ν_J can be increased

$a_B = 30$ Å。因为 J 正比于电子波函数的重叠，它会通过施主之间的"J 栅极"所产生的静电势而变化[13]。如下所示，当 $4J \approx \mu_B B$ 的时候，核之间会产生显著的耦合，这个已经逼近了施主之间的最小要求间距 100 ~ 200 Å。然而实际上的间距可能远大于这个值，因为 J 栅极可以被施加正偏压来减少施主之间的势垒，量子计算机所需要的栅极的尺寸已经接近了当今电子学微加工技术的极限。

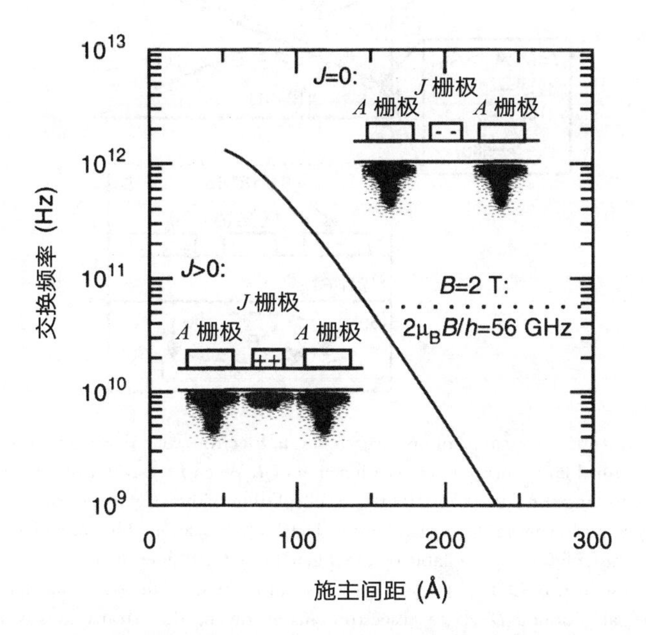

图 3. J 栅极可以改变施主之间的静电势垒 V，以此来增强或者减弱交换耦合，正比于电子波函数的重叠。当 $V=0$ 时，硅的交换频率 $(4J/h)$ 如图所示。

对于双电子系统来说，交换作用使得电子的单重态 $(|\uparrow\downarrow - \downarrow\uparrow\rangle)$ 能量相对于三重态能量降低[30]。($|\uparrow\downarrow\rangle$ 符号被用来代表电子自旋态，$|01\rangle$ 符号表示核自旋态；在 $|\downarrow\downarrow 11\rangle$ 状态下，所有自旋在同一方向。为了简化，归一化的常数被省略)。然而在磁场存在时，当 $J < \mu_B B/2$ 时（如图 4a 所示），$|\downarrow\downarrow\rangle$ 将会是电子的基态。在 $|\downarrow\downarrow\rangle$ 态中，核态的能量可以用微扰理论取到 A 的二阶项来计算。当 $A_1 = A_2 = A$ 时，$|10-01\rangle$ 态的能量相比 $|10+01\rangle$ 态的能量小了：

$$h\nu_J = 2A^2\left(\frac{1}{\mu_B B - 2J} - \frac{1}{\mu_B B}\right) \tag{3}$$

$|11\rangle$ 态在 $|10+01\rangle$ 态上方距离能量 $h\nu_A$ 处，而 $|00\rangle$ 在 $|10-01\rangle$ 下方距离能量 $h\nu_A$ 处，由公式（1）给出。对于 Si:^{31}P 系统，在 $B = 2$ T 且 $4J/h = 30$ GHz 时，公式（3）给出解为 $\nu_J = 75$ kHz。这个核自旋交换频率近似于计算机进行二进制运算的频率（ν_J 可以

by increasing J, but at the expense of also increasing the relaxation rate of the coupled nuclear–electron spin excitations). The speed of single spin operations is determined by the size of B_{ac} and is comparable to 75 kHz when $B_{ac} = 10^{-3}$ T.

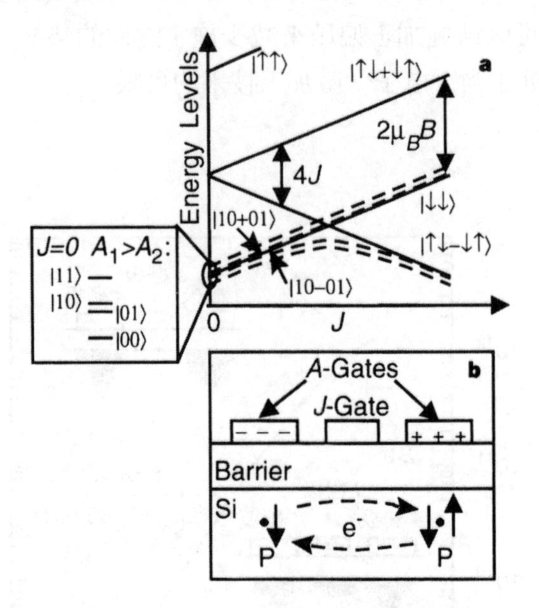

Fig. 4. Two qubit quantum logic and spin measurement. **a**, Electron (solid lines) and lowest energy-coupled electron–nuclear (dashed lines) energy levels as a function of J. When $J < \mu_B B/2$, two qubit computations are performed by controlling the $|10-01\rangle - |10+01\rangle$ level splitting with a J gate. Above $J = \mu_B B/2$, the states of the coupled system evolve into states of differing electron polarization. The state of the nucleus at $J = 0$ with the larger energy splitting (controllable by the A gate bias) determines the final electron spin state after an adiabatic increase in J. **b**, Only $|\uparrow\downarrow - \downarrow\uparrow\rangle$ electrons can make transitions into states in which electrons are bound to the same donor (D^- states). Electron current during these transitions is measurable using capacitive techniques, enabling the underlying spin states of the electrons and nuclei to be determined.

Spin Measurements

Measurement of nuclear spins in the proposed quantum computer is accomplished in a two-step process: distinct nuclear spin states are adiabatically converted into states with different electron polarization, and the electron spin is determined by its effect on the symmetry of the orbital wavefunction of an exchange-coupled two-electron system. A procedure for accomplishing this conversion is shown in Fig. 4. While computation is done when $J < \mu_B B/2$ and the electrons are fully polarized, measurements are made when $J > \mu_B B/2$, and $|\uparrow\downarrow - \downarrow\uparrow\rangle$ states have the lowest energy (Fig. 4a). As the electron levels cross, the $|\downarrow\downarrow\rangle$ and $|\uparrow\downarrow - \downarrow\uparrow\rangle$ states are coupled by hyperfine interactions with the nuclei. During an adiabatic increase in J, the two lower-energy nuclear spin states at $J = 0$ evolve into $|\uparrow\downarrow - \downarrow\uparrow\rangle$ states when $J > \mu_B B/2$, whereas the two higher-energy nuclear states remain $|\downarrow\downarrow\rangle$. If, at $J = 0$, $A_1 > A_2$, the orientation of nuclear spin 1 alone will determine whether the system evolves into the $|\uparrow\downarrow - \downarrow\uparrow\rangle$ or the $|\downarrow\downarrow\rangle$ state during an adiabatic increase in J.

通过增加J来增加，但是也以增加了耦合原子核–电子自旋激发的弛豫速率为代价）。单个自旋的运算速度是由B_{ac}的大小来决定的，当$B_{ac} = 10^{-3}$ T时相当于75 kHz。

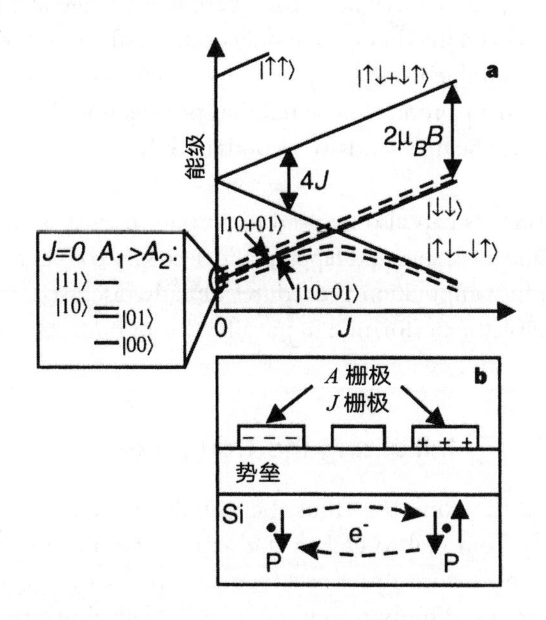

图4. 两个量子比特的逻辑和自旋测量。**a**，电子（实线）和最低能量耦合电子–核自旋（虚线）能级随J的变化。当$J < \mu_B B/2$，两个量子比特计算通过使用J栅极控制$|10-01\rangle - |10+01\rangle$能级分裂来实现。在$J = \mu_B B/2$上方，耦合系统的态演化为不同电子极化的态。原子核的具有很大能量分裂的（受A栅极偏压的控制）在$J = 0$的态决定了在J绝热增加后，电子最终的自旋态。**b**，只有$|\uparrow\downarrow - \downarrow\uparrow\rangle$电子才可以转换成电子与同一个施主结合的态（$D^-$态）。这个转换过程的电子电流是可以通过电容技术来测量的，这就使得电子和原子核的基础自旋态可以被探测。

自旋测量

在本文提出的量子计算机中，核自旋测量是分两步来实现的：不同的原子核自旋态绝热转化为不同电子极化的态，且电子自旋是由其对交换耦合双电子系统的轨道波函数对称性的影响来决定的。图4展示了如何实现这一转换的过程。当计算完成时$J < \mu_B B/2$且电子被充分地极化，测量在$J > \mu_B B/2$时进行，$|\uparrow\downarrow - \downarrow\uparrow\rangle$态有最低的能量（图4a）。当电子能级交叉时，$|\downarrow\downarrow\rangle$态和$|\uparrow\downarrow - \downarrow\uparrow\rangle$态就会通过与原子核产生超精细相互作用耦合。在$J$绝热增加过程中，当$J > \mu_B B/2$时，两个处于$J = 0$的低能量核自旋态演化为$|\uparrow\downarrow - \downarrow\uparrow\rangle$态，然而两个高能级的核自旋态仍然保持$|\downarrow\downarrow\rangle$态。如果在$J = 0$时，$A_1 > A_2$，1号原子核自旋的取向将独自决定系统在$J$的绝热增加过程中是演化为$|\uparrow\downarrow - \downarrow\uparrow\rangle$态或者是$|\downarrow\downarrow\rangle$态。

A method to detect the electron spin state by using electronic means is shown in Fig. 4b. Both electrons can become bound to the same donor (a D^- state) if the A gates above the donors are biased appropriately. In Si:P, the D^- state is always a singlet with a second electron binding energy of 1.7 meV (refs 31, 32). Consequently, a differential voltage applied to the A gates can result in charge motion between the donors that only occurs if the electrons are in a singlet state. This charge motion is measurable using sensitive single-electron capacitance techniques[33]. This approach to spin measurement produces a signal that persists until the electron spin relaxes, a time that, as noted above, can be thousands of seconds in Si:P.

The spin measurement process can also be used to prepare nuclear spins in a prescribed state by first determining the state of a spin and flipping it if necessary so that it ends up in the desired spin state. As with the spin computation procedures already discussed, spin measurement and preparation can in principle be performed in parallel throughout the computer.

Initializing the Computer

Before any computation, the computer must be initialized by calibrating the A gates and the J gates. Fluctuations from cell to cell in the gate biases necessary to perform logical operations are an inevitable consequence of variations in the positions of the donors and in the sizes of the gates. The parameters of each cell, however, can be determined individually using the measurement capabilities of the computer, because the measurement technique discussed here does not require precise knowledge of the J and A couplings. The A-gate voltage at which the underlying nuclear spin is resonant with an applied B_{ac} can be determined using the technique of adiabatic fast passage[34]: when $B_{ac} = 0$, the nuclear spin is measured and the A gate is biased at a voltage known to be off resonance. B_{ac} is then switched on, and the A gate bias is swept through a prescribed voltage interval. B_{ac} is then switched off and the nuclear spin is measured again. The spin will have flipped if, and only if, resonance occurred within the prescribed A-gate voltage range. Testing for spin flips in increasingly small voltage ranges leads to the determination of the resonance voltage. Once adjacent A gates have been calibrated, the J gates can be calibrated in a similar manner by sweeping J-gate biases across resonances of two coupled cells.

This calibration procedure can be performed in parallel on many cells, so calibration is not a fundamental impediment to scaling the computer to large sizes. Calibration voltages can be stored on capacitors located on the Si chip adjacent to the quantum computer. External controlling circuitry would thus need to control only the timing of gate biases, and not their magnitudes.

Spin Decoherence Introduced by Gates

In the quantum computer architecture outlined above, biasing of A gates and J gates enables custom control of the qubits and their mutual interactions. The presence of

一种用来探测电子自旋态的方法是使用电子手段，如图 4b 所示。如果施主上方的 A 栅极被施加适当的偏压，两个电子可以都结合到同一个施主（D^- 态）。在 Si:P 中 D^- 态总是具有 1.7 meV 第二电子结合能的单重态[31,32]。因此，施加一个差分电压于栅极 A 上可以导致只有处于单重态的电子可以在施主间产生电荷移动。这种电荷移动可以通过灵敏的单电子电容技术[33]来测量。这种测量自旋的方法产生的信号持续到电子自旋发生弛豫为止，如前所述，在 Si:P 系统中这种过程可以有几千秒的时间。

自旋的测量过程也可以用来将核自旋初始化到指定的态上：首先确定自旋的态，如果需要的话就翻转它使它停在想要的自旋态上。正如我们已经讨论的自旋计算过程，自旋的测量和准备原则上可以在计算机中并行执行。

初始化计算机

在进行任何计算之前，计算机必须通过定标 A 栅极和 J 栅极来实现初始化。施主位置以及栅极大小的不均匀必然造成不同单元执行逻辑运算所必需的栅极偏压的涨落，不过每个单元的参数都可以通过计算机的测量功能来独自地确定，因为这里讨论的测量技术不需要非常准确地了解 J 栅极和 A 栅极的耦合。A 栅极下面核自旋与外加 B_{ac} 共振时，所需要施加的 A 栅极电压可以通过绝热快速通过技术[34]来确定：当 $B_{ac}=0$ 时，核自旋被测量，且 A 栅极被施加一个能消除共振的偏压。B_{ac} 然后被启动，在指定的电压区间内扫描 A 栅极偏压。然后关闭 B_{ac} 并再次测量核自旋。当且仅当共振在指定的 A 栅极电压范围内出现的时候，自旋才会被翻转。在越来越小的电压范围内测试自旋的翻转将能确定出共振的电压。当邻近的 A 栅极被定标之后，J 栅极可以用同样的方法在两个耦合单元的共振区扫描 J 栅极偏压来定标。

这种定标程序可以在多个单元上面并行进行，所以定标问题不是将计算机扩展到大尺寸的主要阻碍。定标电压可以存储在连接到量子计算机上的硅芯片的电容中。因此，外部控制电路只需要控制栅极偏压的时间，而不是他们的大小。

栅极导致的自旋退相干

在上面概述的量子计算机的设计结构中，对 A 类和 J 类栅极施加偏压可以自定义地控制量子比特以及他们之间的相互作用。然而如果栅极的偏压振荡出期望的

the gates, however, will lead to decoherence of the spins if the gate biases fluctuate away from their desired values. These effects need to be considered to evaluate the performance of any gate-controlled quantum computer. During the computation, the largest source of decoherence is likely to arise from voltage fluctuations on the A gates. (When $J < \mu_B B/2$, modulation of the state energies by the J gates is much smaller than by the A gates. J exceeds $\mu_B B/2$ only during the measurement process, when decoherence will inevitably occur.) The precession frequencies of two spins in phase at $t = 0$ depends on the potentials on their respective A gates. Differential fluctuations of the potentials produce differences in the precession frequency. At some later time $t = t_\phi$, the spins will be 180° out of phase; t_ϕ can be estimated by determining the transition rate between $|10+01\rangle$ (spins in phase) and $|10-01\rangle$ (spins 180° out of phase) of a two-spin system. The hamiltonian that couples these states is $H_\phi = \frac{1}{4}h\Delta(\sigma_z^{1n} - \sigma_z^{2n})$, where Δ is the fluctuating differential precession frequency of the spins. Standard treatment of fluctuating hamiltonians[34] predicts: $t_\phi^{-1} = \pi^2 S_\Delta(\nu_{st})$, where S_Δ is the spectral density of the frequency fluctuations, and ν_{st} is the frequency difference between the $|10-01\rangle$ and $|10+01\rangle$ states. At a particular bias voltage, the A gates have a frequency tuning parameter $\alpha = d\Delta/dV$. Thus:

$$t_\phi^{-1} = \pi^2 \alpha^2(V) S_V(\nu_{st}) \tag{4}$$

where S_V is the spectral density of the gate voltage fluctuations.

S_V for good room temperature electronics is of order 10^{-18} V^2/Hz, comparable to the room temperature Johnson noise of a 50-Ω resistor. The value of α, estimated from Fig. 2, is 10–100 MHz V^{-1}, yielding t_ϕ = 10–1,000 s; α is determined by the size of the donor array cells and cannot readily be reduced (to increase t_ϕ) without reducing the exchange interaction between cells. Because α is a function of the gate bias (Fig. 2), t_ϕ can be increased by minimizing the voltage applied to the A gates.

Although equation (4) is valid for white noise, at low frequencies it is likely that materials-dependent fluctuations ($1/f$ noise) will be the dominant cause of spin dephasing. Consequently, it is difficult to give hard estimates of t_ϕ for the computer. Charge fluctuations within the computer (arising from fluctuating occupancies of traps and surface states, for example) are likely to be particularly important, and minimizing them will place great demands on computer fabrication.

Although materials-dependent fluctuations are difficult to estimate, the low-temperature operations of the computer and the dissipationless nature of quantum computing mean that, in principle, fluctuations can be kept extremely small: using low-temperature electronics to bias the gates (for instance, by using on chip capacitors as discussed above) could produce $t_\phi \approx 10^6$ s. Electronically controlled nuclear spin quantum computers thus have the theoretical capability to perform at least 10^5 to perhaps 10^{10} logical operations during t_ϕ, and can probably meet Preskill's criterion[8] for an error probability of 10^{-6} per qubit operation.

范围，栅极的存在将导致自旋退相干。这些效应应该在任何通过栅极控制来实现的量子计算机的性能评价中被考虑到。在计算过程中，导致退相干的主要来源一般是 A 栅极上的电压振荡。(当 $J<\mu_B B/2$ 时，态能量被 J 栅极调制的幅度远小于被 A 栅极调制的幅度。J 超过 $\mu_B B/2$ 只会发生在测量过程，那时退相干是不可避免的。) 在 $t=0$ 时刻同相位中的两个自旋的进动频率取决于他们各自 A 栅极上的电势。这两个电势的差值波动引起进动频率的相对变化。在稍后的 $t=t_\phi$ 时刻，这两个自旋的相位将会相差 180 度(反相位)；t_ϕ 可以通过这个双自旋系统在 $|10+01\rangle$(自旋同相位)和 $|10-01\rangle$(自旋 180 度反相位)两种状态间的转换速率来确定。这些态之间耦合的哈密顿量为 $H_\phi=\frac{1}{4}h\Delta(\sigma_z^{1n}-\sigma_z^{2n})$，其中 Δ 是不同自旋进动频率差值的振荡。对该振荡哈密顿量进行标准处理[34] 导出：$t_\phi^{-1}=\pi^2 S_\Delta(\nu_{st})$，其中 S_Δ 是频率振荡的频谱密度，ν_{st} 是 $|10-01\rangle$ 和 $|10+01\rangle$ 态之间的频率差。在一个固定的偏压下，A 类栅极有一个频率调谐参数 $\alpha=\mathrm{d}\Delta/\mathrm{d}V$。因此

$$t_\phi^{-1}=\pi^2\alpha^2(V)S_V(\nu_{st}) \tag{4}$$

其中 S_V 是栅极电压涨落的频谱密度。

良好的室温电子元件的 S_V 的量级是 $10^{-18}\ \mathrm{V^2/Hz}$，这和一个 50 Ω 的电阻的室温约翰逊噪音是接近的。从图 2 中估算出的 α 值大概为 10 ~ 100 $\mathrm{MHz\cdot V^{-1}}$，导出 $t_\phi=10\sim1{,}000$ s；α 可以由施主阵列单元的大小来确定，且不能在不减小单元之间的相互作用的情况下被轻易地减小(来增加 t_ϕ)。因为 α 是栅极偏压的函数(图 2)，t_ϕ 可以通过最小化 A 类栅极偏压来增加。

虽然公式 4 对白噪声来说是正确的，在低频率条件下，导致自旋退相的主要原因很可能是依赖于材料的波动($1/f$ 噪音)。因此给出计算机一个准确的预期值 t_ϕ 是非常困难的。计算机内部的电荷涨落(例如由陷阱和表面态的占位数涨落引起)很可能是非常重要的，这就要求在计算机制造中最小化这种电荷涨落。

虽然依赖材料的波动很难预期，在低温条件下的计算机运算以及量子计算无耗散的特征意味着，原则上波动可以被保持得很小：使用低温电子器件来给栅极施加偏压(例如之前讨论中提到的使用芯片电容)可以使得 t_ϕ 约等于 10^6 s。电子器件控制的核自旋量子计算机因此理论上具备了在 t_ϕ 时间内执行 10^5 到 10^{10} 逻辑运算的能力，这样基本上就满足了普雷斯基尔的每个量子比特运算误码率为 10^{-6} 的准则[8]。

Constructing the Computer

Building the computer presented here will obviously be an extraordinary challenge: the materials must be almost completely free of spin ($I \neq 0$ isotopes) and charge impurities to prevent dephasing fluctuations from arising within the computer. Donors must be introduced into the material in an ordered array hundreds of Å beneath the surface. Finally, gates with lateral dimensions and separations ~100 Å must be patterned on the surface, registered to the donors beneath them. Although it is possible that the computer can use SiO_2 as the barrier material (the standard MOS technology used in most current conventional electronics), the need to reduce disorder and fluctuations to a minimum means that heteroepitaxial materials, such as Si/SiGe, may ultimately be preferable to Si/SiO_2.

The most obvious obstacle to building to the quantum computer presented above is the incorporation of the donor array into the Si layer beneath the barrier layer. Currently, semiconductor structures are deposited layer by layer. The δ-doping technique produces donors lying on a plane in the material, with the donors randomly distributed within the plane. The quantum computer envisaged here requires that the donors be placed into an ordered one- or two-dimensional array; furthermore, precisely one donor must be placed into each array cell, making it extremely difficult to create the array by using lithography and ion implantation or by focused deposition. Methods currently under development to place single atoms on surfaces using ultra-high-vacuum scanning tunnelling microscopy[35] or atom optics techniques[36] are likely candidates to be used to position the donor array. A challenge will be to grow high-quality Si layers on the surface subsequent to placement of the donors.

Fabricating large arrays of donors may prove to be difficult, but two-spin devices, which can be used to test the logical operations and measurement techniques presented here, can be made using random doping techniques. Although only a small fraction of such devices will work properly, adjacent conventional Si electronic multiplexing circuitry can be used to examine many devices separately. The relative ease of fabricating such "hybrid"(quantum–conventional) circuits is a particularly attractive feature of Si-based quantum computation.

In a Si-based nuclear spin quantum computer, the highly coherent quantum states necessary for quantum computation are incorporated into a material in which the ability to implement complex computer architectures is well established. The substantial challenges facing the realization of the computer, particularly in fabricating 100-Å-scale gated devices, are similar to those facing the next generation of conventional electronics; consequently, new manufacturing technologies being developed for conventional electronics will bear directly on efforts to develop a quantum computer in Si. Quantum computers sufficiently complex that they can achieve their theoretical potential may thus one day be built using the same technology that is used to produce conventional computers.

(**393**, 133-137; 1998)

构造计算机

构造一个这里讨论的计算机显然将是非常巨大的挑战：选取的材料必须完全没有自旋($I \neq 0$同位素)和电荷杂质，以避免计算机中出现退相位扰动。施主必须作为一个有序的阵列导入材料表面几百埃以下。最后，横向尺寸和间距为100 Å左右的栅极被加工在表面，与其下方的施主对应。虽然使用二氧化硅(大多数当前传统电子器件所使用的标准MOS技术)作为势垒层材料是可行的，但是由于需要最小化无序和扰动，意味着异质外延材料如硅/硅锗(Si/SiGe)最终可能优于硅/二氧化硅(Si/SiO_2)。

构造上文描述的量子计算机最明显的困难是如何将施主阵列注入势垒层下方的硅层中。现阶段半导体结构是逐层沉积制造的。δ掺杂技术得到的施主位于材料中的单层中，其中施主是随机分布在这一层中的。我们这里设想的量子计算机要求施主被放置于有序的一维或者二维的阵列中；另外，施主必须被准确地放置在每个阵列单元中，这就使得通过光刻技术和离子注入或者通过聚焦沉积方法都很难制造出这种阵列。目前正在发展的将单原子放在表面的技术是利用超高真空扫描隧道显微镜[35]或者原子光学技术[36]实现的，这些是可能被用来放置施主阵列的备选方法。接着的一个挑战将是在施主放置处表面上生长高质量的硅层。

制造大的施主阵列也许是非常困难的，但是用来检验此处介绍的逻辑计算和测量技术的双自旋器件可以使用随机掺杂技术来制造。虽然只有一小部分这样的器件可以正常工作，但是相关的传统硅电子学中的复用电路可以用来独立地检测许多器件。制造这种“混合”(量子–传统)电路的相对简易性是硅基量子计算机的独特魅力所在。

在基于硅的核自旋量子计算机中，量子计算所必需的高度相干的量子态被整合到能够实现复杂计算机结构的材料中。实现量子计算机面临的重要挑战，尤其是100 Å尺度的栅极器件制造，与下一代传统电子器件面临的挑战类似；因此，正为传统电子器件研发的新制造技术将与开发硅量子计算机的努力直接相关。足够复杂以至于可以实现其理论潜力的量子计算机，也许会在将来的某一天使用传统计算机类似的技术搭建起来。

(姜克 翻译；杜江峰 审稿)

B. E. Kane
Semiconductor Nanofabrication Facility, School of Physics, University of New South Wales, Sydney 2052, Australia

Received 10 November 1997; accepted 24 February 1998.

References:

1. Steane, A. Quantum computing. *Rep. Prog. Phys.* **61,** 117-173 (1998).
2. Bennett, C. H. Quantum information and computation. *Physics Today* 24-30 (Oct. 1995).
3. Shor, P. W. in *Proc. 35th Annu. Symp. Foundations of Computer Science* (ed. Goldwasser, S.) 124-134 (IEEE Computer Society, Los Alamitos, CA, 1994).
4. Ekert, A. & Jozsa, R. Quantum computation and Shor's factoring algorithm. *Rev. Mod. Phys.* **68,** 733-753 (1996).
5. Grover, L. K. Quantum mechanics helps in searching for a needle in a haystack. *Phys. Rev. Lett.* **79,** 325-328 (1997).
6. Calderbank, A. R. & Shor, P. W. Good quantum error correcting codes exist. *Phys. Rev. A* **54,** 1098-1105 (1996).
7. Steane, A. M. Error correcting codes in quantum theory. *Phys. Rev. Lett.* **77,** 793-797 (1996).
8. Preskill, J. Reliable quantum computers. *Proc. R. Soc. Lond. A* **454,** 385-410 (1998).
9. Lloyd, S. A potentially realizable quantum computer. *Science* **261,** 1569-1571 (1993).
10. DiVincenzo, D. P. Quantum computation. *Science* **270,** 255-261 (1995).
11. Gershenfeld, N. A. & Chuang, I. L. Bulk spin-resonance quantum computation. *Science* **275,** 350-356 (1997).
12. Cory, D. G., Fahmy, A. F. & Havel, T. F. Ensemble quantum computing by NMR spectroscopy. *Proc. Natl Acad. Sci. USA* **94,** 1634-1639 (1997).
13. Loss, D. & DiVincenzo, D. P. Quantum computation with quantum dots. *Phys. Rev. A* **57,** 120-126 (1998).
14. Privman, V., Vagner, I. D. & Kventsel, G. Quantum computation in quantum Hall systems. *Phys. Lett. A* **239,** 141-146 (1998).
15. Slichter, C. P. *Principles of Magnetic Resonance* 3rd edn, Ch 4 (Springer, Berlin, 1990).
16. Dobers, M., Klitzing, K. v., Schneider, J., Weimann, G. & Ploog, K. Electrical detection of nuclear magnetic resonance in GaAs-Al_xGa_{1-x}As heterostructures. *Phys. Rev. Lett.* **61,** 1650-1653 (1988).
17. Stich, B., Greulich-Weber, S. & Spaeth, J.-M. Electrical detection of electron nuclear double resonance in silicon. *Appl. Phys. Lett.* **68,** 1102-1104 (1996).
18. Kane, B. E., Pfeiffer, L. N. & West, K. W. Evidence for an electric-field-induced phase transition in a spin-polarized two-dimensional electron gas. *Phys. Rev. B* **46,** 7264-7267 (1992).
19. Wald, K. W., Kouwenhoven, L. P., McEuen, P. L., van der Vaart, N. C. & Foxon, C. T. Local dynamic nuclear polarization using quantum point contacts. *Phys. Rev. Lett.* **73,** 1011-1014 (1994).
20. Dixon, D. C., Wald, K. R., McEuen, P. L. & Melloch, M. R. Dynamic polarization at the edge of a two-dimensional electron gas. *Phys. Rev. B* **56,** 4743-4750 (1997).
21. *CRC Handbook of Chemistry and Physics* 77th edn 11-38 (CRC Press, Boca Raton, Florida, 1996).
22. Feher, G. Electron spin resonance on donors in silicon. I. Electronic structure of donors by the electron nuclear double resonance technique. *Phys. Rev.* **114,** 1219-1244 (1959).
23. Wilson, D. K. & Feher, G. Electron spin resonance experiments on donors in silicon. III. Investigation of excited states by the application of uniaxial stress and their importance in relaxation processes. *Phys. Rev.* **124,** 1068-1083 (1961).
24. Waugh, J. S. & Slichter, C. P. Mechanism of nuclear spin-lattice relaxation in insulators at very low temperatures. *Phys. Rev. B* **37,** 4337-4339 (1988).
25. Kohn, W. *Solid State Physics* Vol. 5 (eds Seitz, F. & Turnbull, D.) 257-320 (Academic, New York, 1957).
26. DiVincenzo, D. P. Two-bit gates are universal for quantum computation. *Phys. Rev. A* **51,** 1015-1021 (1995).
27. Lloyd, S. Almost any quantum logic gate is universal. *Phys. Rev. Lett.* **75,** 346-349 (1995).
28. Herring, C. & Flicker, M. Asymptotic exchange coupling of two hydrogen atoms. *Phys. Rev.* **134,** A362–A366 (1964).
29. Andres, K., Bhatt, R. N., Goalwin, P., Rice, T. M. & Walstedt, R. E. Low-temperature magnetic susceptibility of Si:P in the nonmetallic region. *Phys. Rev. B* **24,** 244-260 (1981).
30. Ashcroft, N. W. & Mermin, N. D. in *Solid State Physics* Ch. 32 (Saunders College, Philadelphia, 1976).
31. Larsen, D. M. Stress dependence of the binding energy of D^- centers in Si. *Phys. Rev. B* **23,** 5521-5526 (1981).
32. Larsen, D. M. & McCann, S. Y. Variational studies of two- and three-dimensional D^- centers in magnetic fields. *Phys. Rev. B* **46,** 3966-3970 (1992).
33. Ashoori, R. C. Electrons in artificial atoms. *Nature* **379,** 413-419 (1996).
34. Abragam, A. *Principles of Nuclear Magnetism* (Oxford Univ. Press, London, 1961).
35. Lyding, J. W. UHV STM nanofabrication: progress, technology spin-offs, and challenges. *Proc. IEEE* **85,** 589-600 (1997).
36. Adams, C. S., Sigel, J. & Mlynek, J. Atom optics. *Phys. Rep.* **240,** 143-210 (1994).
37. Warren, W. S. The usefulness of NMR quantum computing. *Science* **277,** 1688-1690 (1997).

Acknowledgements. This work has been supported by the Australian Research Council. I thank R. G. Clark for encouragement and E. Hellman for suggesting that the work in ref. 18 could be relevant to quantum computation.

Correspondence should be addressed to the author (e-mail: kane@newt.phys.unsw.edu.au).

Collective Dynamics of "Small-world" Networks

D. J. Watts and S. H. Strogatz

Editor's Note

We live in a world of networks, from human social networks and biological food webs to neural systems, transportation networks and the modern Internet. In this paper, Duncan Watts and Steven Strogatz demonstrate that these networks often share surprising architectural similarities. Most notably, they are "small worlds", in that only a few steps are needed to move along links from any one point to another, even in networks containing an enormous number of elements. Such networks also tend to be highly "clustered", with many redundant paths, making them connected wholes that are resilient to the removal of links. This paper stimulated an explosion of work on complex networks by providing a new conceptual framework for analysing them.

Networks of coupled dynamical systems have been used to model biological oscillators[1-4], Josephson junction arrays[5,6], excitable media[7], neural networks[8-10], spatial games[11], genetic control networks[12] and many other self-organizing systems. Ordinarily, the connection topology is assumed to be either completely regular or completely random. But many biological, technological and social networks lie somewhere between these two extremes. Here we explore simple models of networks that can be tuned through this middle ground: regular networks "rewired" to introduce increasing amounts of disorder. We find that these systems can be highly clustered, like regular lattices, yet have small characteristic path lengths, like random graphs. We call them "small-world" networks, by analogy with the small-world phenomenon[13,14] (popularly known as six degrees of separation[15]). The neural network of the worm *Caenorhabditis elegans*, the power grid of the western United States, and the collaboration graph of film actors are shown to be small-world networks. Models of dynamical systems with small-world coupling display enhanced signal-propagation speed, computational power, and synchronizability. In particular, infectious diseases spread more easily in small-world networks than in regular lattices.

TO interpolate between regular and random networks, we consider the following random rewiring procedure (Fig. 1). Starting from a ring lattice with n vertices and k edges per vertex, we rewire each edge at random with probability p. This construction allows us to "tune" the graph between regularity ($p = 0$) and disorder ($p = 1$), and thereby to probe the intermediate region $0 < p < 1$, about which little is known.

“小世界”网络的集体动力学

沃茨，斯特罗加茨

编者按

我们生活在一个网络世界里：从人类社会网络和生物食物网到神经系统、运输网络和现代互联网。本文中，邓肯·沃茨和史蒂文·斯特罗加茨证明了这些网络通常具有惊人的结构相似性。最值得注意的是，它们是“小世界”，因为即使在包含大量元素的网络中，也只需要几个步骤就可以沿着链路从一个点移动到另一个点。这种网络也倾向于高度“集群化”，许多冗余路径使得它们连接在一起成为一个整体，从而对链路的移除具有恢复力。本文提供了一个新的概念框架来分析复杂网络，促进了复杂网络研究工作的爆炸式增长。

耦合动力系统网络已被广泛应用于生物振荡[1-4]、约瑟夫森结阵列[5,6]、可激发介质[7]、神经网络[8-10]、空间博弈[11]、基因控制网络[12]以及多种其他自组织系统的建模。通常情况下，其拓扑结构被认为或是完全规则的，或是完全随机的。但是许多生物、技术和社会的网络介于这两种极端状况之间。本文阐述了一种处于这种中间态的简单的网络模型：将规则网络“重新连接”来引入更多无序性。我们发现这些系统可以像有序晶格一样被高度集聚，同时具有很小的特征路径长度，类似随机图。通过与小世界现象[13,14]（人们所熟知的六度分隔[15]）的类比，我们称它们为“小世界”网络。秀丽隐杆线虫的神经网络、美国西部的电力网络以及电影演员的合作图都表现为小世界网络。具有小世界耦合的动力系统模型显示了更强的信号传播速度、计算能力和同步性。特别需要指出的是，传染病在小世界网络中传播要比在规则网络中传播更容易。

为了实现规则网络和随机网络之间的中间状态，我们来考虑以下的随机重连过程（图1）。从一个具有 n 个顶点，每个顶点 k 条边的环形晶格网络开始，我们以概率 p 随机地将每条边重新连接。这种构造允许我们在规则（$p=0$）和无序（$p=1$）之间“调谐”，从而探索目前被了解得很少的 $0<p<1$ 的中间区域。

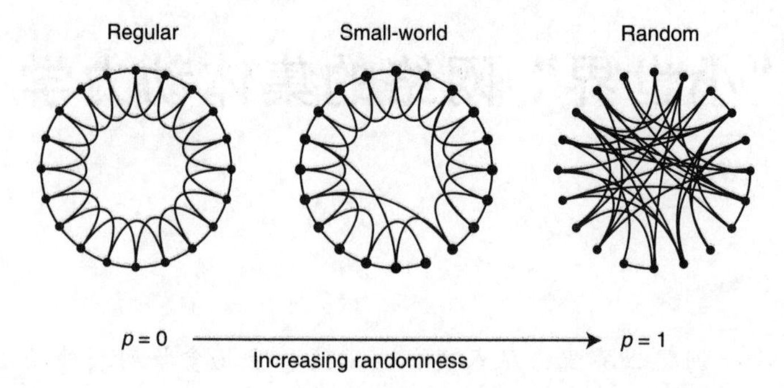

Fig. 1. Random rewiring procedure for interpolating between a regular ring lattice and a random network, without altering the number of vertices or edges in the graph. We start with a ring of n vertices, each connected to its k nearest neighbours by undirected edges. (For clarity, $n = 20$ and $k = 4$ in the schematic examples shown here, but much larger n and k are used in the rest of this Letter.) We choose a vertex and the edge that connects it to its nearest neighbour in a clockwise sense. With probability p, we reconnect this edge to a vertex chosen uniformly at random over the entire ring, with duplicate edges forbidden; otherwise we leave the edge in place. We repeat this process by moving clockwise around the ring, considering each vertex in turn until one lap is completed. Next, we consider the edges that connect vertices to their second-nearest neighbours clockwise. As before, we randomly rewire each of these edges with probability p, and continue this process, circulating around the ring and proceeding outward to more distant neighbours after each lap, until each edge in the original lattice has been considered once. (As there are $nk/2$ edges in the entire graph, the rewiring process stops after $k/2$ laps.) Three realizations of this process are shown, for different values of p. For $p = 0$, the original ring is unchanged; as p increases, the graph becomes increasingly disordered until for $p = 1$, all edges are rewired randomly. One of our main results is that for intermediate values of p, the graph is a small-world network: highly clustered like a regular graph, yet with small characteristic path length, like a random graph. (See Fig. 2.)

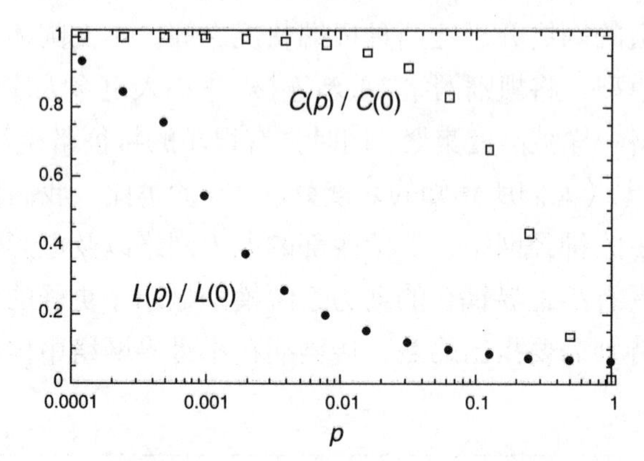

Fig. 2. Characteristic path length $L(p)$ and clustering coefficient $C(p)$ for the family of randomly rewired graphs described in Fig. 1. Here L is defined as the number of edges in the shortest path between two vertices, averaged over all pairs of vertices. The clustering coefficient $C(p)$ is defined as follows. Suppose that a vertex v has k_v neighbours; then at most $k_v (k_v - 1)/2$ edges can exist between them (this occurs when every neighbour of v is connected to every other neighbour of v). Let C_v denote the fraction of these allowable edges that actually exist. Define C as the average of C_v over all v. For friendship networks, these statistics have intuitive meanings: L is the average number of friendships in the shortest chain connecting two people; C_v reflects the extent to which friends of v are also friends of each other; and thus C measures the cliquishness of a typical friendship circle. The data shown in the figure are averages over 20 random realizations of the rewiring process described in Fig. 1, and have been normalized by the values $L(0)$, $C(0)$ for a regular lattice. All the graphs have $n = 1{,}000$ vertices and an average degree of $k = 10$ edges per vertex. We note that a logarithmic horizontal scale has been used to resolve the rapid drop in $L(p)$, corresponding to

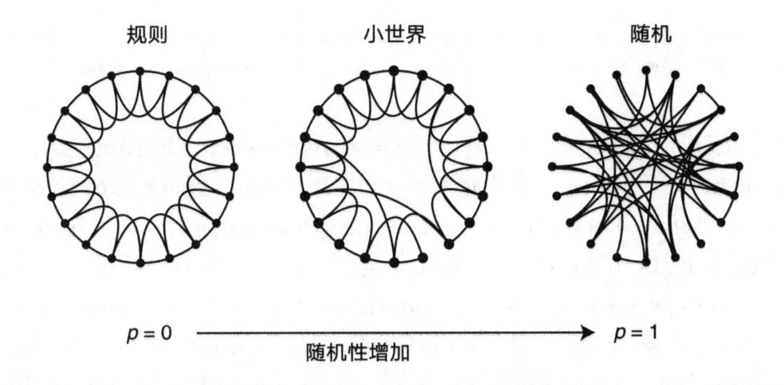

图 1. 随机重新连接来制造处于规则环形晶格网络和随机网络之间的状态，而顶点和边的数量不变。我们从一个具有 n 个顶点的环开始，每个顶点都与最近的 k 个相邻顶点通过无向边相连。(为了表述清晰，图中的例子取 $n = 20$，$k = 4$，但在文中的其他部分我们取了更大的 n 和 k 值。) 我们选取一个顶点以及顺时针方向上连接该顶点和最近邻点的边。以概率 p 将这条边重新连接到整个环上随机选取的任一顶点，不允许有重复连接的边；否则保持该边不动。我们沿环的顺时针方向重复这个过程，按顺序考虑到每个顶点直到一圈完成。接下来，我们考虑连接顶点与其顺时针方向第二近邻点的边。像上次操作一样，我们以概率 p 随机地重新连接这些边，然后我们重复这个过程，沿环循环，并在每一圈之后对连接更远近邻点的边继续该操作，直到每条原始网络中的边都被考虑到为止。(因为整个图中共有 $nk/2$ 条边，所以重新连接过程在 $k/2$ 圈后结束。) 图中表示了不同 p 时，该过程的三种实现方式。$p = 0$ 时，原始环不变；随 p 增大图形无序性开始增加，到 $p = 1$ 时，所有的边都被随机地重新连接。我们得出的一个主要结论是，对于 p 处于中间值时，图为一小世界网络：像规则图一样高度集聚，又像随机图一样具有很小的特征路径长度。(见图 2)

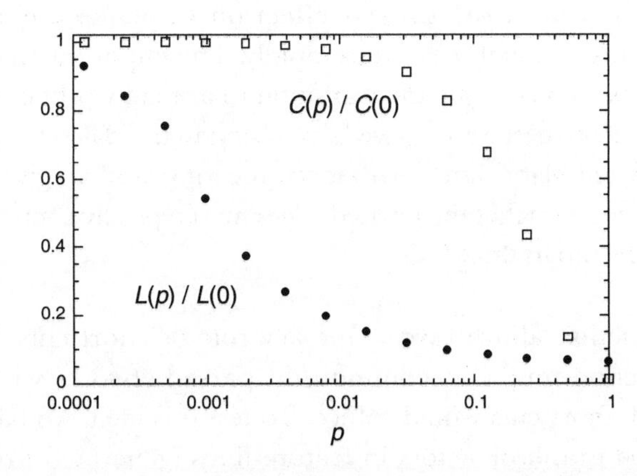

图 2. 特征路径长度 $L(p)$ 与图 1 中所描述的随机重连图形的集聚系数 $C(p)$ 的关系。L 被定义为两顶点之间最短路径上的边的数量，为所有顶点对的平均值。集聚系数 $C(p)$ 按下述定义。假设一顶点 v 有 k_v 个邻点；那么至多有 $k_v(k_v-1)/2$ 条边连接它们(这种情况发生在 v 的每个邻点都与 v 的所有其他邻点相连)。用 C_v 来表示真正存在的边占这些所有被允许出现的边的比例。定义 C 为所有的顶点 v 对应 C_v 的平均值。对于朋友关系网络，这些统计数据有着直观的含义：L 为两个人之间最短链中朋友关系的平均数；C_v 反映 v 的朋友中有多少互相之间也是朋友；因而 C 测量朋友关系圈的小集团性。图中所示的数据为图 1 中所描述的 20 种随机重连过程的平均值，并相对规则晶格网络的 $L(0)$ 和 $C(0)$ 进行了归一化。所有的图形都有 $n = 1,000$ 个顶点，每个顶点平均有 $k = 10$ 条边。我们使用对数横坐标以解决 $L(p)$ 骤降

the onset of the small-world phenomenon. During this drop, $C(p)$ remains almost constant at its value for the regular lattice, indicating that the transition to a small world is almost undetectable at the local level.

We quantify the structural properties of these graphs by their characteristic path length $L(p)$ and clustering coefficient $C(p)$, as defined in Fig. 2 legend. Here $L(p)$ measures the typical separation between two vertices in the graph (a global property), whereas $C(p)$ measures the cliquishness of a typical neighbourhood (a local property). The networks of interest to us have many vertices with sparse connections, but not so sparse that the graph is in danger of becoming disconnected. Specifically, we require $n \gg k \gg \ln(n) \gg 1$, where $k \gg \ln(n)$ guarantees that a random graph will be connected[16]. In this regime, we find that $L \sim n/2k \gg 1$ and $C \sim 3/4$ as $p \rightarrow 0$, while $L \approx L_{\text{random}} \sim \ln(n)/\ln(k)$ and $C \approx C_{\text{random}} \sim k/n \ll 1$ as $p \rightarrow 1$. Thus the regular lattice at $p = 0$ is a highly clustered, large world where L grows linearly with n, whereas the random network at $p = 1$ is a poorly clustered, small world where L grows only logarithmically with n. These limiting cases might lead one to suspect that large C is always associated with large L, and small C with small L.

On the contrary, Fig. 2 reveals that there is a broad interval of p over which $L(p)$ is almost as small as L_{random} yet $C(p) \gg C_{\text{random}}$. These small-world networks result from the immediate drop in $L(p)$ caused by the introduction of a few long-range edges. Such "short cuts" connect vertices that would otherwise be much farther apart than L_{random}. For small p, each short cut has a highly nonlinear effect on L, contracting the distance not just between the pair of vertices that it connects, but between their immediate neighbourhoods, neighbourhoods of neighbourhoods and so on. By contrast, an edge removed from a clustered neighbourhood to make a short cut has, at most, a linear effect on C; hence $C(p)$ remains practically unchanged for small p even though $L(p)$ drops rapidly. The important implication here is that at the local level (as reflected by $C(p)$), the transition to a small world is almost undetectable. To check the robustness of these results, we have tested many different types of initial regular graphs, as well as different algorithms for random rewiring, and all give qualitatively similar results. The only requirement is that the rewired edges must typically connect vertices that would otherwise be much farther apart than L_{random}.

The idealized construction above reveals the key role of short cuts. It suggests that the small-world phenomenon might be common in sparse networks with many vertices, as even a tiny fraction of short cuts would suffice. To test this idea, we have computed L and C for the collaboration graph of actors in feature films (generated from data available at http://us.imdb.com), the electrical power grid of the western United States, and the neural network of the nematode worm *C. elegans*[17]. All three graphs are of scientific interest. The graph of film actors is a surrogate for a social network[18], with the advantage of being much more easily specified. It is also akin to the graph of mathematical collaborations centred, traditionally, on P. Erdös (partial data available at http://www.acs.oakland.edu/~grossman/erdoshp.html). The graph of the power grid is relevant to the efficiency and robustness of power networks[19]. And *C. elegans* is the sole example of a completely mapped neural network.

的问题，对应小世界现象的开始。在这个骤降过程中，规则晶格网络的 $C(p)$ 基本保持不变，也就意味着网络结构向小世界的转变在局域上是无法探测的。

通过它们的特征路径长度 $L(p)$ 和集聚系数 $C(p)$（图 2 注给出了这两项的定义），我们将这些图形的结构属性进行量化。这里 $L(p)$ 测量图形中两顶点之间的分隔（一种全局属性），而 $C(p)$ 测量近邻顶点间的小集团性（一种局域属性）。我们感兴趣的网络有很多顶点但连接很稀疏，但是没有稀疏到存在使图断开的危险。具体地讲，我们要求 $n \gg k \gg \ln(n) \gg 1$，其中 $k \gg \ln(n)$ 确保了随机图会被连接[16]。在这种机制下，我们发现当 $p \to 0$ 时，$L \sim n/2k \gg 1$，$C \sim 3/4$，而当 $p \to 1$ 时，$L \approx L_{\text{随机}} \sim \ln(n)/\ln(k)$，$C \approx C_{\text{随机}} \sim k/n \ll 1$。因此在 $p = 0$ 时的规则晶格网络是一高度群聚、L 随 n 线性增加的大世界，而 $p = 1$ 时的随机网络是一低群聚、L 只随 n 对数增长的小世界。这些带有限制性的情况很可能让人们认为大的 C 永远与大 L 相关，而小 C 与小 L 相关。

相反地，图 2 揭示了在 $L(p)$ 小到与 $L_{\text{随机}}$ 接近，但 $C(p) \gg C_{\text{随机}}$ 时，p 存在一个很宽的跨度。这些小世界网络源于由若干远程边的引入而引起的 $L(p)$ 的骤降。这样的"捷径"连接起了原本要相隔较 $L_{\text{随机}}$ 远很多的顶点。对于小的 p 来说，每一条捷径都产生很强的非线性效应作用于 L，不仅仅缩短了它所连接的顶点之间的距离，而且还缩短了它们与直接邻点之间的距离，以及与邻点的邻点之间的距离，等等。相比之下，从群聚的邻点中移除用于制造捷径的一条边，最多产生线性效应作用于 C；因此对于小的 p 即使 $L(p)$ 迅速减小，$C(p)$ 也仍保持不变。这里的一个很重要的启示是，就局部而言（通过 $C(p)$ 所反应的），向小世界的转变几乎无法探测。为了检验这些结果的稳健性，我们测试了多种不同类型的初始规则图，以及随机重连的不同算法，所有的测试都给出性质类似的结果。唯一的要求是重新连接的边通常必须连接原本比 $L_{\text{随机}}$ 远很多的顶点。

以上理想的构造揭示了捷径的核心作用。它意味着小世界现象在多顶点的稀疏网络中很普遍，很少的捷径就满足要求。为了验证这个想法，我们计算了故事片电影中演员的合作图（数据由 http://us.imdb.com 提供）、美国西部的电力网络以及秀丽隐杆线虫神经网络[17]这三个图的 L 和 C。所有这三个图都引起了科学界的兴趣。电影演员合作图是社会网络的代表[18]，具有很容易被细化定义的优势。它也与传统上的、以保罗·埃尔德什为中心的数学合作图相似（部分数据可以在 http://www.acs.oakland.edu/~grossman/erdoshp.html 上获得）。电力网络的图与电力网络的效率和稳定性相关[19]。而秀丽隐杆线虫是完整的神经网络的唯一例子。

Table 1 shows that all three graphs are small-world networks. These examples were not hand-picked; they were chosen because of their inherent interest and because complete wiring diagrams were available. Thus the small-world phenomenon is not merely a curiosity of social networks[13,14] nor an artefact of an idealized model—it is probably generic for many large, sparse networks found in nature.

Table 1. Empirical examples of small-world networks

	L_{actual}	L_{random}	C_{actual}	C_{random}
Film actors	3.65	2.99	0.79	0.00027
Power grid	18.7	12.4	0.080	0.005
C. elegans	2.65	2.25	0.28	0.05

Characteristic path length L and clustering coefficient C for three real networks, compared to random graphs with the same number of vertices (n) and average number of edges per vertex (k). (Actors: $n = 225{,}226$, $k = 61$. Power grid: $n = 4{,}941$, $k = 2.67$. *C. elegans*: $n = 282$, $k = 14$.) The graphs are defined as follows. Two actors are joined by an edge if they have acted in a film together. We restrict attention to the giant connected component[16] of this graph, which includes ~90% of all actors listed in the Internet Movie Database (available at http://us.imdb.com), as of April 1997. For the power grid, vertices represent generators, transformers and substations, and edges represent high-voltage transmission lines between them. For *C. elegans*, an edge joins two neurons if they are connected by either a synapse or a gap junction. We treat all edges as undirected and unweighted, and all vertices as identical, recognizing that these are crude approximations. All three networks show the small-world phenomenon: $L \gtrsim L_{random}$ but $C \gg C_{random}$.

We now investigate the functional significance of small-world connectivity for dynamical systems. Our test case is a deliberately simplified model for the spread of an infectious disease. The population structure is modelled by the family of graphs described in Fig. 1. At time $t = 0$, a single infective individual is introduced into an otherwise healthy population. Infective individuals are removed permanently (by immunity or death) after a period of sickness that lasts one unit of dimensionless time. During this time, each infective individual can infect each of its healthy neighbours with probability r. On subsequent time steps, the disease spreads along the edges of the graph until it either infects the entire population, or it dies out, having infected some fraction of the population in the process.

Two results emerge. First, the critical infectiousness r_{half}, at which the disease infects half the population, decreases rapidly for small p (Fig. 3a). Second, for a disease that is sufficiently infectious to infect the entire population regardless of its structure, the time $T(p)$ required for global infection resembles the $L(p)$ curve (Fig. 3b). Thus, infectious diseases are predicted to spread much more easily and quickly in a small world; the alarming and less obvious point is how few short cuts are needed to make the world small.

Our model differs in some significant ways from other network models of disease spreading[20-24]. All the models indicate that network structure influences the speed and extent of disease transmission, but our model illuminates the dynamics as an explicit function of structure (Fig. 3), rather than for a few particular topologies, such as random graphs, stars and chains[20-23]. In the work closest to ours, Kretschmar and Morris[24] have shown

表 1 显示所有三个图都是小世界网络。这些例子不是特别挑选出来的；它们之所以被选择是因为它们本身具有吸引力，而且它们的连接完备清晰。因此小世界现象不只是社会网络的奇异现象[13,14]，也不是一个理想模型的人为性质——它很可能是自然中发现的许多大的、稀疏的网络的普遍现象。

表 1. 小世界网络的实际举例

	$L_{真实}$	$L_{随机}$	$C_{真实}$	$C_{随机}$
电影演员	3.65	2.99	0.79	0.00027
电力网络	18.7	12.4	0.080	0.005
秀丽隐杆线虫	2.65	2.25	0.28	0.05

三个真实网络的特征路径长度 L 和集聚系数 C，与具有相同的顶点数(n)及每顶点的平均边数(k)的随机图的比较。(电影演员：$n = 225{,}226$，$k = 61$。电力网络：$n = 4{,}941$，$k = 2.67$。秀丽隐杆线虫：$n = 282$，$k = 14$。)图的定义如下。两演员如果出演过同一电影，他们即被一边连接。我们仅关注其中的最大连通子集[16]，其中包含了截至 1997 年 4 月网络电影数据库(源自 http://us.imdb.com)所列出的所有演员的约 90%。对于电力网络，顶点代表发电机、变压器和变电站，边代表他们之间的高压电线。对于秀丽隐杆线虫，如果两神经元通过突触或者缝隙连接相连，则用边将二者连接。我们假设所有的边无方向无加权，所有的顶点都是相同的，所以这只是一种粗糙的近似。所有三个网络显示出小世界现象：$L \gtrsim L_{随机}$而 $C \gg C_{随机}$。

我们现在来研究以小世界方式连接的动力学系统的功能意义。我们的测试案例是被有意简化的一传染性疾病的传播模型。人口的相互作用结构用图 1 所描述的一系列图给出。在时间 $t = 0$ 时，一感染性个体被引入原本健康的人群。被感染的个体在患病经过一无量纲时间单位后被永久性移除(通过免疫或死亡)。在这段时间内，每个被感染的个体以概率 r 传染每个健康的邻居。在随后的时间段，疾病沿图的边传播直到把疾病传染给所有的人，或者疾病传染一部分人之后消亡。

这里出现了两个结果。首先，考查疾病传染一半人口的临界感染率 $r_{半数}$，在 p 较小时，它迅速减小(图 3a)。其次，无论什么结构，对于足以传染整个人群的疾病，整体感染所需要的时间 $T(p)$ 与 $L(p)$ 曲线相似(图 3b)。因此，传染性疾病可能会更容易、更迅速地在小世界中传播；需要注意但不太明显的一点是最少需要多少捷径可以构造这样的小世界。

我们的模型与其他疾病传播的网络模型[20-24]有很大程度的不同。其他所有的模型都显示网络结构影响疾病传播的速度和程度，但是我们的模型给出了动力学行为作为结构性质的一个明确函数关系(图 3)，而不是只考虑少数的特定拓扑结构，例如随机图、星形网络和链[20-23]。与我们最接近的研究中，克雷奇马尔和莫里斯[24]已

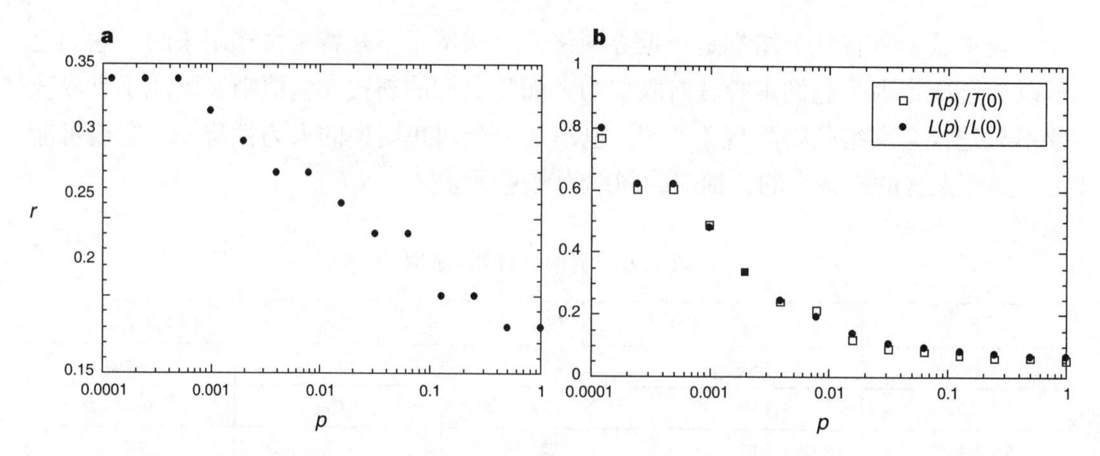

Fig. 3. Simulation results for a simple model of disease spreading. The community structure is given by one realization of the family of randomly rewired graphs used in Fig. 1. **a**, Critical infectiousness r_{half}, at which the disease infects half the population, decreases with p. **b**, The time $T(p)$ required for a maximally infectious disease ($r = 1$) to spread throughout the entire population has essentially the same functional form as the characteristic path length $L(p)$. Even if only a few per cent of the edges in the original lattice are randomly rewired, the time to global infection is nearly as short as for a random graph.

that increases in the number of concurrent partnerships can significantly accelerate the propagation of a sexually-transmitted disease that spreads along the edges of a graph. All their graphs are disconnected because they fix the average number of partners per person at $k = 1$. An increase in the number of concurrent partnerships causes faster spreading by increasing the number of vertices in the graph's largest connected component. In contrast, all our graphs are connected; hence the predicted changes in the spreading dynamics are due to more subtle structural features than changes in connectedness. Moreover, changes in the number of concurrent partners are obvious to an individual, whereas transitions leading to a smaller world are not.

We have also examined the effect of small-world connectivity on three other dynamical systems. In each case, the elements were coupled according to the family of graphs described in Fig. 1. (1) For cellular automata charged with the computational task of density classification[25], we find that a simple "majority-rule" running on a small-world graph can outperform all known human and genetic algorithm-generated rules running on a ring lattice. (2) For the iterated, multi-player "Prisoner's dilemma"[11] played on a graph, we find that as the fraction of short cuts increases, cooperation is less likely to emerge in a population of players using a generalized "tit-for-tat"[26] strategy. The likelihood of cooperative strategies evolving out of an initial cooperative/non-cooperative mix also decreases with increasing p. (3) Small-world networks of coupled phase oscillators synchronize almost as readily as in the mean-field model[2], despite having orders of magnitude fewer edges. This result may be relevant to the observed synchronization of widely separated neurons in the visual cortex[27] if, as seems plausible, the brain has a small-world architecture.

We hope that our work will stimulate further studies of small-world networks. Their distinctive combination of high clustering with short characteristic path length cannot be

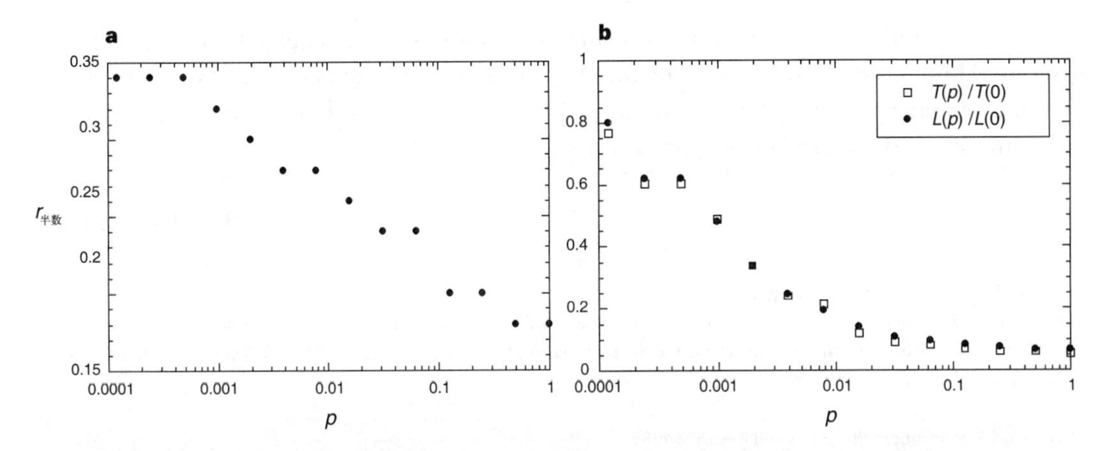

图 3. 一个简单的疾病传播模型的模拟结果。社群结构由图 1 中随机重连图族的一个实现给出。**a**，疾病感染半数人群的临界感染率 $r_{半数}$ 随 p 减小。**b**，最强感染性疾病（$r=1$）传播至整个人群所需的时间 $T(p)$ 具有与特征路径长度 $L(p)$ 相同的函数形式。即使原始晶格网络中只有几个百分比的边被重连，整体感染的时间也与随机图中的一样短。

经指出并行性伴侣关系数目的增加会加速沿图边传播的性疾病的传染。他们所有的图形都是断开的，因为他们设定了每个人的伴侣平均数为 $k=1$。并行性伴侣关系数目的增加会通过增加最大连通集团中顶点的数目引起传播加速。相比而言，我们的图是相连的；因此所预测的传播动态行为上的改变是由于更多的精细结构特征而非连通性的改变。此外，并行性伴侣关系数目的改变对于个体来说影响是明显的，但对整体结构向小世界转变的影响却不明显。

我们也测试了小世界性质在其他三个动态系统中的效应。在每种情况下，元素通过图 1 所描述的系统网络发生耦合。(1) 对于负责密度分类计算的元胞自动机[25]，我们发现一个简单的在小世界图中运行的"多数规则"可以胜过所有已知在环形晶格网络中运行的人为以及遗传算法产生的规则。(2) 对于一个基于网络迭代的、多人的"囚徒困境"[11]，我们发现随捷径数量的增加，运用"以牙还牙"[26] 战略的参与者更不容易产生合作。由原始的合作/非合作组合发展产生的合作战略的可行性也随着 p 的增加而降低。(3) 耦合相位振荡的小世界网络的同步几乎与在平均场模型[2] 中的一样容易，尽管边的数量少了几个量级。这个结果可能与所观测到的视觉皮层中被广泛分离的神经元的同步性[27] 相关，如此看来，大脑似乎也具有小世界体系结构。

我们希望本文可以促进小世界网络更深层次的研究。小世界网络同时具有高聚集度与短特征路径长度的独特特征，不能用传统的近似方法例如基于规则晶格网络

captured by traditional approximations such as those based on regular lattices or random graphs. Although small-world architecture has not received much attention, we suggest that it will probably turn out to be widespread in biological, social and man-made systems, often with important dynamical consequences.

(**393**, 440-442; 1998)

Duncan J. Watts* **& Steven H. Strogatz**
Department of Theoretical and Applied Mechanics, Kimball Hall, Cornell University, Ithaca, New York 14853, USA
* Present address: Paul F. Lazarsfeld Center for the Social Sciences, Columbia University, 812 SIPA Building, 420 W118 St, New York, New York 10027, USA.

Received 27 November 1997; accepted 6 April 1998.

References:

1. Winfree, A. T. *The Geometry of Biological Time* (Springer, New York, 1980).
2. Kuramoto, Y. *Chemical Oscillations, Waves, and Turbulence* (Springer, Berlin, 1984).
3. Strogatz, S. H. & Stewart, I. Coupled oscillators and biological synchronization. *Sci. Am.* **269**(6), 102-109 (1993).
4. Bressloff, P. C., Coombes, S. & De Souza, B. Dynamics of a ring of pulse-coupled oscillators: a group theoretic approach. *Phys. Rev. Lett.* **79**, 2791-2794 (1997).
5. Braiman, Y., Lindner, J. F. & Ditto, W. L. Taming spatiotemporal chaos with disorder. *Nature* **378**, 465-467 (1995).
6. Wiesenfeld, K. New results on frequency-locking dynamics of disordered Josephson arrays. *Physica B* **222**, 315-319 (1996).
7. Gerhardt, M., Schuster, H. & Tyson, J. J. A cellular automaton model of excitable media including curvature and dispersion. *Science* **247**, 1563-1566 (1990).
8. Collins, J. J., Chow, C. C. & Imhoff, T. T. Stochastic resonance without tuning. *Nature* **376**, 236-238 (1995).
9. Hopfield, J. J. & Herz, A. V. M. Rapid local synchronization of action potentials: Toward computation with coupled integrate-and-fire neurons. *Proc. Natl Acad. Sci. USA* **92**, 6655-6662 (1995).
10. Abbott, L. F. & van Vreeswijk, C. Asynchronous states in neural networks of pulse-coupled oscillators. *Phys. Rev. E* **48**(2), 1483-1490 (1993).
11. Nowak, M. A. & May, R. M. Evolutionary games and spatial chaos. *Nature* **359**, 826-829 (1992).
12. Kauffman, S. A. Metabolic stability and epigenesis in randomly constructed genetic nets. *J. Theor. Biol.* **22**, 437-467 (1969).
13. Milgram, S. The small world problem. *Psychol. Today* **2**, 60-67 (1967).
14. Kochen, M. (ed.) *The Small World* (Ablex, Norwood, NJ, 1989).
15. Guare, J. *Six Degrees of Separation: A Play* (Vintage Books, New York, 1990).
16. Bollabás, B. *Random Graphs* (Academic, London, 1985).
17. Achacoso, T. B. & Yamamoto, W. S. *AY's Neuroanatomy of C. elegans for Computation* (CRC Press, Boca Raton, FL, 1992).
18. Wasserman, S. & Faust, K. *Social Network Analysis: Methods and Applications* (Cambridge Univ. Press, 1994).
19. Phadke, A. G. & Thorp, J. S. *Computer Relaying for Power Systems* (Wiley, New York, 1988).
20. Sattenspiel, L. & Simon, C. P. The spread and persistence of infectious diseases in structured populations. *Math. Biosci.* **90**, 341-366 (1988).
21. Longini, I. M. Jr A mathematical model for predicting the geographic spread of new infectious agents. *Math. Biosci.* **90**, 367-383 (1988).
22. Hess, G. Disease in metapopulation models: implications for conservation. *Ecology* **77**, 1617-1632 (1996).
23. Blythe, S. P., Castillo-Chavez, C. & Palmer, J. S. Toward a unified theory of sexual mixing and pair formation. *Math. Biosci.* **107**, 379-405 (1991).
24. Kretschmar, M. & Morris, M. Measures of concurrency in networks and the spread of infectious disease. *Math. Biosci.* **133**, 165-195 (1996).
25. Das, R., Mitchell, M. & Crutchfield, J. P. in *Parallel Problem Solving from Nature* (eds Davido, Y., Schwefel, H.-P. & Männer, R.) 344-353 (Lecture Notes in Computer Science 866, Springer, Berlin, 1994).
26. Axelrod, R. *The Evolution of Cooperation* (Basic Books, New York, 1984).
27. Gray, C. M., König, P., Engel, A. K. & Singer, W. Oscillatory responses in cat visual cortex exhibit intercolumnar synchronization which reflects global stimulus properties. *Nature* **338**, 334-337 (1989).

Acknowledgements. We thank B. Tjaden for providing the film actor data, and J. Thorp and K. Bae for the Western States Power Grid data. This work was supported by the US National Science Foundation (Division of Mathematical Sciences).

Correspondence and requests for materials should be addressed to D.J.W. (e-mail: djw24@columbia.edu).

或者随机图的方法来获得。尽管小世界体系结构还没有引起很大的关注，我们推测它将在生物、社会以及人工系统等广泛存在，并且通常对系统的动力学行为有重要的影响。

（崔宁 翻译；狄增如 审稿）

An Electrophoretic Ink for All-printed Reflective Electronic Displays

B. Comiskey *et al.*

Editor's Note

The printed word retains some advantages over electronic displays. The subliminal flickering of screens tires the eye, and they are fragile, expensive, energy-hungry, and lose contrast in bright ambient light. This has stimulated a quest for "electronic paper" that combines the benefits of printed paper with the convenience and versatility of electronic information systems. Here researchers at MIT's Media Lab report one of the first candidate materials: a layer of transparent microcapsules on a flexible plastic sheet. Each capsule contains white and black pigment particles that can be drawn to the top surface by an electric field. This foundational paper led to a spinoff company, E-Ink, which, in partnership with microelectronics companies, now sells a range of portable display devices.

It has for many years been an ambition of researchers in display media to create a flexible low-cost system that is the electronic analogue of paper. In this context, microparticle-based displays[1-5] have long intrigued researchers. Switchable contrast in such displays is achieved by the electromigration of highly scattering or absorbing microparticles (in the size range 0.1–5 μm), quite distinct from the molecular-scale properties that govern the behaviour of the more familiar liquid-crystal displays[6]. Microparticle-based displays possess intrinsic bistability, exhibit extremely low power d.c. field addressing and have demonstrated high contrast and reflectivity. These features, combined with a near-lambertian viewing characteristic, result in an "ink on paper" look[7]. But such displays have to date suffered from short lifetimes and difficulty in manufacture. Here we report the synthesis of an electrophoretic ink based on the microencapsulation of an electrophoretic dispersion[8]. The use of a microencapsulated electrophoretic medium solves the lifetime issues and permits the fabrication of a bistable electronic display solely by means of printing. This system may satisfy the practical requirements of electronic paper.

PREVIOUS approaches to fabricating particle-based displays have been based on rotating bichromal spheres in glass cavities[1] or elastomeric slabs[2,3], and electrophoresis in glass cavities[4,5]. The advantageous optical and electronic characteristics of microparticle displays, as compared to liquid-crystal displays, result from the ability to electrostatically migrate highly scattering pigments such as titanium dioxide ($n = 2.7$), yielding a difference in optical index of refraction with the surrounding dielectric liquid of $\Delta n = 1.3$ and black pigments with very high absorption and hiding power. This results in a very short scattering

一种可用于全印刷反射式电子显示器的电泳墨水

科米斯基等

编者按

与电子显示相比，印刷的文字仍然具有一定的优势。电子显示器的轻微闪烁容易使眼睛疲劳，另外，它们易碎、昂贵、耗能，在明亮的环境光线下对比度下降。因而刺激了能够兼顾纸质印刷物的优点和电子信息系统的便利性与广泛性的“电子纸张”的需求。麻省理工学院媒体实验室的研究人员在本文中报道了首选材料之一：置于软塑料片上的透明微胶囊层。每个胶囊中都含有在电场中可移向顶面的黑、白两种颜色的微粒。这篇基础性的文章导致一个派生公司——E-Ink 的问世，该公司与微电子公司合作后，一系列便携式显示设备已进入市场。

长期以来，显示媒介领域的研究人员一直有一种追求，即发明一种柔韧的低成本系统，也就是实现纸张的电子模拟。在这种情况下，基于微粒的显示器 [1-5] 早已引起了研究人员们的兴趣。这类显示器中可切换的对比度是通过强散射或强吸收微粒(尺寸范围为 0.1~5 μm)的电迁移来实现的，完全不同于人们熟知的通过分子的性质来调控的液晶显示器 [6]。基于微粒的显示器拥有本征双稳态，表现出极低的直流电场编址，并且具有很高的对比度和反射率。这些特征，再加之近朗伯显示特性，就能产生“纸上墨迹”的视觉效果 [7]。但是这类显示器却存在着寿命短，制作难度大的缺点。本文中我们报道了一种基于电泳分散微胶囊化 [8] 的电泳型墨水的合成。使用微胶囊化的电泳介质解决了寿命问题，并且实现了仅用印刷技术便可制造双稳态电子显示器的要求。该系统可以达到电子纸张应用于实际的要求。

以往生产基于微粒的显示器的方法是以在玻璃腔 [1] 或弹性板 [2,3] 中的双色拧转球，和它们在玻璃腔中的电泳 [4,5] 为基础的。与液晶显示器相比，微粒显示器具有优越的光学和电学特性，基于高散射颜料例如二氧化钛($n = 2.7$)的静电迁移能力，使其和周围介电液体及高吸收和遮盖力黑色颜料之间的光学折射率差异较大($\Delta n = 1.3$)。从而导致散射长度很短(约 1 μm)而相应的反射率和对比度很高，效果

length (~1 μm) and a correspondingly high reflectivity and contrast, similar to that of ink on paper. For comparison, typical index differences in a scattering-mode liquid crystal are less than $\Delta n = 0.25$ (ref. 9).

Despite these favourable attributes, microparticle displays at present suffer from a number of shortcomings. These include, in the case of the bichromal sphere system, difficulty in obtaining complete rotation (and thus complete contrast) due to a fall-off in the dipole force close to normal angles and difficulty of manufacture. In the case of electrophoretic systems, the shortcomings include reduced lifetime due to colloidal instability[10]. Finally, bistable display media of all types, to date, are not capable of being manufactured with simple processes such as printing. (Although recent results have been obtained in ink-jetting of electroluminescent doped polymer films for organic light-emitting structures[11], such emissive non-bistable displays do not meet a key criterion of "electronic paper", namely the persistent display of information with zero power consumption.)

To realize a printable bistable display system, we have synthesized an electrophoretic ink by microencapsulating droplets of an electrophoretic dispersion in individual microcapsules with diameters in the range of 30–300 μm. Figure 1a indicates schematically the operation of a system of microencapsulated differently coloured and charged microparticles, in which one or the other species of particles may be migrated towards the viewer by means of an externally applied electric field. Figure 1b shows cross-sectional photomicrographs of a single microcapsule addressed with positive and negative fields.

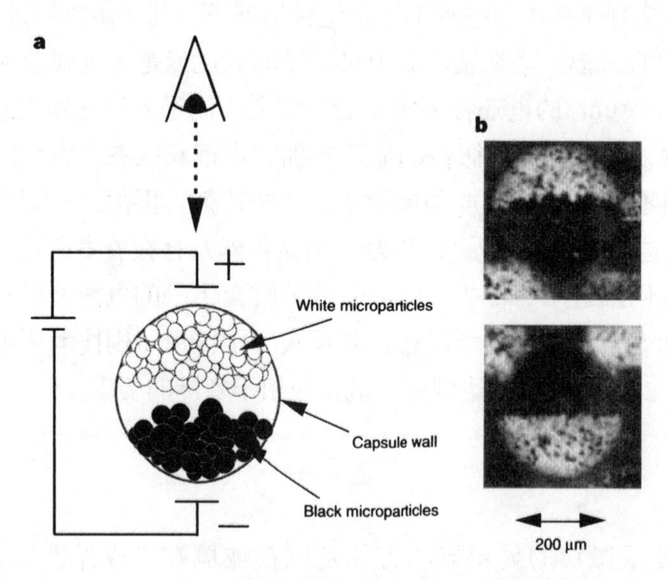

Fig. 1. Electrophoretic microcapsule. **a**, Schematic illustration of microencapsulated electrophoretic image display (white and black microparticles system). The top transparent electrode becomes positively charged, resulting in negatively-charged white microparticles migrating towards it. Oppositely charged black microparticles move towards the bottom electrode. **b**, Photomicrograph of an individual microcapsule addressed with a positive and negative field.

和纸上墨迹相近。相比较而言散射型液晶的典型折射率差值则小于 $\Delta n = 0.25$(参考文献 9)。

尽管具有这些有利的属性，现阶段的微粒显示器仍然存在许多缺点。其中包括，在双色拧转球系统中，由于在接近直角时偶极力下降，使其难以完全拧转(从而达到最大反差)，此外还有制造上的困难。至于电泳系统，缺点包括胶体不稳定性[10]引起的寿命缩短。总之，至今为止，各种类型的双稳态显示媒体都不可能通过如印刷这样的简单过程来生产。(尽管用喷墨法在掺杂的电致发光聚合物薄膜上制备有机发光结构最近已有进展[11]，但这类发光非双稳态显示器并没有达到“电子纸张”的关键指标，即零能耗时信息的持久显示。)

为了实现可印刷的双稳态显示系统，我们合成了一种电泳墨水，通过将电泳分散的液滴微胶囊化，微胶囊的直径范围为 30 ~ 300 μm。图 1a 显示了不同颜色的带电微粒微胶囊化系统的操作示意图，在该系统中一种或另一种粒子可以通过外加电场移向观察者。图 1b 显示了用正负电场编址的单一微胶囊的横截面显微照片。

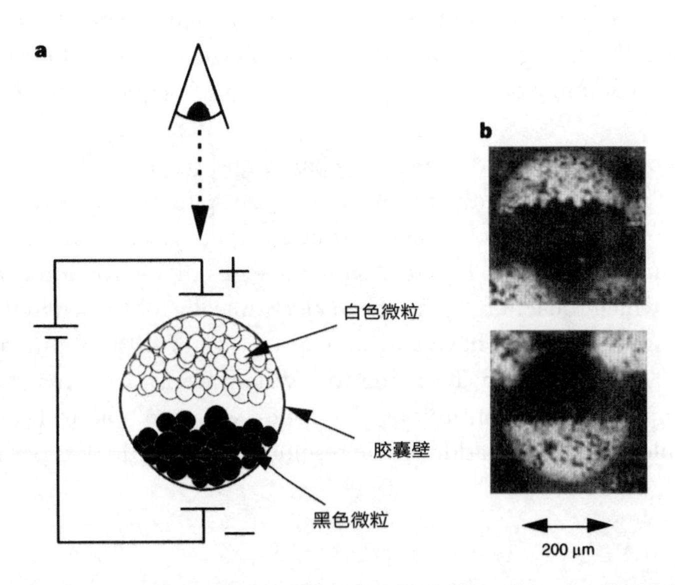

图 1. 电泳微胶囊。**a**，电泳成像显示器微胶囊化的示意图(黑白微粒系统)。上方的透明电极带正电，引起带负电的白色微粒向它迁移。带相反电荷的黑色微粒向下方电极迁移。**b**，用正负电场编址单个微胶囊的显微照片。

The microencapsulated electrophoretic system was synthesized using the following process. The internal phase of the microcapsules was composed of a mixed dispersion of black and white microparticles in a dielectric fluid. To obtain white microparticles, a suspension of rutile titanium dioxide (specific gravity = 4.2) in molten low-molecular-weight polyethylene was atomized. A similar process was used to prepare black microparticles with an inorganic black pigment. The resulting particles were sieved to obtain a dry powder with an average diameter of 5 μm. Alternatively, smaller monodisperse particles, with a diameter less than 1 μm, were prepared chemically. The polyethylene serves to reduce the specific gravity of the particles (typically to ~1.5) and present a modified surface chemistry for charging purposes. The particles in suspension acquire a surface charge due to the electrical double layer[12]. The black and white particles have different zeta potential (and hence different mobility) due to the electroconductivity of the black pigment[13]. These microparticles are then dispersed in a mixture of tetrachloroethylene (specific gravity, 1.6) and an aliphatic hydrocarbon (specific gravity, 0.8) which is specific-gravity-matched to the manufactured particles. In the case where the particles are designed with charges of opposite sign, they are prevented from coagulation by providing a physical polymeric adsorbed layer on the surface of each particle which provides strong inter-particle repulsive forces[10]. Alternatively, a single particle system (white microparticles) dispersed in a dyed (Oil Blue N) dielectric fluid was prepared.

This suspension, which in typical electrophoretic displays would be interposed between two glass electrodes, was then emulsified into an aqueous phase and microencapsulated by means of an *in situ* polycondensation of urea and formaldehyde. This process produces discrete mechanically strong and optically clear microcapsules. The microcapsules are optionally filtered to obtain a desired size range, and are subsequently washed and dried.

Microcapsules (white particle in dye) were prepared and dispersed in a carrier (ultraviolet-curable urethane) and subsequently coated onto a transparent conductive film (indium tin oxide (ITO) on polyester). Rear electrodes printed from a silver-doped polymeric ink were then applied to the display layer. Figure 2 shows a series of microphotographs at different magnifications in which the letter "k" has been electronically addressed in the electronic ink. The sample shown was sieved to have a mean capsule size of 40 ± 10 μm yielding a capsule resolution of ~600 dots per inch. By going to a system using structured top electrodes (or address lines as opposed to a continuous electrode), we were able to fractionally address single microcapsules, yielding an addressable resolution of ~1,200 dots per inch.

微胶囊电泳系统经由以下工序合成。微胶囊的分散相由黑白微粒在介电流体的分散系混合而成。白色微粒是通过金红石型二氧化钛（比重 = 4.2）在熔化的低分子量聚乙烯的悬浮液中雾化而成。黑色微粒可用无机黑色颜料经过类似的过程制备。所得粒子经筛分得到平均直径为 5 μm 的干粉。此外直径小于 1 μm 的较小单分散粒子可用化学方法制备。聚乙烯用来降低粒子的比重（通常降至 1.5 左右）和提供经化学修饰的表面使之能够带电。悬浮液中粒子表面电荷来自双电层作用[12]。黑色颜料的导电性使黑白粒子具有不同的界面电位（进而有不同的迁移率）[13]。这些微粒被分散于四氯乙烯（比重 1.6）和一种脂肪烃（比重 0.8）的混合物中，混合物与所制备粒子的比重相匹配。在粒子按设计分别带有异性电荷的情况下，在粒子表面上通过物理吸附覆盖上一层高聚物，使得粒子间存在强互斥力[10]以避免发生聚沉。此外，还制备了分散在染色（油蓝 N）介电流体中的单粒子系统（白色微粒）。

该悬浮液置于典型电泳显示器的两块玻璃电极之间，在水相中乳化并通过尿素和甲醛的原位缩聚反应实现微胶囊化。该方法可制得高强度、光学性能清晰、分立的微胶囊。筛选后可获得所需尺寸范围的微胶囊，随之进行清洗和干燥。

在载体（紫外光固化的聚氨酯）中制备和分散微胶囊（染料中的白色粒子）后，随即覆以透明导电薄膜（载有氧化铟锡（ITO）的聚酯）。然后，用于显示层的背电极由银掺杂的聚合物墨水印制而成。图 2 显示了不同放大率下的一系列显微照片，为字母“k”在电子墨水中的电子编址。所示样品曾经过筛选，胶囊平均尺寸为 40 ± 10 μm，得到的胶囊分辨率为每英寸约 600 点。通过运用结构化的面电极（或者编址线与连续电极方向相反）系统，我们可对部分单一微胶囊编址，得到的编址分辨率为每英寸约 1,200 点。

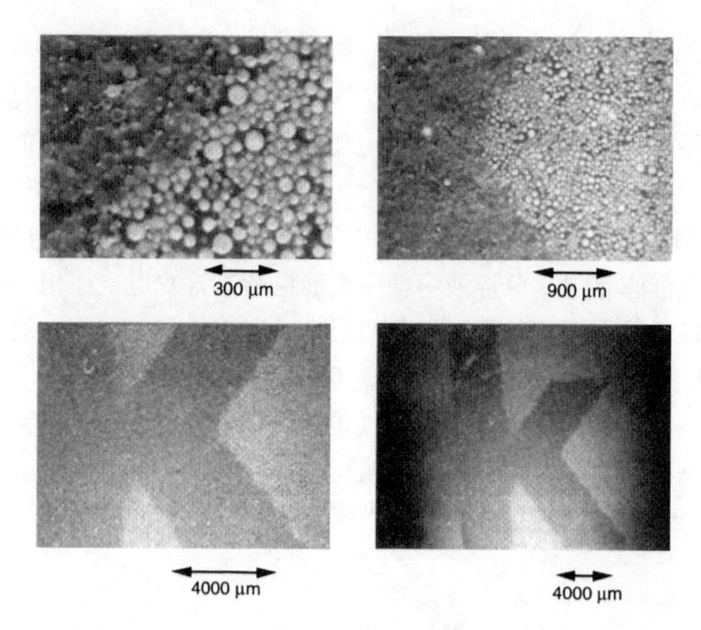

Fig. 2. Electrophoretic ink. Photomicrographs of 200-μm-thick film of electronic ink ("white particles in dye" type) with a capsule diameter of 40 ± 10 μm (top view). The electronically addressed letter "k" is white, other areas are blue.

In addition to making a flexible and printable system, microencapsulating the electrophoretic dispersion has solved a longstanding problem with electrophoretic displays, namely limited lifetime due to particle clustering, agglomeration and lateral migration[10]. This is because the dispersion is physically contained in discrete compartments, and cannot move or agglomerate on a scale larger than the capsule size. In ink samples that we have prepared to date, $> 10^7$ switching cycles have been observed with no degradation in performance. We have measured contrast ratios of 7:1 with 35% reflectance and a near-lambertian viewing characteristic. Using the same measurement system, we measured newspaper at 5:1 contrast and 55% reflectance. We have observed image storage times of several months, after which the material may be addressed readily.

In Fig. 3a we show maximum minus minimum reflected optical power (non-normalized contrast) versus drive frequency for several different applied fields; open symbols indicate data taken from low to high frequency, filled symbols show data from high to low frequency. Figure 3b shows plots of contrast versus applied field for different frequencies. The data in Fig. 3 were obtained for a 200-μm film of electronic ink (of the type "white particles in dye"), with capsules of diameter 40 ± 10 μm on ITO polyester with a silver-ink rear electrode.

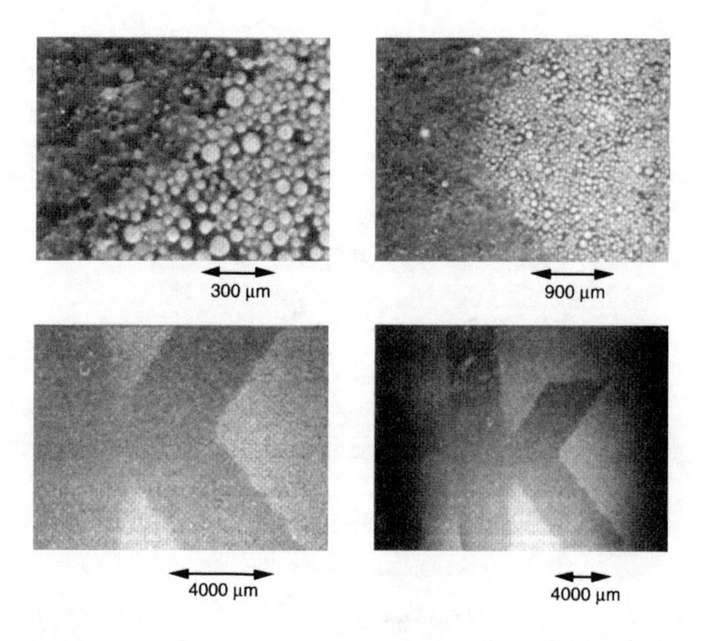

图 2. 电泳墨水。胶囊直径为 40±10 μm，厚度为 200 μm 的电子墨水（“染料中白色粒子”型）薄膜显微照片（俯视图）。电子编址的字母“k”为白色，其他区域为蓝色。

除了制作灵活的可印刷系统之外，微胶囊电泳分散法解决了电泳显示器的一个存在已久的问题，即由于粒子簇集，聚沉，以及横向迁移带来的寿命受限问题[10]。因为分散被物理地限制在分立的空间中，移动和聚集都不可能超出胶囊的尺寸。目前为止，我们制备的墨水样品，经 10^7 次以上的切换循环，没有观测到性能有所下降。测量得到的对比度为 7:1，反射率为 35%，并且观测到近朗伯显示特性。运用相同的测量系统，我们测得报纸的对比度为 5:1，反射率为 55%。而且经几个月的存放后，该材料仍然可以很容易地编址。

图 3a 中展示了反射光强的最大和最小值之差（对比度未归一化）与不同外加场下驱动频率的关系；空心图形代表频率由低至高时采集的数据，实心图形代表频率由高至低时所采集的数据。图 3b 为不同频率下对比度与外加电场的关系。图 3 中的数据源于 200 μm 的电子墨水（“染料中白色粒子”型）薄膜，在具有银墨水背电极的 ITO 聚酯膜上的胶囊直径为 40±10 μm。

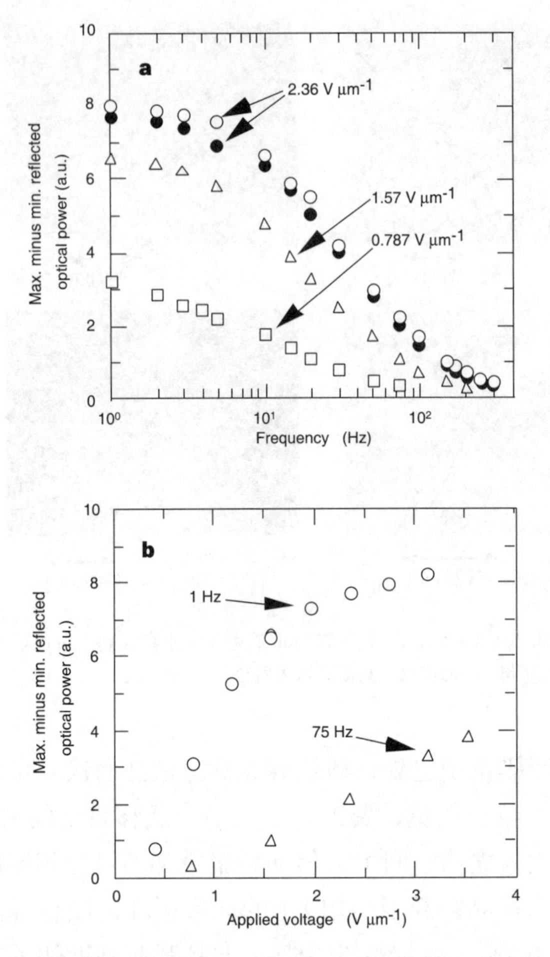

Fig. 3. Properties of a 200-μm-thick film of electronic ink ("white particles in dye" type) with capsule diameter of 40 ± 10 μm. **a**, Maximum minus minimum reflected optical power (contrast) versus drive frequency for several different applied fields (open symbols, data taken from low to high frequency; filled symbols, from high to low frequency). **b**, Plot of contrast versus applied field for different frequencies.

We may calculate the particle mobility, zeta potential and charge per particle as follows. The low-frequency asymptote in Fig. 3a indicates that at sufficiently slow driving frequency there is time for the internal particles to make a full traverse of the capsule, after which there is no further contribution to the contrast. Thus the cusp of the sigmoid in Fig. 3a may be taken to indicate the frequency (5 Hz) at which the particle just traverses a round trip (2×40 μm = 80 μm) in the microcapsule yielding a velocity of $v = 400$ μm s^{-1}. The Reynolds number is given by $\mathrm{Re} = \rho v r \eta^{-1}$ where ρ is the internal fluid density, r is the particle radius, η is the internal fluid viscosity and v is the particle velocity. For the present system we have $\rho = 1.6 \times 10^{-15}$ kg μm^{-3}, $v = 400$ μm s^{-1}, $r = 0.5$ μm, and $\eta = 0.8 \times 10^{-9}$ kg μm^{-1} s^{-1}; and $\mathrm{Re} = 0.0004 \ll 1$ and the flow is laminar[14]. The transient time to establish the laminar flow may be estimated from the Navier–Stokes equation to be $\tau_{ss} = (1/9) r^2 \rho \eta^{-1} = 55$ ns. Thus for time-scales of interest, the particle mobility is given as $\mu = v/E = \epsilon\zeta/6\pi\eta = q/12\pi r\eta$ where ϵ is the dielectric constant of the internal fluid, ζ is the zeta potential of the particles and q is the charge per particle. Thus for our system we have

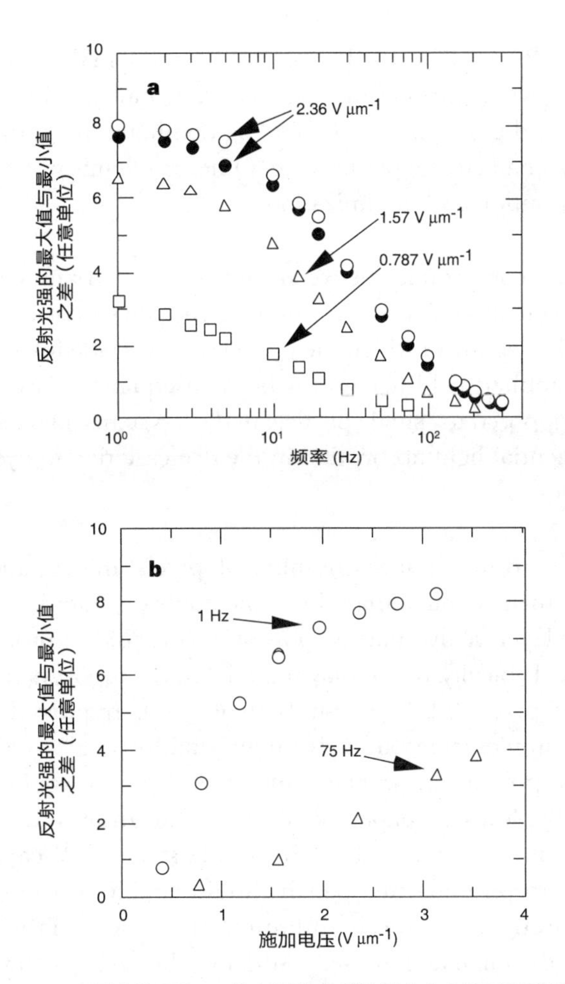

图 3. 胶囊直径为 40 ± 10 μm，厚度为 200 μm 的电子墨水（“染料中白色粒子”型）薄膜的属性。**a**，反射光强的最大值与最小值之差（对比度）与不同外场下驱动频率的关系（空心图形，频率由低至高采集的数据；实心图形，频率由高至低所采集的数据）。**b**，不同频率下对比度与外场的关系。

我们可以通过如下方法计算粒子的迁移率、界面电位和单位粒子的荷电量。图 3a 中低频渐近线意味着当驱动频率足够低时，内部粒子有时间完全横穿胶囊，之后对对比度没有进一步的贡献。因此图 3a 中 S 形的突起点可以视为粒子刚好在胶囊中以 5 Hz 的频率往返一周（2 × 40 μm = 80 μm）时的速率，为 $v = 400\ \mu m \cdot s^{-1}$。雷诺常数由 $Re = \rho v r \eta^{-1}$ 给出，其中 ρ 为内部流体密度，r 为粒子半径，η 为内部流体黏度，v 为粒子速度。对于现在的系统，有 $\rho = 1.6 \times 10^{-15}\ kg \cdot \mu m^{-3}$，$v = 400\ \mu m \cdot s^{-1}$，$r = 0.5\ \mu m$，$\eta = 0.8 \times 10^{-9}\ kg \cdot \mu m^{-1} \cdot s^{-1}$，$Re = 0.0004 \ll 1$，并且流动为层流[14]。确定层流的瞬态时间可以从纳维–斯托克斯方程中估算出来，为 $\tau_{ss} = (1/9)\ r^2 \rho \eta^{-1} = 55$ ns。因此在我们所感兴趣的时间尺度之内，粒子迁移率可以表示为 $\mu = v/E = \epsilon\zeta/6\pi\eta = q/12\pi r\eta$，其中 ϵ 为内部流体的介电常数，ζ 为粒子的界面电位，q 为单位粒子的电量。对于我们的系统，$\mu = 169\ \mu m^2 \cdot V^{-1} \cdot s^{-1}$，$\epsilon = 2.5 \times 8.85 \times 10^{-6}\ kg \cdot \mu m \cdot s^{-2} \cdot V^{-2}$，$\zeta = 120$ mV，

$\mu = 169\ \mu m^2\ V^{-1}\ s^{-1}$, $\epsilon = 2.5 \times 8.85 \times 10^{-6}$ kg $\mu m\ s^{-2}\ V^{-2}$, $\zeta = 120$ mV, and $q = 2.6 \times 10^{-18}$ C = $16e^-$. The very small charge per particle indicates the mechanism for "field-off" bistability. Particle/capsule-wall and particle/particle binding dominate the particle/particle repulsion coming from the very small charge per particle. Other mechanisms for "field-off" bistability include charge redistribution and minimization.

Figure 3a indicates the compromise between switching speed and contrast; Fig. 3b shows the compromise between applied field and contrast. In both sets of curves, the high-contrast asymptote indicates the regime in which the particles have made full traverse of the capsule, and further time or applied field has no further consequence. Two-particle (black/white particles) systems, as opposed to "single particle in dye" systems, relax these trade-offs due to the absence of exponential light absorption in the dye case due to dye interstitially present between white particles.

In order to efficiently address a large number of pixels, matrix addressing is required. Matrix addressing in turn requires that either the contrast material (passive matrix) or an underlying electronic layer (active matrix) possess a threshold in order to prevent crosstalk between address lines. Typically, the display material may support passive matrix addressing for small numbers of pixels[15]. Larger numbers of pixels require the nonlinearity of an electronic element as implemented in active matrix addressing[16]. Active matrix structures, typically formed from polysilicon on glass, are expensive and would obviate many of the advantages of a flexible low-cost paper-like display. Our group has recently demonstrated an all-printed metal–insulator–metal (MIM) diode structure[17] capable of forming the active matrix for electrophoretic ink, which should enable us to drive high-pixel-count sheets of electrophoretic ink while maintaining the ease of fabrication and low-cost structure provided by the ink material itself. Work by other groups may also be of use in this endeavour[18]. We believe that such systems will open up a new field of research in printable microscopic electro-mechanical systems. Coupled with recent advances in printable logic, such technology offers the prospect of fundamentally changing the nature of printing, from printing form to printing function.

(**394**, 253-255; 1998)

Barrett Comiskey, J. D. Albert, Hidekazu Yoshizawa & Joseph Jacobson
Massachusetts Institute of Technology, The Media Laboratory, 20 Ames Street, Cambridge, Massachusetts 02139-4307, USA

Received 8 December 1997; accepted 18 May 1998.

References:

1. Pankove, J. I. *Color Reflection Type Display Panel* (Tech. Note No. 535, RCA Lab., Princeton, NJ, 1962).
2. Sheridon, N. K. & Berkovitz, M. A. The gyricon–a twisting ball display. *Proc. Soc. Information Display* **18(3,4)**, 289-293 (1977).
3. Sheridon, N. K. *et al.* in *Proc. Int. Display Research Conf.* (ed. Jay Morreale) L82–L85 (Toronto, 1997).
4. Ota, I., Honishi, J. & Yoshiyama, M. Electrophoretic image display panel. *Proc. IEEE* **61,** 832-836 (1973).

$q = 2.6 \times 10^{-18}$ C = 16e$^-$。单位粒子带电量极低表示遵守“去场”双稳态机制。粒子/胶囊壁和粒子/粒子间的结合对来自带电量很小的微粒间的粒子/粒子互斥起主导作用。“去场”双稳态的其他机制包括电荷再分布以及最小化。

图 3a 指示出切换速度与对比度之间的制约关系；图 3b 显示了外场与对比度之间的制约关系。在两组曲线中，高对比度渐近线暗示着粒子完全穿越胶囊的机制，随后时间和外场都不再起作用。和“染料中的单粒子系统”不同的是，在双粒子（黑/白粒子）系统中，由于白粒子处于染料之中，而染料对光不发生指数型吸收，因而削弱了这种制约。

大量像素的有效编址要应用矩阵编址。矩阵编址又依次要求对比材料（被动矩阵）或者基本电层（主动矩阵）具有一个阈值以防止编址线之间的串扰。通常，显示材料可以支持少量像素的被动矩阵编址 [15]。大量像素要求电子元件在实现主动矩阵编址时具有非线性 [16]。主动矩阵结构，通常是由在玻璃上的多晶硅形成，昂贵而且不具备低成本的柔韧性类纸张显示器的许多优点。我们的团队近期提出了一种可印刷的金属–绝缘体–金属（MIM）的二极管结构 [17]，可以组成电泳墨水的主动矩阵，应当有可能使驱动电泳墨水的高像素数印刷板成为现实，因为使用了这种墨水材料，保持了制造的简单易行和结构的低成本。其他团队的工作也可能应用到这项研究之中 [18]。我们相信这种系统会开创可印刷微观机电系统的一个新研究领域。结合现阶段其他可印刷技术的思路，这种技术预期将从根本上改变印刷的特性，从印刷形式到印刷功能。

（崔宁 翻译；宋心琦 审稿）

5. Dalisa, A. L. Electrophoretic display technology. *IEEE Trans. Electron Devices* **ED-24(7),** 827-834 (1977).

6. Castellano, J. A. & Harrison, K. J. in *The Physics and Chemistry of Liquid Crystal Devices* (ed. Sprokel, G. J.) 263-288 (Plenum, New York, 1980).

7. Fitzhenry-Ritz, B. Optical properties of electrophoretic image displays. *IEEE Trans. Electron Devices* **ED-28(6),** 726-735 (1981).

8. Comiskey, B., Albert, J. D. & Jacobson, J. Electrophoretic ink: a printable display material. In *Digest of Tech. Papers* 75-76 (Soc. Information Display, Boston, 1997).

9. Okada, M., Hatano, T. & Hashimoto, K. Reflective multicolor display using cholesteric liquid crystals. In *Digest of Tech. Papers* 1019-1026 (Soc. Information Display, Boston, 1997).

10. Murau, P. & Singer, B. The understanding and elimination of some suspension instabilities in an electrophoretic display. *J. Appl. Phys.* **49,** 4820-4829 (1978).

11. Hebner, T. R., Wu, C. C., Marcy, D., Lu, M. H. & Sturm, J. C. Ink-jet printing of doped polymers for organic light emitting devices. *Appl. Phys. Lett.* **72,** 519-521 (1998).

12. Fowkes, F. M., Jinnai, H., Mostafa, M. A., Anderson, F. W. & Moore, R. J. in *Colloids and Surfaces in Reprographic Technology* (eds Hair, M. & Croucher, M. D.) (Am. Chem. Soc., Washington DC, 1982).

13. Claus, C. J. & Mayer, E. F. in *Xerography and Related Processes* (eds Dessauer, J. H. & Clark, H. E.) 341-373 (Focal, New York, 1965).

14. Landau, L. D. & Lifshitz, E. M. *Fluid Mechanics* (Pergamon, New York, 1959).

15. Ota, I., Sato, T., Tanka, S., Yamagami, T. & Takeda, H. Electrophoretic display devices. In *Laser 75* 145-148 (Optoelectronics Conf. Proc., Munich, 1975).

16. Shiffman, R. R. & Parker, R. H. An electrophoretic image display with internal NMOS address logic and display drivers. *Proc. Soc. Information Display* **25.2,** 105-115 (1984).

17. Park, J. & Jacobson, J. in *Proc. Materials Research Soc.* B8.2 (Mater. Res. Soc., Warrendale, PA, 1998).

18. Service, R. F. Patterning electronics on the cheap. *Science* **278,** 383 (1997).

Correspondence and requests for materials should be addressed to J.J. (e-mail: jacobson@media.mit.edu).

Light Speed Reduction to 17 Metres per Second in an Ultracold Atomic Gas

L. V. Hau *et al.*

Editor's Note

Light moves more slowly in physical media than in empty space—but generally only by a little. Yet interference effects associated with quantum physics can in principle be used to slow the speed of light dramatically in a specially designed medium. Here, Lene Vestergaard Hau and colleagues demonstrate this effect in an ultracold gas of sodium atoms. They use a laser beam travelling at right angles to a test beam to prepare the atoms such that a fraction of them are in particular excited electronic states. Light in the test beam, being repeatedly absorbed and re-emitted by these atoms, travels at a speed of only 17 metres per second, some 20 million times slower than light in a vacuum.

Techniques that use quantum interference effects are being actively investigated to manipulate the optical properties of quantum systems[1]. One such example is electromagnetically induced transparency, a quantum effect that permits the propagation of light pulses through an otherwise opaque medium[2-5]. Here we report an experimental demonstration of electromagnetically induced transparency in an ultracold gas of sodium atoms, in which the optical pulses propagate at twenty million times slower than the speed of light in a vacuum. The gas is cooled to nanokelvin temperatures by laser and evaporative cooling[6-10]. The quantum interference controlling the optical properties of the medium is set up by a "coupling" laser beam propagating at a right angle to the pulsed "probe" beam. At nanokelvin temperatures, the variation of refractive index with probe frequency can be made very steep. In conjunction with the high atomic density, this results in the exceptionally low light speeds observed. By cooling the cloud below the transition temperature for Bose–Einstein condensation[11-13] (causing a macroscopic population of alkali atoms in the quantum ground state of the confining potential), we observe even lower pulse propagation velocities ($17\ m\ s^{-1}$) owing to the increased atom density. We report an inferred nonlinear refractive index of $0.18\ cm^2\ W^{-1}$ and find that the system shows exceptionally large optical nonlinearities, which are of potential fundamental and technological interest for quantum optics.

THE experiment is performed with a gas of sodium atoms cooled to nanokelvin temperatures. Our atom cooling set-up is described in some detail in ref. 14. Atoms emitted from a "candlestick" atomic beam source[15] are decelerated in a Zeeman slower and loaded into a magneto-optical trap. In a few seconds we collect a cloud of 10^{10} atoms at a temperature of 1 mK and a density of $6 \times 10^{11}\ cm^{-3}$. The atoms are then polarization

光速在超冷原子气中降低至 17 米每秒

豪等

编者按

光在物质媒介中的传播速度比在真空中的慢，但通常只慢一点点。然而，与量子物理相关的干涉效应原则上可以使光速在特别设计的介质中显著变慢。本文中，莱娜·韦斯特戈·豪和其同事们在超冷钠原子气体中展示了这种效应：使用与测试光束成直角传播的激光束来制备原子，使得其中的一部分处在特定的电子激发态。测试光束中的光，被这些原子反复吸收和重新发射，以每秒 17 米的速度传播，大约为真空中光速的 2,000 万分之一。

基于量子干涉效应的技术被积极研究用来操纵量子系统中的光学性质[1]。其中的一个例子就是电磁感应透明，一种允许光脉冲在其他状况下不透明的介质中传播的量子现象[2-5]。在此，我们报道了在超冷钠原子气中电磁感应透明的实验验证，光脉冲以在真空中光速的 2,000 万分之一的速度在该介质中传播。气体通过激光和蒸发制冷被降温至纳开 (10^{-9} K)[6-10]。控制介质光学性质的量子干涉通过与脉冲“探测”光束成直角传播的“耦合”激光束来建立。在纳开温度下，折射率随探测频率可以发生非常急剧的变化。与高原子密度相结合，就会导致所观察到的极低光速。我们将原子云冷却至玻色–爱因斯坦凝聚临界温度[11-13]以下（引起处于囚禁势量子基态的碱金属原子的宏观布居），由于原子数密度增大，我们观测到更低的脉冲传播速度 ($17\ m \cdot s^{-1}$)。经过测算，本文报道了该体系的非线性折射率系数为 $0.18\ cm^2 \cdot W^{-1}$，发现该系统显示出非常大的光学非线性，这对于量子光学的研究而言具有潜在的基础和技术吸引力。

本实验是在降温至纳开的钠原子中进行的。文献 14 中对我们的原子冷却装置有一些细节性的描述。从“烛台”原子束源[15]发出的原子在塞曼减速器中被减速，然后被加载到一个磁光阱中。若干秒后我们在 1 mK 温度，$6 \times 10^{11}\ cm^{-3}$ 密度下采集到一个由 10^{10} 个原子组成的原子云。这些原子随后被偏振梯度冷却若干毫秒至 50 μK， gradient cooled for a few milliseconds to 50 μK and optically pumped into the $F=1$ ground state with an equal population of the three magnetic sublevels. We then turn all laser beams off and confine the atoms magnetically in the "4 Dee" trap[14]. Only atoms in the $M_F=-1$ state, with magnetic dipole moments directed opposite to the magnetic field direction (picked as the quantization axis), are trapped in the asymmetric harmonic trapping potential.This magnetic filtering results in a sample of atoms that are all in a single atomic state (state $|1\rangle$ in Fig. 1b) which allows adiabatic optical preparation of the atoms, as described below, and minimal heating of the cloud.

Next we evaporatively cool the atoms for 38 s to the transition temperature for Bose–Einstein condensation, T_c. The magnetic fields are then adjusted to adiabatically soften the trap. The resulting trapping potential has a frequency of $f_z=$ 21 Hz along the symmetry (z) axis of the 4 Dee trap, and transverse frequencies $f_x=f_y=69$ Hz. The bias field, parallel to the z axis, is 11 G. When we cool well below T_c, we are left with 1–2 million atoms in the condensate. For these parameters the transition occurs at a temperature of $T_c=$ 435 nK and a peak density in the cloud of 5×10^{12} cm^{-3}.

We now apply a linearly polarized laser beam, the coupling beam, tuned to the transition between the unpopulated hyperfine states $|2\rangle$ and $|3\rangle$ (Fig. 1b).This beam couples states $|2\rangle$ and $|3\rangle$ and creates a quantum interference for a weaker probe laser beam (left circularly polarized) which is tuned to the $|1\rangle\rightarrow|3\rangle$ transition. A stable eigenstate (the "dark state") of the atom in the presence of coupling and probe lasers is a coherent superposition of the two hyperfine ground states $|1\rangle$ and $|2\rangle$. The ratio of the probability amplitudes is such that the contributions to the atomic dipole moment induced by the two lasers exactly cancel. The quantum interference occurs in a narrow interval of probe frequencies, with a width determined by the coupling laser power.

并被光泵抽运至 $F = 1$ 的基态，原子平均布居于三个磁子能级上。然后我们关闭所有的激光束，并用磁场将原子限制在“4 Dee”阱[14]中。只有处于 $M_F = -1$ 态的原子，其磁偶极矩与磁场方向（被选作量子化轴）相反，被捕获于非对称的谐振势阱中。这个磁过滤使所有的样品原子都处于同一原子态（图 1b 中的量子态 $|1\rangle$），这就使下文所述的原子绝热光学制备和对原子云最低限度的加热成为可能。

接下来我们将原子蒸发制冷 38 秒至玻色–爱因斯坦凝聚的临界转变温度 T_c。调整磁场，绝热地减小势阱的梯度。这样所获得的势阱沿 4 Dee 势阱对称轴（z）方向的频率 $f_z = 21$ Hz，横向频率 $f_x = f_y = 69$ Hz。偏置场平行于 z 轴，为 11 G。当温度完全低于 T_c，我们只剩下处于凝聚态的一百万到两百万个原子。在这些参数下，转变发生在 $T_c = 435$ nK，原子云峰值密度为 5×10^{12} cm^{-3}。

我们现在来加一线偏振激光束，即耦合光束，将之调谐至无粒子分布的超精细态 $|2\rangle$ 和 $|3\rangle$ 之间的跃迁（图 1b）。该光束将态 $|2\rangle$ 和 $|3\rangle$ 相耦合，从而对一更弱的调谐在态 $|1\rangle \rightarrow |3\rangle$ 跃迁的探测激光光束（左旋圆偏振）产生了量子干涉。在耦合激光和探测激光同时存在时，原子的稳定本征态（“暗”态）是两个超精细基态 $|1\rangle$ 和 $|2\rangle$ 的相干叠加。概率幅比值被设置为两激光对所激发的原子偶极矩的贡献恰好相抵消。量子干涉发生在探测频率的一狭窄的范围内，宽度由耦合激光功率决定。

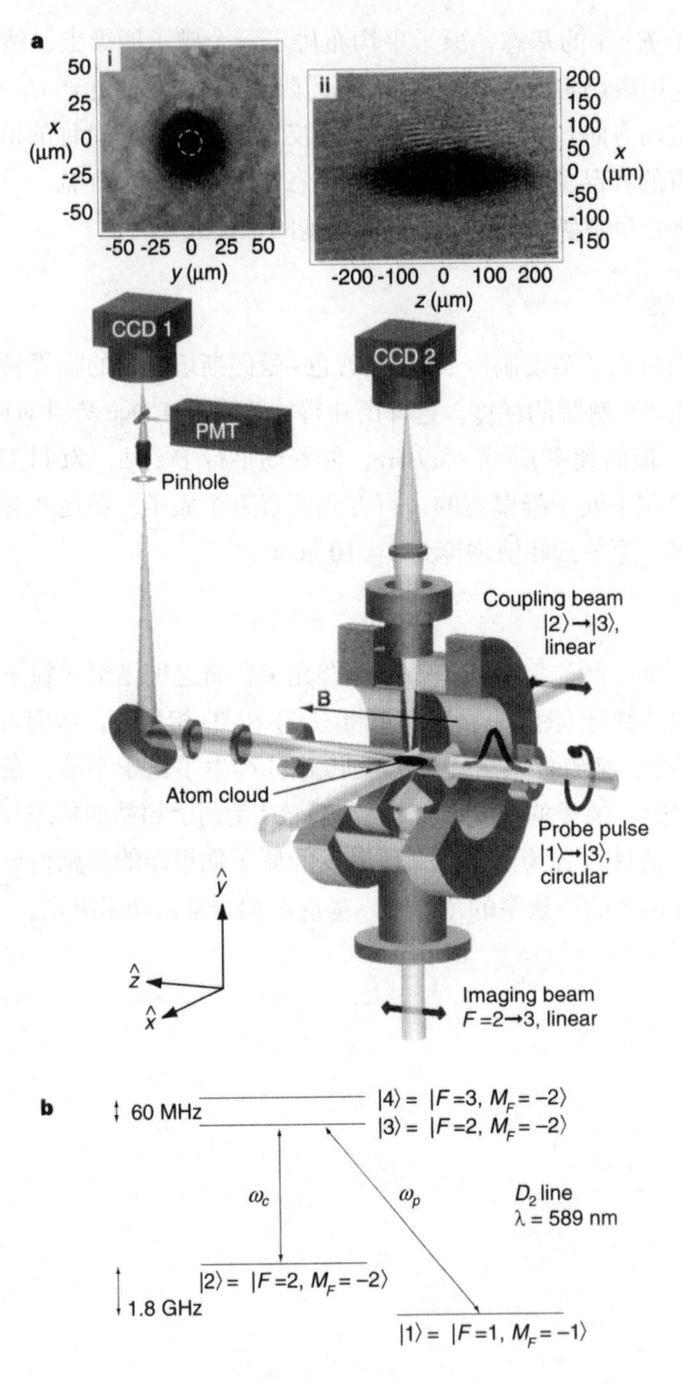

Fig. 1. Experimental set-up. A "coupling" laser beam propagates along the x axis with its linear polarization along the 11-G bias field in the z direction. The "probe" laser pulse propagates along the z axis and is left-circularly polarized. With a flipper mirror in front of the camera CCD 1, we direct this probe beam either to the camera or to the photomultiplier (PMT). For pulse delay measurements, we place a pinhole in an external image plane of the imaging optics and select a small area, 15 μm in diameter, of the probe beam centred on the atom clouds (as indicated by the dashed circle in inset (i)). The pulse delays are measured with the PMT. The imaging beam propagating along the y axis is used

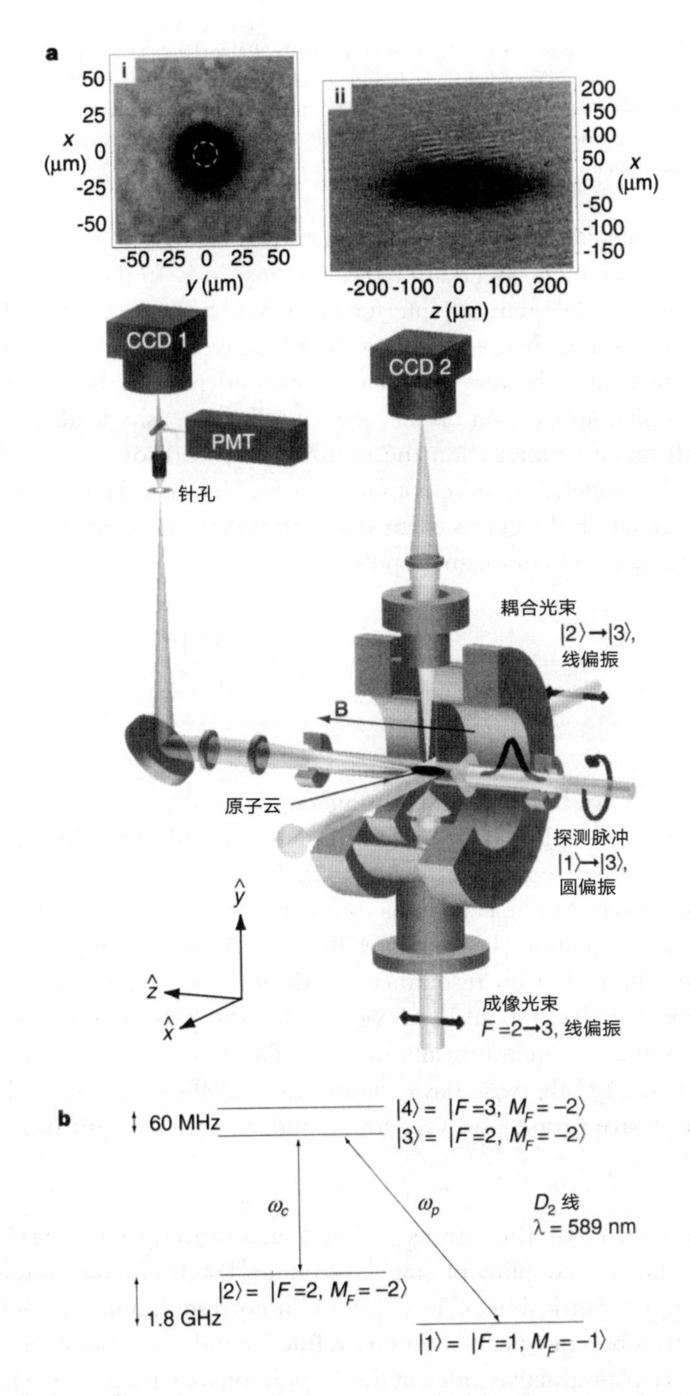

图 1. 实验装置图。一"耦合"激光束沿 x 轴传播，其线偏振方向 z 轴上加有 11 G 的偏置场。"探测" 激光脉冲沿 z 轴传播，为左旋圆偏振。利用摄像机 CCD1 前的转向镜，我们可以引导探测光速进入摄像机或者进入光电倍增管(PMT)。对于脉冲延迟的测量，我们在成像光路中一个外部的成像平面上放置一个针孔，并选择探测光束在原子云中心直径为 15 μm 的小区域(插图(i)中虚线圆所示)。脉冲延迟通过 PMT 测量。沿 y 轴传播的成像光束被用来使原子云在 CCD2 上成像，以获得原子云在沿脉冲传播方向

to image atom clouds onto camera CCD 2 to find the length of the clouds along the pulse propagation direction (z axis) for determination of light speeds. Inset (ii) shows atoms cooled to 450 nK which is 15 nK above T_c. (Note that this imaging beam is never applied at the same time as the probe pulse and coupling laser). The position of a cloud and its diameter in the two transverse directions, x and y, are found with CCD 1. Inset (i) shows an image of a condensate.

Figure 2a shows the calculated transmission of the probe beam as a function of its detuning from resonance for parameters which are typical of this work. In the absence of dephasing of the $|1\rangle \rightarrow |2\rangle$ transition, the quantum interference would be perfect, and at line centre, the transmission would be unity. Figure 2b shows the refractive index for the probe beam as a function of detuning. Due to the very small Doppler broadening of the $|1\rangle \rightarrow |2\rangle$ transition in our nanokelvin samples, application of very low coupling intensity leads to a transparency peak with a width much smaller than the natural line width of the $|1\rangle \rightarrow |3\rangle$ transition. Correspondingly, the dispersion curve is much steeper than can be obtained by any other technique, and this results in the unprecedented low group velocities reported here. The group velocity v_g for a propagating electromagnetic pulse is[16-19]:

$$v_g = \frac{c}{n(\omega_p) + \omega_p \dfrac{dn}{d\omega_p}} \approx \frac{\hbar c \epsilon_0}{2\omega_p} \frac{|\Omega_c|^2}{|\boldsymbol{\mu}_{13}|^2 N} \tag{1}$$

Here $n(\omega_p)$ is the refractive index at probe frequency ω_p (rad s^{-1}), $|\Omega_c|^2$ is the square of the Rabi frequency for the coupling laser and varies linearly with intensity, $\boldsymbol{\mu}_{13}$ is the electric dipole matrix element between states $|1\rangle$ and $|3\rangle$, N is the atomic density, and ϵ_0 is the permittivity of free space. At line centre, the refractive index is unity, and the second term in the denominator of equation (1) dominates the first. An important characteristic of the refractive index profile is that on resonance the dispersion of the group velocity is zero (see ref. 16), that is, $d^2n/d\omega_p^2 = 0$, and to lowest order, the pulse maintains its shape as it propagates. The established quantum interference allows pulse transmission through our atom clouds which would otherwise have transmission coefficients of e^{-110} (below T_c), and creates a steep dispersive profile and very low group velocity for light pulses propagating through the clouds.

We note that the centres of the curves in Fig. 2 are shifted by 0.6 MHz from probe resonance. This is due to a coupling of state $|2\rangle$ to state $|4\rangle$ through the coupling laser field, which results in an a.c. Stark shift of level $|2\rangle$ and a corresponding line shift of the $2 \rightarrow 3$ transition. As the transparency peak and unity refractive index are obtained at two-photon resonance, this leads to a refractive index at the $1 \rightarrow 3$ resonance frequency which is different from unity. The difference is proportional to the a.c. Stark shift and hence to the coupling laser intensity, which is important for predicting the nonlinear refractive index as described below.

（z 轴）上的长度，并以此来确定光速。插图（ii）为降温至 450 nK 的原子，即在临界温度 T_c 以上 15 nK。（注意该成像光束不能与探测光束或者耦合光束同时打开）。原子云的位置和在 x 和 y 两横向上的直径由 CCD1 测得。插图（i）为一凝聚态的成像图。

基于本文典型参数所确定的谐振频率，图 2a 为计算所得的探测光束透射率随其频率失谐的变化曲线。当 $|1\rangle\to|2\rangle$ 跃迁不存在失相时，量子干涉是完美的，在线中心，透射率应该是 1。图 2b 给出了探测光束的折射率与失谐的函数关系。由于我们的纳开温度样品中 $|1\rangle\to|2\rangle$ 跃迁时非常小的多普勒展宽，施加非常小的耦合强度都会导致一个宽度比 $|1\rangle\to|3\rangle$ 跃迁的自然线宽窄很多的透明度峰。相应地，色散曲线比通过其他任何技术所获得的都更陡，这也就导致了本文所述的前所未有的极低群速度。电磁脉冲传播的群速度 v_g 为[16-19]：

$$v_g=\frac{c}{n(\omega_p)+\omega_p\dfrac{dn}{d\omega_p}}\approx\frac{\hbar c\epsilon_0}{2\omega_p}\frac{|\Omega_c|^2}{|\boldsymbol{\mu}_{13}|^2N}\tag{1}$$

其中，$n(\omega_p)$ 为探测频率 $\omega_p(\mathrm{rad\cdot s^{-1}})$ 时的折射率，$|\Omega_c|^2$ 为耦合激光的拉比频率的平方，它随光强线性变化，$\boldsymbol{\mu}_{13}$ 为态 $|1\rangle$ 和 $|3\rangle$ 间的电偶极矩矩阵元，N 为原子数密度，ϵ_0 为自由空间的介电常数。在线中央，折射率为 1，等式（1）分母中的第二项的影响超过第一项，占主导地位。折射率图谱的一个重要特性是共振时群速度色散为 0（参见文献 16），即 $d^2n/d\omega_p^2=0$，就最低阶来说，脉冲传播过程中保持波形不变。所建立的量子干涉允许脉冲传输通过在其他状态下透射系数仅为 e^{-110}（T_c 以下）的原子云，从而产生了极陡的色散图谱和极低的群速度。

我们注意到图 2 中的曲线中心与探测共振频率偏离了 0.6 MHz。这是由于耦合激光场使态 $|2\rangle$ 和 $|4\rangle$ 发生了耦合，从而能级 $|2\rangle$ 有交流斯塔克位移和与之相应的 2→3 跃迁的谱线位移。由于透明度峰值和等于 1 的折射率是在两光子共振时获得的，这就导致 1→3 跃迁共振频率下的折射率不为 1。二者之间的差别与交流斯塔克位移成比例，也因此与耦合激光强度成比例，这点对下文所述的预估非线性折射率至关重要。

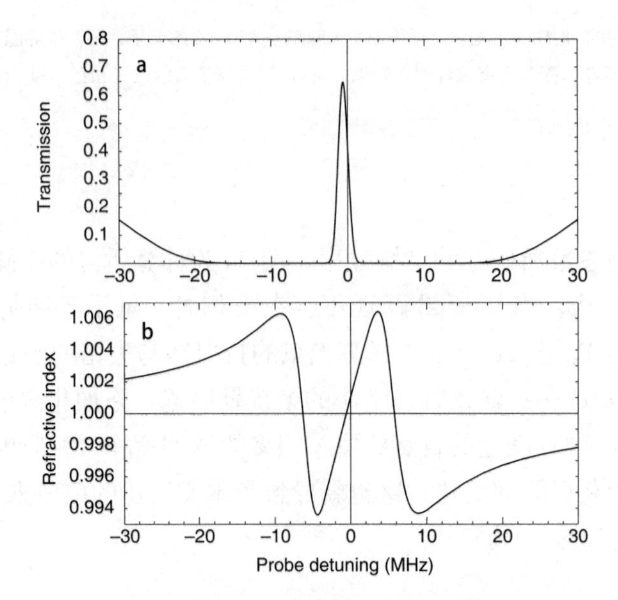

Fig. 2. Effect of probe detuning. **a**, Transmission profile. Calculated probe transmission as a function of detuning from the $|1\rangle \rightarrow |3\rangle$ resonance for an atom cloud cooled to 450 nK, with a peak density of 3.3×10^{12} cm^{-3} and a length of 229 μm (corresponding to the cloud in inset (ii) of Fig. 1a). The coupling laser is resonant with the $|2\rangle \rightarrow |3\rangle$ transition and has a power density of 52 mW cm^{-2}. **b**, Refractive index profile. The calculated refractive index is shown as a function of probe detuning for the same parameters as in **a**. The steepness of the slope at resonance is inversely proportional to the group velocity of transmitted light pulses and is controlled by the coupling laser intensity. Note that as a result of the a.c. Stark shift of the $|2\rangle \rightarrow |3\rangle$ transition, caused by a coupling of states $|2\rangle$ and $|4\rangle$ through the coupling laser field, the centre of the transmission and refractive index profiles is shifted by 0.6 MHz. The shift of the refractive index profile results in the nonlinear refractive index described in the text.

A diagram of the experiment is shown in Fig. 1a. The 2.5-mm-diameter coupling beam propagates along the x axis with its linear polarization parallel to the **B** field. The 0.5-mm-diameter, σ^- polarized probe beam propagates along the z axis. The size and position of the atom cloud in the transverse directions, x and y, are obtained by imaging the transmission profile of the probe beam after the cloud onto a charge-coupled-device (CCD) camera. An image of a condensate is shown as inset (i). A 55 mW cm^{-2} coupling laser beam was present during the 10-μs exposure of the atoms to a 5 mW cm^{-2} probe beam tuned close to resonance. The $f/7$ imaging optics are diffraction-limited to a resolution of 7 μm.

During the pulse delay experiments, a pinhole (placed in an external image plane of the lens system) is used to select only the part of the probe light that has passed through the central 15 μm of the atom cloud where the column density is the greatest. The outline of the pinhole is indicated with the dashed circle in inset (i).

Both coupling and probe beams are derived from the same dye laser. The frequency of the coupling beam is set by an acousto-optic modulator (AOM) to the $|2\rangle \rightarrow |3\rangle$ resonance. Here we take into account both Zeeman shifts and the a.c. Stark shift described above.

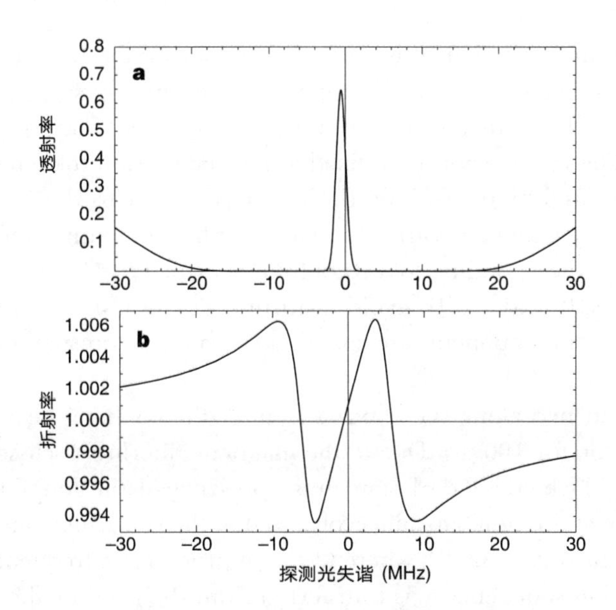

图 2. 探测光失谐的影响。**a**，透射率曲线。计算所得的探测透射率与激光频率相对原子云从 $|1\rangle \rightarrow |3\rangle$ 共振失谐的函数关系，该原子云被降温至 450 nK，峰值密度为 3.3×10^{12} cm^{-3}，长度为 229 μm(对应图 1a 中插图 (ii) 的原子云)。耦合激光与 $|2\rangle \rightarrow |3\rangle$ 的跃迁共振，功率密度为 52 mW · cm^{-2}。**b**，折射率曲线。计算所得的折射率与 **a** 中相同参数下探测光失谐的函数关系。达到共振时斜率的陡度与投射光脉冲的群速度成反比，并且由耦合激光强度所控制。要注意到，通过耦合激光场，态 $|2\rangle$ 和 $|4\rangle$ 产生耦合所造成的 $|2\rangle \rightarrow |3\rangle$ 跃迁的交流斯塔克位移的存在，导致透射率曲线的中心与折射率曲线的中心相差 0.6 MHz。折射率曲线的偏移引起了文中所述的非线性折射率现象。

实验图示如图 1a 所示。直径 2.5 mm 的耦合光束沿 x 轴传播，其线偏振方向平行于磁场 **B**。探测光束直径 0.5 mm、σ^- 偏振，沿 z 轴传播。原子云在横向 x，y 方向的尺寸和位置由探测光束通过原子云后成像在电荷耦合器件 (CCD) 摄像机上的透射率剖面获得。插图 (i) 为一凝聚态的图像。当原子在调谐至与共振频率接近的 5 mW · cm^{-2} 的探测光束中暴露 10 μs 的过程中，另一 55 mW · cm^{-2} 的耦合激光光束同时存在。f/7 成像光学受衍射限制分辨率为 7 μm。

在脉冲延迟的实验中，利用一针孔 (置于透镜系统的外成像面) 来只选择那些通过原子云中心 15 μm 的部分探测光束，这部分原子柱的数密度最大。针孔的轮廓在插图 (i) 中以虚线圆标记出。

耦合光束和探测光束都源于相同的染料激光。耦合激光器的频率由一声光调制器 (AOM) 设定为 $|2\rangle \rightarrow |3\rangle$ 的共振频率。这里我们将塞曼位移和上文所提到的交流斯塔克位移都考虑在内。

The corresponding probe resonance is found by measuring the transmission of the probe beam as a function of its frequency. We apply a fast frequency sweep, across 32 MHz in 50 μs, and determine resonance from the transmission peak. The sweep is controlled by a separate AOM. The frequency is then fixed at resonance, and the temporal shape of the probe pulse is generated by controlling the r.f. drive power to the AOM. The resulting pulse is approximately gaussian with a full-width at half-maximum of 2.5 μs. The peak power is 1 mW cm^{-2} corresponding to a Rabi frequency of $\Omega_p = 0.20\,A$, where the Einstein A coefficient is 6.3×10^7 rad s^{-1}. To avoid distortion of the pulse, it is made of sufficient duration that its Fourier components are contained within the transparency peak.

Probe pulses are launched along the z axis 4 μs after the coupling beam is turned on (the coupling field is left on for 100 μs). Due to the magnetic filtering discussed above, all atoms are initially in state $|1\rangle$ which is a dark state in the presence of the coupling laser only. When the pulse arrives, the atoms adiabatically evolve so that the probability amplitude of state $|2\rangle$ is equal to the ratio $\Omega_p/(\Omega_p^2 + \Omega_c^2)^{1/2}$, where Ω_p is the probe Rabi frequency. To establish the coherent superposition state, energy is transferred from the front of the probe pulse to the atoms and the coupling laser field. At the end of the pulse, the atoms adiabatically return to the original state $|1\rangle$ and the energy returns to the back of the probe pulse with no net energy and momentum transfer to the atomic cloud. Because the refractive index is unity, the electric field is unchanged as the probe pulse enters the medium. As the group velocity is decreased, the total energy density must increase so as to keep constant the power per area. This increase is represented by the energy stored in the atoms and the coupling laser field during pulse propagation through the cloud.

The pulses are recorded with a photomultiplier (3-ns response time) after they penetrate the atom clouds. The output from the photomultiplier is amplified by a 150-MHz-bandwidth amplifier and the waveforms are recorded on a digital scope. With a "flipper" mirror in front of the camera we control whether the probe beam is directed to the camera or to the photomultiplier.

The result of a pulse delay measurement is shown in Fig. 3. The front pulse is a reference pulse obtained with no atoms present. The pulse delayed by 7.05 μs was slowed down in an atom cloud with a length of 229 μm (see Fig. 1a, inset (ii)). The resulting light speed is 32.5 m s^{-1}. We used a coupling laser intensity of 12 mW cm^{-2} corresponding to a Rabi frequency of $\Omega_c = 0.56\,A$. The cloud was cooled to 450 nK (which is 15 nK above T_c), the peak density was 3.3×10^{12} cm^{-3} , and the total number of atoms was 3.8×10^6 . From these numbers we calculate that the pulse transmission coefficient would be e^{-63} in the absence of the coupling laser. The probe pulse was indeed observed to be totally absorbed by the atoms when the coupling beam was left off. Inhomogeneous broadening due to spatially varying Zeeman shifts is negligible (~20 kHz) for the low temperatures and correspondingly small cloud sizes used here.

相应的探测共振频率由测量探测光束的透射率与频率的函数关系得出。我们运用一快速频率扫描，在 50 μs 内扫描 32 MHz，通过透射峰值来确定共振。扫描受另一 AOM 控制。然后将频率固定在共振频率，探测脉冲的时域波形就由控制 AOM 的无线电频率驱动来产生。所得脉冲是一个近高斯分布，半高宽为 2.5 μs。拉比频率 $\Omega_p = 0.20A$ 时的峰值功率为 1 mW · cm^{-2}。这里取爱因斯坦自激系数 A 为 6.3×10^7 rad · s^{-1}。为避免脉冲失真，我们给予其足够的持续时间来使其傅里叶分量包含在透明度峰之中。

在耦合光束打开 4 μs 之后探测脉冲沿 z 轴发射（耦合场保持打开 100 μs）。由于上文所讨论的磁过滤，所有的原子最初都处于 $|1\rangle$ 态，只有耦合激光存在时，这是一个暗态。当接收到光脉冲时，原子绝热演化，导致 $|2\rangle$ 态的概率幅与 $\Omega_p/(\Omega_p^2+\Omega_c^2)^{1/2}$ 相等，其中 Ω_p 是探测光拉比频率。为了实现相干叠加态，能量从探测光脉冲的前沿传递到原子和耦合激光场。在探测光脉冲末端，原子绝热地回到原始态 $|1\rangle$，能量返回到探测脉冲后端，但没有净能量和动量转移到原子云。由于折射率等于 1，在探测脉冲进入介质时电场不变。随着群速度降低，总能量密度必须提高才能保证单位面积功率不变。这种提高就体现在脉冲在原子云中传播时，能量在原子以及耦合激光场的储存。

这些脉冲在穿过原子云后被一光电倍增管（响应时间为 3 ns）所记录。光电倍增管的输出信号被 150 MHz 带宽的放大器放大，波形显示在一数字示波器上。在摄像机前放置一“转向”镜，我们就可以控制探测光束是射入摄像机还是射入光电倍增管中。

探测光脉冲延迟测量的结果如图 3 所示。前面的脉冲是没有原子存在时的参考脉冲。脉冲穿过长度为 229 μm 的原子云时被减慢，延迟了 7.05 μs（见图 1a，插图(ii)）。由此可得光速为 32.5 m · s^{-1}。实验中我们使用的耦合激光强度为 12 mW · cm^{-2}，对应拉比频率为 $\Omega_c = 0.56A$。原子云被冷却至 450 nK（临界温度 T_c 以上 15 nK）。峰值密度为 3.3×10^{12} cm^{-3}，原子总数为 3.8×10^6。根据这些数据我们可以计算得出在没有耦合激光时，脉冲透射系数为 e^{-63}。实验中确实观察到当耦合激光关闭时，探测脉冲完全被原子所吸收。本实验在低温下进行且原子云尺寸相对较小，由空间变化的塞曼位移引起的非均匀展宽（约 20 kHz）可以忽略。

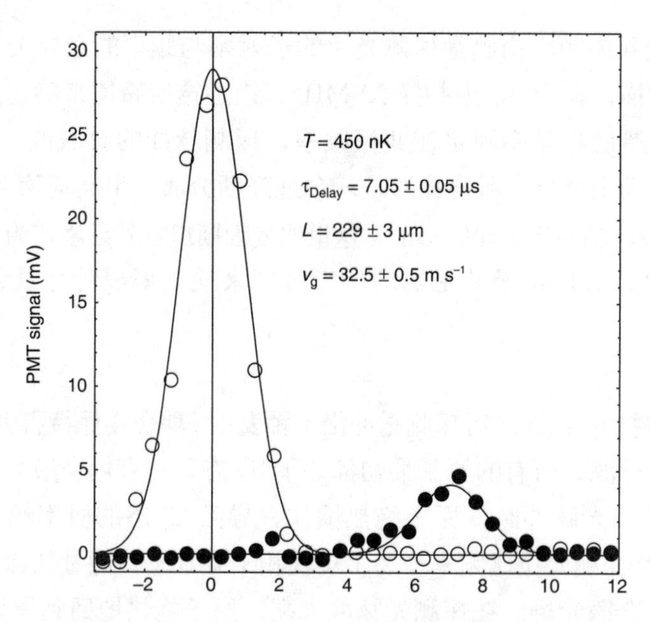

Fig. 3. Pulse delay measurement. The front pulse (open circles) is a reference pulse with no atoms in the system. The other pulse (filled circles) is delayed by 7.05 μs in a 229-μm-long atom cloud (see inset (ii) in Fig. 1a).The corresponding light speed is 32.5 m s^{-1}. The curves represent gaussian fits to the measured pulses.

The size of the atom cloud in the z direction is obtained with another CCD camera. For this purpose, we use a separate 1 mW cm^{-2} laser beam propagating along the vertical y axis and tuned 20 MHz below the $F = 2 \rightarrow 3$ transition. The atoms are pumped to the $F = 2$ ground state for 10 μs before the imaging which is performed with an exposure time of 10 μs. We image the transmission profile of the laser beam after the atom cloud with diffraction-limited $f/5$ optics. An example is shown in Fig. 1a, inset (ii), where the asymmetry of the trap is clear from the cloud's elliptical profile. We note that the imaging laser is never applied at the same time as the coupling laser and probe pulse, and for each recorded pulse or CCD picture a new cloud is loaded.

We measured a series of pulse delays and corresponding cloud sizes for atoms cooled to temperatures between 2.5 μK and 50 nK. From these pairs of numbers we obtain the corresponding propagation velocities (Fig. 4). The open circles are for a coupling power of 52 mW cm^{-2} ($\Omega_c = 1.2\Gamma$). The light speed is inversely proportional to the atom density (equation (1)) which increases with lower temperatures, with an additional density increase when a condensate is formed. The filled circles are for a coupling power of 12 mW cm^{-2}. The lower coupling power is seen to cause a decrease of group velocities in agreement with equation (1). We obtain a light speed of 17 m s^{-1} for pulse propagation in an atom cloud initially prepared as an almost pure Bose–Einstein condensate (condensate fraction is $\geqslant 90\%$). Whether the cloud remains a condensate during and after pulse propagation is an issue that is beyond the scope of this Letter.

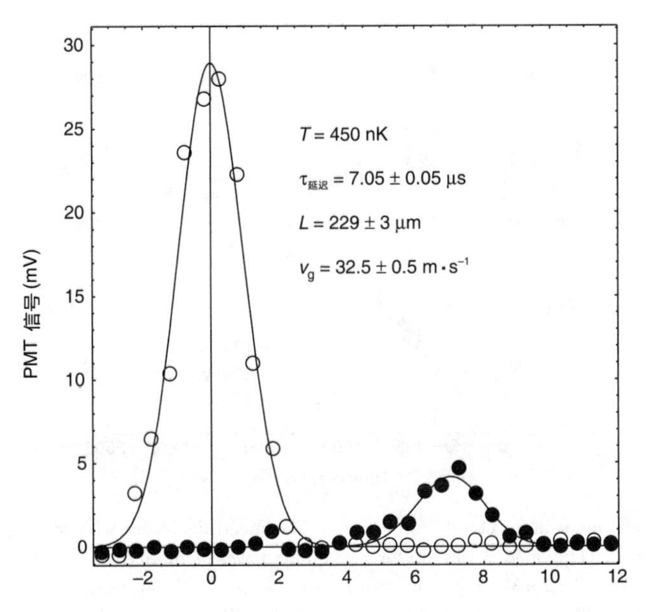

图 3. 脉冲延迟测量。前面的脉冲（空心圆）为系统中没有原子的参考脉冲。另一个脉冲（实心圆）在 229 μm 长的原子云中被延迟 7.05 μs(见图 1a 插图(ii))。相应的光速为 32.5 m · s^{-1}。曲线为测量脉冲的高斯拟合。

沿 z 方向的原子云尺寸通过另一 CCD 摄像机捕获。为了达到这个目的，我们运用了另一独立的 1 mW · cm^{-2} 的激光束沿垂直的 y 轴传播，并调谐到比 $F = 2 \to 3$ 跃迁低 20 MHz。原子在 10 μs 的曝光成像之前，先被泵到 $F = 2$ 的基态 10 μs。我们使用衍射极限 $f/5$ 的光学器件在原子云之后对激光束的纵向透射率剖面进行成像。一个例子如图 1a 插图(ii)所示，其中由云的椭圆轮廓可清楚看见磁阱的不对称性。值得注意的是，成像激光不会与耦合激光及探测脉冲同时施加，并且对于每个记录的脉冲或 CCD 图片，都加载一新的原子云。

我们在原子冷却到温度 2.5 μK 和 50 μK 之间测量了一系列脉冲延迟和相应的云尺寸。从这些数据对中，我们获得了相应的传播速度(图 4)。空心圆对应耦合光强 52 mW · cm^{-2}($\Omega_c = 1.2A$)。光速与原子数密度成反比(等式(1))，而原子数密度随温度降低而增大，当凝聚态形成时原子数密度进一步增加。实心圆对应耦合光强 12 mW · cm^{-2}。与等式(1)一致，可以看到耦合光强的降低引起群速度的减小。在最初制备的几乎纯净的玻色–爱因斯坦凝聚态(凝聚比率大于等于 90%)原子云中，我们获得了 17 m · s^{-1} 的光脉冲传播速度。在脉冲传播过程中和传播后原子云是否保持凝聚不在本文讨论范围之内。

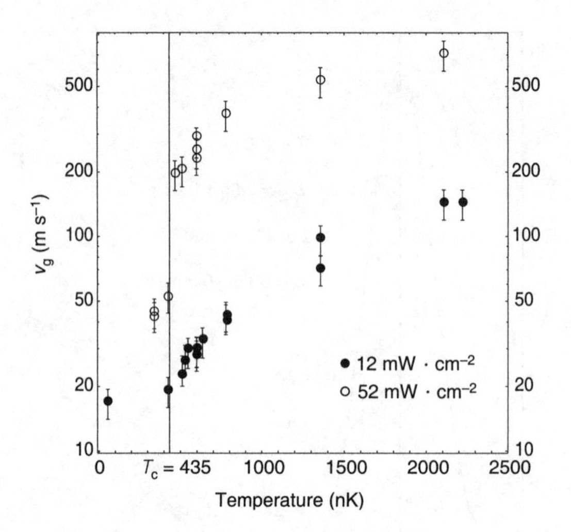

Fig. 4. Light speed versus atom cloud temperature. The speed decreases with temperature due to the atom density increase. The open circles are for a coupling power of 52 mW cm^{-2} and the filled circles are for a coupling power of 12 mW cm^{-2}. The temperature T_c marks the transition temperature for Bose–Einstein condensation. The decrease in group velocity below T_c is due to a density increase of the atom cloud when the condensate is formed. From imaging measurements we obtain a maximum atom density of 8×10^{13} cm^{-3} at a temperature of 200 nK. Here, the dense condensate component constitutes 60% of all atoms, and the total atom density is 16 times larger than the density of a non-condensed cloud at T_c. The light speed measurement at 50 nK is for a cloud with a condensate fraction $\geqslant$ 90%. The finite dephasing rate due to state $|4\rangle$ does not allow pulse penetration of the most dense clouds. This problem could be overcome by tuning the laser to the D_1 line as described in the text.

Transitions from state $|2\rangle$ to state $|4\rangle$, induced by the coupling laser (detuned by 60 MHz from this transition), result in a finite decay rate of the established coherence between states $|1\rangle$ and $|2\rangle$ and limit pulse transmission. The dephasing rate is proportional to the power density of the coupling laser and we expect, and find, that probe pulses have a peak transmission that is independent of coupling intensity and a velocity which reduces linearly with this intensity. The dephasing time is determined from the slope of a semi-log plot of transmission versus pulse delay[19]. At a coupling power of 12 mW cm^{-2}, we measured a dephasing time of 9 μs for atom clouds just above T_c.

Giant Kerr nonlinearities are of interest for areas of quantum optics such as optical squeezing, quantum nondemolition, and studies of nonlocality. It was recently proposed that they may be obtained using electromagnetically induced transparency[20]. Here we report the first (to our knowledge) measurement of such a nonlinearity. The refractive index for zero probe detuning is given by $n = 1 + (n_2 I_c)$ where I_c is the coupling laser intensity, and n_2 the cross phase nonlinear refractive index. As seen from Fig. 2b, the nonlinear term $(n_2 I_c)$ equals the product of the slope of the refractive index at probe resonance and the a.c. Stark shift of the $|2\rangle \rightarrow |3\rangle$ transition caused by the coupling laser. We can then express n_2 by the formula (see equation (1));

$$n_2 = \frac{\Delta\omega_S}{I_c}\frac{dn}{d\omega_p} \approx \frac{1}{2\pi}\frac{\Delta\omega_S}{I_c}\frac{\lambda}{v_g} \qquad (2)$$

where $\Delta\omega_S$ is the a.c. Stark shift, proportional to I_c, and λ the wavelength of the probe

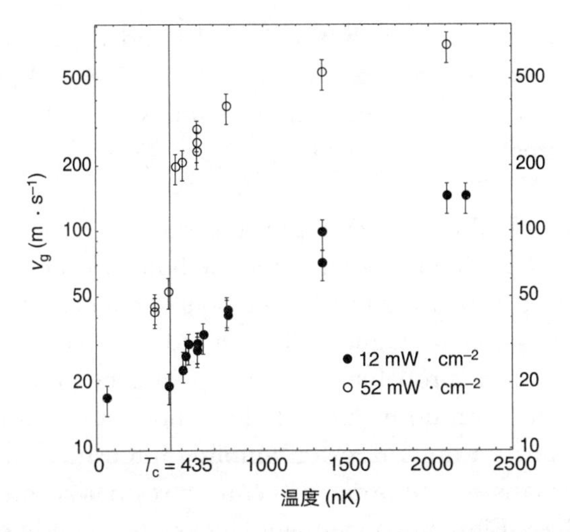

图 4. 光速与原子云温度的关系。由于原子数密度的增大，光速随温度的降低而降低。空心圆对应耦合光强 52 mW · cm^{-2}，实心圆对应耦合光强 12 mW · cm^{-2}。温度 T_c 标记出了玻色–爱因斯坦凝聚的临界转变温度。T_c 以下的群速度减小是由形成凝聚态时原子云密度增大而引起。通过成像测量，我们得到 200 nK 时最大的原子数密度为 8×10^{13} cm^{-3}。密集凝聚态成分占所有原子的 60%，并且总原子数密度是 T_c 时非凝聚态云密度的 16 倍。50 nK 时的光速测量是在凝聚分数 ≥ 90% 的云中进行的。由于态 $|4\rangle$ 造成的有限的退相率不允许脉冲穿过最密集的云。这个问题可以通过文中所述的调谐激光至 D_1 线的方法来解决。

由耦合激光激发的 $|2\rangle\rightarrow|4\rangle$ 态的跃迁（耦合激光与该跃迁之间存在 60 MHz 失谐），使得所建立的 $|1\rangle$ 和 $|2\rangle$ 态的相干具有一个有限衰变率，并限制了探测光脉冲的透射。退相率与耦合激光的功率密度成比例，我们预期并发现，探测脉冲具有一个透射率峰值，该值不受耦合强度以及随此强度线性减小的速度的影响。退相时间由透射率–脉冲延迟的半对数曲线的斜率决定 [19]。在耦合光强为 12 mW · cm^{-2} 时，对于温度刚刚超过 T_c 的原子云，我们测量到退相时间为 9 μs。

巨克尔非线性在量子光学领域具有很大的吸引力，例如光学压缩、量子非破坏性以及非定域性等方面的研究。近期的研究还指出，这些现象很可能通过电磁感应透明来获得 [20]。本文报道了（据我们所知）首次对这样一种非线性的测量。零失谐探测光的折射率表达式为 $n=1+(n_2 I_c)$，其中 I_c 为耦合激光强度，n_2 为交叉相位非线性折射率。正如图 2b 所示，非线性项 $n_2 I_c$ 等于探测光在共振时的折射率曲线斜率与耦合激光激发的 $|2\rangle\rightarrow|3\rangle$ 跃迁的交流斯塔克位移之积。因此 n_2 可以用以下公式表示（见等式（1））：

$$n_2=\frac{\Delta\omega_S}{I_c}\frac{dn}{d\omega_p}\approx\frac{1}{2\pi}\frac{\Delta\omega_S}{I_c}\frac{\lambda}{v_g} \tag{2}$$

其中 $\Delta\omega_S$ 为交流斯塔克位移，与 I_c 成比例，λ 为探测跃迁的光波长。当耦合激光强

transition. We measured an a.c. Stark shift of 1.3×10^6 rad s^{-1} for a coupling laser intensity of 40 mW cm^{-2}. For a measured group velocity of 17 m s^{-1} (Fig. 4), we obtain a nonlinear refractive index of 0.18 cm^2 W^{-1}. This nonlinear index is $\sim 10^6$ times greater than that measured in cold Cs atoms[21].

With a system that avoids the $|1\rangle - |2\rangle$ dephasing rate described above (which can be obtained by tuning to the D_1 line in sodium), the method used here could be developed to yield the collision-induced dephasing rate of the double condensate which is generated in the process of establishing electromagnetically induced transparency (see also refs 22, 23). In that case, the square of the probability amplitude for state $|3\rangle$ could be kept below 10^{-5} during pulse propagation, with no heating of the condensate as a result. With improved frequency stability of our set-up and lower coupling intensities, even lower light speeds would be possible, perhaps of the order of centimetres per second, comparable to the speed of sound in a Bose–Einstein condensate. Under these conditions we expect phonon excitation during light pulse propagation through the condensate. By deliberately tuning another laser beam to the $|2\rangle \rightarrow |4\rangle$ transition, it should be possible to demonstrate optical switching at the single photon level[24]. Finally, we note that during propagation of the atom clouds, light pulses are compressed in the z direction by a ratio of c/v_g. For our experimental parameters, that results in pulses with a spatial extent of only 43 μm.

(**397**, 594-598; 1999)

Lene Vestergaard Hau[*†], S. E. Harris[‡], Zachary Dutton[*†] & Cyrus H. Behroozi[*§]

[*] Rowland Institute for Science, 100 Edwin H. Land Boulevard, Cambridge, Massachusetts 02142, USA

[†] Department of Physics, [§] Division of Engineering and Applied Sciences, Harvard University, Cambridge, Massachusetts 02138, USA

[‡] Edward L. Ginzton Laboratory, Stanford University, Stanford, California 94305, USA

Received 3 November; accepted 21 December 1998.

References:

1. Knight, P. L., Stoicheff, B. & Walls, D. (eds) Highlights in quantum optics. *Phil. Trans. R. Soc. Lond. A* **355**, 2215-2416 (1997).
2. Harris, S. E. Electromagnetically induced transparency. *Phys. Today* **50(7)**, 36-42 (1997).
3. Scully, M. O. & Zubairy, M. S. *Quantum Optics* (Cambridge Univ. Press, 1997).
4. Arimondo, E. in *Progress in Optics* (ed. Wolf, E.) 257-354 (Elsevier Science, Amsterdam, 1996).
5. Bergmann, K., Theuer, H. & Shore, B. W. Coherent population transfer among quantum states of atoms and molecules. *Rev. Mod. Phys.* **70**, 1003-1006 (1998).
6. Chu, S. The manipulation of neutral particles. *Rev. Mod. Phys.* **70**, 685-706 (1998).
7. Cohen-Tannoudjii, C. N. Manipulating atoms with photons. *Rev. Mod. Phys.* **70**, 707-719 (1998).
8. Phillips, W. D. Laser cooling and trapping of neutral atoms. *Rev. Mod. Phys.* **70**, 721-741 (1998).
9. Hess, H. F. Evaporative cooling of magnetically trapped and compressed spin-polarized hydrogen. *Phys. Rev. B* **34**, 3476-3479 (1986).
10. Masuhara, N. *et al.* Evaporative cooling of spin-polarized atomic hydrogen. *Phys. Rev. Lett.* **61**, 935-938 (1988).
11. Anderson, M. H., Ensher, J. R., Matthews, M. R., Wieman, C. E. & Cornell, E. A. Observation of Bose-Einstein condensation in a dilute atomic vapor. *Science* **269**, 198-201 (1995).
12. Davis, K. B. *et al.* Bose-Einstein condensation in a gas of sodium atoms. *Phys. Rev. Lett.* **75**, 3969-3973 (1995).
13. Bradley, C. C., Sackett, C. A. & Hulet, R. G. Bose-Einstein condensation of lithium: observation of limited condensate number. *Phys. Rev. Lett.* **78**, 985-989 (1997).
14. Hau, L. V. *et al.* Near-resonant spatial images of confined Bose-Einstein condensates in a 4-Dee magnetic bottle. *Phys. Rev. A* **58**, R54-R57 (1998).

度为 40 $mW \cdot cm^{-2}$ 时我们测量到一 1.3×10^6 $rad \cdot s^{-1}$ 的交流斯塔克位移。而对于测量所得的 17 $m \cdot s^{-1}$ 的群速度(图 4)，可得非线性折射率为 0.18 $cm^2 \cdot W^{-1}$。该非线性折射率是在冷铯原子中相应值的 10^6 倍 [21]。

应用上文所述的可避免 $|1\rangle-|2\rangle$ 退相率的系统(通过调谐至钠 D_1 线而获得)，可以对这里所使用的方法进一步发展以得出在建立电磁感应透明过程中发生的双凝聚态碰撞诱导失相率(参见文献 22 和 23)。在这种情况下，态 $|3\rangle$ 概率幅的平方在脉冲传播过程中可以保持在 10^{-5} 以下，因此不存在凝聚态加热。由于我们装置频率稳定性的提高以及耦合强度的减小，更低的光速也是可能实现的，可能是几厘米每秒的量级，可与玻色–爱因斯坦凝聚中的音速相比。在此条件下我们预期光脉冲传播通过凝聚体时有声子被激发。通过谨慎调谐另一激光束至 $|2\rangle\rightarrow|4\rangle$ 跃迁，应该有可能在单光子水平上演示光学开关 [24]。最后本文指出，光束在原子云传播过程中，光脉冲在 z 方向被压缩，比率为 c/v_g。对于本实验所采用的参数，这导致脉冲的空间范围仅为 43 μm。

(崔宁 翻译；石锦卫 审稿)

15. Hau, L. V., Golovchenko, J. A. & Burns, M. M. A new atomic beam source: The "candlestick". *Rev. Sci. Instrum.* **65,** 3746-3750 (1994).

16. Harris, S. E., Field, J. E. & Kasapi, A. Dispersive properties of electromagnetically induced transparency. *Phys. Rev. A* **46,** R29-R32 (1992).

17. Grobe, R., Hioe, F. T. & Eberly, J. H. Formation of shape-preserving pulses in a nonlinear adiabatically integrable system. *Phys. Rev. Lett.* **73,** 3183-3186 (1994).

18. Xiao, M., Li, Y.-Q., Jin, S.-Z. & Gea-Banacloche, J. Measurement of dispersive properties of electromagnetically induced transparency in rubidium atoms. *Phys. Rev. Lett.* **74,** 666-669 (1995).

19. Kasapi, A., Jain, M., Yin, G. Y. & Harris, S. E. Electromagnetically induced transparency: propagation dynamics. *Phys. Rev. Lett.* **74,** 2447-2450 (1995).

20. Schmidt, H. & Imamoglu, A. Giant Kerr nonlinearities obtained by electromagnetically induced transparency. *Opt. Lett.* **21,** 1936-1938 (1996).

21. Lambrecht, A., Courty, J. M., Reynaud, S. & Giacobino, E. Cold atoms: A new medium for quantum optics. *Appl. Phys. B* **60,** 129-134 (1995).

22. Hall, D. S., Matthews, M. R., Wieman, C. E. & Cornell, E. A. Measurements of relative phase in two-component Bose-Einstein condensates. *Phys. Rev. Lett.* **81,** 1543-1546 (1998).

23. Ruostekoski, J. & Walls, D. F. Coherent population trapping of Bose-Einstein condensates: detection of phase diffusion. *Eur. Phys. J. D* (submitted).

24. Harris, S. E. & Yamamoto, Y. Photon switching by quantum interference. *Phys. Rev. Lett.* **81,** 3611-3614 (1998).

Acknowledgements. We thank J. A. Golovchenko for discussions and C. Liu for experimental assistance. L.V.H. acknowledges support from the Rowland Institute for Science. S.E.H. is supported by the US Air Force Office of Scientific Research, the US Army Research Office, and the US Office of Naval Research. C.H.B. is supported by an NSF fellowship.

Correspondence and requests for materials should be addressed to L.V.H. (e-mail: hau@rowland.org).

Observation of Coherent Optical Information Storage in an Atomic Medium Using Halted Light Pulses

C. Liu *et al.*

Editor's Note

Laser light shining on an opaque medium can sometimes make it transparent to other light, a phenomenon known as electromagnetically induced transparency. The optical properties of the medium change so dramatically that light pulses can be slowed and compressed by many orders of magnitude. Here Chien Liu and colleagues report on experiments in which they used the effect to bring laser pulses to a complete standstill in a cold gas of sodium atoms held in a magnetic trap. They also show that information may be frozen into the atomic medium for as long as 1 microsecond, and then recovered. Liu and colleagues suggest that this technique could be useful in the development of quantum computers and other quantum information systems.

Electromagnetically induced transparency[1-3] is a quantum interference effect that permits the propagation of light through an otherwise opaque atomic medium; a "coupling" laser is used to create the interference necessary to allow the transmission of resonant pulses from a "probe" laser. This technique has been used[4-6] to slow and spatially compress light pulses by seven orders of magnitude, resulting in their complete localization and containment within an atomic cloud[4]. Here we use electromagnetically induced transparency to bring laser pulses to a complete stop in a magnetically trapped, cold cloud of sodium atoms. Within the spatially localized pulse region, the atoms are in a superposition state determined by the amplitudes and phases of the coupling and probe laser fields. Upon sudden turn-off of the coupling laser, the compressed probe pulse is effectively stopped; coherent information initially contained in the laser fields is "frozen" in the atomic medium for up to 1 ms. The coupling laser is turned back on at a later time and the probe pulse is regenerated: the stored coherence is read out and transferred back into the radiation field. We present a theoretical model that reveals that the system is self-adjusting to minimize dissipative loss during the "read" and "write" operations. We anticipate applications of this phenomenon for quantum information processing.

WITH the coupling and probe lasers used in the experiment, the atoms are accurately modelled as three-level atoms interacting with the two laser fields (Fig. 1a). Under perfect electromagnetically-induced transparency (EIT) conditions (two-photon resonance), a stationary eigenstate exists for the system of a three-level atom and resonant laser fields, where the atom is in a "dark", coherent superposition of states $|1\rangle$ and $|2\rangle$:

运用停止光脉冲来观测原子介质中的相干光学信息存储

刘谦（音译）等

编者按

照射在不透明介质上的激光有时会使介质对其他光线透明，这种现象称为电磁感应透明。介质的光学性质变化非常显著，以至于光脉冲可以被放慢和压缩许多个数量级。本文中，刘谦（音译）与其同事们报道了使用这种效应的实验，在这些实验中，他们使激光脉冲完全停滞在磁阱中的冷钠原子气团中。他们还展示了信息可冻结在原子介质中长达1微秒，然后恢复。刘谦与其同事们认为这种技术可以用于量子计算机和其他量子信息系统的开发。

电磁感应透明[1-3]是一种量子干涉效应，它允许光在其他状态下非透明的原子介质中传播；“耦合”激光被用来产生必要的干涉，以实现从“探测”激光发出的共振脉冲的传输。这种技术[4-6]已经被用于减缓及空间压缩光脉冲达七个数量级之多，从而导致其在原子云内的完全局域化和容纳[4]。本文运用电磁感应透明来使激光脉冲在磁阱捕获的冷钠原子云中完全停止。在空间局域化脉冲区域内，原子处于由耦合激光场和探测激光场的振幅和相位所决定的叠加态。当突然关闭耦合激光时，压缩的探测光束会有效地停止；最初存储在激光场中的相干信息就被“冷冻”在原子介质中，最长可达1 ms。随后耦合激光被重新开启，探测脉冲便再次产生：所存储的相干信息被读取出来并转移回辐射场。本文提出了一个理论模型，揭示了该系统的自调节特征使得“读”和“写”操作过程中耗散损失最小。我们预期这种现象可以被应用于量子信息处理。

通过实验中所使用的耦合和探测激光，原子被精确地建模为与两激光场相互作用的三能级原子（图1a）。在完全电磁感应透明（EIT）条件下（双光子共振），对于三能级原子和共振激光场组成的系统存在一稳定的本征态，其中原子处于态 $|1\rangle$ 和 $|2\rangle$ 的相干叠加“暗”状态：

$$|D\rangle = \frac{\Omega_c|1\rangle - \Omega_p|2\rangle \exp[i(\mathbf{k}_p - \mathbf{k}_c)\cdot\mathbf{r} - i(\omega_p - \omega_c)t]}{\sqrt{\Omega_c^2 + \Omega_p^2}} \quad (1)$$

Here Ω_p and Ω_c are the Rabi frequencies, $\mathbf{k}_p$ and $\mathbf{k}_c$ the wavevectors, and ω_p and ω_c the optical angular frequencies of the probe and coupling lasers, respectively. The Rabi frequencies are defined as $\Omega_{p,c} \equiv e\, \mathbf{E}_{p,c}\cdot\mathbf{r}_{13,23}/\hbar$, where e is the electron charge, $\mathbf{E}_{p,c}$ are the slowly varying envelopes of probe and coupling field amplitudes, and $e\,\mathbf{r}_{13,23}$ are the electric dipole moments of the atomic transitions. The dark state does not couple to the radiatively decaying state $|3\rangle$, which eliminates absorption of the laser fields[1-3].

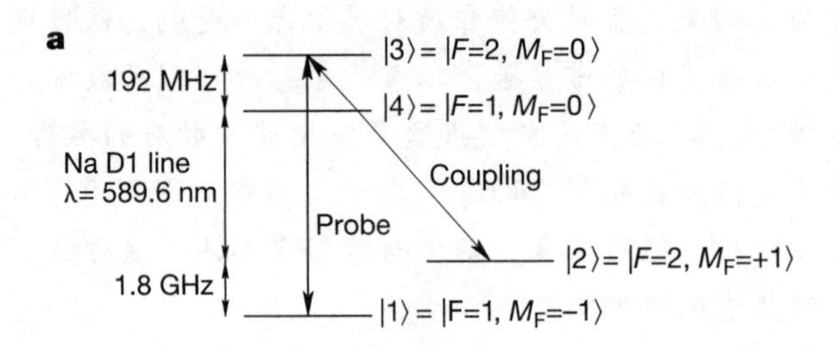

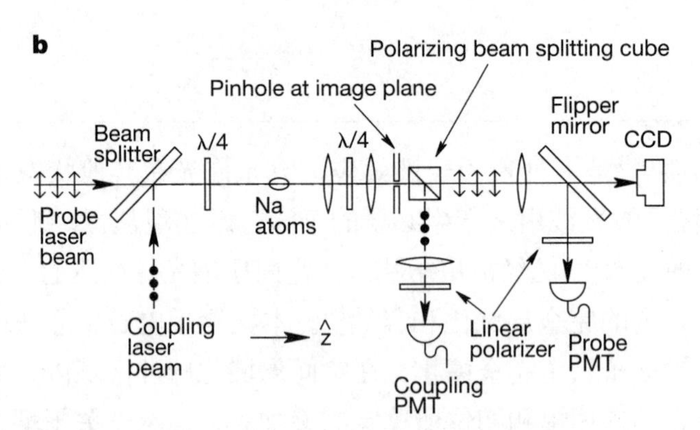

Fig. 1. Experimental set-up and procedure. **a**, States $|1\rangle$, $|2\rangle$ and $|3\rangle$ form the three-level EIT system. The cooled atoms are initially magnetically trapped in state $|1\rangle = |3S, F=1, M_F=-1\rangle$. Stimulated photon exchanges between the probe and coupling laser fields create a "dark" superposition of states $|1\rangle$ and $|2\rangle$, which renders the medium transparent for the resonant probe pulses. **b**, We apply a 2.2-mm diameter, σ^--polarized coupling laser, resonant with the $|3S, F=2, M_F=+1\rangle \rightarrow |3P, F=2, M_F=0\rangle$ transition, and a co-propagating, 1.2-mm diameter σ^+-polarized probe pulse tuned to the $|3S, F=1, M_F=-1\rangle \rightarrow |3P, F=2, M_F=0\rangle$ transition. The two laser beams start out with orthogonal linear polarizations (two-headed arrows and filled circles show the directions of linear polarization of the probe and coupling lasers, respectively). They are combined with a beam splitter, circularly polarized with a quarter-wave plate ($\lambda/4$), and then injected into the atom cloud. After leaving the cloud, the laser beams pass a second quarter-wave plate and regain their original linear polarizations before being separated with a polarizing beam-splitting cube. The atom cloud is imaged first onto an external image plane and then onto a CCD (charge-coupled device) camera. A pinhole is placed in the external image plane and positioned at the centre of the cloud image. With the pinhole and flipper mirror in place, only those portions of the probe and coupling laser beams that have passed through the central region of the cloud are selected and monitored simultaneously by two photomultiplier tubes (PMTs). States $|1\rangle$ and $|2\rangle$ have identical first-order Zeeman shifts so the two-photon resonance is maintained across the trapped atom clouds. Cold atoms and co-propagating lasers eliminate Doppler effects. However, off-resonance transitions to state $|4\rangle$ prevent perfect transmission of the light pulses in this case.

$$|D\rangle = \frac{\Omega_c|1\rangle - \Omega_p|2\rangle \exp[i(\mathbf{k}_p - \mathbf{k}_c)\cdot \mathbf{r} - i(\omega_p - \omega_c)t]}{\sqrt{\Omega_c^2 + \Omega_p^2}} \tag{1}$$

这里 Ω_p 和 Ω_c 分别为探测和耦合激光的拉比频率，$\mathbf{k}_p$ 和 $\mathbf{k}_c$ 为波矢，ω_p 和 ω_c 为光学角频率。拉比频率定义为 $\Omega_{p,c} \equiv e\mathbf{E}_{p,c} \cdot \mathbf{r}_{13,23}/\hbar$,，其中 e 为电子电荷，$\mathbf{E}_{p,c}$ 为探测和耦合场振幅的缓慢变化包络，$e\mathbf{r}_{13,23}$ 为原子跃迁的电偶极矩。暗态不会与辐射衰减态 $|3\rangle$ 耦合，从而抵消了激光场的吸收[1-3]。

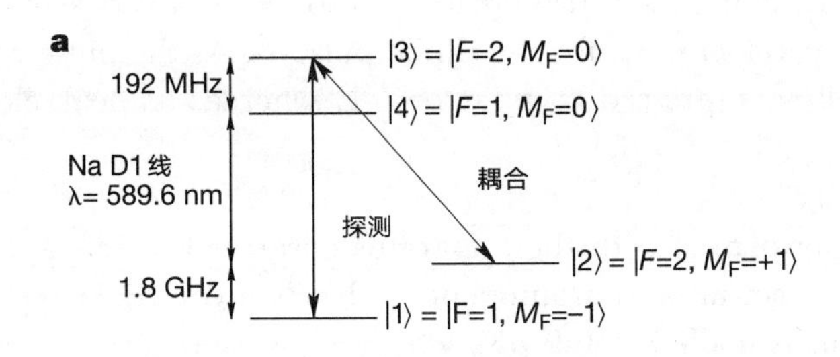

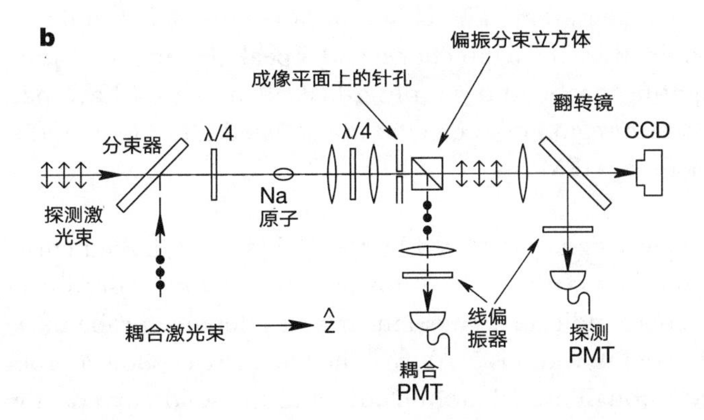

图 1. 实验装置和实验过程。**a**，量子态 $|1\rangle$，$|2\rangle$，$|3\rangle$ 构成一个三能级 EIT 系统。被降温的原子最初被磁阱捕获于态 $|1\rangle = |3S,\ F=1,\ M_F=-1\rangle$。探测和耦合激光场之间的受激光子交换产生了态 $|1\rangle$ 和 $|2\rangle$ 的“暗”叠加态，这使得介质对共振探测脉冲是透明的。**b**，我们施加一个直径 2.2 mm 的 σ^- 偏振耦合激光，频率与 $|3S,\ F=2,\ M_F=+1\rangle \rightarrow |3P,\ F=2,\ M_F=0\rangle$ 态的跃迁共振，还施加一个同向传播的，1.2 mm 直径，频率调谐至 $|3S,\ F=1,\ M_F=-1\rangle \rightarrow |3P,\ F=2,\ M_F=0\rangle$ 跃迁的 σ^+ 偏振探测激光。两激光束以正交的线性偏振方向发出（双箭头和实心圆分别代表探测和耦合光束的偏振方向）。它们通过分束器合成，经四分之一（λ/4）波片形成圆偏振，然后射入原子云。从原子云中射出后，激光束又经过四分之一（λ/4）波片，在被偏振分束立方体分离之前重获原来各自的线偏振态。原子云首先成像在一外置的成像平面，然后成像在 CCD（电荷耦合器件）摄像机上。一针孔被置于外置的成像平面上，位于云图像的中心。通过针孔和翻转镜的适当摆放，只有通过了云中心区域的探测和耦合激光光束被采集并且同时由两个光电倍增管（PMT）监测。量子态 $|1\rangle$ 和 $|2\rangle$ 有相同的一阶塞曼位移，所以双光子共振在被捕获的原子云中保持下来。冷原子和同向传播的激光消除了多普勒效应。但是，在这种情况下，由于存在到量子态 $|4\rangle$ 的非共振跃迁，光脉冲无法获得理想的透射率。

Atoms are prepared (magnetically trapped) in a particular internal quantum state $|1\rangle$ (Fig. 1a). The atom cloud is first illuminated by a coupling laser, resonant with the $|2\rangle$–$|3\rangle$ transition. With only the coupling laser on and all atoms in $|1\rangle$, the system is in a dark state (equation (1) with $\Omega_p = 0$). A probe laser pulse, tuned to the $|1\rangle$–$|3\rangle$ transition and co-propagating with the coupling laser, is subsequently sent through the atomic medium. Atoms within the pulse region are driven into the dark-state superposition of states $|1\rangle$ and $|2\rangle$, determined by the ratio of the instantaneous Rabi frequencies of the laser fields (equation (1)).

The presence of the coupling laser field creates transparency, a very steep refractive index profile, and low group velocity, V_g, for the probe pulse[1-10]. As the pulse enters the atomic medium, it is spatially compressed by a factor c/V_g whereas its peak electric amplitude remains constant during the slow-down[4,7].

The experiment is performed with the apparatus described in refs 4 and 11. Figure 1 shows the new optical set-up and atomic energy levels involved. A typical cloud of 11 million sodium atoms is cooled to 0.9 μK, which is just above the critical temperature for Bose–Einstein condensation. The cloud has a length of 339 μm in the z direction, a width of 55 μm in the transverse directions, and a peak density of 11 μm^{-3}. Those portions of the co-propagating probe and coupling laser beams that have passed through the 15-μm-diameter centre region of the cloud are selected and monitored simultaneously by two photomultiplier tubes (PMTs).

Figure 2a shows typical signals detected by the PMTs. The dashed curve is the measured intensity of the coupling laser, which is turned on a few microseconds before the probe pulse. The open circles indicate a gaussian-shaped reference probe pulse recorded in the absence of atoms (1/e full width is 5.70 μs). The filled circles show a probe pulse measured after it has passed through a cold atom cloud, and the solid curve is a gaussian fit to the data. The delay of this probe pulse, relative to the reference pulse, is 11.8 μs corresponding to a group velocity of 28 $m\ s^{-1}$, a reduction by a factor of 10^7 from its vacuum value. The measured delay agrees with the theoretical prediction of 12.2 μs based on a measured coupling Rabi frequency Ω_c of 2.57 MHz $\times$ 2π and an observed atomic column density of 3,670 μm^{-2}.

原子制备(磁阱捕获)时处于一特定的内量子态 $|1\rangle$ (图 1a)。原子云最初受一耦合激光辐射，与 $|2\rangle$–$|3\rangle$ 跃迁共振。只有耦合激光开启，并且所有的原子都处于 $|1\rangle$ 态时，系统处于暗态中(等式(1)中 $\Omega_p = 0$)。一探测激光脉冲，被调谐至 $|1\rangle$–$|3\rangle$ 跃迁，与耦合激光同向传播，随后被发送穿过原子介质。在脉冲区域的原子被驱动至 $|1\rangle$ 和 $|2\rangle$ 的暗叠加态，该叠加态由激光场的瞬时拉比频率的比率决定(等式(1))。

耦合激光场的存在产生了透明现象，并导致探测脉冲的折射率图谱变得非常陡峭，群速度 V_g 变得很低[1-10]。当脉冲进入原子介质时，其空间尺寸被压缩了 c/V_g 倍，而在减速期间其峰值电子振幅保持恒定[4,7]。

实验装置如参考文献 4 和 11 所述。图 1 给出了新的光学装置以及所涉及的原子能级。一典型的原子云包含 1,100 万钠原子，它被冷却至 0.9 μK，刚好在玻色–爱因斯坦凝聚临界温度以上。原子云在 z 方向长 339 μm，横向宽度 55 μm，峰值密度为 11 μm^{-3}。选择同向传播的探测和耦合激光光束中通过原子云中心直径 15 μm 区域的部分，并用两个光电倍增管(PMT)同时监测。

图 2a 给出了 PMT 探测到的典型信号。虚线表示耦合激光的测量强度，比探测脉冲早几微秒打开。空心圆为没有原子存在时(1/e 全宽为 5.70 μs)所记录的高斯型参考探测脉冲。实心圆为探测脉冲通过一冷原子云后所测得的信号，实线为数据的高斯拟合。这个探测脉冲与参考脉冲相比的延迟为 11.8 μs，对应群速度为 28 $m \cdot s^{-1}$，相当于真空中速度的 $1/10^7$。实验测得的延迟与理论预测值 12.2 μs 相符合。该理论预测值是基于测量所得的耦合拉比频率 Ω_c(2.57 MHz × 2π)以及观测所得的原子柱密度(3,670 μm^{-2})获得的。

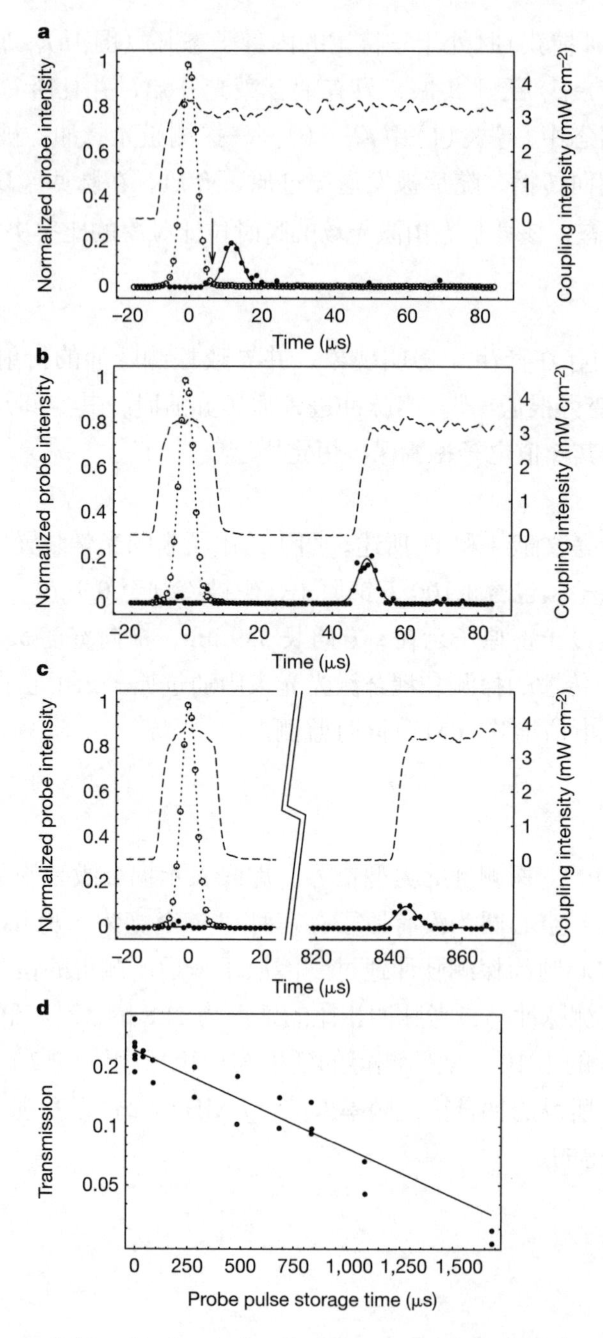

Fig. 2. Measurements of delayed and revived probe pulses. Open circles (fitted to the dotted gaussian curves) show reference pulses obtained as the average of 100 probe pulses recorded in the absence of atoms. Dashed curves and filled circles (fitted to the solid gaussian curves) show simultaneously measured intensities of coupling and probe pulses that have propagated under EIT conditions through a 339-μm-long atom cloud cooled to 0.9 μK. The measured probe intensities are normalized to the peak intensity of the reference pulses (typically, $\Omega_p/\Omega_c = 0.3$ at the peak). **a**, Probe pulse delayed by 11.8 μs. The arrow at 6.3 μs indicates the time when the probe pulse is spatially compressed and contained completely within the atomic cloud. (The intersection of the back edge of the reference pulse and the front edge of the delayed pulse defines a moment when the tail of the probe pulse has just entered the cloud and the leading

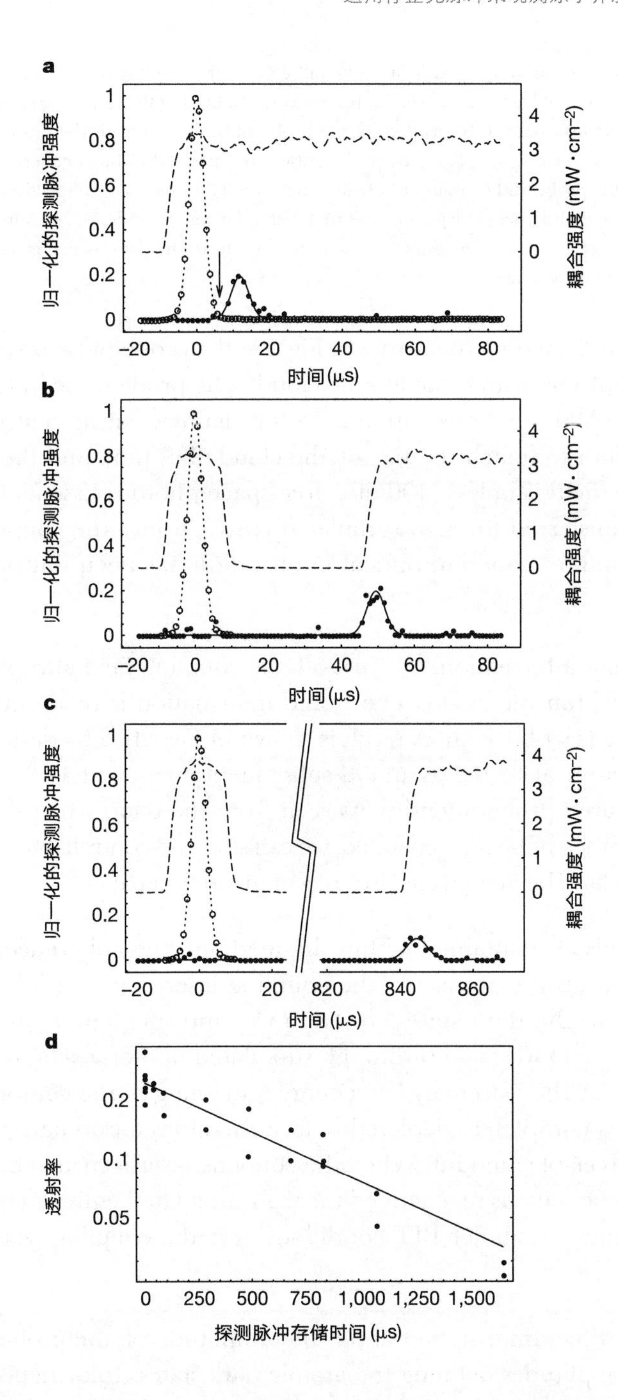

图 2. 探测脉冲延迟和再生的测量。空心圆（高斯拟合用点线表示）对应参考脉冲，为没有原子时所记录的 100 个探测脉冲的平均值。虚线和实心圆（高斯拟合用实线表示）对应在 EIT 条件下同时测量耦合和探测脉冲经由 339 μm 长，降温至 0.9 μK 的原子云传播所得到的强度。测量所得的探测强度相对参考脉冲的峰值强度进行了归一化（一般地，峰值处的 $\Omega_p/\Omega_c = 0.3$）。**a**，探测脉冲延迟 11.8 μs。6.3 μs 处的箭头表示探测脉冲空间上被压缩并被完全包含在原子云中的时刻（参考脉冲的后沿与延迟脉冲的前沿之间的交叉点决定了脉冲尾部刚刚进入云而前沿又恰好要离开的时刻）。**b**，**c**，耦合场在 $t = 6.3$ μs 关闭，

edge is just about to exit.) **b**, **c**, Revival of a probe pulse after the coupling field is turned off at $t = 6.3$ μs and turned back on at $t = 44.3$ μs and $t = 839.3$ μs, respectively. During the time interval when the coupling laser is off, coherent information imprinted by the probe pulse, is stored in the atomic medium. Upon subsequent turn-on of the coupling field, the probe pulse is regenerated through coherent stimulation. The time constants for the probe and coupling PMT amplifiers are 0.3 μs and 3 μs, respectively. The actual turn on/off time for the coupling field is 1 μs, as measured with a fast photodiode. **d**, Measured transmission of the probe pulse energy versus storage time. The solid line is a fit to the data, which gives a 1/e decay time of 0.9 ms for the atomic coherence.

At time $t = 6.3$ μs, indicated by the arrow in Fig. 2a, the probe pulse is spatially compressed and contained completely within the atomic cloud. The probe pulse in free space is 3.4 km long and contains 27,000 photons within a 15-μm diameter at its centre. It is compressed in the atomic medium to match the size of the cloud (339 μm), and the remaining optical energy in the probe field is only 1/400 of a free-space photon. Essentially all of the probe energy has been transferred through stimulated emission into the coupling laser field and the atomic medium, and coherent optical information has been imprinted on the atoms (equation (1)).

To store this coherent information, we turn off the coupling field abruptly when the probe pulse is contained within the cloud. The stored information is read out at a later time by turning the coupling laser back on. A result is shown in Fig. 2b. The dashed curve shows the coupling laser's turn-off at $t = 6.3$ μs and its subsequent turn-on at 44.3 μs. The filled circles represent the measured probe intensity. As seen from the data, when the coupling laser is turned back on the probe pulse is regenerated: we can stop and controllably regenerate the probe pulse. Similar effects have been predicted in a recent theoretical paper[12].

When the probe pulse is contained within the medium, the coherence of the laser fields is already imprinted on the atoms. As the coupling laser is turned off, the probe field is depleted to maintain the dark state (equation (1)) and (negligible) atomic amplitude is transferred from state $|1\rangle$ to state $|2\rangle$ through stimulated photon exchange between the two light fields. Because of the extremely low energy remaining in the compressed probe pulse, as noted above, it is completely depleted before the atomic population amplitudes have changed by an appreciable amount. When the coupling laser is turned back on, the process reverses and the probe pulse is regenerated through stimulated emission into the probe field. It propagates subsequently under EIT conditions as if the coupling beam had never been turned off.

During the storage time, information about the amplitude of the probe field is contained in the population amplitudes defining the atomic dark states. Information about the mode vector of the probe field is contained in the relative phase between different atoms in the macroscopic sample. The use of cold atoms minimizes thermal motion and the associated smearing of the relative phase during the storage time. (We obtain storage times that are up to 50 times larger than the time it takes an atom to travel one laser wavelength. As seen from equation (1), the difference between the wavevectors of the two laser fields determines the wavelength of the periodic phase pattern imprinted on the medium, which is 10^5 times larger than the individual laser wavelengths.)

又分别在 $t = 44.3$ μs 和 $t = 839.3$ μs 重新开启之后探测脉冲的再生。在耦合激光关闭的时间间隔期间内，探测脉冲所携带的相干信息就存储在原子介质中。随着耦合场再次开启，探测脉冲通过相干激励再次产生。探测和耦合 PMT 放大器的时间常数分别为 0.3 μs 和 3 μs。根据快速光电二极管的测量，耦合场的实际开启/关闭时间为 1 μs。**d**，测量所得的探测脉冲能量透射率与存储时间的关系。实线是对数据进行的拟合结果，给出了对于原子相干的 1/e 衰减时间 0.9 ms。

当时间 $t = 6.3$ μs 时，如图 2a 中的箭头所示，探测脉冲空间上被压缩并被完全包含在原子云中。这个探测脉冲在自由空间中长 3.4 km，在其中心 15 μm 直径范围内含有 27,000 个光子。为了与云尺寸(339 μm)相匹配，它在原子介质中被压缩，且探测场所剩余的光能仅为自由空间光子的 1/400。实质上来讲几乎所有的探测能量都通过受激辐射被转移至耦合激光场和原子介质，而相干光学信息也被印记于原子之上(等式(1))。

为了存储这些相干信息，当探测脉冲刚好被原子云包含时我们突然关闭耦合场。所存储的信息在随后耦合激光开启时被读取出来。图 2b 给出了一结果图。虚线显示耦合激光在 $t = 6.3$ μs 时关闭，随后在 44.3 μs 时开启。实心圆表示测得的探测强度。从这些数据中可以看出，当耦合激光再一次被开启时探测脉冲也再次产生，也就是说我们可以停止并控制探测脉冲的再生。类似的现象已经被近期的理论文章所预测[12]。

当探测脉冲被包含在介质中，激光场的相干信息已经被印记在原子上。当耦合激光关闭时，探测场被耗尽以保持暗态(等式(1))，而(可忽略的)原子振幅通过两光场间的受激光子交换从 $|1\rangle$ 态转移到 $|2\rangle$ 态。如上所述，由于被压缩的探测脉冲剩余的能量相当低，在原子布居振幅发生可观的改变之前，能量已经完全被耗尽。当耦合激光被重新开启时，逆过程发生，探测脉冲通过受激辐射到探测场而再生。随后探测脉冲在 EIT 条件下传播，就像耦合光束从未被关闭一样。

在存储时间内，探测场的振幅信息包含在定义原子暗态的布居振幅之中。探测场的模式矢量信息包含在宏观样本中不同原子间的相对相位之中。冷原子的运用使热运动降至最低，也就使存储过程中相应的相对相位拖尾效应达到最小。(我们获得存储时间最长达到一个原子通过一个激光波长所需时间的 50 倍。通过等式(1)可以得出，两激光场波矢的差决定了印记在介质中的周期性相位图样的波长，该波长是单个激光波长的 10^5 倍。)

The regenerated probe pulse in Fig. 2b has the same shape as the "normal" EIT pulse shown in Fig. 2a. Figure 2c shows a case where the optical coherence is stored in the atomic medium for more than 800 μs before it is read out by the coupling laser. Here the amplitude of the revived probe pulse is reduced compared to that of the pulse in Fig. 2b. Figure 2d shows the measured transmission for a series of pulses as a function of their storage time in the atom cloud. The data are consistent with an exponential decay with a 1/e decay time of 0.9 ms, comparable to the calculated mean free time of 0.5 ms between elastic collisions in the atom cloud with a density of 11 μm^{-3}. Further studies of the decoherence mechanisms are planned but are beyond the scope of this Letter.

We have verified experimentally that the probe pulse is regenerated through stimulated rather than spontaneous emission. To do this, we prepared all atoms in state $|2\rangle$ and subsequently turned on the coupling laser alone. The coupling laser was completely absorbed for tens of microseconds without generating any signal in the probe PMT.

In Fig. 3a–c, we show three PMT signal traces recorded under similar conditions except that we vary the intensity, I_{c2}, of the coupling laser when it is turned back on. When I_{c2} is larger than the original coupling intensity, I_{c1}, the amplitude of the revived probe pulse increases and its temporal width decreases (Fig. 3a). For $I_{c2} < I_{c1}$, the opposite occurs (Fig. 3c). These results support our physical picture of the process. The stored atomic coherence dictates the ratio of the Rabi frequencies of the coupling and revived probe fields, as well as the spatial width of the regenerated pulse. In Fig. 3d we show that with a large I_{c2}, the peak intensity of the revived probe pulse exceeds that of the original input pulse by 40%.

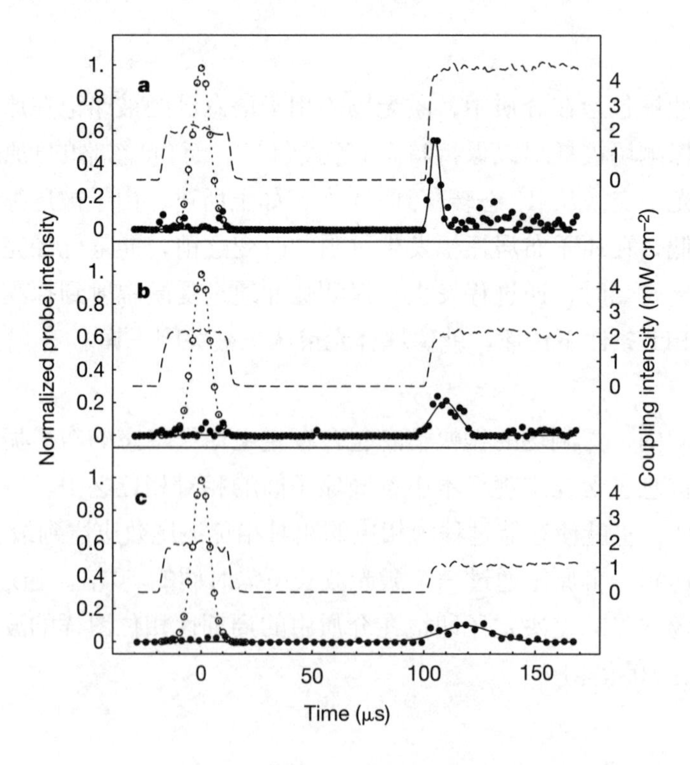

图 2b 中的再生探测脉冲与图 2a 中所示的“正常”EIT 脉冲波形相同。图 2c 显示了光学相干信息在被耦合激光读取前，存储在原子介质中超过 800 μs 的情况。这里的再生探测脉冲振幅与图 2b 相比有所降低。图 2d 为测量所得的一系列脉冲在原子云中的透射率与存储时间的函数关系图。这些数据给出了 0.9 ms 的 1/e 衰减时间，与计算所得的在密度为 11 μm^{-3} 的原子云中弹性碰撞的平均自由时间 0.5 ms 相当。关于退相干机制的进一步研究正在计划中，但不在本文讨论范围之内。

我们已经在实验上验证了探测脉冲的再生是由于受激辐射而非自发辐射。为了实现这个目的，我们准备了所有原子处于 $|2\rangle$ 的态，随后单独开启耦合激光。耦合激光在几十个毫秒内被完全吸收，但没有在 PMT 上产生任何信号。

图 3a ~ 3c 显示了在相似条件下记录的三种 PMT 信号轨迹，不同之处在于我们改变了当耦合激光再次开启时的强度 I_{c2}。当 I_{c2} 比最初的耦合强度 I_{c1} 大时，再生的探测脉冲振幅增大，其时间宽度减小（图 3a）。对于 $I_{c2} < I_{c1}$，情况相反（图 3c）。这些结果支持我们为整个过程建立的物理图像。存储的原子相干信息限定了耦合与再生探测场的拉比频率比率，以及再生脉冲的空间宽度。在图 3d 中，当 I_{c2} 很大时，再生的探测脉冲峰值强度超过初始输入脉冲的 40%。

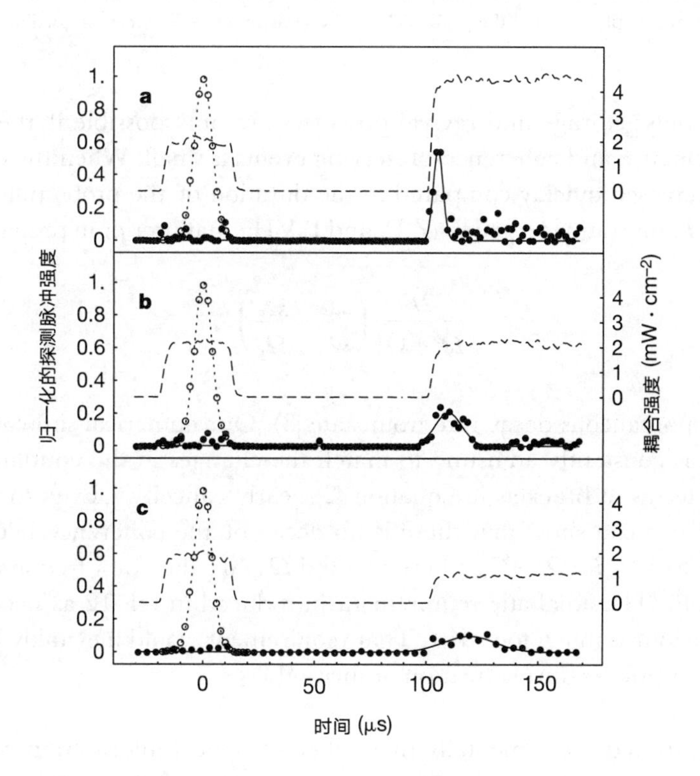

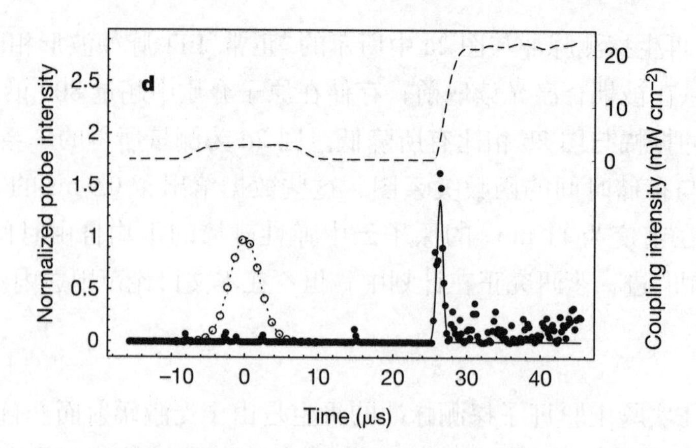

Fig. 3. Measurements of revived probe pulses for varying intensities (I_{c2}) of the second coupling pulse. The intensity (I_{c1}) of the first coupling pulse is held constant. **a–c**, The figures are recorded for I_{c2}/I_{c1} ratios of 2, 1, and 0.5, respectively. A series of data show that the height and the inverse temporal width of the revived pulses are each proportional to I_{c2}. These observations are consistent with our physical picture. Because the atomic coherence dictates the ratio of the Rabi frequencies for the coupling and revived probe fields (equation (1)), the intensity of the regenerated probe pulse is proportional to the intensity of the coupling laser when it is turned back on. Furthermore, the spatial width of the revived pulse is determined by the distribution of the atomic coherence and is thus the same as the spatial extent of the original compressed pulse. The group velocity of the probe pulse under EIT conditions is proportional to the coupling intensity[4,7]. With a larger (smaller) I_{c2}, the revived probe pulse acquires a proportionally larger (smaller) group velocity, which causes its temporal width to be inversely proportional to I_{c2}. Panel **d** shows that the intensity of the revived probe pulse can exceed that of the original input pulse, in this instance by 40%. (The observed peak-to-peak fluctuation of laser intensity is less than 10%.) The energy in the revived probe pulses is the same in all panels **a–d**, owing to the fact that the total stored amplitude of state $|2\rangle$ atoms (available to stimulate photons into the probe field) is the same in all cases. Meanings of lines and symbols as in Fig. 2.

Dissipationless pulse storage and revival processes are only possible if the ratio between the rates of dissipative and coherence-preserving events is small. When the coupling field is increased or decreased quickly compared to the duration of the probe pulse (τ) but slowly compared to $1/\Gamma$, this ratio is equal to (Z.D. and L.V.H., manuscript in preparation)

$$\frac{2\Gamma}{\Omega_c^2+\Omega_p^2}\left(\frac{\dot{\Omega}_p}{\Omega_p}-\frac{\dot{\Omega}_c}{\Omega_c}\right) \tag{2}$$

where Γ is the spontaneous decay rate from state $|3\rangle$. Our numerical simulations show that the probe field is constantly adjusting to match the changes in the coupling field in such a way that the terms in brackets in equation (2) nearly cancel[13,14]. Even for turn-off times faster than $1/\Gamma$, we can show that there is no decay of the coherence between states $|1\rangle$ and $|2\rangle$ as long as $\tau \gg \Gamma/(\Omega_{c0}^2+\Omega_{p0}^2)$; here Ω_{c0} and Ω_{p0} are the Rabi frequencies before the coupling turn-off. (The adiabatic requirement introduced in ref. 12 as necessary for non-dissipative behaviour is much too strict. That requirement would inevitably break down for low coupling laser powers during turn-on or turn-off.)

We have demonstrated experimentally that coherent optical information can be stored in

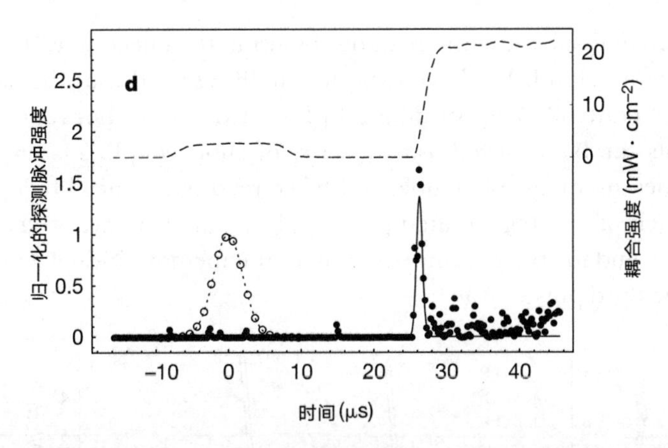

图 3. 不同的第二耦合脉冲强度(I_{c2})下再生探测脉冲的测量。第一耦合脉冲强度(I_{c1})保持恒定。**a ~ c**，记录的是分别对应 I_{c2}/I_{c1} 之比为 2，1 以及 0.5 的图形。一系列数据表明，再生脉冲的高度、时间宽度的倒数均与 I_{c2} 成正比。观察到的现象与我们的物理图像相一致。由于原子相干限定了耦合和再生探测场的拉比频率之比(等式(1))，再生探测脉冲的强度与耦合激光再次开启时的强度成正比。另外，再生脉冲的空间宽度由原子相干的分布决定，因此与最初被压缩的脉冲的空间范围相同。EIT 条件下探测脉冲的群速度与耦合强度成正比[4,7]。对于更大(或更小)的 I_{c2}，再生探测脉冲获得了按比例更大(或更小)的群速度，从而导致其时间宽度与 I_{c2} 成反比。图 **d** 显示再生的探测脉冲强度可以超过初始输入脉冲的强度，图中所示情况超过了 40%。(观察到的激光强度的峰–峰波动在 10% 以内。)再生探测脉冲的能量在所有 **a ~ d** 图中相同，这是由于在所有四种情况下，处于量子态 $|2\rangle$ 的原子(可用于将光子激发到探测场中)的总存储振幅相等。线和符号所对应的含义与图 2 一致。

无耗散的脉冲存储和再生过程只有在耗散率和相干保持事件率之比很小时才可能实现。当耦合场的增大或减小与探测脉冲的持续时间(τ)相比很迅速，而与 $1/\Gamma$ 相比变化缓慢时，这个比值等于(扎卡里·达顿和莱娜·韦斯特戈·豪，稿件准备中)

$$\frac{2\Gamma}{\Omega_{\mathrm{c}}^2+\Omega_{\mathrm{p}}^2}\left(\frac{\dot{\Omega}_{\mathrm{p}}}{\Omega_{\mathrm{p}}}-\frac{\dot{\Omega}_{\mathrm{c}}}{\Omega_{\mathrm{c}}}\right) \tag{2}$$

这里 Γ 为态 $|3\rangle$ 的自发衰减率。我们的数值模拟显示，探测场一直通过使等式(2)的括号内项几乎相消的方式不断调节至与耦合场变化相匹配[13,14]。甚至对于关闭时间小于 $1/\Gamma$ 的情况，我们也可以证明，只要 $\tau \gg \Gamma/(\Omega_{\mathrm{c_0}}^2+\Omega_{\mathrm{p_0}}^2)$，态 $|1\rangle$ 和 $|2\rangle$ 的相干就不存在衰减；这里 $\Omega_{\mathrm{c_0}}$ 和 $\Omega_{\mathrm{p_0}}$ 为在耦合激光关闭前的拉比频率。(文献 12 中提到的对于非耗散行为所必需的绝热要求太过严格。这个条件在低耦合激光电源开启和关闭时就不可避免地被破坏。)

本文实验论证了相干光学信息可以被存储在原子介质中，并随后在磁阱捕获的

an atomic medium and subsequently read out by using the effect of EIT in a magnetically trapped, cooled atom cloud. We have experimentally verified that the storage and read-out processes are controlled by stimulated photon transfers between two laser fields. Multiple read-outs can be achieved using a series of short coupling laser pulses. In Fig. 4a and b we show measurements of double and triple read-outs spaced by up to hundreds of microseconds. Each of the regenerated probe pulses contains part of the contents of the "atomic memory", and for the parameters chosen, the memory is depleted after the second pulse and after the third pulse.

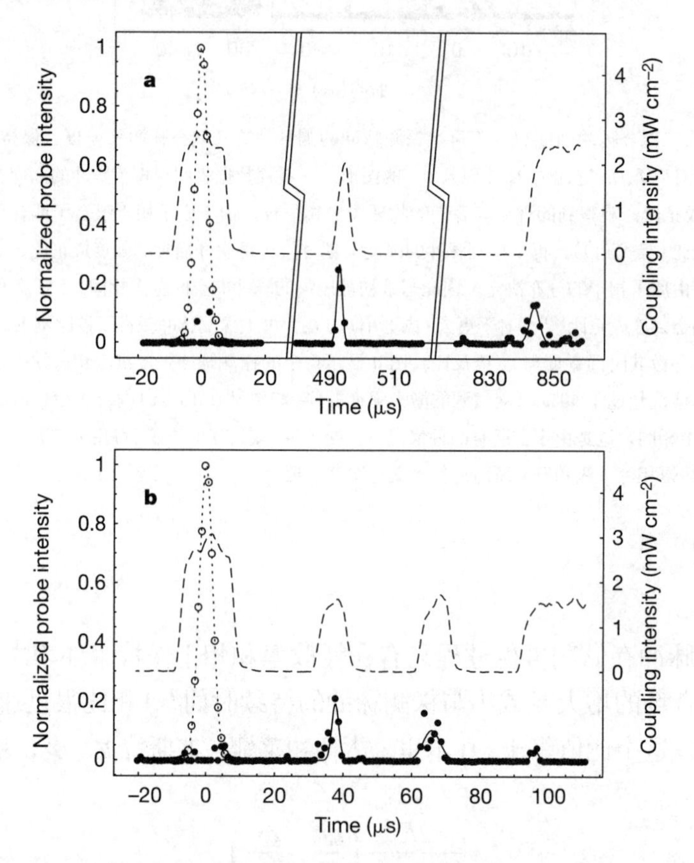

Fig. 4. Measurements of double and triple read-out of the atomic memory. To deplete the atomic memory in these cases, we use two (**a**) and three (**b**) short coupling pulses. The total energy in the two (three) revived probe pulses is measured to be the same as the energy in the single revived probe pulse obtained with a single, long coupling laser pulse (as used in Figs 2 and 3). Meanings of lines and symbols as in Fig. 2.

We believe that this system could be used for quantum information transfer; for example, to inter-convert stationary and flying qubits[15]. By injection of multiple probe pulses into a Bose–Einstein condensate—where we expect that most atomic collisions are coherence-preserving—and with use of controlled atom–atom interactions, quantum information processing may be possible during the storage time.

(**409**, 490-493; 2001)

冷原子云中利用 EIT 效应读取出来。本文通过实验验证了存储和读取过程是由两激光场之间的受激光子转移来控制的。应用一系列短耦合激光脉冲可以实现多重读取。图 4a 和 4b 显示了间隔高达几百个毫秒的二重和三重读取的测量结果。每个再生的探测脉冲包含“原子内存”的部分内容，而对于本实验所选择的参数，内存在第二和第三个脉冲之后被耗尽。

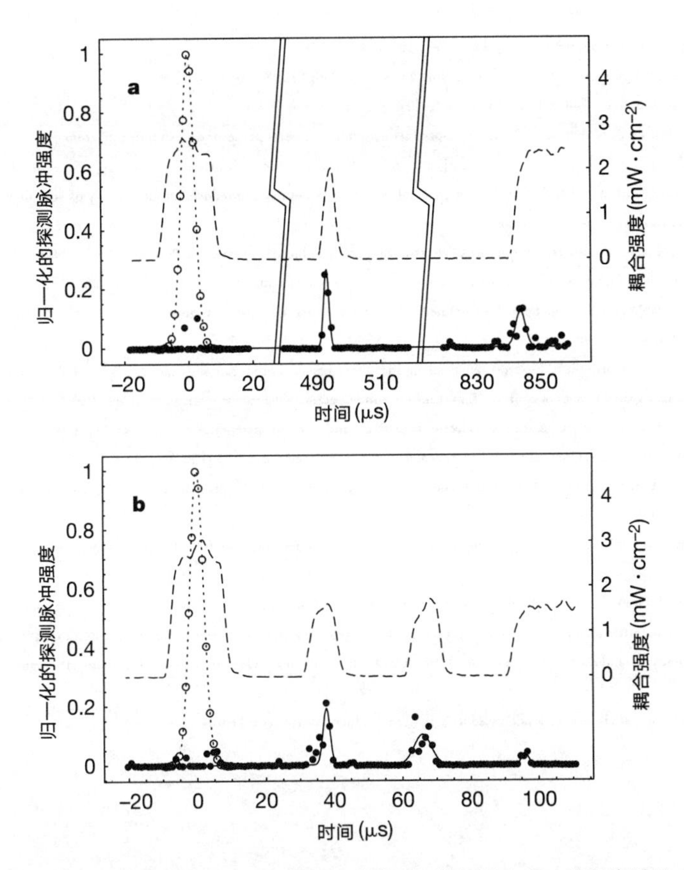

图 4. 原子内存的二重和三重读取测量。为了耗尽这些情况下的原子内存，我们使用了二重(**a**)和三重(**b**)短耦合脉冲。二(三)重再生探测脉冲的总能量与具有单一长耦合激光脉冲的单再生探测脉冲(图 2 和 3 中使用的脉冲)的能量相等。线和符号所对应的含义与图 2 一致。

我们相信该系统可以应用于量子信息传递；例如，用于定态量子比特和飞行量子比特的互相转换[15]。通过注入多重探测脉冲至玻色–爱因斯坦凝聚体(我们预计此条件下多数原子碰撞是可以相干保留的)并运用受控制的原子–原子之间的相互作用，量子信息处理在存储时间之内或许是可能的。

(崔宁 翻译；石锦卫 审稿)

Chien Liu*†, Zachary Dutton*‡, Cyrus H. Behroozi*† & Lene Vestergaard Hau*†‡
* Rowland Institute for Science, 100 Edwin H. Land Boulevard, Cambridge, Massachusetts 02142, USA
† Division of Engineering and Applied Sciences, ‡ Department of Physics, Harvard University, Cambridge, Massachusetts 02138, USA

Received 13 October; accepted 17 November 2000.

References:

1. Harris, S. E. Electromagnetically induced transparency. *Phys. Today* **50**, 36-42 (1997).
2. Scully, M. O. & Zubairy, M. S. *Quantum Optics* (Cambridge Univ. Press, Cambridge, 1997).
3. Arimondo, E. in *Progress in Optics* (ed. Wolf, E.) 257-354 (Elsevier Science, Amsterdam, 1996).
4. Hau, L. V., Harris, S. E., Dutton, Z. & Behroozi, C. H. Light speed reduction to 17 metres per second in an ultracold atomic gas. *Nature* **397**, 594-598 (1999).
5. Kash, M. M. *et al.* Ultraslow group velocity and enhanced nonlinear optical effects in a coherently driven hot atomic gas. *Phys. Rev. Lett.* **82**, 5229-5232 (1999).
6. Budker, D., Kimball, D. F., Rochester, S. M. & Yashchuk, V. V. Nonlinear magneto-optics and reduced group velocity of light in atomic vapor with slow ground state relaxation. *Phys. Rev. Lett.* **83**, 1767-1770 (1999).
7. Harris, S. E., Field, J. E. & Kasapi, A. Dispersive properties of electromagnetically induced transparency. *Phys. Rev. A* **46**, R29-R32 (1992).
8. Grobe, R., Hioe, F. T. & Eberly, J. H. Formation of shape-preserving pulses in a nonlinear adiabatically integrable system. *Phys. Rev. Lett.* **73**, 3183-3186 (1994).
9. Xiao, M., Li, Y.-Q., Jin, S.-Z. & Gea-Banacloche, J. Measurement of dispersive properties of electromagnetically induced transparency in rubidium atoms. *Phys. Rev. Lett.* **74**, 666-669 (1995).
10. Kasapi, A., Jain, M., Yin, G. Y. & Harris, S. E. Electromagnetically induced transparency: propagation dynamics. *Phys. Rev. Lett.* **74**, 2447-2450 (1995).
11. Hau, L. V. *et al.* Near-resonant spatial images of confined Bose-Einstein condensates in a 4-Dee magnetic bottle. *Phys. Rev. A* **58**, R54-R57 (1998).
12. Fleischhauer, M. & Lukin, M. D. Dark-state polaritons in electromagnetically induced transparency. *Phys. Rev. Lett.* **84**, 5094-5097 (2000).
13. Harris, S. E. Normal modes for electromagnetically induced transparency. *Phys. Rev. Lett.* **72**, 52-55 (1994).
14. Fleischhauer, M. & Manak, A. S. Propagation of laser pulses and coherent population transfer in dissipative three-level systems: An adiabatic dressed-state picture. *Phys. Rev. A* **54**, 794-803 (1996).
15. DiVincenzo, D. P. The physical implementation of quantum computation. Preprint quant-ph/0002077 at ⟨http://xxx.lanl.gov⟩ (2000).

Acknowledgements. We thank J. Golovchenko for discussions during which the idea of the rapid turn off and on of the coupling laser first emerged. We also thank M. Burns for critical reading of the manuscript. This work was supported by the Rowland Institute for Science, the Defense Advanced Research Projects Agency, the US Airforce Office of Scientific Research, and the US Army Research Office OSD Multidisciplinary University Research Initiative Program.

Correspondence and requests for materials should be addressed to C.L. (e-mail: chien@deas.harvard.edu).

Experimental Violation of a Bell's Inequality with Efficient Detection

Experimental Violation of a Bell's Inequality with Efficient Detection

M. A. Rowe *et al.*

Editor's Note

By the new millennium, careful tests of John Bell's celebrated "inequalities" had indicated that quantum physics seems to be inconsistent with any local realistic interpretation—that is, with a view in which systems have definite properties independent of other parts of the universe, and no influence can travel faster than light. But argument persisted due to various "loopholes" linked to experimental imperfections, including the limited efficiency of particle detectors. Here Mary Rowe and colleagues report violation of Bell's inequalities in experiments using heavy beryllium ions, for which they were able to detect all particles and thus close one loophole. Nonetheless, not all possible local realistic interpretations of quantum theory have been excluded even now.

Local realism is the idea that objects have definite properties whether or not they are measured, and that measurements of these properties are not affected by events taking place sufficiently far away[1]. Einstein, Podolsky and Rosen[2] used these reasonable assumptions to conclude that quantum mechanics is incomplete. Starting in 1965, Bell and others constructed mathematical inequalities whereby experimental tests could distinguish between quantum mechanics and local realistic theories[1,3-5]. Many experiments[1,6-15] have since been done that are consistent with quantum mechanics and inconsistent with local realism. But these conclusions remain the subject of considerable interest and debate, and experiments are still being refined to overcome "loopholes" that might allow a local realistic interpretation. Here we have measured correlations in the classical properties of massive entangled particles ($^9Be^+$ ions): these correlations violate a form of Bell's inequality. Our measured value of the appropriate Bell's "signal" is 2.25 ± 0.03, whereas a value of 2 is the maximum allowed by local realistic theories of nature. In contrast to previous measurements with massive particles, this violation of Bell's inequality was obtained by use of a complete set of measurements. Moreover, the high detection efficiency of our apparatus eliminates the so-called "detection" loophole.

EARLY experiments to test Bell's inequalities were subject to two primary, although seemingly implausible, loopholes. The first might be termed the locality or "lightcone" loophole, in which the correlations of apparently separate events could result from unknown subluminal signals propagating between different regions of the apparatus. Aspect[16] has given a brief history of this issue, starting with the experiments of ref. 8 and highlighting the strict relativistic separation between measurements reported by the

利用高效检测来实验验证贝尔不等式的违背

罗等

编者按

新的千年到来之前，对约翰·贝尔著名的“不等式”进行的种种仔细的验证表明量子物理学似乎与任何定域实在论的解释都不相吻合，定域实在论认为系统有确定的性质，与世界的其他事物无关，且任何影响都不能超光速传播。但由于存在与实验不完美有关的各种各样的、包括粒子探测器的有效探测效率在内的“漏洞”，争论一直持续存在。本文中，玛丽·罗和其同事们报道了利用重铍离子所做的实验中贝尔不等式的违背。由于他们在实验中可以探测到所有的粒子，从而堵上了一个漏洞。然而，即使是现在，仍不能排除对量子理论的所有可能的定域实在论的解释。

定域实在论的观点认为，无论是否被测量，物体都具有确定的性质，对这些性质的测量并不受足够远发生的事件的影响[1]。爱因斯坦–波多尔斯基–罗森[2]用这些合理的假设，得出量子力学是不完备的结论。从1965年开始，贝尔和其他人构建了一个数学不等式，依靠实验验证能区分量子力学与定域实在论的理论[1,3-5]。已进行的很多实验[1,6-15]与量子力学理论相符，而与定域实在论不符。但这些结论仍然是大家相当感兴趣和有争议的问题，一些实验仍在不断完善以克服那些允许定域实在论解释的“漏洞”。我们现已测量了大质量的纠缠粒子（$^9Be^+$离子）在经典性质下的关联：这些关联违反了贝尔不等式的形式。我们得到的贝尔“信号”测量值是2.25 ± 0.03，然而2是自然界定域实在论所允许的最大值。与过去的重粒子测量相比，这次背离贝尔不等式的结果是使用一组完备测量得出的。此外，我们装置的高检测效率消除了所谓的“检测”漏洞。

验证贝尔不等式的早期实验受限于两个似乎令人难以置信的初级漏洞。第一个漏洞可称为定域性的，或“光锥”漏洞，这个漏洞指的是两个明显分离的事件之间的关联有可能是某些未知的亚光速信号在装置的不同区域之间传播造成的。阿斯佩[16]给出了这个问题的简明综述，综述从文献8的实验开始，并着重介绍了由因斯布鲁克小组[15]报道的测量之间严格的相对论性分离。日内瓦实验[14,17]也报道了类似的结果。

Innsbruck group[15]. Similar results have also been reported for the Geneva experiment[14,17]. The second loophole is usually referred to as the detection loophole. All experiments up to now have had detection efficiencies low enough to allow the possibility that the subensemble of detected events agrees with quantum mechanics even though the entire ensemble satisfies Bell's inequalities. Therefore it must be assumed that the detected events represent the entire ensemble; a fair-sampling hypothesis. Several proposals for closing this loophole have been made[18-24]; we believe the experiment that we report here is the first to do so. Another feature of our experiment is that it uses massive particles. A previous test of Bell's inequality was carried out on protons[25], but the interpretation of the detected events relied on quantum mechanics, as symmetries valid given quantum mechanics were used to extrapolate the data to a complete set of Bell's angles. Here we do not make such assumptions.

A Bell measurement of the type suggested by Clauser, Horne, Shimony and Holt[5] (CHSH) consists of three basic ingredients (Fig. 1a). First is the preparation of a pair of particles in a repeatable starting configuration (the output of the "magic" box in Fig. 1a). Second, a variable classical manipulation is applied independently to each particle; these manipulations are labelled ϕ_1 and ϕ_2. Finally, in the detection phase, a classical property with two possible outcomes is measured for each of the particles. The correlation of these outcomes

$$q(\phi_1, \phi_2) = \frac{N_{\text{same}}(\phi_1, \phi_2) - N_{\text{different}}(\phi_1, \phi_2)}{N_{\text{same}} + N_{\text{different}}} \tag{1}$$

is measured by repeating the experiment many times. Here N_{same} and $N_{\text{different}}$ are the number of measurements where the two results were the same and different, respectively. The CHSH form of Bell's inequalities states that the correlations resulting from local realistic theories must obey:

$$B(\alpha_1, \delta_1, \beta_2, \gamma_2) = |q(\delta_1, \gamma_2) - q(\alpha_1, \gamma_2)| + |q(\delta_1, \beta_2) + q(\alpha_1, \beta_2)| \leqslant 2 \tag{2}$$

where α_1 and δ_1 (β_2 and γ_2) are specific values of ϕ_1 (ϕ_2). For example, in a photon experiment[15], parametric down-conversion prepares a pair of photons in a singlet Einstein–Podolsky–Rosen (EPR) pair. After this, a variable rotation of the photon polarization is applied to each photon. Finally, the photons' polarization states, vertical or horizontal, are determined.

第二个漏洞通常被认为是检测漏洞。至今的所有实验检测的效率很低，以至于允许存在即使整个系综满足贝尔不等式，而检测事件的子系综与量子力学符合的可能性。因此，必须假设检测的事件代表整个系综；这是一个合理的取样假设。现已提出了堵上这个漏洞的数个建议[18-24]；我们相信，本文所述乃是首次针对堵上此漏洞所进行的实验。我们实验的另一个特点是采用大质量粒子。之前的贝尔不等式验证是利用质子进行的[25]，但检测事件的解释依据是量子力学，如量子力学中给出的对称性被用于将数据外推到贝尔角的完备集。此处我们并未采用这些假设。

由克劳塞、霍恩、希莫尼和霍尔特[5](CHSH)提出的这类贝尔测量包括三个基本部分(图 1a)。第一部分是制备在可重复的起始组态中的一对粒子(图 1 a 中"魔盒"的输出)。第二部分是对每个粒子独立地用一个可变的经典操控，这些操控标记为 ϕ_1 和 ϕ_2；最后在检测阶段，对每个粒子的经典性质进行测量，皆有两个可能的结果。这些结果的关联函数为

$$q(\phi_1, \phi_2) = \frac{N_{\text{same}}(\phi_1, \phi_2) - N_{\text{different}}(\phi_1, \phi_2)}{N_{\text{same}} + N_{\text{different}}} \tag{1}$$

式(1)是多次重复实验的测量结果。式中 N_{same} 和 $N_{\text{different}}$ 分别是两个结果相同和不同时的测量数目。贝尔不等式的 CHSH 形式表述为，由定域实在论得出的相关性必须遵从下式：

$$B(\alpha_1, \delta_1, \beta_2, \gamma_2) = |q(\delta_1, \gamma_2) - q(\alpha_1, \gamma_2)| + |q(\delta_1, \beta_2) + q(\alpha_1, \beta_2)| \leqslant 2 \tag{2}$$

式中 α_1 和 δ_1(β_2 和 γ_2)是 ϕ_1(ϕ_2)的特定值。例如，在光子实验中[15]，利用参数下转换技术把一对光子制备在爱因斯坦–波多尔斯基–罗森(EPR)单重态上，然后让每个光子通过一个角度可变的偏振片。最后确定光子的偏振状态：垂直或水平。

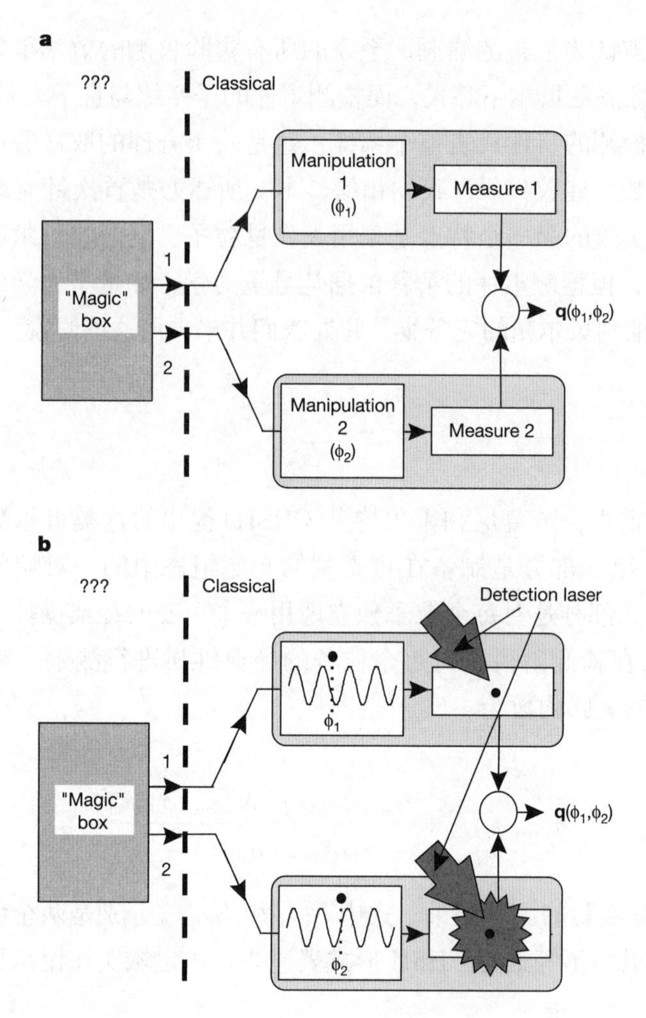

Fig. 1. Illustration of how Bell's inequality experiments work. The idea is that a "magic box" emits a pair of particles. We attempt to determine the joint properties of these particles by applying various classical manipulations to them and observing the correlations of the measurement outputs. **a**, A general CHSH type of Bell's inequality experiment. **b**, Our experiment. The manipulation is a laser wave applied with phases ϕ_1 and ϕ_2 to ion 1 and ion 2 respectively. The measurement is the detection of photons emanating from the ions upon application of a detection laser. Two possible measurement outcomes are possible, detection of few photons (as depicted for ion 1 in the figure) or the detection of many photons (as depicted for ion 2 in the figure).

Our experiment prepares a pair of two-level atomic ions in a repeatable configuration (entangled state). Next, a laser field is applied to the particles; the classical manipulation variables are the phases of this field at each ion's position. Finally, upon application of a detection laser beam, the classical property measured is the number of scattered photons emanating from the particles (which effectively measures their atomic states). Figure 1b shows how our experiment maps onto the general case. Entangled atoms produced in the context of cavity-quantum-electrodynamics[26] could similarly be used to measure Bell's inequalities.

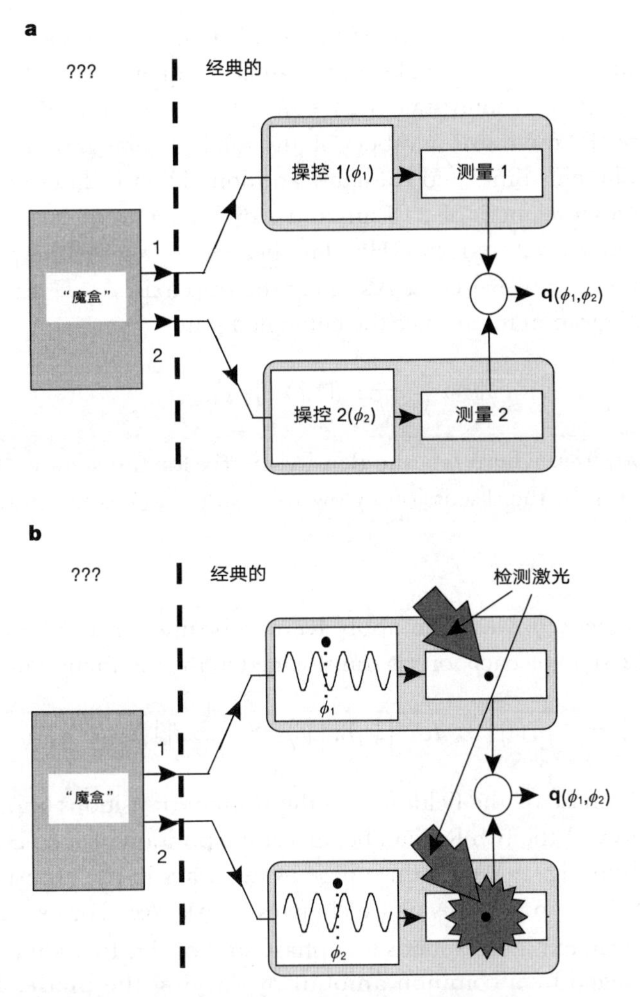

图 1. 贝尔不等式验证实验如何开展的示意图。实验思想是一个“魔盒”发射一对粒子。通过应用各种经典的操控方法，我们试图测定这些粒子的联合性质，并观测其测量输出的关联。**a**，一般的 CHSH 型贝尔不等式实验；**b**，我们的实验。操控是用相位为 ϕ_1 和 ϕ_2 的激光波分别加到离子 1 和 2 上。测量是对离子施加探测激光后，检测从离子发射出来的光子。有两个可能的测量结果，几乎检测不到光子（如图中对离子 1 的描绘）或检测到许多光子（如图中对离子 2 的描绘）。

实验中我们把一对二能级原子离子制备在可重复组态（纠缠态）上。其次，在粒子上加一激光场；经典操控的变量是每个离子所在处这个场的相位。最后，应用检测激光束之后，被测的经典性质是从粒子发射出的散射光子的数量（通过它可有效地测量粒子的原子态）。图 1b 展示了我们的实验与一般实验的对应。在腔量子电动力学意义下产生的纠缠的原子[26]能类似地用于测量贝尔不等式。

The experimental apparatus is as described in ref. 27. Two $^9Be^+$ ions are confined along the axis of a linear Paul trap with an axial centre-of-mass frequency of 5 MHz. We select two resolved levels of the $2S_{1/2}$ ground state, $|\downarrow\rangle \equiv |F=2, m_F=-2\rangle$ and $|\uparrow\rangle \equiv |F=1, m_F=-1\rangle$, where F and m_F are the quantum numbers of the total angular momentum. These states are coupled by a coherent stimulated Raman transition. The two laser beams used to drive the transition have a wavelength of 313 nm and a difference frequency near the hyperfine splitting of the states, $\omega_0 \cong 2\pi \times 1.25$ GHz. The beams are aligned perpendicular to each other, with their difference wavevector $\Delta\mathbf{k}$ along the trap axis. As described in ref. 27, it is possible in this configuration to produce the entangled state

$$|\psi_2\rangle = \frac{1}{\sqrt{2}}(|\uparrow\uparrow\rangle - |\downarrow\downarrow\rangle) \tag{3}$$

The fidelity $F = \langle\psi_2|\rho|\psi_2\rangle$, where ρ is the density matrix for the state we make, was about 88% for the data runs. In the discussion below we assume $|\psi_2\rangle$ as the starting condition for the experiment.

After making the state $|\psi_2\rangle$, we again apply Raman beams for a pulse of short duration (~400 ns) so that the state of each ion j is transformed in the interaction picture as

$$|\uparrow_j\rangle \rightarrow \frac{1}{\sqrt{2}}(|\uparrow_j\rangle - ie^{-i\phi_j}|\downarrow_j\rangle); \; |\downarrow_j\rangle \rightarrow \frac{1}{\sqrt{2}}(|\downarrow_j\rangle - ie^{i\phi_j}|\uparrow_j\rangle) \tag{4}$$

The phase, ϕ_j, is the phase of the field driving the Raman transitions (more specifically, the phase difference between the two Raman beams) at the position of ion j and corresponds to the inputs ϕ_1 and ϕ_2 in Fig. 1. We set this phase in two ways in the experiment. First, as an ion is moved along the trap axis this phase changes by $\Delta\mathbf{k} \cdot \Delta\mathbf{x}_j$. For example, a translation of $\lambda/\sqrt{2}$ along the trap axis corresponds to a phase shift of 2π. In addition, the laser phase on both ions is changed by a common amount by varying the phase, ϕ_s, of the radio-frequency synthesizer that determines the Raman difference frequency. The phase on ion j is therefore

$$\phi_j = \phi_s + \Delta\mathbf{k} \cdot \mathbf{x}_j \tag{5}$$

In the experiment, the axial trap strength is changed so that the ions move about the centre of the trap symmetrically, giving $\Delta\mathbf{x}_1 = -\Delta\mathbf{x}_2$. Therefore the trap strength controls the differential phase, $\Delta\phi \equiv \phi_1 - \phi_2 = \Delta\mathbf{k} \cdot (\mathbf{x}_1 - \mathbf{x}_2)$, and the synthesizer controls the total phase, $\phi_{tot} \equiv \phi_1 + \phi_2 = 2\phi_s$. The calibration of these relations is discussed in the Methods.

The state of an ion, $|\downarrow\rangle$ or $|\uparrow\rangle$, is determined by probing the ion with circularly polarized light from a "detection" laser beam[27]. During this detection pulse, ions in the $|\downarrow\rangle$ or bright state scatter many photons, and on average about 64 of these are detected with a photomultiplier tube, while ions in the $|\uparrow\rangle$ or dark state scatter very few photons. For two ions, three cases can occur: zero ions bright, one ion bright, or two ions bright. In the one-ion-bright case it is not necessary to know which ion is bright because the Bell's

实验装置如文献 27 中所描述。两个 $^9Be^+$ 离子以 5 MHz 的轴向质心频率约束在沿着线性保罗阱的轴上。我们选择 $2S_{1/2}$ 基态的两个可分辨的能级，$|\downarrow\rangle \equiv |F=2, m_F=-2\rangle$ 和 $|\uparrow\rangle \equiv |F=1, m_F=-1\rangle$，式中 F 和 m_F 是总角动量的量子数。这些态经由相干受激拉曼跃迁耦合。用于驱动跃迁的两束激光波长为 313 nm，两者的差频接近态的超精细分裂 $\omega_0 \cong 2\pi \times 1.25$ GHz 的跃迁。两束激光彼此相互垂直，其波矢差 $\Delta\mathbf{k}$ 沿着阱的轴。如文献 27 中所述，在这种组态中可能产生纠缠态

$$|\psi_2\rangle = \frac{1}{\sqrt{2}}(|\uparrow\uparrow\rangle - |\downarrow\downarrow\rangle) \tag{3}$$

对于运行数据，保真度 $F=\langle\psi_2|\rho|\psi_2\rangle$ 约为 88%，式中 ρ 是我们所产生的态的密度矩阵。在以下讨论中，我们假定 $|\psi_2\rangle$ 是实验的起始条件。

在产生了态 $|\psi_2\rangle$ 后，我们再应用拉曼光束作为一个短脉冲(约 400 ns)，因此每个离子 j 的态在相互作用绘景中转变为

$$|\uparrow_j\rangle \to \frac{1}{\sqrt{2}}(|\uparrow_j\rangle - ie^{-i\phi_j}|\downarrow_j\rangle);\ |\downarrow_j\rangle \to \frac{1}{\sqrt{2}}(|\downarrow_j\rangle - ie^{i\phi_j}|\uparrow_j\rangle) \tag{4}$$

相位 ϕ_j 是在离子 j 的位置上驱动拉曼跃迁的场的相位(更明确地说，是两束拉曼光束之间的相位差)，对应于图 1 中的输入 ϕ_1 和 ϕ_2。我们在实验中按两种方式设定其相位。首先，在离子沿着阱轴运动时，相位变化为 $\Delta\mathbf{k}\cdot\Delta\mathbf{x}_j$。例如，沿着阱轴的平移 $\lambda/\sqrt{2}$，相当于 2π 的相移。此外，改变决定拉曼差频的射频合成器的相位 ϕ_s，使加在两个离子上的激光相位发生同量的变化。从而离子 j 的相位是

$$\phi_j = \phi_s + \Delta\mathbf{k}\cdot\mathbf{x}_j \tag{5}$$

在实验中，轴向阱的强度改变，因此离子绕阱的中心对称地运动，给出 $\Delta\mathbf{x}_1 = -\Delta\mathbf{x}_2$。因此阱的强度控制了相位差值，$\Delta\phi \equiv \phi_1 - \phi_2 = \Delta\mathbf{k}\cdot(\mathbf{x}_1 - \mathbf{x}_2)$，合成器控制了总相位 $\phi_{tot} \equiv \phi_1 + \phi_2 = 2\phi_s$。这些关系的标定在方法一节中进行讨论。

一个离子的态，$|\downarrow\rangle$ 或 $|\uparrow\rangle$，用“检测”激光束中的圆偏振光探测离子来确定[27]。在检测脉冲持续时间内，处于 $|\downarrow\rangle$ 态或明态的离子散射许多光子，用光电倍增管检测到平均约 64 个，处于 $|\uparrow\rangle$ 态或暗态的离子散射很少的光子。对两个离子而言可能出现三种情况：无离子处于明态，一个离子处于明态，两个离子处于明态。在一个离子处于明态的情况下，不必知道是哪一个离子，因为贝尔的测量仅要求知道离子

measurement requires only knowledge of whether or not the ions' states are different. Figure 2 shows histograms, each with 20,000 detection measurements. The three cases are distinguished from each other with simple discriminator levels in the number of photons collected with the phototube.

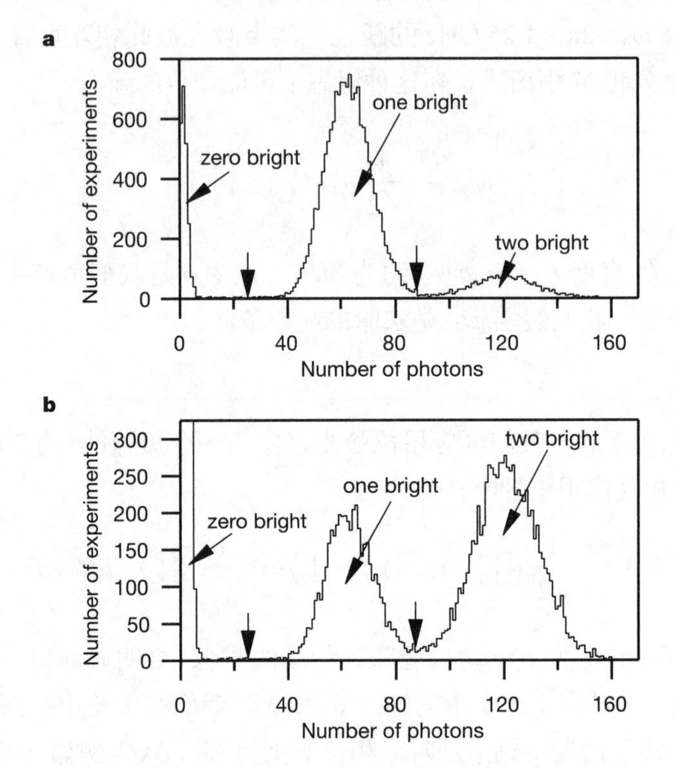

Fig. 2. Typical data histograms comprising the detection measurements of 20,000 experiments taking a total time of about 20 s. In each experiment the population in the $|\uparrow\rangle$ state is first coherently transferred to the $|F=1, M_F=+1\rangle$ to make it even less likely to fluoresce upon application of the detection laser. The detection laser is turned on and the number of fluorescence photons detected by the phototube in 1 ms is recorded. The cut between the one bright and two bright cases is made so that the fractions of two equal distributions which extend past the cut points are equal. The vertical arrows indicate the location of the cut between the 0 (1) bright and 1 (2) bright peaks at 25 (86) counts. **a**, Data histogram with a negative correlation using $\phi_1 = 3\pi/8$ and $\phi_2 = 3\pi/8$. For these data $N_0 \cong 2{,}200$, $N_1 \cong 15{,}500$ and $N_2 \cong 2{,}300$. **b**, Data histogram with a positive correlation using $\phi_1 = 3\pi/8$ and $\phi_2 = -\pi/8$. For these data $N_0 \cong 7{,}700$, $N_1 \cong 4{,}400$ and $N_2 \cong 7{,}900$. The zero bright peak extends vertically to 2,551.

An alternative description of our experiment can be made in the language of spin-one-half magnetic moments in a magnetic field (directed in the $\hat{z}$ direction). The dynamics of the spin system are the same as for our two-level system[28]. Combining the manipulation (equation (4)) and measurement steps, we effectively measure the spin projection of each ion j in the $\hat{r}_j$ direction, where the vector $\hat{r}_j$ is in the $\hat{x}$–$\hat{y}$ plane at an angle ϕ_j to the $\hat{y}$ axis. Although we have used quantum-mechanical language to describe the manipulation and measurement steps, we emphasize that both are procedures completely analogous to the classical rotations of wave-plates and measurements of polarization in an optical apparatus.

态是否不同。图 2 示出了直方图，每个实验重复测量 20,000 次。用光电管收集到的光子进行计数即可区分这三种情况。

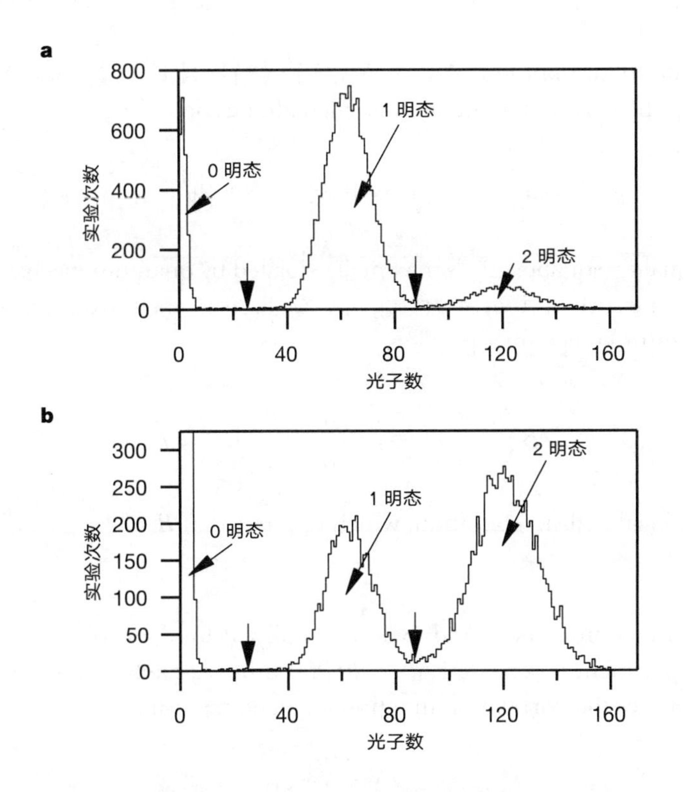

图 2. 包含 20,000 次实验的检测结果的典型数据直方图，所用的总时间约为 20 s。在每次实验中，在 $|\uparrow\rangle$ 态的粒子数，首先相干转移到 $|F=1,\ M_F=+1\rangle$，使得它在检测激光作用下也不发出荧光。打开检测激光，光电管在 1 ms 内记录检测到的荧光光子数。在一个离子处于明态和两个离子处于明态的情况之间进行截断，使两个相等分布扩展到截断点以外的部分所占的比例相等。垂直的箭头表示在 25(86) 计数时，在 0(1) 个离子处于明态与 1(2) 个离子处于明态峰值之间的截断。**a**. 用 $\phi_1=3\pi/8$ 和 $\phi_2=3\pi/8$ 的负相关数据直方图。对于这些数据，$N_0\cong 2{,}200$，$N_1\cong 15{,}500$ 和 $N_2\cong 2{,}300$。**b**. 用 $\phi_1=3\pi/8$ 和 $\phi_2=-\pi/8$ 的正相关数据直方图。对于这些数据，$N_0\cong 7{,}700$，$N_1\cong 4{,}400$ 和 $N_2\cong 7{,}900$。无离子处于明态的峰值将垂直扩展到 2,551。

我们的实验还能用磁场中（在 $\hat{z}$ 方向上）1/2 自旋磁矩的语言这另一种方式进行描述。自旋系统的动力学与我们的二能级系统是相同的[28]。结合操控（式(4)）和测量步骤，我们有效地测量了每个离子 j 在 $\hat{r}_j$ 方向上的自旋投影，其中矢量 $\hat{r}_j$ 在 $\hat{x}$–$\hat{y}$ 平面上与 $\hat{y}$ 呈角度 ϕ_j。虽然，我们已经使用量子力学的语言来描述操控和测量步骤，我们强调两者是与光学装置中波片的经典旋转和偏振的测量完全类似的步骤。

Here we calculate the quantum-mechanical prediction for the correlation function. Our manipulation step transforms the starting state, $|\psi_2\rangle$, to

$$|\psi_2'\rangle = \frac{1}{2\sqrt{2}}\left\{\left(1+e^{i(\phi_1+\phi_2)}\right)\left(|\uparrow\uparrow\rangle - e^{-i(\phi_1+\phi_2)}|\downarrow\downarrow\rangle\right) - i\left(1-e^{i(\phi_1+\phi_2)}\right)\left(e^{-i\phi_2}|\uparrow\downarrow\rangle + e^{-i\phi_1}|\downarrow\uparrow\rangle\right)\right\} \quad (6)$$

Using the measurement operators $\hat{N}_{\text{same}} = N_{\text{tot}}[|\uparrow\uparrow\rangle\langle\uparrow\uparrow| + |\downarrow\downarrow\rangle\langle\downarrow\downarrow|]$ and $\hat{N}_{\text{different}} = N_{\text{tot}}[|\uparrow\downarrow\rangle\langle\uparrow\downarrow| + |\downarrow\uparrow\rangle\langle\downarrow\uparrow|]$, the correlation function is calculated to be

$$q(\phi_1, \phi_2) = \frac{1}{8}\left[2\left|1+e^{i(\phi_1+\phi_2)}\right|^2 - 2\left|1-e^{i(\phi_1+\phi_2)}\right|^2\right] = \cos(\phi_1+\phi_2) \quad (7)$$

The CHSH inequality (equation (2)) is maximally violated by quantum mechanics at certain sets of phase angles. One such set is $\alpha_1 = -(\pi/8)$, $\delta_1 = 3\pi/8$, $\beta_2 = -(\pi/8)$ and $\gamma_2 = 3\pi/8$. With these phase angles quantum mechanics predicts

$$B\left(-\frac{\pi}{8}, \frac{3\pi}{8}, -\frac{\pi}{8}, \frac{3\pi}{8}\right) = 2\sqrt{2} \quad (8)$$

This violates the local realism condition, which requires that $B \leqslant 2$.

The correlation function is measured experimentally at four sets of phase angles, listed in Table 1. The experiment is repeated $N_{\text{tot}} = 20{,}000$ times at each of the four sets of phases. For each set of phases the correlation function is calculated using

$$q = \frac{(N_0 + N_2) - N_1}{N_{\text{tot}}} \quad (9)$$

Here N_0, N_1 and N_2 are the number of events with zero, one and two ions bright, respectively. The correlation values from the four sets of phase angles are combined into the Bell's signal, $B(\alpha_1, \delta_1, \beta_2, \gamma_2)$, using equation (2). The correlation values and resulting Bell's signals from five data runs are given in Table 2.

Table 1. The four sets of phase angles used for the Bell's experiment

Experiment input	ϕ_1	ϕ_2	$\Delta\phi$	ϕ_{tot}
$\alpha_1 \beta_2$	$-\pi/8$	$-\pi/8$	0	$-\pi/4$
$\alpha_1 \gamma_2$	$-\pi/8$	$3\pi/8$	$-\pi/2$	$+\pi/4$
$\delta_1 \beta_2$	$3\pi/8$	$-\pi/8$	$+\pi/2$	$+\pi/4$
$\delta_1 \gamma_2$	$3\pi/8$	$3\pi/8$	0	$+3\pi/4$

这里，我们计算了关联函数的量子力学预测结果。我们的操控步骤将起始态 $|\psi_2\rangle$ 变换为

$$|\psi_2'\rangle = \frac{1}{2\sqrt{2}}\left\{\left(1+e^{i(\phi_1+\phi_2)}\right)\left(|\uparrow\uparrow\rangle - e^{-i(\phi_1+\phi_2)}|\downarrow\downarrow\rangle\right) - i\left(1-e^{i(\phi_1+\phi_2)}\right)\left(e^{-i\phi_2}|\uparrow\downarrow\rangle + e^{-i\phi_1}|\downarrow\uparrow\rangle\right)\right\} \tag{6}$$

应用测量算符 $\hat{N}_{\text{same}} = N_{\text{tot}}[|\uparrow\uparrow\rangle\langle\uparrow\uparrow| + |\downarrow\downarrow\rangle\langle\downarrow\downarrow|]$ 和 $\hat{N}_{\text{different}} = N_{\text{tot}}[|\uparrow\downarrow\rangle\langle\uparrow\downarrow| + |\downarrow\uparrow\rangle\langle\downarrow\uparrow|]$，计算的关联函数为

$$q(\phi_1, \phi_2) = \frac{1}{8}[2|1+e^{i(\phi_1+\phi_2)})|^2 - 2|1-e^{i(\phi_1+\phi_2)}|^2] = \cos(\phi_1+\phi_2) \tag{7}$$

根据量子力学，CHSH 不等式（式(2)）在特定相角组合处达到了最大的背离。一组这样的组合为 $\alpha_1 = -(\pi/8)$，$\delta_1 = 3\pi/8$，$\beta_2 = -(\pi/8)$ 及 $\gamma_2 = 3\pi/8$。用这些相角，量子力学预测

$$B\left(-\frac{\pi}{8}, \frac{3\pi}{8}, -\frac{\pi}{8}, \frac{3\pi}{8}\right) = 2\sqrt{2} \tag{8}$$

这背离了要求 $B \leqslant 2$ 的定域实在论条件。

实验上用列于表 1 的四组相角测量关联函数。四组相角的每一组，实验都重复 $N_{\text{tot}} = 20{,}000$ 次。对每组相位，用下式计算关联函数

$$q = \frac{(N_0+N_2)-N_1}{N_{\text{tot}}} \tag{9}$$

式中 N_0, N_1, N_2 分别是以 0, 1 和 2 个离子处于明态的事件数目。四组相角的相关值已用式(2)代入贝尔信号 $B(\alpha_1, \delta_1, \beta_2, \gamma_2)$ 中。表 2 给出了五次数据运行的相关值和得出的贝尔信号。

表 1. 用于贝尔实验的四组相角

实验输入	ϕ_1	ϕ_2	$\Delta\phi$	ϕ_{tot}
$\alpha_1\,\beta_2$	$-\pi/8$	$-\pi/8$	0	$-\pi/4$
$\alpha_1\,\gamma_2$	$-\pi/8$	$3\pi/8$	$-\pi/2$	$+\pi/4$
$\delta_1\,\beta_2$	$3\pi/8$	$-\pi/8$	$+\pi/2$	$+\pi/4$
$\delta_1\,\gamma_2$	$3\pi/8$	$3\pi/8$	0	$+3\pi/4$

Table 2. Correlation values and resulting Bell's signals for five experimental runs

Run number	$q(\alpha_1, \beta_2)$	$q(\alpha_1, \gamma_2)$	$q(\delta_1, \beta_2)$	$q(\delta_1, \gamma_2)$	$B(\alpha_1, \delta_1, \beta_2, \gamma_2)$
1	0.541	0.539	0.569	−0.573	2.222
2	0.575	0.570	0.530	−0.600	2.275
3	0.551	0.634	0.590	−0.487	2.262
4	0.575	0.561	0.559	−0.551	2.246
5	0.541	0.596	0.537	−0.571	2.245

The experimental angle values were $\alpha_1 = -(\pi/8)$, $\delta_1 = 3\pi/8$, $\beta_2 = -(\pi/8)$, and $\gamma_2 = 3\pi/8$. The statistical errors are 0.006 and 0.012 for the q and B values respectively. The systematic errors (see text) are 0.03 and 0.06 for the q and B values respectively.

So far we have described the experiment in terms of perfect implementation of the phase angles. In the actual experiment, however, α_1, δ_1, β_2 and γ_2 are not quite the same angles both times they occur in the Bell's inequality. In our experiment the dominant reason for this error results from the phase instability of the synthesizer, which can cause the angles to drift appreciably during four minutes, the time required to take a complete set of measurements. This random drift causes a root-mean-squared error for the correlation function of ± 0.03 on this timescale, which propagates to an error of ± 0.06 for the Bell's signal. The error for the Bell's signal from the five combined data sets is then ± 0.03, consistent with the run-to-run variation observed. Averaging the five Bell's signals from Table 2, we arrive at our experimental result, which is

$$B\left(-\frac{\pi}{8}, \frac{3\pi}{8}, -\frac{\pi}{8}, \frac{3\pi}{8}\right) = 2.25 \pm 0.03 \tag{10}$$

If we take into account the imperfections of our experiment (imperfect state fidelity, manipulations, and detection), this value agrees with the prediction of quantum mechanics.

The result above was obtained using the outcomes of every experiment, so that no fair-sampling hypothesis is required. In this case, the issue of detection efficiency is replaced by detection accuracy. The dominant cause of inaccuracy in our state detection comes from the bright state becoming dark because of optical pumping effects. For example, imperfect circular polarization of the detection light allows an ion in the $|\downarrow\rangle$ state to be pumped to $|\uparrow\rangle$, resulting in fewer collected photons from a bright ion. Because of such errors, a bright ion is misidentified 2% of the time as being dark. This imperfect detection accuracy decreases the magnitude of the measured correlations. We estimate that our Bell's signal would be 2.37 with perfect detection accuracy.

We have thus presented experimental results of a Bell's inequality measurement where a measurement outcome was recorded for every experiment. Our detection efficiency was high enough for a Bell's inequality to be violated without requiring the assumption of fair sampling, thereby closing the detection loophole in this experiment. The ions were separated by a distance large enough that no known interaction could affect the results;

表 2. 五次实验运行的相关值和得出的贝尔信号

运行序号	$q(\alpha_1, \beta_2)$	$q(\alpha_1, \gamma_2)$	$q(\delta_1, \beta_2)$	$q(\delta_1, \gamma_2)$	$B(\alpha_1, \delta_1, \beta_2, \gamma_2)$
1	0.541	0.539	0.569	−0.573	2.222
2	0.575	0.570	0.530	−0.600	2.275
3	0.551	0.634	0.590	−0.487	2.262
4	0.575	0.561	0.559	−0.551	2.246
5	0.541	0.596	0.537	−0.571	2.245

实验的角度数值是 $\alpha_1 = -(\pi/8)$，$\delta_1 = 3\pi/8$，$\beta_2 = -(\pi/8)$ 及 $\gamma_2 = 3\pi/8$。对 q 和 B 的统计误差分别为 0.006 和 0.012。对 q 和 B 的系统误差（见正文）分别为 0.03 和 0.06。

在此为止，我们已用理想的相角设定描述了实验。然而，在实际实验中，两次代入贝尔不等式的角 α_1、δ_1、β_2 和 γ_2 并不是完全相同的。在我们的实验中，这个误差主要来自合成器的相位不稳定度，使相角在四分钟内产生了明显的漂移，这正是一整套测量所需要的时间。这项随机漂移对于关联函数来说，在这一时标上产生了 ±0.03 的方均根误差，它传递到贝尔信号的误差为 ±0.06。来自五个数据组的贝尔信号的误差则为 ±0.03，与观测到的每次运行变化是相符的。对表 2 中五个贝尔信号求平均，我们得到的实验结果为

$$B\left(-\frac{\pi}{8}, \frac{3\pi}{8}, -\frac{\pi}{8}, \frac{3\pi}{8}\right) = 2.25 \pm 0.03 \tag{10}$$

如果考虑到我们实验中的不完美的方面（不完美的态保真度、操控和检测），这个结果与量子力学的预测值是一致的。

上述结果是利用每个实验的输出量得出的，因此不要求有合理取样的前提。在这种情况下，检测效率的问题由检测准确度代替。在我们实验中，态检测的不准确的主要原因来自光抽运效应引起的明态变为暗态。例如，检测光不完美的圆偏振允许离子由 $|\downarrow\rangle$ 态抽运到 $|\uparrow\rangle$，从而处于明态的离子收集到较少的光子。由于这类误差的存在，一个离子处于明态而被错判为处于暗态的概率是 2%。这类不完美的检测准确度降低了测量相关性的量级。我们估计本文贝尔信号在完美检测准确度下将是 2.37。

我们已经给出了每次实验记录的贝尔不等式测量的实验结果。我们的检测效率高到足以不要求有合理取样的前提也可以给出贝尔不等式背离的结论，从而在这个实验中堵上了检测漏洞这个问题。离子之间相距很远，远到没有已知的相互作用可

however, the lightcone loophole remains open here. Further details of this experiment will be published elsewhere.

Methods

Phase calibration

The experiment was run with specific phase differences of the Raman laser beam fields at each ion. In order to implement a complete set of laser phases, a calibration of the phase on each ion as a function of axial trap strength was made. We emphasize that the calibration method is classical in nature. Although quantum mechanics guided the choice of calibration method, no quantum mechanics was used to interpret the signal. General arguments are used to describe the signal resulting from a sequence of laser pulses and its dependence on the classical physical parameters of the system, the laser phase at the ion, and the ion's position.

In the calibration procedure, a Ramsey experiment was performed on two ions. The first $\pi/2$ Rabi rotation was performed identically each time. The laser phases at the ions' positions for the second $\pi/2$ Rabi rotation were varied, ϕ_1 for ion 1 and ϕ_2 for ion 2. The detection signal is the total number of photons counted during detection. With an auxiliary one-ion experiment we first established empirically that the individual signal depends only on the laser phase at an individual ion and is $C+A\cos\phi_j$. Here C and A are the offset and amplitude of the one-ion signal. We measure the detector to be linear, so that the detection signal is the sum of the two ions' individual signals. The two-ion signal is therefore

$$C+A\cos\phi_1+C+A\cos\phi_2=2C+2A\cos\left[\frac{1}{2}(\phi_1+\phi_2)\right]\cos\left[\frac{1}{2}(\phi_1-\phi_2)\right] \tag{11}$$

By measuring the fringe amplitude and phase as $\phi_s=(\phi_1+\phi_2)/2$ is swept, we calibrate $\phi_1-\phi_2$ as a function of trap strength and ensure that $\phi_1+\phi_2$ is independent of trap strength.

We use the phase convention that at the ion separation used for the entanglement preparation pulse the maximum of the correlation function is at $\phi_1=\phi_2=0$ (or $\Delta\phi=\phi_{\text{tot}}=0$). Our measurement procedure begins by experimentally finding this condition of $\phi_1=\phi_2=0$ by keeping $\Delta\phi=0$ and scanning the synthesizer phase to find the maximum correlation. The experiment is then adjusted to the phase angles specified above by switching the axial trap strength to set $\Delta\phi$ and incrementing the synthesizer phase to set ϕ_{tot}.

Locality issues

The ions are separated by a distance of approximately 3 μm, which is greater than 100 times the size of the wavepacket of each ion. Although the Coulomb interaction strongly couples the ions' motion, it does not affect the ions' internal states. At this distance, all known relevant interactions are expected to be small. For example, dipole–dipole interactions between the ions slightly modify the light-scattering intensity, but this effect is negligible for the ion–ion separations used[29]. Also, the detection

以影响实验结果；但是，本文还没有解决光锥漏洞的问题。这个实验的更多细节将发表在其他地方。

方　法

相位标定

实验是在每个离子的拉曼激光光束场的特定相位差下运行的。为了实现激光相位的完备集，每个离子上的激光相位被标定为轴向阱强度的函数。我们强调标定方法本质上是经典的。虽然，量子力学影响了标定方法的选择，但并没有使用量子力学来解释信号。一般的理论用于描述从一系列激光脉冲得出的信号，以及它与系统的经典物理参量的相关关系，经典的物理参量包括离子上的激光相位和离子的位置。

在标定过程中，对两个离子进行了拉姆齐实验。每次实验中第一个 $\pi/2$ 的拉比旋转都相同。对于第二个 $\pi/2$ 拉比旋转，离子处的激光相位发生了改变，离子 1 处为 ϕ_1，离子 2 处为 ϕ_2。检测信号是在检测期间记录的总光子数。用一个辅助的单离子实验，我们首先在实践经验上确立了单个信号仅与单个离子处的激光相位有关，信号记为 $C+A\cos\phi_j$。此处的 C 和 A 是单离子信号的偏置和振幅。我们测得检测器是线性的，因此检测信号是两个离子的单独信号之和。从而两个离子的信号为

$$C+A\cos\phi_1+C+A\cos\phi_2=2C+2A\cos\left[\frac{1}{2}(\phi_1+\phi_2)\right]\cos\left[\frac{1}{2}(\phi_1-\phi_2)\right] \tag{11}$$

当扫描 $\phi_s=(\phi_1+\phi_2)/2$ 时，通过测量条纹振幅和相位，我们标定了 $\phi_1-\phi_2$ 作为阱强度的函数，并确保 $\phi_1+\phi_2$ 与阱强度无关。

我们按照惯例，在用于纠缠制备脉冲的离子分离处，关联函数在 $\phi_1=\phi_2=0$(或 $\Delta\phi=\phi_{tot}=0$) 处取得最大值。在测量的开始，我们通过保持 $\Delta\phi=0$，进而用实验方法找到 $\phi_1=\phi_2=0$ 的条件，并扫描合成器的相位来找到关联函数最大值。然后在实验中通过调整轴阱强度来设置 $\Delta\phi$，通过增加合成器相位来设置 ϕ_{tot}，把相角调整到上文给出的相应值。

定域性的争论

离子间相隔的距离约为 3 μm，这是每个离子的波包尺寸的 100 倍。虽然，库仑相互作用与离子的运动发生强烈的耦合，但它并不影响离子的内部状态。在这个距离上，所有已知的相互作用的预期都是很小的。例如，在离子之间的偶极–偶极相互作用稍微改变了光散射强度，但这种影响对于所用的离子之间的距离来说是

solid angle is large enough that Young's interference fringes, if present, are averaged out[30]. Even though all known interactions would cause negligible correlations in the measurement outcomes, the ion separation is not large enough to eliminate the lightcone loophole.

We note that the experiment would be conceptually simpler if, after creating the entangled state, we separated the ions so that the input manipulations and measurements were done individually. However, unless we separated the ions by a distance large enough to overcome the lightcone loophole, this is only a matter of convenience of description and does not change the conclusions that can be drawn from the results.

(**409**, 791-794; 2001)

M. A. Rowe*, D. Kielpinski*, V. Meyer*, C. A. Sackett*, W. M. Itano*, C. Monroe† & D. J. Wineland*

* Time and Frequency Division, National Institute of Standards and Technology, Boulder, Colorado 80305, USA

† Department of Physics, University of Michigan, Ann Arbor, Michigan 48109, USA

Received 25 October; accepted 30 November 2000.

References:

1. Clauser, J. F. & Shimony, A. Bell's theorem: experimental tests and implications. *Rep. Prog. Phys.* **41**, 1883-1927 (1978).
2. Einstein, A., Podolsky, B. & Rosen, N. Can quantum-mechanical description of reality be considered complete? *Phys. Rev.* **47**, 777-780 (1935).
3. Bell, J. S. On the Einstein-Podolsky-Rosen paradox. *Physics* **1**, 195-200 (1965).
4. Bell, J. S. in *Foundations of Quantum Mechanics* (ed. d'Espagnat, B.) 171-181 (Academic, New York, 1971).
5. Clauser, J. F., Horne, M. A., Shimony, A. & Holt, R. A. Proposed experiment to test local hidden-variable theories. *Phys. Rev. Lett.* **23**, 880-884 (1969).
6. Freedman, S. J. & Clauser, J. F. Experimental test of local hidden-variable theories. *Phys. Rev. Lett.* **28**, 938-941 (1972).
7. Fry, E. S. & Thompson, R. C. Experimental test of local hidden-variable theories. *Phys. Rev. Lett.* **37**, 465-468 (1976).
8. Aspect, A., Grangier, P. & Roger, G. Experimental realization of Einstein-Podolsky-Rosen-Bohm *Gedankenexperiment*: a new violation of Bell's inequalities. *Phys. Rev. Lett.* **49**, 91-94 (1982).
9. Aspect, A., Dalibard, J. & Roger, G. Experimental test of Bell's inequalities using time-varying analyzers. *Phys. Rev. Lett.* **49**, 1804-1807 (1982).
10. Ou, Z. Y. & Mandel, L. Violation of Bell's inequality and classical probability in a two-photon correlation experiment. *Phys. Rev. Lett.* **61**, 50-53 (1988).
11. Shih, Y. H. & Alley, C. O. New type of Einstein-Podolsky-Rosen-Bohm experiment using pairs of light quanta produced by optical parametric down conversion. *Phys. Rev. Lett.* **61**, 2921-2924 (1988).
12. Tapster, P. R., Rarity, J. G. & Owens, P. C. M. Violation of Bell's inequality over 4 km of optical fiber. *Phys. Rev. Lett.* **73**, 1923-1926 (1994).
13. Kwiat, P. G., Mattle, K., Weinfurter, H. & Zeilinger, A. New high-intensity source of polarization-entangled photon pairs. *Phys. Rev. Lett.* **75**, 4337-4341 (1995).
14. Tittel, W., Brendel, J., Zbinden, H. & Gisin, N. Violation of Bell inequalities by photons more than 10 km apart. *Phys. Rev. Lett.* **81**, 3563-3566 (1998).
15. Weihs, G. *et al.* Violation of Bell's inequality under strict Einstein locality conditions. *Phys. Rev. Lett.* **81**, 5039-5043 (1998).
16. Aspect, A. Bell's inequality test: more ideal than ever. *Nature* **398**, 189-190 (1999).
17. Gisin, N. & Zbinden, H. Bell inequality and the locality loophole: active versus passive switches. *Phys. Lett. A* **264**, 103-107 (1999).
18. Lo, T. K. & Shimony, A. Proposed molecular test of local hidden-variable theories. *Phys. Rev. A* **23**, 3003-3012 (1981).
19. Kwiat, P. G., Eberhard, P. H., Steinberg, A. M. & Chiao, R. Y. Proposal for a loophole-free Bell inequality experiment. *Phys. Rev. A* **49**, 3209-3220 (1994).
20. Huelga, S. F., Ferrero, M. & Santos, E. Loophole-free test of the Bell inequality. *Phys. Rev. A* **51**, 5008-5011 (1995).
21. Fry, E. S., Walther, T. & Li, S. Proposal for a loophole free test of the Bell inequalities. *Phys. Rev. A* **52**, 4381-4395 (1995).
22. Freyberger, M., Aravind, P. K., Horne, M. A. & Shimony, A. Proposed test of Bell's inequality without a detection loophole by using entangled Rydberg atoms. *Phys. Rev. A* **53**, 1232-1244 (1996).
23. Brif, C. & Mann, A. Testing Bell's inequality with two-level atoms via population spectroscopy. *Europhys. Lett.* **49**, 1-7 (2000).
24. Beige, A., Munro, W. J. & Knight, P. L. A Bell's inequality test with entangled atoms. *Phys. Rev. A* **62**, 052102-1–052102-9 (2000).
25. Lamehi-Rachti, M. & Mittig, W. Quantum mechanics and hidden variables: a test of Bell's inequality by the measurement of the spin correlation in low-energy proton-proton scattering. *Phys. Rev. D* **14**, 2543-2555 (1976).

可以忽略不计的[29]。同时，检测立体角大到足以使杨氏干涉条纹（如存在）被平均掉[30]。即使所有已知的相互作用在测量结果中的相关性可以忽略不计，但离子之间的距离仍不会大到足以消除光锥漏洞。

我们注意到，如果在建立纠缠态后分开离子，使输入操控和测量分别进行，实验将在概念上更为简单。然而，除非我们把离子之间的距离扩大到足以克服光锥漏洞，否则，这只是描述方便与否的问题，而并不会改变从实验结果所得出的结论。

（沈乃澂 翻译；李军刚 审稿）

26. Hagley, E. *et al.* Generation of Einstein-Podolsky-Rosen pairs of atoms. *Phys. Rev. Lett.* **79**, 1-5 (1997).
27. Sackett, C. A. *et al.* Experimental entanglement of four particles. *Nature* **404**, 256-259 (2000).
28. Feynman, R. P., Vernon, F. L. & Hellwarth, R. W. Geometrical representation of the Schrödinger equation for solving maser problems. *J. Appl. Phys.* **28**, 49-52 (1957).
29. Richter, T. Cooperative resonance fluorescence from two atoms experiencing different driving fields. *Optica Acta* **30**, 1769-1780 (1983).
30. Eichmann, U. *et al.* Young's interference experiment with light scattered from two atoms. *Phys. Rev. Lett.* **70**, 2359-2362 (1993).

Acknowledgements. We thank A. Ben-Kish, J. Bollinger, J. Britton, N. Gisin, P. Knight, P. Kwiat and I. Percival for useful discussions and comments on the manuscript. This work was supported by the US National Security Agency (NSA) and the Advanced Research and Development Activity (ARDA), the US Office of Naval Research, and the US Army Research Office. This paper is a contribution of the National Institute of Standards and Technology and is not subject to US copyright.

Correspondence and requests for materials should be addressed to D.J.W. (e-mail: david.wineland@boulder.nist.gov).

Superconductivity at 39 K in Magnesium Diboride

J. Nagamatsu *et al.*

Editor's Note

By 2001, a variety of materials based on copper oxides that superconduct at temperatures over 130 K had been observed. Here physicist Jun Nagamatsu and colleagues reported their observation of superconductivity at a temperature of "just" 39 K, but in a very different material: the metallic compound magnesium diboride. This work established a new record for the superconducting transition temperature in metallic compounds, for which the previous high had been, since 1973, only 23 K. Later work would show that the behaviour of this compound fits the standard Bardeen–Cooper–Schrieffer theory of superconductivity, although with some peculiarities. Due to the low cost of its constituent elements and fabrication, magnesium diboride has become a widely used superconductor for practical applications.

In the light of the tremendous progress that has been made in raising the transition temperature of the copper oxide superconductors (for a review, see ref. 1), it is natural to wonder how high the transition temperature, T_c, can be pushed in other classes of materials. At present, the highest reported values of T_c for non-copper-oxide bulk superconductivity are 33 K in electron-doped $Cs_xRb_yC_{60}$ (ref. 2), and 30 K in $Ba_{1-x}K_xBiO_3$ (ref. 3). (Hole-doped C_{60} was recently found[4] to be superconducting with a T_c as high as 52 K, although the nature of the experiment meant that the supercurrents were confined to the surface of the C_{60} crystal, rather than probing the bulk.) Here we report the discovery of bulk superconductivity in magnesium diboride, MgB_2. Magnetization and resistivity measurements establish a transition temperature of 39 K, which we believe to be the highest yet determined for a non-copper-oxide bulk superconductor.

THE samples were prepared from powdered magnesium (Mg; 99.9%) and powdered amorphous boron (B; 99%) in a dry box. The powders were mixed in an appropriate ratio (Mg:B = 1:2), ground and pressed into pellets. The pellets were heated at 973 K under a high argon pressure, 196 MPa, using a hot isostatic pressing (HIP) furnace (O_2Dr.HIP, Kobelco) for 10 hours. Powder X-ray diffraction was performed by a conventional X-ray spectrometer with a graphite monochromator (RINT-2000, Rigaku). Intensity data were collected with CuKα radiation over a 2θ range from 5° to 80° at a step width of 0.02°.

Figure 1 shows a typical X-ray diffraction pattern of MgB_2 taken at room temperature. All the intense peaks can be indexed assuming an hexagonal unit cell, with $a = 3.086$ Å and $c = 3.524$ Å. Figure 2 shows the crystal structure of MgB_2 (ref. 5), of which the space

二硼化镁在 39 K 时的超导性

永松纯等

编者按

到 2001 年为止，在温度超过 130 K 的多种基于铜氧化物的材料中已观测到超导性。本文中物理学家永松纯及其同事们报道了在温度“仅仅为”39 K 时观测到的超导性，但这是在一种非常不一样的材料中观测到的：金属化合物二硼化镁。这项工作创造了金属化合物超导转变温度的一个新纪录，因为从 1973 年到此之前，这种转变温度最高只有 23 K。后续的工作表明，尽管存在一些特殊性，这种化合物的性质符合标准的 BCS（巴丁–库珀–施里弗）超导理论。由于原料和制造成本低，二硼化镁已被广泛地用于超导体的实际应用中。

鉴于在提高铜氧化物的超导转变温度上所取得的巨大成就（见参考文献 1 中的综述），其他类型材料的超导转变温度 T_c 可以被提高到何种程度自然而然地引起了人们的关注。现有报道中，非铜氧化物体超导性的 T_c 最高值，在电子掺杂的 $Cs_xRb_yC_{60}$ 中为 33 K（参考文献 2），在 $Ba_{1-x}K_xBiO_3$ 中为 30 K（参考文献 3）。（近期发现空穴掺杂的 C_{60} 超导转变温度 T_c 高达 52 K[4]，尽管那次实验的原本目的是为了揭示超导电流仅局限于 C_{60} 晶体表面，而非检测整体。）本文中报告了二硼化镁（MgB_2）中体超导的发现。磁化和电阻测量表明超导转变温度为 39 K，我们相信这是非铜氧化物体超导的至今为止所确定的最高转变温度。

样品是用粉末状镁（Mg; 99.9%）和粉末状的非晶硼（B; 99%）在干燥箱中制备的。两种粉末以适当的比例（Mg : B = 1 : 2）混合，研磨并挤压成圆片。将这些圆片置于热等静压（HIP）炉（O_2Dr.HIP，日本神钢集团）中，在 196 Mpa 的高氩压强和 973 K 的温度下加热 10 小时。粉末 X 射线衍射通过带有石墨单色仪的常规 X 射线光谱仪（RINT-2000，日本理学公司）来进行。强度数据利用 Cu 靶的 Kα 辐射采集，2θ 范围从 5° 到 80°，步长宽度为 0.02°。

图 1 为一种典型的在室温下获取的 MgB_2 的 X 射线衍射图。假定一种六方晶胞，其 $a = 3.086$ Å，$c = 3.524$ Å，可以将所有高强度的峰进行标记。图 2 展示了 MgB_2

group is *P6/mmm* (no.191). As shown in Fig. 2, the boron atoms are arranged in layers, with layers of Mg interleaved between them. The structure of each boron layer is the same as that of a layer in the graphite structure: each boron atom is here equidistant from three other boron atoms. Therefore, MgB_2 is composed of two layers of boron and magnesium along the *c* axis in the hexagonal lattice.

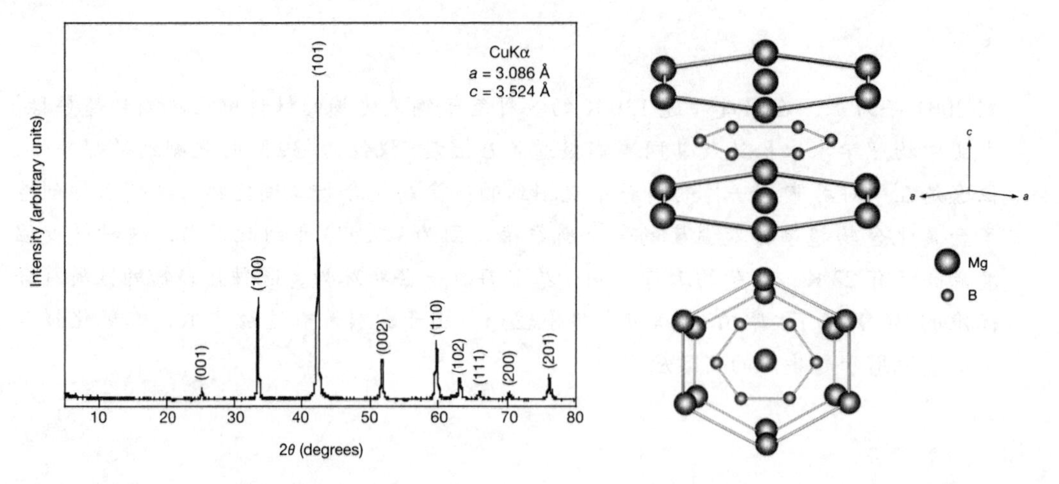

Fig. 1. X-ray diffraction pattern of MgB_2 at room temperature.

Fig. 2. Crystal structure of MgB_2.

Magnetization measurements were also performed with a SQUID magnetometer (MPMSR2, Quantum Design). Figure 3 shows the magnetic susceptibility ($\chi = M/H$, where M is magnetization and H is magnetic field) of MgB_2 as a function of temperature, under conditions of zero field cooling (ZFC) and field cooling (FC) at 10 Oe. The existence of the superconducting phase was then confirmed unambiguously by measuring the Meissner effect on cooling in a magnetic field. The onset of a well-defined Meissner effect was observed at 39 K. A superconducting volume fraction of 49% under a magnetic field of 10 Oe was obtained at 5 K, indicating that the superconductivity is bulk in nature. The standard four-probe technique was used for resistivity measurements.

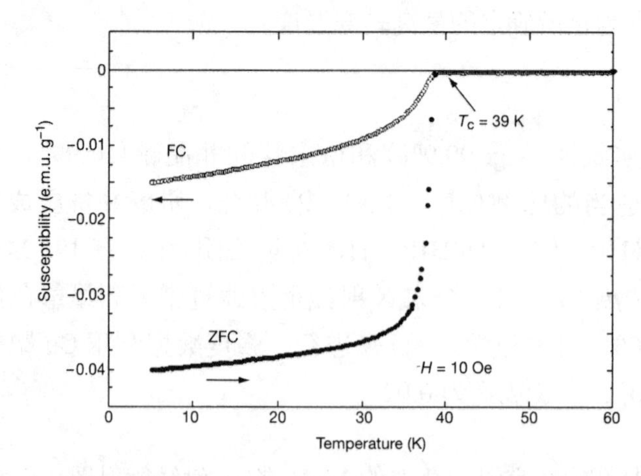

Fig. 3. Magnetic susceptibility χ of MgB_2 as a function of temperature. Data are shown for measurements under conditions of zero field cooling (ZFC) and field cooling (FC) at 10 Oe.

的晶体结构（参考文献 5），空间群为 $P6/mmm$（编号 191）。如图 2 所示，硼原子是按层排列的，镁层交错插入硼层。每一层的硼原子结构与石墨每层的结构相同：其中每个硼原子与其他三个硼原子是等距的。因此，MgB_2 是由两层硼以及一层镁沿六方晶格 c 轴构成的。

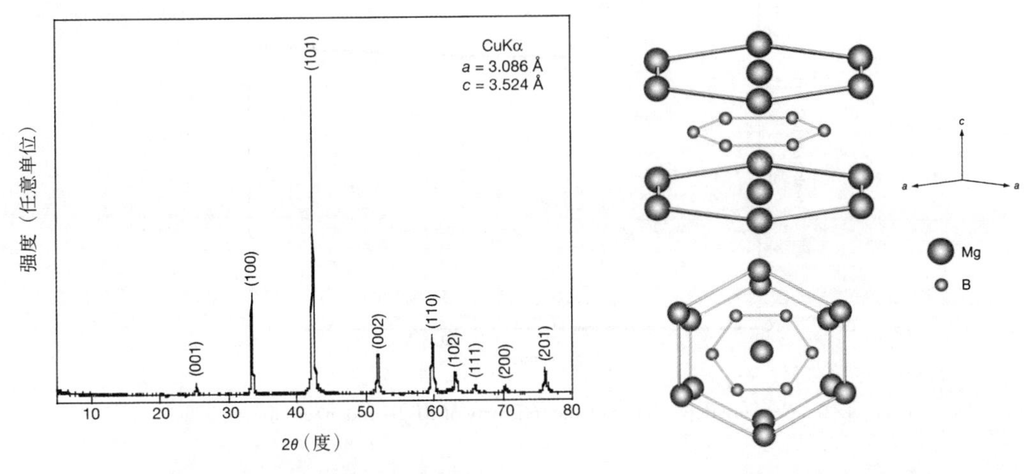

图 1. 室温下 MgB_2 的典型 X 射线衍射图

图 2. MgB_2 的晶体结构

磁化强度测量同样是通过超导量子干涉磁强计（SQUID）完成（MPMSR2, Quantum Design）。图 3 显示了 MgB_2 的磁化率（$\chi = M/H$，其中 M 为磁化强度，H 为磁场）在零场冷却（ZFC）以及在 10 Oe 的场冷却（FC）条件下随温度变化的曲线。通过测量在磁场中冷却状态下的迈斯纳效应，明确地证实了超导相的存在。在温度为 39 K 时，开始观测到明确的迈斯纳效应。在温度为 5 K 时，10 Oe 磁场下超导体积分数为 49%，这就表明本质上是体超导性。在电阻测量中我们使用的是常规的四探针技术。

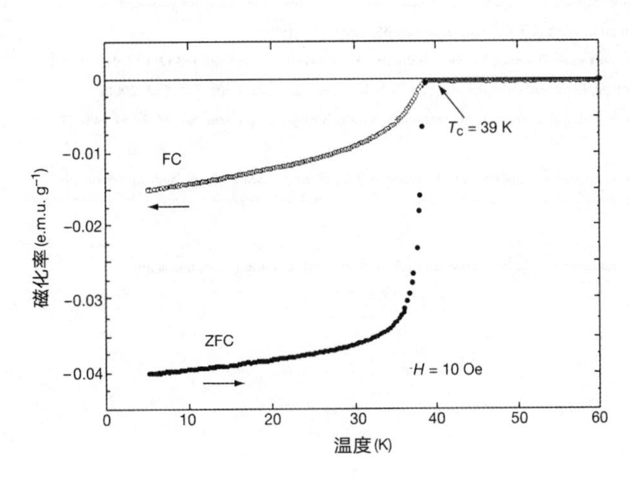

图 3. MgB_2 的磁化率 χ 随温度的变化。数据分别表示在零场冷却（ZFC）以及在 10 Oe 的场冷却（FC）条件下的测量结果。

Figure 4 shows the temperature dependence of the resistivity of MgB_2 under zero magnetic field. The onset and end-point transition temperatures are 39 K and 38 K, respectively, indicating that the superconductivity was truly realized in this system.

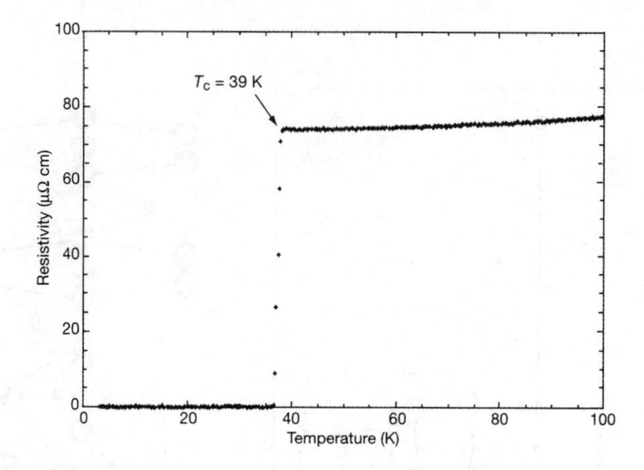

Fig. 4. Temperature dependence of the resistivity of MgB_2 under zero magnetic field.

(**410**, 63-64; 2001)

Jun Nagamatsu*, Norimasa Nakagawa*, Takahiro Muranaka*, Yuji Zenitani* & Jun Akimitsu*†
* Department of Physics, Aoyama-Gakuin University, Chitosedai, Setagaya-ku, Tokyo 157-8572, Japan
† CREST, Japan Science and Technology Corporation, Kawaguchi, Saitama 332-0012, Japan

Received 24 January; accepted 5 February 2001.

References:

1. Takagi, H. in *Proc. Int. Conf. on Materials and Mechanisms of Superconductivity, High Temperature Superconductors VI. Physica C* **341-348**, 3-7 (2000).
2. Tanigaki, K. *et al.* Superconductivity at 33 K in $Cs_xRb_yC_{60}$. *Nature* **352**, 222-223 (1991).
3. Cava, R. V. *et al.* Superconductivity near 30 K without copper: the $Ba_{0.6}K_{0.4}BiO_3$ perovskite. *Nature* **332**, 814-816 (1988).
4. Schön, J. H., Kloc, Ch. & Batlogg, B. Superconductivity at 52 K in hole-doped C_{60}. *Nature* **408**, 549-552 (2000).
5. Jones, M. & Marsh, R. The preparation and structure of magnesium boride, MgB_2. *J. Am. Chem. Soc.* **76**, 1434-1436 (1954).

Acknowledgements. This work was partially supported by a Grant-in-Aid for Science Research from the Ministry of Education, Science, Sports and Culture, Japan and by a grant from CREST.

Correspondence and requests for materials should be addressed to J.A. (e-mail: jun@soliton.phys.aoyama.ac.jp).

图 4 显示了 MgB_2 在零磁场下电阻率随温度的变化关系。转变温度的起止点分别出现在 39 K 和 38 K，这就表明在该系统中真正实现了超导性。

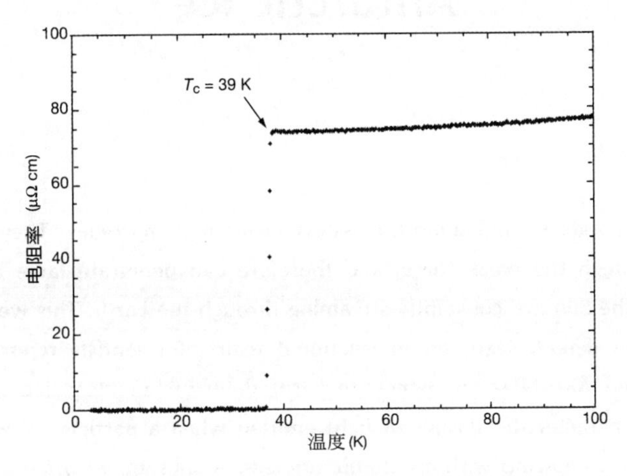

图 4. 零磁场下 MgB_2 的电阻率随温度的变化关系

（崔宁 翻译；韩汝珊 郭建栋 审稿）

Observation of High-energy Neutrinos Using Čerenkov Detectors Embedded Deep in Antarctic Ice

E. Andrés *et al.*

Editor's Note

Neutrinos are chargeless and almost massless elementary particles. They interact with other matter only through the weak force, and therefore can penetrate large amounts of matter—neutrinos from the Sun are constantly streaming through the Earth. This weak interaction makes them difficult to detect. Here an international team of scientists report a proof-of-concept experiment called AMANDA, in which they set detectors in clear ice in the Antarctic and looked for the characteristic flashes of light emitted when a particle called a muon is created by interaction of a neutrino with an atomic nucleus. A subsequent project using a much larger volume of ice as the detector, called IceCube, is now operating near the South Pole.

Neutrinos are elementary particles that carry no electric charge and have little mass. As they interact only weakly with other particles, they can penetrate enormous amounts of matter, and therefore have the potential to directly convey astrophysical information from the edge of the Universe and from deep inside the most cataclysmic high-energy regions[1]. The neutrino's great penetrating power, however, also makes this particle difficult to detect. Underground detectors have observed low-energy neutrinos from the Sun and a nearby supernova[2], as well as neutrinos generated in the Earth's atmosphere. But the very low fluxes of high-energy neutrinos from cosmic sources can be observed only by much larger, expandable detectors in, for example, deep water[3,4] or ice[5]. Here we report the detection of upwardly propagating atmospheric neutrinos by the ice-based Antarctic muon and neutrino detector array (AMANDA). These results establish a technology with which to build a kilometre-scale neutrino observatory necessary for astrophysical observations[1].

HIGH-ENERGY neutrinos must be generated in the same astrophysical sources that produce high-energy cosmic rays[1]. These sources are a matter of speculation, but are thought to reside in shocked or violent environments such as are found in supernova remnants, active galactic nuclei, and gamma-ray bursters. The interaction of any high-energy proton or nucleus with matter or radiation in the source will produce neutrinos, some of which will have line-of-sight trajectories to Earth. AMANDA detects neutrinos with energies above a few tens of GeV by observing the Čerenkov radiation from muons that are produced in neutrino–nucleon interactions in the ice surrounding the detector or

借助南极冰下深处的切伦科夫探测器观测高能中微子

安德烈斯等

编者按

中微子是无电荷且几乎无质量的基本粒子。它们只能通过弱力与其他物质相互作用，因此可以穿透大量物质——来自太阳的中微子就不断地流经地球。这种弱相互作用使中微子难以被检测。本文中，一个国际科学家团队报告了名为 AMANDA 的概念验证型实验，他们将探测器设置在南极透明的冰层中，并寻找由中微子与原子核相互作用而产生的被称为 μ 子的粒子发出的特征性闪光。随后的冰立方 (IceCube) 项目使用更大体量的冰作为探测器，目前已在南极附近运行。

中微子是一种不携带电荷且质量极小的基本粒子。由于这类粒子与其他粒子的相互作用很弱，能够穿透大量的物质，因此很可能可以直接传递来自宇宙边缘以及大部分高能激变区域深处的天体物理信息[1]。然而，中微子极强的穿透力也使得对该粒子的探测变得非常困难。在以前的研究中，研究人员通过地下探测器已经探测到来自太阳和附近超新星的低能中微子[2]，以及地球大气中产生的中微子。但是，来自宇宙源的高能中微子通量很低，只能通过更大的、可扩展的探测器才能观测到，比如放置于深水[3,4]或冰[5]中的探测器。本文将报道通过基于冰的南极 μ 子和中微子探测器阵列 (AMANDA) 对向上传播的大气中微子的探测。这些结果将确立一种技术，可以用于建设千米尺度的中微子天文台，从而用于天体物理的观测[1]。

发射高能宇宙射线的天体物理源也必然产生高能中微子[1]。这些源目前尚属推测，但是一般认为存在于极端剧烈环境中，比如超新星遗迹、活动星系核和伽马射线暴。这些源中，任何高能质子或原子核与物质或辐射相互作用，都会产生中微子，其中一些将具有指向地球的视线径迹。中微子与探测器周围冰层或底下岩床中的核子相互作用可以产生 μ 子。通过探测 μ 子发出的切伦科夫辐射，AMANDA 可以探测到能量高于数十 GeV 的中微子[6]。图 1 给出了我们用于探测切伦科夫辐射的光

in the bedrock below[6]. This Čerenkov light is detected by an array of photomultiplier tubes (Fig. 1), which are buried deep in the ice in order to minimize the downward flux of muons produced in cosmic-ray interactions in the atmosphere. These muons constitute the main background for AMANDA. To ensure that the detected muons are produced by a neutrino, we use the Earth as a filter and look for upwardly propagating muons that perforce must have been produced by a neutrino that passed through the Earth. From the relative arrival times of the Čerenkov photons, measured with a precision of a few nanoseconds, we can reconstruct the track of the muon. The direction of the neutrino and muon are collinear within an angle $\theta_{\nu-\mu} \approx 1.5/\sqrt{E_\nu}$ degrees, where E_ν is measured in TeV, thus enabling us to search for point sources of high energy neutrinos.

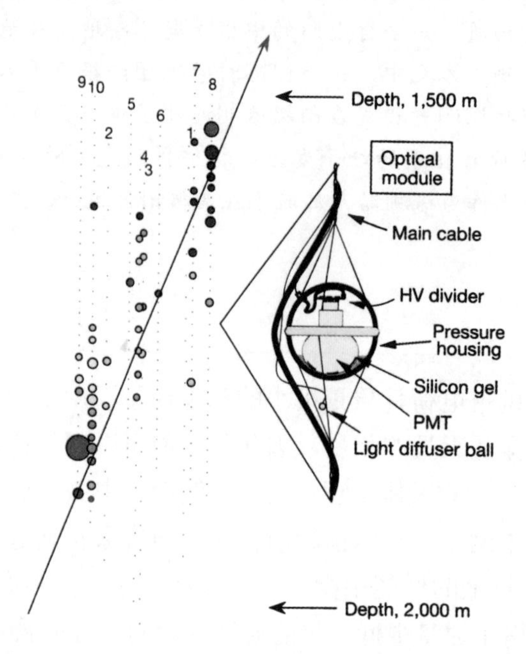

Fig. 1. The AMANDA-B10 detector and a schematic diagram of an optical module. Each dot represents an optical module. The modules are separated by 20 m on the inner strings (1 to 4), and by 10 m on the outer strings (5 to 10). The coloured circles show pulses from the photomultipliers for a particular event; the sizes of the circles indicate the amplitudes of the pulses and the colours correspond to the time of a photon's arrival. Earlier times are in red and later ones in blue. The arrow indicates the reconstructed track of the upwardly propagating muon.

Upwardly propagating atmospheric neutrinos are a well understood source that can be used to verify the detection technique. The results reported here are from analyses of experimental data acquired in 138 days of net operating time during the Antarctic winter of 1997. At that time the detector consisted of 302 optical modules deployed on ten strings at depths of between 1,500 m and 2,000 m (Fig. 1). The instrumented volume is a cylinder of approximately 120 m in diameter and 500 m in height. An optical module consists of an 8-inch photomultiplier tube housed in a glass pressure vessel. A cable provides the high voltage and transmits the anode current signals to the data acquisition electronics at the surface. Figure 1 also shows a representative event that has satisfied the selection criteria for an upwardly moving muon. The effective detection area for muons varies from about 3,000 m^2

电倍增管阵列，它们被埋设于冰层深处，以最大限度地减小大气中宇宙射线相互作用产生的μ子的向下通量。这些大气μ子构成了 AMANDA 的主要背景信号。为了确保探测到的μ子确实产生于中微子，我们将整个地球作为一个大过滤器，探测那些一定是由穿过地球的中微子产生的向上传播的μ子。探测器探测到的切伦科夫光子的相对到达时间的精度在纳秒量级，我们通过记录这些到达时间可以重建μ子的径迹。中微子与μ子的方向在夹角 $\theta_{\nu\text{-}\mu}\approx 1.5$ 度$/(E_{\nu})^{1/2}$ 内共线，其中 E_{ν} 单位为 TeV，这可以使我们寻找到高能中微子点源。

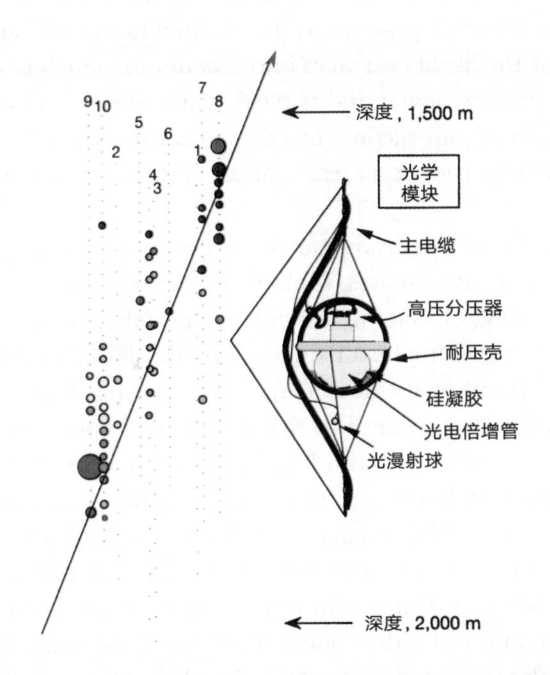

图 1. AMANDA-B10 探测器及其光学模块原理图。图中每个小点代表一个光学模块。在内部串，光学模块之间的距离为 20 m(1 到 4)；在外部串，模块之间的距离为 10 m(5 到 10)。不同颜色的圆圈表示某个事例中光电倍增管产生的脉冲，其中圆圈的尺寸大小代表该脉冲的幅度，颜色对应光子到达的时间。较早的时间用红色表示，较晚的时间用蓝色表示。图中箭头表示斜向上运动的μ子的重建径迹。

向上传播的大气中微子作为源已经为人们所了解，因此可以用大气中微子验证我们的探测技术。我们分析了在 1997 年南极冬季获取的 138 天净运转时间的实验数据，并将结果在本文中予以报告。当时，实验采用的探测器由 302 个光学模块构成，这些模块埋设为 10 串，深度在 1,500 m 到 2,000 m 之间(图 1)。整个探测器设备是一个圆柱体，其直径大约为 120 m，高度约为 500 m。每个光学模块包括一个封装于玻璃质压力容器中的 8 英寸光电倍增管。一根电缆用于提供光电倍增管所需的高电压，并将阳极电流信号传输到位于地面的数据采集电子设备。图 1 中也给出了一个满足向上运动的μ子选择标准的典型事件。根据μ子能量的不同，探测系统的有效

at 100 GeV to about 4×10^4 m^2 for the higher-energy muons ($\geqslant 100$ TeV) that would be produced by neutrinos coming from, for example, the same cosmic sources that produce gamma-ray bursts[7].

Because a knowledge of the optical properties of the ice is essential for track reconstruction, these have been studied extensively[8,9]. The absorption length of blue and ultraviolet light (the relevant wavelengths for our purposes) varies between 85 m and 225 m, depending on depth. The effective scattering length, which combines the mean free path λ with the average scattering angle θ through $\lambda/(1-\langle\cos\theta\rangle)$, varies from 15 m to 40 m. In order to reconstruct the muon tracks we use a maximum-likelihood method, which incorporates the scattering and absorption of photons as determined from calibration measurements. A bayesian formulation of the likelihood takes into account the much larger rate of downward muons relative to the upward signal and is particularly effective in decreasing the chance for a downward muon to be mis-reconstructed as upward. (See refs 6 and 10–13 for more information on optical properties of ice, calibration, and analysis techniques.)

Certain types of events that might appear to be upwardly propagating muons must be considered and eliminated. Rare cases, such as muons that undergo catastrophic energy loss through bremsstrahlung, or that are coincident with other muons, must be investigated. To this end, a series of requirements or quality criteria, based on the characteristic time and spatial pattern of photons associated with a muon track and the response of our detector, are applied to all events that, in a first assessment, appear to be upwardly moving muons. For example, an event that has a large number of optical modules hit by prompt (that is, unscattered) photons, has a high quality. By making these requirements (or "cuts") increasingly selective, we eliminate correspondingly more of the background of false upward events while still retaining a significant fraction of the true upwardly moving muons. Because there is a large space within which the parameters defining these cuts can be optimized, two different and independent analyses of the same set of data have been undertaken. These analyses yielded comparable numbers of upwardly propagating muons (153 in analysis A, 188 in analysis B). Comparison of these results with their respective Monte Carlo simulations shows that they are consistent with each other in terms of the number of events, the number of events in common and, as discussed below, the expected properties of atmospheric neutrinos.

In Fig. 2a, the number of experimental events is compared to simulations of background and signal as a function of the (identical) quality requirements placed on the three types of events: experimental data, simulated upwardly moving muons from atmospheric neutrinos, and a simulated background of downwardly moving cosmic-ray muons. For simplicity in presentation, the levels of the individual cuts have been combined into a single parameter representing the overall event quality. Figure 2b shows ratios of the quantities plotted above. As the quality level is increased, the ratios of simulated background to experimental data, and of experimental data to simulated signal, both continue their rapid decrease, the former toward zero and the latter toward 0.7. Over the same range, the ratio of experimental data to the simulated sum of background and signal remains nearly constant. We conclude that

探测面积范围从 3,000 m^2 到 40,000 m^2，对应从 100 GeV 到高能 μ 子（≥100 TeV）的情形，产生后者的中微子应该来自宇宙源，比如同样产生伽马射线暴的宇宙源[7]。

为了重建中微子的径迹，首先要弄清冰的光学特性。人们对此已进行了广泛的研究[8,9]。根据冰层深度的不同，蓝光和紫外光（我们所用的相关波长）的吸收长度在 85 m 到 225 m 之间变化。由平均自由程 λ 和平均散射角 θ 构成的有效散射长度 $\lambda/(1-\langle\cos\theta\rangle)$ 为 15 ~ 40 m。我们采用最大似然方法重建 μ 子径迹，考虑了通过刻度测量得到的光子的散射和吸收。考虑到向下 μ 子的事例率远远大于向上的信号，对似然值的贝叶斯分析可以有效减小将向下 μ 子错误重建为向上 μ 子的可能性。（参考文献 6 以及 10 ~ 13 介绍了更多关于冰的光学特性、刻度以及分析技术。）

在探测过程中，可能存在某些看似是向上传播的 μ 子的伪事件，对这种情况必须予以考虑并剔除。同时，也必须要考虑那些罕见的情况，比如 μ 子经过韧致辐射损失了大部分能量，或者与其他 μ 子同时出现的情况。为此，一系列基于 μ 子径迹预期产生的光子时间和空间特征图样以及基于探测器的响应的要求或者说质量标准在首次评估中用于筛选那些看似是向上运动的 μ 子事件。比如，瞬发（即未被散射）光子触发大量光学模块，该事件就具有较高的品质。通过不断提高选择的标准（或者说"判选"），我们可以相应地剔除更多的向上 μ 子伪事件背景，同时仍然保留大部分真正向上运动的 μ 子。由于构成筛选标准的各参数具有较大的优化空间，我们对相同的实验数据进行了两种不同且互相独立的分析，分别称为 A 分析和 B 分析。A、B 两种分析结果给出的向上运动的 μ 子数目相当，分别为 153 和 188。这两个结果分别与其相应的蒙特卡罗模拟进行比较，发现二者在事件数目、共同事件数目以及如下所述的大气中微子预期特性方面彼此一致。

如图 2a 所示，我们将实验事件的数量与背景和信号的模拟结果作为（相同）品质判据的函数进行了对比，对以下三类事件——实验数据、模拟的大气中微子产生的向上运动的 μ 子、模拟的向下运动的宇宙射线 μ 子背景——进行了分析。为了简化演示，多个筛选标准合并成了一个代表整体事件品质的参数。图 2b 显示了图 2a 绘制的数量间的比值。随着品质水平的增加，模拟背景与实验数据的比值以及实验数据与模拟信号的比值，同时迅速减小，其中前者趋于 0 而后者趋于 0.7。在上述范围内，实验数据与另两者（模拟的背景和信号）之和的比值几乎保持不变。因此，我们得出以下结论：品质筛选标准可以剔除实验数据中误重建的向下运动的 μ

the quality requirements have reduced the presence of wrongly reconstructed downward muons in the experimental data to a negligible fraction of the signal and that the experimental data behave in the same way as the simulated atmospheric neutrino signal for events that pass the stringent cuts. The estimated uncertainty on the number of events predicted by the signal Monte Carlo simulation, which includes uncertainties in the high-energy atmospheric neutrino flux, the *in situ* sensitivity of the optical modules, and the precise optical properties of the ice, is +40%/−50%. The observed ratio of experiment to simulation (0.7) and the expectation (1.0) therefore agree within the estimated uncertainties.

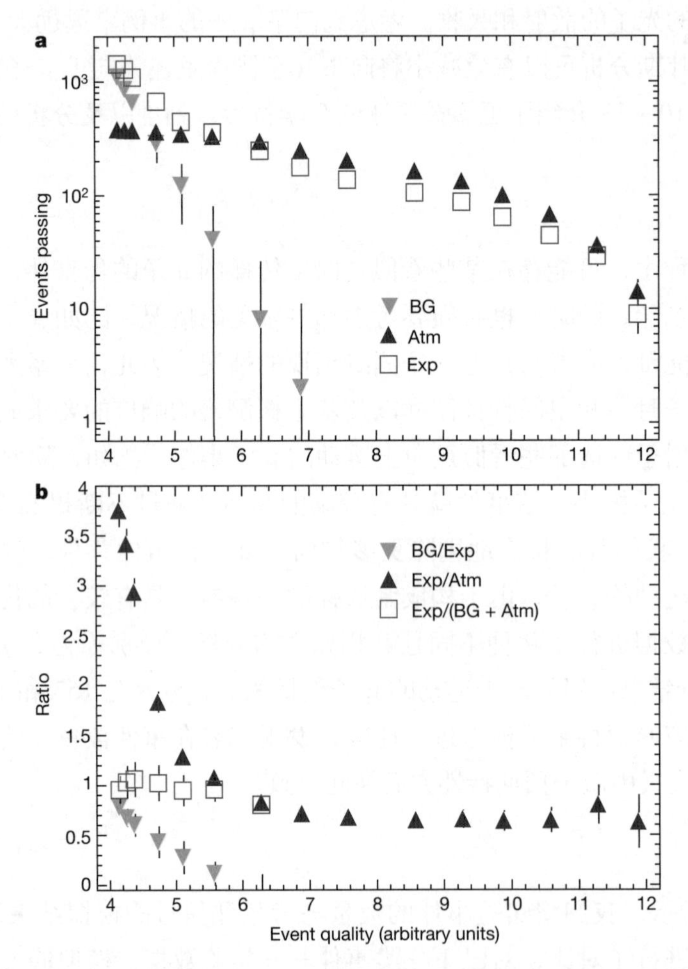

Fig. 2. Experimental data confront expectations. The numbers of reconstructed upwardly moving muon events for the experimental data (Exp) from analysis A are compared to simulations of background cosmic-ray muons (BG) and simulations of atmospheric neutrinos (Atm) as a function of "event quality", a variable indicating the combined severity of the cuts designed to enhance the signal. The comparison begins at a quality level of 4. Cuts were made on a number of parameters including the reconstructed zenith angle (> 100 degrees), maximum likelihood of the reconstruction, topological distributions of the detected photons, and the number of optical modules recording unscattered photons. The optimum levels of the cuts were determined by comparing the relative rejection rates for Monte Carlo simulated neutrino events and background events. **b**, Ratios of the quantities shown in **a**.

子，直至与信号相比达到可以忽略的水平；经过严格筛选后，实验数据与模拟大气中微子信号的行为规律相同。蒙特卡罗模拟预测的信号事件数的估算不确定度为+40%/−50%，该不确定度包括了高能大气中微子通量、光学模块的实地灵敏度以及冰层准确光学特性的不确定度。因此，在上述估算不确定度内，观察到的实验和模拟比值(0.7)与期望值(1.0)相符合。

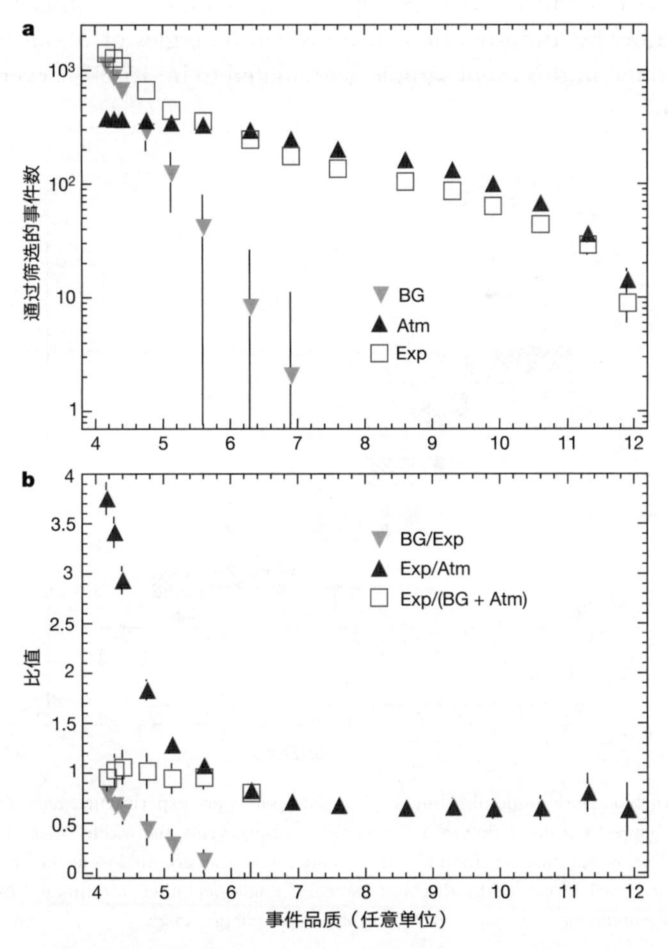

图2. 实验数据与期望值比较。A分析中重建实验数据中向上运动的μ子事件的数目(用Exp表示)，与模拟宇宙射线μ子背景(BG)以及模拟大气中微子(Atm)，作为“事件品质”的函数进行了比较。其中“事件品质”这一变量表示用于增强信号的若干筛选标准的组合的严格程度。图中对比起始于品质水平4。筛选标准由若干参量构成，这些参量包括重建天顶角(>100度)、重建径迹的最大似然值、探测光子的拓扑分布以及记录未散射光子的光学模块数目。通过比较蒙特卡罗模拟中微子事件和背景事件的相对排斥率，可以确定筛选标准的最佳水平。**b**图给出了**a**图中数量的比值。

The shape of the zenith-angle distribution of the 188 events from analysis B is compared to a simulation of the atmospheric neutrino signal in Fig. 3, where the absolute Monte Carlo rate has been normalized to the experimental rate. The variation of the measured rate with zenith angle is reproduced by the simulation to within the statistical uncertainty. We note that the tall geometry of the detector favours the more vertical muons. The arrival directions of the upwardly moving muons observed in both analyses are shown in Fig. 4. A statistical analysis indicates no evidence for point sources in these samples. The agreement between experiment and simulation of atmospheric neutrino signal, as demonstrated in Figs 2 and 3, taken together with comparisons for a number of other variables (to be published elsewhere) leads us to conclude that the upcoming muon events observed by AMANDA are produced mainly by atmospheric neutrinos with energies of about 50 GeV to a few TeV. The background in this event sample is estimated to be 15 ± 7% events, and is due to misreconstructions.

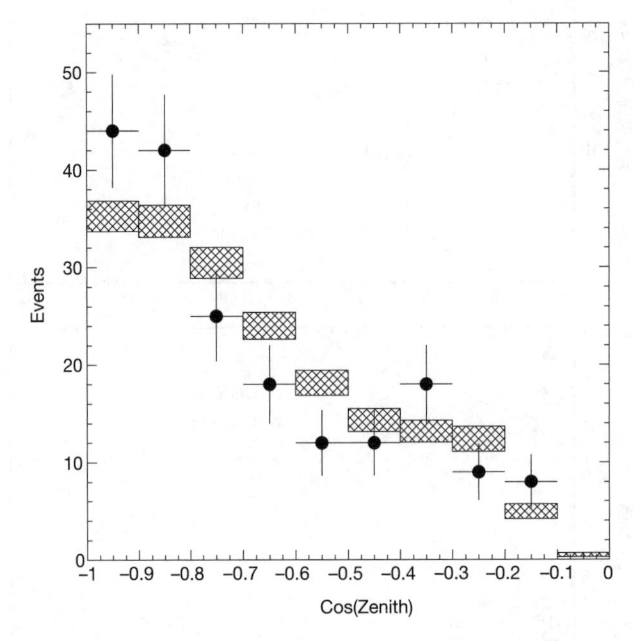

Fig. 3. Reconstructed zenith angle distribution. The data points are experimental data (from analysis B) and the shaded boxes are a simulation of atmospheric neutrino events, the widths of the boxes indicating the error bars. The overall normalization of the simulation has been adjusted to fit the data. The possible effects of neutrino oscillations on the flux and its zenith angle dependence estimated from the Super-Kamiokande measurements[20], are expected to be small in our energy range.

From the consistency of the selected event sample with muons generated by atmospheric neutrinos, and in particular the absence of an excess of high-energy events with a large number of optical modules that had been hit, we can determine an upper limit on a diffuse extraterrestrial neutrino flux. Assuming a hard E^{-2} spectrum characteristic of shockwave acceleration, we expect to reach a sensitivity of order $dN/dE_\nu = 10^{-6} E_\nu^{-2}$ cm^{-2} s^{-1} sr^{-1} GeV^{-1}. This value is low enough to be in the range where a number of models[14-19] predict fluxes, a few of which are larger[15,16]. Most recent estimates are smaller[17-19]. The present level of sensitivity

图 3 给出了 B 分析中 188 个事件的天顶角分布与模拟大气中微子信号的对比，其中蒙特卡罗绝对事例率根据实验值进行了归一化。在统计不确定度之内，测量事例率随天顶角的变化与模拟结果一致。我们注意到，探测器较高的几何尺寸会更利于探测到垂直方向的 μ 子。在 A 和 B 两种分析中，观测到的向上运动的 μ 子的到达方向如图 4 所示。统计分析表明，这些样本尚不足以提供点源存在的证据。图 2 与图 3 已经证明实验与大气中微子信号模拟的一致性；同时对比其他变量(具体研究结果即将发表)的话，我们得出以下结论：AMANDA 观测到的上行 μ 子事件主要是由那些能量介于 50 GeV 到若干 TeV 之间的大气中微子产生。事件样本中的背景事件估计为 15%±7%，这是由于错误的径迹重建造成的。

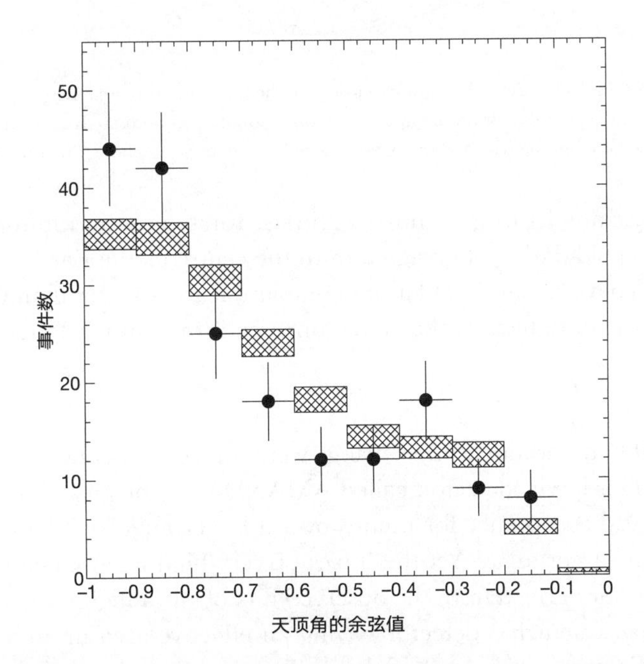

图 3. 重建径迹的天顶角分布。图中数据点是取自分析 B 的实验数据，阴影矩形框表示大气中微子事件的模拟，其中矩形框的宽度代表误差棒。模拟的整体归一化已经经过调整以拟合数据。中微子振荡对通量及通量的天顶角依赖的可能影响可以通过超级神冈实验的测量估算出来[20]，在我们研究的能量范围内，该影响预计很小。

根据筛选的事件样本与大气中微子产生的 μ 子的一致性，特别是触发大量光学模块的高能事件不存在超出的情况下，我们可以确定地外弥散中微子通量的上限。假设激波加速具有 E^{-2} 形式的硬谱特征，那么我们预计灵敏度的量级将达到 $dN/dE_{\nu}=10^{-6}\ E_{\nu}^{-2}\ \mathrm{cm^{-2}\cdot s^{-1}\cdot sr^{-1}\cdot GeV^{-1}}$。该值很小，位于很多模型预测的通量范围之内[14-19]，其中一些通量会更大些[15,16]。不过近期的估算值大多数相对更小[17-19]。我们设想可

and the prospects for improving it through longer exposure times and better determination of muon energy illustrate the ability of large-area detectors to test theoretical models that assume the hadronic origin of TeV photons from active galaxies—models which would be difficult to confirm or exclude without the ability to observe high-energy neutrinos.

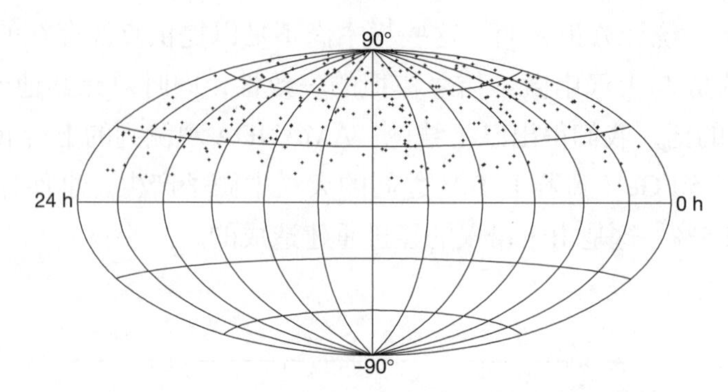

Fig. 4. Distribution in declination and right ascension of the upwardly propagating events on the sky. The 263 events shown here are taken from the upward muons contained in both analysis A and analysis B. The median difference between the true and the reconstructed muon angles is about 3 to 4 degrees.

Searches for neutrinos from gamma-ray bursts, for magnetic monopoles, supernova collapses and for a cold dark matter signal from the centre of the Earth are also in progress and, with only 138 days of data, yield limits comparable to or better than those from smaller underground neutrino detectors that have operated for a much longer period (see refs 10–13).

From 1997 to 1999 an additional nine strings were added in a concentric cylinder around AMANDA-B10. This larger detector, called AMANDA-II, consists of 677 optical modules and has an improved acceptance for muons over a larger angular interval. Data are being taken now with the larger array. Yet the fluxes of very high energy neutrinos predicted by theoretical models[14] or derived from the observed flux of ultra high energy cosmic rays[19] are sufficiently low that a neutrino detector having an effective area up to a square kilometre is required for their observation and study[1,14]. Plans are therefore being made for a much larger detector, IceCube, consisting of 4,800 photomultipliers to be deployed on 80 strings. This proposed neutrino telescope would have an effective area of about 1 km^2, an energy threshold near 100 GeV and a pointing accuracy for muons of better than one degree for high-energy events. In conclusion, the observation of neutrinos by a neutrino telescope deep in the Antarctic ice cap, a goal that was once thought difficult if not impossible, represents an important step toward establishing the field of high-energy neutrino astronomy first envisioned over 40 years ago.

(**410**, 441-443; 2001)

以通过更长时间的曝光以及更好地确定μ子的能量来改善灵敏度，目前的灵敏度水平和这些设想说明了大面积探测器检验理论模型的能力。这些理论模型假设活动星系发出的TeV能量的光子来源于强子，如果没有观测高能中微子能力，证实或排除这些模型将会非常困难。

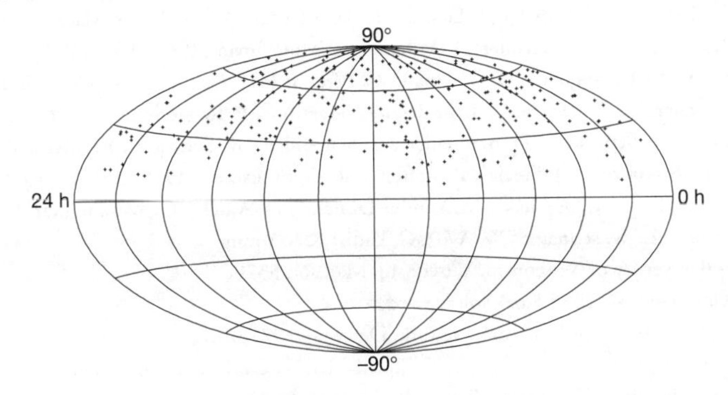

图4. 向上传播的μ子事件在天空中赤纬和赤经上的分布。图中所示的263个事件取自A和B两个分析中包含的共同的上行μ子事件。μ子的真实角度和重建角度之差的中值约为3～4度。

寻找来自伽马射线暴的中微子，寻找磁单极子和超新星坍缩，寻找来自地球中心的冷暗物质，这些工作也在进行之中。而且，仅使用138天的数据，就可给出与那些已长时间运行的小型地下中微子探测器相媲美或更好的上限（见参考文献10～13）。

从1997年到1999年，在AMANDA-B10周围的同心圆柱体中又添加了9串探测器。这个更大的探测器被称为AMANDA-II，由677个光学模块构成，在更大的角度范围内改善了μ子的接收度。目前，这个更巨大的阵列正在采集数据。但是无论是理论模型预言[14]还是由超高能宇宙射线通量观测值推导出的[19]高能中微子通量都是非常低的，以至于为了观测研究这些中微子，探测器的有效面积需要达到一平方千米[1,14]。所以，目前正在计划研制更大的探测器，该探测器被称为冰立方（IceCube），由分布在80根串上的4,800个光电倍增管构成。这一计划中的中微子探测器将具有1 km^2 大小的有效面积，其探测能量阈值接近100 GeV，对于高能事件的μ子的指向精度优于1度。总而言之，通过南极冰盖下的中微子望远镜观测到中微子意味着人们朝40多年前就试图建立的高能中微子天文学研究领域迈出了重要的一步，尽管这一设想一度被认为是极难甚至是无法实现的目标。

（金世超 翻译；曹俊 审稿）

E. Andrés[*], P. Askebjer[†], X. Bai[‡], G. Barouch[*], S. W. Barwick[§], R. C. Bay[‖], K.-H. Becker[¶], L. Bergström[†], D. Bertrand[#], D. Bierenbaum[§], A. Biron[☆], J. Booth[§], O. Botner[**], A. Bouchta[☆], M. M. Boyce[*], S. Carius[††], A. Chen[*], D. Chirkin[‖¶], J. Conrad[**], J. Cooley[*], C. G. S. Costa[#], D. F. Cowen[‡‡], J. Dailing[§], E. Dalberg[†], T. DeYoung[*], P. Desiati[☆], J.-P. Dewulf[#], P. Doksus[*], J. Edsjö[†], P. Ekström[†], B. Erlandsson[†], T. Feser[§§], M. Gaug[☆], A. Goldschmidt[‖‖], A. Goobar[†], L. Gray[*], H. Haase[☆], A. Hallgren[**], F. Halzen[*], K. Hanson[‡‡], R. Hardtke[*], Y. D. He[‖], M. Hellwig[§§], H. Heukenkamp[☆], G. C. Hill[*], P. O. Hulth[†], S. Hundertmark[§], J. Jacobsen[‖‖], V. Kandhadai[*], A. Karle[*], J. Kim[§], B. Koci[*], L. Köpke[§§], M. Kowalski[☆], H. Leich[☆], M. Leuthold[☆], P. Lindahl[††], I. Liubarsky[*], P. Loaiza[**], D. M. Lowder[‖], J. Ludvig[‖‖], J. Madsen[*], P. Marciniewski[**], H. S. Matis[‖‖], A. Mihalyi[‡‡], T. Mikolajski[☆], T. C. Miller[‡], Y. Minaeva[†], P. Miočinović[‖], P. C. Mock[§], R. Morse[*], T. Neunhöffer[§§], F. M. Newcomer[‡‡], P. Niessen[☆], D. R. Nygren[‖‖], H. Ögelman[*], C. Pérez de los Heros[**], R. Porrata[§], P. B. Price[‖], K. Rawlins[*], C. Reed[§], W. Rhode[¶], A. Richards[‖], S. Richter[☆], J. Rodríguez Martino[†], P. Romenesko[*], D. Ross[§], H. Rubinstein[†], H.-G. Sander[§§], T. Scheider[§§], T. Schmidt[☆], D. Schneider[*], E. Schneider[§], R. Schwarz[*], A. Silvestri[¶☆], M. Solarz[‖], G. M. Spiczak[‡], C. Spiering[☆], N. Starinsky[*], D. Steele[*], P. Steffen[☆], R. G. Stokstad[‖‖], O. Streicher[☆], Q. Sun[†], I. Taboada[‡‡], L. Thollander[†], T. Thon[☆], S. Tilav[*], N. Usechak[§], M. Vander Donckt[#], C. Walck[†], C. Weinheimer[§§], C. H. Wiebusch[☆], R. Wischnewski[☆], H. Wissing[☆], K. Woschnagg[‖], W. Wu[§], G. Yodh[§] & S. Young[§]

[*] Department of Physics, University of Wisconsin, Wisconsin, Madison 53706, USA

[†] Fysikum, Stockholm University, S-11385 Stockholm, Sweden

[‡] Bartol Research Institute, University of Delaware, Newark, Delaware 19716, USA

[§] Department of Physics and Astronomy, University of California, Irvine, California 92697, USA

[‖] Department of Physics, University of California, Berkeley, California 94720, USA

[¶] Fachbereich 8 Physik, BUGH Wuppertal, D-42097 Wuppertal, Germany

[#] Brussels Free University, Science Faculty CP230, Boulevard du Triomphe, B-1050 Brussels, Belgium

[☆] DESY-Zeuthen, D-15735 Zeuthen, Germany

[**] Department of Radiation Sciences, Uppsala University, S-75121 Uppsala, Sweden

[††] Department of Technology, Kalmar University, S-39129 Kalmar, Sweden

[‡‡] Department of Physics and Astronomy, University of Pennsylvania, Philadelphia, Pennsylvania 19104, USA

[§§] Institute of Physics, University of Mainz, Staudinger Weg 7, D-55099 Mainz, Germany

[‖‖] Institute for Nuclear and Particle Astrophysics, Lawrence Berkeley National Laboratory, Berkeley, California 94720, USA

Received 15 September 2000; accepted 25 January 2001.

References:

1. Gaisser, T. K., Halzen, F. & Stanev, T. Particle physics with high-energy neutrinos. *Phys. Rep.* **258**, 173-236 (1995).
2. Totsuka, Y. Neutrino astronomy. *Rep. Prog. Phys.* **55**, 377-430 (1992).
3. Roberts, A. The birth of high-energy neutrino astronomy: A personal history of the DUMAND project. *Rev. Mod. Phys.* **64**, 259-312 (1992).
4. Balkanov, V. A. *et al.* An upper limit on the diffuse flux of high energy neutrinos obtained with the Baikal detector NT-96. *Astropart. Phys.* **14**, 61-67 (2000).
5. Lowder, D. M. *et al.* Observation of muons using the polar ice as a Čerenkov detector. *Nature* **353**, 331-333 (1991).
6. Andres, E. *et al.* The AMANDA neutrino telescope: Principle of operation and first results. *Astropart. Phys.* **13**, 1-20 (2000).
7. Waxman, E. & Bahcall, J. N. High-energy neutrinos from cosmological gamma ray burst fireballs. *Phys. Rev. Lett.* **78**, 2992-2295 (1997).
8. Askjeber, P. *et al.* Optical properties of the south pole ice at depths between 0.8 and 1 km. *Science* **267**, 1147-1150 (1995).
9. Price, B. P. Implications of optical properties of ocean, lake, and ice for ultrahigh-energy neutrino detection. *Appl. Opt.* **36**, 1965-1975 (1997).
10. Wischnewski, R. *et al.* in *Proc. 26th Int. Cosmic Ray Conf., Salt Lake City* Vol. 2 (eds Kieda, D., Salamon, M. & Dingus, B.) 229-232 (1999).
11. Dalberg, E. *et al.* in *Proc. 26th Int. Cosmic Ray Conf., Salt Lake City* Vol. 2 (eds Kieda, D., Salamon, M. & Dingus, B.) 348-351 (1999).
12. Bay, R. *et al.* in *Proc. 26th Int. Cosmic Ray Conf., Salt Lake City* Vol. 2 (eds Kieda, D., Salamon, M. & Dingus, B.) 225-228 (1999).
13. Niessen, P. *et al.* in *Proc. 26th Int. Cosmic Ray Conf., Salt Lake City* Vol. 2 (eds Kieda, D., Salamon, M. & Dingus, B.) 344-347 (1999).
14. Learned, J. G. & Mannheim, K. High-energy neutrino astrophysics. *Ann. Rev. Nucl. Sci.* **50**, 679-749 (2000).
15. Szabo, A. P. & Protheroe, R. J. Implications of particle acceleration in active galactic nuclei for cosmic rays and high-energy neutrino astronomy. *Astropart. Phys.* **2**, 375-392 (1994).
16. Stecker, F. W. & Salamon, M. H. High-energy neutrinos from quasars. *Space Sci. Rev.* **75**, 341-355 (1996).
17. Nellen, L., Mannheim, K. & Biermann, P. L. Neutrino production through hadronic cascades in AGN accretion disks. *Phys. Rev. D* **47**, 5270-5274 (1993).
18. Mannheim, K., Protheroe, R. J. & Rachen, J. P. Cosmic ray bound for models of extragalactic neutrino production. *Phys. Rev. D* **63**, 023003-1–023003-16 (2001).
19. Waxman, E. & Bahcall, J. N. High energy neutrinos from astrophysical sources. An upper bound. *Phys. Rev. D* **59**, 023002-1–023002-8 (1999).

20. Fukuda, Y. *et al.* (Super-Kamiokande Collaboration) Measurement of the flux and zenith-angle distribution of upward through-going muons by Super-Kamiokande. *Phys. Rev. Lett.* **82**, 2644-2648 (1999).

Acknowledgements. This research was supported by the following agencies: US National Science Foundation, Office of Polar Programs; US National Science Foundation, Physics Division; University of Wisconsin Alumni Research Foundation; US Department of Energy; Swedish Natural Science Research Council; Swedish Polar Research Secretariat; Knut and Allice Wallenberg Foundation, Sweden; German Ministry for Education and Research; US National Energy Research Scientific Computing Center (supported by the Office of Energy Research of the US Department of Energy); UC-Irvine AENEAS Supercomputer Facility; Deutsche Forschungsgemeinschaft (DFG). D.F.C. acknowledges the support of the NSF CAREER programme and C.P.d.l.H. acknowledges support from the European Union 4th Framework of Training and Mobility of Researchers.

Correspondence and requests for materials should be addressed to F.H. (e-mail: halzen@pheno.physics.wisc.edu).

Upper Limits to Submillimetre-range Forces from Extra Space-time Dimensions

J. C. Long *et al.*

Editor's Note

Gravity is the weakest of the four fundamental forces, being 40 orders of magnitude weaker than the electromagnetic force. In string theory—an attempt to unify the description of particles and forces—this weakness arises because the force is dissipated in six extra spatial dimensions that are "curled up" and therefore invisible to us. It has been suggested that the effect of these curled-up dimensions might become evident at very small length scales, whereby objects would feel more gravitation force at those scales. Here physicist Joshua Long and colleagues describe an experiment that looked for a deviation of the gravitational force from its familiar inverse-square law on length scales of 100 micrometres. They found no such deviation.

String theory is the most promising approach to the long-sought unified description of the four forces of nature and the elementary particles[1], but direct evidence supporting it is lacking. The theory requires six extra spatial dimensions beyond the three that we observe; it is usually supposed that these extra dimensions are curled up into small spaces. This "compactification" induces "moduli" fields, which describe the size and shape of the compact dimensions at each point in space-time. These moduli fields generate forces with strengths comparable to gravity, which according to some recent predictions[2-7] might be detected on length scales of about 100 μm. Here we report a search for gravitational-strength forces using planar oscillators separated by a gap of 108 μm. No new forces are observed, ruling out a substantial portion of the previously allowed parameter space[4] for the strange and gluon moduli forces, and setting a new upper limit on the range of the string dilaton[2,3] and radion[5-7] forces.

THE combined potential energy V due to a modulus force and newtonian gravity may be written:

$$V = -\int d\mathbf{r}_1 \int d\mathbf{r}_2 \, \frac{G\rho_1(\mathbf{r}_1)\rho_2(\mathbf{r}_2)}{r_{12}} \, [1 + \alpha \exp(-r_{12}/\lambda)] \tag{1}$$

The first term is Newton's universal gravitation law, with G the gravitational constant, r_{12} the distance between two points $\mathbf{r}_1$ and $\mathbf{r}_2$ in the test masses, and ρ_1, ρ_2 the mass densities of the two bodies. The second term is a Yukawa potential, with α the strength of the new force relative to gravity, and λ the range. As for any force-mediating field, the range of the modulus force is related to its mass m by $\lambda = \hbar/mc$. Previous tests of Newtonian gravity

来自额外空间维度在亚毫米尺度上力程的上限

朗等

编者按

引力是四种基本力中最弱的，比电磁力要小四十个数量级。在弦论——一个试图统一描述粒子和力的尝试——中，引力的弱性是由于引力向着六个额外的卷曲不可见维度泄漏导致的。已经有一些建议说这些卷曲维度的效应可能在很小的尺度下会变得很明显，在这些尺度上物体将会感受到更大的引力。本文中，物理学家乔舒亚·朗和合作者们介绍了一个实验，这个实验在 100 μm 的尺度上寻找引力偏离通常熟知的平方反比律的迹象。他们发现没有这样的偏离。

弦理论是人们一直在寻找的、统一自然界四种力和基本粒子的最有希望的方法[1]，但是现在仍然缺乏直接支持它的证据。除了我们观察到的三个维度之外，弦理论还要求存在六个额外空间维度。通常假定这些额外的维度卷曲成了很小的空间。这个“紧致化”诱导出了“模”场，用来描述时空中每一点紧致维度的大小和形状。根据最近的预测[2-7]，这些模场产生了和引力大小可比的力，可能能在 100 μm 尺度上被探测到。本文中，我们报告了以下工作：用相隔 108 μm 的两个平面振子来搜寻这一引力强度量级的力。结果未观察到新的力，从而排除了一大块之前对于奇异模力和胶子模力来说是允许的参数空间[4]，对弦伸缩子力[2,3]和径向子力[5-7]的力程给出了一个新的上限。

由模力和牛顿引力结合得到的势能 V 可以写作

$$V = -\int d\mathbf{r}_1 \int d\mathbf{r}_2 \frac{G\rho_1(\mathbf{r}_1)\rho_2(\mathbf{r}_2)}{r_{12}} [1+\alpha\exp(-r_{12}/\lambda)] \tag{1}$$

第一项是牛顿万有引力定律，G 是引力常数，r_{12} 是检验质量位置 $\mathbf{r}_1$ 和 $\mathbf{r}_2$ 两点间的距离，ρ_1，ρ_2 是两个物体的质量密度。第二项是汤川势，α 是新的力相对于引力的强度，λ 是力程。对于任何传递相互作用的场，模力的力程和它的质量 m 存在如下关系 $\lambda = \hbar/mc$。之前关于牛顿引力的检验和对于新的宏观力的搜寻涵盖了从光年到纳

and searches for new macroscopic forces have covered length scales from light-years to nanometres[8-10], and it has been found that new forces of gravitational strength can be excluded for ranges λ from 200 μm to nearly a light-year[8,11,12], but limits on new forces become poor very rapidly below 200 μm (refs 9, 13).

The strange and gluon moduli arise in a scenario where supersymmetry (a hypothesized symmetry incorporated in string theory that relates bosons to fermions) is broken at 10–100 TeV, and all compactification occurs near the ultrahigh Planck scale of 10^{19} GeV where gravity is unified (in these scenarios) with other forces. In this case a range of approximately a millimetre is specifically predicted. The dilaton is a scalar field required for the consistency of string theory whose vacuum value fixes the strength of the interaction between strings. It may be interpreted as the modulus describing the size of the tenth space dimension that is compactified in M-theory to yield string theory in nine space dimensions. (M-theory is the structure that unifies the various types of string theory into a single framework[1].) The strength α associated with the dilaton can be computed, but the range λ at present can only be constrained by experiment; little is known about the mechanism which gives the dilaton mass. The radion results from a very different scenario in which one or more dimensions compactify at TeV energy scales, but here a range λ accessible to our experiment is again predicted.

The special significance of millimetre scales derives from a mass formula of the form $m \approx M^2/M_P$, where M_P is the mass associated with the Planck scale, 10^{19} GeV$/c^2$, and M is a mass of 1–100 TeV$/c^2$, leading to λ in the centimetre to micrometre range. In the strange and gluon moduli case, M is the scale where supersymmetry breaking occurs, whereas in the radion modulus case it corresponds to the size of one or more compact dimensions. This formula also applies to the ADD theory[14] (named for its authors) in which two compact extra dimensions of millimetre size modify gravity itself. In the ADD theory M is the fundamental length scale where all physics is unified while m is a mass corresponding to the size of the large compact extra dimensions. Many scenarios of compactification, symmetry breaking and mass generation are still viable, so although the possibility of observing new forces at millimetre scales is exciting, such experiments cannot currently falsify string theory.

The planar geometry of our source and detector masses (Fig. 1) is chosen to concentrate as much mass density as possible at the length scale of interest. It is approximately null with respect to the $1/r^2$ newtonian background, a helpful feature in the context of a new force search. A cantilever mode (similar to the motion of a diving board) of the tungsten source mass is driven to a tip amplitude of 19 μm at the resonant frequency of the detector mass. The tungsten detector mass is a double torsional oscillator[15]; in the resonant mode of interest the two rectangular sections of the detector counter-rotate about the torsion axis, with most of the amplitude confined to the smaller rectangle under the source mass and shield. Torsional motion of the detector will be driven if a mass coupled force is present between the source and detector. The motions are detected with a capacitive transducer, followed by a preamplifier, filters and a lock-in amplifier. To suppress background forces due to electrostatics and residual gas, a stiff conducting shield is fixed between the test masses. With

米的尺度[8-10]，已有结果表明，在 200 μm 到接近一个光年的尺度范围内不存在引力强度量级的新力[8,11,12]，但是在 200 μm 之下的尺度，对出现新力的限制会快速变弱（参考文献 9 和 13）。

奇异模空间和胶子模空间出现在 10 ~ 100 TeV 下超对称（一个弦论中猜想的将玻色子和费米子联系起来的对称性）破缺的时候，而所有的紧致化过程则是发生在临近超高普朗克尺度（10^{19} GeV）上的（这些情况下的引力和其他力是统一的）。在这种情况下，可以预测出一个大约是毫米级的力程。伸缩子是弦理论自洽性所要求的一个标量场，它的真空值决定了弦之间相互作用的强度。它或许可以被理解成是 M 理论中描述第十个空间维度的尺度的模参数，这个空间维度在 M 理论中被做了紧致化来得到具有九个空间维度的弦论（M 理论是将所有不同的弦论整合成一个框架的理论结构[1]）。和伸缩子相关的强度 α 可以通过计算得到，但是力程 λ 目前只能通过实验来限制，并且我们几乎不知道给出伸缩子质量的机制是什么。径向子是由一个完全不同的机制得到的，这一机制下一个或者多个额外维度的紧致化发生在 TeV 能标尺度上，但这里实验可以得到的力程 λ 也是预测出来的。

毫米尺度特殊的重要性可以通过形如 $m \approx M^2/M_P$ 的质量公式导出，这里 M_P 是和普朗克尺度相关的质量 10^{19} GeV/c^2，M 是 1 ~ 100 TeV/c^2 的质量，导致 λ 的尺度在厘米到微米范围之间。在奇异模空间和胶子模空间中，M 是超对称破缺发生的尺度，而在径向子模空间的情况下，它对应于一个或者多个紧致维度的尺度。这个公式适用于 ADD 理论[14]（根据其提出者们的姓名命名），这一理论中两个毫米尺度的紧致化额外维会修改引力。在 ADD 理论中，M 是一个基本尺度，在这个尺度上所有的物理理论都是统一起来的，而 m 是对应于大的紧致化额外维尺度的质量。在许多紧致化图像中，对称性破缺和质量产生依然是可靠的，因此，尽管在毫米尺度上观察到新力的可能性是令人激动的，但这些实验目前还不能证伪弦理论。

我们选择的源和探测器质量呈现为图 1 所示的平面几何，是为了使其尽可能地在感兴趣的尺度上集中更高的质量密度。它几乎不受 $1/r^2$ 牛顿背景的影响，这在寻找一个新力的时候是一个有帮助的特征。钨制的源的悬臂模式（和跳水板的运动类似）在探测器共振频率的驱动下出现了 19 μm 的尖端振幅。钨探测器是双扭转振子[15]，在我们感兴趣的共振模式下，探测器的两个矩形部分绕着扭转轴反转，绝大部分振幅都被限制在防护屏和源下面的小矩形中。如果源和探测器之间存在一个质量耦合力的话将会驱动探测器的扭转运动。运动用电容传感器探测，跟着是一个前置放大器、过滤器和锁定放大器。为了压低由于静电和剩余气体产生的背景力，一个刚性的导电防护屏固定在两个检验质量之间。对于静止的源，源和探测器的距离

the source at rest, the gap between source and detector is adjusted to 108 μm, and the entire apparatus is placed in a vacuum enclosure and maintained at pressures below 2×10^{-7} torr.

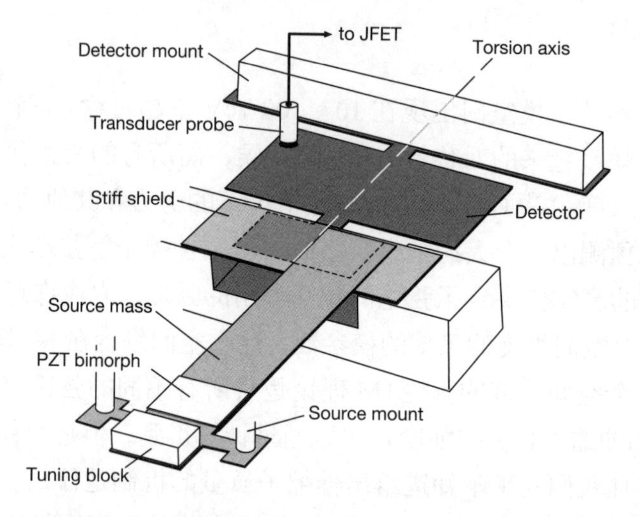

Fig. 1. Major components of the apparatus. The smaller rectangle of the tungsten detector (under the shield) is 11.455 mm wide, 5.080 mm long and 195 μm thick. The detector is annealed at 1,300 °C to increase its mechanical Q to 25,000. In operation the 1,173.085-Hz resonant frequency of the detector is stabilized by actively controlling the detector temperature to 305 K. The tungsten source mass is 35 mm long, 7 mm wide and 305 μm thick. The source mass resonant frequency is tuned to the detector and driven by the PZT (lead zirconate titanate) piezoelectric bimorph. The shield is a 60-μm-thick sapphire plate coated with 100 nm of gold on both sides. The test masses and the shield are supported by three separate five-stage passive vibration isolation stacks[23], each providing approximately 200-dB attenuation at 1 kHz. Mechanical probes are used to directly measure the relative orientation and position of each component, and to measure the source mass amplitude. Detector motions are sensed by a cylindrical capacitive probe supported 100 μm above a rear corner of the large rectangle of the detector mass. The probe is biased at 200 V through a 100 GΩ resistor, and connected to an SK 152 junction field-effect transistor (JFET) through a blocking capacitor. The JFET noise temperature of 100 mK is more than sufficient to detect 305 K thermal motions of the detector mass. The JFET preamplifier is followed by a second preamplifier, filters, and finally a two-phase lock-in amplifier. The total voltage gain from the capacitive probe to the lock-in input is approximately 1,600. A crystal-controlled oscillator provides a reference signal for the lock-in amplifier and drives the source mass PZT through a 1:10 step-up transformer.

Figure 2 shows histograms of the raw data, collected with the source mass drive tuned both on and off the detector resonant frequency. Each plot contains data from one channel of the lock-in amplifier, corresponding to one of two orthogonal phases of the detector motion at the drive frequency. The widths of the off-resonance distributions are due to preamplifier noise, whereas the on-resonance distributions are due to the sum of preamplifier noise and detector thermal motions. The on- and off-resonance means shown in Fig. 3 agree within their standard deviations, indicating the absence of a significant resonant force signal.

调整到 108 μm，整个仪器放在真空罩中，然后将压强控制在 2×10^{-7} torr 以下。

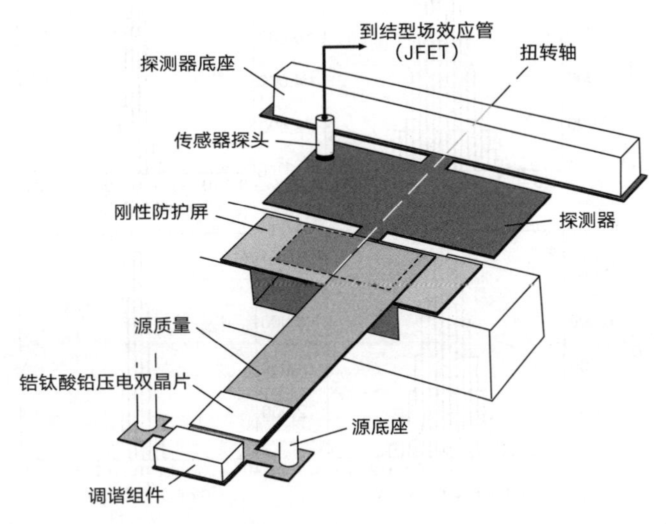

图 1. 仪器的主要构成。钨探测器的小矩形（在防护屏下面）有 11.455 mm 宽，5.080 mm 长，195 μm 厚。探测器在 1,300 ℃下退火以便增加它的机械品质因子 Q 到 25,000。在操作上，探测器 1,173.085 Hz 的共振频率是通过主动地将探测器的温度控制在 305 K 来稳定下来。钨制的源有 35 mm 长，7 mm 宽，305 μm 厚。源的共振频率调制到探测器的共振频率，并由锆钛酸铅（PZT）压电双晶片驱动。防护屏是一个 60 μm 厚的蓝宝石板，两侧镀上了 100 nm 厚的金。检验质量和防护屏由三个分立的五步被动隔振层支撑 [23]，每一个在 1 kHz 时提供了大约 200 dB 的噪音减弱。机械探头用来直接探测每个组分的相对取向和位置，还用来测量源的振幅。探测器的运动通过在探测器大矩形后面的角的上方 100 μm 处的一个圆柱形电容式探头来感知。探头通过一个 100 GΩ 的电阻在 200 V 电压上偏置，并且通过一个隔直流电容器连着一个 SK 152 的结型场效应晶体管（JFET）。JFET 100 mK 的噪声温度足够用来探测 305 K 的探测器的热运动了。JFET 前置放大器接着第二个前置放大器、过滤器，最后是一个两相锁定放大器。从电容式探头到锁定放大器输入的总电压放大了约 1,600 倍。一个晶体调控的振子给锁定放大器提供了一个参考信号，并且通过一个 1∶10 的升压变压器驱动源质量的 PZT。

图 2 展示了原始数据的直方图，当源的驱动频率在探测器的共振频率时和不在探测器的共振频率时都收集了数据。每个图都包含来自锁定放大器的一个通道的信号，对应于在驱动频率下探测器运动的两个正交模式中的一个。非共振分布的展宽是因为前置放大器的噪声，而共振分布的展宽来自前置放大器的噪声和探测器的热运动之和。图 3 中这个共振和非共振的平均值在标准偏差之内是吻合的，表明不存在一个显著的共振力的信号。

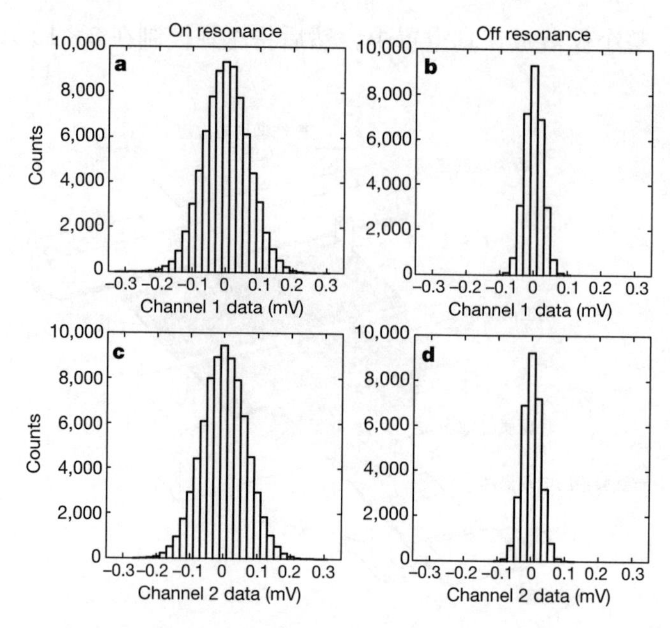

Fig. 2. Distributions of data samples. Data were recorded at 1 Hz with a lock-in bandwidth chosen to include the noise power of the detector thermal oscillations, which was used for calibration. Each data cycle began with five 120-sample diagnostic runs with a direct-current bias of 5–10 V applied to the shield to induce a large test force, transmitted from source to detector via deflections of the shield. The biased runs were recorded at five drive frequencies separated by 15 mHz to cover the detector resonance. The shield was then grounded and the cycle continued with 720 samples with the drive tuned on-resonance (**a, c**) and 288 samples with the drive tuned 2 Hz below the detector resonance (**b, d**). The off-resonance run provided a continuous zero check. A total of 108 such cycles were acquired over five days yielding 77,760 on-resonance samples. The biased diagnostic data show that the source mass amplitude, the detector Q and resonant frequency, and the electronic gain were all stable throughout the data set.

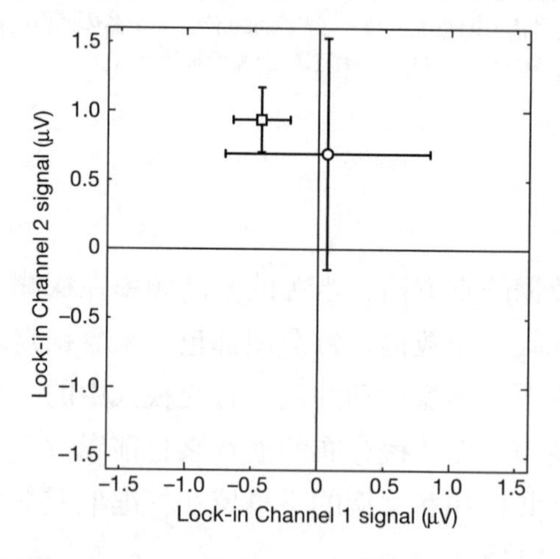

Fig. 3. Means of the off- and on-resonance data samples. The circular point with the larger standard deviations is the on-resonance mean. Correlations between nearby samples have been accounted for in computing the standard deviations shown by error bars. The small offset from the origin is due to leakage of the reference signal internal to the lock-in. Measurements with the shield removed and a bias voltage between the source and detector show that the phase for an attractive force is 189°. The on-resonance mean minus the off-resonance mean at 189° is -0.44 ± 0.82 μV. Based on calibration via the equipartition

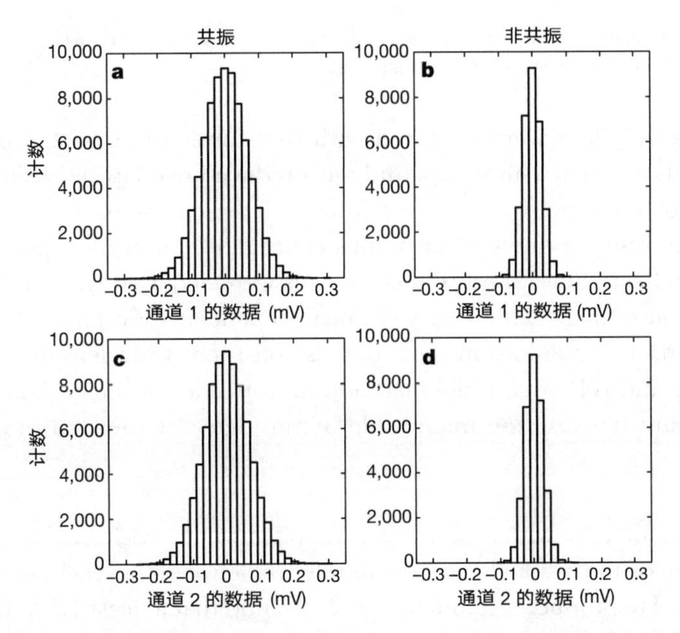

图 2. 数据样本的分布。数据在 1 Hz 处收集，选取的锁定带宽包含了探测器热振动的噪声功率以用于校正。每个数据周期从 120 个样本的 5 次诊断运行开始，在防护屏上应用 5 ~ 10 V 的直流偏压，用来产生一个很大的检测力，然后通过防护屏的偏转在源到探测器之间传递。偏置运行在五个驱动频率下记录，这五个驱动频率之间间隔 15 mHz 从而覆盖探测器共振频率。防护屏之后接地，然后这个周期继续在驱动频率调到共振频率的 720 个样本（**a** 和 **c**）和驱动频率调到探测器共振频率以下 2 Hz 处的 288 个的样本（**b** 和 **d**）中进行。非共振运行提供了一个连续的零检验。五天时间内获得了共 108 个这样的循环，产生了 77,760 个共振样本。偏置诊断的数据显示源质量的振幅、探测器的品质因子 Q 和共振频率以及电子增益在整个数据集上都是稳定的。

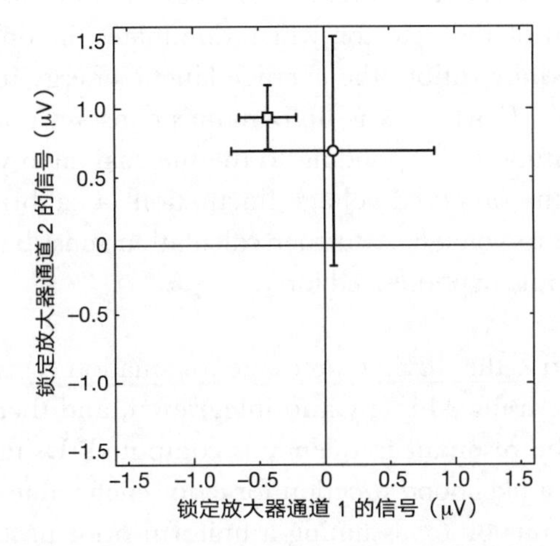

图 3. 非共振和共振数据样本的平均值。具有更大标准偏差的圆点是共振的平均值。邻近样本之间的关联在计算标准偏差时已经考虑进去了，标准偏差用误差棒来体现。原点处的微小偏移来源于锁定放大器内部的参考信号的溢出。去掉防护屏，并在源和探测器之间加上偏置电压进行了探测，结果显示吸引力的相位是 189°。在 189°，共振的平均值减去非共振的平均值是 -0.44 ± 0.82 μV。基于通过能均分定理

theorem this corresponds to a lumped force amplitude at the edge of the detector (smaller rectangle in Fig. 1) of -1.2 ± 2.2 fN, where the negative sign indicates repulsion.

Additional cycles of data were acquired with the source mass on the opposite side of the detector, with a larger, 1-mm gap, and with reduced overlap between the source and detector. No resonant signal was observed in any of these sessions, making it unlikely that the observed null result is due to a fortuitous cancellation of surface potential, magnetic, and/or acoustic effects. Several on-resonance runs were acquired with different transducer probe bias voltage settings. The observed linear dependence on bias voltage of the root-mean-square (r.m.s.) fluctuations in these data is consistent with detector motion due only to thermal noise and rules out additional motion from transducer back-action noise. This check is important because the magnitude of the detector thermal motion is used for calibration.

Data from diagnostic runs with a bias voltage applied to the shield can be used to estimate the minimum size of the residual potential difference between the shield and the (grounded) test masses needed to produce a resonant signal. We find that at least 1.5 V would be needed to generate an effect above detector thermal noise, about an order of magnitude larger than the residual potential difference actually measured between the shield and test masses. The most important magnetic background effect involves eddy currents generated when the source mass moves in an external magnetic field. Fields produced by the source currents create eddy currents in the detector, which then interact with the external field. Studies of this effect with large applied magnetic fields show that the ambient field actually present cannot generate a signal greater than one-fifth of the thermal-noise-limited sensitivity.

The instrument can be calibrated in several ways, but the most accurate method is to use the r.m.s. thermal motion of the detector, which dominates the on-resonance distributions in Fig. 2. According to equipartition, the average kinetic energy in each normal mode of the detector is equal to $\frac{1}{2}kT$; where k is Boltzmann's constant and T is the temperature. The normal mode amplitude corresponding to the thermal energy can be calculated, and by comparing this with the observed voltage fluctuations a calibration can be established relating mode amplitude to voltage. A further calculation must be done to find the mode amplitude resulting from any hypothesized force.

For given values of α and λ the driving force due to equation (1) is computed at 30 values of the source mass phase using Monte Carlo integration, and then the Fourier amplitude of the driving force at the resonant frequency is computed. Using the observed statistics of the data we construct a likelihood function for α for each value of λ and compute 95% confidence-level upper limits on α, assuming a uniform prior probability density function (PDF) for α. This analysis is complicated by the presence of uncertainties in the geometrical and mechanical parameters needed to compute the driven displacement. To include these effects we actually construct a likelihood function of α and the uncertain parameters, and then integrate out the uncertain parameters using prior PDFs based on their experimental uncertainties. The most important of these parameters is the 108-μm equilibrium gap

的校正，这对应于一个在探测器边缘(图 1 中的小矩形)大小为 −1.2 ± 2.2 fN 的集总力振幅，负号表明它是排斥力。

更多的数据周期通过将源放在探测器的另外一侧获得，源和探测器之间有一个比之前大的、大小为 1 mm 的间隙，并且有着更小的交叠。在这些测量中都没有看到任何的共振信号，使得观测到的零结果不太可能源于表面势、磁和(或)声学效应的偶然抵消。设定不同的传感器探头的偏置电压获得了几次共振。在这些数据中观测到的方均根误差的涨落对于偏置电压的线性依赖和只由热噪声产生的探测器运动是自洽的，排除了传感器反作用噪声导致的额外运动。这个检查是重要的，因为探测器热运动的大小是被用来做校正的。

来自诊断运行(在防护屏上施加偏置电压)的数据可以用来估计防护屏和(接地的)检验质量之间的剩余电势差的最小尺度，这个电势差是需要的，它可以产生一个共振信号。我们发现至少需要 1.5 V 的电压来产生一个高于探测器热噪声的效应，1.5 V 大约比测得的在防护屏和检验质量之间的剩余电势差高一个数量级。最重要的磁背景效应是当源在外磁场运动时产生的涡流。源电流产生的场在探测器中产生了涡流，涡流之后会和外部的场相互作用。关于这个大施加磁场带来的效应的研究表明，周围实际存在的场所产生的信号，不能大于热噪声限制给出的灵敏度的五分之一。

这个实验器材可以通过几种方式校正，但是最准确的方法是用探测器的方均根热运动进行校正，它在图 2 的共振分布中占主导。根据能均分定理，每个探测器正则模式的平均动能等于$\frac{1}{2}kT$；k 是玻尔兹曼常数而 T 是温度。对应于热能的正则模式的振幅可以算出来，比较它和观测到的电压涨落，可以通过将模式振幅和电压联系起来建立一个校正。如果要寻找任何假设存在的力所产生的模式振幅，需要进行更进一步的计算。

对于给定的 α 和 λ 的值，由方程(1)给出的驱动力在源相位取的 30 个值下用蒙特卡罗积分计算，然后算出驱动力在共振频率下的傅里叶振幅。用观测到的数据统计，我们对每个 λ 的值构建了一个 α 的似然函数，然后在假定了一个 α 的均匀先验概率密度分布函数(PDF)的基础上，在 95% 的置信度上计算了 α 的上限。由于计算驱动位移时所需要的几何和动力学参数存在不确定性，这个分析会变得复杂。为了包括这些效应，我们实际上构建了一个 α 和这些不确定参数的似然函数，然后基于实验不确定性，利用先验概率密度分布函数积掉了不确定的参数。这些参数中最重要的是在源和探测器之间具有 6 μm 不确定性的 108 μm 平衡间隙。分析方法的进一

between source and detector which has a 6-μm uncertainty. Further details of our analysis methods are given elsewhere[16].

Our results are shown in Fig. 4, together with other experimental limits and the theoretical predictions. Our limit is the strongest available between 10 and 100 μm. When the gluon and strange moduli forces were first proposed, the areas of their allowed regions in the (α, λ) parameter space were 4.4 and 5.2 square decades respectively. By now they are nearly excluded with only 0.75 and 1.4 square decades still available. Our limit on the dilaton range for $\alpha = 2{,}000$ is $\lambda < 23$ μm (corresponding to $m > 8.6$ meV), a factor of two better than the previous limit. For the radion modulus we set an upper limit on λ of 88 μm, close to the value of 40 μm estimated in the theory (for one extra dimension). For the ADD theory with two large extra compact dimensions, we do not quite reach the limit on the size of these dimensions already set in refs 11 and 12.

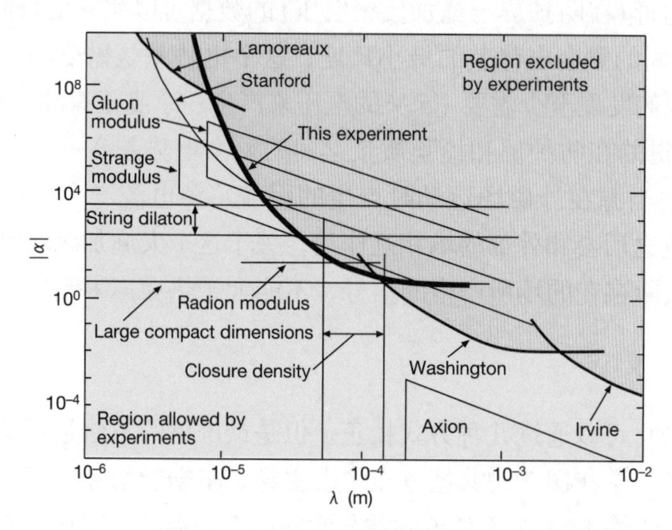

Fig. 4. Current limits on new gravitational strength forces between 1 μm and 1 cm. Our result is a 95% confidence-level upper limit on the Yukawa strength α as a function of range λ (solid bold curve). It is shown together with limits from previous experiments (Lamoreaux[24], Washington[12], Irvine[25]) and theoretical predictions newly constrained (gluon modulus[4], strange modulus[4], string dilaton[2], radion modulus[7]). An unpublished limit from the Stanford experiment[26] is also shown; it is derived in the presence of a background force. The dilaton strength is somewhat model-dependent and there is a range of values reported in the literature[2]. We have chosen the region $200 < \alpha < 3{,}000$, which includes most values. Also shown are predictions from the ADD theory with two large compact extra dimensions[14,27,28], axion mediated forces[20,21], and the λ region corresponding to a cosmological energy density between 1.0 and 0.1 times the closure density[17]. For the moduli, dilaton, and ADD theories, the upper bounds on λ of the regions shown are set at the experimental limits that were known at the time the theories were proposed.

Besides forces from extra dimensions, two other ideas have suggested new weak forces at submillimetre scales. The cosmological energy density needed to close the universe, if converted to a length by taking its inverse fourth root (in natural units where $\hbar = c = 1$), corresponds to about 100 μm. This fact has led to repeated attempts[17-19] to address difficulties connected with the very small observed size of Einstein's cosmological constant by introducing new forces with a range near 100 μm. Our result is the best upper bound

步细节在其他论文中给出[16]。

图 4 展示了我们的结果，其中也包括其他实验的限制和理论的预言。我们的限制在 10 μm 到 100 μm 范围内是已知的方法中最强的。当胶子模力和奇异模力第一次被提出的时候，它们在 (α, λ) 参数空间中所允许的区域分别是 4.4 和 5.2 个单位。现在它们除了 0.75 和 1.4 个允许的单位外，其余的都被排除了。关于 $\alpha = 2{,}000$ 的伸缩子的范围，我们的限制是 $\lambda < 23$ μm(对应于 $m > 8.6$ MeV)，比之前的极限好上一倍。对于径向子的模空间，我们给出了一个 $\lambda = 88$ μm 的上限，接近理论(一个额外维)预言的 40 μm 的值。对于有两个大的额外紧致维度的 ADD 理论，我们并没有达到文献 11 和 12 中给出的关于这些维度的大小的极限。

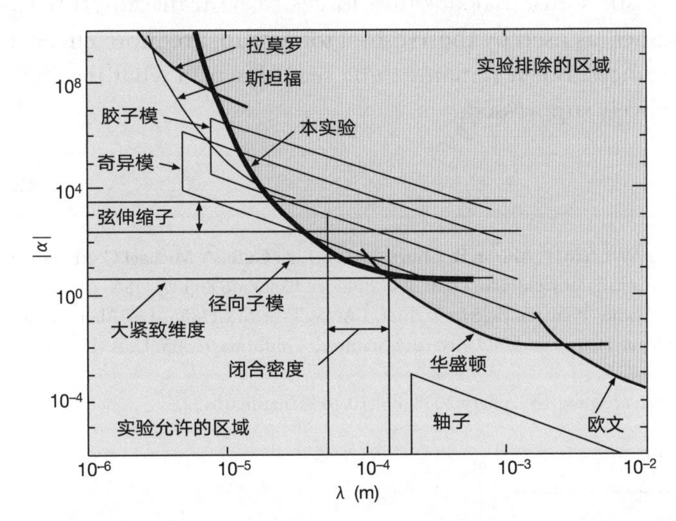

图 4. 目前 1 μm 到 1 cm 之间的新引力强度的限制。我们的结果是一个在 95% 置信度上汤川强度 α 函数的上限，它是关于力程 λ 的函数(粗体实线)。之前的实验(拉莫罗[24]，华盛顿[12]，欧文[25])给出的极限以及理论预言给出的新极限也一并展示在图中(胶子模[4]，奇异模[4]，弦伸缩子[2]，径向子模[7])。一个来自斯坦福的未发表的实验[26]也展示在图中，它是在存在背景力的前提下得到的。伸缩子的强度某种意义上是依赖于模型，在文献中有一个取值范围[2]。我们选择区域 $200 < \alpha < 3{,}000$，它包含了大部分的值。图中也展示了分别由具有两个大的紧致额外维度的 ADD 理论[14,27,28]，轴子传递的力[20,21]以及宇宙学能量密度在 1.0 到 0.1 倍闭合能量密度之间的 λ 区域[17]所给出的预言结果。对于模空间、伸缩子和 ADD 理论，展示的 λ 区域的上界设置在理论提出时人们已知的实验所能达到的极限。

除了来自额外维的力，还有其他两个想法可以在亚毫米尺度给出新的弱的相互作用。用来闭合宇宙的宇宙学(临界)能量密度，如果通过做它的负四次方根来得到一个长度(在自然单位制下 $\hbar = c = 1$)，结果大概是 100 μm。这个事实使得人们不断尝试[17-19]通过引入一个力程大约 100 μm 的力来解决爱因斯坦的宇宙学常数的观测值很小所带来的困难。我们的结果是这个区域内能得到的最好的上界，但是我们还

on α in this region, but we have not quite reached gravitational sensitivity. Finally, the oldest of these predictions, still out of reach, is the very feeble axion-mediated force[20,21]. The axion is a field intended to explain why the violation of charge-parity symmetry is so small in quantum chromodynamics, the theory of the strong nuclear force.

Experiments of the sort reported here constrain string-inspired scenarios by setting very restrictive limits on predicted submillimetre forces. Of course, the actual observation of any new force would be a major advance. Because several theoretical scenarios point especially to these length scales, it is an important goal for the future to reach gravitational strength at even shorter distances, perhaps down to 10 μm. Experiments attempting to reach such distances will confront rapidly increasing background forces, especially electrostatic forces arising from the spatially non-uniform surface potentials of metals[22]. Electric fields due to surface potentials can be shielded with good conductors, but because of the finite stiffness of any shield they still cause background forces to be transmitted between test masses. Stretched membranes (as used by the Washington group) are more effective than stiff plates at the shortest distances, but it remains to be seen down to what distance the background forces can be effectively suppressed.

(**421**, 922-925; 2003)

Joshua C. Long*†, Hilton W. Chan*†, Allison B. Churnside*, Eric A. Gulbis*, Michael C. M. Varney* & John C. Price*

* Physics Department, University of Colorado, UCB 390, Boulder, Colorado 80309, USA

† Present addresses: Los Alamos Neutron Science Center, LANSCE-3, MS-H855, Los Alamos, New Mexico 87545, USA (J.C.L.); and Physics Department, Stanford University, Stanford, California 94305, USA (H.W.C.)

Received 21 October 2002; accepted 13 January 2003; doi:10.1038/nature01432.

References:

1. Greene, B. *The Elegant Universe: Superstrings, Hidden Dimensions, and the Quest for the Ultimate Theory* (Norton, New York, 1999).
2. Kaplan, D. B. & Wise, M. B. Couplings of a light dilaton and violations of the equivalence principle. *J. High Energy Phys.* **8**, 37 (2000).
3. Taylor, T. R. & Veneziano, G. Dilaton couplings at large distances. *Phys. Lett. B* **213**, 450-454 (1988).
4. Dimopoulos, S. & Giudice, G. Macroscopic forces from supersymmetry. *Phys. Lett. B* **379**, 105-114 (1996).
5. Antoniadis, I. A possible new dimension at a few TeV. *Phys. Lett. B* **246**, 377-384 (1990).
6. Antoniadis, I., Dimopoulos, S. & Dvali, G. Millimeter-range forces in superstring theories with weak- scale compactification. *Nucl. Phys. B* **516**, 70-82 (1998).
7. Chacko, Z. & Perazzi, E. Extra dimensions at the weak scale and deviations from Newtonian gravity. Preprint hep-ph/0210254 available at ⟨arXiv.org⟩ (2002).
8. Fischbach, E. & Talmadge, C. *The Search for Non-Newtonian Gravity* (Springer, New York, 1999).
9. Bordag, M., Mohideen, U. & Mostepanenko, V. M. New Developments in the Casimir effect. *Phys. Rep.* **353**, 1-205 (2001).
10. Long, J. C., Chan, H. W. & Price, J. C. Experimental status of gravitational-strength forces in the sub-centimeter regime. *Nucl. Phys. B* **539**, 23-34 (1999).
11. Hoyle, C. D. *et al.* Sub-millimeter tests of the gravitational inverse-square law: A search for "large" extra dimensions. *Phys. Rev. Lett.* **86**, 1418-1421 (2001).
12. Adelberger, E. G. Sub-mm tests of the gravitational inverse-square law. Preprint hep-ex/0202008 available at ⟨arXiv.org⟩ (2002).
13. Fischbach, E., Krause, D. E., Mostepanenko, V. M. & Novello, M. New constraints on ultrashort-ranged Yukawa interactions from atomic force microscopy. *Phys. Rev. D* **64**, 075010 (2001).
14. Arkani-Hamed, N., Dimopoulos, S. & Dvali, G. The hierarchy problem and new dimensions at a millimeter. *Phys. Lett. B* **429**, 263-272 (1998).
15. Kleiman, R. N., Kaminsky, G. K., Reppy, J. D., Pindak, R. & Bishop, D. J. Single-crystal silicon high-Q torsional oscillators. *Rev. Sci. Instrum.* **56**, 2088-2091 (1985).
16. Long, J. C. *et al.* New experimental limits on macroscopic forces below 100 microns. Preprint hep-ph/0210004 available at ⟨arXiv.org⟩ (2002).
17. Beane, S. R. On the importance of testing gravity at distances less than 1 cm. *Gen. Rel. Grav.* **29**, 945-951 (1997).

没有达到引力的灵敏度。最后，这些预言中最古老的一个，也是目前依然探测不到的，是很微弱的轴子传递的相互作用[20,21]。轴子的引入是为了解释为什么在量子色动力学（强核力的理论）中电荷–宇称对称性的破缺是如此的微小。

本文报道的这类实验对预测在亚毫米尺度上存在额外力的类弦理论给出了一个很强约束性的极限。当然，能够观测到任何新的力都将是一个巨大的进步。因为一些理论特别地指向了这些尺度，未来在更小的距离（可能低至 10 μm 以下）上到达引力的强度将是一个重要的目标。试图到达这个距离的实验将会面临更快速增长的背景力，尤其是由于金属表面空间不均匀的势所产生的静电力[22]。表面势产生的电场可以被良导体屏蔽，但是由于任何防护屏都有有限的刚性系数，它们依然会造成表面力在检验质量之间传递。延展的膜（即华盛顿课题组所使用的材料）在小距离下比刚性的盘更加有效，但是低至多大的距离下背景力可以被有效地压低还不能确定。

（安宇森 翻译；蔡荣根 审稿）

18. Sundrum, R. Towards an effective particle-string resolution of the cosmological constant problem. *J High Energy Phys.* **7**, 1 (1999).

19. Schmidhuber, C. Old puzzles. Preprint hep-th/0207203 available at ⟨arXiv.org⟩ (2002).

20. Moody, J. E. & Wilczek, F. New macroscopic forces? *Phys. Rev. D* **30**, 130-138 (1984).

21. Rosenberg, L. J. & van Bibber, K. A. Searches for invisible axions. *Phys. Rep.* **325**, 1-39 (2000).

22. Price, J. C. in *Proc. Int. Symp. on Experimental Gravitational Physics* (eds Michelson, P., En-ke, H. & Pizzella, G.) 436-439 (World Scientific, Singapore, 1988).

23. Chan, H. W., Long, J. C. & Price, J. C. Taber vibration isolator for vacuum and cryogenic applications. *Rev. Sci. Instrum.* **70**, 2742-2750 (1999).

24. Lamoreaux, S. K. Demonstration of the Casimir force in the 0.6 to 6 μm range. *Phys. Rev. Lett.* **78**, 5-8 (1997).

25. Hoskins, J. K., Newman, R. D., Spero, R. & Shultz, J. Experimental tests of the gravitational inverse-square law for mass separations from 2 to 105 cm. *Phys. Rev. D* **32**, 3084-3095 (1985).

26. Chiaverini, J., Smullin, S. J., Geraci, A. A., Weld, D. M. & Kapitulnik, A. New experimental constraints on non-Newtonian forces below 100 microns. Preprint hep-ph/0209325 available at ⟨arXiv.org⟩ (2002).

27. Floratos, E. G. & Leontaris, G. K. Low scale unification, Newton's law and extra dimensions. *Phys. Lett. B* **465**, 95-100 (1999).

28. Kehagias, A. & Sfetsos, K. Deviations from the $1/r^2$ Newton law due to extra dimensions. *Phys. Lett. B* **472**, 39-44 (2000).

Acknowledgements. We thank E. Lagae for work in the laboratory, and C. Briggs, T. Buxkemper, L. Czaia, H. Green, S. Gustafson and H. Rohner of the University of Colorado and JILA instrument shops for technical assistance. We also gratefully acknowledge discussions with S. de Alwis, B. Dobrescu and S. Dimopoulos. This work is supported by grants from the US National Science Foundation.

Competing interests statement. The authors declare that they have no competing financial interests.

Correspondence and requests for materials should be addressed to J.C.P. (e-mail: john.price@colorado.edu).

A Precision Measurement of the Mass of the Top Quark

DØ Collaboration*

Editor's Note

Although the standard model of particle physics is extremely successful, it does not in itself explain the masses of subatomic particles. Instead, they are hypothesized to arise from the interactions of particles with the so-called Higgs boson. Here an international team called the D0 collaboration, working at the Tevatron supercollider at Fermilab in Illinois, report a high-precision measurement of the mass of the top quark, which helps to constrain the range of allowable masses for the Higgs boson. The results suggest that it is more massive than was previously thought, so that more energy will be required to make it. The Higgs boson was finally observed in the Large Hadron Collider at CERN in Geneva in 2012. Its mass is lighter than was expected.

The standard model of particle physics contains parameters—such as particle masses—whose origins are still unknown and which cannot be predicted, but whose values are constrained through their interactions. In particular, the masses of the top quark (M_t) and W boson (M_W)[1] constrain the mass of the long-hypothesized, but thus far not observed, Higgs boson. A precise measurement of M_t can therefore indicate where to look for the Higgs, and indeed whether the hypothesis of a standard model Higgs is consistent with experimental data. As top quarks are produced in pairs and decay in only about 10^{-24} s into various final states, reconstructing their masses from their decay products is very challenging. Here we report a technique that extracts more information from each top-quark event and yields a greatly improved precision (of ± 5.3 GeV/c^2) when compared to previous measurements[2]. When our new result is combined with our published measurement in a complementary decay mode[3] and with the only other measurements available[2], the new world average for M_t becomes[4] 178.0 ± 4.3 GeV/c^2. As a result, the most likely Higgs mass increases from the experimentally excluded[5] value[6] of 96 to 117 GeV/c^2, which is beyond current experimental sensitivity. The upper limit on the Higgs mass at the 95% confidence level is raised from 219 to 251 GeV/c^2.

THE discovery of the top quark in 1995 served as one of the major confirmations of the validity of the standard model (SM)[7,8]. Of its many parameters, the mass of the top quark, in particular, reflects some of the most crucial aspects of the SM. This is because, in principle, the top quark is point-like and should be massless; yet, through its

* A full list of participants and affiliations is given in the original paper.

顶夸克质量的精确测量

DØ 项目合作组 *

编者按

尽管粒子物理的标准模型是非常成功的，但模型本身解释不了亚原子粒子的质量问题，亚原子粒子的质量被假定来源于粒子与所谓的希格斯玻色子之间的相互作用。本文中，一个名为 D0 合作组的国际团队通过伊利诺伊州费米实验室的万亿电子伏特加速器 (Tevatron) 得到了顶夸克质量的精确测定，由此有助于限定希格斯玻色子允许质量的范围。结果表明，希格斯玻色子的质量比之前预期的要大，因而需要更大的能量才能产生希格斯玻色子。2012 年，位于瑞士日内瓦的欧洲核子研究组织 (CERN) 的大型强子对撞机终于观测到希格斯玻色子，但其质量比预期的要轻。

粒子物理标准模型包含了像粒子质量这一类的参数，这类参数起源不清，无法预言，但是其量值可以通过它们之间的相互作用来限定。特别是一直被预言存在但是从未被观测到的希格斯玻色子的质量，可以由顶夸克的质量 (M_t) 和 W 玻色子的质量 (M_W) [1] 进行限定。因而，M_t 的精确测量可以更好地指导我们寻找希格斯子，甚至可验证标准模型预言的希格斯子是否与实验数据相符。由于顶夸克成对产生，并在 10^{-24} 秒后就衰变为各种末态，通过衰变的产物来重建其质量是非常困难的。本文提出的技术可以从每一个顶夸克对事例中获取更多信息并得到比之前测量值 [2] 更高的精度 (±5.3 GeV/c^2)。我们最新的结果结合之前发表的互补衰变模式 (全轻衰变模式) 的测量值 [3] 和目前其余已知的结果 [2]，得到的新的全球 M_t 平均值是 178.0±4.3 GeV/c^2 [4]。由此，最有可能的希格斯子质量的值从已经被实验 [5,6] 排除的 96 GeV/c^2 提高到 117 GeV/c^2，这比当前的实验灵敏度要高。在 95% 置信度水平，希格斯子的质量上限从 219 GeV/c^2 提高到 251 GeV/c^2。

1995 年顶夸克的发现是对标准模型 (SM) 的重要支持之一 [7,8]。在众多参数中，顶夸克的质量尤其反映了标准模型的一些根本特征。这是因为本质上顶夸克是一个无质量的点粒子，然而通过与预言的希格斯场的相互作用，顶夸克获得的物理质量大致相当于一个金原子核质量。这个质量如此之大，使得顶夸克 (和 W 玻色子一起)

---

* 合作组的参与者及所属单位的名单请参见原文。

interactions with the hypothesized Higgs field, the physical mass of the top quark appears to be about the mass of a gold nucleus. Because it is so heavy, the top quark (along with the W boson) provides an unusually sensitive tool for investigating the Higgs field. M_W is known to a precision of 0.05%, while the uncertainty on M_t is at the 3% level[1]. Improvements in both measurements are required to restrict further the allowed range of mass for the Higgs; however, given the large uncertainty in M_t, an improvement in its precision is particularly important. As has been pointed out recently[9,10], a potential problem for the SM is that, on the basis of the currently accepted value for M_t, the most likely value of the Higgs mass[6] lies in a range that has already been excluded by experiment[5]. Precise knowledge of the Higgs mass is crucial for our understanding of the SM and any possible new physics beyond it. For example, in a large class of supersymmetric models (theoretically preferred solutions to the deficiencies of the SM), the Higgs mass has to be less than about 135 GeV/c^2. Although, unlike the SM, supersymmetry predicts more than one Higgs boson, the properties of the lightest one are expected to be essentially the same as those for the SM Higgs boson. Thus, if the SM-like Higgs is heavier than about 135 GeV/c^2, it would disfavour a large class of supersymmetric models. In addition, some of the current limits on supersymmetric particles from LEP[11] are extremely sensitive to M_t. In fact, for M_t greater than 179 GeV/c^2, the bounds on one of the major supersymmetry parameters, $\tan\beta$, which relates the properties of the SM-like Higgs boson and its heavier partners, would disappear completely[12]. Hence, in addition to the impact on searches for the Higgs boson, other important consequences call for improved precision on M_t, and this goal is the main subject of this paper.

The DØ experiment at the Fermilab Tevatron has studied a sample of $t\bar{t}$ events produced in proton–antiproton ($p\bar{p}$) interactions[13]. The total energy of 1.8 TeV released in a head-on collision of a 900-GeV p and a 900-GeV $\bar{p}$ is almost as large as the rest energy of ten gold nuclei. Each top (antitop) quark decays almost immediately into a bottom $b(\bar{b})$ quark and a W^+ (W^-) boson, and we have reexamined those events in which one of the W bosons decays into a charged lepton (electron or muon) and a neutrino, and the other W into a quark and an antiquark (see Fig. 1). These events and their selection criteria are identical to those used to extract the mass of the top quark in our previous publication, and correspond to an integrated luminosity of 125 events per pb. (That is, given the production cross-section of the $t\bar{t}$ in $p\bar{p}$ collisions at 1.8 TeV of 5.7 pb, as measured by DØ[14], these data correspond to approximately 700 produced $t\bar{t}$ pairs, a fraction of which is fully detected in various possible decay modes. Approximately 30% of these correspond to the lepton + jets topology categorized in Fig. 2, where "jet" refers to products of the fragmentation of a quark into a collimated group of particles that are emitted along the quark's original direction.) The main background processes correspond to multijet production (20%), where one of the jets is reconstructed incorrectly as a lepton, and the W + jets production with leptonic W decays (80%), which has the same topology as the $t\bar{t}$ signal.

成为一个超常灵敏的研究希格斯场的手段。M_W 的测量值仅有 0.05% 的不确定度，而 M_t 的测量值有 3% 的不确定度[1]。要进一步缩小希格斯子质量的可能范围，需要改进这两个量值的精确度。但因为 M_t 的测量值不确定度大，所以对其精度的改进就格外地重要。正如最近[9,10]已指出的，标准模型的一个潜在问题在于，根据当前公认的 M_t 测量值，最有可能的希格斯子质量[6]落在已被实验排除的区间[5]。准确地认识希格斯子的质量，对于我们理解标准模型或者标准模型以外的其他可能的新物理理论，都是非常重要的。比如说，在一大类超对称模型中（解决标准模型缺陷的首选理论），希格斯子质量必须小于约 135 GeV/c^2。和标准模型不同的是，超对称性预言的希格斯子不止一个，但是其中最轻的那一个的性质基本上与标准模型中的希格斯玻色子相同。因此，如果类标准模型希格斯玻色子的质量超过 135 GeV/c^2，这一大类超对称模型就不再被看好。另外，从大型正负电子对撞机（LEP）[11]得到的一些超对称粒子当前的限制对 M_t 非常敏感。实际上，当 M_t 大于 179 GeV/c^2 时，一个联系类标准模型希格斯玻色子及其更重的伴子的主要的超对称参数（$\tan\beta$）的限制将完全消失[12]。除了对希格斯玻色子的寻找有影响，改进 M_t 的测量精度还有其他重要意义，这个目标也是本文的主要研究问题。

DØ 项目在费米实验室的万亿电子伏特加速器（Tevatron）上研究了用质子–反质子（$p\bar{p}$）对撞产生的顶夸克–反顶夸克（$t\bar{t}$）事例[13]。900 GeV 的质子（p）和 900 GeV 的反质子（$\bar{p}$）对撞释放了 1.8 TeV 的总能量，几乎相当于十个金原子核的静止质量。每个顶（反顶）夸克几乎立即衰变为一个底（反底）夸克 $b(\bar{b})$ 和一个 $W^+(W^-)$ 玻色子。我们重新检查了那些一个 W 玻色子衰变为一个带电轻子（电子或者 μ 子）和一个中微子，而另一个 W 玻色子衰变为一个夸克和一个反夸克的事例（图 1）。挑选这些事例所用选择条件及实验数据样本与我们之前发表的获取顶夸克质量的工作中所用的选择条件和数据样本相同，即对应每 pb 有 125 个事例积分亮度的实验数据。（就是说，与 DØ 实验中测得的结果[14]一致，假定 $p\bar{p}$ 对撞中的 $t\bar{t}$ 在能量为 1.8 TeV 时产生截面为 5.7 皮靶，这些数据大概对应于 700 个 $t\bar{t}$ 对，其中只有部分通过各种可能的衰变模式检测到了。大概有 30% $t\bar{t}$ 衰变模式对应于图 2 归类的轻子 + 喷注拓扑，其中“喷注”这个词表示夸克分裂产生的成群粒子沿夸克原来的方向平行前进）。主要的背景过程包括多喷注产生（20%，其中的一个喷注被错误地重建为轻子）和与 $t\bar{t}$ 信号有相同拓扑结构的轻子型 W 衰变的 W+ 喷注产生（80%）。

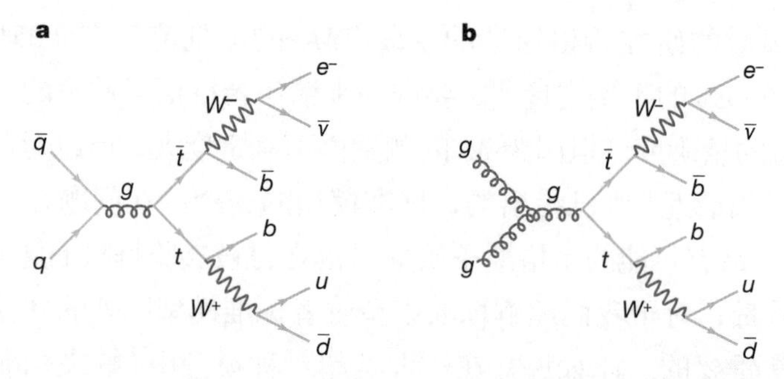

Fig. 1. Feynman diagrams for $t\bar{t}$ production in $p\bar{p}$ collisions, with subsequent decays into an electron, neutrino, and quarks. Quark–antiquark production (**a**) is dominant, but gluon fusion (**b**) contributes ~10% to the cross-section. This particular final state ($e\nu u\bar{d}b\bar{b}$) is one of the channels used in the analysis.

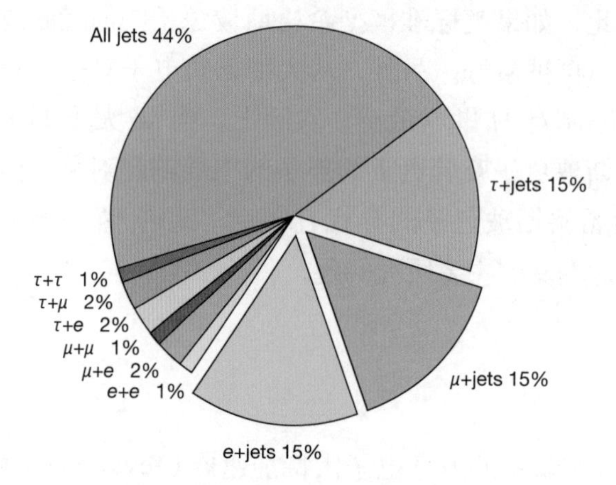

Fig. 2. Relative importance of various $t\bar{t}$ decay modes. The "lepton+jets" channel used in this analysis corresponds to the two offset slices of the pie-chart and amounts to 30% of all the $t\bar{t}$ decays

The previous DØ measurement of M_t in this lepton + jets channel is $M_t = 173.3 \pm 5.6$ (stat) ± 5.5 (syst) GeV/c^2, and is based on 91 candidate events. Information pertaining to the older analysis and the DØ detector can be found elsewhere[13,15].

The new method of M_t measurement is similar to one suggested previously (ref. 16 and references therein, and ref. 17) for $t\bar{t}$ dilepton decay channels (where both W bosons decay leptonically), and used in previous mass analyses of dilepton events[3], and akin to an approach suggested for the measurement of the mass of the W boson at LEP[18-20]. The critical differences from previous analyses in the lepton + jets decay channel lie in: (1) the assignment of more weight to events that are well measured or more likely to correspond to $t\bar{t}$ signal, and (2) the handling of the combinations of final-state objects (lepton, jets and imbalance in transverse momentum, the latter being a signature for an undetected neutrino) and their identification with top-quark decay products in an event (such as from ambiguity in choosing jets that correspond to b or $\bar{b}$ quarks from the decays of the t and $\bar{t}$ quarks). Also, because leading-order matrix elements were used to calculate the event weights, only events with exactly four jets are kept in this analysis, resulting in a candidate sample of 71 events. Although we are left with fewer events, the new method for extracting M_t provides

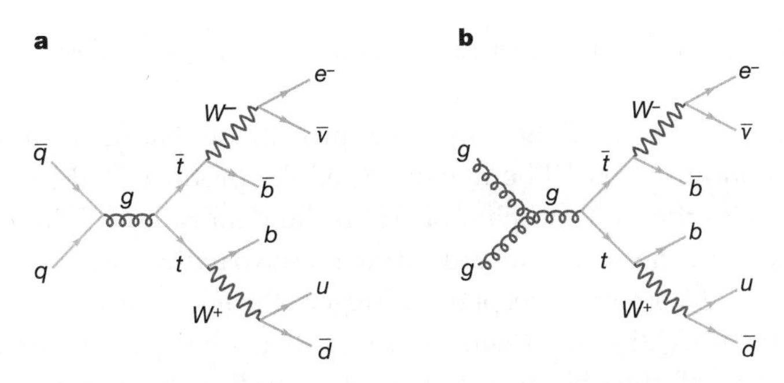

图 1. $p\bar{p}$(正反质子)对撞中 $t\bar{t}$(正反顶夸克)产生的费曼图。产生的顶夸克随即衰变为一个电子、中微子和数个夸克。(a) 图所示的夸克–反夸克产生过程是主要的,(b) 图中的胶子聚变过程贡献大约 10% 的截面。这个末态($e\nu u\bar{d}b\bar{b}$)是分析中用到的通道之一。

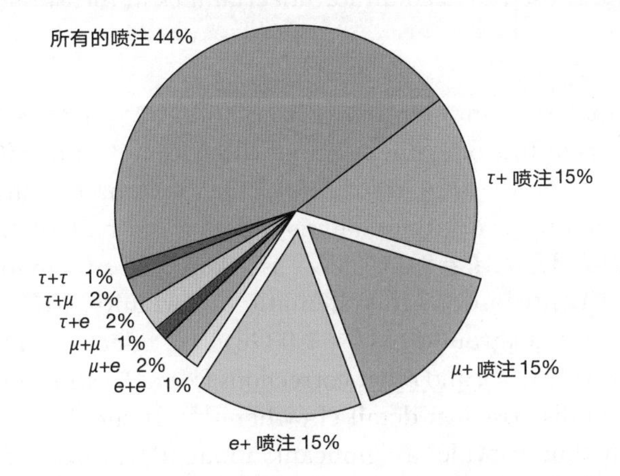

图 2. 各个 $t\bar{t}$ 衰变模式的比例。本分析中用到的“轻子 + 喷注”衰变道对应于扇面图中分割出的两片,总共占 $t\bar{t}$ 衰变的 30%。

以前在通过轻子 + 喷注衰变道测量 M_t 值的 DØ 实验中,根据 91 个候选事例得到的量值是 $M_t = 173.3 \pm 5.6$(统计不确定度)± 5.5(系统不确定度)GeV/c^2。以前的分析以及 DØ 探测器的相关信息可以在相关文献 [13,15] 中找到。

测量 M_t 的新方法类似于之前文献(参考文献 16 以及其中引用的文献,参考文献 17)提出的关于 $t\bar{t}$ 双轻子衰变道(其中两个 W 玻色子的衰变都是轻子型的)中的测量方法,该方法曾被应用于之前双轻子事例的质量分析 [3],与之前建议的在 LEP 上测量 W 玻色子质量的方法也是相似的 [18-20]。轻子 + 喷注衰变道的分析与之前的分析关键的不同之处在于:(1) 测量得比较精确的事例或者更可能对应于 $t\bar{t}$ 信号的事例这点被赋予了更大的权重,(2) 对末态产物(轻子、喷注和不平衡的横向动量,最后一项表明了一个未探测到的中微子的存在)的各种组合的处理以及在事例中用顶夸克衰变产物对它们的确证(比如解决喷注是对应于 t 夸克衰变中的 b 夸克还是 $\bar{t}$ 夸克衰变中的 $\bar{b}$ 夸克的不确定性问题)。另外,由于使用领头阶矩阵元计算事例权重,分析中只保留了恰好有 4 个喷注的事例,这样得到的候选样本有 71 个事例。虽然事例数

substantial improvement in both statistical and systematic uncertainties.

We calculate as a function of M_t the differential probability that the measured variables in any event correspond to signal. The maximum of the product of these individual event probabilities provides the best estimate of M_t in the data sample. The impact of biases from imperfections in the detector and event-reconstruction algorithms is taken into account in two ways. Geometric acceptance, trigger efficiencies, event selection, and so on enter through a multiplicative acceptance function that is independent of M_t. Because the angular directions of all the objects in the event, as well as the electron momentum, are measured with high precision, their measured values are used directly in the calculation of the probability that any event corresponds to $t\bar{t}$ or background production. The known momentum resolution is used to account for uncertainties in measurements of jet energies and muon momenta.

As in the previous analysis[13], momentum conservation in γ+jet events is used to check that the energies of jets in the experiment agree with Monte Carlo (MC) simulations. This calibration has an uncertainty $\delta E = (0.025E + 0.5\ \text{GeV})$. Consequently, all jet energies in our sample are rescaled by $\pm\delta E$, the analysis redone, and half of the difference in the two rescaled results for M_t ($\delta M_t = 3.3\ \text{GeV}/c^2$) is taken as a systematic uncertainty from this source. All other contributions to systematic uncertainty: MC modelling of signal ($\delta M_t = 1.1\ \text{GeV}/c^2$) and background ($\delta M_t = 1.0\ \text{GeV}/c^2$), effect of calorimeter noise and event pile-up ($\delta M_t = 1.3\ \text{GeV}/c^2$), and other corrections from M_t extraction ($\delta M_t = 0.6\ \text{GeV}/c^2$) are much smaller, and discussed in detail elsewhere[21,22]. It should be noted that the new mass measurement method provides a significant (about 40%, from ± 5.5 to $\pm 3.9\ \text{GeV}/c^2$) reduction in systematic uncertainty, which is ultimately dominated by the measurement of jet energies. For details on the new analysis, see the Methods.

The final result is $M_t = 180.1 \pm 3.6\ (\text{stat}) \pm 3.9\ (\text{syst})\ \text{GeV}/c^2$. The improvement in statistical uncertainty over our previous measurement is equivalent to collecting a factor of 2.4 as much data. Combining the statistical and systematic uncertainties in quadrature, we obtain $M_t = 180.1 \pm 5.3\ \text{GeV}/c^2$, which is consistent with our previous measurement in the same channel (at about 1.4 standard deviations), and has a precision comparable to all previous M_t measurements combined[1].

The new measurement can be combined with that obtained for the dilepton sample that was also collected at DØ during run I (ref. 3) ($M_t = 168.4 \pm 12.3\ (\text{stat}) \pm 3.6\ (\text{syst})\ \text{GeV}/c^2$), to yield the new DØ average for the mass of the top quark:

$$M_t = 179.0 \pm 5.1\ \text{GeV}/c^2 \tag{1}$$

Combining this with measurements from the CDF experiment[2] provides a new "world average" (based on all measurements available) for the mass of the top quark[4]:

$$M_t = 178.0 \pm 4.3\ \text{GeV}/c^2 \tag{2}$$

少了，但是提取 M_t 的新方法在统计不确定度和系统不确定度上都有实质性的改进。

我们计算了以 M_t 及事例观测量作为变量的函数，并将该函数对应为信号的微分概率。这些独立事例概率的最大乘积提供了数据样本中 M_t 的最佳估计值。我们从两个方面考虑了探测器和事例重建算法的不完美引起的偏差影响。几何接收率、触发效率、事例选择等被包含在一个不依赖 M_t 的多元接收函数中。由于事例中所有对象的角度方向和电子动量的测量精度都很高，它们的测量值被直接用来计算任何对应于 $t\bar{t}$ 事例或者背景事例的产生概率。使用已知的动量分辨率来解释测量喷注能量和 μ 子动量时的不确定度。

和以前的分析 [13] 一样，γ+ 喷注事例中的动量守恒被用来检查喷注的能量是否与蒙特卡罗（MC）模拟结果相符合。这个校准的不确定度 $\delta E = (0.025E + 0.5\ \text{GeV})$。因此，我们的样本中所有喷注能量都以 $\pm\delta E$ 重新标度，两次重新标度和分析后的 M_t（$\delta M_t = 3.3\ \text{GeV}/c^2$）结果之差的一半被作为喷注能量测量不确定度引起的系统性不确定度。对系统不确定度有贡献的其他因素包括：信号和背景的蒙特卡罗建模（$\delta M_t = 1.1\ \text{GeV}/c^2$ 和 $\delta M_t = 1.0\ \text{GeV}/c^2$），量能器噪声和事例堆积的影响（$\delta M_t = 1.3\ \text{GeV}/c^2$）以及其他源于 M_t 提取的修正（$\delta M_t = 0.6\ \text{GeV}/c^2$）。以上这些因素的贡献要小得多，有另文 [21,22] 具体讨论。这里需要指出，新的质量测量方法明显减小了主要由测量喷注能量导致的系统不确定度（减小了约 40%，从 $\pm 5.5\ \text{GeV}/c^2$ 减到 $\pm 3.9\ \text{GeV}/c^2$）。关于新分析方法的细节，请参看本文后面的**方法**。

最后的结果是 $M_t = 180.1 \pm 3.6$（统计误差）± 3.9（系统误差）GeV/c^2。和之前的测量结果比较，新方法在统计不确定度上的改进相当于将数据量增加至原来的 2.4 倍。将统计和系统不确定度求均方根，我们得到 $M_t = 180.1 \pm 5.3\ \text{GeV}/c^2$，这与我们之前在相同衰变道的测量结果相符（在大约 1.4 个标准差偏离下），与之前所有的 M_t 测量的联合精度相当 [1]。

将新的测量结果与通过第 I 轮 DØ 实验（参考文献 3）中搜集到的双轻子样本得到的结果（$M_t = 168.4 \pm 12.3$（统计误差）± 3.6（系统误差）GeV/c^2）相结合，得出顶夸克质量的新的 DØ 平均：

$$M_t = 179.0 \pm 5.1\ \text{GeV}/c^2 \tag{1}$$

把这一结果和 CDF 实验的测量结果 [2] 进一步结合起来，得到以我们最近的结果为主导的顶夸克质量新的“全球平均”（基于目前所有已知的测量结果）[4]：

$$M_t = 178.0 \pm 4.3\ \text{GeV}/c^2 \tag{2}$$

dominated by our new measurement. This new world average shifts the best-fit value of the expected Higgs mass from 96 GeV/c^2 to 117 GeV/c^2 (see Fig. 3), which is now outside the experimentally excluded region, yet accessible in the current run of the Tevatron and at future runs at the Large Hadron Collider (LHC), at present under construction at CERN. (The upper limit on the Higgs mass at the 95% confidence level changes from 219 GeV/c^2 to 251 GeV/c^2.) Figure 3 shows the effect of using only the new DØ top mass for fits to the Higgs mass, and indicates a best value of 123 GeV/c^2 and the upper limit of 277 GeV/c^2 at the 95% confidence level. It should be noted that the horizontal scale in Fig. 3 is logarithmic, and the limits on the Higgs boson mass are therefore asymmetric.

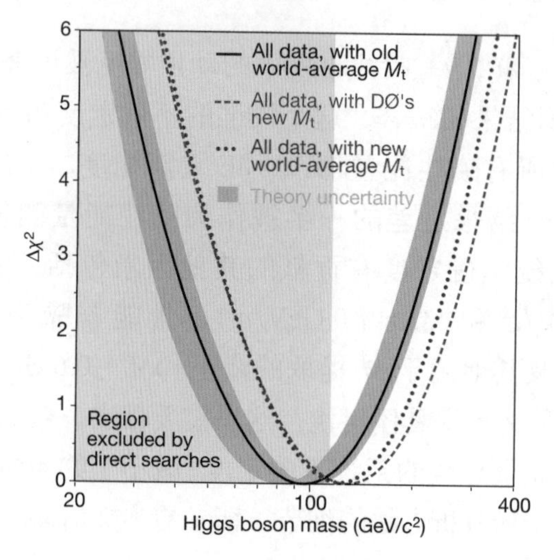

Fig. 3. Current experimental constraints on the mass of the Higgs boson. The χ^2 for a global fit to electroweak data[6] is shown as a function of the Higgs mass. The solid line corresponds to the result for the previous world-averaged $M_t = 174.3 \pm 5.1$ GeV/c^2, with the blue band indicating the impact of theoretical uncertainty. The dotted line shows the result for the new world-average M_t of 178.0 ± 4.3 GeV/c^2, whereas the dashed line corresponds to using only the new DØ average of 179.0 ± 5.1 GeV/c^2. The yellow-shaded area on the left indicates the region of Higgs masses excluded by experiment (>114.4 GeV/c^2 at the 95% confidence level[5]). The improved M_t measurement shifts the most likely value of the Higgs mass above the experimentally excluded range.

The new method is already being applied to data being collected by the CDF and DØ experiments at the new run of the Fermilab Tevatron and should provide even higher precision on the determination of M_t, equivalent to more than a doubling of the data sample, relative to using the conventional method. An ultimate precision of about 2 GeV/c^2 on the mass of the top quark is expected to be reached in several years of Tevatron operation. Further improvement may eventually come from the LHC.

Methods

The probability density as a function of M_t can be written as a convolution of the calculable cross-

这个新的全球平均将预言的希格斯子质量的最佳拟合值从 96 GeV/c^2 提高到 117 GeV/c^2(见图 3)，现在这个值已经落在了实验排除的区间之外，但可以在 Tevatron 上目前正运行的实验中以及 CERN 目前正在建设的大型强子对撞机 (LHC) 上将开展的实验中达到。(在 95% 置信度水平下，希格斯子质量的上限值从 219 GeV/c^2 变为 251 GeV/c^2)。图 3 显示如果只采用最新的 DØ 顶夸克质量来拟合希格斯子质量，获得的最佳的结果为 123 GeV/c^2，在 95% 置信度水平下最佳上限值为 277 GeV/c^2。注意图 3 中水平坐标是对数尺度，因而希格斯玻色子质量的上下限值是不对称的。

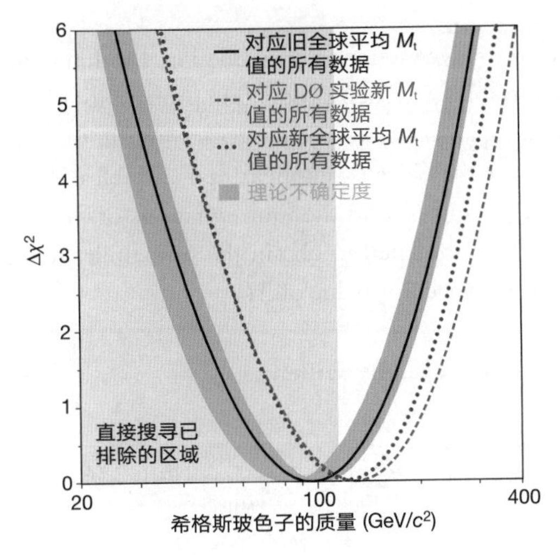

图 3. 当前实验得出的对希格斯玻色子质量的限定。图中显示电弱数据[6]全局拟合的 χ^2 值随希格斯子质量的变化。实线对应于之前的全球平均 $M_t = 174.3 \pm 5.1$ GeV/c^2 的结果，蓝色带状区域表示理论不确定度的影响。点线表示新的全球平均 $M_t = 178.0 \pm 4.3$ GeV/c^2 的结果，短横线对应只采用 DØ 实验新平均值 179.0 ± 5.1 GeV/c^2 的结果。左边黄色阴影区域表示实验排除的希格斯质量范围 (> 114.4 GeV/c^2，95% 置信水平下[5])。改进后的 M_t 测量值使得最有可能的希格斯子质量移至实验排除的区间之外。

新方法已经被用来处理费米实验室 Tevatron 上新一轮 CDF 实验和 DØ 实验中所收集到的数据，以期得到更精确的 M_t 值，以达到相当于常规方法两倍多数据样本的精度。可以预期经过未来几年在 Tevatron 上的实验，顶夸克质量的精度最终能够达到 2 GeV/c^2。而更进一步的精度提升最终将有赖于 LHC 的投入使用。

方　法

作为 M_t 函数的概率密度可以表示为可计算的截面与所有来自测量分辨率的效应的

section and any effects from measurement resolution

$$P(x, M_t) = \frac{1}{\sigma(M_t)} \int d\sigma(y, M_t) dq_1 dq_2 f(q_1) f(q_2) W(y, x) \quad (3)$$

where $W(y, x)$, our general transfer function, is the normalized probability for the measured set of variables x to arise from a set of nascent (partonic) variables y, $d\sigma(y, M_t)$ is the partonic theoretical differential cross-section, $f(q)$ are parton distribution functions that reflect the probability of finding any specific interacting quark (antiquark) with momentum q within the proton (antiproton), and $\sigma(M_t)$ is the total cross-section for producing $t\bar{t}$. The integral in equation (3) sums over all possible parton states, leading to what is observed in the detector.

The acceptance of the detector is given in terms of a function $A(x)$ that relates the probability $P_m(x, M_t)$ of measuring the observed variables x to their production probability $P(x, M_t)$: $P_m(x, M_t) = A(x)P(x, M_t)$. Effects from energy resolution, and so on are taken into account in the transfer function $W(y, x)$. The integrations in equation (3) over the eleven well-measured variables (three components of charged-lepton momentum and eight jet angles) and the four equations of energy-momentum conservation leave five integrals that must be performed to obtain the probability that any event represents $t\bar{t}$ (or background) production for some specified value of M_t.

The probability for a $t\bar{t}$ interpretation can be written as:

$$P_{t\bar{t}} = \frac{1}{12\sigma_{t\bar{t}}} \int d^5\Omega \sum_{\text{perm.},\nu} |M_{t\bar{t}}|^2 \frac{f(q_1)f(q_2)}{|q_1||q_2|} \Phi_6 W_{\text{jets}}(E_{\text{part}}, E_{\text{jet}})$$

where Ω represent a set of five integration variables, $M_{t\bar{t}}$ is the leading-order matrix element for $t\bar{t}$ production[23,24], $f(q_1)$ and $f(q_2)$ are the CTEQ4M parton distribution functions for the incident quarks[25], Φ_6 is the phase-space factor for the six-object final state, and the sum is over all 12 permutations of the jets and all possible neutrino solutions. $W_{\text{jets}}(E_{\text{part}}, E_{\text{jet}})$ corresponds to a function that maps parton-level energies E_{part} to energies measured in the detector E_{jet} and is based on MC studies. A similar expression, using a matrix element for W+jets production (the dominant background source) that is independent of M_t, is used to calculate the probability for a background interpretation, P_{bkg}.

Studies of samples of HERWIG (ref. 26; we used version 5.1) MC events indicate that the new method is capable of providing almost a factor-of-two reduction in the statistical uncertainty on the extracted M_t. These studies also reveal that there is a systematic shift in the extracted M_t that depends on the amount of background there is in the data. To minimize this effect, a selection is introduced, based on the probability that an event represents background. The specific value of the P_{bkg} cut-off is based on MC studies carried out before applying the method to data, and, for $M_t = 175$ GeV/c^2, retains 71% of the signal and 30% of the background. A total of 22 data events out of our 71 candidates pass this selection.

The final likelihood as a function of M_t is written as:

卷积

$$P(x, M_t) = \frac{1}{\sigma(M_t)} \int d\sigma(y, M_t) dq_1 dq_2 f(q_1) f(q_2) W(y, x) \tag{3}$$

其中广义传递函数 $W(y, x)$ 表示从一组部分子水平的变量 y 中产生探测器水平变量 x 的归一化概率；$d\sigma(y, M_t)$ 是部分子水平的理论微分截面；$f(q)$ 是部分子的分布函数，表示质子（反质子）内发现处于特定相互作用中、动量为 q 的夸克（反夸克）的概率，$\sigma(M_t)$ 是产生 $t\bar{t}$ 的总截面。式（3）中的积分要对所有可能的部分子态进行求和，从而得到探测器水平的观测结果。

探测器的接受特性由函数 $A(x)$ 表达。这个函数将观测到的探测器水平变量 x 的概率 $P_m(x, M_t)$ 与其产生概率 $P(x, M_t)$ 联系起来：$P_m(x, M_t) = A(x)P(x, M_t)$。能量分辨率等效应包含在传递函数 $W(y, x)$ 中。式（3）中对 11 个精确测量的变量（带电轻子动量的 3 个分量以及 8 个喷注角度）的积分，加上四个能量–动量守恒关系，使得要完成 5 次积分才能得到特定 M_t 质量值下任何可代表 $t\bar{t}$（或者背景）产生事例的概率。

$t\bar{t}$ 事例判读的概率可以表示为

$$P_{t\bar{t}} = \frac{1}{12\sigma_{t\bar{t}}} \int d^5\Omega \sum_{\text{perm.},\nu} |M_{t\bar{t}}|^2 \frac{f(q_1)f(q_2)}{|q_1||q_2|} \Phi_6 W_{\text{jets}}(E_{\text{part}}, E_{\text{jet}})$$

其中 Ω 代表一组积分变量（5 个），对于 $t\bar{t}$ 产生事例，$M_{t\bar{t}}$ 是其领头阶矩阵元[23,24]。$f(q_1)$ 和 $f(q_2)$ 是入射夸克的 CTEQ4M 部分子分布函数[25]，Φ_6 是 6 体末态的相空间因子，求和运算针对喷注的全部 12 种排列和所有可能的中微子解进行。$W_{\text{jets}}(E_{\text{part}}, E_{\text{jet}})$ 对应于一个基于蒙特卡罗模拟研究的、将部分子水平的能量 E_{part} 映射到探测器测量到的能量 E_{jet} 的函数。利用一个不依赖 M_t 的 W+喷注产生（主要的背景来源）的矩阵元，一个类似表达式被用来计算背景判读概率 P_{bkg}。

对 HERWIG（参考文献 26，我们用的是 5.1 版本）蒙特卡罗事例样本的研究表明，新方法能够使得到的 M_t 统计不确定度降低将近一半。这些研究也揭示，获取的 M_t 值存在一个系统漂移，这一漂移依赖于数据中的背景数目。为了尽量减小这一效应，我们根据一个事例代表背景的概率引入了一个遴选程序。应用本方法处理数据之前，根据蒙特卡罗模拟，确定了具体的 P_{bkg} 截断值。对于 $M_t = 175\ \text{GeV}/c^2$，71% 的信号和 30% 的背景都得到保留。71 个候选事例中，有 22 个通过了遴选。

最终作为 M_t 函数的似然函数表达式如下：

$$\ln L(M_t) = \sum_{i=1}^{n} \ln[c_1 P_{t\bar{t}}(x_i, M_t) + c_2 P_{bkg}(x_i)] - N \int A(x)[c_1 P_{t\bar{t}}(x, M_t) + c_2 P_{bkg}(x)] \mathrm{d}x$$

The integration is performed using MC methods. The best value of M_t (when L is at its maximum L_{max}) represents the most likely mass of the top quark in the final N-event sample, and the parameters c_i reflect the amounts of signal and background. MC studies show that there is a downward shift of 0.5 GeV/c^2 in the extracted mass, and this correction is applied to the result. Reasonable changes in the cut-off on P_{bkg} do not have a significant impact on M_t.

Figure 4 shows the value of $L(M_t)/L_{max}$ as a function of M_t for the 22 events that pass all selection criteria, after correction for the 0.5 GeV/c^2 bias in mass. The likelihood is maximized with respect to the parameters c_i at each mass point. The gaussian fit in the figure yields $M_t = 180.1$ GeV/c^2, with a statistical uncertainty of $\delta M_t = 3.6$ GeV/c^2. The systematic uncertainty, dominated by the measurement of jet energies, as discussed above, amounts to $\delta M_t = 3.9$ GeV/c^2. When added in quadrature to the statistical uncertainty from the fit, it yields the overall uncertainty on the new M_t measurement of ± 5.3 GeV/c^2.

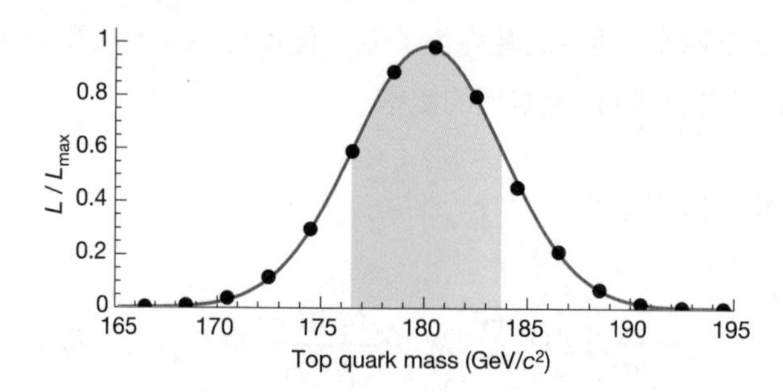

Fig. 4. Determination of the mass of the top quark using the maximum-likelihood method. The points represent the likelihood of the fit used to extract M_t divided by it maximum value, as a function of M_t (after a correction for a -0.5 GeV/c^2 mass bias, see text). The solid line shows a gaussian fit to the points. The maximum likelihood corresponds to a mass of 180.1 GeV/c^2, which is the new DØ measurement of M_t in the lepton+jets channel. The shaded band corresponds to the range of ± 1 standard deviation, and indicates the ± 3.6 GeV/c^2 statistical uncertainty of the fit.

(**429**, 638-642; 2004)

Received 23 January; accepted 21 April 2004; doi:10.1038/nature02589.

References:

1. Hagiwara, K. *et al.* Review of particle physics. *Phys. Rev. D* **66**, 010001 (2002).

2. Affolder, T. *et al.* (CDF Collaboration). Measurement of the top quark mass with the Collider Detector at Fermilab. *Phys. Rev. D* **63**, 032003 (2001).

3. Abbott, B. *et al.* (DØ Collaboration). Measurement of the top quark mass in the dilepton channel. *Phys. Rev. D* **60**, 052001 (1999).

4. The CDF Collaboration, the DØ Collaboration, and the TEVATRON Electro-Weak Working Group. Combination of CDF and DØ Results on the Top-Quark Mass. Preprint at http://www.arXiv.org/hep-ex/0404010 (2004).

5. Barate, R. *et al.* (ALEPH Collaboration, DELPHI Collaboration, L3 Collaboration, OPAL Collaboration, and LEP Working Group for Higgs boson searches). Search for the standard model Higgs boson at LEP. *Phys. Lett. B* **565**, 61-75 (2003).

6. The LEP Collaborations ALEPH, DELPHI, L3, and OPAL, the LEP Electroweak Working Group, and the SLD Heavy Flavour Group. A combination of

$$\ln L(M_t)=\sum_{i=1}^{n}\ln[c_1P_{t\bar{t}}(x_i,M_t)+c_2P_{bkg}(x_i)]-N\int A(x)[c_1P_{t\bar{t}}(x,M_t)+c_2P_{bkg}(x)]dx$$

积分运算是通过蒙特卡罗方法完成的。M_t 的最佳值（当 L 等于其最大值 L_{max}）代表了最终的 N 事例样本中最有可能的顶夸克质量值，参数 c_i 反映信号和背景的数量。蒙特卡罗模拟研究表明获取的质量值有一个向下的 0.5 GeV/c^2 的漂移，由此我们对结果进行了修正。在合理范围内 P_{bkg} 截断值的变动对 M_t 没有显著的影响。

图 4 显示了进行 0.5 GeV/c^2 的质量偏差的修正之后，通过所有遴选标准的 22 个事例的 $L(M_t)/L_{max}$ 比值随 M_t 的变化。在图中每一个质量数据点，对参数 c_i 取最大似然。图中所示数据的高斯拟合得到 M_t = 180.1 GeV/c^2，统计不确定度 δM_t = 3.6 GeV/c^2。如上文所述，系统不确定度主要受喷注能量的测量的影响，大小为 δM_t = 3.9 GeV/c^2。系统不确定度与拟合得到的统计不确定度求均方根，得到新的 M_t 测量值的总不确定度为 ±5.3 GeV/c^2。

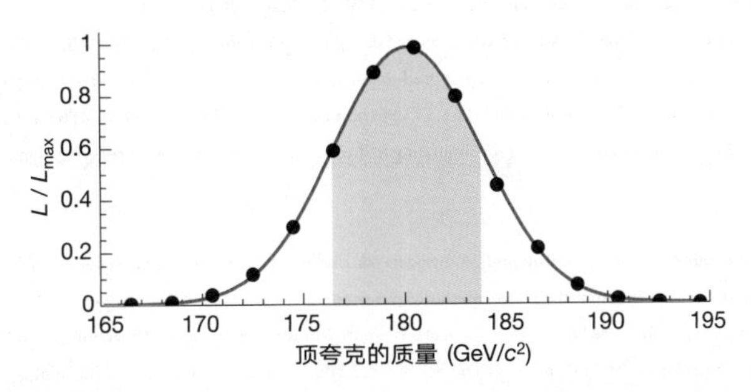

图 4. 用最大似然法确定顶夸克的质量。数据点表示用以获取 M_t 的拟合的似然值与其最大值的比，该比是 M_t 的函数（已进行 0.5 GeV/c^2 的质量偏差的修正，见正文）。实线表示对数据点的高斯拟合结果。最大似然值对应质量为 180.1 GeV/c^2，这也是新的轻子 + 喷注衰变道中 M_t 的 DØ 实验测定值。带颜色的区域对应正负一个标准差区间，表示 ±3.6 GeV/c^2 的拟合统计不确度。

（何钧 翻译；曹庆宏 审稿）

preliminary electroweak measurements and constraints on the standard model. Preprint at http://www.arXiv.org/hep-ex/0312023 (2003).

7. Abe, F. *et al.* (CDF Collaboration). Observation of top quark production in $p\bar{p}$ collisions with the Collider Detector at Fermilab. *Phys. Rev. Lett.* **74**, 2626-2631 (1995).
8. Abachi, S. *et al.* (DØ Collaboration). Observation of the top quark. *Phys. Rev. Lett.* **74**, 2632-2637 (1995).
9. Ellis, J. The 115 GeV Higgs odyssey. *Comments Nucl. Part. Phys. A* **2**, 89-103 (2002).
10. Chanowitz, M. S. Electroweak data and the Higgs boson mass: a case for new physics. *Phys. Rev. D* **66**, 073002 (2002).
11. The LEP Collaborations ALEPH, DELPHI, L3 & OPAL, the LEP Higgs Working Group. Searches for the neutral Higgs bosons of the MSSM: preliminary combined results using LEP data collected at energies up to 209 GeV. Preprint at http://www.arXiv.org/hep-ex/0107030 (2001).
12. Degrassi, G., Heinemeyer, S., Hollik, W., Slavich, P. & Weiglein, G. Towards high-precision predictions for the MSSM Higgs sector. *Eur. Phys. J. C* **28**, 133-143 (2003).
13. Abbott, B. *et al.* (DØ Collaboration). Direct measurement of the top quark mass by the DØ collaboration. *Phys. Rev. D* **58**, 052001 (1998).
14. Abazov, V. M. *et al.* (DØ Collaboration). $t\bar{t}$ production cross section in $p\bar{p}$ collisions at $\sqrt{s}$ = 1.8 TeV. *Phys. Rev. D* **67**, 012004 (2003).
15. Abachi, S. *et al.* (DØ Collaboration). The DØ detector. *Nucl. Instrum. Methods A* **338**, 185-253 (1994).
16. Dalitz, R. H. & Goldstein, G. R. Test of analysis method for top–antitop production and decay events. *Proc. R. Soc. Lond. A* **445**, 2803-2834 (1999).
17. Kondo, K. *et al.* Dynamical likelihood method for reconstruction of events with missing momentum. 3: Analysis of a CDF high p_T $e\mu$ event as $t\bar{t}$ production. *J. Phys. Soc. Jpn* **62**, 1177-1182(1993).
18. Berends, F. A., Papadopoulos, C. G. & Pittau, R. On the determination of M_W and TGCs in W-pair production using the best measured kinematical variables. *Phys. Lett. B* **417**, 385-389 (1998).
19. Abreu, P. *et al.* (DELPHI Collaboration). Measurement of the W pair cross-section and of the W Mass in e^+e^- interactions at 172 GeV. *Eur. Phys. J. C* **2**, 581-595 (1998).
20. Juste, A. Measurement of the W mass in e^+e^- annihilation PhD thesis, 1-160, Univ. Autonoma de Barcelona (1998).
21. Estrada, J. Maximal use of kinematic information for extracting the top quark mass in single-lepton $t\bar{t}$ events PhD thesis, 1-132, Univ. Rochester (2001).
22. Canelli, F. Helicity of the W boson in single-lepton $t\bar{t}$ events PhD thesis. 1-241, Univ. Rochester (2003).
23. Mahlon, G. & Parke, S. Angular correlations in top quark pair production and decay at hadron colliders. *Phys. Rev. D* **53**, 4886-4896 (1996).
24. Mahlon, G. & Parke, S. Maximizing spin correlations in top quark pair production at the Tevatron. *Phys. Lett. B* **411**, 173-179 (1997).
25. Lai, H. L. *et al.* (CTEQ Collaboration). Global QCD analysis and the CTEQ parton distributions. *Phys. Rev. D* **51**, 4763-4782 (1995).
26. Marchesini, G. *et al.* HERWIG: a Monte Carlo event generator for simulating hadron emission reactions with interfering gluons. *Comput. Phys. Commun.* **67**, 465-508 (1992).

Acknowledgements. We are grateful to our colleagues A. Quadt and M. Mulders for reading of the manuscript and comments. We thank the staffs at Fermilab and collaborating institutions, and acknowledge support from the Department of Energy and National Science Foundation (USA), Commissariat à L'Energie Atomique and CNRS/Institut National de Physique Nucléaire et de Physique des Particules (France), Ministry for Science and Technology and Ministry for Atomic Energy (Russia), CAPES, CNPq and FAPERJ (Brazil), Departments of Atomic Energy and Science and Education (India), Colciencias (Colombia), CONACyT (Mexico), Ministry of Education and KOSEF (South Korea), CONICET and UBACyT (Argentina), The Foundation for Fundamental Research on Matter (The Netherlands), PPARC (UK), Ministry of Education (Czech Republic), the A. P. Sloan Foundation, and the Research Corporation.

Authors' contributions. We wish to note the great number of contributions made by the late Harry Melanson to the DØ experiment, through his steady and inspirational leadership of the physics, reconstruction and algorithm efforts.

Competing interests statement. The authors declare that they have no competing financial interests.

Correspondence and requests for materials should be addressed to J. Estrada (estrada@fnal.gov).

Experimental Investigation of Geologically Produced Antineutrinos with KamLAND

T. Araki *et al.*

Editor's Note

The KamLAND anti-neutrino detector at the Kamioka mine in Japan was the first detector sensitive enough to detect neutrino oscillations—the spontaneous transformation of a neutrino from one type to another—by looking at anti-neutrinos from nuclear reactors. As Takeo Araki and colleagues note here, the detector was also sensitive enough to detect neutrinos produced in the Earth's interior from radioactive decay of uranium or thorium. Previous estimates of the energy released by such decays suggested that it accounted for fully half of the total energy dissipated inside the planet, making radioactivity a significant contributor to the heat of the deep earth. The results here broadly agree with the earlier figures, and put an improved upper limit on its value.

The detection of electron antineutrinos produced by natural radioactivity in the Earth could yield important geophysical information. The Kamioka liquid scintillator antineutrino detector (KamLAND) has the sensitivity to detect electron antineutrinos produced by the decay of ^{238}U and ^{232}Th within the Earth. Earth composition models suggest that the radiogenic power from these isotope decays is 16 TW, approximately half of the total measured heat dissipation rate from the Earth. Here we present results from a search for geoneutrinos with KamLAND. Assuming a Th/U mass concentration ratio of 3.9, the 90 per cent confidence interval for the total number of geoneutrinos detected is 4.5 to 54.2. This result is consistent with the central value of 19 predicted by geophysical models. Although our present data have limited statistical power, they nevertheless provide by direct means an upper limit (60 TW) for the radiogenic power of U and Th in the Earth, a quantity that is currently poorly constrained.

THE Kamioka liquid scintillator antineutrino detector (KamLAND) has demonstrated neutrino oscillation using electron antineutrinos ($\bar{\nu}_e$s) with energies of a few MeV from nuclear reactors[1,2]. Additionally, KamLAND is the first detector sensitive enough to measure $\bar{\nu}_e$s produced in the Earth from the ^{238}U and ^{232}Th decay chains. Using $\bar{\nu}_e$s to study processes inside the Earth was first suggested by Eder[3] and Marx[4], and has been reviewed a number of times[5-10]. As $\bar{\nu}_e$s produced from the ^{238}U and ^{232}Th decay chains have exceedingly small interaction cross-sections, they propagate undisturbed in the Earth's interior, and their measurement near the Earth's surface can be used to gain information on their sources. Although the detection of $\bar{\nu}_e$s from ^{40}K decay would also be of great interest in geophysics, with possible applications in the interpretation of geo-magnetism, their energies are too

KamLAND 对地质来源的反中微子的实验研究

荒木岳夫等

编者按

KamLAND 反中微子探测器位于日本的神冈矿井下。它是第一个足够灵敏到通过监测来自反应堆的反中微子而探测到中微子振荡（即一种类型的中微子自发转变为另一种类型）的探测器。如荒木岳夫及其同事所指出的，该探测器也足够灵敏到可探测地球内部铀和钍放射性衰变产生的中微子。之前的估算表明，这些衰变产生的能量约占地球释放的总能量的一半，这使得放射性成为地球深部热能的一个主要来源。本文的结果与之前的数据大致吻合，并给出了更好的上限值。

探测地球中天然放射性产生的电子反中微子可以提供重要的地球物理信息。神冈液体闪烁体反中微子探测器（简称 KamLAND）可以探测到地球中放射元素 ^{238}U 和 ^{232}Th 衰变产生的电子反中微子（地球中微子）。地球组成模型表明，这些同位素衰变的放射性生热为 16 TW，大约占地球总散失热率测量值的一半。本文介绍了 KamLAND 探测地球中微子的结果。假设 Th/U 的质量丰度比为 3.9，探测到 90% 置信区间内地球中微子总数为 4.5 到 54.2 个。该结果与地球物理模型预测的中心值（19 个）一致。尽管现有数据统计量有限，这些数据仍然以直接的方式给出了此前知之甚少的地球中 U 和 Th 放射性生热的上限（60 TW）。

神冈液体闪烁体反中微子探测器（KamLAND）通过对来自核反应堆的几兆电子伏的电子反中微子（$\overline{\nu}_e$）的探测，证明了反应堆中微子振荡现象[1,2]。另外，KamLAND 也是第一个足够灵敏并探测到地球中 ^{238}U 和 ^{232}Th 衰变链产生的 $\overline{\nu}_e$ 的探测器。埃德[3]和马克思[4]第一次提出可利用 $\overline{\nu}_e$ 来测量地球内部性质，之后该观点被引述了数次[5-10]。^{238}U 和 ^{232}Th 衰变链产生的 $\overline{\nu}_e$ 具有极其小的相互作用截面，因此它们不受干扰地在地球内部传播，在地球表面附近对 $\overline{\nu}_e$ 进行测量可获得其产生来源的信息。尽管探测 ^{40}K 衰变产生的 $\overline{\nu}_e$ 对地球物理学也具有重要意义，可能用以解释地磁

low to be detected with KamLAND. The antineutrino flux above our detection threshold from other long-lived isotopes is expected to be negligible.

The Radiogenic Earth

The total power dissipated from the Earth (heat flow) has been measured with thermal techniques[11] to be 44.2 ± 1.0 TW. Despite this small quoted error, a more recent evaluation[12] of the same data (assuming much lower hydrothermal heat flow near mid-ocean ridges) has led to a lower figure of 31 ± 1 TW. On the basis of studies of chondritic meteorites[13] the calculated radiogenic power is thought to be 19 TW, 84% of which is produced by ^{238}U and ^{232}Th decay. Some models of mantle convection suggest that radiogenic power is a larger fraction of the total power[14,15].

^{238}U and ^{232}Th decay via a series of well-established α and β^- processes[16] terminating in the stable isotopes ^{206}Pb and ^{208}Pb, respectively. Each β^- decay produces a daughter nucleus, an electron and a $\bar{\nu}_e$. The $\bar{\nu}_e$ energy distribution is well established[17], and includes a correction for the electromagnetic interaction between the electron and the charge distribution of the daughter nucleus. Figure 1 shows the expected $\bar{\nu}_e$ distribution, $\mathrm{d}n(E_\nu)/\mathrm{d}E_\nu$, as a function of $\bar{\nu}_e$ energy, E_ν, for the ^{238}U and ^{232}Th decay chains.

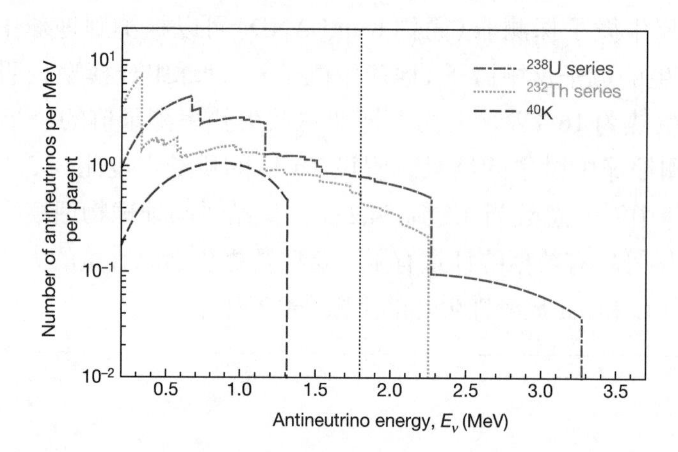

Fig. 1. The expected ^{238}U, ^{232}Th and ^{40}K decay chain electron antineutrino energy distributions. KamLAND can only detect electron antineutrinos to the right of the vertical dotted black line; hence it is insensitive to ^{40}K electron antineutrinos.

Ignoring the negligible neutrino absorption, the expected $\bar{\nu}_e$ flux at a position $\mathbf{r}$ for each isotope is given by:

$$\frac{\mathrm{d}\phi(E_\nu, \mathbf{r})}{\mathrm{d}E_\nu} = A\frac{\mathrm{d}n(E_\nu)}{\mathrm{d}E_\nu}\int_{V_\oplus}\mathrm{d}^3\mathbf{r}'\frac{a(\mathbf{r}')\rho(\mathbf{r}')P(E_\nu, |\mathbf{r}-\mathbf{r}'|)}{4\pi|\mathbf{r}-\mathbf{r}'|^2} \quad (1)$$

where A is the decay rate per unit mass, the integral is over the volume of the Earth, $a(\mathbf{r}')$ is the isotope mass per unit rock mass, $\rho(\mathbf{r}')$ is the rock density, and $P(E_\nu, |\mathbf{r}-\mathbf{r}'|)$ is the $\bar{\nu}_e$

现象，但是它们的能量太低，不能被 KamLAND 探测到。来自其他长寿命同位素且能量高于我们探测阈值的反中微子被认为可忽略不计。

地球的放射能

地球释放的总能量（热流）已通过热学方法[11]测得，为 44.2 ± 1.0 TW。尽管以上结果给出的误差很小，最近利用相同数据的推算[12]（假定洋中脊附近热液的热流低得多）给出一个更低的值 31 ± 1 TW。基于对球粒陨石的研究[13]，计算出的放射性生热约为 19 TW，其中 84% 来自 ^{238}U 和 ^{232}Th 衰变。一些地幔对流模型给出的放射性生热占总能量的比例更高[14,15]。

^{238}U 和 ^{232}Th 经过一系列已知的 α 衰变和 β^- 衰变后[16]，最终分别变成稳定同位素 ^{206}Pb 和 ^{208}Pb。每次 β^- 衰变产生一个子核、一个电子和一个 $\bar{\nu}_e$。这个 $\bar{\nu}_e$ 的能谱已经比较精确[17]，包括对电子和子核电荷分布之间电磁相互作用的修正。图 1 是来自 ^{238}U 和 ^{232}Th 衰变链的预期 $\bar{\nu}_e$ 分布，即 $dn(E_\nu)/dE_\nu$，该结果是 $\bar{\nu}_e$ 能量 (E_ν) 的函数。

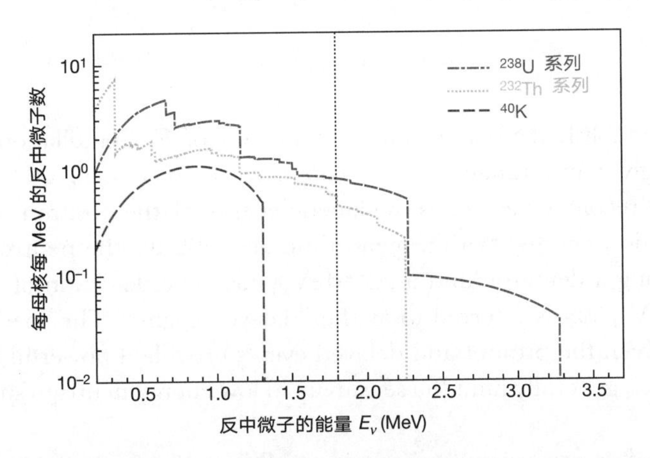

图 1. 从 ^{238}U、^{232}Th 和 ^{40}K 衰变链产生的电子反中微子的预期能量分布。KamLAND 只能探测到黑竖点虚线右侧部分的电子反中微子，因此对 ^{40}K 电子反中微子不敏感。

忽略微不足道的中微子吸收，每种同位素在位置 **r** 处的 $\bar{\nu}_e$ 预期通量如下：

$$\frac{d\phi(E_\nu, \mathbf{r})}{dE_\nu} = A\frac{dn(E_\nu)}{dE_\nu}\int_{V_\oplus} d^3\mathbf{r}'\frac{a(\mathbf{r}')\rho(\mathbf{r}')P(E_\nu, |\mathbf{r}-\mathbf{r}'|)}{4\pi|\mathbf{r}-\mathbf{r}'|^2} \tag{1}$$

其中 A 是单位质量的衰变率，积分针对整个地球的体积进行，$a(\mathbf{r}')$ 是每单位岩体质量的同位素质量，$\rho(\mathbf{r}')$ 是岩石密度，$P(E_\nu, |\mathbf{r}-\mathbf{r}'|)$ 是 $\bar{\nu}_e$ 穿过 $|\mathbf{r}-\mathbf{r}'|$ 距离后的“存活”

"survival" probability after travelling a distance $|\mathbf{r}-\mathbf{r}'|$. This probability derives from the now accepted phenomenon of neutrino oscillation, and can be written, for two neutrino flavours as[18]

$$P(E_\nu, L) \cong 1-\sin^2 2\theta_{12}\sin^2\left(\frac{1.27\Delta m_{12}^2[\mathrm{eV}^2]L[\mathrm{m}]}{E_\nu[\mathrm{MeV}]}\right) \tag{2}$$

where $L = |\mathbf{r}-\mathbf{r}'|$. The neutrino oscillation parameters $\Delta m_{12}^2 = 7.9^{+0.6}_{-0.5} \times 10^{-5}$ eV^2 and $\sin^2 2\theta_{12} = 0.82 \pm 0.07$ are also determined with KamLAND[2] using reactor $\overline{\nu}_e$s with energies above those of geoneutrinos, combined with solar neutrino experiments[19]. Corrections from three flavour neutrino oscillation ($< 5\%$) and "matter effects"[20] ($\sim 1\%$) are ignored. For typical geoneutrino energies, the approximation $P(E_\nu, |L|) = 1-0.5\sin^2 2\theta_{12}$ only affects the accuracy of the integral in equation (1) at 1% owing to the distributed $\overline{\nu}_e$ production points. This approximation, used in this paper, neglects energy spectrum distortions.

Geoneutrino Detection

KamLAND is located in the Kamioka mine, 1,000 m below the summit of Mt Ikenoyama, Gifu prefecture, Japan (36° 25′ 36″ N, 137° 18′ 43″ E). It detects electron antineutrinos in $\sim$1 kton of liquid scintillator via neutron inverse β-decay,

$$\overline{\nu}_e + p \rightarrow e^+ + n \tag{3}$$

which has a well-established cross-section[21] as a function of E_ν. Scintillation light from the e^+, "prompt event", gives an estimate of the incident $\overline{\nu}_e$ energy, $E_\nu \approx E_{e+} + 0.8$ MeV (neglecting the small neutron recoil), where E_{e+} is the kinetic energy of the positron plus the electron–positron annihilation energy. With a mean time of $\sim$200 μs, the neutron is captured by a proton, producing a deuteron and a 2.2 MeV γ-ray. The detection of scintillation light from this 2.2 MeV γ-ray is referred to as the "delayed event". The spatial and temporal coincidences between the prompt and delayed events provide a powerful tool for reducing backgrounds, which generally limit the sensitivity in low energy neutrino studies.

A reference model[22] is constructed using seismic data to divide the Earth into continental crust, oceanic crust, mantle, core and sediment. Some of these regions are further sub-divided, with each sub-region having different U and Th concentrations. This model assumes that U and Th are absent from the core. The expected geoneutrino flux at KamLAND, including a suppression factor of 0.59 due to neutrino oscillations, is 2.34×10^6 cm^{-2} s^{-1} and 1.98×10^6 cm^{-2} s^{-1} from the ^{238}U and ^{232}Th decay chains, respectively. Including the detection cross-section, the number of geoneutrinos expected at KamLAND from ^{238}U and ^{232}Th decay is 3.85×10^{-31} $\overline{\nu}_e$ per target proton per year, 79% of which is due to ^{238}U. Figure 2 shows that a large fraction of the expected geoneutrino flux originates in the area surrounding KamLAND. The effect of local geology was studied extensively in the context of the reference model[22] and was found to produce less than a 10% error on the total expected flux.

概率。该概率从已被证实的中微子振荡现象中得来。对两种中微子味道，该概率可写为[18]

$$P(E_\nu, L) \simeq 1-\sin^2 2\theta_{12}\sin^2\left(\frac{1.27\Delta m_{12}^2[\mathrm{eV}^2]L[\mathrm{m}]}{E_\nu[\mathrm{MeV}]}\right) \tag{2}$$

其中 $L=|\mathbf{r}-\mathbf{r}'|$。利用能量高于地球中微子的反应堆 $\bar{\nu}_e$ 的 KamLAND 实验[2]，并结合太阳中微子实验[19]，确定了中微子振荡参数的大小：$\Delta m_{12}^2=7.9_{-0.5}^{+0.6}\times10^{-5}\ \mathrm{eV}^2$，$\sin^2 2\theta_{12}=0.82\pm0.07$。三味中微子振荡效应（$<5\%$）和“物质效应”[20]（$\sim1\%$）的修正可忽略。对典型的地球中微子能量，取 $P(E_\nu,|L|)=1-0.5\sin^2 2\theta_{12}$ 的近似对公式（1）积分精度的影响只有 1%，主要来自分散的 $\bar{\nu}_e$ 产生地点。本文中使用这种近似，忽略了能谱的变形。

地球中微子探测

KamLAND 实验位于日本岐阜县池野山地下 1,000 米处的神冈矿井中（北纬 36° 25′ 36″，东经 137° 18′ 43″）。它通过中子反 β 衰变过程，在约 1 千吨的液体闪烁体中探测电子反中微子，

$$\bar{\nu}_e+p\rightarrow e^+ + n \tag{3}$$

该过程的反应截面非常清楚[21]，是 E_ν 的函数。来自 e^+ 的闪烁光被称为“快信号”，它给出入射 $\bar{\nu}_e$ 的能量的估计值，$E_\nu \approx E_{e^+}+0.8\ \mathrm{MeV}$（忽略了小的中子反冲）。此处 E_{e^+} 是正电子动能加上电子–正电子的湮灭能。经过平均约 200 μs 的时间，中子被质子俘获，产生一个氘核和一个 2.2 MeV 的 γ 光子。探测到的 2.2 MeV γ 光子的闪烁光被称为“慢信号”。快慢信号在时间和位置上的符合提供了一个减少本底的强大工具。在低能中微子研究中，本底通常会制约实验灵敏度。

利用地震数据构建的一个参考模型[22]将地球划分为陆壳、洋壳、地幔、地核和沉积层。部分上述区域可进一步细分，每个子区域有不同的 U 和 Th 丰度。这个模型假定地核里不含 U 和 Th。预期 KamLAND 实验站点来自 ^{238}U 和 ^{232}Th 衰变链的地球中微子通量分别为 $2.34\times10^6\ \mathrm{cm}^{-2}\cdot\mathrm{s}^{-1}$ 和 $1.98\times10^6\ \mathrm{cm}^{-2}\cdot\mathrm{s}^{-1}$，其中考虑了来自中微子振荡的平均压低因子 0.59。考虑探测截面后，KamLAND 预期能捕获的来自 ^{238}U 和 ^{232}Th 衰变链的地球中微子事例率为 3.85×10^{-31} 个 $\bar{\nu}_e$ 每靶质子每年，其中 79% 来自 ^{238}U。从图 2 中可以看出，预期的地球中微子通量中，很大一部分在 KamLAND 附近区域产生。在参考模型中细致地研究了当地的地质效应[22]，发现导致的误差小于总预期通量的 10%。

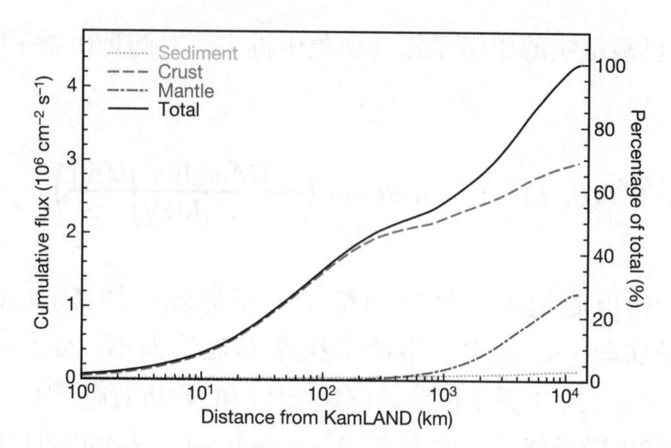

Fig. 2. The expected total ^{238}U and ^{232}Th geoneutrino flux within a given distance from KamLAND[22]. Approximately 25% and 50% of the total flux originates within 50 km and 500 km of KamLAND, respectively. The line representing the crust includes both the continental and the almost negligible oceanic contribution.

The data presented here are based on a total detector live-time of 749.1 ± 0.5 d after basic cuts to ensure the reliability of the data. The number of target protons is estimated at $(3.46 \pm 0.17) \times 10^{31}$ on the basis of target proton density and a spherical fiducial scintillator volume with 5 m radius, resulting in a total exposure of $(7.09 \pm 0.35) \times 10^{31}$ target proton years. The overall efficiency for detecting geoneutrino candidates with energies between 1.7 and 3.4 MeV in the fiducial volume is estimated to be 0.687 ± 0.007. The energy range reaches below the inverse β-decay threshold owing to the detector energy resolution.

Backgrounds for geoneutrino candidates are dominated by $\bar{\nu}_e$s from nuclear reactors in the vicinity of the detector, and by α-particle induced neutron backgrounds due to radioactive contamination within the detector. Reactor $\bar{\nu}_e$s reach substantially higher energies, as shown in Fig. 3. Therefore, the oscillation parameters in ref. 2 were determined by analysing $\bar{\nu}_e$s with energies greater than 3.4 MeV, where there is no signal from the geoneutrinos. Using these parameters, the number of nuclear reactor $\bar{\nu}_e$ background events used by the "rate only" analysis discussed below is determined to be 80.4 ± 7.2.

The α-particle-induced neutron background is due to the $^{13}C(\alpha,n)^{16}O$ reaction where the α-particle is produced in ^{210}Po decay with a kinetic energy of 5.3 MeV. The ^{210}Po is produced by the decay of ^{210}Pb, which has a half-life of 22 yr. The ^{210}Pb resulted from the decay of ^{222}Rn contamination, and is distributed throughout the detector. The neutrons in the $^{13}C(\alpha,n)^{16}O$ reaction are produced with kinetic energy up to 7.3 MeV. Owing to scintillation light quenching for high ionization density, only about one-third of this energy is converted into "visible" energy as the neutrons thermalize. The thermal neutrons are captured by protons with a mean capture time of ~200 μs, producing a delayed signal identical to that from neutron inverse β-decay. The number of ^{13}C nuclei in the fiducial volume is determined from the measured $^{13}C/^{12}C$ ratio in the KamLAND scintillator. On the basis of the $^{13}C(\alpha,n)^{16}O$ reaction cross-section[23], the α-particle energy loss in the scintillator[24], and the number of ^{210}Po decays, the total number of neutrons produced is

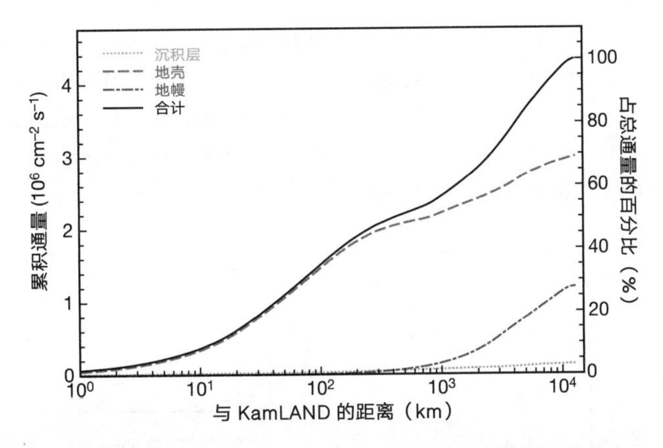

图 2. 在 KamLAND 不同距离范围内产生的 ^{238}U 和 ^{232}Th 地球中微子通量 [22]。大约 25% 和 50% 的总通量来自 KamLAND 附近 50 km 和 500 km 范围内。图中地壳的贡献包括陆壳和几乎可忽略的洋壳的贡献。

本文中的数据经过了一些基本挑选，以确保数据的可靠性，总的探测器有效时间为 749.1 ± 0.5 天。基于平均靶质子密度和半径为 5 m 的闪烁体球形有效体积，总的靶质子数估计为 $(3.46 \pm 0.17) \times 10^{31}$ 个，因此总的曝光量为 $(7.09 \pm 0.35) \times 10^{31}$ 靶质子 · 年。选取有效体积内能量在 1.7 MeV 到 3.4 MeV 之间的地球中微子候选者，总探测效率为 0.687 ± 0.007。能量范围取到反 β 衰变的阈值以下是因为探测器存在能量分辨精度。

地球中微子候选者中的本底主要来自探测器附近核反应堆产生的电子反中微子，以及探测器中放射性杂质产生的 α 粒子引发的中子本底。如图 3 所示，反应堆中微子可达到的能量比地球中微子要高得多。因此参考文献 2 里的振荡参数通过分析能量大于 3.4 MeV 的 $\bar{\nu}_e$ 得到，此能量区间内无地球中微子信号。利用这些参数，得到反应堆中微子本底数为 80.4 ± 7.2，下面将要介绍的“事例率”分析法中用到此值。

α 粒子引发的中子本底来自 $^{13}C(\alpha,n)^{16}O$ 反应，其中 α 粒子来自 ^{210}Po 衰变，动能为 5.3 MeV。^{210}Po 来自半衰期为 22 年的 ^{210}Pb 衰变。^{210}Pb 来自 ^{222}Rn 衰变，而 ^{222}Rn 污染分布在整个探测器内。$^{13}C(\alpha,n)^{16}O$ 反应产生的中子最大动能为 7.3 MeV。由于闪烁光在高电离密度下的淬灭效应，当中子慢化时，只有约三分之一的能量转化为“可见”能量。热中子被质子所俘获，平均俘获时间约为 200 μs，产生一个与中子反 β 衰变相同的慢信号。^{13}C 原子核在探测器有效体积内的数量通过测量 KamLAND 闪烁体的 $^{13}C/^{12}C$ 比值得到。基于 $^{13}C(\alpha,n)^{16}O$ 的反应截面 [23]、α 粒子在闪烁体里的能量损失 [24] 以及 ^{210}Po 衰变数，预期产生的总中子数为 93 ± 22，误差主要由 $^{13}C(\alpha,n)^{16}O$

expected to be 93 ± 22. This error is dominated by estimated 20% and 14% uncertainties in the total $^{13}C(\alpha,n)^{16}O$ reaction cross-section and the number of ^{210}Po decays, respectively. The neutron energy distribution is calculated using the measured neutron angular distributions in the centre of mass frame[25,26]. Including the efficiency for passing the $\bar{\nu}_e$ candidate cuts, the number of (α,n) background events is estimated to be 42 ± 11.

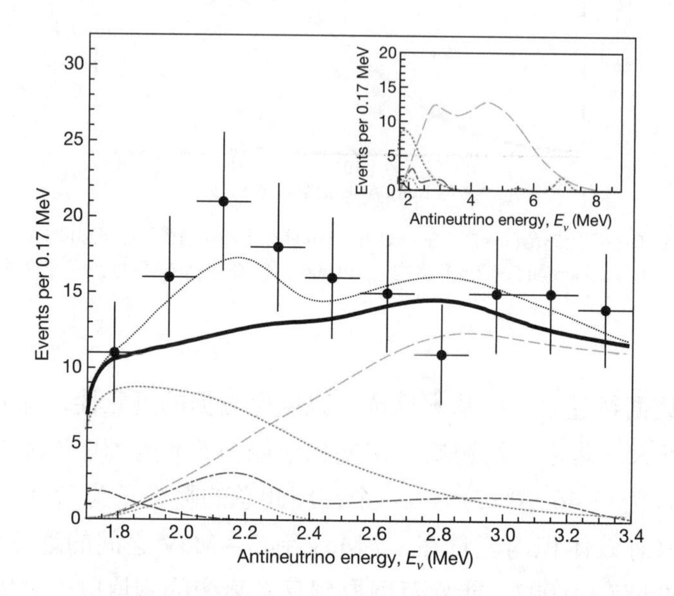

Fig. 3. $\bar{\nu}_e$ energy spectra in KamLAND. Main panel, experimental points together with the total expectation (thin dotted black line). Also shown are the total expected spectrum excluding the geoneutrino signal (thick solid black line), the expected signals from ^{238}U (dot-dashed red line) and ^{232}Th (dotted green line) geoneutrinos, and the backgrounds due to reactor $\bar{\nu}_e$ (dashed light blue line), $^{13}C(\alpha,n)^{16}O$ reactions (dotted brown line), and random coincidences (dashed purple line). Inset, expected spectra extended to higher energy. The geoneutrino spectra are calculated from our reference model, which assumes 16 TW radiogenic power from ^{238}U and ^{232}Th. The error bars represent $\pm$ 1 standard deviation intervals.

There is a small contribution to the background from random coincidences, $\bar{\nu}_e$s from the β^- decay of long lived nuclear reactor fission products, and radioactive isotopes produced by cosmic rays. Using an out-of-time coincidence cut from 10 ms to 20 s, the random coincidence background is estimated to be 2.38 ± 0.01 events. Using the expected $\bar{\nu}_e$ energy spectrum[27] for long lived nuclear reactor fission products, the corresponding background is estimated to be 1.9 ± 0.2 events. The most significant background due to radioactive isotopes produced by cosmic rays is from the β^- decay $^9Li \rightarrow 2\alpha + n + e^- + \bar{\nu}_e$, which has a neutron in the final state. On the basis of events correlated with cosmic rays, the estimated number of background events caused by radioactive ^9Li is 0.30 ± 0.05. Other backgrounds considered and found to be negligible include spontaneous fission, neutron emitters and correlated decays in the radioactive background decay chains, fast neutrons from cosmic ray interactions, (γ,n) reactions and solar ν_e induced break-up of ^2H. The total background is estimated to be 127 ± 13 events (1σ error).

The total number of observed $\bar{\nu}_e$ candidates is 152, with their energy distribution shown in

反应截面的 20% 不确定度和 ^{210}Po 衰变数的 14% 不确定度决定。通过在质心系下测量的中子角分布计算得到中子的能量分布 [25,26]，再考虑 $\bar{\nu}_e$ 事例挑选的效率，估算出 (α,n) 本底大约有 42 ± 11 个。

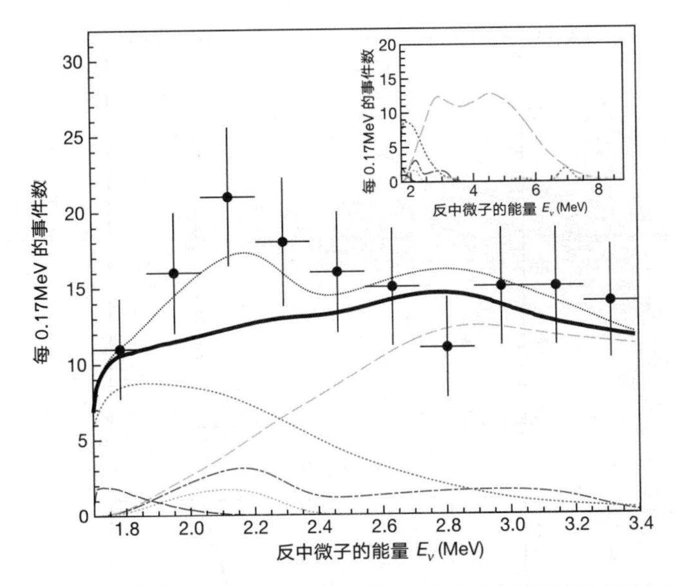

图 3. KamLAND 测得的电子反中微子能谱。主图为实验数据点和总的预期能谱（细黑点线）。同时显示了除地球中微子信号外所有预期能谱的总和（粗黑实线）、^{238}U 地球中微子预期信号（红点划线）、^{232}Th 地球中微子预期信号（绿点线）、反应堆中微子本底（浅蓝虚线）、$^{13}C(\alpha,n)^{16}O$ 反应本底（棕点线）以及随机符合的本底（紫虚线）。小插图显示了延伸到更高能量下的预期能谱。地球中微子能谱从我们的参考模型计算而来，其中假定了 ^{238}U 和 ^{232}Th 共产生 16 TW 的放射性热能。误差棒代表 ±1 倍标准偏差。

其他较小的本底还包括随机符合本底、长寿命的反应堆裂变产物发生 β^- 衰变产生的 $\bar{\nu}_e$ 以及宇宙线产生的放射性同位素。根据信号时间窗以外的 10 ms 到 20 s 之间的符合情况来估算，约有 2.38 ± 0.01 个随机符合事例。根据长寿命的反应堆裂变产物的预期 $\bar{\nu}_e$ 能谱 [27]，相应的本底估计为 1.9 ± 0.2 个事例。宇宙线产生的最显著的放射性同位素本底来自末态有一个中子的 β^- 衰变 $^9Li\rightarrow2\alpha+n+e^-+\bar{\nu}_e$。基于与宇宙线关联的事例，由放射性的 ^9Li 产生的预期本底事例为 0.30 ± 0.05 个。其他考虑过但发现可忽略不计的本底还包括自发裂变、放射性本底衰变链中的中子发射和关联衰变、宇宙线产生的快中子、(γ,n) 反应以及太阳中微子引发的 ^2H 分解。总本底估计为 127 ± 13 个事例（1σ 误差）。

观测到的 $\bar{\nu}_e$ 候选者总数为 152 个，其能量分布显示在图 3 中。考虑到地球中微

Fig. 3. Including the geoneutrino detection systematic errors, parts of which are correlated with the background estimation errors, a "rate only" analysis gives 25^{+19}_{-18} geoneutrino candidates from the ^{238}U and ^{232}Th decay chains. Dividing by the detection efficiency, live-time, and number of target protons, the total geoneutrino detected rate obtained is $5.1^{+3.9}_{-3.6} \times 10^{-31}$ $\bar{\nu}_e$ per target proton per year.

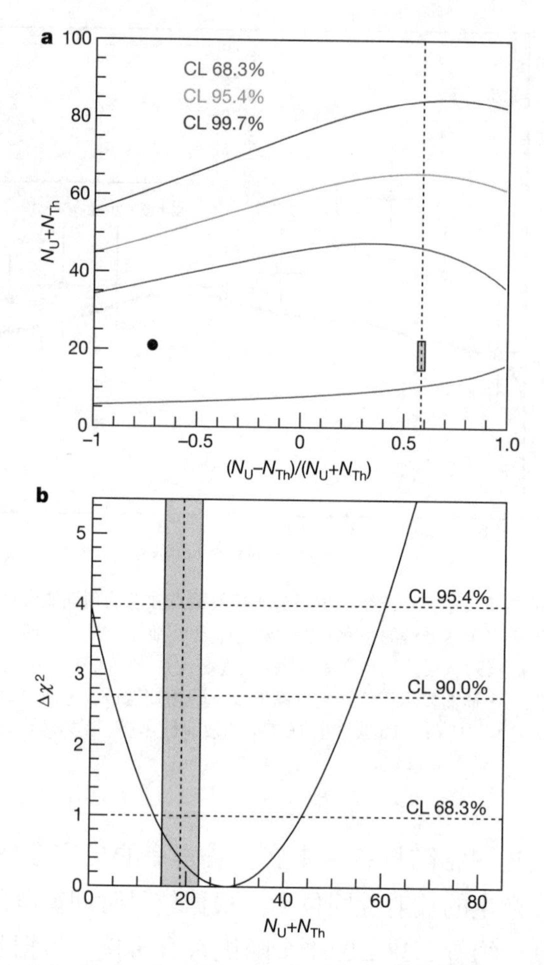

Fig. 4. Confidence intervals for the number of geoneutrinos detected. Panel **a** shows the 68.3% confidence level (CL; red), 95.4% CL (green) and 99.7% CL (blue) contours for detected ^{238}U and ^{232}Th geoneutrinos. The small shaded area represents the prediction from the geophysical model. The vertical dashed line represents the value of $(N_U-N_{Th})/(N_U+N_{Th})$ assuming the mass ratio, Th/U = 3.9, derived from chondritic meteorites, and accounting for the ^{238}U and ^{232}Th decay rates and the $\bar{\nu}_e$ detection efficiencies in KamLAND. The dot represents our best fit point, favouring 3 ^{238}U geoneutrinos and 18 ^{232}Th geoneutrinos. Panel **b** shows $\Delta\chi^2$ as a function of the total number of ^{238}U and ^{232}Th geoneutrino candidates, fixing the normalized difference to the chondritic meteorites constraint. The grey band gives the value of N_U+N_{Th} predicted by the geophysical model.

We also perform an un-binned maximum likelihood analysis of the $\bar{\nu}_e$ energy spectrum between 1.7 and 3.4 MeV, using the known shape of the signal and background spectra. As the neutrino oscillation parameters do not significantly affect the expected shape of the

子探测的系统误差（其中一部分与本底估算误差相关联），采用事例率分析法，^{238}U 和 ^{232}Th 衰变链产生的地球中微子数为 25^{+19}_{-18}。该事例数除以探测器效率、取数时间以及靶质子数，探测到的地球中微子总事例率为每年每靶质子 $5.1^{+3.9}_{-3.6}\times10^{-31}$ 个 $\bar{\nu}_e$。

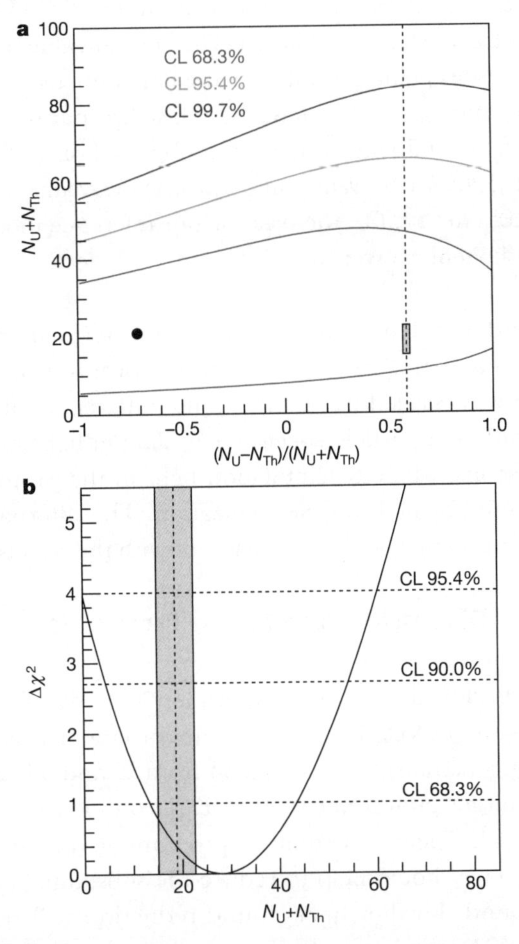

图 4. 探测到的地球中微子事例数的置信区间。图 **a** 是探测到的 ^{238}U 和 ^{232}Th 地球中微子数分别在 68.3% 置信度（红色）、95.4% 置信度（绿色）和 99.7% 置信度（蓝色）下的等值线。小的阴影区是地球物理模型的预期，竖虚线表示基于从球粒陨石推出的 Th/U 质量比 3.9，并考虑了 ^{238}U 和 ^{232}Th 衰变率以及 KamLAND 对 $\bar{\nu}_e$ 的探测效率等情况下的 (N_U-N_{Th}) 与 (N_U+N_{Th}) 的比值。黑点为最佳拟合点，倾向于 ^{238}U 地球中微子数为 3 个，^{232}Th 地球中微子数为 18 个。图 **b** 是 $\Delta\chi^2$ 随 ^{238}U 和 ^{232}Th 地球中微子总数的函数分布，其中归一化后的差别固定到球粒陨石的限制值。灰色区域带是地球物理模型预期的 N_U+N_{Th} 值。

我们也利用已知的信号能谱和本底能谱的形状，对 1.7 ~ 3.4 MeV 区间的 $\bar{\nu}_e$ 能谱进行了一项不分能量区间的最大似然分析。中微子振荡参数对预期的地球中微子能谱形状影响很小，因此采用了无振荡的谱形。而反应堆中微子本底的谱形则包括

geoneutrino signal, the un-oscillated shape is assumed. However, the oscillation parameters are included in the reactor background shape. Figure 4a shows the confidence intervals for the number of observed ^{238}U and ^{232}Th geoneutrinos. Based on a study of chondritic meteorites[28], the Th/U mass ratio in the Earth is believed to be between 3.7 and 4.1, and is known better than either absolute concentration. Assuming a Th/U mass ratio of 3.9, we estimate the 90% confidence interval for the total number of ^{238}U and ^{232}Th geoneutrino candidates to be 4.5 to 54.2, as shown in Fig. 4b. The central value of 28.0 is consistent with the "rate only" analysis. At this point, the value of the fit parameters are $\Delta m_{12}^2 = 7.8 \times 10^{-5}$ eV^2, $\sin^2 2\theta_{12} = 0.82$, $p_\alpha = 1.0$, and $q_\alpha = 1.0$, where these last two parameters are defined in the Methods section. The 99% confidence upper limit obtained on the total detected ^{238}U and ^{232}Th geoneutrino rate is 1.45×10^{-30} $\bar{\nu}_e$ per target proton per year, corresponding to a flux at KamLAND of 1.62×10^7 cm^{-2} s^{-1}. On the basis of our reference model, this corresponds to an upper limit on the radiogenic power from ^{238}U and ^{232}Th decay of 60 TW.

As a cross-check, an independent analysis[29] has been performed using a partial data set, including detection efficiency, of 2.6×10^{31} target proton years. In this analysis, the ^{13}C(α,n)^{16}O background was verified using the minute differences in the time structures of scintillation light from different particle species. Scintillation light in the prompt part of $\bar{\nu}_e$ events is caused by positrons, whereas scintillation light in the prompt part of ^{13}C(α,n)^{16}O background events is caused by neutron thermalization. This alternative analysis produced a slightly larger geoneutrino signal, which is consistent with the results presented here.

Discussion and Future Prospects

In conclusion, we have performed the first experimental study of antineutrinos from the Earth's interior using KamLAND. The present measurement is consistent with current geophysical models, and constrains the $\bar{\nu}_e$ emission from U and Th in the planet to be less than 1.45×10^{-30} $\bar{\nu}_e$ per target proton per year at 99% confidence limits, corresponding to a flux of 1.62×10^7 cm^{-2} s^{-1}. There is currently a programme underway to reduce the ^{210}Pb content of the detector. This should help to reduce the substantial systematic error due to the ^{13}C(α,n)^{16}O background. Further background reduction will require a new detector location, far away from nuclear reactors. The reported investigation of geoneutrinos should pave the way to future and more accurate measurements, which may provide a new window for the exploration of the Earth.

Methods

As shown in Fig. 5, KamLAND[1] consists of 1 kton of ultrapure liquid scintillator contained in a transparent nylon/EVOH (ethylene vinyl alcohol copolymer) composite film balloon suspended in non-scintillating oil. Charged particles deposit their kinetic energy in the scintillator; some of this energy is converted into scintillation light. The scintillation light is then detected by an array of 1,325 17-inch-diameter photomultiplier tubes (PMTs) and 554 20-inch-diameter PMTs mounted on the inner surface of an 18-m-diameter spherical stainless-steel containment vessel. A 3.2-

了振荡参数。图 4a 显示了观测到的 ^{238}U 和 ^{232}Th 地球中微子数的置信区间。基于对球粒陨石的研究 [28]，地球上 Th/U 的质量比被认为在 3.7 ~ 4.1 之间，相比于 Th 和 U 各自的绝对丰度，人们了解的 Th/U 质量比是更为准确的。假设 Th/U 的质量比为 3.9，我们估计 90% 置信区间内 ^{238}U 和 ^{232}Th 地球中微子总数为 4.5 到 54.2，如图 4b 所示。中心值 28.0 与前面事例率分析方法的结果一致。此处拟合参数的值分别为 $\Delta m_{12}^2 = 7.8 \times 10^{-5}$ eV^2、$\sin^2 2\theta_{12} = 0.82$、$p_\alpha = 1.0$、$q_\alpha = 1.0$，其中最后两个参数的定义在下面的方法章节中介绍。探测到的 ^{238}U 和 ^{232}Th 地球中微子总事例率的 99% 置信度上限为 1.45×10^{-30} 个 $\bar{\nu}_e$ 每靶质子每年，对应到达 KamLAND 的地球中微子通量为 1.62×10^7 cm$^{-2} \cdot$ s^{-1}。基于我们的参考模型，这相应于 ^{238}U 和 ^{232}Th 衰变的放射性功率上限为 60 TW。

作为验证，使用总量为 2.6×10^{31} 个靶质子 · 年的部分数据（包含探测器效率）进行了独立分析 [29]。在分析中，^{13}C$(\alpha,n)^{16}$O 本底通过不同粒子种类产生闪烁光的微小时间分布差异得到了验证。$\bar{\nu}_e$ 事例的快信号是由正电子产生的闪烁光，而 ^{13}C$(\alpha,n)^{16}$O 本底事例的快信号是中子慢化产生的闪烁光。此分析得到稍微大一点的地球中微子信号，与本文给出的结果一致。

讨论和前景

综上所述，我们利用 KamLAND 探测器，对来自地球内部的反中微子进行了首次实验研究。测量结果与现有的地球物理模型一致：在 99% 的置信限，限制了地球中 U 和 Th 发射的 $\bar{\nu}_e$ 数小于每靶质子每年 1.45×10^{-30} 个，对应于通量 1.62×10^7 cm$^{-2} \cdot$ s^{-1}。目前正在执行计划以减少探测器中 ^{210}Pb 的含量，这将帮助降低由 ^{13}C$(\alpha,n)^{16}$O 本底带来的较大的系统误差。进一步降低本底需要一个新的远离核反应堆的实验点。本文报告的对地球中微子的研究应该为未来更准确的测量铺平了道路，这可能为探索地球提供了一个新的窗口。

方　法

如图 5 所示，KamLAND[1] 由装在透明的尼龙/EVOH（乙烯–乙烯醇共聚物）复合膜气球里的 1 千吨超纯净的液体闪烁体组成，气球悬浮在不产生闪烁光的矿物油中。带电粒子在液体闪烁体里沉积能量，这些能量的一部分转化为闪烁光，被分布在四周的 1,325 个直径为 17 英寸的光电倍增管（PMT）和 554 个直径为 20 英寸的 PMT 探测到。PMT 安装在直径为 18 米的球形不锈钢容器的内表面。包围着球形容器的是装有 225 个直径为 20 英寸的 PMT

kton water-Cherenkov detector with 225 20-inch-diameter PMTs surrounds the containment sphere. This outer detector tags cosmic-ray muons and absorbs γ-rays and neutrons from the surrounding rock.

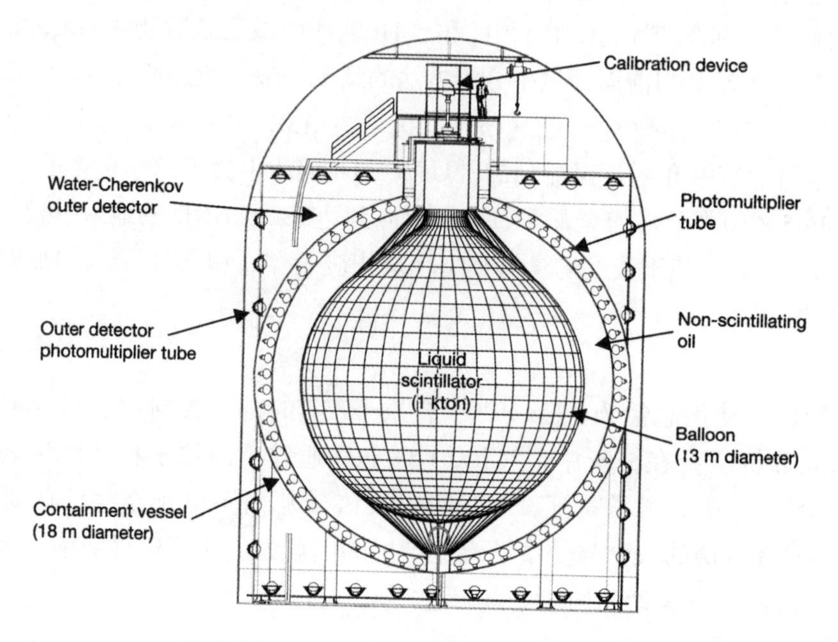

Fig. 5. Schematic diagram of the KamLAND detector.

The arrival times of photons at the PMTs allow us to determine the location of particle interactions inside the detector, and the amount of detected light after correcting for spatial variation of the detector response allows us to determine the particle's energy. The event location and energy determination is calibrated with γ-ray sources deployed vertically down the centre of the detector. To be classified as a $\overline{\nu}_e$ candidate, the time coincidence between the prompt and delayed events (ΔT) is required to satisfy $0.5\ \mu s < \Delta T < 500\ \mu s$. The position of the prompt ($\mathbf{r}_p$) and delayed ($\mathbf{r}_d$) events with respect to the centre of the detector are required to satisfy $|\mathbf{r}_p| < 5$ m; $|\mathbf{r}_d| < 5$ m and $|\mathbf{r}_p - \mathbf{r}_d| < 1.0$ m: The energy of the electron antineutrino is required to satisfy $1.7\ \text{MeV} < E_\nu < 3.4\ \text{MeV}$ and the energy of the delayed event (E_d) is required to satisfy $1.8\ \text{MeV} < E_d < 2.6\ \text{MeV}$.

Given N_U and N_{Th} geoneutrinos detected from the ^{238}U and ^{232}Th decay chains, the expected energy distribution of the candidates is

$$\frac{d\overline{N}(E_\nu)}{dE_\nu} = N_U \frac{dP_U(E_\nu)}{dE_\nu} + N_{Th} \frac{dP_{Th}(E_\nu)}{dE_\nu} + \frac{dN_r(E_\nu; \Delta m_{12}^2, \sin^2 2\theta_{12})}{dE_\nu} + P_\alpha \frac{dN_\alpha(E_\nu/q_\alpha)}{dE_\nu} + \sum_k \frac{dN_k(E_\nu)}{dE_\nu} \tag{4}$$

where $dP_U(E_\nu)/dE_\nu$ and $dP_{Th}(E_\nu)/dE_\nu$ are the normalized expected geoneutrino spectra from ^{238}U and ^{232}Th decay chains. The third term on the right hand side of equation (4) is the energy spectrum of the expected $\overline{\nu}_e$ reactor background, which is a function of the neutrino oscillation parameters Δm_{12}^2

的 3.2 千吨水切伦科夫探测器，这个外部探测器用来标记宇宙线 μ 子并吸收来自周围岩石的 γ 射线和中子。

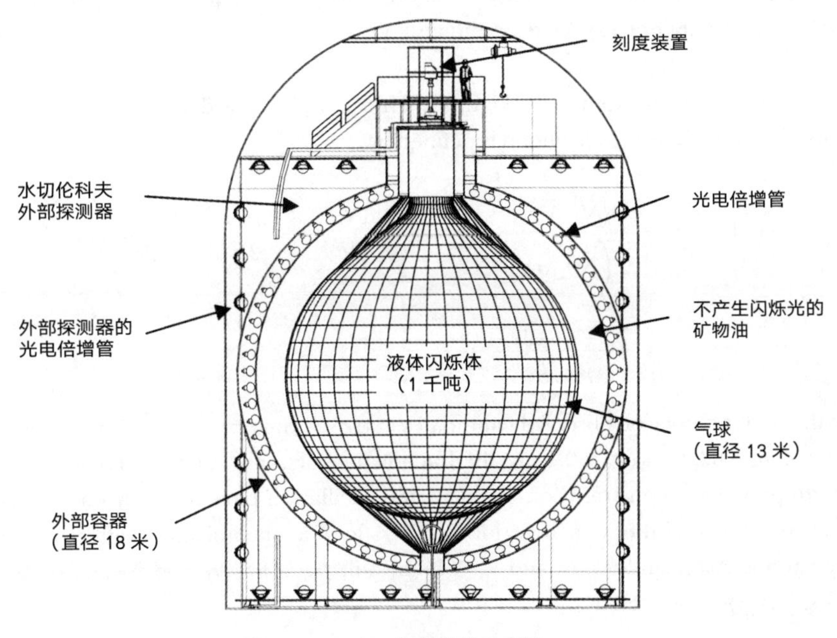

图 5. KamLAND 探测器的示意图

根据光子到达 PMT 的时间分布可确定探测器内粒子发生相互作用的位置。修正探测器响应随位置的变化后，可通过探测到的总光子数确定粒子的能量。事例的位置和能量通过垂伸至探测器中心的 γ 源进行刻度。对 $\overline{\nu}_e$ 事例的挑选，要求满足以下条件：快慢信号之间的时间符合 (ΔT) 满足 $0.5\ \mu s < \Delta T < 500\ \mu s$；相对于探测器中心的快信号 $(\mathbf{r}_p)$ 和慢信号 $(\mathbf{r}_d)$ 的位置分别满足 $|\mathbf{r}_p| < 5$ m，$|\mathbf{r}_d| < 5$ m 以及 $|\mathbf{r}_p - \mathbf{r}_d| < 1.0$ m；电子反中微子的能量满足 $1.7\ \text{MeV} < E_\nu < 3.4\ \text{MeV}$；慢信号能量 E_d 满足 $1.8\ \text{MeV} < E_d < 2.6\ \text{MeV}$。

假定有 N_U 和 N_{Th} 个探测到的地球中微子分别来自 ^{238}U 和 ^{232}Th 衰变链，预期探测到的全部 $\overline{\nu}_e$ 候选者的能量分布为：

$$\begin{aligned}\frac{d\overline{N}(E_\nu)}{dE_\nu} &= N_U\frac{dP_U(E_\nu)}{dE_\nu} + N_{Th}\frac{dP_{Th}(E_\nu)}{dE_\nu} + \frac{dN_r(E_\nu;\ \Delta m_{12}^2,\ \sin^2 2\theta_{12})}{dE_\nu} \\ &\quad + P_\alpha\frac{dN_\alpha(E_\nu/q_\alpha)}{dE_\nu} + \sum_k\frac{dN_k(E_\nu)}{dE_\nu}\end{aligned} \tag{4}$$

其中，$dP_U(E_\nu)/dE_\nu$ 和 $dP_{Th}(E_\nu)/dE_\nu$ 分别是 ^{238}U 和 ^{232}Th 衰变链产生的归一化后的预期地球中微子能谱，等式右边第三项是反应堆中微子本底的能谱，它是中微子振荡参数 Δm_{12}^2 和 $\sin^2 2\theta_{12}$ 的函数。$dN_\alpha(E_\nu/q_\alpha)/dE_\nu$ 是 $^{13}C(\alpha,n)^{16}O$ 本底经过能量和事例率的缩放因子 q_α 和 p_α 修正后的

and $\sin^2 2\theta_{12}$. $dN_\alpha(E_\nu/q_\alpha)/dE_\nu$ is the energy spectrum of the expected $^{13}C(\alpha,n)^{16}O$ background with energy and rate scaling factors q_α and p_α, respectively. The sum is over the other known backgrounds where $dN_k(E_\nu)/dE_\nu$ is the expected energy spectrum of the background. All expected spectra include energy smearing due to the detector energy resolution. Integrating equation (4) between 1.7 and 3.4 MeV gives the total number of expected candidates, $\overline{N}$.

The number of geoneutrinos from the ^{238}U and ^{232}Th decay chains is determined from an unbinned maximum likelihood fit. The log likelihood is defined by

$$\log L = -\frac{(N-\overline{N})^2}{2\left(\overline{N}+\sigma_{\overline{N}}^2\right)} + \sum_{i=1}^{N}\log\frac{1}{\overline{N}}\frac{d\overline{N}(E_i)}{dE_\nu} - \frac{(p_\alpha-1)^2}{2\sigma_p^2} - \frac{(q_\alpha-1)^2}{2\sigma_q^2} - \frac{\chi^2(\Delta m_{12}^2, \sin^2 2\theta_{12})}{2} \quad (5)$$

where N is the total number of observed candidates and $\sigma_{\overline{N}}$ is the error on $\overline{N}$. E_i is the energy of the ith $\overline{\nu}_e$ candidate. $\sigma_p = 0.24$ and $\sigma_q = 0.1$ are the fractional errors on q_α and p_α, respectively. The term $\chi^2(\Delta m^2, \sin^2 2\theta)$ provides a constraint on the neutrino oscillation parameters from the KamLAND reactor measurements and the solar neutrino results[30]. $\log L$ is maximized at different values of N_U and N_{Th} by varying Δm_{12}^2, $\sin^2 2\theta_{12}$, p_α and q_α. The best fit point for N_U and N_{Th} corresponds to the maximum $\log L$. A $\Delta\chi^2$ is defined by

$$\Delta\chi^2 = 2(\log L_{max} - \log L) \quad (6)$$

where $\log L_{max}$ is the $\log L$ at the best fit point. The confidence intervals are calculated from this $\Delta\chi^2$.

(**436**, 499-503; 2005)

T. Araki[1], S. Enomoto[1], K. Furuno[1], Y. Gando[1], K. Ichimura[1], H. Ikeda[1], K. Inoue[1], Y. Kishimoto[1], M. Koga[1], Y. Koseki[1], T. Maeda[1], T. Mitsui[1], M. Motoki[1], K. Nakajima[1], H. Ogawa[1], M. Ogawa[1], K. Owada[1], J.-S. Ricol[1], I. Shimizu[1], J. Shirai[1], F. Suekane[1], A. Suzuki[1], K. Tada[1], S. Takeuchi[1], K. Tamae[1], Y. Tsuda[1], H. Watanabe[1], J. Busenitz[2], T. Classen[2], Z. Djurcic[2], G. Keefer[2], D. Leonard[2], A. Piepke[2], E. Yakushev[2], B. E. Berger[3], Y. D. Chan[3], M. P. Decowski[3], D. A. Dwyer[3], S. J. Freedman[3], B. K. Fujikawa[3], J. Goldman[3], F. Gray[3], K. M. Heeger[3], L. Hsu[3], K. T. Lesko[3], K.-B. Luk[3], H. Murayama[3], T. O'Donnell[3], A. W. P. Poon[3], H. M. Steiner[3], L. A. Winslow[3], C. Mauger[4], R. D. McKeown[4], P. Vogel[4], C. E. Lane[5], T. Miletic[5], G. Guillian[6], J. G. Learned[6], J. Maricic[6], S. Matsuno[6], S. Pakvasa[6], G. A. Horton-Smith[7], S. Dazeley[8], S. Hatakeyama[8], A. Rojas[8], R. Svoboda[8], B. D. Dieterle[9], J. Detwiler[10], G. Gratta[10], K. Ishii[10], N. Tolich[10], Y. Uchida[10], M. Batygov[11], W. Bugg[11], Y. Efremenko[11], Y. Kamyshkov[11], A. Kozlov[11], Y. Nakamura[11], H. J. Karwowski[12], D. M. Markoff[12], K. Nakamura[12], R. M. Rohm[12], W. Tornow[12], R. Wendell[12], M.-J. Chen[13], Y.-F. Wang[13] & F. Piquemal[14]

[1] Research Center for Neutrino Science, Tohoku University, Sendai 980-8578, Japan

[2] Department of Physics and Astronomy, University of Alabama, Tuscaloosa, Alabama 35487, USA

[3] Physics Department, University of California at Berkeley and Lawrence Berkeley National Laboratory, Berkeley, California 94720, USA

[4] W. K. Kellogg Radiation Laboratory, California Institute of Technology, Pasadena, California 91125, USA

[5] Physics Department, Drexel University, Philadelphia, Pennsylvania 19104, USA

[6] Department of Physics and Astronomy, University of Hawaii at Manoa, Honolulu, Hawaii 96822, USA

[7] Department of Physics, Kansas State University, Manhattan, Kansas 66506, USA

[8] Department of Physics and Astronomy, Louisiana State University, Baton Rouge, Louisiana 70803, USA

[9] Physics Department, University of New Mexico, Albuquerque, New Mexico 87131, USA

预期能谱；最后一项是对已知其他所有本底能谱 $dN_k(E_\nu)/dE_\nu$ 的求和。所有的能谱都包含了由探测器能量分辨引起的能量弥散。对公式(4)在 1.7 ~ 3.4 MeV 能量区间内进行积分给出总的预期 $\overline{\nu}_e$ 事例数 $\overline{N}$。

^{238}U 和 ^{232}Th 衰变链产生的地球中微子数目由不分能量区间的最大似然法拟合得到。对数似然函数的定义如下，

$$\log L = -\frac{(N-\overline{N})^2}{2\left(\overline{N}+\sigma_{\overline{N}}^2\right)} + \sum_{i=1}^{N}\log\frac{1}{\overline{N}}\frac{d\overline{N}(E_i)}{dE_\nu} - \frac{(p_\alpha-1)^2}{2\sigma_p^2} - \frac{(q_\alpha-1)^2}{2\sigma_q^2} - \frac{\chi^2(\Delta m_{12}^2, \sin^2 2\theta_{12})}{2} \tag{5}$$

其中，N 是总的观测事例数，$\sigma_{\overline{N}}$ 是 $\overline{N}$ 的误差，E_i 是第 i 个 $\overline{\nu}_e$ 事例的能量，$\sigma_p = 0.24$ 和 $\sigma_q = 0.1$ 分别是 p_α 和 q_α 的相对误差。$\chi^2(\Delta m^2, \sin^2 2\theta)$ 项对 KamLAND 反应堆中微子测量和太阳中微子测量 [30] 得到的振荡参数提供了一个约束。通过改变 Δm_{12}^2、$\sin^2 2\theta_{12}$、p_α 和 q_α 这四个参数，将 $\log L$ 在不同 $N_{\rm U}$ 和 $N_{\rm Th}$ 值处最大化，最大的 $\log L$ 即是 $N_{\rm U}$ 和 $N_{\rm Th}$ 的最佳拟合点。$\Delta\chi^2$ 的定义如下

$$\Delta\chi^2 = 2(\log L_{\max} - \log L) \tag{6}$$

其中 $\log L_{\max}$ 是 $\log L$ 在最佳拟合点的值，而置信区间由 $\Delta\chi^2$ 计算得到。

（韩然 翻译；曹俊 审稿）

[10] Physics Department, Stanford University, Stanford, California 94305, USA
[11] Department of Physics and Astronomy, University of Tennessee, Knoxville, Tennessee 37996, USA
[12] Physics Department, Duke University, Durham, North Carolina 27008, USA, and Physics Department, North Carolina State, Raleigh, North Carolina 27695, USA, and Physics Department, University of North Carolina, Chapel Hill, North Carolina 27599, USA
[13] Institute of High Energy Physics, Beijing 100039, China
[14] CEN Bordeaux-Gradignan, IN2P3-CNRS and University Bordeaux I, F-33175 Gradignan Cedex, France

Received 25 May; accepted 4 July 2005.

References:

1. Eguchi, K. *et al.* First results from KamLAND: Evidence for reactor antineutrino disappearance. *Phys. Rev. Lett.* **90**, 021802 (2003).
2. Araki, T. *et al.* Measurement of neutrino oscillation with KamLAND: Evidence of spectral distortion. *Phys. Rev. Lett.* **94**, 081801 (2005).
3. Eder, G. Terrestrial neutrinos. *Nucl. Phys.* **78**, 657-662 (1966).
4. Marx, G. Geophysics by neutrinos. *Czech. J. Phys. B* **19**, 1471-1479 (1969).
5. Avilez, C., Marx, G. & Fuentes, B. Earth as a source of antineutrinos. *Phys. Rev. D* **23**, 1116-1117 (1981).
6. Krauss, L. M., Glashow, S. L. & Schramm, D. N. Antineutrino astronomy and geophysics. *Nature* **310**, 191-198 (1984).
7. Kobayashi, M. & Fukao, Y. The Earth as an antineutrino star. *Geophys. Res. Lett.* **18**, 633-636 (1991).
8. Raghavan, R. S. *et al.* Measuring the global radioactivity in the Earth by multidetector antineutrino spectroscopy. *Phys. Rev. Lett.* **80**, 635-638 (1998).
9. Rothschild, C. G., Chen, M. C. & Calaprice, F. P. Antineutrino geophysics with liquid scintillator detectors. *Geophys. Res. Lett.* **25**, 1083-1086 (1998).
10. Mantovani, F., Carmignani, L., Fiorentini, G. & Lissia, M. Antineutrinos from Earth: A reference model and its uncertainties. *Phys. Rev. D* **69**, 013001 (2004).
11. Pollack, H. N., Hurter, S. J. & Johnson, J. R. Heat flow from the Earth's interior: analysis of the global data set. *Rev. Geophys.* **31**, 267-280 (1993).
12. Hofmeister, A. M. & Criss, R. E. Earth's heat flux revised and linked to chemistry. *Tectonophysics* **395**, 159-177 (2005).
13. McDonough, W. F. & Sun, S.-s. The composition of the Earth. *Chem. Geol.* **120**, 223-253 (1995).
14. Jackson, M. J. & Pollack, H. N. On the sensitivity of parameterized convection to the rate of decay of internal heat sources. *J. Geophys. Res.* **89**, 10103-10108 (1984).
15. Richter, F. M. Regionalized models for the thermal evolution of the Earth. *Earth Planet. Sci. Lett.* **68**, 471-484 (1984).
16. Firestone, R. B. *Table of Isotopes* 8th edn (John Wiley, New York, 1996).
17. Behrens, H. & Jänecke, J. *Landolt-Börnstein - Group I, Elementary Particles, Nuclei and Atoms* Vol. 4 (Springer, Berlin, 1969).
18. McKeown, R. D. & Vogel, P. Neutrino masses and oscillations: triumphs and challenges. *Phys. Rep.* **394**, 315-356 (2004).
19. Ahmed, S. N. *et al.* Measurement of the total active ^{8}B solar neutrino flux at the Sudbury Neutrino Observatory with enhanced neutral current sensitivity. *Phys. Rev. Lett.* **92**, 181301 (2004).
20. Wolfenstein, L. Neutrino oscillations in matter. *Phys. Rev. D* **17**, 2369-2374 (1978).
21. Vogel, P. & Beacom, J. F. Angular distribution of neutron inverse beta decay, $\bar{\nu}_e + p \to e^+ + n$. *Phys. Rev. D* **60**, 053003 (1999).
22. Enomoto, S. *Neutrino Geophysics and Observation of Geo-neutrinos at KamLAND.* Thesis, Tohoku Univ. (2005); available at ⟨http://www.awa.tohoku.ac.jp/KamLAND/publications/Sanshiro_thesis.pdf⟩.
23. JENDL Japanese Evaluated Nuclear Data Library. ⟨http://wwwndc.tokai.jaeri.go.jp/jendl/jendl.html⟩ (2004).
24. Apostolakis, J. Geant—Detector description and simulation tool. ⟨http://wwwasd.web.cern.ch/wwwasd/geant/index.html⟩ (2003).
25. Walton, R. B., Clement, J. D. & Borlei, F. Interaction of neutrons with oxygen and a study of the $C^{13}(\alpha,n)O^{16}$ reaction. *Phys. Rev.* **107**, 1065-1075 (1957).
26. Kerr, G. W., Morris, J. M. & Risser, J. R. Energy levels of ^{17}O from $^{13}C(\alpha, \alpha_0)^{13}C$ and $^{13}C(\alpha,n)^{16}O$. *Nucl. Phys. A* **110**, 637-656 (1968).
27. Kopeikin, V. I. *et al.* Inverse beta decay in a nonequilibrium antineutrino flux from a nuclear reactor. *Phys. Atom. Nuclei* **64**, 849-854 (2001).
28. Rocholl, A. & Jochum, K. P. Th, U and other trace elements in carbonaceous chondrites: Implications for the terrestrial and solar-system Th/U ratios. *Earth Planet. Sci. Lett.* **117**, 265-278 (1993).
29. Tolich, N. *Experimental Study of Terrestrial Electron Anti-neutrinos with KamLAND* Thesis, Stanford Univ. (2005); available at ⟨http://www.awa.tohoku.ac.jp/KamLAND/publications/Nikolai_thesis.pdf⟩.
30. KamLAND collaboration. Data release accompanying the 2nd KamLAND reactor result. ⟨http://www.awa.tohoku.ac.jp/KamLAND/datarelease/2ndresult.html⟩ (2005).

Acknowledgements. We thank E. Ohtani and N. Sleep for advice and guidance. The KamLAND experiment is supported by the COE program of the Japanese Ministry of Education, Culture, Sports, Science, and Technology, and by the United States Department of Energy. The reactor data were provided courtesy of the following associations in Japan: Hokkaido, Tohoku, Tokyo, Hokuriku, Chubu, Kansai, Chugoku, Shikoku and Kyushu Electric Power Companies, Japan Atomic Power Co. and Japan Nuclear Cycle Development Institute. Kamioka Mining and Smelting Company provided services for activity at the experimental site.

Author Information. Reprints and permissions information is available at npg.nature.com/reprintsandpermissions. The authors declare no competing financial interests. Correspondence and requests for materials should be addressed to S.E. (sanshiro@awa.tohoku.ac.jp) or N.T. (nrtolich@lbl.gov).